Test Yourself

Take a quiz at the *Microbiology: A Systems Approach* Online Learning Center to gauge your mastery of chapter content.

Each chapter quiz is specially constructed to test your comprehension of key concepts. Immediate feedback explains incorrect responses. You can even e-mail your quiz results to your professor.

Results Reporter

Out of 25 questions, you answered 18 correctly, for a final grade of 72%.

18 correct (72%)
7 incorrect (28%)
0 unanswered (0%)

Your Results:

The correct answer for each question is indicated by a ✓.

1 INCORRECT If an antimicrobial agent is bacteriostatic,

- ●A)complete killing of bacteria occurs with its use.
- ✓ ○B)bacteria will resume growth upon removal of the agent.
- ○C)no viable organisms remain after its use.
- ○D)it will kill only spores.
- ○E)it will kill bacteria and inhibit fungi.

Feedback: Incorrect. The correct answer is bacteria will resume growth upon removal of the agent. Choice A is incorrect because bacteriostatic means an agent is capable of inhibiting the growth of bacteria.

Routing Information

Date: Wed Nov 10 13:58:23 EST 2004

My name: _____
Section ID: _____

Email these results to:

	Email address:	Format:
Me:	_____	Text ▾
My Instructor:	_____	Text ▾
My TA:	_____	Text ▾
Other:	_____	Text ▾

E-Mail The Results

Course Tools

Powerful tools are available through the Online Learning Center to supplement your textbook. Additional case studies, web-based exercises, and links to relevant and dynamic websites provide content to enhance your studies.

As you study, a 24-hour tutorial service is only an e-mail away.

This "homework hotline" offers you the opportunity to discuss text questions with our Microbiology consultant.

Send an e-mail by filling out the form below, and then hitting the Submit button.

McGraw-Hill Higher Education E-mail Privacy Notice

Contacting us via e-mail reveals your e-mail address and any other information you include. We will use this information to help us fulfill your order or respond to your inquiry. McGraw-Hill Higher Education does not share the information it collects about customers and prospects from its web sites with other companies within the family of The McGraw-Hill Companies or with organizations outside the family of The McGraw-Hill Companies.

If you would like to confirm the accuracy of the information we have collected from you, or if you have questions about the uses of this information, please email a request to mhhe_webmaster@mcgraw-hill.com.

The McGraw-Hill Companies are leaders in protecting the private information requested from their customers. If you would like more information about The McGraw-Hill Companies Customer Privacy Policy, click here.

Name _____

Email address _____

Your comments:

_____

Submit

Internet Exercises

Microscopy is an essential tool that enables scientists to view specimens too small for the unaided eye. View images of microscopes from history at http://micro.magnet.fsu.edu/primer/museum/index.html.

Access the database of images at http://www.cellsalive.com/ to view prokaryotic, eukaryotic and viral specimens. What characteristics of microscopes allow us to discern fine details of small objects? What limits the use of the light microscope? Compare the light micrographs with electron micrographs at Dennis Kunkel Microscopy, Inc. (http://www.denniskunkel.com/) and (http://www.booklabs.wisc.edu/uwmr/newuwmr/fesem.html). What feature of the electron microscope permits enhanced clarity and imaging of specimens too small for the light microscope?

Case 2

A 27-year old white female presented at the walking clinic of her local physician on August 15. On physical exam, the patient had a fever of 38.5C. She appeared fatigued, had tender joints, and complained of a headache, a stiff neck and a backache. The physician noticed a circular "rash" about 5 inches in diameter, with a bright red leading edge and a dim center in the form of a "bull's eye". The physician noted an irregular heart beat. The patient complained of lack of ability to concentrate.

The patient gave the following history. She is a graduate student in the wildlife program at the university in town. She was in the field for three weeks in Wisconsin during the months of May and June. She tracks small mammals in the field and studies their behavior. It had been a warm, wet spring and she complained of a large number of biting flies, mosquitoes and ticks in the area. She felt well until about 2 weeks after returning to her home. Since that time, many of her symptoms had progressed. She finally found that she could take it no more.

1. What is your best diagnosis of this case?
2. What features are critical to your diagnosis?
3. What further steps should be taken to clear up the problem?

Visit www.mhhe.com/cowan1

Microbiology
A SYSTEMS APPROACH

Marjorie Kelly Cowan
Miami University

Kathleen Park Talaro
Pasadena City College

Boston   Burr Ridge, IL   Dubuque, IA   Madison, WI   New York   San Francisco   St. Louis
Bangkok   Bogotá   Caracas   Kuala Lumpur   Lisbon   London   Madrid   Mexico City
Milan   Montreal   New Delhi   Santiago   Seoul   Singapore   Sydney   Taipei   Toronto

Higher Education

MICROBIOLOGY: A SYSTEMS APPROACH

1 2 3 4 5 6 7 8 9 0 QPV/QPV 0 9 8 7 6 5

ISBN 0–07–291804–7

Editorial Director: *Kent A. Peterson*
Publisher: *Colin H. Wheatley*
Developmental Editor: *Rose Koos*
Managing Developmental Editor: *Patricia Hesse*
Marketing Manager: *Tami Petsche*
Senior Project Manager: *Jayne Klein*
Senior Production Supervisor: *Laura Fuller*
Senior Media Project Manager: *Jodi K. Banowetz*
Lead Media Technology Producer: *John J. Theobald*
Designer: *Laurie B. Janssen*
Cover Designer: *Rokusek Design*
(USE) Cover Image: *Dr. Steve Patterson/Science Photo Library*
Senior Photo Research Coordinator: *John C. Leland*
Photo Research: *David Tietz*
Supplement Producer: *Brenda A. Ernzen*
Compositor: *The GTS Companies/Los Angeles, CA Campus*
Typeface: *10/12 Times Roman*
Printer: *Quebecor World Versailles Inc.*

About the Cover: The cover image is a colored scanning electron micrograph (SEM) showing coronavirus particles (yellow) on the surface of a culture cell (blue). Coronaviruses are responsible for causing common colds and gastroenteritis. The SARS virus is also a coronavirus.

The credits section for this book begins on page C-1 and is considered an extension of the copyright page.

Library of Congress Cataloging-in-Publication Data

Cowan, M. Kelly.
    Microbiology : a systems approach / Marjorie M. Kelly Cowan, Kathleen Park Talaro.—1st ed.
        p.   cm.
    Includes index.
    ISBN 0–07–291804–7 (alk. paper)
    1. Microbiology. I. Talaro, Kathleen P. II. Title.

QR41.2.C69   2006
579—dc22

                                        2004057885
                                        CIP

www.mhhe.com

About the Authors

Kelly Cowan has been a microbiologist at Miami University since 1993. She received her Ph.D. at the University of Louisville, and later worked at the University of Maryland Center of Marine Biotechnology and the University of Groningen in The Netherlands.

Her first love is teaching—both doing it and studying how to do it better. She teaches nursing microbiology and non-majors microbiology every year. She is a member of the Undergraduate Education Committee of the American Society for Microbiology, and past president of the Ohio Branch of the American Society for Microbiology. In 1997 she won a Celebration of Teaching Award sponsored by the Greater Cincinnati Consortium of Colleges and

Universities. Since 2003 she has served as the Chief Academic Officer of Miami University Middletown.

Kelly has published (with her students) twenty-four research articles stemming from her work on bacterial adhesion mechanisms and plant-derived antimicrobial compounds. She holds two patents for strategies to block microbial attachment. Kelly also travels extensively to present her research, and to talk to other professors about teaching.

When she's not teaching, researching or traveling, she's listening to live music at home played by her two sons Taylor (16) and Sam (13), whose musical tastes run from Robert Johnson to the Rolling Stones to Sum41.

Kathleen Park Talaro is a microbiologist, author, illustrator, photographer, and educator at Pasadena City College. She began her college education at Idaho State University in Pocatello. There, she found a niche that fit her particular abilities and interests, spending part of her time as a scientific illustrator and part as a biology lab assistant. After graduation with a B.S. in biology, she entered graduate school at Arizona State University, majoring in physiological ecology. During her graduate studies she participated in two research expeditions to British Columbia with the Scripps Institution of Oceanography. Kathy continued to expand her background, first finishing a Master's degree at Occidental College and later taking additional specialized coursework in microbiology at California Institute of Technology and California State University.

If there is one continuing theme reverberating through Kathy's experiences, it is the love of education and teaching. She has been teaching allied health microbiology and majors biology courses for nearly 30 years. Kathy finds great joy in watching her students develop their early awareness of microorganisms—when they first come face-to-face with the reality of them on their hands, in the air, in their food, and, of course, nearly everywhere.

Kathy is a member of the American Society for Microbiology and the American Association for the Advancement of Science. She keeps active in self-study and research, and continues to attend workshops and conferences to remain current in her field. Kathy has also been active in science outreach programs by teaching Saturday workshops in microbiology and DNA technology to high school and junior high students.

We dedicate this book to all public health workers
who devote their lives to bringing the advances and medicines
enjoyed by the industrialized world to all humans.

Brief Contents

Contents

CHAPTER 4

Procaryotic Profiles: The Bacteria and Archaea   89

INSIGHT 4.1   *Discovery*
Biofilms—The Glue of Life   96

INSIGHT 4.2   *Discovery*
The Gram Stain: A Grand Stain   98

INSIGHT 4.3   *Discovery*
Redefining Bacterial Size   113

CHAPTER 5

Eucaryotic Cells and Microorganisms   119

INSIGHT 5.1   *Historical*
The Extraordinary Emergence of Eucaryotic Cells   121

INSIGHT 5.2   *Discovery*
The Many Faces of Fungi   133

CHAPTER 11

Physical and Chemical Control of Microbes 315

CHAPTER 12

Drugs, Microbes, Host—The Elements of Chemotherapy 347

CHAPTER 13

Microbe-Human Interactions: Infection and Disease 383

INSIGHT 13.1   *Discovery*

Life Without Flora   389

INSIGHT 13.2   *Medical*

Laboratory Biosafety Levels and Classes of Pathogens   391

INSIGHT 13.3   *Medical*

The Classic Stages of Clinical Infections   396

INSIGHT 13.4   *Medical*

A Quick Guide to the Terminology of Infection and Disease   398

INSIGHT 13.5   *Historical*

The History of Human Guinea Pigs   413

CHAPTER **14**

Nonspecific Host Defenses   417

INSIGHT 14.1   *Medical*

When Inflammation Gets Out of Hand   432

INSIGHT 14.2   *Medical*

The Dynamics of Inflammatory Mediators   435

INSIGHT 14.3   *Medical*

Some Facts About Fever   436

CHAPTER **15**

Specific Immunity and Immunization   445

CHAPTER **16**

Disorders in Immunity   483

CHAPTER **17**

Diagnosing Infections   515

INSIGHT 17.1   *Discovery*

The Uncultured   517

INSIGHT 17.2   *Medical*

When Positive Is Negative: How to Interpret Serological Test Results   526

CHAPTER 18

Infectious Diseases Affecting the Skin and Eyes   539

INSIGHT 18.1   *Medical*

The Skin Predators: *Staphylococcus* and *Streptococcus*   550

INSIGHT 18.2   *Historical*

Smallpox: An Ancient Scourge Revisited   556

INSIGHT 18.3   *Medical*

Naming Skin Lesions   559

CHAPTER 19

Infectious Diseases Affecting the Nervous System   577

INSIGHT 19.1   *Discovery*

Baby Food and Meningitis   588

INSIGHT 19.2   *Medical*

A Long Way from Egypt: West Nile Virus in the United States   590

INSIGHT 19.3   *Historical*

Polio   599

INSIGHT 19.4   *Discovery*

Botox: No Wrinkles. No Headaches. No Worries?   605

CHAPTER 20

Infectious Diseases Affecting the Cardiovascular and Lymphatic Systems   613

INSIGHT 20.1   *Medical*

Atherosclerosis   616

INSIGHT 20.2   *Medical*

The Arthropod Vectors of Infectious Disease   626

INSIGHT 20.3   *Medical*

AIDS-Defining Illnesses (ADIs)   638

CHAPTER **21**

**Infectious Diseases Affecting the Respiratory
System   653**

INSIGHT 21.1   *Medical*

Fungal Lung Diseases   669

INSIGHT 21.2   *Discovery*

Bioterror in the Lungs   678

CHAPTER **22**

**Infectious Diseases Affecting the
Gastrointestinal Tract   687**

INSIGHT 22.1   *Medical*

Stools: To Culture or Not to Culture?   700

INSIGHT 22.2   *Discovery*

**A Little Water, Some Sugar and Salt, Save
Millions of Lives   706**

INSIGHT 22.3   *Medical*

Microbes Have Fingerprints, Too   712

INSIGHT 22.4   *Discovery*

Treating Inflammatory Bowel Disease with Worms?   721

Preface

It's not difficult, or even necessary these days, to convince anyone that microbiology is an important subject to study. In recent years anthrax, West Nile virus, SARS, and of course, HIV have already done that job. What can be difficult is helping students understand and apply the challenging concepts of microbiology, and make sense of the good and the bad that come from microorganisms.

With more than 30 years of combined college teaching experience, we have tested numerous teaching strategies to find those that help students succeed in their study of microbiology. We learn more about the subject matter and more about our students with every class we teach, and each time we are impressed by how much our students already know, how hard they are willing to work, and their life aspirations. Our presentation of the concepts in this book has been greatly influenced by our students' insights, feedback, and accomplishments. In addition we have incorporated into the textbook our tried and proven teaching strategies as is seen through our unique organization of the disease chapters, conversational writing style, vivid artwork, and effective pedagogy.

This textbook is also a product of a unique synergy between us as co-authors. Kelly brings to the mix her experience and success in research and teaching, while Kathy is a celebrated author who has gained the respect of instructors and students for her distinctive, easy-to-understand writing style. The end result is timely content and microbiology research embedded in an accessible presentation.

We humbly present this book in the hope that using it will turn readers on to the wonderful world of microbiology, and the fascinating—and sometimes heartbreaking—world of infectious disease.

What Sets This Book Apart?

Distinctive Organization of Infectious Disease Chapters

Following the tradition of microbiology textbooks, the first 16 chapters of *Microbiology: A Systems Approach* provide the basics about microorganisms: what they are, the methods used to study them, human attempts to control them, and our bodies' defenses against them. For chapters 17–23, we have developed an unequaled level of organization in our presentation of the infectious disease material.

Exclusive Chapter   Chapter 17, "Diagnosing Infections," is unique among microbiology textbooks: it brings together in one place the methods used to diagnose infectious diseases. It starts with collecting samples from the patient, and details the biochemical, serological, and molecular methods used to identify causative microbes.

Highly Organized Disease Chapters   Like other books, chapters 18–23 present the diseases according to the human organ systems. However, the organization of the material within each of these chapters has been taken to a new level.

The traditional organ system approach makes sense (to anyone who has experienced an infection!), but still leaves organizational threads hanging. Within a given organ system chapter, diseases are discussed in random order, and there is often no consistent pattern to what is said about each disease.

This book improves upon that approach by organizing the infectious agents according to the symptoms or condition they cause, instead of in a random order. For example, in the respiratory disease chapter, there is a major heading called "Community-Acquired Pneumonia"—a condition that can be caused by several different microbes. Each of those microbes is discussed under that heading, in a systematic manner. At the end of the section, the microbes are summarized in a **Checkpoint table** called "Community-Acquired Pneumonia." Conditions with only one possible cause, such as pertussis, also end with a Checkpoint table that includes the single causative agent.

✔ CHECKPOINT 21.6	Pertussis (Whooping Cough)
Causative Organism(s)	*Bordetella pertussis*
Most Common Modes of Transmission	Droplet contact
Virulence Factors	FHA (adhesion), pertussis toxin and tracheal cytotoxin, endotoxin
Culture/Diagnosis	Grown on B-G, charcoal or potato-glycerol agar; diagnosis can be made on symptoms
Prevention	Acellular vaccine (DTaP), erythromycin or trimethoprim; sulfamethoxazole for contacts
Treatment	Mainly supportive; erythromycin to decrease communicability

This approach is refreshingly logical, systematic, and intuitive, as it encourages clinical and critical modes of thinking in students—the type of thinking they will be using if their eventual careers are in health care. Students learn to examine multiple possibilities for a given condition, and grow accustomed to looking for commonalities and differences among the various organisms that cause a given condition. In addition, they learn to consider the kinds of conditions that are caused by only one microbe.

Along with the higher level of organization offered in this book, students are provided with key pedagogical tools at the end of each disease chapter to reinforce and tie together the information they've just learned. Each disease chapter ends with a **summary figure**—a "glass body" that highlights the affected organs discussed in the chapter—and a **taxonomic list of organisms.** The distinctive summary figure lists the diseases that were presented in the chapter, with the microbes that could cause them *color-coded by type of microorganism.* The taxonomic list of organisms is presented in tabular form so students can see the diversity of microbes causing diseases in that system, and appreciate their taxonomic positions.

In summary, the disease presentation in this book makes the world of infectious diseases come together for the student. It presents the information within a consistent organizational structure (known to facilitate learning) and embeds it within a structure that teaches clinical, and critical modes of thinking.

An Engaging Writing Style, Praised by Reviewers

Our goal was to achieve a precise balance in writing, so students will easily comprehend the material without compromising the level of presentation. One of the key strengths of this text comes from our efforts in making difficult concepts understandable, as well as intriguing and exciting for students. We use this consistent, direct approach throughout the text—in the narrative, the illustrations, and throughout the pedagogical aids. Analogies, case studies, and real-world examples also help students relate microbiology to their world.

> "I rate this book as the highest, for readability, in a college microbiology textbook. I found the writing style to be clear and straightforward without sacrificing completeness." Todd Herman, UCLA

> "I found this textbook very easy to read. I think the sentence structure and vocabulary provide a more conversational style of text than other textbooks that I have reviewed or used. In my experience, students prefer this style of writing and are more likely to really read the text and understand it." Valerie A. Watson, West Virginia University

A Vivid Art Program That Explains Itself

Kathy Talaro brings her experience as a teacher, microbiologist, *and* illustrator to this text. Her insight and expertise provide an inimitable blend of scientific accuracy and aesthetics. Vivid, multi-dimensional illustrations complement self-contained, concept-specific narrative; it is not necessary to read page content surrounding artwork to grasp concepts being illustrated. Development of the artwork in this manner further enhances learning and helps to build a solid foundation of understanding.

A special Art Consultant Board, composed of experienced instructors, also worked closely with us through the development of this book to ensure accuracy and effectiveness in the art.

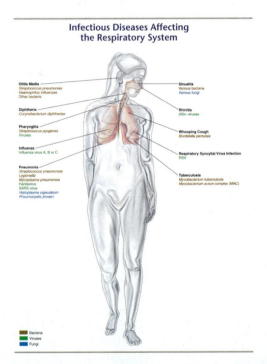

Infectious Diseases Affecting the Respiratory System

Taxonomic Organization of Microorganisms Causing Disease in the Respiratory Tract		
Microorganism	**Disease**	**Chapter Location**
Gram-Positive Bacteria		
Streptococcus pneumoniae	Otitis media, pneumonia	Otitis media, p. 657
		Pneumonia, p. 675
S. pyogenes	Pharyngitis	Pharyngitis, p. 658
Corynebacterium diphtheriae	Diphtheria	Diphtheria, p. 662
Gram-Negative Bacteria		
Haemophilus influenzae	Otitis media	Otitis media, p. 657
Bordetella pertussis	Whooping cough	Whooping cough, p. 664
Mycobacterium tuberculosis, * *M. avium* complex	Tuberculosis	Tuberculosis, p. 668
Legionella spp.	Pneumonia	Pneumonia, p. 676
Other Bacteria		
Mycoplasma pneumoniae	Pneumonia	Pneumonia, p. 677
RNA Viruses		
Respiratory syncytial virus	RSV disease	RSV disease, p. 665
Influenza virus A, B, and C	Influenza	Influenza, p. 666
Hantavirus	Hantavirus pulmonary syndrome	Pneumonia, p. 677
SARS-associated coronavirus	SARS	Pneumonia, p. 678
Fungi		
Pneumocystis jiroveci	*Pneumocystis* pneumonia	Pneumonia, p. 681
Histoplasma capsulatum	Histoplasmosis	Pneumonia, p. 679

*There is some debate about the gram status of the genus *Mycobacterium*; it is generally not considered gram positive or gram negative.

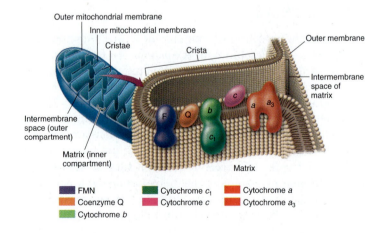

Pedagogy Designed for the Way Students Learn

Microbiology: A Systems Approach makes learning easier through its carefully crafted pedagogical system. Following is a closer look at some of the key features that our students have taught us are useful.

- All chapters open with **In the News** mysteries to solve. These real-world case studies help students appreciate and understand how microbiology impacts our lives on a daily basis. The solutions appear later in the chapter, after the necessary elements have been presented.

- A **Chapter Overview** at the beginning of each chapter provides students with a framework from which to begin their study of a chapter.
- In chapters 1–16 and 24, major sections of the chapter are followed by **Checkpoints** that repeat and summarize the concepts of that section. In the disease chapters (18–23) the Checkpoints are in the form of the disease tables described earlier.
- **Insight** readings allow students to delve into material that goes beyond the chapter concepts and consider the application of those concepts. The Insight readings are divided into four categories: Discovery, Historical, Medical, and Microbiology.
- All chapters end with a **summary,** and a comprehensive array of **end-of-chapter questions.** The questions are not just multiple-choice, but also critical thinking questions that often have no correct answer. Considering and answering these questions, and even better, discussing them with fellow students, can make the difference between temporary (or limited) learning and true knowledge of the concepts.

Teaching Supplements

McGraw-Hill offers various tools and technology products to support *Microbiology: A Systems Approach.* Instructors can obtain teaching aids by calling the Customer Service Department, at 800-338-3987, or contacting their local McGraw-Hill sales representative.

Digital Content Manager CD-ROM

This cross-platform CD-ROM is a multimedia collection of visual resources that allows instructors to utilize artwork from the text in multiple formats and create *customized* classroom presentations, visually based tests and quizzes, dynamic course website content, or attractive printed support material. The assets on this CD-ROM are organized by chapter within the following easy-to-use folders:

Art Library Full-color digital files of all the illustrations in the book, plus the same art saved in black and white versions, can be readily incorporated into lecture presentations, exams, or custom-made classroom materials. These images are also pre-inserted into blank PowerPoint slides for ease of use.

Photo Library Digital files of all photographs from the text can be reproduced for multiple classroom uses.

Table Library Every table that appears in the text is provided in digital form.

PowerPoint Lecture Outline Library Ready-made presentations that combine art and lecture notes are provided for each of the 24 chapters of the text. These lecture outlines can be used as they are, or can be tailored to reflect preferred lecture topics and sequences.

Animation Library More than 50 full-color animations are available to harness the visual impact of processes in motion. Import these dynamic files into classroom presentations or online course materials.

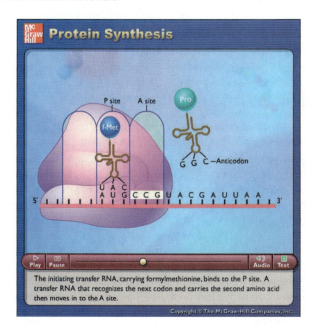

TextEdit Art Library Every illustration from the textbook is provided in PowerPoint. Instructors may revise, move, or delete labels to create customized presentations and exams.

Active Art Library Key figures are saved in manipulable layers that can be isolated and customized to meet the needs of the lecture environment.

Video Library 36 dynamic motion sequences have been produced to bring microorganisms to life for students.

Instructor's Testing and Resource CD-ROM

This cross-platform CD-ROM features a computerized test bank utilizing McGraw-Hill's EZ Test, a flexible and easy-to-use electronic testing program. The program allows instructors to create tests from book-specific items. It accommodates a wide range of question types, and instructors may add their own questions. Multiple versions of the test can be created, and any test can be exported for use with course management systems such as WebCT, BlackBoard or PageOut. The program is available for Windows and Macintosh environments. Word files of the test bank are included for those who prefer to work outside of the test-generator software. The Instructor's Manual is also available on this CD-ROM in both Word and PDF formats.

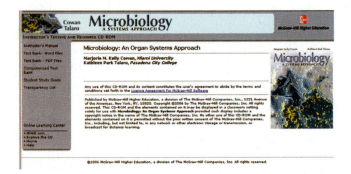

Transparencies

This set of 300 overhead transparencies includes key, full-color figures from the textbook for classroom projection.

Online Learning Center

http://www.mhhe.com/cowan1

The Cowan/Talaro Online Learning Center provides a vast array of resources to enhance the teaching and learning experience. Moreover, these resources are easily loaded into course management systems such as WebCT or Blackboard. Instructors can contact their McGraw-Hill representative for more details.

eInstruction

The classroom performance system (CPS) utilizes wireless technology to bring interactivity into the classroom or lecture hall. Instructors and students receive immediate feedback through wireless response pads that are easy to use and engage students. eInstruction can assist instructors by:

- Taking attendance
- Administering quizzes and tests
- Creating a lecture with intermittent questions
- Using the CPS grade book to manage lectures and student comprehension
- Integrating interactivity into their PowerPoint presentations

Course Management Systems

Text-specific content is available for the most popular course management systems, including **WebCT, Blackboard, eCollege, and McGraw-Hill's own PageOut.** Instructors can quickly and easily set up a course website that includes text-specific quizzes, additional readings, interactives, and animations that support the McGraw-Hill text they are using. The content is taken from the Online Learning Center that accompanies the textbook. McGraw-Hill provides this service to adopters at no additional charge to students, and also provides customer support ranging from general technical support to staff training.

Learning Supplements

Students may order supplemental study materials by contacting their local bookstore.

Online Learning Center

http://www.mhhe.com/cowan1

Students can visit this book-specific website to find a variety of resources to enhance their learning. The resources support each chapter in the textbook, and some of the features include:

- **Self-quizzing**
- **Animations** of key processes
- Electronic **flashcards** to review key vocabulary
- Additional **clinical case presentations**
- **Internet exercises** to encourage use of the Internet as a resource to gather and evaluate information

Student Study Art Notebook

This handy study aid, a bound and printed notebook containing all of the artwork from the text, allows students to jot notes during lecture and complete self-tests to identify and work through microbiological processes.

Student Study Guide

A valuable student resource, written by Nancy Boury, the Student Study Guide goes beyond the standard multiple-choice and true-false self-quizzing. The author has provided a wealth of study assets, including key concepts, vocabulary review, self-tests, and more.

Microbes in Motion CD-ROM

Microbes in Motion is a cross-platform, interactive CD-ROM that brings microbiology to life through interactive video, audio, animations, and hyperlinking. It is an easy-to-use tutorial that is ideal for self-quizzing, class preparation, or review of microbiological concepts.

HyperClinic CD-ROM

Students will have fun with this interactive CD-ROM while learning valuable concepts and gaining practical experience in clinical microbiology. Packed with more than 100 case studies and over 200 pathogens supported with audio, video, and interactive screens, students will gain confidence as they take on the role of the professionals.

Acknowledgments

We have collected a stack of papers that seems to be five feet high; this stack is the written record of the careful consideration that our reviewers gave to early versions of this book. We are deeply grateful to every one of the expert instructors listed here, who took the time to comb through each chapter and point out errors or confusing wording or organization. Each of them has brought to bear on this book what their students have taught them. A wealth of experience, knowledge, and dedication to students is represented in this list of names! We sincerely thank them for the vast improvements they brought to this book.

Reviewers

Elizabeth Wheeler Alm, *Central Michigan University*

Penny P. Antley, *University of Louisiana at Lafayette*

Marcie L. Baer, *Shippensburg University*

Gail Baker, *Laguardia Community College*

A. Clyde Blauer, *Snow College*

Clifford W. Bond, *Montana State University*

Kathryn Brooks, *Michigan State University*

Linda D. Bruslind, *Oregon State University*

D. Kim Burnham, *Oklahoma State University*

Robert M. Carey, *Pima Community College*

Daniel E. Cayton, *El Centro College*

Naowarat Cheeptham, *The University College of the Cariboo*

Jim Collins, *University of Arizona*

Judith A. Coston, *Bossier Parish Community College*

Sarah Crawford, *Connecticut State University*

John R. Dankert, *University of Louisiana at Lafayette*

Janet M. Decker, *University of Arizona*

Charles J. Dick, *Pasco-Hernando Community College*

Bob F. Drake, *Southwest Tennessee Community College*

David Drake, *University of Iowa*

Larry E. Eason, *Pasco-Hernando Community College*

Angela M. Edwards, *Trident Technical College*

David L. Elmendorf, *University of Central Oklahoma*

Teresa G. Fischer, *Indian River Community College*

Katherine Foreman, *Moraine Valley Community College*

Pamela B. Fouché, *Walters State Community College*

S. Marvin Friedman, *Hunter College*

Kathryn Germain, *Southwest Tennessee Community College*

Sandra Gibbons, *Moraine Valley Community College*

Judy Gnarpe, *University of Alberta*

Indhu Gopal, *Carolina's College of Health Sciences*

Thomas Gorczyca, *Northern Essex Community College*

Brinda Govindan, *San Francisco State University*

Judy Haber, *California State University, Fresno*

Richard Hanke, *Rose State College*

Todd Herman, *University of California Los Angeles*

James B. Jensen, *Brigham Young University*

Gilbert H. John, *Oklahoma State University*

Richard D. Karp, *University of Cincinnati*

George Keller, *Samford University*

Scott S. Kinnes, *Azusa Pacific University*

Dennis J. Kitz, *Southern Illinois University, Edwardsville*

Kenneth S. Landreth, *West Virginia University*

Jeff G. Leid, *Northern Arizona University*

Shawn Lester, *Montgomery College*

Roger Lightner, *University of Arkansas, Fort Smith*

William Lorowitz, *Weber State University*

Bernard MacLennan, *University College of Cape Breton*

A. Charles McBride, *Ivy Tech State College*

Colleen McDermott, *University of Wisconsin, Oshkosh*

Robert J. McDonough, *Georgia Perimeter College*

Marvita D. McGuire, *Tulsa Community College*

Sherry Meeks, *University of Central Oklahoma*

Gloria R. Mihalik, *Lynn University*

Fernando Monroy, *Northern Arizona University*

Pamela Moolenaar-Wirsiy, *Georgia Perimeter College*

David W. Morris, *George Washington University*

Karen Nakaoka, *Weber State University*

Murad Odeh, *South Texas Community College*

Natalie Osterhoudt, *Broward Community College*

Gregory E. Paquette, *University of Rhode Island*

Jack Pennington, *St. Louis Community College, Forest Park*

Beverly Perry, *Houston Community College*

Indiren Pillay, *Southwest Tennessee Community College*

Nirmala V. Prabhu, *Edison College*

Judith A. Prask, *Montgomery Community College*

Davis W. Pritchett, *University of Louisiana, Monroe*

David Quincey, *Bournemouth University*

S.N. Rajagopal, *University of Wisconsin, La Crosse*

Laurie L. Richardson, *Florida International University*

Luis A. Rodriguez, *San Antonio College*

Lisa Rutledge, *Columbia State Community College*

Sarmad Saman, *Massachusetts Bay Community College*

Todd Sandrin, *University of Wisconsin, Oshkosh*

Gene M. Scalarone, *Idaho State University*

Virginia Schurman, *Community College of Baltimore County, Essex*

Teri Shors, *University of Wisconsin, Oshkosh*

Edward Simon, *Purdue University*

Kevin Sorensen, *Snow College*

Angela L. Spence, *Southwest Missouri State University*

Timothy A. Steele, *Des Moines University*

Gail A. Stewart, *Camden County College*

Steven J. Thurlow, *Jackson Community College*

Michael Troyan, *Pennsylvania State University*

Jonathan Van Hamme, *The University College of the Cariboo*

Randy L. Wade, *Los Angeles Harbor College*

Musau WaKabongo, *Des Moines University*

Valerie A. Watson, *West Virginia University*

Dwight D. Wray, *Brigham Young University*

Text Consultant Panel

The following individuals provided us with invaluable feedback throughout the development of this textbook. Their candid comments helped us fine-tune the breadth and depth of the content, and we are grateful for their input.

William Boyko, *Sinclair Community College*

Genie Brackenridge, *Waubonsee Community College*

Kathryn Brooks, *Michigan State University*

Jim Collins, *University of Arizona*

Rita Connolly, *Camden County College*

Bob F. Drake, *Southwest Tennessee Community College*

S. Marvin Friedman, *Hunter College*

Jeff G. Leid, *Northern Arizona University*

William Lorowitz, *Weber State University*

Sherry Meeks, *University of Central Oklahoma*

Pamela Moolenaar-Wirsiy, *Georgia Perimeter College*

Beverly Perry, *Houston Community College*

Davis W. Pritchett, *University of Louisiana, Monroe*

Luis A. Rodriguez, *San Antonio College*

Michael Troyan, *Pennsylvania State University*

Van Wheat, *South Texas Community College*

Art Consultant Panel

We are appreciative of the thoughtful and honest feedback given to us by the art consultant panel. The individuals listed below reviewed specific figures as they were being developed and provided invaluable comments on presentation, color, style, and clarity. Their feedback has enhanced the quality and effectiveness of the art program.

Kathryn Brooks, *Michigan State University*

Kathryn Germain, *Southwest Tennessee Community College*

Todd Herman, *University of California Los Angeles*

Paulette Royt, *George Mason University*

Gail A. Stewart, *Camden County College*

Dwight D. Wray, *Brigham Young University*

Case Study Contributors

The 24 case studies that open each of the chapters in the book were put together by the authors and a team of instructors. Members of the team (besides ourselves) and the cases they contributed, are listed here.

Linda D. Bruslind, *Oregon State University*, Chapter 16

Jim Collins, *University of Arizona*, Chapter 11

Janet M. Decker, *University of Arizona*, Chapter 13

Pamela B. Fouché, *Walters State Community College*, Chapter 7

Judy Gnarpe, *University of Alberta*, Chapter 15

Dawn Janich, *Community College of Philadelphia*, Chapter 10

Karen Nakaoka, *Weber State University*, Chapters 4, 8, 9, and 14

Murad Odeh, *South Texas Community College*, Chapter 12

Todd Sandrin, *University of Wisconsin, Oshkosh*, Chapter 24

Teri Shors, *University of Wisconsin, Oshkosh*, Chapters 1, 2, 5, and 6

Timothy A. Steele, *Des Moines University*, Chapter 17

Valerie A. Watson, *West Virginia University*, Chapter 3

A Note of Thanks from Kelly Cowan

It must be obvious by now that writing a textbook is a group activity. Thank you, Kathy, for giving me the chance to apprentice with you. I am also grateful to Pat Reidy and Colin Wheatley for championing this project at McGraw-Hill. Pat Hesse and her team at McGraw-Hill—Jayne Klein, Rose Koos, Laurie Janssen, Wayne Harms, and the indomitable Tami Petsche—were truly the "Little Engine That Could," and are responsible for the book coming to fruition. Great thanks are due Indiren Pillay and Teri Shors for their extensive work on the end-of-chapter materials. Nick Nelson and Suzanne Evans, former students of mine who taught me about life as well as learning, proofread the text and the art and did a fabulous job. Their involvement in the project allowed me to sleep a little more soundly. In a much broader sense, I am personally indebted to four important mentors in my career: Ron Doyle at the University of Louisville, Madilyn Fletcher (then at) the University of Maryland, Henk Busscher at the University of Groningen, and Anne Morris Hooke at Miami University.

Finally, I want to thank my family support team, Paul, Taylor, and Sam. They gave me love, understanding, coffee—and the computer every time I asked for it.

A Note of Thanks from Kathy Talaro

I, too, owe a debt to the McGraw-Hill team, and to Kelly Cowan, for sharing her dynamic and relevant approach to microbiology. She has shaped the textbook into a unique and effective instrument that will be a valued addition to the genre. On a personal note, I wish to recognize my family for supporting and grounding me over the past many years. To the two David B.'s and Nicole: many thanks for putting up with all of that "microbiology talk" over dinner and for keeping me laughing when I most needed it.

Unique Systems-Based Approach Enhances Comprehension

Cowan/Talaro takes a *unique* approach to diseases by **consistently** covering multiple causative agents of a particular disease in the same section and summarizing this information in Checkpoint tables. The causative agents are categorized in a logical manner based on the presenting symptoms in the patient. Through this approach, students study how diseases affect patients—the way future healthcare professionals will encounter the material on the job.

considered one of the most infectious of all bacteria. The term "lawnmower" tularemia refers to tularemia acquired while performing grass-mowing or brush-cutting chores. Cases of tularemia have appeared in people who have accidentally run over dead rabbits while lawn mowing, presumably from inhaling aerosolized bacteria.

After an incubation period ranging from a few days to 3 weeks, acute symptoms of headache, backache, fever, chills, malaise, and weakness appear. Further clinical manifestations are tied to the portal of entry. They include ulcerative skin lesions, swollen lymph glands, conjunctival inflammation, sore throat, intestinal disruption, and pulmonary involvement. The death rate in the most serious forms of disease is 10%, but proper treatment with gentamycin or tetracycline reduces mortality to almost zero. Because the intracellular persistence of *F. tularensis* can lead to relapses, antimicrobial therapy must not be discontinued prematurely. Protection is available in the form of a live attenuated vaccine. Laboratory workers and other occupationally exposed personnel must wear gloves, masks, and eyewear.

CHECKPOINT 20.4	Tularemia
Causative Organism(s)	*Francisella tularensis*
Most Common Modes of Transmission	Vector, biological; also direct contact with body fluids from infected animal; airborne
Virulence Factors	Intracellular growth
Culture/Diagnosis	Culture dangerous to lab workers and not reliable; serology most often used
Prevention	Live attenuated vaccine for high-risk individuals
Treatment	Gentamycin or tetracycline

Infectious Mononucleosis

This lymphatic system disease, which is often simply called "mono" or the "kissing disease," can be caused by a number of bacteria or viruses, but the vast majority of cases are caused by the **Epstein-Barr virus (EBV)**, and most of the remainder are caused by cytomegalovirus (CMV). Both of these viruses are in the herpes family.

Signs and Symptoms

The symptoms of mononucleosis are sore throat, high fever, and cervical lymphadenopathy, which develop after a long incubation period (30 to 50 days). Many patients also have a gray-white exudate in the throat, a skin rash, and enlarged spleen and liver. A notable sign of mononucleosis is sudden leukocytosis, consisting initially of infected B cells and later T cells. Fatigue is a hallmark of the disease. Patients remain fatigued for a period of weeks. During that time, they are advised not to engage in strenuous activity due to the possibility of injuring their enlarged spleen (or liver).

Eventually, the strong, cell-mediated immune response is decisive in controlling the infection and preventing complications. But after recovery, people usually remain chronically infected with EBV and CMV.

Epstein-Barr Virus

Although "mono" was first described more than a century ago, its most frequent cause was finally discovered through a series of accidental events starting in 1958, when Michael Burkitt discovered an unusual malignant tumor in African children (Burkitt's lymphoma) that appeared to be infectious. Later, Michael Epstein and Yvonne Barr cultured a virus from tumors that showed typical herpesvirus morphology. Evidence that the two diseases had a common cause was provided when a laboratory technician accidentally acquired mononucleosis while working with the Burkitt's lymphoma virus. The Epstein-Barr virus shares morphological and antigenic features with other herpesviruses, and in addition, it contains a circular form of DNA that is readily spliced into the host cell DNA.

Scientists have long suspected a link between chronic EBV infection and illnesses such as chronic fatigue syndrome, but the connection is still controversial. In 2003, a report in the *New England Journal of Medicine* presented strong evidence that chronic EBV infection was necessary, although probably not sufficient, to cause certain forms of Hodgkin's lymphoma.

Pathogenesis and Virulence Factors The latency of the virus, and its ability to splice its DNA into host cell DNA, make it an extremely versatile virus that can avoid the host's immune response.

Transmission and Epidemiology More than 90% of the world's population is infected with EBV. In general, the virus causes no noticeable symptoms, but the time of life when the virus is first encountered seems to matter. In the case of EBV, infection during the teen years seems to result in disease, whereas infection before or after this period is usually asymptomatic. You will soon see that infection with CMV during the fetal period can lead to severe disease.

Direct oral contact and contamination with saliva are the principal modes of transmission, although transfer through blood transfusions, sexual contact, and organ transplants is possible.

Culture and Diagnosis A differential blood count that shows excess lymphocytes, reduced neutrophils, and large, atypical lymphocytes with lobulated nuclei and vacuolated cytoplasm is suggestive of EBV infection (figure 20.6). A test called the "Monospot test" detects *heterophile antibodies*—which are antibodies that are not directed against EBV but are seen when a person has an EBV infection. This test is not reliable in children younger than age 4, in which case a specific EBV antigen/antibody test is conducted.

Summary Checkpoint Tables

Following the textual discussion of each disease, a table summarizes the characteristics of agents that can cause that disease.

Consistent, Clinical Presentation of Diseases

For each disease, the discussion begins with an introduction to the disease and its signs and symptoms. Next, the causative agent or agents of that disease are presented and the following areas are discussed:

- pathogenesis and virulence factors
- transmission and epidemiology
- culture and diagnosis
- prevention and treatment

Prevention and Treatment The usual treatments for infectious mononucleosis are directed at symptomatic relief of fever and sore throat. Hospitalization is rarely needed. Occasionally, rupture of the spleen necessitates immediate surgery to remove it.

Cytomegalovirus

Cytomegalovirus (CMV) is also a herpesvirus. It is generally distinguished by its ability to produce giant (megalo) cells (cyto) with nuclear and cytoplasmic inclusion bodies. Like other herpesviruses, both EBV and CMV have a tendency to become latent in host cells. Infections are likely to be permanent. The viruses do not reemerge the way herpes simplex viruses do, unless a patient becomes severely immunocompromised. (CMV ocular symptoms are a common complication of AIDS—affecting up to 40% of all AIDS patients.)

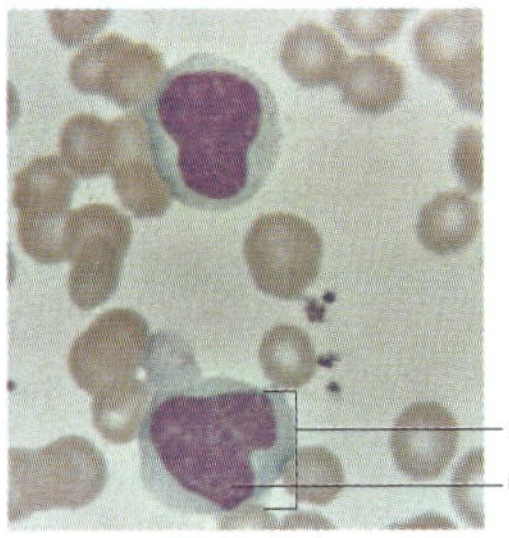

Pathogenesis and Virulence Factors The ability of the virus to fuse cells and its latency both contribute to its virulence.

Transmission and Epidemiology Like EBV, CMV is ubiquitous in humans. Unlike EBV, CMV generally causes disease only in fetuses, newborns, and immunodeficient adults. Although not covered here, CMV infection of fetuses affects up to 5,000 babies a year and can cause long-term neurological and sensory disturbances.

CMV is transmitted in saliva, respiratory mucus, milk, urine, semen, cervical secretions, and feces. Transmission usually involves intimate contact such as sex, vaginal birth, transplacental infection, blood transfusion, and organ transplantation.

Culture and Diagnosis During CMV mononucleosis, the virus can be isolated from virtually all organs as well as from epithelial tissue. Cell enlargement and prominent inclusions in the cytoplasm and nucleus are suggestive of CMV. The virus can be cultured and tested with monoclonal antibody against a CMV protein called *early nuclear antigen*. Direct ELISA tests and DNA probe analysis are also useful in diagnosis. Testing serum for antibodies may fail to diagnose infection in neonates and in the immunocompromised, so it is less reliable.

Prevention and Treatment Drug therapy is generally reserved for serious disease in immunosuppressed patients, and not for CMV mononucleosis. The three main drugs are ganciclovir, valacyclovir, and foscarnet, which have toxic side effects and cannot be administered for long periods. The development of a vaccine is hampered by the lack of an animal that can be infected with human cytomegalovirus. One crucial concern is whether vaccine-stimulated antibodies would be protective, since patients already seropositive can become naturally reinfected. Despite these odds, clinical trials began in late 2003 for an experimental CMV vaccine.

FIGURE 20.6 Evidence of Epstein-Barr infection in the blood smear of a patient with infectious mononucleosis. Note the abnormally large lymphocytes containing indented nuclei with light discolorations.

(Labels: Lymphocyte, Nucleus)

CHECKPOINT 20.5	Infectious Mononucleosis	
Causative Organism(s)	Epstein-Barr virus (EBV)	Cytomegalovirus (CMV)
Most Common Modes of Transmission	Direct, indirect contact, parenteral	Direct, indirect contact, parenteral, vertical
Virulence Factors	Latency, ability to incorporate into host DNA	Latency, ability to fuse cells
Culture/Diagnosis	Differential blood count, Monospot test for heterophile antibody, specific ELISA	Virus isolation and growth, ELISA or PCR tests
Prevention	–	Vaccine in trials
Treatment	Supportive	Only for immunosuppressed patients, not usually for mononucleosis
Distinctive Features	Most common in teens	More common in adults, dangerous to fetus

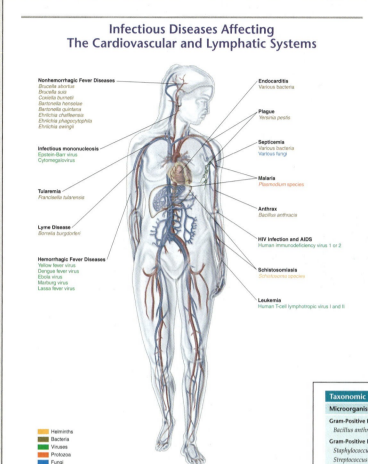

Infectious Diseases Affecting The Cardiovascular and Lymphatic Systems

Nonhemorrhagic Fever Diseases
Brucella abortus
Brucella suis
Coxiella burnetii
Bartonella henselae
Bartonella quintana
Ehrlichia chaffeensis
Ehrlichia phagocytophila
Ehrlichia ewingii

Infectious mononucleosis
Epstein-Barr virus
Cytomegalovirus

Tularemia
Francisella tularensis

Lyme Disease
Borrelia burgdorferi

Hemorrhagic Fever Diseases
Yellow fever virus
Dengue fever virus
Ebola virus
Marburg virus
Lassa fever virus

Endocarditis
Various bacteria

Plague
Yersinia pestis

Septicemia
Various bacteria
Various fungi

Malaria
Plasmodium species

Anthrax
Bacillus anthracis

HIV Infection and AIDS
Human immunodeficiency virus 1 or 2

Schistosomiasis
Schistosoma species

Leukemia
Human T-cell lymphotropic virus I and II

Helminths
Bacteria
Viruses
Protozoa
Fungi

Summary Figures

After the diseases of a particular body system have been discussed, students are invited to study the **summary figure** at the end of the chapter—a "glass body" that highlights the affected organs and lists the diseases that were presented in the chapter. In addition, the microbes that could cause the diseases are *color-coded by type of microorganism.*

This summary figure, along with the checkpoint tables, provide an excellent set of study tools.

Taxonomic List of Organisms

A **taxonomic list of organisms** is also presented at the end of each disease chapter, so students can see the diversity of microbes causing diseases in that system.

Taxonomic Organization of Microorganisms Causing Disease in the Cardiovascular and Lymphatic System		
Microorganism	**Disease**	**Chapter Location**
Gram-Positive Endospore-Forming Bacteria		
Bacillus anthracis	Anthrax	Anthrax, p. 635
Gram-Positive Bacteria		
Staphylococcus aureus	Acute endocarditis	Endocarditis, p. 617
Streptococcus pyogenes	Acute endocarditis	Endocarditis, p. 617
Streptococcus pneumoniae	Acute endocarditis	Endocarditis, p. 617
Gram-Negative Bacteria		
Yersinia pestis	Plague	Plague, p. 619
Francisella tularensis	Tularemia	Tularemia, p. 621
Borelia burgdorferi	Lyme disease	Lyme disease, p. 624
Brucella abortus, B. suis	Brucellosis	Nonhemorrhagic fever diseases, p.628
Coxiella burnetii	Q fever	Nonhemorrhagic fever diseases, p. 628
Bartonella henselae	Cat-scratch disease	Nonhemorrhagic fever diseases, p. 629
Bartonella quintana	Trench fever	Nonhemorrhagic fever diseases, p. 629
Ehrlichia chaffeensis, E. phagocytophila, E. ewingii	Ehrlichiosis	Nonhemorrhagic fever diseases, p. 629
Neisseria gonorrhoeae	Acute endocarditis	Endocarditis, p. 617
Rickettsia rickettsii	Rocky mountain spotted fever	Nonhemorrhagic fever diseases, p. 630
DNA Viruses		
Epstein-Barr virus	Infectious mononucleosis	Infectious mononucleosis, p. 622
Cytomegalovirus	Infectious mononucleosis	Infectious mononucleosis, p. 622
RNA Viruses		
Yellow fever virus	Yellow fever	Hemorrhagic fevers, p. 626
Dengue fever virus	Dengue fever	Hemorrhagic fevers, p. 627
Ebola and Marburg viruses	Ebola and Marburg hemorrhagic fevers	Hemorrhagic fevers, p. 627
Lassa fever virus	Lassa fever	Hemorrhagic fevers, p. 627
Retroviruses		
Human immunodeficiency virus 1 and 2	HIV infection and AIDS	HIV infection and AIDS, p. 636
Human T-cell lymphotropic virus I	Adult T-cell leukemia	Leukemias, p. 645
Human T-cell lymphotropic virus II	Hairy-cell leukemia (?)	Leukemias, p. 645
Protozoa		
Plasmodium falciparum, P. vivax, P. ovale, P. malariae	Malaria	Malaria, p. 631

Instructional Art Program Clarifies Concepts

Cowan/Talaro's *Microbiology: A Systems Approach* provides visually powerful artwork that paints conceptual pictures for students. The art combines vivid colors, multi-dimensionality, and self-contained, concept-specific narrative to help students study the challenging concepts of microbiology from a visual perspective—a proven study technique.

Process Figures

Cowan/Talaro illustrates many difficult microbiology concepts in steps that students find easy to follow. Each step is clearly illustrated in the figure and correlated to accompanying narrative to benefit all types of learners.

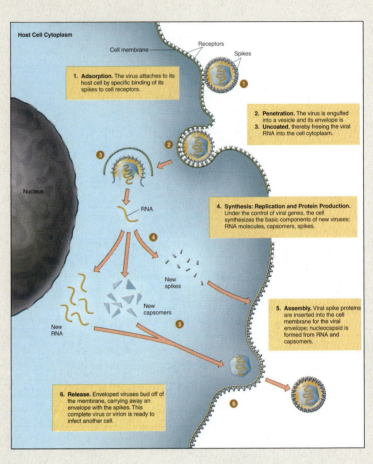

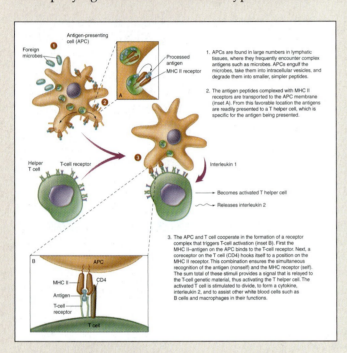

Combination Figures

Line art combined with photos gives students two perspectives: the realism of photos and the explanatory clarity of line drawings.

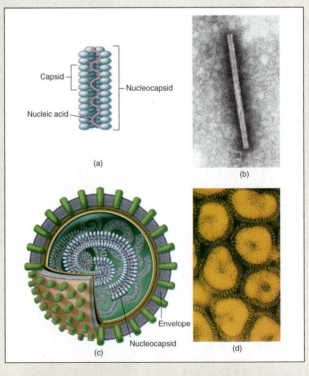

> *"The artwork, tables, and figures for this text are well done, giving the students great visual help in understanding the concepts."*
>
> Karen G. Nakaoka, Weber State University

Clinical Photos

Color photos of individuals affected by disease provide students with a real-life, clinical view of how microorganisms manifest themselves in the human body.

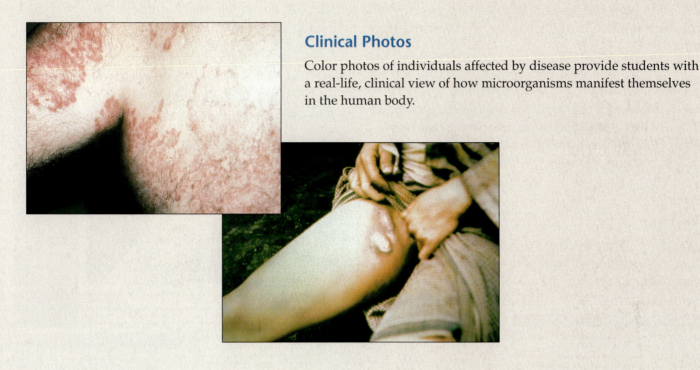

Overview Figures

Many challenging concepts of microbiology consist of numerous interrelated activities. Cowan/Talaro visually summarizes these concepts to help students piece the activities together for a complete, conceptual picture.

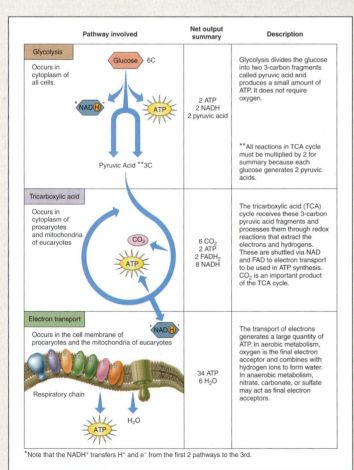

Pedagogical Aids Promote Systematic Learning

Cowan/Talaro organizes each chapter with consistent pedagogical tools. Such tools enable students to develop a consistent learning strategy and enhance their understanding of the concepts.

In the News

All chapters open with a real-world case study to help students appreciate and understand how microbiology impacts lives on a daily basis. The solution to the case study appears later in the chapter, after necessary elements have been presented.

Checkpoints

Major sections of all chapters end with a summary of the significant concepts covered. In the disease chapters (18-23) the Checkpoints take the form of tables that summarize the characteristics of the infectious agent(s) discussed.

✔ CHECKPOINT

- Bioenergetics describes metabolism in terms of production, utilization, and transfer of energy by cells.
- Catabolic pathways release energy through three pathways: glycolysis, the tricarboxylic acid cycle, and the respiratory electron transport system.
- Cellular respiration is described by the nature of the final electron acceptor. Aerobic respiration implies that O_2 is the final electron acceptor. Anaerobic respiration implies that some other molecule is the final electron acceptor. If the final electron acceptor is an organic molecule, the anaerobic process is considered fermentation.
- Carbohydrates are preferred cell energy sources because they are superior hydrogen (electron) donors.
- Glycolysis is the catabolic process by which glucose is oxidized and converted into two molecules of pyruvic acid, with a net gain of 2 ATP. The formation of ATP is via substrate-level phosphorylation.
- The tricarboxylic acid cycle processes the 3-carbon pyruvic acid and generates three CO_2 molecules. The electrons it releases are transferred to redox carriers for energy harvesting. It also generates 2 ATPs.
- The electron transport chain generates free energy through sequential redox reactions collectively called oxidative phosphorylation. This energy is used to generate up to 38 ATP for each glucose molecule catabolized.

Insight Readings

Current, real-world readings allow students to consider applications of the concepts they are studying. The Insight readings are divided into four categories: Discovery, Historical, Medical, and Microbiology.

Microbial Metabolism
The Chemical Crossroads of Life

CHAPTER 8

IN THE NEWS

A 65-year-old female was admitted to the hospital with labored breathing, hiccups, neck and jaw stiffness, and abdominal pain, with generalized rigidity of her abdominal muscles. Although the patient had no fever, she had an elevated white blood cell (WBC) count. One week earlier, she had been treated for a knee laceration, at which time her wound was cleansed and sutured. At that time, she was given a tetanus toxoid injection and antibiotic therapy with cephalexin.

Muscle spasms progressed, and the patient was transferred to the ICU. She was unable to speak because of rigidity of the muscles of the face, neck, and jaw. The patient was given tetanus immune globulin. Because of continued breathing difficulties, she was intubated, given mechanical ventilation and metronidazole. Lorazepam to control the spasms, and fentanyl for pain management of the muscle spasms, were administered. The sutures from the knee were removed and the wound debrided. Coagulase-negative *Staphylococcus* grew from the wound exudate. After 12 days in the ICU, she was given a tracheostomy. Over time, the patient... tor. On day 46, she was transferred to a rehabilitation...

- What disease do you suspect this patient had?
- Was the coagulase-negative Staphylococcus, whic... causing muscle weakness and difficulty breathing?
- Was the environment in this infected wound anae...

CHAPTER OVERVIEW

- Cells are constantly involved in an orderly activity called metabolism that encompasses all of their chemical and energy transactions.
- Enzymes are essential metabolic participants that drive cell reactions.

IN THE NEWS *(Continued from page 213)*

The case described at the beginning of the chapter was caused by *Clostridium tetani*. Even though an obligately anaerobic organism was not cultured from the wound, the wound evidently, was infected with *Clostridium tetani* as a result of the traumatic laceration. The conditions in the wound were anaerobic enough to allow for growth of these bacilli. It is even probable that the coagulase-negative *Staphylococcus* consumed enough oxygen to allow for the tetanus bacilli's growth, resulting in the production of tetanus toxin, which causes muscle spasms.

Fortunately, this is a rare disease in developed areas of the world in which the tetanus toxoid vaccine is widely used. When tetanus does occur in this setting, it usually occurs in people who are either not immunized or in the elderly who have not received adequate boosters of the vaccine.

See: Bunch, T. J. et al. 2002. Respiratory failure in tetanus: Case report. Chest 122:1488–1492.

INSIGHT 6.3   *Discovery*

Artificial Viruses Created!

Newspapers are filled with stories of the debate over the ethics of creating life through cloning techniques. Dolly the cloned sheep and the cattle, swine, and goats that have followed in her footsteps have raised ethical questions about scientists "playing God," when they harvest genetic material from an animal and create an identical organism from it, as is the case with cloning.

Meanwhile, in a much less publicized event in 2002, scientists at the State University of New York at Stony Brook succeeded in artificially creating a virus that is virtually identical to natural poliovirus. They used DNA nucleotides they bought "off the shelf" and put them together according to the published poliovirus sequence. They then added an enzyme that would transcribe the DNA sequence into the RNA genome used by poliovirus. They ended up with a virus that was nearly identical to poliovirus (see illustration), with a similar capsid as well as a similar ability to infect host cells and reproduce itself.

The creation of the virus was greeted with controversy, particularly because poliovirus is potentially devastating to human health. The scientists, who were working on a biowarfare defense project funded by the Department of Defense, argued that they were demonstrating what could be accomplished if information and chemicals fell into the wrong hands.

In 2003, another lab in Rockville, Maryland, manufactured a "working" bacteriophage, a harmless virus called phi X. Their hope is to create microorganisms from which they can harness energy—for use as a renewable energy source.

Both of these viruses have tiny genomes compared with higher organisms: phi X has 5,400 nucleotide base pairs, and the RNA genome of polio is only 7,500 bases long. Similar duplication of complex cells and organisms is not yet possible. (Even the single-celled bacteria typically have millions of base pairs in their DNA, and the human genome consists of 3 billion base pairs.) But the prospect of harmful misuse of the new technology has prompted scientific experts to team with national security and bioethics experts to discuss the pros and cons of the new technology, and ways to ensure its acceptable uses.

"The general style of writing, the pedagogic constructs used. . . and linked boxed texts, are all to be applauded."

David Quincey, Bournemouth University

Chapter Summary with Key Terms

A brief outline of the main chapter concepts is provided for students, and important terms are highlighted.

Multiple-Choice Questions

Students can assess their knowledge by answering this set of questions.

Concept Questions

Suggested as a "writing-to-learn" experience, students are asked to address the concept questions with one- or two-paragraph answers.

Critical Thinking Questions

Using the facts and concepts they just studied, students must reason and problem-solve to answer the critical-thinking questions. Such questions do not have a single correct answer, and thus, open the discussion.

Internet Search Topics

Opportunities for further research into the concepts just covered are outlined at the end of each chapter, and students are directed to the Cowan/Talaro Online Learning Center to further their studies.

"I like the end-of-chapter questions, especially the critical thinking which goes beyond what one finds in most texts."

Janet M. Decker, University of Arizona

Chapter Summary With Key Terms

9.1 Introduction to Genetics and Genes: Unlocking the Secrets of Heredity

A. **Genetics** is the study of **heredity** and can be studied at the level of the organism, **genome, chromosome, gene,** and **DNA.** Genes provide the information needed to construct **proteins,** which have structural or catalytic functions in the cell.

B. **DNA** is a long molecule in the form of a double helix. Each strand of the helix consists of a string of **nucleotides** which form hydrogen bonds with their counterparts on the other strand. **Adenine** base pairs with **thymine** while **guanine** base pairs with **cytosine.** The order of the nucleotides specifies which amino acids will be used to construct proteins during the process of translation.

C. DNA **replication** is **semiconservative** and requires the participation of several enzymes.

9.2 Applications of the DNA Code: Transcription and Translation

A. DNA is used to produce RNA **(transcription)** and RNA is then used to produce protein **(translation).**

B. **RNA:** Unlike DNA, RNA is single stranded, contains **uracil** instead of thymine and **ribose** instead of **deoxyribose.**

1. Major forms of RNA found in the cell include **mRNA, tRNA,** and **rRNA.**
2. The genetic information contained in DNA is copied to produce an RNA molecule. **Codons** in the mRNA pair with **anticodons** in the tRNA to specify what amino acids to assemble on the ribosome during translation.

C. *Transcription and Translation:* Transcription occurs when **RNA polymerase** copies the template strand of a segment of DNA. RNA is always made in the 5' to 3' direction. Translation occurs when the mRNA is used to direct the synthesis of proteins on the ribosome. Codons in the mRNA pair with anticodons in the tRNA to assemble a string of amino acids. This occurs until a stop codon is reached.

D. *Eucaryotic Gene Expression:* Eucaryotic genes are composed of **exons** (expressed sequences) and **introns** (intervening sequences). The introns must be removed and the exons spliced together to create the final mRNA.

E. The genetics of viruses is quite diverse.

1. Genomes of viruses are found in many physical forms not seen in cells, including dsDNA, ssDNA, dsRNA, and ssRNA.
2. DNA viruses tend to replicate in the nucleus while RNA viruses replicate in the cytoplasm. Retroviruses synthesize dsDNA from ssRNA.

9.3 Genetic Regulation of Protein Synthesis and Metabolism

Protein synthesis is regulated through gene induction or repression, as controlled by an **operon.** Operons consist of several structural genes controlled by a common regulatory element.

A. *Inducible operons* such as the lactose operon are normally off but can be turned on in the presence of lactose.

B. *Repressible operons* are normally on but can be turned off when their end product is present.

C. Many antibiotics prevent growth by interfering with transcription or translation.

9.4 Mutations: Changes in the

A. Permanent changes ... are known as **muta...** **spontaneous** or **ind...**

B. **Point mutations** ent... and are categorized ... **back-mutations,** bas... nucleotide(s).

C. Many mutations, pa... mismatched bases ar... be corrected using e...

D. The **Ames test** meas... by determining the a... mutations in bacteria...

9.5 DNA Recombination E...

A. Intermicrobial trans... permit gene sharing... recombination inclu... and **transduction.**

B. **Transposons** are DN... to different places w... consequence genera... chromosome structu...

Multiple-Choice Questions

1. What is the smallest unit of heredity?
   - a. chromosome
   - b. gene
   - c. codon
   - d. nucleotide

2. A nucleotide contains which of the following?
   - a. 5 C sugar
   - b. nitrogen base
   - c. phosphate
   - d. b and c only
   - e. all of these

3. The nitrogen bases in DNA are bonded to the
   - a. phosphate
   - b. deoxyribose
   - c. ribose
   - d. hydrogen

4. DNA replication is semicons... will become half of the ...
   - a. RNA, DNA
   - b. template, finished

5. In DNA, adenine is the com... cytosine is the complement...
   - a. guanine, thymine
   - b. uracil, guanine

---

6. The base pairs are held together primarily by
   - a. covalent bonds
   - b. hydrogen bonds
   - c. ionic bonds
   - d. gyrases

7. Why must the lagging strand of DNA be replicated in short pieces?
   - a. because of limited space
   - b. otherwise, the helix will become distorted
   - c. the DNA polymerase can synthesize in only one direction
   - d. to make proofreading of code easier

8. Messenger RNA is formed by _____ of a gene on the DNA template strand.
   - a. transcription
   - b. replication
   - c. translation
   - d. transformation

9. Transfer RNA is the molecule that
   - a. contributes to the structure of ribosomes
   - b. adapts the genetic code to protein structure
   - c. transfers the DNA code to mRNA
   - d. provides the master code for amino acids

10. As a general rule, the template strand on DNA will always begin with a/an
    - a. TAC
    - b. AUG
    - c. ATG
    - d. UAC

11. The *lac* operon is usually in the _____ position and is activated by a/an _____ molecule.
    - a. on, repressor
    - b. off, inducer
    - c. on, inducer
    - d. off, repressor

12. For mutations to have an effect on populations of microbes, they must be
    - a. inheritable
    - b. permanent
    - c. beneficial
    - d. a and b
    - e. all of the above

13. Which of the following characteristics is *not* true of a plasmid?
    - a. It is a circular piece of DNA.
    - b. It is required for normal cell function.
    - c. It is found in bacteria.
    - d. It can be transferred from cell to cell.

14. Which genes can be transferred by all three methods of intermicrobial transfer?
    - a. capsule production
    - b. toxin production
    - c. F factor
    - d. drug resistance

15. Which of the following would occur through specialized transduction?
    - a. acquisition of Hfr plasmid
    - b. transfer of genes for toxin production
    - c. transfer of genes for capsule formation
    - d. transfer of a plasmid with genes for degrading pesticides

16. Multiple Matching. Fill in the blanks with all the letters of the words below that apply.
    - _____ genetic transfer that occurs after the donor is dead
    - _____ carries the codon
    - _____ carries the anticodon
    - _____ a process synonymous with mRNA synthesis
    - _____ bacteriophages participate in this transfer
    - _____ duplication of the DNA molecule
    - _____ process in which transcribed DNA code is deciphered into a polypeptide
    - _____ involves plasmids
    - a. replication
    - b. tRNA
    - f. mRNA

Concept Questions

These questions are suggested as a *writing-to-learn* expe... For each question, compose a one- or two-paragraph an... includes the factual information needed to completely a... question.

1. Compare the genetic material of eucaryotes, bacteri... viruses in terms of general structure, size, and mode... replication.

2. Briefly describe how DNA is packaged to fit inside a...

3. Describe what is meant by the antiparallel arrangem... DNA.

4. On paper, replicate the following segment of DNA:

   5' ATCGGCTACGTTCAC3'
   3' TAGCCGATGCAAGTG5'

   a. Show the direction of replication of the new strand... explain what the lagging and leading strands are...
   b. Explain how this is semiconservative replication... new strands identical to the original segment of ...

---

12. What are the functions of start and stop codons? Give examples of them.

13. The following sequence represents triplets on DNA:

    TAC CAG ATA CAC TCC CCT GCG ACT

    a. Give the mRNA codons and tRNA anticodons that correspond with this sequence, and then give the sequence of amino acids in the polypeptide.
    b. Provide another mRNA strand that can be used to synthesize this same protein.
    c. Looking at figure 9.14, give the type and order of the amino acids in the peptide.

14. a. Summarize how bacterial and eucaryotic cells differ in gene structure, transcription, and translation.
    b. Discuss the roles of exons and introns.

15. a. What is an operon? Describe the functions of regulators, promoters, and operators.
    b. Compare and contrast the *lac* operon with a repressible operon system.

16. Explain the ideas behind the Ames test and what it is used for.

17. Describe the principal types of mutations. Give an example of a mutation that is beneficial and one that is lethal or harmful.

18. a. Compare conjugation, transformation, and transduction on the basis of general method, nature of donor, and nature of recipient.
    b. Explain the differences between general and specialized transduction, using drawings.

19. By means of a flowchart, show the possible jumps that a transposon can make. Show the involvement of viruses in its movement.

Critical Thinking Questions

Critical thinking is the ability to reason and solve problems using facts and concepts. These questions can be approached from a number of angles, and in most cases, they do not have a single correct answer.

1. Knowing that retroviruses operate on the principle of reversing the direction of transcription from RNA to DNA, propose a drug that might possibly interfere with their replication.

2. Using the piece of DNA in concept question 13, show a deletion, an insertion, a substitution, and nonsense mutations. Which ones are frameshift mutations? Are any of your mutations nonsense? Missense? (Use the universal code to determine this.)

3. Using figure 9.14 and table 9.4, go through the steps in mutation of a codon followed by its transcription and translation that will give the end result in silent, missense, and nonsense mutations. Why is a change in the RNA code alone not really a mutation?

4. Explain the principle of "wobble" and find four amino acids that are encoded by wobble bases (figure 9.14). Suggest some benefits of this phenomenon to microorganisms.

5. The enzymes required to carry out transcription and translation are themselves produced through these same processes. Speculate which may have come first in evolution—proteins or nucleic acids—and explain your choice.

6. Why can one not reliably predict the sequence of nucleotides on mRNA or DNA by observing the amino acid sequence of proteins?

7. Speculate on the manner in which transposons may be involved in cancer.

8. Explain what is meant by the expression: phenotype = genotype + environment.

Try this

A simple test you can do to demonstrate the coiling of DNA in bacteria is to open a large elastic band, stretch it taut, and twist it. First it will form a loose helix, then a tighter helix, and finally, to relieve stress, it will twist back upon itself. Further twisting will result in a series of knotlike bodies; this is how bacterial DNA is condensed.

Internet Search Topics

1. Do an Internet search under the heading "DNA music." Explore several websites, discovering how this music is made and listening to some examples.

2. Find information on introns. Explain at least five current theories as to their possible functions.

3. Go to the Online Learning Center for chapter 9 of this text at http://www.mhhe.com/cowan1. Access the URLs listed under Internet Search Topics and research the following:

   Log on to one or more of the listed URLs and locate animations, three-dimensional graphics, and interactive tutorials that help you to visualize replication, transcription, and translation.

The Main Themes of Microbiology

IN THE NEWS

Severe acute respiratory syndrome (SARS) is a newly identified respiratory infection caused by a novel coronavirus. The SARS pandemic is believed to have originated in the Guangdong Province of China during the fall of 2002. A SARS patient from this region traveled to Hong Kong on February 15th, 2003, and may have infected several guests at a hotel where he resided. One of the hotel guests was a resident of Hong Kong. By February 24th, the hotel resident came down with a fever, chills, dry cough, runny nose, and malaise. Over the next several days, his symptoms worsened to pneumonia, leading to his hospitalization at the Prince of Wales Hospital in Hong Kong.

The Prince of Wales Hospital is a large medical teaching hospital of the Chinese University of Hong Kong. By March 12th, a large-scale outbreak of SARS occurred inside of the hospital. During the initial outbreak, March 15th through 25th, 2003, 44% of the SARS cases (68 of 156) admitted to the Prince of Wales Hospital were hospital workers. SARS is a contagious disease that spreads from person to person primarily through contact with respiratory droplets containing the SARS virus. Chinese University researchers and the Hong Kong Hospital Authority conducted studies to determine why hospital workers were so vulnerable to SARS at this hospital.

► *Can you think of what factors contributed to the increased rates of SARS transmission seen among hospital workers?*

► *What precautions would you take in caring for SARS patients?*

CHAPTER OVERVIEW

► Microorganisms, also called microbes, are organisms that require a microscope to be readily observed.

► In terms of numbers and range of distribution, microbes are the dominant organisms on earth.

► Major groups of microorganisms include bacteria, algae, protozoa, fungi, parasitic worms, and viruses.

► Microbiology involves study in numerous areas involving cell structure, function, genetics, immunology, biochemistry, epidemiology, and ecology.

► Microorganisms are essential to the operation of the earth's ecosystems, as photosynthesizers, decomposers, and recyclers.

▶ Humans use the versatility of microbes to make improvements in industrial production, agriculture, medicine, and environmental protection.

▶ The beneficial qualities of microbes are in contrast to the many infectious diseases they cause.

▶ Microorganisms are the oldest organisms, having evolved over 3.5 billion years of earth's history to the modern varieties we now observe.

▶ Microbiologists use the scientific method to develop theories and explanations for microbial phenomena.

▶ The history of microbiology is marked by numerous significant discoveries and events in microscopy, culture techniques, and other methods of handling or controlling microbes.

▶ Microbes are classified into groups according to evolutionary relationships, provided with standard scientific names, and identified by specific characteristics.

1.1 The Scope of Microbiology

Microbiology is a specialized area of biology that deals with living things ordinarily too small to be seen without magnification. Such **microscopic** organisms are collectively referred to as **microorganisms** (my"-kroh-or'-gun-izms), **microbes,** or several other terms, depending upon the purpose. Some people call them germs or bugs in reference to their role in infection and disease, but those terms have other biological meanings and perhaps place undue emphasis on the disagreeable reputation of microorganisms. There are several major groups of microorganisms that we'll be studying. They are **bacteria, viruses, fungi, protozoa, algae,** and **helminths** (parasitic worms). They all have different biological characteristics. The nature of microorganisms makes them both very easy and very difficult to study. Easy, because they reproduce so rapidly and we can quickly grow large populations in the laboratory. Difficult, because we can't see them directly. We rely on a variety of indirect means of analyzing them in addition to using microscopes.

Microbiology is one of the largest and most complex of the biological sciences because it includes many diverse biological disciplines. Microbiologists study every aspect of microbes—their genetics, their physiology, their characteristics that may cause disease or lead to benefits, the way they interact with the environment, the way they interact with mammalian hosts, and their uses in industry and agriculture.

Here are some descriptions of different branches of study within microbiology.

Immunology studies the complex web of immune chemicals and cells that are produced in response to infection. It also concerns itself with the study of *hypersensitivity*, inappropriate immune responses that can be harmful to the human host. *Allergy* is one example of hypersensitivity (see chapter 16).

Public health microbiology and epidemiology aim to monitor and control the spread of diseases in communities. The principal U.S. and global institutions involved in this concern are the United States Public Health Service (USPHS) with its main agency, the Centers for Disease Control and Prevention (CDC) located in Atlanta, Georgia, and the World Health Organization (WHO), the medical limb

of the United Nations. The CDC collects information on disease from around the United States and publishes it in a weekly newsletter called the *Morbidity and Mortality Weekly Report*. The CDC website, www.cdc.gov, is also one of the most reliable sources of information about diseases you will study in this book.

Food microbiology, dairy microbiology, and aquatic microbiology examine the ecological and practical roles of microbes in food and water (see chapter 24).

Agricultural microbiology is concerned with the relationships between microbes and crops, with an emphasis on improving yields and combating plant diseases.

Biotechnology includes any process in which humans use the metabolism of living things to arrive at a desired product, ranging from bread making to gene therapy. It is a tool used in industrial microbiology, which is concerned with the uses of microbes to produce or harvest large quantities of substances such as beer, vitamins, amino acids, drugs, and enzymes (see chapters 10 and 24).

Genetic engineering and recombinant DNA technology involve techniques that deliberately alter the genetic makeup of organisms to mass-produce human hormones and other drugs, create totally new substances, and develop organisms with unique methods of synthesis and adaptation. This is the most powerful and rapidly growing area in modern microbiology (see chapter 10).

Each of the major disciplines in microbiology contains numerous subdivisions or specialties that in turn deal with a specific subject area or field. In fact, many areas of this science have become so specialized that it is not uncommon for a microbiologist to spend his or her whole life concentrating on a single group or type of microbe, biochemical process, or disease. On the other hand, rarely is one person a single type of microbiologist, and most can be classified in several ways. There are, for instance, bacterial physiologists who study industrial processes, molecular biologists who focus on the genetics of viruses, epidemiologists who are also nurses, and dentists who specialize in the microbiology of gum disease.

Studies in microbiology have led to greater understanding of many general biological principles. For example, the

FIGURE 1.1   Evolutionary timeline.
The first bacteria appeared approximately 3.5 billion years ago. They were the only form of life for half of the earth's history.

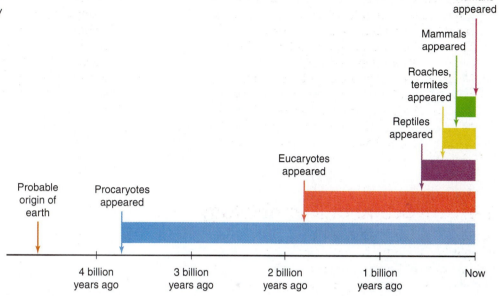

1.2 The Impact of Microbes on Earth: Small Organisms with a Giant Effect

The most important knowledge that should emerge from a microbiology course is the profound influence microorganisms have on all aspects of the earth and its residents. For billions of years, microbes have extensively shaped the development of the earth's habitats and the evolution of other life forms. It is understandable that scientists searching for life on other planets first look for signs of microorganisms.

Bacterial-type organisms have been on this planet for about 3.5 billion years, according to the fossil record. It appears that they were the only living inhabitants on earth for almost 2 billion years. At that time (about 1.8 billion years ago) a more complex type of single-celled organism arose, of a **eucaryotic** (yoo"-kar-ee-ah'-tik) cell type. Eu-cary means *true nucleus,* which gives you a hint that those first inhabitants, the bacteria, had no true nucleus. For that reason they are called **procaryotes** (proh"-kar-ee-otes) (pre-nucleus). The early eucaryotes were the precursors of the cell type that eventually formed multicellular animals, including humans. But you can see from **figure 1.1** how long that took! On the scale pictured in the figure, humans seem to have just appeared. The bacteria preceded even the earliest animals by about 3 billion years. This is a good indication that humans are not likely to, nor should we try to, eliminate bacteria from our environment. They've survived and adapted to many catastrophic changes over the course of their geologic history.

Another indication of the huge influence bacteria exert is how **ubiquitous** they are. Microbes can be found nearly everywhere, from deep in the earth's crust, to the polar ice caps and oceans, to the bodies of plants and animals. Being mostly invisible, the actions of microorganisms are usually not as obvious or familiar as those of larger plants and animals. They make up for their small size by occurring in large numbers and living in places that many other organisms cannot survive. Above all, they play central roles in the earth's landscape that are essential to life.

Microbial Involvement in Energy and Nutrient Flow

Microbes are deeply involved in the flow of energy and food through the earth's ecosystems.[1] Most people are aware that plants carry out **photosynthesis,** which is the light-fueled conversion of carbon dioxide to organic material, accompanied by the formation of oxygen. But microorganisms were photosynthesizing long before the first plants appeared. In fact, they were responsible for changing the atmosphere of the earth from one without oxygen, to one with oxygen. Today photosynthetic microorganisms (including algae) account for more than 50% of the earth's photosynthesis, contributing the majority of the oxygen to the atmosphere **(figure 1.2a).**

Another process that helps keep the earth in balance is the process of biological **decomposition** and nutrient recycling. Decomposition involves the breakdown of dead matter and wastes into simple compounds that can be directed back into the natural cycles of living things **(figure 1.2b).** If it were not for multitudes of bacteria and fungi, many chemical elements would become locked up and unavailable to

study of microorganisms established universal concepts concerning the chemistry of life (see chapters 2 and 8), systems of inheritance (see chapter 9), and the global cycles of nutrients, minerals, and gases (see chapter 24).

---

1. Ecosystems are any interactions that occur between living organisms and their environment.

(a) Summer pond with a thick mat of algae—a rich photosynthetic community.

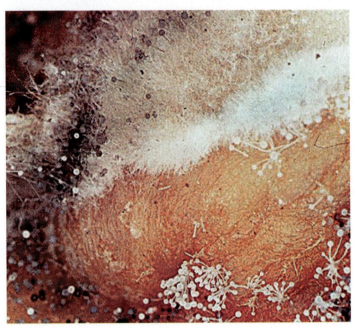

(b) An orange being decomposed by a common soil fungus.

FIGURE 1.2   Microbial habitats.

organisms. In the long-term scheme of things, microorganisms are the main forces that drive the structure and content of the soil, water, and atmosphere. For example:

- The very temperature of the earth is regulated by "greenhouse gases," such as carbon dioxide and methane, that create an insulation layer in the atmosphere and help retain heat. Much of this gas is produced by microbes living in the environment and the digestive tracts of animals.
- Recent estimates propose that, based on weight and numbers, up to 50% of all organisms exist within and beneath the earth's crust in sediments, rocks, and even volcanoes. It is increasingly evident that this enormous underground community of microbes is a significant influence on weathering, mineral extraction, and soil formation.
- Bacteria and fungi live in complex associations with plants that assist the plants in obtaining nutrients and water and may protect them against disease. Microbes form similar interrelationships with animals, notably in the stomach of cattle, where a rich assortment of bacteria digest the complex carbohydrates of the animals' diets.

1.3 Human Use of Microorganisms

Microorganisms clearly have monumental importance to the earth's operation. It is this very same diversity and versatility that also makes them excellent candidates for solving human problems. By accident or choice, humans have been using microorganisms for thousands of years to improve life and even to shape civilizations. Yeasts, a type of microscopic fungi, cause bread to rise and ferment sugar into alcohol to make wine and beers. These and other "home" uses of microbes have been in use for thousands of years. For example,

historical records show that households in ancient Egypt kept moldy loaves of bread to apply directly to wounds and lesions. When humans manipulate microorganisms to make products in an industrial setting, it is called biotechnology. For example, some specialized bacteria have unique capacities to mine precious metals **(figure 1.3a)** or to produce enzymes that are used in laundry detergents.

Genetic engineering is a newer area of biotechnology that manipulates the genetics of microbes, plants, and animals for the purpose of creating new products and genetically modified organisms. One powerful technique for designing new organisms is termed **recombinant DNA.** This technology makes it possible to deliberately alter DNA[2] and to switch genetic material from one organism to another. Bacteria and fungi were some of the first organisms to be genetically engineered, because they are so adaptable to changes in their genetic makeup. Recombinant DNA technology has unlimited potential in terms of medical, industrial, and agricultural uses. Microbes can be engineered to synthesize desirable proteins such as drugs, hormones, and enzymes **(figure 1.3b).**

Among the genetically unique organisms that have been designed by bioengineers are bacteria that contain a natural pesticide, yeast that produce human insulin, pigs that produce human hemoglobin, and plants that do not ripen too rapidly. The techniques also pave the way for characterizing human genetic material and diseases.

Another way of tapping into the unlimited potential of microorganisms is the relatively new science of **bioremediation** (by'-oh-ree-mee-dee-ay"-shun). This process involves

---

2. DNA, or deoxyribonucleic acid, the chemical substance that comprises the genetic material of organisms.

(a) An aerial view of a copper mine looks like a giant quilt pattern. The colored patches are bacteria in various stages of extracting metals from the ore.

(b) Microbes as synthesizers. A large complex fermentor manufactures drugs and enzymes using microbial metabolism.

(c) A bioremediation platform placed in a river for the purpose of detoxifying water containing industrial pollutants.

FIGURE 1.3 **Microbes at work.**

the introduction of microbes into the environment to restore stability or to clean up toxic pollutants. Bioremediation is required to control the massive levels of pollution from industry and modern living. Microbes have a surprising capacity to break down chemicals that would be harmful to other organisms. Agencies and companies have developed microbes to handle oil spills and detoxify sites contaminated with heavy metals, pesticides, and other chemical wastes **(figure 1.3c)**. The solid waste disposal industry is interested in developing methods for degrading the tons of garbage in landfills, especially human-made plastics and paper products. One form of bioremediation that has been in use for some time is the treatment of water and sewage. Since clean freshwater supplies are dwindling worldwide, it will become even more important to find ways to reclaim polluted water.

1.4 Infectious Diseases and the Human Condition

One of the most fascinating aspects of the microorganisms with which we share the earth is that, despite all of the benefits they provide, they also contribute significantly to human misery as **pathogens** (path′-oh-jenz). The vast majority of microorganisms that associate with humans cause no harm. In fact, they provide many benefits to their human hosts. There is little doubt that a diverse microbial flora living in and on humans is an important part of human ecology. However, humanity is also plagued by nearly 2,000 different microbes that can cause various types of disease. Infectious diseases still devastate human populations worldwide, despite significant strides in understanding and treating them. The most recent estimates from the World Health Organization (WHO) point to a total of 10 billion new infections across the world every year. (There are more infections than people because many people acquire more than one infection.) Infectious diseases are also among the most common causes of death in much of humanity, and they still kill a significant percentage of the U.S. population. **Table 1.1** depicts the 10 top causes of death per year (by all causes, infectious and noninfectious) in the United States and worldwide. The worldwide death toll from infections is about 13 million people per year. In **figure 1.4** you can see the top infectious causes of death displayed in a different way. Note that many of these infections are treatable with drugs or preventable with vaccines. Those hardest hit are residents in countries where access to adequate medical care is lacking. One-third of the earth's inhabitants live on less than $1 per day, are malnourished, and are not fully immunized.

Malaria, which kills more than a million people every year worldwide, is caused by a microorganism transmitted by mosquitoes (see chapter 20). Currently the most effective way for citizens of developing countries to avoid infection with malaria is to sleep under a bed net, since the mosquitoes are most active in the evening. Yet even this inexpensive solution

TABLE 1.1	Top Causes of Death—All Diseases		
United States	**No. of Deaths**	**Worldwide**	**No. of Deaths**
1. Heart disease	725,000	1. Heart disease	11.1 million
2. Cancer	550,000	2. Cancer	7.1 million
3. Stroke	167,000	3. Stroke	5.5 million
4. Chronic lower-respiratory disease	124,000	4. Respiratory infections*	3.9 million
5. Unintentional injury (accidents)	97,000	5. Chronic lower-respiratory disease	3.6 million
6. Diabetes	68,000	6. Accidents	3.5 million
7. Influenza and pneumonia	63,000	7. HIV/AIDS	2.9 million
8. Alzheimer's disease	45,000	8. Perinatal conditions	2.5 million
9. Kidney problems	35,000	9. Diarrheal diseases	2.0 million
10. Septicemia (bloodstream infection)	30,000	10. Tuberculosis	1.6 million

*Diseases in red are those most clearly caused by microorganisms.

Data adapted from The World Health Report 2002 (World Health Organization).

is beyond the reach of many. Mothers in Southeast Asia and elsewhere have to make nightly decisions about which of their children will sleep under the single family bed net, since a second one, priced at about $3 to $5, is too expensive for them.

IN THE NEWS (Continued from page 1)

During the initial outbreak of SARS at the Prince of Wales Hospital in Hong Kong, hospital workers were confronted with a new infectious disease caused by a virus. SARS was transmitted quickly among hospital workers. There were concerns that the new coronavirus was spreading through small aerosols or contact with contaminated surfaces in the hospital environment. The epidemiological investigation led by the Chinese University and Hong Kong Hospital Authority determined that hospital workers did not take special protective measures when in contact with SARS patients during the initial outbreak of the disease. Personal protection such as wearing masks, goggles, caps, and gowns was inadequate and workers had less than two hours of infection control training. Many did not understand infection control procedures and used personal protection equipment inconsistently.

The study revealed that 40% to 50% of hospital workers experienced difficulties with their masks fitting properly, fogging of protective goggles, and general compliance problems. This case provides an example of the consequences of inadequate infection control measures. Proper training and implementation of infection control measures reduces the risks of breakthrough transmission of the SARS virus.

See: Lau, J. T. F. et al. 2004. SARS transmission among hospital workers in Hong Kong. Emerg. Infect. Dis.

Adding to the overload of infectious diseases, we are also witnessing an increase in the number of new (emerging) and older (reemerging) diseases. SARS, AIDS, hepatitis C, and viral encephalitis are examples of recently identified diseases that cause severe mortality and morbidity and are currently on the rise. To somewhat balance this trend, there have

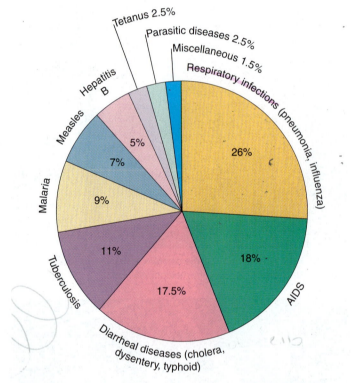

FIGURE 1.4 Worldwide infectious disease statistics.
This figure depicts the ten most common infectious causes of death.

also been some advances in eradication of diseases such as polio, measles, leprosy, and certain parasitic worms. The WHO is currently on a global push to vaccinate children against the most common childhood diseases.

One of the most eye-opening discoveries in recent years is that many diseases which used to be considered noninfectious probably do involve microbial infection. The most famous of these is gastric ulcers, now known to be caused by a bacterium called *Helicobacter*. But there are more. A connection has been established between certain cancers and viruses, between diabetes and the Coxsackie virus, and between

INSIGHT 1.1   *Historical*

The More Things Change . . .

In 1967, the surgeon general of the United States delivered a speech to Congress: "It is time to close the book on infectious diseases," he said. "The war against pestilence is over."

In 1998, Surgeon General David Satcher had a different message. The *Miami Herald* reported his speech with this headline: "Infectious Diseases a Rising Peril; Death Rates in U.S. Up 58% Since 1980."

The middle of the last century was a time of great confidence in science and medicine. With the introduction of antibiotics in the 1940s, and a lengthening list of vaccines that prevented the most frightening diseases, Americans felt that it was only a matter of time before diseases caused by microorganisms (i.e., infectious diseases) would be completely manageable. The nation's attention turned to the so-called chronic diseases, such as heart disease, cancer, and stroke.

So what happened to change the optimism of the 1960s to the warning expressed in the speech from 1998? Dr. Satcher explained it this way: "Organisms changed and people changed." First, we are becoming more susceptible to infectious disease precisely because of advances in medicine. People are living longer. Sicker people are staying alive much longer than in the

United States Surgeon General David Satcher in 1998.

United States Surgeon General Luther Terry addressing press conference in 1964.

past. Older and sicker people have heightened susceptibility to what we might call garden-variety microbes. Second, the population has become more mobile. Travelers can crisscross the globe in a matter of hours, taking their microbes with them and introducing them into new "naive" populations. Third, there are growing numbers of microbes that truly are new (or at least, new to us). The conditions they cause are called **emerging diseases.** Changes in agricultural practices and encroachment of humans on wild habitats are just two probable causes of emerging diseases. Fourth, microorganisms have demonstrated their formidable capacity to respond and adapt to our attempts to control them, most spectacularly by becoming resistant to the effects of our miracle drugs.

And there's one more thing: Evidence is mounting that many conditions formerly thought to be caused by genetics or lifestyle, such as heart disease and cancer, can often be at least partially caused by microorganisms.

Microbes never stop surprising us—in their ability to harm but also to help us. The best way to keep up is to learn as much as you can about them. This book is a good place to start.

---

schizophrenia and a virus called the borna agent. Diseases as disparate as multiple sclerosis, obsessive compulsive disorder, and coronary artery disease have been linked to chronic infections with microorganisms. It seems that the golden age of microbiological discovery, during which all of the "obvious" diseases were characterized, and cures or preventions were devised for them, should more accurately be referred to as the *first* golden age. We're now discovering the subtler side of microorganisms. Their roles in quiet but slowly destructive diseases are now well known. These include female infertility caused by *Chlamydia* infection, and malignancies such as liver cancer (hepatitis viruses) and cervical cancer

(human papillomavirus). Most scientists expect that, in time, many chronic conditions will be found to have some association with microbial agents.

As mentioned earlier, another important development in infectious disease trends is the increasing number of patients with weakened defenses that are kept alive for extended periods. They are subject to infections by common microbes that are not pathogenic to healthy people. There is also an increase in microbes that are resistant to drugs. It appears that even with the most modern technology available to us, microbes still have the "last word," as the great French microbiologist Louis Pasteur observed **(Insight 1.1).**

✔ CHECKPOINT

- Microorganisms are defined as "living organisms too small to be seen with the naked eye." Among the members of this huge group of organisms are bacteria, fungi, protozoa, algae, viruses, and parasitic worms.
- Microorganisms live nearly everywhere and influence many biological and physical activities on earth.
- There are many kinds of relationships between microorganisms and humans; most are beneficial, but some are harmful.
- The scope of microbiology is incredibly diverse. It includes basic microbial research, research on infectious diseases, study of prevention and treatment of disease, environmental functions of microorganisms, and industrial use of microorganisms for commercial, agricultural, and medical purposes.
- In the last 120 years, microbiologists have identified the causative agents for many infectious diseases. In addition, they have discovered distinct connections between microorganisms and diseases whose causes were previously unknown.
- Microorganisms: We have to learn to live with them because we cannot live without them.

1.5 The General Characteristics of Microorganisms

Cellular Organization

As discussed earlier, two basic cell lines appeared during evolutionary history. These lines, termed **procaryotic cells** and **eucaryotic cells,** differ primarily in the complexity of their cell structure **(figure 1.5a).**

In general, procaryotic cells are smaller than eucaryotic cells, and they lack special structures such as a nucleus and **organelles.** Organelles are small membrane-bound cell struc-

tures that perform specific functions in eucaryotic cells. These two cell types and the organisms that possess them (called procaryotes and eucaryotes) are covered in more detail in chapters 4 and 5.

All procaryotes are microorganisms, but only some eucaryotes are microorganisms. The bodies of most microorganisms consist of either a single cell or just a few cells **(figure 1.6).** Because of their role in disease, certain animals such as helminth worms and insects, many of which can be seen with the naked eye, are also considered in the study of microorganisms. Even in its seeming simplicity, the microscopic world is every bit as complex and diverse as the macroscopic one. There is no doubt that microorganisms also outnumber macroscopic organisms by a factor of several million.

A Note on Viruses

Viruses are subject to intense study by microbiologists. They are not cells. They are small particles that exist at a level of complexity somewhere between large molecules and cells (see **figure 1.5b**). Viruses are much simpler than cells; they are composed essentially of a small amount of hereditary material wrapped up in a protein covering. Some biologists refer to viruses as parasitic particles; others consider them to be very primitive organisms. One thing is certain—they are highly dependent on a host cell's machinery for their activities.

Microbial Dimensions: How Small Is Small?

When we say that microbes are too small to be seen with the unaided eye, what sorts of dimensions are we talking

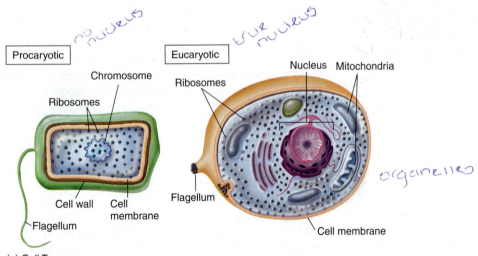

(a) Cell Types

Microbial cells are of the small, relatively simple procaryotic variety (left) or the larger, more complex eucaryotic type (right). (Not to scale)

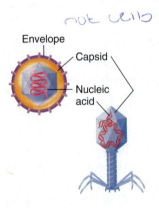

(b) Virus Types

Viruses are tiny particles, not cells, that consist of genetic material surrounded by a protective covering. Shown here are a human virus (top) and bacterial virus (bottom). (Not to scale)

FIGURE 1.5   Cell structure.

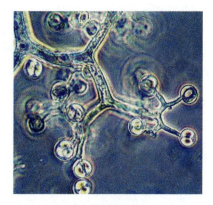

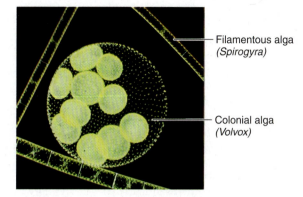

Bacterium: *E. coli*

Fungus: *Thamnidium*

Algae: *Volvox* and *Spirogyra*

Filamentous alga (*Spirogyra*)

Colonial alga (*Volvox*)

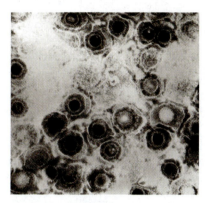

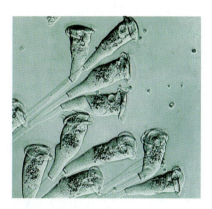

Virus: *Herpes simplex*

Protozoan: *Vorticella*

Helminth: Head (scolex) of *Taenia solium*

FIGURE 1.6 **The six types of microorganisms.**
(Organisms are not shown at the same magnifications.)

about? This concept is best visualized by comparing microbial groups with the larger organisms of the macroscopic world and also with the molecules and atoms of the molecular world **(figure 1.7).** Whereas the dimensions of macroscopic organisms are usually given in centimeters (cm) and meters (m), those of most microorganisms fall within the range of micrometers (μm) and sometimes, nanometers (nm) and millimeters (mm). The size range of most microbes extends from the smallest viruses, measuring around 20 nm and actually not much bigger than a large molecule, to protozoans measuring 3 to 4 mm and visible with the naked eye.

Lifestyles of Microorganisms

The majority of microorganisms live a free existence in habitats such as soil and water, where they are relatively harmless and often beneficial. A free-living organism can derive all required foods and other factors directly from the nonliving environment. Some microorganisms require interactions with other organisms. Sometimes these microbes are termed **parasites.** They are harbored and nourished by other living organisms, called **hosts.** A parasite's actions cause damage to its host through infection and disease. Although parasites cause important diseases, they make up only a small proportion of microbes.

✔ CHECKPOINT

- Excluding the viruses, there are two types of microorganisms: procaryotes, which are small and lack a nucleus and organelles, and eucaryotes, which are larger and have both a nucleus and organelles.
- Viruses are not cellular and are therefore sometimes called particles rather than organisms. They are included in microbiology because of their small size and close relationship with cells.
- Most microorganisms are measured in micrometers, with two exceptions. The helminths are measured in millimeters, and the viruses are measured in nanometers.
- Contrary to popular belief, most microorganisms are harmless, free-living species that perform vital functions in both the environment and larger organisms. Comparatively few species are agents of disease.

1.6 The Historical Foundations of Microbiology

If not for the extensive interest, curiosity, and devotion of thousands of microbiologists over the last 300 years, we would know little about the microscopic realm that surrounds us. Many of the discoveries in this science have resulted from the prior work of men and women who toiled long hours in dimly lit laboratories with the crudest of tools.

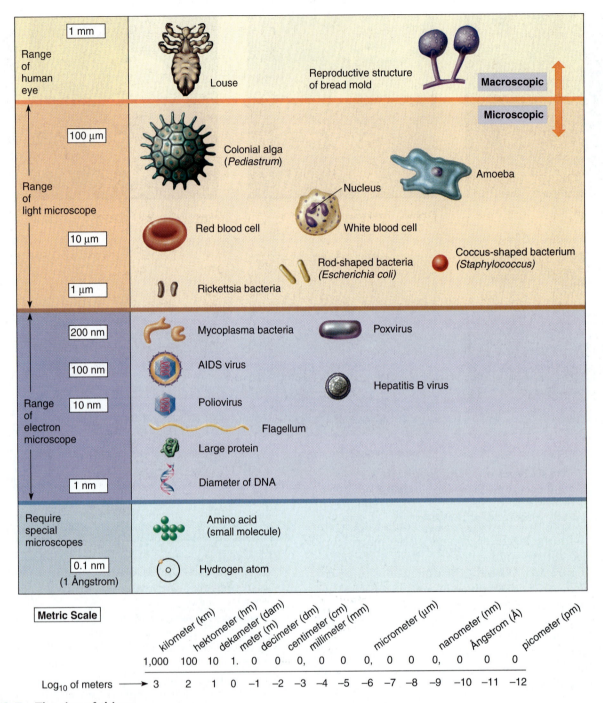

FIGURE 1.7   The size of things.
Common measurements encountered in microbiology and a scale of comparison from the macroscopic to the microscopic, molecular, and atomic. Most microbes encountered in our studies will fall between 100 µm and 10 nm in overall dimensions. The microbes shown are more or less to scale within size zone but not between size zones.

Each additional insight, whether large or small, has added to our current knowledge of living things and processes. This section will summarize the prominent discoveries made in the past 300 years: microscopy, the rise of the scientific method, and the development of medical microbiology, including the germ theory and the origins of modern microbiological techniques. See table B.1 in appendix B, which summarizes some of the pivotal events in microbiology, from its earliest beginnings to the present.

The Development of the Microscope: "Seeing Is Believing"

It is likely that from the very earliest history, humans noticed that when certain foods spoiled they became inedible or caused illness, and yet other "spoiled" foods did no harm and even had enhanced flavor. Indeed, several centuries ago, there was already a sense that diseases such as the black plague and smallpox were caused by some sort of transmis-

FIGURE 1.8 **An oil painting of Antonie van Leeuwenhoek (1632–1723) sitting in his laboratory.**
J. R. Porter and C. Dobell have commented on the unique qualities Leeuwenhoek brought to his craft: "He was one of the most original and curious men who ever lived. It is difficult to compare him with anybody because he belonged to a genus of which he was the type and only species, and when he died his line became extinct."

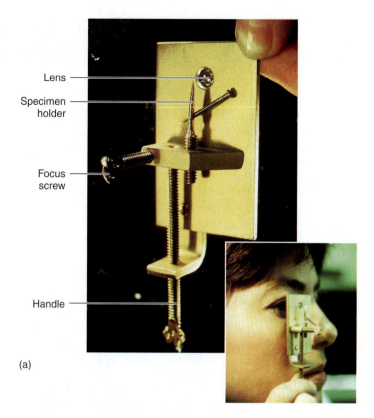

(a)

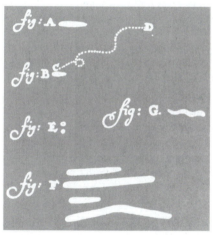

(b)

FIGURE 1.9 **Leeuwenhoek's microscope.**
(a) A brass replica of a Leeuwenhoek microscope and how it is held. **(b)** Examples of bacteria drawn by Leeuwenhoek.

sible matter. But the causes of such phenomena were vague and obscure because the technology to study them was lacking. Consequently, they remained cloaked in mystery and regarded with superstition—a trend that led even well-educated scientists to believe in spontaneous generation **(Insight 1.2).** True awareness of the widespread distribution of microorganisms and some of their characteristics was finally made possible by the development of the first microscopes. These devices revealed microbes as discrete entities sharing many of the characteristics of larger, visible plants and animals. Several early scientists fashioned magnifying lenses, but their microscopes lacked the optical clarity needed for examining bacteria and other small, single-celled organisms. The most careful and exacting observations awaited the clever single-lens microscope hand-fashioned by Antonie van Leeuwenhoek, a Dutch linen merchant and self-made microbiologist **(figure 1.8).**

Paintings of historical figures like the one of Leeuwenhoek in figure 1.8 don't always convey a meaningful feeling for the event or person depicted. Imagine a dusty shop in Holland in the late 1600s. Ladies in traditional Dutch garb came in and out, choosing among the bolts of linens for their draperies and upholstery. Between customers, Leeuwenhoek retired to the workbench in the back of his shop, grinding glass lenses to ever-finer specifications. He could see with increasing clarity the threads in his fabrics. Eventually he became interested in things other than thread counts. He took rainwater from a clay pot and smeared it on his specimen holder, and peered at it through his finest lens. He found "animals appearing to me ten thousand times less than those which may be perceived in the water with the naked eye."

He didn't stop there. He scraped the plaque from his teeth, and from the teeth of some volunteers who had never cleaned their teeth in their lives, and took a good close look at that. He recorded: "In the said matter there were many very little living animalcules, very prettily a-moving. . . . Moreover, the other animalcules were in such enormous numbers, that all the water . . . seemed to be alive." Leeuwenhoek started sending his observations to the Royal Society of London, and eventually he was recognized as a scientist of great merit.

Leeuwenhoek constructed more than 250 small, powerful microscopes that could magnify up to 300 times **(figure 1.9).**

INSIGHT 1.2   *Historical*

The Fall of Superstition and the Rise of Microbiology

For thousands of years, people believed that certain living things arose from vital forces present in nonliving or decomposing matter. This ancient belief, known as **spontaneous generation,** was continually reinforced as people observed that meat left out in the open soon "produced" maggots, that mushrooms appeared on rotting wood, that rats and mice emerged from piles of litter, and other similar phenomena. Though some of these early ideas seem quaint and ridiculous in light of modern knowledge, we must remember that, at the time, mysteries in life were accepted, and the scientific method was not widely practiced.

Even after single-celled organisms were discovered during the mid-1600s, the idea of spontaneous generation continued to exist. Some scientists assumed that microscopic beings were an early stage in the development of more complex ones.

Over the subsequent 200 years, scientists waged an experimental battle over the two hypotheses that could explain the origin of simple life forms. Some tenaciously clung to the idea of **abiogenesis** (ah-bee"-oh-jen-uh-sis), which embraced spontaneous generation. On the other side were advocates of **biogenesis** saying that living things arise only from others of their same kind. There were serious proponents on both sides, and each side put forth what appeared on the surface to be plausible explanations of why their evidence was more correct. Gradually, the abiogenesis hypothesis was abandoned, as convincing evidence for biogenesis continued to mount. The following series of experiments were among the most important in finally tipping the balance.

Among the important variables to be considered in challenging the hypotheses were the effects of nutrients, air, and heat and the presence of preexisting life forms in the environment. One of the first people to test the spontaneous generation theory was Francesco Redi of Italy. He conducted a simple experiment in which he placed meat in a jar and covered it with fine gauze. Flies gathering at the jar were blocked from entering and thus laid their eggs on the outside of the gauze. The maggots subsequently developed without access to the meat, indicating that maggots were

the offspring of flies and did not arise from some "vital force" in the meat. This and related experiments laid to rest the idea that more complex animals such as insects and mice developed through abiogenesis, but it did not convince many scientists of the day that simpler organisms could not arise in that way.

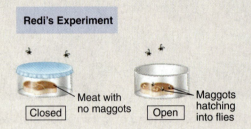

Redi's Experiment

Meat with no maggots — Closed
Maggots hatching into flies — Open

The Frenchman Louis Jablot reasoned that even microscopic organisms must have parents, and his experiments with infusions (dried hay steeped in water) supported that hypothesis. He divided an infusion that had been boiled to destroy any living things into two containers: a heated container that was closed to the air and a heated container that was freely open to the air. Only the open vessel developed microorganisms, which he presumed had

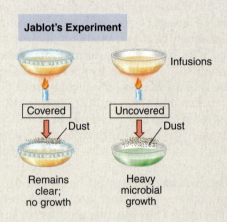

Jablot's Experiment

Infusions

Covered — Dust
Uncovered — Dust

Remains clear; no growth
Heavy microbial growth

Considering that he had no formal training in science and that he was the first person ever to faithfully record this strange new world, his descriptions of bacteria and protozoa (which he called "animalcules") were astute and precise. Because of Leeuwenhoek's extraordinary contributions to microbiology, he is known as the father of bacteriology and protozoology.

From the time of Leeuwenhoek, microscopes became more complex and improved with the addition of refined lenses, a condenser, finer focusing devices, and built-in light sources. The prototype of the modern compound microscope, in use from about the mid-1800s, was capable of magnifications of 1,000 times or more. Even our modern laboratory microscopes are not greatly different in basic structure and function from those early microscopes. The

technical characteristics of microscopes and microscopy are a major focus of chapter 3.

The Establishment of the Scientific Method

A serious impediment to the development of true scientific reasoning and testing was the tendency of early scientists to explain natural phenomena by a mixture of belief, superstition, and argument. The development of an experimental system that answered questions objectively and was not based on prejudice marked the beginning of true scientific thinking. These ideas gradually crept into the consciousness of the scientific community during the 1600s. The general approach taken by scientists to explain a certain natural

entered in air laden with dust. Regrettably, the validation of biogenesis was temporarily set back by John Needham, an Englishman who did similar experiments using mutton gravy. His results were in conflict with Jablot's because both his heated and unheated test containers teemed with microbes. Unfortunately, his experiments were done before the realization that heat-resistant endospores were not killed by mere boiling. Apparently Jablot had been lucky; his infusions had no endospores.

Additional experiments further defended biogenesis. Franz Shultze and Theodor Schwann of Germany felt sure that air was the source of microbes and sought to prove this by passing air through strong chemicals or hot glass tubes into heat-treated infusions in flasks. When the infusions again remained devoid of living things, the supporters of abiogenesis claimed that the treatment of the air had made it harmful to the spontaneous development of life.

Shultze and Schwann's Test

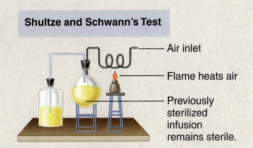

- Air inlet
- Flame heats air
- Previously sterilized infusion remains sterile.

Then, in the mid-1800s, the acclaimed microbiologist Louis Pasteur entered the arena. He had recently been studying the roles of microorganisms in the fermentation of beer and wine, and it was clear to him that these processes were brought about by the activities of microbes introduced into the beverage from air, fruits, and grains. The methods he used to discount abiogenesis were simple yet brilliant.

To further clarify that air and dust were the source of microbes, Pasteur filled flasks with broth and fashioned their openings into elongate, swan-neck–shaped tubes. The flasks' openings were freely open to the air but were curved so that gravity would cause any airborne dust particles to deposit in the lower part of the necks. He heated the flasks to sterilize the broth and then incubated them. As long as the flask remained intact, the broth remained sterile, but if the neck was broken off so that dust fell directly down into the container, microbial growth immediately commenced.

Pasteur summed up his findings, "For I have kept from them, and am still keeping from them, that one thing which is above the power of man to make; I have kept from them the germs that float in the air, I have kept from them life."

Pasteur's Experiment

Microbes being destroyed

Vigorous heat is applied.

Broth free of live cells (sterile)

Neck on second sterile flask is broken; growth occurs.

Neck intact; airborne microbes are trapped at base, and broth is sterile.

phenomenon is called the **scientific method.** A primary aim of this method is to formulate a **hypothesis,** a tentative explanation to account for what has been observed or measured. A good hypothesis should be in the form of a statement. It must be capable of being either supported or discredited by careful, systematic observation or experimentation. For example, the statement that "microorganisms cause diseases" can be experimentally determined by the tools of science, but the statement that "diseases are caused by evil spirits" cannot.

There are various ways to apply the scientific method, but probably the most common is called the **deductive approach.** In the deductive approach, a scientist constructs a hypothesis, tests its validity by outlining particular events that are predicted by the hypothesis, and then performs experiments to test for those events **(figure 1.10).** The deductive process states: "*If the hypothesis is valid, then certain specific events can be expected to occur.*"

A lengthy process of experimentation, analysis, and testing eventually leads to conclusions that either support or refute the hypothesis. If experiments do not uphold the hypothesis—that is, if it is found to be flawed—the hypothesis or some part of it is rejected; it is either discarded or modified to fit the results of the experiment (see **figure 1.10b**). If the hypothesis is supported by the results from the experiment, it is not (or should not be) immediately accepted as fact. It then must be tested and retested. Indeed, this is an important guideline in the acceptance of a hypothesis. The

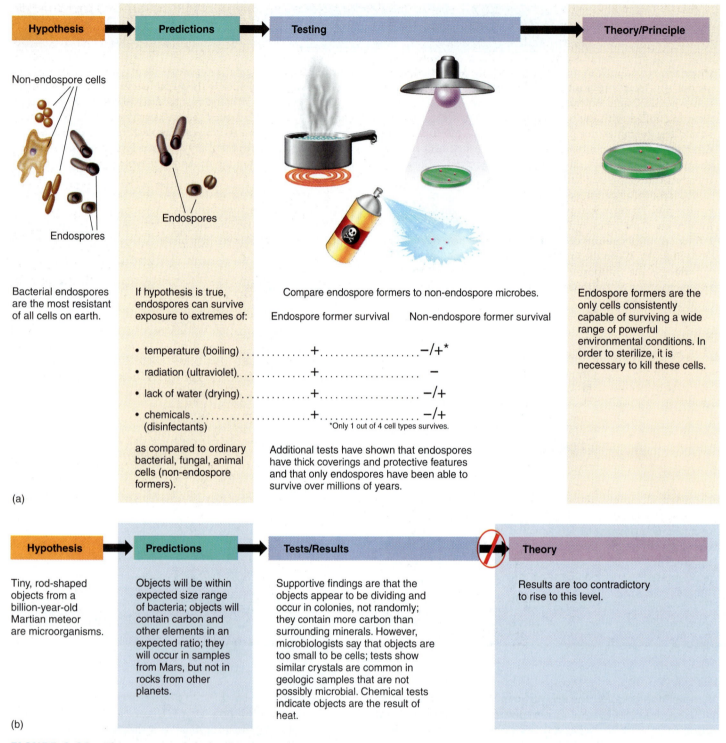

(a)

Hypothesis	Predictions	Testing	Theory/Principle

Hypothesis: Bacterial endospores are the most resistant of all cells on earth.

Non-endospore cells

Endospores

Predictions: If hypothesis is true, endospores can survive exposure to extremes of:

Endospores

- temperature (boiling)
- radiation (ultraviolet)
- lack of water (drying)
- chemicals (disinfectants)

as compared to ordinary bacterial, fungal, animal cells (non-endospore formers).

Testing: Compare endospore formers to non-endospore microbes.

	Endospore former survival	Non-endospore former survival
temperature (boiling)	+	–/+*
radiation (ultraviolet)	+	–
lack of water (drying)	+	–/+
chemicals (disinfectants)	+	–/+

*Only 1 out of 4 cell types survives.

Additional tests have shown that endospores have thick coverings and protective features and that only endospores have been able to survive over millions of years.

Theory/Principle: Endospore formers are the only cells consistently capable of surviving a wide range of powerful environmental conditions. In order to sterilize, it is necessary to kill these cells.

(b)

Hypothesis	Predictions	Tests/Results	Theory

Hypothesis: Tiny, rod-shaped objects from a billion-year-old Martian meteor are microorganisms.

Predictions: Objects will be within expected size range of bacteria; objects will contain carbon and other elements in an expected ratio; they will occur in samples from Mars, but not in rocks from other planets.

Tests/Results: Supportive findings are that the objects appear to be dividing and occur in colonies, not randomly; they contain more carbon than surrounding minerals. However, microbiologists say that objects are too small to be cells; tests show similar crystals are common in geologic samples that are not possibly microbial. Chemical tests indicate objects are the result of heat.

Theory: Results are too contradictory to rise to this level.

FIGURE 1.10   The pattern of deductive reasoning.
The deductive process starts with a general hypothesis that predicts specific expectations. **(a)** This example is based on a well-established principle. **(b)** This example is based on a new hypothesis that has not stood up to critical testing.

results of the experiment must be published and then repeated by other investigators.

In time, as each hypothesis is supported by a growing body of data and survives rigorous scrutiny, it moves to the next level of acceptance—the **theory.** A theory is a collection of statements, propositions, or concepts that explains or accounts for a natural event. A theory is not the result of a sin-gle experiment repeated over and over again, but is an entire body of ideas that expresses or explains many aspects of a phenomenon. It is not a fuzzy or weak speculation, as is sometimes the popular notion, but a viable declaration that has stood the test of time and has yet to be disproved by serious scientific endeavors. Often, theories develop and progress through decades of research and are added to and modified by new findings. At some point, evidence of the accuracy and predictability of a theory is so compelling that the next level of confidence is reached and the theory becomes a law, or principle. For example, although we still refer to the germ *theory* of disease, so little question remains that microbes can cause disease that it has clearly passed into the realm of law.

Science and its hypotheses and theories must progress along with technology. As advances in instrumentation allow new, more detailed views of living phenomena, old theories may be reexamined and altered and new ones proposed. But scientists do not take the stance that theories or even "laws" are ever absolutely proved. The characteristics that make scientists most effective in their work are curiosity, open-mindedness, skepticism, creativity, cooperation, and readiness to revise their views of natural processes as new discoveries are made. The events described in Insights 1.2 and **1.3** provide important examples.

The Development of Medical Microbiology

Early experiments on the sources of microorganisms led to the profound realization that microbes are everywhere: Not only are air and dust full of them, but the entire surface of the earth, its waters, and all objects are inhabited by them. This discovery led to immediate applications in medicine. Thus the seeds of medical microbiology were sown in the mid to latter half of the nineteenth century with the introduction of the germ theory of disease and the resulting use of sterile, aseptic, and pure culture techniques.

The Discovery of Spores and Sterilization

Following Pasteur's inventive work with infusions (see Insight 1.2), it was not long before English physicist John Tyndall provided the initial evidence that some of the microbes in dust and air have very high heat resistance and that particularly vigorous treatment is required to destroy them. Later, the discovery and detailed description of heat-resistant bacterial endospores by Ferdinand Cohn, a German botanist, clarified the reason that heat would sometimes fail to completely eliminate all microorganisms. The modern sense of the word **sterile,** meaning completely free of all life forms including spores and viruses, was established from that point on (see chapter 11). The capacity to sterilize objects and materials is an absolutely essential part of microbiology, medicine, dentistry, and some industries.

The Development of Aseptic Techniques

From earliest history, humans experienced a vague sense that "unseen forces" or "poisonous vapors" emanating from decomposing matter could cause disease. As the study of microbiology became more scientific and the invisible was made visible, the fear of such mysterious vapors was replaced by the knowledge and sometimes even the fear of "germs." About 120 years ago, the first studies by Robert Koch clearly linked a microscopic organism with a specific disease. Since that time, microbiologists have conducted a continuous search for disease-causing agents.

At the same time that abiogenesis was being hotly debated, a few budding microbiologists began to suspect that microorganisms could cause not only spoilage and decay but also infectious diseases. It occurred to these rugged individualists that even the human body itself was a source of infection. Dr. Oliver Wendell Holmes, an American physician, observed that mothers who gave birth at home experienced fewer infections than did mothers who gave birth in the hospital, and the Hungarian Dr. Ignaz Semmelweis showed quite clearly that women became infected in the maternity ward after examinations by physicians coming directly from the autopsy room. The English surgeon Joseph Lister took notice of these observations and was the first to introduce **aseptic** (ay-sep'-tik) **techniques** aimed at reducing microbes in a medical setting and preventing wound infections. Lister's concept of asepsis was much more limited than our modern precautions. It mainly involved disinfecting the hands and the air with strong antiseptic chemicals, such as phenol, prior to surgery. It is hard for us to believe, but as recently as the late 1800s surgeons wore street clothes in the operating room and had little idea that hand washing was important. Lister's techniques and the application of heat for sterilization became the bases for microbial control by physical and chemical methods, which are still in use today.

The Discovery of Pathogens and the Germ Theory of Disease

Two ingenious founders of microbiology, Louis Pasteur of France **(figure 1.11)** and Robert Koch of Germany **(figure 1.12),** introduced techniques that are still used today. Pasteur

FIGURE 1.11 **Louis Pasteur (1822–1895), one of the founders of microbiology.**
Few microbiologists can match the scope and impact of his contributions to the science of microbiology.

FIGURE 1.12   Robert Koch looking through a microscope with colleague Richard Pfeiffer looking on.
Robert Koch won the Nobel Prize for Physiology or Medicine in 1905 for his work on *M. tuberculosis*. Richard Pfeiffer discovered *Haemophilus influenzae* and was a pioneer in typhoid vaccination.

✔ CHECKPOINT

- Our current understanding of microbiology is the cumulative work of thousands of microbiologists, many of whom literally gave their lives to advance knowledge in this field.
- The microscope made it possible to see microorganisms and thus to identify their widespread presence, particularly as agents of disease.
- Antonie van Leeuwenhoek is considered the father of bacteriology and protozoology because he was the first person to produce precise, correct descriptions of these organisms.
- The theory of spontaneous generation of living organisms from "vital forces" in the air was disproved once and for all by Louis Pasteur.
- The scientific method is a process by which scientists seek to explain natural phenomena. It is characterized by specific procedures that either support or discredit an initial hypothesis.
- Knowledge acquired through the scientific method is rigorously tested by repeated experiments by many scientists to verify its validity. A collection of valid hypotheses is called a theory. A theory supported by much data collected over time is called a law.
- Scientific truth changes through time as new research brings new information. Scientists must be able and willing to change theory in response to new data.
- Medical microbiologists developed the germ theory of disease and introduced the critically important concept of aseptic technique to control the spread of disease agents.
- Koch's postulates are the cornerstone of the germ theory of disease. They are still used today to pinpoint the causative agent of a specific disease.
- Louis Pasteur and Robert Koch were the leading microbiologists during the golden age of microbiology (1875–1900); each had his own research institute.

made enormous contributions to our understanding of the microbial role in wine and beer formation. He invented pasteurization and completed some of the first studies showing that human diseases could arise from infection. These studies, supported by the work of other scientists, became known as the **germ theory of disease.** Pasteur's contemporary, Koch, established *Koch's postulates,* a series of proofs that verified the germ theory and could establish whether an organism was pathogenic and which disease it caused (see chapter 13). About 1875, Koch used this experimental system to show that anthrax was caused by a bacterium called *Bacillus anthracis.* So useful were his postulates that the causative agents of 20 other diseases were discovered between 1875 and 1900, and even today, they are the standard for identifying pathogens.

Numerous exciting technologies emerged from Koch's prolific and probing laboratory work. During this golden age of the 1880s, he realized that study of the microbial world would require separating microbes from each other and growing them in culture. It is not an overstatement to say that he and his colleagues invented most of the techniques that are described in chapter 3: inoculation, isolation, media, maintenance of pure cultures, and preparation of specimens for microscopic examination. Other highlights in this era of discovery are presented in later chapters on microbial control (see chapter 11) and vaccination (see chapter 15).

1.7 Taxonomy: Organizing, Classifying, and Naming Microorganisms

Students just beginning their microbiology studies are often dismayed by the seemingly endless array of new, unusual, and sometimes confusing names for microorganisms. Learning microbial **nomenclature** is very much like learning a new language, and occasionally its demands may be a bit overwhelming. But paying attention to proper microbial names is just like following a baseball game or a theater production: You cannot tell the players apart without a program! Your understanding and appreciation of microorganisms will be greatly improved by learning a few general rules about how they are named.

The formal system for organizing, classifying, and naming living things is **taxonomy.** This science originated more than 250 years ago when Carl von Linné (also known as Linnaeus; 1701–1778), a Swedish botanist, laid down the basic rules for taxonomic categories, or **taxa.** Von Linné realized early on that a system for recognizing and defining the

INSIGHT 1.3 — *Discovery*

The Serendipity of the Scientific Method: Discovering Drugs

Discoveries in science are not always determined by the strict formulation and testing of a formal hypothesis. Quite often, they involve serendipity* and the luck of being in the right place and time, followed by a curiosity and willingness to change the direction of an experiment. This is especially true in the field of drug discoveries. The first antibiotic, penicillin, was discovered in the late 1920s by Dr. Alexander Fleming, who found a mold colony growing on a culture of bacteria that was wiping out the bacteria. He isolated the active ingredient that eventually launched the era of antibiotics. The search for new drugs to treat infections and cancer has been a continuous focus since that time. Even though the detailed science of testing a drug and working out its chemical structure and action require sophisticated scientific technology, the first and most important part of discovery often lies in a keen eye and an open mind.

In 1987, Dr. Michael Zasloff, a physician and molecular biologist, was doing research in gene expression using African clawed frogs as a source of eggs. After performing surgery on the frogs and routinely placing them back in a nonsterile aquarium, he was surprised to notice that most of the time the frogs did not get infected or die. If the animal had been a mammal such as a mouse, it would probably not have survived the nonsterile surgery. This led him to conclude that the frog's skin must provide some form of natural protection. He observed that when the skin was stimulated by injury or irritants, it formed a thick white coating in a few moments that reminded him of a self-made "bandage" over the wound. He took a section of skin and extracted the components that were responsible for killing the microbes. His tests showed that they were small proteins called peptides, which he named

*serendipity: making useful discoveries by accident.

magainins, after the Hebrew word for shield. Within 6 months of these findings, Dr. Zasloff made the decision to completely change the subject of his research and started up a new biotechnology company (Magainin Pharmaceuticals) to explore the therapeutic potential for magainins as well as other frog peptides.

The initial tests on this new class of drugs would indicate that they do indeed destroy a variety of bacteria as well as fungi, protozoa, and viruses. Although they are toxic to human cells too, this makes them a possible candidate for cancer treatment. Currently the drugs are being synthesized and tested in the lab for effectiveness and safety. Dr. Zasloff's intriguing observation and subsequent experiments had the impact of opening up a whole new area of biology: isolating antimicrobic peptides from multicellular organisms. Additional studies have shown that these compounds are widespread among amphibians, fish, birds, mammals, and plants. A number of companies are involved in developing applications for animal peptides. This discovery has been well timed, since resistance among microorganisms to traditional drugs is a continuing problem.

An African clawed frog responding to an irritant on its back first forms spots and then a thick opaque blotch of protective chemicals.

properties of living things would prevent chaos in scientific studies by providing each organism with a unique name and an exact "slot" in which to catalogue it. This classification would then serve as a means for future identification of that same organism and permit workers in many biological fields to know if they were indeed discussing the same organism. The von Linné system has served well in categorizing the 2 million or more different types of organisms that have been discovered since that time.

The primary concerns of taxonomy are classification, nomenclature, and identification. These three areas are interrelated and play a vital role in keeping a dynamic inventory of the extensive array of living things. *Classification* is the orderly arrangement of organisms into groups, preferably in a format that shows evolutionary relationships. *Nomenclature* is the process of assigning names to the various taxonomic rankings of each microbial species. *Identification* is the process of discovering and recording the traits of organisms so that

they may be placed in an overall taxonomic scheme. A survey of some general methods of identification appears in chapter 3.

The Levels of Classification

The main taxa, or groups, in a classification scheme are organized into several descending ranks, beginning with **domain,** which is a giant, all-inclusive category based on a unique cell type, and ending with a **species,** the smallest and most specific taxon. All the members of a domain share only one or few general characteristics, whereas members of a species are essentially the same kind of organism—that is, they share the majority of their characteristics. The taxa between the top and bottom levels are, in descending order: **kingdom, phylum** or **division,**[3] **class, order, family,** and

3. The term *phylum* is used for protozoa and animals; the term *division* is used for bacteria, algae, plants, and fungi.

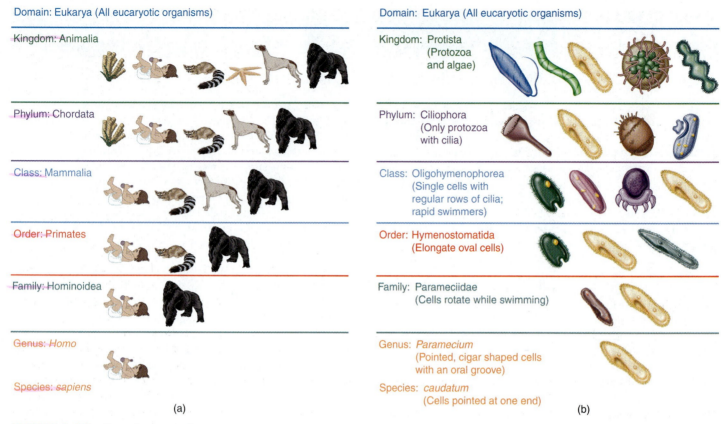

FIGURE 1.13   Sample taxonomy.
Two organisms belonging to the Eukarya domain, traced through their taxonomic series. **(a)** Modern humans, *Homo sapiens*. **(b)** A common protozoan, *Paramecium caudatum*.

genus. Thus, each domain can be subdivided into a series of kingdoms, each kingdom is made up of several phyla, each phylum contains several classes, and so on. Because taxonomic schemes are to some extent artificial, certain groups of organisms do not exactly fit into the eight taxa. In that case, additional levels can be imposed immediately above (super) or below (sub) a taxon, giving us such categories as superphylum and subclass.

To illustrate the fine points of this system, we compare the taxonomic breakdowns of a human and a protozoan **(figure 1.13)**. Humans and protozoa belong to the same domain (Eukarya) but are placed in different kingdoms. To emphasize just how broad the category kingdom is, ponder the fact that we belong to the same kingdom as jellyfish. Of the several phyla within this kingdom, humans belong to the Phylum Chordata, but even a phylum is rather all-inclusive, considering that humans share it with other vertebrates as well as with creatures called sea squirts. The next level, Class Mammalia, narrows the field considerably by grouping only those vertebrates that have hair and suckle their young. Humans belong to the Order Primates, a group that also includes apes, monkeys, and lemurs. Next comes the Family Hominoidea, containing only humans and apes. The final levels are our genus, *Homo* (all races of modern and ancient humans), and our species, *sapiens* (meaning wise). Notice that for both the human and the protozoan, the categories become less inclusive

and the individual members more closely related. Other examples of classification schemes are provided in sections of chapters 4 and 5 and in several later chapters.

We need to remember that all taxonomic **hierarchies** are based on the judgment of scientists with certain expertise in a particular group of organisms and that not all other experts may agree with the system being used. Consequently, no taxa are permanent to any degree; they are constantly being revised and refined as new information becomes available or new viewpoints become prevalent. Because this text does not aim to emphasize details of taxonomy, we will usually be concerned with only the most general (kingdom, phylum) and specific (genus, species) levels.

Assigning Specific Names

Many larger organisms are known by a common name suggested by certain dominant features. For example, a bird species might be called a red-headed blackbird or a flowering species a black-eyed Susan. Some species of microorganisms (especially pathogens) are also called by informal names, such as the gonococcus (*Neisseria gonorrhoeae*) or the tubercle bacillus (*Mycobacterium tuberculosis*), but this is not the usual practice. If we were to adopt common names such as the "little yellow coccus" or the "club-shaped diphtheria bacterium," the terminology would become even more cumbersome and challenging than scientific names. Even worse, common names are notorious for varying from region to region, even within the same country. A decided advantage of standardized nomenclature is that it provides a universal language, thereby enabling scientists from all countries on the earth to freely exchange information.

The method of assigning the **scientific, or specific name** is called the **binomial (two-name) system of nomenclature.** The scientific name is always a combination of the generic (genus) name followed by the species name. The generic part of the scientific name is capitalized, and the species part begins with a lowercase letter. Both should be italicized (or underlined if italics are not available), as follows:

Staphylococcus aureus

Because other taxonomic levels are not italicized and consist of only one word, one can always recognize a scientific name. An organism's scientific name is sometimes abbreviated to save space, as in *S. aureus*, but only if the genus name has already been stated. The source for nomenclature is usually Latin or Greek. If other languages such as English or French are used, the endings of these words are revised to have Latin endings. In general, the name first applied to a species will be the one that takes precedence over all others. An international group oversees the naming of every new organism discovered, making sure that standard procedures have been followed and that there is not already an earlier name for the organism or another organism with that same name. The inspiration for names is extremely varied and often rather imaginative. Some species have been named in honor of a microbiologist who originally discovered the microbe or who has made outstanding contributions to the field. Other names may designate a characteristic of the microbe (shape, color), a location where it was found, or a disease it causes. Some examples of specific names, their pronunciations, and their origins are:

- *Staphylococcus aureus* (staf'-i-lo-kok'-us ah'-ree-us) Gr. *staphule*, bunch of grapes, *kokkus*, berry, and Gr. *aureus*, golden. A common bacterial pathogen of humans.
- *Campylobacter jejuni* (cam'-peh-loh-bak-ter jee-joo'-neye) Gr. *kampylos*, curved, *bakterion*, little rod, and *jejunum*, a section of intestine. One of the most important causes of intestinal infection worldwide.
- *Lactobacillus sanfrancisco* (lak"-toh-bass-ill'-us san-fran-siss'-koh) L. *lacto*, milk, and *bacillus*, little rod. A bacterial species used to make sourdough bread.
- *Vampirovibrio chlorellavorus* (vam-py'-roh-vib-ree-oh klor-ell-ah'-vor-us) F. *vampire*; L. *vibrio*, curved cell; *Chlorella*, a genus of green algae; and *vorus*, to devour. A small, curved bacterium that sucks out the cell juices of *Chlorella*.
- *Giardia lamblia* (jee-ar'-dee-uh lam'-blee-uh) for Alfred Giard, a French microbiologist, and Vilem Lambl, a Bohemian physician, both of whom worked on the organism, a protozoan that causes a severe intestinal infection.

Here's a helpful hint: These names may seem difficult to pronounce and the temptation is to simply "slur over them." But when you encounter the name of a microorganism in the chapters ahead it will be extremely useful to take the time to sound them out and repeat them until they seem familiar. You are much more likely to remember them that way—and they are less likely to end up in a tangled heap with all of the new language you will be learning.

The Origin and Evolution of Microorganisms

As we indicated earlier, *taxonomy*, the classification of biological species, is a system used to organize all of the forms of life. In biology today there are different methods for deciding on taxonomic categories, but they all rely on the degree of relatedness among organisms. The natural relatedness between groups of living things is called their *phylogeny*. So, biologists use phylogenetic relationships to create a system of taxonomy.

To understand the relatedness among organisms, we must understand some fundamentals of evolution. **Evolution** is an important theme that underlies all of biology, including microbiology. Put simply, evolution states that living things change gradually through hundreds of millions of years and that these evolvements result in various types of structural and functional changes through many generations. The process of evolution is selective: Those changes that most favor the survival of a particular organism or group of organisms tend to be retained, and those that are less beneficial to survival tend to be lost. Space does not permit a detailed analysis of evolutionary theories, but the occurrence of evolution is supported by a tremendous amount of evidence from the fossil record and from the study of **morphology** (structure), **physiology** (function), and **genetics** (inheritance). Evolution accounts for the millions of different species on the earth and their adaptation to its many and diverse habitats.

Evolution is founded on two preconceptions: (1) that all new species originate from preexisting species and (2) that closely related organisms have similar features because they evolved from common ancestral forms. Usually, evolution progresses toward greater complexity, and evolutionary stages range from simple, primitive forms that are close to an ancestral organism to more complex, advanced forms. Although we use the terms *primitive* and *advanced* to denote the degree of change from the original set of ancestral traits, it is very important to realize that all species presently residing on the earth are modern, but some have arisen more recently in evolutionary history than others.

The phylogeny, or evolutionary relatedness, of organisms is often represented by a diagram of a tree. The trunk of the tree represents the main ancestral lines and the branches show offshoots into specialized groups of organisms. This sort of arrangement places the more ancient groups at the bottom and the more recent ones at the top. The branches

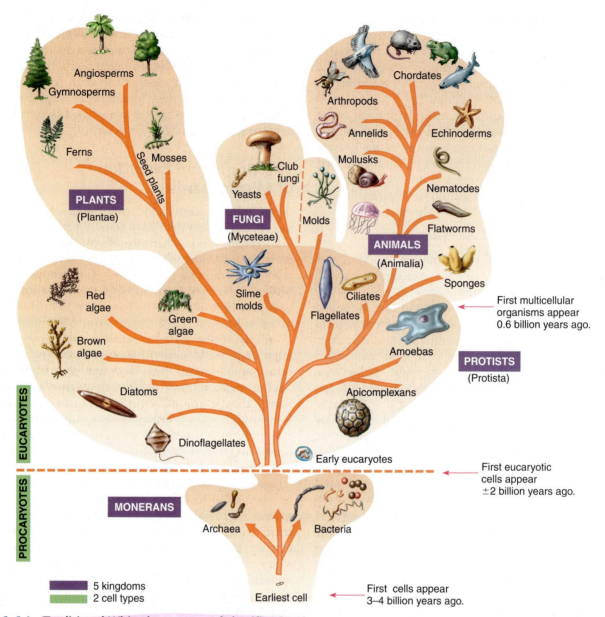

FIGURE 1.14 **Traditional Whittaker system of classification.**
In this system kingdoms are based on cell structure and type, the nature of body organization, and nutritional type. Bacteria and Archaea (monerans) are made of procaryotic cells and are unicellular. Protists are made of eucaryotic cells and are mostly unicellular. They can be photosynthetic (algae), or they can feed on other organisms (protozoa). Fungi are eucaryotic cells and are unicellular or multicellular; they have cell walls and are not photosynthetic. Plants have eucaryotic cells, are multicellular, have cell walls, and are photosynthetic. Animals have eucaryotic cells, are multicellular, do not have cell walls, and derive nutrients from other organisms.
After Dolphin, *Biology Lab Manual,* 4th ed., Fig. 14.1, p. 177, McGraw-Hill Companies.

may also indicate origins, how closely related various organisms are, and an approximate timescale for evolutionary history (**figures 1.14** and **1.15**).

Systems of Presenting a Universal Tree of Life

The first phylogenetic trees of life were constructed on the basis of just two kingdoms (plants and animals). In time, it became clear that certain organisms did not truly fit either of those categories, so a third kingdom for simpler organisms that lacked tissue differentiation (protists) was recognized. Eventually, when significant differences became evident even among the protists, a fourth kingdom was proposed for the bacteria. Robert Whittaker built on this work and during the period of 1959–1969 added a fifth kingdom for fungi. Whittaker's five-kingdom system quickly became the standard.

From that time until very recently, phylogenetic trees of life have looked like the one in figure 1.14. The relationships

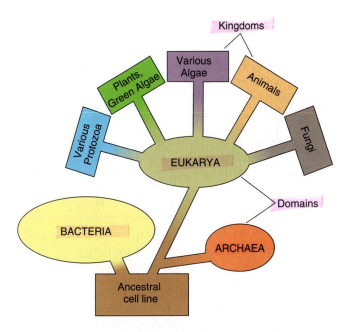

3 cell types, showing relationship with domains and kingdoms

FIGURE 1.15 **Woese-Fox system.**
A system for representing the origins of cell lines and major taxonomic groups as proposed by Carl Woese and colleagues. They propose three distinct cell lines placed in superkingdoms called domains. The first primitive cells, called progenotes, were ancestors of both lines of procaryotes (Domains Bacteria and Archaea), and the Archaea emerged from the same cell line as eucaryotes (Domain Eukarya). Some of the traditional kingdoms are still present with this system (see figure 1.14). Protozoa and some algal groups (called various algae here) are lumped into general categories.

that were considered in constructing the tree were those based on structural similarities and differences, such as bacterial and eucaryotic cellular organization, and the way the organisms got their nutrition. By 1959 these methods indicated that there were five major subdivisions, or kingdoms: the monera, fungi, protists, plants, and animals. Within these kingdoms were two major cell types, the procaryotic and eucaryotic. The system has proved very useful.

Recently, newer methods for determining phylogeny have led to the development of a differently shaped tree—with important implications for our understanding of evolutionary relatedness. The new techniques are those of *molecular biology*, defined as the study of genes—both their structure and function—at the molecular level. Molecular biological methods have demonstrated that certain types of molecules in cells, called small ribosomal ribonucleic acid (rRNA), provide a "living record" of the evolutionary history of an organism. Analysis of this molecule in procaryotic and eucaryotic cells indicates that certain unusual cells called **archaea** (originally archaebacteria) are so different from the other two groups that they should be included in a separate superkingdom. Many archaea are characterized by their abil-

ity to live in extreme environments, such as hot springs or highly salty environments. Under the microscope they resemble bacteria, but molecular biology has revealed that the cells of archaea, though procaryotic in nature, are actually more closely related to eucaryotic cells than to bacterial cells (see table 4.6). To reflect these relationships, Carl Woese and George Fox have proposed a system that assigns all organisms to one of three domains, each described by a different type of cell (see figure 1.15). The procaryotic cell types are placed in the Domains **Archaea** and **Bacteria.** Eucaryotes are all placed in the Domain **Eukarya.** It is believed that these three superkingdoms arose from an ancestor most similar to the archaea. This new system is still undergoing analysis, and it somewhat complicates the presentation of organisms in that it disposes of some traditional groups, although many of the traditional kingdoms still work within this framework. The original Kingdom Protista is now a collection of protozoa and algae that exist in several separate kingdoms (discussed in chapter 5).

This new scheme does not greatly affect our presentation of most microbes, because we discuss them at the genus or species level. But be aware that biological taxonomy, and more importantly, our view of how organisms evolved on earth, is in a period of transition. Keep in mind that our methods of classification reflect our current understanding and are constantly changing as new information is uncovered.

Please note that viruses are *not* included in any of the classification or evolutionary schemes, because they are not cells and their position cannot be given with any confidence. Their special taxonomy is discussed in chapter 6.

✔ **CHECKPOINT**

- Taxonomy is the formal filing system scientists use to classify living organisms. It puts every organism in its place and makes a place for every living organism.
- The taxonomic system has three primary functions: classification, nomenclature, and identification of species.
- The eight major taxa, or groups, in the taxonomic system are (in descending order): domain, kingdom, phylum or division, class, order, family, genus, and species.
- The binomial system of nomenclature describes each living organism by two names: genus and species.
- Taxonomy groups organisms by phylogenetic similarity, which in turn is based on evolutionary similarities in morphology, physiology, and genetics.
- Evolutionary patterns show a treelike branching from simple, primitive life forms to complex, advanced life forms.
- The Whittaker five-kingdom classification system places all bacteria in the Kingdom Procaryotae and subdivides the eucaryotes into Kingdom Protista, Myceteae, Animalia, and Plantae.
- The Woese-Fox classification system places all eucaryotes in the Domain (Superkingdom) Eukarya and subdivides the procaryotes into the two Domains Archaea and Bacteria.

Chapter Summary With Key Terms

1.1 The Scope of Microbiology

A. **Microbiology** is the study of **bacteria, viruses, fungi, protozoa,** and **algae,** which are collectively called **microorganisms,** or microbes. In general, microorganisms are **microscopic** and, unlike **macroscopic** organisms, which are readily visible, they require magnification to be adequately observed or studied.

B. The simplicity, growth rate, and adaptability of microbes are some of the reasons that microbiology is so diverse and has branched out into many subsciences and applications. Important subsciences include **immunology, epidemiology,** public health, food, dairy, aquatic, and industrial microbiology.

1.2 The Impact of Microbes on Earth: Small Organisms with a Giant Effect

Microbes live in most of the world's habitats and are indispensable for normal, balanced life on earth. They play many roles in the functioning of the earth's ecosystems.

A. Microbes are **ubiquitous.**

B. Eucaryotes, which contain nuclei, arose from procaryotes, which do not contain nuclei.

C. Microbes are involved in nutrient production and energy flow. Algae and certain bacteria trap the sun's energy to produce food through **photosynthesis.**

D. Other microbes are responsible for the breakdown and recycling of nutrients through **decomposition.** Microbes are essential to the maintenance of the air, soil, and water.

1.3 Human Use of Microorganisms

Microbes have been called upon to solve environmental, agricultural, and medical problems.

A. **Biotechnology** applies the power of microbes toward the manufacture of industrial products, foods, and drugs.

B. Microbes form the basis of **genetic engineering** and **recombinant DNA** technology, which alter genetic material to produce new products and modified life forms.

C. In **bioremediation,** microbes are used to clean up pollutants and wastes in natural environments.

1.4 Infectious Diseases and the Human Condition

A. Nearly 2,000 microbes are **pathogens** that cause infectious diseases. Infectious diseases result in high levels of mortality and morbidity (illness). Many infections are emerging, meaning that they are newly identified pathogens gaining greater prominence. Many older diseases are also increasing.

B. Some diseases previously thought to be non-infectious may involve microbial infections (e.g., *Helicobacter,* causing gastric ulcers, and Coxsackie viruses, causing diabetes).

C. An increasing number of individuals have weak immune systems, which makes them more susceptible to infectious diseases.

1.5 The General Characteristics of Microorganisms

A. Microbial cells are either the small, relatively simple, non-nucleated procaryotic variety or the larger, more complex eucaryotic type that contain a nucleus and **organelles.**

B. **Viruses** are microorganisms, but are not cells. They are smaller in size and infect their procaryotic or eucaryotic hosts in order to reproduce themselves.

C. **Parasites** are free-living microorganisms that cause damage to their **hosts** through infection and disease.

1.6 The Historical Foundations of Microbiology

A. Microbiology as a science is about 200 years old. Hundreds of contributors have provided discoveries and knowledge to enrich our understanding.

B. With his simple microscope, Leeuwenhoek discovered organisms he called animalcules. As a consequence of his findings and the rise of the **scientific method,** the notion of **spontaneous generation,** or **abiogenesis,** was eventually abandoned for **biogenesis.** The scientific method develops rational **hypotheses** and **theories** that can be tested. Theories that withstand repeated scrutiny become law in time.

C. Early microbiology blossomed with the conceptual developments of **sterilization, aseptic techniques,** and the **germ theory of disease.**

1.7 Taxonomy: Organizing, Classifying, and Naming Microorganisms

A. **Taxonomy** is a hierarchical scheme for the classification, identification, and **nomenclature** of organisms, which are grouped in categories called **taxa,** based on features ranging from general to specific.

B. Starting with the broadest category, the taxa are **domain, kingdom, phylum** (or **division**), **class, order, family, genus,** and **species.** Organisms are assigned **binomial scientific names** consisting of their genus and species names.

C. The latest classification scheme for living things is based on the genetic structure of their ribosomes. The Woese-Fox system recognizes often three domains: **Archaea,** simple procaryotes that often live in extreme environments; **Bacteria,** typical procaryotes; and **Eukarya,** all types of eucaryotic organisms.

D. An alternative classification scheme uses a five-kingdom organization: Kingdom Procaryotae (Monera), containing the **eubacteria** and the archaea; Kingdom Protista, containing primitive unicellular microbes such as algae and protozoa; Kingdom Myceteae, containing the fungi; Kingdom Animalia, containing animals; and Kingdom Plantae, containing plants.

Multiple-Choice Questions

Select the correct answer from the answers provided. For questions with blanks, choose the combination of answers that most accurately completes the statement.

1. Which of the following is not considered a microorganism?
   a. alga
   b. bacterium
   c. protozoan
   d. mushroom

2. An area of microbiology that is concerned with the occurrence of disease in human populations is
   a. immunology
   b. parasitology
   c. epidemiology
   d. bioremediation

3. Which process involves the deliberate alteration of an organism's genetic material?
   a. bioremediation
   b. biotechnology
   c. decomposition
   d. recombinant DNA

4. A prominent difference between procaryotic and eucaryotic cells is the
   a. larger size of procaryotes
   b. lack of pigmentation in eucaryotes
   c. presence of a nucleus in eucaryotes
   d. presence of a cell wall in procaryotes

5. Which of the following parts was absent from Leeuwenhoek's microscopes?
   a. focusing screw
   b. lens
   c. specimen holder
   d. condenser

6. Abiogenesis refers to the
   a. spontaneous generation of organisms from nonliving matter
   b. development of life forms from preexisting life forms
   c. development of aseptic technique
   d. germ theory of disease

7. A hypothesis can be defined as
   a. a belief based on knowledge
   b. knowledge based on belief
   c. a scientific explanation that is subject to testing
   d. a theory that has been thoroughly tested

8. Which early microbiologist was most responsible for developing sterile laboratory techniques?
   a. Louis Pasteur
   b. Robert Koch
   c. Carl von Linné
   d. John Tyndall

9. Which scientist is most responsible for finally laying the theory of spontaneous generation to rest?
   a. Joseph Lister
   b. Robert Koch
   c. Francesco Redi
   d. Louis Pasteur

10. When a hypothesis has been thoroughly supported by long-term study and data, it is considered
    a. a law
    b. a speculation
    c. a theory
    d. proved

11. Which is the correct order of the taxonomic categories, going from most specific to most general?
    a. domain, kingdom, phylum, class, order, family, genus, species
    b. division, domain, kingdom, class, family, genus, species
    c. species, genus, family, order, class, phylum, kingdom, domain
    d. species, family, class, order, phylum, kingdom

12. By definition, organisms in the same _____ are more closely related than are those in the same _____.
    a. order, family
    b. class, phylum
    c. family, genus
    d. phylum, division

13. Which of the following are procaryotic?
    a. bacteria
    b. archaea
    c. protists
    d. both a and b

14. Order the following items by size, using numbers: 1 = smallest and 8 = largest.
    _3_ AIDS virus
    _7_ amoeba
    _4_ rickettsia
    _2_ protein
    _8_ worm
    _5_ coccus-shaped bacterium
    _6_ white blood cell
    _1_ atom

15. Which of the following is not an emerging infectious disease?
    a. SARS
    b. hepatitis C
    c. mononucleosis
    d. AIDS

16. How would you classify a virus?
    a. procaryotic
    b. eucaryotic
    c. neither a nor b

Concept Questions

These questions are suggested as a *writing-to-learn* experience. For each question, compose a one- or two-paragraph answer that includes the factual information needed to completely address the question.

1. Explain the important contributions microorganisms make in the earth's ecosystems.

2. Describe five different ways in which humans exploit microorganisms for our benefit.

3. Identify the groups of microorganisms included in the scope of microbiology, and explain the criteria for including these groups in the field.

4. Why was the abandonment of the spontaneous generation theory so significant? Using the scientific method, describe the steps you would take to test the theory of spontaneous generation.

5. a. Differentiate between a hypothesis and a theory.
   b. Is the germ theory of disease really a law, and why?

6. a. Differentiate between taxonomy, classification, and nomenclature.
   b. What is the basis for a phylogenetic system of classification?
   c. What is a binomial system of nomenclature, and why is it used?
   d. Give the correct order of taxa, going from most general to most specific. Create a mnemonic (memory) device for recalling the order.

7. Compare the new domain system with the five-kingdom system. Does the newer system change the basic idea of procaryotes and eucaryotes? What is the third cell type?

8. Evolution accounts for the millions of different species on the earth and their adaptation to its many and diverse habitats. Explain this. Cite examples in your answer.

Critical Thinking Questions

Critical thinking is the ability to reason and solve problems using facts and concepts. These questions can be approached from a number of angles, and in most cases, they do not have a single correct answer.

1. What do you suppose the world would be like if there were cures for all infectious diseases and a means to destroy all microbes? What characteristics of microbes would prevent this from happening?

2. a. Where do you suppose the "new" infectious diseases come from?
   b. Name some factors that could cause older diseases to show an increase in the number of cases.
   c. Comment on the sensational ways that some tabloid media portray infectious diseases to the public.

3. Look up each disease shown on figure 1.4 in the index and see which ones could be prevented by vaccines or cured with drugs. Are there other ways (besides vaccines) to prevent any of these?

4. Correctly label the types of microorganisms in the drawing on the right, using basic characteristics featured in the chapter. (Organisms are not to scale.)

5. What events, discoveries, or inventions were probably the most significant in the development of microbiology and why?

6. Can you develop a scientific hypothesis and means of testing the cause of stomach ulcers? (Is it caused by an infection? By too much acid? By a genetic disorder?)

7. Where do you suppose viruses came from? Why do they require the host's cellular machinery?

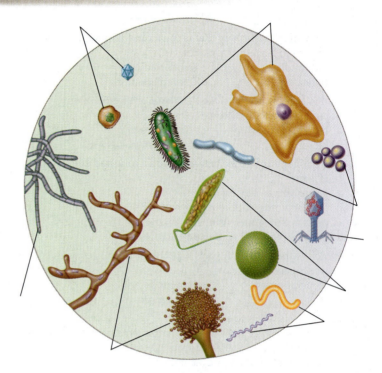

8. Construct the scientific name of a newly discovered species of bacterium, using your name, a pet's name, a place, or a unique characteristic. Be sure to use proper notation and endings.

9. Archaea are often found in hot, sulfuric, acidic, salty habitats, much like the early earth's conditions. Speculate on the origins of life, especially as it relates to the archaea.

Internet Search Topics

1. Using a search engine on the World Wide Web, search for the phrase *emerging diseases*. Adding terms like WHO and CDC will refine your search and take you to several appropriate websites. List the top 10 emerging diseases in the United States and worldwide.

2. Go to the student Online Learning Center for chapter 1 of this text at http://www.mhhe.com/cowan1. Access the URLs listed under Internet Search Topics and research the following:
   a. Explore the "trees of life." Compare the main relationships among the three major domains.
   b. Observe the comparative sizes of microbes arrayed on the head of a pin.
   c. Look at the discussion of biology prefixes and suffixes. A little time spent here could make the rest of your microbiology studies much smoother.

The Chemistry of Biology

IN THE NEWS

In 1996, *The Lancet,* a major British medical research journal most often read by physicians and medical researchers, asked for assistance in treating a 29-year-old man suffering from an unusual malady: severe body odor for 5 years. In 1991, the man pricked his finger with a chicken bone while dressing chickens at work. A physician examined him in a hospital. His finger was abnormally red in color, but not swollen. It also had a distinct odor. The patient was treated with several prolonged courses of antibiotics, hyperbaric oxygen, and ultraviolet light without any benefits. Surgery was performed on his finger in order to detect any pus-containing bacteria or soft tissue damage. None was detected. Skin biopsy samples were performed and found to be normal.

Eventually, the skin coloring of his finger returned to normal. His odor continued. The smell emanating from his affected arm "could be detected across a large room, and when confined to a smaller examination room became almost intolerable." Five years later the man still suffered from this odor. He experienced depression and social isolation.

The following bacteria were repeatedly isolated from samples of skin scrapings taken from his finger and arm: *Clostridium novyi, Clostridium cochlearium,* and *Clostridium malenominatum.* All antibiotics had failed to eradicate these bacteria. At the time, these bacteria were not known to grow on the skin or cause disease. All of them had occasionally been found in chickens. These organisms had become part of the patient's normal microbial skin flora.

▶ *What is the definition of normal microbial flora?*

▶ *Could the three different species of* Clostridium *bacteria be responsible for the man's body odor?*

CHAPTER OVERVIEW

▶ The understanding of living cells and processes is enhanced by a knowledge of chemistry.

▶ The structure and function of all matter in the universe is based on atoms.

▶ Atoms have unique structures and properties that allow chemical reactions to occur.

▶ Atoms contain protons, neutrons, and electrons in combinations to form elements.

▶ Living things are composed of approximately 25 different elements.

▶ Elements interact to form bonds that result in molecules and compounds with different characteristics than the elements that form them.

▶ Atoms and molecules undergo chemical reactions such as oxidation/reduction, ionization, and dissolution.

▶ The properties of carbon have been critical in forming macromolecules of life such as proteins, fats, carbohydrates, and nucleic acids.

▶ The structure and shape of a macromolecule dictate its functions.

▶ Cells carry out fundamental activities of life, such as growth, metabolism, reproduction, synthesis, and transport, that are all essentially chemical reactions on a grand scale.

2.1 Atoms, Bonds, and Molecules: Fundamental Building Blocks

The universe is composed of an infinite variety of substances existing in the gaseous, liquid, and solid states. All such tangible materials that occupy space and have mass are called **matter.** The organization of matter—whether air, rocks, or bacteria—begins with individual building blocks called atoms. An **atom** is defined as a tiny particle that cannot be subdivided into smaller substances without losing its properties. Even in a science dealing with very small things, an atom's minute size is striking; for example, an oxygen atom is only 0.0000000013 mm (0.0013 nm) in diameter, and one million of them in a cluster would barely be visible to the naked eye.

Although scientists have not directly observed the detailed structure of an atom, the exact composition of atoms has been well established by extensive physical analysis using sophisticated instruments. In general, an atom derives its properties from a combination of subatomic particles called **protons** (p^+), which are positively charged; **neutrons** (n^0), which have no charge (are neutral); and **electrons** (e^-), which are negatively charged. The relatively larger protons and neutrons make up a central core, or **nucleus,**[1] that is surrounded by one or more electrons **(figure 2.1).** The nucleus makes up the larger mass (weight) of the atom,

---

1. Be careful not to confuse the nucleus of an atom with the nucleus of a cell (discussed later).

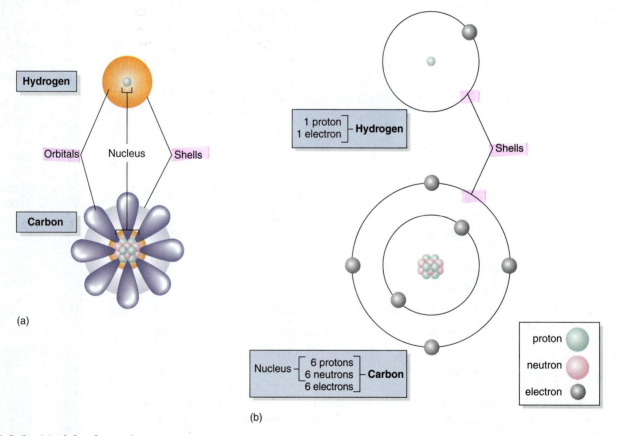

(a)

(b)

FIGURE 2.1   Models of atomic structure.
(a) Three-dimensional models of hydrogen and carbon that approximate their actual structure. The nucleus is surrounded by electrons in orbitals that occur in levels called shells. Hydrogen has just one shell and one orbital. Carbon has two shells and four orbitals; the shape of the outermost orbitals is paired lobes rather than circles or spheres. **(b)** Simple models of the same atoms make it easier to show the numbers and arrangements of shells and electrons, and the numbers of protons and neutrons in the nucleus. (Not to accurate scale.)

whereas the electron region accounts for the greater volume. To get a perspective on proportions, consider this: If an atom were the size of a football stadium, the nucleus would be about the size of a marble! The stability of atomic structure is largely maintained by: (1) the mutual attraction of the protons and electrons (opposite charges attract each other) and (2) the exact balance of proton number and electron number, which causes the opposing charges to cancel each other out. At least in theory, then, isolated intact atoms do not carry a charge.

Different Types of Atoms: Elements and Their Properties

All atoms share the same fundamental structure. All protons are identical, all neutrons are identical, and all electrons are identical. But when these subatomic particles come together in specific, varied combinations, unique types of atoms called **elements** result. Each element has a characteristic atomic structure and predictable chemical behavior. To date, 115 elements, both naturally occurring and artificially produced by physicists, have been described. By convention, an element is assigned a distinctive name with an abbreviated shorthand symbol. The elements are often depicted in a periodic table. **Table 2.1** lists some of the elements common to biological systems, their atomic characteristics, and some of the natural and applied roles they play.

The Major Elements of Life and Their Primary Characteristics

The unique properties of each element result from the numbers of protons, neutrons, and electrons it contains, and each element can be identified by certain physical measurements.

Each element is assigned an **atomic number (AN)** based on the number of protons it has. The atomic number is a valuable measurement because an element's proton number does not vary, and knowing it automatically tells you the usual number of electrons (recall that a neutral atom has an equal number of protons and electrons). Another useful measurement is the **mass[2] number (MN)**, equal to the number of protons and neutrons. If one knows the mass number and the atomic number, it is possible to determine the numbers of neutrons by subtraction. Hydrogen is a unique element because its common form has only one proton, one electron, and no neutron, making it the only element with the same atomic and mass number.

Isotopes are variant forms of the same element that differ in the number of neutrons and thus have different mass numbers. These multiple forms occur naturally in certain proportions. Carbon, for example, exists primarily as carbon 12 with 6 neutrons (MN=12); but a small amount (about 1%)

is carbon 13 with 7 neutrons and carbon 14 with 8 neutrons. Although isotopes have virtually the same chemical properties, some of them have unstable nuclei that spontaneously release energy in the form of radiation. Such *radioactive isotopes* play a role in a number of research and medical applications. Because they emit detectable signs, they can be used to trace the position of key atoms or molecules in chemical reactions, they are tools in diagnosis and treatment, and they are even applied in sterilization procedures (see ionizing radiation in chapter 11). Another application of isotopes is in dating fossils and other ancient materials **(Insight 2.1).** An element's **atomic weight** is the average of the mass numbers of all its isotopic forms (table 2.1).

Electron Orbitals and Shells

The structure of an atom can be envisioned as a central nucleus surrounded by a "cloud" of electrons that constantly rotate about the nucleus in pathways (see figure 2.1). The pathways, called **orbitals,** are not actual objects or exact locations, but represent volumes of space in which an electron is likely to be found. Electrons occupy energy shells, proceeding from the lower-level energy electrons nearest the nucleus to the higher-energy electrons in the farthest orbitals.

Electrons fill the orbitals and shells in *pairs,* starting with the shell nearest the nucleus. The first shell contains one orbital and a maximum of 2 electrons; the second shell has four orbitals and up to 8 electrons; the third shell with 9 orbitals can hold up to 18 electrons; and the fourth shell with 16 orbitals contains up to 32 electrons. The number of orbitals and shells and how completely they are filled depends on the numbers of electrons, so that each element will have a unique pattern. For example, helium (AN=2) has only a filled first shell of 2 e^-; oxygen (AN=8) has a filled first shell and a partially filled second shell of 6 e^-; and magnesium (AN=12) has a filled first shell, a filled second one, and a third shell that fills only one orbital, so is nearly empty. As we will see, the chemical properties of an element are controlled mainly by the distribution of electrons in the outermost shell. Figure 2.1 and **figure 2.2** present various simplified models of atomic structure and electron maps.

---

2. Mass refers to the amount of matter that a particle contains. The proton and neutron have almost exactly the same mass, which is about 1.7×10^{-24} g, or 1 dalton.

✔ CHECKPOINT

- Protons (p^+) and neutrons (n^0) make up the nucleus of an atom. Electrons (e^-) orbit the nucleus.
- All elements are composed of atoms but differ in the numbers of protons, neutrons, and electrons they possess.
- Elements are identified by *atomic weight*, or *mass*, or by *atomic number.*
- Isotopes are varieties of one element that contain the same number of protons but different numbers of neutrons.
- The number of electrons in an element's outermost orbital (compared with the total number possible) determines its chemical properties and reactivity.

TABLE 2.1   The Major Elements of Life and Their Primary Characteristics

Element	Atomic Symbol*	Atomic Number	Atomic Weight	Ionized Form**	Significance in Microbiology
Calcium	Ca	20	40.1	Ca^{++}	Part of outer covering of certain shelled amoebas; stored within bacterial spores
Carbon	C	6	12.0	—	Principal structural component of biological molecules
Carbon•	C14	6	14.0	—	Radioactive isotope used in dating fossils
Chlorine	Cl	17	35.5	Cl^-	Component of disinfectants; used in water purification
Cobalt	Co	27	58.9	Co^{++}, Co^{+++}	Trace element needed by some bacteria to synthesize vitamins
Cobalt•	Co60	27	60	—	An emitter of gamma rays; used in food sterilization; used to treat cancer
Copper	Cu	29	63.5	Cu^+, Cu^{++}	Necessary to the function of some enzymes; Cu salts are used to treat fungal and worm infections
Hydrogen	H	1	1	H^+	Necessary component of water and many organic molecules; H_2 gas released by bacterial metabolism
Hydrogen•	H3	1	3	—	Tritium has 2 neutrons; radioactive; used in clinical laboratory procedures
Iodine	I	53	126.9	I^-	A component of antiseptics and disinfectants; contained in a reagent of the Gram stain
Iodine•	I131, I125	53	131, 125		Radioactive isotopes for diagnosis and treatment of cancers
Iron	Fe	26	55.8	Fe^{++}, Fe^{+++}	Necessary component of respiratory enzymes; some microbes require it to produce toxin
Magnesium	Mg	12	24.3	Mg^{++}	A trace element needed for some enzymes; component of chlorophyll pigment
Manganese	Mn	25	54.9	Mn^{++}, Mn^{+++}	Trace element for certain respiratory enzymes
Nitrogen	N	7	14.0	—	Component of all proteins and nucleic acids; the major atmospheric gas
Oxygen	O	8	16.0	—	An essential component of many organic molecules; molecule used in metabolism by many organisms
Phosphorus	P	15	31	—	A component of ATP, nucleic acids, cell membranes; stored in granules in cells
Phosphorus•	P32	15	32	—	Radioactive isotope used as a diagnostic and therapeutic agent
Potassium	K	19	39.1	K^+	Required for normal ribosome function and protein synthesis; essential for cell membrane permeability
Sodium	Na	11	23.0	Na^+	Necessary for transport; maintains osmotic pressure; used in food preservation
Sulfur	S	16	32.1	—	Important component of proteins; makes disulfide bonds; storage element in many bacteria
Zinc	Zn	30	65.4	Zn^{++}	An enzyme cofactor; required for protein synthesis and cell division; important in regulating DNA

*Based on the Latin name of the element. The first letter is always capitalized; if there is a second letter, it is always lowercased.

**A dash indicates an element that is usually found in combination with other elements, rather than as an ion.

INSIGHT 2.1   *Microbiology*

Searching for Ancient Life with Isotopes

Determining the age of the earth and the historical time frame of living things has long been a priority of biologists. Much evidence comes from fossils, geologic sediments, and genetic studies, yet there has always been a need for an exacting scientific reference for tracing samples back in time, possibly even to the beginnings of the earth itself. One very precise solution to this problem comes from patterns that exist in isotopes. The isotopes of an element have the same basic chemical structure, but over billions of years, they have come to vary slightly in the number of neutrons. For example, carbon has three isotopes: C12, predominantly found in living things; C13, a less common form associated with nonliving matter; and C14, a radioactive isotope. All isotopes exist in relatively predictable proportions in the earth, solar system, and even universe, so that any variations from the expected ratios would indicate some other factor besides random change.

Isotope chemists use giant machines called microprobes to analyze the atomic structures in fossils and rock samples. These amazing machines can rapidly sort and measure the types and amounts of isotopes, which reflect a sample's age and possibly its origins. The accuracy of this method is such that it can be used like an "atomic clock." It was recently used to verify the dateline for the origins of the first life forms, using 3.85-billion-year-old sediment samples from Greenland. Testing indicated that the content of C12 in the samples was substantially higher than the amount in inorganic rocks, and it was concluded that living cells must have accumulated the C12. This finding shows that the origin of life was 400 million years earlier than the previous estimates.

In a separate study, some ancient Martian meteorites were probed to determine if certain microscopic rods could be some form of microbes (see figure 1.10*b*). By measuring the ratios of oxygen isotopes in carbonate ions (CO_3^{2-}), chemists were able to detect significant fluctuations in the isotopes from different parts of the same meteorite. Such differences would most likely be caused by huge variations in temperature or other extreme environments that are incompatible with life. From this evidence, they concluded that the tiny rods were not Martian microbes.

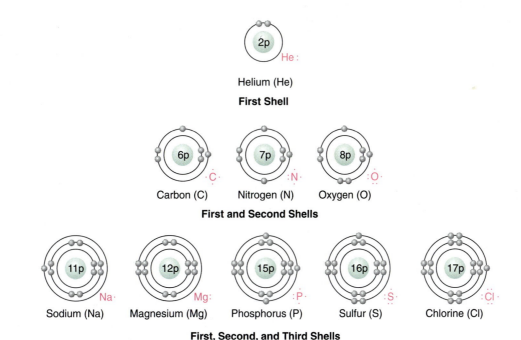

FIGURE 2.2 **Electron orbitals and shells.**
Models of several elements show how the shells are filled by electrons as the atomic numbers increase (numbers noted inside nuclei). Electrons tend to appear in pairs, but certain elements have incompletely filled outer shells. Chemists depict elements in shorthand form (red Lewis structures) that indicate only the valence electrons, since these are the electrons involved in chemical bonds.

Bonds and Molecules

Most elements do not exist naturally in pure, uncombined form but are bound together as molecules and compounds. A **molecule** is a distinct chemical substance that results from the combination of two or more atoms. Some molecules such as oxygen (O_2) and nitrogen gas (N_2) consist of atoms of the same element. Molecules that are combinations of two or more *different* elements are termed **compounds.** Compounds such as water (H_2O) and biological molecules (proteins, sugars, fats) are the predominant substances in living systems. When atoms bind together in molecules, they lose the properties of the atom and take on the properties of the combined substance. In the same way that an atom has an atomic weight, a molecule has a molecular weight (MW), which is calculated from the sum of all of the atomic weights of the atoms it contains.

The **chemical bonds** of molecules and compounds result when two or more atoms share, donate (lose), or accept (gain) electrons **(figure 2.3).** The number of electrons in the outermost shell of an element is known as its **valence.** The valence determines the degree of reactivity and the types of bonds an element can make. Elements with a filled outer orbital are relatively stable because they have no extra electrons to share with or donate to other atoms. For example, helium has one filled shell, with no tendency either to give up electrons or to take them from other elements, making it a stable, inert (nonreactive) gas. Elements with partially filled outer orbitals are less stable and are more apt to form some sort of bond. Many chemical reactions are based on the tendency of atoms with unfilled outer shells to gain greater stability by achieving, or at least approximating, a filled outer shell. For example, an atom such as oxygen that can accept 2 additional electrons will bond readily with atoms (such as hydrogen) that can share or donate electrons. We explore some additional examples of the basic types of bonding in the following section.

In addition to reactivity, the number of electrons in the outer shell also dictates the number of chemical bonds an atom can make. For instance, hydrogen can bind with one other atom, oxygen can bind with up to two other atoms, and carbon can bind with four.

Covalent Bonds and Polarity: Molecules with Shared Electrons

Covalent (cooperative valence) **bonds** form between atoms with valences that suit them to sharing electrons rather than to donating or receiving them. A simple example is hydrogen gas (H_2), which consists of two hydrogen atoms. A hydrogen atom has only a single electron, but when two of them combine, each will bring its electron to orbit about both nuclei, thereby approaching a filled orbital (2 electrons) for both atoms and thus creating a single **covalent** bond **(figure 2.4a).** Covalent bonding also occurs in oxygen gas (O_2), but with a difference. Because each atom has 2 electrons to share in this molecule, the combination creates two pairs of shared electrons, also known as a double covalent bond **(figure 2.4b).** The majority of the molecules associated with living things are composed of single and double covalent bonds between the most common biological elements (carbon, hydrogen, oxygen, nitrogen, sulfur, and phosphorus), which are discussed in more depth in chapter 7. A slightly more complex pattern of covalent bonding is shown for methane gas (CH_4) in **figure 2.4c.**

Other effects of bonding result in differences in polarity. When atoms of different electronegativity[3] form covalent bonds, the electrons are not shared equally and may be pulled more toward one atom than another. This pull causes one end of a molecule to assume a partial negative charge and the other end to assume a partial positive charge. A molecule with such an asymmetrical distribution of charges is termed **polar** and has positive and

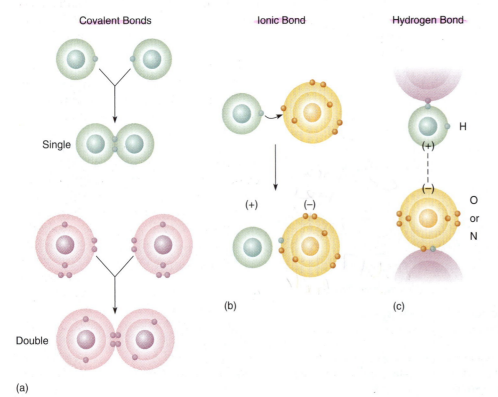

(a)

FIGURE 2.3   General representation of three types of bonding.
(a) Covalent bonds, both single and double. **(b)** Ionic bond. **(c)** Hydrogen bond. Note that hydrogen bonds are represented in models and formulas by dotted lines, as shown in **(c).**

3. Electronegativity—the ability to attract electrons.

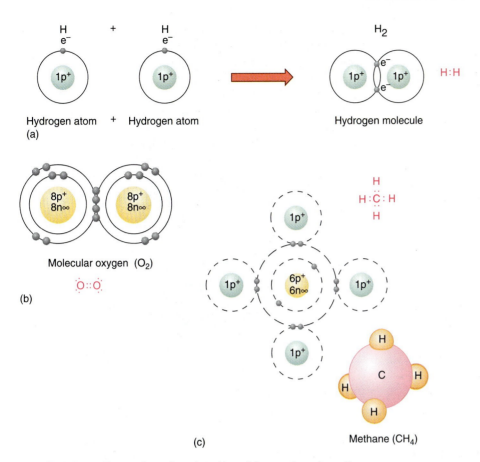

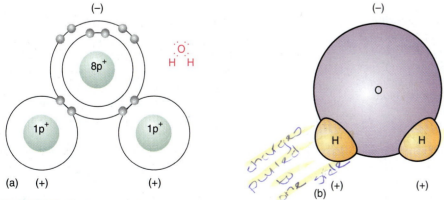

FIGURE 2.4 **Examples of molecules with covalent bonding.**
(a) A hydrogen molecule is formed when two hydrogen atoms share their electrons and form a single bond. **(b)** In a double bond, the outer orbitals of two oxygen atoms overlap and permit the sharing of 4 electrons (one pair from each) and the saturation of the outer orbital for both. **(c)** Simple, working, and three-dimensional models of methane. Note that carbon has 4 electrons to share and hydrogens each have one, thereby completing the shells for all atoms in the compound, and creating 4 single bonds.

FIGURE 2.5 **Polar molecule.**
(a) Simple model and **(b)** a three-dimensional model of a water molecule indicate the polarity, or unequal distribution, of electrical charge, which is caused by the pull of the shared electrons toward the oxygen side of the molecule.

negative poles. Observe the water molecule shown in **figure 2.5** and note that, because the oxygen atom is larger and has more protons than the hydrogen atoms, it will tend to draw the shared electrons with greater force toward its nucleus. This unequal force causes the oxygen part of the molecule to express a negative charge (due to the electrons' being attracted there) and the hydrogens to express a positive charge (due to the protons). The polar nature of water plays an extensive role in a number of biological reactions, which are discussed later. Polarity is a significant property of many large molecules in living systems and greatly influences both their reactivity and their structure.

When covalent bonds are formed between atoms that have the same or similar electronegativity, the electrons are shared equally between the two atoms. Because of this balanced distribution, no part of the molecule has a greater attraction for the electrons. This sort of electrically neutral molecule is termed **nonpolar.**

Ionic Bonds: Electron Transfer Among Atoms

In reactions that form **ionic bonds,** electrons are transferred completely from one atom to another and are not shared. These reactions invariably occur between atoms with valences that complement each other, meaning that one atom has an unfilled shell that will readily accept electrons and the other atom has an unfilled shell that will readily lose electrons. A striking example is the reaction that occurs between sodium (Na) and chlorine (Cl). Elemental sodium is a soft, lustrous metal so reactive that it can burn flesh, and molecular chlorine is a very poisonous yellow gas. But when the two are combined, they form sodium chloride[4] (NaCl)—the familiar nontoxic table salt—a compound with properties quite different from either parent element (**figure 2.6**).

How does this transformation occur? Sodium has 11 electrons (2 in shell one, 8 in shell two, and only 1 in shell three), so it is 7 short of having a complete outer

4. In general, when a salt is formed, the ending of the name of the negatively charged ion is changed to -ide.

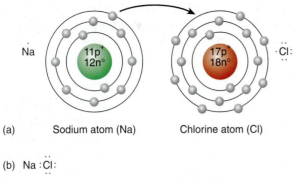

(a) Sodium atom (Na) Chlorine atom (Cl)

(b) Na :Cl:

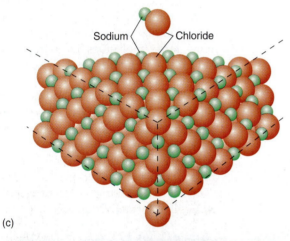

(c)

(d)

FIGURE 2.6 Ionic bonding between sodium and chlorine.
(a) When the two elements are placed together, sodium loses its single outer orbital electron to chlorine, thereby filling chlorine's outer shell. **(b)** Simple model of ionic bonding. **(c)** Sodium and chloride ions form large molecules, or crystals, in which the two atoms alternate in a definite, regular, geometric pattern. **(d)** Note the cubic nature of NaCl crystals at the macroscopic level.

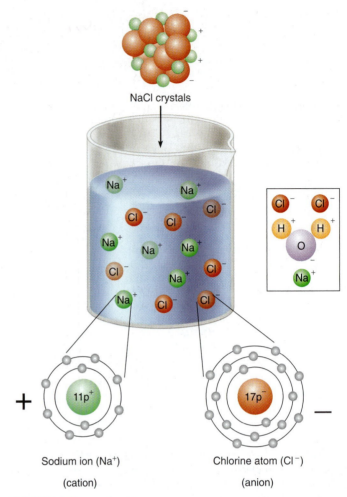

FIGURE 2.7 Ionization.
When NaCl in the crystalline form is added to water, the ions are released from the crystal as separate charged particles (cations and anions) into solution. (See also figure 2.11.) In this solution, Cl^- ions are attracted to the hydrogen component of water, and Na^+ ions are attracted to the oxygen (box).

shell. Chlorine has 17 electrons (2 in shell one, 8 in shell two, and 7 in shell three), making it 1 short of a complete outer shell. These two atoms are very reactive with one another, because a sodium atom will readily donate its single electron and a chlorine atom will avidly receive it. (The reaction is slightly more involved than a single sodium atom's combining with a single chloride atom **(Insight 2.2)**, but this complexity does not detract from the fundamental reaction as described here.) The outcome of this reaction is not many single, isolated molecules of NaCl but rather a solid crystal complex that interlinks millions of sodium and chloride ions **(figure 2.6c and d).**

Ionization: Formation of Charged Particles Molecules with intact ionic bonds are electrically neutral, but they can produce charged particles when dissolved in a liquid called a solvent. This phenomenon, called **ionization,** occurs when the ionic bond is broken and the atoms dissociate (separate) into unattached, charged particles called **ions** **(figure 2.7).** To illustrate what imparts a charge to ions, let us look again at the reaction between sodium and chlorine. When a sodium atom reacts with chlorine and loses one electron, the sodium is left with one more proton than electrons. This imbalance produces a positively charged sodium ion (Na^+). Chlorine, on the other hand, has gained one electron and now has one more electron than protons, producing a negatively charged ion (Cl^-). Positively charged ions are termed **cations,** and negatively charged ions are termed **anions.** (A good mnemonic device is to think of the "t" in cation as a plus (+) sign and the first "n" in anion as a negative (−) sign.) Substances such as salts, acids, and bases that release ions when dissolved in water are termed **electrolytes** because their charges enable them

INSIGHT 2.2   *Discovery*

Redox: Electron Transfer and Oxidation-Reduction Reactions

The metabolic work of cells, such as synthesis, movement, and digestion, revolves around energy exchanges and transfers. The management of energy in cells is almost exclusively dependent on chemical rather than physical reactions because most cells are far too delicate to operate with heat, radiation, and other more potent forms of energy. The outer-shell electrons are readily portable and easily manipulated sources of energy. It is in fact the movement of electrons from molecule to molecule that accounts for most energy exchanges in cells. Fundamentally, then, a cell must have a supply of atoms that can gain or lose electrons if they are to carry out life processes.

The phenomenon in which electrons are transferred from one atom or molecule to another is termed an **oxidation** and **reduction** (shortened to **redox**) **reaction.** Although the term *oxidation* was originally adopted for reactions involving the addition of oxygen, the term oxidation can include any reaction causing electron release, regardless of the involvement of oxygen. By comparison, reduction is any reaction that causes an atom to receive electrons. All redox reactions occur in pairs. To analyze the phenomenon, let us again review the production of NaCl, but from a different standpoint. Although it is true that these atoms form ionic bonds, the chemical combination of the two is also a type of redox reaction.

When these two atoms react to form sodium chloride, a sodium atom gives up an electron to a chlorine atom. During this reaction, sodium is oxidized because it loses an electron, and

chlorine is reduced because it gains an electron. To take this definition further, an atom or molecule, such as sodium, that can donate electrons and thereby reduce another molecule is a reducing agent; one that can receive extra electrons and thereby oxidize another molecule is an oxidizing agent. You may find this concept easier to keep straight if you think of redox agents as partners: The one that gives its electrons away is oxidized; the partner that receives the electrons is reduced. (A mnemonic device to keep track of this is *LEO* says *GER: Lose* Electrons *Oxidized; Gain* Electrons *Reduced.*)

Redox reactions are essential to many of the biochemical processes discussed in chapter 8. In cellular metabolism, electrons alone can be transferred from one molecule to another as described here, but sometimes oxidation and reduction occur with the transfer of hydrogen atoms (which are a proton and an electron) from one compound to another.

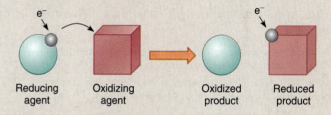

Simplified diagram of the exchange of electrons during an oxidation-reduction reaction.

to conduct an electrical current. Owing to the general rule that particles of like charge repel each other and those of opposite charge attract each other, we can expect ions to interact electrostatically with other ions and polar molecules. Such interactions are important in many cellular chemical reactions, in the formation of solutions, and in the reactions microorganisms have with dyes. The transfer of electrons from one molecule to another constitutes a significant mechanism by which biological systems store and release energy.

Hydrogen Bonding   Some types of bonding do not involve sharing, losing, or gaining electrons, but instead are due to attractive forces between nearby molecules or atoms. One such bond is a **hydrogen bond,** a weak type of bond that forms between a hydrogen covalently bonded to one molecule and an oxygen or nitrogen atom on the same molecule or on a different molecule. Because hydrogen in a covalent bond tends to be positively charged, it will attract a nearby negatively charged atom and form an easily disrupted bridge with it. This type of bonding is usually represented in molecular models with a dotted line. A simple example of hydrogen bonding occurs between water molecules **(figure 2.8).** More extensive hydrogen bonding is partly responsible for the structure and stability of proteins and nucleic acids, as you will see later on.

IN THE NEWS   *(Continued from page 25)*

The opening chapter case was about a man whose normal skin flora had been changed by the colonization of three unusual microbes from the same bacterial genus (*Clostridium*) after being pricked by a chicken bone. *Clostridium* infections in humans are rare. Two examples are *Clostridium tetani,* the cause of tetanus, and *Clostridium perfringens,* the cause of gas gangrene. Using laboratory techniques the man's odor was determined to be caused by compounds produced by the clostridial bacteria during their metabolism. Metabolism is a series of chemical reactions that provide energy in a form the organism can use for its own purposes. In simplified terms, the energy comes from an electron that is donated by an atom at the beginning of the process and accepted by another atom at the end of the process. One of the odor-causing chemical compounds produced was *N*-butyric acid (or normal-butyric acid). *N*-butyric acid is a fatty acid commonly found in rancid butter or vomit and has an unpleasant odor. In this case, the *N*-butyric acid was produced by the *Clostridium* species through the anaerobic fermentation of carbohydrates. Carbohydrates are a basic class of organic compound that are often used by organisms to store or consume energy. The chemistry of anaerobic respiration is discussed in chapter 8.

N-butyric acids have been shown to inhibit the growth of other bacteria. This may help to explain the altered indigenous flora of the

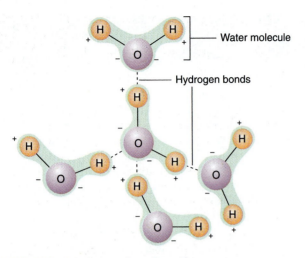

FIGURE 2.8   Hydrogen bonding in water.
Because of the polarity of water molecules, the negatively charged oxygen end of one water molecule is weakly attracted to the positively charged hydrogen end of an adjacent water molecule.

patient, allowing for the overgrowth of three different species of *Clostridium.*

See: Mills, C. M., Llewelyn, M. B., Kelly, D. R., and Holt, P. 1996. A man who pricked his finger and smelled putrid for 5 years. Case report. Lancet 348:1282.

Araki, Y., et al. 2002. Oral administration of a product derived from Clostridium butyricum in rats. Int. J. Mol. Med. 9:53–57.

Chemical Shorthand: Formulas, Models, and Equations   The atomic content of molecules can be represented by a few convenient formulas. We have already been exposed to the molecular formula, which concisely gives the atomic symbols and the number of the elements involved in subscript (CO_2, H_2O). More complex molecules such as glucose ($C_6H_{12}O_6$) can also be symbolized this way, but this formula is not unique, since fructose and galactose also share it. Molecular formulas are useful, but they only summarize the atoms in a compound; they do not show the position of bonds between atoms. For this purpose, chemists use structural formulas illustrating the relationships of the atoms and the number and types of bonds **(figure 2.9).** Other structural models present the three-dimensional appearance of a molecule, illustrating the orientation of atoms (differentiated by color codes) and the molecule's overall shape **(figure 2.10).**

The printed page tends to make molecules appear static, but this picture is far from correct, because molecules are capable of changing through chemical reactions. For ease in tracing chemical exchanges between atoms or molecules, and to derive some sense of the dynamic character of reactions, chemists use shorthand equations containing symbols, numbers, and arrows to simplify or summarize the major characteristics of a reaction. Molecules entering or starting a reaction are called **reactants,** and substances left by a reaction are called products. In most instances, summary chemical reactions do not give the details of the exchange, in order to keep the expression simple and to save space.

In a *synthesis reaction,* the reactants bond together in a manner that produces an entirely new molecule (reactant A plus reactant B yields product AB). An example is the production of sulfur dioxide, a by-product of burning sulfur fuels and an important component of smog:

$$S + O_2 \rightarrow SO_2$$

Some synthesis reactions are not such simple combinations. When water is synthesized, for example, the reaction does not really involve one oxygen atom combining with two hydrogen atoms, because elemental oxygen exists as O_2 and elemental hydrogen exists as H_2. A more accurate equation for this reaction is:

$$2H_2 + O_2 \rightarrow 2H_2O$$

The equation for reactions must be balanced—that is, the number of atoms on one side of the arrow must equal the number on the other side to reflect all of the participants in the reaction. To arrive at the total number of atoms in the reaction, multiply the prefix number by the subscript number; if no number is given, it is assumed to be 1.

In *decomposition reactions,* the bonds on a single reactant molecule are permanently broken to release two or more product molecules. One example is the resulting molecules when large nutrient molecules are digested into smaller units; a simpler example can be shown for the common chemical hydrogen peroxide:

$$2H_2O_2 \rightarrow 2H_2O + O_2$$

During *exchange reactions,* the reactants trade portions between each other and release products that are combinations of the two. This type of reaction occurs between acids and bases when they form water and a salt:

$$AB + XY \rightleftharpoons AX + BY$$

The reactions in biological systems can be reversible, meaning that reactants and products can be converted back and forth. These reversible reactions are symbolized with a double arrow, each pointing in opposite directions, as in the exchange reaction above. Whether a reaction is reversible depends on the proportions of these compounds, the difference in energy state of the reactants and products, and the presence of **catalysts** (substances that increase the rate of a reaction). Additional reactants coming from another reaction can also be indicated by arrows that enter or leave at the main arrow:

$$X + Y \xrightarrow{\quad CD \searrow C \quad} XYD$$

Solutions: Homogeneous Mixtures of Molecules

A **solution** is a mixture of one or more substances called **solutes** uniformly dispersed in a dissolving medium called a **solvent.** An important characteristic of a solution is that the solute cannot be separated by filtration or ordinary settling.

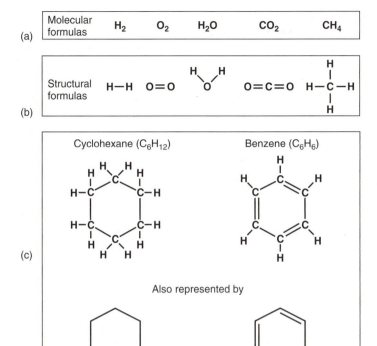

FIGURE 2.9 Comparison of molecular and structural formulas.

(a) Molecular formulas provide a brief summary of the elements in a compound. **(b)** Structural formulas clarify the exact relationships of the atoms in the molecule, depicting single bonds by a single line and double bonds by two lines. **(c)** In structural formulas of organic compounds, cyclic or ringed compounds may be completely labeled, or **(d)** they may be presented in a shorthand form in which carbons are assumed to be at the angles and attached to hydrogens. See figure 2.14 for structural formulas of three sugars with the same molecular formula, $C_6H_{12}O_6$.

FIGURE 2.10 Three-dimensional, or space-filling, models of (a) water, (b) carbon dioxide, and (c) glucose.
The red atoms are oxygen, the white ones hydrogen, and the black ones carbon.

The solute can be gaseous, liquid, or solid, and the solvent is usually a liquid. Examples of solutions are salt or sugar dissolved in water and iodine dissolved in alcohol. In general, a solvent will dissolve a solute only if it has similar electrical characteristics as indicated by the rule of solubility, expressed simply as "like dissolves like." For example, water is a polar molecule and will readily dissolve an ionic solute such as NaCl, yet a nonpolar solvent such as benzene will not dissolve NaCl.

Water is the most common solvent in natural systems, having several characteristics that suit it to this role. The polarity of the water molecule causes it to form hydrogen bonds with other water molecules, but it can also interact readily with charged or polar molecules. When an ionic solute such as NaCl crystals is added to water, it is dissolved, thereby releasing Na^+ and Cl^- into solution. Dissolution occurs because Na^+ is attracted to the negative pole of the water molecule and Cl^- is attracted to the positive pole; in this way, they are drawn away from the crystal separately into solution. As it leaves, each ion becomes **hydrated,** which means that it is surrounded by a sphere of water molecules **(figure 2.11).** Molecules such as salt or sugar that attract water to their surface are termed **hydrophilic.** Nonpolar molecules, such as benzene, that repel water are considered **hydrophobic.** A third class of molecules, such as the phospholipids in cell membranes, are considered **amphipathic** because they have both hydrophilic and hydrophobic properties.

Because most biological activities take place in aqueous (water-based) solutions, the concentration of these solutions can be very important (see chapter 7). The **concentration** of a solution expresses the amount of solute dissolved in a certain amount of solvent. It can be calculated by weight, volume, or percentage. A common way to calculate percentage of concentration is to use the weight of the solute, measured in grams (g), dissolved in a specified volume of solvent, measured in milliliters (ml). For example, dissolving 3 g of NaCl in 100 ml of water produces a 3% solution; dissolving 30 g in 100 ml produces a 30% solution; and dissolving 3 g in 1,000 ml (1 liter) produces a 0.3% solution. A solution with a small amount of solute and a relatively greater amount of solvent (0.3%) is considered dilute or weak. On the other hand, a solution containing significant percentages of solute (30%) is considered concentrated or strong.

A common way to express concentration of biological solutions is by its molar concentration, or *molarity* (M). A standard molar solution is obtained by dissolving one *mole,* defined as the molecular weight of the compound in grams, in 1 L (1,000 ml) of solution. To make a 1 M solution of sodium chloride, we would dissolve 58 g of NaCl to give

FIGURE 2.11
Hydration spheres formed around ions in solution.
In this example, a sodium cation attracts the negatively charged region of water molecules, and a chloride anion attracts the positively charged region of water molecules. In both cases, the ions become covered with spherical layers of specific numbers and arrangements of water molecules.

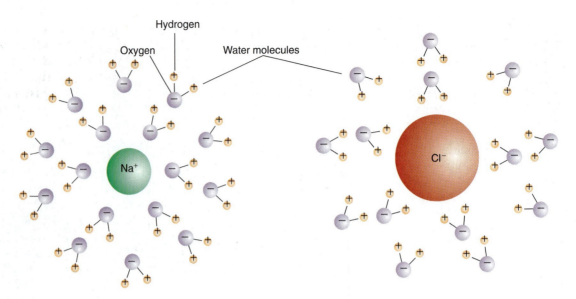

1 L of solution; a 0.1 M solution would require 5.8 g of NaCl in 1 L of solution.

Acidity, Alkalinity, and the pH Scale

Another factor with far-reaching impact on living things is the concentration of acidic or basic solutions in their environment. To understand how solutions develop acidity or basicity, we must look again at the behavior of water molecules. Hydrogens and oxygen tend to remain bonded by covalent bonds, but in certain instances, a single hydrogen can break away as the ionic form (H^+), leaving the remainder of the molecule in the form of an OH^- ion. The H^+ ion is positively charged because it is essentially a hydrogen ion that has lost its electron; the OH^- is negatively charged because it remains in possession of that electron. Ionization of water is constantly occurring, but in pure water containing no other ions, H^+ and OH^- are produced in equal amounts, and the solution remains neutral. By one definition, a solution is considered **acidic** when a component dissolved in water (acid) releases excess hydrogen ions[5] (H^+); a solution is **basic** when a component releases excess hydroxyl ions (OH^-), so that there is no longer a balance between the two ions.

To measure the acid and base concentrations of solutions, scientists use the **pH** scale, a graduated numerical scale that ranges from 0 (the most acidic) to 14 (the most basic). This scale is a useful standard for rating relative acidity and basicity; use **figure 2.12** to familiarize yourself with the pH readings of some common substances. It is not an arbitrary scale but actually a mathematical derivation based on the negative logarithm (reviewed in appendix B) of the concentration of H^+ ions in moles per liter (symbolized as [H^+]) in a solution, represented as:

$$pH = -\log[H^+]$$

Acidic solutions have a greater concentration of H^+ than OH^-, starting with pH 0, which contains 1.0 moles H^+/1. Each of the subsequent whole-number readings in the scale changes in [H^+] by a tenfold reduction, so that pH 1 contains [0.1 moles H^+/1], pH 2 contains [0.01 moles H^+/1], and so on, continuing in the same manner up to pH 14, which contains [0.00000000000001 moles H^+/1]. These same concentrations can be represented more manageably by exponents: pH 2 has a [H^+] of 10^{-2} moles, and pH 14 has a [H^+] of 10^{-14} moles **(table 2.2)**. It is evident that the pH units are derived from the exponent itself. Even though the basis for the pH scale is [H^+], it is important to note that, as the [H^+] in a solution decreases, the [OH^-] increases in direct proportion. At midpoint—pH 7, or neutrality—the concentrations are exactly equal and neither predominates, this being the pH of pure water previously mentioned.

In summary, the pH scale can be used to rate or determine the degree of acidity or basicity (also called alkalinity) of a solution. On this scale, a pH below 7 is acidic, and the lower the pH, the greater the acidity; a pH above 7 is basic, and the higher the pH, the greater the basicity. Incidentally, although pHs are given here in even whole numbers, more often, a pH reading exists in decimal form; for example, pH 4.5 or 6.8 (acidic) and pH 7.4 or 10.2 (basic). Because of the damaging effects of very concentrated acids or bases, most cells operate best under neutral, weakly acidic, or weakly basic conditions (see chapter 7).

Aqueous solutions containing both acids and bases may be involved in **neutralization** reactions, which give rise to water and other neutral by-products. For example, when equal molar solutions of hydrochloric acid (HCl) and sodium hydroxide (NaOH, a base) are mixed, the reaction proceeds as follows:

$$HCl + NaOH \rightarrow H_2O + NaCl$$

Here the acid and base ionize to H^+ and OH^- ions, which form water, and other ions, Na^+ and Cl^-, which form sodium chloride. Any product other than water that arises when acids and bases react is called a salt. Many of the

---

5. Actually, it forms a hydronium ion (H_3O^+), but for simplicity's sake, we will use the notation of H^+.

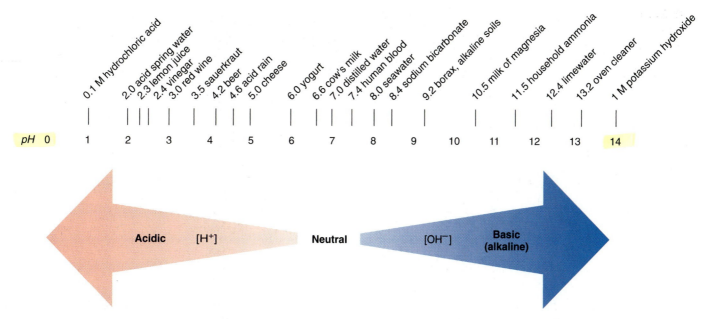

FIGURE 2.12   The pH scale.
Shown are the relative degrees of acidity and basicity and the approximate pH readings for various substances.

TABLE 2.2	Hydrogen Ion and Hydroxyl Ion Concentrations at a Given pH		
Moles/L of Hydrogen Ions	**Logarithm**	**pH**	**Moles/L of OH⁻**
1.0	10^{-0}	0	10^{-14}
0.1	10^{-1}	1	10^{-13}
0.01	10^{-2}	2	10^{-12}
0.001	10^{-3}	3	10^{-11}
0.0001	10^{-4}	4	10^{-10}
0.00001	10^{-5}	5	10^{-9}
0.000001	10^{-6}	6	10^{-8}
0.0000001	10^{-7}	7	10^{-7}
0.00000001	10^{-8}	8	10^{-6}
0.000000001	10^{-9}	9	10^{-5}
0.0000000001	10^{-10}	10	10^{-4}
0.00000000001	10^{-11}	11	10^{-3}
0.000000000001	10^{-12}	12	10^{-2}
0.0000000000001	10^{-13}	13	10^{-1}
0.00000000000001	10^{-14}	14	10^{-0}

organic acids (such as lactic and succinic acids) that function in **metabolism** are available as the acid and the salt form (such as lactate, succinate), depending on the conditions in the cell (see chapter 8).

The Chemistry of Carbon and Organic Compounds

So far, our main focus has been on the characteristics of atoms, ions, and small, simple substances that play diverse roles in the structure and function of living things. These substances are often lumped together in a category called **inorganic chemicals.**

A chemical is usually inorganic if it does not contain both carbon and hydrogen. Examples of inorganic chemicals include NaCl (sodium chloride), $Mg_3(PO_4)_2$ (magnesium phosphate), $CaCO_3$ (calcium carbonate), and CO_2 (carbon dioxide). In reality, however, most of the chemical reactions and structures of living things occur at the level of more complex molecules, termed **organic chemicals.** These are carbon compounds with a basic framework of the element carbon bonded to other atoms. Organic molecules vary in complexity from the simplest, methane (CH_4; see figure 2.4c), which has a molecular weight of 16, to certain antibody molecules (produced by an immune reaction) that have a molecular weight of nearly 1,000,000 and are among the most complex molecules on earth.

The role of carbon as the fundamental element of life can best be understood if we look at its chemistry and bonding patterns. The valence of carbon makes it an ideal atomic building block to form the backbone of organic molecules; it has 4 electrons in its outer orbital to be shared with other atoms (including other carbons) through covalent bonding. As a result, it can form stable chains containing thousands of carbon atoms and still has bonding sites available for forming covalent bonds with numerous other atoms. The bonds that carbon forms are linear, branched, or ringed, and it can form four single bonds, two double bonds, or one triple bond **(figure 2.13).** The atoms with which carbon is most often associated in organic compounds are hydrogen, oxygen, nitrogen, sulfur, and phosphorus.

Functional Groups of Organic Compounds

One important advantage of carbon's serving as the molecular skeleton for living things is that it is free to bind with an unending array of other molecules. These special molecular groups or accessory molecules that bind to organic compounds

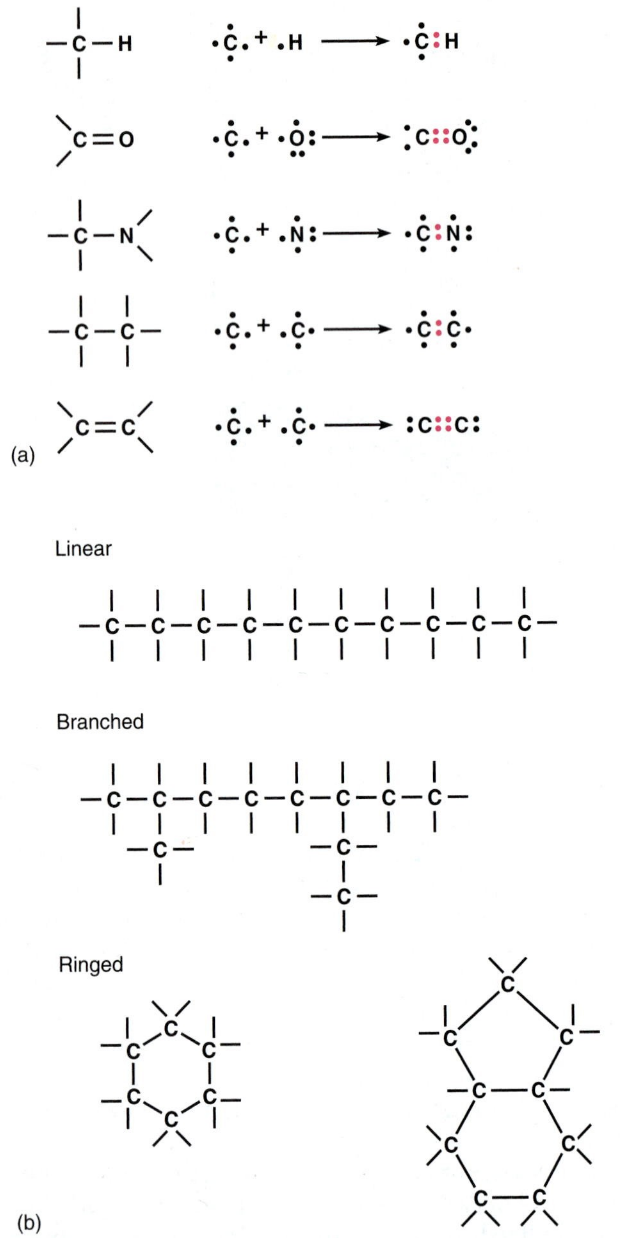

(a)

Linear

Branched

Ringed

(b)

FIGURE 2.13   **The versatility of bonding in carbon.**
In most compounds, each carbon makes a total of four bonds.
(a) Both single and double bonds can be made with other carbons,
oxygen, and nitrogen; single bonds are made with hydrogen. Simple
electron models show how the electrons are shared in these bonds.
(b) Multiple bonding of carbons can give rise to long chains,
branched compounds, and ringed compounds, many of which are
extraordinarily large and complex.

are called **functional groups.** Functional groups help define
the chemical class of certain groups of organic compounds
and confer unique reactive properties on the whole molecule
(table 2.3). Because each type of functional group behaves in a
distinctive manner, reactions of an organic compound can be
predicted by knowing the kind of functional group or groups
it carries. Many synthesis, decomposition, and transfer reac-

TABLE 2.3	Representative Functional Groups and Classes of Organic Compounds	
Formula of Functional Group	**Name**	**Class of Compounds**
$R^* - O - H$	Hydroxyl	Alcohols, carbohydrates
$R - C \overset{O}{\underset{OH}{=}}$	Carboxyl	Fatty acids, proteins, organic acids
$R - C(H)(H) - NH_2$	Amino	Proteins, nucleic acids
$R - C \overset{O}{\underset{O-R}{=}}$	Ester	Lipids
$R - C(H)(H) - SH$	Sulfhydryl	Cysteine (amino acid), proteins
$R - C \overset{O}{\underset{H}{=}}$	Carbonyl, terminal end	Aldehydes, polysaccharides
$R - C(=O) - C -$	Carbonyl, internal	Ketones, polysaccharides
$R - O - P(=O)(OH) - OH$	Phosphate	DNA, RNA, ATP

tions rely upon functional groups such as R—OH or R—NH$_2$.
The —R designation on a molecule is shorthand for residue,
and its placement in a formula indicates that the group attached at that site varies from one compound to another.

✓ **CHECKPOINT**

■ Covalent bonds are chemical bonds in which electrons are
shared between atoms. Equally distributed electrons form
nonpolar covalent bonds, whereas unequally distributed
electrons form polar covalent bonds.

TABLE 2.4	Macromolecules and Their Functions	
Macromolecule	**Description/Basic Structure**	**Examples/Notes**
Carbohydrates		
Monosaccharides	3- to 7-carbon sugars	Glucose, fructose / Sugars involved in metabolic reactions; building block of disaccharides and polysaccharides
Disaccharides	Two monosaccharides	Maltose (malt sugar) / Composed of two glucoses; an important breakdown product of starch
		Lactose (milk sugar) / Composed of glucose and galactose
		Sucrose (table sugar) / Composed of glucose and fructose
Polysaccharides	Chains of monosaccharides	Starch, cellulose, glycogen / Cell wall, food storage
Lipids		
Triglycerides	Fatty acids + glycerol	Fats, oils / Major component of cell membranes; storage
Phospholipids	Fatty acids + glycerol + phosphate	Membranes
Waxes	Fatty acids, alcohols	Mycolic acid / Cell wall of mycobacteria
Steroids	Ringed structure	Cholesterol, ergosterol / Membranes of eucaryotes and some bacteria
Proteins		
	Amino acids	Enzymes; part of cell membrane, cell wall, ribosomes, antibodies / Metabolic reactions; structural components
Nucleic acids		
	Pentose sugar + phosphate + nitrogenous base	
	Purines: adenine, guanine	
	Pyrimidines: cytosine, thymine, uracil	
Deoxyribonucleic acid (DNA)	Contains deoxyribose sugar and thymine, not uracil	Chromosomes; genetic material of viruses / Inheritance
Ribonucleic acid (RNA)	Contains ribose sugar and uracil, not thymine	Ribosomes; mRNA, tRNA / Expression of genetic traits

- Ionic bonds are chemical bonds resulting from opposite charges. The outer electron shell either donates or receives electrons from another atom so that the outer shell of each atom is completely filled.
- Hydrogen bonds are weak chemical bonds that form between covalently bonded hydrogens and either oxygens or nitrogens on different molecules.
- Chemical equations express the chemical exchanges between atoms or molecules. Some arrangements contain more energy than others, and chemical reactions such as synthesis or decomposition may require or release the difference in energy.
- Solutions are mixtures of solutes and solvents that cannot be separated by filtration or settling.
- The pH, ranging from a highly *acidic* solution to highly *basic* solution, refers to the concentration of hydrogen ions. It is expressed as a number from 0 to 14.
- Biologists define organic molecules as those containing both carbon and hydrogen.
- Carbon is the backbone of biological compounds because of its ability to form single, double, or triple covalent bonds with itself and many different elements.
- Functional (R) groups are specific arrangements of organic molecules that confer distinct properties, including chemical reactivity, to organic compounds.

2.2 Macromolecules: Superstructures of Life

The compounds of life fall into the realm of **biochemistry.** Biochemicals are organic compounds produced by (or components of) living things, and they include four main families: carbohydrates, lipids, proteins, and nucleic acids **(table 2.4).** The compounds in these groups are assembled from smaller molecular subunits, or building blocks, and because they are often very large compounds, they are termed **macromolecules.** All macromolecules except lipids are formed by polymerization, a process in which repeating subunits termed **monomers** are bound into chains of various lengths termed **polymers.** For example, proteins (polymers) are composed of a chain of amino acids (monomers). The large size and complex, three-dimensional shape of macromolecules enables them to function as structural components, molecular messengers, energy sources, enzymes (biochemical catalysts), nutrient stores, and sources of genetic information. In the following section and in later chapters, we consider numerous concepts relating to the roles of macromolecules in cells. Table 2.4 will also be a useful reference when you study metabolism in chapter 8.

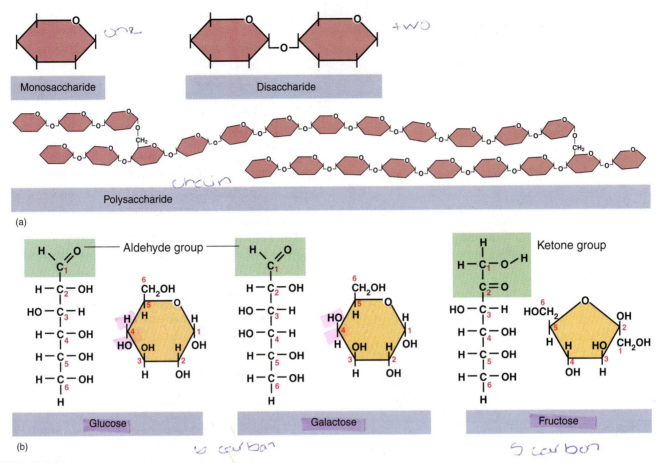

FIGURE 2.14   Common classes of carbohydrates.
(a) Major saccharide groups, named for the number of sugar units each contains. (b) Three hexoses with the same molecular formula and different structural formulas. Both linear and ring models are given. The linear form emphasizes aldehyde and ketone groups, although in solution the sugars exist in the ring form. Note that the carbons are numbered so as to keep track of reactions within and between monosaccharides.

Carbohydrates: Sugars and Polysaccharides

The term **carbohydrate** originates from the way that most members of this chemical class resemble combinations of carbon and water. Although carbohydrates can be generally represented by the formula $(CH_2O)_n$, in which n indicates the number of units of this combination of atoms, some carbohydrates contain additional atoms of sulfur or nitrogen. In molecular configuration, the carbons form chains or rings with two or more hydroxyl groups and either an aldehyde or a ketone group, giving them the technical designation of *polyhydroxy aldehydes* or *ketones* (figure 2.14).

Carbohydrates exist in a great variety of configurations. The common term sugar (**saccharide**) refers to a simple carbohydrate such as a monosaccharide or a disaccharide that has a sweet taste. A **monosaccharide** is a simple polyhydroxy aldehyde or ketone molecule containing from 3 to 7 carbons; a **disaccharide** is a combination of two monosaccharides; and a **polysaccharide** is a polymer of five or more

monosaccharides bound in linear or branched chain patterns (see figure 2.14). Monosaccharides and disaccharides are specified by combining a prefix that describes some characteristic of the sugar with the suffix **-ose.** For example, **hexoses** are composed of 6 carbons, and **pentoses** contain 5 carbons. **Glucose** (Gr. sweet) is the most common and universally important hexose; **fructose** is named for fruit (one of its sources); and xylose, a pentose, derives its name from the Greek word for wood. Disaccharides are named similarly: **lactose** (L. milk) is an important component of milk; **maltose** means malt sugar; and **sucrose** (Fr. sugar) is common table sugar or cane sugar.

The Nature of Carbohydrate Bonds

The subunits of disaccharides and polysaccharides are linked by means of **glycosidic bonds,** in which carbons (each is assigned a number) on adjacent sugar units are bonded to the same oxygen atom like links in a chain (**figure 2.15).** For example, maltose is formed when the number 1 carbon on a glucose bonds to the oxygen on the number 4

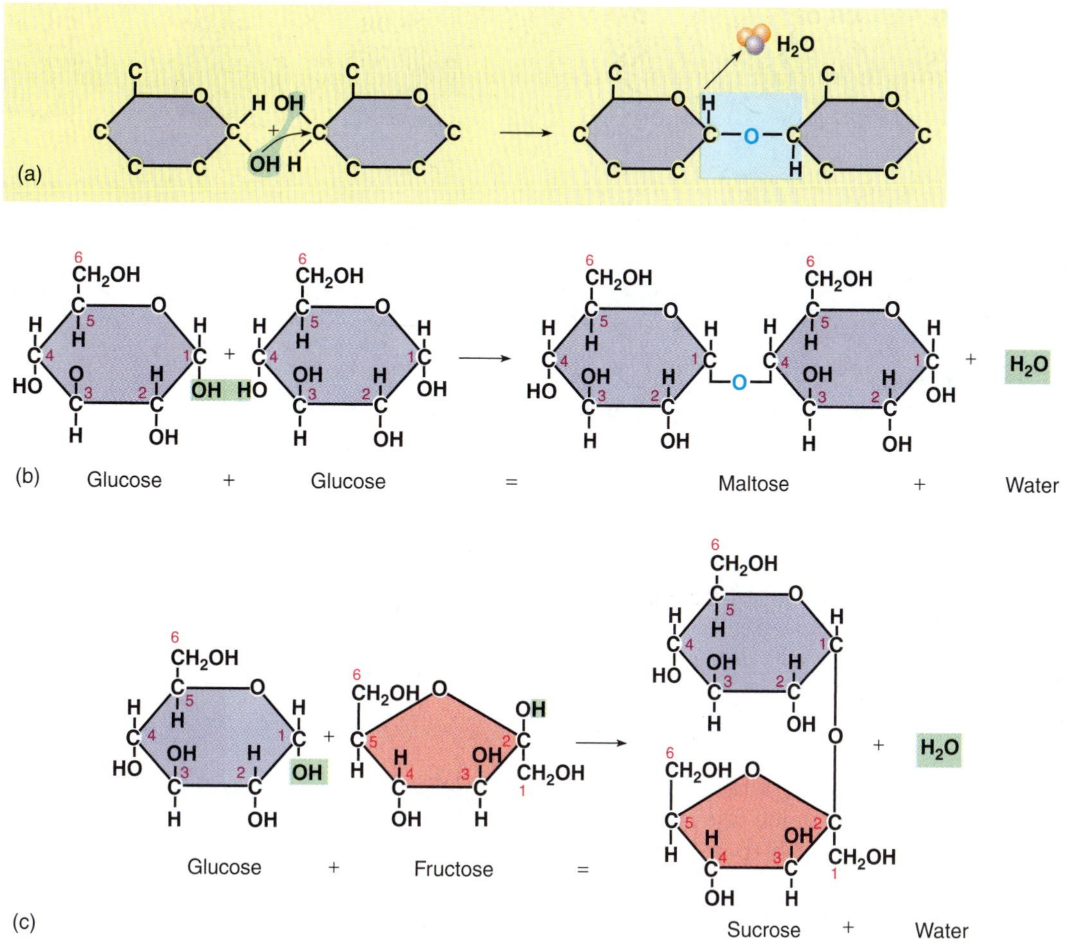

FIGURE 2.15 Glycosidic bond.
(a) General scheme in the formation of a glycosidic bond by dehydration synthesis. (b) Formation of the 1,4 bond between two α glucoses to produce maltose and water. (c) Formation of the 1,2 bond between glucose and fructose to produce sucrose and water.

carbon on a second glucose; sucrose is formed when glucose and fructose bind oxygen between their number 1 and number 2 carbons; and lactose is formed when glucose and galactose connect by their number 1 and number 4 carbons. In order to form this bond, one carbon gives up its OH group and the other (the one contributing the oxygen to the bond) loses the H from its OH group. Because a water molecule is produced, this reaction is known as **dehydration synthesis,** a process common to most polymerization reactions (see proteins, page 45). Three polysaccharides (starch, cellulose, and glycogen) are structurally and biochemically distinct, even though all are polymers of the same monosaccharide—glucose. The basis for their differences lies primarily in the exact way the glucoses are bound together, which greatly affects the characteristics of the end product (**figure 2.16**). The synthesis and breakage of each type of bond requires a specialized catalyst called an enzyme (see chapter 8).

The Functions of Polysaccharides

Polysaccharides typically contribute to structural support and protection and serve as nutrient and energy stores. The cell walls in plants and many microscopic algae derive their strength and rigidity from **cellulose,** a long, fibrous polymer (**figure 2.16a**). Because of this role, cellulose is probably one of the most common organic substances on the earth, yet it is digestible only by certain bacteria, fungi, and protozoa. These microbes, called decomposers, play an essential role in breaking down and recycling plant materials (see figure 7.2). Some bacteria secrete slime layers of a glucose polymer called *dextran*. This substance causes a sticky layer to develop on teeth that leads to plaque, described later in chapter 22.

Other structural polysaccharides can be conjugated (chemically bonded) to amino acids, nitrogen bases, lipids, or proteins. **Agar,** an indispensable polysaccharide in preparing solid culture media, is a natural component of certain seaweeds. It is a complex polymer of galactose and sulfur-containing

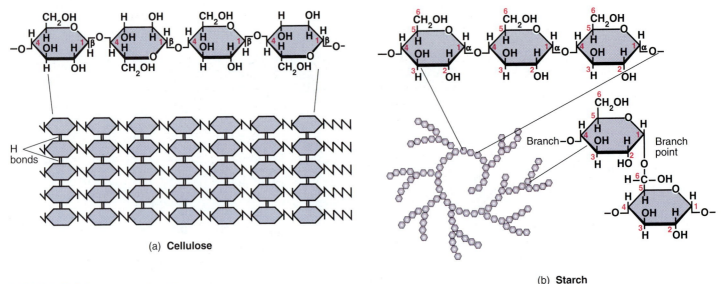

FIGURE 2.16 Polysaccharides.
(a) Cellulose is composed of β glucose bonded in 1,4 bonds that produce linear, lengthy chains of polysaccharides that are H-bonded along their length. This is the typical structure of wood and cotton fibers. (b) Starch is also composed of glucose polymers, in this case α glucose. The main structure is amylose bonded in a 1,4 pattern, with side branches of amylopectin bonded by 1,6 bonds. The entire molecule is compact and granular.

carbohydrates. The exoskeletons of certain fungi contain **chitin** (ky-tun), a polymer of glucosamine (a sugar with an amino functional group). **Peptidoglycan** (pep-tih-doh-gly′-kan) is one special class of compounds in which polysaccharides (glycans) are linked to peptide fragments (a short chain of amino acids). This molecule provides the main source of structural support to the bacterial cell wall. The cell wall of gram-negative bacteria also contains **lipopolysaccharide,** a complex of lipid and polysaccharide responsible for symptoms such as fever and shock (see chapters 4 and 13).

The outer surface of many cells has a delicate "sugar coating" composed of polysaccharides bound in various ways to proteins (the combination is called mucoprotein or glycoprotein). This structure, called the **glycocalyx,** functions in attachment to other cells or as a site for *receptors*—surface molecules that receive and respond to external stimuli. Small sugar molecules account for the differences in human blood types, and carbohydrates are a component of large protein molecules called antibodies. Some viruses have glycoproteins on their surface with which they bind to and invade their host cells.

Polysaccharides are usually stored by cells in the form of glucose polymers such as starch **(figure 2.16b)** or **glycogen,** but only organisms with the appropriate digestive enzymes can break them down and use them as a nutrient source. Because a water molecule is required for breaking the bond between two glucose molecules, digestion is also termed **hydrolysis.** Starch is the primary storage food of green plants, microscopic algae, and some fungi; glycogen (animal starch) is a stored carbohydrate for animals and certain groups of bacteria and protozoa.

Lipids: Fats, Phospholipids, and Waxes

The term **lipid,** derived from the Greek word *lipos,* meaning fat, is not a chemical designation, but an operational term for a variety of substances that are not soluble in polar solvents such as water (recall that oil and water do not mix) but will dissolve in nonpolar solvents such as benzene and chloroform. This property occurs because the substances we call lipids contain relatively long or complex C—H (hydrocarbon) chains that are nonpolar and thus hydrophobic. The main groups of compounds classified as lipids are triglycerides, phospholipids, steroids, and waxes.

Important storage lipids are the **triglycerides,** a category that includes fats and oils. Triglycerides are composed of a single molecule of glycerol bound to three fatty acids **(figure 2.17). Glycerol** is a 3-carbon alcohol[6] with three OH groups that serve as binding sites, and fatty acids are long-chain hydrocarbon molecules with a carboxyl group (COOH) at one end that is free to bind to the glycerol. The bond that forms between the —OH group and the —COOH is defined as an **ester bond.** The hydrocarbon portion of a fatty acid can vary in length from 4 to 24 carbons and, depending on the fat, it may be saturated or unsaturated. If all carbons in the chain are single-bonded to 2 other carbons and 2 hydrogens, the fat is saturated; if there is at least one C=C double bond in the chain, it is unsaturated. The structure of fatty acids is what gives fats and oils (liquid fats) their greasy, insoluble nature. In general, solid fats (such as

6. Alcohols are hydrocarbons containing OH groups.

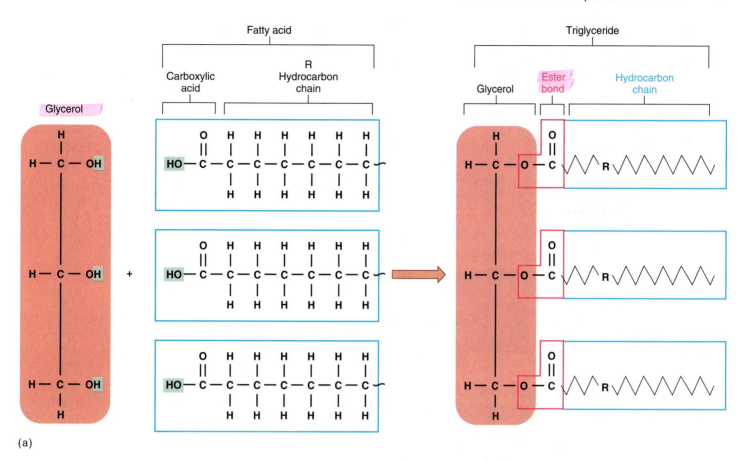

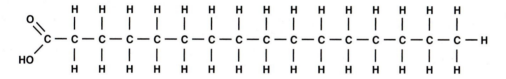

FIGURE 2.17 **Synthesis and structure of a triglyceride.**
(a) Because a water molecule is released at each ester bond, this is another form of dehydration synthesis. The jagged lines and R symbol represent the hydrocarbon chains of the fatty acids, which are commonly very long. **(b)** Structural formulas for a saturated and unsaturated fatty acid.

beef tallow) are more saturated, and oils (or liquid fats) are more unsaturated. In most cells, triglycerides are stored in long-term concentrated form as droplets or globules. When the ester linkage is acted on by digestive enzymes called lipases, the fatty acids and glycerol are freed to be used in metabolism. Fatty acids are a superior source of energy, yielding twice as much per gram as other storage molecules (starch). Soaps are K^+ or Na^+ salts of fatty acids whose qualities make them excellent grease removers and cleaners (see chapter 11).

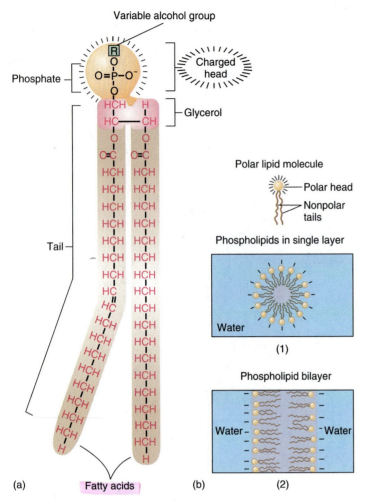

FIGURE 2.18   Phospholipids—membrane molecules.
(a) A model of a single molecule of a phospholipid. The phosphate-alcohol head lends a charge to one end of the molecule; its long, trailing hydrocarbon chain is uncharged. **(b)** The behavior of phospholipids in water-based solutions causes them to become arranged **(1)** in single layers called micelles, with the charged head oriented toward the water phase and the hydrophobic nonpolar tail buried away from the water phase, or **(2)** in double-layered phospholipid systems with the hydrophobic tails sandwiched between two hydrophilic layers.

TABLE 2.5	Twenty Amino Acids and Their Abbreviations	
Acid	**Abbreviation**	**Characteristic of R Groups***
Alanine	Ala	NP
Arginine	Arg	+
Asparagine	Asn	P
Aspartic acid	Asp	−
Cysteine	Cys	P
Glutamic acid	Glu	−
Glutamine	Gln	P
Glycine	Gly	P
Histidine	His	+
Isoleucine	Ile	NP
Leucine	Leu	NP
Lysine	Lys	+
Methionine	Met	NP
Phenylalanine	Phe	NP
Proline	Pro	NP
Serine	Ser	P
Threonine	Thr	P
Tryptophan	Trp	NP
Tyrosine	Tyr	P
Valine	Val	NP

*NP, nonpolar; P, polar; +, positively charged; −, negatively charged.

Membrane Lipids

A class of lipids that serves as a major structural component of cell membranes is the **phospholipids.** Although phospholipids also contain glycerol and fatty acids, they have some significant differences from triglycerides. Phospholipids contain only two fatty acids attached to the glycerol, and the third glycerol binding site holds a phosphate group. The phosphate is in turn bonded to an alcohol, which varies from one phospholipid to another **(figure 2.18a).** These lipids have a hydrophilic region from the charge on the phosphoric acid–alcohol "head" of the molecule and a hydrophobic region that corresponds to the long, uncharged "tail" (formed by the fatty acids). When exposed to an aqueous solution, the charged heads are attracted to the water phase, and the nonpolar tails are repelled from the water phase **(figure 2.18b).** This property causes lipids to naturally assume single and double layers (bilayers), which contribute to their biological significance in membranes. When two single layers of polar lipids come together to form a double layer, the outer hydrophilic face of each single layer will orient itself toward the solution, and the hydrophobic portions will become immersed in the core of the bilayer. The structure of lipid bilayers confers characteristics on membranes such as selective permeability and fluid nature **(Insight 2.3).**

Miscellaneous Lipids

Steroids are complex ringed compounds commonly found in cell membranes and animal hormones. The best known of these is the sterol (meaning a steroid with an OH group) called **cholesterol (figure 2.19).** Cholesterol reinforces the structure of the cell membrane in animal cells and in an unusual group of cell-wall-deficient bacteria called the mycoplasmas (see chapter 4). The cell membranes of fungi also contain a sterol, called ergosterol. *Prostaglandins* are fatty acid derivatives found in trace amounts that function in inflammatory and allergic reactions, blood clotting, and smooth muscle contraction. Chemically, a *wax* is an ester formed between a long-chain alcohol and a saturated fatty acid. The resulting material is typically pliable and soft when warmed but hard and water-resistant when cold (paraffin, for example). Among living things, fur, feathers, fruits, leaves, human skin, and insect exoskeletons are naturally water-

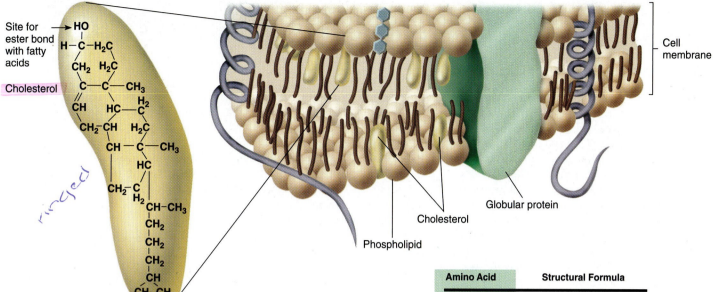

Site for ester bond with fatty acids

Cholesterol

ringed

Phospholipid

Cholesterol

Globular protein

Cell membrane

FIGURE 2.19 **Formula for cholesterol, an alcoholic steroid that is inserted in some membranes.**
Cholesterol can become esterified with fatty acids at its OH group, imparting a polar quality similar to that of phospholipids.

proofed with a coating of wax. Bacteria that cause tuberculosis and leprosy produce a wax (wax D) that repels ordinary laboratory stains and contributes to their pathogenicity.

Proteins: Shapers of Life

The predominant organic molecules in cells are **proteins,** a fitting term adopted from the Greek word *proteios,* meaning first or prime. To a large extent, the structure, behavior, and unique qualities of each living thing are a consequence of the proteins they contain. To best explain the origin of the special properties and versatility of proteins, we must examine their general structure. The building blocks of proteins are **amino acids,** which exist in 20 different naturally occurring forms **(table 2.5).** Various combinations of these amino acids account for the nearly infinite variety of proteins. Amino acids have a basic skeleton consisting of a carbon (called the α carbon) linked to an amino group (NH_2), a carboxyl group (COOH), a hydrogen atom (H), and a variable R group. The variations among the amino acids occur at the R group, which is different in each amino acid and imparts the unique characteristics to the molecule and to the proteins that contain it **(figure 2.20).** A covalent bond called a **peptide bond** forms between the amino group on one amino acid and the carboxyl group on another

FIGURE 2.20 **Structural formulas of selected amino acids.**
The basic structure common to all amino acids is shown in blue type and the variable group, or R group, is placed in a colored box. Note the variations in structure of this reactive component.

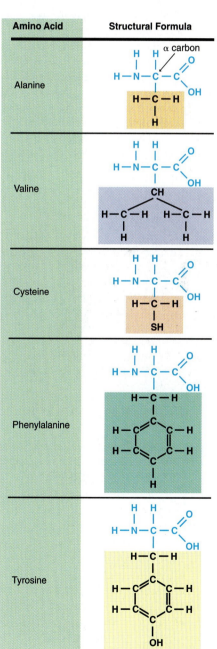

Amino Acid	Structural Formula
Alanine	
Valine	
Cysteine	
Phenylalanine	
Tyrosine	

INSIGHT 2.3   *Discovery*

Membranes: Cellular Skins

The word **membrane** appears frequently in descriptions of cells in this chapter and in chapters 4 and 5. The word itself describes any lining or covering, including such multicellular structures as the mucous membranes of the body. From the perspective of a single cell, however, a membrane is a thin, double-layered sheet composed of lipids such as phospholipids and sterols (averaging about 40% of membrane content) and protein molecules (averaging about 60%). The primary role of membranes is as a cell membrane that completely encases the cytoplasm. Membranes are also components of eucaryotic organelles such as nuclei, mitochondria, and chloroplasts, and they appear in internal pockets of certain procaryotic cells. Even some viruses, which are not cells at all, can have a membranous protective covering.

Cell membranes are so thin—on the average, just 0.0070 μm (7 nm) thick—that they cannot actually be seen with an optical microscope. Even at magnifications made possible by electron microscopy (500,000×), very little of the precise architecture can be visualized, and a cross-sectional view has the appearance of railroad tracks. Following detailed microscopic and chemical analysis, S. J. Singer and C. K. Nicholson proposed a simple and elegant theory for membrane structure called the **fluid mosaic model.**

According to this theory, a membrane is a continuous bilayer formed by lipids that are oriented with the polar lipid heads toward the outside and the nonpolar tails toward the center of the membrane. Embedded at numerous sites in this bilayer are various-sized globular proteins. Some proteins are situated only at the surface; others extend fully through the entire membrane. The configuration of the inner and outer sides of the membrane can be quite different because of the variations in protein shape and position.

Membranes are dynamic and constantly changing because the lipid phase is in motion and many proteins can migrate freely about, somewhat as icebergs do in the ocean. This fluidity is essential to such activities as engulfment of food and discharge or secretion by cells. The structure of the lipid phase provides an impenetrable barrier to many substances. This property accounts for the selective permeability and capacity to regulate transport of molecules. It also serves to segregate activities within the cell's cytoplasm. Membrane proteins function in receiving molecular signals (receptors), in binding and transporting nutrients, and in acting as enzymes, topics to be discussed in chapters 7 and 8.

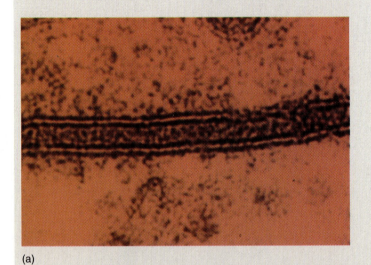

(a)

(b)

(a) Extreme magnification of a cross section of a cell membrane, which appears as double tracks. **(b)** A generalized version of the fluid mosaic model of a cell membrane indicates a bilayer of lipids with globular proteins embedded to some degree in the lipid matrix. This structure explains many characteristics of membranes, including flexibility, solubility, permeability, and transport.

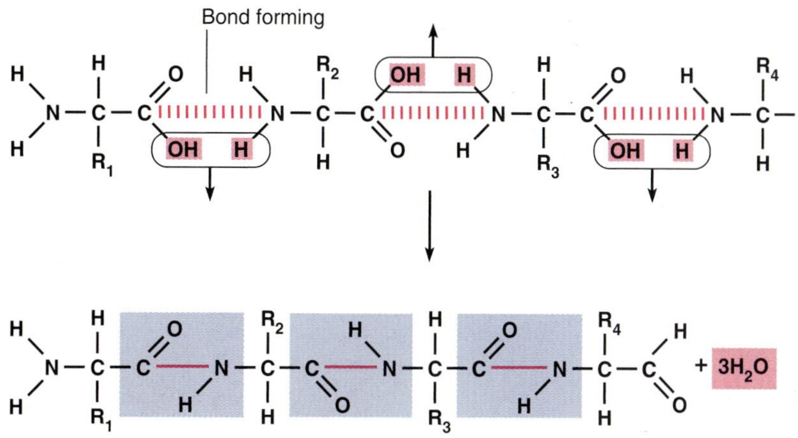

FIGURE 2.21   The formation of peptide bonds in a tetrapeptide.

amino acid. As a result of peptide bond formation, it is possible to produce molecules varying in length from two amino acids to chains containing thousands of them.

Various terms are used to denote the nature of compounds containing peptide bonds. **Peptide** usually refers to a molecule composed of short chains of amino acids, such as a dipeptide (two amino acids), a tripeptide (three), and a tetrapeptide (four) **(figure 2.21).** A **polypeptide** contains an unspecified number of amino acids, but usually has more than 20, and is often a smaller subunit of a protein. A protein is the largest of this class of compounds and usually contains a minimum of 50 amino acids. It is common for the terms *polypeptide* and *protein* to be used interchangeably, though not all polypeptides are large enough to be considered proteins. In chapter 9 we see that protein synthesis is not just a random connection of amino acids; it is directed by information provided in DNA.

Protein Structure and Diversity

The reason that proteins are so varied and specific is that they do not function in the form of a simple straight chain of amino acids (called the primary structure). A protein has a natural tendency to assume more complex levels of organization, called the secondary, tertiary, and quaternary structures **(figure 2.22).** The **primary (1°) structure** is more correctly described as the type, number, and order of amino acids in the chain, which varies extensively from protein to protein. The **secondary (2°) structure** arises when various functional groups exposed on the outer surface of the molecule interact by forming hydrogen bonds. This interaction causes the amino acid chain to twist into a coiled configuration called the α *helix* or to fold into an accordion pattern called a β-*pleated sheet.* Some proteins contain both types of secondary configurations. Proteins at the secondary level undergo a third degree of torsion called the **tertiary (3°) structure** created by additional bonds between functional groups **(figure 2.22c).** In proteins with the sulfur-containing amino acid **cysteine,** considerable tertiary stability is achieved through covalent disulfide bonds between sulfur atoms on two different parts of the molecule. Some complex proteins assume a **quaternary (4°) structure,** in

which more than one polypeptide forms a large, multiunit protein. This is typical of antibodies (see chapter 15) and some enzymes that act in cell synthesis.

The most important outcome of intrachain[7] bonding and folding is that each different type of protein develops a unique shape, and its surface displays a distinctive pattern of pockets and bulges. As a result, a protein can react only with molecules that complement or fit its particular surface features like a lock and key. Such a degree of specificity can provide the functional diversity required for many thousands of different cellular activities. **Enzymes** serve as the catalysts for all chemical reactions in cells, and nearly every reaction requires a different enzyme (see chapter 8). **Antibodies** are complex glycoproteins with specific regions of attachment for bacteria, viruses, and other microorganisms; certain bacterial toxins (poisonous products) react with only one specific organ or tissue; and proteins embedded in the cell membrane have reactive sites restricted to a certain nutrient. Some proteins function as *receptors* to receive stimuli from the environment. The functional three-dimensional form of a protein is termed the *native state,* and if it is disrupted by some means, the protein is said to be *denatured.* Such agents as heat, acid, alcohol, and some disinfectants disrupt (and thus denature) the stabilizing intrachain bonds and cause the molecule to become nonfunctional, as described in chapter 11.

The Nucleic Acids: A Cell Computer and Its Programs

The nucleic acids, **deoxyribonucleic acid (DNA)** and **ribonucleic acid (RNA),** were originally isolated from the cell nucleus. Shortly thereafter, they were also found in other parts of nucleated cells, in cells with no nuclei (bacteria), and in viruses. The universal occurrence of nucleic acids in all known cells and viruses emphasizes their important roles as informational molecules. DNA, the master computer of cells, contains

---

7. **Intra**chain means within the chain; **inter**chain would be between two chains.

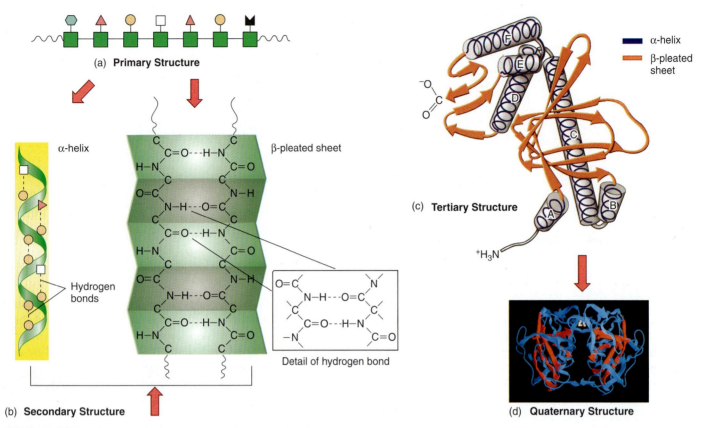

FIGURE 2.22 **Stages in the formation of a functioning protein.**
(a) Its primary structure is a series of amino acids bound in a chain. **(b)** Its secondary structure develops when the chain forms hydrogen bonds that fold it into one of several configurations such as an α-helix or β-pleated sheet. Some proteins have several configurations in the same molecule. **(c)** A protein's tertiary structure is due to further folding of the molecule into a three-dimensional mass that is stabilized by hydrogen, ionic, and disulfide bonds between functional groups. The letters and arrows denote the order and direction of the folded chain. **(d)** The quaternary structure exists only in proteins that consist of more than one polypeptide chain. Shown here is a computer model of the nitrogenase iron protein, with the two polypeptide chains (right and left) arranged symmetrically.

a special coded genetic program with detailed and specific instructions for each organism's heredity. It transfers the details of its program to RNA, operator molecules responsible for carrying out DNA's instructions and translating the DNA program into proteins that can perform life functions. For now, let us briefly consider the structure and some functions of DNA, RNA, and a close relative, adenosine triphosphate (ATP).

Both nucleic acids are polymers of repeating units called **nucleotides,** each of which is composed of three smaller units: a **nitrogen base,** a pentose (5-carbon) sugar, and a *phosphate* **(figure 2.23a).** The nitrogen base is a cyclic compound that comes in two forms: *purines* (two rings) and *pyrimidines* (one ring). There are two types of purines—**adenine (A)** and **guanine (G)**—and three types of pyrimidines—**thymine (T), cytosine (C),** and **uracil (U) (figure 2.24).** A characteristic that differentiates DNA from RNA is that DNA contains all of the nitrogen bases except uracil, and RNA contains all of the nitrogen bases except thymine. The nitrogen base is covalently bonded to the sugar *ribose* in RNA and *de-*

oxyribose (because it has one less oxygen than ribose) in DNA. Phosphate (PO_4^{3-}), a derivative of phosphoric acid (H_3PO_4), provides the final covalent bridge that connects sugars in series. Thus, the backbone of a nucleic acid strand is a chain of alternating phosphate-sugar-phosphate-sugar molecules, and the nitrogen bases branch off the side of this backbone (see **figure 2.23b,c**).

The Double Helix of DNA

DNA is a huge molecule formed by two very long polynucleotide strands linked along their length by hydrogen bonds between complementary pairs of nitrogen bases. The pairing of the nitrogen bases occurs according to a predictable pattern: Adenine ordinarily pairs with thymine, and cytosine with guanine. The bases are attracted in this way because each pair shares oxygen, nitrogen, and hydrogen atoms exactly positioned to align perfectly for hydrogen bonds **(figure 2.25).**

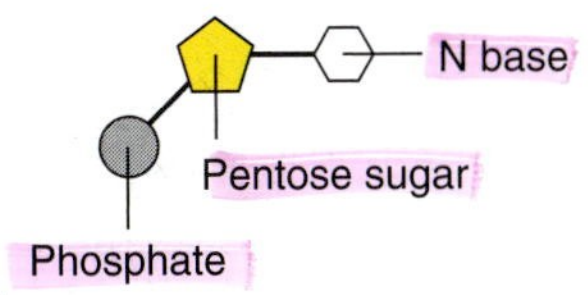

(a)

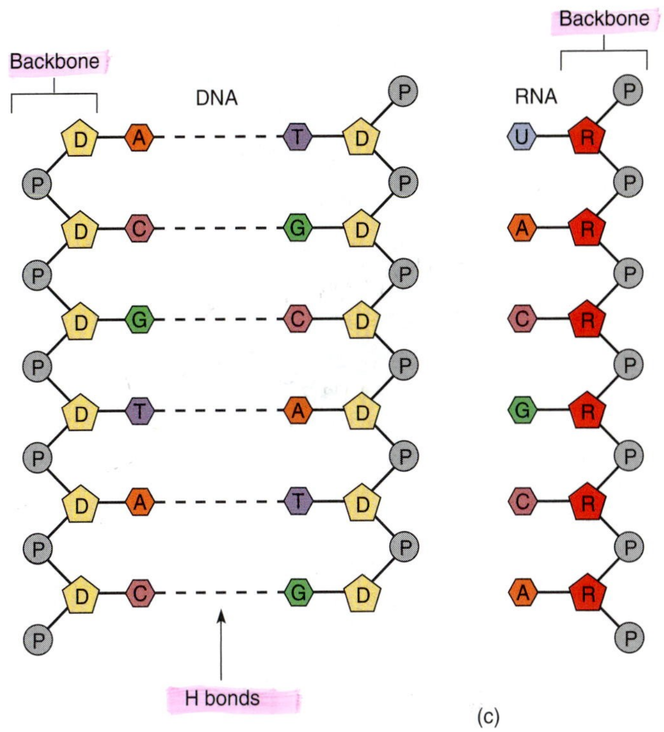

(b) (c)

FIGURE 2.23 The general structure of nucleic acids. **(a)** A nucleotide, composed of a phosphate, a pentose sugar, and a nitrogen base (either A, T, U, C, or G), is the monomer of both DNA and RNA. **(b)** In DNA, the polymer is composed of alternating deoxyribose (D) and phosphate (P) with nitrogen bases (A, T, C, G) attached to the deoxyribose. DNA almost always exists in pairs of strands, oriented so that the bases are paired across the central axis of the molecule. **(c)** In RNA, the polymer is composed of alternating ribose (R) and phosphate (P) attached to nitrogen bases (A, U, C, G), but it is only a single strand.

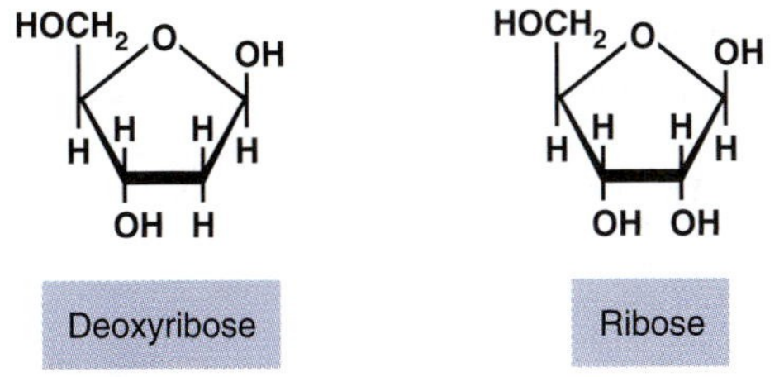

(a) **Pentose Sugars**

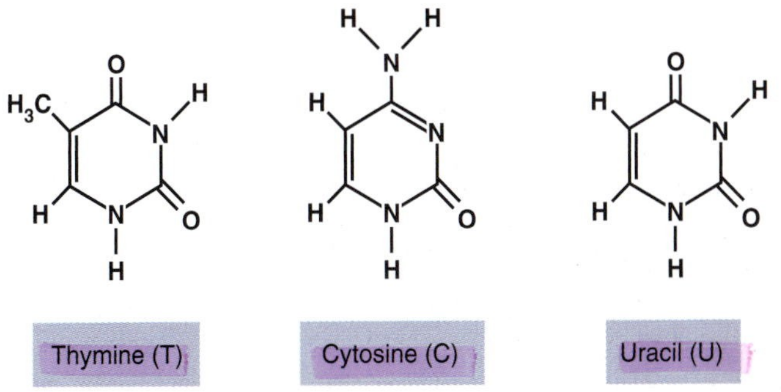

(c) **Pyrimidines**

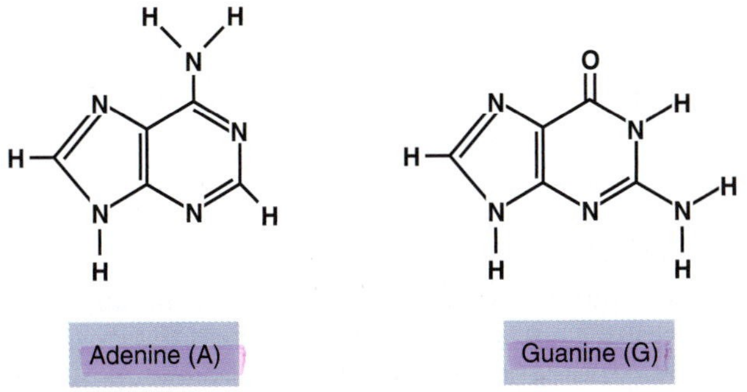

(b) **Purines**

FIGURE 2.24 The sugars and nitrogen bases that make up DNA and RNA. **(a)** DNA contains deoxyribose, and RNA contains ribose. **(b)** A and G purines are found in both DNA and RNA. **(c)** C pyrimidine is found in both DNA and RNA, but T is found only in DNA, and U is found only in RNA.

FIGURE 2.25 **A structural representation of the double helix of DNA.**
Shown are the details of hydrogen bonds between the nitrogen bases of the two strands.

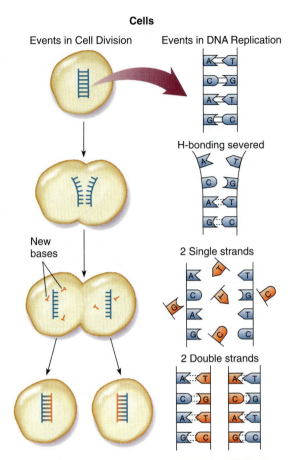

FIGURE 2.26 **Simplified view of DNA replication in cells.**
The DNA in the cell's chromosome must be duplicated as the cell is dividing. This duplication is accomplished through the separation of the double DNA strand into two single strands. New strands are then synthesized using the original strands as guides to assemble the correct new complementary bases.

For ease in understanding the structure of DNA, it is sometimes compared to a ladder, with the sugar-phosphate backbone representing the rails and the paired nitrogen bases representing the steps. Owing to the manner of nucleotide pairing and stacking of the bases, the actual configuration of DNA is a *double helix* that looks somewhat like a spiral staircase. As is true of protein, the structure of DNA is intimately related to its function. DNA molecules are usually extremely long, a feature that satisfies a requirement for storing genetic information in the sequence of base pairs the molecule contains. The hydrogen bonds between pairs can be disrupted when DNA is being copied, and the fixed complementary base pairing is essential to maintain the genetic code.

Making New DNA: Passing on the Genetic Message

The biological properties of cells and viruses are ultimately programmed by a master code composed of nucleic acids. This code is in the form of DNA in all cells and many viruses; other viruses are based on RNA alone. Regardless of the exact genetic program, both cells and viruses will continue to exist only if they can duplicate their genetic material and pass it on to subsequent generations. **Figure 2.26** summarizes the main steps in this process and how it differs between cells and viruses.

During its division cycle, the cell has a mechanism for making a copy of its DNA by **replication,** using the original strand as a pattern (figure 2.26). Note that replication is guided by the double-stranded nature of DNA and the precise pairing of bases that create the master code. Replication

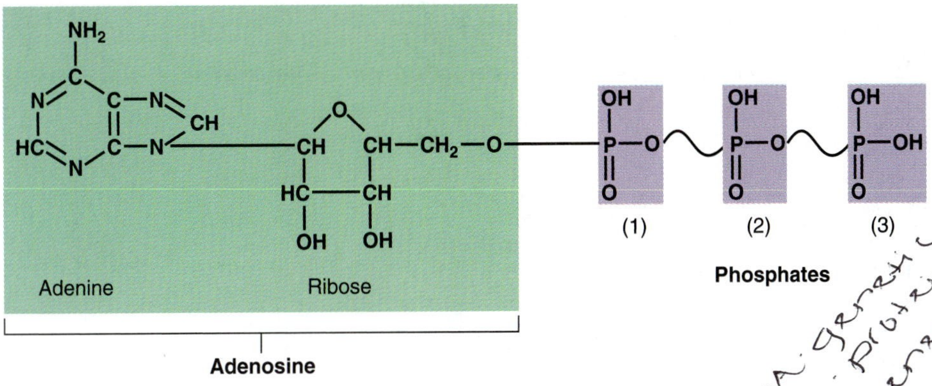

FIGURE 2.27 **The structural formula of an ATP molecule, the chemical form of energy transfer in cells.** The wavy lines that connect the phosphates represent bonds that release large amounts of energy when broken. Within the cell, several of the hydroxyl groups on the phosphates would be negatively charged oxygens.

requires the separation of the double strand into two single strands by an enzyme that helps to split the hydrogen bonds along the length of the molecule. This event exposes the base code and makes it available for copying. Free nucleotides are used to synthesize matching strands that complement the bases in the code by adhering to the pairing requirements of A–T and C–G. The end result is two separate double strands with the same order of bases as the original molecule.

RNA: Organizers of Protein Synthesis

Like DNA, RNA consists of a long chain of nucleotides. However, RNA is a single strand containing ribose sugar instead of deoxyribose and uracil instead of thymine (see figure 2.23). Several functional types of RNA are formed using the DNA template through a replication-like process. Three major types of RNA are important for protein synthesis. Messenger RNA (mRNA) is a copy of a gene from DNA that provides the order and type of amino acids in a protein; transfer RNA (tRNA) is a carrier that delivers the correct amino acids for protein assembly; and ribosomal RNA (rRNA) is a major component of ribosomes (described in chapter 4). More information on these important processes is presented in chapter 9.

ATP: The Energy Molecule of Cells

A relative of RNA involved in an entirely different cell activity is **adenosine triphosphate (ATP).** ATP is a nucleotide containing adenine, ribose, and three phosphates rather than just one **(figure 2.27).** It belongs to a category of high-energy compounds (also including guanosine triphosphate, GTP) that give off energy when the bond is broken between the second and third (outermost) phosphate. The presence of these high-energy bonds makes it possible for ATP to release and store energy for cellular chemical reactions. Breakage of the bond of the terminal phosphate releases energy to do cellular work and also generates adenosine diphosphate (ADP). ADP can be converted back to ATP when the third phosphate is restored, thereby serving as an energy depot. Carriers for oxidation-reduction activities (nicotinamide adenine dinucleotide [NAD], for instance) are also derivatives of nucleotides (see chapter 8).

✔ CHECKPOINT

- Macromolecules are very large organic molecules (polymers) built up by polymerization of smaller molecular subunits (monomers).
- Carbohydrates are biological molecules whose polymers are monomers linked together by glycosidic bonds. Their main functions are protection and support (in organisms with cell walls) and also nutrient and energy stores.
- Lipids are biological molecules such as fats that are insoluble in water and contain special ester linkages. Their main functions are cell components, cell secretions, and nutrient and energy stores.
- Proteins are biological molecules whose polymers are chains of amino acid monomers linked together by peptide bonds.
- Proteins are called the "shapers of life" because of the many biological roles they play in cell structure and cell metabolism.
- Protein structure determines protein function. Structure and shape is dictated by amino acid composition and by the pH and temperature of the protein's immediate environment.
- Nucleic acids are biological molecules whose polymers are chains of nucleotide monomers linked together by phosphate–pentose sugar covalent bonds. Double-stranded nucleic acids are linked together by hydrogen bonds. Nucleic acids are information molecules that direct cell metabolism and reproduction. Nucleotides such as ATP also serve as energy transfer molecules in cells.

2.3 Cells: Where Chemicals Come to Life

As we proceed in this chemical survey from the level of simple molecules to increasingly complex levels of macromolecules, at some point we cross a line from the realm of lifeless molecules and arrive at the fundamental unit of life called a **cell.**[8] A cell is indeed a huge aggregate of carbon, hydrogen,

---

8. The word *cell* was originally coined from an Old English term meaning "small room" because of the way plant cells looked to early microscopists.

TABLE 2.6   A General Comparison of Procaryotic and Eucaryotic Cells and Viruses*

Function or Structure	Characteristic	Procaryotic Cells	Eucaryotic Cells	Viruses**
Genetics	Nucleic acids	+	+	+
	Chromosomes	+	+	−
	True nucleus	−	+	−
	Nuclear envelope	−	+	−
Reproduction	Mitosis	−	+	−
	Production of sex cells	+/−	+	−
	Binary fission	+	+	−
Biosynthesis	Independent	+	+	−
	Golgi apparatus	−	+	
	Endoplasmic reticulum	−	+	
	Ribosomes	+***	+	
Respiration	Enzymes	+	+	−
	Mitochondria	−	+	−
Photosynthesis	Pigments	+/−	+/−	−
	Chloroplasts	−	+/−	−
Motility/locomotor structures	Flagella	+/−***	+/−	−
	Cilia	−	+/−	−
Shape/protection	Membrane	+	+	+/−
	Cell wall	+***	+/−	− (have capsids instead)
	Capsule	+/−	+/−	−
Complexity of function		+	+	+/−
Size (in general)		0.5–3 μm****	2–100 μm	< 0.2 μm

handwritten annotations: "organelles" (bracketing Golgi apparatus, Endoplasmic reticulum, Ribosomes), "proteins" (next to Enzymes), "organelles" (next to Mitochondria)

*+ means most members of the group exhibit this characteristic; − means most lack it; +/− means some members have it and some do not.

**Viruses cannot participate in metabolic or genetic activity outside their host cells.

***The procaryotic type is functionally similar to the eucaryotic, but structurally unique.

****Much smaller and much larger bacteria exist; see Insight 4.3.

oxygen, nitrogen, and many other atoms, and it follows the basic laws of chemistry and physics, but it is much more. The combination of these atoms produces characteristics, reactions, and products that can only be described as *living*.

Fundamental Characteristics of Cells

The bodies of living things such as bacteria and protozoa consist of only a single cell, whereas those of animals and plants contain trillions of cells. Regardless of the organism, all cells have a few common characteristics. They tend to be spherical, polygonal, cubical, or cylindrical, and their protoplasm (internal cell contents) is encased in a cell or cytoplasmic membrane (see Insight 2.3). They have chromosomes containing DNA and ribosomes for protein synthesis, and they are exceedingly complex in function. Aside from these few similarities, most cell types fall into one of two fundamentally different lines (discussed in chapter 1): the small, seemingly simple procaryotic cells and the larger, structurally more complicated eucaryotic cells.

Eucaryotic cells are found in animals, plants, fungi, and protists. They contain a number of complex internal parts called organelles that perform useful functions for the cell involving growth, nutrition, or metabolism. By convention, organelles are defined as cell components that perform specific functions and are enclosed by membranes. Organelles also partition the eucaryotic cell into smaller compartments. The most visible organelle is the nucleus, a roughly ball-shaped mass surrounded by a double membrane that contains the DNA of the cell. Other organelles include the Golgi apparatus, endoplasmic reticulum, vacuoles, and mitochondria.

Procaryotic cells are possessed only by the bacteria and archaea. Sometimes it may seem that procaryotes are the microbial "have-nots" because, for the sake of comparison, they are described by what they lack. They have no nucleus or other organelles. This apparent simplicity is misleading, because the fine structure of procaryotes is complex. Overall, procaryotic cells can engage in nearly every activity that eucaryotic cells can, and many can function in ways that eucaryotes cannot. **Table 2.6** compares features of procaryotic and eucaryotic cells and viruses.

Processes That Define Life

To lay the groundwork for a detailed coverage of cells in chapters 4 and 5, this section provides an overview of cell structure and function and introduces the primary characteristics of life. The biological activities or properties that help define and characterize cells as living entities are:

1. growth;
2. reproduction and heredity;

3. metabolism, including cell synthesis and the release of energy;
4. movement and/or irritability;
5. cell support, protection, and storage mechanisms; and
6. the capacity to transport substances into and out of the cell.

Although eucaryotic cells have specific organelles to perform these functions, procaryotic cells must rely on a few simple, multipurpose cell components. As indicated in chapter 1, viruses are not cells, are not generally considered living things, and show certain signs of life only when they invade a host cell. Table 2.6 indicates their relative simplicity compared with cells.

Reproduction: Bearing Offspring

A cell's **genome** (jee'-nohm), its complete set of genetic material, is composed of elongate strands of DNA. The DNA is packed into discrete bodies called **chromosomes.** In eucaryotic cells, the chromosomes are located within a nuclear membrane.[9] Procaryotic DNA occurs in a special type of circular chromosome that is not enclosed by a membrane of any sort.

Living things devote a portion of their life cycle to producing offspring that will carry on their particular genetic line for many generations. In *sexual reproduction,* offspring are produced through the union of sex cells from two parents. In *asexual reproduction,* offspring originate through the division of a single parent cell into two daughter cells. Sexual reproduction occurs in most eucaryotes, and eucaryotic cells also reproduce asexually by several processes. One is a type of cell division called *binary fission,* a simple process in which the cell splits equally in two. Many eucaryotic cells engage in **mitosis** (my-toh'-sis), an orderly division of chromosomes that usually accompanies cell division (discussed in chapter 5). In contrast, procaryotic cells reproduce primarily by binary fission. They have no mitotic apparatus, nor do they reproduce by typical sexual means.

Metabolism: Chemical and Physical Life Processes

Cells synthesize proteins using hundreds of tiny particles called **ribosomes.** In eucaryotes, ribosomes are dispersed throughout the cell or inserted into membranous sacs known as the **endoplasmic reticulum.** Procaryotes have smaller ribosomes scattered throughout the protoplasm, since they lack an endoplasmic reticulum. Eucaryotes generate energy by chemical reactions in the **mitochondria,** whereas procaryotes use their cell membrane for this purpose. Photosynthetic microorganisms (algae and some bacteria) trap solar energy by means of pigments and convert it to chemical energy in the cell. Algae (eucaryotes) have compact, membranous bundles called **chloroplasts,** which contain the pigment and perform the photosynthetic reactions. Photosynthetic reactions and pigments of procaryotes do not occur in chloroplasts, but in specialized areas of the cell membrane. These characteristics are described in more detail in chapters 4 and 5.

Irritability or Motility

All cells have the capacity to respond to chemical, mechanical, or light stimuli. This quality, called **irritability,** helps cells adapt to the environment and obtain nutrients. Although not present in all cells, true **motility,** or self-propulsion, is a notable sign of life. Eucaryotic cells move by one of the following locomotor organelles: cilia, which are short, hairlike appendages; flagella, which are longer, whiplike appendages; or pseudopods, fingerlike extensions of the cell membrane. Motile procaryotes move by means of unusual, propeller-like flagella unique to bacteria or by special fibrils that produce a gliding form of motility. They have no cilia or pseudopods.

Protection and Storage

Many cells are supported and protected by rigid cell walls, which prevent them from rupturing while also providing support and shape. Among eucaryotes, cell walls occur in plants, microscopic algae, and fungi, but not in animals or protozoa. The majority of procaryotes have cell walls, but they differ in composition from the eucaryotic varieties. As protection against depleted nutrient sources, many microbes store nutrients intracellularly. Eucaryotes store nutrients in membranous sacs called vacuoles, and procaryotes concentrate them in crystals called granules or inclusions.

Transport: Movement of Nutrients and Wastes

Cell survival depends on drawing nutrients from the external environment and expelling waste and other metabolic products from the internal environment. This two-directional transport is accomplished in both eucaryotes and procaryotes by the cell membrane. This membrane, described in Insight 2.3, has a very similar structure in both eucaryotic and procaryotic cells. Eucaryotes have an additional organelle, the **Golgi apparatus,** that assists in sorting and packaging molecules for transport and removal from the cell.

As mentioned, you will learn more about these life processes in subsequent chapters. First, however, we need to review some of the methods by which microbiologists study these organisms. These techniques are the topic of chapter 3.

✔ **CHECKPOINT**

- As the atom is the fundamental unit of matter, so is the cell the fundamental unit of life.
- All true cells contain biological molecules that carry out the processes that define life: metabolism and reproduction. These two basic processes are supported by the functions of irritability and motility, protection, storage, and transport.
- The cell membrane is critically important to all cells because it controls the interchange between the cell and its environment.

---

9. Eucaryotic cells also carry a different sort of chromosome within their mitochondria.

Chapter Summary With Key Terms

2.1 Atoms, Bonds, and Molecules: Fundamental Building Blocks

A. Atomic Structure and Elements

1. All **matter** in the universe is composed of minute particles called **atoms**—the simplest form of matter not divisible into a simpler substance by chemical means. Atoms are composed of smaller particles called **protons, neutrons,** and **electrons.**

2. Atoms that differ in numbers of the protons, neutrons, and electrons are elements. **Elements** can be described by **mass number (MN)** and **atomic number (AN),** and each is known by a distinct name and symbol. Elements may exist in variant forms called **isotopes.**

B. Bonds and Molecules

1. Atoms interact to form **chemical bonds** and **molecules.** If the atoms combining to make a molecule are different elements, then the substance is termed a **compound.**

2. The type of bond is dictated by the electron makeup of the outer orbitals **(valence)** of the atoms. Bond types include:

   a. **Covalent bonds,** with shared electrons. The molecule shares the electrons; the balance of charge will be **polar** if unequal or **nonpolar** if equally shared/electrically neutral.

   b. **Ionic bonds,** where electrons are transferred to an atom that can come closer to filling up the outer orbital. Dissociation of these compounds leads to the formation of charged **cations** and **anions.**

   c. **Hydrogen bonds** involve weak covalent bonds between hydrogen and nearby electronegative oxygens and nitrogens.

C. Solutions, Acids, Bases, and pH

1. A **solution** is a combination of a solid, liquid, or gaseous chemical (the **solute**) dissolved in a liquid medium (the **solvent**). Water is the most common solvent in natural systems.

2. Ionization of water leads to the release of hydrogen ions (H^+) and hydroxyl (OH^-) ions. The **pH** scale expresses the concentration of H^+ such that a pH of less than 7.0 is considered **acidic,** and a pH of more than that, indicating fewer H^+, is considered **basic.**

2.2 Macromolecules: Superstructures of Life

A. Biochemistry studies those molecules that are found in living things. These are based on **organic** compounds, which usually consist of carbon and hydrogen covalently bonded in various combinations. **Inorganic** compounds do not contain both carbon and hydrogen in combination.

B. **Macromolecules** are very large compounds and are generally assembled from single units called **monomers** by polymerization.

C. Macromolecules of life fall into basic categories of **carbohydrates, lipids, proteins,** and nucleic acids.

1. **Carbohydrates** are composed of carbon, hydrogen, and oxygen and contain aldehyde or ketone groups.

   a. **Monosaccharides** such as glucose are the simplest carbohydrates with 3 to 7 carbons; these are the monomers of carbohydrates.

   b. **Disaccharides** such as lactose consist of two monosaccharides joined by **glycosidic bonds. Polysaccharides** such as starch and **peptidoglycan** are chains of five or more monosaccharides.

2. **Lipids** contain long hydrocarbon chains and are not soluble in polar solvents such as water due to their nonpolar, hydrophobic character. Examples are **triglycerides, phospholipids,** sterols and waxes.

3. **Proteins** are highly complex macromolecules that are crucial in most, if not all, life processes.

   a. **Amino acids** are the basic building blocks of proteins. They all share a basic structure of an amino group, a carboxyl group, an R group, and hydrogen bonded to a carbon atom. There are 20 different R groups, which define the basic set of 20 amino acids, found in all of life.

   b. A **peptide** is a short chain of amino acids bound by **peptide bonds:** a **protein** contains more than 50 amino acids.

   c. The structure of a protein is very important to the function it has. This is described by the **primary structure** (the chain of amino acids), the **secondary structure** (formation of helices and sheets due to hydrogen bonding within the chain), **tertiary structure** (cross-links, especially disulfide bonds, between secondary structures), and **quaternary structure** (formation of multisubunit proteins). The incredible variation in shapes is the basis for the diverse roles proteins play as **enzymes, antibodies,** receptors, and structural components.

4. Nucleic acids

   a. **Nucleotides** are the building blocks of nucleic acids. They are composed of a **nitrogen base, a pentose** sugar, and phosphate, Nitrogen bases are ringed compounds: **adenine (A), guanine (G), cytosine (C), thymine (T),** and **uracil (U).** Pentose sugars may be deoxyribose or ribose.

   b. **Deoxyribonucleic acid (DNA)** is a polymer of nucleotides that occurs as a double-stranded helix with hydrogen bonding in pairs between the helices. It has all of the bases except uracil, and the pentose sugar is deoxyribose. DNA is the master code for a cell's life processes and must be transmitted to the offspring through **replication.**

   c. **Ribonucleic acid (RNA)** is a polymer of nucleotides where the sugar is **ribose** and the **uracil** is used instead of thymine. It is almost always found single stranded and is used to express the DNA code into proteins.

   d. **Adenosine triphosphate (ATP)** is a nucleotide involved in the transfer and storage of energy in cells.

2.3 Cells: Where Chemicals Come to Life
  A. All living things are composed of **cells,** which are aggregates of macromolecules that carry out living processes.
  B. Cells can be divided into two basic types: procaryotes and eucaryotes.
  C. Cells show the basic essentials characteristics of life. Parts of cells and macromolecules do not show these characteristics.

Multiple-Choice Questions

1. The smallest unit of matter with unique characteristics is
   a. an electron
   b. a molecule
   c. an atom
   d. a proton

2. The __x__ charge of a proton is exactly balanced by the ____ charge of a (an) __e__.
   a. negative, positive, electron
   b. positive, neutral, neutron
   c. positive, negative, electron
   d. neutral, negative, electron

3. Electrons move around the nucleus of an atom in pathways called
   a. shells
   b. orbitals
   c. circles
   d. rings

4. Which part of an element does not vary in number?
   a. electron
   b. neutron
   c. proton
   d. all of these vary

5. If a substance contains two or more elements of different types, it is considered
   a. a compound
   b. a monomer
   c. a molecule
   d. organic

6. Bonds in which atoms share electrons are defined as ____ bonds.
   a. hydrogen
   b. ionic
   c. double
   d. covalent

7. Hydrogen bonds can form between ____ adjacent to each other.
   a. two hydrogen atoms
   b. two oxygen atoms
   c. a hydrogen atom and an oxygen atom
   d. negative charges

8. An atom that can donate electrons during a reaction is called
   a. an oxidizing agent
   b. a reducing agent
   c. an ionic agent
   d. an electrolyte

9. In a solution of NaCl and water, NaCl is the ____ and water is the ____.
   a. acid, base
   b. base, acid
   c. solute, solvent
   d. solvent, solute

10. A solution with a pH of 2 ____ than a solution with a pH of 8.
    a. has less H^+
    b. has more H^+
    c. has more OH^-
    d. is less concentrated

11. Fructose is a type of
    a. disaccharide
    b. monosaccharide
    c. polysaccharide
    d. amino acid

12. Bond formation in polysaccharides and polypeptides is accompanied by the removal of a
    a. hydrogen atom
    b. hydroxyl ion
    c. carbon atom
    d. water molecule

13. The monomer unit of polysaccharides such as starch and cellulose is
    a. fructose
    b. glucose
    c. ribose
    d. lactose

14. A phospholipid contains
    a. three fatty acids bound to glycerol
    b. three fatty acids, a glycerol, and a phosphate
    c. two fatty acids and a phosphate bound to glycerol
    d. three cholesterol molecules bound to glycerol

15. Proteins are synthesized by linking amino acids with ____ bonds.
    a. disulfide
    b. glycosidic
    c. peptide
    d. ester

16. The amino acid that accounts for disulfide bonds in the tertiary structure of proteins is
    a. tyrosine
    b. glycine
    c. cysteine
    d. serine

17. DNA is a hereditary molecule that is composed of
    a. deoxyribose, phosphate, and nitrogen bases
    b. deoxyribose, a pentose, and nucleic acids
    c. sugar, proteins, and thymine
    d. adenine, phosphate, and ribose

18. What is meant by DNA replication?
    a. duplication of the sugar-phosphate backbone
    b. matching of base pairs
    c. formation of the double helix
    d. the exact copying of the DNA code into two new molecules

19. Proteins can function as
    a. enzymes
    b. receptors
    c. antibodies
    d. a, b, and c

20. RNA plays an important role in what biological process?
    a. replication
    b. protein synthesis
    c. lipid metabolism
    d. water transport

Concept Questions

These questions are suggested as a *writing-to-learn* experience. For each question, compose a one- or two-paragraph answer that includes the factual information needed to completely address the question.

1. How are the concepts of an atom and an element related? What causes elements to differ?

2. a. How are mass number and atomic number derived? What is the atomic weight?
   b. Using data in table 2.1, give the electron number of nitrogen, sulfur, calcium, phosphorus, and iron.
   c. What is distinctive about isotopes of elements, and why are they important?

3. a. How is the concept of molecules and compounds related?
   b. Compute the molecular weight of oxygen and methane.

4. a. Why is an isolated atom neutral?
   b. Describe the concept of the atomic nucleus, electron orbitals, and shells.
   c. What causes atoms to form chemical bonds?
   d. Why do some elements not bond readily?
   e. Draw the atomic structure of magnesium and predict what kinds of bonds it will make.

5. Distinguish between the general reactions in covalent, ionic, and hydrogen bonds.

6. a. Which kinds of elements tend to make covalent bonds?
   b. Distinguish between a single and a double bond.
   c. What is polarity?
   d. Why are some covalent molecules polar and others nonpolar?
   e. What is an important consequence of the polarity of water?

7. a. Which kinds of elements tend to make ionic bonds?
   b. Exactly what causes the charges to form on atoms in ionic bonds?
   c. Verify the proton and electron numbers for Na^+ and Cl^-.
   d. Differentiate between an anion and a cation.
   e. What kind of ion would you expect magnesium to make, on the basis of its valence?

8. Differentiate between an oxidizing agent and a reducing agent.

9. Why are hydrogen bonds relatively weak?

10. a. Compare the three basic types of chemical formulas.
    b. Review the types of chemical reactions and the general ways they can be expressed in equations.

11. a. Define solution, solvent, and solute.
    b. What properties of water make it an effective biological solvent, and how does a molecule like NaCl become dissolved in it?

c. How is the concentration of a solution determined?
d. What is molarity? Tell how to make a 1M solution of $Mg_3(PO_4)_2$ and a 0.1 M solution of $CaSO_4$.

12. a. What determines whether a substance is an acid or a base?
    b. Briefly outline the pH scale.
    c. How can a neutral salt be formed from acids and bases?

13. a. What atoms must be present in a molecule for it to be considered organic?
    b. What characteristics of carbon make it ideal for the formation of organic compounds?
    c. What are functional groups?
    d. Differentiate between a monomer and a polymer.
    e. How are polymers formed?
    f. Name several inorganic compounds.

14. a. What characterizes the carbohydrates?
    b. Differentiate between mono-, di-, and polysaccharides, and give examples of each.
    c. What is a glycosidic bond?
    d. What are some of the functions of polysaccharides in cells?

15. a. Draw simple structural molecules of triglycerides and phospholipids to compare their differences and similarities.
    b. What is an ester bond?
    c. How are saturated and unsaturated fatty acids different?
    d. What characteristic of phospholipids makes them essential components of cell membranes?
    e. Why is the hydrophilic end of phospholipids attracted to water?

16. a. Describe the basic structure of an amino acid.
    b. What makes the amino acids distinctive, and how many of them are there?
    c. What is a peptide bond?
    d. Differentiate between a peptide, a polypeptide, and a protein.
    e. Explain what causes the various levels of structure of a protein molecule.
    f. What functions do proteins perform in a cell?

17. a. Describe a nucleotide and a polynucleotide, and compare and contrast the general structure of DNA and RNA.
    b. Name the two purines and the three pyrimidines.
    c. Why is DNA called a double helix?
    d. What is the function of RNA?
    e. What is ATP, and what is its function in cells?

18. What biological activities define life? How will these activities differ in procaryotes versus eucaryotes?

Critical Thinking Questions

Critical thinking is the ability to reason and solve problems using facts and concepts. These questions can be approached from a number of angles, and in most cases, they do not have a single correct answer.

1. The "octet rule" in chemistry helps predict the tendency of atoms to acquire or donate electrons from the outer shell. It says that those with fewer than 4 tend to donate electrons and those with more than 4 tend to accept additional electrons; those with exactly 4 can do both. Using this rule, determine what category each of the following elements falls into: N, S, C, P, O, H, Ca, Fe, and Mg. (You will need to work out the valence of the atoms.)

2. Predict the kinds of bonds that occur in ammonium (NH_3), phosphate (PO_4), disulfide (S—S), and magnesium chloride ($MgCl_2$). (Use simple models such as those in figure 2.4.)

3. Work out the following problems:
   a. What is the number of protons in helium?
   b. Will an H bond form between $H_3C—CH{=}O$ and H_2O? Why or why not?
   c. Draw the following molecules and determine which are polar: Cl_2, NH_3, CH_4.
   d. What is the pH of a solution with a concentration of 0.00001 moles/ml (M) of H^+?
   e. What is the pH of a solution with a concentration of 0.00001 moles/ml (M) of OH^-?

4. a. Describe how hydration spheres are formed around cations and anions.
   b. What kind of substances will be expected to be hydrophilic and hydrophobic, and what makes them so?

   c. Distinguish between polar and ionic compounds, using your own words.

5. In what way are carbon-based compounds like children's Tinker Toys or Lego blocks?

6. Is galactose an aldehyde or a ketone sugar?

7. a. How many water molecules are released when a triglyceride is formed?
   b. How many peptide bonds are in a tetrapeptide?

8. Looking at figure 2.25, can you see why adenine forms hydrogen bonds with thymine and why cytosine forms them with guanine?

9. Saturated fats are solid at room temperature and unsaturated fats are not. Is butter an example of a saturated or unsaturated fat? Is olive oil an example of a saturated or unsaturated fat? What are trans-fatty acids? Why is there currently a dietary trans-fatty acid debate?

Internet Search Topics

1. Use a search engine to explore the topic of isotopes and dating ancient rocks. How can isotopes be used to determine if rocks contain evidence of life?

2. Go to the Online Learning Center for chapter 2 of this text at http://www.mhhe.com/cowan1. Make a search for basic information on elements, using one or both of the websites listed in the Science Zone. Click on the icons for C, H, N, O, P, S and list the source, biological importance, and other useful information about these elements.

Tools of the Laboratory

The Methods for Studying Microorganisms

A 94-year-old woman went to her local hospital emergency department in mid-November 2001 complaining of a 5-day history of weakness, fever, nonproductive cough, and generalized myalgia (muscle aches). Otherwise, for a person her age she was fairly healthy, although she did suffer from chronic obstructive pulmonary disease, hypertension, and chronic kidney failure.

On physical examination, her heart rate was above normal and she had a fever of 102.3°F (39.1°C). The rest of her physical examination was normal. Initial laboratory studies (blood cell count, blood chemistries, and chest X ray) were also normal except for the chemical urine testing. This finding along with the fever suggested an infec-

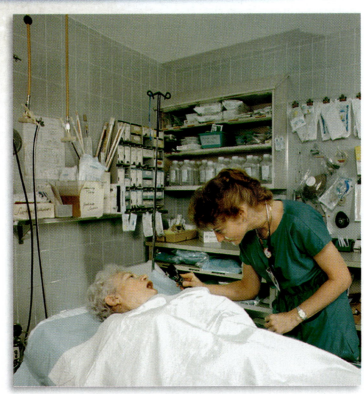

tion, so the patient was admitted to the hospital. Samples of blood and urine were sent to the microbiology laboratory and set up appropriately.

The next day, microscopic evaluation of the urine culture revealed rod-shaped bacteria that stained red, and the blood culture revealed rods that stained purple. (The liquid blood culture was then transferred to appropriate solid media.) This finding in the blood was unusual, so a sample culture was sent to the state health department laboratory. Antibiotic therapy was adjusted, yet the patient's condition deteriorated. Her most serious symptoms localized to her chest, and she was transferred to the intensive care unit. Four days after admission, the health department announced that the bacteria found in the patient's blood were *Bacillus anthracis.* She was suffering from inhalation anthrax. Further testing showed these bacteria to be of the same strain that had been involved in the recent bioterrorist attack. Despite treatment, the patient died on the fifth day after admission.

▶ *What techniques and equipment are used when the bacteria are observed as being purple and red? How are these findings reported?*

▶ *What are the stages of processing a blood sample?*

CHAPTER OVERVIEW

▶ Microbes are managed and characterized with the Five I's—inoculation, incubation, isolation, inspection, and identification.

▶ Cultures are made by removing a sample from a desired source and placing it in containers of media.

▶ Media can be varied in chemical and physical form and functional purposes, depending on the intention.

▶ Growth and isolation of microbes leads to pure cultures that permit the study and testing of single species.

▶ Cultures can be used to provide information on microbial morphology, biochemistry, and genetic characteristics.

▶ Unknown, invisible samples can become known and visible.

▶ The microscope is a powerful tool for magnifying and resolving cells and their parts.

▶ Microscopes exist in several forms, using light, radiation, and electrons to form images.

▶ Specimens and cultures are prepared for study in fresh (live) or fixed (dead) form.

▶ Staining procedures highlight cells and allow them to be described and identified.

3.1 Methods of Culturing Microorganisms—The Five I's

Biologists studying large organisms such as animals and plants can, for the most part, immediately see and differentiate their experimental subjects from the surrounding environment and from one another. In fact, they can use their senses of sight, smell, hearing, and even touch to detect and evaluate identifying characteristics and to keep track of growth and developmental changes. Because microbiologists cannot rely as much as other scientists on senses other than sight, they are confronted by some unique problems. First, most habitats (such as the soil and the human mouth) harbor microbes in complex associations, so it is often necessary to separate the species from one another. Second, to maintain and keep track of such small research subjects, microbiologists usually have to grow them under artificial conditions. A third difficulty in working with microbes is that they are invisible and widely distributed, and undesirable ones can be introduced into an experiment and cause misleading results. These impediments motivated the development of techniques to control microbes and their growth, primarily sterile, aseptic, and pure culture techniques.[1]

Microbiologists use five basic techniques to manipulate, grow, examine, and characterize microorganisms in the laboratory: inoculation, incubation, isolation, inspection, and identification (the Five I's; **figure 3.1**). Some or all of these procedures are performed by microbiologists, whether the beginning laboratory student, the researcher attempting to isolate drug-producing bacteria from soil, or the clinical microbiologist working with a specimen from a patient's infection. These procedures make it possible to handle and maintain microorganisms as discrete entities whose detailed biology can be studied and recorded.

Inoculation: Producing a Culture

To cultivate, or **culture,** microorganisms, one introduces a tiny sample (the inoculum) into a container of nutrient medium (pl. media), which provides an environment in which they multiply. This process is called **inoculation.** Any instrument used for sampling and inoculation must initially be *sterile* (see footnote 1). The observable growth that appears in or on the medium is known as a culture. The nature of the sample being cultured depends on the objectives of the analysis. Clinical specimens for determining the cause of an infectious disease are obtained from body fluids (blood, cerebrospinal fluid), discharges (sputum, urine, feces), or diseased tissue. Other samples subject to microbiological analysis are soil, water, sewage, foods, air, and inanimate objects. Procedures for proper specimen collection are discussed in chapter 17.

Isolation: Separating One Species from Another

Certain **isolation** techniques are based on the concept that if an individual bacterial cell is separated from other cells and provided adequate space on a nutrient surface, it will grow into a discrete mound of cells called a **colony (figure 3.2).** If it was formed from a single cell, a colony consists of just that one species and no other. Proper isolation requires that a small number of cells be inoculated into a relatively large volume or over an expansive area of medium. It generally requires the following materials: a medium that has a relatively firm surface (see agar in "Physical States of Media," page 64), a Petri plate (a clear, flat dish with a cover), and inoculating tools. In the streak plate method, a small droplet of culture or sample is spread over the surface of the medium according to a pattern that gradually thins out the sample and separates the cells spatially over several sections of the plate **(figure 3.3a,b).** Because of its ease and effectiveness, the streak plate is the method of choice for most applications.

In the loop dilution, or pour plate, technique, the sample is inoculated serially into a series of cooled but still liquid agar tubes so as to dilute the number of cells in each successive tube in the series **(figure 3.3c,d).** Inoculated tubes are then plated out (poured) into sterile Petri plates and are allowed to solidify (harden). The end result (usually in the second or third plate) is that the number of cells per volume is so decreased that cells have ample space to

---

1. *Sterile* means the complete absence of viable microbes: *aseptic* refers to prevention of infection; *pure culture* refers to growth of a single species of microbe.

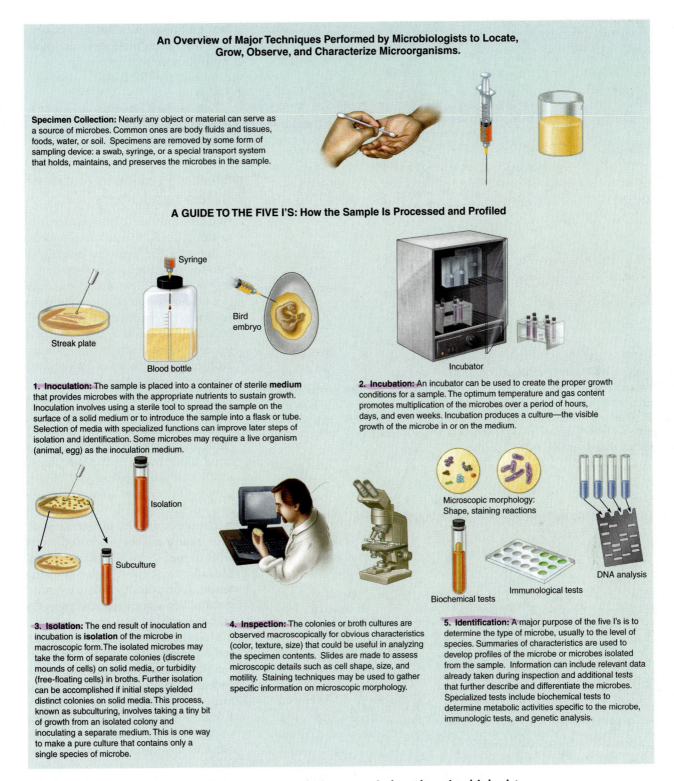

An Overview of Major Techniques Performed by Microbiologists to Locate, Grow, Observe, and Characterize Microorganisms.

Specimen Collection: Nearly any object or material can serve as a source of microbes. Common ones are body fluids and tissues, foods, water, or soil. Specimens are removed by some form of sampling device: a swab, syringe, or a special transport system that holds, maintains, and preserves the microbes in the sample.

A GUIDE TO THE FIVE I'S: How the Sample Is Processed and Profiled

Syringe

Streak plate

Blood bottle

Bird embryo

Incubator

1. Inoculation: The sample is placed into a container of sterile **medium** that provides microbes with the appropriate nutrients to sustain growth. Inoculation involves using a sterile tool to spread the sample on the surface of a solid medium or to introduce the sample into a flask or tube. Selection of media with specialized functions can improve later steps of isolation and identification. Some microbes may require a live organism (animal, egg) as the inoculation medium.

2. Incubation: An incubator can be used to create the proper growth conditions for a sample. The optimum temperature and gas content promotes multiplication of the microbes over a period of hours, days, and even weeks. Incubation produces a culture—the visible growth of the microbe in or on the medium.

Isolation

Subculture

Microscopic morphology: Shape, staining reactions

Biochemical tests

Immunological tests

DNA analysis

3. Isolation: The end result of inoculation and incubation is **isolation** of the microbe in macroscopic form. The isolated microbes may take the form of separate colonies (discrete mounds of cells) on solid media, or turbidity (free-floating cells) in broths. Further isolation can be accomplished if initial steps yielded distinct colonies on solid media. This process, known as subculturing, involves taking a tiny bit of growth from an isolated colony and inoculating a separate medium. This is one way to make a pure culture that contains only a single species of microbe.

4. Inspection: The colonies or broth cultures are observed macroscopically for obvious characteristics (color, texture, size) that could be useful in analyzing the specimen contents. Slides are made to assess microscopic details such as cell shape, size, and motility. Staining techniques may be used to gather specific information on microscopic morphology.

5. Identification: A major purpose of the five I's is to determine the type of microbe, usually to the level of species. Summaries of characteristics are used to develop profiles of the microbe or microbes isolated from the sample. Information can include relevant data already taken during inspection and additional tests that further describe and differentiate the microbes. Specialized tests include biochemical tests to determine metabolic activities specific to the microbe, immunologic tests, and genetic analysis.

FIGURE 3.1 **A summary of the general laboratory techniques carried out by microbiologists.**
It is not necessary to perform all the steps shown or to perform them exactly in this order, but all microbiologists participate in at least some of these activities. In some cases, one may proceed right from the sample to inspection, and in others, only inoculation and incubation on special media are required.

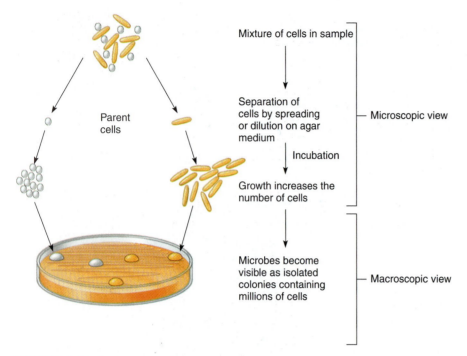

Mixture of cells in sample — Microscopic view

Parent cells

Separation of cells by spreading or dilution on agar medium

Incubation

Growth increases the number of cells

Microbes become visible as isolated colonies containing millions of cells — Macroscopic view

FIGURE 3.2   Isolation technique.
Stages in the formation of an isolated colony, showing the microscopic events and the macroscopic result. Separation techniques such as streaking can be used to isolate single cells. After numerous cell divisions, a macroscopic mound of cells, or a colony, will be formed. This is a relatively simple yet successful way to separate different types of bacteria in a mixed sample.

TABLE 3.1	Three Categories of Media Classification	
Physical State (Medium's Normal Consistency)	**Chemical Composition (Type of Chemicals Medium Contains)**	**Functional Type (Purpose of Medium)***
1. Liquid	1. Synthetic (chemically defined)	1. General purpose
2. Semisolid		2. Enriched
3. Solid (can be converted to liquid)	2. Nonsynthetic (not chemically defined)	3. Selective
		4. Differential
4. Solid (cannot be liquefied)		5. Anaerobic growth
		6. Specimen transport
		7. Assay
		8. Enumeration

*Some media can serve more than one function. For example, a medium such as brain-heart infusion is general purpose and enriched; mannitol salt agar is both selective and differential; and blood agar is both enriched and differential.

grow into separate colonies. One difference between this and the streak plate method is that in this technique some of the colonies will develop deep in the medium itself and not just on the surface.

With the spread plate technique, a small volume of liquid, diluted sample is pipetted onto the surface of the medium and spread around evenly by a sterile spreading tool (sometimes called a "hockey stick"). Like the streak plate, cells are pushed into separate areas on the surface so that they can form individual colonies **(figure 3.3e,f)**.

Before we continue to cover information on the Five I's, we will take a side trip to look at media in more detail.

Media: Providing Nutrients in the Laboratory

A major stimulus to the rise of microbiology in the late 1800s was the development of techniques for growing microbes out of their natural habitats and in pure form in the laboratory. This milestone enabled the close examination of a microbe and its morphology, physiology, and genetics. It was evident from the very first that for successful cultivation, each microorganism had to be provided with all of its required nutrients in an artificial medium.

Some microbes require only a very few simple inorganic compounds for growth; others need a complex list of specific inorganic and organic compounds. This tremendous diversity is evident in the types of media that can be prepared. At least 500 different types of media are used in culturing and identifying microorganisms. Culture media are contained in test tubes, flasks, or Petri plates, and they are inoculated by such tools as loops, needles, pipettes, and swabs. Media are extremely varied in nutrient content and consistency and can be specially formulated for a particular purpose. Culturing microbes that cannot grow on artificial media (all viruses and certain bacteria) requires cell cultures or host animals **(Insight 3.1).**

For an experiment to be properly controlled, sterile technique is necessary. This means that the inoculation must start with a sterile medium and inoculating tools with sterile tips must be used. Measures must be taken to prevent introduction of nonsterile materials, such as room air and fingers, directly into the media.

Types of Media

Media can be classified according to three properties **(table 3.1):**

1. physical state,
2. chemical composition, and
3. functional type.

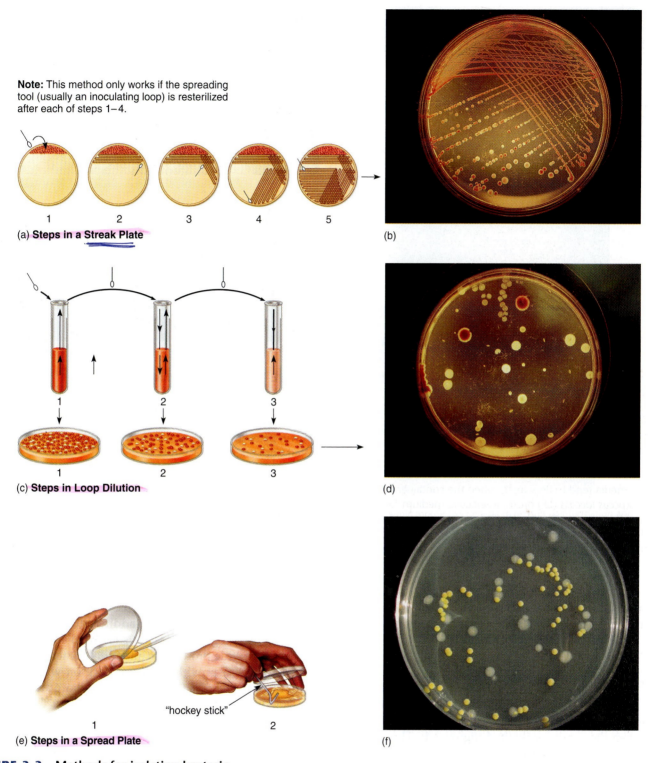

Note: This method only works if the spreading tool (usually an inoculating loop) is resterilized after each of steps 1–4.

1   2   3   4   5

(a) **Steps in a Streak Plate**

(b)

1   2   3

1   2   3

(c) **Steps in Loop Dilution**

(d)

"hockey stick"

1   2

(e) **Steps in a Spread Plate**

(f)

FIGURE 3.3   Methods for isolating bacteria.
(a) Steps in a quadrant streak plate and **(b)** resulting isolated colonies of bacteria. **(c)** Steps in the loop dilution method and **(d)** the appearance of plate 3. **(e)** Spread plate and **(f)** its result.

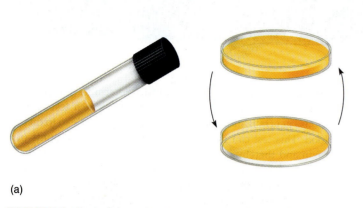

(a)

(b)

FIGURE 3.6   Solid media that are reversible to liquids.
(a) Media containing 1–5% agar are solid enough to remain in place when containers are tilted or inverted. They are liquefiable by heat but generally not by bacterial enzymes. **(b)** Nutrient gelatin contains enough gelatin (12%) to take on a solid consistency. The top tube shows it as a solid. The bottom tube indicates a result when microbial enzymes digest the gelatin and liquefy it.

Complex, or nonsynthetic, media contain at least one ingredient that is *not* chemically definable—not a simple, pure compound and not representable by an exact chemical formula. Most of these substances are extracts of animals, plants, or yeasts, including such materials as ground-up cells, tissues, and secretions. Examples are blood, serum, and meat extracts or infusions. Other nonsynthetic ingredients are milk, yeast extract, soybean digests, and peptone. Peptone is a partially digested protein, rich in amino acids, that is often used as a carbon and nitrogen source. Nutrient broth, blood agar, and MacConkey agar, though different in function and appearance, are all complex nonsynthetic media.

They present a rich mixture of nutrients for microbes that have complex nutritional needs.

A specific example can be used to compare the differences between a synthetic medium and a nonsynthetic one. Both synthetic *Euglena* medium (table 3.2) and nonsynthetic nutrient broth contain amino acids. But *Euglena* medium has three known amino acids in known amounts, whereas nutrient broth contains amino acids (from peptone) in variable types and amounts. Pure inorganic salts and organic acids are added in precise quantities for *Euglena*, whereas those components are provided by undefined beef extract in nutrient broth.

Media to Suit Every Function

Microbiologists have many types of media at their disposal, with new ones being devised all the time. Depending upon what is added, a microbiologist can fine-tune a medium for nearly any purpose. Until recently, microbiologists knew of only a few species of bacteria or fungi that could not be cultivated artificially. Newer DNA detection technologies have shown us just how wrong we were; it is now thought that there are many times more microbes that we don't know how to cultivate in the lab than those that we do. Previous discovery and identification of microorganisms relied on our ability to grow them. Now we can detect a single bacterium in its natural habitat.

General-purpose media are designed to grow as broad a spectrum of microbes as possible. As a rule, they are nonsynthetic and contain a mixture of nutrients that could support the growth of a variety of microbial life. Examples include nutrient agar and broth, brain-heart infusion, and trypticase soy agar (TSA). TSA contains partially digested milk protein (casein), soybean digest, NaCl, and agar.

An **enriched medium** contains complex organic substances such as blood, serum, hemoglobin, or special **growth factors** (specific vitamins, amino acids) that certain species

TABLE 3.2	Medium for the Growth and Maintenance of the Green Alga *Euglena*
Glutamic acid (aa)	6 g
Aspartic acid (aa)	4 g
Glycine (aa)	5 g
Sucrose (c)	30 g
Malic acid (oa)	2 g
Succinic acid (oa)	1.04 g
Boric acid	1.14 mg
Thiamine hydrochloride (v)	12 mg
Monopotassium phosphate	0.6 g
Magnesium sulfate	0.8 g
Calcium carbonate	0.16 g
Ammonium carbonate	0.72 g
Ferric chloride	60 mg
Zinc sulfate	40 mg
Manganese sulfate	6 mg
Copper sulfate	0.62 mg
Cobalt sulfate	5 mg
Ammonium molybdate	1.34 mg

Note: These ingredients are dissolved in 1,000 ml of water.
aa, amino acid; c, carbohydrate; oa, organic acid; v, vitamin; g, gram; mg, milligram.

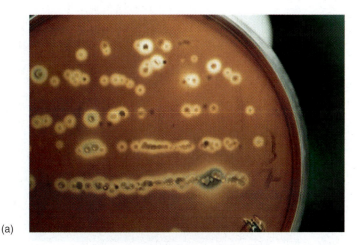

(a)

(b)

FIGURE 3.7   Examples of enriched media.
(a) Blood agar plate growing bacteria from the human throat. Note that this medium also differentiates among different colonies by their appearance. **(b)** Chocolate agar, a medium that gets its brown color from heated blood, not from chocolate. It is commonly used to culture the fastidious gonococcus *Neisseria gonorrhoeae.*

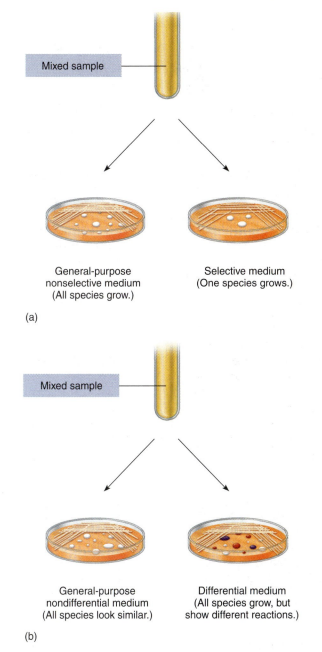

Mixed sample

General-purpose
nonselective medium
(All species grow.)

Selective medium
(One species grows.)

(a)

Mixed sample

General-purpose
nondifferential medium
(All species look similar.)

Differential medium
(All species grow, but
show different reactions.)

(b)

✗FIGURE 3.8   Comparison of selective and differential media with general-purpose media.
(a) The same mixed sample containing three different species is streaked onto plates of general-purpose nonselective medium and selective medium. Note the results. **(b)** Another mixed sample containing three different species is streaked onto plates of general-purpose nondifferential medium and differential medium. Note the results.

must have in order to grow. Bacteria that require growth factors and complex nutrients are termed **fastidious.** Blood agar, which is made by adding sterile sheep, horse, or rabbit blood to a sterile agar base **(figure 3.7a)** is widely employed to grow fastidious streptococci and other pathogens. Pathogenic *Neisseria* (one species causes gonorrhea) are grown on Thayer-Martin medium or chocolate agar, which is made by heating blood agar **(figure 3.7b).**

Selective and Differential Media.   Some of the cleverest and most inventive media recipes belong to the categories of selective and differential media **(figure 3.8).** These media are

designed for special microbial groups, and they have extensive applications in isolation and identification. They can permit, in a single step, the preliminary identification of a genus or even a species.

A **selective medium (table 3.3)** contains one or more agents that inhibit the growth of a certain microbe or microbes

TABLE 3.3	Selective Media, Agents, and Functions	
Medium	**Selective Agent**	**Used For**
Mueller tellurite	Potassium tellurite	Isolation of *Corynebacterium diphtheriae*
Enterococcus faecalis broth	Sodium azide, tetrazolium	Isolation of fecal enterococci
Phenylethanol agar	Phenylethanol chloride	Isolation of staphylococci and streptococci
Tomato juice agar	Tomato juice, acid	Isolation of lactobacilli from saliva
MacConkey agar	Bile, crystal violet	Isolation of gram-negative enterics
Salmonella/Shigella (SS) agar	Bile, citrate, brilliant green	Isolation of *Salmonella* and *Shigella*
Lowenstein-Jensen	Malachite green dye	Isolation and maintenance of *Mycobacteria*
Sabouraud's agar	pH of 5.6 (acid)	Isolation of fungi—inhibits bacteria

(a)

(b)

FIGURE 3.9 Examples of media that are both selective and differential.

(a) Mannitol salt agar is used to isolate members of the genus *Staphylococcus*. It is selective for this genus because it can grow in the presence of 7.5% sodium chloride, whereas many other species are inhibited by this high concentration. It contains a dye that also differentiates those species of *Staphylococcus* that produce acid from mannitol and turn the phenol red dye to a bright yellow. (b) MacConkey agar differentiates between lactose-fermenting bacteria (indicated by a pink-red reaction in the center of the colony) and lactose-negative bacteria (indicated by an off-white colony with no dye reaction).

(call them A, B, and C) but not others (D) and thereby encourages, or *selects*, microbe D and allows it to grow. Selective media are very important in primary isolation of a specific type of microorganism from samples containing dozens of different species—for example, feces, saliva, skin, water, and soil. They hasten isolation by suppressing the unwanted background organisms and favoring growth of the desired ones.

Mannitol salt agar (MSA) **(figure 3.9a)** contains a high concentration of NaCl (7.5%) that is quite inhibitory to most human pathogens. One exception is the genus *Staphylococcus*, which grows well in this medium and consequently can be amplified in mixed samples. Bile salts, a component of feces, inhibit most gram-positive bacteria while permitting many gram-negative rods to grow. Media for isolating intestinal pathogens (MacConkey agar, Hektoen enteric [HE] agar) contain bile salts as a selective agent **(figure 3.9b).** Dyes such as methylene blue and crystal violet also inhibit certain gram-positive bacteria. Other agents that have selective properties are antimicrobial drugs and acid. Some selective media contain strongly inhibitory agents to favor the growth of a pathogen that would otherwise be overlooked because of its low numbers in a specimen. Selenite and brilliant green dye are used in media to isolate *Salmonella* from feces, and sodium azide is used to isolate enterococci from water and food (see EF broth, figure 3.4b).

Differential media grow several types of microorganisms but are designed to display visible differences among those microorganisms. Differentiation shows up as variations in colony size or color, in media color changes, or in the formation of gas bubbles and precipitates **(table 3.4).** These variations come from the type of chemicals these media con-

tain and the ways that microbes react to them. For example, when microbe X metabolizes a certain substance not used by organism Y, then X will cause a visible change in the medium and Y will not. The simplest differential media show two reaction types such as the use or nonuse of a particular nutrient or a color change in some colonies but not in others. Some media are sufficiently complex to show three or four different reactions **(figure 3.10).** A single medium can be both selective and differential, owing to different ingredients in its composition. MacConkey agar, for example, appears in table 3.3 (selective media) and table 3.4 (differential media).

Dyes can be used as differential agents because many of them are pH indicators that change color in response to the production of an acid or a base. For example, MacConkey agar contains neutral red, a dye that is yellow when neutral and

TABLE 3.4 Differential Media

Medium	Substances That Facilitate Differentiation	Differentiates Between
Blood agar	Intact red blood cells	Types of hemolysis
Mannitol salt agar	Mannitol, phenol red, and 7.5% NaCl	Species of *Staphylococcus* NaCl also inhibits the salt-sensitive species
Hektoen enteric (HE) agar	Brom thymol blue, acid fuchsin, sucrose, salicin, thiosulfate, ferric ammonium citrate, and bile	*Salmonella, Shigella,* other lactose fermenters from nonfermenters Dyes and bile also inhibit gram-positive bacteria
MacConkey agar	Lactose, neutral red	Bacteria that ferment lactose (lowering the pH) from those that do not
Urea broth	Urea, phenol red	Bacteria that hydrolyze urea to ammonia
Sulfur indole motility (SIM)	Thiosulfate, iron	H_2S gas producers from nonproducers
Triple-sugar iron agar (TSIA)	Triple sugars, iron, and phenol red dye	Fermentation of sugars, H_2S production
XLD agar	Lysine, xylose, iron, thiosulfate, phenol red	*Enterobacter, Escherichia, Proteus, Providencia, Salmonella,* and *Shigella*
Birdseed agar	Seeds from thistle plant	*Cryptococcus neoformans* and other fungi

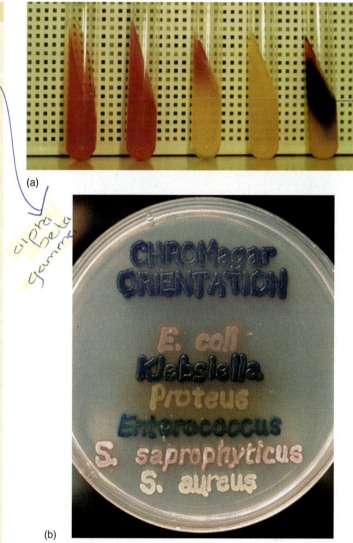

(a)

(b)

FIGURE 3.10 Media that differentiate characteristics.
(a) Triple-sugar iron agar (TSIA) in a slant tube. This medium contains three fermentable carbohydrates, phenol red to indicate pH changes, and a chemical (iron) that indicates H_2S gas production. Reactions (from left to right) are: no growth; growth with no acid production; acid production in the bottom (butt) only; acid production all through the medium; and acid production in the butt with H_2S gas formation (black). **(b)** A state-of-the-art medium developed for culturing and identifying the most common urinary pathogens. CHROMagar Orientation™ uses color-forming reactions to distinguish at least seven species and permits rapid identification and treatment. In the example, the bacteria were streaked so as to spell their own names.

pink or red when acidic. A common intestinal bacterium such as *Escherichia coli* that gives off acid when it metabolizes the lactose in the medium develops red to pink colonies, and one like *Salmonella* that does not give off acid remains its natural color (off-white). Spirit blue agar is used to detect the hydrolysis (digestion) of fats by lipase enzyme. Positive hydrolysis is indicated by the dark blue color that develops in colonies.

Miscellaneous Media A reducing medium contains a substance (thioglycollic acid or cystine) that absorbs oxygen or slows the penetration of oxygen in a medium, thus reducing its availability. Reducing media are important for growing anaerobic bacteria or for determining oxygen requirements of isolates (described in chapter 7). Carbohydrate fermentation media contain sugars that can be fermented (converted to acids) and a pH indicator to show this reaction (figure 3.9a and **figure 3.11**). Media for other biochemical reactions that

provide the basis for identifying bacteria and fungi are presented in chapter 17.

Transport media are used to maintain and preserve specimens that have to be held for a period of time before clinical analysis or to sustain delicate species that die rapidly if not held under stable conditions. Stuart's and Amie's transport media contain salts, buffers, and absorbants to prevent cell destruction by enzymes, pH changes, and toxic substances, but will

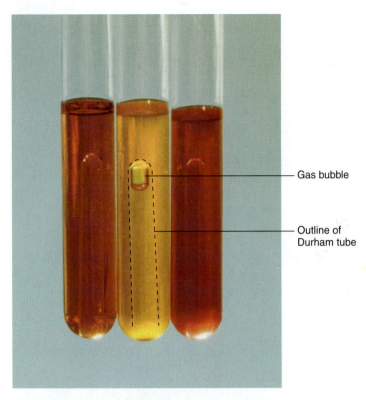

Gas bubble

Outline of
Durham tube

FIGURE 3.11   Carbohydrate fermentation in broths.
This medium is designed to show fermentation (acid production)
and gas formation by means of a small, inverted Durham tube for
collecting gas bubbles. The tube on the left is an uninoculated
negative control; the center tube is positive for acid (yellow) and
gas (open space); the tube on the right shows growth but neither
acid nor gas.

not support growth. Assay media are used by technologists to
test the effectiveness of antimicrobial drugs (see chapter 12)
and by drug manufacturers to assess the effect of disinfectants,
antiseptics, cosmetics, and preservatives on the growth of microorganisms. Enumeration media are used by industrial and
environmental microbiologists to count the numbers of organisms in milk, water, food, soil, and other samples.

✔ **CHECKPOINT**

- Most microorganisms can be cultured on artificial media, but
  some can be cultured only in living tissue or in cells.
- Artificial media are classified by their *physical state* as either
  liquid, semisolid, liquefiable solid, or nonliquefiable solid.
- Artificial media are classified by their *chemical composition* as
  either *synthetic* or *nonsynthetic,* depending on whether the exact chemical composition is known.
- Artificial media are classified by their *function* as either
  general-purpose media or media with one or more specific
  purposes. Enriched, selective, differential, transport, assay,
  and enumerating media are all examples of media designed
  for specific purposes.

Incubation, Inspection, and Identification

Once a container of medium has been inoculated, it is **incubated,** which means it is placed in a temperature-controlled
chamber (incubator) to encourage multiplication. Although
microbes have adapted to growth at temperatures ranging
from freezing to boiling, the usual temperatures used in laboratory propagation fall between 20° and 40°C. Incubators can
also control the content of atmospheric gases such as oxygen
and carbon dioxide that may be required for the growth of
certain microbes. During the incubation period (ranging from
a day to several weeks), the microbe multiplies and produces
growth that is observable macroscopically. Microbial growth
in a liquid medium materializes as cloudiness, sediment,
scum, or color. A common manifestation of growth on solid
media is the appearance of colonies, especially in bacteria and
fungi. Colonies are actually large masses of piled-up cells (see
chapter 4).

In some ways, culturing microbes is analogous to gar-
dening. Cultures are formed by "seeding" tiny plots (media)
with microbial cells. Extreme care is taken to exclude weeds
(contaminants). A **pure culture** is a container of medium that
grows only a single known species or type of microorganism
(figure 3.12a). This type of culture is most frequently used for
laboratory study, because it allows the systematic examination and control of one microorganism by itself. Instead of
the term *pure culture,* some microbiologists prefer the term
axenic, meaning that the culture is free of other living things
except for the one being studied. A standard method for
preparing a pure culture is to **subculture,** or make a second-
level culture from a well-isolated colony. A tiny bit of cells is
transferred into a separate container of media and incubated
(see figure 3.1, step 3).

A **mixed culture (figure 3.12b)** is a container that holds
two or more *identified,* easily differentiated species of microorganisms, not unlike a garden plot containing both carrots and
onions. A **contaminated culture (figure 3.12c)** was once pure
or mixed (and thus a known entity) but has since had **contaminants** (unwanted microbes of uncertain identity) introduced into it, like weeds into a garden. Because contaminants
have the potential for causing disruption, constant vigilance
is required to exclude them from microbiology laboratories,
as you will no doubt witness from your own experience.

How does one determine what sorts of microorganisms
have been isolated in cultures? Certainly microscopic appearance can be valuable in differentiating the smaller, simpler
procaryotic cells from the larger, more complex eucaryotic cells.
Appearance can be especially useful in identifying eucaryotic
microorganisms to the level of genus or species because of their
distinctive morphological features; however, bacteria are generally not identifiable by these methods because very different
species may appear quite similar. For them, we must include
other techniques, some of which characterize their cellular metabolism. These methods, called biochemical tests, can determine fundamental chemical characteristics such as nutrient
requirements, products given off during growth, presence of
enzymes, and mechanisms for deriving energy.

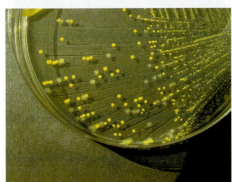

FIGURE 3.12 **Various conditions of cultures.**
(a) Three tubes containing pure cultures of *Escherichia coli* (white), *Micrococcus luteus* (yellow), and *Serratia marcescens* (red). **(b)** A mixed culture of *M. luteus* and *E. coli* readily differentiated by their colors. **(c)** This plate of *S. marcescens* was overexposed to room air, and it has developed a large, white colony. Because this intruder is not desirable and not identified, the culture is now contaminated.

Several modern analytical and diagnostic tools that focus on genetic characteristics can detect microbes based on their DNA. Identification can also be accomplished by testing the isolate against known antibodies (immunological testing). In the case of certain pathogens, further information on a microbe is obtained by inoculating a suitable laboratory animal. A profile is prepared by compiling physiological testing results with both macroscopic and microscopic traits. The profile then becomes the raw material used in final identification. In chapter 17, we present more detailed examples of identification methods.

Maintenance and Disposal of Cultures

In most medical laboratories, the cultures and specimens constitute a potential hazard and require immediate and proper disposal. Both steam sterilizing (see autoclave, chapter 11) and incineration (burning) are used to destroy microorganisms. On the other hand, many teaching and research laboratories maintain a line of stock cultures that represent "living catalogues" for study and experimentation. The largest culture collection can be found at the American Type Culture Collection in Manassas, Virginia, which maintains a voluminous array of frozen and freeze-dried fungal, bacterial, viral, and algal cultures.

> ### ✔ CHECKPOINT
>
> - The Five I's—inoculation, incubation, isolation, inspection, and identification—summarize the kinds of laboratory procedures used in microbiology.
> - Following *inoculation*, cultures are *incubated* at a specified temperature to encourage growth.
> - *Isolated colonies* that originate from single cells are composed of large numbers of cells piled up together.
> - A culture may exist in one of the following forms: A *pure culture* contains only one species or type of microorganism. A *mixed culture* contains two or more known species. A *contaminated culture* contains both known and unknown (unwanted) microorganisms.
> - During *inspection*, the cultures are examined and evaluated macroscopically and microscopically.
> - Microorganisms are *identified* in terms of their macroscopic or immunological morphology; their microscopic morphology; their biochemical reactions; and their genetic characteristics.
> - Microbial cultures are usually disposed of in two ways: steam sterilization or incineration.

3.2 The Microscope: Window on an Invisible Realm

Imagine Leeuwenhoek's excitement and wonder when he first viewed a drop of rainwater and glimpsed an amazing microscopic world teeming with unearthly creatures. Beginning microbiology students still experience this sensation, and even experienced microbiologists remember their first view. The microbial existence is indeed another world, but it would remain largely uncharted without an essential tool: the microscope. Your efforts in exploring microbes will be more meaningful if you understand some essentials of **microscopy** and specimen preparation.

Magnification and Microscope Design

The two key characteristics of a reliable microscope are magnification, or the ability enlarge objects, and resolving power, or the ability to show detail.

A discovery by early microscopists that spurred the advancement of microbiology was that a clear, glass sphere could act as a lens to magnify small objects. Magnification in most microscopes results from a complex interaction between visible light waves and the curvature of the lens. When a beam or ray of light transmitted through air strikes and passes through the convex surface of glass, it experiences some degree of **refraction,** defined as the bending or change in the angle of the light ray as it passes through a medium such as a lens. The greater the difference in the composition of the two substances the light passes between, the more pronounced is the refraction. When an object is placed a certain distance from the spherical lens and illuminated with light, an optical replica, or image, of it is formed by the refracted light. Depending upon the size and curvature of the lens, the image appears enlarged to a particular degree, which is called its power of magnification and is usually identified with a number combined with × (read "times"). This behavior of light is evident if one looks through an everyday object such as a glass ball or a magnifying glass **(figure 3.13).** It is basic to the function of all optical, or light, microscopes, though many of them have additional features that define, refine, and increase the size of the image.

The first microscopes were simple, meaning they contained just a single magnifying lens and a few working parts. Examples of this type of microscope are a magnifying glass, a hand lens, and Leeuwenhoek's basic little tool shown earlier in figure 1.9*a*. Among the refinements that led to the development of today's compound microscope were the addition of a second magnifying lens system, a lamp in the base to give off visible light and illuminate the specimen, and a special lens called the condenser that converges or focuses the rays of light to a single point on the object. The fundamental parts of a modern compound light microscope are illustrated in **figure 3.14.**

FIGURE 3.13   **Effects of magnification.**
Demonstration of the magnification and image-forming capacity of clear glass "lenses." Given a proper source of illumination, this magnifying glass and crystal ball magnify a ruler two to three times.

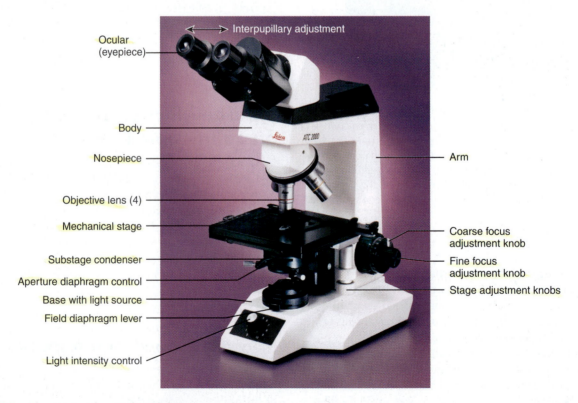

FIGURE 3.14   **The parts of a student laboratory microscope.**
This microscope is a compound light microscope with two oculars (called binocular). It has four objective lenses, a mechanical stage to move the specimen, a condenser, an iris diaphragm, and a built-in lamp.

Principles of Light Microscopy

To be most effective, a microscope should provide adequate magnification, resolution, and clarity of image. Magnification of the object or specimen by a compound microscope occurs in two phases. The first lens in this system (the one closest to the specimen) is the objective lens, and the second (the one closest to the eye) is the ocular lens, or eyepiece (figure 3.15). The objective forms the initial image of the specimen, called the **real image**. When this image is projected up through the microscope body to the plane of the eyepiece, the ocular lens forms a second image, the **virtual image**. The virtual image is the one that will be received by the eye and converted to a retinal and visual image. The

magnifying power of the objective alone usually ranges from 4× to 100×, and the power of the ocular alone ranges from 10× to 20×. The total power of magnification of the final image formed by the combined lenses is a product of the separate powers of the two lenses:

Power of objective	×	Usual Power of ocular	=	Total magnification
10× low power objective		10×	=	100×
40× high dry objective		10×	=	400×
100× oil immersion objective		10×	=	1,000×

Microscopes are equipped with a nosepiece holding three or more objectives that can be rotated into position as needed. The power of the ocular usually remains constant for a given microscope. Depending on the power of the ocular, the total magnification of standard light microscopes can vary from 40× with the lowest power objective (called the scanning objective) to 2,000× with the highest power objective (the oil immersion objective).

Resolution: Distinguishing Magnified Objects Clearly   As important as magnification is for visualizing tiny objects or cells, an additional optical property is essential for seeing clearly. That property is resolution, or **resolving power.** Resolution is the capacity of an optical system to distinguish or separate two adjacent objects or points from one another. For example, at a certain fixed distance, the lens in the human eye can resolve two small objects as separate points just as long as the two objects are no closer than 0.2 mm apart. The eye examination given by optometrists is in fact a test of the resolving power of the human eye for various-sized letters read at a distance of 20 feet. Because microorganisms are extremely small and usually very close together, they will not be seen with clarity or any degree of detail unless the microscope's lenses can resolve them.

A simple equation in the form of a fraction expresses the main determining factors in resolution:

$$\text{Resolving power (R.P.)} = \frac{\text{Wavelength of light in nm}}{2 \times \text{Numerical aperture of objective lens}}$$

From this equation it is evident that the resolving power is a function of the wavelength of light that forms the image, along with certain characteristics of the objective. The light source for optical microscopes consists of a band of colored wavelengths in the visible spectrum. The shortest visible wavelengths are in the violet-blue portion of the spectrum (400 nm), and the longest are in the red portion (750 nm). Because the wavelength must pass between the objects that are being resolved, shorter wavelengths (in the 400–500 nm range) will provide better resolution (figure 3.16). Some microscopes have a special blue filter placed over the lamp to limit the longer wavelengths of light from entering the specimen.

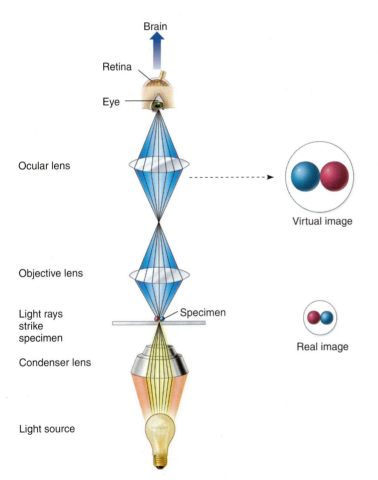

FIGURE 3.15   The pathway of light and the two stages in magnification of a compound microscope.
As light passes through the condenser, it forms a solid beam that is focused on the specimen. Light leaving the specimen that enters the objective lens is refracted so that an enlarged primary image, the real image, is formed. One does not see this image, but its degree of magnification is represented by the lower circle. The real image is projected through the ocular, and a second image, the virtual image, is formed by a similar process. The virtual image is the final magnified image that is received by the retina and perceived by the brain. Notice that the lens systems cause the image to be reversed.

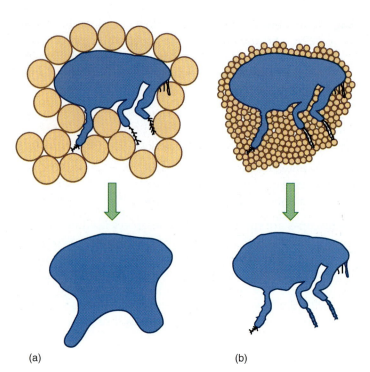

(a)   (b)

FIGURE 3.16 **Effect of wavelength on resolution.**
A simple model demonstrates how the wavelength influences
the resolving power of a microscope. Here an outline of a flea
represents the object being illuminated, and two different-sized
circles represent the wavelengths of light. In **(a)**, the longer waves
are too large to penetrate between the finer spaces and produce a
fuzzy, undetailed image. In **(b)**, shorter waves are small enough to
enter small spaces and produce a much more detailed image that
is recognizable as a flea.

The other factor influencing resolution is the **numerical
aperture,** a mathematical constant that describes the relative
efficiency of a lens in bending light rays. Without going into
the mathematical derivation of this constant, it is sufficient
to say that each objective has a fixed numerical aperture
reading that is determined by the microscope design and
ranges from 0.1 in the lowest power lens to approximately
1.25 in the highest power (oil immersion) lens. The most important thing to remember is that a higher numerical aperture number will provide better resolution. In order for the
oil immersion lens to arrive at its maximum resolving capacity, a drop of oil must be inserted between the tip of the
lens and the specimen on the glass slide. Because oil has the
same optical qualities as glass, it prevents refractive loss that
normally occurs as peripheral light passes from the slide
into the air; this property effectively increases the numerical
aperture **(figure 3.17).** There is an absolute limitation to resolution in optical microscopes, which can be demonstrated
by calculating the resolution of the oil immersion lens using
a blue-green wavelength of light:

$$\text{R.P.} = \frac{500 \text{ nm}}{2 \times 1.25}$$

$$= 200 \text{ nm (or } 0.2 \text{ } \mu\text{m)}$$

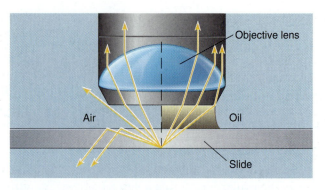

FIGURE 3.17 **Workings of an oil immersion lens.**
To maximize its resolving power, an oil immersion lens (the one
with highest magnification) must have a drop of oil placed at its
tip. This forms a continuous medium to transmit a beam of light
from the condenser to the objective and effectively increase the
numerical aperture. Without oil, some of the peripheral light that
passes through the specimen is scattered into the air or onto the
glass slide; this scattering decreases resolution.

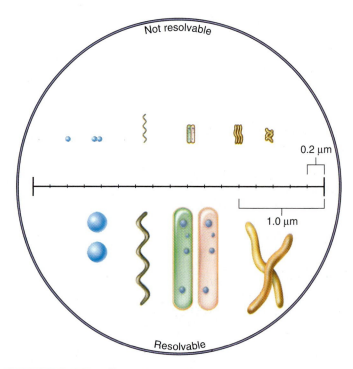

FIGURE 3.18 **Effect of magnification.**
Comparison of cells that would not be resolvable versus those that
would be resolvable under oil immersion at 1,000× magnification.
Note that in addition to differentiating two adjacent things, good
resolution also means being able to observe an object clearly.

In practical terms, this means that the oil immersion lens
can resolve any cell or cell part as long as it is at least 0.2 μm
in diameter, and that it can resolve two adjacent objects as
long as they are at least 0.2 μm apart **(figure 3.18).** In general,
organisms that are 0.5 μm or more in diameter are readily

seen. This includes fungi and protozoa and some of their internal structures, and most bacteria. However, a few bacteria and most viruses are far too small to be resolved by the optical microscope and require electron microscopy (discussed later in this chapter). In summary then, the factor that most limits the clarity of a microscope's image is its resolving power. Even if a light microscope were designed to magnify several thousand times, its resolving power could not be increased, and the image it produced would simply be enlarged and fuzzy.

Other constraints to the formation of a clear image are the quality of the lens and light source and the lack of contrast in the specimen. No matter how carefully a lens is constructed, flaws remain. A typical problem is spherical aberration, a distortion in the image caused by irregularities in the lens, which creates a curved, rather than flat, image (see figure 3.13). Another is chromatic aberration, a rainbow-like image that is caused by the lens acting as a prism and separating visible light into its colored bands. Brightness and direction of illumination also affect image formation.

Because too much light can reduce contrast and burn out the image, an adjustable iris diaphragm on most microscopes controls the amount of light entering the condenser. The lack of contrast in cell components is compensated for by using special lenses (the phase-contrast microscope) and by adding dyes.

Variations on the Optical Microscope

Optical microscopes that use visible light can be described by the nature of their *field*, meaning the circular area viewed through the ocular lens. With special adaptations in lenses, condensers, and light sources, four special types of microscopes can be described: bright-field, dark-field, phase-contrast, and interference. A fifth type of optical microscope, the fluorescence microscope, uses ultraviolet radiation as the illuminating source, and another, the confocal microscope, uses a laser beam. Each of these microscopes is adapted for viewing specimens in a particular way, as described in the next sections and summarized in **table 3.5**.

TABLE 3.5	Comparisons of Types of Microscopy		
Microscope	**Maximum Practical Magnification**	**Resolution**	**Important Features**
Visible light as source of illumination			
Bright-field	2,000×	0.2 μm (200 nm)	Common multipurpose microscope for live and preserved stained specimens; specimen is dark, field is white; provides fair cellular detail
Dark-field	2,000×	0.2 μm	Best for observing live, unstained specimens; specimen is bright, field is black; provides outline of specimen with reduced internal cellular detail
Phase-contrast	2,000×	0.2 μm	Used for live specimens; specimen is contrasted against gray background; excellent for internal cellular detail
Differential interference	2,000×	0.2 μm	Provides brightly colored, highly contrasting, three-dimensional images of live specimens
Ultraviolet rays as source of illumination			
Fluorescent	2,000×	0.2 μm	Specimens stained with fluorescent dyes or combined with fluorescent antibodies emit visible light; specificity makes this microscope an excellent diagnostic tool
Confocal	2,000×	0.2 μm	Specimens stained with fluorescent dyes are scanned by laser beam, multiple images (optical sections) are combined into three-dimensional image by a computer; unstained specimens can be viewed using light reflected from specimen
Electron beam forms image of specimen			
Transmission electron microscope (TEM)	100,000×	0.5 nm	Sections of specimen are viewed under very high magnification; finest detailed structure of cells and viruses is shown; used only on preserved material
Scanning electron microscope (SEM)	650,000×	10 nm	Scans and magnifies external surface of specimen; produces striking three-dimensional image

Bright-Field Microscopy

The bright-field microscope is the most widely used type of light microscope. Although we ordinarily view objects like the words on this page with light reflected off the surface, a bright-field microscope forms its image when light is transmitted through the specimen. The specimen, being denser and more opaque than its surroundings, absorbs some of this light, and the rest of the light is transmitted directly up through the ocular into the field. As a result, the specimen will produce an image that is darker than the surrounding brightly illuminated field. The bright-field microscope is a multipurpose instrument that can be used for both live, unstained material and preserved, stained material. The bright-field image is compared with that of other microscopes in **figure 3.19.**

Dark-Field Microscopy

A bright-field microscope can be adapted as a dark-field microscope by adding a special disc called a *stop* to the condenser. The stop blocks all light from entering the objective lens except peripheral light that is reflected off the sides of the specimen itself. The resulting image is a particularly striking one: brightly illuminated specimens surrounded by a dark (black) field **(figure 3.19b).** Some of Leeuwenhoek's more successful microscopes probably operated with dark-field illumination. The most effective use of dark-field microscopy is to visualize living cells that would be distorted by drying or heat or cannot be stained with the usual methods. It can outline the organism's shape and permit rapid recognition of swimming cells that might appear in dental and other infections, but it does not reveal fine internal details.

Phase-Contrast and Interference Microscopy

If similar objects made of clear glass, ice, cellophane, or plastic are immersed in the same container of water, an observer would have difficulty telling them apart because they have similar optical properties. Internal components of a live, unstained cell also lack contrast and can be difficult to distinguish. But cell structures do differ slightly in density, enough that they can alter the light that passes through them in subtle ways. The phase-contrast microscope has been constructed to take advantage of this characteristic. This microscope contains devices that transform the subtle changes in light waves passing through the specimen into differences in light intensity. For example, denser cell parts such as organelles alter the pathway of light more than less dense regions (the cytoplasm). Light patterns coming from these regions will vary in contrast. The amount of internal detail visible by this method is greater than by either bright-field or dark-field methods. The phase-contrast microscope is most useful for observing intracellular structures such as bacterial spores, granules, and organelles, as well as the

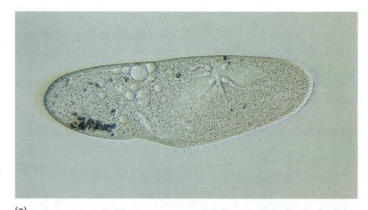

(a)

(b)

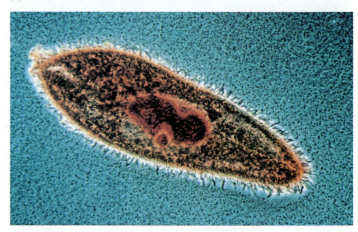

(c)

FIGURE 3.19   Three views of a basic cell.
A live cell of *Paramecium* viewed with **(a)** bright-field (400×), **(b)** dark-field (400×), and **(c)** phase-contrast (400×). Note the difference in the appearance of the field and the degree of detail shown by each method of microscopy. Only in phase-contrast are the cilia (fine hairs) on the cells noticeable. Can you see the nucleus? The oral groove?

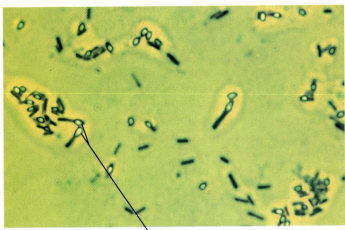

Bacterial spores

(a)

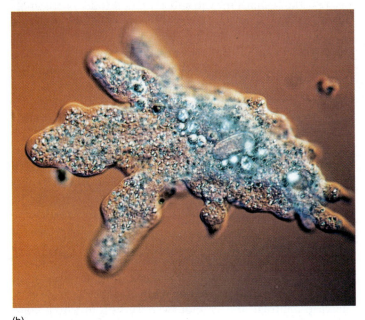

(b)

FIGURE 3.20 Visualizing internal structures.
(a) Phase-contrast micrograph of a bacterium containing spores. The relative density of the spores causes them to appear as bright, shiny objects against the darker cell parts (600×). **(b)** Differential interference micrograph of *Amoeba proteus,* a common protozoan. Note the outstanding internal detail, the depth of field, and the bright colors, which are not natural (160×).

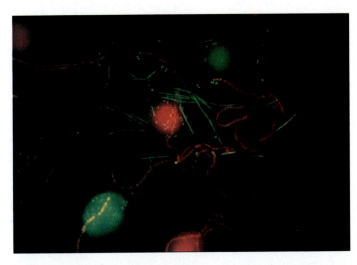

FIGURE 3.21 Fluorescent staining on a fresh sample of cheek scrapings from the oral cavity.
Cheek epithelial cells are the larger unfocused red or green cells. Bacteria appearing here are streptococci (tiny spheres in long chains) and filamentous rods. This particular staining technique also indicates whether cells are alive or dead; live cells fluoresce green, and dead cells fluoresce red.

locomotor structures of eucaryotic cells (**figure 3.19c** and **figure 3.20a**).

Like the phase-contrast microscope, the differential interference contrast (DIC) microscope provides a detailed view of unstained, live specimens by manipulating the light. But this microscope has additional refinements, including two prisms that add contrasting colors to the image and two beams of light rather than a single one. DIC microscopes pro-duce extremely well-defined images that are vividly colored and appear three-dimensional (**figure 3.20b**).

Fluorescence Microscopy

The fluorescence microscope is a specially modified compound microscope furnished with an ultraviolet (UV) radiation source and a filter that protects the viewer's eye from injury by these dangerous rays. The name of this type of microscopy originates from the use of certain dyes (acridine, fluorescein) and minerals that show **fluorescence.** The dyes emit visible light when bombarded by shorter ultraviolet rays. For an image to be formed, the specimen must first be coated or placed in contact with a source of fluorescence. Subsequent illumination by ultraviolet radiation causes the specimen to give off light that will form its own image, usually an intense yellow, orange, or red against a black field.

Fluorescence microscopy has its most useful applications in diagnosing infections caused by specific bacteria, protozoans, and viruses. A staining technique with fluorescent dyes is commonly used to detect *Mycobacterium tuberculosis* (the agent of tuberculosis) in patients' specimens (see figure 21.20). In a number of diagnostic procedures, fluorescent dyes are affixed to specific antibodies. These *fluorescent antibodies* can be used to detect the causative agents in such diseases as syphilis, chlamydiosis, trichomoniasis, herpes, and influenza. A newer technology using fluorescent nucleic acid stains can differentiate between live and dead cells in mixtures (**figure 3.21**). A fluorescence

FIGURE 3.22 **Confocal microscopy of a basic cell.**
This *Paramecium* was stained with fluorescent dyes and visualized by a scanning confocal microscope.

microscope can be handy for locating microbes in complex mixtures because only those cells targeted by the technique will fluoresce.

Most optical microscopes have difficulty forming a clear image of cells at higher magnifications, because cells are often too thick for conventional lenses to focus all levels of the cell simultaneously. This is especially true of larger cells with complex internal structures. A newer type of microscope that overcomes this impediment is called the *scanning confocal microscope*. This microscope uses a laser beam of light to scan various depths in the specimen and deliver a sharp image focusing on just a single plane. It is thus able to capture a highly focused view at any level, ranging from the surface to the middle of the cell. It is most often used on fluorescently stained specimens, but it can also be used to visualize live unstained cells and tissues **(figure 3.22).**

Electron Microscopy

If conventional light microscopes are our windows on the microscopic world, then the electron microscope (EM) is our window on the tiniest details of that world. Although this microscope was originally conceived and developed for studying nonbiological materials such as metals and small electronics parts, biologists immediately recognized the importance of the tool and began to use it in the early 1930s. One of the most impressive features of the electron microscope is the resolution it provides.

Unlike light microscopes, the electron microscope forms an image with a beam of electrons that can be made to travel in wavelike patterns when accelerated to high speeds. These waves are 100,000 times shorter than the waves of visible light. Because resolving power is a function of wavelength, electrons have tremendous power to resolve minute structures. Indeed, it is possible to resolve atoms with an electron microscope, though the practical resolution for biological applications is approximately 0.5 nm. Because the resolution is so substantial, it follows that magnification can also be extremely high—usually between 5,000× and 1,000,000× for biological specimens and up to 5,000,000× in some applications. Its capacity for magnification and resolution makes the EM an invaluable tool for seeing the finest structure of cells and viruses. If not for electron microscopes, our understanding of biological structure and function would still be in its early theoretical stages.

In fundamental ways, the electron microscope is a derivative of the compound microscope. It employs components analogous to, but not necessarily the same as, those in light microscopy **(figure 3.23).** For instance, it magnifies in stages by means of two lens systems, and it has a condensing lens, a specimen holder, and focusing apparatus. Otherwise, the two types have numerous differences **(table 3.6).** An electron gun aims its beam through a vacuum to ring-shaped electromagnets that focus this beam on the specimen. Specimens must be pretreated with chemicals or dyes to increase contrast and cannot be observed in a live state. The enlarged image is displayed on a viewing screen or photographed for further study rather than being observed directly through an eyepiece. Because images produced by electrons lack color, electron micrographs (a micrograph is a photograph of a microscopic object) are always shades of black, gray, and white. The color-enhanced micrographs used in this and other textbooks have computer-added color.

Two general forms of EM are the transmission electron microscope (TEM) and the scanning electron microscope (SEM) (see table 3.5). Transmission electron microscopes are the method of choice for viewing the detailed structure of cells and viruses. This microscope produces its image by transmitting electrons through the specimen. Because electrons cannot readily penetrate thick preparations, the specimen must be sectioned into extremely thin slices (20–100 nm thick) and stained or coated with metals that will increase image contrast. The darkest areas of TEM micrographs represent the thicker (denser) parts, and the lighter areas indicate the more transparent and less dense

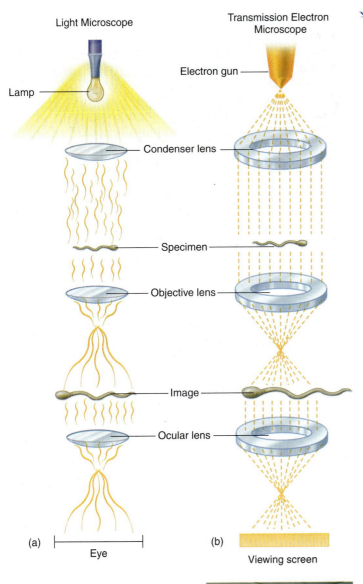

Light Microscope

Lamp

Condenser lens

Specimen

Objective lens

Image

Ocular lens

(a)

Eye

Transmission Electron Microscope

Electron gun

Condenser lens

Specimen

Objective lens

Image

Ocular lens

(b)

Viewing screen

Final image can be displayed on fluorescent screen or photographed. (c)

FIGURE 3.23 Comparison of two microscopes.
(a) the light microscope and **(b)** one type of electron microscope (EM; transmission type). These diagrams are highly simplified, especially for the electron microscope, to indicate the common components. Note that the EM's image pathway is actually upside down compared with that of a light microscope. **(c)** The EM is a larger machine with far more complicated working parts than most light microscopes.

TABLE 3.6	Comparison of Light Microscopes and Electron Microscopes	
Characteristic	**Light or Optical**	**Electron (Transmission)**
Useful magnification	2,000×	1,000,000× or more
Maximum resolution	200 nm	0.5 nm
Image produced by	Light rays	Electron beam
Image focused by	Glass objective lens	Electromagnetic objective lenses
Image viewed through	Glass ocular lens	Fluorescent screen
Specimen placed on	Glass slide	Copper mesh
Specimen may be alive	Yes	No
Specimen requires special stains or treatment	Not always	Yes
Colored images possible	Yes	No

parts **(figure 3.24)**. The TEM can also be used to produce negative images and shadow casts of whole microbes.

The scanning electron microscope provides some of the most dramatic and realistic images in existence. This instrument is designed to create an extremely detailed three-dimensional view of all kinds of objects—from plaque on teeth to tapeworm heads. To produce its images, the SEM does not transmit electrons; it bombards the surface of a whole, metal-coated specimen with electrons while scanning back and forth over it. A shower of electrons deflected from the surface is picked up with great fidelity by a sophisticated detector, and the electron pattern is displayed as an image on a television screen. The contours of the specimens resolved with scanning electron micrography are very revealing and often surprising. Areas that look smooth and flat with the light microscope display intriguing surface features with the SEM **(figure 3.25)**. Improved technology has continued to refine electron microscopes and to develop variations on the basic plan. One of the most inventive relatives of the EM is the scanning probe microscope **(Insight 3.2)**.

Preparing Specimens for Optical Microscopes

A specimen for optical microscopy is generally prepared by mounting a sample on a suitable glass slide that sits on the stage between the condenser and the objective lens. The manner in which a slide specimen, or mount, is prepared depends upon: (1) the condition of the specimen, either in a living or preserved state; (2) the aims of the examiner, whether to observe overall structure, identify the microorganisms,

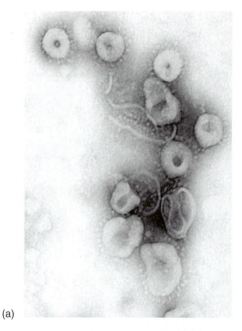

(a)

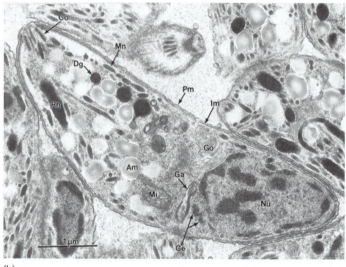

(b)

FIGURE 3.24   Transmission Electron Micrographs.
(a) A sample from the respiratory tract reveals coronaviruses (corona for the crownlike envelope) that cause infectious bronchitis (100,000×). A new form of this virus is responsible for severe acute respiratory syndrome (SARS) in humans. **(b)** A section through an infectious stage of *Toxoplasma gondii,* the cause of toxoplasmosis. Labels indicate fine structures such as cell membrane (Pm), Golgi complex (Go), nucleus (Nu), mitochondrion (Mi), centrioles (Ce), and granules (Am, Dg).

or see movement; and (3) the type of microscopy available, whether it is bright-field, dark-field, phase-contrast, or fluorescence.

Fresh, Living Preparations

Live samples of microorganisms are placed in wet mounts or in hanging drop mounts so that they can be observed as near

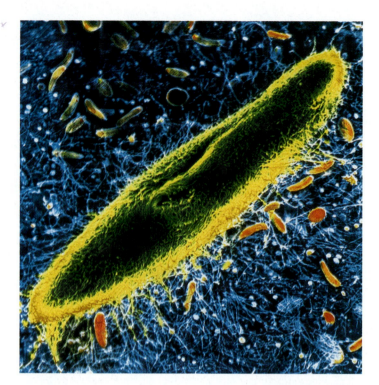

FIGURE 3.25   A false-color scanning electron micrograph (SEM) of *Paramecium,* covered in masses of fine hairs (100×). These are actually its locomotor and feeding structures—the cilia. Cells in the surrounding medium are bacteria that serve as the protozoan's "movable feast." Compare this with figure 3.19 to appreciate the outstanding three-dimensional detail shown by an SEM.

to their natural state as possible. The cells are suspended in a suitable fluid (water, broth, saline) that temporarily maintains viability and provides space and a medium for locomotion. A wet mount consists of a drop or two of the culture placed on a slide and overlaid with a cover glass. Although this type of mount is quick and easy to prepare, it has certain disadvantages. The cover glass can damage larger cells, and the slide is very susceptible to drying and can contaminate the handler's fingers. A more satisfactory alternative is the hanging drop preparation made with a special concave (depression) slide, a Vaseline adhesive or sealant, and a coverslip from which a tiny drop of sample is suspended. These types of short-term mounts provide a true assessment of the size, shape, arrangement, color, and motility of cells. Greater cellular detail can be observed with phase-contrast or interference microscopy.

Fixed, Stained Smears

A more permanent mount for long-term study can be obtained by preparing fixed, stained specimens. The smear technique, developed by Robert Koch more than 100 years ago, consists of spreading a thin film made from a liquid suspension of cells on a slide and air-drying it. Next, the air-dried smear is usually heated gently by a process called

INSIGHT 3.2   Discovery

The Evolution in Resolution: Probing Microscopes

In the past, chemists, physicists, and biologists had to rely on indirect methods to provide information on the structures of the smallest molecules. But technological advances have created a new generation of microscopes that "see" atomic structure by actually feeling it. *Scanning probe microscopes* operate with a minute needle tapered to a tip that can be as narrow as a single atom! This probe scans over the exposed surface of a material and records an image of its outer texture. These revolutionary microscopes have such profound resolution that they have the potential to image single atoms (but not subatomic structure yet) and to magnify 100 million times. The scanning tunneling microscope (STM) was the first of these microscopes. It uses a tungsten probe that hovers near the surface of an object and follows its topography while simultaneously giving off an electrical signal of its pathway, which is then imaged on a screen. The STM is used primarily for detecting defects on the surfaces of electrical conductors and computer chips composed of silicon, but it has also provided the first incredible close-up views of DNA, the genetic material (see Insight 9.1).

Another exciting new variant is the atomic force microscope (AFM), which gently forces a diamond and metal probe down onto the surface of a specimen like a needle on a record. As it moves along the surface, any deflection of the metal probe is detected by a sensitive device that relays the information to an imager. The AFM is very useful in viewing the detailed functions of biological molecules such as antibodies and enzymes.

These powerful new microscopes, along with tools that can move and position atoms, have spawned a field called *nanotechnology*—the science of the "small." Scientists in this area use physics, chemistry, biology, and engineering to explore and manipulate small molecules and atoms. Working at these dimensions, they hope to create tiny molecular tools to miniaturize computers and other electronic devices. In the future, it may be possible to use microstructures to deliver drugs, analyze DNA, and treat disease.

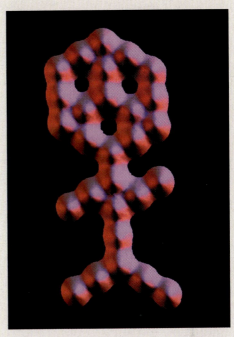

"Carbon monoxide man." An atomic force microscope image of a single carbon monoxide molecule, with its single carbon and oxygen atoms magnified several million times.

---

heat fixation that simultaneously kills the specimen and secures it to the slide. Another important action of fixation is to preserve various cellular components in a natural state with minimal distortion. Fixation of some microbial cells is performed with chemicals such as alcohol and formalin.

Like images on undeveloped photographic film, the unstained cells of a fixed smear are quite indistinct, no matter how great the magnification or how fine the resolving power of the microscope. The process of "developing" a smear to create contrast and make inconspicuous features stand out requires staining techniques. Staining is any procedure that applies colored chemicals called dyes to specimens. Dyes impart a color to cells or cell parts by becoming affixed to them through a chemical reaction. In general, they are classified as basic (cationic) dyes, which have a positive charge, or acidic (anionic) dyes, which have a negative charge. Because

chemicals of opposite charge are attracted to each other, cell parts that are negatively charged will attract basic dyes and those that are positively charged will attract acidic dyes (table 3.7). Many cells, especially those of bacteria, have numerous negatively charged acidic substances and thus stain more readily with basic dyes. Acidic dyes, on the other hand, tend to be repelled by cells, so they are good for negative staining (discussed in the next section).

Negative Versus Positive Staining   Two basic types of staining technique are used, depending upon how a dye reacts with the specimen (summarized in table 3.7). Most procedures involve a **positive stain,** in which the dye actually sticks to the specimen and gives it color. A **negative stain,** on the other hand, is just the reverse (like a photographic negative). The dye does not stick to the specimen but settles around its outer boundary, forming a silhouette. In a sense,

TABLE 3.7 Comparison of Positive and Negative Stains

	Positive Staining	Negative Staining
Appearance of cell	Colored by dye	Clear and colorless
	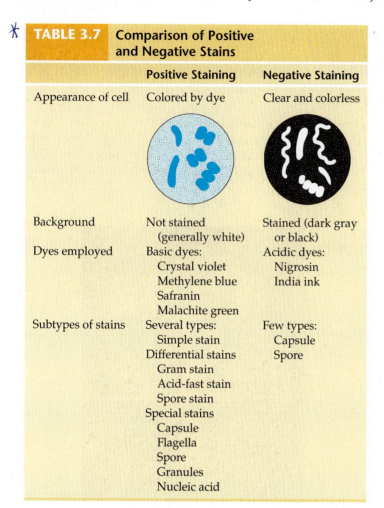	
Background	Not stained (generally white)	Stained (dark gray or black)

Dyes employed	Basic dyes:
	    Crystal violet
	    Methylene blue
	    Safranin
	    Malachite green
		Acidic dyes:
		    Nigrosin
		    India ink

Subtypes of stains	Several types:
	    Simple stain
	    Differential stains
	        Gram stain
	        Acid-fast stain
	        Spore stain
	    Special stains
	        Capsule
	        Flagella
	        Spore
	        Granules
	        Nucleic acid
		Few types:
		    Capsule
		    Spore

Most simple staining techniques take advantage of the ready binding of bacterial cells to dyes like malachite green, crystal violet, basic fuchsin, and safranin. Simple stains cause all cells in a smear to appear more or less the same color, regardless of type, but they can still reveal bacterial characteristics such as shape, size, and arrangement.

Types of Differential Stains   A satisfactory differential stain uses differently colored dyes to clearly contrast two cell types or cell parts. Common combinations are red and purple, red and green, or pink and blue. Differential stains can also pinpoint other characteristics, such as the size, shape, and arrangement of cells. Typical examples include Gram, acid-fast, and endospore stains. Some staining techniques (spore, capsule) fall into more than one category.

Gram staining, a century-old method named for its developer, Hans Christian Gram, remains the most universal diagnostic staining technique for bacteria. It permits ready differentiation of major categories based upon the color reaction of the cells: *gram-positive*, which stain purple, and *gram-negative*, which stain pink (red). The Gram stain is the basis of several important bacteriological topics, including bacterial taxonomy, cell wall structure, and identification and diagnosis of infection; in some cases, it even guides the selection of the correct drug for an infection. Gram staining is discussed in greater detail in Insight 4.2.

negative staining "stains" the glass slide to produce a dark background around the cells. Nigrosin (blue-black) and India ink (a black suspension of carbon particles) are the dyes most commonly used for negative staining. The cells themselves do not stain because these dyes are negatively charged and are repelled by the negatively charged surface of the cells. The value of negative staining is its relative simplicity and the reduced shrinkage or distortion of cells, as the smear is not heat fixed. A quick assessment can thus be made regarding cellular size, shape, and arrangement. Negative staining is also used to accentuate the capsule that surrounds certain bacteria and yeasts **(figure 3.26).**

Simple Versus Differential Staining   Positive staining methods are classified as simple, differential, or special (figure 3.26). Whereas **simple stains** require only a single dye and an uncomplicated procedure, **differential stains** use two different-colored dyes, called the *primary dye* and the *counterstain*, to distinguish between cell types or parts. These staining techniques tend to be more complex and sometimes require additional chemical reagents to produce the desired reaction.

IN THE NEWS (Continued from page 59)

The Gram stain is used to visualize and differentiate bacteria into broad categories. Purple-stained bacteria are called gram positive, and red, gram negative. These classifications relate information about the cell wall structure of each. Because bacteria are so small, the highest magnification lens (100×) on a bright-field compound microscope is used to distinguish cells and determine their color and shape. The optic properties of immersion oil placed between the glass slide and this lens combine to resolve objects as small as 0.2 μm into view.

The blood sample is processed as follows: (1) inoculum—the blood is aseptically obtained and placed into liquid medium; (2) incubation—the specimen is left overnight in appropriate conditions (this is an ongoing process); (3) inspection—a Gram stain is prepared and viewed; (4) isolation—the culture growing in the liquid is transferred with proper technique to differential and selective solid media; (5) identification—the announcement of *Bacillus anthracis*. After isolation, further biotesting is performed to identify this bacterium. To further identify this particular strain as identical to those in other anthrax cases, a DNA study is also performed.

See: CDC. 2001. Update: Investigation of bioterrorism-related inhalational anthrax—Connecticut, 2001. MMWR 50:1049–1051.

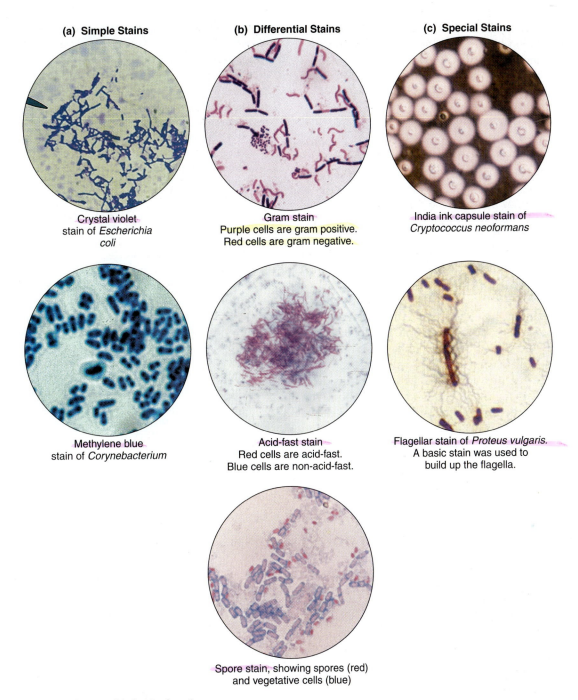

(a) Simple Stains

Crystal violet
stain of *Escherichia coli*

Methylene blue
stain of *Corynebacterium*

(b) Differential Stains

Gram stain
Purple cells are gram positive.
Red cells are gram negative.

Acid-fast stain
Red cells are acid-fast.
Blue cells are non-acid-fast.

Spore stain, showing spores (red)
and vegetative cells (blue)

(c) Special Stains

India ink capsule stain of
Cryptococcus neoformans

Flagellar stain of *Proteus vulgaris.*
A basic stain was used to
build up the flagella.

FIGURE 3.26 **Types of microbiological stains.**
(a) Simple stains. **(b)** Differential stains: Gram, acid-fast, and spore. **(c)** Special stains: capsule and flagellar.

The **acid-fast stain,** like the Gram stain, is an important diagnostic stain that differentiates acid-fast bacteria (pink) from nonacid-fast bacteria (blue). This stain originated as a specific method to detect *Mycobacterium tuberculosis* in specimens. It was determined that these bacterial cells have a particularly impervious outer wall that holds fast (tightly or tenaciously) to the dye (carbol fuchsin) even when washed

with a solution containing acid or acid alcohol. This stain is used for other medically important mycobacteria such as the Hansen's disease (leprosy) bacillus and for *Nocardia*, an agent of lung or skin infections.

The endospore stain (spore stain) is similar to the acid-fast method in that a dye is forced by heat into resistant bodies called spores or endospores (their formation and significance are discussed in chapter 4). This stain is designed to distinguish between spores and the cells that they come from (so-called **vegetative cells**). Of significance in medical microbiology are the gram-positive, spore-forming members of the genus *Bacillus* (the cause of anthrax) and *Clostridium* (the cause of botulism and tetanus)—dramatic diseases of universal fascination that we consider in later chapters.

Special stains are used to emphasize certain cell parts that are not revealed by conventional staining methods. Capsule staining is a method of observing the microbial capsule, an unstructured protective layer surrounding the cells of some bacteria and fungi. Because the capsule does not react with most stains, it is often negatively stained with India ink, or it may be demonstrated by special positive stains. The fact that not all microbes exhibit capsules is a useful feature for identifying pathogens. One example is *Cryptococcus*, which causes a serious fungal meningitis in AIDS patients (see chapter 19).

Flagellar staining is a method of revealing flagella, the tiny, slender filaments used by bacteria for locomotion. Because the width of bacterial flagella lies beyond the resolving power of the light microscope, in order to be seen, they must be enlarged by depositing a coating on the outside of the filament and then staining it. This stain works best with fresh, young cultures, because flagella are delicate and can be lost or damaged on older cells. Their presence, number, and arrangement on a cell are taxonomically useful.

✔ CHECKPOINT

- Magnification, resolving power, lens quality, and illumination source all influence the clarity of specimens viewed through the optical microscope.
- The maximum resolving power of the optical microscope is 200 nm, or 0.2 μm. This is sufficient to see the internal structures of eucaryotes and the morphology of most bacteria.
- There are six types of optical microscopes. Four types use visible light for illumination: bright-field, dark-field, phase-contrast, and interference microscopes. The fluorescence microscope uses UV light for illumination, but it has the same resolving power as the other optical microscopes. The confocal microscope can use UV light or visible light reflected from specimens.
- Electron microscopes (EM) use electrons, not light waves, as an illumination source to provide high magnification (5,000× to 1,000,000×) and high resolution (0.5 nm). Electron microscopes can visualize cell ultrastructure (TEM) and three-dimensional images of cell and virus surface features (SEM).
- Specimens viewed through optical microscopes can be either alive or dead, depending on the type of specimen preparation, but all EM specimens are dead because they must be viewed in a vacuum.
- Stains are important diagnostic tools in microbiology because they can be designed to differentiate cell shape, structure, and biochemical composition of the specimens being viewed.

Chapter Summary With Key Terms

3.1   Methods of Culturing Microorganisms—The Five I's

A. Microbiology as a science is very dependent on a number of specialized laboratory techniques. Laboratory steps routinely employed in microbiology are inoculation, incubation, isolation, inspection, and identification.

1. Initially, a specimen must be collected from a source, whether environmental or a patient.
2. **Inoculation** of a **medium** is the first step in obtaining a **culture** of the microorganisms present.
3. **Isolation** of the microorganisms, so that each microbial cell present is separated from the others and forms discrete **colonies**, is aided by inoculation techniques such as streak plates, pour plates, and spread plates.
4. **Incubation** of the medium with the microbes under the right conditions allows growth to visible colonies. Generally, isolated colonies would be **subcultured** for further testing at this point. The goal is a **pure culture** in most cases, or a **mixed culture. Contaminated cultures** can ruin correct analysis and study.
5. **Inspection** begins with macroscopic characteristics of the colonies, and continues with microscopic analysis.
6. **Identification** correlates the various morphological, physiological, genetic, and serological traits as

needed to be able to pinpoint the actual species or even strain of microbe.

B. Media: Providing Nutrients in the Laboratory
1. Artificial media allow the growth and isolation of microorganisms in the laboratory, and can be classified by their physical state, chemical composition, and functional types. The nutritional requirements of microorganisms in the laboratory may be simple or complex.
2. Physical types of media include those that are **liquid,** such as broths and milk, those that are **semisolid,** and those that are solid. Solid media may be liquefiable, containing a solidifying agent such as **agar** or gelatin.
3. Chemical composition of a medium may be completely chemically defined, thus synthetic. Nonsynthetic, or complex, media contain ingredients that are not completely definable.
4. Functional types of media serve different purposes, often allowing biochemical tests to be performed at the same time. Types include general-purpose, **enriched, selective, differential,** anaerobic (reducing), assay, and enumeration **media.** Transport media are important for conveying certain clinical specimens to the laboratory.

5. In certain instances, microorganisms have to be grown in cell cultures or host animals.
6. Cultures are maintained by large collection facilities such as the American Type Culture Collection located in Manassas, Virginia.

3.2 The Microscope: Window on an Invisible Realm

A. Optical, or light, microscopy depends on lenses that **refract** light rays, drawing the rays to a focus to produce a magnified image.
   1. A simple microscope consists of a single magnifying lens, whereas a compound microscope relies on two lenses: the ocular lens and the objective lens.
   2. The total power of magnification is calculated from the product of the ocular and objective magnifying powers.
   3. Resolution, or the **resolving power,** is a measure of a microscope's capacity to make clear images of very small objects. Resolution is improved with shorter wavelengths of illumination and with a higher **numerical aperture** of the lens. Light microscopes are limited to magnifications around 2,000× by the resolution.
   4. Modifications in the lighting or the lens system give rise to the bright-field, dark-field, phase-contrast, interference, and fluorescence microscopes.
B. Electron microscopy depends on electromagnets that serve as lenses to focus electron beams. A transmission electron microscope (TEM) projects the electrons through prepared sections of the specimen, providing detailed structural images of cells, cell parts, and viruses. A scanning electron microscope (SEM) is more like dark-field microscopy, bouncing the electrons off the surface of the specimen to detectors.
C. Specimen preparation in optical microscopy is governed by the condition of the specimen, the purpose of the inspection, and the type of microscope being used.
   1. Wet mounts and hanging drop mounts permit examination of the characteristics of live cells, such as motility, shape, and arrangement.
   2. Fixed mounts are made by drying and heating a film of the specimen called a smear. This is then stained using dyes to permit visualization of cells or cell parts.
D. Staining uses either basic (cationic) dyes with positive charges or acidic (anionic) dyes with negative charges. The surfaces of microbes are negatively charged and attract basic dyes. This is the basis of positive staining. In **negative staining,** the microbe repels the dye and it stains the background. Dyes may be used alone and in combination.
   1. Simple stains use just one dye, and highlight cell morphology.
   2. Differential stains require a primary dye and a contrasting counterstain in order to distinguish cell types or parts. Important differential stains include the **Gram stain, acid-fast stain,** and the endospore stain.
   3. Special stains are designed to bring out distinctive characteristics. Examples include capsule stains and flagellar stains.

Multiple-Choice Questions

1. Which of the following is not one of the Five I's?
   a. inspection   d. incubation
   b. identification   e. inoculation
   c. induction

2. The term *culture* refers to the ____ growth of microorganisms in ____.
   a. rapid, an incubator   c. microscopic, the body
   b. macroscopic, media   d. artificial, colonies

3. A mixed culture is
   a. the same as a contaminated culture
   b. one that has been adequately stirred
   c. one that contains two or more known species
   d. a pond sample containing algae and protozoa

4. Agar is superior to gelatin as a solidifying agent because agar
   a. does not melt at room temperature
   b. solidifies at 75°C
   c. is not usually decomposed by microorganisms
   d. both a and c

5. The process that most accounts for magnification is
   a. a condenser
   b. refraction of light rays
   c. illumination
   d. resolution

6. A subculture is a
   a. colony growing beneath the media surface
   b. culture made from a contaminant
   c. culture made in an embryo
   d. culture made from an isolated colony

7. Resolution is ____ with a longer wavelength of light.
   a. improved   c. not changed
   b. worsened   d. not possible

8. A real image is produced by the
   a. ocular   c. condenser
   b. objective   d. eye

9. A microscope that has a total magnification of 1,500× when using the oil immersion objective has an ocular of what power?
   a. 150×   c. 15×
   b. 1.5×   d. 30×

10. The specimen for an electron microscope is always
    a. stained with dyes   c. killed
    b. sliced into thin sections   d. viewed directly

11. Motility is best observed with a
    a. hanging drop preparation
    b. negative stain
    c. streak plate
    d. flagellar stain

12. Bacteria tend to stain more readily with cationic (positively charged) dyes because bacteria
    a. contain large amounts of alkaline substances
    b. contain large amounts of acidic substances
    c. are neutral
    d. have thick cell walls

13. The primary difference between a TEM and SEM is in
    a. magnification capability
    b. colored versus black-and-white images
    c. preparation of the specimen
    d. type of lenses

14. **Multiple Matching.** For each type of medium, select all descriptions that fit. For media that fit more than one description, briefly explain why this is the case.

    abf mannitol salt agar   a. selective medium
    _df_ chocolate agar   b. differential medium
    abf MacConkey agar   c. chemically defined (synthetic)
    _ef_ nutrient broth   medium

    _df_ Sabouraud's agar   d. enriched medium
    bef triple-sugar iron agar   e. general-purpose medium
    _ac_ *Euglena* agar   f. complex medium
    bef SIM medium   g. transport medium

15. A fastidious organism must be grown on what type of medium?
    a. general-purpose medium
    b. differential medium
    c. synthetic medium
    d. enriched medium

16. What type of medium is used to maintain and preserve specimens before clinical analysis?
    a. selective medium
    b. transport medium
    c. enriched medium
    d. differential medium

Concept Questions

These questions are suggested as a *writing-to-learn* experience. For each question, compose a one- or two-paragraph answer that includes the factual information needed to completely address the question.

1. a. Describe briefly what is involved in the Five I's.
   b. Name three basic differences between inoculation and contamination.

2. a. Name two ways that pure, mixed, and contaminated cultures are similar and two ways that they differ from each other.
   b. What must be done to avoid contamination?

3. a. Explain what is involved in isolating microorganisms and why it is necessary to do this.
   b. Compare and contrast three common laboratory techniques for separating bacteria in a mixed sample.
   c. Describe how an isolated colony forms.
   d. Explain why an isolated colony and a pure culture are not the same thing.

4. a. Explain the two principal functions of dyes in media.
   b. Differentiate among the ingredients and functions of enriched, selective, and differential media.

5. Differentiate between microscopic and macroscopic methods of observing microorganisms, citing a specific example of each method.

6. a. Contrast the concepts of magnification, refraction, and resolution.
   b. Briefly explain how an image is made and magnified.
   c. Trace the pathway of light from its source to the eye, explaining what happens as it passes through the major parts of the microscope.

7. a. On the basis of the formula for resolving power, explain why a smaller R.P. value is preferred to a larger one.
   b. What does it mean in practical terms if the resolving power is 1.0 μm?

c. How does a value greater than 1.0 μm compare? (Is it better or worse?)
d. How does a value less than 1.0 μm compare?
e. What can be done to a microscope to improve resolution?

8. Compare bright-field, dark-field, phase-contrast, and fluorescence microscopy as to field appearance, specimen appearance, light source, and uses.

9. a. Compare and contrast the optical compound microscope with the electron microscope.
   b. Why is the resolution so superior in the electron microscope?
   c. What will you never see in an unretouched electron micrograph?
   d. Compare the way that the image is formed in the TEM and SEM.

10. Evaluate the following preparations in terms of showing microbial size, shape, motility, and differentiation: spore stain, negative stain, simple stain, hanging drop slide, and Gram stain.

11. a. Itemize the various staining methods, and briefly characterize each.
    b. For a stain to be considered a differential stain, what must it do?
    c. Explain what happens in positive staining to cause the reaction in the cell.
    d. Explain what happens in negative staining that causes the final result.

12. a. Why are some bacteria difficult to grow in the laboratory? Relate this to what you know so far about metabolism.
    b. Why are viruses hard to cultivate in the laboratory?

Critical Thinking Questions

Critical thinking is the ability to reason and solve problems using facts and concepts. These questions can be approached from a number of angles, and in most cases, they do not have a single correct answer.

1. Describe the steps you would take to isolate, cultivate, and identify a microbial pathogen from a urine sample. (Hint: Look at the Five I's.)

2. A certain medium has the following composition:

Glucose	15 g
Yeast extract	5 g
Peptone	5 g
KH_2PO_4	2 g
Distilled water	1,000 ml

   a. To what chemical category does this medium belong?
   b. How could you convert *Euglena* agar (table 3.2) into a nonsynthetic medium?

3. a. Name four categories that blood agar fits into.
   b. Name four differential reactions that TSIA shows.
   c. Can you tell what functional kind of medium *Enterococcus faecalis* medium is?

4. a. What kind of medium might you make to selectively grow a bacterium that lives in the ocean?
   b. One that lives in the human stomach?
   c. What characteristic of dyes makes them useful in differential media?
   d. Why are intestinal bacteria able to grow on media containing bile?

5. a. When buying a microscope, what features are most important to check for?
   b. What is probably true of a $20 microscope that claims to magnify 1,000×?

6. How can one obtain 2,000× magnification with a 100× objective?

7. a. In what ways are dark-field microscopy and negative staining alike?
   b. How is the dark-field microscope like the scanning electron microscope?

8. Biotechnology companies have engineered hundreds of different types of mice, rats, pigs, goats, cattle, and rabbits to have genetic diseases similar to diseases of humans or to synthesize drugs and other biochemical products. They have patented these animals and sell them to researchers for study and experimentation.
   a. What do you think of creating new life forms just for experimentation?
   b. Comment on the benefits, safety, and ethics of this trend.

9. This is a test of your living optical system's resolving power. Prop your book against a wall about 20 inches away and determine the line in the illustration below that is no longer resolvable by your eye. See if you can determine your actual resolving power, using a millimeter ruler.

So, Naturalists observe,

a flea has smaller

fleas that on him prey;

and these have smaller still

to bite 'em; and so proceed,

ad infinitum.

Source: Poem by Jonathan Swift.

10. Some human pathogenic bacteria are resistant to most antibiotics. How would you prove a bacterium is resistant to antibiotics using laboratory culture techniques?

Internet Search Topics

1. Search through several websites using the keywords "electron micrograph." Find examples of TEM and SEM micrographs and their applications in science and technology.

2. Search using the words "laboratory identification of anthrax" to make an outline of the basic techniques used in analysis of the microbe, under the headings of the Five I's.

3. Go to the Online Learning Center for chapter 3 of this text at http://www.mhhe.com/cowan1. Access the URLs listed under Internet Search Topics and research the following:

   a. Explore the website listed, which contains a broad base of information and images on microscopes and microscopy. Visit the photo gallery to compare different types of microscope images.

   b. Use the interactive website listed to see clearly how the numerical aperture changes with magnification.

Procaryotic Profiles

The Bacteria and Archaea

From April 3rd to April 24th, 2001, nine cases of pneumonia occurred in elderly residents (median age of 86 years) living at a long-term care facility in New Jersey. Seven of the nine patients had *Streptococcus pneumoniae* isolated from blood cultures, with capsular serotyping revealing that all isolates were serotype 14 and of the same clonal group. Seven of the nine patients also lived in the same wing of the nursing home. The two patients that were culture negative did contain gram-positive diplococci in their sputum and had chest X rays consistent with pneumonia. Epidemiological studies of the patients and controls revealed that all who developed pneumonia had no documented record of vaccination with the pneumococcal polysaccharide vaccine (PPV). In contrast, about 50% of the controls were vaccinated with PPV. Even though other risk factors were assessed, the lack of vaccination with PPV was the only one strongly associated with illness. Unfortunately, despite treatment, four of the nine patients with pneumonia died.

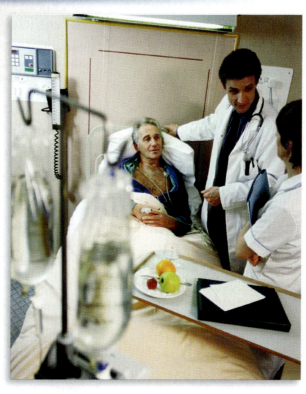

Once the outbreak was recognized, PPV was offered to those 55 residents who had not yet been vaccinated: 37 of these were vaccinated, whereas the other 18 were either ineligible or refused the vaccine. Other control measures included refusal to admit patients without a history of PPV vaccine.

▶ *What special advantage does the capsule confer on the pathogen* Streptococcus pneumoniae?

▶ *Why are those who have been vaccinated against* Streptococcus pneumoniae *more resistant to infection by this agent?*

CHAPTER OVERVIEW

▶ Procaryotic cells are the smallest, simplest, and most abundant cells on earth.

▶ Representative procaryotes include bacteria and archaea, both of which lack a nucleus and organelles but are functionally complex.

▶ The structure of bacterial cells is compact and capable of adaptations to a multitude of habitats.

▶ The cell is encased in an envelope that protects, supports, and regulates transport.

▶ Bacteria have special structures for motility and adhesion in the environment.

▶ Bacterial cells contain genetic material in one or a few chromosomes, and ribosomes for synthesizing proteins.

► Bacteria have the capacity for reproduction, nutrient storage, dormancy, and resistance to adverse conditions.

► Shape, size, and arrangement of bacterial cells are extremely varied.

► Bacterial taxonomy and classification is based on their structure, metabolism, and genetics.

► Archaea are procaryotes related to eucaryotic cells that possess unique biochemistry and genetics.

4.1 Procaryotic Form and Function

The evolutionary history of procaryotic cells extends back at least 3.8 billion years. It is now generally thought that the very first cells to appear on the earth were a type of archaea possibly related to modern forms that live on sulfur compounds in geothermal ocean vents. The fact that these organisms have endured for so long in such a variety of habitats indicates a cellular structure and function that are amazingly versatile and adaptable. The general cellular organization of a procaryotic cell can be represented with this flowchart:

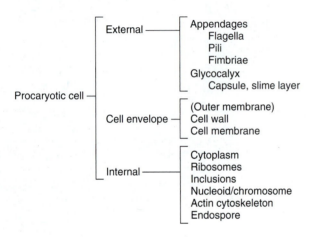

All bacterial cells invariably have a cell membrane, cytoplasm, ribosomes, and one (or a few) chromosome(s); the majority have a cell wall and some form of surface coating or glycocalyx. Specific structures that are found in some, but not all, bacteria are flagella, pili, fimbriae, capsules, slime layers, inclusions, an actin cytoskeleton, and endospores.

The Structure of a Generalized Procaryotic Cell

Bacterial cells appear featureless and two-dimensional when viewed with an ordinary microscope. Not until they are subjected to the scrutiny of the electron microscope and biochemical studies does their intricate and functionally complex nature become evident. The descriptions of procaryotic structure, except where otherwise noted, refer to the **bacteria**, a category of procaryotes with peptidoglycan in their cell walls. **Figure 4.1** presents a three-dimensional anatomical view of a generalized (rod-shaped) bacterial cell. As we survey the principal anatomical features of this cell, we will perform a microscopic dissection of sorts, following a course that begins with the outer cell structures and proceeds to the internal contents.

4.2 External Structures

Appendages: Cell Extensions

Several discrete types of accessory structures sprout from the surface of bacteria. These elongate **appendages** are common but are not present on all species. Appendages can be divided into two major groups: those that provide motility (flagella and axial filaments), and those that provide attachments or channels (fimbriae and pili).

Flagella—Bacterial Propellers

The procaryotic **flagellum** (flah-jel'-em), an appendage of truly amazing construction, is certainly unique in the biological world. The primary function of flagella is to confer **motility,** or self-propulsion—that is, the capacity of a cell to swim freely through an aqueous habitat. The extreme thinness of a bacterial flagellum necessitates high magnification to reveal its special architecture, which has three distinct parts: the filament, the hook (sheath), and the basal body **(figure 4.2).** The **filament,** a helical structure composed of proteins, is approximately 20 nm in diameter and varies from 1 to 70 μm in length. It is inserted into a curved, tubular hook. The hook is anchored to the cell by the basal body, a stack of rings firmly anchored through the cell wall, to the cell membrane and the outer membrane. This arrangement permits the hook with its filament to rotate 360°, rather than undulating back and forth like a whip as was once thought.

One can generalize that all spirilla, about half of the bacilli, and a small number of cocci are flagellated (these bacterial shapes are shown in figure 4.22). Flagella vary both in number and arrangement according to two general patterns: (1) In a *polar* arrangement, the flagella are attached at one or both ends of the cell. Three subtypes of this pattern are: **monotrichous** (mah"-noh-trik'-us), with a single flagellum; **lophotrichous** (lo"-foh), with small bunches or tufts of flagella emerging from the same site; and **amphitrichous** (am"-fee), with flagella at both poles of the cell. (2) In a **peritrichous** (per"-ee) arrangement, flagella are dispersed randomly over the surface of the cell **(figure 4.3).**

The presence of motility is one piece of information used in the laboratory identification or diagnosis of pathogens. Special stains or electron microscope preparations must be used to see arrangement, since flagella are too minute to be seen in live preparations with a light microscope. Often it is sufficient to know simply whether a bacterial species is motile. One way to detect motility is to stab a tiny mass of cells into a soft (semisolid) medium in a test tube. Growth spreading rapidly through the entire medium is indicative of motility. Alternatively, cells can be observed microscopically with a hanging

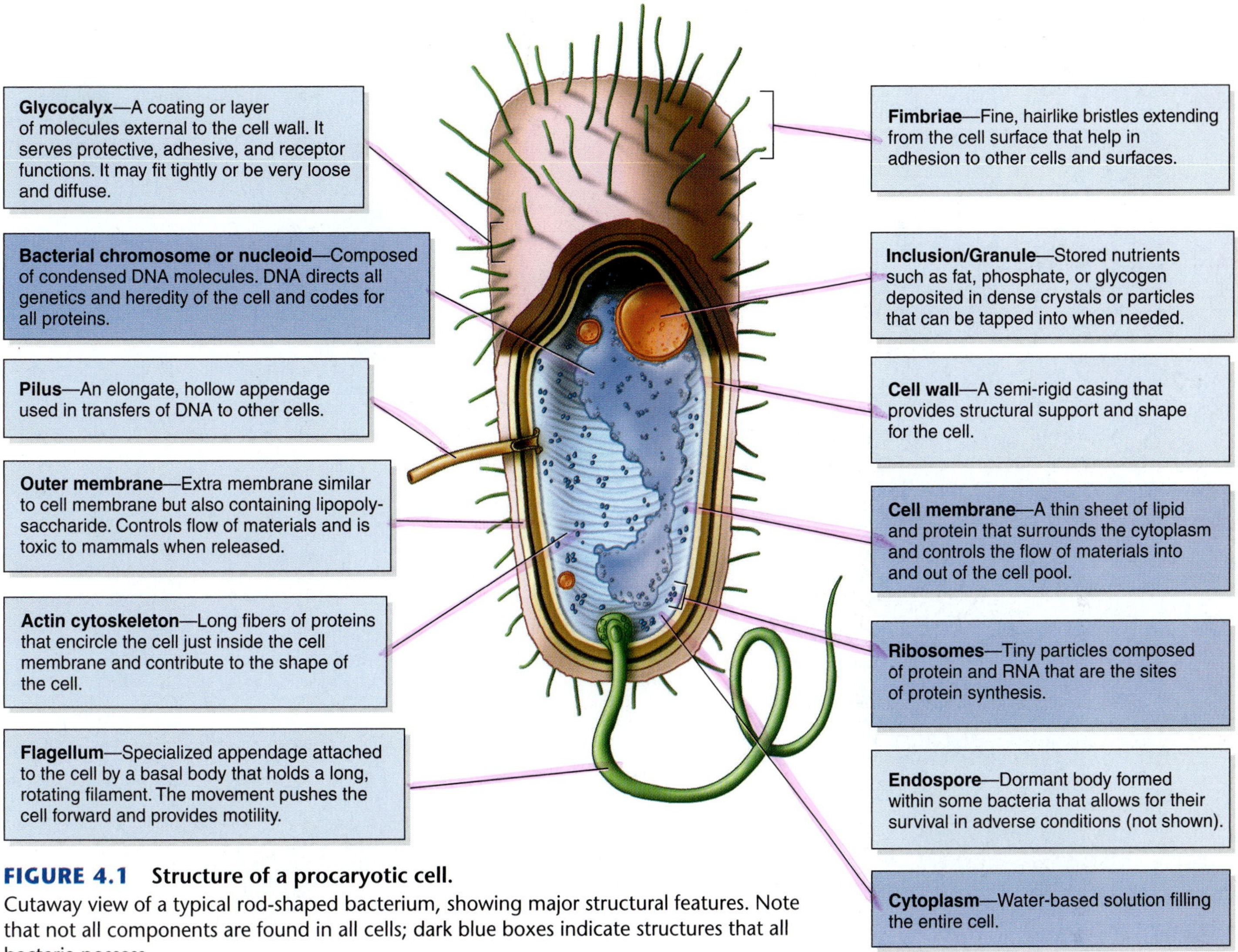

Glycocalyx—A coating or layer of molecules external to the cell wall. It serves protective, adhesive, and receptor functions. It may fit tightly or be very loose and diffuse.

Bacterial chromosome or nucleoid—Composed of condensed DNA molecules. DNA directs all genetics and heredity of the cell and codes for all proteins.

Pilus—An elongate, hollow appendage used in transfers of DNA to other cells.

Outer membrane—Extra membrane similar to cell membrane but also containing lipopolysaccharide. Controls flow of materials and is toxic to mammals when released.

Actin cytoskeleton—Long fibers of proteins that encircle the cell just inside the cell membrane and contribute to the shape of the cell.

Flagellum—Specialized appendage attached to the cell by a basal body that holds a long, rotating filament. The movement pushes the cell forward and provides motility.

Fimbriae—Fine, hairlike bristles extending from the cell surface that help in adhesion to other cells and surfaces.

Inclusion/Granule—Stored nutrients such as fat, phosphate, or glycogen deposited in dense crystals or particles that can be tapped into when needed.

Cell wall—A semi-rigid casing that provides structural support and shape for the cell.

Cell membrane—A thin sheet of lipid and protein that surrounds the cytoplasm and controls the flow of materials into and out of the cell pool.

Ribosomes—Tiny particles composed of protein and RNA that are the sites of protein synthesis.

Endospore—Dormant body formed within some bacteria that allows for their survival in adverse conditions (not shown).

Cytoplasm—Water-based solution filling the entire cell.

FIGURE 4.1   Structure of a procaryotic cell.
Cutaway view of a typical rod-shaped bacterium, showing major structural features. Note that not all components are found in all cells; dark blue boxes indicate structures that all bacteria possess.

drop slide. A truly motile cell will flit, dart, or wobble around the field, making some progress, whereas one that is nonmotile jiggles about in one place but makes no progress.

Fine Points of Flagellar Function   Flagellated bacteria can perform some rather sophisticated feats. They can detect and move in response to chemical signals—a type of behavior called **chemotaxis** (ke″-moh-tak′-sis). Positive chemotaxis is movement of a cell in the direction of a favorable chemical stimulus (usually a nutrient); negative chemotaxis is movement away from a repellent (potentially harmful) compound.

The flagellum is effective in guiding bacteria through the environment primarily because the system for detecting chemicals is linked to the mechanisms that drive the flagellum. Located in the cell membrane are clusters of receptors[1] that bind specific molecules coming from the immediate environment. The attachment of sufficient numbers of these molecules transmits signals to the flagellum and sets it into rotary motion. If several flagella are present, they become aligned and rotate as a group **(figure 4.4)**. As a flagellum

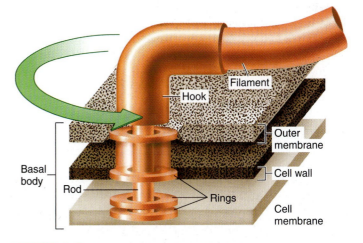

FIGURE 4.2   Details of the basal body in a gram-negative cell.
The hook, rings, and rod function together as a tiny device that rotates the filament 360°.

---

1. Cell surface molecules that bind specifically with other molecules.

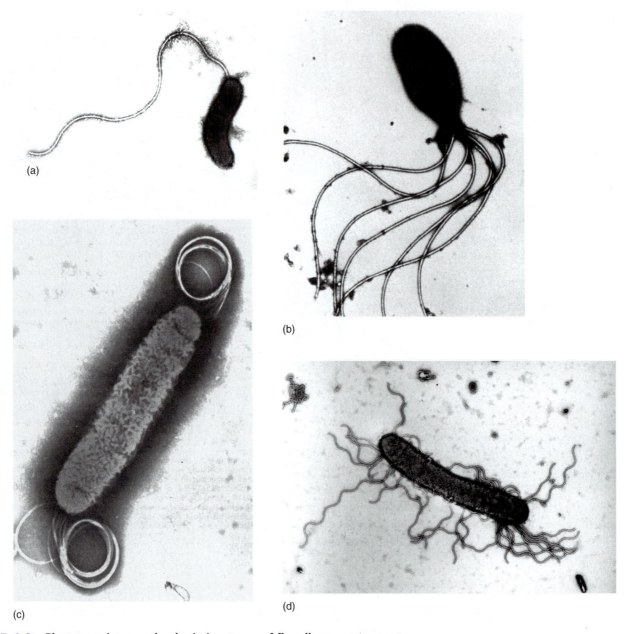

FIGURE 4.3 **Electron micrographs depicting types of flagellar arrangements.**
(a) Monotrichous flagellum on the predatory bacterium *Bdellovibrio*. **(b)** Lophotrichous flagella on *Vibrio fischeri*, a common marine bacterium (23,000×). **(c)** Unusual flagella on *Aquaspirillum* are amphitrichous (and lophotrichous) in arrangement and coil up into tight loops. **(d)** An unidentified bacterium discovered inside *Paramecium* cells exhibits peritrichous flagella.
(b) From Reichelt and Baumann, *Arch. Microbiol.* 94:283–330. © Springer-Verlag, 1973.

FIGURE 4.4 **The operation of flagella and the mode of locomotion in bacteria with polar and peritrichous flagella.**
(a) In general, when a polar flagellum rotates in a counterclockwise direction, the cell swims forward. When the flagellum reverses direction and rotates clockwise, the cell stops and tumbles. **(b)** In peritrichous forms, all flagella sweep toward one end of the cell and rotate as a single group. During tumbles, the flagella lose coordination.

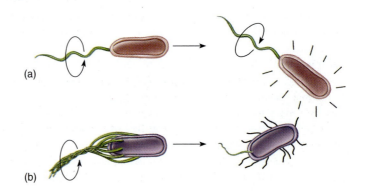

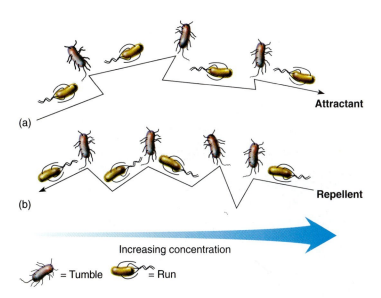

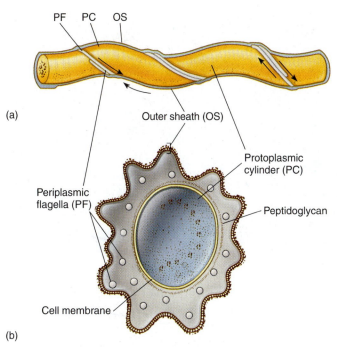

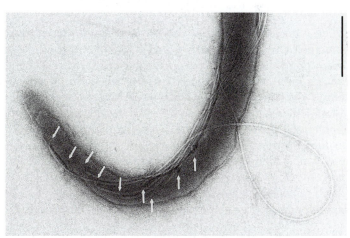

FIGURE 4.5 **Chemotaxis in bacteria.**
The cell shows a primitive mechanism for progressing **(a)** toward positive stimuli and **(b)** away from irritants by swimming in straight runs or by tumbling. Bacterial runs allow straight, undisturbed progress toward the stimulus, whereas tumbles interrupt progress to allow the bacterium to redirect itself away from the stimulus after sampling the environment.

rotates counterclockwise, the cell itself swims in a smooth linear direction toward the stimulus; this action is called a *run*. Runs are interrupted at various intervals by *tumbles*, during which the flagellum reverses direction and causes the cell to stop and change its course. It is believed that attractant molecules inhibit tumbles and permit progress toward the stimulus. Repellents cause numerous tumbles, allowing the bacterium to redirect itself away from the stimulus **(figure 4.5).** Some photosynthetic bacteria exhibit *phototaxis*, a type of movement in response to light rather than chemicals.

Periplasmic Flagella

Corkscrew-shaped bacteria called **spirochetes** (spy'-roh-keet) show an unusual, wriggly mode of locomotion caused by two or more long, coiled threads, the periplasmic flagella or *axial filaments*. A periplasmic flagellum is a type of internal flagellum that is enclosed in the space between the cell wall and the cell membrane **(figure 4.6).** The filaments curl closely around the spirochete coils yet are free to contract and impart a twisting or flexing motion to the cell. This form of locomotion must be seen in live cells such as the spirochete of syphilis to be truly appreciated.

Appendages for Attachment and Mating

The structures termed **pilus** (pil-us) and **fimbria** (fim'-bree-ah) both refer to bacterial surface appendages that provide some type of adhesion, but not locomotion.

Fimbriae are small, bristlelike fibers sprouting off the surface of many bacterial cells **(figure 4.7).** Their exact composition varies, but most of them contain protein. Fimbriae have an inherent tendency to stick to each other and to sur-

FIGURE 4.6 **The orientation of periplasmic flagella on the spirochete cell.**
(a) Longitudinal section. **(b)** Cross section. Contraction of the filaments imparts a spinning and undulating pattern of locomotion. **(c)** Electron micrograph captures the details of periplasmic flagella and their insertion points (arrows) in *Borrelia burgdorferi.* One flagellum has escaped the outer sheath, probably during preparation for EM. (Bar = 0.2 μm)

faces. They may be responsible for the mutual clinging of cells that leads to biofilms and other thick aggregates of cells on the surface of liquids and for the microbial colonization of inanimate solids such as rocks and glass **(Insight 4.1).** Some pathogens can colonize and infect host tissues because of a tight adhesion between their fimbriae and epithelial cells **(figure 4.7b).** For example, the gonococcus (agent of gonorrhea) colonizes the genitourinary tract, and *Escherichia coli* colonizes the intestine by this means. Mutant forms of these pathogens that lack fimbriae are unable to cause infections.

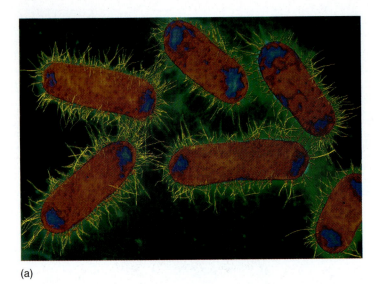

(a)

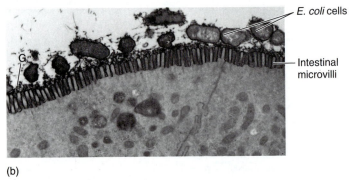

(b)

FIGURE 4.7   Form and function of bacterial fimbriae.
(a) Several cells of pathogenic *Escherichia coli* covered with numerous stiff fibers called fimbriae (30,000×). Note also the dark blue granules, which are the chromosomes. **(b)** A row of *E. coli* cells tightly adheres by their fimbriae to the surface of intestinal cells (12,000×). This is how the bacterium clings and gains access to the body during an infection. (G = glycocalyx)

A pilus (also called a *sex pilus*) is an elongate, rigid tubular structure made of a special protein, *pilin*. So far, true pili have been found only on gram-negative bacteria, where they are utilized in a "mating" process between cells called **conjugation,**[2] which involves partial transfer of DNA from one cell to another **(figure 4.8).** A pilus from the donor cell unites with a recipient cell thereby providing a cytoplasmic connection for making the transfer. Production of pili is controlled genetically, and conjugation takes place only between compatible gram-negative cells. Conjugation in gram-positive bacteria does occur, but involves aggregation proteins rather than sex pili. The roles of pili and conjugation are further explored in chapter 9.

The Bacterial Surface Coating, or Glycocalyx

The bacterial cell surface is frequently exposed to severe environmental conditions. The **glycocalyx** develops as a coating of macromolecules to protect the cell and, in some cases, help it adhere to its environment. Glycocalyces differ among bacteria in thickness, organization, and chemical composition. Some bacteria are covered with a loose shield called a slime layer that evidently protects them from loss of water and nutrients **(figure 4.9a).** Other bacteria produce **capsules** of repeating polysaccharide units, of protein, or of both **(figures 4.9b and 4.10).** A capsule is bound more tightly to the cell than a slime layer is, and it has a thicker, gummy consistency that gives a prominently sticky (mucoid) character to the colonies of most encapsulated bacteria.

Specialized Functions of the Glycocalyx   Capsules are formed by many pathogenic bacteria, such as *Streptococcus pneumoniae* (a cause of pneumonia, an infection of the lung), *Haemophilus influenzae* (one cause of meningitis), and *Bacil-*

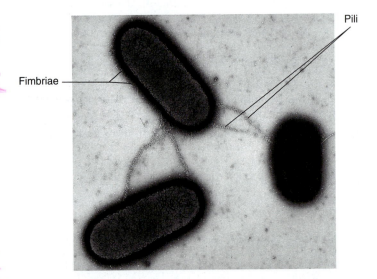

FIGURE 4.8   Three bacteria in the process of conjugating.
Clearly evident are the sex pili forming mutual conjugation bridges between a donor (upper cell) and two recipients (two lower cells). (Fimbriae can also be seen on the donor cell.)

lus anthracis (the cause of anthrax). Encapsulated bacterial cells generally have greater pathogenicity because capsules protect the bacteria against white blood cells called phagocytes. Phagocytes are a natural body defense that can engulf and destroy foreign cells through phagocytosis, thus preventing infection. A capsular coating blocks the mechanisms that phagocytes use to attach to and engulf bacteria. By escaping phagocytosis, the bacteria are free to multiply and infect body tissues. Encapsulated bacteria that mutate to nonencapsulated forms usually lose their pathogenicity.

Other types of glycocalyces can be important in formation of biofilms. The thick, white plaque that forms on teeth comes in part from the surface slimes produced by certain

---

2. Although the term *mating* is sometimes used for this process, it is not a form of sexual reproduction.

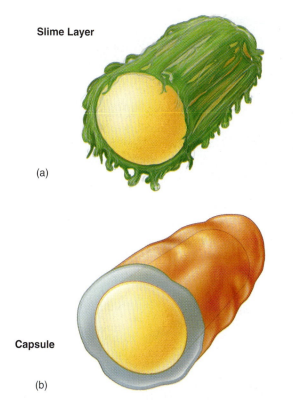

Slime Layer

(a)

Capsule

(b)

FIGURE 4.9 **Bacterial cells sectioned to show the types of glycocalyces.**
(a) The slime layer is a loose structure that is easily washed off.
(b) The capsule is a thick, structured layer that is not readily removed.

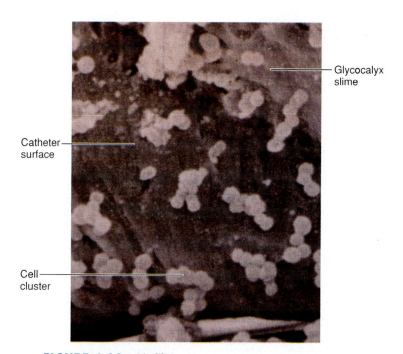

Glycocalyx slime

Catheter surface

Cell cluster

FIGURE 4.11 **Biofilm.**
Scanning electron micrograph of *Staphylococcus aureus* cells attached to a catheter by a slime secretion.

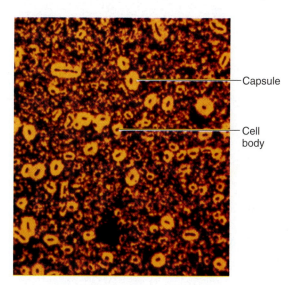

Capsule

Cell body

FIGURE 4.10 **Encapsulated bacteria.**
Staining reveals the microscopic appearance of a large, well-developed capsule.

streptococci in the oral cavity. This slime protects them from being dislodged from the teeth and provides a niche for other oral bacteria that, in time, can lead to dental disease. The glycocalyx of some bacteria is so highly adherent that it is responsible for persistent colonization of nonliving materials such as plastic catheters, intrauterine devices, and metal pacemakers that are in common medical use **(figure 4.11).**

IN THE NEWS *(Continued from page 89)*

The outbreak of pneumococcal pneumonia described at the beginning of the chapter points out that the presence of certain bacterial structures, such as a capsule, enhances virulence. Studies have shown that since the capsule allows the bacterium to resist host phagocytosis, encapsulated strains of *Streptococcus pneumoniae* are virulent, whereas those with no capsule are not. In fact, those individuals who have antibodies specific for the polysaccharide capsule of the *Streptococcus pneumoniae* strain will be resistant to attack by that strain. This knowledge has been used to make a vaccine for adults, using 23 types of polysaccharide capsular antigens, which, when injected, will elicit specific antibodies that protect from the most common strains causing pneumococcal pneumonia. The serum antibodies that arise after vaccination specifically coat the bacterial capsule and allow for uptake of the bacteria by the host phagocytes. Efficacy of the vaccine is shown in studies in which incidence of pneumococcal disease in the elderly is reduced in those vaccinated. This disease is significant, as the Centers for Disease Control and Prevention (CDC) estimates that about a half million cases occur each year, resulting in about 40,000 deaths in the United States.

As was the case with this outbreak, the highest mortality rate (30–40%) occurs in the elderly or in those with underlying medical conditions. CDC estimates that about half of these deaths could be prevented through use of the pneumococcal vaccine.

See: CDC. 2001. Outbreak of pneumococcal pneumonia among unvaccinated residents of a nursing home—New Jersey, April 2001. MMWR 50:707–710.

CDC. 1997. Prevention of pneumococcal disease: Recommendations of the Advisory Committee on Immunization Practices (ACIP). MMWR 46, No. RR-09.

INSIGHT 4.1   *Discovery*

Biofilms—The Glue of Life

Being aware of the widespread existence of microorganisms on earth, we should not be surprised that, when left undisturbed, they gather in masses, cling to various surfaces, and capture available moisture and nutrients. The formation of these living layers, called **biofilms,** is actually a universal phenomenon that all of us have observed. Consider the scum that builds up in toilet bowls and shower stalls in a short time if they are not cleaned; or the algae that collect on the walls of swimming pools; and, more intimately—the constant deposition of plaque on teeth. Microbes making biofilms is a primeval tendency that has been occurring for billions of years as a way to create stable habitats with adequate access to food, water, atmosphere, and other essential factors. Biofilms are often cooperative associations among several microbial groups (bacteria, fungi, algae, and protozoa) as well as plants and animals.

Substrates are most likely to accept a biofilm if they are moist and have developed a thin layer of organic material such as polysaccharides or glycoproteins on their exposed surface (see figure at right). This depositing process occurs within a few minutes to hours, making a slightly sticky texture that attracts primary colonists, usually bacteria. These early cells attach (adsorb to) and begin to multiply on the surface. As they grow, various secreted substances in their glycocalyx (receptors, fimbriae, slime layers, capsules) increase the binding of cells to the surface and thicken the biofilm. As the biofilm evolves, it undergoes specific adaptations to the habitat in which it forms. In many cases, the earliest colonists contribute nutrients and create microhabitats that serve as a matrix for other microbes to attach and grow into the film, forming complete communities. The biofilm varies in thickness and complexity, depending upon where it occurs and how long it keeps developing. Complexity ranges from single cell layers to thick microbial mats with dozens of dynamic interactive layers.

Biofilms are a profoundly important force in the development of terrestrial and aquatic environments. They dwell permanently in bedrock and the earth's sediments, where they play an essential role in recycling elements, leaching minerals, and soil formation. Biofilms associated with plant roots promote the mutual exchange of nutrients between the microbes and roots. Invasive biofilms can wreak havoc with human-made structures such as cooling towers, storage tanks, air conditioners, and even stone buildings.

Biofilms also have serious medical implications. Most healthy human tissues do not accrue these thick layers of microbial life. Normal flora are generally limited to single-cell associations with skin and mucous membranes. But biofilms accumulate on damaged tissues (such as rheumatic heart valves), hard tissues (teeth), and foreign materials (catheters, IUDs, artificial hip joints). Microbes in a biofilm are extremely difficult to eradicate with antimicrobials. Previously it was assumed that the drugs had

difficulty penetrating the viscous biofilm matrix. Now scientists have discovered that bacteria in biofilms turn on different genes when they are in a biofilm than when they are "free-floating." This altered gene expression gives the bacteria a different set of characteristics, often making them impervious to antibiotics.

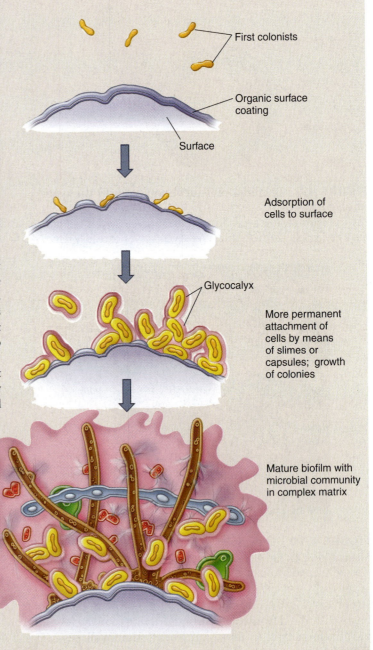

First colonists

Organic surface coating

Surface

Adsorption of cells to surface

Glycocalyx

More permanent attachment of cells by means of slimes or capsules; growth of colonies

Mature biofilm with microbial community in complex matrix

4.3 The Cell Envelope: The Boundary Layer of Bacteria

The majority of bacteria have a chemically complex external covering, termed the cell envelope, that lies outside of the cytoplasm. It is composed of two or three basic layers: the cell wall, the cell membrane, and, in some bacteria, the outer membrane. The layers of the envelope are stacked one upon another and are often tightly bonded together like the outer husk and casings of a coconut. Although each envelope layer performs a distinct function, together they act as a single protective unit.

Differences in Cell Envelope Structure

More than a hundred years ago, long before the detailed anatomy of bacteria was even remotely known, a Danish physician named Hans Christian Gram developed a staining technique, the **Gram stain,** that delineates two generally different groups of bacteria **(Insight 4.2).** The two major groups shown by this technique are the gram-positive bacteria and the gram-negative bacteria. Because the Gram stain does not actually reveal the nature of these physical differences, we must turn to the electron microscope and to biochemical analysis.

The extent of the differences between gram-positive and gram-negative bacteria is evident in the physical appearance of their cell envelopes **(figure 4.12).** In gram-positive cells, a microscopic section resembles an open-faced sandwich with two layers: the thick cell wall, composed primarily of peptidoglycan (defined in the next section), and the cell membrane. A similar section of a gram-negative cell envelope shows a complete sandwich with three layers: an outer membrane, a thin cell wall, and the cell membrane.

Moving from outside to in, the outer membrane (if present) lies just under the glycocalyx. Next comes the cell wall. Finally, the innermost layer is always the cell membrane. Since only some bacteria have an outer membrane, we'll discuss the cell wall first.

Structure of the Cell Wall

The cell wall accounts for a number of important bacterial characteristics. In general, it helps determine the shape of a bacterium, and it also provides the kind of strong structural support necessary to keep a bacterium from bursting or collapsing because of changes in osmotic pressure. In this way, the cell wall functions like a bicycle tire that maintains the necessary shape and prevents the more delicate inner tube from bursting when it is expanded.

The cell walls of most bacteria gain their relatively rigid quality from a unique macromolecule called **peptidoglycan (PG).** This compound is composed of a repeating framework of long *glycan* chains cross-linked by short peptide fragments to provide a strong but flexible support framework

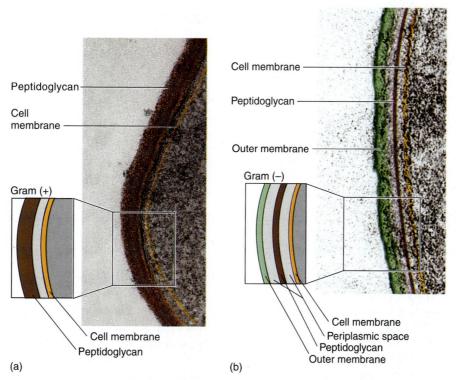

FIGURE 4.12   **A comparison of the envelopes of gram-positive and gram-negative cells.**
(a) A photomicrograph of a gram-positive cell wall/membrane and an artist's interpretation of its open-faced sandwich–style layering with two layers. **(b)** A photomicrograph of a gram-negative cell wall/membrane and an artist's interpretation of its complete sandwich–style layering with three distinct layers.

INSIGHT 4.2   *Discovery*

The Gram Stain: A Grand Stain

In 1884, Hans Christian Gram discovered a staining technique that could be used to make bacteria in infectious specimens more visible. His technique consisted of timed, sequential applications of crystal violet (the primary dye), Gram's iodine (IKI, the mordant), an alcohol rinse (decolorizer), and a contrasting counterstain. The initial counterstain used was yellow or brown and was later replaced by the red dye, safranin. Since that substitution, bacteria that stained purple are called gram-positive, and those that stained red are called gram-negative.

Although these staining reactions involve an attraction of the cell to a charged dye (see chapter 3), it is important to note that the terms *gram-positive* and *gram-negative* are not used to indicate the electrical charge of cells or dyes but whether or not a cell retains the primary dye-iodine complex after decolorization. There is nothing specific in the reaction of gram-positive cells to the primary dye or in the reaction of gram-negative cells to the counterstain. The different results in the Gram stain are due to differences in the structure of the cell wall and how it reacts to the series of reagents applied to the cells.

In the first step, crystal violet is added to the cells in a smear and stains them all the same purple color. The second and key differentiating step is the addition of the mordant—Gram's iodine. The mordant is a stabilizer that causes the dye to form large crystals in the peptidoglycan meshwork of the cell wall. Because the peptidoglycan layer in gram-positive cells is thicker, the entrapment of the dye is far more extensive in them than in gram-negative cells. Application of alcohol in the third step dissolves lipids in the outer membrane and removes the dye from the peptidoglycan layer and the gram-negative cells. By contrast, the crystals of dye tightly embedded in the peptidoglycan of gram-positive bacteria are relatively inaccessible and resistant to removal. Because gram-negative bacteria are colorless

after decolorization, their presence is demonstrated by applying the counterstain safranin in the final step.

This century-old staining method remains the universal basis for bacterial classification and identification. It permits differentiation of four major categories based upon color reaction and shape: gram-positive rods, gram-positive cocci, gram-negative rods, and gram-negative cocci (see table 4.4). The Gram stain can also be a practical aid in diagnosing infection and in guiding drug treatment. For example, gram staining a fresh urine or throat specimen can help pinpoint the possible cause of infection, and in some cases it is possible to begin drug therapy on the basis of this stain. Even in this day of elaborate and expensive medical technology, the Gram stain remains an important and unbeatable first tool in diagnosis.

Step	Microscopic Appearance of Cell		Chemical Reaction in Cell Wall (very magnified view)	
	Gram (+)	Gram (−)	Gram (+)	Gram (−)
1. Crystal violet			Both cell walls affix the dye	
2. Gram's iodine			Dye crystals trapped in wall	No effect of iodine
3. Alcohol			Crystals remain in cell wall	Outer membrane weakened; wall loses dye
4. Safranin (red dye)			Red dye has no effect	Red dye stains the colorless cell

(figure 4.13). The amount and exact composition of peptidoglycan varies among the major bacterial groups.

Because many bacteria live in aqueous habitats with a low solute concentration, they are constantly absorbing excess water by osmosis. Were it not for the strength and relative rigidity of the peptidoglycan in the cell wall, they would rupture from internal pressure. Understanding this function of the cell wall has been a tremendous boon to the drug industry. Several types of drugs used to treat infection (penicillin, cephalosporins) are effective because they target the peptide cross-links in the peptidoglycan, thereby disrupting

its integrity. With their cell walls incomplete or missing, such cells have very little protection from **lysis** (ly'-sis). Lysozyme, an enzyme contained in tears and saliva, provides a natural defense against certain bacteria by hydrolyzing the bonds in the glycan chains and causing the wall to break down. (Chapter 11 discusses the actions of antimicrobial chemical agents.)

The Gram-Positive Cell Wall

The bulk of the gram-positive cell wall is a thick, homogeneous sheath of peptidoglycan ranging from 20 to 80 nm in thickness.

(a) The peptidoglycan of a cell wall can be seen as a crisscross network pattern similar to a chain-link fence, forming a single massive molecule that molds the outer structure of the cell into a tight box.

The Gram-Negative Cell Wall

The gram-negative wall is a single, thin (1–3 nm) sheet of peptidoglycan. Although it acts as a somewhat rigid protective structure as previously described, its thinness gives gram-negative bacteria a relatively greater flexibility and sensitivity to lysis. A well-developed *periplasmic space* surrounds the peptidoglycan (see figure 4.14). This space is an important reaction site for a large and varied pool of substances that enter and leave the cell.

(b) An idealized view of the molecular pattern of peptidoglycan. It contains alternating glycans (G and M) bound together in long strands. The G stands for N-acetyl glucosamine, and the M stands for N-acetyl muramic acid. A muramic acid molecule binds to an adjoining muramic acid on a parallel chain by means of a cross-linkage of peptides.

Glycan chains

Peptide cross-links

Nontypical Cell Walls

Several bacterial groups lack the cell wall structure of gram-positive or gram-negative bacteria, and some bacteria have no cell wall at all. Although these exceptional forms can stain positive or negative in the Gram stain, examination of their fine structure and chemistry shows that they do not really fit the descriptions for typical gram-negative or -positive cells. For example, the cells of *Mycobacterium* and *Nocardia* contain peptidoglycan and stain gram-positive, but the bulk of their cell wall is composed of unique types of lipids. One of these is a very-long-chain fatty acid called *mycolic acid*, or cord factor, that contributes to the pathogenicity of this group (see chapter 21). The thick, waxy nature imparted to the cell wall by these lipids is also responsible for a high degree of resistance to certain chemicals and dyes. Such resistance is the basis for the **acid-fast stain** used to diagnose tuberculosis and leprosy. In this stain, hot carbol fuchsin dye becomes tenaciously attached (is held fast) to these cells so that an acid-alcohol solution will not remove the dye (see chapter 3).

(c) A detailed view of the links between the muramic acids. Tetrapeptide chains branching off the muramic acids connect by interbridges also composed of amino acids. The types of amino acids in the interbridge can vary and it may be lacking entirely (gram-negative cells). It is this linkage that provides rigid yet flexible support to the cell and that may be targeted by drugs like penicillin. ✳

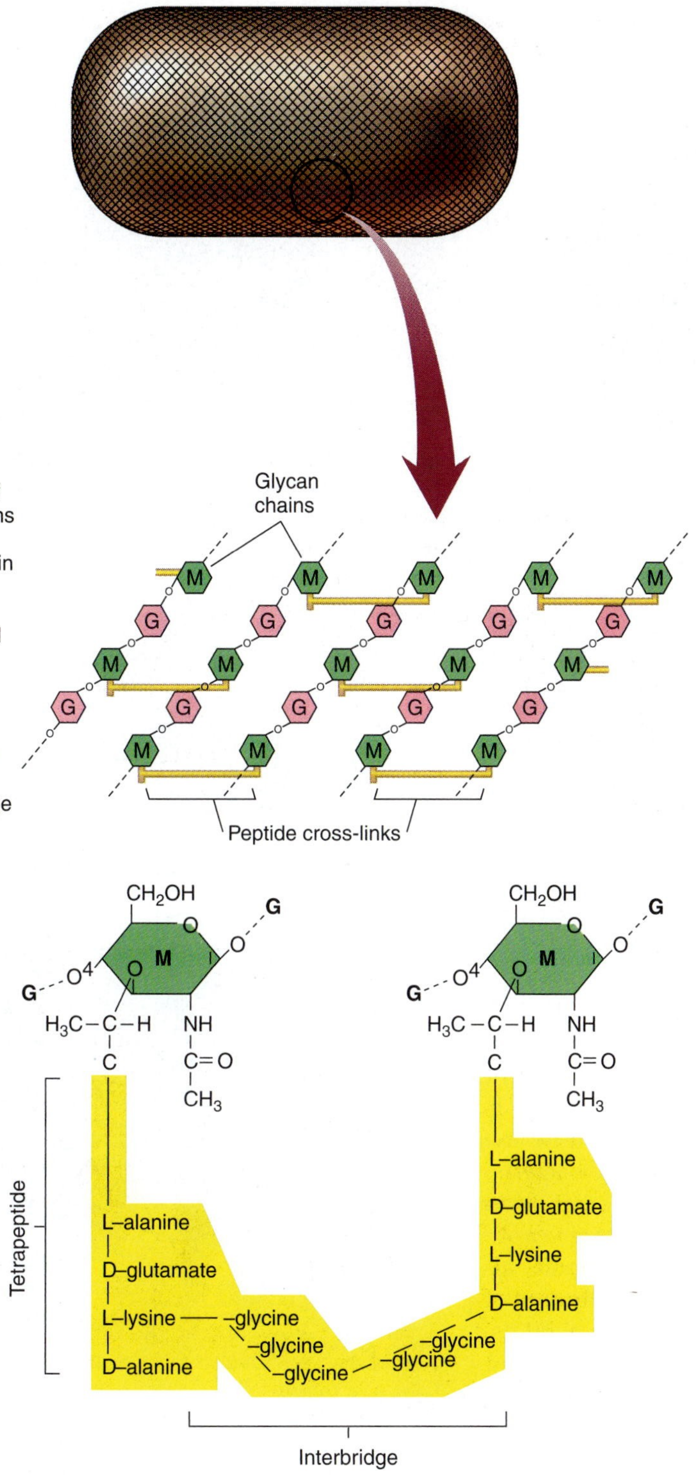

CH$_2$OH

G

M

O

G — O⁴

H$_3$C — C — H   NH

C   C=O

CH$_3$

Tetrapeptide

L-alanine

D-glutamate

L-lysine — glycine

D-alanine — glycine

glycine

glycine

CH$_2$OH

G

M

O

G — O⁴

H$_3$C — C — H   NH

C   C=O

CH$_3$

L-alanine

D-glutamate

L-lysine

D-alanine

glycine

glycine

Interbridge

FIGURE 4.13   Structure of peptidoglycan in the cell wall.

Because they are from a more ancient and primitive line of procaryotes, the archaea exhibit unusual and chemically distinct cell walls. In some, the walls are composed almost entirely of polysaccharides, and in others, the walls are pure protein; but as a group, they all lack the true peptidoglycan structure described previously. Since a few archaea and all mycoplasmas (next section) lack a cell wall entirely, their cell membrane must serve the dual functions of support as well as transport.

It also contains tightly bound acidic polysaccharides, including teichoic acid and lipoteichoic acid (**figure 4.14**). Teichoic acid is a polymer of ribitol or glycerol and phosphate embedded in the peptidoglycan sheath. Lipoteichoic acid is similar in structure but is attached to the lipids in the plasma membrane. These molecules appear to function in cell wall maintenance and enlargement during cell division, and they also contribute to the acidic charge on the cell surface.

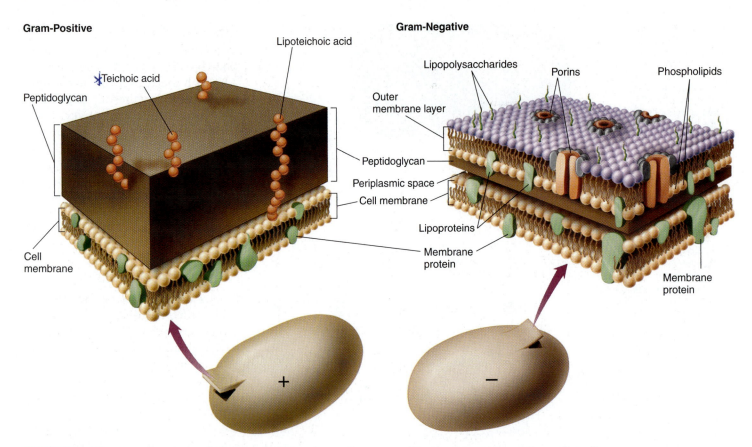

FIGURE 4.14   A comparison of the detailed structure of gram-positive and gram-negative cell walls.

Mycoplasmas and Other Cell-Wall-Deficient Bacteria

Mycoplasmas are bacteria that naturally lack a cell wall. Although other bacteria require an intact cell wall to prevent the bursting of the cell, the mycoplasma cell membrane is stabilized by sterols and is resistant to lysis. These extremely tiny, pleomorphic cells are very small bacteria, ranging from 0.1 to 0.5 μm in size. They range in shape from filamentous to coccus or doughnut-shaped. They are *not* obligate parasites and can be grown on artificial media, although added sterols are required for the cell membranes of some species. Mycoplasmas are found in many habitats, including plants, soil, and animals. The most important medical species is *Mycoplasma pneumoniae* (**figure 4.15**), which adheres to the epithelial cells in the lung and causes an atypical form of pneumonia in humans (described in chapter 21).

Some bacteria that ordinarily have a cell wall can lose it during part of their life cycle. These wall-deficient forms are referred to as **L forms** or L-phase variants (for the Lister Institute, where they were discovered). L forms arise naturally from a mutation in the wall-forming genes, or they can be induced artificially by treatment with a chemical such as lysozyme or penicillin that disrupts the cell wall. When a gram-positive cell is exposed to either of these two chemicals, it will lose the cell wall completely and become a **protoplast**, a fragile cell bounded only by a membrane that is highly susceptible to lysis (**figure 4.16a**). A gram-negative cell exposed

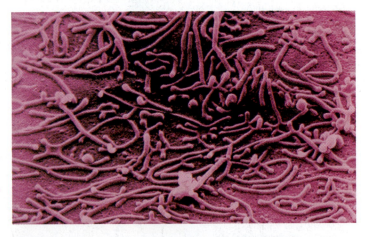

FIGURE 4.15   Scanning electron micrograph of *Mycoplasma pneumoniae* (62,000×).
Cells like these that naturally lack a cell wall exhibit extreme variation in shape.

to these same substances loses it peptidoglycan but retains its outer membrane, leaving a less fragile but nevertheless weakened **spheroplast** (**figure 4.16b**). Evidence points to a role for L forms in certain infections.

The Gram-Negative Outer Membrane

The outer membrane is somewhat similar in construction to the cell membrane, except that it contains specialized types

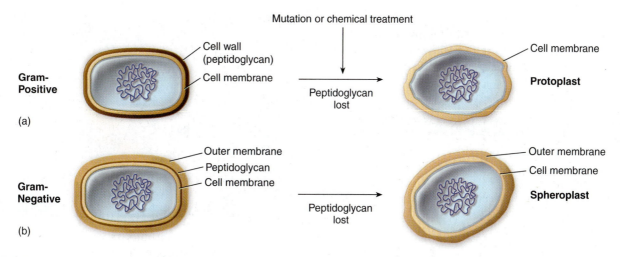

FIGURE 4.16 **The conversion of walled bacterial cells to L forms:** **(a)** gram-positive bacteria; **(b)** gram-negative bacteria.

of polysaccharides and proteins. The uppermost layer of the OM contains *lipopolysaccharide* (LPS). The polysaccharide chains extending off the surface function as antigens and receptors. The lipid portion of LPS has been referred to as *endotoxin* because it stimulates fever and shock reactions in gram-negative infections such as meningitis and typhoid fever. The innermost layer of the OM is a phospholipid layer anchored by means of lipoproteins to the peptidoglycan layer below. The outer membrane serves as a partial chemical sieve by allowing only relatively small molecules to penetrate. Access is provided by special membrane channels formed by *porin proteins* that completely span the outer membrane. The size of these porins can be altered so as to block the entrance of harmful chemicals, making them one defense of gram-negative bacteria against certain antibiotics (see figure 4.14).

Cell Membrane Structure

Appearing just beneath the cell wall is the cell, or cytoplasmic, membrane, a very thin (5–10 nm), flexible sheet molded completely around the cytoplasm. Its general composition was described in chapter 2 as a lipid bilayer with proteins embedded to varying degrees (see Insight 2.3). Bacterial cell membranes have this typical structure, containing primarily phospholipids (making up about 30–40% of the membrane mass) and proteins (contributing 60–70%). Major exceptions to this description are the membranes of mycoplasmas, which contain high amounts of sterols—rigid lipids that stabilize and reinforce the membrane—and the membranes of archaea, which contain unique branched hydrocarbons rather than fatty acids.

Photosynthetic procaryotes such as cyanobacteria contain dense stacks of internal membranes that carry the photosynthetic pigments, which we describe later on.

Functions of the Cell Membrane

Because bacteria have none of the eucaryotic organelles, the cell membrane provides a site for functions such as energy reactions, nutrient processing, and synthesis. A major action of the cell membrane is to regulate *transport,* that is, the passage of nutrients into the cell and the discharge of wastes. Although water and small uncharged molecules can diffuse across the membrane unaided, the membrane is a *selectively permeable* structure with special carrier mechanisms for passage of most molecules (see chapter 7). The glycocalyx and cell wall can bar the passage of large molecules, but they are not the primary transport apparatus. The cell membrane is also involved in *secretion,* or the discharge of a metabolic product into the extracellular environment.

The membranes of procaryotes are an important site for a number of metabolic activities. Most enzymes of respiration and ATP synthesis reside in the cell membrane since procaryotes lack mitochondria (see chapter 8). Enzyme structures located in the cell membrane also help synthesize structural macromolecules to be incorporated into the cell envelope and appendages. Other products (enzymes and toxins) are secreted by the membrane into the extracellular environment.

Practical Considerations of Differences in Cell Envelope Structure

Variations in cell envelope anatomy contribute to several other differences between the two cell types. The outer membrane contributes an extra barrier in gram-negative bacteria that makes them more impervious to some antimicrobial chemicals such as dyes and disinfectants, so they are generally more difficult to inhibit or kill than are gram-positive bacteria. One exception is for alcohol-based compounds, which can dissolve the lipids in the outer membrane and disturb its integrity. Treating infections caused by gram-negative bacteria often requires different drugs from gram-positive infections, especially drugs that can cross the outer membrane.

The cell envelope or its parts can interact with human tissues and contribute to disease. Proteins attached to the outer portion of the cell wall of several gram-positive species, including *Corynebacterium diphtheriae* (the agent of diphtheria) and *Streptococcus pyogenes* (the cause of strep throat), also have toxic properties. The lipids in the cell walls of certain *Mycobacterium* species are harmful to human cells as well. Because most macromolecules in the cell walls are foreign to humans, they stimulate antibody production by the immune system (see chapter 15).

✔ **CHECKPOINT**

- Bacteria are the oldest form of cellular life. They are also the most widely dispersed, occupying every conceivable microclimate on the planet.
- The external structures of bacteria include appendages (flagella, fimbriae, and pili) and the glycocalyx.
- Flagella vary in number and arrangement as well as in the type and rate of motion they produce.
- The cell envelope is the complex boundary structure surrounding a bacterial cell. In gram-negative bacteria, the envelope consists of an outer membrane, the cell wall, and the cell membrane. Gram-positive bacteria have only the cell wall and cell membrane.
- In a Gram stain, gram-positive bacteria retain the crystal violet and stain purple. Gram-negative bacteria lose the crystal violet and stain red from the safranin counterstain.
- Gram-positive bacteria have thick cell walls of peptidoglycan and acidic polysaccharides such as teichoic acid, and they have a thin periplasmic space. The cell walls of gram-negative bacteria are thinner and have a wide periplasmic space.
- The outer membrane of gram-negative cells contains lipopolysaccharide (LPS). LPS is toxic to mammalian hosts.
- The bacterial cell membrane is typically composed of phospholipids and proteins, and it performs many metabolic functions as well as transport activities.

4.4 Bacterial Internal Structure

Contents of the Cell Cytoplasm

Encased by the cell membrane is a dense, gelatinous solution referred to as **cytoplasm,** which is another prominent site for many of the cell's biochemical and synthetic activities. Its major component is water (70–80%), which serves as a solvent for the cell pool, a complex mixture of nutrients including sugars, amino acids, and salts. The components of this pool serve as building blocks for cell synthesis or as sources of energy. The cytoplasm also contains larger, discrete cell masses such as the chromatin body, ribosomes, mesosomes, granules, and actin strands that act as a cytoskeleton in bacteria that have them.

Bacterial Chromosomes and Plasmids: The Sources of Genetic Information

The hereditary material of most bacteria exists in the form of a single circular strand of DNA designated as the **bacterial chromosome.** (Some bacteria have multiple chromosomes.) By definition, bacteria do not have a nucleus; that is, their DNA is not enclosed by a nuclear membrane but instead is

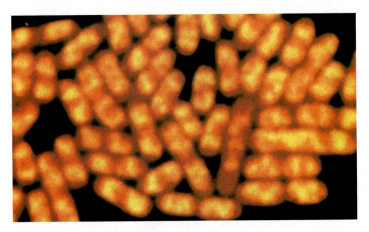

FIGURE 4.17 **Chromosome structure.**
Fluorescent staining highlights the chromosomes of the bacterial pathogen *Salmonella enteriditis*. The cytoplasm is orange, and the chromosome fluoresces bright yellow.

aggregated in a dense area of the cell called the **nucleoid.** The chromosome is actually an extremely long molecule of DNA that is tightly coiled around special basic protein molecules so as to fit inside the cell compartment. Arranged along its length are genetic units (genes) that carry information required for bacterial maintenance and growth. When exposed to special stains or observed with an electron microscope, chromosomes have a granular or fibrous appearance **(figure 4.17).**

Although the chromosome is the minimal genetic requirement for bacterial survival, many bacteria contain other, nonessential pieces of DNA called **plasmids.** These tiny strands exist as separate double-stranded circles of DNA, although at times they can become integrated into the chromosome. During conjugation, they may be duplicated and passed on to related nearby bacteria. During bacterial reproduction they are duplicated and passed on to offspring. They are not essential to bacterial growth and metabolism, but they often confer protective traits such as resisting drugs and producing toxins and enzymes (see chapter 9). Because they can be readily manipulated in the laboratory and transferred from one bacterial cell to another, plasmids are an important agent in modern genetic engineering techniques.

Ribosomes: Sites of Protein Synthesis

A bacterial cell contains thousands of tiny **ribosomes** which are made of RNA and protein. When viewed even by very high magnification, ribosomes show up as fine, spherical specks dispersed throughout the cytoplasm that often occur in chains (polysomes). Many are also attached to the cell membrane. Chemically, a ribosome is a combination of a special type of RNA called ribosomal RNA, or rRNA (about 60%), and protein (40%). One method of characterizing ribosomes is by S, or Svedberg,[3] units, which rate the molecular sizes of various cell parts that have been spun down and separated

---

3. Named in honor of T. Svedberg, the Swedish chemist who developed the ultracentrifuge in 1926.

by molecular weight and shape in a centrifuge. Heavier, more compact structures sediment faster and are assigned a higher S rating. Combining this method of analysis with high-resolution electron micrography has revealed that the procaryotic ribosome, which has an overall rating of 70S, is actually composed of two smaller subunits **(figure 4.18).** They fit together to form a miniature platform upon which protein synthesis is performed. We examine the more detailed functions of ribosomes in chapter 9.

Inclusions, or Granules: Storage Bodies

Most bacteria are exposed to severe shifts in the availability of food. During periods of nutrient abundance, some can compensate by laying down nutrients intracellularly in **inclusion bodies,** or **inclusions,** of varying size, number, and content. As the environmental source of these nutrients becomes depleted, the bacterial cell can mobilize its own storehouse as required. Some inclusion bodies enclose condensed, energy-rich organic substances, such as glycogen and poly β-hydroxybutyrate (PHB), within special single-layered membranes **(figure 4.19).** A unique type of inclusion found in some aquatic bacteria are gas vesicles that provide buoyancy and flotation. Other inclusions, also called granules, contain crystals of inorganic compounds and are not enclosed by membranes. Sulfur granules of photosynthetic bacteria and polyphosphate granules of *Corynebacterium* and *Mycobacterium,* described later, are of this type. The latter represent an important source of building blocks for nucleic acid and ATP synthesis. They have been termed **metachromatic granules** because they stain a contrasting color (red, purple) in the presence of methylene blue dye.

Perhaps the most unique cell granule is not involved in cell nutrition but rather in cell orientation. Magnetotactic bacteria contain crystalline particles of iron oxide (magnetosomes) that have magnetic properties. Evidently the bacteria use these granules to be pulled by the polar and gravitational fields into deeper habitats with a lower oxygen content.

The Actin Cytoskeleton

Until very recently, scientists thought that the shape of all bacteria was completely determined by the peptidoglycan layer (cell wall). Although this is true of some bacteria, particularly the cocci, other bacteria produce long polymers of a protein called **actin,** arranged in helical ribbons around the cell just under the cell membrane **(figure 4.20).** These fibers appear to confer cell shape, perhaps by influencing the way peptidoglycan is manufactured. The fibers have been found in rod-shaped and spiral bacteria.

Bacterial Endospores: An Extremely Resistant Stage

Ample evidence indicates that the anatomy of bacteria helps them adjust rather well to adverse habitats. But of all microbial structures, nothing can compare to the bacterial **endo-**

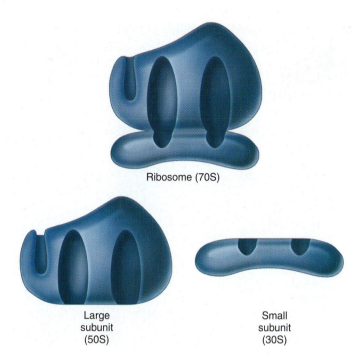

Ribosome (70S)

Large subunit (50S)

Small subunit (30S)

FIGURE 4.18 **A model of a procaryotic ribosome, showing the small (30S) and large (50S) subunits, both separate and joined.**

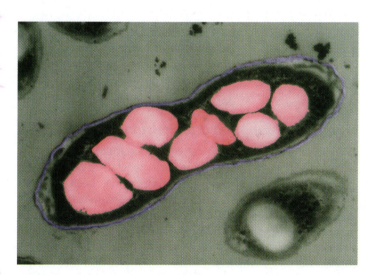

FIGURE 4.19 **An example of a storage inclusion in a bacterial cell (32,500×).**
Substances such as polyhydroxybutyrate can be stored in an insoluble, concentrated form that provides an ample, long-term supply of that nutrient.

spore (or simply spore) for withstanding hostile conditions and facilitating survival.

Endospores are dormant bodies produced by the bacteria *Bacillus, Clostridium,* and *Sporosarcina.* These bacteria have a two-phase life cycle—a vegetative cell and an endospore **(figure 4.21).** The vegetative cell is a metabolically active and growing entity that can be induced by environmental conditions to undergo spore formation, or **sporulation.** Once formed, the spore exists in an inert, resting condition that shows up

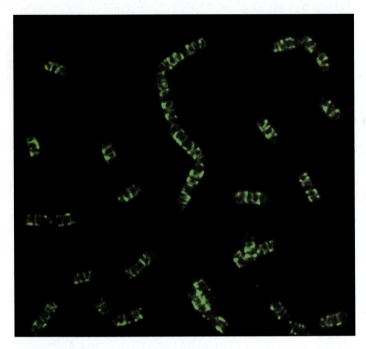

FIGURE 4.20 **Bacterial cytoskeleton.** The actin fibers are fluorescently stained.

prominently in a spore or Gram stain **(table 4.1)**. Features of spores, including size, shape, and position in the vegetative cell, are somewhat useful in identifying some species. Both gram-positive and gram-negative bacteria can form endospores, but the medically relevant ones are all gram-positive.

Endospore Formation and Resistance

The depletion of nutrients, especially an adequate carbon or nitrogen source, is the stimulus for a vegetative cell to begin endospore formation. Once this stimulus has been received by the vegetative cell, it undergoes a conversion to a committed sporulating cell called a **sporangium.** Complete transformation of a vegetative cell into a sporangium and then into an endospore requires 6 to 8 hours in most spore-forming species. Table 4.1 illustrates some major physical and chemical events in this process. Bacterial endospores are the hardiest of all life forms, capable of withstanding extremes in heat, drying, freezing, radiation, and chemicals that would readily kill vegetative cells. Their survival under such harsh conditions is due to several factors. The heat resistance of spores has been linked to their high content of calcium and *dipicolinic acid,* although the exact role of these chemicals is not yet clear. We know, for instance, that heat destroys cells by inactivating proteins and DNA and that this process requires a certain amount of water in the protoplasm. Because the deposition of calcium dipicolinate in the endospore removes water and leaves the endospore very dehydrated, it is less vulnerable to the effects of heat. It is also metabolically inactive and highly resistant to damage from further drying. The thick, impervious cortex and spore coats also protect against radiation and chemicals (table 4.1). The longevity of bacterial spores verges

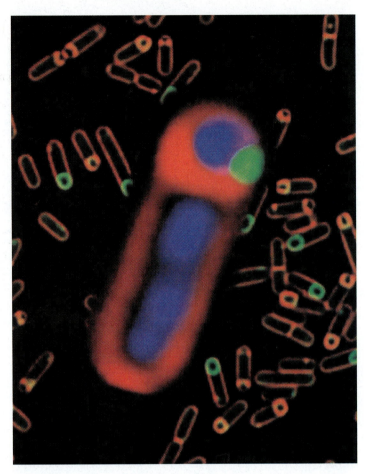

FIGURE 4.21 These biological "safety pins" are actually stages in endospore formation of *Bacillus subtilis,* stained with fluorescent proteins. The large red and blue cell is a vegetative cell in the early stages of sporulating. The developing spores are shown in green and orange.

on immortality. One record describes the isolation of viable endospores from a fossilized bee that was 25 million years old. More recently, microbiologists unearthed a viable endospore from a 250-million-year-old salt crystal. Initial analysis of this ancient microbe indicates it is a species of *Bacillus* that is genetically different from known species.

A NOTE ON TERMINOLOGY

The word *spore* can have more than one usage in microbiology. It is a generic term that refers to any tiny compact cells that are produced by vegetative or reproductive structures of microorganisms. Spores can be quite variable in origin, form, and function. The bacterial type discussed here is called an endospore, because it is produced inside a cell. It functions in *survival,* not in reproduction, because no increase in cell numbers is involved in its formation. In contrast, the fungi produce many different types of spores for both survival and reproduction (see chapter 5).

TABLE 4.1		General Stages in Endospore Formation	
Stage		**State of Cell**	**Process/Event**
1		Vegetative cell	Cell in early stage of binary fission doubles chromosome.
2		Vegetative cell becomes **sporangium** in preparation for sporulation	One chromosome and a small bit of cytoplasm are walled off as a protoplast at one end of the cell. This core contains the minimum structures and chemicals necessary for guiding life processes. During this time, the sporangium remains active in synthesizing compounds required for spore formation.
3		Sporangium	The protoplast is engulfed by the sporangium to continue the formation of various protective layers around it.
4		Sporangium with prospore	Special peptidoglycan is laid down to form a cortex around the spore protoplast, now called the prospore; calcium and dipicolinic acid are deposited; core becomes dehydrated and metabolically inactive.
5		Sporangium with prospore	Three heavy and impervious protein spore coats are added.
6		Mature endospore	Endospore becomes thicker, and heat resistance is complete; sporangium is no longer functional and begins to deteriorate.
7		Free spore	Complete lysis of sporangium frees spore; it can remain dormant yet viable for thousands of years.
8		Germination	Addition of nutrients and water reverses the dormancy. The spore then swells and liberates a young vegetative cell.
9		Vegetative cell	Restored vegetative cell.

Fluorescent stain of *Bacillus subtilis.*

TEM of cross-section of free endospore.

The Germination of Endospores

After lying in a state of inactivity for an indefinite time, endospores can be revitalized when favorable conditions arise. The breaking of dormancy, or germination, happens in the presence of water and a specific chemical or environmental stimulus (germination agent). Once initiated, it proceeds to completion quite rapidly ($1\frac{1}{2}$ hours). Although the specific germination agent varies among species, it is generally a small organic molecule such as an amino acid or an inorganic salt. This agent stimulates the formation of hydrolytic (digestive) enzymes by the endospore membranes. These enzymes digest the cortex and expose the core to water. As the core rehydrates and takes up nutrients, it begins to grow out of the endospore coats. In time, it reverts to a fully active vegetative cell, resuming the vegetative cycle.

Medical Significance of Bacterial Spores

Although the majority of spore-forming bacteria are relatively harmless, several bacterial pathogens are sporeformers. In fact, some aspects of the diseases they cause are related to the persistence and resistance of their spores. *Bacillus anthracis* is the agent of anthrax; its persistence in endospore form makes it an ideal candidate for bioterrorism. The genus *Clostridium* includes even more pathogens, including *C. tetani*, the cause of tetanus (lockjaw), and *C. perfringens*, the cause of gas gangrene. When the spores of these species are embedded in a wound that contains dead tissue, they can germinate, grow, and release potent toxins. Another toxin-forming species, *C. botulinum*, is the agent of botulism, a deadly form of food poisoning. (Each of these disease conditions is discussed in the infectious disease chapters, according to the organ systems they affect.)

Because they inhabit the soil and dust, endospores are a constant intruder where sterility and cleanliness are important. They resist ordinary cleaning methods that use boiling water, soaps, and disinfectants, and they frequently contaminate cultures and media. Hospitals and clinics must take precautions to guard against the potential harmful effects of endospores in wounds. Endospore destruction is a particular concern of the food-canning industry. Several endospore-forming species cause food spoilage or poisoning. Ordinary boiling (100°C) will usually not destroy such spores, so canning is carried out in pressurized steam at 120°C for 20 to 30 minutes. Such rigorous conditions will ensure that the food is sterile and free from viable bacteria.

✔ CHECKPOINT

- The cytoplasm of bacterial cells serves as a solvent for materials used in all cell functions.
- The genetic material of bacteria is DNA. Genes are arranged on large, circular chromosomes. Additional genes are carried on plasmids.

- Bacterial ribosomes are dispersed in the cytoplasm in chains (polysomes) and are also embedded in the cell membrane.
- Bacteria may store nutrients in their cytoplasm in structures called inclusions. Inclusions vary in structure and the materials that are stored.
- Some bacteria manufacture long actin filaments that help determine their cellular shape.
- A few families of bacteria produce dormant bodies called endospores, which are the hardiest of all life forms, surviving for hundreds or thousands of years.
- The genera *Bacillus* and *Clostridium* are sporeformers, and both contain deadly pathogens.

4.5 Bacterial Shapes, Arrangements, and Sizes

For the most part, bacteria function as independent single-celled, or unicellular, organisms. Although it is true that an individual bacterial cell can live attached to others in colonies or other such groupings, each one is fully capable of carrying out all necessary life activities, such as reproduction, metabolism, and nutrient processing (unlike the more specialized cells of a multicellular organism).

Bacteria exhibit considerable variety in shape, size, and colonial arrangement. It is convenient to describe most bacteria by one of three general shapes as dictated by the configuration of the cell wall (**figure 4.22**). If the cell is spherical or

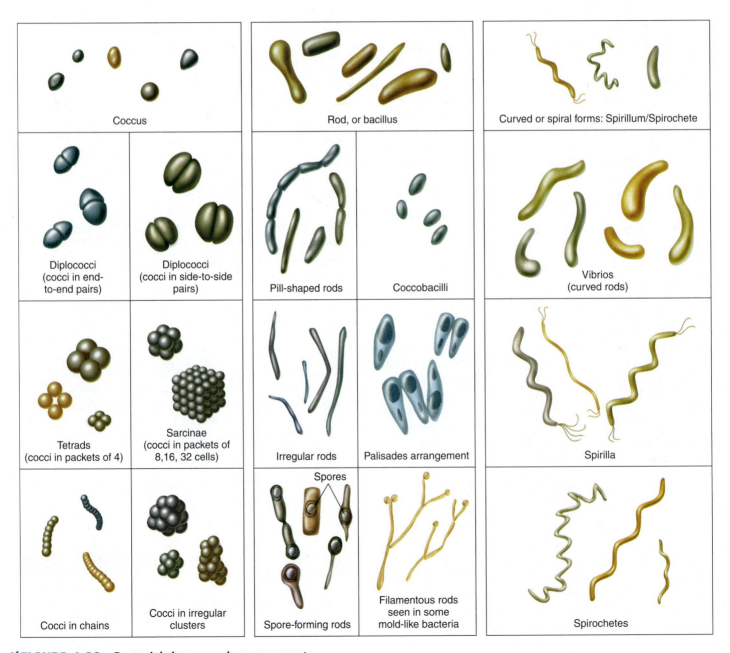

Coccus

Rod, or bacillus

Curved or spiral forms: Spirillum/Spirochete

Diplococci (cocci in end-to-end pairs)

Diplococci (cocci in side-to-side pairs)

Pill-shaped rods

Coccobacilli

Vibrios (curved rods)

Tetrads (cocci in packets of 4)

Sarcinae (cocci in packets of 8, 16, 32 cells)

Irregular rods

Palisades arrangement

Spirilla

Cocci in chains

Cocci in irregular clusters

Spores

Spore-forming rods

Filamentous rods seen in some mold-like bacteria

Spirochetes

FIGURE 4.22   Bacterial shapes and arrangements.
May not be shown to exact scale.

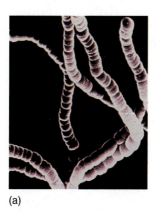
(a)

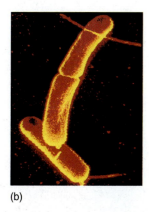
(b)

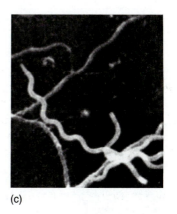
(c)

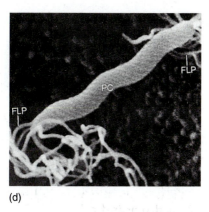
(d)

FIGURE 4.23   **SEM photographs of basic bacterial shapes reveal their three dimensions and surface features.**
(a) Cocci in chains. **(b)** A rod-shaped bacterium (*Escherichia coli*) in a diplobacillus arrangement. **(c)** A spirochete (*Borrelia burgdorferi*, the cause of Lyme disease) is a long, thin cell with irregular coils and no external flagella. **(d)** A spirillum is thicker with a few even coils (PC) and external flagella (FLP). Can you tell what the flagellar arrangement is?

ball-shaped, the bacterium is described as a **coccus** (kok'-us). Cocci can be perfect spheres, but they also can exist as oval, bean-shaped, or even pointed variants. A cell that is cylindrical (longer than wide) is termed a rod, or **bacillus** (bah-sil'-lus). There is also a genus named *Bacillus*. As might be expected, rods are also quite varied in their actual form. Depending on the bacterial species, they can be blocky, spindle-shaped, round-ended, long and threadlike (filamentous), or even clubbed or drumstick-shaped. When a rod is short and plump, it is called a **coccobacillus;** if it is gently curved, it is a **vibrio** (vib'-ree-oh). A bacterium having the shape of a curviform or spiral-shaped cylinder is called a **spirillum** (spy-ril'-em), a rigid helix, twisted twice or more along its axis (like a corkscrew). Another spiral cell mentioned earlier in conjunction with periplasmic flagella is the spirochete, a more flexible form that resembles a spring. Refer to **table 4.2** for a comparison of other features of the two helical bacterial forms. Because bacterial cells look two-dimensional and flat with traditional staining and microscope techniques, they are seen to best advantage with a scanning electron microscope that emphasizes their striking three-dimensional forms **(figure 4.23).**

It is common for cells of a single species to vary to some extent in shape and size. This phenomenon, called **pleomorphism (figure 4.24),** is due to individual variations in cell wall

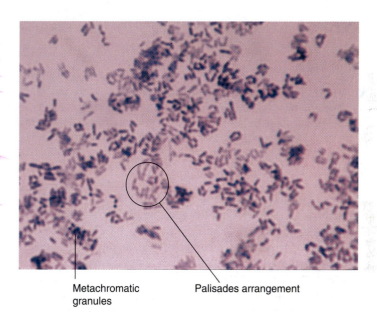

Metachromatic granules   Palisades arrangement

FIGURE 4.24   **Pleomorphism in *Corynebacterium*.**
Cells occur in a great variety of shapes and sizes (800×). This genus typically exhibits an unusual formation called a palisades arrangement, but some cells have other appearances. Close examination will also reveal darkly stained granules inside the cells.

TABLE 4.2	**Comparison of the Two Spiral-Shaped Bacteria**				
	Overall Appearance	**Mode of Locomotion**	**Number of Helical Turns**	**Gram Reaction (Cell Wall Type)**	**Examples of Important Types**
Spirilla	Rigid helix	Polar flagella; cells swim by rotating around like corkscrews; do not flex 1 to several flagella; can be in tufts	Varies from 1 to 20	Gram-negative	Most are harmless; one species, *Spirillum minor*, causes rat bite fever
Spirochetes	Flexible helix	Periplasmic flagella within sheath; cells flex; can swim by rotation or by creeping on surfaces 2 to 100 periplasmic flagella	Varies from 3 to 70	Gram-negative	*Treponema pallidum*, cause of syphilis; *Borrelia* and *Leptospira*, important pathogens

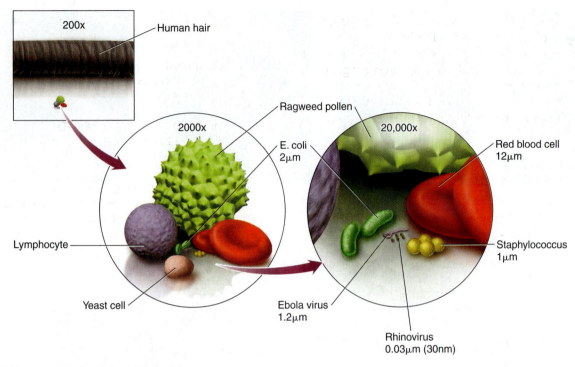

★FIGURE 4.25   The dimensions of bacteria.

The sizes of bacteria range from those just barely visible with light microscopy (0.2 μm) to those measuring a thousand times that size. Cocci measure anywhere from 0.5 to 3.0 μm in diameter; bacilli range from 0.2 to 2.0 μm in diameter and from 0.5 to 20 μm in length; vibrios and spirilla vary from 0.2 to 2.0 μm in diameter and from 0.5 to 100 μm in length. Spirochetes range from 0.1 to 3.0 μm in diameter and from 0.5 to 250 μm in length. Note the range of sizes as compared with eucaryotic cells and viruses. Comparisons are given as average sizes.

structure caused by nutritional or slight hereditary differences. For example, although the cells of *Corynebacterium diphtheriae* are generally considered rod-shaped, in culture they display variations such as club-shaped, swollen, curved, filamentous, and coccoid. Pleomorphism reaches an extreme in the mycoplasmas, which entirely lack cell walls and thus display extreme variations in shape (see figure 4.15).

The cells of bacteria can also be categorized according to arrangement, or style of grouping (see figure 4.22). The main factors influencing the arrangement of a particular cell type are its pattern of division and how the cells remain attached afterward. The greatest variety in arrangement occurs in cocci, which can be single, in pairs (diplococci), in **tetrads** (groups of four), in irregular clusters (both staphylococci and micrococci), or in chains of a few to hundreds of cells (streptococci). An even more complex grouping is a cubical packet of eight, sixteen, or more cells called a **sarcina** (sar'-sih-nah). These different coccal groupings are the result of the division of a coccus in a single plane, in two perpendicular planes, or in several intersecting planes; after division, the resultant daughter cells remain attached.

Bacilli are less varied in arrangement because they divide only in the transverse plane (perpendicular to the axis). They occur either as single cells, as a pair of cells with their ends attached (diplobacilli), or as a chain of several cells (streptobacilli). A **palisades** (pal'-ih-saydz) arrangement, typical of the corynebacteria, is formed when the cells of a chain remain partially attached by a small hinge region at the ends. The cells tend to fold (snap) back upon each other, forming a row of cells oriented side by side (see figure 4.23). The reaction can be compared to the behavior of boxcars on a jackknifed train, and the result looks superficially like an irregular picket fence. Spirilla are occasionally found in short chains, but spirochetes rarely remain attached after division. Comparative sizes of typical cells are presented in **figure 4.25.**

✓ CHECKPOINT

- Most bacteria have one of three general shapes: coccus (round), bacillus (rod), or spiral, based on the configuration of the cell wall. Two types of spiral cells are spirochetes and spirilla.
- Shape and arrangement of cells are key means of describing bacteria. Arrangements of cells are based on the number of planes in which a given species divides.
- Cocci can divide in many planes to form pairs, chains, packets, or clusters. Bacilli divide only in the transverse plane. If they remain attached, they form chains or palisades.

Classification Systems in the Procaryotae

Classification systems serve both practical and academic purposes. They aid in differentiating and identifying unknown species in medical and applied microbiology. They are also useful in organizing bacteria and as a means of studying their relationships and origins. Since classification was started around 200 years ago, several thousand species of bacteria and archaea have been identified, named, and catalogued.

For years scientists have had intense interest in tracing the origins of and evolutionary relationships among bacteria, but doing so has not been an easy task. One of the questions that has plagued taxonomists is, What characteristics are the most indicative of closeness in ancestry? Early bacteriologists found it convenient to classify bacteria according to shape, variations in arrangement, growth characteristics, and habitat. However, as more species were discovered and as techniques for studying their biochemistry were developed, it soon became clear that similarities in cell shape, arrangement, and staining reactions do not automatically indicate relatedness. Even though the gram-negative rods look alike, there are hundreds of different species, with highly significant differences in biochemistry and genetics. If we attempted to classify them on the basis of Gram stain and shape alone, we could not assign them to a more specific level than class. Increasingly, classification schemes are turning to genetic and molecular traits that cannot be visualized under a microscope or in culture.

One of the most viable indicators of evolutionary relatedness and affiliation is comparison of the sequence of nitrogen bases in ribosomal RNA, a major component of ribosomes. Ribosomes have the same function (protein synthesis) in all cells, and they tend to remain more or less stable in their nucleic acid content over long periods. Thus, any major differences in the sequence, or "signature," of the rRNA is likely to indicate some distance in ancestry. This technique is powerful at two levels: It is effective for differentiating general group differences (it was used to separate the three superkingdoms of life discussed in chapter 1), and it can be fine-tuned to identify at the species level (for example in *Mycobacterium* and *Legionella*). Elements of these and other identification methods are presented in more detail in chapter 17.

The definitive published source for bacterial classification, called *Bergey's Manual*, has been in print continuously since 1923. The basis for the early classification in *Bergey's* was the **phenotypic** traits of bacteria, such as their shape, cultural behavior, and biochemical reactions. These traits are still used extensively by clinical microbiologists or researchers who need to quickly identify unknown bacteria. As methods for RNA and DNA analysis became available, this information was used to supplement the phenotypic information. The current version of the publication, called *Bergey's Manual of Systematic Bacteriology*, presents a comprehensive view of bacterial relatedness, combining phenotypic information with rRNA sequencing information to classify bacteria; it is a huge five-volume set. (We need to remember that all bacterial classification systems are in a state of constant flux; no system is ever finished.)

With the explosion of information about evolutionary relatedness among bacteria, the need for a *Bergey's Manual* that contained easily accessible information for identifying unknown bacteria became apparent. Now there is a separate book, called *Bergey's Manual of Determinative Bacteriology*,

TABLE 4.3	Major Taxonomic Groups of Bacteria per *Bergey's Manual*
Division I. Gracilicutes: Gram-Negative Bacteria	
Class I.	Scotobacteria: Gram-negative non-photosynthetic bacteria
Class II.	Anoxyphotobacteria: Gram-negative photosynthetic bacteria that do not produce oxygen (purple and green bacteria)
Class III.	Oxyphotobacteria: Gram-negative photosynthetic bacteria that evolve oxygen (cyanobacteria)
Division II. Firmicutes: Gram-Positive Bacteria	
Class I.	Firmibacteria: Gram-positive rods or cocci (examples in table 4.4)
Class II.	Thallobacteria: Gram-positive branching cells (the actinomycetes)
Division III. Tenericutes	
Class I.	Mollicutes: Bacteria lacking a cell wall (the mycoplasmas)
Division IV. Mendosicutes	
Class I.	Archaebacteria: Procaryotes with atypical compounds in the cell wall and membranes

Source: Data from *Bergey's Manual of Determinative Bacteriology*, 9th ed. Williams & Wilkins Company, Baltimore, 1994.

based entirely on phenotypic characteristics. It is utilitarian in focus, categorizing bacteria by traits commonly assayed in clinical, teaching, and research labs **(table 4.3).** It is widely used by microbiologists who need to identify bacteria but need not know their evolutionary backgrounds. This phenotypic classification is more useful for students of medical microbiology, as well.

Taxonomic Scheme

Bergey's Manual of Determinative Bacteriology organizes the Kingdom Procaryotae into four major divisions. These somewhat natural divisions are based upon the nature of the cell wall. The **Gracilicutes** (gras"-ih-lik'-yoo-teez) have gramnegative cell walls and thus are thin-skinned; the **Firmicutes** have gram-positive cell walls that are thick and strong; the **Tenericutes** (ten"-er-ik'-yoo-teez) lack a cell wall and thus are soft; and the **Mendosicutes** (men-doh-sik'-yoo-teez) are the archaea (also called archaebacteria), primitive procaryotes with unusual cell walls and nutritional habits. The first two divisions contain the greatest number of species. The 200 or so species that cause human and animal diseases can be found in four classes: the Scotobacteria, Firmibacteria, Thallobacteria, and Mollicutes. The system used in *Bergey's Manual* further organizes bacteria into subcategories such as classes, orders, and families, but these are not available for all groups.

Diagnostic Scheme

As mentioned earlier, many medical microbiologists prefer an informal working system that outlines the major families and genera. **Table 4.4** is an example of an adaptation of the

TABLE 4.4   Medically Important Families and Genera of Bacteria, with Notes on Some Diseases*

I. Bacteria with gram-positive cell wall structure

Cocci in clusters or packets that are aerobic or facultative
Family Micrococcaceae: *Staphylococcus* (members cause boils, skin infections)

Cocci in pairs and chains that are facultative
Family Streptococcaceae: *Streptococcus* (species cause strep throat, dental caries)

Anaerobic cocci in pairs, tetrads, irregular clusters
Family Peptococcaceae: *Peptococcus, Peptostreptococcus* (involved in wound infections)

Spore-forming rods
Family Bacillaceae: *Bacillus* (anthrax), *Clostridium* (tetanus, gas gangrene, botulism)

Non-spore-forming rods
Family Lactobacillaceae: *Lactobacillus, Listeria* (milk-borne disease), *Erysipelothrix* (erysipeloid)
Family Propionibacteriaceae: *Propionibacterium* (involved in acne)

Family Corynebacteriaceae: *Corynebacterium* (diphtheria)

Family Mycobacteriaceae: *Mycobacterium* (tuberculosis, leprosy)

Family Nocardiaceae: *Nocardia* (lung abscesses)

Family Actinomycetaceae: *Actinomyces* (lumpy jaw), *Bifidobacterium*

Family Streptomycetaceae: *Streptomyces* (important source of antibiotics)

II. Bacteria with gram-negative cell wall structure

Aerobic cocci
Neisseria (gonorrhea, meningitis), *Branhamella*
Aerobic coccobacilli
Moraxella, Acinetobacter
Anaerobic cocci
Family Veillonellaceae
Veillonella (dental disease)
Miscellaneous rods
Brucella (undulant fever), *Bordetella* (whooping cough), *Francisella* (tularemia)
Aerobic rods
Family Pseudomonadaceae: *Pseudomonas* (pneumonia, burn infections)
Miscellaneous: *Legionella* (Legionnaires' disease)
Facultative or anaerobic rods and vibrios
Family Enterobacteriaceae: *Escherichia, Edwardsiella, Citrobacter, Salmonella* (typhoid fever), *Shigella*
(dysentery), *Klebsiella, Enterobacter, Serratia, Proteus, Yersinia* (one species causes plague)

Family Vibronaceae: *Vibrio* (cholera, food infection), *Campylobacter, Aeromonas*

Miscellaneous genera: *Chromobacterium, Flavobacterium, Haemophilus* (meningitis), *Pasteurella,*
Cardiobacterium, Streptobacillus
Anaerobic rods
Family Bacteroidaceae: *Bacteroides, Fusobacterium* (anaerobic wound and dental infections)
Helical and curviform bacteria
Family Spirochaetaceae: *Treponema* (syphilis), *Borrelia* (Lyme disease), *Leptospira* (kidney infection)
Obligate intracellular bacteria
Family Rickettsiaceae: *Rickettsia* (Rocky Mountain spotted fever), *Coxiella* (Q fever)
Family Bartonellaceae: *Bartonella* (trench fever, cat scratch disease)
Family Chlamydiaceae: *Chlamydia* (sexually transmitted infection)

III. Bacteria with no cell walls

Family Mycoplasmataceae: *Mycoplasma* (pneumonia), *Ureaplasma* (urinary infection)

*Details of pathogens and diseases in chapters 18 through 23.

phenotypic method of classification that might be used in clinical microbiology. This system is more applicable for diagnosis because it is restricted to bacterial disease agents, depends less on nomenclature, and is based on readily accessible morphological and physiological tests rather than on phylogenetic relationships. It also divides the bacteria into gram-positive, gram-negative, and those without cell walls and then subgroups them according to cell shape, arrangement, and certain physiological traits such as oxygen usage: *Aerobic* bacteria use oxygen in metabolism; *anaerobic* bacteria do not use oxygen in metabolism; and facultative bacteria may or may not use oxygen. Further tests not listed on the table would be required to separate closely related genera and species. Many of these are included in later chapters on specific bacterial groups.

Species and Subspecies in Bacteria

Among most organisms, the species level is a distinct, readily defined, and natural taxonomic category. In animals, for instance, a species is a distinct type of organism that can produce viable offspring only when it mates with others of its own kind. This definition does not work for bacteria primarily because they do not exhibit a typical mode of sexual reproduction. They can accept genetic information from unrelated forms, and they can also alter their genetic makeup by a variety of mechanisms. Thus, it is necessary to hedge a bit when we define a bacterial species. Theoretically, it is a collection of bacterial cells, all of which share an overall similar pattern of traits, in contrast to other groups whose pattern differs significantly. Although the boundaries that separate two closely related species in a genus are in some cases very arbitrary, this definition still serves as a method to separate the bacteria into various kinds that can be cultured and studied. As additional information on bacterial genomes is discovered, it may be possible to define species according to specific combinations of genetic codes found only in a particular isolated culture.

Individual members of given species can show variations, as well. Therefore more categories within species exist, but they are not well defined. Microbiologists use terms like *subspecies*, *strain*, or *type* to designate bacteria of the same species that have differing characteristics. *Serotype* refers to representatives of a species that stimulate a distinct pattern of antibody (serum) responses in their hosts, owing to distinct surface molecules.

> ### ✔ CHECKPOINT
>
> - Bacteria are formally classified by phylogenetic relationships and phenotypic characteristics.
> - Medical identification of pathogens uses an informal system of classification based on Gram stain, morphology, biochemical reactions, and metabolic requirements.
> - A bacterial species is loosely defined as a collection of bacterial cells that shares an overall similar pattern of traits different from other groups of bacteria.
> - Variant forms within a species (subspecies) include strains and types.

4.6 Survey of Procaryotic Groups with Unusual Characteristics

The bacterial world is so diverse that we cannot do complete justice to it in this introductory chapter. This variety extends into all areas of bacterial biology, including nutrition, mode of life, and behavior. Certain types of bacteria exhibit such unusual qualities that they deserve special mention. In this minisurvey, we will consider some medically important groups and some more remarkable representatives of bacteria living free in the environment that are ecologically important. Many of the bacteria mentioned here do not have the morphology typical of bacteria discussed previously, and in a few cases, they are vividly different **(Insight 4.3)**.

Unusual Forms of Medically Significant Bacteria

Most bacteria are free-living or parasitic forms that can metabolize and reproduce by independent means. Two groups of bacteria—the rickettsias and chlamydias—have adapted to life inside their host cells, where they are considered **obligate intracellular parasites.**

Rickettsias

Rickettsias[4] are distinctive, very tiny, gram-negative bacteria **(figure 4.26)**. Although they have a somewhat typical bacterial morphology, they are atypical in their life cycle and other adaptations. Most are pathogens that alternate between a mammalian host and blood-sucking arthropods,[5] such as fleas, lice, or ticks. Rickettsias cannot survive or multiply outside a host cell and cannot carry out metabolism completely on their own, so they are closely attached to their hosts. Several important human diseases are caused by rickettsias. Among these are Rocky Mountain spotted fever, caused by *Rickettsia rickettsii* (transmitted by ticks), and endemic typhus, caused by *Rickettsia typhi* (transmitted by lice).

Chlamydias

Bacteria of the genus *Chlamydia* are similar to the rickettsias in that they require host cells for growth and metabolism, but they are not closely related and are not transmitted by arthropods. Because of their tiny size and obligately parasitic lifestyle, they were at one time considered a type of virus. Species that carry the greatest medical impact are *Chlamydia trachomatis*, the cause of both a severe eye infection (trachoma) that can lead to blindness and one of the most common sexually transmitted diseases; and *Chlamydia pneumoniae*, an agent in lung infections.

Diseases caused by rickettsias and by *Chlamydia* species are described in more detail in the infectious disease chapters according to the organ systems they affect.

---

4. Named for Howard Ricketts, a physician who first worked with these organisms and later lost his life to typhus.

5. An arthropod is an invertebrate with jointed legs, such as an insect, tick, or spider.

Nucleus

FIGURE 4
rickettsia
Its mass gr
the nucleus

FIGURE 4.28 **Behavior of purple sulfur bacteria.**
Floating purple mats are huge masses of purple sulfur bacteria
blooming in the Baltic Sea. Photosynthetic bacteria can have
significant effects on the ecology of certain habitats.

FIGURE 4.29 **Myxobacterium.**
A photograph of an actual mature fruiting body of a myxobacterium.
Source: ASM News 63 (August 1997): 425. Photographer David Graham.

superkingdom (the Domain Archaea). We include them in
this chapter because they are procaryotic in general structure
and they do share many bacterial characteristics. But evidence is accumulating that they are actually more closely related to Domain Eukarya than to bacteria. For example,
archaea and eucaryotes share a number of ribosomal RNA sequences that are not found in bacteria, and their protein synthesis and ribosomal subunit structures are similar. **Table 4.5**
outlines selected points of comparison of the three domains.

that are deep enough for the anaerobic conditions they require yet where their pigment can still absorb wavelengths
of light **(figure 4.28).** These bacteria are named for their predominant colors, but they can also develop brown, pink,
purple, blue, and orange coloration. Both groups utilize sulfur compounds (H_2S, S) in their metabolism.

Gliding, Fruiting Bacteria

The gliding bacteria are a mixed collection of gram-
negative bacteria that live in water and soil. The
name is derived from the tendency of members to
glide over moist surfaces. The gliding property evidently involves rotation of filaments or fibers just
under the outer membrane of the cell wall. They do
not have flagella. Several morphological forms exist,
including slender rods, long filaments, cocci, and
some miniature, tree-shaped fruiting bodies. Probably the most intriguing and exceptional members of
this group are the slime bacteria, or myxobacteria
(figure 4.29). What sets the myxobacteria apart from
other bacteria are the complexity and advancement of
their life cycle. During this cycle, the vegetative cells
swarm together and differentiate into a many-celled,
colored structure called the fruiting body. The fruiting body is a survival structure that makes spores by
a method very similar to that of certain fungi. These
fruiting structures are often large enough to be seen
with the unaided eye on tree bark and plant debris.

Archaea: The Other Procaryotes

The discovery and characterization of novel procaryotic cells that have unusual anatomy, physiology, and genetics changed our views of microbial
taxonomy and classification (see chapter 1). These
single-celled, simple organisms, called **archaea,** are
now considered a third cell type in a separate

Free-Liv

Photosy

The nutrit
they deriv
thetic bact
special lig
sunlight t
inorganic
thetic bact
synthesis
as sulfur g

Cyanobe

The cyand
years and
further st
gram-nega
structure.
and they
mentous
packets su
specialized
membrane
chlorophy
ure 4.27a).
to float on
and cysts
able by pl
bacteria ir

✱

TABLE 4.5	Comparison of Three Cellular Domains		
Characteristic	**Bacteria**	**Archaea**	**Eukarya**
Cell type	Procaryotic	Procaryotic	Eucaryotic
Chromosomes	Single, or few, circular	Single, circular	Several, linear
Types of ribosomes	70S	70S but structure is similar to 80S	80S
Contains unique ribosomal RNA signature sequences	+	+	+
Number of sequences shared with Eukarya	1	3	(all)
Protein synthesis similar to Eukarya	−	+	
Presence of peptidoglycan in cell wall	+	−	−
Cell membrane lipids	Fatty acids with ester linkages	Long-chain, branched hydrocarbons with ether linkages	Fatty acids with ester linkages
Sterols in membrane	− (some exceptions)	−	+

(a)

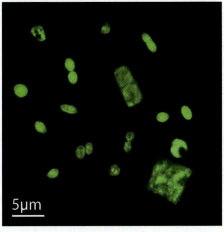

5μm

(b)

FIGURE 4.30 **Halophiles around the world.**
(a) A solar evaporation pond in Owens Lake, California, is extremely high in salt and mineral content. The archea that dominate in this hot, saline habitat produce brilliant red pigments with which they absorb light to drive cell synthesis. (b) A sample taken from a saltern in Australia viewed by fluorescent microscopy (1,000×). Note the range of cell shapes (cocci, rods, and square) found in this community.

Among the ways that the archaea differ significantly from other cell types are that certain genetic sequences are found only in their rRNA, and that they have unique membrane lipids and cell wall construction. It is clear that the archaea are the most primitive of all life forms and are most closely related to the first cells that originated on the earth 4 billion years ago. The early earth is thought to have contained a hot, anaerobic "soup" with sulfuric gases and salts in abundance. The modern archaea still live in the remaining habitats on the earth that have these same ancient conditions—the most extreme habitats in nature. It is for this reason that they are often called extremophiles, meaning that they "love" extreme conditions in the environment.

Metabolically, the archaea exhibit nearly incredible adaptations to what would be deadly conditions for other organisms. These hardy microbes have adapted to multiple combinations of heat, salt, acid, pH, pressure, and atmosphere. Included in this group are methane producers, hyperthermophiles, extreme halophiles, and sulfur reducers.

Members of the group called *methanogens* can convert CO_2 and H_2 into methane gas (CH_4) through unusual and complex pathways. These archaea are common inhabitants of anaerobic swamp mud, the bottom sediments of lakes and oceans, and even the digestive systems of animals. The gas they produce collects in swamps and may become a source of fuel. Methane may also contribute to the "greenhouse effect," which maintains the earth's temperature and can contribute to global warming (see chapter 24).

Other types of archaea—the extreme halophiles—require salt to grow and may have such a high salt tolerance that they can multiply in sodium chloride solutions (36% NaCl) that would destroy most cells. They exist in the saltiest places on the earth—inland seas, salt lakes, salt mines, and salted fish. They are not particularly common in the ocean because the salt content is not high enough. Many of the "halobacteria" use a red pigment to synthesize ATP in the presence of light. These pigments are responsible for "red herrings," the color of the Red Sea, and the red color of salt ponds **(figure 4.30)**.

Archaea adapted to growth at very low temperatures are called *psychrophilic* (loving cold temperatures); those growing at very high temperatures are *hyperthermophilic* (loving high temperatures). Hyperthermophiles flourish at temperatures between 80° and 105°C and cannot grow at 50°C. They live in volcanic waters and soils and submarine vents and are also often salt- and acid-tolerant as well. One member, *Thermoplasma*, lives in hot, acidic habitats in the waste piles around coal mines that regularly sustain a pH of 1 and a temperature of nearly 60°C. Researchers sampling sulfur vents in the deep ocean discovered thermophilic archaea flourishing at temperatures up to 250°C—150° above the temperature of boiling water! Not only were these archaea growing prolifically at this high temperature, but they were also living at 265 atmospheres of pressure. (On the earth's surface, pressure is about one atmosphere.) For additional discussion of the unusual adaptations of archaea, see chapter 7.

✔ CHECKPOINT

- The rickettsias are a group of bacteria that are intracellular parasites, dependent on their eucaryote host for energy and nutrients. Most are pathogens that alternate between arthropods and mammalian hosts.
- The chlamydias are also small, intracellular parasites that infect humans, mammals, and birds. They do not require arthropod vectors.
- Many bacteria are free-living, rather than parasitic. The photosynthetic bacteria and gliding bacteria encompass many subgroups that colonize specialized habitats, not other living organisms.
- Archaea are another type of procaryotic cell that constitute the third domain of life. They exhibit unusual biochemistry and genetics that make them different from bacteria. Many members are adapted to extreme habitats with low or high temperature, salt, pressure, or acid.

Chapter Summary With Key Terms

4.1 Procaryotic Form and Function
General Features of Procaryotes
A. Procaryotes consist of two major groups, the bacteria and the archaea. Life on earth would not be possible without them.
B. Procaryotic cells lack the membrane-surrounded organelles and nuclear compartment of eucaryotic cells but are still complex in their structure and function. All procaryotes have a cell membrane, cytoplasm, ribosomes, and a chromosome.

4.2 External Structures
Appendages: Cell Extensions
Some bacteria have projections that extend from the cell. **Flagella** (and internal **axial filaments** found in spirochetes) are used for motility. **Fimbriae** function in adhering to the environment; **pili** provide a means for genetic exchange. The **glycocalyx** may be a slime layer or a capsule.

4.3 The Cell Envelope: The Boundary Layer of Bacteria
A. Most procaryotes are surrounded by a protective envelope that consists of either two or three parts: the **cytoplasmic membrane** and the **cell wall (peptidoglycan)** are present in almost all bacteria; the **outer membrane** is an additional layer present only in gram-negative bacteria.
B. The Gram stain differentiates two types of cells on the basis of their cell envelopes; gram-positive bacteria have a cytoplasmic membrane and a thick cell wall, whereas gram-negative bacteria have a cytoplasmic membrane, a thin cell wall, and an additional outer membrane.

4.4 Bacterial Internal Structure
The cell cytoplasm is a watery substance that holds some or all of the following internal structures in bacteria: the **chromosome**(s) condensed in the **nucleoid; ribosomes** which serve as the sites of protein synthesis and are 70S in size; extra genetic information in the form of **plasmids;** storage structures known as **inclusions;** an **actin cytoskeleton** which helps give the bacterium its shape; and in some bacteria an **endospore** which is a highly resistant structure for survival. Bacterial endospores are not involved in reproduction.

4.5 Bacterial Shapes, Arrangements, and Sizes
A. Most bacteria are unicellular and are found in a great variety of shapes, arrangements and sizes. General shapes include **cocci, bacilli,** and helical forms such as **spirilla** and **spirochetes.** Some show great variation within the species in shape and size and are **pleomorphic.** Other variations include **coccobacilli, vibrios,** and filamentous forms.
B. Procaryotes divide by binary fission and do not utilize mitosis. Various arrangements result from cell division and are termed **diplococci,** streptococci, staphylococci, **tetrads,** and **sarcina** for cocci; bacilli may form pairs, chains, or **palisades.**
C. An important taxonomic system is standardized by *Bergey's Manual of Determinative Bacteriology,* which divides procaryotes into four major groups:
    1. **Gracilicutes:** Bacteria with gram-negative cell walls.
    2. **Firmicutes:** Bacteria with gram-positive cell walls.
    3. **Tenericutes:** Bacteria without cell walls.
    4. **Mendosicutes:** Archaebacteria (archae).
D. Bacterial species may be divided into strains and types.

4.6 Survey of Procaryotic Groups with Unusual Characteristics
Several groups of bacteria are so different that they have not always fit well in classification schemes.
A. Medically important bacteria: **Rickettsias** and chlamydias are within the gram-negative group but are small **obligate intracellular parasites** that replicate within cells of the hosts they invade.
B. Nonpathogenic bacterial groups: The majority of bacterial species are free-living and not involved in disease. Unusual groups include photosynthetic bacteria such as cyanobacteria, which provide oxygen to the environment, and the green and purple bacteria.
C. Archaea, the other major procaryote group: Archaea share many characteristics of procaryotes but do have some differences with bacteria in certain genetic aspects and some cell components. Many are adapted to extreme environments, as may have been found originally on earth. They are not considered medically important, but are of ecological and potential economic importance.

Multiple-Choice Questions

1. Which of the following is not found in all bacterial cells?
   a. cell membrane   c. ribosomes
   b. a nucleoid   d. actin cytoskeleton

2. The major locomotor structures in bacteria are
   a. flagella   c. fimbriae
   b. pili   d. cilia

3. Pili are tubular shafts in _____ bacteria that serve as a means of _____.
   a. gram-positive, genetic exchange
   b. gram-positive, attachment
   c. gram-negative, genetic exchange
   d. gram-negative, protection

4. An example of a glycocalyx is
   a. a capsule   c. outer membrane
   b. pili   d. a cell wall

5. Which of the following is a primary bacterial cell wall function?
   a. transport   c. support
   b. motility   d. adhesion

6. Which of the following is present in both gram-positive and gram-negative cell walls?
   a. an outer membrane   c. teichoic acid
   b. peptidoglycan   d. lipopolysaccharides

7. Metachromatic granules are concentrated crystals of _____ that are found in _____.
   a. fat, *Mycobacterium*
   b. dipicolinic acid, *Bacillus*
   c. sulfur, *Thiobacillus*
   d. PO₄, *Corynebacterium*

8. Bacterial endospores function in
   a. reproduction   c. protein synthesis
   b. survival   d. storage

9. A bacterial arrangement in packets of eight cells is described as a _____.
   a. micrococcus   c. tetrad
   b. diplococcus   d. sarcina

10. The major difference between a spirochete and a spirillum is
    a. presence of flagella   c. the nature of motility
    b. the presence of twists d. size

11. Which division of bacteria has a gram-positive cell wall?
    a. Gracilicutes   c. Firmicutes
    b. Archaea   d. Tenericutes

12. To which division of bacteria do cyanobacteria belong?
    a. Tenericutes   c. Firmicutes
    b. Gracilicutes   d. Mendosicutes

13. Which stain is used to distinguish differences between the cell walls of medically important bacteria?
    a. simple stain
    b. acridine orange stain
    c. Gram stain
    d. negative stain

Concept Questions

These questions are suggested as a *writing-to-learn* experience. For each question, compose a one- or two-paragraph answer that includes the factual information needed to completely address the question.

1. a. Name several general characteristics that could be used to define the procaryotes.
   b. Do any other microbial groups besides bacteria have procaryotic cells?
   c. What does it mean to say that bacteria are ubiquitous? In what habitats are they found? Give some general means by which bacteria derive nutrients.

2. a. Describe the structure of a flagellum and how it operates. What are the four main types of flagellar arrangement?
   b. How does the flagellum dictate the behavior of a motile bacterium? Differentiate between flagella and periplasmic flagella.
   c. List some direct and indirect ways that one can determine bacterial motility.

3. a. Explain the position of the glycocalyx.
   b. What are the functions of slime layers and capsules?
   c. How is the presence of a slime layer evident even at the level of a colony?

4. Differentiate between pili and fimbriae.

5. a. Compare the cell envelopes of gram-positive and gram-negative bacteria.
   b. What function does peptidoglycan serve?
   c. To which part of the cell envelope does it belong?
   d. Give a simple description of its structure.
   e. What happens to a cell that has its peptidoglycan disrupted or removed?
   f. What functions does the LPS layer serve?

6. a. What is the Gram stain?
   b. What is there in the structure of bacteria that causes some to stain purple and others to stain red?
   c. How does the precise structure of the cell walls differ in gram-positive and gram-negative bacteria?
   d. What other properties besides staining are different in gram-positive and gram-negative bacteria?

e. What is the periplasmic space, and how does it function?
   f. What characteristics does the outer membrane confer on gram-negative bacteria?

7. List five functions that the cell membrane performs in bacteria.

8. a. Compare the composition of the bacterial chromosome (nucleoid) and plasmids.
   b. What are the functions of each?

9. a. What is unique about the structure of bacterial ribosomes?
   b. How do they function?
   c. Where are they located?

10. a. Compare and contrast the structure and function of inclusions and granules.
    b. What are metachromatic granules, and what do they contain?

11. a. Describe the vegetative stage of a bacterial cell.
    b. Describe the structure of an endospore, and explain its function.
    c. Describe the endospore-forming cycle.
    d. Explain why an endospore is not considered a reproductive body.
    e. Why are endospores so difficult to destroy?

12. a. Draw the three bacterial shapes.
    b. How are spirochetes and spirilla different?
    c. What is a vibrio? A coccobacillus?
    d. What is pleomorphism?
    e. What is the difference between the use of the term bacillus and the name *Bacillus? Staphylococcus* and staphylococcus?

13. a. Rank the size ranges in bacteria according to shape.
    b. Rank the bacteria in relationship to viruses and eucaryotic cell size.
    c. Use the size bars to measure the cells in figure 4.30 and Insight 4.3.

14. a. What characteristics are used to classify bacteria?
    b. What are the most useful characteristics for categorizing bacteria into families?

15. a. How is the species level in bacteria defined?
    b. Name at least three ways bacteria are grouped below the species level.

16. a. Describe at least two circumstances that give rise to L forms.
    b. How do L forms survive?
    c. In what ways are they important?

17. Name several ways in which bacteria are medically and ecologically important.

18. a. Explain the characteristics of archaea that indicate that they constitute a unique domain of living things that is neither bacterial nor eucaryotic.
    b. What leads microbiologists to believe the archaea are more closely related to eucaryotes than to bacteria?
    c. What is meant by the term *extremophile?* Describe some archaeal adaptations to extreme habitats.

Critical Thinking Questions

Critical thinking is the ability to reason and solve problems using facts and concepts. These questions can be approached from a number of angles, and in most cases, they do not have a single correct answer.

1. What would happen if one stained a gram-positive cell only with safranin? A gram-negative cell only with crystal violet? What would happen to the two types if the mordant were omitted?

2. What is required to kill endospores? How do you suppose archaeologists were able to date some spores as being thousands (or millions) of years old?

3. Using clay, demonstrate how cocci can divide in several planes and show the outcome of this division. Show how the arrangements of bacilli occur, including palisades.

4. Using a corkscrew and a spring to compare the flexibility and locomotion of spirilla and spirochetes, explain which cell type is represented by each object.

5. Under the microscope, you see a rod-shaped cell that is swimming rapidly forward.
   a. What do you automatically know about that bacterium's structure?
   b. How would a bacterium use its flagellum for phototaxis?
   c. Can you think of another function of flagella besides locomotion?

6. a. Name a bacterium that has no cell walls.
   b. How is it protected from osmotic destruction?

7. a. Name a bacterium that is aerobic, gram positive, and spore-forming.
   b. What habitat would you expect this species to occupy?

8. a. Name an acid-fast bacterium.
   b. What characteristics make this bacterium different from other gram-positive bacteria?

9. a. Name two main groups of obligate intracellular parasitic bacteria.
   b. Why can't these groups live independently?

10. a. Name a bacterium that contains sulfur granules.
    b. What is the advantage in storing these granules?

11. a. Name a bacterium that uses chlorophyll to photosynthesize.
    b. Describe the two major groups of photosynthetic bacteria.
    c. How are they similar?
    d. How are they different?

12. a. What are some possible adaptations that the giant bacterium *Thiomargarita* has had to make because of its large size?
    b. If a regular bacterium were the size of an elephant, estimate the size of a nanobe at that scale.

13. Propose a hypothesis to explain how bacteria and archaea could have, together, given rise to eucaryotes.

14. Explain or illustrate exactly what will happen to the cell wall if the synthesis of the interbridge is blocked by penicillin. What if the glycan is hydrolyzed by lysozyme?

15. Ask your lab instructor to help you make a biofilm and examine it under the microscope. One possible technique is to suspend a glass slide in an aquarium for a few weeks, then carefully air-dry, fix, and Gram stain it. Observe the diversity of cell types.

Internet Search Topics

1. Go to a search engine and type in "Martian Microbes." Look for papers and information that support or reject the idea that fossil structures discovered in an ancient meteor from Mars could be bacteria. What are some of the reasons that microbiologists are skeptical of this possibility?

2. Search the Internet for information on nanobacteria. Give convincing reasons why these are or are not real organisms.

3. Go to the Online Learning Center for chapter 4 of this text at http://www.mhhe.com/cowan1. Access the URLs listed under Internet Search Topics and research the following:

   Go to the Cells Alive website as listed. Click on "Microbiology" and go to the "Dividing Bacteria" and "Bacterial Motility" options to observe short clips on these topics.

Eucaryotic Cells and Microorganisms

IN THE NEWS

During June of 2000, several children in Delaware, Ohio, were hospitalized at Grady Memorial General Hospital (GMH) after experiencing watery diarrhea, abdominal cramps, vomiting, and loss of appetite. Dr. McDermott, a new gastroenterologist at GMH, who also had a strong interest in infectious diseases, was asked to examine the children. Their illness lasted from 1 to 44 days and nearly half of them complained of intermittent bouts of diarrhea. By July 20th, over 150 individuals—mainly children and young adults between the ages of 20 and 40—experienced similar signs or symptoms. Dr. McDermott suspected that their illness was due to a microbial infection and queried the Delaware City County Health Department (DCCHD) to investigate this mysterious outbreak further.

Dr. McDermott helped the DCCHD team in surveying individuals hospitalized for intermittent diarrhea. They questioned individuals about recent travel, their sources of drinking water, visits to pools and lakes, swimming behaviors, contact with sick persons or young animals, and day-care attendance. The DCCHD's investigation reported that the outbreaks were linked to a swimming pool located at a private club in central Ohio. The swimming pool was closed on July 28th. A total of 700 clinical cases among residents of Delaware County and three neighboring counties were identified during the entire span of the outbreak that began late June and continued through September. At least five fecal accidents were observed during that time period at the pool. Only one of these accidents was of diarrheal origin. Outbreaks of gastrointestinal distress associated with recreational water activities have increased in recent years, with most being caused by the organism in this case.

- ▶ *Do you know what microorganism might be the cause of the outbreak?*
- ▶ *How can a single fecal accident contaminate an entire pool and cause so many clinical cases of gastrointestinal distress?*

CHAPTER OVERVIEW

▶ Eucaryotic cells are large complex cells divided into separate compartments by membrane-bound components called organelles.

▶ Major organelles—the nucleus, mitochondria, chloroplasts, endoplasmic reticulum, Golgi apparatus, and locomotor appendages—each serve an essential function to the cell, such as heredity, production of energy, synthesis, transport, and movement.

▶ Eucaryotic cells are found in fungi, protozoa, algae, plants, and animals, and they exhibit single-celled, colonial, and multicellular body plans.

▶ Fungi are eucaryotes that feed on organic substrates, have cell walls, reproduce asexually and sexually by spores, and exist in macroscopic or microscopic forms.

▶ Most fungi are free-living decomposers that are beneficial to biological communities; some may cause infections in animals and plants.

▶ Microscopic fungi include yeasts with spherical budding cells and molds with elongate filamentous hyphae in mycelia.

▶ Algae are aquatic photosynthetic protists with rigid cell walls and chloroplasts containing chlorophyll and other pigments.

▶ Algae belong to several groups based on their type of pigments, cell wall, stored food materials, and body plan.

▶ Protozoa are protists that feed by engulfing other cells, lack a cell wall, usually have some type of locomotor organelle, and may form dormant cysts.

▶ Subgroups of protozoa differ in their organelles of motility (flagella, cilia, pseudopods, nonmotile).

▶ Most protozoa are free-living aquatic cells that feed on bacteria and algae, and a few are animal parasites.

▶ The infective helminths are flatworms and roundworms that have greatly modified body organs so as to favor their parasitic lifestyle.

5.1 The History of Eucaryotes

Evidence from paleontology indicates that the first eucaryotic cells appeared on the earth approximately 2 billion years ago. Some fossilized cells that look remarkably like modern-day algae or protozoa appear in shale sediments from China, Russia, and Australia that date from 850 million to 950 million years ago **(figure 5.1).** Biologists have discovered convincing evidence to suggest that the eucaryotic cell evolved from procaryotic organisms by a process of intracellular **symbiosis** (sim-beye-oh'-sis) **(Insight 5.1).** It now seems clear that some of the **organelles** that distinguish eucaryotic cells originated from procaryotic cells that became trapped inside them. The structure of these first eucaryotic cells was so versatile that eucaryotic microorganisms soon spread out into available habitats and adopted greatly diverse styles of living.

The first primitive eucaryotes were probably single-celled and independent, but, over time, some forms began to aggregate, forming colonies. With further evolution, some of the cells within colonies became *specialized*, or adapted to perform a particular function advantageous to the whole

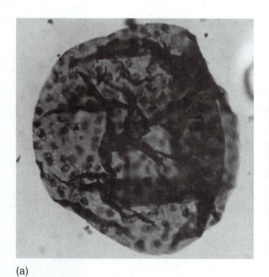
(a)

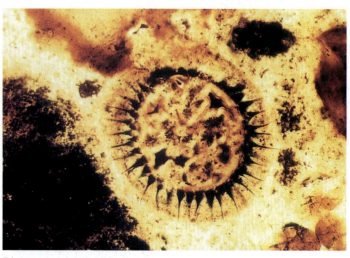
(b)

FIGURE 5.1 **Ancient eucaryotic protists caught up in fossilized rocks.**
(a) An alga-like cell found in Siberian shale deposits and dated from 850 million to 950 million years ago. **(b)** A large, disc-like cell bearing a crown of spines is from Chinese rock dated 590 million to 610 million years ago.

INSIGHT 5.1   *Historical*

The Extraordinary Emergence of Eucaryotic Cells

For years, biologists have grappled with the problem of how a cell as complex as the eucaryotic cell originated. One of the most fascinating explanations is that of **endosymbiosis,** which proposes that eucaryotic cells arose when a much larger procaryotic cell engulfed smaller bacterial cells that began to live and re-produce inside the procaryotic cell rather than being de-stroyed. As the smaller cells took up permanent residence, they came to perform specialized functions for the larger cell, such as food synthesis and oxygen utilization, that enhanced the cell's versatility and survival. Over time, when the cells evolv-ed into a single functioning entity, the relationship became obligatory. Although the theory of endosymbiosis has been greeted with some controversy, this phenomenon has been demonstrated in the laboratory with amoebas infected with bacteria that gradually became dependent upon the bacteria for survival.

The biologist most responsible for validation of the theory of endosymbiosis is Dr. Lynn Margulis. Using modern molec-ular techniques, she has accumulated convincing evidence of the relationships between the organelles of modern eucaryotic cells and the structure of bacteria. In many ways, the mito-chondrion of eucaryotic cells is something like a tiny cell within a cell. It is capable of independent division, contains a circular chromosome that has bacterial DNA sequences, and has ribosomes that are clearly procaryotic. Mitochondria also have bacterial membranes and can be inhibited by drugs that affect only bacteria. This link is so well established that recent phylogenetic trees of bacteria show mitochondria and chloro-plasts as two of the bacterial branches.

Chloroplasts likely arose when endosymbiotic cyanobac-teria provided their host cells with a built-in feeding mecha-nism. Evidence is seen in a modern flagellated protist that harbors specialized chloroplasts with cyanobacterial chloro-phyll and thylakoids. Margulis also has convincing evidence that eucaryotic cilia and flagella are the consequence of en-dosymbiosis between spiral bacteria and the cell membrane of early eucaryotic cells.

It is tempting to envision the development of the eu-caryotic cell through a fusion of archaeal and bacterial cells. The archaea would have served as a source of ribosomes and certain aspects of protein synthesis, and bacteria would have given rise to mitochondria and chloroplasts. The first eukarya probably resembled modern protists such as *Giardia* or *Plagiopyla*. Researchers are pursuing further genetic analy-sis of these microbes and their organelles to add further sup-port to the theory.

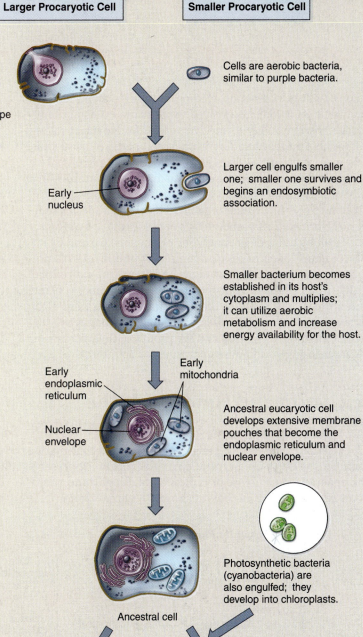

Larger Procaryotic Cell

Smaller Procaryotic Cell

Cell would have flexible membrane and internal extensions that could surround the nucleoid, forming a simple envelope that becomes the early nucleus.

Cells are aerobic bacteria, similar to purple bacteria.

Early nucleus

Larger cell engulfs smaller one; smaller one survives and begins an endosymbiotic association.

Smaller bacterium becomes established in its host's cytoplasm and multiplies; it can utilize aerobic metabolism and increase energy availability for the host.

Early endoplasmic reticulum

Early mitochondria

Nuclear envelope

Ancestral eucaryotic cell develops extensive membrane pouches that become the endoplasmic reticulum and nuclear envelope.

Photosynthetic bacteria (cyanobacteria) are also engulfed; they develop into chloroplasts.

Ancestral cell

Chloroplast

Protozoa, fungi, animals

Algae, higher plants

TABLE 5.1	Eucaryotic Organisms Studied in Microbiology	
Always Unicellular	**May Be Unicellular or Multicellular**	**Always Multicellular**
Protozoa	Fungi Algae	Helminths (have unicellular egg or larval forms)

colony, such as locomotion, feeding, or reproduction. Complex multicellular organisms evolved as individual cells in the organism lost the ability to survive apart from the intact colony. Although a multicellular organism is composed of many cells, it is more than just a disorganized assemblage of cells like a colony. Rather, it is composed of distinct groups of cells that cannot exist independently of the rest of the body. The cell groupings of multicellular organisms that have a specific function are termed *tissues*, and groups of tissues make up *organs*.

Looking at modern eucaryotic organisms, we find examples of many levels of cellular complexity **(table 5.1)**. All protozoa, as well as numerous algae and fungi, are unicellular. Truly multicellular organisms are found only among plants and animals and some of the fungi (mushrooms) and algae (seaweeds). Only certain eucaryotes are traditionally studied by microbiologists—primarily the protozoa, the microscopic algae and fungi, and animal parasites, or helminths.

5.2 Form and Function of the Eucaryotic Cell: External Structures

The cells of eucaryotic organisms are so varied that no one member can serve as a typical example. **Figure 5.2** presents the generalized structure of typical algal, fungal, and protozoan cells. The outline below shows the organization of a eucaryotic cell. Compare this outline to the one found on page 90 in chapter 4.

In general, eucaryotic microbial cells have a cytoplasmic membrane, nucleus, mitochondria, endoplasmic reticulum, Golgi apparatus, vacuoles, cytoskeleton, and glycocalyx. A cell wall, locomotor appendages, and chloroplasts, are found only in some groups. In the following sections, we cover the microscopic structure and functions of the eucaryotic cell. As with the procaryotes, we begin on the outside and proceed inward through the cell.

Locomotor Appendages: Cilia and Flagella

Motility allows a microorganism to locate life-sustaining nutrients and to migrate toward positive stimuli such as sunlight; it also permits avoidance of harmful substances and stimuli. Locomotion by means of flagella or cilia is common in protozoa, many algae, and a few fungal and animal cells.

Although they share the same name, eucaryotic flagella are much different from those of procaryotes. The eucaryotic flagellum is thicker (by a factor of 10), structurally more complex, and covered by an extension of the cell membrane. A single flagellum is a long, sheathed cylinder containing regularly spaced hollow tubules—microtubules—that extend along its entire length **(figure 5.3b)**. A cross section reveals nine pairs of closely attached microtubules surrounding a single central pair. This scheme, called the 9 + 2 arrangement, is a universal pattern of flagella and cilia **(figure 5.3a)**. During locomotion, the adjacent microtubules slide past each other, whipping the flagellum back and forth. Although details of this process are too complex to discuss here, it involves expenditure of energy and a coordinating mechanism in the cell membrane. Flagella can move the cell by pushing it forward like a fishtail or by pulling it by a lashing or twirling motion **(figure 5.3c)**. The placement and number of

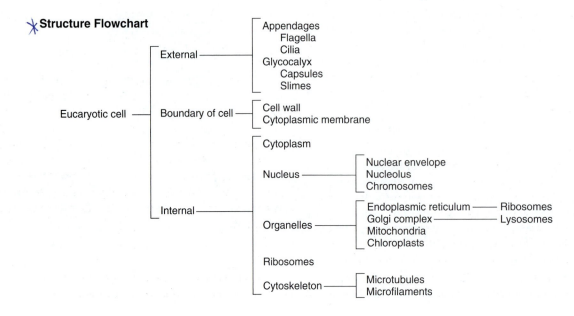

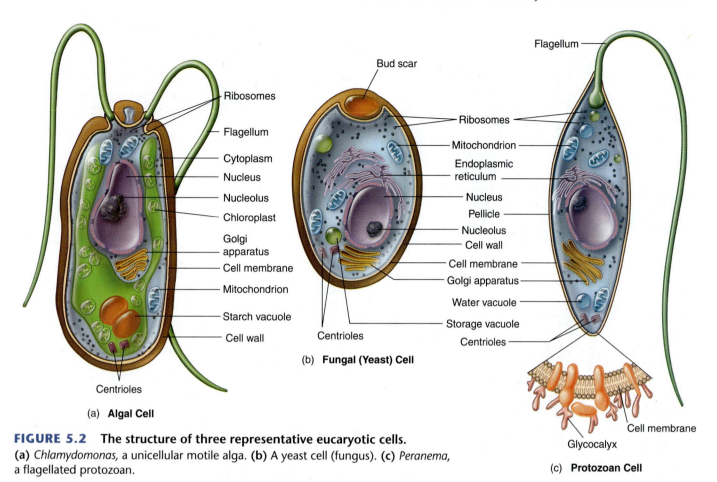

FIGURE 5.2 **The structure of three representative eucaryotic cells.**
(a) *Chlamydomonas,* a unicellular motile alga. **(b)** A yeast cell (fungus). **(c)** *Peranema,* a flagellated protozoan.

flagella can be useful in identifying flagellated protozoa and certain algae.

Cilia are very similar in overall architecture to flagella, but they are shorter and more numerous (some cells have several thousand). They are found only on a single group of protozoa and certain animal cells. In the ciliated protozoa, the cilia occur in rows over the cell surface, where they beat back and forth in regular oarlike strokes (see **figure 5.4**). Such protozoa are among the fastest of all motile cells. The fastest ciliated protozoon can swim up to 2,500 μm/s—a meter and a half per minute! On some cells, cilia also function as feeding and filtering structures.

The Glycocalyx

Most eucaryotic cells have a **glycocalyx,** an outermost boundary that comes into direct contact with the environment (see figure 5.2c). This structure is usually composed of polysaccharides and appears as a network of fibers, a slime layer, or a capsule much like the glycocalyx of procaryotes. Because of its positioning, the glycocalyx contributes to protection, adherence of cells to surfaces, and reception of signals from other cells and from the environment. The nature of the layer beneath the glycocalyx varies among the several eucaryotic groups. Fungi and most algae have a thick, rigid cell wall surrounding a cell membrane, whereas protozoa, a few algae, and all animal cells lack a cell wall and have only a cell membrane.

Form and Function of the Eucaryotic Cell: Boundary Structures

The Cell Wall

The cell walls of the fungi and algae are rigid and provide structural support and shape, but they are different in chemical composition from procaryotic cell walls. Fungal cell walls have a thick, inner layer of polysaccharide fibers composed of chitin or cellulose and a thin outer layer of mixed glycans **(figure 5.5).** The cell walls of algae are quite varied in chemical composition. Substances commonly found among various algal groups are cellulose, pectin,[1] mannans,[2] and minerals such as silicon dioxide and calcium carbonate.

---

1. A polysaccharide composed of galacturonic acid subunits.

2. A polymer of the sugar known as mannose.

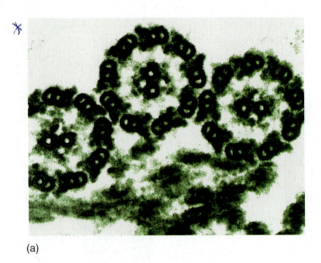

(a)

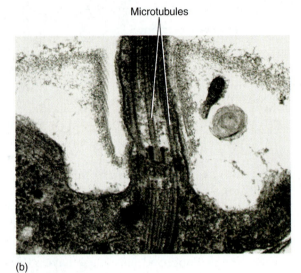

Microtubules

(b)

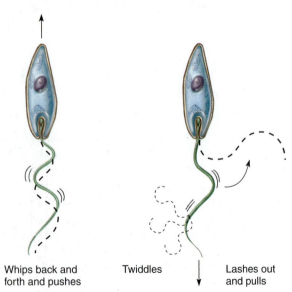

(c)   Whips back and   Twiddles   Lashes out
       forth and pushes   and pulls

FIGURE 5.3   The structures of microtubules.

(a) A cross section that reveals the typical 9 + 2 arrangement found in both flagella and cilia. **(b)** Longitudinal section through a flagellum, showing microtubules. **(c)** Locomotor patterns seen in flagellates.

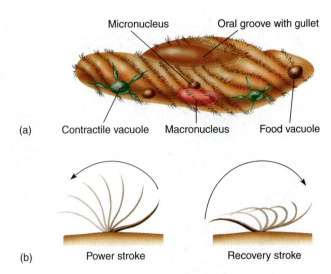

Micronucleus   Oral groove with gullet

(a)   Contractile vacuole   Macronucleus   Food vacuole

(b)   Power stroke   Recovery stroke

FIGURE 5.4   Structure and locomotion in ciliates.
(a) The structure of a typical representative, *Paramecium*.
(b) Cilia beat in coordinated waves, driving the cell forward and backward. View of a single cilium shows that it has a pattern of movement like a swimmer, with a power forward stroke and a repositioning stroke.

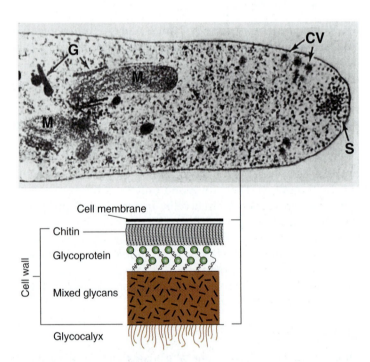

Cell membrane

Chitin

Glycoprotein

Cell wall

Mixed glycans

Glycocalyx

FIGURE 5.5   Boundary structure.
Cross section through the tip of a fungal cell to show the general structure of the cell wall in relationship to the cell membrane and glycocalyx. Top: Photomicrograph (S, growing tip; CV, coated vesicles; G, Golgi apparatus; M, mitochondrion). Bottom: The cell wall is a thick, rigid structure composed of complex layers of polysaccharides and proteins.

The Cytoplasmic Membrane

The cytoplasmic (cell) membrane of eucaryotic cells is a typical bilayer of phospholipids in which protein molecules are embedded. In addition to phospholipids, eucaryotic membranes also contain *sterols* of various kinds. Sterols are different from phospholipids in both structure and behavior, as you may recall from chapter 2. Their relative rigidity confers stability on eucaryotic membranes. This strengthening feature is extremely important in cells that lack a cell wall. Cytoplasmic membranes of eucaryotes are functionally similar to those of procaryotes, serving as selectively permeable barriers in transport. Unlike procaryotes, eucaryotic cells also contain a number of individual membrane-bound organelles that are extensive enough to account for 60% to 80% of their volume.

✔ CHECKPOINT

- Eucaryotes are cells with a nucleus and organelles compartmentalized by membranes. They might have originated from procaryote ancestors about 2 billion years ago. Eucaryotic cell structure enabled eucaryotes to diversify from single cells into a huge variety of complex multicellular forms.

- The cell structures common to most eucaryotes are the cell membrane, nucleus, vacuoles, mitochondria, endoplasmic reticulum, Golgi apparatus, and a cytoskeleton. Cell walls, chloroplasts, and locomotor organs are present in some eucaryote groups.

- Microscopic eucaryotes use locomotor organs such as flagella or cilia for moving themselves or their food.

- The glycocalyx is the outermost boundary of most eucaryotic cells. Its functions are protection, adherence, and reception of chemical signals from the environment or from other organisms. The glycocalyx is supported by either a cell wall or a cell membrane.

- The cytoplasmic (cell) membrane of eucaryotes is similar in function to that of procaryotes, but it differs in composition, possessing sterols as additional stabilizing agents.

5.3 Form and Function of the Eucaryotic Cell: Internal Structures

The Nucleus: The Control Center

The nucleus is a compact sphere that is the most prominent organelle of eucaryotic cells. It is separated from the cell cytoplasm by an external boundary called a nuclear envelope. The envelope has a unique architecture. It is composed of two parallel membranes separated by a narrow space, and it is perforated with small, regularly spaced openings, or pores, formed at sites where the two membranes unite (**figure 5.6**). The nuclear pores are passageways through which macromolecules migrate from the nucleus to the cytoplasm and vice versa. The nucleus contains a matrix called the nucleoplasm and a granular mass, the **nucleolus,** that can stain

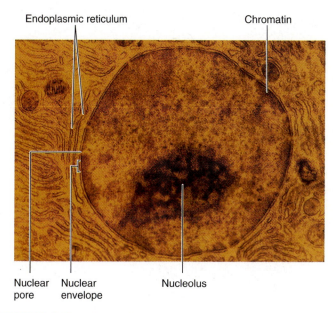

FIGURE 5.6   The nucleus.
Electron micrograph section of an interphase nucleus, showing its most prominent features.

more intensely than the immediate surroundings because of its RNA content. The nucleolus is the site for ribosomal RNA synthesis and a collection area for ribosomal subunits. The subunits are transported through the nuclear pores into the cytoplasm for final assembly into ribosomes.

A prominent feature of the nucleoplasm in stained preparations is a network of dark fibers known as **chromatin** because of its attraction for dyes. Analysis has shown that chromatin actually comprises the eucaryotic **chromosomes,** large units of genetic information in the cell. The chromosomes in the nucleus of most cells are not readily visible because they are long, linear DNA molecules bound in varying degrees to **histone** proteins, and they are far too fine to be resolved as distinct structures without extremely high magnification. During **mitosis,** however, when the duplicated chromosomes are separated equally into daughter cells, the chromosomes themselves become readily visible as discrete bodies (**figure 5.7**). This appearance arises when the DNA becomes highly condensed by forming coils and supercoils around the histones to prevent the chromosomes from tangling as they are separated into new cells. This process is described in more detail in chapter 9.

The nucleus, as you've just seen, contains instructions in the form of DNA. Elaborate processes have evolved for transcription and duplication of this genetic material. Much of the protein synthesis and other work of the cell takes place outside the nucleus in the cell's other organelles.

Endoplasmic Reticulum: A Passageway in the Cell

The **endoplasmic reticulum (ER)** is a microscopic series of tunnels used in transport and storage. Two kinds of endoplasmic reticulum are the **rough endoplasmic reticulum (RER)**

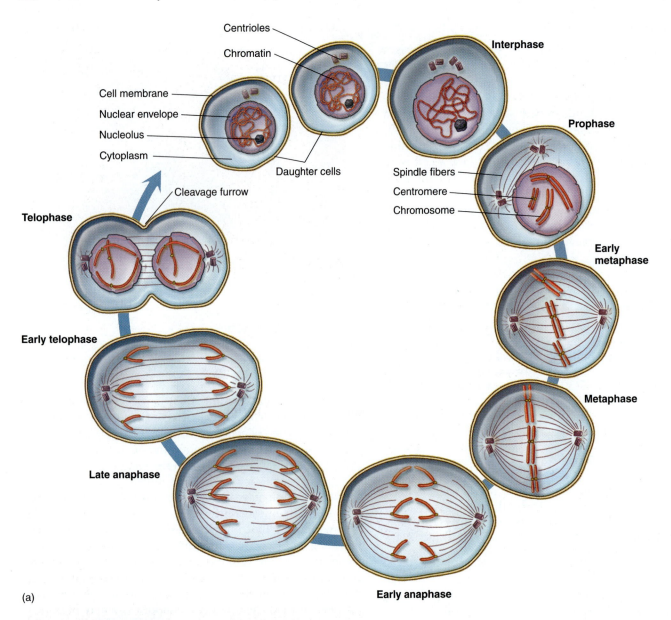

(a)

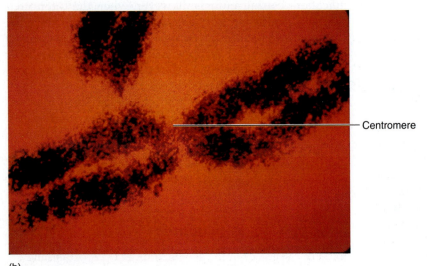

(b)

FIGURE 5.7 **Changes in the cell and nucleus that accompany mitosis in a eucaryotic cell such as a yeast.**

(a) Before mitosis (at interphase), chromosomes are visible only as chromatin. As mitosis proceeds (early prophase), chromosomes take on a fine, threadlike appearance as they condense, and the nuclear membrane and nucleolus are temporarily disrupted. **(b)** By metaphase, the chromosomes are fully visible as X-shaped structures. The shape is due to duplicated chromosomes attached at a central point, the centromere. Spindle fibers attach to these and facilitate the separation of individual chromosomes during metaphase. Later phases serve in the completion of chromosomal separation and division of the cell proper into daughter cells.

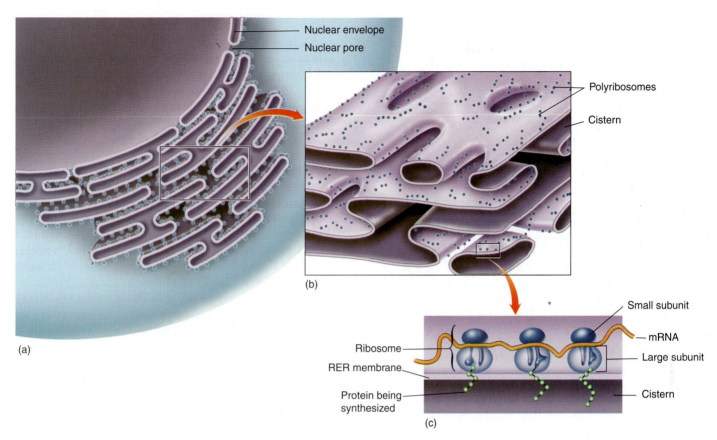

Nuclear envelope
Nuclear pore

Polyribosomes

Cistern

(a)

(b)

Small subunit

mRNA

Ribosome

Large subunit

RER membrane

Protein being
synthesized

Cistern

(c)

FIGURE 5.8 **The origin and detailed structure of the rough endoplasmic reticulum (RER).**
(a) Schematic view of the origin of the RER from the outer membrane of the nuclear envelope. **(b)** Three-dimensional projection of the RER. **(c)** Detail of the orientation of a ribosome on the RER membrane.

(figure 5.8) and the **smooth endoplasmic reticulum (SER).** Electron micrographs show that the RER originates from the outer membrane of the nuclear envelope and extends in a continuous network through the cytoplasm, even out to the cell membrane. This architecture permits the spaces in the RER, or cisternae, to transport materials from the nucleus to the cytoplasm and ultimately to the cell's exterior. The RER appears rough because of large numbers of ribosomes partly attached to its membrane surface. Proteins synthesized on the ribosomes are shunted into the cavity of the reticulum and held there for later packaging and transport. In contrast to the RER, the SER is a closed tubular network without ribosomes that functions in nutrient processing and in synthesis and storage of nonprotein macromolecules such as lipids.

Golgi Apparatus: A Packaging Machine

The **Golgi**[3] (gol′-jee) **apparatus,** also called the Golgi **complex** or body, is the site in the cell in which proteins are modified and then sent to their final destinations. It is a discrete organelle consisting of a stack of several flattened, disc-shaped sacs called cisternae. These sacs have outer limiting membranes and cavities like those of the endoplasmic reticulum, but they do not form a continuous network **(figure 5.9).** This organelle is always closely associated with the endoplasmic reticulum both in its location and function. At a site where it meets the Golgi apparatus, the endoplasmic reticulum buds off tiny membrane-bound packets of protein called *transitional vesicles* that are picked up by the forming face of the Golgi apparatus. Once in the complex itself, the proteins are often modified by the addition of polysaccharides and lipids. The final action of this apparatus is to pinch off finished *condensing vesicles* that will be conveyed to organelles such as lysosomes or transported outside the cell as secretory vesicles **(figure 5.10).**

Nucleus, Endoplasmic Reticulum, and Golgi Apparatus: Nature's Assembly Line

As the keeper of the eucaryotic genetic code, the nucleus ultimately governs and regulates all cell activities. But, because the nucleus remains fixed in a specific cellular site, it must direct these activities through a structural and chemical network (figure 5.10). This network includes ribosomes, which originate in the nucleus, and the rough endoplasmic reticulum, which is continuously connected with the nuclear envelope. Initially, a segment of the genetic code of DNA containing the instructions for producing a protein is copied into RNA and passed out through the nuclear pores directly to the ribosomes

3. Named for C. Golgi, an Italian histologist who first described the apparatus in 1898.

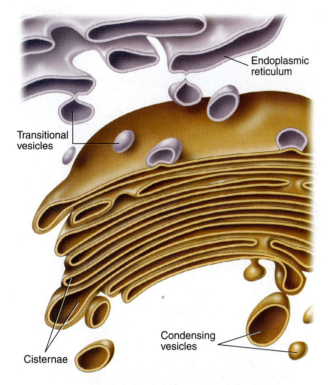

FIGURE 5.9   **Detail of the Golgi apparatus.**
The flattened layers are cisternae. Vesicles enter the upper surface
and leave the lower surface.

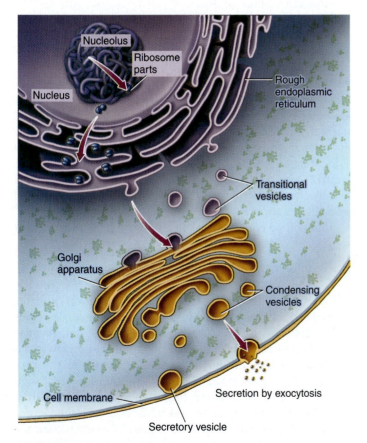

FIGURE 5.10   **The transport process.**
The cooperation of organelles in protein synthesis and transport:
nucleus → RER → Golgi apparatus → vesicles → secretion.

on the endoplasmic reticulum. Here, specific proteins are synthesized from the RNA code and deposited in the lumen (space) of the endoplasmic reticulum. After being transported to the Golgi apparatus, the protein products are chemically modified and packaged into vesicles that can be used by the cell in a variety of ways. Some of the vesicles contain enzymes to digest food inside the cell; other vesicles are secreted to digest materials outside the cell, and yet others are important in the enlargement and repair of the cell wall and membrane.

A **lysosome** is one type of vesicle originating from the Golgi apparatus that contains a variety of enzymes. Lysosomes are involved in intracellular digestion of food particles and in protection against invading microorganisms. They also participate in digestion and removal of cell debris in damaged tissue. Other types of vesicles include **vacuoles** (vak'-yoo-ohl), which are membrane-bound sacs containing fluids or solid particles to be digested, excreted, or stored. They are formed in phagocytic cells (certain white blood cells and protozoa) in response to food and other substances that have been engulfed. The contents of a food vacuole are digested through the merger of the vacuole with a lysosome. This merged structure is called a phagosome **(figure 5.11)**. Other types of vacuoles are used in storing reserve food such as fats and glycogen. Protozoa living in freshwater habitats regulate osmotic pressure by means of contractile vacuoles, which regularly expel excess water that has diffused into the cell (described later).

Mitochondria: Energy Generators of the Cell

Although the nucleus is the cell's control center, none of the cellular activities it commands could proceed without a constant supply of energy, the bulk of which is generated in most eucaryotes by **mitochondria** (my"-toh-kon'-dree-uh). When viewed with light microscopy, mitochondria appear as round or elongated particles scattered throughout the cytoplasm. The internal ultrastructure reveals that a single mitochondrion consists of a smooth, continuous outer membrane that forms the external contour, and an inner, folded membrane nestled neatly within the outer membrane **(figure 5.12a)**. The folds on the inner membrane, called **cristae** (kris'-te), may be tubular, like fingers, or folded into shelflike bands.

The cristae membranes hold the enzymes and electron carriers of aerobic respiration. This is an oxygen-using process that extracts chemical energy contained in nutrient molecules and stores it in the form of high-energy molecules, or ATP. More detailed functions of mitochondria are covered in chapter 8. The spaces around the cristae are filled with a chemically complex fluid called the **matrix,** which holds ribosomes, DNA, and the pool of enzymes and other compounds involved in the metabolic cycle. Mitochondria (along with chloroplasts) are unique among organelles in that they divide independently of the cell, contain circular strands of DNA, and have procaryotic-sized 70S ribosomes. These findings have prompted some intriguing speculations on their evolutionary origins (see Insight 5.1).

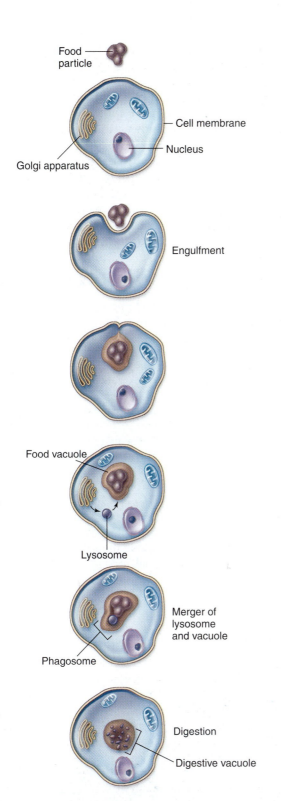

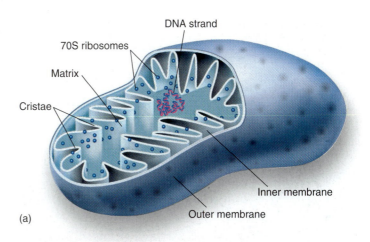

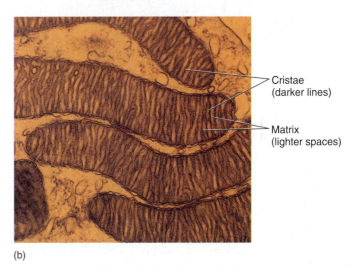

(a)

(b)

FIGURE 5.12   General structure of a mitochondrion.
(a) A three-dimensional projection. **(b)** An electron micrograph.
In most cells, mitochondria are elliptical or spherical, although in
certain fungi, algae, and protozoa, they are long and filament-like.

**FIGURE 5.11   The origin and action of lysosomes in
phagocytosis.**

Chloroplasts: Photosynthesis Machines

Chloroplasts are remarkable organelles found in algae and
plant cells that are capable of converting the energy of sunlight
into chemical energy through photosynthesis. The photosyn-

thetic role of chloroplasts makes them the primary producers
of organic nutrients upon which all other organisms (except
certain bacteria) ultimately depend. Another important photosynthetic product of chloroplasts is oxygen gas. Although
chloroplasts resemble mitochondria, chloroplasts are larger,
contain special pigments, and are much more varied in shape.

There are differences among various algal chloroplasts,
but most are generally composed of two membranes, one enclosing the other. The smooth, outer membrane completely
covers an inner membrane folded into small, disclike sacs
called **thylakoids** that are stacked upon one another into
grana. These structures carry the green pigment chlorophyll
and sometimes additional pigments as well. Surrounding the
thylakoids is a ground substance called the **stroma (figure 5.13).** The role of the photosynthetic pigments is to absorb
and transform solar energy into chemical energy, which is
then used during reactions in the stroma to synthesize carbohydrates. We further explore some important aspects of photosynthesis in chapters 7 and 24.

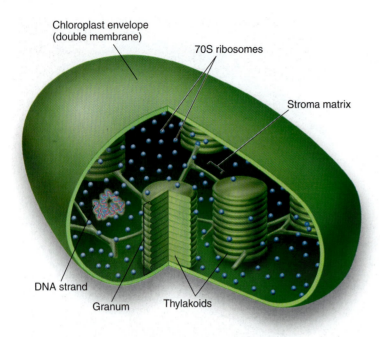

Chloroplast envelope
(double membrane)

70S ribosomes

Stroma matrix

DNA strand

Granum

Thylakoids

FIGURE 5.13   Detail of an algal chloroplast.

Ribosomes: Protein Synthesizers

In an electron micrograph of a eucaryotic cell, ribosomes are numerous, tiny particles that give a "dotted" appearance to the cytoplasm. Ribosomes are distributed in two ways: Some are scattered freely in the cytoplasm and cytoskeleton; others are intimately associated with the rough endoplasmic reticulum as previously described. Multiple ribosomes are often found arranged in short chains called polyribosomes (polysomes). The basic structure of eucaryotic ribosomes is similar to that of procaryotic ribosomes, described in chapter 4. Both are composed of large and small subunits of ribonucleoprotein (see figure 5.8). By contrast, however, the eucaryotic ribosome (except in the mitochondrion) is the larger 80S variety that is a combination of 60S and 40S subunits. As in the procaryotes, eucaryotic ribosomes are the staging areas for protein synthesis.

The Cytoskeleton: A Support Network

The cytoplasm of a eucaryotic cell is criss-crossed by a flexible framework of molecules called the cytoskeleton (**figure 5.14**). This framework appears to have several functions, such as anchoring organelles, providing support, and permitting shape changes and movement in some cells. The two main types of cytoskeletal elements are *microfilaments* and *microtubules*. **Microfilaments** are thin protein strands that attach to the cell membrane and form a network through the cytoplasm. Some microfilaments are responsible for movements

of the cytoplasm, often made evident by the streaming of organelles around the cell in a cyclic pattern. Other microfilaments are active in *amoeboid motion,* a type of movement typical of cells such as amoebas and phagocytes that produces extensions of the cell membrane (pseudopods) into which the cytoplasm flows. **Microtubules** are long, hollow tubes that maintain the shape of eucaryotic cells without walls and transport substances from one part of a cell to another. The spindle fibers that play an essential role in mitosis are actually microtubules that attach to chromosomes and separate them into daughter cells. As indicated earlier, microtubules are also responsible for the movement of cilia and flagella.

✔ CHECKPOINT

- The genome of eucaryotes is located in the nucleus, a spherical structure surrounded by a double membrane. The nucleus contains the nucleolus, the site of ribosome synthesis. DNA is organized into chromosomes in the nucleus.
- The endoplasmic reticulum (ER) is an internal network of membranous passageways extending throughout the cell.
- The Golgi apparatus is a packaging center that receives materials from the ER and then forms vesicles around them for storage or for transport to the cell membrane for secretion.
- The mitochondria generate energy in the form of ATP to be used in numerous cellular activities.
- Chloroplasts, membranous packets found in plants and algae, are used in photosynthesis.
- Ribosomes are the sites for protein synthesis present in both eucaryotes and procaryotes.
- The cytoskeleton maintains the shape of cells and produces movement of cytoplasm within the cell, movement of chromosomes at cell division, and, in some groups, movement of the cell as a unit.

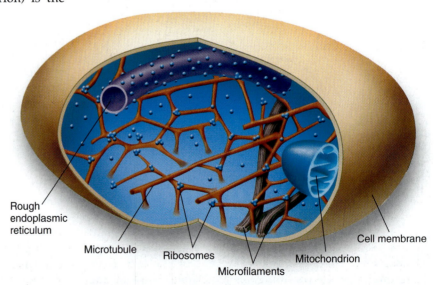

Rough
endoplasmic
reticulum

Microtubule   Ribosomes

Microfilaments

Mitochondrion

Cell membrane

FIGURE 5.14   A model of the cytoskeleton.
Depicted is the relationship between microtubules, microfilaments, and organelles.

Survey of Eucaryotic Microorganisms

With the general structure of the eucaryotic cell in mind, let us next examine the amazingly wide range of adaptations that this cell type has undergone. The following sections contain a general survey of the principal eucaryotic microorganisms—fungi, algae, protozoa, and parasitic worms—while also introducing elements of their structure, life history, classification, identification, and importance.

5.4 The Kingdom of the Fungi

The position of the **fungi** in the biological world has been debated for many years. Although they were originally classified with the green plants (along with algae and bacteria), they were later separated from plants and placed in a group with algae and protozoa (the Protista). Even at that time, however, many microbiologists were struck by several unique qualities of fungi that warranted their being placed into their own separate kingdom, and eventually they were.

The Kingdom Fungi, or Myceteae, is large and filled with forms of great variety and complexity. For practical purposes, the approximately 100,000 species of fungi can be divided into two groups: the *macroscopic fungi* (mushrooms, puffballs, gill fungi) and the *microscopic fungi* (molds, yeasts). Although the majority of fungi are either unicellular or colonial, a few complex forms such as mushrooms and puffballs are considered multicellular. Cells of the microscopic fungi exist in two basic morphological types: yeasts and hyphae. A yeast cell is distinguished by its round to oval shape and by its mode of asexual reproduction. It grows swellings on its surface called buds which then become separate cells. **Hyphae** (hy'-fee) are long, threadlike cells found in the bodies of filamentous fungi, or molds (**figure 5.15**). Some species form a **pseudohypha,** a chain of yeasts formed when buds remain attached in a row (**figure 5.16**). Because of its manner of formation, it is not a true hypha like that of molds. While some fungal cells exist only in a yeast form and others occur primarily as hyphae, a few, called **dimorphic,** can take either form, depending upon growth conditions, such as changing temperature. This variability in growth form is particularly characteristic of some pathogenic molds.

Fungal Nutrition

All fungi are **heterotrophic.** They acquire nutrients from a wide variety of organic materials called **substrates (figure 5.17).** Most fungi are **saprobes,** meaning that they obtain these substrates from the remnants of dead plants and animals in soil or aquatic habitats. Fungi can also be parasites on the bodies of living animals or plants, although very few fungi absolutely require a living host. In general, the fungus penetrates the substrate and secretes enzymes that reduce it to small molecules that can be absorbed by the cells. Fungi have enzymes for digesting an incredible array of substances, including feathers, hair, cellulose, petroleum products, wood, and rubber. It has been said that every naturally

(a)

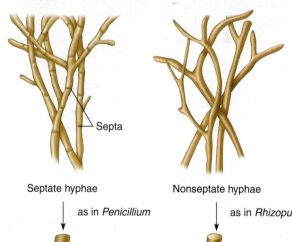

Septum

(b)

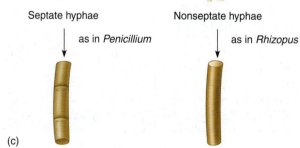

Septa

Septate hyphae   Nonseptate hyphae

as in *Penicillium*   as in *Rhizopus*

(c)

✳FIGURE 5.15 *Diplodia maydis,* **a pathogenic fungus of corn plants.**

(a) Scanning electron micrograph of a single colony showing its filamentous texture (24×). (b) Close-up of hyphal structure (1,200×). (c) Basic structural types of hyphae.

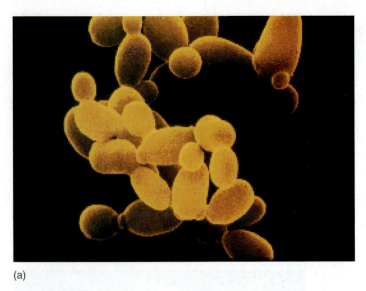

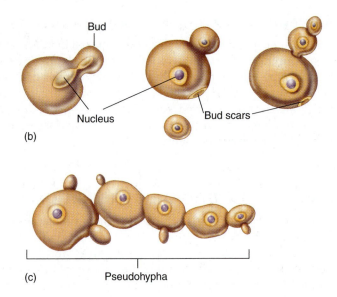

FIGURE 5.16 Microscopic morphology of yeasts.
(a) Scanning electron micrograph of the brewer's, or baker's, yeast *Saccharomyces cerevisiae* (21,000×). (b) Formation and release of yeast buds. (c) Formation of pseudohypha (a chain of budding yeast cells).

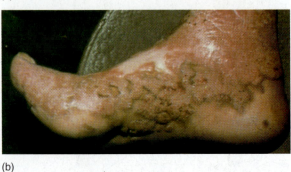

FIGURE 5.17 Nutritional sources (substrates) for fungi.
(a) A fungal mycelium growing on raspberries. The fine hyphal filaments and black sporangia are typical of *Rhizopus*. (b) The skin of the foot infected by a soil fungus, *Fonsecaea pedrosoi*.

occurring organic material on the earth can be attacked by some type of fungus. Fungi are often found in nutritionally poor or adverse environments. Various fungi thrive in substrates with high salt or sugar content, at relatively high temperatures, and even in snow and glaciers. Their medical and agricultural impact is extensive. A number of species cause **mycoses** (fungal infections) in animals, and thousands of species are important plant pathogens. Fungal toxins may cause disease in humans, and airborne fungi are a frequent cause of allergies and other medical conditions **(Insight 5.2).**

Organization of Microscopic Fungi

The cells of most microscopic fungi grow in loose associations or colonies. The colonies of yeasts are much like those of bacteria in that they have a soft, uniform texture and appearance. The colonies of filamentous fungi are noted for the striking cottony, hairy, or velvety textures that arise from their microscopic organization and morphology. The woven, intertwining mass of hyphae that makes up the body or colony of a mold is called a **mycelium.**

Although hyphae contain the usual eucaryotic organelles, they also have some unique organizational features. In most fungi, the hyphae are divided into segments by cross walls, or **septa,** a condition called septate (figure 5.15*c*). The nature of the septa varies from solid partitions with no communication between the compartments to partial walls with small pores that allow the flow of organelles and nutrients between adjacent compartments. Nonseptate hyphae consist of one long, continuous cell *not* divided into individual compartments by cross walls. With this construction, the cytoplasm and organelles move freely from one region to another, and each hyphal element can have several nuclei.

INSIGHT 5.2   *Discovery*

The Many Faces of Fungi

The Mother of All Fungi

Far from being microscopic, one of the largest organisms in the world is a massive basidiomycete growing in a forest in eastern Oregon. The mycelium of a single colony of *Armillaria ostoye* covers an area of 2,200 acres and stretches 3.5 miles across and 3 feet into the ground. Most of the fungus lies hidden beneath the ground, where it periodically fruits into edible mushrooms. Experts with the forest service took random samples over a large area and found that the mycelium is from genetically identical stock. They believe this fungus was started from a single mating pair of sexual hyphae around 2,400 years ago. Another colossal species from Washington State covers 1,500 acres and weighs around 100 tons. Both fungi penetrate the root structure of trees and spread slowly from tree to tree, sapping their nutrients and gradually destroying whole forests.

Armillaria (honey) mushrooms sprouting from the base of a tree in Oregon provide only a tiny hint of the massive fungus from which they arose.

A Fungus in Your Future

Biologists are developing some rather imaginative uses for fungi as a way of controlling both the life and death of plants. Dutch and Canadian researchers studying ways to control a devastating fungus infection of elm trees (Dutch elm disease) have come up with a brand new use of an old method—they actually vaccinate the trees. Ordinarily, the disease fungus invades the plant vessels and chokes off the flow of water. The natural tendency of the elm to defend itself by surrounding and inactivating the fungus is too slow to save it from death. But treating the elm trees before they get infected helps them develop an immunity to the disease. Plants are vaccinated somewhat like humans and animals: nonpathogenic spores or proteins from fungi are injected into the tree over a period of time. So far it appears that the symptoms of disease and the degree of damage can be significantly reduced. This may be the start of a whole new way to control fungal pests.

At the other extreme, government biologists working for narcotic control agencies have unveiled a recent plan to use fungi to kill unwanted plants. The main targets would be plants grown to produce illegal drugs like cocaine and heroin in the hopes of cutting down on these drugs right at the source. A fungus infection (*Fusarium*) that wiped out 30% of the coca crop in Peru dramatically demonstrated how effective this might be. Since then, at least two other fungi that could destroy opium poppies and marijuana plants have been isolated.

Purposefully releasing plant pathogens such as *Fusarium* into the environment has stirred a great deal of controversy. Critics in South America emphasize that even if the fungus appears specific to a particular plant, there is too much potential for it to switch hosts to food and ornamental plants and wreak havoc with the ecosystem. United States biologists who support the plan of using fungal control agents say that it is not as dangerous as massive spraying with pesticides, and that extensive laboratory tests have proved that the species of fungi being used will be very specific to the illegal drug plants and will not affect close relatives. Limited field tests will be started in the near future, paid for by a billion-dollar fund created by the U.S. government as part of its war on drugs.

Fungi, Fungi, Everywhere

The importance of fungi in the ecological structure of the earth is well founded. They are essential contributors to complex environments such as soil, and they play numerous beneficial roles as decomposers of organic debris and as partners to plants. Fungi also have great practical importance due to their metabolic versatility. They are productive sources of drugs (penicillin) to treat human infections and other diseases, and they are used in industry to ferment foods and synthesize organic chemicals.

The fact that they are so widespread also means that they frequently share human living quarters, especially in locations that provide ample moisture and nutrients. Often their presence is harmless and limited to a film of mildew on shower stalls or other moist environments. In some cases, depending on the amount of contamination and the type of mold, these indoor fungi can also give rise to various medical problems. Such common air contaminants as *Penicillium, Aspergillus, Cladosporium,* and *Stachybotrys* (see figure 5.25) all have the capacity to give off airborne spores and toxins that, when inhaled, cause a whole spectrum of symptoms sometimes referred to as "sick building syndrome." The usual source of harmful fungi is the presence of chronically water-damaged walls, ceilings, and other building materials that have come to harbor these fungi. People exposed to these houses or buildings report symptoms that range from skin rash, flulike reactions, sore throat, and headaches to fatigue, diarrhea, allergies, and immune suppression. Recent reports of sick buildings have been on the rise, affecting thousands of people, and some deaths have been reported in small children. The control of indoor fungi requires correcting the moisture problem, removing the contaminated materials, and decontaminating the living spaces. Mycologists are currently studying the mechanisms of toxic effects with an aim to develop better diagnosis and treatment.

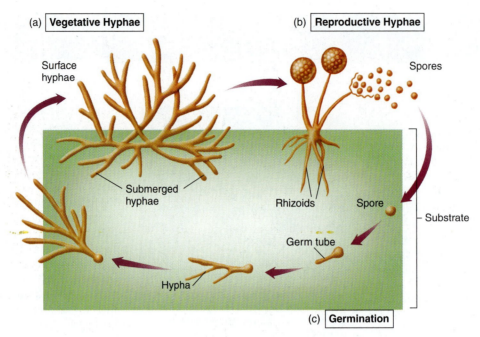

(a) **Vegetative Hyphae**

Surface hyphae

Submerged hyphae

(b) **Reproductive Hyphae**

Spores

Rhizoids   Spore

Substrate

Germ tube

Hypha

(c) **Germination**

FIGURE 5.18 **Functional types of hyphae using the mold *Rhizopus* as an example.**
(a) Vegetative hyphae are those surface and submerged filaments that digest, absorb, and distribute nutrients from the substrate. This species also has special anchoring structures called rhizoids. **(b)** Later, as the mold matures, it sprouts reproductive hyphae that produce asexual spores. **(c)** During the asexual life cycle, the free mold spores settle on a substrate and send out germ tubes that elongate into hyphae. Through continued growth and branching, an extensive mycelium is produced. So prolific are the fungi that a single colony of mold can easily contain 5,000 spore-bearing structures. If each of these released 2,000 single spores and if every spore were able to germinate, we would soon find ourselves in a sea of mycelia. Most spores do not germinate, but enough are successful to keep the numbers of fungi and their spores very high in most habitats.

Hyphae can also be classified according to their particular function. Vegetative hyphae (mycelia) are responsible for the visible mass of growth that appears on the surface of a substrate and penetrates it to digest and absorb nutrients. During the development of a fungal colony, the vegetative hyphae give rise to structures called reproductive, or fertile, hyphae which branch off vegetative mycelium. These hyphae are responsible for the production of fungal reproductive bodies called **spores.** Other specializations of hyphae are illustrated in **figure 5.18.**

Reproductive Strategies and Spore Formation

Fungi have many complex and successful reproductive strategies. Most can propagate by the simple outward growth of existing hyphae or by fragmentation, in which a separated piece of mycelium can generate a whole new colony. But the primary reproductive mode of fungi involves the production of various types of spores. Do not confuse fungal spores with the more resistant, nonreproductive bacterial spores. Fungal spores are responsible not only for multiplication but also for survival, producing genetic variation, and dissemination. Because of their compactness and relatively light weight, spores are dispersed widely through the environment by air, water, and living things. Upon encountering a favorable substrate, a

spore will germinate and produce a new fungus colony in a very short time (figure 5.18).

The fungi exhibit such a marked diversity in spores that they are largely classified and identified by their spores and spore-forming structures. Although there are some elaborate systems for naming and classifying spores, we will present only a basic overview of the principal types. The most general subdivision is based on the way the spores arise. Asexual spores are the products of mitotic division of a single parent cell, and sexual spores are formed through a process involving the fusing of two parental nuclei followed by meiosis.

Asexual Spore Formation

On the basis of the nature of the reproductive hypha and the manner in which the spores originate, there are two subtypes of asexual spore (figure 5.19):

1. **Sporangiospores (figure 5.19a)** are formed by successive cleavages within a saclike head called a **sporangium,** which is attached to a stalk, the sporangiophore. These spores are initially enclosed but are released when the sporangium ruptures.

2. **Conidia** (conidiospores) are free spores not enclosed by a spore-bearing sac **(figure 5.19b)**. They develop either by the pinching off of the tip of a special fertile hypha or by the segmentation of a preexisting vegetative hypha.

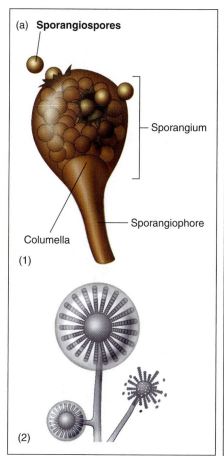

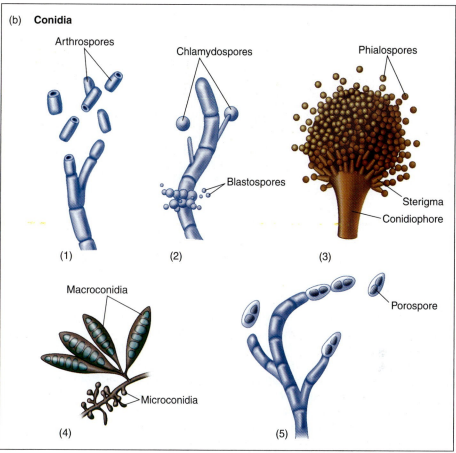

FIGURE 5.19   **Types of asexual mold spores.**
(a) Sporangiospores: (*1*) *Absidia*, (*2*) *Syncephalastrum*, **(b)** Conidia: (*1*) arthrospores (e.g., *Coccidioides*), (*2*) chlamydospores and blastospores (e.g., *Candida albicans*), (*3*) phialospores (e.g., *Aspergillus*), (*4*) macroconidia and microconidia (e.g., *Microsporum*), and (*5*) porospores (e.g., *Alternaria*).

Conidia are the most common asexual spores, and they occur in these forms:

arthrospore (ar'-thro-spor) Gr. *arthron*, joint. A rectangular spore formed when a septate hypha fragments at the cross walls.

chlamydospore (klam-ih'-doh-spor) Gr. *chlamys*, cloak. A spherical conidium formed by the thickening of a hyphal cell. It is released when the surrounding hypha fractures, and it serves as a survival or resting cell.

blastospore. A spore produced by budding from a parent cell that is a yeast or another conidium; also called a bud.

phialospore (fy'-ah-lo-spor) Gr. *phialos*, a vessel. A conidium that is budded from the mouth of a vase-shaped spore-bearing cell called a phialide or *sterigma*, leaving a small collar.

microconidium and *macroconidium*. The smaller and larger conidia formed by the same fungus under varying conditions. Microconidia are one-celled, and macroconidia have two or more cells.

porospore. A conidium that grows out through small pores in the spore-bearing cell; some are composed of several cells.

Sexual Spore Formation

If fungi can propagate themselves successfully with millions of asexual spores, what is the function of their sexual spores? The answer lies in important variations that occur when fungi of different genetic makeup combine their genetic material. Just as in plants and animals, this linking of genes from two parents creates offspring with combinations of genes different from that of either parent. The offspring from such a union can have slight variations in form and function that are potentially advantageous in the adaptation and survival of their species.

The majority of fungi produce sexual spores at some point. The nature of this process varies from the simple fusion of fertile hyphae of two different strains to a complex union of differentiated male and female structures and the development of special fruiting structures. We will consider the three most common sexual spores: zygospores, ascospores, and basidiospores. These spore types provide an important basis for classifying the major fungal divisions.

Zygospores are sturdy diploid spores formed when hyphae of two opposite strains (called the plus and minus strains) fuse and create a diploid zygote that swells and becomes covered by strong, spiny walls **(figure 5.20)**. When its wall is

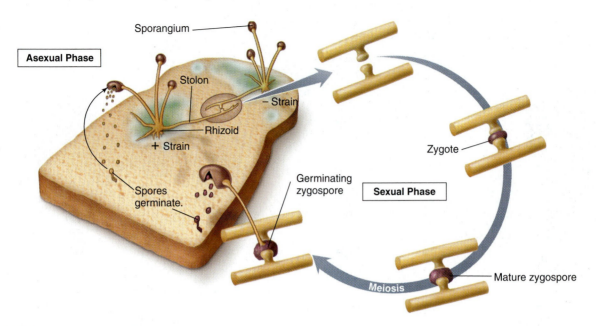

FIGURE 5.20 **Formation of zygospores in *Rhizopus stolonifer*.**
Sexual reproduction occurs when two mating strains of hyphae grow together, fuse, and form a mature zygospore. Germination of the zygospore involves meiotic division and production of a haploid sporangium that looks just like the asexual one.

disrupted, and moisture and nutrient conditions are suitable, the zygospore germinates and forms a mycelium that gives rise to a sporangium. Meiosis of diploid cells of the sporangium results in haploid nuclei that develop into sporangiospores. Both the sporangia and the sporangiospores that arise from sexual processes are outwardly identical to the asexual type, but because the spores arose from the union of two separate fungal parents, they are not genetically identical.

In general, haploid spores called **ascospores** are created inside a special fungal sac, or **ascus** (pl. asci) **(figure 5.21).** Although details can vary among types of fungi, the ascus and ascospores are formed when two different strains or sexes join together to produce offspring. In many species, the male sexual organ fuses with the female sexual organ. The end result is a number of terminal cells, each containing a diploid nucleus. Through differentiation, each of these cells enlarges to form an ascus, and its diploid nucleus undergoes meiosis (often followed by mitosis) to form four to eight haploid nuclei that will mature into ascospores. A ripe ascus breaks open and releases the ascospores. Some species form an elaborate fruiting body to hold the asci.

Basidiospores (bah-sid'-ee-oh-sporz) are haploid sexual spores formed on the outside of a club-shaped cell called a **basidium (figure 5.22).** In general, spore formation follows the same pattern of two mating types coming together, fusing, and forming terminal cells with diploid nuclei. Each of these cells becomes a basidium, and its nucleus produces, through meiosis, four haploid nuclei. These nuclei are extruded through the top of the basidium, where they develop into basidiospores. Notice the location of the basidia along the gills in mushrooms, which are often dark from the spores they contain. It may be a surprise to discover that the fleshy part of a mushroom is actually a fruiting body designed to protect and help disseminate its sexual spores.

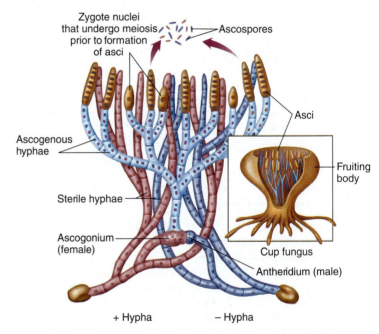

FIGURE 5.21 **Production of ascospores in a cup fungus.**
Inset shows the cup-shaped fruiting body that houses the asci.

Fungal Classification

It is often difficult for microbiologists to assign logical and useful classification schemes to microorganisms that also reflect their evolutionary relationships. This difficulty is due to the fact that the organisms do not always perfectly fit the neat categories made for them, and even experts cannot always agree on the nature of the categories. The fungi are no exception, and there are several ways to classify them. For our purposes, we will adopt a classification scheme with a

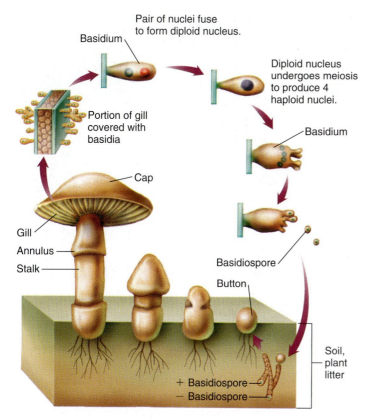

FIGURE 5.22   Formation of basidiospores in a mushroom.

Labels in figure:
Pair of nuclei fuse to form diploid nucleus.
Basidium
Diploid nucleus undergoes meiosis to produce 4 haploid nuclei.
Portion of gill covered with basidia
Basidium
Cap
Basidiospore
Gill
Annulus
Stalk
Button
Soil, plant litter
+ Basidiospore
− Basidiospore

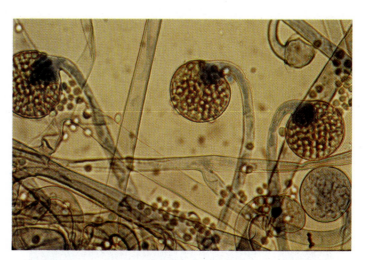

FIGURE 5.23   A representative Zygomycota, *Circinella.* Note the sporangia, sporangiospores, and nonseptate hyphae.

medical mycology emphasis, in which the Kingdom Fungi is subdivided into two subkingdoms, the Amastigomycota and the Mastigomycota. The Amastigomycota are common inhabitants of terrestrial habitats; several of them are human pathogens. Some of the characteristics used to separate them into subgroups are the type of sexual reproduction and their hyphal structure. The Mastigomycota are primitive filamentous fungi that live primarily in water and may cause disease in potatoes and grapes.

The next section outlines the four divisions of Amastigomycota, including major characteristics and important members.

Amastigomycota: Fungi That Produce Sexual and Asexual Spores (Perfect)

Division I—Zygomycota (also Phycomycetes)   Sexual spores: zygospores; asexual spores: mostly sporangiospores, some conidia. Hyphae are usually nonseptate. If septate, the septa are complete. Most species are free-living saprobes; some are animal parasites. Can be obnoxious contaminants in the laboratory and on food and vegetables. Examples are mostly molds: *Rhizopus,* a black bread mold; *Mucor; Syncephalastrum; Circinella* (**figure 5.23,** asexual phase only).

Division II—Ascomycota (also Ascomycetes)   Sexual spores: produce ascospores in asci; asexual spores: many types of conidia, formed at the tips of conidiophores. Hyphae with porous septa. Many important species. Examples: *Histoplasma,* the cause

of Ohio Valley fever; *Microsporum,* one cause of ringworm (a common name for certain fungal skin infections that often grow in a ringed pattern); *Penicillium,* one source of antibiotics (**figure 5.24,** asexual phase only); and *Saccharomyces,* a yeast used in making bread and beer. Most of the species in this division are either molds or yeasts; it also includes many human and plant pathogens, such as *Pneumocystis (carinii) jiroveci,* a pathogen of AIDS patients.

Division III—Basidiomycota (also Basidiomycetes)   Sexual reproduction by means of basidia and basidiospores; asexual spores: conidia. Incompletely septate hyphae. Some plant parasites and one human pathogen. Fleshy fruiting bodies are common. Examples: mushrooms, puffballs, bracket fungi, and plant pathogens called rusts and smuts. The one human pathogen, the yeast *Cryptococcus neoformans,* causes an invasive systemic infection in several organs, including the skin, brain, and lungs (see chapter 19).

Amastigomycota: Fungi That Produce Only Asexual Spores (Imperfect)

From the beginnings of fungal classification, any fungus that lacked a sexual state was called "imperfect" and was placed in a catchall category, the Fungi Imperfecti, or Deuteromycota. A species would remain classified in that category until its sexual state was described (if ever). Gradually, many species of Fungi Imperfecti were found to make sexual spores, and they were assigned to the taxonomic grouping that best fit those spores. This system created a quandary, because several very important species, especially pathogens, had to be regrouped and renamed. In many cases, the older names were so well entrenched in the literature that it was easier to retain them and assign a new generic name for their sexual stage.

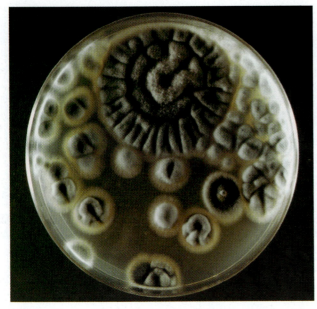

(a)

(b)

FIGURE 5.24   A common Ascomycota, *Penicillium*.
(a) Macroscopic view of a typical blue-green colony. (b) Microscopic view shows the brush arrangement of phialospores (220×).

Consequently, *Blastomyces* and *Histoplasma* are also known as *Ajellomyces* (sexual phase).

   Division IV—Deuteromycota   Asexual spores: conidia of various types. Hyphae septate. Majority are yeasts or molds, some dimorphic. Saprobes and a few animal and plant parasites. Examples: Several human pathogens were originally placed in this group, especially the imperfect states of *Blastomyces* and *Microsporum*. Other species are *Coccidioides immitis,* the cause of valley fever; *Candida albicans,* the cause of various yeast infections; *Cladosporium,* a common mildew fungus; and *Stachybotrys,* a toxic mold **(figure 5.25).**

Fungal Identification and Cultivation

Fungi are identified in medical specimens by first being isolated on special types of media and then being observed

FIGURE 5.25   Mycelium and spores of a representative of Deuteromycota called *Stachybotrys.*
This black mold is implicated in building contamination that leads to toxic diseases (see Insight 5.2).

macroscopically and microscopically. Examples of media for cultivating fungi are cornmeal, blood, and Sabouraud's agar. The latter medium is useful in isolating fungi from mixed samples because of its low pH, which inhibits the growth of bacteria but not of most fungi. Because the fungi are classified into general groups by the presence and type of sexual spores, it would seem logical to identify them in the same way, but sexual spores are rarely if ever demonstrated in the laboratory setting. As a result, the asexual spore-forming structures and spores are usually used to identify organisms to the level of genus and species. Other characteristics that contribute to identification are hyphal type, colony texture and pigmentation, physiological characteristics, and genetic makeup.

The Roles of Fungi in Nature and Industry

Nearly all fungi are free-living and do not require a host to complete their life cycles. Even among those fungi that are pathogenic, most human infection occurs through accidental contact with an environmental source such as soil, water, or dust. Humans are generally quite resistant to fungal infection, except for two main types of fungal pathogens: the primary pathogens, which can infect even healthy persons, and the opportunistic pathogens, which attack persons who are already weakened in some way.

   Mycoses (fungal infections) vary in the way the agent enters the body and the degree of tissue involvement **(table 5.2).** The list of opportunistic fungal pathogens has been increasing in the past few years because of newer medical techniques that keep immunocompromised patients alive. Even so-called harmless species found in the air and dust around us may be able to cause opportunistic infections in patients who already have AIDS, cancer, or diabetes (see Insight 21.1 in chapter 21).

TABLE 5.2	Major Fungal Infections of Humans	
Degree of Tissue Involvement and Area Affected	**Name of Infection**	**Name of Causative Fungus**
Superficial (not deeply invasive)		
Outer epidermis	Tinea versicolor	*Malassezia furfur*
Epidermis, hair, and dermis can be attacked	Dermatophytosis, also called tinea or ringworm of the scalp, body, feet (athlete's foot), toenails	*Microsporum, Trichophyton,* and *Epidermophyton*
Mucous membranes, skin, nails	Candidiasis, or yeast infection	*Candida albicans*
Systemic (deep; organism enters lungs; can invade other organs)		
Lung	Coccidioidomycosis (San Joaquin Valley fever)	*Coccidioides immitis*
	North American blastomycosis (Chicago disease)	*Blastomyces dermatitidis*
	Histoplasmosis (Ohio Valley fever)	*Histoplasma capsulatum*
	Cryptococcosis (torulosis)	*Cryptococcus neoformans*
Lung, skin	Paracoccidioidomycosis (South American blastomycosis)	*Paracoccidioides brasiliensis*

Fungi are involved in other medical conditions besides infections (see Insight 5.2). Fungal cell walls give off chemical substances that can cause allergies. The toxins produced by poisonous mushrooms can induce neurological disturbances and even death. The mold *Aspergillus flavus* synthesizes a potentially lethal poison called aflatoxin, which is the cause of a disease in domestic animals that have eaten grain contaminated with the mold and is also a cause of liver cancer in humans.

Fungi pose an ever-present economic hindrance to the agricultural industry. A number of species are pathogenic to field plants such as corn and grain, and fungi also rot fresh produce during shipping and storage. It has been estimated that as much as 40% of the yearly fruit crop is consumed not by humans but by fungi. On the beneficial side, however, fungi play an essential role in decomposing organic matter and returning essential minerals to the soil. They form stable associations with plant roots (mycorrhizae) that increase the ability of the roots to absorb water and nutrients. Industry has tapped the biochemical potential of fungi to produce large quantities of antibiotics, alcohol, organic acids, and vitamins. Some fungi are eaten or used to impart flavorings to food. The yeast *Saccharomyces* produces the alcohol in beer and wine and the gas that causes bread to rise. Blue cheese, soy sauce, and cured meats derive their unique flavors from the actions of fungi (see chapter 24).

5.5 The Protists

The algae and protozoa have been traditionally combined into the Kingdom Protista. The two major taxonomic categories of this kingdom are Subkingdom Algae and Subkingdom Protozoa. Although these general types of microbes are now known to occupy several kingdoms, it is still useful to retain the concept of a protist as any unicellular or colonial organism that lacks true tissues. We will only briefly mention algae, as they do not cause human infections for the most part.

The Algae: Photosynthetic Protists

The **algae** are a group of photosynthetic organisms usually recognized by their larger members, such as seaweeds and kelps. In addition to being beautifully colored and diverse in appearance, they vary in length from a few micrometers to 100 meters. Algae occur in unicellular, colonial, and filamentous forms, and the larger forms can possess tissues and simple organs. **Figure 5.26** depicts various types of algae. Algal cells as a group exhibit all of the eucaryotic organelles. The most noticeable of these are the chloroplasts, which contain, in addition to the green pigment chlorophyll, a number of other pigments that create the yellow, red, and brown coloration of some groups.

Algae are widespread inhabitants of fresh and marine waters. They are one of the main components of the large floating community of microscopic organisms called **plankton.** In this capacity, they play an essential role in the aquatic food web and produce most of the earth's oxygen. Other algal habitats include the surface of soil, rocks, and plants, and several species are even hardy enough to live in hot springs or snowbanks.

Animal tissues would be rather inhospitable to algae, so algae are rarely infectious. One exception is *Prototheca*, an unusual non-photosynthetic alga, which has been associated with skin and subcutaneous infections in humans and animals.

The primary medical threat from algae is due to a type of food poisoning caused by the toxins of certain marine algae. During particular seasons of the year, the overgrowth of these motile algae imparts a brilliant red color to the water, which is referred to as a "red tide." When intertidal animals feed, their bodies accumulate toxins given off by the algae that can persist for several months. Paralytic shellfish poisoning is caused by eating exposed clams or other invertebrates. It is marked by severe neurological symptoms and can be fatal. Ciguatera is a serious intoxication caused by algal toxins that have accumulated in fish such as bass and mackerel. Cooking does not destroy the toxin, and there is no antidote.

Several episodes of a severe infection caused by *Pfiesteria piscicida*, a toxic algal form, have been reported over the past several years in the United States. The disease was first reported in fish and was later transmitted to humans. This newly identified species occurs in at least 20 forms, including spores, cysts, and amoebas (see **figure 5.26c**), that can release

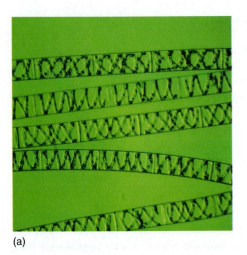

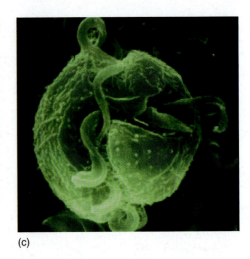

(a)   (b)   (c)

FIGURE 5.26 **Representative microscopic algae.**
(a) *Spirogyra*, a colonial filamentous form with spiral chloroplasts. **(b)** A strew of beautiful algae called diatoms shows the intricate and varied structure of their silica cell wall. **(c)** *Pfiesteria piscicida.* Although it is free-living, it is known to parasitize fish and release potent toxins that kill fish and sicken humans.

potent toxins. Both fish and humans develop neurological symptoms and bloody skin lesions. The cause of the epidemic has been traced to nutrient-rich agricultural runoff water that promoted the sudden "bloom" of *Pfiesteria*. These microbes first attacked and killed millions of fish and later people whose occupations exposed them to fish and contaminated water.

Biology of the Protozoa

If a poll were taken to choose the most engrossing and vivid group of microorganisms, many biologists would choose the protozoa. Although their name comes from the Greek for "first animals," they are far from being simple, primitive organisms. The protozoa constitute a very large group (about 65,000 species) of creatures that although single-celled, have startling properties when it comes to movement, feeding, and behavior. Although most members of this group are harmless, free-living inhabitants of water and soil, a few species are parasites collectively responsible for hundreds of millions of infections of humans each year. Before we consider a few examples of important pathogens, let us examine some general aspects of protozoan biology.

Protozoan Form and Function

Most protozoan cells are single cells containing the major eucaryotic organelles except chloroplasts. Their organelles can be highly specialized for feeding, reproduction, and locomotion. The cytoplasm is usually divided into a clear outer layer called the **ectoplasm** and a granular inner region called the *endoplasm*. Ectoplasm is involved in locomotion, feeding, and protection. Endoplasm houses the nucleus, mitochondria, and food and contractile vacuoles. Some ciliates and flagellates[4] even have organelles that work somewhat like a prim-

itive nervous system to coordinate movement. Because protozoa lack a cell wall, they have a certain amount of flexibility. Their outer boundary is a cell membrane that regulates the movement of food, wastes, and secretions. Cell shape can remain constant (as in most ciliates) or can change constantly (as in amoebas). Certain amoebas (foraminiferans) encase themselves in hard shells made of calcium carbonate. The size of most protozoan cells falls within the range of 3 to 300 μm. Some notable exceptions are giant amoebas and ciliates that are large enough (3–4 mm in length) to be seen swimming in pond water.

Nutritional and Habitat Range Protozoa are heterotrophic and usually require their food in a complex organic form. Free-living species scavenge dead plant or animal debris and even graze on live cells of bacteria and algae. Some species have special feeding structures such as oral grooves, which carry food particles into a passageway or gullet that packages the captured food into vacuoles for digestion. A remarkable feeding adaptation can be seen in the ciliate *Didinium*, which can easily devour another microbe that is nearly its size. Some protozoa absorb food directly through the cell membrane. Parasitic species live on the fluids of their host, such as plasma and digestive juices, or they can actively feed on tissues.

Although protozoa have adapted to a wide range of habitats, their main limiting factor is the availability of moisture. Their predominant habitats are fresh and marine water, soil, plants, and animals. Even extremes in temperature and pH are not a barrier to their existence; hardy species are found in hot springs, ice, and habitats with low or high pH. Many protozoa can convert to a resistant, dormant stage called a cyst.

Styles of Locomotion Except for one group (the Apicomplexa), protozoa are motile by means of **pseudopods** ("false foot"), **flagella,** or **cilia.** A few species have both pseudopods (also called pseudopodia) and flagella. Some unusual protozoa move by a gliding or twisting movement that does not appear

---

4. The terms *ciliate* and *flagellate* are common names of protozoan groups that move by means of cilia and flagella.

to involve any of these locomotor structures. Pseudopods are blunt, branched, or long and pointed, depending on the particular species. The flowing action of the pseudopods results in amoeboid motion, and pseudopods also serve as feeding structures in many amoebas. The structure and behavior of flagella and cilia were discussed in the first section of this chapter. Flagella vary in number from one to several, and in certain species they are attached along the length of the cell by an extension of the cytoplasmic membrane called the *undulating membrane* (see figure 5.28). In most ciliates, the cilia are distributed over the entire surface of the cell in characteristic patterns. Because of the tremendous variety in ciliary arrangements and functions, ciliates are among the most diverse and awesome cells in the biological world. In certain protozoa, cilia line the oral groove and function in feeding; in others, they fuse together to form stiff props that serve as primitive rows of walking legs.

Life Cycles and Reproduction Most protozoa are recognized by a motile feeding stage called the **trophozoite** that requires ample food and moisture to remain active. A large number of species are also capable of entering into a dormant, resting stage called a **cyst** when conditions in the environment become unfavorable for growth and feeding. During *encystment*, the trophozoite cell rounds up into a sphere, and its ectoplasm secretes a tough, thick cuticle around the cell membrane **(figure 5.27).** Because cysts are more resistant than ordinary cells to heat, drying, and chemicals, they can survive adverse periods. They can be dispersed by air currents and may even be an important factor in the spread of diseases such as amoebic dysentery. If provided with moisture and nutrients, a cyst breaks open and releases the active trophozoite.

The life cycles of protozoans vary from simple to complex. Several protozoan groups exist only in the trophozoite state. Many alternate between a trophozoite and a cyst stage, depending on the conditions of the habitat. The life cycle of a parasitic protozoan dictates its mode of transmission to other hosts. For example, the flagellate *Trichomonas vaginalis* causes a common sexually transmitted disease. Because it does not form cysts, it is more delicate, and must be transmitted by intimate contact between sexual partners. In contrast, intestinal pathogens such as *Entamoeba histolytica* and *Giardia lamblia* form cysts and are readily transmitted in contaminated water and foods.

All protozoa reproduce by relatively simple, asexual methods, usually mitotic cell division. Several parasitic species, including the agents of malaria and toxoplasmosis, reproduce asexually inside a host cell by multiple fission. Sexual reproduction also occurs during the life cycle of most protozoa. Ciliates participate in **conjugation,** a form of genetic exchange in which members of two different mating types

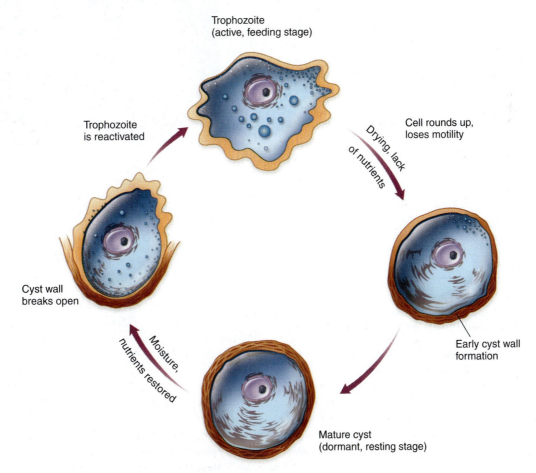

Trophozoite
(active, feeding stage)

Trophozoite
is reactivated

Cell rounds up,
loses motility

Drying, lack
of nutrients

Cyst wall
breaks open

Moisture,
nutrients restored

Early cyst wall
formation

Mature cyst
(dormant, resting stage)

※ **FIGURE 5.27** **The general life cycle exhibited by many protozoa.**
All protozoa have a trophozoite form, but not all produce cysts.

fuse temporarily and exchange micronuclei. This process of sexual recombination yields new and different genetic combinations that can be advantageous in evolution.

IN THE NEWS *(Continued from page 119)*

(Continued from page 119)

The disease in the opening case study was **cryptosporidiosis.** It is caused by a single-celled protozoan parasite named *Cryptosporidium parvum. C. parvum* undergoes a complex life cycle. During one stage in the life cycle, thick-walled oocysts that are 3 to 5 μm in size are formed. It was these oocysts that were detected in water samples from the pool. When Dr. McDermott educated the patients about this disease, she emphasized that all it takes is the ingestion of 1 to 10 oocysts to cause disease. This is referred to as a very "low infectious dose." It is why one fecal accident can sufficiently contaminate an entire swimming pool in which individuals may accidentally swallow only one or two mouthfuls of contaminated water. The oocysts are extremely resistant to disinfectants and the recommended concentrations of chlorine used in swimming pools. Because of their small size, they may not be removed efficiently by pool filters.

Cryptosporidium infected over 370,000 individuals in Milwaukee, Wisconsin, in 1993. This was the largest waterborne outbreak in U.S. history. During that epidemic, the public water supply was contaminated with human sewage containing *C. parvum* oocysts. Besides pool-associated and contaminated drinking water illness, there have been reports of *C. parvum* oocysts in oysters intended for human consumption that were harvested from sites where there were high levels of fecal contamination due to wastewater outfalls and cattle farms. Oysters remove oocysts from contaminated waters and retain them on their gills and within their body.

See: CDC. 2001. Protracted outbreaks of cryptosporidiosis associated with swimming pool use—Ohio and Nebraska, 2000. MMWR 50:406–410.

Fayer, R. et. al. 1999. Cryptosporidium parvum in oysters from commercial harvesting sites in the Chesapeake Bay. Emerg. Infect. Dis. 5:706–710.

Classification of Selected Medically Important Protozoa

Taxonomists have not escaped problems classifying protozoa. They, too, are very diverse and frequently frustrate attempts to generalize or place them in neat groupings. We will use a simple system of four groups, based on method of motility, mode of reproduction, and stages in the life cycle, summarized here.

The Mastigophora (Flagellated)   Motility is primarily by flagella alone or by both flagellar and amoeboid motion. Single nucleus. Sexual reproduction, when present, by syngamy; division by longitudinal fission. Several parasitic forms lack mitochondria and Golgi apparatus. Most species form cysts and are free-living; the group also includes several parasites. Some species are found in loose aggregates or colonies, but most are solitary. Members include: *Trypanosoma* and

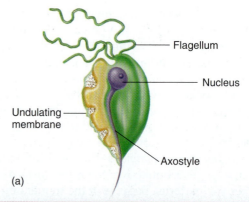

(a)

(b)

FIGURE 5.28   **The structure of a typical mastigophoran,** *Trichomonas vaginalis.*
This genital tract pathogen is shown in **(a)** a drawing and **(b)** a scanning electron micrograph.

Leishmania, important blood pathogens spread by insect vectors; *Giardia,* an intestinal parasite spread in water contaminated with feces; *Trichomonas,* a parasite of the reproductive tract of humans spread by sexual contact **(figure 5.28).**

The Sarcodina (Amoebas)   Cell form is primarily an amoeba **(figure 5.29).** Major locomotor organelles are pseudopods, although some species have flagellated reproductive states. Asexual reproduction by fission. Two groups have an external shell; mostly uninucleate; usually encyst. Most amoebas are free-living and not infectious; *Entamoeba* is a pathogen or parasite of humans; shelled amoebas called foraminifera and radiolarians are responsible for chalk deposits in the ocean.

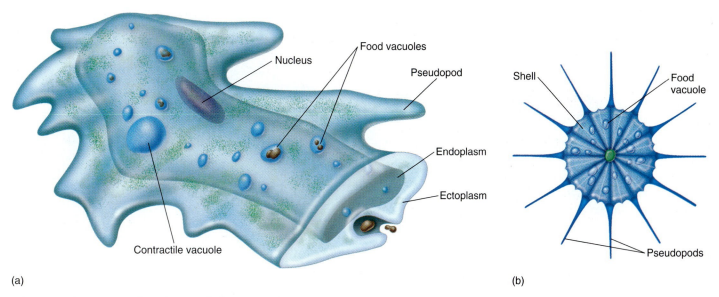

(a)

(b)

FIGURE 5.29   Examples of sarcodinians.
(a) The structure of an amoeba. **(b)** Radiolarian, a shelled amoeba with long, pointed pseudopods. Pseudopods are used in both movement and feeding.

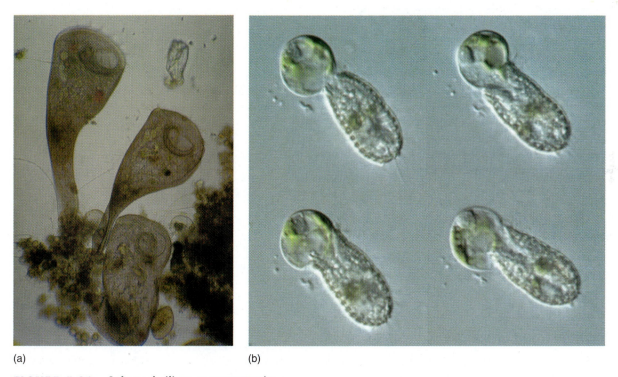

(a)

(b)

FIGURE 5.30   Selected ciliate representatives.
(a) Large, funnel-shaped *Stentor* with a rotating row of cilia around its oral cavity. Currents produced by the cilia sweep food particles into the gullet. **(b)** Stages in the process of *Coleps* feeding on an alga (round cell). The predaceous ciliate gradually pulls its prey into a large oral groove.

The Ciliophora (Ciliated) Trophozoites are motile by cilia; some have cilia in tufts for feeding and attachment; most develop cysts; have both macronuclei and micronuclei; division by transverse fission; most have a definite mouth and feeding organelle; show relatively advanced behavior **(figure 5.30).** The majority of ciliates are free-living and harmless.

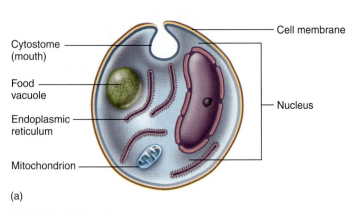

Cytostome (mouth)

Food vacuole

Endoplasmic reticulum

Mitochondrion

Cell membrane

Nucleus

(a)

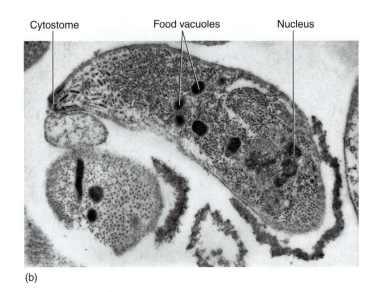

Cytostome

Food vacuoles

Nucleus

(b)

FIGURE 5.31   Sporozoan protozoan.
(a) General cell structure. Note the lack of specialized locomotor organelles. **(b)** Scanning electron micrograph of the sporozoite of *Cryptosporidium*, an intestinal parasite of humans and other mammals.

The Apicomplexa (Sporozoa)   Motility is absent in most cells except male gametes. Life cycles are complex, with well-developed asexual and sexual stages. Sporozoa produce special sporelike cells called **sporozoites (figure 5.31)** following sexual reproduction, which are important in transmission of infections; most form thick-walled zygotes called oocysts; entire group is parasitic. *Plasmodium,* the most prevalent protozoan parasite, causes 100 million to 300 million cases of malaria each year worldwide. It is an intracellular parasite with a complex cycle alternating between humans and mosquitoes. *Toxoplasma gondii* causes an acute infection (toxoplasmosis) in humans, which is acquired from cats and other animals.

Just as with the procaryotes and other eucaryotes, protozoans that cause disease produce symptoms in different organ systems. These diseases are covered in chapters 18 through 23.

Protozoan Identification and Cultivation

The unique appearance of most protozoa makes it possible for a knowledgeable person to identify them to the level of genus and often species by microscopic morphology alone. Characteristics to consider in identification include the shape and size of the cell; the type, number, and distribution of locomotor structures; the presence of special organelles or cysts; and the number of nuclei. Medical specimens taken from blood, sputum, cerebrospinal fluid, feces, or the vagina are smeared directly onto a slide and observed with or without special stains. Occasionally, protozoa are cultivated on artificial media or in laboratory animals for further identification or study.

TABLE 5.3	Major Pathogenic Protozoa, Infections, and Primary Sources
Protozoan/Disease	**Reservoir/Source**
Amoeboid Protozoa	
Amoebiasis: *Entamoeba histolytica*	Human/water and food
Brain infection: *Naegleria, Acanthamoeba*	Free-living in water
Ciliated Protozoa	
Balantidiosis: *Balantidium coli*	Zoonotic in pigs
Flagellated Protozoa	
Giardiasis: *Giardia lamblia*	Zoonotic/water and food
Trichomoniasis: *T. hominis, T. vaginalis*	Human
Hemoflagellates	
Trypanosomiasis: *Trypanosoma brucei, T. cruzi*	Zoonotic/ vector-borne
Leishmaniasis: *Leishmania donovani, L. tropica, L. brasiliensis*	Zoonotic/ vector-borne
Apicomplexan Protozoa	
Malaria: *Plasmodium vivax, P. falciparum, P. malariae*	Human/vector-borne
Toxoplasmosis: *Toxoplasma gondii*	Zoonotic/vector-borne
Cryptosporidiosis: *Cryptosporidium*	Free-living/water, food
Cyclosporiasis: *Cyclospora cayetanensis*	Water/fresh produce

Important Protozoan Pathogens

Although protozoan infections are very common, they are actually caused by only a small number of species often restricted geographically to the tropics and subtropics **(table 5.3).** In this survey, we look at examples from two protozoan groups that illustrate some of the main features of protozoan diseases.

Protozoa are traditionally studied along with the helminths in the science of parasitology. Although a parasite is more accurately defined as an organism that obtains food and other requirements at the expense of a host, the term *parasite* is often used to denote protozoan and helminth pathogens.

Pathogenic Flagellates: Trypanosomes Trypanosomes are protozoa belonging to the genus *Trypanosoma* (try"-pan-oh-soh'-mah). The two most important representatives are *T. brucei* and *T. cruzi*, species that are closely related but geographically restricted. *Trypanosoma brucei* occurs in Africa, where it causes approximately 35,000 new cases of sleeping sickness each year (see chapter 19). *Trypanosoma cruzi*, the cause of Chagas disease,[5] is endemic to South and Central America, where it infects several million people a year. Both species have long, crescent-shaped cells with a single flagellum that is sometimes attached to the cell body by an undulating membrane. Both occur in the blood during infection and are transmitted by blood-sucking vectors. We will use *T. cruzi* to illustrate the phases of a trypanosomal life cycle and to demonstrate the complexity of parasitic relationships.

The trypanosome of Chagas disease relies on the close relationship of a warm-blooded mammal and an insect that feeds on mammalian blood. The mammalian hosts are numerous, including dogs, cats, opossums, armadillos, and foxes. The vector is the *reduviid* (ree-doo'-vee-id) *bug*, an insect that is sometimes called the "kissing bug" because of its habit of biting its host at the corner of the mouth. Transmission occurs from bug to mammal and from mammal to bug, but usually not from mammal to mammal, except across the placenta during pregnancy. The general phases of this cycle are presented in **figure 5.32.**

The trypanosome trophozoite multiplies in the intestinal tract of the reduviid bug and is harbored in the feces. The bug seeks a host and bites the mucous membranes, usually of the eye, nose, or lips. As it fills with blood, the bug soils the bite with feces containing the trypanosome. Ironically, the victims themselves inadvertently contribute to the entry of the microbe by scratching the bite wound. The trypanosomes ultimately become established and multiply in muscle and white blood cells. Periodically, these parasitized cells rupture, releasing large numbers of new trophozoites into the blood. Eventually, the trypanosome can spread to many systems, including the lymphoid organs, heart, liver, and brain. Manifestations of the resultant disease range from mild to very severe, and include fever, inflammation, and heart and brain damage. In many cases, the disease has an extended course and can cause death.

Infective Amoebas: *Entamoeba* Several species of amoebas cause disease in humans, but probably the most common disease is amoebiasis, or amoebic dysentery, caused by *Enta-*

<hr>

5. Named for Carlos Chagas, the discoverer of *T. cruzi*.

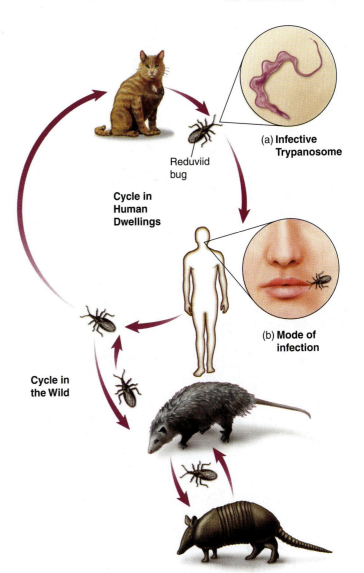

(a) **Infective Trypanosome**

Reduviid bug

Cycle in Human Dwellings

(b) **Mode of infection**

Cycle in the Wild

FIGURE 5.32 **Cycle of transmission in Chagas disease.** Trypanosomes (inset *a*) are transmitted among mammalian hosts and human hosts by means of a bite from the kissing bug (inset *b*).

moeba histolytica (see chapter 22). This microbe is widely distributed in the world, from northern zones to the tropics, and is nearly always associated with humans. Amoebic dysentery is the fourth most common protozoan infection in the world. This microbe has a life cycle quite different from the trypanosomes in that it does not involve multiple hosts and a blood-sucking vector. It lives part of its cycle as a trophozoite and part as a cyst. Because the cyst is the more resistant form and can survive in water and soil for several weeks, it is the more important stage for transmission. The primary way that people become infected is by ingesting food or water contaminated with human feces.

Figure 5.33 shows the major features of the amoebic dysentery cycle, starting with the ingestion of cysts. The viable, heavy-walled cyst passes through the stomach unharmed. Once inside the small intestine, the cyst germinates into a large multinucleate amoeba that subsequently divides to form

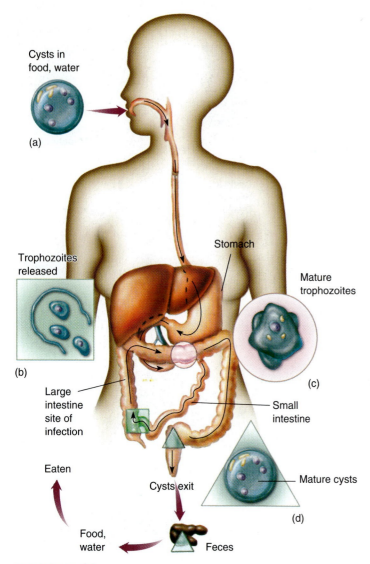

Cysts in food, water

(a)

Stomach

Trophozoites released

Mature trophozoites

(b)

(c)

Large intestine site of infection

Small intestine

Eaten

Mature cysts

Cysts exit

(d)

Food, water

Feces

FIGURE 5.33 **Stages in the infection and transmission of amoebic dysentery.**
Arrows show the route of infection; insets show the appearance of *Entamoeba histolytica.* **(a)** Cysts are eaten. **(b)** Trophozoites (amoebas) emerge from cysts. **(c)** Trophozoites invade the large intestinal wall. **(d)** Mature cysts are released in the feces, and may be spread through contaminated food and water.

small amoebas (the trophozoite stage). These trophozoites migrate to the large intestine and begin to feed and grow. From this site, they can penetrate the lining of the intestine and invade the liver, lungs, and skin. Common symptoms include gastrointestinal disturbances such as nausea, vomiting, and diarrhea, leading to weight loss and dehydration. Untreated cases with extensive damage to the organs experience a high death rate. The cycle is completed in the infected human when certain trophozoites in the feces begin to form cysts, which then pass out of the body with fecal matter. Knowledge of the amoebic cycle and role of cysts has been helpful in controlling the disease. Important preventive measures include sewage treatment, curtailing the use of human feces as fertilizers, and adequate sanitation of food and water.

5.6 The Parasitic Helminths

Tapeworms, flukes, and roundworms are collectively called helminths, from the Greek word meaning worm. Adult animals are usually large enough to be seen with the naked eye, and they range from the longest tapeworms, measuring up to about 25 m in length, to roundworms less than 1 mm in length. Nevertheless, they are included among microorganisms because of their infective abilities and because the microscope is necessary to identify their eggs and larvae.

On the basis of morphological form, the two major groups of parasitic helminths are the flatworms (Phylum Platyhelminthes), with a very thin, often segmented body plan (**figure 5.34**), and the roundworms (Phylum Aschelminthes, also called **nematodes**), with an elongate, cylindrical, unsegmented body (**figure 5.35**). The flatworm group is subdivided into the **cestodes,** or tapeworms, named for their long, ribbonlike arrangement, and the **trematodes,** or flukes, characterized by flat, ovoid bodies. Not all flatworms and roundworms are parasites by nature; many live free in soil and water. Because most disease-causing helminths spend part of their lives in the gastrointestinal tract, they are discussed in chapter 22.

General Worm Morphology

All helminths are multicellular animals equipped to some degree with organs and organ systems. In parasitic helminths, the most developed organs are those of the reproductive tract, with some degree of reduction in the digestive, excretory, nervous, and muscular systems. In particular groups, such as the cestodes, reproduction is so dominant that the worms are reduced to little more than a series of flattened sacs filled with ovaries, testes, and eggs (see **figure 5.34***a*). Not all worms have such extreme adaptations as cestodes, but most have a highly developed reproductive potential, thick cuticles for protection, and mouth glands for breaking down the host's tissue.

Life Cycles and Reproduction

The complete life cycle of helminths includes the fertilized egg (embryo), larval, and adult stages. In the majority of helminths, adults derive nutrients and reproduce sexually in a host's body. In nematodes, the sexes are separate and usually different in appearance; in trematodes, the sexes can be either separate or **hermaphroditic,** meaning that male and female sex organs are in the same worm; cestodes are generally hermaphroditic. For a parasite's continued survival as a species, it must complete the life cycle by transmitting an infective form, usually an egg or larva, to the body of another host, either of the same or a different species. By convention, the host in which larval development occurs is the intermediate (secondary) host, and adulthood and mating occur in the **definitive (final) host.** A transport host is an intermediate host that experiences no parasitic development but is an essential link in the completion of the cycle.

In general, sources for human infection are contaminated food, soil, and water or infected animals, and routes of infection are by oral intake or penetration of unbroken skin.

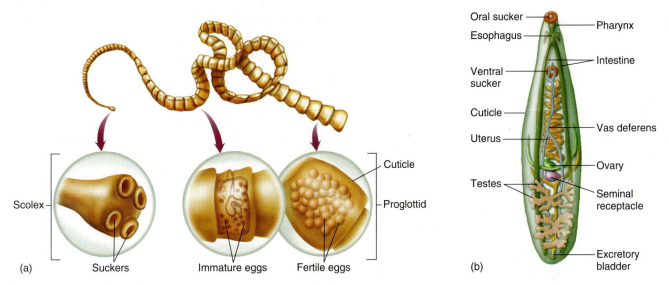

FIGURE 5.34   **Parasitic flatworms.**
(a) A cestode (beef tapeworm), showing the scolex; long, tapelike body; and magnified views of immature and mature proglottids (body segments). **(b)** The structure of a trematode (liver fluke). Note the suckers that attach to host tissue and the dominance of reproductive and digestive organs.

Humans are the definitive hosts for many of the parasites listed in **table 5.4,** and in about half the diseases, they are also the sole biological reservoir. In other cases, animals or insect vectors serve as reservoirs or are required to complete worm development. In the majority of helminth infections, the worms must leave their host to complete the entire life cycle.

Fertilized eggs are usually released to the environment and are provided with a protective shell and extra food to aid their development into larvae. Even so, most eggs and larvae are vulnerable to heat, cold, drying, and predators and are destroyed or unable to reach a new host. To counteract this formidable mortality rate, certain worms have adapted a reproductive capacity that borders on the incredible: A single female *Ascaris*[6] can lay 200,000 eggs a day, and a large female can contain over 25,000,000 eggs at varying stages of development! If only a tiny number of these eggs makes it to another host, the parasite will have been successful in completing its life cycle.

A Helminth Cycle: The Pinworm

To illustrate a helminth cycle in humans, we will use the example of a roundworm, *Enterobius vermicularis*, the pinworm or seatworm. This worm causes a very common infestation of the large intestine (see figure 5.35). Worms range from 2 to 12 mm long and have a tapered, curved cylinder shape. The condition they cause, enterobiasis, is usually a simple, uncomplicated infection that does not spread beyond the intestine.

A cycle starts when a person swallows microscopic eggs picked up from another infected person by direct contact or by touching articles that person has touched. The eggs hatch in the intestine and then release larvae that mature into adult worms within about one month. Male and female worms

6. *Ascaris* is a genus of parasitic intestinal roundworms.

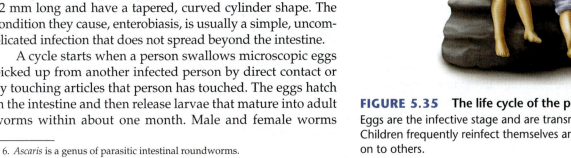

FIGURE 5.35   **The life cycle of the pinworm, a roundworm.**
Eggs are the infective stage and are transmitted by unclean hands. Children frequently reinfect themselves and also pass the parasite on to others.

TABLE 5.4   Major Helminths of Humans and Their Modes of Transmission

Classification	Common Name of Disease or Worm	Life Cycle Requirement	Spread to Humans By
Nematodes (Roundworms)			
Intestinal Nematodes			Ingestion
Infective in egg (embryo) stage			
Trichuris trichiura	Whipworm	Humans	Fecal pollution of soil with eggs
Ascaris lumbricoides	Ascariasis	Humans	Fecal pollution of soil with eggs
Enterobius vermicularis	Pinworm	Humans	Close contact
Infective in larval stage			
Necator americanus	New World hookworm	Humans	Fecal pollution of soil with eggs
Ancylostoma duodenale	Old World hookworm	Humans	Fecal pollution of soil with eggs
Strongyloides stercoralis	Threadworm	Humans; may live free	Fecal pollution of soil with eggs
Trichinella spiralis	Trichina worm	Pigs, wild mammals	Consumption of meat containing larvae
Tissue Nematodes			Burrowing of larva into tissue
Onchocerca volvulus	River blindness	Humans, black flies	Fly bite
Dracunculus medinensis	Guinea worm	Humans and *Cyclops* (an aquatic invertebrate)	Ingestion of water containing *Cyclops*
Trematodes			
Schistosoma japonicum	Blood fluke	Humans and snails	Ingestion of fresh water containing larval stage
S. mansoni	Blood fluke	Humans and snails	Ingestion of fresh water containing larval stage
S. haematobium	Blood fluke	Humans and snails	Ingestion of fresh water containing larval stage
Opisthorchis sinensis	Chinese liver fluke	Humans, snails, fish	Consumption of fish
Fasciola hepatica	Sheep liver fluke	Herbivores (sheep, cattle)	Consumption of water and water plants
Cestodes			
T. solium	Pork tapeworm	Humans, swine	Consumption of undercooked or raw pork
Diphyllobothrium latum	Fish tapeworm	Humans, fish	Consumption of undercooked or raw fish
Hymenolepis nana	Dwarf tapeworm	Humans	Oral-fecal; close contact

mate, and the female migrates out to the anus to deposit eggs, which cause intense itchiness that is relieved by scratching. Herein lies a significant means of dispersal: Scratching contaminates the fingers, which, in turn, transfer eggs to bedclothes and other inanimate objects. This person becomes a host and a source of eggs and can spread them to others in addition to reinfesting himself. Enterobiasis occurs most often among families and in other close living situations. Its distribution is worldwide among all socioeconomic groups, but it seems to attack younger people more frequently than older ones.

Helminth Classification and Identification

The helminths are classified according to their shape; their size; the degree of development of various organs; the presence of hooks, suckers, or other special structures; the mode of reproduction; the kinds of hosts; and the appearance of eggs and larvae. They are identified in the laboratory by microscopic detection of the adult worm or its larvae and eggs, which often have distinctive shapes or external and internal structures. Occasionally, they are cultured in order to verify all of the life stages.

Distribution and Importance of Parasitic Worms

About 50 species of helminths parasitize humans (table 5.4). They are distributed in all areas of the world that support human life. Some worms are restricted to a given geographic region, and many have a higher incidence in tropical areas. This knowledge must be tempered with the realization that jet-age travel, along with human migration, is gradually changing the patterns of worm infections, especially of those species that do not require alternate hosts or special climatic conditions for development. The yearly estimate of worldwide cases numbers in the billions, and these are not confined to developing countries. A conservative estimate places 50,000,000 helminth infections in North America alone. The primary targets are malnourished children.

You have now learned about the variety of organisms that microbiologists study and classify. And as you've seen, many such organisms are capable of causing disease. In chapter 6, you'll learn about the "not-quite-organisms" that can cause disease, namely, viruses.

✔ CHECKPOINT

- The eucaryotic microorganisms include the Fungi (Myceteae), the Protista (algae and protozoa), and the Helminths (Kingdom Animalia).
- The Kingdom Fungi (Myceteae) is composed of nonphotosynthetic haploid species with cell walls. The fungi are either saprobes or parasites, and may be unicellular, colonial, or multicellular. Forms include yeasts (unicellular budding cells) and molds (filamentous cells called hyphae). Their primary means of reproduction involves asexual and sexual spores.

- The protists are mostly unicellular or colonial eucaryotes that lack specialized tissues. There are two major organism types: the Algae and the Protozoa. Algae are photosynthetic organisms that contain chloroplasts with chlorophyll and other pigments. Protozoa are heterotrophs that usually display some form of locomotion. Most are single-celled trophozoites, and many produce a resistant stage, or cyst.
- The Kingdom Animalia has only one group that contains members that are (sometimes) microscopic. These are the helminths or worms. Parasitic members include flatworms and roundworms that are able to invade and reproduce in human tissues.

Chapter Summary With Key Terms

5.1 The History of Eucaryotes

5.2 and 5.3 Form and Function of the Eucaryotic Cell: External and Internal Structures

A. Eucaryotic cells are complex and compartmentalized into individual organelles.

B. Major organelles and other structural features include: Appendages (cilia, flagella), **glycocalyx,** cell wall, cytoplasmic (or cell) membrane, **organelles** (nucleus, **nucleolus, endoplasmic reticulum, Golgi complex, mitochondria,** chloroplasts), ribosomes, cytoskeleton (**microfilaments,** microtubules). A review comparing the major differences between eucaryotic and procaryotic cells is provided in table 2.6, page 52.

5.4 The Kingdom of the Fungi

Common names of the macroscopic fungi are mushrooms, bracket fungi, and puffballs. Microscopic fungi are known as yeasts and molds.

A. *Overall Morphology:* At the cellular (microscopic) level, fungi are typical eucaryotic cells, with thick cell walls. Yeasts are single cells that form buds and **pseudohyphae. Hyphae** are long, tubular filaments that can be septate or nonseptate and grow in a network called a **mycelium;** hyphae are characteristic of the filamentous fungi called molds.

B. *Nutritional Mode/Distribution:* All are **heterotrophic.** The majority are harmless **saprobes** living off organic **substrates** such as dead animal and plant tissues. A few are **parasites,** living on the tissues of other organisms, but none are obligate. Distribution is extremely widespread in many habitats.

C. *Reproduction:* Primarily through **spores** formed on special reproductive hyphae. In asexual reproduction, spores are formed through budding, partitioning of a hypha, or in special sporogenous structures; examples are **conidia** and **sporangiospores.** In sexual reproduction, spores are formed following fusion of male and female strains and the formation of a sexual structure; sexual spores are one basis for classification.

D. *Major Groups:* The four main divisions among the terrestrial fungi, given with sexual spore type, are Zygomycota (**zygospores**), Ascomycota (**ascospores**),

Basidiomycota (**basidiospores**), and Amastigomycota (only asexual spores).

E. *Importance:* Fungi are essential decomposers of plant and animal detritis in the environment. Economically beneficial as sources of antibiotics; used in making foods and in genetic studies. Adverse impacts include: decomposition of fruits and vegetables; human infections, or **mycoses;** some produce substances that are toxic if eaten.

5.5 The Protists

A. The Algae
Include photosynthetic kelps and seaweeds.
1. *Overall Morphology:* Are unicellular, colonial, filamentous or larger forms.
2. *Nutritional Mode/Distribution:* Photosynthetic; fresh and marine water habitats; main component of **plankton.**
3. *Importance:* Provide the basis of the food web in most aquatic habitats. Certain algae produce neurotoxins that are harmful to humans and animals.

B. The Protozoa
Include large single-celled organisms; a few are pathogens.
1. *Overall Morphology:* Most are unicellular; lack a cell wall. The cytoplasm is divided into ectoplasm and endoplasm. Many convert to a resistant, dormant stage called a **cyst.**
2. *Nutritional Mode/Distribution:* All are heterotrophic. Most are free-living in a moist habitat (water, soil); feed by engulfing other microorganisms and organic matter.
3. *Reproduction:* Asexual by binary fission and **mitosis,** budding; sexual by fusion of free-swimming gametes, conjugation.
4. *Major Groups:* Protozoa are subdivided into four groups based upon mode of locomotion and type of reproduction: Mastigophora, the flagellates, motile by flagella; Sarcodina, the amoebas, motile by pseudopods; Ciliophora, the ciliates, motile by cilia; Apicomplexa, motility not well developed; produce unique reproductive structures.

5. *Importance:* Ecologically important in food webs and decomposing organic matter. Medical significance: hundreds of millions of people are afflicted with one of the many protozoan infections (malaria, trypanosomiasis, amoebiasis). Can be spread from host to host by insect vectors.

5.6 The Parasitic Helminths

Includes three categories: roundworms, tapeworms, and flukes.

A. *Overall Morphology:* Animal cells; multicellular; individual organs specialized for reproduction, digestion, movement, protection, though some of these are reduced.

B. *Reproductive Mode:* Includes embryo, larval, and adult stages. Majority reproduce sexually. Sexes may be hermaphroditic.

C. *Epidemiology:* Developing countries in the tropics hardest hit by helminth infections; transmitted via ingestion of larvae or eggs in food; from soil or water. They afflict billions of humans.

Multiple-Choice Questions

1. Both flagella and cilia are found primarily in
   a. algae
   b. protozoa
   c. fungi
   d. both b and c

2. Features of the nuclear envelope include
   a. ribosomes
   b. a double membrane structure
   c. pores that allow communication with the cytoplasm
   d. b and c
   e. all of these

3. The cell wall is found in which eucaryotes?
   a. fungi   c. protozoa
   b. algae   d. a and b

4. What is embedded in rough endoplasmic reticulum?
   a. ribosomes
   b. Golgi apparatus
   c. chromatin
   d. vesicles

5. Yeasts are _____ fungi, and molds are _____ fungi.
   a. macroscopic, microscopic
   b. unicellular, filamentous
   c. motile, nonmotile
   d. water, terrestrial

6. In general, fungi derive nutrients through
   a. photosynthesis
   b. engulfing bacteria
   c. digesting organic substrates
   d. parasitism

7. A hypha divided into compartments by cross walls is called
   a. nonseptate
   b. imperfect
   c. septate
   d. perfect

8. Algae generally contain some type of
   a. spore
   b. chlorophyll
   c. locomotor organelle
   d. toxin

9. Almost all protozoa have a
   a. locomotor organelle   c. pellicle
   b. cyst stage   d. trophozoite stage

10. The protozoan trophozoite is the
    a. active feeding stage
    b. inactive dormant stage
    c. infective stage
    d. spore-forming stage

11. All mature sporozoa are
    a. parasitic   c. carried by vectors
    b. nonmotile   d. both a and b

12. Parasitic helminths reproduce with
    a. spores
    b. eggs and sperm
    c. mitosis
    d. cysts
    e. all of these

13. Mitochondria likely originated from
    a. archaea
    b. invaginations of the cell membrane
    c. purple bacteria
    d. cyanobacteria

14. **Single Matching.** Select the description that best fits the word in the left column.
    _____ diatom   a. the cause of malaria
    _____ *Rhizopus*   b. single-celled alga with silica in its cell wall
    _____ *Histoplasma*   c. fungal cause of Ohio Valley fever
    _____ *Cryptococcus* d. the cause of amoebic dysentery
    _____ euglenid   e. genus of black bread mold
    _____ dinoflagellate f. helminth worm involved in pinworm infection
    _____ *Trichomonas*   g. motile flagellated alga with eyespots
    _____ *Entamoeba*   h. a yeast that infects the lungs
    _____ *Plasmodium*   i. flagellated protozoan genus that causes an STD
    _____ *Enterobius*   j. alga that causes red tides

15. Human fungal infections involve and affect what areas of the human body?
    a. skin   c. lungs
    b. mucous membranes   d. a, b, and c

16. Most helminth infections
    a. are localized to one site in the body
    b. spread through major systems of the body
    c. develop within the spleen
    d. develop within the liver

Concept Questions

These questions are suggested as a *writing-to-learn* experience. For each question, compose a one- or two-paragraph answer that includes the factual information needed to completely address the question.

1. Construct a chart that reviews the major similarities and differences between procaryotic and eucaryotic cells.

2. a. Which kingdoms of the five-kingdom system contain eucaryotic microorganisms? How do unicellular, colonial, and multicellular organisms differ from each other?
   b. Give examples of each type.

3. a. Describe the anatomy and functions of each of the major eucaryotic organelles.
   b. How are flagella and cilia similar? How are they different?
   c. Compare and contrast the smooth ER, the rough ER, and the Golgi apparatus in structure and function.

4. Trace the synthesis of cell products, their processing, and their packaging through the organelle network.

5. Describe some of the ways that organisms use lysosomes.

6. For what reasons would a cell need a "skeleton"?

7. a. Differentiate between the yeast and hypha types of fungal cell.
   b. What is a mold?
   c. What does it mean if a fungus is dimorphic?

8. a. How does a fungus feed?
   b. Where would one expect to find fungi?

9. a. Describe the functional types of hyphae.
   b. Describe the two main types of asexual fungal spores and how they are formed.
   c. What are some types of conidia?
   d. What is the reproductive potential of molds in terms of spore production?
   e. How do mold spores differ from procaryotic spores?

10. a. Explain the importance of sexual spores to fungi.
    b. Describe the three main types of sexual spores, and construct a simple diagram to show how each is formed.

11. How are fungi classified? Give an example of a member of each fungus division and describe its structure and importance.

12. What is a mycosis? What kind of mycosis is athlete's foot? What kind is coccidioidomycosis?

13. What is a working definition of a "protist"?

14. a. Describe the principal characteristics of algae that separate them from protozoa.
    b. How are algae important?
    c. What causes the many colors in the algae?
    d. Are there any algae of medical importance?

15. a. Explain the general characteristics of the protozoan life cycle.
    b. Describe the protozoan adaptations for feeding.
    c. Describe protozoan reproductive processes.

16. a. Briefly outline the characteristics of the four protozoan groups.
    b. What is an important pathogen in each group?

17. a. Which protozoan group is the most complex in structure and behavior?
    b. In life cycle?
    c. What characteristics set the apicomplexa apart from the other protozoan groups?

18. a. Construct a chart that compares the four groups of eucaryotic microorganisms (fungi, algae, protozoa, helminths) in cellular structure.
    b. Indicate whether each group has a cell wall, chloroplasts, motility, or some other distinguishing feature.
    c. Include also the manner of nutrition and body plan (unicellular, colonial, filamentous, or multicellular) for each group.

19. Discuss the adaptations of parasitic worms to their lifestyles, and explain why these adaptations are necessary or advantageous to the worms' survival.

Critical Thinking Questions

Critical thinking is the ability to reason and solve problems using facts and concepts. These questions can be approached from a number of angles, and in most cases, they do not have a single correct answer.

1. Suggest some ways that one would go about determining if mitochondria and chloroplasts are a modified procaryotic cell.

2. Give the common name of a eucaryotic microbe that is unicellular, walled, non-photosynthetic, nonmotile, and bud-forming.

3. Give the common name of a microbe that is unicellular, nonwalled, motile with flagella, and has chloroplasts.

4. Which group of microbes has long, thin pseudopods and is encased in a hard shell?

5. What general type of multicellular parasite is composed primarily of thin sacs of reproductive organs?

6. a. Name two parasites that are transmitted in the cyst form.
   b. How must a non-cyst-forming pathogenic protozoan be transmitted? Why?

7. You just found an old container of food in the back in your refrigerator. You open it and see a mass of multicolored fuzz. As a budding microbiologist, describe how you would determine what types of organisms are growing on the food.

8. Explain what factors could cause opportunistic mycoses to be a growing medical problem.

9. a. How are bacterial endospores and cysts of protozoa alike?
   b. How do they differ?

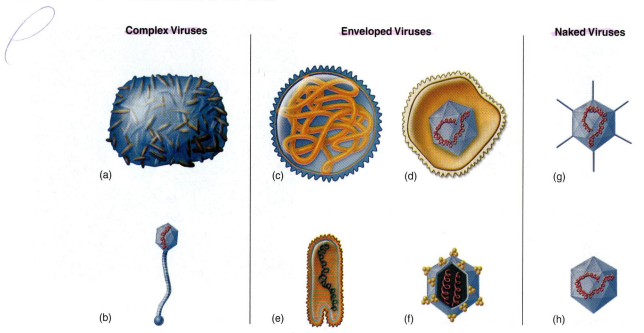

FIGURE 6.9 Morphology of viruses.
Complex viruses: (a) poxvirus, a large DNA virus; **(b)** flexible-tailed bacteriophage. **Enveloped viruses: (c)** mumps virus, an enveloped RNA virus with a helical nucleocapsid; **(d)** herpesvirus, an enveloped DNA virus with an icosahedral nucleocapsid; **(e)** rhabdovirus, a helical RNA virus with a bullet-shaped envelope; **(f)** HIV, an RNA retrovirus with an icosahedral capsid. **Naked viruses: (g)** adenovirus, a DNA virus with fibers on the capsid; **(h)** papillomavirus, a DNA virus that causes warts.

The Viral Envelope

When **enveloped viruses** (mostly animal) are released from the host cell, they take with them a bit of its membrane system in the form of an envelope, as described later on. Some viruses bud off the cell membrane; others leave via the nuclear envelope or the endoplasmic reticulum. Whichever avenue of escape, the viral envelope differs significantly from the host's membranes. In the envelope, some or all of the regular membrane proteins are replaced with special viral proteins. Some proteins form a binding layer between the envelope and capsid of the virus, and glycoproteins (proteins bound to a carbohydrate) remain exposed on the outside of the envelope. These protruding molecules, called **spikes** or peplomers, are essential for the attachment of viruses to the next host cell. Because the envelope is more supple than the capsid, enveloped viruses are pleomorphic and range from spherical to filamentous in shape.

Functions of the Viral Capsid/Envelope

The outermost covering of a virus is indispensable to viral function because it protects the nucleic acid from the effects of various enzymes and chemicals when the virus is outside the host cell. For example, the capsids of enteric (intestinal) viruses such as polio and hepatitis A are resistant to the acid- and protein-digesting enzymes of the gastrointestinal tract. Capsids and envelopes are also responsible for helping to introduce the viral DNA or RNA into a suitable host cell, first by binding to the cell surface

and then by assisting in penetration of the viral nucleic acid (to be discussed in more detail later in the chapter). In addition, parts of viral capsids and envelopes stimulate the immune system to produce antibodies that can neutralize viruses and protect the host's cells against future infections (see chapter 15).

Complex Viruses: Atypical Viruses

Two special groups of viruses, termed complex viruses (**figure 6.10),** are more intricate in structure than the helical, icosahedral, naked, or enveloped viruses just described. The poxviruses (including the agent of smallpox) are very large DNA viruses that lack a regular capsid and have in its place several layers of lipoproteins and coarse surface fibrils. Some members of another group of very complex viruses, the **bacteriophages** (bak-teer'-ee-oh-fay") have a polyhedral head, a helical tail, and fibers for attachment to the host cell. Their mode of multiplication is covered in a later section of this chapter. Figure 6.9 summarizes the morphological types of some common viruses.

Nucleic Acids: At the Core of a Virus

The sum total of the genetic information carried by an organism is known as its **genome.** So far, one biological constant is that the genetic information of living cells is carried by nucleic acids (DNA, RNA). Viruses, although neither alive nor cells, are no exception to this rule, but there is a significant difference. Unlike cells, which contain both DNA

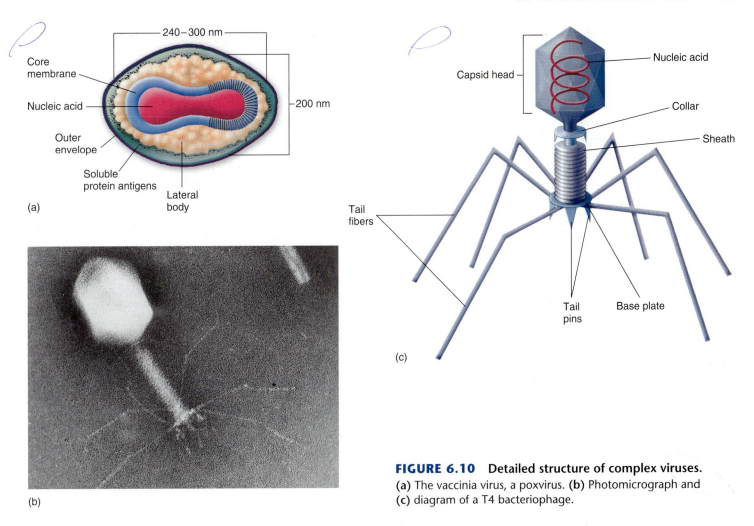

FIGURE 6.10 **Detailed structure of complex viruses.**
(a) The vaccinia virus, a poxvirus. (b) Photomicrograph and
(c) diagram of a T4 bacteriophage.

and RNA, viruses contain either DNA or RNA *but not both*. Because viruses must pack into a tiny space all of the genes necessary to instruct the host cell to make new viruses, the number of viral genes is quite small compared with that of a cell. It varies from four genes in hepatitis B virus to hundreds of genes in some herpesviruses. Viruses possess only the genes needed to invade host cells and redirect their activity. By comparison, the bacterium *Escherichia coli* has approximately 4,000 genes, and a human cell has approximately 30,000–40,000 genes. These additional genes allow cells to carry out the complex metabolic activity necessary for independent life.

In chapter 2 you learned that DNA usually exists as a double-stranded molecule and that RNA is single-stranded. Although most viruses follow this same pattern, a few exhibit distinctive and exceptional forms. Notable examples are the parvoviruses, which contain single-stranded DNA, and reoviruses (a cause of respiratory and intestinal tract infections), which contain double-stranded RNA. In fact, viruses exhibit wide variety in how their RNA or DNA is configured. DNA viruses can have single-stranded (ss) or double-stranded (ds) DNA; the dsDNA can be arranged linearly or

in ds circles. RNA viruses can be double-stranded but are more often single-stranded. You will learn in chapter 9 that all proteins are made by "translating" the nucleic acid code on a single strand of RNA into an amino acid sequence. Single stranded RNA genomes that are ready for immediate translation into proteins are called *positive-sense* RNA. Other RNA genomes have to be converted into the proper form to be made into proteins, and these are called *negative-sense* RNA. RNA genomes may also be *segmented*, meaning that the individual genes exist on separate pieces of RNA. The influenza virus (an orthomyxovirus) is an example of this. A special type of RNA virus is called a *retrovirus*. We'll discuss it later. **Tables 6.2** and **6.3** summarize the structures of some medically relevant DNA and RNA viruses.

In all cases, these tiny strands of genetic material carry the blueprint for viral structure and functions. In a very real sense, viruses are genetic parasites because they cannot multiply until their nucleic acid has reached the internal habitat of the host cell. At the minimum, they must carry genes for synthesizing the viral capsid and genetic material, for regulating the actions of the host, and for packaging the mature virus.

Other Substances in the Virus Particle

In addition to the protein of the capsid, the proteins and lipids of envelopes, and the nucleic acid of the core, viruses can contain enzymes for specific operations within their host cell. They may come with pre-formed enzymes that are required for viral replication. Examples include *polymerases* (pol-im'-ur-ace) that synthesize DNA and RNA and replicases that copy RNA. The AIDS virus comes equipped with *reverse transcriptase* for synthesizing DNA from RNA. However, viruses completely lack the genes for synthesis of metabolic enzymes. As we shall see, this deficiency has little consequence, because viruses have adapted to completely take over their hosts' metabolic resources. Some viruses can actually carry away substances from their host cell. For instance, arenaviruses pack along host ribosomes, and retroviruses "borrow" the host's tRNA molecules.

TABLE 6.2 Medically Relevant DNA Virus Groups

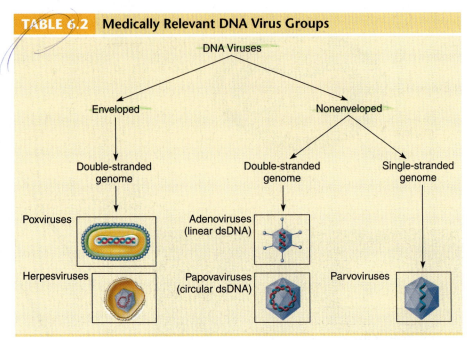

Adapted from: *Poxviridae* from Buller et al., National Institute of Allergy & Infectious Disease, Department of Health & Human Services.

TABLE 6.3 Medically Relevant RNA Viruses

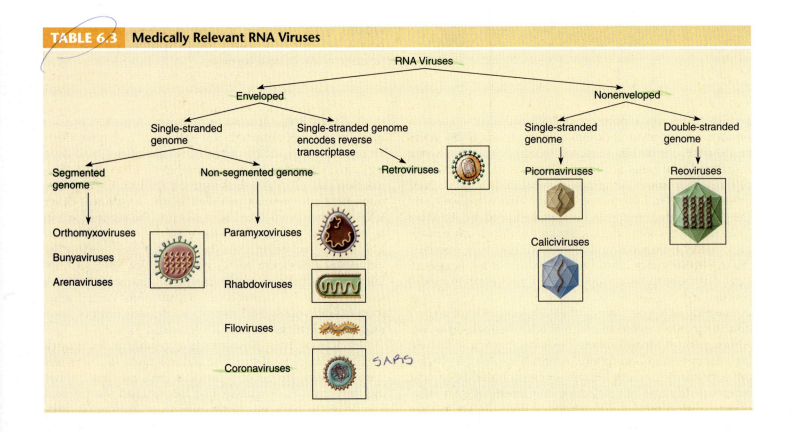

TABLE 6.4	Examples from the Three Orders of Viruses			
Genome Type	**Order**	**Family**	**Genus**	**Species**
dsDNA	Caudovirales	Poxviridae	Orthopoxvirus	Vaccinia virus
neg (ss)RNA	Mononegavirales	Paramyxoviridae	Morbillivirus	Measles virus
pos (ss)RNA	Nidovirales	Togaviridae	Rubivirus	Rubella virus

Adapted from van Regenmortel, M., editor, et al. 2000. *Virus Taxonomy. Seventh Report of the International Committee on Taxonomy of Viruses.* New York: Academic Press.

6.4 How Viruses Are Classified and Named

Although viruses are not classified as members of the kingdoms discussed in chapter 1, they are diverse enough to require their own classification scheme to aid in their study and identification. In an informal and general way, we have already begun classifying viruses—as animal, plant, or bacterial viruses; enveloped or naked viruses; DNA or RNA viruses; and helical or icosahedral viruses. These introductory categories are certainly useful in organization and description, but the study of specific viruses requires a more standardized method of nomenclature. For many years, the animal viruses were classified mainly on the basis of their hosts and the kind of diseases they caused. Newer systems for naming viruses also take into account the actual nature of the virus particles themselves, with only partial emphasis on host and disease. The main criteria presently used to group viruses are structure, chemical composition, and similarities in genetic makeup.

In 2000 the International Committee on the Taxonomy of Viruses issued their latest report on the classification of viruses. They listed 3 orders, 63 families, and 263 genera of viruses. Previous to 2000 there had been only a single recognized order of viruses. Examples of each of the three orders of viruses are presented in **table 6.4.** Note the naming conventions—that is, virus families are written with "-viridae" on the end of the name, and genera end with "-virus."

Historically, some virologists had created an informal *species* naming system that mirrors the species names in higher organisms, using genus and species epithets such as *Measles morbillivirus.* This has not been an official designation, however. The species category has created a lot of controversy within the virology community, with many scientists arguing that non-organisms such as viruses can never be speciated. Others argue that viruses are too changeable, and thus fine distinctions used for deciding on species classifications will quickly disappear. Over the past decade, virologists have largely accepted the concept of viral species, defining them as consisting of members that have a number of properties in common but have some variation in their properties. In other words, a virus is placed in a species on the basis of a collection of properties. For viruses that infect humans, species may be defined based on relatively minor differences in host range, pathogenicity, or antigenicity. The important thing to remember is that viral species designations, in the words of one preeminent viral taxonomist, are "fuzzy sets with hazy boundaries."[3]

Because the use of standardized species names has not been widely accepted, the genus or common English vernacular names (for example, poliovirus and rabies virus) predominate in discussions of specific viruses in this text. **Table 6.5** illustrates the naming system for important viruses and the diseases they cause.

6.5 Modes of Viral Multiplication

Viruses are closely associated with their hosts. In addition to providing the viral habitat, the host cell is absolutely necessary for viral multiplication. The process of viral multiplication is an extraordinary biological phenomenon. Viruses have often been aptly described as minute parasites that seize control of the synthetic and genetic machinery of cells. The nature of this cycle dictates viral pathogenicity, transmission, the responses of the immune defenses, and human measures to control viral infections. From these perspectives, we cannot overemphasize the importance of a working knowledge of the relationship between viruses and their host cells.

Multiplication Cycles in Animal Viruses

The general phases in the life cycle of animal viruses are **adsorption, penetration, uncoating, synthesis, assembly,** and **release** from the host cell. The length of the entire multiplication cycle varies from 8 hours in polioviruses to 36 hours in herpesviruses. See **figure 6.11** for the major phases of one type of animal virus.

Adsorption and Host Range

Invasion begins when the virus encounters a susceptible host cell and adsorbs specifically to receptor sites on the cell membrane. The membrane receptors that viruses attach to are usually glycoproteins the cell requires for its normal function. For example, the rabies virus affixes to the acetylcholine receptor of nerve cells, and the human immunodeficiency virus (HIV or AIDS virus) attaches to the CD4 protein on

---

3. van Regenmortel, M. H. V., and Mahy, B. W. J. Emerging issues in virus taxonomy. *Emerg. Infect. Dis.* [serial online] 2004 Jan [*date cited*]. Available from www.cdc.gov/ncidod/EID/vol10no1/03-0279.htm

TABLE 6.5	Important Human Virus Families, Genera, Common Names, and Types of Diseases			
	Family	**Genus of Virus**	**Common Name of Genus Members**	**Name of Disease**
DNA Viruses				
	Poxviridae	Orthopoxvirus	Variola and vaccinia	Smallpox, cowpox
	Herpesviridae	Simplexvirus	Herpes simplex (HSV) 1 virus	Fever blister, cold sores
			Herpes simplex (HSV) 2 virus	Genital herpes
		Varicellovirus	Varicella zoster virus (VZV)	Chickenpox, shingles
		Cytomegalovirus	Human cytomegalovirus (CMV)	CMV infections
	Adenoviridae	Mastadenovirus	Human adenoviruses	Adenovirus infection
	Papovaviridae	Papillomavirus	Human papillomavirus (HPV)	Several types of warts
		Polyomavirus	JC virus (JCV)	Progressive multifocal leukoencephalopathy (PML)
	Hepadnaviridae	Hepadnavirus	Hepatitis B virus (HBV or Dane particle)	Serum hepatitis
	Parvoviridae	Erythrovirus	Parvovirus B19	Erythema infectiosum
RNA Viruses				
	Picornaviridae	Enterovirus	Poliovirus	Poliomyelitis
			Coxsackievirus	Hand-foot-mouth disease
		Hepatovirus	Hepatitis A virus (HAV)	Short-term hepatitis
		Rhinovirus	Human rhinovirus	Common cold, bronchitis
	Calciviridae	Calicivirus	Norwalk virus	Viral diarrhea, Norwalk virus syndrome
	Togaviridae	Alphavirus	Eastern equine encephalitis virus	Eastern equine encephalitis (EEE)
			Western equine encephalitis virus	Western equine encephalitis (WEE)
			Yellow fever virus	Yellow fever
			St. Louis encephalitis virus	St. Louis encephalitis
		Rubivirus	Rubella virus	Rubella (German measles)
	Flaviviridae	Flavivirus	Dengue fever virus	Dengue fever
			West Nile fever virus	West Nile fever
	Bunyaviridae	Bunyavirus	Bunyamwera viruses	California encephalitis
		Hantavirus	Sin Nombre virus	Respiratory distress syndrome
		Phlebovirus	Rift Valley fever virus	Rift Valley fever
		Nairovirus	Crimean–Congo hemorrhagic fever virus (CCHF)	Crimean–Congo hemorrhagic fever
	Filoviridae	Filovirus	Ebola, Marburg virus	Ebola fever
	Reoviridae	Coltivirus	Colorado tick fever virus	Colorado tick fever
		Rotavirus	Human rotavirus	Rotavirus gastroenteritis
	Orthomyxoviridae	Influenza virus	Influenza virus, type A (Asian, Hong Kong, and swine influenza viruses)	Influenza or "flu"
	Paramyxoviridae	Paramyxovirus	Parainfluenza virus, types 1–5	Parainfluenza
			Mumps virus	Mumps
		Morbillivirus	Measles virus	Measles (red)
		Pneumovirus	Respiratory syncytial virus (RSV)	Common cold syndrome
	Rhabdoviridae	Lyssavirus	Rabies virus	Rabies (hydrophobia)
	Retroviridae	Oncornavirus	Human T-cell leukemia virus (HTLV)	T-cell leukemia
		Lentivirus	HIV (human immunodeficiency viruses 1 and 2)	Acquired immunodeficiency syndrome (AIDS)
	Arenaviridae	Arenavirus	Lassa virus	Lassa fever
	Coronaviridae	Coronavirus	Infectious bronchitis virus (IBV)	Bronchitis
			Enteric corona virus	Coronavirus enteritis
			SARS virus	Severe acute respiratory syndrome

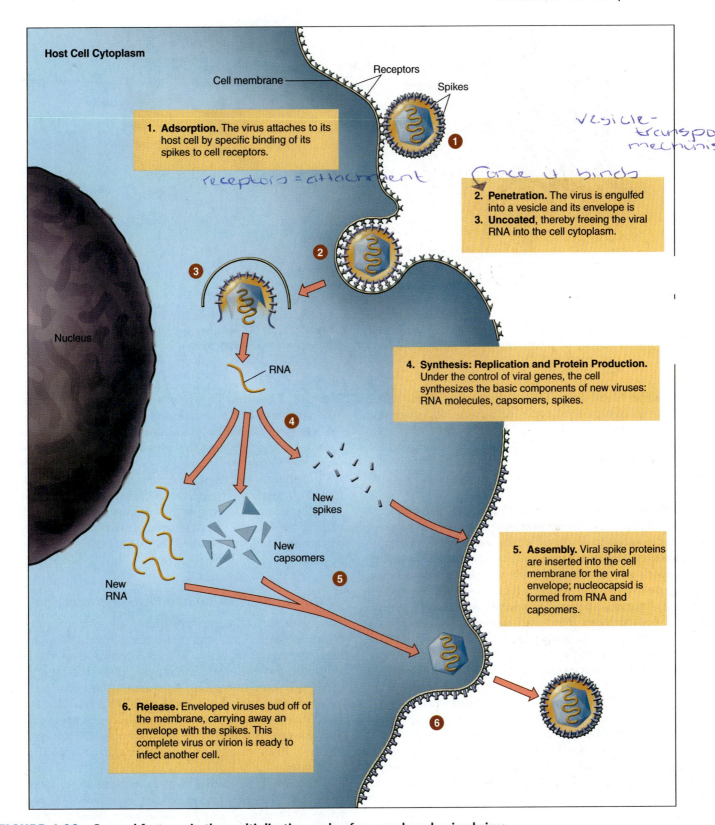

Host Cell Cytoplasm

Cell membrane

Receptors

Spikes

1. **Adsorption.** The virus attaches to its host cell by specific binding of its spikes to cell receptors.

receptors = attachment [handwritten]

Vesicle-transport mechanism [handwritten]

once it binds [handwritten]

2. **Penetration.** The virus is engulfed into a vesicle and its envelope is
3. **Uncoated,** thereby freeing the viral RNA into the cell cytoplasm.

Nucleus

RNA

4. **Synthesis: Replication and Protein Production.** Under the control of viral genes, the cell synthesizes the basic components of new viruses: RNA molecules, capsomers, spikes.

New spikes

New capsomers

New RNA

5. **Assembly.** Viral spike proteins are inserted into the cell membrane for the viral envelope; nucleocapsid is formed from RNA and capsomers.

6. **Release.** Enveloped viruses bud off of the membrane, carrying away an envelope with the spikes. This complete virus or virion is ready to infect another cell.

FIGURE 6.11   **General features in the multiplication cycle of an enveloped animal virus.**
Using an RNA virus (rubella virus), the major events are outlined, although other viruses will vary in exact details of the cycle.

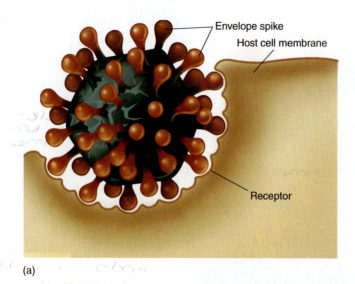

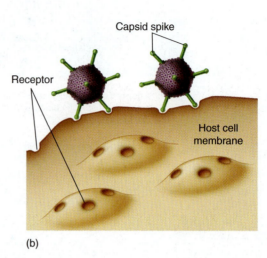

FIGURE 6.12 **The mode by which animal viruses adsorb to the host cell membrane.**
(a) An enveloped coronavirus with prominent spikes. The configuration of the spike has a complementary fit for cell receptors. The process in which the virus lands on the cell and plugs into receptors is termed docking. **(b)** An adenovirus has a naked capsid that adheres to its host cell by nestling surface molecules on its capsid into the receptors on the host cell's membrane.

certain white blood cells. The mode of attachment varies between the two general types of viruses. In enveloped forms such as influenza virus and HIV, glycoprotein spikes bind to the cell membrane receptors. Viruses with naked nucleocapsids (adenovirus, for example) use molecules on their capsids that adhere to cell membrane receptors **(figure 6.12).**

Because a virus can invade its host cell only through making an exact fit with a specific host molecule, the range of hosts it can infect in a natural setting is limited. This limitation, known as the **host range,** may be as restricted as hepatitis B, which infects only liver cells of humans; intermediate like the poliovirus, which infects intestinal and nerve cells of primates (humans, apes, and monkeys); or as broad as the rabies virus, which can infect various cells of all mammals. Cells that lack compatible virus receptors are resistant to adsorption and invasion by that virus. This explains why, for example, human liver cells are not infected by the canine hepatitis virus and dog liver cells cannot host the human hepatitis A virus. It also explains why viruses usually have tissue specificities called *tropisms* (troh'-pizmz) for certain cells in the body. The hepatitis B virus targets the liver, and the mumps virus targets salivary glands. However, the fact that many viruses can be manipulated to infect cells that they would not infect naturally makes it possible to cultivate them in the laboratory.

Penetration/Uncoating of Animal Viruses

Animal viruses exhibit some impressive mechanisms for entering a host cell. The flexible cell membrane of the host is penetrated by the whole virus or its nucleic acid **(figure 6.13).** In penetration by **endocytosis (figure 6.13a),** the entire virus

is engulfed by the cell and enclosed in a vacuole or vesicle. When enzymes in the vacuole dissolve the envelope and capsid, the virus is said to be **uncoated,** a process that releases the viral nucleic acid into the cytoplasm. The exact manner of uncoating varies, but in most cases, the virus fuses with the wall of the vesicle. Another means of entry involves direct fusion of the viral envelope with the host cell membrane (as in influenza and mumps viruses) **(figure 6.13b).** In this form of penetration, the envelope merges directly with the cell membrane, thereby liberating the nucleocapsid into the cell's interior.

Synthesis: Replication and Protein Production

The synthetic and replicative phases of animal viruses are highly regulated and extremely complex at the molecular level. Free viral nucleic acid exerts control over the host's synthetic and metabolic machinery. How this control proceeds will vary, depending on whether the virus is a DNA or an RNA virus. In general, the DNA viruses (except poxviruses) enter the host cell's nucleus and are replicated and assembled there. With few exceptions (such as retroviruses), RNA viruses are replicated and assembled in the cytoplasm.

The details of animal virus replication are discussed in **Insight 6.2.** Here we provide a brief overview of the process, using RNA viruses as a model. Almost immediately upon entry, the viral nucleic acid alters the genetic expression of the host and instructs it to synthesize the building blocks for new viruses. First, the RNA of the virus becomes a message for synthesizing viral proteins (translation). The viruses with positive-sense RNA molecules already contain the correct message for translation into proteins. Viruses with

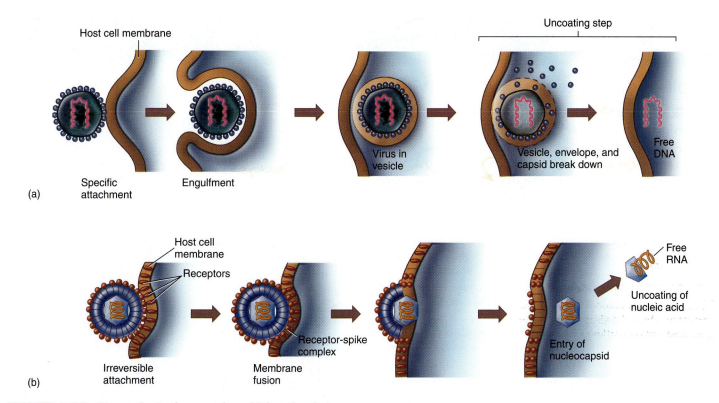

(a)

Host cell membrane

Specific attachment — Engulfment — Virus in vesicle — Uncoating step — Vesicle, envelope, and capsid break down — Free DNA

(b)

Host cell membrane

Receptors

Irreversible attachment — Membrane fusion — Receptor-spike complex — Entry of nucleocapsid — Free RNA — Uncoating of nucleic acid

FIGURE 6.13   Two principal means by which animal viruses penetrate.
(a) Endocytosis (engulfment) and uncoating of a herpesvirus. **(b)** Fusion of the cell membrane with the viral envelope (mumps virus).

negative-sense RNA molecules must first be converted into a positive sense message. Some viruses come equipped with the necessary enzymes for synthesis of viral components; others utilize those of the host. In the next phase, new RNA is synthesized using host nucleotides. Proteins for the capsid, spikes, and viral enzymes are synthesized on the host's ribosomes using its amino acids.

Assembly of Animal Viruses: Host Cell As Factory

Toward the end of the cycle, mature virus particles are constructed from the growing pool of parts. In most instances, the capsid is first laid down as an empty shell that will serve as a receptacle for the nucleic acid strand. Electron micrographs taken during this time show cells with masses of viruses, often in crystalline packets **(figure 6.14).** One important event leading to the release of enveloped viruses is the insertion of viral spikes into the host's cell membrane so they can be picked up as the virus buds off with its envelope, as discussed earlier.

Release of Mature Viruses

To complete the cycle, assembled viruses leave their host in one of two ways. Nonenveloped and complex viruses that reach maturation in the cell nucleus or cytoplasm are released when the cell lyses or ruptures. Enveloped viruses

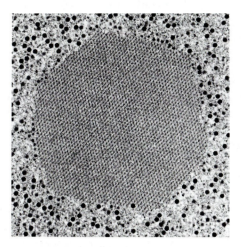

FIGURE 6.14   Nucleus of a cell, containing a crystalline mass of adenovirus (35,000×).

are liberated by **budding** or **exocytosis**[4] from the membranes of the cytoplasm, nucleus, endoplasmic reticulum, or vesicles. During this process, the nucleocapsid binds to the membrane, which curves completely around it and forms a small pouch. Pinching off the pouch releases the

---

4. For enveloped viruses, these terms are interchangeable. They mean the release of a virus from an animal cell by enclosing it in a portion of membrane derived from the cell.

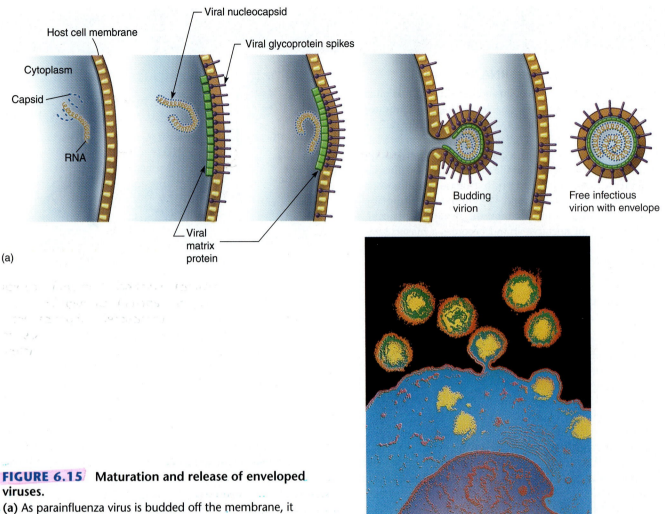

Host cell membrane

Cytoplasm

Capsid

RNA

Viral nucleocapsid

Viral glycoprotein spikes

Viral matrix protein

(a)

Budding virion

Free infectious virion with envelope

(b)

FIGURE 6.15 **Maturation and release of enveloped viruses.**
(a) As parainfluenza virus is budded off the membrane, it simultaneously picks up an envelope and spikes. **(b)** AIDS viruses (HIV) leave their host T cell by budding off its surface.

virus with its envelope **(figure 6.15).** Budding of enveloped viruses causes them to be shed gradually, without the sudden destruction of the cell. Regardless of how the virus leaves, most active viral infections are ultimately lethal to the cell because of accumulated damage. Lethal damages include a permanent shutdown of metabolism and genetic expression, destruction of cell membrane and organelles, toxicity of virus components, and release of lysosomes.

The number of viruses released by infected cells is variable, controlled by factors such as the size of the virus and the health of the host cell. About 3,000 to 4,000 virions are released from a single cell infected with poxviruses, whereas a poliovirus-infected cell can release over 100,000 virions. If even a small number of these virions happens to meet another susceptible cell and infect it, the potential for rapid viral proliferation is immense.

Damage to the Host Cell and Persistent Infections

The short- and long-term effects of viral infections on animal cells are well documented. **Cytopathic** (sy"-toh-path'-ik) **effects** (CPEs) are defined as virus-induced damage to the cell that alters its microscopic appearance. Individual cells can become disoriented, undergo gross changes in shape or size, or develop intracellular changes **(figure 6.16a).** It is common to note *inclusion bodies,* or compacted masses of viruses or damaged cell organelles, in the nucleus and cytoplasm **(figure 6.16b).** Examination of cells and tissues for cytopathic effects is an important part of the diagnosis of viral infections. **Table 6.6** summarizes some prominent cytopathic effects associated with specific viruses. One very common CPE is the fusion of multiple host cells into single large cells containing multiple nuclei. These **syncytia** are a result of some viruses' ability to fuse membranes. One virus (respiratory syncytial virus) is even named for this effect.

Although accumulated damage from a virus infection kills most host cells, some cells maintain a carrier relationship, in which the cell harbors the virus and is not immediately lysed. These so-called *persistent infections* can last from a few weeks to the remainder of the host's life. One of the more serious complications occurs with the measles virus. It may remain hidden in brain cells for many years, causing progressive damage and loss of function. Several viruses remain in a

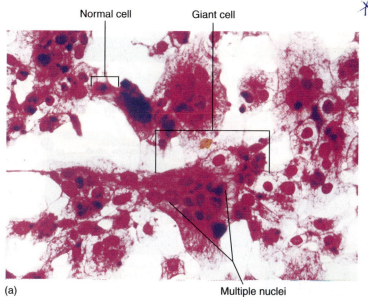

Normal cell   Giant cell

(a)

Multiple nuclei

Inclusion bodies

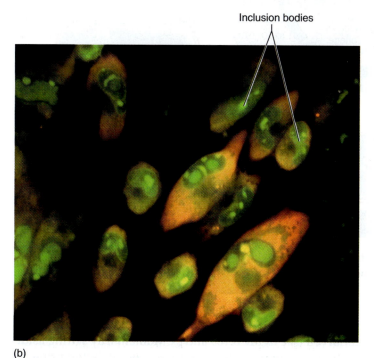

(b)

FIGURE 6.16 **Cytopathic changes in cells and cell cultures infected by viruses.**
(a) Human epithelial cells infected by herpes simplex virus demonstrate multinucleate giant cells. **(b)** Fluorescent-stained human cells infected with cytomegalovirus. Note the inclusion bodies (arrows). Note also that both viruses disrupt the cohesive junctions between cells.

chronic latent state,[5] periodically becoming reactivated. Examples of this are herpes simplex viruses (cold sores and genital herpes) and herpes zoster virus (chickenpox and shingles). Both viruses can go into latency in nerve cells and later emerge under the influence of various stimuli to cause recur-

TABLE 6.6	Cytopathic Changes in Selected Virus-Infected Animal Cells
Virus	**Response in Animal Cell**
Smallpox virus	Cells round up; inclusions appear in cytoplasm
Herpes simplex	Cells fuse to form multinucleated syncytia; nuclear inclusions (see figure 6.16)
Adenovirus	Clumping of cells; nuclear inclusions
Poliovirus	Cell lysis; no inclusions
Reovirus	Cell enlargement; vacuoles and inclusions in cytoplasm
Influenza virus	Cells round up; no inclusions
Rabies virus	No change in cell shape; cytoplasmic inclusions (Negri bodies)
Measles virus	Syncytia form (multinucleate)

rent symptoms. Specific damage that occurs in viral diseases is covered more completely in chapters 18 through 23.

Some animal viruses enter their host cell and permanently alter its genetic material, leading to cancer. These viruses are termed *oncogenic,* and their effect on the cell is called **transformation.** A startling feature of these viruses is that their nucleic acid is consolidated into the host DNA. Transformed cells have an increased rate of growth; alterations in chromosomes; changes in the cell's surface molecules; and the capacity to divide for an indefinite period, unlike normal animal cells. Mammalian viruses capable of initiating tumors are called **oncoviruses.** Some of these are DNA viruses such as papillomavirus (genital warts are associated with cervical cancer), herpesviruses (Epstein-Barr virus causes Burkitt's lymphoma), and hepatitis B virus. Two viruses related to HIV—HTLV I and II[6]—are involved in human cancers. These findings have spurred a great deal of speculation on the possible involvement of viruses in cancers whose cause is still unknown. Additional information on the connection between viruses and cancer is found in chapters 9 and 20.

CHECKPOINT

■ Virus size range is from 20 nm to 450 nm (diameter). Viruses are composed of an outer protein capsid enclosing either DNA or RNA plus a variety of enzymes. Some viruses also exhibit an envelope around the capsid.

■ Viruses go through a multiplication cycle that generally involves adsorption, penetration (sometimes followed by uncoating), viral synthesis and assembly, and viral release by lysis or budding.

■ These events turn the host cell into a factory solely for making and shedding new viruses. This results in the ultimate destruction of the cell.

■ Animal viruses can cause acute infections or can persist in host tissues as chronic latent infections that can reactivate periodically throughout the host's life. Some persistent animal viruses are oncogenic.

---

5. Meaning that they exist in an inactive state over long periods.

6. Human T-cell lymphotropic viruses: cause types of leukemia.

lytic cycle. The lysogenic phase is depicted as part of figure 6.17. Lysogeny is a less deadly form of parasitism than the full lytic cycle and is thought to be an advancement that allows the virus to spread without killing the host. Many bacteria that infect humans are lysogenized by phages. And sometimes that is very bad news for the human: Occasionally phage genes in the bacterial chromosome cause the production of toxins or enzymes that cause pathology in the human. When a bacterium acquires a new trait from its temperate phage, it is called **lysogenic conversion**. The phenomenon was first discovered in the 1950s in the bacterium that causes diphtheria, *Corynebacterium diphtheriae*. The diphtheria toxin responsible for the deadly nature of the disease is a bacteriophage product. *C. diphtheriae* without the phage are harmless. Other bacteria that are made virulent by their prophages are *Vibrio cholerae*, the agent of cholera, and *Clostridium botulinum*, the cause of botulism. On page 171 we described a similar relationship that exists between certain animal viruses and human cells.

The cycle of bacterial and animal viruses (see figures 6.11 and 6.17) illustrates general features of viral multiplication in a very concrete and memorable way. The two cycles are compared in **table 6.7**. It is fascinating to realize that viruses are capable of lying "dormant" in their host cells, possibly becoming active at some later time. Because of the intimate

association between the genetic material of the virus and host, phages occasionally serve as transporters of bacterial genes from one bacterium to another and consequently can play a profound role in bacterial genetics. This phenomenon, called transduction, is one way that genes for toxin production and drug resistance are transferred between bacteria (see chapters 9 and 12).

> ### ✔ CHECKPOINT
>
> - Bacteriophages vary significantly from animal viruses in their methods of adsorption, penetration, site of replication, and method of exit from host cells.
> - Lysogeny is a condition in which viral DNA is inserted into the bacterial chromosome and remains inactive for an extended period. It is replicated right along with the chromosome every time the bacterium divides.
> - Some bacteria express virulence traits that are coded for by the bacteriophage DNA in their chromosomes. This phenomenon is called lysogenic conversion.

6.6 Techniques in Cultivating and Identifying Animal Viruses

One problem hampering earlier animal virologists was their inability to propagate specific viruses routinely in pure culture and in sufficient quantities for their studies. Virtually all of the pioneering attempts at cultivation had to be performed in an organism that was the usual host for the virus. But this method had its limitations. How could researchers have ever traced the stages of viral multiplication if they had been restricted to the natural host, especially in the case of human viruses? Fortunately, systems of cultivation with broader applications were developed, including *in vivo* (in vee'-voh) inoculation of laboratory-bred animals and embryonic bird tissues and *in vitro* (in vee'-troh) cell (or tissue) culture methods. Such use of substitute host systems permits greater control, uniformity, and wide-scale harvesting of viruses.

The primary purposes of viral cultivation are:

1. to isolate and identify viruses in clinical specimens;
2. to prepare viruses for vaccines; and
3. to do detailed research on viral structure, multiplication cycles, genetics, and effects on host cells.

Using Live Animal Inoculation

Specially bred strains of white mice, rats, hamsters, guinea pigs, and rabbits are the usual choices for animal cultivation of viruses. Invertebrates (insects) or nonhuman primates are occasionally used as well. Because viruses can exhibit some host specificity, certain animals can propagate a given virus more readily than others. Depending on the particular experiment, tests can be performed on adult, juvenile, or newborn animals. The animal is exposed to the virus by injection of a viral preparation or specimen into the brain, blood, muscle, body cavity, skin, or footpads.

TABLE 6.7	Comparison of Bacteriophage and Animal Virus Multiplication	
	Bacteriophage	**Animal Virus**
Adsorption	Precise attachment of special tail fibers to cell wall	Attachment of capsid or envelope to cell surface receptors
Penetration	Injection of nucleic acid through cell wall; no uncoating of nucleic acid	Whole virus is engulfed and uncoated, or virus surface fuses with cell membrane, nucleic acid is released
Synthesis and Assembly	Occurs in cytoplasm Cessation of host synthesis Viral DNA or RNA is replicated and begins to function Viral components synthesized	Occurs in cytoplasm and nucleus Cessation of host synthesis Viral DNA or RNA is replicated and begins to function Viral components synthesized
Viral Persistence	Lysogeny	Latency, chronic infection, cancer
Release from Host Cell	Cell lyses when viral enzymes weaken it	Some cells lyse; enveloped viruses bud off host cell membrane
Cell Destruction	Immediate	Immediate or delayed

Using Bird Embryos

An embryo is an early developmental stage of animals marked by rapid differentiation of cells. Birds undergo their embryonic period within the closed protective case of an egg, which makes an incubating bird egg a nearly perfect system for viral propagation. It is an intact and self-supporting unit, complete with its own sterile environment and nourishment. Furthermore, it furnishes several embryonic tissues that readily support viral multiplication.

Chicken, duck, and turkey eggs are the most common choices for inoculation. The egg must be injected through the shell, usually by drilling a hole or making a small window. Rigorous sterile techniques must be used to prevent contamination by bacteria and fungi from the air and the outer surface of the shell. The exact tissue that is inoculated is guided by the type of virus being cultivated and the goals of the experiment **(figure 6.21)**.

Viruses multiplying in embryos may or may not cause effects visible to the naked eye. The signs of viral growth include death of the embryo, defects in embryonic development, and localized areas of damage in the membranes, resulting in discrete, opaque spots called pocks (a variant of *pox*). If a virus does not produce overt changes in the developing embryonic tissue, virologists have other methods of detection. Embryonic fluids and tissues can be prepared for direct examination with an electron microscope. Certain viruses can also be detected by their ability to agglutinate red blood cells (form big clumps) or by their reaction with an antibody of known specificity that will affix to its corresponding virus, if it is present.

Using Cell (Tissue) Culture Techniques

The most important early discovery that led to easier cultivation of viruses in the laboratory was the development of a simple and effective way to grow populations of isolated animal cells in culture. These types of in vitro cultivation systems are termed cell culture or tissue culture. (Although these terms are used interchangeably, cell culture is probably a more accurate description.) So prominent is this method that most viruses are propagated in some sort of cell culture, and much of the virologist's work involves developing and maintaining these cultures. Animal cell cultures are grown in sterile chambers with special media that contain the correct nutrients required by animal cells to survive. The cultured cells grow in the form of a *monolayer,* a single, confluent sheet of cells that supports viral multiplication and permits close inspection of the culture for signs of infection **(figure 6.22)**.

Cultures of animal cells usually exist in the primary or continuous form. *Primary cell cultures* are prepared by placing freshly isolated animal tissue in a growth medium. The cells undergo a series of mitotic divisions to produce a monolayer. Embryonic, fetal, adult, and even cancerous tissues have served as sources of primary cultures. A primary culture retains several characteristics of the original tissue from which it was derived, but this original line generally has a limited existence. Eventually, it will die out or mutate into a

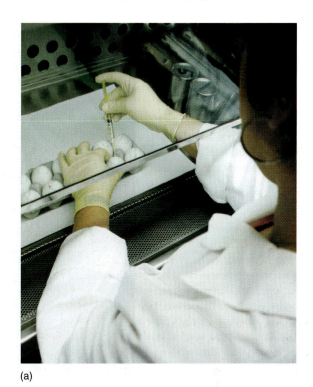

(a)

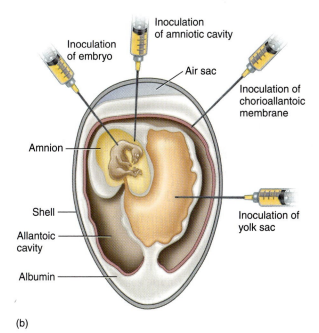

(b)

FIGURE 6.21 **Cultivating animal viruses in a developing bird embryo.**

(a) A technician inoculates fertilized chicken eggs with viruses in the first stage of preparing vaccines. This process requires the highest levels of sterile and aseptic precautions. Influenza vaccine is prepared this way. (b) The shell is perforated using sterile techniques, and a virus preparation is injected into a site selected to grow the viruses. Targets include the allantoic cavity, a fluid-filled sac that functions in embryonic waste removal; the amniotic cavity, a sac that cushions and protects the embryo itself; the chorioallantoic membrane, which functions in embryonic gas exchange; the yolk sac, a membrane that mobilizes yolk for the nourishment of the embryo; and the embryo itself.

INSIGHT 6.4   *Medical*

Uncommon Facts About the Common Cold

The common cold touches the lives of humans more than any other viral infection, afflicting at least half the population every year and accounting for millions of hours of absenteeism from work and school. The reason for its widespread distribution is not that it is more virulent or transmissible than other infections, but that symptoms of colds are linked to hundreds of different viruses and viral strains. Among the known causative viruses, in order of importance, are rhinoviruses (which cause about half of all colds), paramyxoviruses, enteroviruses, coronaviruses, reoviruses, and adenoviruses. A given cold can be caused by a single virus type or it can result from a mixed infection.

The name implies a relationship with cold weather or drafts. But studies in which human volunteers with wet heads or feet chilled or exposed to moist, frigid air have failed to support such a link. Most colds occur in the late autumn, winter, and early spring—all periods of colder weather—but this seasonal connection has more to do with being confined in closed spaces with carriers than with temperature.

The most significant single factor in the spread of colds is contamination of hands with mucous secretions. The portal of entry is the mucous membranes of the nose and eyes. The most common symptom is a nasal discharge, and the least common is fever, except in infants and children.

Finding a cure for the common cold has been a long-standing goal of medical science. This quest is not motivated by the clinical nature of a cold, which is really a rather benign infection. A more likely reason to search for a "magic cold bullet" is productivity in the workplace and schools, as well as the potential profits from a truly effective cold drug.

One nonspecific approach has been to destroy the virus outright and halt its spread. Special facial tissues impregnated with mild acid have been marketed for use during the cold season. After years of controversy, a recent study has finally shown that taking megadoses of vitamin C at the onset of a cold can be beneficial. Zinc lozenges may also help retard the onset of cold symptoms.

The important role of natural interferon in controlling many cold viruses has led to the testing and marketing of a nasal spray containing recombinant interferon. Another company is currently developing a novel therapy based on monoclonal antibodies. These antibodies are raised to the site on the human cell (receptor) to which the rhinovirus attaches. In theory, these antibodies should occupy the cell receptor, competitively inhibit viral attachment, and prevent infection. Experiments in chimpanzees and humans showed that, when administered intranasally, this antibody preparation delayed the onset of symptoms and reduced their severity.

IN THE NEWS (Continued from page 153)

The disease in question was avian influenza, or bird flu. It was caused by avian strains of influenza type A viruses. The avian influenza A viruses typically do not infect humans. The first avian influenza virus was isolated from terns in South Africa in 1961. These viruses circulate among wild birds. The disease is very contagious among the birds and can be deadly among domesticated birds such as chickens. In recent years, outbreaks of avian influenza have been occurring among poultry populations in countries throughout Asia. Few countries, however, experience human cases. In 1997, a highly pathogenic strain infected humans directly from domesticated birds for the first time.

In a 2003–2004 outbreak, the bird flu emerged on poultry farms in the United States (for the first time) and Asia. Human cases—such as the one in this case study—occurred in Thailand and Vietnam. As of February 9, 2004, Thailand reported 23 laboratory confirmed human cases of avian influenza with a 78% mortality rate. There was a mass culling of chickens. Most of the human cases resulted from contact with infected poultry or contaminated surfaces. The virus is shed from birds in saliva, nasal secretions, and feces.

The Centers for Disease Control and Prevention (CDC) and World Health Organization (WHO) are involved in investigative activities related to the outbreaks and in developing rapid detection kits and a vaccine against the circulating avian influenza strains.

See: CDC. 2004. Cases of influenza A (H5N1)—Thailand, 2004. MMWR 53:100–103.
www.cdc.gov/flu/avian/outbreak.htm
www.who.int/csr/disease/avian_influenza/avian_faqs/en/

6.8 Other Noncellular Infectious Agents

Not all noncellular infectious agents have typical viral morphology. One group of unusual forms, even smaller and simpler than viruses, is implicated in chronic, persistent diseases in humans and animals. These diseases are called spongiform encephalopathies because the brain tissue removed from affected animals resembles a sponge. The infection has a long period of latency (usually several years) before the first clinical signs appear. Signs range from mental derangement to loss of muscle control. The diseases are progressive and universally fatal.

A common feature of these conditions is the deposition of distinct protein fibrils in the brain tissue. Researchers have hypothesized that these fibrils are the agents of the disease and have named them **prions** (pree'-onz).

Creutzfeldt-Jakob disease afflicts the central nervous system of humans and causes gradual degeneration and death. Cases in which medical workers developed the disease after handling autopsy specimens seem to indicate that it is transmissible, but by an unknown mechanism. Several animals (sheep, mink, elk) are victims of similar transmissible diseases. Bovine spongiform encephalopathy, or "mad cow disease," was recently the subject of fears and a crisis in Europe when researchers found evidence that the disease could be acquired by humans who consumed contaminated beef. This was the first incidence of prion disease transmission from animals to humans. Several hundred Europeans developed symptoms of a variant form of Creutzfeldt-Jakob disease, leading to strict governmental controls on exporting cattle and beef products. In 2003, isolated cows with BSE were found in Canada and in the United States. Extreme precautionary measures have been taken to protect North American consumers. (This disease is described in more detail in chapter 19.)

The exact mode of prion infection is currently being analyzed. The fact that prions are composed primarily of protein (no nucleic acid) has certainly revolutionized our ideas of what can constitute an infectious agent. One of the most compelling questions is just how a prion could be replicated, since all other infectious agents require some nucleic acid.

Other fascinating virus-like agents in human disease are defective forms called satellite viruses that are actually dependent on other viruses for replication. Two remarkable examples are the adeno-associated virus (AAV), which can replicate only in cells infected with adenovirus, and the delta agent, a naked strand of RNA that is expressed only in the presence of the hepatitis B virus and can worsen the severity of liver damage.

Plants are also parasitized by virus-like agents called **viroids** that differ from ordinary viruses by being very small (about one-tenth the size of an average virus) and being composed of only naked strands of RNA, lacking a capsid or any other type of coating. Viroids are significant pathogens in several economically important plants, including tomatoes, potatoes, cucumbers, citrus trees, and chrysanthemums.

6.9 Treatment of Animal Viral Infections

The nature of viruses has at times been a major impediment to effective therapy. Because viruses are not bacteria, antibiotics aimed at disrupting procaryotic cells do not work on them. On the other hand, many antiviral drugs block virus replication by targeting the function of host cells, and can cause severe side effects. Antiviral drugs are designed to target one of the steps in the viral life cycle you learned about earlier in this chapter. Azidothymide (AZT), a drug used to treat AIDS, targets the synthesis stage. A newer class of HIV drugs, the protease inhibitors, interrupts the assembly phase of the viral life cycle. Another compound that shows some potential for treating and preventing viral infections is a naturally occurring human cell product called *interferon* (see chapters 12 and 14). Vaccines that stimulate immunity are an extremely valuable tool but are available for only a limited number of viral diseases (see chapter 16).

We have completed our survey of procaryotes, eucaryotes, and viruses and have described characteristics of different representatives of these three groups. Chapters 7 and 8 explore how microorganisms maintain themselves, beginning with nutrition (chapter 7) and then looking into microbial metabolism (chapter 8).

✔ CHECKPOINT

- Viruses are easily responsible for several billion infections each year. It is conceivable that many chronic diseases of unknown cause will eventually be connected to viral agents.
- Other noncellular agents of disease are the prions, which are not viruses at all, but protein fibers; viroids, extremely small lengths of protein-coated nucleic acid; and satellite viruses, which require larger viruses to cause disease.
- Viral infections are difficult to treat because the drugs that attack the viral replication cycle also cause serious side effects in the host.

Chapter Summary With Key Terms

6.1 The Search for the Elusive Viruses
Viruses, being much smaller than bacteria, fungi, and protozoa, had to be indirectly studied until the 20th century when they were finally seen with an electron microscope.

6.2 The Position of Viruses in the Biological Spectrum
Scientists don't agree about whether viruses are living or not. They are obligate intracellular parasites.

6.3 The General Structure of Viruses
A. Viruses are infectious particles and not cells; they lack organelles and locomotion of any kind; are large, complex molecules; can be crystalline in form. A virus particle is composed of a nucleic acid core (DNA or RNA, not both) surrounded by a geometric protein shell, or **capsid;** the combination is called a **nucleocapsid;** capsid is **helical** or **icosahedral** in configuration; many

are covered by a membranous envelope containing viral protein **spikes;** complex viruses have additional external and internal structures.

B. *Shapes/Sizes:* Icosahedral, helical, spherical, and cylindrical shaped. Smallest infectious forms range from the largest poxvirus (0.45 mm or 450 nm) to the smallest viruses (0.02 mm or 20 nm).

C. *Nutritional and Other Requirements:* Lack enzymes for processing food or generating energy; are tied entirely to the host cell for all needs **(obligate intracellular parasites).**

D. Viruses are known to parasitize all types of cells, including bacteria, algae, fungi, protozoa, animals, and plants. Each viral type is limited in its **host range** to a single species or group, mostly due to specificity of adsorption of virus to specific host receptors.

6.4 How Viruses Are Classified and Named

A. The two major types of viruses are *DNA* and *RNA viruses*. These are further subdivided into families, depending on shape and size of capsid, presence or absence of an envelope, whether double- or single-stranded nucleic acid, and antigenic similarities.

B. The International Committee on the Taxonomy of Viruses oversees naming and classification of viruses. Viruses are classified into orders, families, and genera. These groupings are based on virus structure, chemical composition, and genetic makeup.

6.5 Modes of Viral Multiplication

A. *Multiplication Cycle: Animal Cells*
1. The life cycle steps of an animal virus are adsorption, penetration/uncoating, synthesis and assembly, and release from the host cell.
2. Some animal viruses cause chronic and persistent infections.
3. Viruses that alter host genetic material may cause *oncogenic* effects.

B. *Multiplication Cycle: Bacteriophages*
1. Bacteriophages are viruses that attack bacteria. They penetrate by injecting their nucleic acid and are released as virulent phage upon **lysis** of the cell.

2. Some viruses go into a latent, or **lysogenic,** phase in which they integrate into the DNA of the host cell and later may be active and produce a lytic infection.

6.6 Techniques in Cultivating and Identifying Animal Viruses

A. The need for an intracellular habitat makes it necessary to grow viruses in living cells, either in the intact host animal, in bird embryos, or in isolated cultures of host cells (cell culture).

B. *Identification:* Viruses are identified by means of **cytopathic effects** (CPE) in host cells, direct examination of viruses or their components in samples, analyzing blood for antibodies against viruses, performing genetic analysis of samples to detect virus nucleic acid, growing viruses in culture, and symptoms.

6.7 Medical Importance of Viruses

A. *Medical:* Viruses attach to specific target hosts or cells. They cause a variety of infectious diseases, ranging from mild respiratory illness (common cold) to destructive and potentially fatal conditions (rabies, AIDS). Some viruses can cause birth defects and cancer in humans and other animals.

B. *Research:* Because of their simplicity, viruses have become an invaluable tool for studying basic genetic principles. Current research is also focused on the possible connection of viruses to chronic afflictions of unknown causes, such as type I diabetes and multiple sclerosis.

6.8 Other Noncellular Infectious Agents

A. Spongiform encephalopathies are chronic persistent neurological diseases caused by **prions.**

B. Examples of neurological diseases include "mad cow disease" and Creutzfeldt-Jakob disease.

C. Other noncellular infectious agents include satellite viruses and viroids.

6.9 Treatment of Animal Viral Infections

Viral infections are difficult to treat because the drugs that attack viral replication also cause serious side effects in the host.

Multiple-Choice Questions

1. A virus is a tiny infectious
   a. cell
   b. living thing
   c. particle
   d. nucleic acid

2. Viruses are known to infect
   a. plants
   b. bacteria
   c. fungi
   d. all organisms

3. The capsid is composed of protein subunits called
   a. spikes
   b. protomers
   c. virions
   d. capsomers

4. The envelope of an animal virus is derived from the ____ of its host cell.
   a. cell wall
   b. membrane
   c. glycocalyx
   d. receptors

5. The nucleic acid of a virus is
   a. DNA only
   b. RNA only
   c. both DNA and RNA
   d. either DNA or RNA

6. The general steps in a viral multiplication cycle are
   a. adsorption, penetration, synthesis, assembly, and release
   b. endocytosis, uncoating, replication, assembly, and budding
   c. adsorption, uncoating, duplication, assembly, and lysis
   d. endocytosis, penetration, replication, maturation, and exocytosis

7. A prophage is an early stage in the development of a/an
   a. bacterial virus
   b. poxvirus
   c. lytic virus
   d. enveloped virus

8. The nucleic acid of animal viruses enters the host cell through
   a. translocation
   b. fusion
   c. endocytosis
   d. all of these

9. In general, RNA viruses multiply in the cell _____, and DNA viruses multiply in the cell _____.
   a. nucleus, cytoplasm
   b. cytoplasm, nucleus
   c. vesicles, ribosomes
   d. endoplasmic reticulum, nucleolus

10. Enveloped viruses carry surface receptors called
    a. buds
    b. spikes
    c. fibers
    d. sheaths

11. Viruses that persist in the cell and cause recurrent disease are considered
    a. oncogenic
    b. cytopathic
    c. latent
    d. resistant

12. Viruses cannot be cultivated in
    a. tissue culture
    b. bird embryos
    c. live mammals
    d. blood agar

13. Clear patches in cell cultures that indicate sites of virus infection are called
    a. plaques
    b. pocks
    c. colonies
    d. prions

14. Label the parts of this virus. Identify the capsid, nucleic acid, and other features of this virus. Can you identify it?

15. Circle the viral infections from this list: cholera, rabies, plague, cold sores, whooping cough, tetanus, genital warts, gonorrhea, mumps, Rocky Mountain spotted fever, syphilis, rubella, rat bite fever.

Concept Questions

These questions are suggested as a *writing-to-learn* experience. For each question, compose a one- or two-paragraph answer that includes the factual information needed to completely address the question.

1. a. Describe 10 *unique* characteristics of viruses (can include structure, behavior, multiplication).
   b. After consulting table 6.1, what additional statements can you make about viruses, especially as compared with cells?

2. a. Explain what it means to be an obligate intracellular parasite.
   b. What is another way to describe the sort of parasitism exhibited by viruses?

3. a. Characterize viruses according to size range.
   b. What does it mean to say that they are ultramicroscopic?
   c. That they are filterable?

4. a. Describe the general structure of viruses.
   b. What is the capsid, and what is its function?
   c. How are the two types of capsids constructed?
   d. What is a nucleocapsid?
   e. Give examples of viruses with the two capsid types.
   f. What is an enveloped virus, and how does the envelope arise?
   g. Give an example of a common enveloped human virus.
   h. What are spikes, how are they formed, and what is their function?

5. a. What dictates the host range of animal viruses?
   b. What are two ways that animal viruses penetrate the host cell?
   c. What is uncoating?
   d. Describe the two ways that animal viruses leave their host cell.

6. a. Describe several cytopathic effects of viruses.
   b. What causes the appearance of the host cell?
   c. How might it be used to diagnose viral infection?

7. a. What does it mean for a virus to be persistent or latent, and how are these events important?
   b. Briefly describe the action of an oncogenic virus.

8. a. What are bacteriophages and what is their structure?
   b. What is a tobacco mosaic virus?
   c. How are the poxviruses different from other animal viruses?

9. a. Since viruses lack metabolic enzymes, how can they synthesize necessary components?
   b. What are some enzymes with which the virus is equipped?

10. a. How are viruses classified? What are virus families?
    b. How are generic and common names used?
    c. Look at table 6.5 and count the total number of different viral diseases. How many are caused by DNA viruses? How many are RNA-virus diseases?

11. a. Compare and contrast the main phases in the lytic multiplication cycle in bacteriophages and animal viruses.
    b. When is a virus a virion?
    c. What is necessary for adsorption?
    d. Why is penetration so different in the two groups?
    e. What is eclipse?
    f. In simple terms, what does the virus nucleic acid do once it gets into the cell?
    g. What processes are involved in assembly?

12. a. What is a prophage or temperate phage?
    b. What is lysogeny?

13. a. Describe the three main techniques for cultivating viruses.
    b. What are the advantages of using cell culture?
    c. The disadvantages of using cell culture?
    d. What is a disadvantage of using live intact animals or embryos?
    e. What is a cell line? A monolayer?
    f. How are plaques formed?

14. a. What is the principal effect of the agent of Creutzfeldt-Jakob disease?
    b. How is the proposed agent different from viruses?
    c. What are viroids?

15. Why are virus diseases more difficult to treat than bacterial diseases?

Critical Thinking Questions

Critical thinking is the ability to reason and solve problems using facts and concepts. These questions can be approached from a number of angles, and in most cases, they do not have a single correct answer.

1. a. What characteristics of viruses could be used to characterize them as life forms?
   b. What makes them more similar to lifeless molecules?

2. a. Comment on the possible origin of viruses. Is it not curious that the human cell welcomes a virus in and hospitably removes its coat as if it were an old acquaintance?
   b. How do spikes play a part in the action of the host cell?

3. a. If viruses that normally form envelopes were prevented from budding, would they still be infectious?
   b. If the RNA of an influenza virus were injected into a cell by itself, could it cause a lytic infection?

4. The end result of most viral infections is death of the host cell.
   a. If this is the case, how can we account for such differences in the damage that viruses do (compare the effects of the cold virus with those of the rabies virus)?
   b. Describe the adaptation of viruses that does not immediately kill the host cell and explain what its function might be.

5. a. Given that DNA viruses can actually be carried in the DNA of the host cell's chromosomes, comment on what this phenomenon means in terms of inheritance in the offspring.
   b. Discuss the connection between viruses and cancers, giving possible mechanisms for viruses that cause cancer.

6. HIV attacks only specific types of human cells, such as certain white blood cells and nerve cells. Can you explain why a virus can enter some types of human cells but not others?

7. a. Consult table 6.5 to determine which viral diseases you have had and which ones you have been vaccinated against.
   b. Which viruses would you investigate as possible oncoviruses?

8. One early problem in cultivating HIV was the lack of a cell line that would sustain indefinitely *in vitro*, but eventually one was developed. What do you expect were the stages in developing this cell line?

9. a. If you were involved in developing an antiviral drug, what would be some important considerations? (Can a drug "kill" a virus?)
   b. How could multiplication be blocked?

10. a. Is there such a thing as a "good virus"? Explain why or why not. Consider both bacteriophages and viruses of eucaryotic organisms.

11. Why is an embryonic or fetal viral infection so harmful?

12. How are computer viruses analogous to real viruses?

13. Discuss some advantages and disadvantages of bacteriophage therapy in treating bacterial infections.

Internet Search Topics

Go to the Online Learning Center for chapter 6 of this text at http://www.mhhe.com/cowan1. Access the URLs listed under Internet Search Topics and research the following:

1. Explore the excellent websites listed for viruses.

   Click on Principles of Virus Architecture and Virus Images and Tutorials.

2. Look up emerging viral diseases and make note of the newest viruses that have arisen since 1999. What kinds of diseases do they cause, and where did they possibly originate from?

3. Find websites that discuss prions and prion-based diseases. What possible way do the prions replicate?

Elements of Microbial Nutrition, Ecology, and Growth

IN THE NEWS

In June 2003, frantic parents rushed a 3-month-old female infant to the emergency room of a regional medical center in rural Tennessee. On initial examination by a triage nurse, "Baby Caroline" appeared listless with unfocused eyes and labored breathing. Her parents reported that, over the past 72 hours, the infant had grown increasingly irritable and had cried weakly and seemed unable to nurse properly. Further questioning revealed that Baby Caroline had had no bowel movements for 3 days. Within 48 hours of admission, she developed flaccid paralysis and ex-

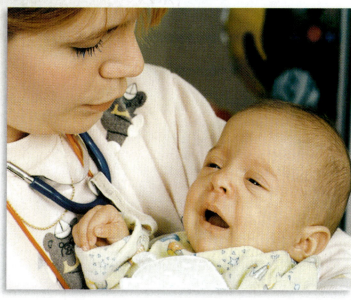

perienced respiratory failure. The child received supportive therapy, including the use of a ventilator and administration of antitoxin. Full recovery occurred in about 4 weeks.

Epidemiologists called in to determine the source of the disease examined the child's home. Baby Caroline's parents stated that they were feeding her a leading brand of powdered infant formula prepared with tap water. A week or so previously, Baby Caroline started to refuse the formula, so her mother sweetened it with fresh honey from the family apiary. Additional questioning revealed that a 2-year-old sibling often "borrowed" the baby's pacifier and played with it in the soil of the backyard. Baby Caroline's mother admitted that she had, on a few occasions, simply retrieved the pacifier and wiped it with tissue before returning it to the infant.

▶ *Based on the information given here, what is the diagnosis of Baby Caroline's illness?*

▶ *What culture methods could an epidemiologist use to determine the source of the causative agents of the disease?*

CHAPTER OVERVIEW

▶ Microbes exist in every known natural habitat on earth.

▶ Microbes show enormous capacity to adapt to environmental factors.

▶ Factors that have the greatest impact on microbes are nutrients, temperature, pH, amount of available water, atmospheric gases, light, pressure, and other organisms.

▶ Nutrition involves absorbing required chemicals from the environment for use in metabolism.

▶ Autotrophs can exist solely on inorganic nutrients, while heterotrophs require both inorganic and organic nutrients.

▶ Energy sources for microbes may come from light or chemicals.

▶ Microbes can thrive at cold, moderate, or hot temperatures.

▶ Oxygen and carbon dioxide are primary gases used in metabolism.

▶ The water content of the cell versus its environment dictates the osmotic adaptations of cells.

▶ Transport of materials by cells across cell membranes involves movement by passive and active mechanisms.

▶ Microbes interact in a variety of ways with one another and with other organisms that share their habitats.

▶ The pattern of population growth in simple microbes is to double the number of cells in each generation.

▶ Growth rate is limited by availability of nutrients and buildup of waste products.

7.1 Microbial Nutrition

Nutrition is a process by which chemical substances called **nutrients** are acquired from the environment and used in cellular activities such as metabolism and growth. With respect to nutrition, microbes are not really so different from humans **(Insight 7.1).** Bacteria living in mud on a diet of inorganic sulfur or protozoa digesting wood in a termite's intestine seem to show radical adaptations, but even these organisms require a constant influx of certain substances from their habitat. In general, all living things require a source of elements such as carbon, hydrogen, oxygen, phosphorus, potassium, nitrogen, sulfur, calcium, iron, sodium, chlorine, magnesium, and certain other elements. But the ultimate source of a particular element, its chemical form, and how much of it the microbe needs are all points of variation between different types of organisms. Any substance, whether in elemental or molecular form, that must be provided to an organism is called an **essential nutrient.** Once absorbed, nutrients are processed and transformed into the chemicals of the cell.

Two categories of essential nutrients are **macronutrients** and **micronutrients.** Macronutrients are required in relatively large quantities and play principal roles in cell structure and metabolism. Examples of macronutrients are carbon, hydrogen, and oxygen. Micronutrients, or **trace elements,** such as manganese, zinc, and nickel are present in much smaller amounts and are involved in enzyme function and maintenance of protein structure. What constitutes a micronutrient can vary from one microbe to another and often must be determined in the laboratory. This determination is made by deliberately omitting the substance in question from a growth medium to see if the microbe can grow in its absence.

Another way to categorize nutrients is according to their carbon content. An inorganic nutrient is an atom or simple molecule that contains a combination of atoms other than carbon and hydrogen. The natural reservoirs of inorganic compounds are mineral deposits in the crust of the earth, bodies of water, and the atmosphere. Examples include metals and their salts (magnesium sulfate, ferric nitrate, sodium phosphate), gases (oxygen, carbon dioxide), and water **(table 7.1).** In contrast, the molecules of organic nutrients contain carbon and hydrogen atoms and are usually the products of living things. They range from the simplest organic molecule, methane (CH_4), to large polymers (carbohydrates, lipids, proteins, and nucleic acids). The source of nutrients is extremely varied: Some microbes obtain their nutrients entirely from in-

TABLE 7.1	Principal Inorganic Reservoirs of Elements
Element	**Inorganic Environmental Reservoir**
Carbon	CO_2 in air; CO_3^{2-} in rocks and sediments
Oxygen	O_2 in air, certain oxides, water
Nitrogen	N_2 in air; NO_3^-, NO_2^-, NH_4^+ in soil and water
Hydrogen	Water, H_2 gas, mineral deposits
Phosphorus	Mineral deposits (PO_4^{3-}, H_3PO_4)
Sulfur	Mineral deposits, volcanic sediments (SO_4^{2-}, H_2S, S^0)
Potassium	Mineral deposits, the ocean (KCl, K_3PO_4)
Sodium	Mineral deposits, the ocean (NaCl, NaSi)
Calcium	Mineral deposits, the ocean ($CaCO_3$, $CaCl_2$)
Magnesium	Mineral deposits, geologic sediments ($MgSO_4$)
Chloride	The ocean (NaCl, NH_4Cl)
Iron	Mineral deposits, geologic sediments ($FeSO_4$)
Manganese, molybdenum, cobalt, nickel, zinc, copper, other micronutrients	Various geologic sediments

organic sources, and others require a combination of organic and inorganic sources. Parasites capable of invading and living on the human body derive all essential nutrients from host tissues, tissue fluids, secretions, and wastes.

Chemical Analysis of Microbial Cytoplasm

Examining the chemical composition of a bacterial cell can indicate its nutritional requirements. **Table 7.2** lists the major contents of the intestinal bacterium *Escherichia coli*. Some of these components are absorbed in a ready-to-use form, and others must be synthesized by the cell from simple nutrients. Several important features of cell composition can be summarized as follows:

- Water content is the highest of all the components (70%).
- Proteins are the next most prevalent chemical.
- About 97% of the dry cell weight is composed of organic compounds.

INSIGHT 7.1   *Discovery*

Dining with an Amoeba

An amoeba gorging itself on bacteria could be compared to a person eating a bowl of vegetable soup, because its nutrient needs are fundamentally similar to that of a human. Most food is a complex substance that contains many different types of nutrients. Some smaller molecules such as sugars can be absorbed directly by the cell; larger food debris and molecules must first be ingested and broken down into a size that can be absorbed. As nutrients are taken in, they add to a dynamic pool of inorganic and organic compounds dissolved in the cytoplasm. This pool will provide raw materials to be assimilated into the organism's own specialized proteins, carbohydrates, lipids, and other macromolecules used in growth and metabolism.

Food particles are phagocytosed into a vacuole that fuses with a lysosome containing digestive enzymes (E). Smaller subunits of digested macromolecules are transported out of the vacuole into the cell pool and are used in the anabolic and catabolic activities of the cell.

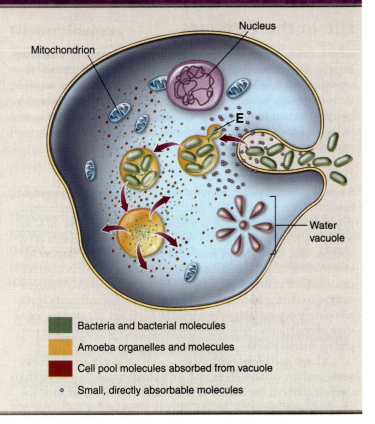

- Bacteria and bacterial molecules
- Amoeba organelles and molecules
- Cell pool molecules absorbed from vacuole
- Small, directly absorbable molecules

TABLE 7.2	Analysis of the Chemical Composition of an *Escherichia coli* Cell			
	% Total Weight	**% Dry Weight**		**% Dry Weight**
Organic Compounds			**Elements**	
Proteins	15	50	Carbon (C)	50
Nucleic acids			Oxygen (O)	20
RNA	6	20	Nitrogen (N)	14
DNA	1	3	Hydrogen (H)	8
Carbohydrates	3	10	Phosphorus (P)	3
Lipids	2	Not determined	Sulfur (S)	1
Miscellaneous	2	Not determined	Potassium (K)	1
			Sodium (Na)	1
Inorganic Compounds			Calcium (Ca)	0.5
Water	70		Magnesium (Mg)	0.5
All others	1	3	Chlorine (Cl)	0.5
			Iron (Fe)	0.2
			Manganese (Mn), zinc (Zn), molybdenum (Mo), copper (Cu), cobalt (Co), zinc (Zn)	0.3

- About 96% of the cell is composed of six elements (represented by CHONPS).
- Chemical elements are needed in the overall scheme of cell growth, but most of them are available to the cell as compounds and not as pure elements (table 7.2).

- A cell as "simple" as *E. coli* contains on the order of 5,000 different compounds, yet it needs to absorb only a few types of nutrients to synthesize this great diversity. These include $(NH_4)_2SO_4$, $FeCl_2$, $NaCl$, trace elements, glucose, KH_2PO_4, $MgSO_4$, $CaHPO_4$, and water.

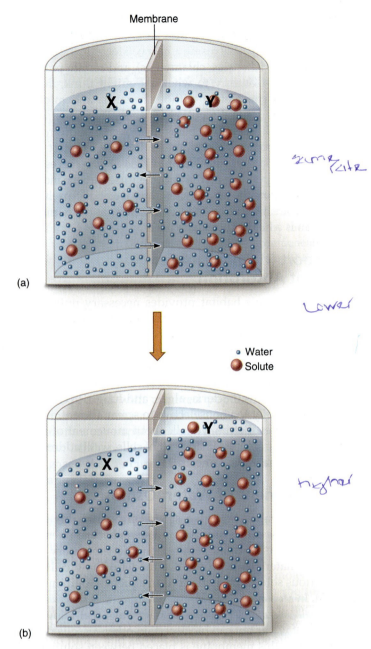

FIGURE 7.3 **Osmosis, the diffusion of water through a selectively permeable membrane.**

(a) A membrane has pores that allow the ready passage of water but not large solute molecules from one side to another. Placement of this membrane between solutions of different solute concentrations (X = less concentrated and Y = more concentrated) results in a diffusion gradient for water. Water molecules undergo diffusion and move across the membrane pores in both directions. Because there is more water in solution X, the opportunity for a water molecule to successfully hit and go through a pore is greater for X than for Y. The result will be a net movement of water from X to Y. **(b)** The level of solution on the Y side rises as water continues to diffuse in. This process will continue until equilibration occurs and the rate of diffusion of water is equal on both sides.

of the cell membrane **(figure 7.4)**. Such systems can be compared using the terms *isotonic*, *hypotonic*, and *hypertonic*. (The root *-tonic* means "tension." *Iso-* means "the same," *hypo-* means "less," and *hyper-* means "over" or "more.")

Under **isotonic** conditions, the environment is equal in solute concentration to the cell's internal environment, and because diffusion of water proceeds at the same rate in both directions, there is no net change in cell volume. Isotonic solutions are generally the most stable environments for cells, because they are already in an osmotic steady-state with the cell. Parasites living in host tissues are most likely to be living in isotonic habitats.

Under **hypotonic** conditions, the solute concentration of the external environment is lower than that of the cell's internal environment. Pure water provides the most hypotonic environment for cells because it has no solute. The net direction of osmosis is from the hypotonic solution into the cell, and cells without walls swell and can burst.

A slightly hypotonic environment can be quite favorable for bacterial cells. The constant slight tendency for water to flow into the cell keeps the cell membrane fully extended and the cytoplasm full. This is the optimum condition for the many processes occurring in and on the membrane. Slight hypotonicity is tolerated quite well by bacteria because of their rigid cell walls.

Hypertonic[1] conditions are also out of balance with the tonicity of the cell's cytoplasm, but in this case, the environment has a higher solute concentration than the cytoplasm. Because a hypertonic environment will force water to diffuse out of a cell, it is said to have high *osmotic pressure* or potential. The growth-limiting effect of hypertonic solutions on microbes is the principle behind using concentrated salt and sugar solutions as preservatives for food, such as in salted hams.

Adaptations to Osmotic Variations in the Environment

Let us now see how specific microbes have adapted osmotically to their environments. In general, isotonic conditions pose little stress on cells, so survival depends on counteracting the adverse effects of hypertonic and hypotonic environments.

A bacterium and an amoeba living in fresh pond water are examples of cells that live in constantly hypotonic conditions. The rate of water diffusing across the cell membrane into the cytoplasm is rapid and constant, and the cells would die without a way to adapt. As just mentioned, the majority of bacterial cells compensate by having a cell wall that protects them from bursting even as the cytoplasmic membrane becomes *turgid* (ter′-jid) from pressure. The amoeba's adaptation is an anatomical and physiological one that requires the constant expenditure of energy. It has a water, or contractile, vacuole that siphons excess water back out into the habitat like a tiny pump.

A microbe living in a high-salt environment (hypertonic) has the opposite problem and must either restrict its loss of

---

1. It will help you to recall these osmotic conditions if you remember that the prefixes iso-, hypo-, and hyper- refer to the environment *outside* of the cell.

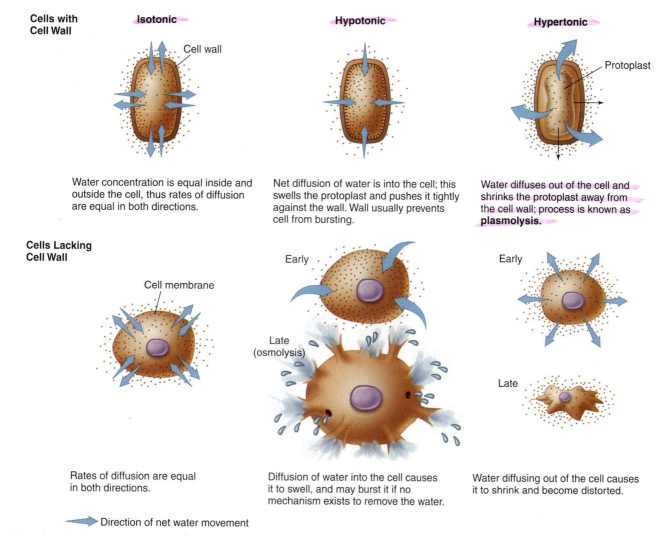

FIGURE 7.4 Cell responses to solutions of differing osmotic content.

water to the environment or increase the salinity of its internal environment. Halobacteria living in the Great Salt Lake and the Dead Sea actually absorb salt to make their cells isotonic with the environment, thus they have a physiological need for a high-salt concentration in their habitats (see halophiles on page 201).

The Movement of Molecules: Diffusion and Transport

The driving force of transport is atomic and molecular movement—the natural tendency of atoms and molecules to be in constant random motion. The existence of this motion is evident in Brownian movement of small particles suspended in liquid. It can be demonstrated by a variety of simple observations. A drop of perfume released into one part of a room is soon smelled in another part, or a lump of sugar in a cup of tea spreads through the whole cup without stirring. This phenomenon of molecular movement, in which atoms or molecules move in a gradient from an area of higher density or concentration to an area of lower density or concentration, is **diffusion (figure 7.5).**

Diffusion

All molecules, regardless of being in a solid, liquid, or gas, are in continuous movement, and as the temperature increases, the molecular movement becomes faster. This is called "thermal" movement. In any solution, including cytoplasm, these moving molecules cannot travel very far without having collisions with other molecules and, therefore, will bounce off each other like millions of pool balls every second. As a result of each collision, the directions of the colliding molecules are altered and the direction of any one molecule is unpredictable and is therefore "random." If we start with a solution in which the solute, or dissolved substance, is more concentrated in one area than another then the random thermal movement of molecules in this solution will eventually distribute the molecules from the area of higher concentration to the area of lower concentration, thus evenly distributing the molecules. This net movement of molecules down their concentration gradient by random thermal motion is known as diffusion. Diffusion of molecules across the cell membrane is largely determined by the concentration gradient and permeability of the substance.

How Molecules Diffuse in Aqueous Solutions

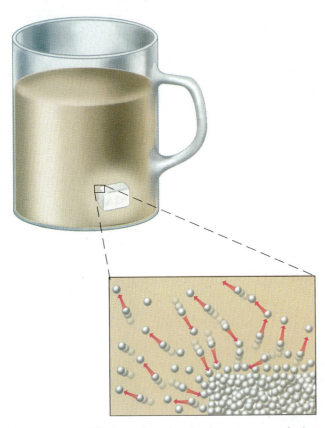

FIGURE 7.5   Diffusion of molecules in aqueous solutions.
A high concentration of sugar exists in the cube at the bottom of
the liquid. An imaginary molecular view of this area shows that
sugar molecules are in a constant state of motion. Those at the
edge of the cube diffuse from the concentrated area into more
dilute regions. As diffusion continues, the sugar will spread evenly
throughout the aqueous phase, and eventually there will be no
gradient. At that point, the system is said to be in equilibrium.

So far, the discussion of passive or simple diffusion has
not included the added complexity of membranes or cell
walls, which hinder simple diffusion by adding a physical barrier. Therefore, simple diffusion is limited to small nonpolar
molecules like oxygen or lipid soluble molecules that may
pass through the membranes. It is imperative that a cell be able
to move polar molecules and ions across the plasma membrane, and given the greatly decreased permeability of these
chemicals simple diffusion will not allow this movement.
Therefore the concept of **facilitated diffusion** must be introduced **(figure 7.6).** This type of mediated transport mechanism
utilizes a carrier protein that will bind a specific substance.
This binding changes the conformation of the carrier proteins
so that the substance is moved across the membrane. Once
the substance is transported, the carrier protein resumes its
original shape and is ready to transport again. These carrier
proteins exhibit **specificity,** which means that they bind and
transport only a single type of molecule. For example, a carrier
protein that transports sodium will not bind glucose. A second
characteristic exhibited by facilitated diffusion is **saturation.**
The rate of transport of a substance is limited by the number of

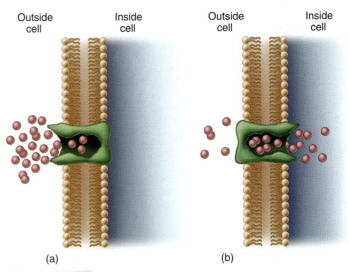

(a)   (b)

FIGURE 7.6   Facilitated diffusion.
Facilitated diffusion involves the attachment of a molecule to a
specific protein carrier. **(a)** Bonding of the molecule causes a
conformational change in the protein that facilitates the molecule's
passage across the membrane. **(b)** The membrane receptor opens
into the cell and releases the molecule.

binding sites on the transport proteins. As the substance's concentration increases so does the rate of transport until the
concentration of the transported substance is such that all of
the transporters' binding sites are occupied. Then the rate of
transport reaches a steady state and cannot move faster despite further increases in the substance's concentration. A third
characteristic of these carrier proteins is that they exhibit competition. This is when two molecules of similar shape can bind
to the same binding site on a carrier protein. The chemical with
the higher binding affinity, or the chemical in the higher concentration, will be transported at a greater rate.

Neither simple diffusion nor facilitated diffusion require
energy, since molecules are moving down a concentration
gradient.

Active Transport: Bringing in Molecules Against a Gradient

Free-living microbes exist under relatively nutrient-starved
conditions and cannot rely completely on slow and rather inefficient passive transport mechanisms. To ensure a constant
supply of nutrients and other required substances, microbes
must capture those that are in extremely short supply and actively transport them into the cell. Features inherent in **active
transport** systems are

1. the transport of nutrients against the diffusion gradient
    or in the same direction as the natural gradient but at a
    rate faster than by diffusion alone,
2. the presence of specific membrane proteins (permeases
    and pumps; **figure 7.7a**), and
3. the expenditure of energy. Examples of substances trans-
    ported actively are monosaccharides, amino acids, or-
    ganic acids, phosphates, and metal ions.

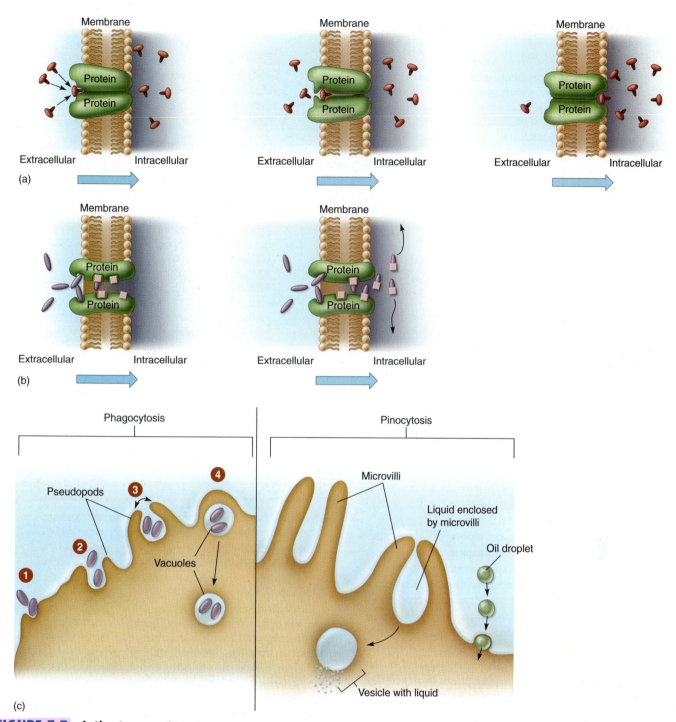

FIGURE 7.7 Active transport.

In active transport mechanisms, energy is expended to transport the molecule across the cell membrane. **(a)** Carrier-mediated active transport. The membrane proteins (permeases) have attachment sites for essential nutrient molecules. As these molecules bind to the permease, they are pumped into the cell's interior through special membrane protein channels. Microbes have these systems for transporting various ions (sodium, iron) and small organic molecules. **(b)** In group translocation, the molecule is actively captured, but along the route of transport, it is chemically altered. By coupling transport with synthesis, the cell conserves energy. **(c)** Endocytosis (phagocytosis and pinocytosis). Solid particles are phagocytosed by large cell extensions called pseudopods, and fluids and/or dissolved substances are pinocytosed into vesicles by very fine cell protrusions called microvilli. Oil droplets fuse with the membrane and are released directly into the cell.

TABLE 7.4	Summary of Transport Processes in Cells			
General Process	**Nature of Transport**	**Examples**	**Description**	**Qualities**
Passive	Energy expenditure is not required. Substances exist in a gradient and move from areas of higher concentration toward areas of lower concentration in the gradient.	Diffusion	A fundamental property of atoms and molecules that exist in a state of random motion	Nonspecific Brownian movement
		Facilitated diffusion	Molecule binds to a receptor in membrane and is carried across to other side	Molecule specific; transports both ways
Active	Energy expenditure is required. Molecules need not exist in a gradient. Rate of transport is increased. Transport may occur against a concentration gradient.	Carrier-mediated active transport	Atoms or molecules are pumped into or out of the cell by specialized receptors. Driven by ATP or the proton motive force	Transports simple sugars, amino acids, inorganic ions (Na^+, K^+)
		Group translocation	Molecule is moved across membrane and simultaneously converted to a metabolically useful substance.	Alternate system for transporting nutrients (sugars, amino acids)
		Bulk transport	Mass transport of large particles, cells, and liquids by engulfment and vesicle formation	Includes endocytosis, phagocytosis, pinocytosis

Some freshwater algae have such efficient active transport systems that an essential nutrient can be found in intracellular concentrations 200 times that of the habitat.

An important type of active transport involves specialized pumps, which can rapidly carry ions such as K^+, Na^+, and H^+ across the membrane. This behavior is particularly important in mitochondrial ATP formation and protein synthesis, as described in chapter 8. Another type of active transport, **group translocation,** couples the transport of a nutrient with its conversion to a substance that is immediately useful inside the cell **(figure 7.7b).** This method is used by certain bacteria to transport sugars (glucose, fructose) while simultaneously adding molecules such as phosphate that prepare them for the next stage in metabolism.

Endocytosis: Eating and Drinking by Cells

Some eucaryotic cells transport large molecules, particles, liquids, or even other cells across the cell membrane. Because the cell usually expends energy to carry out this transport, it is also a form of active transport. The substances transported do not pass physically through the membrane but are carried into the cell by **endocytosis.** First the cell encloses the substance in its membrane, simultaneously forming a vacuole and engulfing it **(figure 7.7c).** Amoebas and certain white blood cells ingest whole cells or large solid matter by a type of endocytosis called **phagocytosis.** Liquids, such as oils or molecules in solution, enter the cell through **pinocytosis.** The mechanisms for transport of molecules into cells are summarized in **table 7.4.**

✔ CHECKPOINT

- Nutrition is a process by which all living organisms obtain substances from their environment to convert to metabolic uses.
- Although the chemical form of nutrients varies widely, all organisms require six bioelements—carbon, hydrogen, oxygen, nitrogen, phosphorus, and sulfur—to survive, grow, and reproduce.
- Nutrients are categorized by the amount required (macronutrients or micronutrients), by chemical structure (organic or inorganic), and by their importance to the organism's survival (essential or nonessential).
- Microorganisms are classified both by the chemical form of their nutrients and the energy sources they utilize.
- Nutrient requirements of microorganisms determine their respective niches in the food webs of major ecosystems.
- Nutrients are transported into microorganisms by two kinds of processes: active transport that expends energy and passive transport that occurs independently of energy input.
- The molecular size and concentration of a nutrient determine the method of transport.

7.2 Environmental Factors That Influence Microbes

Microbes are exposed to a wide variety of environmental factors in addition to nutrients. Microbial ecology focuses on ways that microorganisms deal with or adapt to such factors as heat, cold, gases, acid, radiation, osmotic and hydrostatic pressures, and even other microbes. Adaptation is a complex adjustment in biochemistry or genetics that enables long-term survival and growth. For most microbes, environmental factors fundamentally affect the function of metabolic enzymes. Thus, survival in a changing environment is largely a matter of whether the enzyme systems of microorganisms can adapt to alterations in their habitat. Incidentally, one must be careful to differentiate between growth in a given condition and tolerance, which implies survival without growth.

Temperature Adaptations

Microbial cells are unable to control their temperature and therefore assume the ambient temperature of their natural habitats. Their survival is dependent on adapting to whatever temperature variations are encountered in that habitat. The range of temperatures for microbial growth can be expressed as three *cardinal temperatures*. The **minimum temperature** is the lowest temperature that permits a microbe's continued growth and metabolism; below this temperature, its activities are inhibited. The **maximum temperature** is the highest temperature at which growth and metabolism can proceed. If the temperature rises slightly above maximum, growth will stop, but if it continues to rise beyond that point, the enzymes and nucleic acids will eventually become permanently inactivated (otherwise known as denaturation), and the cell will die. This is why heat works so well as an agent in microbial control. The **optimum temperature** covers a small range, intermediate between the minimum and maximum, which promotes the fastest rate of growth and metabolism (rarely is the optimum a single point).

Depending on their natural habitats, some microbes have a narrow cardinal range, others a broad one. Some strict parasites will not grow if the temperature varies more than a few degrees below or above the host's body temperature. For instance, the typhus rickettsia multiplies only in the range of 32°C to 38°C, and rhinoviruses (one cause of the common cold) multiply successfully only in tissues that are slightly below normal body temperature (33°C to 35°C). Other organisms are not so limited. Strains of *Staphylococcus aureus* grow within the range of 6°C to 46°C, and the intestinal bacterium *Enterococcus faecalis* grows within the range of 0°C to 44°C.

Another way to express temperature adaptation is to describe whether an organism grows optimally in a cold, moderate, or hot temperature range. The terms used for these ecological groups are *psychrophile, mesophile,* and *thermophile* **(figure 7.8)**, respectively.

A **psychrophile** (sy'-kroh-fyl) is a microorganism that has an optimum temperature below 15°C and

is capable of growth at 0°C. It is obligate with respect to cold and generally cannot grow above 20°C. Laboratory work with true psychrophiles can be a real challenge. Inoculations have to be done in a cold room because room temperature can be lethal to the organisms. Unlike most laboratory cultures, storage in the refrigerator incubates, rather than inhibits, them. As one might predict, the habitats of psychrophilic bacteria, fungi, and algae are lakes and rivers, snowfields **(figure 7.9)**, polar ice, and the deep ocean. Rarely, if ever, are they pathogenic. True psychrophiles must be distinguished from *psychrotrophs* or *facultative psychrophiles* that grow slowly in cold but have an optimum temperature above 20°C. Bacteria such as *Staphylococcus aureus* and *Listeria monocytogenes* are a concern because they can grow in refrigerated food and cause food-borne illness.

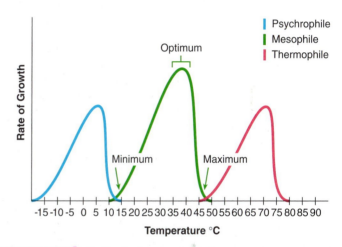

FIGURE 7.8 **Ecological groups by temperature of adaptation.**

Psychrophiles can grow at or near 0°C and have an optimum below 15°C. As a group, mesophiles can grow between 10°C and 50°C, but their optima usually fall between 20°C and 40°C. Generally speaking, thermophiles require temperatures above 45°C and grow optimally between this temperature and 80°C. Note that the extremes of the ranges can overlap to an extent.

(a)

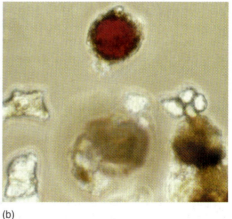
(b)

FIGURE 7.9 **Red snow.**

(a) An early summer snowbank provides a perfect habitat for psychrophilic photosynthetic organisms like *Chlamydomonas nivalis*. **(b)** Microscopic view of this snow alga (actually classified as a "green" alga although a red pigment dominates at this stage of its life cycle).

INSIGHT 7.4   *Discovery*

Cashing in on "Hot" Microbes

The smoldering thermal springs in Yellowstone National Park are more than just one of the geologic wonders of the world. They are also a hotbed of some of the most unusual microorganisms in the world. The thermophiles thriving at temperatures near the boiling point are the focus of serious interest from the scientific community. For many years, biologists have been intrigued that any living organism could function at such high temperatures. Such questions as these come to mind: Why don't they melt and disintegrate, why don't their proteins coagulate, and how can their DNA possibly remain intact?

One of the earliest thermophiles to be isolated was *Thermus aquaticus*. It was discovered by Thomas Brock in Yellowstone's Mushroom Pool in 1965 and was registered with the American Type Culture Collection. Interested researchers studied this species and discovered that it has extremely heat-stable proteins and nucleic acids, and its cell membrane does not break down readily at high temperatures. Later, an extremely heat-stable DNA-replicating enzyme was isolated from the species.

What followed is a riveting example of how pure research for the sake of understanding and discovery also offered up a key ingredient in a multimillion-dollar process. Once an enzyme was discovered that was capable of copying DNA at very high temperatures (65°C to 72°C), researchers were able to develop a technique called the polymerase chain reaction (PCR), which could amplify a single piece of DNA into hundreds of thousands of identical copies. The enzyme, called **Taq polymerase** (from *Thermus aquaticus*), revolutionized forensic science, microbial ecology, and medical diagnosis. (Kary Mullis, who recognized the utility of Taq and developed the PCR technique in 1983, won the Nobel Prize in Chemistry for it in 1993.)

Spurred by this remarkable success story, biotechnology companies have descended on Yellowstone, which contains over 10,000

Biotechnology researchers harvesting samples in Yellowstone National Park.

hot springs, geysers, and hot habitats. These industries are looking to unusual bacteria and archaea as a source of "extremozymes," enzymes that operate under high temperatures and acidity. Many other organisms with useful enzymes have been discovered. Some provide applications in the dairy, brewing, and baking industries for high-temperature processing and fermentations. Others are being considered for waste treatment and bioremediation.

This quest has also brought attention to questions such as: Who owns these microbes, and can their enzymes be patented? In the year 2000, the Park Service secured a legal ruling that allows them to share in the profits from companies and to add that money to their operating budget. The U.S. Supreme Court has also ruled that a microbe isolated from natural habitats cannot be patented. Only the technology that uses the microbe can be patented.

The majority of medically significant microorganisms are **mesophiles** (mez'-oh-fylz), organisms that grow at intermediate temperatures. Although an individual species can grow at the extremes of 10°C or 50°C, the optimum growth temperatures (optima) of most mesophiles fall into the range of 20°C to 40°C. Organisms in this group inhabit animals and plants as well as soil and water in temperate, subtropical, and tropical regions. Most human pathogens have optima somewhere between 30°C and 40°C (human body temperature is 37°C). *Thermoduric* microbes, which can survive short exposure to high temperatures but are normally mesophiles, are common contaminants of heated or pasteurized foods (see chapter 11). Examples include heat-resistant cysts such as *Giardia* or sporeformers such as *Bacillus* and *Clostridium*.

A **thermophile** (thur'-moh-fyl) is a microbe that grows optimally at temperatures greater than 45°C. Such heat-loving microbes live in soil and water associated with volcanic activity, in compost piles, and in habitats directly exposed to the sun. Thermophiles vary in heat requirements, with a

general range of growth of 45°C to 80°C. Most eucaryotic forms cannot survive above 60°C, but a few thermophilic bacteria, called hyperthermophiles, grow between 80°C and 120°C (currently thought to be the temperature limit endured by enzymes and cell structures). Strict thermophiles are so heat-tolerant that researchers may use an autoclave to isolate them in culture. Currently, there is intense interest in thermal microorganisms on the part of biotechnology companies **(Insight 7.4)**.

Gas Requirements

The atmospheric gases that most influence microbial growth are O_2 and CO_2. Of these, oxygen gas has the greatest impact on microbial adaptation. Not only is it an important respiratory gas, but it is also a powerful oxidizing agent that exists in many toxic forms. In general, microbes fall into one of three categories: those that use oxygen and can detoxify it; those that can neither use oxygen nor detoxify it; and those that do not use oxygen but can detoxify it.

How Microbes Process Oxygen

As oxygen enters into cellular reactions, it is transformed into several toxic products. Singlet oxygen ($^1O_2$) is an extremely reactive molecule produced by both living and nonliving processes. Notably, it is one of the substances produced by phagocytes to kill invading bacteria (see chapter 14). The buildup of singlet oxygen and the oxidation of membrane lipids and other molecules can damage and destroy a cell. The highly reactive superoxide ion (O_2^-), peroxide (H_2O_2), and hydroxyl radicals (OH^-) are other destructive metabolic by-products of oxygen. To protect themselves against damage, most cells have developed enzymes that go about the business of scavenging and neutralizing these chemicals. The complete conversion of superoxide ion into harmless oxygen requires a two-step process and at least two enzymes:

$$\text{Step 1.} \quad 2O_2^- + 2H^+ \xrightarrow{\text{Superoxide}\atop\text{dismutase}} H_2O_2 \text{ (hydrogen peroxide)} + O_2$$

$$\text{Step 2.} \quad 2H_2O_2 \xrightarrow{\text{Catalase}} 2H_2O + O_2$$

In this series of reactions (essential for aerobic organisms), the superoxide ion is first converted to hydrogen peroxide and normal oxygen by the action of an enzyme called superoxide dismutase. Because hydrogen peroxide is also toxic to cells (it is used as a disinfectant and antiseptic), it must be degraded by the enzyme catalase into water and oxygen. If a microbe is not capable of dealing with toxic oxygen by these or similar mechanisms, it is forced to live in habitats free of oxygen.

With respect to oxygen requirements, several general categories are recognized. An **aerobe** (air'-ohb) (aerobic organism) can use gaseous oxygen in its metabolism and possesses the enzymes needed to process toxic oxygen products. An organism that cannot grow without oxygen is an **obligate aerobe.** Most fungi and protozoa, as well as many bacteria (genera *Micrococcus* and *Bacillus*), have strict requirements for oxygen in their metabolism.

A **facultative anaerobe** is an aerobe that does not require oxygen for its metabolism and is capable of growth in the absence of it. This type of organism metabolizes by aerobic respiration when oxygen is present, but in its absence, it adopts an anaerobic mode of metabolism such as fermentation. Facultative anaerobes usually possess catalase and superoxide dismutase. A large number of bacterial pathogens fall into this group (for example, gram-negative intestinal bacteria and staphylococci). A **microaerophile** (myk"-roh-air'-oh-fyl) does not grow at normal atmospheric concentrations of oxygen but requires a small amount of it in metabolism. Most organisms in this category live in a habitat (soil, water, or the human body) that provides small amounts of oxygen but is not directly exposed to the atmosphere.

An **anaerobe** (anaerobic microorganism) lacks the metabolic enzyme systems for using oxygen in respiration. Because **strict,** or **obligate, anaerobes** also lack the enzymes for processing toxic oxygen, they cannot tolerate any free oxygen in the immediate environment and will die if exposed to it. Strict anaerobes live in highly reduced habitats, such as deep muds, lakes, oceans, and soil. Even though human cells use oxygen and oxygen is found in the blood and tissues, some body sites present anaerobic pockets or microhabitats where colonization or infection can occur. One region that is an important site for anaerobic infections is the oral cavity. Dental caries are partly due to the complex actions of aerobic and anaerobic bacteria, and most gingival infections consist of similar mixtures of oral bacteria that have invaded damaged gum tissues (see chapter 22). Another common site for anaerobic infections is the large intestine, a relatively oxygen-free habitat that harbors a rich assortment of strictly anaerobic bacteria. Anaerobic infections can occur following abdominal surgery and traumatic injuries (gas gangrene and tetanus). Growing anaerobic bacteria usually requires special media, methods of incubation, and handling chambers that exclude oxygen (**figure 7.10a**).

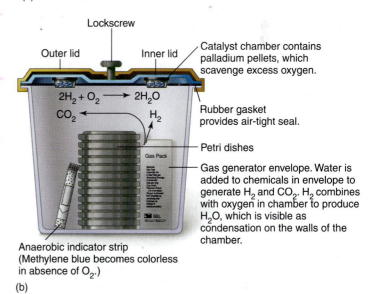

(a)

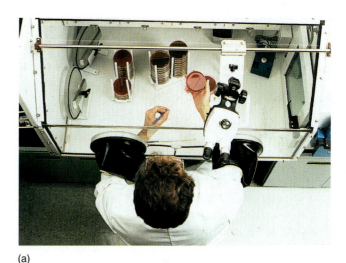

(b)

FIGURE 7.10 **Culturing techniques for anaerobes.**
(a) A special anaerobic environmental chamber makes it possible to handle strict anaerobes without exposing them to air. It also has provisions for incubation and inspection in a completely O_2-free system. **(b)** The anaerobic, or CO_2, incubator system. To create an anaerobic environment, a packet is activated to produce hydrogen gas, and the chamber is sealed tightly. The gas reacts with available oxygen to produce water. Carbon dioxide can also be added to the system for growth of organisms needing high concentrations of it.

Aerotolerant anaerobes do not utilize oxygen but can survive and grow to a limited extent in its presence. These anaerobes are not harmed by oxygen, mainly because they possess alternate mechanisms for breaking down peroxides and superoxide. Certain lactobacilli and streptococci use manganese ions or peroxidases to perform this task.

Determining the oxygen requirements of a microbe from a biochemical standpoint can be a very time-consuming process. Often it is illuminating to perform culture tests with reducing media (those that contain an oxygen-absorbing chemical). One such technique demonstrates oxygen requirements by the location of growth in a tube of fluid thioglycollate (**figure 7.11**).

Although all microbes require some carbon dioxide in their metabolism, *capnophiles* grow best at a higher CO_2 tension than is normally present in the atmosphere. This becomes important in the initial isolation of some pathogens from clinical specimens, notably *Neisseria* (gonorrhea, meningitis), *Brucella* (undulant fever), and *Streptococcus pneumoniae*. Incubation is carried out in a CO_2 incubator that provides 3% to 10% CO_2 (see **figure 7.10***b*).

IN THE NEWS   *(Continued from page 183)*

Baby Caroline suffered from infant botulism. Botulism is a neuroparalytic disease that in adults results from ingestion of or contamination of a wound by preformed toxin of *Clostridium botulinum* or related species. In babies, the disease occurs differently. Infant botulism, often called "floppy baby syndrome," may occur when a toxin is produced in and absorbed from the gastrointestinal tract. In infants, this intoxication begins with ingestion of *Clostridium* spores. In the intestine, spores germinate and the organisms produce a toxin that the child absorbs. Substances such as honey, syrup, and soil may contain the spores.

Binding of the neurotoxin results in loss of muscle tone. The result is a flaccid paralysis ("floppy baby"), which may lead to respiratory and cardiac failure. Since an infant's intestinal flora are not well established, and the immune system is immature, defense mechanisms that would stop the growth of the *Clostridium* may fail. Therefore, infants may develop this form of the disease, whereas symptoms in adults usually require consumption of preformed toxins in items such as improperly home-canned food.

To determine the source of the *Clostridium* spores, one might culture honey, prepared formula, and garden soil. Since organisms of the genus *Clostridium* are anaerobic, samples would be cultured in the absence of oxygen using culture media that select for and allow presumptive identification of these organisms. One suitable medium would be sulfite polymyxin sulfadiazine (SPS) medium on which clostridia produce blackened colonies. All cultures should be performed by state or national labs due to the danger associated with the bacterium.

See: CDC. 2003. Infant botulism—New York City, 2001–2002. MMWR 52:21–24.

FIGURE 7.11   **Use of thioglycollate broth to demonstrate oxygen requirements.**

Thioglycollate is a reducing agent that allows anaerobic bacteria to grow in tubes exposed to air. Oxygen concentration is highest at the top of the tube. When a series of tubes is inoculated with bacteria that differ in O_2 requirements, the relative position of growth provides some indication of their adaptations to oxygen use. Tube 1 (on the left): aerobic *(Pseudomonas aeruginosa)*; Tube 2: facultative *(Staphylococcus aureus)*; Tube 3: facultative *(Escherichia coli)*; Tube 4: obligate anaerobe *(Clostridium butyricum)*.

Effects of pH

Microbial growth and survival are also influenced by the pH of the habitat. The pH was defined in chapter 2 as the degree of acidity or alkalinity (basicity) of a solution. It is expressed by the pH scale, a series of numbers ranging from 0 to 14. The pH of pure water (7.0) is neutral, neither acidic nor basic. As the pH value decreases toward 0, the acidity increases, and as the pH increases toward 14, the alkalinity increases. The majority of organisms live or grow in habitats between pH 6 and 8 because strong acids and bases can be highly damaging to enzymes and other cellular substances.

A few microorganisms live at pH extremes. Obligate *acidophiles* include *Euglena mutabilis*, an alga that grows in acid pools between 0 and 1.0 pH, and *Thermoplasma*, an archaea that lacks a cell wall, lives in hot coal piles at a pH of 1 to 2, and will lyse if exposed to pH 7. Because many molds and yeasts tolerate moderate acid, they are the most common spoilage agents of pickled foods. Alkalinophiles live in hot pools and soils that contain high levels of basic minerals (up to pH 10.0). Bacteria that decompose urine create alkaline conditions, since ammonium (NH_4^+) can be produced when urea (a component of urine) is digested. Metabolism of urea is one way that *Proteus* spp. can neutralize the acidity of the urine to colonize and infect the urinary system.

Osmotic Pressure

Although most microbes exist under hypotonic or isotonic conditions, a few, called osmophiles, live in habitats with a high solute concentration. One common type of osmophile prefers high concentrations of salt; these organisms are called **halophiles** (hay'-loh-fylz). Obligate *halophiles* such as *Halobacterium* and *Halococcus* inhabit salt lakes, ponds, and other hypersaline habitats. They grow optimally in solutions of 25% NaCl but require at least 9% NaCl (combined with other salts) for growth. These archaea have significant modifications in their cell walls and membranes and will lyse in hypotonic habitats. *Facultative halophiles* are remarkably resistant to salt, even though they do not normally reside in high-salt environments. For example, *Staphylococcus aureus* can grow on NaCl media ranging from 0.1% up to 20%. Although it is common to use high concentrations of salt and sugar to preserve food (jellies, syrups, and brines), many bacteria and fungi actually thrive under these conditions and are common spoilage agents.

Miscellaneous Environmental Factors

Various forms of electromagnetic radiation (ultraviolet, infrared, visible light) stream constantly onto the earth from the sun. Some microbes (phototrophs) can use visible light rays as an energy source, but non-photosynthetic microbes tend to be damaged by the toxic oxygen products produced by contact with light. Some microbial species produce yellow carotenoid pigments to protect against the damaging effects of light by absorbing and dismantling toxic oxygen. Other types of radiation that can damage microbes are ultraviolet and ionizing rays (X rays and cosmic rays). In chapter 11, you will see just how these types of energy are applied in microbial control.

Descent into the ocean depths subjects organisms to increasing hydrostatic pressure. Deep-sea microbes called **barophiles** exist under pressures that range from a few times to over 1,000 times the pressure of the atmosphere. These bacteria are so strictly adapted to high pressures that they will rupture when exposed to normal atmospheric pressure.

Because of the high water content of cytoplasm, all cells require water from their environment to sustain growth and metabolism. Water is the solvent for cell chemicals, and it is needed for enzyme function and digestion of macromolecules. A certain amount of water on the external surface of the cell is required for the diffusion of nutrients and wastes. Even in apparently dry habitats, such as sand or dry soil, the particles retain a thin layer of water usable by microorganisms. Only dormant, dehydrated cell stages (for example, spores and cysts) tolerate extreme drying because of the inactivity of their enzymes.

Ecological Associations Among Microorganisms

Up to now, we have considered the importance of nonliving environmental influences on the growth of microorganisms. Another profound influence comes from other organisms that share (or sometimes are) their habitats. In all but the rarest instances, microbes live in shared habitats, which give rise to complex and fascinating associations. Some associations are between similar or dissimilar types of microbes; others involve multicellular organisms such as animals or plants. Interactions can have beneficial, harmful, or no particular effects on the organisms involved; they can be obligatory or nonobligatory to the members; and they often involve nutritional interactions. This outline provides an overview of the major types of microbial associations:

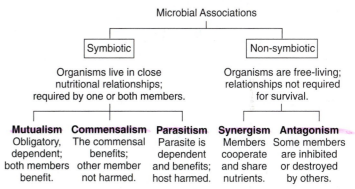

A general term used to denote a situation in which two organisms live together in a close partnership is **symbiosis,**[2] and the members are termed *symbionts*. Three main types of symbiosis occur. **Mutualism** exists when organisms live in an obligatory but mutually beneficial relationship. This association is rather common in nature because of the survival value it has for the members involved. **Insight 7.5** gives several examples to illustrate this concept. In other symbiotic relationships the relationship tends to be unequal, meaning it benefits one member and not the other, and it can be obligatory.

In a relationship known as **commensalism,** the member called the commensal receives benefits, while its coinhabitant is neither harmed nor benefited. A classic commensal interaction between microorganisms called **satellitism** arises when one member provides nutritional or protective factors needed by the other **(figure 7.12).** Some microbes can break down a substance that would be toxic or inhibitory to another microbe. Relationships between humans and resident commensals that derive nutrients from the body are discussed in a later section.

In an earlier section, we introduced the concept of **parasitism** as a relationship in which the host organism provides the parasitic microbe with nutrients and a habitat. Multiplication of the parasite usually harms the host to some extent. As this relationship evolves, the host may even develop tolerance for or dependence on a parasite, at which point we call the relationship commensalism or mutualism.

---

2. Note that *symbiosis* is a neutral term and does not by itself imply benefit or detriment.

INSIGHT 7.5

Life Together: Mutualism

A tremendous variety of mutualistic partnerships occur in nature. These associations gradually evolve over millions of years as the participating members come to rely on some critical substance or habitat that they share. One of the earliest such associations is thought to have resulted in eucaryotic cells (see Insight 5.1).

Protozoan cells often receive growth factors from symbiotic bacteria and algae that, in turn, are nurtured by the protozoan cell. One peculiar ciliate propels itself by affixing symbiotic bacteria to its cell membrane to act as "oars." These relationships become so obligatory that some amoebas and ciliates require mutualistic bacteria for survival. This kind of relationship is especially striking in the complex mutualism of termites, which harbor protozoans specialized to live only inside them. The protozoans, in turn, contain endosymbiotic bacteria. Wood eaten by the termite gets processed by the protozoan and bacterial enzymes, and all three organisms thrive.

Symbiosis Between Microbes and Animals

Microorganisms carry on symbiotic relationships with animals as diverse as sponges, worms, and mammals. Bacteria and protozoa are essential in the operation of the rumen (a complex, four-chambered stomach) of cud-chewing mammals. These mammals produce no enzymes of their own to break down the cellulose that

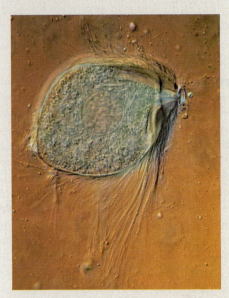

Termites are thought to be responsible for wood damage; however, it is the termite's endosymbiont (the protozoan pictured above) that provides the enzymes for digesting wood.

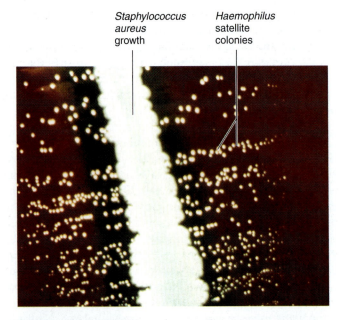

Staphylococcus aureus growth

Haemophilus satellite colonies

FIGURE 7.12 Satellitism, a type of commensalism between two microbes.
In this example, *Staphylococcus aureus* provides growth factors to *Haemophilus influenzae,* which grows as tiny satellite colonies near the streak of *Staphylococcus.* By itself, *Haemophilus* could not grow on blood agar. The *Staphylococcus* gives off several nutrients such as vitamins and amino acids that diffuse out to the *Haemophilus,* thereby promoting its growth.

Synergism is an interrelationship between two or more free-living organisms that benefits them but is not necessary for their survival. Together, the participants cooperate to produce a result that none of them could do alone. This form of shared metabolism can be viewed by this reaction:

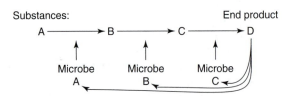

Substances: End product

Each microbe performs a specific action on a chemical in the series. The end product is useful to all three microbes.

An example of synergism is observed in the exchange between soil bacteria and plant roots (see chapter 24). The plant provides various growth factors, and the bacteria help fertilize the plant by supplying it with minerals. In synergistic infections, a combination of organisms can produce tissue damage that a single organism would not cause alone. Gum disease, dental caries, and gas gangrene involve mixed infections by bacteria interacting synergistically.

Antagonism is an association between free-living species that arises when members of a community compete. In this interaction, one microbe secretes chemical substances into the surrounding environment that inhibit or destroy another microbe in the same habitat. The first microbe

is a major part of their diet, but the microbial population harbored in their rumens does. The complex food materials are digested through several stages, during which time the animal regurgitates and chews the partially digested plant matter (the cud) and occasionally burps methane produced by the microbial symbionts.

Thermal Vent Symbionts

Another fascinating symbiotic relationship has been found in the deep hydrothermal vents in the seafloor, where geologic forces spread the crustal plates and release heat and gas. These vents are a focus of tremendous biological and geologic activity. Discoveries first made in the late 1970s demonstrated that the source of energy in this community is not the

A view of a vent community based on mutualism and chemoautotrophy. The giant tube worm *Riftia* houses bacteria in its specialized feeding organ, the trophosome. Raw materials in the form of dissolved inorganic molecules are provided to the bacteria through the worm's circulation. With these, the bacteria produce usable organic food that is absorbed by the worm.

sun, because the vents are too deep for light to penetrate (2,600 m). Instead, this ecosystem is based on a massive chemoautotrophic bacterial population that oxidizes the abundant hydrogen sulfide (H_2S)

gas given off by the volcanic activity there. As the bottom of the food web, these bacteria serve as the primary producers of nutrients that service a broad spectrum of specialized animals.

may gain a competitive advantage by increasing the space and nutrients available to it. Interactions of this type are common in the soil, where mixed communities often compete for space and food. *Antibiosis*—the production of inhibitory compounds such as antibiotics, or bacteriocins—is actually a form of antagonism. Hundreds of naturally occurring antibiotics have been isolated from bacteria and fungi and used as drugs to control diseases (see chapter 12). Bacteriocins are another class of antimicrobial proteins that are toxic to bacteria other than the ones that produced them.

Interrelationships Between Microbes and Humans

The human body is a rich habitat for symbiotic bacteria, fungi, and a few protozoa. Microbes that normally live on the skin, in the alimentary tract, and in other sites are called the *normal microbial flora* (see chapter 13). These residents participate in commensal, parasitic, and synergistic relationships with their human hosts. For example, *Escherichia coli* living symbiotically in the intestine produce vitamin K, and species of symbiotic *Lactobacillus* residing in the vagina help maintain an acidic environment that protects against infection by other microorganisms. Hundreds of commensal species "make a living" on the body without either harming or benefiting it. For example, many bacteria and yeasts reside in the

outer dead regions of the skin; oral microbes feed on the constant flow of nutrients in the mouth; and billions of bacteria live on the wastes in the large intestine. Because the normal flora and the body are in a constant state of change, these relationships are not absolute, and a commensal can convert to a parasite by invading body tissues and causing disease.

✔ CHECKPOINT

- The environmental factors that control microbial growth are temperature, pH, moisture, radiation, gases, and other microorganisms.
- Environmental factors control microbial growth by their influence on microbial enzymes.
- Three cardinal temperatures for a microorganism describe its temperature range and the temperature at which it grows best. These are the minimum temperature, the maximum temperature, and the optimum temperature.
- Microorganisms are classified by their temperature requirements as psychrophiles, mesophiles, or thermophiles.
- Most eucaryotic microorganisms are aerobic, while bacteria vary widely in their oxygen requirements from obligately aerobic to anaerobic.
- Microorganisms live in association with other species that range from mutually beneficial symbiosis to parasitism and antagonism.

7.3 The Study of Microbial Growth

When microbes are provided with nutrients and the required environmental factors, they become metabolically active and grow. Growth takes place on two levels. On one level, a cell synthesizes new cell components and increases its size; on the other level, the number of cells in the population increases. This capacity for multiplication, increasing the size of the population by cell division, has tremendous importance in microbial control, infectious disease, and biotechnology. In the following sections, we will focus primarily on the characteristics of bacterial growth that are generally representative of single-celled microorganisms.

The Basis of Population Growth: Binary Fission

The division of a bacterial cell occurs mainly through **binary,** or **transverse, fission;** *binary* means that one cell becomes two, and *transverse* refers to the division plane forming across the width of the cell. During binary fission, the parent cell enlarges, duplicates its chromosome, and forms a central transverse septum that divides the cell into two daughter cells. This process is repeated at intervals by each new daughter cell in turn, and with each successive round of division, the population increases. The stages in this continuous process are shown in greater detail in **figures 7.13** and **7.14.**

The Rate of Population Growth

The time required for a complete fission cycle—from parent cell to two new daughter cells—is called the **generation,** or **doubling, time.** The term *generation* has a similar meaning as it does in humans. It is the period between an individual's birth and the time of producing offspring. In bacteria, each new fission cycle or generation increases the population by a factor of 2, or doubles it. Thus, the initial parent stage consists of 1 cell, the first generation consists of 2 cells, the second 4, the third 8, then 16, 32, 64, and so on. As long as the environment remains favorable, this doubling effect can continue at a constant rate. With the passing of each generation, the population will double, over and over again.

The length of the generation time is a measure of the growth rate of an organism. Compared with the growth rates of most other living things, bacteria are notoriously rapid. The average generation time is 30 to 60 minutes under optimum conditions. The shortest generation times average 5 to 10 minutes, and longer generation times require days. For example, *Mycobacterium leprae,* the cause of Hansen's disease, has a generation time of 10 to 30 days—as long as some animals. Environmental bacteria commonly have generation time measured in months. Most pathogens have relatively short doubling times. *Salmonella enteritidis* and *Staphylococcus aureus,* bacteria that cause food-borne illness, double in 20 to 30 minutes, which is why leaving food at room temperature

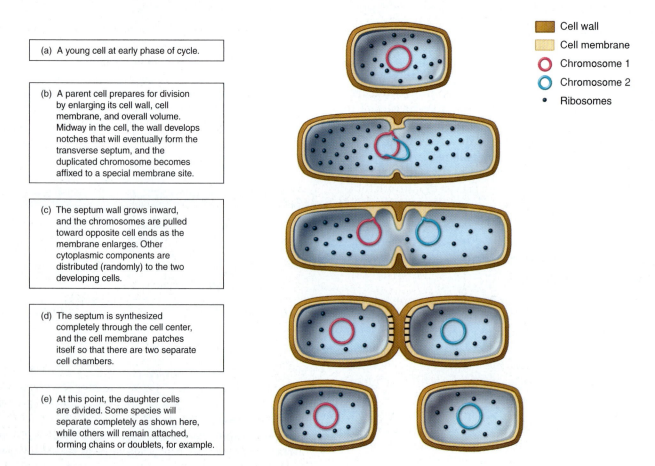

(a) A young cell at early phase of cycle.

(b) A parent cell prepares for division by enlarging its cell wall, cell membrane, and overall volume. Midway in the cell, the wall develops notches that will eventually form the transverse septum, and the duplicated chromosome becomes affixed to a special membrane site.

(c) The septum wall grows inward, and the chromosomes are pulled toward opposite cell ends as the membrane enlarges. Other cytoplasmic components are distributed (randomly) to the two developing cells.

(d) The septum is synthesized completely through the cell center, and the cell membrane patches itself so that there are two separate cell chambers.

(e) At this point, the daughter cells are divided. Some species will separate completely as shown here, while others will remain attached, forming chains or doublets, for example.

- Cell wall
- Cell membrane
- Chromosome 1
- Chromosome 2
- Ribosomes

FIGURE 7.13   **Steps in binary fission of a rod-shaped bacterium.**

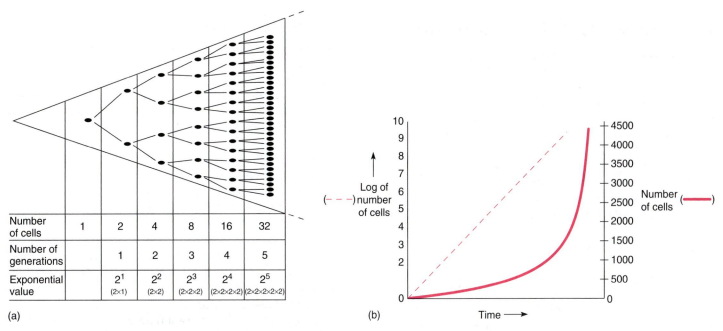

FIGURE 7.14 **The mathematics of population growth.**
(a) Starting with a single cell, if each product of reproduction goes on to divide by binary fission, the population doubles with each new cell division or generation. This process can be represented by logarithms (2 raised to an exponent) or by simple numbers. **(b)** Plotting the logarithm of the cells produces a straight line indicative of exponential growth, whereas plotting the cell numbers arithmetically gives a curved slope.

even for a short period has caused many a person to be suddenly stricken with an attack of food-borne disease. In a few hours, a population of these bacteria can easily grow from a small number of cells to several million.

Figure 7.14 shows several quantitative characteristics of growth: (A) The cell population size can be represented by the number 2 with an exponent ($2^1, 2^2, 2^3, 2^4$); (B) the exponent increases by one in each generation; and (C) the number of the exponent is also the number of the generation. This growth pattern is termed **exponential.** Because these populations often contain very large numbers of cells, it is useful to express them by means of exponents or logarithms (see appendix A). The data from a growing bacterial population are graphed by plotting the number of cells as a function of time. The cell number can be represented logarithmically or arithmetically. Plotting the logarithm number over time provides a straight line indicative of exponential growth. Plotting the data arithmetically gives a constantly curved slope. In general, logarithmic graphs are preferred because an accurate cell number is easier to read, especially during early growth phases.

Predicting the number of cells that will arise during a long growth period (yielding millions of cells) is based on a relatively simple concept. One could use the method of addition 2 + 2 = 4; 4 + 4 = 8; 8 + 8 = 16; 16 + 16 = 32, and so on, or a method of multiplication (for example, $2^5 = 2 \times 2 \times 2 \times 2 \times 2$), but it is easy to see that for 20 or 30 generations, this calculation could be very tedious. An easier way to calculate the size of a population over time is to use an equation such as:

$$N_f = (N_i)2^n$$

In this equation, N_f is the total number of cells in the population at some point in the growth phase, N_i is the starting number, the exponent n denotes the generation number, and 2^n represents the number of cells in that generation. If we know any two of the values, the other values can be calculated. Let us use the example of *Staphylococcus aureus* to calculate how many cells (N_f) will be present in an egg salad sandwich after it sits in a warm car for 4 hours. We will assume that N_i is 10 (number of cells deposited in the sandwich while it was being prepared). To derive n, we need to divide 4 hours (240 minutes) by the generation time (we will use 20 minutes). This calculation comes out to 12, so 2^n is equal to 2^{12}. Using a calculator, we find that 2^{12} is 4,096.

Final number (N_f) = 10 × 4,096
= 40,960 bacterial cells in the sandwich

This same equation, with modifications, is used to determine the generation time, a more complex calculation that requires knowing the number of cells at the beginning and end of a growth period. Such data are obtained through actual testing by a method discussed in the following section.

The Population Growth Curve

In reality, a population of bacteria does not maintain its potential growth rate and does not double endlessly, because in most systems numerous factors prevent the cells from continuously dividing at their maximum rate. Quantitative laboratory studies indicate that a population typically displays a predictable pattern, or **growth curve,** over time. The method traditionally used to observe the population growth pattern is a viable count technique, in which the total number of live cells is counted over a given time period. In brief, this method entails

1. placing a tiny number of cells into a sterile liquid medium;
2. incubating this culture over a period of several hours;
3. sampling the broth at regular intervals during incubation;
4. plating each sample onto solid media; and
5. counting the number of colonies present after incubation.

Insight 7.6 gives the details of this process.

Stages in the Normal Growth Curve

The system of batch culturing described in Insight 7.6 is *closed*, meaning that nutrients and space are finite and there is no mechanism for the removal of waste products. Data from an entire growth period of 3 to 4 days typically produce a curve with a series of phases termed the lag phase, the exponential growth (log) phase, the stationary phase, and the death phase **(figure 7.15).**

The **lag phase** is a relatively "flat" period on the graph when the population appears not to be growing or is growing at less than the exponential rate. Growth lags primarily because

1. the newly inoculated cells require a period of adjustment, enlargement, and synthesis;
2. the cells are not yet multiplying at their maximum rate; and
3. the population of cells is so sparse or dilute that the sampling misses them.

The length of the lag period varies somewhat from one population to another. It is important to note that even though the population of cells is not increasing (growing), individual cells are metabolically active as they increase their contents and prepare to divide.

The cells reach the maximum rate of cell division during the **exponential growth (logarithmic or log) phase,** a period during which the curve increases geometrically. This phase will continue as long as cells have adequate nutrients and the environment is favorable.

At the **stationary growth phase,** the population enters a survival mode in which cells stop growing or grow slowly. The curve levels off because the rate of cell inhibition or death balances out the rate of multiplication. The decline in the growth rate is caused by depleted nutrients and oxygen plus excretion of organic acids and other biochemical pollutants into the growth medium, due to the increased density of cells.

As the limiting factors intensify, cells begin to die at an exponential rate (literally perishing in their own wastes), and they are unable to multiply. The curve now dips downward as the **death phase** begins. The speed with which death occurs depends on the relative resistance of the species and how toxic the conditions are, but it is usually slower than the exponential growth phase. Viable cells often remain many weeks and months after this phase has begun. In the laboratory, refrigeration is used to slow the progression of the death phase so that cultures will remain viable as long as possible.

Practical Importance of the Growth Curve

The tendency for populations to exhibit phases of rapid growth, slow growth, and death has important implications in microbial control, infection, food microbiology, and culture technology. Antimicrobial agents such as heat and disinfectants rapidly accelerate the death phase in all populations, but microbes in the exponential growth phase are more vulnerable to these agents than are those that have entered the stationary phase. In general, actively growing cells are more vulnerable to conditions that disrupt cell metabolism and binary fission.

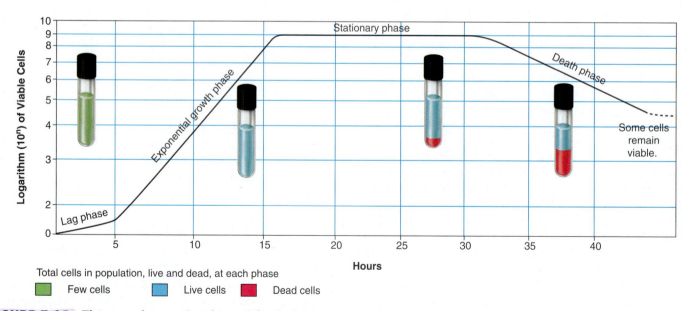

Total cells in population, live and dead, at each phase

🟩 Few cells   🟦 Live cells   🟥 Dead cells

FIGURE 7.15 **The growth curve in a bacterial culture.**
On this graph, the number of viable cells expressed as a logarithm (log) is plotted against time. See text for discussion of the various phases. Note that with a generation time of 30 minutes, the population has risen from 10 (10^1) cells to 1,000,000,000 (10^9) cells in only 16 hours.

INSIGHT 7.6   *Microbiology*

Steps in a Viable Plate Count—Batch Culture Method

A growing population is established by inoculating a flask containing a known quantity of sterile liquid medium with a few cells of a pure culture. The flask is incubated at that bacterium's optimum temperature and timed. The population size at any point in the growth cycle is quantified by removing a tiny measured sample of the culture from the growth chamber and plating it out on a solid medium to develop isolated colonies. This procedure is repeated at evenly spaced intervals (i.e., every hour for 24 hours).

Evaluating the samples involves a common and important principle in microbiology: One colony on the plate represents one cell or colony-forming unit (CFU) from the original sample. Because the CFU of some bacteria is actually composed of several cells (consider the clustered arrangement of *Staphylococcus*, for

instance), using a colony count can underestimate the exact population size to an extent. This is not a serious problem because, in such bacteria, the CFU is the smallest unit of colony formation and dispersal. Multiplication of the number of colonies in a single sample by the container's volume gives a fair estimate of the total population size (number of cells) at any given point. The growth curve is determined by graphing the number for each sample in sequence for the whole incubation period (see figure 7.15).

Because of the scarcity of cells in the early stages of growth, some samples can give a zero reading even if there are viable cells in the culture. The sampling itself can remove enough viable cells to alter the tabulations, but since the purpose is to compare relative trends in growth, these factors do not significantly change the overall pattern.

	60 min	120 min	180 min	240 min	300 min	360 min	420 min	480 min	540 min	600 min
Plates are incubated, colonies are counted	None									
Number of colonies (CFU) per 0.1 ml	<1*	1	3	7	13	23	45	80	135	230
Total estimated cell population in flask	<5,000	5,000	15,000	35,000	65,000	115,000	225,000	400,000	675,000	1,150,000

* Only means that too few cells are present to be assayed.

The viable plate count, a technique for determining population size and rate of growth.

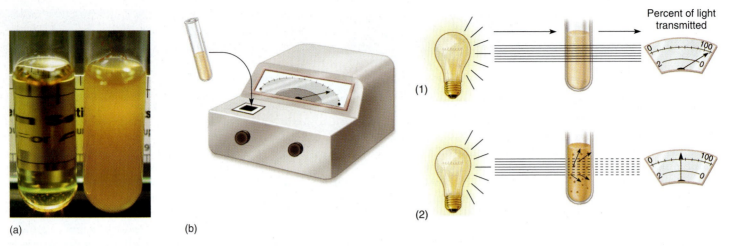

FIGURE 7.16   **Turbidity measurements as indicators of growth.**
(a) Holding a broth to the light is one method of checking for gross differences in cloudiness (turbidity). The broth on the left is transparent, indicating little or no growth; the broth on the right is cloudy and opaque, indicating heavy growth. **(b)** The eye is not sensitive enough to pick up fine degrees in turbidity; more sensitive measurements can be made with a spectrophotometer. (*1*) A tube with no growth will allow light to easily pass. Therefore more light will reach the photodetector and give a higher transmittance value. (*2*) In a tube with growth, the cells scatter the light, resulting in less light reaching the photodetector and, therefore, gives a lower transmittance valve.

Growth patterns in microorganisms can account for the stages of infection (see chapter 13). A person shedding bacteria in the early and middle stages of an infection is more likely to spread it to others than is a person in the late stages. The course of an infection is also influenced by the relatively faster rate of multiplication of the microbe, which can overwhelm the slower growth rate of the host's own cellular defenses.

Understanding the stages of cell growth is crucial for work with cultures. Sometimes a culture that has reached the stationary phase is incubated under the mistaken impression that enough nutrients are present for the culture to multiply. In most cases, it is unwise to continue incubating a culture beyond the stationary phase, because doing so will reduce the number of viable cells and the culture could die out completely. It is also preferable to do stains (an exception is the spore stain) and motility tests on young cultures, because the cells will show their natural size and correct reaction, and motile cells will have functioning flagella.

For certain research or industrial applications, closed batch culturing with its four phases is inefficient. The alternative is an automatic growth chamber called the **chemostat,** or continuous culture system. This device can admit a steady stream of new nutrients and siphon off used media and old bacterial cells, thereby stabilizing the growth rate and cell number. The chemostat is very similar to the industrial fermenters used to produce vitamins and antibiotics (see chapter 24). It has the advantage of maintaining the culture in a biochemically active state and preventing it from entering the death phase.

Other Methods of Analyzing Population Growth

Microbiologists have developed several alternative ways of analyzing bacterial growth qualitatively and quantitatively. One of the simplest methods for estimating the size of a population is through turbidometry. This technique relies on the simple observation that a tube of clear nutrient solution becomes cloudy, or **turbid,** as microbes grow in it. In general, the greater the turbidity, the larger the population size, which can be measured by means of sensitive instruments **(figure 7.16).**

Enumeration of Bacteria

Turbidity readings are useful for evaluating relative amounts of growth, but if a more quantitative evaluation is required, the viable colony count described in Insight 7.6 or some other enumeration (counting) procedure is necessary. The **direct,** or **total, cell count** involves counting the number of cells in a sample microscopically **(figure 7.17).** This technique, very similar to that used in blood cell counts, employs a special microscope slide (cytometer) calibrated to accept a tiny sample that is spread over a premeasured grid. The cell count from a cytometer can be used to estimate the total number of cells in a larger sample (for instance, of milk or water). One inherent inaccuracy in this method as well as in spectrophotometry is that no distinction can be made between dead and live cells, both of which are included in the count.

Counting can be automated by sensitive devices such as the *Coulter counter,* which electronically scans a culture as it passes

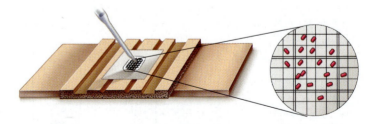

FIGURE 7.17   **Direct microscopic count of bacteria.**
A small sample is placed on the grid under a cover glass. Individual cells, both living and dead, are counted. This number can be used to calculate the total count of a sample.

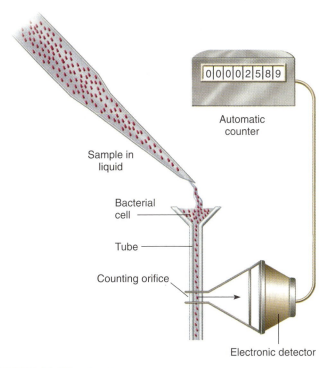

FIGURE 7.18 **Coulter counter.**
As cells pass through this device, they trigger an electronic sensor that tallies their numbers.

through a tiny pipette. As each cell flows by, it is detected and registered on an electronic sensor **(figure 7.18).** A *flow cytometer* works on a similar principle, but in addition to counting, it can measure cell size and even differentiate between live and dead cells. When used in conjunction with fluorescent dyes and anti-

bodies to tag cells, it has been used to differentiate between gram-positive and gram-negative bacteria. It is being adapted for use as a rapid method to identify pathogens in patient specimens and to differentiate blood cells. More sophisticated forms of the flow cytometer can actually sort cells of different types into separate compartments of a collecting device.

Although flow cytometry can be used to count bacteria in natural samples without the need for culturing them, it requires fluorescent labeling of the cells you are interested in detecting, which is not always possible.

A variation of the polymerase chain reaction (PCR) (see Insight 7.4), called real-time PCR, allows scientists to quantify bacteria and other microorganisms that are present in environmental or tissue samples without isolating them and without culturing them.

✔ CHECKPOINT

- Microbial growth refers both to increase in cell size and increase in number of cells in a population.
- The generation time is a measure of the growth rate of a microbial population. It varies in length according to environmental conditions.
- Microbial cultures in a nutrient-limited batch environment exhibit four distinct stages of growth: the lag phase, the exponential growth (log) phase, the stationary phase, and the death phase.
- Microbial cell populations show distinct phases of growth in response to changing nutrient and waste conditions.
- Population growth can be quantified by measuring turbidity, colony counts, and direct cell counts. Other techniques can be used to count bacteria without growing them.

Chapter Summary With Key Terms

7.1 Microbial Nutrition
Nutrition consists of taking in chemical substances (nutrients) and assimilating and extracting energy from them.

A. Substances required for survival are **essential nutrients**—usually containing the elements (C, H, N, O, P, S, Na, Cl, K, Ca, Fe, Mg). Essential nutrients are considered **macronutrients** (required in larger amounts) or **micronutrients** (trace elements required in smaller amounts—Zn, Mn, Cu). Nutrients are classed as either inorganic or organic.

B. A **growth factor** is an organic nutrient (amino acid and vitamin) that cannot be synthesized and must be provided.

C. *Nutritional Types*
1. An **autotroph** depends on carbon dioxide for its carbon needs. An autotroph that derives energy from light is a **photoautotroph;** one that extracts energy from inorganic substances is a **chemoautotroph.** **Methanogens** are chemoautotrophs that produce methane.

2. A **heterotroph** acquires carbon from organic molecules. A **saprobe** is a decomposer that feeds upon dead organic matter, and a **parasite** feeds from a live host and usually causes harm. Parasites that cause damage or death are called **pathogens.**

D. *Transport Mechanisms*
1. A microbial cell must take in nutrients from its surroundings by transporting them across the cell membrane.

2. **Osmosis** is diffusion of water through a selectively permeable membrane.

3. Osmotic changes that affect cells are **hypotonic** solutions, which contain a lower solute concentration, and **hypertonic** solutions, which contain a higher solute concentration. **Isotonic** solutions have the same solute concentration as the inside of the cell. A **halophile** thrives in hypertonic surroundings, and an **obligate halophile** requires a salt concentration of at least 15%, but grows optimally in 25%.

▶ Enzymes derive some of their special characteristics from cofactors such as vitamins, and they show sensitivity to environmental factors.

▶ Enzymes are involved in activities that synthesize, digest, oxidize, and reduce compounds, and convert one substance to another.

▶ Enzymes are regulated by several mechanisms that alter the structure or synthesis of the enzyme.

▶ The energy of living systems resides in the atomic structure of chemicals that can be acted upon and changed.

▶ Cell energetics involves the release of energy that powers the formation of bonds.

▶ The energy of electrons is transferred from one molecule to another in coupled redox reactions.

▶ Electrons are transferred from substrates such as glucose to coenzyme carriers and ultimately captured in high-energy adenosine triphosphate (ATP).

▶ Cell pathways involved in extracting energy from fuels are glycolysis, the tricarboxylic acid cycle, and electron transport.

▶ The molecules used in aerobic respiration are glucose and oxygen, and the products are CO_2, H_2O, and ATP.

▶ Microbes participate in alternate pathways such as fermentation and anaerobic respiration.

▶ Cells manage their metabolites through linked pathways that have numerous functions and can proceed in more than one direction.

8.1 The Metabolism of Microbes

Metabolism, from the Greek term *metaballein*, meaning change, pertains to all chemical reactions and physical workings of the cell. Although metabolism entails thousands of different reactions, most of them fall into one of two general categories. **Anabolism,** sometimes also called *biosynthesis,* is any process that results in synthesis of cell molecules and structures. It is a building and bond-making process that forms larger macromolecules from smaller ones, and it usually requires the input of energy. **Catabolism** is the opposite of anabolism. Catabolic reactions are degradative; they break the bonds of larger molecules into smaller molecules, and often release energy. The linking of anabolism to catabolism ensures the efficient completion of many thousands of cellular processes.

In summary, metabolism performs these functions:

1. degrades macromolecules into smaller molecules and releases energy (catabolism),
2. releases energy in the form of ATP or heat,
3. assembles smaller molecules into larger macromolecules needed for the cell; in this process ATP is utilized to form bonds (anabolism) **(figure 8.1).**

It has built-in controls for reducing or stopping a process that is not in demand and other controls for storing excess nutrients. The metabolic workings of the cell are indeed intricate and complex, but they are also elegant and efficient. It is this very organization that sustains life.

Enzymes: Catalyzing the Chemical Reactions of Life

A microbial cell could be viewed as a microscopic factory, complete with basic building materials, a source of energy, and a "blueprint" for running its extensive network of meta-bolic reactions. But the chemical reactions of life, even when highly organized and complex, cannot proceed without a special class of proteins called **enzymes.**[1] Enzymes are a remarkable example of **catalysts,** chemicals that increase the rate of a chemical reaction without becoming part of the products or being consumed in the reaction. Do not make the mistake of thinking that an enzyme creates a reaction. Because of the great energy of some molecules, a reaction could occur spontaneously at some point even without an enzyme, but at a very slow rate. A study of the enzyme urease shows that it increases the rate of the breakdown of urea by a factor of 100 trillion as compared to an uncatalyzed reaction. Because most uncatalyzed metabolic reactions do not occur fast enough to sustain cell processes, enzymes, which speed up the rate of reactions, are indispensable to life. Other major characteristics of enzymes are summarized in **table 8.1.**

How Do Enzymes Work?

We have said that an enzyme speeds up the rate of a metabolic reaction, but just how does it do this? During a chemical reaction, reactants are converted to products by bond formation or breakage. A certain amount of energy is required to initiate every such reaction, which limits its rate. This resistance to a reaction, which must be overcome for a reaction to proceed, is measurable and is called the **energy of activation** or activation energy **(Insight 8.1).** In the laboratory, overcoming this initial resistance can be achieved by

1. increasing thermal energy (heating) to increase molecular velocity,
2. increasing the concentration of reactants to increase the rate of molecular collisions, or
3. adding a catalyst.

---

1. A few RNA molecules also function on enzymes. See Insight 8.2.

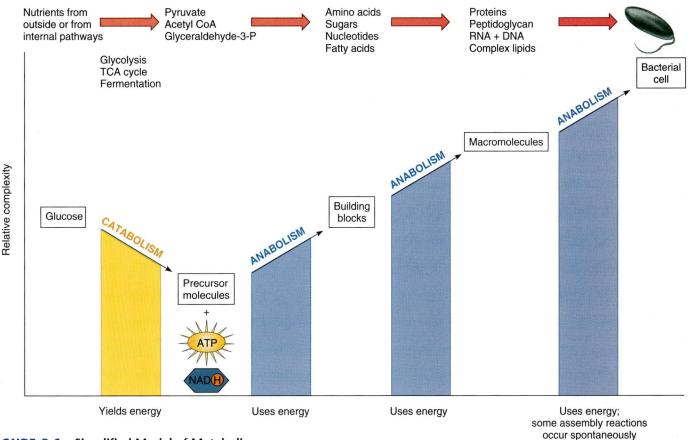

FIGURE 8.1   Simplified Model of Metabolism.
Cellular reactions fall into two major categories. Catabolism involves the breakdown of complex organic molecules to extract energy and form simpler end products. Anabolism uses the energy to synthesize necessary macromolecules and cell structures from precursors.

TABLE 8.1	Checklist of Enzyme Characteristics

- Most are composed of protein; may require cofactors
- Act as organic catalysts to speed up the rate of cellular reactions
- Lower the activation energy required for a chemical reaction to proceed (Insight 8.1)
- Have unique characteristics such as shape, specificity, and function
- Enable metabolic reactions to proceed at a speed compatible with life
- Provide an active site for target molecules called substrates
- Are much larger in size than their substrates
- Associate closely with substrates but do not become integrated into the reaction products
- Are not used up or permanently changed by the reaction
- Can be recycled, thus function in extremely low concentrations
- Are greatly affected by temperature and pH
- Can be regulated by feedback and genetic mechanisms

In most living systems, the first two alternatives are not feasible, because elevating the temperature is potentially harmful and higher concentrations of reactants are not practical. This

leaves only the action of catalysts, and enzymes fill this need efficiently and potently.

At the molecular level, an enzyme promotes a reaction by serving as a physical site upon which the reactant molecules, called **substrates,** can be positioned for various interactions. The enzyme is much larger in size than its substrate, and it presents a unique active site that fits only that particular substrate. Although an enzyme binds to the substrate and participates directly in changes to the substrate, it does not become a part of the products, is not used up by the reaction, and can function over and over again. Enzyme speed, defined as the number of substrate molecules converted per enzyme per second, is well documented. Speeds range from several million for catalase to a thousand for lactate dehydrogenase. To further visualize the roles of enzymes in metabolism, we must next look at their structure.

Enzyme Structure

The primary structure of all enzymes is protein (with some exceptions—see **Insight 8.2**), and they can be classified as simple or conjugated. Simple enzymes consist of protein

the removal of carbon dioxide from organic acids in several metabolic pathways.

The Role of Microbial Enzymes in Disease   Many pathogens secrete unique exoenzymes that help them avoid host defenses or promote their multiplication in tissues. Because these enzymes contribute to pathogenicity, they are referred to as virulence factors, or toxins in some cases. *Streptococcus pyogenes* (a cause of throat and skin infections) produces a streptokinase that digests blood clots and apparently assists in invasion of wounds. Another exoenzyme from this bacterium is called streptolysin.[2] In mammalian hosts, streptolysin damages blood cells and tissues. It is also responsible for lysing red blood cells used in blood agar dishes, and this trait is used for identifying the bacteria growing in culture (see chapter 21). *Pseudomonas aeruginosa*, a respiratory and skin pathogen, produces elastase and collagenase, which digest elastin and collagen, two proteins found in connective tissue. These increase the severity of certain lung diseases and burn infections. *Clostridium perfringens*, an agent of gas gangrene, synthesizes lecithinase C, a lipase that profoundly damages cell membranes and accounts for the tissue death associated with this disease. Not all enzymes digest tissues; some, such as penicillinase, inactivate penicillin and thereby protect a microbe from its effects.

The Sensitivity of Enzymes to Their Environment

The activity of an enzyme is highly influenced by the cell's environment. In general, enzymes operate only under the natural temperature, pH, and osmotic pressure of an organism's habitat. When enzymes are subjected to changes in these normal conditions, they tend to be chemically unstable, or **labile.** Low temperatures inhibit catalysis, and high temperatures denature the apoenzyme. **Denaturation** is a process by which the weak bonds that collectively maintain the native shape of the apoenzyme are broken. This disruption causes extreme distortion of the enzyme's shape and prevents the substrate from attaching to the active site (described in chapter 11). Such nonfunctional enzymes block metabolic reactions and thereby can lead to cell death. Low or high pH or certain chemicals (heavy metals, alcohol) are also denaturing agents.

Regulation of Enzymatic Activity and Metabolic Pathways

Metabolic reactions proceed in a systematic, highly regulated manner that maximizes the use of available nutrients and energy. The cell responds to environmental conditions by adopting those metabolic reactions that most favor growth and survival. Because enzymes are critical to these reactions, the regulation of metabolism is largely the regulation of enzymes by an elaborate system of checks and balances. Let us take a look at some general features of metabolic pathways.

Metabolic Pathways

Metabolic reactions rarely consist of a single action or step. More often, they occur in a multistep series or pathway, with each step catalyzed by an enzyme. An individual reaction is shown in various ways, depending on the purpose at hand (**figure 8.9**). The product of one reaction is often the reactant (substrate) for the next, forming a linear chain of reactions. Many pathways have branches that provide alternate methods for nutrient processing. Others take a cyclic form, in which the starting molecule is regenerated to initiate another turn of the cycle. Pathways generally do not stand alone; they are interconnected and merge at many sites.

Every pathway has one or more enzyme pacemakers (usually the slowest enzyme in the series) that sets the rate of a pathway's progression. These enzymes respond to various

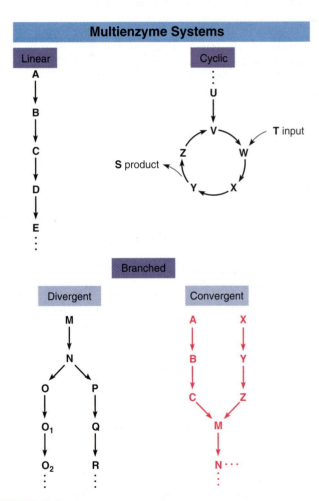

FIGURE 8.9   Patterns of metabolism.

In general, metabolic pathways consist of a linked series of individual chemical reactions that produce intermediary metabolites and lead to a final product. These pathways occur in several patterns, including linear, cyclic, and branched. Anabolic pathways involved in biosynthesis result in a more complex molecule, each step adding on a functional group, whereas catabolic pathways involve the dismantling of molecules and can generate energy. Virtually every reaction in a series involves a specific enzyme.

---

2. Even though most enzyme names end in *-ase* (see Insight 8.3), not all do.

control signals and, in so doing, determine whether a pathway proceeds. Regulation of pacemaker enzymes proceeds on two fundamental levels. Either the enzyme itself is directly inhibited or activated, or the amount of the enzyme in the system is altered (decreased or increased). Factors that affect the enzyme directly provide a means for the system to be finely controlled or tuned, whereas regulation at the genetic level (enzyme synthesis) provides a slower, less sensitive control.

Direct Controls on the Action of Enzymes

The bacterial cell has many ways of directly influencing the activity of its enzymes. It can inhibit enzyme activity by supplying a molecule that resembles the enzyme's normal substrate. The "mimic" can then occupy the enzyme's active site, preventing the actual substrate from binding there. Since the mimic cannot actually be acted on by the enzyme, or function in the way the product would have, the enzyme is effectively shut down. This form of inhibition is called **competitive inhibition,** since the mimic is competing with the substrate for the binding site **(figure 8.10).** (In chapter 12 you will see that some antibiotics use the same strategy of competing with enzymatic active sites to shut down metabolic processes.)

Another form of inhibition can occur with special types of enzymes that have two binding sites—the active site and another area called the regulatory site (figure 8.10). These enzymes are regulated by the binding of molecules other than the substrate in their regulatory sites. Often the regulatory molecule is the product of the enzymatic reaction itself. This provides a negative feedback mechanism that can slow down enzymatic activity once a certain concentration of product is produced. This is **noncompetitive inhibition,** since the regulator molecule does not bind in the same site as the substrate.

Controls on Enzyme Synthesis

Controlling enzymes by controlling their synthesis is another effective mechanism, because enzymes do not last indefinitely. Some wear out, some are deliberately degraded, and others are diluted with each cell division. For catalysis to continue, enzymes eventually must be replaced. This cycle works into the scheme of the cell, where replacement of enzymes can be regulated according to cell demand. The mechanisms of this system are genetic in nature; that is, they require regulation of DNA and the protein synthesis machinery, topics we shall encounter once again in chapter 9.

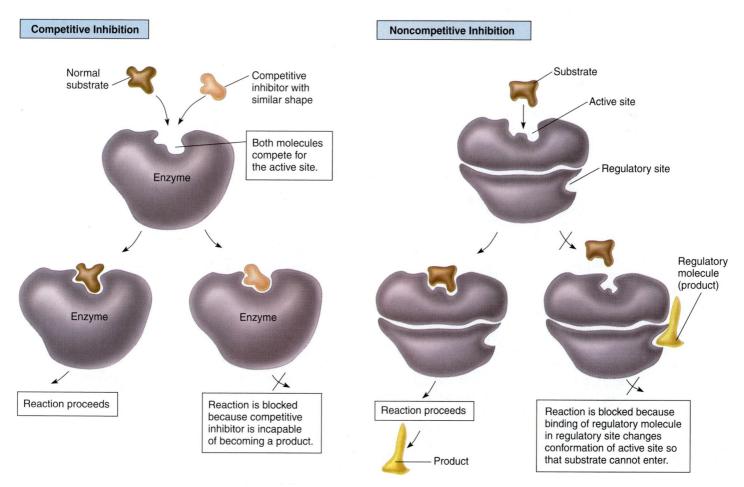

Competitive Inhibition

Normal substrate

Competitive inhibitor with similar shape

Both molecules compete for the active site.

Enzyme

Enzyme

Enzyme

Reaction proceeds

Reaction is blocked because competitive inhibitor is incapable of becoming a product.

Noncompetitive Inhibition

Substrate

Active site

Regulatory site

Regulatory molecule (product)

Reaction proceeds

Reaction is blocked because binding of regulatory molecule in regulatory site changes conformation of active site so that substrate cannot enter.

Product

FIGURE 8.10   **Examples of two common control mechanisms for enzymes.**

Enzyme repression is a means to stop further synthesis of an enzyme somewhere along its pathway. As the level of the end product from a given enzymatic reaction has built to excess, the genetic apparatus responsible for replacing these enzymes is automatically suppressed (**figure 8.11**). The response time is longer than for feedback inhibition, but its effects are more enduring.

A response that resembles the inverse of enzyme repression is **enzyme induction.** In this process, enzymes appear (are induced) only when suitable substrates are present—that is, the synthesis of enzyme is induced by its substrate. Both mechanisms are important genetic control systems in bacteria.

A classic model of enzyme induction occurs in the response of *Escherichia coli* to certain sugars. For example, if a particular strain of *E. coli* is inoculated into a medium whose principal carbon source is lactose, it will produce the enzyme lactase to hydrolyze it into glucose and galactose. If the bacterium is subsequently inoculated into a medium containing only sucrose as a carbon source, it will cease synthesizing lactase and begin synthesizing sucrase. This response enables the organism to adapt to a variety of nutrients, and it also prevents a microbe from wasting energy, making enzymes for which no substrates are present.

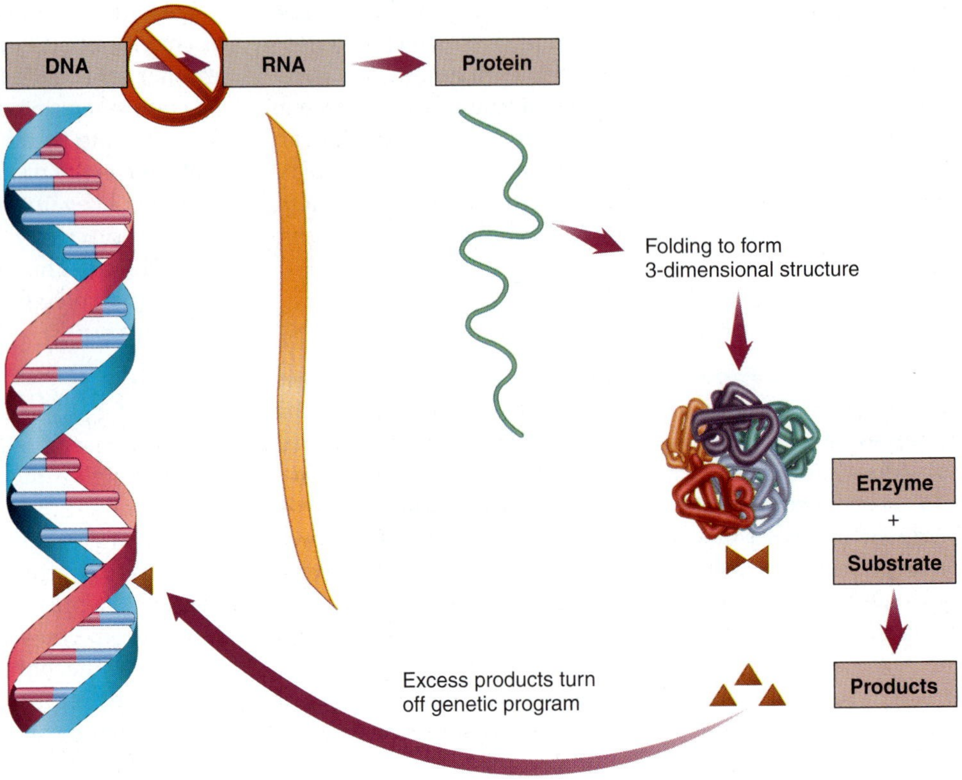

FIGURE 8.11   **One type of genetic control of enzyme synthesis: enzyme repression.**
The enzyme is synthesized continuously until enough product has been made, at which time the excess product reacts with a site on DNA that regulates the enzyme's synthesis, thereby inhibiting further enzyme production.

- To function effectively, enzymes require specific conditions of temperature, pH, and osmotic pressure.
- Enzyme activity is regulated by processes that affect the enzymes directly, or by altering the amount of enzyme produced during protein synthesis.

✔ CHECKPOINT

- Metabolism includes all the biochemical reactions that occur in the cell. It is a self-regulating complex of interdependent processes that encompasses many thousands of chemical reactions.
- Anabolism is the energy-requiring subset of metabolic reactions, which synthesize large molecules from smaller ones.
- Catabolism is the energy-releasing subset of metabolic reactions, which degrade or break down large molecules into smaller ones.
- Enzymes are proteins or RNA that catalyze all biochemical reactions by forming enzyme-substrate complexes. The binding of the substrate by an enzyme makes possible both bond-forming and bond-breaking reactions, depending on the pathway involved. Enzymes may utilize cofactors as carriers and activators.
- Enzymes are classified and named according to the kinds of reactions they catalyze.

8.2 The Pursuit and Utilization of Energy

Energy is defined as the capacity to do work or to cause change. In order to carry out the work of an array of metabolic processes, cells require constant input and expenditure of some form of usable energy. Energy commonly exists in various forms:

1. thermal (heat) energy from molecular motion,
2. radiant (wave) energy from visible light or other rays,
3. electrical energy from a flow of electrons,
4. mechanical energy from a physical change in position,
5. atomic energy from reactions in the nucleus of an atom, and
6. chemical energy present in the bonds of molecules.

Cells are far too fragile to rely in any constant way on thermal or atomic energy for cell transactions. Except for certain

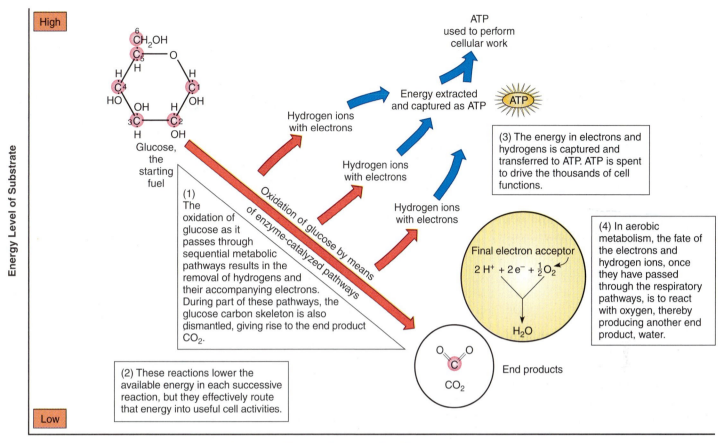

FIGURE 8.12 **A simplified model that summarizes the cell's energy machine.**
The central events of cell energetics include the release of energy during the systematic dismantling of a fuel such as glucose. This is achieved by the shuttling of hydrogens and electrons to sites in the cell where their energy can be transferred to ATP. The final products are CO_2 and H_2O molecules.

chemoautotrophic bacteria, the ultimate source of energy is the sun, but only photosynthetic autotrophs can tap this source directly. This capacity of photosynthesis to convert solar energy into chemical energy provides both a nutritional and an energy basis for all heterotrophic living things. For the most part, only chemical energy can routinely operate cell transactions, and chemical reactions are the universal basis of cellular energetics.

In this section we will further explore the role of energy as it is related to the oxidation of fuels, redox carriers, and the generation of ATP.

Cell Energetics

Cells manage energy in the form of chemical reactions that change molecules. This often involves activities such as the making or breaking of bonds and the transfer of electrons. Not all cellular reactions are equal with respect to energy. Some release energy, and others require it to proceed. For example, a reaction that proceeds as follows:

$$X + Y \xrightarrow{\text{Enzyme}} Z + \text{Energy}$$

releases energy as it goes forward. This type of reaction is termed **exergonic** (ex-er-gon'-ik). Energy of this type is considered free—it is available for doing cellular work. Energy transactions such as the following:

$$\text{Energy} + A + B \xrightarrow{\text{Enzyme}} C$$

are called **endergonic** (en-der-gon'-ik), because they are driven forward with the addition of energy. Exergonic and endergonic reactions are coupled, so that released energy is immediately put to use.

Summaries of metabolism might make it seem that cells "create" energy from nutrients, but they do not. What they actually do is extract chemical energy already present in nutrient fuels and apply that energy toward useful work in the cell, much like a gasoline engine releases energy as it burns fuel. The engine does not actually produce energy, but it converts the potential energy of the fuel to the free energy of work.

The processes by which cells handle energy are well understood, and most of them are explainable in basic biochemical terms. At the simplest level, cells possess specialized enzyme systems that trap the energy present in the bonds of nutrients as they are progressively broken **(figure 8.12).**

During exergonic reactions, energy released by electrons is stored in certain high-energy phosphate molecules such as ATP. As we shall see, the ability of ATP to temporarily store and release the energy of chemical bonds fuels endergonic cell reactions. Before discussing ATP, let us examine the process behind electron transfer: redox reactions.

A Closer Look at Biological Oxidation and Reduction

We stated earlier that biological systems often extract energy through **redox reactions.** Such reactions always occur in pairs, with an electron donor and an electron acceptor, which constitute a *conjugate pair,* or *redox pair.* The reaction can be represented as follows:

Electron donor + Electron acceptor $\longrightarrow$ Electron donor + Electron acceptor
•e⁻   •e⁻
(Reduced)   (Oxidized)   (Oxidized)   (Reduced)

(Review Insight 2.2 for detailed figure.)

This process salvages electrons along with their inherent energy, and it changes the energy balance, leaving the previously reduced compound with less energy than the now oxidized one. The released energy can be captured to **phosphorylate** (add an inorganic phosphate) to ADP or to some other compound. This process stores the energy in a high-energy molecule (ATP, for example). In many cases, the cell does not handle electrons as discrete entities but rather as parts of an atom such as hydrogen. For simplicity's sake, we will continue to use the term *electron transfer,* but keep in mind that hydrogens are often involved in the transfer process. The removal of hydrogens (a hydrogen atom consists of a single proton and a single electron) from a compound during a redox reaction is called dehydrogenation. The job of handling these protons and electrons falls to one or more carriers, which function as short-term repositories for the electrons until they can be transferred. As we shall see, dehydrogenations are an essential supplier of electrons for the respiratory electron transport system.

Electron Carriers: Molecular Shuttles

Electron carriers resemble shuttles that are alternately loaded and unloaded, repeatedly accepting and releasing electrons and hydrogens to facilitate the transfer of redox energy. Most carriers are coenzymes that transfer both electrons and hydrogens, but some transfer electrons only. The most common carrier is NAD (nicotinamide adenine dinucleotide), which carries hydrogens (and a pair of electrons) from dehydrogenation reactions **(figure 8.13).** Reduced NAD can be represented in various ways. Because 2 hydrogens are removed, the actual carrier state is NADH + H⁺, but this is somewhat cumbersome, so we will represent it with the shorter NADH. In catabolic pathways, electrons are ex-

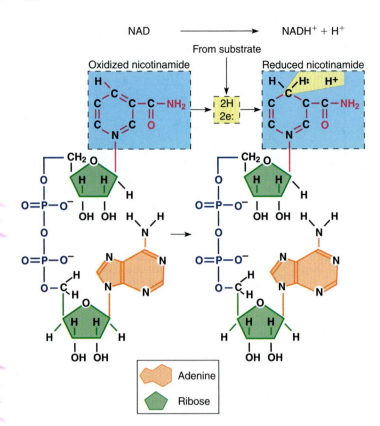

FIGURE 8.13 **Details of NAD reduction.**
The coenzyme NAD contains the vitamin nicotinamide (niacin) and the purine adenine attached to double ribose phosphate molecules (a dinucleotide). The principal site of action is on the nicotinamide (boxed area). Hydrogens and electrons donated by a substrate interact with a carbon on the top of the ring. One hydrogen bonds there, carrying two electrons (H:), and the other hydrogen is carried in solution as H⁺ (a proton).

tracted and carried through a series of redox reactions until the final electron acceptor at the end of a particular pathway is reached (see figure 8.12). In aerobic metabolism, this acceptor is molecular oxygen; in anaerobic metabolism, it is some other inorganic or organic compound. Other common redox carriers are FAD, NADP (NAD phosphate), coenzyme A, and the compounds of the respiratory chain, which are fixed into membranes.

Adenosine Triphosphate: Metabolic Money

In what ways do cells extract chemical energy from electrons, store it, and then tap the storage sources? To answer these questions, we must look more closely at the powerhouse molecule, adenosine triphosphate. ATP has also been described as metabolic money because it can be earned, banked, saved, spent, and exchanged. As a temporary energy repository, ATP provides a connection between energy-yielding catabolism and all other cellular activities that require energy. Some clues to its energy-storing properties lie in its unique molecular structure.

The Molecular Structure of ATP

ATP is a three-part molecule consisting of a nitrogen base (adenine) linked to a 5-carbon sugar (ribose), with a chain of three phosphate groups bonded to the ribose **(figure 8.14).** The type, arrangement, and especially the proximity of atoms in ATP combine to form a compatible but unstable high-energy molecule. The high energy of ATP originates in the orientation of the phosphate groups, which are relatively bulky and carry negative charges. The proximity of these repelling electrostatic charges imposes a strain that is most acute on the bonds between the last two phosphate groups. The strain on the phosphate bonds accounts for the energetic quality of ATP, because removal of the terminal phosphates releases free energy.

Breaking the bonds between two successive phosphates of ATP yields adenosine diphosphate (ADP), which is then converted to adenosine monophosphate (AMP). It is worthwhile noting how economically a cell manages its pool of adenosine nucleotides. AMP derivatives help form the backbone of RNA and are also a major component of certain coenzymes (NAD, FAD, and coenzyme A).

The Metabolic Role of ATP

ATP is the primary energy currency of the cell and when it is used in a chemical reaction it must then be replaced. Therefore, ATP utilization and replenishment is an ongoing cycle. In many instances, the energy released during ATP hydrolysis powers biosynthesis by activating individual subunits before they are enzymatically linked together. ATP is also used to prepare a molecule for catabolism such as the phosphorylation of a 6-carbon sugar during the early stages of glycolysis **(figure 8.15).**

When ATP is utilized by the removal of the terminal phosphate to release energy plus ADP, ATP then needs to be replenished. The reversal of this process, that is, adding the terminal phosphate to ADP, will replenish ATP.

$$ATP \leftrightarrows ADP + P_i + Energy$$

In heterotrophs, the energy infusion that regenerates a high-energy phosphate comes from certain steps of catabolic pathways, in which nutrients such as carbohydrates are degraded and yield energy. ATP is formed when substrates or electron carriers provide a high-energy phosphate that becomes bonded to ADP. Some ATP molecules are formed through *substrate-level phosphorylation*. In substrate-level phosphorylation, ATP is formed by transfer of a phosphate group from a phosphorylated compound (substrate) directly to ADP to yield ATP **(figure 8.16).**

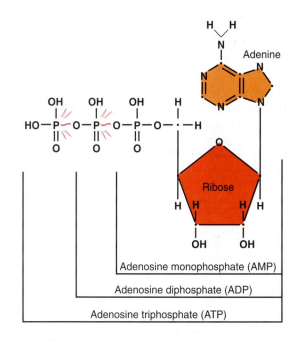

~ Bond that releases
   energy when broken

FIGURE 8.14 **The structure of adenosine triphosphate (ATP) and its partner compounds, ADP and AMP.**

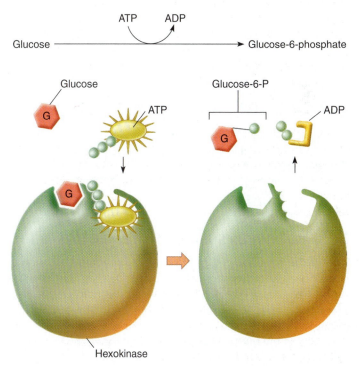

FIGURE 8.15 **An example of phosphorylation of glucose by ATP.**
The first step in catabolizing glucose is the addition of a phosphate from ATP by an enzyme called hexokinase. This use of high-energy phosphate as an activator is a recurring feature of many metabolic pathways.

Substrate

FIGURE 8.16   **ATP formation by substrate-level phosphorylation.**
The inorganic phosphate and the substrates form a bond with high potential energy. In a reaction catalyzed enzymatically, the phosphate is transferred to ADP, thereby producing ATP.

Other ATPs are formed through *oxidative phosphorylation,* a series of redox reactions occurring during the final phase of the respiratory pathway. Phototrophic organisms have a system of *photophosphorylation,* in which the ATP is formed through a series of sunlight-driven reactions (discussed in more detail in chapter 24).

✔ CHECKPOINT

- All metabolic processes require the constant input and expenditure of some form of usable energy. Chemical energy is the currency that runs the metabolic processes of the cell.

- Chemical energy is obtained from the electrons of nutrient molecules through catabolism. It is used to perform the cellular "work" of biosynthesis, movement, membrane transport, and growth.

- Energy is extracted from nutrient molecules by redox reactions. A redox pair of substances passes electrons and hydrogens between them. The donor substance loses electrons, becoming oxidized. The acceptor substance gains electrons, becoming reduced.

- ATP is an important energy molecule of the cell. It donates free energy to anabolic reactions and is continuously regenerated by three phosphorylation processes: substrate-level phosphorylation, oxidative phosphorylation, and (in certain organisms) photophosphorylation.

8.3 Pathways of Bioenergetics

One of the scientific community's greatest achievements was deciphering the biochemical pathways of cells. Initial work with bacteria and yeasts, followed by studies with animal and plant cells, clearly demonstrated metabolic similarities and strongly supported the concept of the universality of metabolism. The study of the production and use of energy by cells is called **bioenergetics,** including catabolic routes that degrade nutrients and anabolic routes that are involved in cell synthesis. Although these pathways are interconnected and interdependent, anabolic pathways are not simply reversals of catabolic ones. At first glance it might seem more economical to use identical pathways, but having different enzymes and divided pathways allows anabolism and catabolism to proceed

simultaneously without interference. For simplicity, we shall focus our discussion on the most common catabolic pathways that will illustrate general principles of other pathways as well.

Catabolism: An Overview of Nutrient Breakdown and Energy Release

The primary catabolism of fuels (such as glucose) that results in energy release in many organisms proceeds through a series of three coupled pathways:

1. **glycolysis** (gly-kol'-ih-sis), also called the Embden-Meyerhof-Parnas (EMP) pathway;
2. the **tricarboxylic acid cycle (TCA),** also known as the citric acid or Krebs cycle;[3] and
3. the **respiratory chain** (electron transport and oxidative phosphorylation).

Each segment of the pathway is responsible for a specific set of actions on various products of glucose. The interconnections of these pathways in aerobic respiration are represented in **figure 8.17,** and their reactions are summarized in **table 8.3.** In the following sections, we will observe each of these pathways in greater detail.

Energy Strategies in Microorganisms

Nutrient processing is extremely varied, especially in bacteria, yet in most cases it is based on three basic catabolic pathways. In previous discussions, microorganisms were categorized according to their requirement for oxygen gas, and this requirement is related directly to their mechanisms of energy release. As we shall see, **aerobic respiration** is a series of reactions (glycolysis, the TCA cycle, and the respiratory chain) that converts glucose to CO_2 and gives off energy (review figure 8.12). It relies on free oxygen as the final acceptor for electrons and hydrogens and produces a relatively large amount of ATP. Aerobic respiration is characteristic of many bacteria, fungi, protozoa, and animals, and it is the system we will emphasize here. Facultative and aerotolerant anaerobes may use only the glycolysis scheme to incompletely oxidize or ferment glucose. In this case, oxygen is not required, organic compounds are the final electron acceptors, and a relatively small amount of ATP is produced. While the growth of aerobic bacteria is usually limited by the availability of substrates, the growth of anaerobes is likely to be stopped when final electron acceptors run out. Some strictly anaerobic microorganisms metabolize by means of **anaerobic respiration.** This system involves the same three pathways as aerobic respiration,

---

3. The EMP pathway is named for the biochemists who first outlined its steps. TCA refers to the involvement of several organic acids containing three carboxylic acid groups, citric acid being the first tricarboxylic acid formed; Krebs is in honor of Sir Hans Krebs who, with F. A. Lipmann, delineated this pathway, an achievement for which they won the Nobel Prize in Physiology or Medicine in 1953.

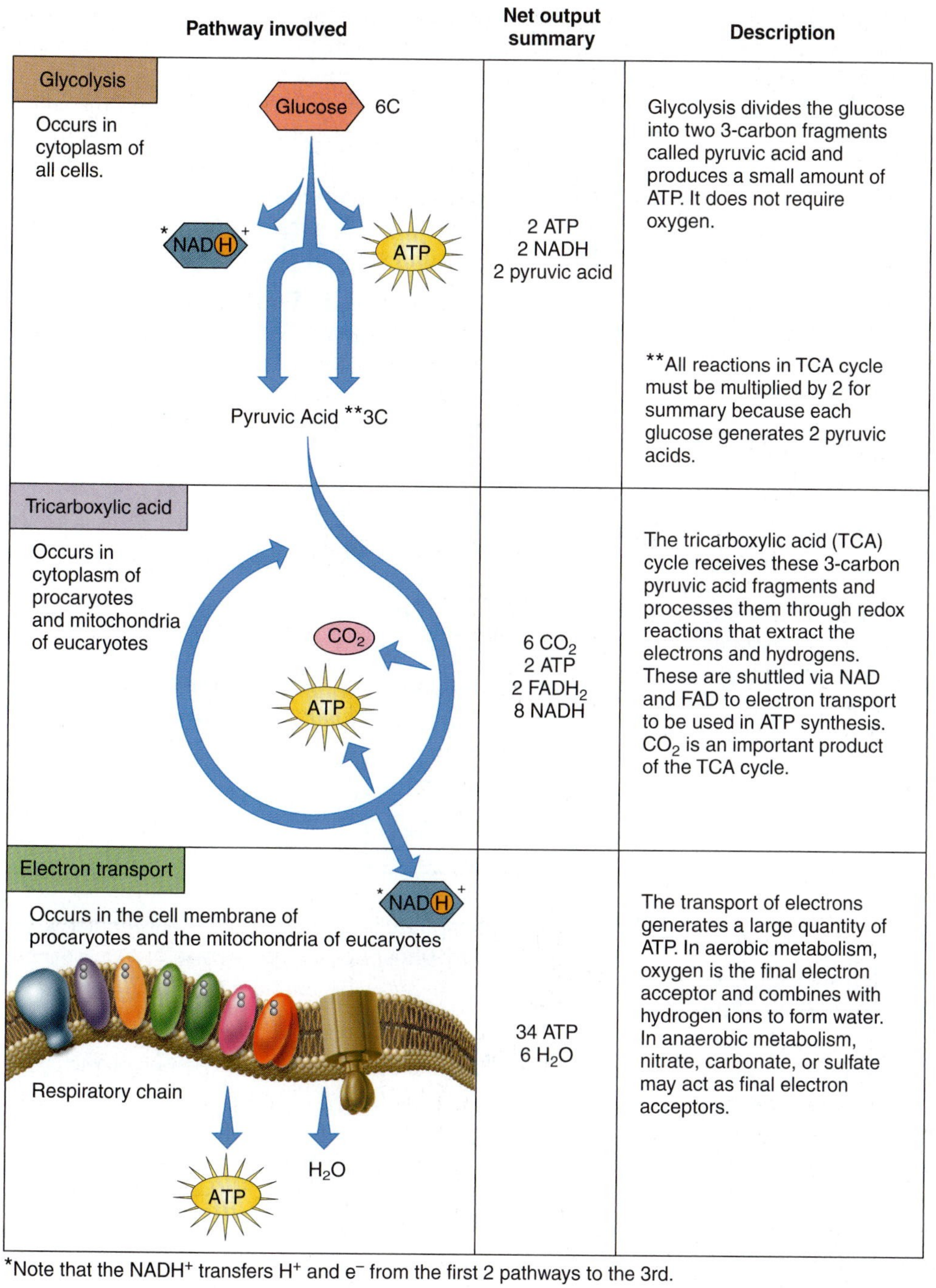

Pathway involved	Net output summary	Description
Glycolysis Occurs in cytoplasm of all cells. Glucose 6C *NAD$\textcircled{H}^+$ ATP Pyruvic Acid **3C	2 ATP 2 NADH 2 pyruvic acid	Glycolysis divides the glucose into two 3-carbon fragments called pyruvic acid and produces a small amount of ATP. It does not require oxygen. **All reactions in TCA cycle must be multiplied by 2 for summary because each glucose generates 2 pyruvic acids.
Tricarboxylic acid Occurs in cytoplasm of procaryotes and mitochondria of eucaryotes CO_2 ATP	6 CO_2 2 ATP 2 $FADH_2$ 8 NADH	The tricarboxylic acid (TCA) cycle receives these 3-carbon pyruvic acid fragments and processes them through redox reactions that extract the electrons and hydrogens. These are shuttled via NAD and FAD to electron transport to be used in ATP synthesis. CO_2 is an important product of the TCA cycle.
Electron transport Occurs in the cell membrane of procaryotes and the mitochondria of eucaryotes *NAD$\textcircled{H}^+$ Respiratory chain ATP H_2O	34 ATP 6 H_2O	The transport of electrons generates a large quantity of ATP. In aerobic metabolism, oxygen is the final electron acceptor and combines with hydrogen ions to form water. In anaerobic metabolism, nitrate, carbonate, or sulfate may act as final electron acceptors.

*Note that the NADH$^+$ transfers H$^+$ and e$^-$ from the first 2 pathways to the 3rd.

FIGURE 8.17 **Overview of the flow, location, and products of pathways in aerobic respiration.** Glucose is degraded through a gradual stepwise process to carbon dioxide and water, while simultaneously extracting energy as ATP and NADH.

TABLE 8.3 **Metabolic Strategies Among Heterotrophic Microorganisms Growing on Glucose**

Scheme	Pathways Involved	Final Electron Acceptor	Products	Chief Microbe Type
Aerobic respiration	Glycolysis, TCA cycle, electron transport	O_2	ATP, CO_2, H_2O	Aerobes; facultative anaerobes
Anaerobic metabolism				
Fermentative	Glycolysis	Organic molecules	ATP, CO_2, ethanol, lactic acid*	Facultative, aerotolerant, strict anaerobes
Respiration	Glycolysis, TCA cycle, electron transport	Various inorganic ions (NO_3^-, SO_4^{2-}, CO_3^{3-})	CO_2, ATP, organic acids, H_2S, CH_4, N_2	Anaerobes; some facultatives

* The products of microbial fermentations are extremely varied and include organic acids, alcohols, and gases.

but it does not use molecular oxygen as the final electron acceptor but instead, NO_3^-, SO_4^{2-}, CO_3^{3-}, and others are utilized as the final electron acceptor. Aspects of fermentation and anaerobic respiration are covered in subsequent sections of this chapter.

Aerobic Respiration

Aerobic respiration is a series of enzyme-catalyzed reactions in which electrons are transferred from fuel molecules such as glucose to oxygen as a final electron acceptor. This pathway is the principal energy-yielding scheme for aerobic heterotrophs, and it provides both ATP and metabolic intermediates for many other pathways in the cell, including those of protein, lipid, and carbohydrate synthesis (figure 8.17).

Aerobic respiration in microorganisms can be summarized by an equation:

$$\text{Glucose } (C_6H_{12}O_6) + 6\,O_2 + 38\,ADP + 38\,P_i \rightarrow$$

$$6\,CO_2 + 6\,H_2O + 38\,ATP$$

The seeming simplicity of the equation for aerobic respiration conceals its complexity. Fortunately, we do not have to present all of the details to address some important concepts concerning its reactants and products, as follows:

1. the steps in the oxidation of glucose,
2. the involvement of coenzyme carriers and the final electron acceptor,
3. where and how ATP originates,
4. where carbon dioxide originates,
5. where oxygen is required, and
6. where water originates.

Glucose: The Starting Compound Carbohydrates such as glucose are good fuels because these compounds are readily oxidized; that is, they are superior hydrogen and electron donors. The enzymatic withdrawal of hydrogen from them also removes electrons that can be used in energy transfers. The end products of the conversion of these carbon compounds are energy-rich ATP and energy-poor carbon dioxide and water. Polysaccharides (starch, glycogen) and disaccharides (maltose, lactose) are stored sources of glucose for the respiratory pathways. Although we use glucose as the main starting compound, other hexoses (fructose, galactose) and fatty acid subunits can enter the pathways of aerobic respiration as well, as we will see in a later section of this chapter.

Glycolysis: The Starting Lineup

A process called glycolysis enzymatically converts glucose through several steps into pyruvic acid. Depending on the organism and the conditions, it may be only the first phase of aerobic respiration, or it may serve as the primary metabolic pathway (fermentation). Glycolysis provides a significant means to synthesize a small amount of ATP anaerobically and also to generate pyruvic acid, an essential intermediary metabolite.

Steps in the Glycolytic Pathway Glycolysis, or the EMP pathway, proceeds along nine linear steps, starting with glucose and ending with pyruvic acid. The first portion of glycolysis involves activation of the substrate, and the steps following involve oxidation reactions of the glucose fragments, the synthesis of ATP, and the formation of pyruvic acid. *Although each step of metabolism is catalyzed by a specific enzyme, we will not mention it for most reactions.* The following outline lists the principal steps of glycolysis.

1. **Glucose** is phosphorylated by means of an **ATP** acting with the enzyme hexokinase. The product is glucose-6-phosphate. (Numbers in chemical names refer to the position of the phosphate on the carbon skeleton.) This is a way of "priming" the system and keeping the glucose inside the cell.

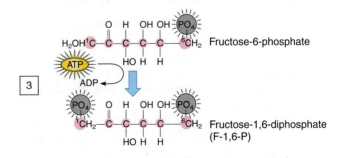

2. Glucose-6-phosphate is converted to its isomer, fructose-6-phosphate, by phosphoglucoisomerase.

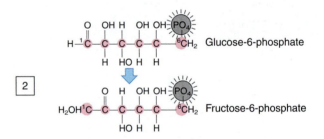

3. Another ATP is spent in phosphorylating the first carbon of fructose-6-phosphate, which yields fructose-1,6-diphosphate.

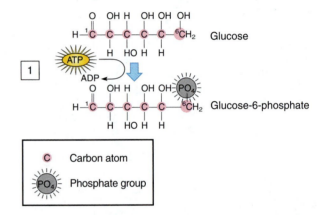

Up to this point, no energy has been released, no oxidation-reduction has occurred, and, in fact, 2 ATPs have been used. In addition, the molecules remain in the 6-carbon state.

4. Now doubly activated, fructose-1,6-diphosphate is split into two 3-carbon fragments: glyceraldehyde-3-phosphate (G-3-P) and dihydroxyacetone phosphate (DHAP). These molecules are isomers, and DHAP is enzymatically converted to G-3-P, which is the more reactive form for subsequent reactions.

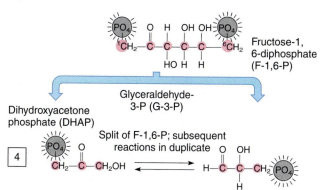

The effect of the splitting of fructose diphosphate is to double every subsequent reaction, because where there was once a single molecule, there are now two to be fed into the remaining pathways.

5. Each molecule of glyceraldehyde-3-phosphate becomes involved in the single oxidation-reduction reaction of glycolysis, a reaction that sets the scene for ATP synthesis. Two reactions occur simultaneously and are catalyzed by the same enzyme, called glyceraldehyde-3-phosphate dehydrogenase. The coenzyme **NAD** picks up hydrogen from G-3-P, forming **NADH.** This step is accompanied by the addition of an inorganic phosphate (PO_4^{3-}) to form an unstable bond on the third carbon of the G-3-P substrate. The product of these reactions is diphosphoglyceric acid.

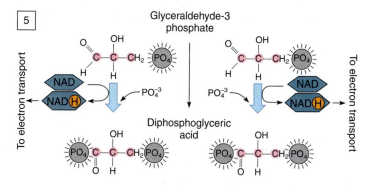

In aerobic organisms, the NADH formed during this step will undergo further reactions in the electron transport system, where the final H acceptor will be oxygen, and an additional 3 ATPs will be generated per NADH. In organisms that ferment glucose anaerobically, the NADH will be oxidized back to NAD, and the hydrogen acceptor will be an organic compound (described later in this chapter).

6. One of the high-energy phosphates of diphosphoglyceric acid is donated to ADP via substrate-level phosphorylation resulting in a molecule of ATP. The product of this reaction is 3-phosphoglyceric acid.

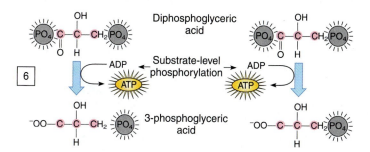

7,8. During this phase, a substrate for the synthesis of a second ATP is produced in two substeps. First, 3-phosphoglyceric acid is converted to 2-phosphoglyceric acid through the shift of a phosphate from the third to the second carbon. Then, the removal of a water molecule from 2-phosphoglyceric acid produces phosphoenolpyruvic acid and generates another high-energy phosphate.

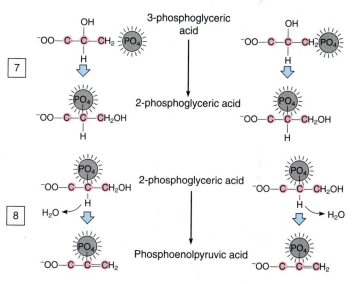

9. In the final reaction of glycolysis, phosphoenolpyruvic acid gives up its high-energy phosphate to form a second ATP, again via substrate-level phosphorylation. This reaction, catalyzed by pyruvate kinase, also produces pyruvic acid (pyruvate), a compound with many roles in metabolism.

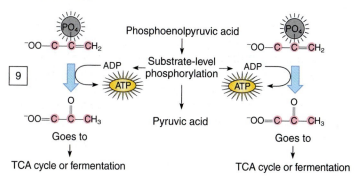

The 2 ATPs formed during steps 6 and 9 are examples of substrate-level phosphorylation, in that the high-energy phosphate is transferred directly from a substrate to ADP. These reactions are catalyzed by kinases, special types of transferase that can phosphorylate a substrate. Because both molecules that arose in step 4 undergo these reactions, an overall total of **4 ATPs** is generated in the partial oxidation of a glucose to 2 pyruvic acids. However, 2 ATPs were expended for steps 1 and 3, so the net number of ATPs available to the cell from these reactions is **2**. For a summary review refer to **figure 8.18.**

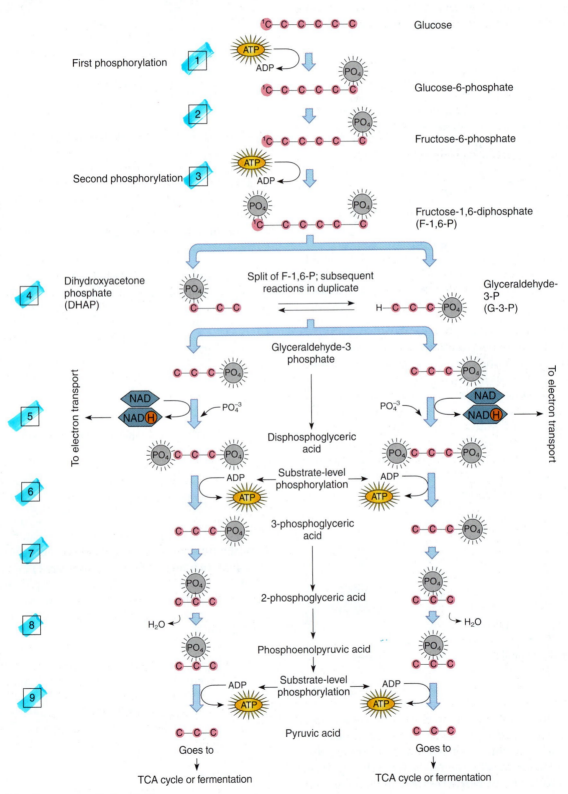

FIGURE 8.18 Summary of glycolysis.

Pyruvic Acid—A Central Metabolite

Pyruvic acid occupies an important position in several pathways, and different organisms handle it in different ways **(figure 8.19)**. In strictly aerobic organisms and some anaerobes, pyruvic acid enters the TCA cycle for further processing and energy release. Facultative anaerobes can adopt a fermentative metabolism, in which pyruvic acid is further reduced into acids or other products.

The Tricarboxylic Acid Cycle— A Carbon and Energy Wheel

As you have seen, the oxidation of glucose yields a comparatively small amount of energy and gives off pyruvic acid. Pyruvic acid is still energy-rich, containing a number of extractable hydrogens and electrons to power ATP synthesis, but this can be achieved only through the work of the second and third phases of respiration, in which pyruvic acid is converted to CO_2 and H_2O. In the following section, we will examine the initial phase of this process, that takes place in the mitochondrial matrix in eucaryotes and in the cytoplasm of bacteria.

The Processing of Pyruvic Acid

To connect the glycolysis pathway to the tricarboxylic acid cycle, for either aerobic or anaerobic respiration, the pyruvic acid is first converted to a starting compound for that cycle **(figure 8.20)**. This step involves the first oxidation-reduction reaction of this phase of respiration, and it also releases the first carbon dioxide molecule. It involves a cluster of enzymes and coenzyme A that participate in the dehydrogenation (oxidation) of pyruvic acid, the reduction of NAD to NADH, and the decarboxylation of pyruvic acid to a 2-carbon acetyl group. The acetyl group remains attached to coenzyme A, forming acetyl coenzyme A (acetyl CoA) that feeds into the TCA cycle. (Pyruvate dehydrogenase, the enzyme complex that makes this reaction possible, is huge. In *E. coli*, it contains four vitamins and has a molecular weight of 6 million—larger than a ribosome!)

The NADH formed during this reaction will be shuttled into electron transport and used to generate ATP via oxidative phosphorylation; its formation is one of five dehydrogenations associated with the conversion reaction and the TCA cycle and accounts for the greater output of ATP in aerobic respiration. **Keep in mind that all reactions described actually happen twice for each glucose because of the two pyruvates that are released during glycolysis.**

An important feature of this pathway is that acetyl groups from the breakdown of certain fats can enter the pathway at this same point, as discussed later on. The acetyl groups are subsequently fed into the tricarboxylic acid cycle and combined with a 4-carbon oxaloacetate molecule to form citric acid, a 6-carbon molecule. Although this seems a cumbersome way to dismantle such a small molecule, it is necessary in biological systems for extracting larger amounts of energy from the remaining fragment of the acetyl groups.

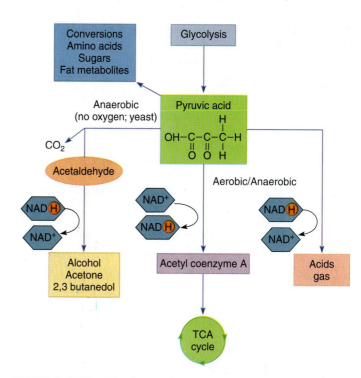

FIGURE 8.19   The fates of pyruvic acid (pyruvate). This metabolite is an important hub in the processing of nutrients by microbes. It may be fermented anaerobically to several end products or oxidized completely to CO_2 and H_2O through the TCA cycle and the electron transport system. It can also serve as a source of raw material for synthesizing amino acids and carbohydrates.

Steps in the TCA Cycle

As you learned earlier, a cyclic pathway is one in which the starting compound is regenerated at the end. The tricarboxylic acid cycle has eight steps, beginning with citric acid formation and ending with oxaloacetic acid[4] (figure 8.20). As we take a single spin around the TCA cycle, it will be helpful to keep track of

1. the numbers of carbons (#C) of each substrate and product,
2. reactions where CO_2 is generated,
3. the involvement of the electron carriers NAD and FAD, and
4. the site of ATP synthesis.

The reactions in the TCA cycle are:

1. Oxaloacetic acid (oxaloacetate; 4C) reacts with the acetyl group (2C) on acetyl CoA, thereby forming citric acid (citrate; 6C) and releasing coenzyme A so it can join with another acetyl group.
2. Citric acid is converted to its isomer, isocitric acid (isocitrate; 6C), to prepare this substrate for the decarboxylation and dehydrogenation of the next step.

---

4. In biochemistry, the terms used for organic acids appear as either the acid form (oxaloacetic acid) or its salt (oxaloacetate).

FIGURE 8.20 **The reactions of a single turn of the TCA cycle.**
Each glucose will produce two spins. Note that this is an enlarged, more detailed view of the middle phase depicted in figure 8.17.
Source: Adapted from Purves and Orions.

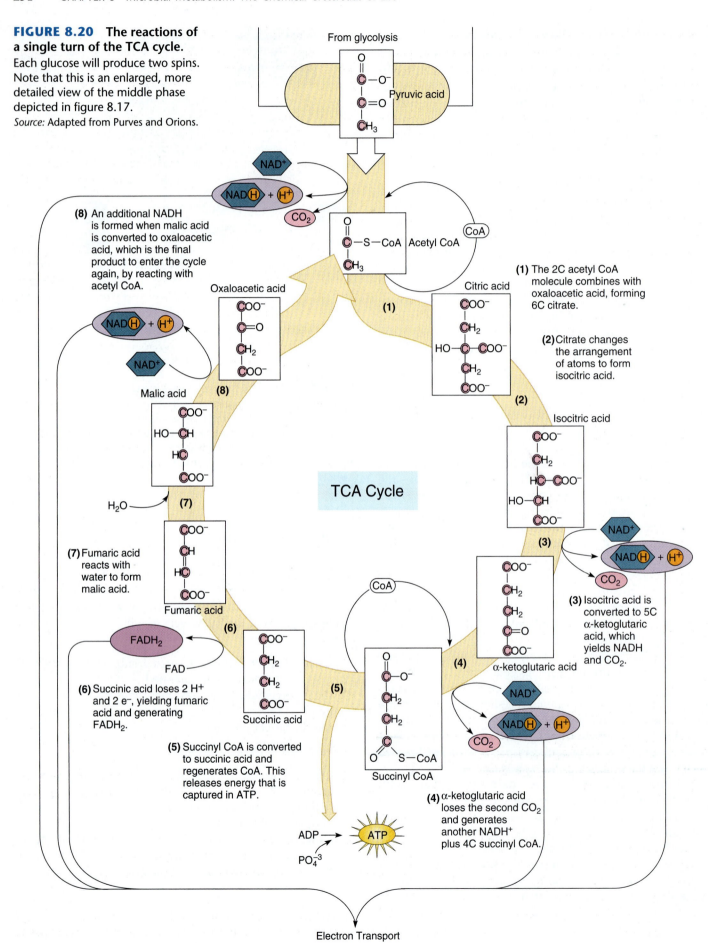

From glycolysis

Pyruvic acid

NAD^+

$NAD \textcircled{H} + \textcircled{H}^+$

CO_2

Acetyl CoA

CoA

(8) An additional NADH is formed when malic acid is converted to oxaloacetic acid, which is the final product to enter the cycle again, by reacting with acetyl CoA.

Oxaloacetic acid

$NAD \textcircled{H} + \textcircled{H}^+$

NAD^+

(1) The 2C acetyl CoA molecule combines with oxaloacetic acid, forming 6C citrate.

Citric acid

(1)

(2) Citrate changes the arrangement of atoms to form isocitric acid.

Malic acid

(8)

(2)

Isocitric acid

TCA Cycle

H_2O

(7)

NAD^+

(3)

$NAD \textcircled{H} + \textcircled{H}^+$

CO_2

(7) Fumaric acid reacts with water to form malic acid.

Fumaric acid

(3) Isocitric acid is converted to 5C α-ketoglutaric acid, which yields NADH and CO_2.

CoA

$FADH_2$

FAD

(6)

NAD^+

(6) Succinic acid loses 2 H^+ and 2 e^-, yielding fumaric acid and generating $FADH_2$.

Succinic acid

(5)

(4)

α-ketoglutaric acid

$NAD \textcircled{H} + \textcircled{H}^+$

CO_2

Succinyl CoA

(5) Succinyl CoA is converted to succinic acid and regenerates CoA. This releases energy that is captured in ATP.

$ADP \longrightarrow$

ATP

PO_4^{-3}

(4) α-ketoglutaric acid loses the second CO_2 and generates another $NADH^+$ plus 4C succinyl CoA.

Electron Transport

3. Isocritic acid is acted upon by an enzyme complex including NAD or NADP (depending on the organism), in a reaction that generates NADH or NADPH, splits off a carbon dioxide, and leaves α-ketoglutaric acid (α-ketoglutarate; 5C).

4. Alpha-ketoglutaric acid serves as a substrate for the last decarboxylation reaction and yet another redox reaction involving coenzyme A and yielding NADH. The product is the high-energy compound succinyl CoA (4C).

At this point, the cycle has completed the formation of 3 CO_2 molecules that balance out the original 3-carbon pyruvic acid that began the TCA. The remaining steps are needed not only to regenerate the oxaloacetic acid to start the cycle again but also to extract more energy from the intermediate compounds leading to oxaloacetic acid.

5. Succinyl CoA is the source of the one substrate level phosphorylation in the TCA cycle. In most microbes, it proceeds with the formation of ATP. The product of this reaction is succinic acid (succinate; 4C).

6. Succinic acid next becomes dehydrogenated, but in this case, the electron and H^+ acceptor is flavin adenine dinucleotide (FAD). The enzyme that catalyzes this reaction, succinyl dehydrogenase, is found in the bacterial cell membrane and mitochondrial crista of eucaryotic cells. $FADH_2$ then directly enters the electron transport system. Fumaric acid (fumarate; 4C) is the product of this reaction.

7. The addition of water to fumaric acid (called hydration) results in malic acid (malate; 4C). This is one of the few reactions in respiration that directly incorporates water.

8. Malic acid is dehydrogenated (with formation of a final NADH), and oxaloacetic acid is formed. This step brings the cycle back to its original starting position, where oxaloacetic acid can react with acetyl coenzyme A.

The Respiratory Chain: Electron Transport and Oxidative Phosphorylation

We now come to the energy chain, which is the final "processing mill" for electrons and hydrogen and the major generator of ATP. Overall, the electron transport system (ETS) consists of a chain of special redox carriers that receive electrons from reduced carriers (NADH, $FADH_2$) generated by glycolysis and the TCA cycle and shuttle them in a sequential and orderly fashion (see the third part of figure 8.17). The flow of electrons down this chain is highly energetic and gives off ATP at various points. The step that finalizes the transport process is the acceptance of electrons and hydrogen by oxygen, producing water. Some variability exists from one organism to another, but the principal compounds that carry out these

complex reactions are NADH dehydrogenase, flavoproteins, coenzyme Q (ubiquinone), and **cytochromes** (sy'-toh-krohm). The cytochromes contain a tightly bound metal atom at their center that is actively involved in accepting electrons and donating them to the next carrier in the series. The highly compartmentalized structure of the respiratory chain is an important factor in its function. Note in **figure 8.21** that the electron transport carriers and enzymes are embedded in the inner mitochondrial membranes in eucaryotes. The equivalent structure for housing them in bacteria is the cell membrane.

Elements of Electron Transport: The Energy Cascade

The principal questions about the electron transport system are: How are the electrons passed from one carrier to another in the series? How is this progression coupled to ATP synthesis? and, Where and how is oxygen utilized? Although the biochemical details of this process are rather complicated, the basic reactions consist of a number of redox reactions now familiar to us. In general, the seven carrier compounds and their enzymes are arranged in linear sequence and are reduced and oxidized in turn.

The sequence of electron carriers in the respiratory chain of most aerobic organisms is

1. NADH dehydrogenase, which is closely associated in a complex with the adjacent carrier, which is
2. flavin mononucleotide (FMN);
3. coenzyme Q;
4. cytochrome b;
5. cytochrome c_1;
6. cytochrome c; and
7. cytochromes a and a_3, which are complexed together.

Conveyance of the NADHs from glycolysis and the TCA cycle to the first carrier sets in motion the remaining six steps.

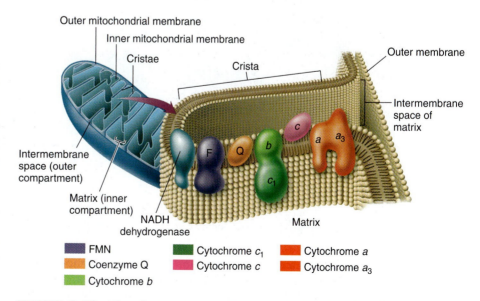

Outer mitochondrial membrane
Inner mitochondrial membrane
Cristae
Crista
Outer membrane
Intermembrane space of matrix
Intermembrane space (outer compartment)
Matrix (inner compartment)
NADH dehydrogenase
Matrix

	FMN		Cytochrome c_1		Cytochrome a
Coenzyme Q		Cytochrome c		Cytochrome a_3
Cytochrome b			

FIGURE 8.21 **The electron transport system in the mitochondrion.**
Location of the electron transport scheme with respect to cristae membranes and matrix.

With each redox exchange, the energy level of the reactants is lessened. The released energy is captured and used by the **ATP synthase** complex, stationed along the cristae in close association with the ETS carriers. Each NADH that enters electron transport can give rise to 3 ATPs. This coupling of ATP synthesis to electron transport is termed **oxidative phosphorylation.** Since $FADH_2$ from the TCA cycle enters the cycle after the NAD and FMN complex reactions, it has less energy to release, and 2 ATPs result from its processing.

The Formation of ATP and Chemiosmosis

What biochemical processes are involved in coupling electron transport to the production of ATP? We will first look at the system in eucaryotes, which have the components of electron transport embedded in a precise sequence on mitochondrial membranes. They are stationed between the inner mitochondrial matrix and the outer intermembrane space **(figure 8.22).** According to a widely accepted concept called **chemiosmosis,** as the electron transport carriers shuttle electrons, they actively pump hydrogen ions (protons) into the outer compartment of the mitochondrion. This process sets up a concentration gradient of hydrogen ions called the *proton motive force (PMF).* The PMF consists of a difference in charge between the outer membrane compartment (+) and the inner membrane compartment (−) **(figure 8.23a).**

Separating the charge has the effect of a battery, which can temporarily store potential energy. This charge will be maintained by the impermeability of the inner cristae membranes to H^+. The only site where H^+ can diffuse into the inner compartment is at the ATP synthase complex, which sets the stage for the final processing of H^+ leading to ATP synthesis.

ATP synthase is a complex enzyme composed of two large units, F_0 and F_1 **(figure 8.23b).** It is embedded in the membrane but part of it rotates like a motor and traps chemical energy. As the H^+ ions flow through the F_0 center of the enzyme by diffusion, the F_1 compartments pull in ADP and P_i.

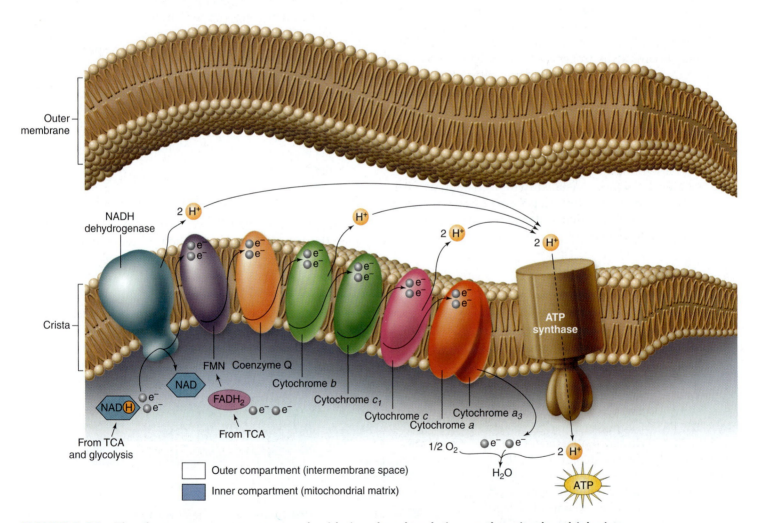

FIGURE 8.22 **The electron transport system and oxidative phosphorylation on the mitochondrial crista.**
Starting at NADH dehydrogenase, electrons brought in from the TCA cycle by NADH are passed along the chain of electron transport carriers. Each adjacent pair of transport molecules undergoes a redox reaction. Coupled to the transport of electrons is the simultaneous active transport of H^+ into the outer compartment by specific carriers. These processes set the scene for ATP synthesis and final H^+ and e^- acceptance by oxygen (see figure 8.23 for details).

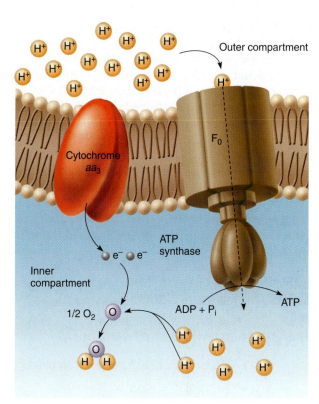

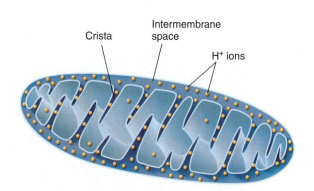

(a) As the carriers in the mitochondrial cristae transport electrons, they also actively pump H⁺ ions (protons) to the intermembrane space, producing a chemical and charge gradient between the outer and inner mitochondrial compartments.

(b) The distribution of electric potential and the concentration gradient of protons across the membrane drive the synthesis of ATP by ATP synthase. The rotation of this enzyme couples diffusion of H^+ to the inner compartment with the bonding of ADP and P_i. The final event of electron transport is the reaction of the electrons with the H^+ and O_2 to form metabolic H_2O. This step is catalyzed by cytochrome oxidase (cytochrome aa_3)

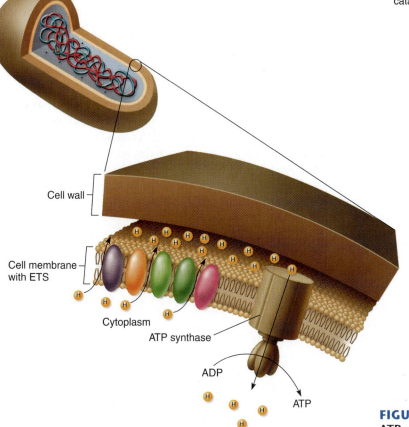

(c) Enlarged view of bacterial cell envelope to show the relationship of electron transport and ATP synthesis. Bacteria have the ETS and ATP synthase stationed in the cell membrane. ETS carriers transport H⁺ and electrons from the cytoplasm to the exterior of the membrane. Here, it is collected to create a gradient just as it occurs in mitochondria.

FIGURE 8.23 Chemiosmosis—the force behind ATP synthesis.

Rotation causes a three-dimensional change in the enzyme that bonds these two molecules, thereby releasing ATP into the inner compartment (figure 8.23b). The enzyme is then rotated back to the start position and will continue the process.

Bacterial ATP synthesis occurs by means of this same overall process. However, bacteria have the ETS stationed in the cell membrane, and the direction of the proton movement is from the cytoplasm to the periplasmic space in gram-negative bacteria, and to the area occupied by the cell wall in gram-positives (figure 8.23c). This difference will affect the amount of ATP produced (discussed in the next section). In both cell types, the chemiosmotic theory has been supported by tests showing that oxidative phosphorylation is blocked if the mitochondrial or bacterial cell membranes are disrupted.

Potential Yield of ATPs from Oxidative Phosphorylation

The total of five NADHs (four from the TCA cycle and one from glycolysis) can be used to synthesize:

15 ATPs for ETS (5×3 per electron pair)

and

$15 \times 2 = 30$ ATPs per glucose

The single FADH produced during the TCA cycle results in:

2 ATPs per electron pair

and

$2 \times 2 = 4$ ATPs per glucose

Table 8.4 summarizes the total of ATP and other products for the entire aerobic pathway. These totals are the potential yields possible but may not be fulfilled by many organisms.

Summary of Aerobic Respiration

Originally, we presented a summary equation for respiration. We are now in a position to tabulate the input and output of this equation at various points in the pathways and sum up the final ATP. Close examination of table 8.4 will review several important facets of aerobic respiration:

1. The total possible yield of ATP is 40: 4 from glycolysis, 2 from the TCA cycle, and 34 from electron transport. However, because 2 ATPs were expended in early glycolysis, this leaves a maximum of **38 ATPs.**

The actual totals may be lower in certain eucaryotic cells because energy is expended in transporting the NADH produced during glycolysis across the mitochondrial membrane. Certain aerobic bacteria come closest to achieving the full total of 38 because they lack mitochondria and thus do not have to use ATP in transport of NADH across the outer mitochondrial membrane.

2. Six carbon dioxide molecules are generated during the TCA cycle.
3. Six oxygen molecules are consumed during electron transport.
4. Six water molecules are produced in electron transport and 2 in glycolysis, but because 2 are used in the TCA cycle, this leaves a net number of 6.

The Terminal Step

The terminal step, during which oxygen accepts the electrons, is catalyzed by cytochrome aa_3, also called cytochrome oxidase. This large enzyme complex is specifically adapted to receive electrons from cytochrome c, pick up hydrogens from the solution, and react with oxygen to form a molecule of water (figure 8.23b). This reaction, though in actuality more complex, is summarized as follows:

$$2\,H^+ + 2\,e^- + \tfrac{1}{2}O_2 \rightarrow H_2O$$

Most eucaryotic aerobes have a fully functioning cytochrome system, but bacteria exhibit wide-ranging variations in this part of the system. Some species lack one or more of the redox steps; others have several alternative electron transport schemes. Because many bacteria lack cytochrome c oxidase, this variation can be used to differentiate among certain genera of bacteria. An oxidase detection test can be used to help identify members of the genera *Neisseria*

TABLE 8.4	Summary of Aerobic Respiration for One Glucose Molecule						
	Glycolysis*	Net Output	TCA Cycle*	Net Output	Respiratory Chain	Net Output	Total Net Output per Glucose
ATP produced	$2 \times 2 =$	2	$1 \times 2 =$	2	$17 \times 2 =$	34	$40 - 2$ (used) $= 38$**
ATP used	2		0		0		
NADH produced	$1 \times 2 =$	2	$4 \times 2 =$	8	0		10
FADH produced	0		$1 \times 2 =$	2	0		2
CO_2 produced	0		$3 \times 2 =$	6	0		6
O_2 used	0		0		$3 \times 2 =$	6	
H_2O produced	2		0		$3 \times 2 =$	6	$8 - 2$ (used) $=$ 6
H_2O used	0		2		0		

*Products are multiplied by 2 because the first figure represents the amount for only one trip through the pathway, and two molecules make this trip for each glucose.

**This amount can vary among microbes.

and *Pseudomonas* and some species of *Bacillus*. Another variation in the cytochrome system is evident in certain bacteria (*Klebsiella*, *Enterobacter*) that can grow even in the presence of cyanide because they lack cytochrome oxidase. Cyanide will cause rapid death in humans and other eucaryotes because it blocks cytochrome oxidase, thereby completely terminating aerobic respiration, but it is harmless to these bacteria.

A potential side reaction of the respiratory chain in aerobic organisms is the incomplete reduction of oxygen to superoxide ion (O_2^-) and hydrogen peroxide (H_2O_2). As mentioned in chapter 7, these toxic oxygen products can be very damaging to cells. Aerobes have neutralizing enzymes to deal with these products, including *superoxide dismutase* and *catalase*. One exception is the genus *Streptococcus*, which can grow well in oxygen yet lacks both cytochromes and catalase. The tolerance of these organisms to oxygen can be explained by the neutralizing effects of a special peroxidase. The lack of cytochromes, catalase, and peroxidases in anaerobes as a rule limits their ability to process free oxygen and contributes to its toxic effects on them.

Alternate Catabolic Pathways

Certain bacteria follow a different pathway in carbohydrate catabolism. The phosphogluconate pathway (also called the hexose monophosphate shunt) provides ways to anaerobically oxidize glucose and other hexoses, to release ATP, to produce large amounts of NADPH, and to process pentoses (5-carbon sugars). This pathway, common in heterolactic fermentative bacteria, yields various end products, including lactic acid, ethanol, and carbon dioxide. Furthermore, it is a significant intermediate source of pentoses for nucleic acid synthesis.

Anaerobic Respiration

Some bacteria have evolved an anaerobic respiratory system that functions like the aerobic cytochrome system except that it utilizes oxygen-containing ions, rather than free oxygen, as the final electron acceptor. Of these, the nitrate (NO_3^-) and nitrite (NO_2^-) reduction systems are best known. The reaction in species such as *Escherichia coli* is represented as:

$$\text{Nitrate reductase}$$
$$\downarrow$$
$$NO_3^- + NADH \rightarrow NO_2^- + H_2O + NAD^+$$

The enzyme nitrate reductase catalyzes the removal of oxygen from nitrate, leaving nitrite and water as products. A test for this reaction is one of the physiological tests used in identifying bacteria.

Some species of *Pseudomonas* and *Bacillus* possess enzymes that can further reduce nitrite to nitric oxide (NO), nitrous oxide (N_2O), and even nitrogen gas (N_2). This process, called **denitrification**, is a very important step in recycling nitrogen in the biosphere. Other oxygen-containing nutrients reduced anaerobically by various bacteria are carbonates and sulfates. None of the anaerobic pathways produce as much ATP as aerobic respiration.

IN THE NEWS *(Continued from page 213)*

The case described at the beginning of the chapter was caused by *Clostridium tetani*. Even though an obligately anaerobic organism was not cultured from the wound, the wound evidently, was infected with *Clostridium tetani* as a result of the traumatic laceration. The conditions in the wound were anaerobic enough to allow for growth of these bacilli. It is even probable that the coagulase-negative *Staphylococcus* consumed enough oxygen to allow for the tetanus bacilli's growth, resulting in the production of tetanus toxin, which causes muscle spasms.

Fortunately, this is a rare disease in developed areas of the world in which the tetanus toxoid vaccine is widely used. When tetanus does occur in this setting, it usually occurs in people who are either not immunized or in the elderly who have not received adequate boosters of the vaccine.

See: Bunch, T. J. et al. 2002. Respiratory failure in tetanus: Case report. Chest 122:1488–1492.

✔ CHECKPOINT

- Bioenergetics describes metabolism in terms of production, utilization, and transfer of energy by cells.
- Catabolic pathways release energy through three pathways: glycolysis, the tricarboxylic acid cycle, and the respiratory electron transport system.
- Cellular respiration is described by the nature of the final electron acceptor. Aerobic respiration implies that O_2 is the final electron acceptor. Anaerobic respiration implies that some other molecule is the final electron acceptor. If the final electron acceptor is an organic molecule, the anaerobic process is considered fermentation.
- Carbohydrates are preferred cell energy sources because they are superior hydrogen (electron) donors.
- Glycolysis is the catabolic process by which glucose is oxidized and converted into two molecules of pyruvic acid, with a net gain of 2 ATP. The formation of ATP is via substrate-level phosphorylation.
- The tricarboxylic acid cycle processes the 3-carbon pyruvic acid and generates three CO_2 molecules. The electrons it releases are transferred to redox carriers for energy harvesting. It also generates 2 ATPs.
- The electron transport chain generates free energy through sequential redox reactions collectively called oxidative phosphorylation. This energy is used to generate up to 38 ATP for each glucose molecule catabolized.

The Importance of Fermentation

Of all the results of pyruvate metabolism, probably the most varied is fermentation. Technically speaking, **fermentation** is the incomplete oxidation of glucose or other carbohydrates in the absence of oxygen. This process uses organic compounds as the terminal electron acceptors and yields a small amount of ATP.

Over time, the term *fermentation* has acquired several looser connotations. Originally, Pasteur called the microbial action of yeast during wine production *ferments,* and to this day, biochemists use the term in reference to the production of ethyl alcohol by yeasts acting on glucose and other carbohydrates. Fermentation is also what bacteriologists call the formation of acid, gas, and other products by the action of various bacteria on pyruvic acid. The process is a common metabolic strategy among bacteria. Industrial processes that produce chemicals on a massive scale through the actions of microbes are also called fermentations (see chapter 24). Each of these usages is acceptable for one application or another.

It may seem that fermentation would yield only meager amounts of energy (2 ATPs maximum per glucose) and that would slow down growth. What actually happens, however, is that many bacteria can grow as fast as they would in the presence of oxygen. This rapid growth is made possible by an increase in the rate of glycolysis. From another standpoint, fermentation permits independence from molecular oxygen and allows colonization of anaerobic environments. It also enables microorganisms with a versatile metabolism to adapt to variations in the availability of oxygen. For them, fermentation provides a means to grow even when oxygen levels are too low for aerobic respiration.

Bacteria that digest cellulose in the rumens of cattle are largely fermentative. After initially hydrolyzing cellulose to glucose, they ferment the glucose to organic acids, which are then absorbed as the bovine's principal energy source. Even human muscle cells can undergo a form of fermentation that permits short periods of activity after the oxygen supply in the muscle has been exhausted. Muscle cells convert pyruvic acid into lactic acid, which allows anaerobic production of ATP to proceed for a time. But this cannot go on indefinitely, and after a few minutes, the accumulated lactic acid causes muscle fatigue.

Products of Fermentation in Microorganisms

Alcoholic beverages (wine, beer, whiskey) are perhaps the most prominent among fermentation products; others are solvents (acetone, butanol), organic acids (lactic, acetic), dairy products, and many other foods. Derivatives of proteins, nucleic acids, and other organic compounds are fermented to produce vitamins, antibiotics, and even hormones such as hydrocortisone.

Fermentation products can be grouped into two general categories: alcoholic fermentation products and acidic fermentation products **(figure 8.24). Alcoholic fermentation** occurs in yeast or bacterial species that have metabolic pathways for converting pyruvic acid to ethanol. This process involves a decarboxylation of pyruvic acid to acetaldehyde, followed by a reduction of the acetaldehyde to ethanol. In oxidizing the NADH formed during glycolysis, NAD is regenerated, thereby allowing the glycolytic pathway to continue. These processes are crucial in the production of beer

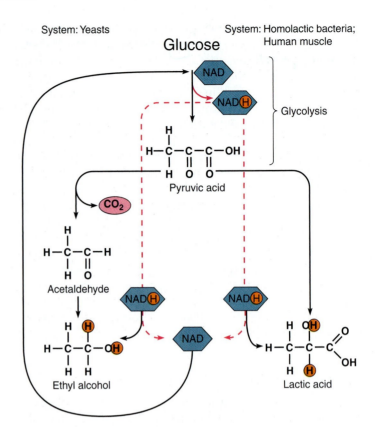

FIGURE 8.24 **The chemistry of fermentation systems that produce acid and alcohol.**
In both cases, the final electron acceptor is an organic compound. In yeasts, pyruvic acid is decarboxylated to acetaldehyde, and the NADH given off in the glycolytic pathway reduces acetaldehyde to ethyl alcohol. In homolactic fermentative bacteria, pyruvic acid is reduced by NADH to lactic acid. Both systems regenerate NAD to feed back into glycolysis or other cycles.

and wine, though the actual techniques for arriving at the desired amount of ethanol and the prevention of unwanted side reactions are important tricks of the brewer's trade (see **Insight 8.4**). Note that the products of alcoholic fermentation are not only ethanol but also CO_2, a gas that accounts for the bubbles in champagne and beer (and the rising of bread dough).

Alcohols other than ethanol can be produced during bacterial fermentation pathways. Certain clostridia produce butanol and isopropanol through a complex series of reactions. Although this process was once an important source of alcohols for industrial use, it has been largely replaced by a nonmicrobial petroleum process.

The pathways of **acidic fermentation** are extremely varied. Lactic acid bacteria ferment pyruvate in the same way that humans do—by reducing it to lactic acid. If the product of this fermentation is mainly lactic acid, as in certain species of *Streptococcus* and *Lactobacillus,* it is termed *homolactic.* The souring of milk is due largely to the production of this acid by bacteria. When glucose is fermented to a mixture of lactic acid, acetic acid, and carbon dioxide, as is the case with

INSIGHT 8.4   *Historical*

Pasteur and the Wine-to-Vinegar Connection

The microbiology of alcoholic fermentation was greatly clarified by Louis Pasteur after French wine makers hired him to uncover the causes of periodic spoilage in wines. Especially troublesome was the conversion of wine to vinegar and the resultant sour flavor. Up to that time, wine formation had been considered strictly a chemical process. After extensively studying beer making and wine grapes, Pasteur concluded that wine, both fine and not-so-fine, was the result of microbial action on the juices of the grape and that wine "disease" was caused by contaminating organisms that produced undesirable products such as acid. Although he did not know it at the time, the bacterial contaminants responsible for the acidity of the spoiled wines were likely to be *Acetobacter* or *Gluconobacter* introduced by the grapes, air, or wine-making apparatus. These common gram-negative genera further oxidized

ethanol to acetic acid and are presently used in commercial vinegar production. The following formula shows how this is accomplished:

$$H-\underset{\underset{H}{|}}{\overset{\overset{H}{|}}{C}}-\underset{\underset{H}{|}}{\overset{\overset{H}{|}}{C}}-OH \longrightarrow H-\underset{\underset{H}{|}}{\overset{\overset{H}{|}}{C}}-C\underset{OH}{\overset{O}{\diagup}} + 2H^{+}$$

Ethanol   Acetic acid

Pasteur's far-reaching solution to the problem is still with us today—mild heating, or *pasteurization*, of the grape juice to destroy the contaminants, followed by inoculation of the juice with a pure yeast culture. The topic of wine making is explored further in chapter 24.

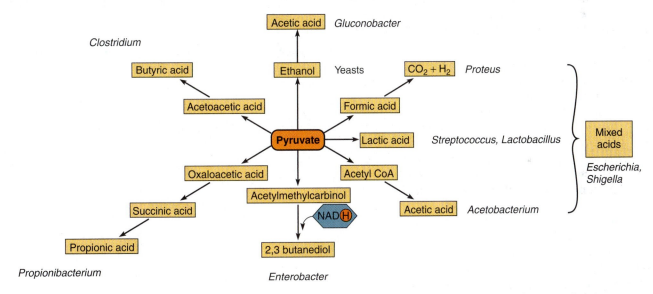

FIGURE 8.25   Miscellaneous products of pyruvate fermentation and the bacteria involved in their production.

Leuconostoc and other species of *Lactobacillus*, the process is termed *heterolactic fermentation*.

Many members of the family Enterobacteriaceae (*Escherichia*, *Shigella*, and *Salmonella*) possess enzyme systems for converting pyruvic acid to several acids simultaneously **(figure 8.25). Mixed acid fermentation** produces a combination of acetic, lactic, succinic, and formic acids, and it lowers the pH of a medium to about 4.0. *Propionibacterium* produces primarily propionic acid, which gives the characteristic flavor to Swiss cheese while fermentation gas (CO_2) produces the holes. Some members also further decompose formic acid completely to carbon dioxide and hydrogen gases. Because enteric bacteria commonly occupy the intestine, this

fermentative activity accounts for the accumulation of some types of gas—primarily CO_2 and H_2—in the intestine. Some bacteria reduce the organic acids and produce the neutral end product 2,3-butanediol **(Insight 8.5)**.

We have provided only a brief survey of fermentation products, but it is worth noting that microbes can be harnessed to synthesize a variety of other substances by varying the raw materials provided them. In fact, so broad is the meaning of the word *fermentation* that the large-scale industrial syntheses by microorganisms often utilize entirely different mechanisms from those described here, and they even occur aerobically, particularly in antibiotic, hormone, vitamin, and amino acid production (see chapter 24).

INSIGHT 8.5   *Discovery*

Fermentation and Biochemical Testing

The knowledge and understanding of the fermentation products of given species are important not only in industrial production but also in identifying bacteria by biochemical tests. Fermentation patterns in enteric bacteria, for example, are an important identification tool. Specimens are grown in media containing various carbohydrates, and the production of acid or acid and gas is noted (see figure 3.11). For instance, *Escherichia* ferments the milk sugar lactose, whereas *Shigella* and *Proteus* do not. On the basis of its gas production during glucose fermentation, *Escherichia* can be further differentiated from *Shigella*, which ferments glucose but does not generate gas. Other enteric bacteria are separated on the basis of whether glucose is fermented to mixed acids or to 2,3-butanediol. *Escherichia coli* produces mixed acids, whereas *Enterobacter* and *Serratia* form primarily 2,3-butanediol. These are part of a test known as IMViC, described in chapter 17. The M refers to the methyl red test for mixed acids, and the V refers to the Voges-Proskauer test for the butanediol pathway.

8.4 Biosynthesis and the Crossing Pathways of Metabolism

Our discussion now turns from catabolism and energy extraction to anabolic functions and biosynthesis. In this section we present aspects of intermediary metabolism, including amphibolic pathways, the synthesis of simple molecules, and the synthesis of macromolecules.

The Frugality of the Cell— Waste Not, Want Not

It must be obvious by now that cells have mechanisms for careful management of carbon compounds. Rather than being dead ends, most catabolic pathways contain strategic molecular intermediates (metabolites) that can be diverted into anabolic pathways. In this way, a given molecule can serve multiple purposes, and the maximum benefit can be derived from all nutrients and metabolites of the cell pool. The property of a system to integrate catabolic and anabolic pathways to improve cell efficiency is termed **amphibolism** (am-fee-bol'-izm).

At this point in the chapter, you can appreciate a more complex view of metabolism than that presented at the beginning, in figure 8.1. **Figure 8.26** demonstrates the amphibolic nature of intermediary metabolism. The pathways of glucose catabolism are an especially rich "metabolic marketplace." The principal sites of amphibolic interaction occur during glycolysis (glyceraldehyde-3-phosphate and pyruvic acid) and the TCA cycle (acetyl coenzyme A and various organic acids).

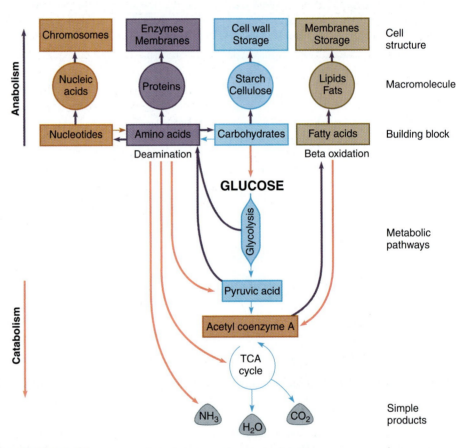

FIGURE 8.26   **An amphibolic view of metabolism.**
Intermediate compounds such as pyruvic acid and acetyl coenzyme A serve an amphibolic function. With comparatively small modifications, these compounds can be converted into other compounds and enter a different pathway. Note that catabolism of glucose (center) furnishes numerous intermediates for anabolic pathways that synthesize amino acids, fats, nucleic acids, and carbohydrates. These building blocks can serve in further synthesis of larger molecules to construct various cell components.

Amphibolic Sources of Cellular Building Blocks

Glyceraldehyde-3-phosphate can be diverted away from glycolysis and converted into precursors for amino acid, carbohydrate, and triglyceride (fat) synthesis. (A precursor molecule is a compound that is the source of another compound.) Earlier we noted the numerous directions that pyruvic acid catabolism can take. In terms of synthesis, pyruvate also plays a pivotal role in providing intermediates for amino acids. In the event of an inadequate glucose supply, pyruvate serves as the starting point in glucose synthesis from various metabolic intermediates, a process called **gluconeogenesis** (gloo"-koh-nee"-oh-gen'-uh-sis).

The acetyl group that starts the TCA cycle is another extremely versatile metabolite that can be fed into a number of synthetic pathways. This 2-carbon fragment can be converted as a single unit into one of several amino acids, or a number of these fragments can be condensed into hydrocarbon chains that are important building blocks for fatty acid and lipid synthesis. Note that the reverse is also true—fats can be degraded to acetyl and thereby enter the TCA cycle at acetyl coenzyme A. This aerobic process, called **beta oxidation,** can provide a large amount of energy. Oxidation of a 6-carbon fatty acid yields 50 ATP, compared with 38 for a 6-carbon sugar.

Two metabolites of carbohydrate catabolism that the TCA cycle produces, oxaloacetic acid and α-ketoglutaric acid, are essential intermediates in the synthesis of certain amino acids. This occurs through **amination,** the addition of an amino group to a carbon skeleton **(figure 8.27a).** A certain core group of amino acids can then be used to synthesize others. Amino acids and carbohydrates can be interchanged through **transamination (figure 8.27b).**

Pathways that synthesize the nitrogen bases (purines, pyrimidines), which are components of DNA and RNA, originate in amino acids and so can be dependent on intermediates from the TCA cycle as well. Because the coenzymes NAD, NADP, FAD, and others contain purines and pyrimidines similar to the nucleic acids, their synthetic pathways are also dependent on amino acids. During times of carbohydrate deprivation, organisms can likewise convert amino acids to intermediates of the TCA cycle by **deamination** (removal of an amino group) and thereby derive energy from proteins. Deamination results in the formation of nitrogen waste products such as ammonium ions or urea **(figure 8.27c).**

Formation of Macromolecules

Monosaccharides, amino acids, fatty acids, nitrogen bases, and vitamins—the building blocks that make up the various macromolecules and organelles of the cell—come from two possible sources. They can enter the cell from the outside as nutrients, or they can be synthesized through various cellular pathways. The degree to which an organism can synthesize its own building blocks (simple molecules) is determined by its genetic makeup, a factor that varies tremendously from

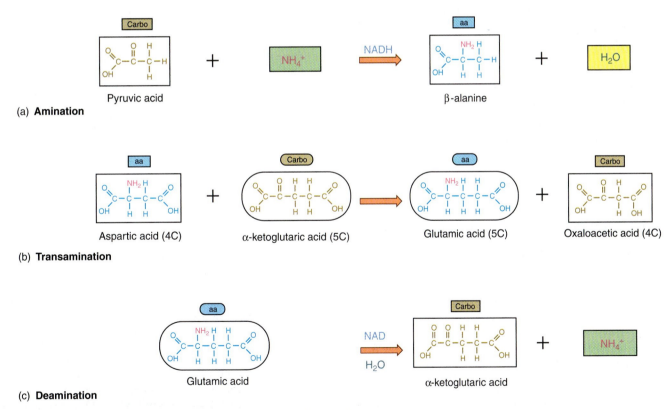

FIGURE 8.27   Reactions that produce and convert amino acids.
All of them require energy as ATP or NAD and specialized enzymes. **(a)** Through amination (the addition of an ammonium molecule [amino group]), a carbohydrate can be converted to an amino acid. **(b)** Through transamination (transfer of an amino group from an amino acid to a carbohydrate fragment), metabolic intermediates can be converted to amino acids that are in low supply. **(c)** Through deamination (removal of an amino group), an amino acid can be converted to a useful intermediate of carbohydrate catabolism. This is how proteins are used to derive energy. Ammonium is one waste product.

3. An enzyme _____ the activation energy required for a chemical reaction.
   a. increases   c. lowers
   b. converts   d. catalyzes

4. An enzyme
   a. becomes part of the final products
   b. is nonspecific for substrate
   c. is consumed by the reaction
   d. is heat and pH labile

5. An apoenzyme is where the _____ is located.
   a. cofactor   e. redox reaction
   b. coenzyme   d. active site

6. Many coenzymes are
   a. metals   c. proteins
   b. vitamins   d. substrates

7. To digest cellulose in its environment, a fungus produces a/an
   a. endoenzyme   c. catalase
   b. exoenzyme   d. polymerase

8. Energy in biological systems is primarily
   a. electrical   c. radiant
   b. chemical   d. mechanical

9. Energy is carried from catabolic to anabolic reactions in the form of _____.
   a. ADP
   b. high-energy ATP bonds
   c. coenzymes
   d. inorganic phosphate

10. Exergonic reactions
    a. release potential energy
    b. consume energy
    c. form bonds
    d. occur only outside the cell

11. A reduced compound is
    a. NAD   c. NADH
    b. FAD   d. ADP

12. Most oxidation reactions in microbial bioenergetics involve the
    a. removal of electrons and hydrogens
    b. addition of electrons and hydrogens
    c. addition of oxygen
    d. removal of oxygen

13. A product or products of glycolysis is/are
    a. ATP   c. CO_2
    b. H_2O d. both a and b

14. Fermentation of a glucose molecule has the potential to produce a net number of _____ ATPs.
    a. 4   c. 40
    b. 2   d. 0

15. Complete oxidation of glucose in aerobic respiration can yield a net output of _____ ATP.
    a. 40   c. 38
    b. 6   d. 2

16. The compound that enters the TCA cycle from glycolysis is
    a. citric acid   c. pyruvic acid
    b. oxaloacetic acid   d. acetyl coenzyme A

17. The $FADH_2$ formed during the TCA cycle enters the electron transport system at which site?
    a. NADH dehydrogenase
    b. cytochrome
    c. coenzyme Q
    d. ATP synthase

18. ATP synthase complexes can generate _____ ATPs for each NADH that enters electron transport.
    a. 1   c. 3
    b. 2   d. 4

19. **Matching.** Match the process a, b, or c with the metabolic events in the list.
    a. glycolysis
    b. TCA (Krebs) cycle
    c. electron transport/oxidative phosphorylation
    _c_ H^+ and e^- are delivered to O_2 as the final acceptor.
    _a_ Pyruvic acid is formed.
    _b_ GTP is formed.
    _c_ H_2O is produced.
    _b_ CO_2 is formed.
    _a_ Fructose diphosphate is split into two 3-carbon fragments.
    _c_ $NADH^+$ is oxidized.
    _c_ ATP synthase is active.

Concept Questions

These questions are suggested as a *writing-to-learn* experience. For each question, compose a one- or two-paragraph answer that includes the factual information needed to completely address the question.

1. a. Describe the chemistry of enzymes and explain how the apoenzyme forms.
   b. Show diagrammatically the interaction of holoenzyme and its substrate and general products that can be formed from a reaction.

2. Differentiate among the chemical composition and functions of various cofactors. Provide examples of each type.

3. a. Two steps in glycolysis are catalyzed by allosteric enzymes. These are: (1) step 2, catabolized by phosphoglucoisomerase, and (2) step 9, catabolized by pyruvate kinase. Suggest what metabolic products might regulate these enzymes.
   b. How might one place these regulators in figure 8.18?

4. Explain how oxidation of a substrate proceeds without oxygen.

5. In the following redox pairs, which compound is reduced and which is oxidized?
   a. NAD and NADH
   b. $FADH_2$ and FAD

c. lactic acid and pyruvic acid
d. NO_3^- and NO_2^-
e. Ethanol and acetaldehyde

6. a. Describe the roles played by ATP and NAD in metabolism.
   b. What particular features of their structure lend them to these functions?

7. Discuss the relationship of:
   a. anabolism to catabolism
   b. ATP to ADP
   c. glycolysis to fermentation
   d. electron transport to oxidative phosphorylation

8. a. What is meant by the concept of the "final electron acceptor"?
   b. What are the final electron acceptors in aerobic, anaerobic, and fermentative metabolism?

9. Name the major ways that substrate-level phosphorylation is different from oxidative phosphorylation.

10. Compare and contrast the location of glycolysis, TCA cycle, and electron transport in procaryotic and eucaryotic cells.

11. a. Outline the basic steps in glycolysis, indicating where ATP is used and given off.
    b. Where does NADH originate, and what is its fate in an aerobe?
    c. What is the fate of NADH in a fermentative organism?

12. a. What is the source of ATP in the TCA cycle?
    b. How many ATPs could be formed from the original glucose molecule?
    c. How does the total of ATPs generated differ between bacteria and many eucaryotes? What causes this difference?

13. Name the sources of oxygen in bacteria that use anaerobic respiration.

14. a. Summarize the chemiosmotic theory of ATP formation.
    b. What is unique about the actions of ATP synthase?

15. How are aerobic and anaerobic respiration different?

16. Compare the general equation for aerobic metabolism with table 8.4 and verify that all figures balance.

17. Water is one of the end products of aerobic respiration. Where in the metabolic cycles is it formed, and where is it used?

Critical Thinking Questions

Critical thinking is the ability to reason and solve problems using facts and concepts. These questions can be approached from a number of angles, and in most cases, they do not have a single correct answer.

1. Using the simplified chart below, fill in a summary of the major starting compounds required and the products given off by each phase of metabolism. Use arrows to pinpoint approximately where the reactions take place.

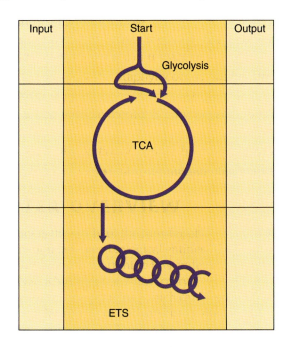

2. Use the graph below to diagram the energetics of a chemical reaction, with and without an enzyme. Be sure to position reactants and products at appropriate points and to indicate the stages in the reaction and the energy levels.

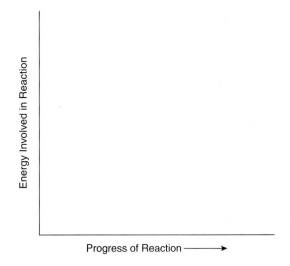

3. Using the concept of fermentation, describe the microbial (biochemical) mechanisms that cause milk to sour.

4. Explain how it is possible for certain microbes to survive and grow in the presence of cyanide, which would kill many other organisms.

5. Suggest the advantages of having metabolic pathways staged in specific membrane or organelle locations rather than being free in the cytoplasm.

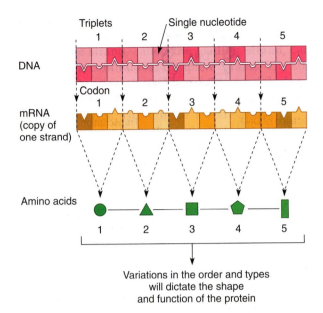

FIGURE 9.10 **Simplified view of the DNA-protein relationship.**
The DNA molecule is a continuous chain of base pairs, but the sequence must be interpreted in groups of three base pairs (a triplet). Each triplet as copied into mRNA codons will translate into one amino acid; consequently, the ratio of base pairs to amino acids is 3:1.

TABLE 9.2		Types of Ribonucleic Acid		
RNA Type	**Contains Codes For**		**Function in Cell**	**Translated**
Messenger (mRNA)	Sequence of amino acids in protein		Carries the DNA master code to the ribosome	Yes
Transfer (tRNA)	A cloverleaf tRNA to carry amino acids		Brings amino acids to ribosome during translation	No
Ribosomal (rRNA)	Several large structural rRNA molecules		Forms the major part of a ribosome and participates in protein synthesis	No
Micro (miRNA) and small interfering (siRNA)	Regulatory RNAs		Regulation of gene expression and coiling of chromatin	No
Primer	An RNA that can begin DNA replication		Primes DNA	No
Ribozymes	RNA enzymes, parts of splicer enzymes		Remove introns from other RNAs in eucaryotes	No

DNA strand **(figure 9.10).** Thus, one gene differs from another in its composition of triplets. An equally important part of this concept is that each triplet represents a code for a particular amino acid. When the triplet code is transcribed and translated, it dictates the type and order of amino acids in a polypeptide (protein) chain.

The final key points that connect DNA and protein function are:

1. A protein's primary structure—the order and type of amino acids in the chain—determines its characteristic shape and function.
2. Proteins ultimately determine phenotype, the expression of all aspects of cell function and structure. Put more simply, living things are what their proteins make them.
3. DNA is mainly a blueprint that tells the cell which kinds of proteins to make and how to make them.

The Major Participants in Transcription and Translation

Transcription, the formation of RNA using DNA as a template, and translation, the synthesis of proteins using RNA as a template, are highly complex. A number of components participate: most prominently, messenger RNA, transfer RNA, ribosomes, several types of enzymes, and a storehouse of raw materials. After first examining each of these components,

we shall see how they come together in the assembly line of the cell.

RNAs: Tools in the Cell's Assembly Line

Ribonucleic acid is an encoded molecule like DNA, but its general structure is different in several ways:

1. It is a single-stranded molecule that exists in helical form. This single strand can assume secondary and tertiary levels of complexity due to bonds within the molecule, leading to specialized forms of RNA (tRNA and rRNA—figure 9.9).
2. RNA contains **uracil,** instead of thymine, as the complementary base-pairing mate for adenine. This does not change the inherent DNA code in any way because the uracil still follows the pairing rules.
3. Although RNA, like DNA, contains a backbone that consists of alternating sugar and phosphate molecules, the sugar in RNA is **ribose** rather than deoxyribose.

The many functional types of RNA range from small regulatory pieces to large structural ones **(table 9.2 and Insight 9.3).** All types of RNA are formed through transcription of a DNA gene, but only mRNA is further translated into another type of molecule (protein).

Messenger RNA: Carrying DNA's Message

Messenger RNA (mRNA) is a transcript (copy) of a structural gene or genes in the DNA. It is synthesized by a process similar to synthesis of the leading strand during DNA replication, and the complementary base-pairing rules ensure that the code will be faithfully copied in the mRNA transcript. The message of this transcribed strand is later read as a series of triplets called **codons (figure 9.11b),** and the length of the mRNA molecule varies from about 100 nucleotides to several thousand. The details of transcription and the function of mRNA in translation will be covered shortly.

Transfer RNA: The Key to Translation

Transfer RNA (tRNA) is also a copy of a specific region of DNA; however, it differs from mRNA. It is uniform in length, being 75 to 95 nucleotides long, and it contains sequences of bases that form hydrogen bonds with complementary sections of the same tRNA strand. At these points, the molecule bends back upon itself into several *hairpin loops,* giving the molecule a secondary *cloverleaf* structure that folds even further into a complex, three-dimensional helix **(figure 9.11a).** This compact molecule is an adaptor that converts RNA language into protein language. The bottom loop of the cloverleaf exposes a triplet, the **anticodon,** that both designates the specificity of the tRNA and complements mRNA's codons. At the opposite end of the molecule is a binding site for the amino acid that is specific for that tRNA's anticodon. For each of the 20 amino acids, there is at least one specialized type of tRNA to carry it. Binding of an amino acid to its specific tRNA, a process known as "charging" the tRNA, takes place in two enzyme-driven steps: First an ATP activates the amino acid, and then this group binds to the acceptor end of the tRNA. Because tRNA is the molecule that will convert the master code on mRNA into a protein, the accuracy of this step is crucial.

The Ribosome: A Mobile Molecular Factory for Translation

The procaryotic (70S) ribosome is a particle composed of tightly packaged ribosomal RNA **(rRNA)** and protein. The rRNA component of the ribosome is also a long polynucleotide

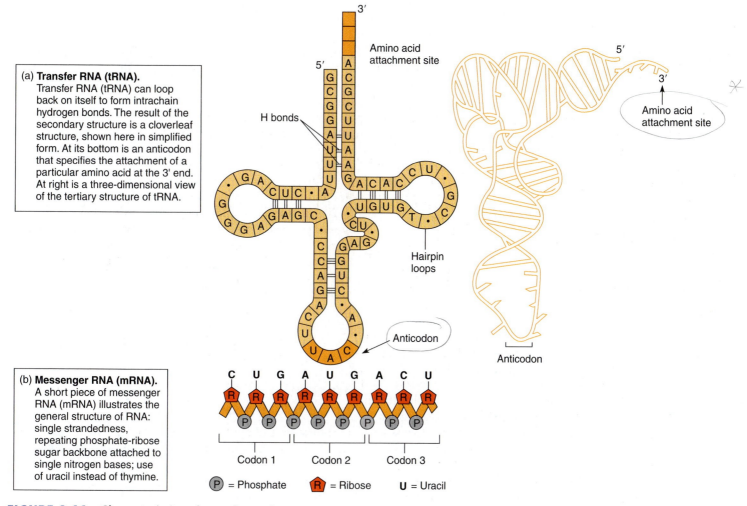

(a) Transfer RNA (tRNA).
Transfer RNA (tRNA) can loop back on itself to form intrachain hydrogen bonds. The result of the secondary structure is a cloverleaf structure, shown here in simplified form. At its bottom is an anticodon that specifies the attachment of a particular amino acid at the 3' end. At right is a three-dimensional view of the tertiary structure of tRNA.

(b) Messenger RNA (mRNA).
A short piece of messenger RNA (mRNA) illustrates the general structure of RNA: single strandedness, repeating phosphate-ribose sugar backbone attached to single nitrogen bases; use of uracil instead of thymine.

Amino acid attachment site

H bonds

Hairpin loops

Anticodon

Amino acid attachment site

Anticodon

Codon 1   Codon 2   Codon 3

P = Phosphate   R = Ribose   U = Uracil

FIGURE 9.11   Characteristics of transfer and messenger RNA.

INSIGHT 9.3 — Discovery

Small RNAs: An Old Dog Shows Off Some New (?) Tricks

Since the earliest days of molecular biology, RNA has been an overlooked worker of the cell, quietly ferrying the information in DNA to ribosomes to direct the formation of proteins. Current research however is showing a new, dynamic role for RNA in the cell that may forever change the reputation of this humble molecule.

Short lengths of RNA, ranging in size from 21 to 28 nucleotides, seem to have the ability to control the expression of certain genes. Experiments in late 2001 identified an enzyme that produced two types of RNAs, dubbed micro RNAs (miRNAs) and small interfering RNAs (siRNAs), by cleaving a larger, double stranded RNA precursor. Once produced, these small RNA molecules direct a cellular enzyme complex called RISC to degrade specific mRNA molecules, silencing the expression of these genes. The process, known as RNA interference, or RNAi, is thought to protect cells from foreign DNA that could otherwise damage the integrity of the genome.

A second type of regulation seems to occur when small RNAs alter the structure of chromosomes. As DNA and proteins coil together to form chromatin, small RNAs direct how tightly or loosely the chromatin is constructed (see Insight 9.1). Just as a closed book cannot be read, DNA sequences contained within tightly coiled chromatin are generally inaccessible to the cell, silencing the expression of those genes. RNAi seems to be the next step toward unraveling the mechanism behind this well-known but poorly understood phenomenon.

One of the most promising roles for RNAi lies in its use as a laboratory tool; enabling researchers to easily regulate the expression of genes to study their effects. Currently, such research is carried out by creating "gene knockouts," a time- and labor-intensive process.

Our knowledge of the full role of small RNAs in the cell is just beginning. In the meantime, scientists will keep studying small RNAs while being mindful of the old adage, "Good things come in small packages."

---

molecule. It forms complex three-dimensional figures that contribute to the structure and function of ribosomes (see figure 9.9). The interactions of proteins and rRNA create the two subunits of the ribosome that engage in final translation of the genetic code (see figure 9.13). A metabolically active bacterial cell can accommodate up to 20,000 of these minuscule factories—all actively engaged in reading the genetic program, taking in raw materials, and producing proteins at an impressive rate.

Transcription: The First Stage of Gene Expression

During transcription, the DNA code is converted to RNA through several stages, directed by a huge and very complex enzyme system, **RNA polymerase (figure 9.12).** Only one strand of the DNA—the *template strand*—contains meaningful instructions for synthesis of a functioning polypeptide. The nontranscribed strand is called the *coding strand*. The strand of DNA that serves as a template varies from one gene to another.

Transcription is initiated when RNA polymerase recognizes a segment of the DNA called the *promoter region.* This region consists of two sequences of DNA just prior to the beginning of the gene to be transcribed. The first sequence, which occurs approximately 35 bases prior to the start of transcription, is tightly bound by RNA polymerase. Tran-

scription is allowed to begin when the DNA helix begins to unwind at the second sequence, which is located about 10 bases prior to the start of transcription. As the DNA helix unwinds, the polymerase advances and begins synthesizing an RNA molecule complementary to the template strand of DNA. The nucleotide sequence of promoters differs only slightly from gene to gene, with all promoters being rich in adenine and thymine.

During elongation, which proceeds in the 5′ to 3′ direction (with regard to the growing RNA molecule), the mRNA is assembled by the addition of nucleotides that are complementary to the DNA template. Remember that uracil (U) is placed as adenine's complement. As elongation continues, the part of DNA already transcribed is rewound into its original helical form. At termination, the polymerases recognize another code that signals the separation and release of the mRNA strand, also called the **transcript.** How long is the mRNA? The very smallest mRNA might consist of 100 bases; an average-sized mRNA might consist of 1,200 bases; and a large one, of several thousand.

Translation: The Second Stage of Gene Expression

In translation, all of the elements needed to synthesize a protein, from the mRNA to the amino acids, are brought together

(a) Overall view of a gene. Each gene contains a specific promoter region and a leader sequence for guiding the beginning of transcription. This is followed by the region of the gene that codes for a polypeptide and ends with a series of terminal sequences that stop translation.

(b) DNA is unwound at the promoter by RNA polymerase. Only one strand of DNA, called the template strand, is copied by the RNA polymerase. This strand runs in the 3' to 5' direction.

(c) As the RNA polymerase moves along the strand, it adds complementary nucleotides as dictated by the DNA template, forming the single-stranded mRNA that reads in the 5' to 3' direction.

(d) The polymerase continues transcribing until it reaches a termination site and the mRNA transcript is released for translation. Note that the section of the DNA that has been transcribed is rewound into its original configuration.

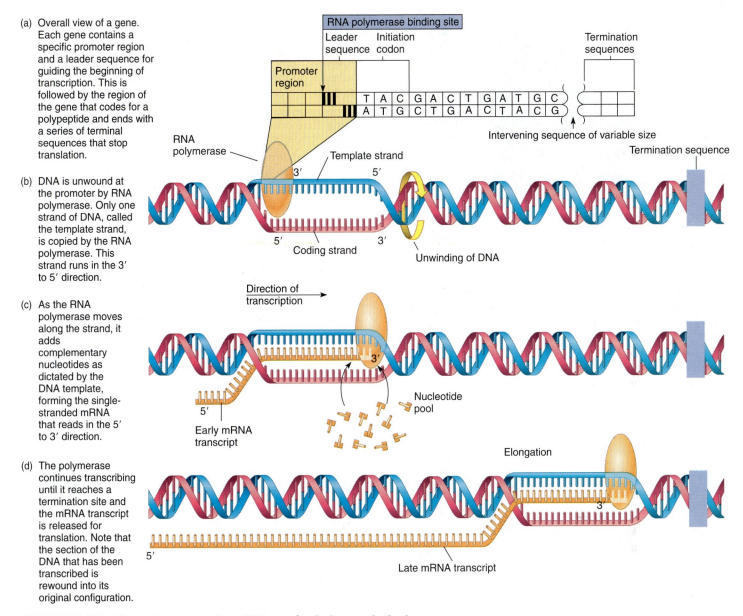

FIGURE 9.12   **The major events in mRNA synthesis (transcription).**

on the ribosomes **(figure 9.13).** The process occurs in five stages: initiation, elongation, termination, and protein folding and processing.

Initiation of Translation

The mRNA molecule leaves the DNA transcription site and is transported to ribosomes in the cytoplasm. Ribosomal subunits are specifically adapted to assembling and forming sites to hold the mRNA and tRNAs. The ribosome thus recognizes these molecules and stabilizes reactions between them. The small subunit binds to the 5' end of the mRNA, and the large subunit supplies enzymes for making peptide bonds on the protein. The ribosome begins to scan the mRNA by moving in the 5' to 3' direction along the mRNA. The first codon it encounters is the START codon, which is almost always AUG (and rarely, GUG).

With the mRNA message in place on the assembled ribosome, the next step in translation involves entrance of tRNAs with their amino acids. The pool of cytoplasm around the region contains a complete array of tRNAs, previously charged by having the correct amino acid attached. The step in which the complementary tRNA meets with the mRNA code is guided by the two sites on the large subunit of the ribosome called the P site (left) and the A site (right).[1] Think of these sites as shallow depressions in the larger subunit of the ribosome, each of which accommodates a tRNA. The ribosome also has an exit or E site where used tRNAs are released.

---

1. P stands for peptide site; A stands for aminoacyl (amino acid) site; E stands for exit site.

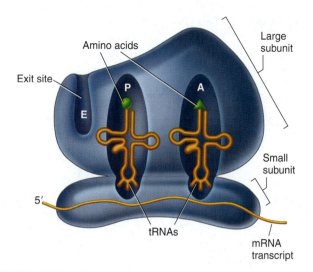

FIGURE 9.13 **The "players" in translation.**
A ribosome serves as the stage for protein synthesis. Assembly of the small and large subunits results in specific sites for holding the mRNA and two tRNAs with their amino acids.

The Master Genetic Code: The Message in Messenger RNA

By convention, the master genetic code is represented by the mRNA codons and the amino acids they specify **(figure 9.14)**. Except in a very few cases, this code is universal, whether for procaryotes, eucaryotes, or viruses. It is worth noting that once the triplet code on mRNA is known, the original DNA sequence, the complementary tRNA code, and the types of amino acids in the protein are automatically known **(figure 9.15)**. However, one cannot predict (backwards) from protein structure what the exact mRNA codons are because of a factor called **redundancy**,[2] meaning that a particular amino acid can be coded for by more than a single codon.

In figure 9.14, the mRNA codons and their corresponding amino acid specificities are given. Because there are 64 different triplet codes[3] and only 20 different amino acids, it is not surprising that some amino acids are represented by several codons. For example, leucine and serine can each be represented by any of six different triplets, and only tryptophan and methionine are represented by a single codon. In such

---

2. This property is also called "degeneracy" by some books.

3. $64 = 4^3$ (the 4 different codons in all possible combinations of 3).

	Second Base Position				
	U	**C**	**A**	**G**	
U	UUU } Phenylalanine UUC UUA } Leucine UUG	UCU UCC Serine UCA UCG	UAU } Tyrosine UAC UAA } STOP** UAG	UGU } Cysteine UGC UGA STOP** UGG Tryptophan	U C A G
C	CUU CUC Leucine CUA CUG	CCU CCC Proline CCA CCG	CAU } Histidine CAC CAA } Glutamine CAG	CGU CGC Arginine CGA CGG	U C A G
A	AUU AUC Isoleucine AUA AUG START Methionine*	ACU ACC Threonine ACA ACG	AAU } Asparagine AAC AAA } Lysine AAG	AGU } Serine AGC AGA } Arginine AGG	U C A G
G	GUU GUC Valine GUA GUG	GCU GCC Alanine GCA GCG	GAU } Aspartic acid GAC GAA } Glutamic acid GAG	GGU GGC Glycine GGA GGG	U C A G

First Base Position (left axis) — *Third Base Position* (right axis)

* This codon initiates translation.
**For these codons, which give the orders to stop translation, there are no corresponding tRNAs and no amino acids.

FIGURE 9.14 **The genetic code: codons of mRNA that specify a given amino acid.**
The master code for translation is found in the mRNA codons.

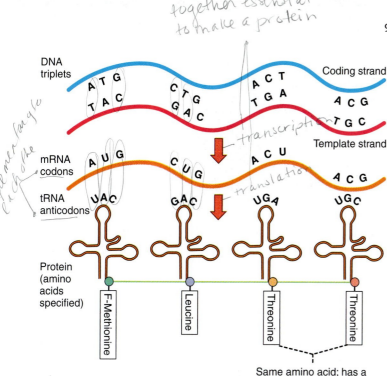

(handwritten annotations on figure:) together essential to make a protein; transcription; translation; Complementary each other

DNA triplets — Coding strand

A T G / T A C   C T G / G A C   A C T / T G A   A C G / T G C — Template strand

mRNA codons — A U G   C U G   A C U   A C G

tRNA anticodons — U A C   G A C   U G A   U G C

Protein (amino acids specified) — F-Methionine   Leucine   Threonine   Threonine

Same amino acid; has a different codon and anticodon

FIGURE 9.15   Interpreting the DNA code.
If the DNA sequence is known, the mRNA codon can be surmised. If a codon is known, the anticodon and, finally, the amino acid sequence can be determined. The reverse is not possible (determining the exact codon or anticodon from amino acid sequence) due to the redundancy of the code.

codons as leucine, only the first two nucleotides are required to encode the correct amino acid, and the third nucleotide does not change its sense. This property, called *wobble,* is thought to permit some variation or mutation without harming the message.

The Beginning of Protein Synthesis

With mRNA serving as the guide, the stage is finally set for actual protein assembly. The correct tRNA (labeled 1 on **figure 9.16**) enters the P site and binds to the **start codon (AUG)** presented by the mRNA. Rules of pairing dictate that the anticodon of this tRNA must be complementary to the mRNA codon AUG, thus the tRNA with anticodon UAC will first occupy site P. It happens that the amino acid carried by the initiator tRNA in bacteria is *formyl methionine* (fMet; see figure 9.14), though in many cases, it may not remain a permanent part of the finished protein.

Continuation and Completion of Protein Synthesis: Elongation and Termination

While reviewing the dynamic process of protein assembly, you will want to remain aware that the ribosome shifts its "reading frame" to the right along the mRNA from one codon to the next. This brings the next codon into place on the ribosome and makes a space for the next tRNA to enter the A position. A peptide bond is formed between the amino

acids on the adjacent tRNAs, and the polypeptide grows in length.

Elongation begins with the filling of the A site by a second tRNA (2 on **figure 9.16**). The identity of this tRNA and its amino acid is dictated by the second mRNA codon.

The entry of tRNA 2 into the A site brings the two adjacent tRNAs in favorable proximity for a peptide bond to form between the amino acids (aa) they carry. The fMet is transferred from the first tRNA to aa 2, resulting in two coupled amino acids called a dipeptide (**figure 9.16b).**

For the next step to proceed, some room must be made on the ribosome, and the next codon in sequence must be brought into position for reading. This process is accomplished by *translocation,* the enzyme-directed shifting of the ribosome to the right along the mRNA strand, which causes the blank tRNA (1) to be discharged from the ribosome (**figure 9.16c)** at the E site. This also shifts the tRNA holding the dipeptide into P position. Site A is temporarily left empty. The tRNA that has been released is now free to drift off into the cytoplasm and become recharged with an amino acid for later additions to this or another protein.

The stage is now set for the insertion of tRNA 3 at site A as directed by the third mRNA codon (**figure 9.16d).** This insertion is followed once again by peptide bond formation between the dipeptide and aa 3 (making a tripeptide), splitting of the peptide from tRNA 2, and translocation. This releases tRNA 2, shifts mRNA to the next position, moves tRNA 3 to position P, and opens position A for the next tRNA (4). From this point on, peptide elongation proceeds repetitively by this same series of actions out to the end of the mRNA.

The termination of protein synthesis is not simply a matter of reaching the last codon on mRNA. It is brought about by the presence of at least one special codon occurring just after the codon for the last amino acid. Termination codons— UAA, UAG, and UGA—are codons for which there is no corresponding tRNA. Although they are often called **nonsense codons,** they carry a necessary and useful message: *Stop here.* When this codon is reached, a special enzyme breaks the bond between the final tRNA and the finished polypeptide chain, releasing it from the ribosome.

Before newly made proteins can carry out their structural or enzymatic roles, they often require finishing touches. Even before the peptide chain is released from the ribosome, it begins folding upon itself to achieve its biologically active tertiary conformation. Other alterations, called *posttranslational* modifications, may be necessary. Some proteins must have the starting amino acid (formyl methionine) clipped off; proteins destined to become complex enzymes have cofactors added; and some join with other completed proteins to form quaternary levels of structure.

The operation of transcription and translation is machinelike in its precision. Protein synthesis in bacteria is both efficient and rapid. At 37°C, 12 to 17 amino acids per second are added to a growing peptide chain. An average protein consisting of about 400 amino acids requires less than half a minute for complete synthesis. Further efficiency is gained when the translation of mRNA starts while

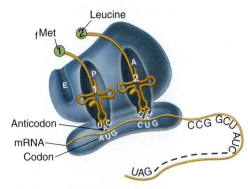

(a) Entrance of tRNAs 1 and 2

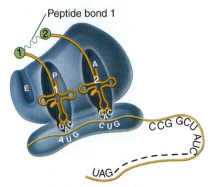

(b) Formation of peptide bond

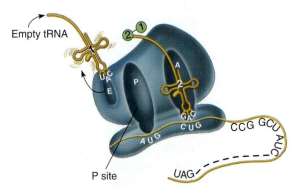

(c) Discharge of tRNA 1 at E site

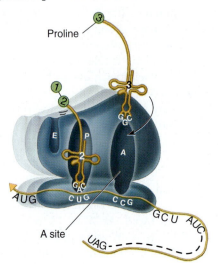

(d) First translocation; tRNA 2 shifts into
    P site; enter tRNA 3 by ribosome

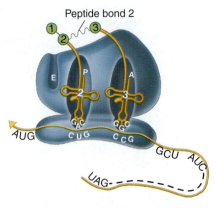

(e) Formation of peptide bond

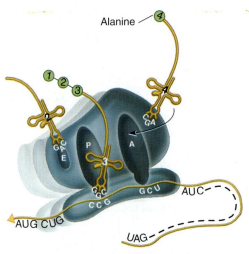

(f) Discharge of tRNA 2; second
    translocation; enter tRNA 4

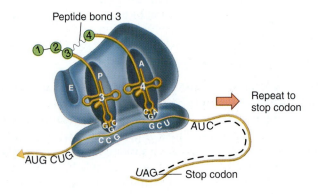

(g) Formation of peptide bond

FIGURE 9.16   **The events in protein synthesis.**

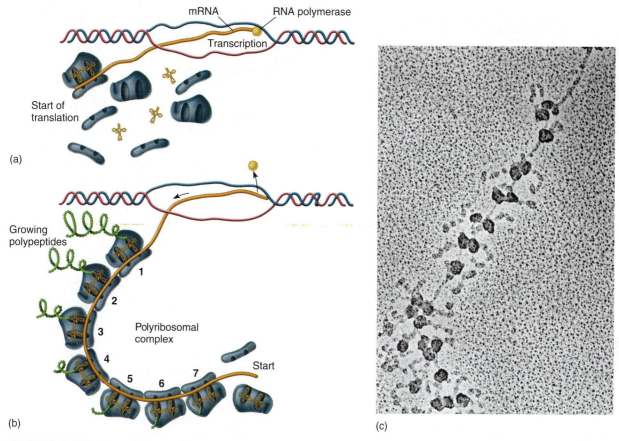

FIGURE 9.17   **Speeding up the protein assembly line in bacteria.**

(a) The mRNA transcript encounters ribosomal parts immediately as it leaves the DNA. **(b)** The ribosomal factories assemble along the mRNA in a chain, each ribosome reading the message and translating it into protein. Many products will thus be well along the synthetic pathway before transcription has even terminated. **(c)** Photomicrograph of a polyribosomal complex in action. Note that the protein "tails" vary in length depending on the stage of translation.

transcription is still occurring **(figure 9.17).** A single mRNA is long enough to be fed through more than one ribosome simultaneously. This permits the synthesis of hundreds of protein molecules from the same mRNA transcript arrayed along a chain of ribosomes. This **polyribosomal complex** is indeed an assembly line for mass production of proteins. Protein synthesis consumes an enormous amount of energy. Nearly 1,200 ATPs are required just for synthesis of an average-sized protein.

Eucaryotic Transcription and Translation: Similar Yet Different

Eucaryotes and procaryotes share many similarities in protein synthesis. The start codon in eucaryotes is also AUG, but it codes for a different form of methionine. Another difference is that eucaryotic mRNAs code for just one protein, unlike bacterial mRNAs, which often contain information from several genes in series.

There are a few differences between procaryotic and eucaryotic gene expression. The presence of the DNA in a separate compartment (the nucleus) means that eucaryotic transcription and translation cannot be simultaneous. The

mRNA transcript must pass through pores in the nuclear membrane and be carried to the ribosomes in the cytoplasm for translation.

We have given the simplified definition of a gene that works well for procaryotes, but most eucaryotic genes are not colinear[4]—meaning that they do *not* exist as an uninterrupted series of triplets coding for a protein. A eucaryotic gene contains the code for a protein, but located along the gene are one to several intervening sequences of bases, called **introns,** that do not code for protein. Introns are interspersed between coding regions, called **exons,** that will be translated into protein **(figure 9.18).** We can use words as examples. A short section of colinear procaryotic gene might read TOM SAW OUR DOG DIG OUT; a eucaryotic gene that codes for the same portion would read TOM SAW XZKP FPL OUR DOG QZWVP DIG OUT. The recognizable words are the exons, and the nonsense letters represent the introns.

This unusual genetic architecture, sometimes called a split gene, requires further processing before translation.

---

4. Colinearity means that the base sequence can be read directly into a series of amino acids.

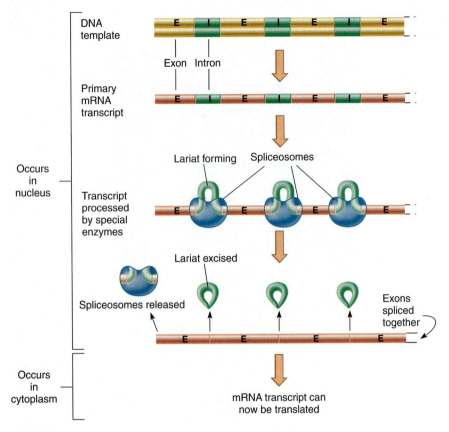

FIGURE 9.18   The split gene of eucaryotes.
Eucaryotic genes have an additional complicating factor in their translation. Their coding sequences, or exons (E), are interrupted at intervals by segments called introns (I) that are not part of that protein's code. Introns are transcribed but not translated, which necessitates their removal by RNA splicing enzymes before translation.

Transcription of the entire gene with both exons and introns occurs first, producing a pre-mRNA. Next, a type of RNA and protein called a *spliceosome* recognizes the exon-intron junctions and enzymatically cuts through them. The action of this splicer enzyme loops the introns into lariat-shaped pieces, excises them, and joins the exons end to end. By this means, a strand of mRNA with no intron material is produced. This completed mRNA strand can then proceed to the cytoplasm to be translated.

At first glance, this system seems to be a cumbersome way to make a transcript, and the value of this extra genetic baggage is still the subject of much debate. Several different types of introns have been discovered, some of which do code for cell substances. Many introns have been found to code for an enzyme called reverse transcriptase, which can convert RNA into DNA. Other introns are translated into endonucleases, enzymes that can snip DNA and allow insertions and deletions into the sequence. Some experts hypothesize that introns represent a "sink" for extra bits of genetic material that could be available for splicing into existing genes, thus promoting genetic change and evolution. There is also evidence that introns serve as genetic regulators and in ribosomal assembly.

The Genetics of Animal Viruses

The genetic nature of viruses was described in chapter 6. Viruses essentially consist of one or more pieces of DNA or RNA enclosed in a protective coating. Above all, they are genetic parasites that require access to their host cell's genetic and metabolic machinery to be replicated, transcribed, and translated, and they also have the potential for genetically changing the cells. Because they contain only those genes needed for the production of new viruses, the genomes of viruses tend to be very compact and economical. In fact, this very simplicity makes them excellent subjects for the study of gene function.

The genetics of viruses is quite diverse. In many viruses, the nucleic acid is linear in form; in others, it is circular. The genome of most viruses exists in a single molecule, though in a few, it is segmented into several smaller molecules. Most viruses contain normal double-stranded (ds) DNA or single-stranded (ss) RNA, but other patterns exist. There are ssDNA viruses, dsRNA viruses, and retroviruses, which work backward by making dsDNA from ssRNA. In some instances, viral genes overlap one another, and in a few DNA viruses, both strands contain a translatable message.

A few generalities can be stated about viral genetics. In all cases, the viral nucleic acid penetrates the cell and is introduced into the host's gene-processing machinery at some point. In successful infection, an invading virus instructs the host's machinery to synthesize large numbers of new virus particles by a mechanism specific to a particular group. With few exceptions, replication of the DNA molecule of DNA animal viruses occurs in the nucleus, where the cell's DNA replication machinery lies and the genome of RNA viruses is replicated in the cytoplasm. In all viruses, viral mRNA is translated into viral proteins on host cell ribosomes using host tRNA.

✔ **CHECKPOINT**

- Information in DNA is converted to proteins by the process of transcription and translation. These proteins may be structural or functional in nature. Structural proteins contribute to the architecture of the cell while functional proteins (enzymes) control an organism's metabolic activities.

- The DNA code occurs in groups of three bases; this code is copied onto RNA as codons; the message determines the types of amino acids in a protein. This code is universal in all cells and viruses.

■ The processes of transcription and translation are similar but not identical for procaryotes and eucaryotes. Eucaryotes transcribe DNA in the nucleus, remove its introns, and translate it in the cytoplasm. Bacteria transcribe and translate simultaneously because the DNA is not sequestered in a nucleus and the bacterial DNA is free of introns.

■ The genetic material of viruses can be either DNA or RNA occurring in single, double, positive-, or negative-sense strands. Viral DNA contains just enough information to force a cell to become a virus factory.

9.3 Genetic Regulation of Protein Synthesis and Metabolism

In chapter 8 we surveyed the metabolic reactions in cells and the enzymes involved in those reactions. At that time, we mentioned that some enzymes are regulated, and that one form of regulation occurs at the genetic level. This control mechanism ensures that genes are active only when their products are required. In this way, enzymes will be produced as they are needed and prevent the waste of energy and materials in dead-end synthesis. Genetic function in procaryotes is regulated by a specific collection of genes called an **operon.** Operons consist of a coordinated set of genes, all of which are regulated as a single unit. Operons are described as either inducible or repressible. The category each operon falls into is determined by how transcription is affected by the environment surrounding the cell. Many catabolic operons are inducible, meaning that the operon is turned on (induced) by the substrate of the enzyme for which the structural genes code. In this way, the enzymes needed to metabolize a nutrient (lactose, for example) are only produced when that nutrient is present in the environment. Repressible operons often contain genes coding for anabolic enzymes, such as those used to synthesize amino acids. In the case of these operons, several genes in series are turned off (repressed) by the product synthesized by the enzyme.

The Lactose Operon: A Model for Inducible Gene Regulation in Bacteria

The best understood cell system for explaining control through genetic induction is the **lactose (lac) operon.** The system, first described in 1961 by François Jacob and Jacques Monod, accounts for the regulation of lactose metabolism in *Escherichia coli.* Many other operons with similar modes of action have since been identified, and together they furnish convincing evidence that the environment of a cell can have great impact on gene expression.

The lactose operon has three important features **(figure 9.19):**

1. the **regulator,** composed of the gene that codes for a protein capable of repressing the operon (a **repressor**);

2. the *control locus,* composed of two areas, the **promoter** (recognized by RNA polymerase) and the **operator,** a sequence that acts as an on/off switch for transcription; and

3. the *structural locus,* made up of three genes, each coding for a different enzyme needed to catabolize lactose.

One of the enzymes, β-galactosidase, hydrolyzes the lactose into its monosaccharides; another, permease, brings lactose across the cell membrane.

The operon is an efficient strategy that permits genes for a particular metabolic pathway to be induced or repressed in unison by a single regulatory element. The enzymes of the *lac* operon are of the inducible sort mentioned in chapter 8. The promoter, operator, and structural components lie adjacent to one another, but the regulator can be at a distant site.

In inducible systems like the *lac* operon, the operon is normally in an *off* mode and does not initiate enzyme synthesis when the appropriate substrate is absent **(figure 9.19a).** How is the operon maintained in this mode? The key is in the repressor protein that is coded by the regulatory gene. This relatively large molecule is **allosteric,** meaning it has two binding sites, one for the operator and another for lactose. In the absence of lactose, this repressor binds with the operator locus, thereby blocking the transcription of the structural genes lying downstream. Think of the repressor as a lock on the operator, and if the operator is locked, the structural genes cannot be transcribed. Importantly, the regulator gene lies upstream (to the left) of the operator region and is transcribed constitutively since it is not controlled in tandem with the operon.

If lactose is added to the cell's environment, it triggers several events that turn the operon *on.* The binding of lactose to the repressor protein causes a conformational change in the repressor that dislodges it from the operator segment **(figure 9.19b).** With the operator opened up, RNA polymerase can now bind to the promoter. The structural genes are transcribed in a single unbroken transcript coding for all three enzymes. (During translation, however, each protein is synthesized separately.) Because lactose is ultimately responsible for stimulating protein synthesis, it is called an *inducer.*

As lactose is depleted, further enzyme synthesis is not necessary, so the order of events reverses. At this point there is no longer sufficient lactose to inhibit the repressor, hence the repressor is again free to attach to the operator. The operator is locked, and transcription of the structural genes and enzyme synthesis related to lactose both stop.

A fine but important point about the *lac* operon is that it functions only in the absence of glucose or if the cell's energy needs are not being met by the available glucose. Glucose is the preferred carbon source because it can be used immediately in growth, and does not require induction of an operon. When glucose is present, a second regulatory system ensures that the *lac* operon is inactive, regardless of lactose levels in the environment.

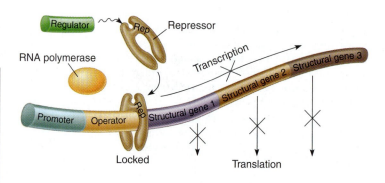

(a) Operon Off. In the absence of lactose, a repressor protein (the product of a regulatory gene located elsewhere on the bacterial chromosome) attaches to the operator of the operon. This effectively locks the operator and prevents any transcription of structural genes downstream (to its right). Suppression of transcription (and consequently, of translation) prevents the unnecessary synthesis of enzymes for processing lactose.

(b) Operon On. Upon entering the cell, the substrate (lactose) becomes a genetic inducer by attaching to the repressor, which loses its grip and falls away. The RNA polymerase is now free to bind to the promoter and initiate transcription, and the enzymes produced by translation of the mRNA perform the necessary reactions on their lactose substrate.

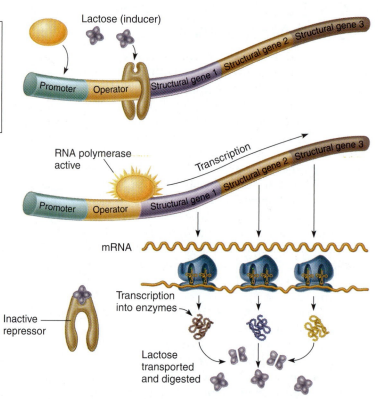

FIGURE 9.19 **The lactose operon in bacteria: how inducible genes are controlled by substrate.**

A Repressible Operon

Bacterial systems for synthesis of amino acids, purines and pyrimidines, and many other processes work on a slightly different principle—that of repression. Similar factors such as repressor proteins, operators, and a series of structural genes exist for this operon, but with some important differences. Unlike the *lac* operon, this operon is normally in the *on* mode and will be turned *off* only when this nutrient is no longer required. The excess nutrient plays role as a **corepressor** needed to block the action of the operon.

A growing cell that needs the amino acid arginine (arg) effectively illustrates the operation of a repressible operon. Under these conditions, the *arg* operon is set to *on*, and arginine is being actively synthesized through the action of the operon's enzymatic products **(figure 9.20a)**. In an active cell, the arginine will be used immediately, and the repressor will remain inactive (unable to bind the operator) because there is too little free arginine to activate it. As the cell's metabolism begins to slow down, however, the synthesized arginine will no longer be used up and will accumulate. The free arginine is then available to act as a corepressor by attaching to the repressor. This reaction changes the shape of the repressor, making it capable of binding to the operator. Transcription stops; arginine is no longer synthesized **(figure 9.20b)**.

Analogous gene control mechanisms in eucaryotic cells are not as well understood, but it is known that gene function can be altered by intrinsic regulatory segments similar to operons. Some molecules, called transcription factors, insert on the grooves of the DNA molecule and enhance transcription of specific genes. Examples include zinc "fingers" and leucine "zippers." These transcription factors can regulate gene expression in response to environmental stimuli such as nutri-

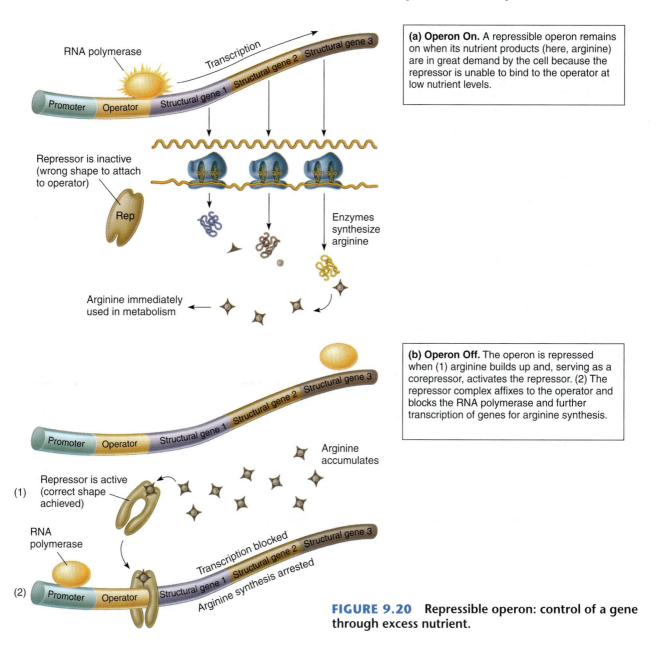

(a) Operon On. A repressible operon remains on when its nutrient products (here, arginine) are in great demand by the cell because the repressor is unable to bind to the operator at low nutrient levels.

(b) Operon Off. The operon is repressed when (1) arginine builds up and, serving as a corepressor, activates the repressor. (2) The repressor complex affixes to the operator and blocks the RNA polymerase and further transcription of genes for arginine synthesis.

FIGURE 9.20 **Repressible operon: control of a gene through excess nutrient.**

ents, toxin levels, or even temperature. Eucaryotic genes are also regulated during growth and development, leading to the hundreds of different tissue types found in higher multicellular organisms.

Antibiotics That Affect Transcription and Translation

Naturally occurring cell nutrients are not the only agents capable of modifying gene expression. Some infection therapy is based on the concept that certain drugs react with DNA, RNA, or ribosomes and thereby alter genetic expression (see chapter 12). Treatment with such drugs is based on an important premise: that growth of the infectious agent will be inhibited by blocking its protein-synthesizing machinery selectively, without disrupting the cell synthesis of the patient receiving the therapy.

Drugs that inhibit protein synthesis exert their influence on transcription or translation. For example, the rifamycins used in therapy for tuberculosis bind to RNA polymerase, blocking the initiation step of transcription, and are selectively more active against bacterial RNA polymerase than the corresponding eucaryotic enzyme. Actinomycin D binds to bacterial DNA and halts mRNA chain elongation, but it also binds to human DNA. For this reason, it is very toxic and never used to treat bacterial infections, though it can be applied in tumor treatment.

The ribosome is a frequent target of antibiotics that inhibit ribosomal function and ultimately protein synthesis. The value and safety of these antibiotics again depend upon the differential susceptibility of procaryotic and eucaryotic ribosomes. One problem with drugs that selectively disrupt procaryotic ribosomes is that the mitochondria of humans contain a procaryotic type of ribosome, and these drugs may inhibit the function of the host's mitochondria. One group of antibiotics (including erythromycin and spectinomycin) prevents translation by interfering with the attachment of mRNA to ribosomes. Chloramphenicol, lincomycin, and tetracycline bind to the ribosome in a way that blocks the elongation of the polypeptide, and aminoglycosides (such as streptomycin) inhibit peptide initiation and elongation. It is interesting to note that these drugs have served as important tools to explore genetic events because they can arrest specific stages in these processes.

✓ CHECKPOINT

- Gene expression must be orchestrated to coordinate the organism's needs with nutritional resources. Genes can be turned "on" and "off" by specific molecules, which expose or hide their nucleotide codes for transcribing proteins. Most, but not all, of these proteins are enzymes.

- Operons are collections of genes in bacteria that code for products with a coordinated function. They include genes for operational and structural components of the cell. Nutrients can combine with regulator gene products to turn a set of structural genes on (inducible genes) or off (repressible genes). The *lac* (lactose) operon is an example of an inducible operon. The *arg* (arginine) operon is an example of a repressible operon.

- The rifamycins, tetracyclines, and aminoglycosides are classes of antibiotics that are effective because they interfere with transcription and translation processes in microorganisms.

9.4 Mutations: Changes in the Genetic Code

As precise and predictable as the rules of genetic expression seem, permanent changes do occur in the genetic code. Indeed, genetic change is the driving force of evolution. In microorganisms such changes may become evident in altered gene expression, such as the appearance or disappearance of anatomical or physiological traits. For example, a pigmented bacterium can lose its ability to form pigment, or a strain of the malarial parasite can develop resistance to a drug. When phenotypic changes are due to changes in the genotype, it is called a **mutation.** On a strictly molecular level, a mutation is an alteration in the nitrogen base sequence of DNA. It can involve the loss of base pairs, the addition of base pairs, or a rearrangement in the order of base pairs. Do not confuse this with genetic recombination, in which microbes transfer whole segments of genetic information between themselves.

A microorganism that exhibits a natural, nonmutated characteristic is known as a **wild type,** or wild strain. If a microorganism bears a mutation, it is called a **mutant strain.** Mutant strains can show variance in morphology, nutritional characteristics, genetic control mechanisms, resistance to chemicals, temperature preference, and nearly any type of enzymatic function. Mutant strains are very useful for track-

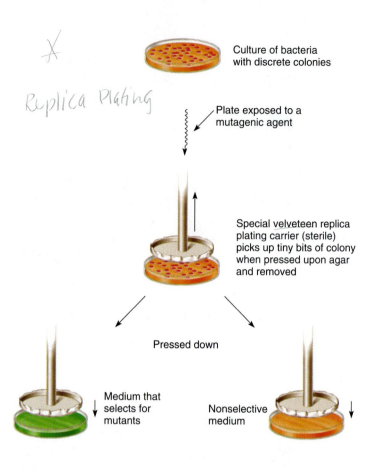

Colonies of mutant strains

Mixture of wild-type and mutant strains

FIGURE 9.21   The general basis of replica plating. This method was developed by Joshua Lederberg for detecting and isolating mutant strains of microorganisms.

ing genetic events, unraveling genetic organization, and pinpointing genetic markers. A classic method of detecting mutant strains involves addition of various nutrients to a culture to screen for its use of that nutrient. For example, in a culture of a wild-type bacterium that is lactose-positive (meaning it has the necessary enzymes for fermenting this sugar), a small number of mutant cells have become lactose-negative, having lost the capacity to ferment this sugar. If the culture is plated on a medium containing indicators for fermentation, each colony can be observed for its fermentation reaction, and the negative strain isolated. Another standard method of detecting and isolating microbial mutants is by replica plating **(figure 9.21).**

Causes of Mutations

A mutation is described as spontaneous or induced, depending upon its origin. A **spontaneous mutation** is a random change in the DNA arising from errors in replication that occur randomly. The frequency of spontaneous mutations has been measured for a number of organisms. Mutation rates vary tremendously, from one mutation in 10^5 replications (a high rate) to one mutation in 10^{10} replications (a low rate). The rapid rate of bacterial reproduction allows these mutations to be observed more readily in bacteria than in most eucaryotes.

Induced mutations result from exposure to known **mutagens,** which are primarily physical or chemical agents that interact with DNA in a disruptive manner **(table 9.3).** The carefully controlled use of mutagens has proved a useful way to induce mutant strains of microorganisms for study.

Chemical mutagenic agents act in a variety of ways to change the DNA. Agents such as acridine dyes insert completely across the DNA helices between adjacent bases to produce mutations that distort the helix. Analogs[5] of the nitrogen bases (5-bromodeoxyuridine and 2-aminopurine, for example) are chemical mimics of natural bases that are incorporated into DNA during replication. Addition of these abnormal bases leads to mistakes in base-pairing. Many chemical mutagens are also carcinogens, or cancer-causing agents (see the discussion of the Ames test in a later section of this chapter).

Physical agents that alter DNA are primarily types of radiation. High-energy gamma rays and X rays introduce major physical changes into DNA, and it accumulates breaks that may not be repairable. Ultraviolet (UV) radiation induces abnormal bonds between adjacent pyrimidines that prevent normal replication. Exposure to large doses of radiation can be fatal, which is why radiation is so effective in microbial control; it can also be carcinogenic in animals. (The use of UV to control microorganisms is described further in chapter 11.)

Categories of Mutations

Mutations range from large mutations, in which large genetic sequences are gained or lost, to small ones that affect only a single base on a gene. These latter mutations, which involve addition, deletion, or substitution of single bases, are called **point mutations.**

To understand how a change in DNA influences the cell, remember that the DNA code appears in a particular order of triplets (three bases) that is transcribed into mRNA codons, each of which specifies an amino acid. A permanent alteration in the DNA that is copied faithfully into mRNA and translated can change the structure of the protein. A change in a protein can likewise change the morphology and physiology of a cell. Most mutations have a harmful effect on the cell, leading to cell dysfunction or death; these are called lethal mutations. Neutral mutations produce neither adverse nor helpful changes. A small number of mutations are beneficial in that they provide the cell with a useful change in structure or physiology.

---

5. An analog is a chemical structured very similarly to another chemical except for minor differences in functional groups.

TABLE 9.3	Selected Mutagenic Agents and Their Effects
Agent	**Effect**
Chemical	
Nitrous acid, bisulfite	Removes an amino group from some bases
Ethidium bromide	Inserts between the paired bases
Acridine dyes	Cause frameshifts due to insertion between base pairs
Nitrogen base analogs	Compete with natural bases for sites on replicating DNA
Radiation	
Ionizing (gamma rays, X rays)	Form free radicals that cause single or double breaks in DNA
Ultraviolet	Causes cross-links between adjacent pyrimidines

Any change in the code that leads to placement of a different amino acid is called a **missense mutation.** A missense mutation can do one of the following:

1. create a faulty, nonfunctional (or less functional) protein,
2. produce a protein which functions in a different manner, or
3. cause no significant alteration in protein function.

A **nonsense mutation,** on the other hand, changes a normal codon into a stop codon that does not code for an amino acid and stops the production of the protein wherever it occurs. A nonsense mutation almost always results in a nonfunctional protein. A **silent mutation** alters a base but does not change the amino acid and thus has no effect. For example, because of the redundancy of the code, ACU, ACC, ACG, and ACA all code for threonine, so a mutation that changes only the last base will not alter the sense of the message in any way. A **back-mutation** occurs when a gene that has undergone mutation reverses (mutates back) to its original base composition.

Mutations also occur when one or more bases are inserted into or deleted from a newly synthesized DNA strand. This type of mutation, known as a **frameshift,** is so named because the reading frame of the mRNA has been changed. Frameshift mutations nearly always result in a nonfunctional protein because every amino acid after the mutation is different from what was coded for in the original DNA. Also note that insertion or deletion of bases in multiples of three (3, 6, 9, etc.) results in the addition or deletion of amino acids but does not disturb the reading frame. The effects of all of these types of mutations can be seen in **table 9.4.**

Repair of Mutations

Earlier we indicated that DNA has a proofreading mechanism to repair mistakes in replication that might otherwise

TABLE 9.4	Classification of Major Types of Mutations	
		Example
(a)	Wild type (original, non-mutated sequence)	THE BIG BAD DOG ATE THE FAT RED CAT
Substitution mutations		
(b)	Missense	THE BIG BAD DOG ATE THE FIT RED CAT
(c)	Nonsense	THE BIG BAD (stop)
Frameshift mutations		
(d)	Insertion	THE BIG BAB DDO GAT ETH EFA TRE DCA T
(e)	Deletion	THE BIG BDD OGA TET HEF ATR EDC AT

Categories of mutations based on type of DNA alteration
(a) The wild-type sequence of a gene is the DNA sequence found in most organisms and is generally considered the "normal" sequence. (b) A missense mutation causes a different amino acid to be incorporated into a protein. Effects range from unnoticeable to severe, based on how different the two amino acids are. (c) A nonsense mutation converts a codon to a stop codon, resulting in premature termination of protein synthesis. Effects of this type of mutation are almost always severe. (d, e) Insertion and deletion mutations cause a change in the reading frame of the mRNA, resulting in a protein in which every amino acid after the mutation is affected. Because of this, frameshift mutations almost always result in a nonfunctional protein.

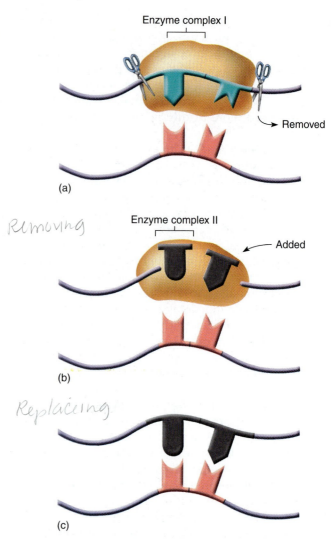

FIGURE 9.22 **Excision repair of mutation by enzymes.**
(a) The first enzyme complex recognizes one or several incorrect bases and removes them. **(b)** The second complex (DNA polymerase I and ligase) places correct bases and seals the gaps. **(c)** Repaired DNA.

become permanent (see page 258). Because mutations are potentially life-threatening, the cell has additional systems for finding and repairing DNA that has been damaged by various mutagenic agents and processes. Most ordinary DNA damage is resolved by enzymatic systems specialized for finding and fixing such defects.

DNA that has been damaged by ultraviolet radiation can be restored by photoactivation or light repair. This repair mechanism requires visible light and a light-sensitive enzyme, DNA photolyase, which can detect and attach to the damaged areas (sites of abnormal pyrimidine binding). Ultraviolet repair mechanisms are successful only for a relatively small number of UV mutations. Cells cannot repair severe, widespread damage and will die. In humans, the genetic disease *xeroderma pigmentosa* is due to nonfunctioning genes for the enzyme photolyase. Persons suffering from this rare disorder develop severe skin cancers; this relation provides strong evidence for a link between cancer and mutations.

Mutations can be excised by a series of enzymes that remove the incorrect bases and add the correct ones. This process is known as *excision repair*. First, enzymes break the bonds between the bases and the sugar-phosphate strand at the site of the error. A different enzyme subsequently removes the defective bases one at a time, leaving a gap that will be filled in by

DNA polymerase I and ligase (**figure 9.22**). A repair system can also locate mismatched bases that were missed during proofreading: for example, C mistakenly paired with A, or G with T. The base must be replaced soon after the mismatch is made, or it will not be recognized by the repair enzymes.

The Ames Test

New agricultural, industrial, and medicinal chemicals are constantly being added to the environment, and exposure to them is widespread. The discovery that many such compounds are mutagenic and that up to 83% of these mutagens are linked to cancer is significant. Although animal testing has been a standard method of detecting chemicals with carcinogenic potential, a more rapid screening system called the **Ames test**[6] is also commonly used. In this ingenious test, the

6. Named for its creator, Bruce Ames.

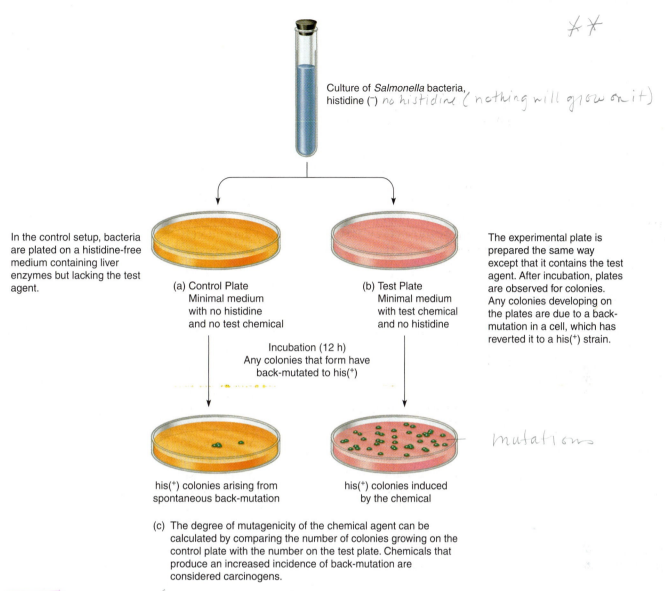

Culture of *Salmonella* bacteria, histidine (⁻) *no histidine (nothing will grow on it)*

In the control setup, bacteria are plated on a histidine-free medium containing liver enzymes but lacking the test agent.

The experimental plate is prepared the same way except that it contains the test agent. After incubation, plates are observed for colonies. Any colonies developing on the plates are due to a back-mutation in a cell, which has reverted it to a his(⁺) strain.

(a) Control Plate
Minimal medium with no histidine and no test chemical

(b) Test Plate
Minimal medium with test chemical and no histidine

Incubation (12 h)
Any colonies that form have back-mutated to his(⁺)

his(⁺) colonies arising from spontaneous back-mutation

his(⁺) colonies induced by the chemical

mutations

(c) The degree of mutagenicity of the chemical agent can be calculated by comparing the number of colonies growing on the control plate with the number on the test plate. Chemicals that produce an increased incidence of back-mutation are considered carcinogens.

FIGURE 9.23 The Ames test. *for carcinogens*
This test is based on a strain of *Salmonella typhimurium* that cannot synthesize histidine his(⁻). It lacks the enzymes to repair DNA so that mutations show up readily, and has leaky cell walls that permit the ready entrance of chemicals. Many potential carcinogens (benzanthracene and aflatoxin, for example) are mutagenic agents only after being acted on by mammalian liver enzymes, so an extract of these enzymes is added to the test medium.

experimental subjects are bacteria whose gene expression and mutation rate can be readily observed and monitored. The premise is that any chemical capable of mutating bacterial DNA can similarly mutate mammalian (and thus human) DNA and is therefore potentially hazardous.

One indicator organism in the Ames test is a mutant strain of *Salmonella typhimurium*[7] that has lost the ability to synthesize the amino acid histidine, a defect highly susceptible to back-mutation because the strain also lacks DNA repair mechanisms. Mutations that cause reversion to the wild strain, which is capable of synthesizing histidine, occur spontaneously at a low rate. A test agent is considered a mutagen if it enhances the rate of back-mutation beyond levels that would occur spontaneously. One variation on this testing procedure is outlined in **figure 9.23.** The Ames test has proved invaluable for screening an assortment of environmental and dietary chemicals for mutagenicity and carcinogenicity without resorting to animal studies.

Positive and Negative Effects of Mutations

Many mutations are not repaired. How the cell copes with them depends on the nature of the mutation and the strategies available to that organism. Mutations are permanent and heritable and will be passed on to the offspring of organisms and new viruses and become a long-term part of the gene

---

7. *S. typhimurium* inhabits the intestine of poultry and causes food poisoning in humans. It is used extensively in genetic studies of bacteria.

pool. Most mutations are harmful to organisms; others provide adaptive advantages.

If a mutation leading to a nonfunctional protein occurs in a gene for which there is only a single copy, as in haploid or simple organisms, the cell will probably die. This happens when certain mutant strains of *E. coli* acquire mutations in the genes needed to repair damage by UV radiation. Mutations of the human genome affecting the action of a single protein (mostly enzymes) are responsible for more than 3,500 diseases (see chapter 10).

Although most spontaneous mutations are not beneficial, a small number contribute to the success of the individual and the population by creating variant strains with alternate ways of expressing a trait. Microbes are not "aware" of this advantage and do not direct these changes; they simply respond to the environment they encounter. Those organisms with beneficial mutations can more readily adapt, survive, and reproduce. In the long-range view, mutations and the variations they produce are the raw materials for change in the population and, thus, for evolution.

Mutations that create variants occur frequently enough that any population contains mutant strains for a number of characteristics, but as long as the environment is stable, these mutants will never comprise more than a tiny percentage of the population. When the environment changes, however, it can become hostile for the survival of certain individuals, and only those microbes bearing protective mutations will be equipped to survive in the new environment. In this way, the environment naturally selects certain mutant strains that will reproduce, give rise to subsequent generations, and in time, be the dominant strain in the population. Through these means, any change that confers an advantage during selection pressure will be retained by the population. One of the clearest models for this sort of selection and adaptation is acquired drug resistance in bacteria (see chapter 12). Bacteria have also developed a mechanism for increasing their adaptive capacity through genetic exchange, called genetic recombination.

9.5 DNA Recombination Events

Genetic recombination through sexual reproduction is an important means of genetic variation in eucaryotes. Although bacteria have no exact equivalent to sexual reproduction, they exhibit a primitive means for sharing or recombining parts of their genome. An event in which one bacterium donates DNA to another bacterium is a type of genetic transfer termed **recombination,** the end result of which is a new strain different from both the donor and the original recipient strain. Recombination in bacteria depends in part on the fact that bacteria contain extrachromosomal DNA—that is, plasmids—and are adept at interchanging genes. Genetic exchanges have tremendous effects on the genetic diversity of bacteria, and unlike spontaneous mutations, are generally beneficial to them. They provide additional genes for resistance to drugs and metabolic poisons, new nutritional and metabolic capabilities, and increased virulence and adaptation to the environment.

In general, any organism that contains (and expresses) genes that originated in another organism is called a **recombinant.**

Transmission of Genetic Material in Bacteria

DNA transfer between bacterial cells typically involves small pieces of DNA in the form of plasmids or chromosomal fragments. Plasmids are small, circular pieces of DNA that contain their own origin of replication and therefore can replicate independently of the bacterial chromosome. Plasmids are found in many bacteria (as well as some fungi) and typically contain, at most, only a few dozen genes. Although plasmids are not necessary for bacterial survival they often carry useful traits, such as antibiotic resistance. Chromosomal fragments that have escaped from a lysed bacterial cell are also commonly involved in the transfer of genetic information between cells. An important difference between plasmids and fragments is that while a plasmid has its own origin of replication and is stably replicated and inherited, chromosomal fragments must integrate themselves into the bacterial chromosome in order to be replicated and eventually passed to progeny cells. The process of genetic recombination is rare in nature, but its frequency can be increased in the laboratory, where the ability to shuffle genes between organisms is highly prized.

Depending upon the mode of transmission, the means of genetic recombination in bacteria is called conjugation, transformation, or transduction. **Conjugation** requires the attachment of two related species and the formation of a bridge that can transport DNA. **Transformation** entails the transfer of naked DNA and requires no special vehicle. **Transduction** is DNA transfer mediated through the action of a bacterial virus **(table 9.5).**

Conjugation: Bacterial "Sex"

Conjugation is a mode of genetic exchange in which a plasmid or other genetic material is transferred by a donor to a recipient cell via a direct connection **(figure 9.24).** Both gram-negative and gram-positive cells can conjugate. In gram-negative cells, the donor has a plasmid **(fertility,** or **F′ factor)** that allows the synthesis of a conjugative **pilus.** The recipient cell is a related species or genus that has a recognition site on its surface. A cell's role in conjugation is denoted by F$^+$ for the cell that has the F plasmid and by F$^-$ for the cell that lacks it. Contact is made when a pilus grows out from the F$^+$ cell, attaches to the surface of the F$^-$ cell, contracts, and draws the two cells together **(figure 9.24a;** see also figure 4.8). In both gram-positive and gram-negative cells, an opening is created between the connected cells, and the replicated DNA passes across from one cell to the other **(figure 9.24b).** Conjugation is a conservative process, in that the donor bacterium generally retains a copy of the genetic material being transferred.

TABLE 9.5	Types of Intermicrobial Exchange		
Mode	**Factors Involved**	**Direct or Indirect***	**Examples of Genes Transferred**
Conjugation	Donor cell with pilus Fertility plasmid in donor Both donor and recipient alive Bridge forms between cells to transfer DNA	Direct	Drug resistance; resistance to metals; toxin production; enzymes; adherence molecules; degradation of toxic substances; uptake of iron
Transformation	Free donor DNA (fragment) Live, competent recipient cell	Indirect	Polysaccharide capsule; unlimited with cloning techniques
Transduction	Donor is lysed bacterial cell Defective bacteriophage is carrier of donor DNA Live recipient cell of same species as donor	Indirect	Toxins; enzymes for sugar fermentation; drug resistance

*Direct means the donor and recipient are in contact during exchange; indirect means they are not.

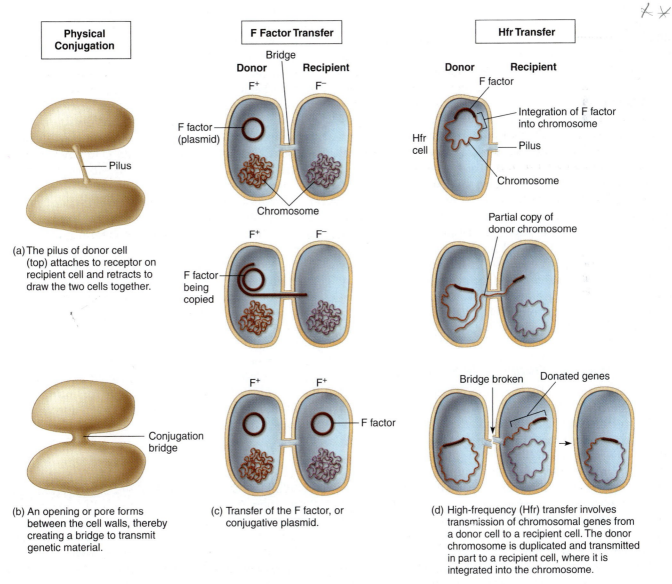

Physical Conjugation

(a) The pilus of donor cell (top) attaches to receptor on recipient cell and retracts to draw the two cells together.

Pilus

(b) An opening or pore forms between the cell walls, thereby creating a bridge to transmit genetic material.

Conjugation bridge

F Factor Transfer

Bridge
Donor   Recipient
F⁺   F⁻

F factor (plasmid)

Chromosome

F⁺   F⁻

F factor being copied

F⁺   F⁺

F factor

(c) Transfer of the F factor, or conjugative plasmid.

Hfr Transfer

Donor   Recipient
F factor

Hfr cell

Integration of F factor into chromosome

Pilus

Chromosome

Partial copy of donor chromosome

Bridge broken   Donated genes

(d) High-frequency (Hfr) transfer involves transmission of chromosomal genes from a donor cell to a recipient cell. The donor chromosome is duplicated and transmitted in part to a recipient cell, where it is integrated into the chromosome.

FIGURE 9.24 Conjugation: genetic transmission through direct contact between two cells.

There are hundreds of conjugative plasmids with some variations in their properties. One of the best understood plasmids is the F factor in *E. coli*, which exhibits these patterns of transfer:

1. The donor (F$^+$) cell makes a copy of its F factor and transmits this to a recipient (F$^-$) cell. The F$^-$ cell is thereby changed into an F$^+$ cell capable of producing a pilus and conjugating with other cells (**figure 9.24c**). No additional donor genes are transferred at this time.
2. In high-frequency recombination (Hfr) donors, the fertility factor has been integrated into the F$^+$ donor chromosome.

The term *high-frequency recombination* was adopted to denote that a cell with an integrated F factor transmits its chromosomal genes at a higher frequency than other cells.

The F factor can direct a more comprehensive transfer of part of the donor chromosome to a recipient cell. This transfer occurs through duplication of the DNA by means of the rolling circle mechanism. One strand of DNA is retained by the donor, and the other strand is transported across to the recipient cell (**figure 9.24d**). The F factor may not be transferred during this process. The transfer of an entire chromosome takes about 100 minutes, but the pilus bridge between cells is ordinarily broken before this time, and rarely is the entire genome of the donor cell transferred.

Conjugation has great biomedical importance. Special **resistance (R) plasmids,** or **factors,** that bear genes for resisting antibiotics and other drugs are commonly shared among bacteria through conjugation. Transfer of R factors can confer multiple resistance to antibiotics such as tetracycline, chloramphenicol, streptomycin, sulfonamides, and penicillin. This phenomenon is discussed further in chapter 12. Other types of R factors carry genetic codes for resistance to heavy metals (nickel and mercury) or for synthesizing virulence factors (toxins, enzymes, and adhesion molecules) that increase the pathogenicity of the bacterial strain. Conjugation studies have also provided an excellent way to map the bacterial chromosome.

Transformation: Capturing DNA from Solution

One of the cornerstone discoveries in microbial genetics was made in the late 1920s by the English biochemist Frederick Griffith working with *Streptococcus pneumoniae* and laboratory mice. The pneumococcus exists in two major strains based on the presence of the capsule, colonial morphology, and pathogenicity. Encapsulated strains bear a smooth (S) colonial appearance and are virulent; strains lacking a capsule have a rough (R) appearance and are nonvirulent. (Recall from chapter 4 that the capsule protects a bacterium from the phagocytic host defenses.) To set the groundwork, Griffith showed that when mice were injected with a live, virulent (S) strain, they soon died (**figure 9.25a**). Mice injected with a live, nonvirulent (R) strain remained alive and healthy (**figure 9.25b**). Next he tried a variation on this theme. First, he heat-killed an S strain and injected it into mice, which remained healthy (**figure 9.25c**). Then came the

ultimate test: Griffith injected both dead S cells and live R cells into mice, with the result that the mice died from pneumococcal blood infection (**figure 9.25d**). If killed bacterial cells do not come back to life and the nonvirulent live strain was harmless, why did the mice die? Although he did not know it at the time, Griffith had demonstrated that dead S cells, while passing through the body of the mouse, broke open and released some of their DNA (by chance, that part containing the genes for making a capsule). A few of the live R cells subsequently picked up this loose DNA and were transformed by it into virulent, capsule-forming strains.

Later studies supported the concept that a chromosome released by a lysed cell breaks into fragments small enough to be accepted by a recipient cell and that DNA, even from a dead cell, retains its genetic code. This nonspecific acceptance by a bacterial cell of small fragments of soluble DNA from the surrounding environment is termed *transformation*. Transformation is apparently facilitated by special DNA-binding proteins on the cell wall that capture DNA from the surrounding medium. Cells that are capable of accepting genetic material through this means are termed *competent*. The new DNA is processed by the cell membrane and transported into the cytoplasm, where some of it is inserted into the bacterial chromosome. Transformation is a natural event found in several groups of gram-positive and gram-negative bacterial species. In addition to genes coding for the capsule, bacteria also exchange genes for antibiotic resistance and bacteriocin synthesis in this way.

Because transformation requires no special appendages, and the donor and recipient cells do not have to be in direct contact, the process is useful for certain types of recombinant DNA technology. With this technique, foreign genes from a completely unrelated organism are inserted into a plasmid, which is then introduced into a competent bacterial cell through transformation. These recombinations can be carried out in a test tube, and human genes can be experimented upon and even expressed outside the human body by placing them in a microbial cell. This same phenomenon in eucaryotic cells, termed *transfection*, is an essential aspect of genetically engineered yeasts, plants, and mice, and it has been proposed as a future technique for curing genetic diseases in humans. These topics are covered in more detail in chapter 10.

IN THE NEWS (Continued from page 249)

The chapter opener described the first known case of infection with VRSA. Analysis of this VRSA strain isolated from the infected exit site of the patient's catheter indicated the possibility that this resistant *S. aureus* strain had acquired the gene for vancomycin resistance from the resistant strain of *E. faecalis* (VRE), isolated from the patient's chronic foot ulcer. The fact that the patient was treated with vancomycin before the VRSA infection allowed for selection of vancomycin-resistant organisms. Although

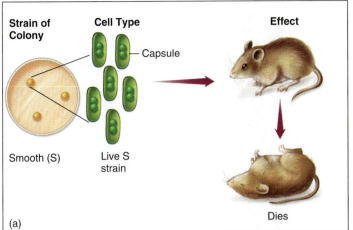

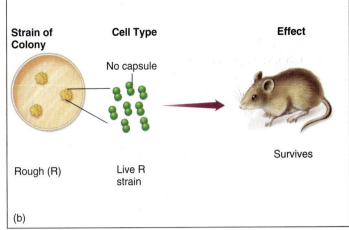

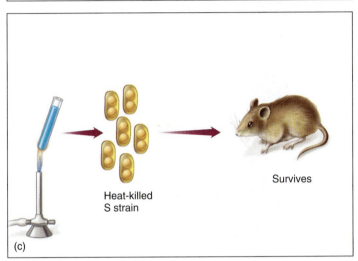

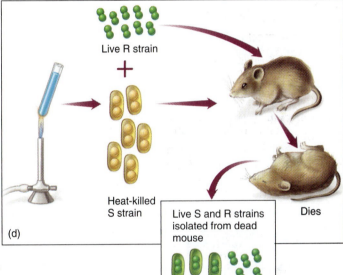

FIGURE 9.25 **Griffith's classic experiment in transformation.**
In essence, this experiment proved that DNA released from a killed cell can be acquired by a live cell. The cell receiving this new DNA is genetically transformed—in this case, from a nonvirulent strain to a virulent one.

prior studies showed that the transfer of the *van*A gene by conjugation of *Enterococcus* species with *S. aureus* was possible *in vitro* (under laboratory conditions), this case provided good evidence that this gene transfer could occur in a living patient.

See: Staphylococcus aureus *resistant to vancomycin—United States, 2002.* MMWR *51:565–566.*

Noble, W. C., Virani, Z., and Cree, R. G. 1992. Co-transfer of vancomycin and other resistance genes from Enterococcus faecalis *NCTC 12201 to* Staphylococcus aureus. *FEMS Microbiol. Lett.* 93:195–198.

Transduction: The Case of the Piggyback DNA

Bacteriophages (bacterial viruses) have been previously described as destructive bacterial parasites. Viruses can in fact serve as genetic vectors (an entity that can bring foreign DNA into a cell). The process by which a bacteriophage serves as the carrier of DNA from a donor cell to a recipient cell is transduction. Although it occurs naturally in a broad spectrum of bacteria, the participating bacteria in a single

transduction event must be the same species because of the specificity of viruses for host cells.

There are two versions of transduction. In *generalized transduction* **(figure 9.26),** random fragments of disintegrating host DNA are taken up by the phage during assembly. Virtually any gene from the bacterium can be transmitted through this means. In *specialized transduction* **(figure 9.27),** a highly specific part of the host genome is regularly incorporated into the virus. This specificity is explained by the prior existence of a temperate prophage inserted in a fixed site on the bacterial chromosome. When activated, the prophage DNA separates from the bacterial chromosome, carrying a small segment of host genes with it. During a lytic cycle, these specific viral-host gene combinations are incorporated into the viral particles and carried to another bacterial cell.

Several cases of specialized transduction have biomedical importance. The virulent strains of bacteria such as *Corynebacterium diphtheriae, Clostridium* spp., and *Streptococcus pyogenes* all produce toxins with profound physiological effects, whereas nonvirulent strains do not produce toxins. It turns

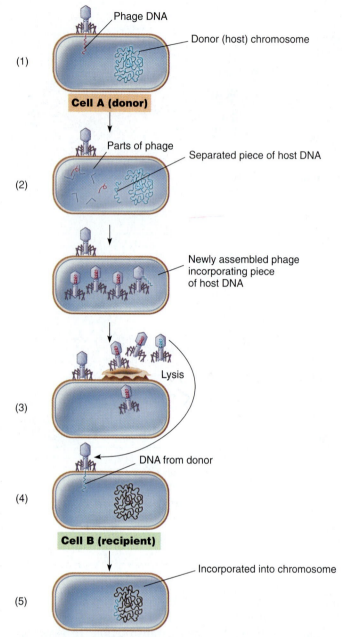

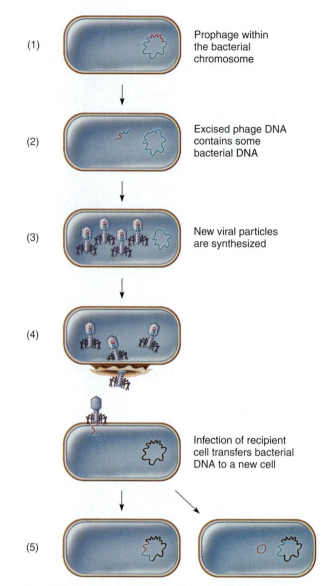

Recombination results in two possible outcomes.

FIGURE 9.26 Generalized transduction: genetic transfer by means of a virus carrier.
(1) A phage infects cell A (the donor cell) by normal means.
(2) During replication and assembly, a phage particle incorporates a segment of bacterial DNA by mistake. **(3)** Cell A then lyses and releases the mature phages, including the genetically altered one.
(4) The altered phage adsorbs to and penetrates another host cell (cell B), injecting the DNA from cell A rather than viral nucleic acid. **(5)** Cell B receives this donated DNA, which recombines with its own DNA. Because the virus is defective (biologically inactive as a virus), it is unable to complete a lytic cycle. The transduced cell survives and can use this new genetic material.

FIGURE 9.27 Specialized transduction: transfer of specific genetic material by means of a virus carrier.
(1) Specialized transduction begins with a cell that contains a prophage (a viral genome integrated into the host cell chromosome). **(2)** Rarely, the virus enters a lytic cycle and, as it excises itself from its host cell, inadvertently includes some bacterial DNA. **(3)** Replication and assembly results in production of a chimeric virus, containing some bacterial DNA. **(4)** Release of the recombinant virus and subsequent infection of a new host results in transfer of bacterial DNA between cells. **(5)** Recombination can occur between the bacterial chromosome and the virus DNA, resulting in either bacterial DNA or a combination of viral and bacterial DNA being incorporated into the bacterial chromosome.

out that toxicity arises from the expression of bacteriophage genes that have been introduced by transduction. Only those bacteria infected with a temperate phage are toxin formers. (Details of toxin action are discussed in the organ system–specific disease chapters.) Other instances of transduction are seen in staphylococcal transfer of drug resistance and in the transmission of gene regulators in gram-negative rods (*Escherichia, Salmonella*).

Transposons: "This Gene Is Jumpin'"   One type of genetic transferral of great interest involves transposable elements, or **transposons.** Transposons have the distinction of shifting from one part of the genome to another and so are termed "jumping genes." When the idea of their existence in corn plants was first postulated by the geneticist Barbara McClintock, it was greeted with some skepticism, because it had long been believed that the location of a given gene was set and that genes did not or could not move around. Now it is evident that jumping genes are widespread among procaryotic and eucaryotic cells and viruses.

All transposons share the general characteristic of traveling from one location to another on the genome—from one chromosomal site to another, from a chromosome to a plasmid, or from a plasmid to a chromosome **(figure 9.28).** Because transposons can occur in plasmids, they can also be transmitted from one cell to another in bacteria and a few eucaryotes. Some transposons replicate themselves before jumping to the next location, and others simply move without replicating first.

Transposons contain DNA that codes for the enzymes needed to remove and reintegrate the transposon at another site in the genome. Flanking the coding region of the DNA are sequences called inverted repeats, which mark the point at which the transposon is removed or reinserted into the genome. The smallest transposons consist of only these two genetic sequences and are often referred to as *insertion elements*. A type of transposon called *retrotransposon* can transcribe DNA into RNA, and then back into DNA for insertion in a new location. Other transposons contain additional genes that provide traits such as antibiotic resistance or toxin production.

The overall effect of transposons—to scramble the genetic language—can be beneficial or adverse, depending upon such variables as where insertion occurs in a chromosome, what kinds of genes are relocated, and the type of cell involved. In bacteria, transposons are known to be involved in

1. changes in traits such as colony morphology, pigmentation, and antigenic characteristics;
2. replacement of damaged DNA; and
3. the intermicrobial transfer of drug resistance (in bacteria).

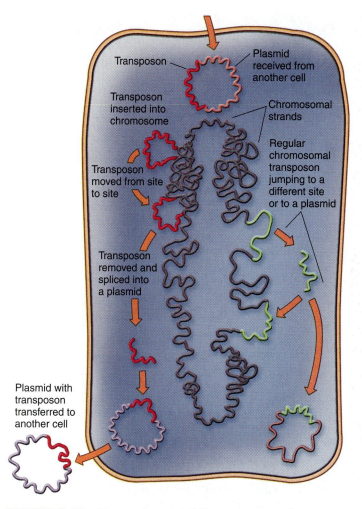

FIGURE 9.28   **Transposons: shifting segments of the genome.**
Potential mechanisms in the movement of transposons in bacterial cells. Some transposons are plasmids that are shifted from site to site; others are regular parts of chromosomes that are moved out of one site and spliced into another.

CHECKPOINT

- Changes in the genetic code can occur by two means: mutation and recombination. Mutation means a change in the nucleotide sequence of the organism's genome.
- Recombination means the addition of genes from an outside source, such as a virus or another cell.

- Mutations can be either spontaneous or induced by exposure to some external mutagenic agent.
- All cells have enzymes that repair damaged DNA. When the degree of damage exceeds the ability of the enzymes to make repairs, mutations occur.
- Mutation-induced changes in DNA nucleotide sequencing range from a single nucleotide to addition or deletion of large sections of genetic material.
- The Ames test is used to identify potential carcinogens on the basis of their ability to cause back-mutations in bacteria.
- Genetic recombination occurs in eucaryotes through sexual reproduction. In bacteria, recombination occurs through the processes of transformation, conjugation, and transduction.
- Transposons are genes that can relocate from one part of the genome to another, causing rearrangement of genetic material. Such rearrangements have either beneficial or harmful consequences for the organism involved.

Chapter Summary With Key Terms

9.1 Introduction to Genetics and Genes: Unlocking the Secrets of Heredity

A. **Genetics** is the study of **heredity** and can be studied at the level of the organism, **genome, chromosome, gene,** and **DNA.** Genes provide the information needed to construct **proteins**, which have structural or catalytic functions in the cell.

B. **DNA** is a long molecule in the form of a double helix. Each strand of the helix consists of a string of **nucleotides** which form hydrogen bonds with their counterparts on the other strand. **Adenine** base pairs with **thymine** while **guanine** base pairs with **cytosine.** The order of the nucleotides specifies which amino acids will be used to construct proteins during the process of translation.

C. DNA **replication** is **semiconservative** and requires the participation of several enzymes.

9.2 Applications of the DNA Code: Transcription and Translation

A. DNA is used to produce RNA **(transcription)** and RNA is then used to produce protein **(translation).**

B. RNA: Unlike DNA, RNA is single stranded, contains **uracil** instead of thymine and **ribose** instead of **deoxyribose.**
1. Major forms of RNA found in the cell include **mRNA, tRNA,** and **rRNA.**
2. The genetic information contained in DNA is copied to produce an RNA molecule. **Codons** in the mRNA pair with **anticodons** in the tRNA to specify what amino acids to assemble on the ribosome during translation.

C. *Transcription and Translation:* Transcription occurs when **RNA polymerase** copies the template strand of a segment of DNA. RNA is always made in the 5′ to 3′ direction. Translation occurs when the mRNA is used to direct the synthesis of proteins on the ribosome. Codons in the mRNA pair with anticodons in the tRNA to assemble a string of amino acids. This occurs until a stop codon is reached.

D. *Eucaryotic Gene Expression:* Eucaryotic genes are composed of **exons** (expressed sequences) and **introns** (intervening sequences). The introns must be removed and the exons spliced together to create the final mRNA.

E. The genetics of viruses is quite diverse.
1. Genomes of viruses are found in many physical forms not seen in cells, including dsDNA, ssDNA, dsDNA, and ssDNA.
2. DNA viruses tend to replicate in the nucleus while RNA viruses replicate in the cytoplasm. Retroviruses synthesize dsDNA from ssRNA.

9.3 Genetic Regulation of Protein Synthesis and Metabolism

Protein synthesis is regulated through gene induction or repression, as controlled by an **operon.** Operons consist of several structural genes controlled by a common regulatory element.

A. *Inducible operons* such as the lactose operon are normally off but can be turned on by a lactose inducer.

B. *Repressible operons* are usually on but can be turned off when their end product is no longer needed.

C. Many antibiotics prevent bacterial growth by interfering with transcription or translation.

9.4 Mutations: Changes in the Genetic Code

A. Permanent changes in the genome of a microorganism are known as **mutations.** Mutations may be **spontaneous** or **induced.**

B. **Point mutations** entail a change in one or a few bases and are categorized as **missense-, nonsense-, silent-,** or **back-mutations,** based on the effect of the change in nucleotide(s).

C. Many mutations, particularly those involving mismatched bases or damage from ultraviolet light, can be corrected using enzymes found in the cell.

D. The **Ames test** measures the mutagenicity of chemicals by determining the ability of a chemical to induce mutations in bacteria.

9.5 DNA Recombination Events

A. Intermicrobial transfer and genetic recombination permit gene sharing between bacteria. Major types of recombination include **conjugation, transformation,** and **transduction.**

B. **Transposons** are DNA sequences that regularly move to different places within the genome of a cell, as a consequence generating mutations and variations in chromosome structure.

Multiple-Choice Questions

1. What is the smallest unit of heredity?
   a. chromosome
   b. gene
   c. codon
   d. nucleotide

2. A nucleotide contains which of the following?
   a. 5 C sugar
   b. nitrogen base
   c. phosphate
   d. b and c only
   e. all of these

3. The nitrogen bases in DNA are bonded to the
   a. phosphate
   b. deoxyribose
   c. ribose
   d. hydrogen

4. DNA replication is semiconservative because the _____ strand will become half of the _____ molecule.
   a. RNA, DNA
   b. template, finished
   c. sense, mRNA
   d. codon, anticodon

5. In DNA, adenine is the complementary base for _____, and cytosine is the complement for _____.
   a. guanine, thymine
   b. uracil, guanine
   c. thymine, guanine
   d. thymine, uracil

6. The base pairs are held together primarily by
   a. covalent bonds
   c. ionic bonds
   b. hydrogen bonds
   d. gyrases

7. Why must the lagging strand of DNA be replicated in short pieces?
   a. because of limited space
   b. otherwise, the helix will become distorted
   c. the DNA polymerase can synthesize in only one direction
   d. to make proofreading of code easier

8. Messenger RNA is formed by _____ of a gene on the DNA template strand.
   a. transcription
   c. translation
   b. replication
   d. transformation

9. Transfer RNA is the molecule that
   a. contributes to the structure of ribosomes
   b. adapts the genetic code to protein structure
   c. transfers the DNA code to mRNA
   d. provides the master code for amino acids

10. As a general rule, the template strand on DNA will always begin with
    a. TAC
    c. ATG
    b. AUG
    d. UAC

11. The *lac* operon is usually in the _____ position and is activated by a/an _____ molecule.
    a. on, repressor
    c. on, inducer
    b. off, inducer
    d. off, repressor

12. For mutations to have an effect on populations of microbes, they must be
    a. inheritable
    d. a and b
    b. permanent
    e. all of the above
    c. beneficial

13. Which of the following characteristics is *not* true of a plasmid?
    a. It is a circular piece of DNA.
    b. It is required for normal cell function.
    c. It is found in bacteria.
    d. It can be transferred from cell to cell.

14. Which genes can be transferred by all three methods of intermicrobial transfer?
    a. capsule production
    c. F factor
    b. toxin production
    d. drug resistance

15. Which of the following would occur through specialized transduction?
    a. acquisition of Hfr plasmid
    b. transfer of genes for toxin production
    c. transfer of genes for capsule formation
    d. transfer of a plasmid with genes for degrading pesticides

16. **Multiple Matching.** Fill in the blanks with all the letters of the words below that apply.
    e/h genetic transfer that occurs after the donor is dead
    f carries the codon
    b carries the anticodon
    g a process synonymous with mRNA synthesis
    e bacteriophages participate in this transfer
    a duplication of the DNA molecule
    i process in which transcribed DNA code is deciphered into a polypeptide
    c/e/h involves plasmids
    a. replication
    f. mRNA
    b. tRNA
    g. transcription
    c. conjugation
    h. transformation
    d. ribosome
    i. translation
    e. transduction
    j. none of these

Concept Questions

These questions are suggested as a *writing-to-learn* experience. For each question, compose a one- or two-paragraph answer that includes the factual information needed to completely address the question.

1. Compare the genetic material of eucaryotes, bacteria, and viruses in terms of general structure, size, and mode of replication.

2. Briefly describe how DNA is packaged to fit inside a cell.

3. Describe what is meant by the antiparallel arrangement of DNA.

4. On paper, replicate the following segment of DNA:

   5′ A T C G G C T A C G T T C A C 3′
   3′ T A G C C G A T G C A A G T G 5′

   a. Show the direction of replication of the new strands and explain what the lagging and leading strands are.
   b. Explain how this is semiconservative replication. Are the new strands identical to the original segment of DNA?

5. Name several characteristics of DNA structure that enable it to be replicated with such great fidelity generation after generation.

6. Explain the following relationship: DNA formats RNA, which makes protein.

7. What message does a gene provide? How is the language of the gene expressed?

8. If a protein is 3,300 amino acids long, how many nucleotide pairs long is the gene sequence that codes for it?

9. Compare the structure and functions of DNA and RNA.

10. a. Where does transcription begin?
    b. What are the template and coding strands of DNA?
    c. Why is only one strand transcribed, and is the same strand of DNA always transcribed?

11. Compare and contrast the actions of DNA and RNA polymerase.

12. What are the functions of start and stop codons? Give examples of them.

13. The following sequence represents triplets on DNA:

    TAC CAG ATA CAC TCC CCT GCG ACT

    a. Give the mRNA codons and tRNA anticodons that correspond with this sequence, and then give the sequence of amino acids in the polypeptide.
    b. Provide another mRNA strand that can be used to synthesize this same protein.
    c. Looking at figure 9.14, give the type and order of the amino acids in the peptide.

14. a. Summarize how bacterial and eucaryotic cells differ in gene structure, transcription, and translation.
    b. Discuss the roles of exons and introns.

15. a. What is an operon? Describe the functions of regulators, promoters, and operators.
    b. Compare and contrast the *lac* operon with a repressible operon system.

16. Explain the ideas behind the Ames test and what it is used for.

17. Describe the principal types of mutations. Give an example of a mutation that is beneficial and one that is lethal or harmful.

18. a. Compare conjugation, transformation, and transduction on the basis of general method, nature of donor, and nature of recipient.
    b. Explain the differences between general and specialized transduction, using drawings.

19. By means of a flowchart, show the possible jumps that a transposon can make. Show the involvement of viruses in its movement.

Critical Thinking Questions

Critical thinking is the ability to reason and solve problems using facts and concepts. These questions can be approached from a number of angles, and in most cases, they do not have a single correct answer.

1. Knowing that retroviruses operate on the principle of reversing the direction of transcription from RNA to DNA, propose a drug that might possibly interfere with their replication.

2. Using the piece of DNA in concept question 13, show a deletion, an insertion, a substitution, and nonsense mutations. Which ones are frameshift mutations? Are any of your mutations nonsense? Missense? (Use the universal code to determine this.)

3. Using figure 9.14 and table 9.4, go through the steps in mutation of a codon followed by its transcription and translation that will give the end result in silent, missense, and nonsense mutations. Why is a change in the RNA code alone not really a mutation?

4. Explain the principle of "wobble" and find four amino acids that are encoded by wobble bases (figure 9.14). Suggest some benefits of this phenomenon to microorganisms.

5. The enzymes required to carry out transcription and translation are themselves produced through these same processes. Speculate which may have come first in evolution—proteins or nucleic acids—and explain your choice.

6. Why can one not reliably predict the sequence of nucleotides on mRNA or DNA by observing the amino acid sequence of proteins?

7. Speculate on the manner in which transposons may be involved in cancer.

8. Explain what is meant by the expression: phenotype = genotype + environment.

Try this
A simple test you can do to demonstrate the coiling of DNA in bacteria is to open a large elastic band, stretch it taut, and twist it. First it will form a loose helix, then a tighter helix, and finally, to relieve stress, it will twist back upon itself. Further twisting will result in a series of knotlike bodies; this is how bacterial DNA is condensed.

Internet Search Topics

1. Do an Internet search under the heading "DNA music." Explore several websites, discovering how this music is made and listening to some examples.

2. Find information on introns. Explain at least five current theories as to their possible functions.

3. Go to the Online Learning Center for chapter 9 of this text at http://www.mhhe.com/cowan1. Access the URLs listed under Internet Search Topics.

   Log on to one or more of the listed URLs and locate animations, three-dimensional graphics, and interactive tutorials that help you to visualize replication, transcription, and translation.

Genetic Engineering
A Revolution in Molecular Biology

Japan, March 20, 1995. Five men carrying innocent-looking plastic bags boarded five separate subway trains bound for the Kasumigaseki Station in Tokyo. A few stops before the station, they quietly placed their bags on the floor and jabbed them with their umbrellas. Within an hour, over 5,000 commuters were exposed to sarin nerve gas. Twelve commuters died. Considering it was peak rush hour, it is amazing so few were killed. The Japanese emergency responders were able to identify the poison as sarin gas within hours.

What Is Sarin?

Sarin gas is one member of the organophosphate family. Organophosphates inhibit acetylcholinesterase activity throughout the body. As a result, acetylcholine builds up at nerve endings of cholinergic synapses, causing effects ranging from constricted pupils and profuse sweating to asphyxiation due to the complete constriction of respiratory muscles.

Organophosphates were first developed in 1936. As World War II ended and the cold war began, the superpowers embarked on development of more lethal forms of this compound to use as the chemical weapons commonly referred to as nerve agents. During this same period, milder forms were also developed and used as insecticide. Most developed nations have banned the use of organophosphate insecticides due to their toxicity to nontarget species including humans. Scientists and environmentalists alike recognized the need for developing a method of detoxification and cleanup for these compounds.

How can an entire subway station be cleaned so that no sarin remains? At least two subway workers died as a result of misguided efforts to clean up the poison in the 1995 incident; they attempted to soak up the gas (turned liquid) with newspapers. There are chemical solvents that will clean the sarin from the station, but they are almost as toxic as the nerve gas. What is needed is a nontoxic method to degrade the sarin into safe and chemically stable compounds.

As far back as 1972, a species of *Flavobacterium* was discovered that degrades organophosphates. Since then, several more species from different genera and geographical regions have been isolated that also have enzymes that degrade organophosphate. Unfortunately these organisms have proven difficult to culture in sufficient quantities for bioremediation.

▶ *Can you think of how the organophosphate-degrading activity from these bacteria can be used, even if they themselves can't grow in the area needing to be cleaned up?*

CHAPTER OVERVIEW

▶ **Genetic engineering**[1] deals with the ability of scientists to manipulate DNA through the use of an expanding repertoire of recombinant DNA techniques. These techniques allow DNA to be cut, separated by size, and even sequenced to determine the actual composition and order of nucleotides.

▶ Biotechnology, the use of an organism's biochemical processes to create a product, has been greatly advanced by the introduction of recombinant DNA technologies because organisms can now be genetically modified to accomplish goals that were previously impossible, such as bacteria that have been engineered to produce human insulin.

▶ Differences in DNA between organisms are being exploited so that they can be used to identify people, animals, and microbes, revolutionizing fields such as police work and epidemiology.

▶ Genes can be probed to predict the likelihood of a particular genetic disease long before the disease strikes (and becomes less treatable). In a growing number of cases, missing or mutated genes are being replaced to correct inherited defects.

▶ Recombinant DNA technology is used to create genetically modified food that may have increased nutritional properties, be easier to grow, or may even vaccinate the person eating it against a host of diseases.

▶ Recombinant DNA techniques have given humans greater power over our own and other species than we've ever had before. Such power is not without risks, and **bioethics**[2] is an important part of any discussion of bioengineering.

10.1 Basic Elements and Applications of Genetic Engineering

In chapter 9 we looked at the ways in which microorganisms duplicate, exchange, and use their genetic information. In scientific parlance this is called *basic science*, because no product or application is directly derived from it. Human beings being what they are, however, it is never long before basic knowledge is used to derive *applied science*, or useful products and applications that owe their invention to the basic research that preceded them. As an example, basic science into the workings of the electron has led to the development of television, computers, and cell phones. None of these staples of modern life were envisioned when early physicists were deciphering the nature of subatomic particles, but without the knowledge of how electrons worked, our ability to harness them for our own uses would never have materialized.

The same scenario can be seen with regard to genetics. The knowledge of how DNA was manipulated within the cell to carry out the goals of a microbe allowed scientists to utilize these processes to accomplish goals more to the liking of human beings. Contrary to being a new idea, the methods of genetic manipulation we will review are simply more efficient ways of accomplishing goals that humans have had for thousands of years.

Examples of human goals that have been more efficiently attained through the use of modern genetic technologies can be seen in each of these scenarios:

1. A farmer mates his two largest pigs in the hopes of producing larger offspring. Unfortunately he quite often ends up with small or unhealthy animals due to other genes that are transferred during mating. Genetic manipulation allows for the transfer of specific genes, so that only advantageous traits are selected.

2. Courts have, for thousands of years, relied on a description of a person's phenotype (eye color, hair color, etc.) as a means of identification. By remembering that a phenotype is the product of a particular sequence of DNA, you can quickly see how looking at someone's DNA (perhaps from a drop of blood) gives a clue as to his or her identification.

3. Lastly, many diseases are the result of a missing or dysfunctional protein, and we have generally treated the disease by replacing the protein as best we can, usually resulting in only temporary relief and limited success. Examples include insulin-dependent diabetes, adenosine deaminase deficiency, and blood clotting disorders. Genetic engineering offers the promise that someday soon fixing the underlying mutation responsible for the lack of a particular protein can treat these diseases far more successfully than we've been able to do in the past.

Information on genetic engineering and its biotechnological applications is growing at such an expanding rate that some new discovery or product is disclosed almost on a daily basis. To keep this subject somewhat manageable, we will present essential concepts and applications, organized under the following six topics:

- Tools and Techniques of Genetic Engineering
- Methods in Recombinant DNA Technology
- Biochemical Products of Recombinant DNA Technology
- Genetically Modified Organisms
- Genetic Treatments
- Genome Analysis

---

1. Sometimes genetic engineering is called bioengineering—defined as the direct, deliberate modification of an organism's genome. Biotechnology is a more encompassing term that includes the use of DNA, genes, or genetically altered organisms in commercial production.

2. A field that relates biological issues to human conduct and moral judgment.

10.2 Tools and Techniques of Genetic Engineering

DNA: An Amazing Molecule

All of the intrinsic properties of DNA hold true whether the DNA is in a bacterium or a test tube. For example, the enzyme helicase is able to unwind the two strands of the double helix just as easily in the lab as it does in a bacterial cell. But in the laboratory we can take advantage of our knowledge of DNA chemistry to make helicase unnecessary. It turns out that when DNA is heated to just below boiling (90°C to 95°C), the two strands separate, revealing the information contained in their bases. With the nucleotides exposed, DNA can be easily identified, replicated, or transcribed. If heat-denatured DNA is then slowly cooled, complementary nucleotides will hydrogen bond with one another, and the strands will renature, or regain their familiar double-stranded form **(figure 10.1a)**. As we shall see, this process is a necessary feature of the polymerase chain reaction and nucleic acid probes described later.

Enzymes for Dicing, Splicing, and Reversing Nucleic Acids

The polynucleotide strands of DNA can also be clipped crosswise at selected positions by means of enzymes called **restriction endonucleases**.[3] These enzymes originate in bacterial cells. They recognize foreign DNA and are capable of breaking the phosphodiester bonds between adjacent nucleotides on both strands of DNA, leading to a break in the DNA strand. In the bacterial cell, this ability protects against the incompatible DNA of bacteriophages or plasmids. In the biotechnologist's lab, the enzymes can be used to cleave DNA at desired sites and are a must for the techniques of recombinant DNA technology.

So far, hundreds of restriction endonucleases have been discovered in bacteria. They function to protect bacteria from foreign DNA such as that from phages. Each type has a known sequence of 4 to 10 base pairs as its target, so sites of cutting can be finely controlled. These enzymes have the unique property of recognizing and clipping at base sequences called **palindromes (figure 10.1b)**. Palindromes are sequences of DNA that are identical when read from the 5′ to 3′ direction on one strand and the 5′ to 3′ direction on the other strand.

Endonucleases are named by combining the first letter of the bacterial genus, the first two letters of the species, and the endonuclease number. Thus, *Eco*RI is the first endonuclease found in *Escherichia coli* (in the R strain), and *Hind*III is the third endonuclease discovered in *Haemophilus influenzae* Type d (figure 10.1b).

Endonucleases are used in the laboratory to cut DNA into smaller pieces for further study as well as to remove and insert it during recombinant DNA techniques, described in a subsequent section. Endonucleases such as *Hae*III make

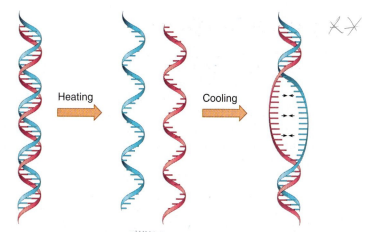

(a) DNA heating and cooling. DNA responds to heat by denaturing—losing its hydrogen bonding, and thereby separating into its two strands. When cooled, the two strands rejoin at complementary sites. The two strands need not be from the same organisms as long as they have matching sites.

Endonuclease	*Eco*RI	*Hind*III	*Hae*III
Cutting pattern	G A A T T C / C T T A A G	A A G C T T / T T C G A A	G G C C / C C G G

(b) Examples of palindromes and cutting patterns.

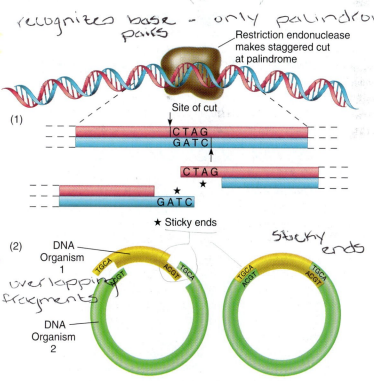

(c) Action of restriction endonucleases. (1) A restriction endonuclease recognizes and cleaves DNA at the site of a specific palindromic sequence. Cleavage can produce staggered tails called <u>sticky ends</u> that accept complementary tails for gene splicing. (2) The sticky ends can be used to join DNA from different organisms by cutting it with the same restriction enzyme, ensuring that all fragments have complementary ends.

FIGURE 10.1   Some useful properties of DNA.

---

3. The meaning of restriction is that the enzymes do not act upon the bacterium's own DNA; an endonuclease nicks DNA internally, not at the ends.

straight, blunt cuts on DNA. But more often the enzymes make staggered symmetrical cuts that leave short tails called "sticky ends." Such adhesive tails will base-pair with complementary tails on other DNA fragments or plasmids. This effect makes it possible to splice genes into specific sites.

The pieces of DNA produced by restriction endonucleases are termed *restriction fragments*. Because DNA sequences vary, even among members of the same species, differences in the cutting pattern of specific restriction endonucleases give rise to restriction fragments of differing lengths, known as *restriction fragment length polymorphisms* (RFLPs). RFLPs allow the direct comparison of the DNA of two different organisms at a specific site, which, as we will see, has many uses.

Another enzyme, called a **ligase**, is necessary to seal the sticky ends together by rejoining the phosphate-sugar bonds cut by endonucleases. Its main application is in final splicing of genes into plasmids and chromosomes.

An enzyme called **reverse transcriptase** is best known for its role in the replication of the AIDS virus and other retroviruses. It also provides geneticists with a valuable tool for converting RNA into DNA. Copies called **complementary DNA, or cDNA,** can be made from messenger, transfer, ribosomal, and other forms of RNA. The technique provides a valuable means of synthesizing eucaryotic genes from mRNA transcripts. The advantage is that the synthesized gene will be free of the intervening sequences (introns) that can complicate the management of eucaryotic genes in genetic engineering. Complementary DNA can also be used to analyze the nucleotide sequence of RNAs, such as those found in ribosomes and transfer RNAs.

Analysis of DNA

One way to produce a readable pattern of DNA fragments is through **gel electrophoresis.** In this technique, samples are placed in compartments (wells) in a soft agar gel and subjected to an electrical current. The phosphate groups in DNA give the entire molecule an overall negative charge, which causes the DNA to move toward the positive pole in the gel. The rate of movement is based primarily on the size of the fragments. The larger fragments move more slowly and remain nearer the top of the gel, whereas the smaller fragments migrate faster and are positioned farther from the wells. The positions of DNA fragments are determined by staining the DNA fragments in the gel **(figure 10.2).** Electrophoresis patterns can be quite distinctive and are very useful in characterizing DNA fragments and comparing the degree

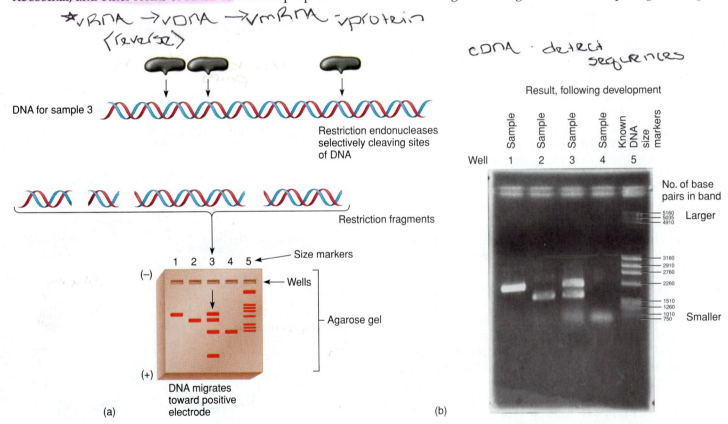

FIGURE 10.2   Revealing the patterns of DNA with electrophoresis.
(a) After cleavage into fragments, DNA is loaded into wells on one end of an agarose gel. When an electrical current is passed through the gel (from the negative pole to the positive pole), the DNA, being negatively charged, migrates toward the positive pole. The larger fragments, measured in numbers of base pairs, migrate more slowly and remain nearer the wells than the smaller (shorter) fragments. **(b)** An actual developed and stained gel reveals a separation pattern of the fragments of DNA. The size of a given DNA band can be determined by comparing it to a known set of markers (lane 5) called a ladder. This method can be used for general screening of DNA or for genetic fingerprints.

of genetic similarities among samples as in a genetic fingerprint (discussed later).

Nucleic Acid Hybridization and Probes

Two different nucleic acids can **hybridize** by uniting at their complementary sites. All different combinations are possible: Single-stranded DNA can unite with other single-stranded DNA or RNA, and RNA can hybridize with other RNA. This property has been the inspiration for specially formulated oligonucleotide tracers called **gene probes.** These probes consist of a short stretch of DNA of a known sequence that will base-pair with a stretch of DNA with a complementary sequence, if one exists in the test sample. Hybridization probes have practical value because they can detect specific nucleotide sequences in unknown samples. So that areas of hybridization can be visualized, the probes carry reporter molecules such as radioactive labels, which are isotopes that emit radiation, or luminescent labels, which give off visible light. Reactions can be revealed by placing photographic film in contact with the test reaction. Fluorescent probes contain dyes that can be visualized with ultraviolet radiation, and enzyme-linked probes react with substrate to release colored dyes.

When probes hybridize with an unknown sample of DNA or RNA, they tag the precise area and degree of hybridization and help determine the nature of nucleic acid present in a sample. In a method called the **Southern blot,**[4] DNA fragments are first separated by electrophoresis and then denatured and transferred to a special filter. A DNA probe is then incubated with the sample, and wherever this probe encounters the segment for which it is complementary, it will attach and form a hybrid. Development of the hybridization pattern will show up as one or more bands **(figure 10.3).** This

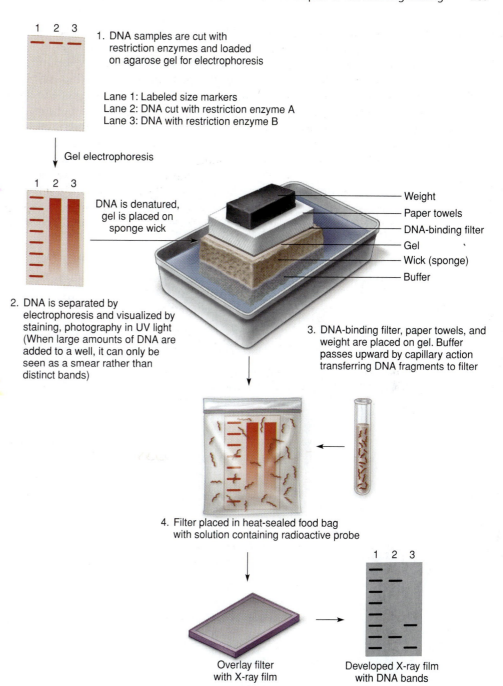

1. DNA samples are cut with restriction enzymes and loaded on agarose gel for electrophoresis

Lane 1: Labeled size markers
Lane 2: DNA cut with restriction enzyme A
Lane 3: DNA with restriction enzyme B

Gel electrophoresis

DNA is denatured, gel is placed on sponge wick

Weight
Paper towels
DNA-binding filter
Gel
Wick (sponge)
Buffer

2. DNA is separated by electrophoresis and visualized by staining, photography in UV light (When large amounts of DNA are added to a well, it can only be seen as a smear rather than distinct bands)

3. DNA-binding filter, paper towels, and weight are placed on gel. Buffer passes upward by capillary action transferring DNA fragments to filter

4. Filter placed in heat-sealed food bag with solution containing radioactive probe

Overlay filter with X-ray film

Developed X-ray film with DNA bands

5. Filter is washed to remove excess probe, dried, film is exposed to produce photographic image of DNA bands

FIGURE 10.3 **Conducting a Southern blot hybridization test.**
The five-step process reveals the hybridized fragments on the X-ray film.

method is a sensitive and specific way to isolate fragments from a complex mixture and to find specific gene sequences on DNA. Southern blot is also one of the important first steps for preparing isolated genes.

Probes are commonly used for diagnosing the cause of an infection from a patient's specimen and identifying a culture of an unknown bacterium or virus. The method for doing this was first outlined in chapter 4. A simple and rapid

---

4. Named for its developer, E. M. Southern. The "northern" blot is a similar method used to analyze RNA, while the western blot detects proteins. There is a reason scientists became scientists and not comedians.

method called a hybridization test does not require electrophoresis. DNA from a test sample is isolated, denatured, placed on an absorbent filter, and combined with a microbe-specific probe **(figure 10.4)**. The blot is then developed and observed for areas of hybridization. Commercially available diagnostic kits are now on the market for identifying intestinal pathogens such as *E. coli*, *Salmonella*, *Campylobacter*, *Shigella*, *Clostridium difficile*, rotaviruses, and adenoviruses.

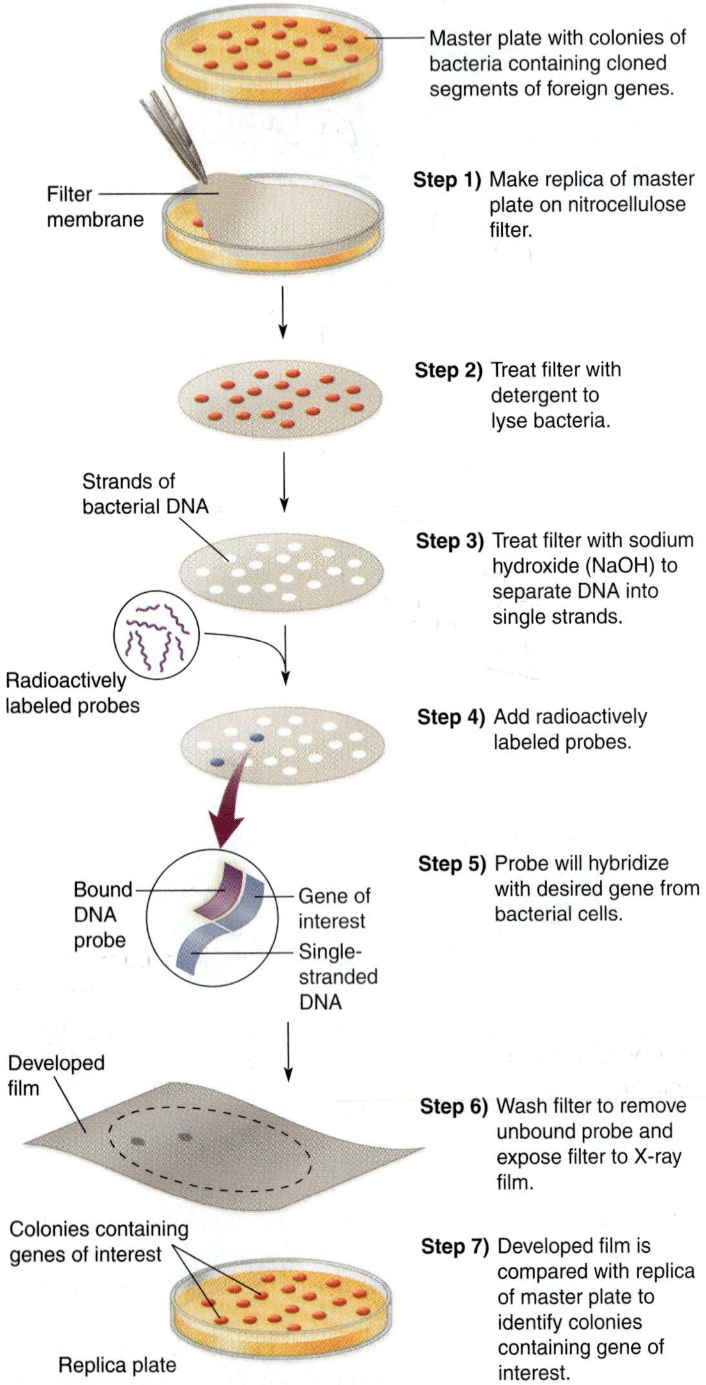

Master plate with colonies of bacteria containing cloned segments of foreign genes.

Filter membrane

Step 1) Make replica of master plate on nitrocellulose filter.

Step 2) Treat filter with detergent to lyse bacteria.

Strands of bacterial DNA

Step 3) Treat filter with sodium hydroxide (NaOH) to separate DNA into single strands.

Radioactively labeled probes

Step 4) Add radioactively labeled probes.

Bound DNA probe

Gene of interest

Single-stranded DNA

Step 5) Probe will hybridize with desired gene from bacterial cells.

Developed film

Step 6) Wash filter to remove unbound probe and expose filter to X-ray film.

Colonies containing genes of interest

Step 7) Developed film is compared with replica of master plate to identify colonies containing gene of interest.

Replica plate

FIGURE 10.4 **A hybridization test relies on the action of microbe-specific probes to identify an unknown bacteria or virus.**

Other bacterial probes exist for *Mycobacterium*, *Legionella*, *Mycoplasma, and Chlamydia;* viral probes are available for herpes simplex and zoster, papilloma (genital warts), hepatitis A and B, and AIDS. DNA probes have also been developed for human genetic markers and some types of cancer.

With another method, called *fluorescent in situ hybridization* (FISH), probes are applied to intact cells and observed microscopically for the presence and location of specific genetic marker sequences on genes. It is a very effective way to locate genes on chromosomes. In situ techniques can also be used to identify unknown bacteria living in natural habitats without having to culture them, and they can be used to detect RNA in cells and tissues.

Methods Used to Size, Synthesize, and Sequence DNA

The relative sizes of nucleic acids are usually denoted by the number of base pairs (bp) or nucleotides they contain. For example, the palindromic sequences recognized by endonucleases are usually 4 to 10 bp in length, an average gene in *E. coli* is approximately 1,300 bp, or 1.3 kilobases (kb), and its entire genome is approximately 4,700,000 bp, 4,700 kb, or 4.7 megabases (Mb). The DNA of the human mitochondrion contains 16 kb, and the Epstein-Barr virus (a cause of infectious mononucleosis) has 172 kb. Humans have approximately 3.1 billion base pairs (Bb) arrayed along 46 chromosomes. The recently completed Human Genome Project had as its goal the elucidation of the entire human genome (see **Insight 10.1**).

DNA Sequencing: Determining the Exact Genetic Code   Analysis of DNA by its size, restriction patterns, and hybridization characteristics is instructive, but the most detailed information comes from determining the actual order and types of bases in DNA. This process, called **DNA sequencing**, provides the identity and order of nucleotides for all types of DNA, including genomic, cDNA, artificial chromosomes, plasmids, and cloned genes. The most common sequencing technique was developed by Frederick Sanger and is based on the synthesis and analysis of a complementary strand of DNA in a test tube **(figure 10.5)**.

Because most DNA being sequenced is very long, it is made more manageable by cutting it into a large number of shorter fragments and separated. The test strands, typically several hundred nucleotides long, are then denatured to expose single strands that will serve as templates to synthesize complementary strands. The fragments are divided into four separate tubes that contain primers to set the start point for the synthesis to begin on one of the strands. The primer is labeled with a fluorescent or radioactive tag, which allows it to be detected. The nucleotides will attach at the 3′ end of the primer, using the template strands as a guide, essentially the same as regular DNA synthesis as performed in a cell.

The tubes are incubated with the necessary DNA polymerase and all four of the regular nucleotides needed to carry out the process of elongating the complementary strand. Each tube also contains a single type (A, T, G, or C) of

(1) Isolated unknown DNA fragment.

Original DNA to be sequenced

(2) DNA is denatured to produce single template strand.

(3) Strand is labeled with specific primer molecule.

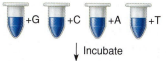

(4) DNA polymerase and regular nucleotide mixture (ATP, CTP, GTP, and TTP) are added; ddG, ddA, ddC, and ddT are placed in separate reaction tubes with the regular nucleotides. The dd nucleotides are labeled with some type of tracer, which allows them to be visualized.

+G   +C   +A   +T

↓ Incubate

(5) Newly replicated strands are terminated at the point of addition of a dd nucleotide.

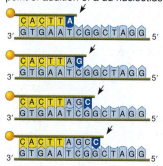

(6) A schematic view of how all possible positions on the fragment are occupied by a labeled nucleotide.

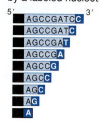

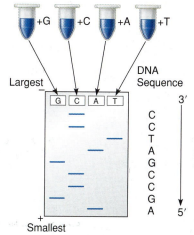

(7) Running the reaction tubes in four separate gel lanes separates them by size and nucleotide type. Reading from bottom to top, one base at a time, provides the correct DNA sequence.

FIGURE 10.5   Steps in a Sanger DNA sequence technique.

dideoxynucleotide (dd), which is critical to the sequencing process. Dideoxynucleotides have no oxygen bound to the 3' carbon of ribose. This oxygen atom is needed for the chemical attachment of the next nucleotide in the growing DNA strand. As the reaction proceeds, strands elongate by adding normal nucleotides, but a small percentage of fragments will randomly incorporate the complementary dd nucleotide and be terminated. Eventually, all possible positions in the sequence will incorporate a terminal dideoxynucleotide, thus producing a series of strands that reflect the correct sequence.

The reaction products are placed into four wells (G, C, A, T) of a polyacrylamide gel, which is sensitive enough to separate strands that differ by only a single nucleotide in length. Electrophoresis separates the fragments in order according to both size and lane, and only the fragments carrying the labeled nucleotides will be readable on the gel. The gel indicates the comparative orientation of the bases, which graduate in size from the smallest fragments that terminate early and migrate farthest (bottom of gel) to successively longer fragments (moving stepwise to the top). Reading the

order of first appearance of a given gel fragment in the G, C, A, or T lanes provides the correct sequence of bases of the complementary strand, and it allows one to infer the sequence of the template strand as well. This method of sequencing is remarkably accurate, with only about one mistake in every 1,000 bases. All of these steps have been automated so that most researchers need only insert a sample into a machine and wait for the printout of the results.

Automation of this process was absolutely necessary for sequencing the entire genomes of humans, mice, and other organisms (see Insight 10.1).

Polymerase Chain Reaction: A Molecular Xerox Machine for DNA

Some of the techniques used to analyze DNA and RNA are limited by the small amounts of test nucleic acid available. This problem was largely solved by the invention of a simple, versatile way to amplify DNA called the **polymerase chain reaction (PCR).** This technique rapidly increases the amount

INSIGHT 10.1   *Discovery*

Okay, the Genome's Sequenced—What's Next?

In early 2001 a press conference was held to announce that the human genome had been sequenced. The 3.1 billion base pairs that make up the DNA found in (nearly) every human cell had been identified and put in the proper order. Champagne corks popped, balloons fell, bands played, reporters reported on the significance of the occasion and . . . nothing else seemed to change. What, you may ask, has the Human Genome Project done for me lately? Here for your perusal are a few FAQs.

Q: Who sequenced the genome?

A: Francis Collins was the head of the publicly funded Human Genome Project (HGP), while Craig Venter was the head of Celera Genomics, a private company that developed a new, more powerful method of DNA sequencing and competed with the HGP. A compromise was finally reached whereby both groups took credit for sequencing the genome.

Q: How big was this project?

A: The Human Genome Project was first discussed in the mid-1980s and got under way in 1990. Although the project was to have taken at least 15 years, advances in technology led to its being completed in just over 10. The 3.1 billion base pairs of DNA code for only about 30,000–40,000 genes, not the 100,000 or so that the genome was thought to contain only a few years ago.

Q: Will I ever see any benefits from this project?

A: Absolutely. Sequencing the genome was only the first step. Knowing what proteins are produced in the body, what they do, and how they interact with one another are essential to understanding the workings of the human body, in both health and disease. By knowing the genetic signatures of different diseases, we will be able to design extraordinarily sensitive diagnostic tests that detect not only a disease but particular subtypes of each malady. With a precise genetic identification of, for example, a tumor, treatment can be tailored to be as effective as possible, while dramatically reducing side effects.

Q: What's next?

A: While everyone was still digesting the human genome (being the equivalent of a million-page book, it's a long read), the mouse genome was completed (in August 2002). The thinking is that areas within genes that are similar in mouse and human (known as regions of homology) are most important for the function of the gene and are the most likely targets for potential drugs or genetic treatments.

Q: Where could I go to check out a really cool website on the human genome?

A: www.ornl.gov/hgmis/

J. Craig Venter, left, and Francis Collins celebrate mapping of the human genome.

---

of DNA in a sample without the need for making cultures or carrying out complex purification techniques. It is so sensitive that it holds the potential to detect cancer from a single cell or to diagnose an infection from a single gene copy. It is comparable to being able to pluck a single DNA "needle" out of a "haystack" of other molecules and make unlimited copies of the DNA. The rapid rate of PCR makes it possible to replicate a target DNA from a few copies to billions of copies in a few hours. It can amplify DNA fragments that consist of a few base pairs to whole genomes containing several million base pairs.

To understand the idea behind PCR, it will be instructive to review figure 9.6, which describes synthesis of DNA as it occurs naturally in cells. The PCR method uses essentially the same events, with the opening up of the double strand, using the exposed strands as templates, the addition of primers, and the actions of a DNA polymerase.

Initiating the reaction requires a few specialized ingredients **(figure 10.6).** The **primers** are synthetic oligonucleotides (short DNA strands) of a known sequence of 15 to 30 bases that serve as landmarks to indicate where DNA amplification will begin. These take the place of RNA primers that would normally be synthesized by primase (in the cell). Depending upon the purposes and what is known about the DNA being replicated, the primers can be random, attaching to any sequence they may fit, or they may be highly specific

(a) In cycle 1, the DNA to be amplified is denatured, primed, and replicated by a polymerase that can function at high temperature. The two resulting strands then serve as templates for a second cycle of denaturation, priming, and synthesis.*

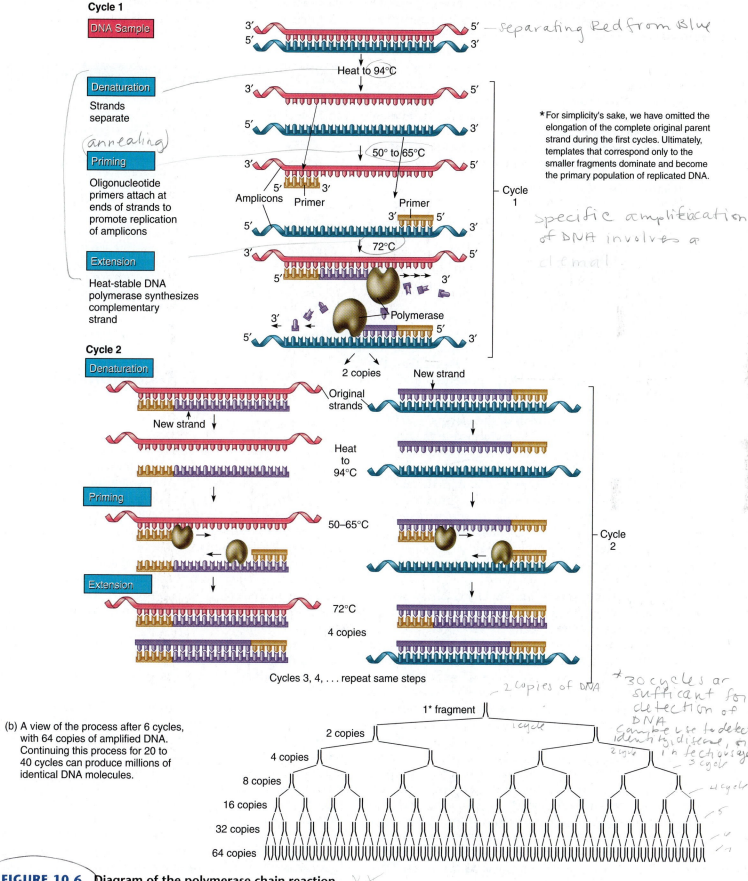

Cycle 1

DNA Sample

— separating Red from Blue

Heat to 94°C

Denaturation

Strands separate

(annealing)

Priming

Oligonucleotide primers attach at ends of strands to promote replication of amplicons

50° to 65°C

Amplicons Primer Primer

Extension

Heat-stable DNA polymerase synthesizes complementary strand

72°C

Polymerase

*For simplicity's sake, we have omitted the elongation of the complete original parent strand during the first cycles. Ultimately, templates that correspond only to the smaller fragments dominate and become the primary population of replicated DNA.

Cycle 1

specific amplification of DNA involves a clonal.

Cycle 2

Denaturation

2 copies   New strand

Original strands

New strand

Heat to 94°C

Priming

50–65°C

Extension

72°C

4 copies

Cycle 2

Cycles 3, 4, . . . repeat same steps

2 copies of DNA

(b) A view of the process after 6 cycles, with 64 copies of amplified DNA. Continuing this process for 20 to 40 cycles can produce millions of identical DNA molecules.

1* fragment

2 copies

4 copies

8 copies

16 copies

32 copies

64 copies

30 cycles ar sufficant for detection of DNA sample use to detect identity, disease, or infectious ag

1 cycle
2 cycle
3 cycle
4 cycle
5

FIGURE 10.6 Diagram of the polymerase chain reaction.

and chosen to amplify a known gene. To keep the DNA strands separated, processing must be carried out at a relatively high temperature. This necessitates the use of special **DNA polymerases** isolated from thermophilic bacteria. Examples of these unique enzymes are Taq polymerase obtained from *Thermus aquaticus* and Vent polymerase from *Thermococcus litoralis*. Enzymes isolated from these thermophilic organisms remain active at the elevated temperatures used in PCR. Another useful component of PCR is a machine called a thermal cycler that automatically initiates the cyclic temperature changes.

The PCR technique operates by repetitive cycling of three basic steps: denaturation, priming, and extension.

1. **Denaturation.** The first step involves heating target DNA to 94°C to separate it into two strands. Next, the system is cooled to between 50°C and 65°C, depending on the exact nucleotide sequence of the primer.
2. **Priming.** Primers are added in a concentration that favors binding to the complementary strand of test DNA. This reaction prepares the two DNA strands, now called **amplicons,** for synthesis.
3. **Extension.** In the third phase, which proceeds at 72°C, DNA polymerase and raw materials in the form of nucleotides are added. Beginning at the free end of the primers on both strands, the polymerases extend the molecule by adding appropriate nucleotides and produce two complete strands of DNA.

It is through cyclic repetition of these steps that DNA becomes amplified. When the DNAs formed in the first cycle are denatured, they become amplicons to be primed and extended in the second cycle. Each subsequent cycle converts the new DNAs to amplicons and doubles the number of copies. The number of cycles required to produce a million molecules is 20, but the process is usually carried out to 30 or 40 cycles. One significant advantage of this technique has been its natural adaptability to automation. A PCR machine can perform 20 cycles on nearly 100 samples in 2 or 3 hours.

Once the PCR is complete, the amplified DNA can be analyzed by any of the techniques discussed earlier. PCR can be adapted to analyze RNA by initially converting an RNA sample to DNA with reverse transcriptase. This cDNA can then be amplified by PCR in the usual manner. It is by such means that ribosomal RNA and messenger RNA are readied for sequencing. The polymerase chain reaction has found prominence as a powerful workhorse of molecular biology, medicine, and biotechnology. It often plays an essential role in gene mapping, the study of genetic defects and cancer, forensics, taxonomy, and evolutionary studies.

For all of its advantages, PCR has some problems. A serious concern is the introduction and amplification of non-target DNA from the surrounding environment, such as a skin cell from the technician carrying out the PCR reaction rather than material from the sample DNA that was supposed to be amplified. Such contamination can be minimized by using equipment and rooms dedicated for DNA

analysis and maintained with the utmost degree of cleanliness. Problems with contaminants can also be reduced by using gene-specific primers and treating samples with special enzymes that can degrade the contaminating DNA before it is amplified.

> ✔ **CHECKPOINT**
>
> - The genetic revolution has produced a wide variety of industrial technologies that translate and radically alter the blueprints of life. The potential of biotechnology promises not only improved quality of life and enhanced economic opportunity but also serious ethical dilemmas. Increased public understanding is essential for developing appropriate guidelines for responsible use of these revolutionary techniques.
> - Genetic engineering utilizes a wide range of methods that physically manipulate DNA for purposes of visualization, sequencing, hybridizing, and identifying specific sequences. The tools of genetic engineering include specialized enzymes, gel electrophoresis, DNA sequencing machines, and gene probes. The polymerase chain reaction (PCR) technique amplifies small amounts of DNA into larger quantities for further analysis.

10.3 Methods in Recombinant DNA Technology: How to Imitate Nature

The primary intent of **recombinant DNA technology** is to deliberately remove genetic material from one organism and combine it with that of a different organism. Its origins can be traced to 1970, when microbiologists first began to duplicate the clever tricks bacteria do naturally with bits of extra DNA such as plasmids, transposons, and proviruses. As mentioned earlier, humans have been trying to artificially influence genetic transmission of traits for centuries. The discovery that bacteria can readily accept, replicate, and express foreign DNA made them powerful agents for studying the genes of other organisms in isolation. The practical applications of this work were soon realized by biotechnologists. Bacteria could be genetically engineered to mass-produce substances such as hormones, enzymes, and vaccines that were difficult to synthesize by the usual industrial methods.

Figure 10.7 provides an overview of the recombinant DNA procedure. An important objective of this technique is to form genetic **clones.** Cloning involves the removal of a selected gene from an animal, plant, or microorganism (the genetic donor) followed by its propagation in a different host organism. Cloning requires that the desired donor gene first be selected, excised by restriction endonucleases, and isolated. The gene is next inserted into a **vector** (usually a plasmid or a virus) that will insert the DNA into a **cloning host.** The cloning host is usually a bacterium or a yeast that can replicate the gene and translate it into the protein

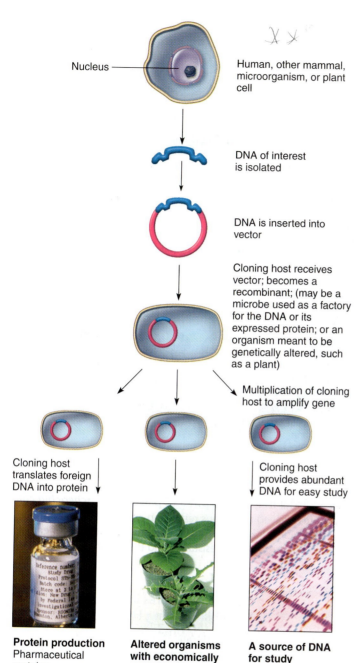

Nucleus

Human, other mammal, microorganism, or plant cell

DNA of interest is isolated

DNA is inserted into vector

Cloning host receives vector; becomes a recombinant; (may be a microbe used as a factory for the DNA or its expressed protein; or an organism meant to be genetically altered, such as a plant)

Multiplication of cloning host to amplify gene

Cloning host translates foreign DNA into protein

Cloning host provides abundant DNA for easy study

Protein production
Pharmaceutical proteins
• insulin
• human growth hormone
Vaccines
• hepatitis B

Altered organisms with economically useful traits
Transgenic plants
• pest resistance
• herbicide resistance
• improved nutritional value

A source of DNA for study
Gene regulation
Gene function
Nucleotide sequencing

FIGURE 10.7   Methods and applications of genetic technology.
Practical applications of genetic engineering include the development of pharmaceuticals, genetically modified organisms, and forensic techniques.

product for which it codes. In the next section, we examine the elements of gene isolation, vectors, and cloning hosts and show how they participate in a complete recombinant DNA procedure.

Technical Aspects of Recombinant DNA and Gene Cloning

The first hurdles in cloning a target gene are to locate its exact site on the genetic donor's chromosome and to isolate it. Some of the first isolated genes came from viruses, which have extremely small and manageable genomes, and later genes were isolated from bacteria and yeasts. At first, the complexity of the human genome made locating specific genes very difficult. However, this process has been greatly improved by tools for excising, mapping, and identifying genes. Among the most common strategies for obtaining genes in an isolated state are:

1. The DNA is removed from cells and separated into fragments by endonucleases. Each fragment is then inserted into a vector and cloned. The cloned fragments undergo Southern blotting and are probed to identify desired sequences. This is a long and tedious process, because each fragment of DNA must be examined for the cloned gene. For example, if the human genome were separated into 20 kb fragments, there would be at least 150,000 clones to test. Alternatively,

2. A gene can be synthesized from isolated mRNA transcripts using reverse transcriptase (cDNA), or

3. A gene can be amplified using PCR in many cases.

Although gene cloning and isolation can be very laborious, a fortunate outcome is that, once isolated, genes can be maintained in a cloning host and vector just like a microbial pure culture. *Genomic libraries,* which are collections of isolated genes, now exist for millions of genes from hundreds of organisms and viruses.

Characteristics of Cloning Vectors

A good recombinant vector has two indispensable qualities: It must be capable of carrying a significant piece of the donor DNA, and it must be readily accepted by the cloning host. Plasmids are excellent vectors because they are small, well characterized, easy to manipulate, and they can be transferred into appropriate host cells through transformation. Bacteriophages also serve well because they have the natural ability to inject DNA into bacterial hosts through transduction. A common vector in early work was an *E. coli* plasmid that carries genetic markers for resistance to antibiotics, although it is restricted by the relatively small amount of foreign DNA it can accept. A modified phage vector, the *Charon*[5] phage, is missing large sections of its genome, so it can carry a fairly large segment of foreign DNA. The simple plasmids and bacteriophages that were a staple of early recombinant DNA methodologies evolved into newer, more advanced vectors. Today, thousands of unique cloning vectors are available commercially. Although every vector has characteristics that make it ideal for a specific project, all vectors

5. Named for the mythical boatman in Hades who carried souls across the River Styx.

base-pair, and a ligase makes the final cova-lent seals. The resultant gene and plasmid combination is called a **recombinant.**

Following this procedure, the recombi-nant is introduced by transformation into the cloning host, a special laboratory strain of *E. coli* that lacks any extra plasmids that could complicate the expression of the gene. Because the recombinant plasmid enters only some of the cloning host cells, it is nec-essary to locate these recombinant clones. Cultures are plated out on medium contain-ing ampicillin, and only those clones that carry the plasmid with ampicillin resistance can form colonies (**figure 10.10**). These re-combinant colonies are selected from the plates and cultured. As the cells multiply, the plasmid is replicated along with the cell's chromosome. In a few hours of growth, there can be billions of cells, each containing the interferon gene. Once the gene has been successfully cloned and tested, this step does not have to be repeated—the recombi-nant strain can be maintained in culture for production purposes.

The bacteria's ability to express the eu-caryotic gene is ensured, because the plas-mid has been modified with the necessary transcription and translation recognition se-quences. As the *E. coli* culture grows it tran-scribes and translates the interferon gene, synthesizes the peptide, and secretes it into the growth medium. At the end of the process, the cloning cells and other chemical and microbial impurities are removed from the medium. Final processing to excise a terminal amino acid from the peptide yields the interferon product in a relatively pure form (see figure 10.9). The scale of this procedure can range from test tube size to gigantic industrial vats that can manufac-ture thousands of gallons of product (see chapter 24).

Although the process we have presented here produces interferon, some variation of it can be used to mass produce a variety of hormones, enzymes, and agricultural products such as pesticides.

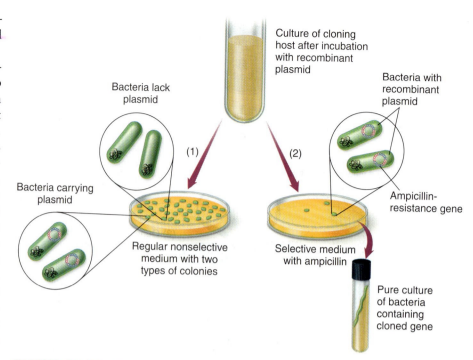

FIGURE 10.10 **One method for screening clones of bacteria that have been transformed with the donor gene.**
(**1**) Plating the culture on nonselective medium will not separate the transformed cells from normal cells, which lack the plasmid. (**2**) Plating on selective medium containing ampicillin will permit only cells containing the plasmid to multiply. Colonies growing on this medium that carry the cloned gene can be used to make a culture for gene libraries, industrial production, and other processes. (Plasmids are shown disproportionate to size of cell.)

10.4 Biochemical Products of Recombinant DNA Technology

Recombinant DNA technology is used by pharmaceutical companies to manufacture medications that cannot be man-ufactured by any other means. Diseases such as diabetes and dwarfism, caused by the lack of an essential hormone, are treated by replacing the missing hormone. Insulin of animal origin was once the only form available to treat diabetes, even though such animal products can cause allergic reac-tions in sensitive individuals. In contrast, dwarfism cannot be treated with animal growth hormones, so originally the only source of human growth hormone (HGH) was the pitu-itaries of cadavers. At one time, not enough HGH was avail-able to treat the thousands of children in need. Another serious problem with using natural human products is the potential for infection. For example, infectious agents such as the prion responsible for Creutzfeld-Jakob disease, described in chapter 19, can be transmitted in this manner. Similarly, clotting factor VIII, a protein needed for blood to clot prop-erly, is missing in persons suffering from hemophilia A. Persons lacking factor VIII have historically received peri-odic injections of the missing protein, which had been col-lected from blood plasma. While the donated protein alleviated the symptoms of factor VIII deficiency, a tragic

✔ **CHECKPOINT**

- Recombinant DNA techniques combine DNA from different sources to produce microorganism "factories" that produce hormones, enzymes, and vaccines on an industrial scale. Cloning is the process by which genes are removed from the original host and duplicated for transfer into a cloning host by means of cloning vectors.
- Plasmids, bacteriophages, and cosmids are types of cloning vectors used to transfer recombinant DNA into a cloning host. Cloning hosts are simply organisms that readily accept recombinant DNA, grow easily, and synthesize large quanti-ties of specific gene product.

TABLE 10.2	Current Protein Products from Recombinant DNA Technology

Immune Treatments

Interferons—peptides used to treat some types of cancer, multiple sclerosis, and viral infections such as hepatitis and genital warts

Interleukins—types of cytokines that regulate the immune function of white blood cells; used in cancer treatment

Orthoclone—an immune suppressant in transplant patients

Macrophage-colony-stimulating factor (GM-CSF)—used to stimulate bone marrow activity after bone marrow grafts

Tumor necrosis factor (TNF)—used to treat cancer *↳ prevent rotting*

Granulocyte-colony-stimulating factor (Neupogen)—developed for treating cancer patients suffering from low neutrophil counts

Hormones

Erythropoietin (EPO)—a peptide that stimulates bone marrow used to treat some forms of anemia

Tissue plasminogen activating factor (tPA)—can dissolve potentially dangerous blood clots

Hemoglobin A—form of artificial blood to be used in place of real blood for transfusions

Factor VIII—needed as replacement blood-clotting factor in type A hemophilia

Relaxin—an aid to childbirth

Human growth hormone (HGH)—stimulates growth in children with dwarfism; prevents wasting syndrome

Enzymes

rH DNase (pulmozyme)—a treatment that can break down the thick lung secretions of cystic fibrosis

Antitrypsin—replacement therapy to benefit emphysema patients

PEG-SOD—a form of superoxide dismutase that minimizes damage to brain tissue after severe trauma

Vaccines

Vaccines for hepatitis B and *Haemophilus influenzae* Type b meningitis

Experimental malaria and AIDS vaccines based on recombinant surface antigens

Miscellaneous

Bovine growth hormone or bovine somatotropin (BST)—given to cows to increase milk production   *40% more milk*

Apolipoprotein—to deter the development of fatty deposits in the arteries and to prevent strokes and heart attacks

Spider silk—a light, tough fabric for parachutes and bulletproof vests

---

side effect was seen in the early 1980s, when a large percentage of the patients receiving plasma-derived factor VIII contracted HIV infections as a result of being exposed to the virus through the donated plasma.

Recombinant technology changed the outcome of these and many other conditions by enabling large-scale manufacture of life-saving hormones and enzymes of human origin. Recombinant human insulin can now be prescribed for diabetics, and recombinant HGH can now be administered to children with dwarfism and Turner syndrome. HGH is also used to prevent the wasting syndrome that occurs in AIDS and cancer patients. In all of these applications, recombinant DNA technology has led to both a safer product and one that can be manufactured in quantities previously unfathomable. Other protein-based hormones, enzymes, and vaccines produced through recombinant DNA are summarized in **table 10.2.**

Nucleic acid products also have a number of medical applications. A new development in vaccine formulation involves using microbial DNA as a stimulus for the immune system. So far, animal tests using DNA vaccines for AIDS and influenza indicate that this may be a breakthrough in vaccine design. Recombinant DNA could also be used to produce DNA-based drugs for the types of gene and antisense therapy discussed in the genetic treatments section.

10.5 Genetically Modified Organisms

Recombinant organisms produced through the introduction of foreign genes are called *transgenic* or genetically modified organisms (GMOs). Foreign genes have been inserted into a variety of microbes, plants, and animals through recombinant DNA techniques developed especially for them. Transgenic "designer" organisms are available for a variety of biotechnological applications. Because they are unique life forms that would never have otherwise occurred, they can be patented.

Recombinant Microbes: Modified Bacteria and Viruses

One of the first practical applications of recombinant DNA in agriculture was to create a genetically altered strain of the bacterium *Pseudomonas syringae*. The wild strain ordinarily contains an ice formation gene that promotes ice or frost formation on moist plant surfaces. Genetic alteration of the frost gene using recombinant plasmids created a different strain that could prevent ice crystals from forming. A commercial product called Frostban has been successfully applied to stop frost damage in strawberry and potato crops.

This **antisense RNA** has bases that are complementary to the sense strand of mRNA in the area surrounding the initiation site. When the antisense RNA binds to its particular mRNA, the now double-stranded RNA is inaccessible to the ribosome, resulting in a loss of translation of that mRNA.

Note that the term *antisense* as used in molecular genetics does *not* mean "no sense." It is used to describe any nucleic acid strand with a base sequence that is complementary to the "sense" or translatable strand. For example, DNA contains a template strand that is transcribed and a matching strand that is not usually transcribed. We can also apply this terminology to RNA. Messenger RNA is considered the translatable strand, and a strand with its complementary sequence of nucleotides would be the antisense strand. To illustrate: If an mRNA sequence read AUGCGAGAC, then an antisense RNA strand for it would read UACGCUCUG.

As this process was tested in the laboratory, two significant differences between procaryotic and eucaryotic systems emerged. The first was that single-stranded DNA was usually used as the antisense agent, as it was easier to manufacture than RNA. The second was that, for some genes, once the antisense strand bound to the mRNA, not only was the hybrid RNA not translated, it was unable even to leave the nucleus.

When an adequate dose of **antisense DNA** is delivered across the cell membrane into the cytoplasm and nucleus, it binds to specific sites on any mRNAs that are the targets of therapy. As a result, the reading of that mRNA transcript on ribosomes will be blocked, and the gene product will not be synthesized (**figure 10.14a**).

Early clinical trials of antisense therapeutics have been promising, with most antisense therapies dramatically lowering the rate of synthesis of the targeted protein. Because the

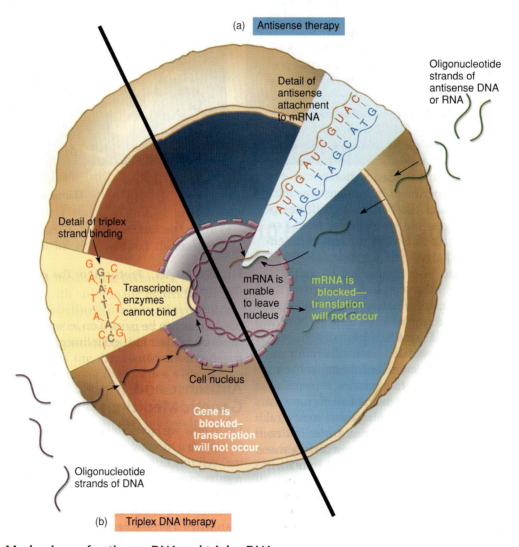

FIGURE 10.14 **Mechanisms of antisense DNA and triplex DNA.**
These genetic medicines are designed to prevent the expression of an undesirable gene or virus. Both use oligonucleotides of known sequence that can bind with genetic targets that are the cause of disease. The site of action is in the nucleus. **(a)** When antisense strands enter the nucleus, they bind to complementary areas on an existing mRNA molecule. Such bound RNA will be unavailable for translation on the cell's ribosomes, and no protein will be made. **(b)** Triplex DNA strands attach by base-pairing to regulatory sites and other sites on the chromosome itself. This action masks the points of attachment for enzymes of transcription, and mRNA cannot be made. Without mRNA, there will be no translation or protein formed.

antisense agent can be targeted to a specific sequence of DNA, side effects are rare. Most concern has been centered on the difficulty of getting high enough levels of the drug into the nucleus of cells and the fact that these drugs must be taken for life. In 1998 Vitravene, an antisense DNA used to treat cytomegalovirus retinitis (blindness brought on by infection with cytomegalovirus in immunocompromised persons), became the first antisense therapy approved for use in humans.

Triplex DNA: Adding a Third Strand

Although the usual structure of DNA is a helix composed of two strands, researchers have known for years that DNA can exist in other naturally and artificially induced formats. One of these unusual variations, *triplex DNA,* is a triple helix formed when a third strand of DNA inserts into the major groove of the molecule. The inserted strand forms hydrogen bonds with the purine bases on one of the adjacent strands (see **figure 10.14***b*). It seems logical that the template of DNA that has an extra strand wedged into its structure can be relatively inaccessible to normal transcription. Indeed, it is theorized that this may be another natural means for cells to control their genetic expression.

The potential therapeutic applications of triplex DNA are still in their infancy. So far, oligonucleotides have been synthesized to form triplex DNA that interacts with regulatory sequences in genes. It can interrupt the action of transcription factors and prevent RNA polymerase from forming mRNA transcripts. Studies performed on cell cultures revealed that the genes coding for oncogenes, viruses, and the receptor for interleukin-2 (an immune stimulant) could be blocked with triplex DNA. In the clinical trials to come, this therapy will no doubt be plagued with some of the same problems mentioned for the antisense approach.

10.7 Genome Analysis: Maps, Fingerprints, and Family Trees

An old geneticist's joke begins: "How do you describe a cat?" The answer, "GATCCT. . . ," is becoming more and more a reality. As was mentioned earlier, DNA technology has allowed us to accomplish many age-old goals by new and improved means. By remembering that phenotypes, whether human, bacterial, or even viral, are the result of specific sequences of DNA, we can easily see how DNA can be used to differentiate among organisms in the same way that observing any of these phenotypes can. Additionally, DNA can be used to "see" the phenotype of an organism no longer present, as when a criminal is identified by DNA extracted from a strand of hair left behind. Finally, possession of a particular sequence of DNA may indicate an increased risk of a genetic disease. Detection of this piece of DNA (known as a marker) can identify a person as being at increased risk for cancer or Alzheimer's disease long before symptoms arise. The ability to

detect diseases before symptoms arise is especially important for diseases such as cancer, where early treatment is sometimes the difference between life and death. With examples like this in mind, let's look at several ways in which new DNA technology is allowing us to accomplish goals in ways that were only dreamed of a few years ago.

Genome Mapping and Screening: An Atlas of the Genome

We have seen a variety of methods for accessing the genomes of organisms, but the most useful information comes from knowing the sequential makeup of the genetic material. Genetic engineers find it very informative to know the **locus,** or exact position, of a particular gene on a chromosome. They also seek information on the types and numbers of **alleles,** which are sites that vary from one individual to another. The process of determining location of loci and other qualities of genomic DNA is called **mapping.** Maps vary in resolution and applications. *Linkage maps* show the relative proximity and order of genes on a chromosome and are relatively low resolution because only a few exact locations are mapped. *Physical maps* are more detailed arrays that depict not only the relative positions of distinct sections of DNA, but give the numerical size in base pairs. This technology uses restriction fragments and fluorescent hybridization probes to visualize the position of a particular selected site along a segment of DNA.

By far the most detailed maps are *sequence maps,* which are produced by the sequencers we discussed earlier. They give an exact order of bases in a plasmid, chromosome, or entire genome. Because this level of resolution is most promising for understanding the nature of the genes, what they code for, and their functions, this form of map dominates the research programs.

Genome sequencing projects have been highly successful, so much so that the genomes of over 1,000 organisms, viruses, and organelles have been fully sequenced. As of 2004, this includes around 1,200 viruses, 155 procaryotes (bacteria and archaea), and hundreds of eucaryotes (including human, mouse, *C. elegans* (a nematode worm), yeast, fruit fly, *Arabidopsis* (a small flowering plant), and rice. One of the remarkable discoveries in this huge enterprise has been how similar the genomes of relatively unrelated organisms are. Humans share around 80% of their DNA codes with mice, about 60% with rice, and even 30% with the worm *C. elegans.*

The new ease with which researchers can sequence genomes was illustrated in the spring of 2003, when a previously unknown virus began causing severe acute respiratory syndrome (SARS) in Southeast Asia. Just 2 weeks after the first virus was isolated from patients, the genome was sequenced and made available to scientists rushing to create a diagnostic test, understand its virulence, and design a vaccine.

Although sequencing provides the ultimate genetic map, it does not automatically identify the exact genes and alleles. Analyzing and storing this massive amount of new data requires specialized computers. Because it is information of

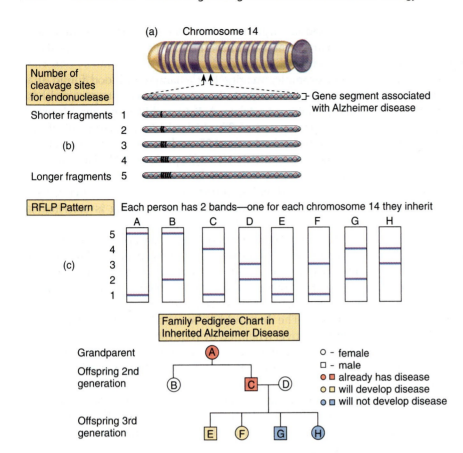

FIGURE 10.16 Pedigree analysis based on genetic screening for familial type Alzheimer's disease.

(a) The site for this disease is on chromosome 14. **(b)** Gene mapping has determined that human chromosomes have one of five different variants of this segment, detectable as different cleavage patterns. **(c)** A Southern blot using the restriction fragments (RFLPs) is performed on a family with three generations. Each person will have two bands that correspond with each of the chromosomes they have inherited. The grandmother (A) with pattern 1,5 and her son (C) with pattern 1,4 both already have the disease. Patterns for his sister (B), his wife (D), and their four children (E, F, G, and H) are also shown. Comparing the band patterns of A and B with those of other family members makes it possible to predict which children will inherit the gene for Alzheimer disease. Because the band shared by A and C is 1, we know that this is the variant associated with Alzheimer's disease. Children E (1,2) and F (1,3) will develop the disease, and children G (2,4) and H (3,4) will not. It is also evident that this form of Alzheimer disease is inherited as a dominant trait.

Because even fossilized DNA can remain partially intact, ancient DNA is being studied for comparative and evolutionary purposes. Anthropologists have traced the migrations of ancient people by analyzing the mitochondrial DNA found in bone fragments. Mitochondrial DNA (mtDNA) is less subject to degradation than is chromosomal DNA and can be used as an evolutionary time clock. A group of evolutionary biologists have been able to recover Neanderthal mtDNA and use it to show that this group of ancient primates could not have been an ancestor of modern humans. Biologists studying canine origins have also been able to show that the wolf is the ancestral form for modern canines.

Knowing the complete nucleotide sequence of the human genome is really only half the battle. With very few exceptions all cells in an organism contain the same DNA, so knowing the sequence of that DNA, while certainly helpful, is of very little use when comparing two cells or tissues from the same organism. Recall that genes are expressed in response to both internal needs and external stimuli, and while the DNA content of a cell is static, the mRNA (and hence protein) content at any given time provides scientists with a profile of genes currently being expressed in the cell. What truly distinguishes a liver cell from a kidney cell or a healthy cell from a diseased cell are the genes expressed in each.

Twin advances in biology and electronics have allowed biologists to view the expression of genes in any given cell using a technique called DNA *microarray analysis*. Prior to the advent of this technology, scientists were able to track

the expression of, at most, a few genes at a time. Given the complex interrelationship that exists between genes in a typical cell, tracking two or three genes was of very little practical use. Microarrays are able to track the expression of thousands of genes at once and are able to do so in a single efficient experiment. Microarrays consist of a glass "chip"[6] (occasionally a silicon chip or nylon membrane is used) onto which have been bound sequences from tens of thousands of different genes. A solution containing fluorescently labeled cDNA, representing all of the mRNA molecules in a cell at a given time, is added to the chip. The labeled cDNA is allowed to bind to any complementary DNA bound to the chip. Bound cDNA is then detected by exciting the fluorescent tag on the cDNA with a laser and recording the fluorescence with a detector linked to a computer. The computer can then interpret this data to determine what mRNAs are present in the cell under a variety of conditions (**figure 10.17**).

Possible uses of microarrays include developing extraordinarily sensitive diagnostic tests that search for a specific pattern of gene expression. As an example, being able to identify a patient's cancer as one of many subtypes (rather than just, for instance, breast cancer) will allow pharmacologists and doctors to treat each cancer with the drug that will be most effective. Again we see that genetic technology can be a very effective way to reach long-held goals.

---

6. It's actually just a glass slide, but all microarray setups have come to be called chips.

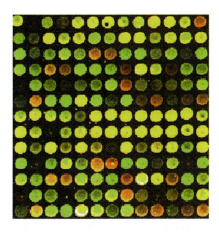

FIGURE 10.17   Gene expression analysis using microarrays.
The microarray seen here consists of oligonucleotides bound to a glass slide. The gene expression of two cells (for example, one healthy and one diseased) can be compared by labeling the cDNA from each cell with either a red or green fluorescent label and allowing the cDNAs to bind to the microarray. A laser is used to excite the bound cDNAs, while a detector records those spots that fluoresce. The color of each spot reveals whether the DNA on the microarray bound to cDNAs are present in the healthy sample, the diseased sample, both samples, or neither sample.

✔ CHECKPOINT

- Gene therapy is the replacement of faulty host genes with functional genes by use of cloning vectors such as viruses and liposomes. This type of transfection can be used to treat genetic disorders and acquired disease. Antisense DNA and triplex DNA are used to block expression of undesirable host genes in plants and animals as well as those of intracellular parasites.
- DNA technology has advanced understanding of basic genetic principles that have significant applications in a wide range of disciplines, particularly medicine, evolution, forensics, and anthropology.
- The Human Genome Project and other genome sequencing projects have revolutionized our understanding of organisms and led to two new biological disciplines, genomics and bioinformatics.
- DNA fingerprinting is a technique by which individuals are identified for purposes of medical diagnosis, genetic ancestry, and forensics.
- Mitochondrial DNA analysis is being used to trace evolutionary origins in animals and plants.
- Microarray analysis can determine what genes are transcribed in a given tissue. It is used to identify and devise treatments for diseases based on the genetic profile of the disease.

Chapter Summary With Key Terms

10.1   Basic Elements and Applications of Genetic Engineering
Genetic engineering refers to the manipulation of an organism's genome and is often used in conjunction with biotechnology, the use of an organism's biochemical and metabolic pathways for industrial production of proteins.

10.2   Tools and Techniques of Genetic Engineering
A. *Amazing DNA*
  1. **Restriction endonucleases** are used to cut DNA at specific recognition sites, resulting in small segments of DNA which can be more easily managed in the laboratory.
  2. The enzyme **ligase** is used to covalently join DNA molecules, whereas **reverse transcriptase** is used to obtain a DNA copy of a particular mRNA.
  3. Short strands of DNA called oligonucleotides are used either as probes for specific DNA sequences or as primers for DNA polymerase.
B. *Nucleic Acid Hybridization and Probes*
  1. The two strands of DNA can be separated (denatured) by heating. These single-stranded nucleic acids will bind, or **hybridize,** to their complementary sequence, allowing them to be used as **gene,** or **hybridization,** probes.
    a. In the Southern blot method, DNA isolated from an organism is probed to identify specific DNA sequences.
    b. A technique originally developed by Frederick Sanger allows for the determination of the exact sequence of a stretch of DNA. The process is called **DNA sequencing.**

  2. The **polymerase chain reaction** is a technique used to amplify a specific segment of DNA several million times in just a few hours.

10.3   Methods in Recombinant DNA Technology: How to Imitate Nature
Technical Aspects of Recombinant DNA Technology
A. Recombinant DNA methods have as their goal the transfer of DNA from one organism to another. This commonly involves inserting the DNA into a **vector** (usually a plasmid or virus) and transferring the recombinant vector to a **cloning host** (usually a bacterium or yeast) for expression.
  1. Examples of cloning vectors include plasmids, bacteriophages, and cosmids, as well as the larger bacterial and yeast artificial chromosomes (BACs and YACs).
  2. Cloning hosts are fast-growing, nonpathogenic organisms that are genetically well understood. Examples include *E. coli* and *Saccharomyces cerevisiae.*
B. Recombinant organisms are created by cutting both the donor DNA and the plasmid DNA with restriction endonucleases and then joining them together with ligase. The vector is then inserted into a cloning host by transformation, and those cells containing a vector are selected by their resistance to antibiotics.
C. When the appropriate recombinant cell has been identified, it is grown on an industrial scale. The cloning host expresses the gene it contains, and the resulting protein is excreted from the cell where it can be recovered from the media and purified.

▶ Microbicidal agents kill microbes by inflicting irreversible damage to the cell. Microbistatic agents temporarily inhibit the reproduction of microbes but do not inflict irreversible damage. Mechanical antimicrobial agents physically remove microbes from materials but do not necessarily kill or inhibit them.

▶ Heat is the most important physical agent in microbial control and can be delivered in both moist (steam sterilization, pasteurization) and dry (incinerators, Bunsen burners) forms.

▶ Radiation exposes materials to high-energy waves that can enter and damage microbes. Examples are ionizing and ultraviolet radiation.

▶ Chemical antimicrobials are available for every level of microbial treatment, from low-level disinfectants to high-level sterilants. Antimicrobial chemicals include halogens, alcohols, phenolics, peroxides, heavy metals, detergents, and aldehydes.

11.1 Controlling Microorganisms

Much of the time in our daily existence, we take for granted tap water that is drinkable, food that is not spoiled, shelves full of products to eradicate "germs," and drugs to treat infections. Controlling our degree of exposure to potentially harmful microbes is a monumental concern in our lives, and it has a long and eventful history (Insight 11.1).

General Considerations in Microbial Control

The methods of microbial control belong to the general category of *decontamination* procedures, in that they destroy or remove contaminants. In microbiology, contaminants are microbes present at a given place and time that are undesirable or unwanted. Most decontamination methods employ either physical agents, such as heat or radiation, or chemical agents such as disinfectants and antiseptics. This separation is convenient, even though the categories overlap in some cases; for instance, radiation can cause damaging chemicals to form, or chemicals can generate heat. A flowchart (figure 11.1) summarizes the major applications and aims in microbial control.

Relative Resistance of Microbial Forms

The primary targets of microbial control are microorganisms capable of causing infection or spoilage that are constantly present in the external environment and on the human body. This targeted population is rarely simple or uniform; in fact, it often contains mixtures of microbes with

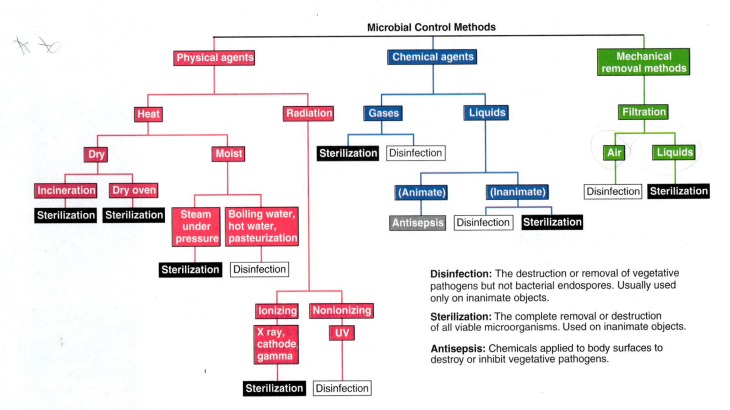

FIGURE 11.1 Microbial control methods.

INSIGHT 11.1   *Historical*

Microbial Control in Ancient Times

No one knows for sure when humans first applied methods that could control microorganisms, but perhaps the discovery and use of fire in prehistoric times was the starting point. We do know that records describing simple measures to control decay and disease appear from civilizations that existed several thousand years ago. We know, too, that these ancient people had no concept that germs caused disease, but they did have a mixture of religious beliefs, skills in observing natural phenomena, and possibly, a bit of luck. This combination led them to carry out simple and sometimes rather hazardous measures that contributed to the control of microorganisms.

Salting, smoking, pickling, and drying foods and exposing food, clothing, and bedding to sunlight were prevalent practices among early civilizations. The Egyptians showed surprising sophistication and understanding of decomposition by embalming the bodies of their dead with strong salts and pungent oils. They introduced filtration of wine and water as well. The Greeks and Romans burned clothing and corpses during epidemics, and they

Illustration of protective clothing used by doctors in the 1700s to avoid exposure to plague victims. The beaklike portion of the hood contained volatile perfumes to protect against foul odors and possibly inhaling "bad air."

stored water in copper and silver containers. The armies of Alexander the Great reportedly boiled their drinking water and buried their wastes. Burning sulfur to fumigate houses and applying sulfur as a skin ointment also date approximately from this era.

During the great plague pandemic of the Middle Ages, it was commonplace to bury corpses in mass graves, burn the clothing of plague victims, and ignite aromatic woods in the houses of the sick in the belief that fumes would combat the disease. In a desperate search for some sort of protection, survivors wore peculiar garments and anointed their bodies with herbs, strong perfume, and vinegar. These attempts may sound foolish and antiquated, but it now appears that they may have had some benefits. Burning wood releases formaldehyde, which could have acted as a disinfectant; herbs, perfume, and vinegar contain mild antimicrobial substances. Each of these early methods, although somewhat crude, laid the foundations for microbial control methods that are still in use today.

---

extreme differences in resistance and harmfulness. Contaminants that can have far-reaching effects if not adequately controlled include bacterial vegetative cells and endospores, fungal hyphae and spores, yeasts, protozoan trophozoites and cysts, worms, viruses, and prions. This scheme compares the general resistance these forms have to physical and chemical methods of control:

Highest resistance

Bacterial endospores, prions

Moderate resistance

Protozoan cysts; some fungal sexual spores (zygospores); some viruses. In general, naked viruses are more resistant than enveloped forms. Among the most resistant viruses are the hepatitis B virus and the poliovirus. Bacteria with more resistant vegetative cells are *Mycobacterium tuberculosis*, *Staphylococcus aureus*, and *Pseudomonas* species.

Least resistance

Most bacterial vegetative cells; fungal spores (other than zygospores) and hyphae; enveloped viruses; yeasts; and protozoan trophozoites

Actual comparative figures on the requirements for destroying various groups of microorganisms are shown in **table 11.1.** Bacterial endospores have traditionally been considered the most resistant microbial entities, being as much

TABLE 11.1	Relative Resistance of Bacterial Endospores and Vegetative Cells to Control Agents		
Method	Endospores*	Vegetative Forms*	Relative Resistance**
Heat (moist)	120°C	80°C	1.5×
Radiation (X-ray) dosage	4000 Grays	1000 Grays	4×
Sterilizing gas (ethylene oxide)	1,200 mg/1	700 mg/1	1.7×
Sporicidal liquid (2% glutaraldehyde)	3 h	10 min	18×

*Values are based on methods (concentration, exposure time, intensity) that are required to destroy the most resistant pathogens in each group.

**The greater resistance of spores versus vegetative cells given as an average figure.

as 18 times harder to destroy than their counterpart vegetative cells. Because of their resistance to microbial control, their destruction is the goal of *sterilization* (see definition in following section) because any process that kills endospores will invariably kill all less resistant microbial forms. Other methods of control (disinfection, antisepsis) act primarily upon microbes that are less hardy than endospores.

A NOTE ABOUT PRIONS

Scientists are just beginning to understand that prions are in a class of their own when it comes to "sterilization" procedures. This chapter defines "sterile" as the absence of all viable microbial life—but none of the procedures described in this chapter are in themselves sufficient to destroy prions. Prions are extraordinarily resistant to heat and chemicals. If instruments or other objects become contaminated with these unique agents, they must either be discarded as biohazards, or, if this is not possible, a combination of chemicals and heat must be applied in accordance with CDC guidelines. The guidelines themselves are constantly evolving as new information becomes available. One U.S. company has marketed a detergent that it says is effective in destroying prions when used in dishwashing devices, but its efficacy has not been extensively tested in "the field." In the meantime, this chapter will discuss sterilization using bacterial endospores as the toughest form of microbial life. When tissues, fluids, or instruments are suspected of containing prions, consultation with infection control experts and/or the CDC is recommended when determining effective sterilization conditions. Chapter 19 describes prions in detail.

Terminology and Methods of Microbial Control

Through the years, a growing terminology has emerged for describing and defining measures that control microbes. To complicate matters, the everyday use of some of these terms can at times be vague and inexact. For example, occasionally one may be directed to "sterilize" or "disinfect" a patient's skin, even though this usage does not fit the technical definition of either term. To lay the groundwork for the concepts in microbial control to follow, we present here a series of concepts, definitions, and usages in antimicrobial control.

Sterilization

Sterilization is a process that destroys or removes all viable microorganisms, including viruses. Any material that has been subjected to this process is said to be **sterile.** These terms should be used only in the strictest sense for methods that have been proved to sterilize. An object cannot be slightly sterile or almost sterile—it is either sterile or not sterile. Control methods that sterilize are generally reserved for inanimate objects, because sterilizing parts of the human body would call for such harsh treatment that it would be highly dangerous and impractical. As we shall see in chapter 13, many internal parts of the body—the brain, muscles, and liver, for example—are naturally free of microbes.

Sterilized products—surgical instruments, syringes, and commercially packaged foods, just to name a few—are essential to human well-being. Although most sterilization is performed with a physical agent such as heat, a few chemicals called *sterilants* can be classified as sterilizing agents because of their ability to destroy spores.

At times, sterilization is neither practicable nor necessary, and only certain groups of microbes need to be controlled. Some antimicrobial agents eliminate only the susceptible vegetative states of microorganisms but do not destroy the more resistant endospore and cyst stages. Keep in mind that the destruction of spores is not always a necessity, because most of the infectious diseases of humans and animals are caused by non-spore-forming microbes.

Microbicidal Agents

The root *-cide,* meaning to kill, can be combined with other terms to define an antimicrobial agent aimed at destroying a certain group of microorganisms. For example, a **bactericide** is a chemical that destroys bacteria except for those in the endospore stage. It may or may not be effective on other microbial groups. A fungicide is a chemical that can kill fungal spores, hyphae, and yeasts. A virucide is any chemical known to inactivate viruses, especially on living tissue. A sporicide is an agent capable of destroying bacterial endospores. A sporicidal agent can also be a sterilant because it can destroy the most resistant of all microbes.

Agents That Cause Microbistasis

The Greek words *stasis* and *static* mean to stand still. They can be used in combination with various prefixes to denote a condition in which microbes are temporarily prevented from multiplying but are not killed outright. Although killing or permanently inactivating microorganisms is the usual goal of microbial control, microbistasis does have meaningful applications. **Bacteriostatic** agents prevent the growth of bacteria on tissues or on objects in the environment, and *fungistatic* chemicals inhibit fungal growth. Materials used to control microorganisms in the body (antiseptics and drugs) have microbistatic effects because many microbicidal compounds can be highly toxic to human cells.

Germicides, Disinfection, Antisepsis

A **germicide,** also called a *microbicide,* is any chemical agent that kills pathogenic microorganisms. A germicide can be used on inanimate (nonliving) materials or on living tissue, but it ordinarily cannot kill resistant microbial cells. Any physical or chemical agent that kills "germs" is said to have germicidal properties.

The related term, **disinfection,** refers to the use of a physical process or a chemical agent (a disinfectant) to destroy vegetative pathogens but not bacterial endospores. It is important to note that disinfectants are normally used only on inanimate objects because, in the concentrations required to be effective, they can be toxic to human and other animal tissue. Disinfection processes also remove the harmful products of microorganisms (toxins) from materials. Examples of disinfection include applying a solution of 5% bleach to an examining table, boiling food utensils used by a sick person, and immersing thermometers in an iodine solution between uses.

In modern usage, **sepsis** is defined as the growth of microorganisms in the blood and other tissues. The term **asepsis** refers to any practice that prevents the entry of infectious agents into sterile tissues and thus prevents infection. Aseptic techniques commonly practiced in health care range from sterile methods that exclude all microbes to **antisepsis**. In antisepsis, chemical agents called **antiseptics** are applied directly to exposed body surfaces (skin and mucous membranes), wounds, and surgical incisions to destroy or inhibit vegetative pathogens. Examples of antisepsis include preparing the skin before surgical incisions with iodine compounds, swabbing an open root canal with hydrogen peroxide, and ordinary handwashing with a germicidal soap.

Methods That Reduce the Numbers of Microorganisms

Several applications in commerce and medicine do not require actual sterilization, disinfection, or antisepsis but are based on reducing the levels of microorganisms (the microbial load) so that the possibility of infection or spoilage is greatly decreased. Restaurants, dairies, breweries, and other food industries consistently handle large numbers of soiled utensils that could readily become sources of infection and spoilage. These industries must keep microbial levels to a minimum during preparation and processing. **Sanitization** is any cleansing technique that mechanically removes microorganisms as well as other debris to reduce contamination to safe levels. A sanitizer is a compound such as soap or detergent used to perform this task.

Cooking utensils, dishes, bottles, cans, and used clothing that have been washed and dried may not be completely free of microbes, but they are considered safe for normal use (sanitary). Air sanitization with ultraviolet lamps reduces airborne microbes in hospital rooms, veterinary clinics, and laboratory installations. It is important to note that some sanitizing processes (such as dishwashing machines) may be rigorous enough to sterilize objects, but this is not true of all sanitization methods. Also note that sanitization is often preferable to sterilization. In a restaurant, for example, you could be given a sterile fork with someone else's old food on it and a sterile glass with lipstick on the rim. On top of this, realize that the costs associated with sterilization would lead to the advent of the $50 fast-food meal. In a situation such as this, the advantage of being sanitary as opposed to sterile can be clearly seen.

It is often necessary to reduce the numbers of microbes on the human skin through **degermation**. This process usually involves scrubbing the skin or immersing it in chemicals, or both. It also emulsifies oils that lie on the outer cutaneous layer and mechanically removes potential pathogens on the outer layers of the skin. Examples of degerming procedures are the surgical handscrub, the application of alcohol wipes to the skin, and the cleansing of a wound with germicidal soap and water. The concepts of antisepsis and degermation clearly overlap, since a degerming procedure can simultaneously be antiseptic, and vice versa.

What Is Microbial Death?

Death is a phenomenon that involves the permanent termination of an organism's vital processes. Signs of life in complex organisms such as animals are self-evident, and death is made clear by loss of nervous function, respiration, or heartbeat. In contrast, death in microscopic organisms that are composed of just one or a few cells is often hard to detect, because they reveal no conspicuous vital signs to begin with. Lethal agents (such as radiation and chemicals) do not necessarily alter the overt appearance of microbial cells. Even the loss of movement in a motile microbe cannot be used to indicate death. This fact has made it necessary to develop special qualifications that define and delineate microbial death.

The destructive effects of chemical or physical agents occur at the level of a single cell. As the cell is continuously exposed to an agent such as intense heat or toxic chemicals, various cell structures become dysfunctional, and the entire cell can sustain irreversible damage. At present, the most practical way to detect this damage is to determine if a microbial cell can still reproduce when exposed to a suitable environment. If the microbe has sustained metabolic or structural damage to such an extent that it can no longer reproduce, even under ideal environmental conditions, then it is no longer viable. The permanent loss of reproductive capability, even under optimum growth conditions, has become the accepted microbiological definition of death.

Factors That Affect Death Rate

The ability to define microbial death has tremendous theoretical and practical importance. It allows medicine and industry to test the conditions required to destroy microorganisms, to pinpoint the ways that antimicrobial agents kill cells, and to establish standards of sterilization and disinfection in these fields. Hundreds of testing procedures have been developed for evaluating physical and chemical agents.

The cells of a culture show marked variation in susceptibility to a given microbicidal agent. Death of the whole population is not instantaneous but begins when a certain threshold of microbicidal agent (some combination of time and concentration) is met. Death continues in a logarithmic manner as the time or concentration of the agent is increased **(figure 11.2)**. Because many microbicidal agents target the cell's metabolic processes, active cells (younger, rapidly dividing) tend to die more quickly than those that are less metabolically active (older, inactive). Eventually, a point is reached at which survival of any cells is highly unlikely; this point is equivalent to sterilization.

The effectiveness of a particular agent is governed by several factors besides time. These additional factors influence the action of antimicrobial agents:

1. The number of microorganisms **(figure 11.2b)**. A higher load of contaminants requires more time to destroy.
2. The nature of the microorganisms in the population **(figure 11.2c)**. In most actual circumstances of disinfection and sterilization, the target population is not a single species of microbe but a mixture of bacteria, fungi, spores, and viruses, presenting a broad spectrum of microbial resistance.
3. The temperature and pH of the environment.

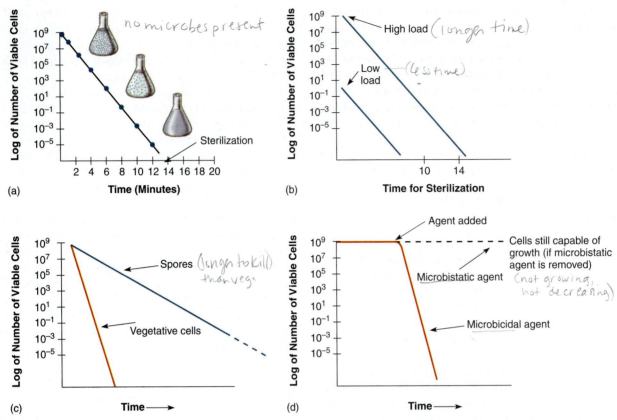

✕ **FIGURE 11.2**   **Factors that influence the rate at which microbes are killed by antimicrobial agents.**
(a) Length of exposure to the agent. During exposure to a chemical or physical agent, all cells of a microbial population, even a pure culture, do not die simultaneously. Over time, the number of viable organisms remaining in the population decreases logarithmically, giving a straight-line relationship on a graph. The point at which the number of survivors is infinitesimally small is considered sterilization. **(b)** Effect of the microbial load. **(c)** Relative resistance of spores versus vegetative forms. **(d)** Action of the agent, whether microbicidal or microbistatic.

4. The concentration (dosage, intensity) of the agent. For example, UV radiation is most effective at 260 nm and most disinfectants are more active at higher concentrations.
5. The mode of action of the agent **(figure 11.2d).** How does it kill or inhibit the microorganism?
6. The presence of solvents, interfering organic matter, and inhibitors. Saliva, blood, and feces can inhibit the actions of disinfectants and even of heat.

The influence of these factors will be discussed in greater detail in subsequent sections.

How Antimicrobial Agents Work: Their Modes of Action

An antimicrobial agent's adverse effect on cells is known as its *mode* (or *mechanism*) *of action*. Agents affect one or more cellular targets, inflicting damage progressively until the cell is no longer able to survive. Antimicrobials have a range of cellular targets, with the agents that are least selective in their targeting tending to be effective against the widest range of microbes (examples include heat and radiation). More selective agents (drugs, for example) tend to target only a single cellular component and are much more restricted as to the microbes they are effective against.

The cellular targets of physical and chemical agents fall into four general categories:

1. the cell wall,
2. the cell membrane,
3. cellular synthetic processes (DNA, RNA), and
4. proteins.

The Effects of Agents on the Cell Wall

The cell wall maintains the structural integrity of bacterial and fungal cells. Several types of chemical agents damage the cell wall by blocking its synthesis, digesting it, or breaking down its surface. A cell deprived of a functioning cell wall becomes fragile and is lysed very easily. Examples of this mode of action include some antimicrobial drugs (penicillins) that interfere with the synthesis of the cell wall in bacteria (described in chapter 12). Detergents and alcohol can also disrupt cell walls, especially in gram-negative bacteria.

How Agents Affect the Cell Membrane

All microorganisms have a cell membrane composed of lipids and proteins, and even some viruses have an outer membranous envelope. As we learned in previous chapters, a cell's membrane provides a two-way system of transport. If this membrane is disrupted, a cell loses its selective permeability and can neither prevent the loss of vital molecules nor bar the entry of damaging chemicals. Loss of those abilities leads to cell death. Detergents called **surfactants** (sir-fak'-tunt) work as microbicidal agents because they lower the surface tension of cell membranes. Surfactants are polar molecules with hydrophilic and hydrophobic regions that can physically bind to the lipid layer and penetrate the internal hydrophobic region of membranes. In effect, this process "opens up" the once tight interface, leaving leaky spots that allow injurious chemicals to seep into the cell and important ions to seep out (figure 11.3).

Agents That Affect Protein and Nucleic Acid Synthesis

Microbial life depends upon an orderly and continuous supply of proteins to function as enzymes and structural molecules. As we saw in chapter 9, these proteins are synthesized via the ribosomes through a complex process called translation. For example, the antibiotic chloramphenicol binds to the ribosomes of bacteria in a way that stops peptide bonds from forming. In its presence, many bacterial cells are inhibited from forming proteins required in growth and metabolism and are thus inhibited from multiplying. Most of the agents that block protein synthesis are drugs used in antimicrobial therapy. These drugs will be discussed in greater detail in chapter 12.

The nucleic acids are likewise necessary for the continued functioning of microbes. DNA must be regularly replicated and transcribed in growing cells, and any agent that either impedes these processes or changes the genetic code is potentially antimicrobial. Some agents bind irreversibly to DNA, preventing both transcription and translation; others are mutagenic agents. Gamma, ultraviolet, or X radiation causes mutations that result in permanent inactivation of DNA. Chemicals such as formaldehyde and ethylene oxide also interfere with DNA and RNA function.

Agents That Alter Protein Function

A microbial cell contains large quantities of proteins that function properly only if they remain in a normal three-dimensional configuration called the *native state*. The antimicrobial properties of some agents arise from their capacity to disrupt, or **denature**, proteins. In general, denaturation occurs when the bonds that maintain the secondary and tertiary structure of the protein are broken. Breaking these bonds will cause the protein to unfold or create random, irregular loops and coils (figure 11.4). One

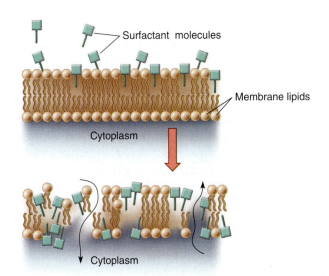

FIGURE 11.3 **Mode of action of surfactants on the cell membrane.**
Surfactants inserting in the lipid bilayer disrupt it and create abnormal channels that alter permeability and cause leakage both into and out of the cell.

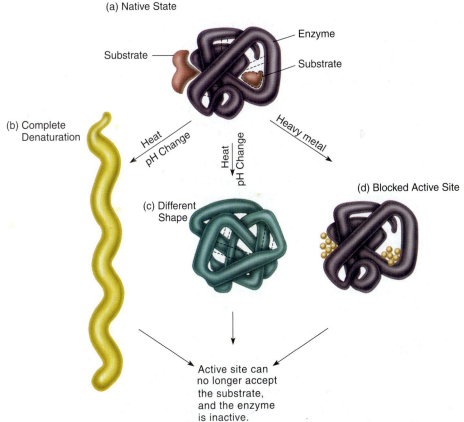

FIGURE 11.4 **Modes of action affecting protein function.**
(a) The native (functional) state is maintained by bonds that create active sites to fit the substrate. Some agents denature the protein by breaking all or some secondary and tertiary bonds. Results are (b) complete unfolding or (c) random bonding and incorrect folding. (d) Some agents react with functional groups on the active site and interfere with bonding.

way that proteins can be denatured is through coagulation by moist heat (the same reaction seen in the irreversible solidification of the white of an egg when boiled). Chemicals such as strong organic solvents (alcohols, acids) and phenolics also coagulate proteins. Other antimicrobial agents, such as metallic ions, attach to the active site of the protein and prevent it from interacting with its correct substrate. Regardless of the exact mechanism, such losses in normal protein function can promptly arrest metabolism. Most antimicrobials of this type are nonselective as to the microbes they affect.

Practical Concerns in Microbial Control

Numerous considerations govern the selection of a workable method of microbial control. These are among the most pressing concerns:

1. Does the application require sterilization, or is disinfection adequate? In other words, must spores be destroyed, or is it necessary to destroy only vegetative pathogens?
2. Is the item to be reused or permanently discarded? If it will be discarded, then the quickest and least expensive method should be chosen.
3. If it will be reused, can the item withstand heat, pressure, radiation, or chemicals?
4. Is the control method suitable for a given application? (For example, ultraviolet radiation is a good sporicidal agent, but it will not penetrate solid materials.) Or, in the case of a chemical, will it leave an undesirable residue?
5. Will the agent penetrate to the necessary extent?
6. Is the method cost- and labor-efficient, and is it safe?

A remarkable variety of substances can require sterilization. They range from durable solids such as rubber to sensitive liquids such as serum, and even to entire office buildings, as seen in 2001 when the Hart Senate Office Building was contaminated with *Bacillus anthracis* endospores. Hundreds of situations requiring sterilization confront the network of persons involved in health care, whether technician, nurse, doctor, or manufacturer, and no universal method works well in every case.

Considerations such as cost, effectiveness, and method of disposal are all important. For example, the disposable plastic items such as catheters and syringes that are used in invasive medical procedures have the potential for infecting the tissues. These must be sterilized during manufacture by a nonheating method (gas or radiation), because heat can damage delicate plastics. After these items have been used, it is often necessary to destroy or decontaminate them before they are discarded because of the potential risk to the handler (from needlesticks). Steam sterilization, which is quick and sure, is a sensible choice at this point, because it does not matter if the plastic is destroyed. Health care workers are held to very high standards of infection prevention.

✔ CHECKPOINT

- Microbial control methods involve the use of physical and chemical agents to eliminate or reduce the numbers of microorganisms from a specific environment so as to prevent the spread of infectious agents, retard spoilage, and keep commercial products safe.
- The population of microbes that cause spoilage or infection varies widely in species composition, resistance, and harmfulness, so microbial control methods must be adjusted to fit individual situations.
- The type of microbial control is indicated by the terminology used. Sterilization agents destroy all viable organisms, including viruses. Antisepsis, disinfection, and sanitization agents reduce the numbers of viable microbes to a specified level.
- Antimicrobial agents are described according to their ability to destroy or inhibit microbial growth. Microbicidal agents cause microbial death. They are described by what they are *-cidal* for: sporocides, bactericides, fungicides, viricides.
- An antiseptic agent is applied to living tissue to destroy or inhibit microbial growth.
- A disinfectant agent is used on inanimate objects to destroy vegetative pathogens but not bacterial endospores.
- Sanitization reduces microbial numbers on inanimate objects to safe levels by physical or chemical means.
- Degermation refers to the process of mechanically removing microbes from the skin.
- Microbial death is defined as the permanent loss of reproductive capability in microorganisms.
- Antimicrobial agents attack specific cell sites to cause microbial death or damage. Any given antimicrobial agent attacks one of four major cell targets: the cell wall, the cell membrane, biosynthesis pathways for DNA or RNA, or protein (enzyme) function.

11.2 Methods of Physical Control

Microorganisms have adapted to the tremendous diversity of habitats the earth provides, even severe conditions of temperature, moisture, pressure, and light. For microbes that normally withstand such extreme physical conditions, our attempts at control would probably have little effect. Fortunately for us, we are most interested in controlling microbes that flourish in the same environment in which humans live. The vast majority of these microbes are readily controlled by abrupt changes in their environment. Most prominent among antimicrobial physical agents is heat. Other less widely used agents include radiation, filtration, ultrasonic waves, and even cold. The following sections will examine some of these methods and explore their practical applications in medicine, commerce, and the home.

Heat As an Agent of Microbial Control

A sudden departure from a microbe's temperature of adaptation is likely to have a detrimental effect on it. As a rule, elevated temperatures (exceeding the maximum growth temperature) are microbicidal, whereas lower temperatures (below the minimum growth temperature) are microbistatic. The two physical states of heat used in microbial control are moist and dry. *Moist heat* occurs in the form of hot water, boiling water, or steam (vaporized water). In practice, the temperature of moist heat usually ranges from 60°C to 135°C. As we shall see, the temperature of steam can be regulated by adjusting its pressure in a closed container. The expression *dry heat* denotes air with a low moisture content that has been heated by a flame or electric heating coil. In practice, the temperature of dry heat ranges from 160°C to several thousand degrees Celsius.

Mode of Action and Relative Effectiveness of Heat

Moist heat and dry heat differ in their modes of action as well as in their efficiency. Moist heat operates at lower temperatures and shorter exposure times to achieve the same effectiveness as dry heat **(table 11.2).** Although many cellular structures are damaged by moist heat, its most microbicidal effect is the coagulation and denaturation of proteins, which quickly and permanently halts cellular metabolism.

Dry heat dehydrates the cell, removing the water necessary for metabolic reactions, and it also denatures proteins. However, the lack of water actually increases the stability of some protein conformations, necessitating the use of higher temperatures when dry heat is employed as a method of microbial control. At very high temperatures, dry heat oxidizes cells, burning them to ashes. This method is the one used in the laboratory when a loop is flamed or in industry when medical waste is incinerated.

Heat Resistance and Thermal Death of Spores and Vegetative Cells

Bacterial endospores exhibit the greatest resistance, and vegetative states of bacteria and fungi are the least resistant to both moist and dry heat. Destruction of spores usually requires temperatures above boiling **(table 11.3),** although resistance varies widely.

Vegetative cells also vary in their sensitivity to heat, though not to the same extent as spores **(table 11.4).** Among bacteria, the death times with moist heat range from 50°C for 3 minutes (*Neisseria gonorrhoeae*) to 60°C for 60 minutes (*Staphylococcus aureus*). It is worth noting that vegetative cells of sporeformers are just as susceptible as vegetative cells of non-sporeformers, and that pathogens are neither more nor less susceptible than nonpathogens. Other microbes, including fungi, protozoa, and worms, are rather similar in their sensitivity to heat. Viruses are surprisingly resistant to heat, with a tolerance range extending from 55°C for 2 to 5 minutes (adenoviruses) to 60°C for 600 minutes (hepatitis A virus). For practical purposes, all non-heat-resistant forms of bacteria, yeasts, molds, protozoa, worms, and viruses are destroyed by exposure to 80°C for 20 minutes.

Practical Concerns in the Use of Heat: Thermal Death Measurements

Adequate sterilization requires that both temperature and length of exposure be considered. As a general rule, higher

TABLE 11.2	Comparison of Times and Temperatures to Achieve Sterilization with Moist and Dry Heat	
	Temperature	Time to Sterilize
Moist heat	121°C	15 min
	125°C	10 min
	134°C	3 min
Dry heat	121°C	600 min
	140°C	180 min
	160°C	120 min
	170°C	60 min

(handwritten annotations: "autoclave"; "does not allow pen. or water")

TABLE 11.3	Thermal Death Times of Various Endospores	
Organism	Temperature	Time of Exposure to Kill Spores
Moist heat		
Bacillus subtilis	121°C	1 min
B. stearothermophilis	121°C	12 min
Clostridium botulinum	120°C	10 min
C. tetani	105°C	10 min
Dry heat		
Bacillus subtilis	121°C	120 min
B. stearothermophilis	140°C	5 min
Clostridium botulinum	120°C	120 min
C. tetani	100°C	60 min

TABLE 11.4	Average Thermal Death Times of Vegetative Stages of Microorganisms	
Microbial Type	Temperature	Time (Min)
Non-spore-forming bacteria	58°C	28
Non-spore-forming bacteria	61°C	18
Vegetative stage of spore-forming bacteria	58°C	19
Fungal spores	76°C	22
Yeasts	59°C	19
Viruses		
Nonenveloped	57°C	29
Enveloped	54°C	22
Protozoan trophozoites	46°C	16
Protozoan cysts	60°C	6
Worm eggs	54°C	3
Worm larvae	60°C	10

temperatures allow shorter exposure times, and lower temperatures require longer exposure times. A combination of these two variables constitutes the **thermal death time,** or TDT, defined as the shortest length of time required to kill all test microbes at a specified temperature. The TDT has been experimentally determined for the microbial species that are common or important contaminants in various heat-treated materials. Another way to compare the susceptibility of microbes to heat is the **thermal death point** (TDP), defined as the lowest temperature required to kill all microbes in a sample in 10 minutes.

Many perishable substances are processed with moist heat. Some of these products are intended to remain on the shelf at room temperature for several months or even years. The chosen heat treatment must render the product free of agents of spoilage or disease. At the same time, the quality of the product and the speed and cost of processing must be considered. For example, in the commercial preparation of canned green beans, one of the cannery's greatest concerns is to prevent growth of the agent of botulism. From several possible TDTs (that is, combinations of time and temperature) for *Clostridium botulinum* spores, the cannery must choose one that kills all spores but does not turn the beans to mush. Out of these many considerations emerges an optimal TDT for a given processing method. Commercial canneries heat low-acid foods at 121°C for 30 minutes, a treatment that sterilizes these foods. Because of such strict controls in canneries, cases of botulism due to commercially canned foods are rare.

Common Methods of Moist Heat Control

The four ways that moist heat is employed to control microbes are

1. steam under pressure;
2. nonpressurized steam;
3. boiling water; and
4. pasteurization.

Sterilization with Steam Under Pressure   At sea level, normal atmospheric pressure is 15 pounds per square inch (psi), or 1 atmosphere. At this pressure water will boil (change from a liquid to a gas) at 100°C, and the resultant steam will remain at exactly that temperature, which is unfortunately too low to reliably kill all microbes. In order to raise the temperature of steam, the pressure at which it is generated must be increased. As the pressure is increased, the temperature at which water boils and the temperature of the steam produced both rise. For example, at a pressure of 20 psi (5 psi above normal), the temperature of steam is 109°C. As the temperature is increased to 10 psi above normal, the steam's temperature rises to 115°C, and at 15 psi above normal (a total of 2 atmospheres), it will be 121°C. It is not the pressure by itself that is killing microbes but the increased temperature it produces.

Such pressure-temperature combinations can be achieved only with a special device that can subject pure steam to pressures greater than 1 atmosphere. Health and commercial industries use an **autoclave** for this purpose, and a comparable home appliance is the pressure cooker. Autoclaves have a fundamentally similar plan: a cylindrical metal chamber with an airtight door on one end and racks to hold materials **(figure 11.5)**. Its construction includes a complex network of valves, pressure and temperature gauges, and ducts for regulating and measuring pressure and conducting the steam into the chamber. Sterilization is achieved when the steam condenses against the objects in the chamber and gradually raises their temperature.

Experience has shown that the most efficient pressure-temperature combination for achieving sterilization is 15 psi, which yields 121°C. It is possible to use higher pressure to reach higher temperatures (for instance, increasing the pressure to 30 psi raises the temperature to 132°C), but doing so will not significantly reduce the exposure time and can harm the items being sterilized. It is important to avoid overpacking or haphazardly loading the chamber, which prevents steam from circulating freely around the contents and impedes the full contact that is necessary. The duration of the process is adjusted according to the bulkiness of the items in the load (thick bundles of material or large flasks of liquid) and how full the chamber is. The range of holding times varies from 10 minutes for light loads to 40 minutes for heavy or bulky ones; the average time is 20 minutes.

The autoclave is a superior choice to sterilize heat-resistant materials such as glassware, cloth (surgical dressings), rubber (gloves), metallic instruments, liquids, paper, some media, and some heat-resistant plastics. If the items are heat-sensitive (plastic Petri plates) but will be discarded, the autoclave is still a good choice. However, the autoclave is ineffective for sterilizing substances that repel moisture (oils, waxes, powders).

Intermittent Sterilization   Selected substances that cannot withstand the high temperature of the autoclave can be subjected to *intermittent sterilization*, also called **tyndallization**.[1] This technique requires a chamber to hold the materials and a reservoir for boiling water. Items in the chamber are exposed to free-flowing steam for 30 to 60 minutes. This temperature is not sufficient to reliably kill spores, so a single exposure will not suffice. On the assumption that surviving spores will germinate into less resistant vegetative cells, the items are incubated at appropriate temperatures for 23 to 24 hours, and then again subjected to steam treatment. This cycle is repeated for 3 days in a row. Because the temperature never gets above 100°C, highly resistant spores that do not germinate may survive even after 3 days of this treatment.

Intermittent sterilization is used most often to process heat-sensitive culture media, such as those containing sera, egg, or carbohydrates (which can break down at higher temperatures) and some canned foods. It is probably not effective in sterilizing items such as instruments and dressings that provide no environment for spore germination, but it certainly can disinfect them.

---

1. Named for the British physicist John Tyndall, who did early experiments with sterilizing procedures.

Pasteurization: Disinfection of Beverages Fresh beverages such as milk, fruit juices, beer, and wine are easily contaminated during collection and processing. Because microbes have the potential for spoiling these foods or causing illness, heat is frequently used to reduce the microbial load and destroy pathogens. **Pasteurization** is a technique in which heat is applied to liquids to kill potential agents of infection and spoilage, while at the same time retaining the liquid's flavor and food value.

Ordinary pasteurization techniques require special heat exchangers that expose the liquid to 71.6°C for 15 seconds (flash method) or to 63°C to 66°C for 30 minutes (batch method). The first method is preferable because it is less likely to change flavor and nutrient content, and it is more effective against certain resistant pathogens such as *Coxiella* and *Mycobacterium*. Although these treatments inactivate most viruses and destroy the vegetative stages of 97% to 99% of bacteria and fungi, they do not kill endospores or **thermoduric** microbes (mostly nonpathogenic lactobacilli, micrococci, and yeasts). Milk is not sterile after regular pasteurization. In fact, it can contain 20,000 microbes per milliliter or more, which explains why even an unopened carton of milk will eventually spoil. Newer techniques can also produce *sterile milk* that has a storage life of 3 months. This milk is processed with ultrahigh temperature (UHT)—134°C for 1 to 2 seconds.

One important aim in pasteurization is to prevent the transmission of milk-borne diseases from infected cows or milk handlers. The primary targets of pasteurization are non-spore-forming pathogens: *Salmonella* species (a common cause of food infection), *Campylobacter jejuni* (acute intestinal infection), *Listeria monocytogenes* (listeriosis), *Brucella* species (undulant fever), *Coxiella burnetii* (Q fever), *Mycobacterium bovis*, *M. tuberculosis*, and several enteric viruses.

Pasteurization also has the advantage of extending milk storage time, and it can also be used by some wineries and breweries to stop fermentation and destroy contaminants.

Boiling Water: Disinfection A simple boiling water bath or chamber can quickly decontaminate items in the clinic and home. Because a single processing at 100°C will not kill all resistant cells, this method can be relied on only for disinfection and not for sterilization. Exposing materials to boiling water for 30 minutes will kill most non-spore-forming pathogens, including resistant species such as the tubercle bacillus and staphylococci. Probably the greatest disadvantage with this method is that the items can be easily recontaminated when removed from the water. Boiling is also a recommended method of disinfecting unsafe drinking water. In the home, boiling water is a fairly reliable way to sanitize and disinfect

(a)

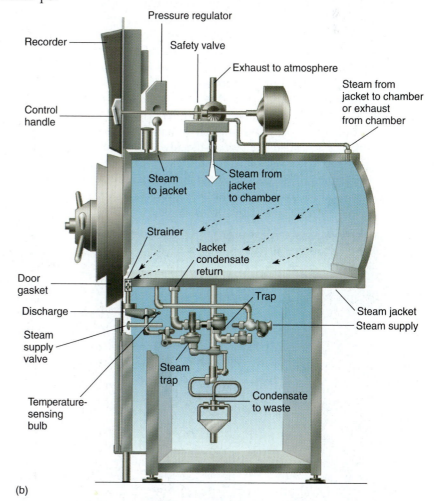
(b)

FIGURE 11.5 **Steam sterilization with the autoclave.**
(a) A large automatic autoclave used in sterilization by drug companies.
(b) Cutaway section, showing autoclave components.

(b) From John J. Perkins, *Principles and Methods of Sterilization in Health Science,* 2nd ed., 1969. Courtesy of Charles C Thomas, Publisher, Springfield, Illinois.

materials for babies, food preparation, and utensils, bedding, and clothing from the sickroom.

Dry Heat: Hot Air and Incineration

Dry heat is not as versatile or as widely used as moist heat, but it has several important sterilization applications. The temperatures and times employed in dry heat vary according to the particular method, but in general, they are greater than with moist heat. **Incineration** in a flame or electric heating coil is perhaps the most rigorous of all heat treatments. The flame of a Bunsen burner reaches 1,870°C at its hottest point, and furnaces/incinerators operate at temperatures of 800°C to 6,500°C. Direct exposure to such intense heat ignites and reduces microbes and other substances to ashes and gas.

Incineration of microbial samples on inoculating loops and needles using a Bunsen burner is a very common practice in the microbiology laboratory. This method is fast and effective, but it is also limited to metals and heat-resistant glass materials. This method also presents hazards to the operator (an open flame) and to the environment (contaminants on needle or loop often spatter when placed in flame). Tabletop infrared incinerators **(figure 11.6)** have replaced Bunsen burners in many labs for these reasons. Large incinerators are regularly employed in hospitals and research labs for complete destruction and disposal of infectious materials such as syringes, needles, cultural materials, dressings, bandages, bedding, animal carcasses, and pathology samples.

The hot-air oven provides another means of dry-heat sterilization. The so-called *dry oven* is usually electric (occasionally gas) and has coils that radiate heat within an enclosed compartment. Heated, circulated air transfers its heat to the materials in the oven. Sterilization requires exposure to

150°C to 180°C for 2 to 4 hours, which ensures thorough heating of the objects and destruction of spores.

The dry oven is used in laboratories and clinics for heat-resistant items that do not sterilize well with moist heat. Substances appropriate for dry ovens are glassware, metallic instruments, powders, and oils that steam does not penetrate well. This method is not suitable for plastics, cotton, and paper, which may burn at the high temperatures, or for liquids, which will evaporate. Another limitation is the time required for it to work.

The Effects of Cold and Desiccation

The principal benefit of cold treatment is to slow growth of cultures and microbes in food during processing and storage. *It must be emphasized that cold merely retards the activities of most microbes.* Although it is true that some microbes are killed by cold temperatures, most are not adversely affected by gradual cooling, long-term refrigeration, or deep-freezing. In fact, freezing temperatures, ranging from −70°C to −135°C, provide an environment that can preserve cultures of bacteria, viruses, and fungi for long periods. Some psychrophiles grow very slowly even at freezing temperatures and can continue to secrete toxic products. Ignorance of these facts is probably responsible for numerous cases of food poisoning from frozen foods that have been defrosted at room temperature and then inadequately cooked. Pathogens able to survive several months in the refrigerator are *Staphylococcus aureus*, *Clostridium* species (sporeformers), *Streptococcus* species, and several types of yeasts, molds, and viruses. Outbreaks of *Salmonella* food infection traced backed to refrigerated foods such as ice cream, eggs, and Tiramisu, are testimony to the inability of freezing temperatures to reliably kill pathogens.

Vegetative cells directly exposed to normal room air gradually become dehydrated, or **desiccated.** Delicate pathogens such as *Streptococcus pneumoniae*, the spirochete of syphilis, and *Neisseria gonorrhoeae* can die after a few hours of air-drying, but many others are not killed and some are even preserved. Endospores of *Bacillus* and *Clostridium* are viable for millions of years under extremely arid conditions. Staphylococci and streptococci in dried secretions, and the tubercle bacillus surrounded by sputum, can remain viable in air and dust for lengthy periods. Many viruses (especially nonenveloped) and fungal spores can also withstand long periods of desiccation. Desiccation can be a valuable way to preserve foods because it greatly reduces the amount of water available to support microbial growth.

It is interesting to note that a combination of freezing and drying—**lyophilization** (ly-off"-il-ih-za'-shun)—is a common method of preserving microorganisms and other cells in a viable state for many years. Pure cultures are frozen instantaneously and exposed to a vacuum that rapidly removes the water (it goes right from the frozen state into the vapor state). This method avoids the formation of ice crystals that would damage the cells. Although not all cells survive this process, enough of them do to permit future reconstitution of that culture.

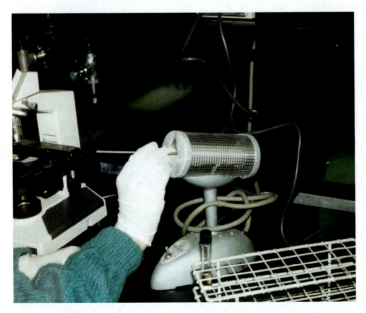

FIGURE 11.6   Dry heat incineration.
Infrared incinerator with shield to prevent spattering of microbial samples during flaming.

As a general rule, chilling, freezing, and desiccation should not be construed as methods of disinfection or sterilization because their antimicrobial effects are erratic and uncertain, and one cannot be sure that pathogens subjected to them have been killed.

Radiation as a Microbial Control Agent

Another way in which energy can serve as an antimicrobial agent is through the use of radiation. **Radiation** is defined as energy emitted from atomic activities and dispersed at high velocity through matter or space. Although radiation exists in many states and can be described and characterized in various ways, we will consider only those types suitable for microbial control: gamma rays, X rays, and ultraviolet radiation.

Modes of Action of Ionizing Versus Nonionizing Radiation

The actual physical effects of radiation on microbes can be understood by visualizing the process of **irradiation,** or bombardment with radiation, at the cellular level **(figure 11.7).** When a cell is bombarded by certain waves or particles, its molecules absorb some of the available energy, leading to one of two consequences: (1) If the radiation ejects orbital electrons from an atom, it causes ions to form; this type of radiation is termed **ionizing radiation.** One of the most sensitive targets for ionizing radiation is DNA, which will undergo mutations on a broad scale. Secondary lethal effects appear to be chemical changes in organelles and the production of toxic substances. Gamma rays, X rays, and high-speed electrons are all ionizing in their effects. (2) **Nonionizing radiation,** best exemplified by UV, excites atoms by raising them to a higher energy state, but it does not ionize them. This atomic excitation, in turn, leads to the formation of abnormal bonds within molecules such as DNA and is thus a source of mutations.

Ionizing Radiation: Gamma Rays, X Rays, and Cathode Rays

Over the past several years, ionizing radiation has become safer and more economical to use, and its applications have mushroomed. It is a highly effective alternative for sterilizing materials that are sensitive to heat or chemicals. Because it sterilizes in the absence of heat, irradiation is a type of **cold (or low-temperature) sterilization.**[2] Devices that emit ionizing rays include gamma-ray machines containing radioactive cobalt, X-ray machines similar to those used in medical diagnosis, and cathode-ray machines that operate like the vacuum tube in a television set. Items are placed in these machines and irradiated for a short time with a carefully chosen dosage. The dosage of radiation is measured in Grays (which

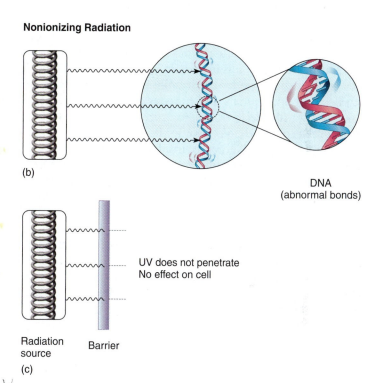

Ionizing Radiation

Radiation source   Barrier   Cell   DNA (breakages)

(a)

Nonionizing Radiation

(b)   DNA (abnormal bonds)

UV does not penetrate
No effect on cell

Radiation source   Barrier

(c)

FIGURE 11.7   Cellular effects of irradiation.
(a) Ionizing radiation can penetrate a solid barrier, bombard a cell, enter it, and dislodge electrons from molecules. Breakage of DNA creates massive mutations. **(b)** Nonionizing radiation enters a cell, strikes molecules, and excites them. The effect on DNA is mutation by formation of abnormal bonds. **(c)** A solid barrier cannot be penetrated by nonionizing radiation.

has replaced the older term, rads). Depending on the application, exposure ranges from 5 to 50 kiloGrays (kGray; a kiloGray is equal to 1000 Grays). Although all ionizing radiations can penetrate liquids and most solid materials, gamma rays are most penetrating, X rays are intermediate, and cathode rays least penetrating.

Applications of Ionizing Radiation

Foods have been subject to irradiation in limited circumstances for more than 50 years. From flour to pork and

---

2. This is a possibly confusing use of the word "cold." In this context it only means the absence of heat. Beer manufacturers have sometimes used this terminology as well. When they say that their product is "cold-filtered," they mean that it has been freed of contaminants via filtration, i.e., in the absence of heat.

ground beef, to fruits and vegetables, radiation is used to kill not only bacterial pathogens but also insects and worms and even to inhibit the sprouting of white potatoes. As soon as radiation is mentioned, however, consumer concern arises that food may be made less nutritious, unpalatable, or even unsafe by subjecting it to ionizing radiation. But irradiated food has been extensively studied, and each of these concerns has been addressed.

Irradiation may lead to a small decrease in the amount of thiamine (vitamin B1) in food, but this change is small enough to be inconsequential. The irradiation process does produce short-lived free radical oxidants, which disappear almost immediately (this same type of chemical intermediate is produced through cooking as well). Certain foods do not irradiate well and are not good candidates for this type of antimicrobial control. The white of eggs becomes milky and liquid, grapefruit gets mushy, and alfalfa seeds do not germinate properly. Lastly, it is important to remember that food is not made radioactive by the irradiation process, and many studies, in both animals and humans, have concluded that there are no ill effects from eating irradiated food. In fact, NASA relies on irradiated meat for its astronauts.

While the potential costs of irradiation will always be debated, the potential benefits are enormous. It has been estimated that irradiation of 50% of the meat and poultry in the United States would result in 900,000 fewer cases of infection, 8,500 fewer hospitalizations, and 350 fewer deaths each year. Radiation is currently approved in the United States for the reduction of bacterial pathogens such as *E. coli* and *Salmonella* in beef and chicken, reduction of *Trichinella* worms in pork, and the reduction of insects on fruits and vegetables. Officials of the United Nations and World Health Organization are proponents of food irradiation. An additional benefit of irradiation is that microbes responsible for food spoilage are killed along with pathogens, leading to an increased shelf life. In any event, no irradiated food can be sold to consumers without clear labeling that this method has been used **(figure 11.8)**. See chapter 24 for further discussion about irradiation of food.

Sterilizing medical products with ionizing radiation is a rapidly expanding field. Drugs, vaccines, medical instruments (especially plastics), syringes, surgical gloves, tissues such as bone and skin, and heart valves for grafting all lend themselves to this mode of sterilization. Since the anthrax attacks of 2001, mail delivered to certain Washington, D.C., zip codes has been irradiated with ionizing radiation. Its main advantages include speed, high penetrating power (it can sterilize materials through outer packages and wrappings), and the absence of heat. Its main disadvantages are potential dangers to radiation machine operators from exposure to radiation and possible damage to some materials.

Nonionizing Radiation: Ultraviolet Rays

Ultraviolet radiation (UV) ranges in wavelength from approximately 100 nm to 400 nm. It is most lethal from 240 nm to 280 nm (with a peak at 260 nm). In everyday practice, the source of UV radiation is the germicidal lamp, which generates radiation at 254 nm. Owing to its lower energy state, UV radiation is not as penetrating as ionizing radiation. Because UV radiation passes readily through air, slightly through liquids, and only poorly through solids, the object to be disinfected must be directly exposed to it for full effect.

As UV radiation passes through a cell, it is initially absorbed by DNA. Specific molecular damage occurs on the pyrimidine bases (thymine and cytosine), which form abnormal linkages with each other called **pyrimidine dimers (figure 11.9)**. These bonds occur between adjacent bases on the same DNA strand and interfere with normal DNA replication and transcription. The results are inhibition of growth

FIGURE 11.8
Sterilization with ionizing radiation.
(a) This irradiation machine uses radioactive cobalt 60 as a gamma radiation source to sterilize fruits, vegetables, meats, fish, and spices. Although this method has stirred some controversy regarding its safety, it is gaining in acceptance because of its ability to increase shelf-life and reduce food-borne infections. **(b)** Regulations dictate that this universal symbol for irradiation must be affixed to all irradiated materials.

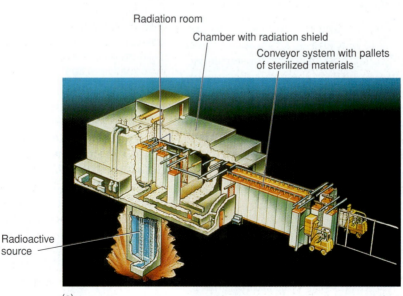

Radiation room
Chamber with radiation shield
Conveyor system with pallets of sterilized materials
Radioactive source

(a)

(b)

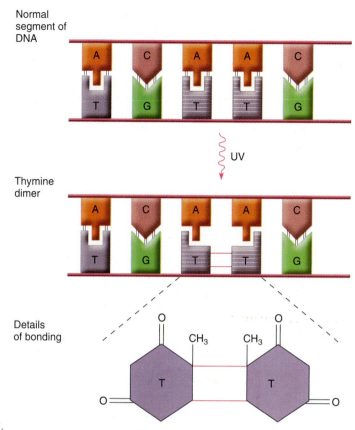

Normal segment of DNA

Thymine dimer

UV

Details of bonding

FIGURE 11.9 **Formation of pyrimidine dimers by the action of ultraviolet (UV) radiation.**
This shows what occurs when two adjacent thymine bases on one strand of DNA are induced by UV rays to bond laterally with each other. The result is a thymine dimer shown in greater detail. Dimers can also occur between adjacent cytosines and thymine and cytosine bases. If they are not repaired, dimers can prevent that segment of DNA from being correctly replicated or transcribed. Massive dimerization is lethal to cells.

FIGURE 11.10 **An ultraviolet (UV) treatment system for disinfection of water.**
Water flows through racks of UV lamps and is exposed to 254 nm UV radiation. This system has a capacity of several million gallons per day and can be used as an alternative to chlorination.

posed directly to a lamp. This method can be used to treat drinking water **(figure 11.10)** and to purify other liquids (milk and fruit juices) as an alternative to heat. Ultraviolet treatment has proved effective in freeing vaccines and plasma from contaminants. The surfaces of solid, nonporous materials such as walls and floors, as well as meat, nuts, tissues for grafting, and drugs, have been successfully disinfected with UV.

One major disadvantage of UV is its poor powers of penetration through solid materials such as glass, metal, cloth, plastic, and even paper. Another drawback to UV is the damaging effect of overexposure on human tissues, including sunburn, retinal damage, cancer, and skin wrinkling.

and cellular death. In addition to altering DNA directly, UV radiation also disrupts cells by generating toxic photochemical products called free radicals. These highly reactive molecules interfere with essential cell processes by binding to DNA, RNA, and proteins. Ultraviolet rays are a powerful tool for destroying fungal cells and spores, bacterial vegetative cells, protozoa, and viruses. Bacterial spores are about 10 times more resistant to radiation than are vegetative cells, but they can be killed by increasing the time of exposure.

Applications of UV Radiation UV radiation is usually directed at disinfection rather than sterilization. Germicidal lamps can cut down on the concentration of airborne microbes as much as 99%. They are used in hospital rooms, operating rooms, schools, food preparation areas, and dental offices. Ultraviolet disinfection of air has proved effective in reducing postoperative infections, preventing the transmission of infections by respiratory droplets, and curtailing the growth of microbes in food-processing plants and slaughterhouses.

Ultraviolet irradiation of liquids requires special equipment to spread the liquid into a thin, flowing film that is ex-

Sterilization by Filtration: Techniques for Removing Microbes

Filtration is an effective method to remove microbes from air and liquids. In practice, a fluid is strained through a filter with openings large enough for the fluid to pass through but too small for microorganisms to pass through **(figure 11.11)**.

Most modern microbiological filters are thin membranes of cellulose acetate, polycarbonate, and a variety of plastic materials (Teflon, nylon) whose pore size can be carefully controlled and standardized. Ordinary substances such as charcoal, diatomaceous earth, or unglazed porcelain are also used in some applications. Viewed microscopically, most filters are perforated by very precise, uniform pores **(figure 11.11b)**. The pore diameters vary from coarse (8 μm) to ultrafine (0.02 μm), permitting selection of the minimum particle size to be trapped. Those with even smaller pore diameters permit true sterilization by removing viruses, and some will even remove large proteins. A sterile liquid filtrate is typically produced by suctioning the liquid through a sterile filter into a presterilized container. These filters are also used to

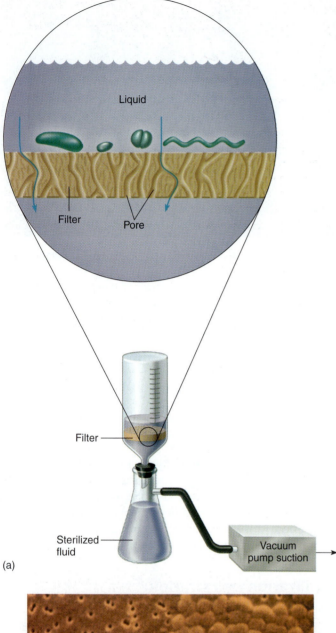

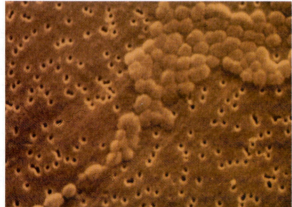

FIGURE 11.11 **Membrane filtration.**
(a) Vacuum assembly for achieving filtration of liquids through suction. Inset shows filter as seen in cross section, with tiny passageways (pores) too small for the microbial cells to enter but large enough for liquid to pass through. **(b)** Scanning electron micrograph of filter, showing relative size of pores and bacteria trapped on its surface (5,900×).

separate mixtures of microorganisms and to enumerate bacteria in water analysis (see chapter 24).

Applications of Filtration Sterilization

Filtration sterilization is used to prepare liquids that cannot withstand heat, including serum and other blood products, vaccines, drugs, IV fluids, enzymes, and media. Filtration has been employed as an alternative method for sterilizing milk and beer without altering their flavor. It is also an important step in water purification. Its use extends to filtering out particulate impurities (crystals, fibers, and so on) that can cause severe reactions in the body. It has the disadvantage of not removing soluble molecules (toxins) that can cause disease.

Filtration is also an efficient means of removing airborne contaminants that are a common source of infection and spoilage. High-efficiency particulate air (HEPA) filters are widely used to provide a flow of sterile air to hospital rooms and sterile rooms. A vacuum with a HEPA filter was even used to remove anthrax spores from the Senate offices most heavily contaminated after the terrorist attack in late 2001 (see Insight 11.4).

✔ CHECKPOINT

- Physical methods of microbial control include heat, cold, radiation, drying, and filtration.
- Heat is the most widely used method of microbial control. It is used in combination with water (moist heat) or as dry heat (oven, flames).
- The thermal death time (TDT) is the shortest length of time required to kill all microbes at a specific temperature. The TDT is longest for spore-forming bacteria and certain viruses.
- The thermal death point (TDP) is the lowest temperature at which all microbes are killed in a specified length of time (10 minutes).
- Autoclaving, or steam sterilization, is the process by which steam is heated under pressure to sterilize a wide range of materials in a comparatively short time (minutes to hours). It is effective for most materials except water-resistant substances such as oils, waxes, and powders.
- Boiling water and pasteurization of beverages disinfect but do not sterilize materials.
- Dry heat is microbicidal under specified times and temperatures. Flame heat, or incineration, is microbicidal. It is used when total destruction of microbes and materials is desired.
- Chilling, freezing, and desiccation are microbistatic but not microbicidal. They are not considered true methods of disinfection because they are not consistent in their effectiveness.
- Ionizing radiation (cold sterilization) by gamma rays and X rays is used to sterilize medical products, meats, and spices. It damages DNA and cell organelles by producing disruptive ions.
- Ultraviolet light, or nonionizing radiation, has limited penetrating ability. It is therefore restricted to disinfecting air and certain liquids.
- Sterilization by filtration removes microbes from heat-sensitive liquids and circulating air. The pore size of the filter determines what kinds of microbes are removed.

INSIGHT 11.2

Microbiology

Pathogen Paranoia: "The Only Good Microbe Is a Dead Microbe"

The sensational publicity over outbreaks of infections such as influenza, anthrax, and microbial food poisoning has monumentally influenced the public view of microorganisms. Thousands of articles have sprinkled the news services over the past 5 years. On the positive side, this glut of information has improved people's awareness of the importance of microorganisms. And, certainly, such knowledge can be seen as beneficial when it leads to well-reasoned and sensible choices, such as using greater care in handwashing, food handling, and personal hygiene. But sometimes a little knowledge can be dangerous. The trend also seems to have escalated into an obsessive fear of "germs" lurking around every corner and a fixation on eliminating microbes from the environment and the human body.

As might be expected, commercial industries have found a way to capitalize on those fears. Every year, the number of products that incorporate antibacterial or germicidal "protection" increases dramatically. A widespread array of cleansers and commonplace materials have already had antimicrobial chemicals added. First it was hand soaps and dishwashing detergents, and eventually the list grew to include shampoos, laundry aids, hand lotions, foot pads for shoes, deodorants, sponges and scrub pads, kitty litter, acne medication, cutting boards, garbage bags, toys, and toothpaste.

One chemical agent routinely added to these products is a phenolic called *triclosan* (Irgasan). This substance is fairly mild and nontoxic and does indeed kill most pathogenic bacteria. However, it does not reliably destroy viruses or fungi and has been linked to cases of skin rashes due to hypersensitivity.

One unfortunate result of the negative news on microbes is how it fosters the feeling that all microbes are harmful. We must not forget that most human beings manage to remain healthy despite the fact that they live in continual intimate contact with microorganisms. We really do not have to be preoccupied with microbes every minute or feel overly concerned that the things we touch, drink, or eat are sterile, as long as they are somewhat clean and free of pathogens. For most of us, resistance to infection is well maintained by our numerous host defenses.

Medical experts are concerned that the widespread overuse of these antibacterial chemicals could favor the survival and growth of resistant strains of bacteria. A study reported in 2000 that many pathogens such as *Mycobacterium tuberculosis* and *Pseudomonas* are naturally resistant to triclosan, and that *E. coli* and *Staphylococcus aureus* have already demonstrated decreased sensitivity to it. The widespread use of this chemical may actually select for "super microbes" that survive ordinary disinfection. Another outcome of overuse of environmental germicides is to reduce the natural contact with microbes that is required to maintain the normal resident flora and stimulate immunities. Constant use of these agents could shift the balance in the normal flora of the body by killing off harmless or beneficial microbes. And there's one more thing. More and more studies are showing that when some bacteria become resistant to antibacterial agents, including triclosan, they simultaneously become resistant to antibiotics, such as tetracycline and erythromycin.

Infectious disease specialists urge a happy medium approach. Instead of filling the home with questionable germicidal products, they encourage cleaning with traditional soaps and detergents, reserving more potent products to reduce the spread of infection among household members.

The molecular structure of triclosan, also known as Irgasan and Ster-Zac, a phenol-based chemical that destroys bacteria by disrupting cell walls and membranes.

11.3 Chemical Agents in Microbial Control

Chemical control of microbes probably emerged as a serious science in the early 1800s, when physicians used chloride of lime and iodine solutions to treat wounds and to wash their hands before surgery. At the present time, approximately 10,000 different antimicrobial chemical agents are manufactured; probably 1,000 of them are used routinely in the health care arena and the home. A genuine need exists to avoid infection and spoilage, but the abundance of products available to "kill germs," "disinfect," "antisepticize," "clean and sanitize," "deodorize," "fight plaque," and "purify the air" indicates a preoccupation with eliminating microbes from the environment that, at times, seems excessive **(Insight 11.2).**

Antimicrobial chemicals occur in the liquid, gaseous, or even solid state, and they range from disinfectants and antiseptics to sterilants and preservatives (chemicals that inhibit the deterioration of substances). For the sake of convenience (and sometimes safety) many solid or gaseous antimicrobial chemicals are dissolved in water, alcohol, or a mixture of the two to produce a liquid solution. Solutions containing pure water as the solvent are termed *aqueous*, whereas those dissolved in pure alcohol or water-alcohol mixtures are termed **tinctures.**

Choosing a Microbicidal Chemical

The choice and appropriate use of antimicrobial chemical agents is of constant concern in medicine and dentistry. Although actual clinical practices of chemical decontamination

TABLE 11.5   Qualities of Chemical Agents Used in Health Care

Agent	Target Microbes	Level of Activity	Toxicity	Comments
Chlorine	Sporicidal (slowly)	Intermediate	Gas is highly toxic; solution irritates skin	Inactivated by organics; unstable in sunlight
Iodine	Sporicidal (slowly)	Intermediate	Can irritate tissue; toxic if ingested	Iodophors* are milder forms
Phenolics	Some bacteria, viruses, fungi	Low to intermediate	Can be absorbed by skin; can cause CNS damage	Poor solubility; expensive
Alcohols	Most bacteria, viruses, fungi	Intermediate	Toxic if ingested; a mild irritant; dries skin	Flammable, fast-acting
Hydrogen peroxide,* stabilized	Sporicidal	High	Toxic to eyes; toxic if ingested	Improved stability; works well in organic matter
Quaternary ammonium compounds	Some bactericidal, virucidal, fungicidal activity	Low	Irritating to mucous membranes; poisonous if taken internally	Weak solutions can support microbial growth; easily inactivated
Soaps	Certain very sensitive species	Very low	Nontoxic; few if any toxic effects	Used for removing soil, oils, debris
Mercurials	Weakly microbistatic	Low	Highly toxic if ingested, inhaled, absorbed	Easily inactivated
Silver nitrate	Bactericidal	Low	Toxic, irritating	Discolors skin
Glutaraldehyde*	Sporicidal	High	Can irritate skin; toxic if absorbed	Not inactivated by organic matter; unstable
Formaldehyde	Sporicidal	Intermediate to high	Very irritating; fumes damaging, carcinogenic	Slow rate of action; limited applications
Ethylene oxide gas*	Sporicidal	High	Very dangerous to eyes, lungs; carcinogenic	Explosive in pure state; good penetration; materials must be aerated
Dyes	Weakly bactericidal, fungicidal	Low	Low toxicity	Stains materials, skin
Chlorhexidine*	Most bacteria, some viruses, fungi	Low to intermediate	Low toxicity	Fast-acting, mild, has residual effects

*These chemicals approach the ideal by having many of the following characteristics: broad spectrum, low toxicity, fast action, penetrating abilities, residual effects, stability, potency in organic matter, and solubility.

vary widely, some desirable qualities in a germicide have been identified, including:

1. rapid action even in low concentrations,
2. solubility in water or alcohol and long-term stability,
3. broad-spectrum microbicidal action without being toxic to human and animal tissues,
4. penetration of inanimate surfaces to sustain a cumulative or persistent action,
5. resistance to becoming inactivated by organic matter,
6. noncorrosive or nonstaining properties,
7. sanitizing and deodorizing properties, and
8. affordability and ready availability.

As yet, no chemical can completely fulfill all of those requirements, but glutaraldehyde and hydrogen peroxide approach this ideal. At the same time, we should question the rather overinflated claims made about certain commercial agents such as mouthwashes and disinfectant air sprays.

Germicides are evaluated in terms of their effectiveness in destroying microbes in medical and dental settings. The three levels of chemical decontamination procedures are *high, intermediate,* and *low* **(table 11.5).** High-level germicides kill endospores, and, if properly used, are sterilants. Materials that necessitate high-level control are medical devices—for example, catheters, heart-lung equipment, and implants—that are not heat-sterilizable and are intended to enter body tissues during medical procedures. Intermediate-level germicides kill fungal (but not bacterial) spores, resistant pathogens such as the tubercle bacillus, and viruses. They are used to disinfect items (respiratory equipment, thermometers) that come into intimate contact with the mucous membranes but are noninvasive. Low levels of disinfection eliminate only vegetative bacteria, vegetative fungal cells, and some viruses. They are used to clean materials such as electrodes, straps, and furniture that touch the skin surfaces but not the mucous membranes.

Factors That Affect the Germicidal Activity of Chemicals

Factors that control the effect of a germicide include the nature of the microorganisms being treated, the nature of the material being treated, the degree of contamination, the time

TABLE 11.6	Required Concentrations and Times for Chemical Destruction of Selected Microbes	
Organism	**Concentration**	**Time**
Agent: Aqueous Iodine		
Staphylococcus aureus	2%	2 min
Escherichia coli	2%	1.5 min
Enteric viruses	2%	10 min
Agent: Chlorine		
Mycobacterium tuberculosis	50 ppm	50 sec
Entamoeba cysts (protozoa)	0.1 ppm	150 min
Hepatitis A virus	3 ppm	30 min
Agent: Phenol		
Staphylococcus aureus	1:85 dil	10 min
Escherichia coli	1:75 dil	10 min
Agent: Ethyl Alcohol		
Staphylococcus aureus	70%	10 min
Escherichia coli	70%	2 min
Poliovirus	70%	10 min
Agent: Hydrogen Peroxide		
Staphylococcus aureus	3%	12.5 sec
Neisseria gonorrhoeae	3%	0.3 sec
Herpes simplex virus	3%	12.8 sec
Agent: Quaternary Ammonium Compound		
Staphylococcus aureus	450 ppm	10 min
Salmonella typhi	300 ppm	10 min
Agent: Silver Ions		
Staphylococcus aureus	8 µg/ml	48 h
Escherichia coli	2 mg/ml	48 h
Candida albicans (yeast)	14 mg/ml	48 h
Agent: Glutaraldehyde		
Staphylococcus aureus	2%	<1 min
Mycobacterium tuberculosis	2%	<10 min
Herpes simplex virus	2%	<10 min
Agent: Ethylene Oxide Gas		
Streptococcus faecalis	500 mg/l	2–4 min
Influenza virus	10,000 mg/l	25 h
Agent: Chlorhexidine		
Staphylococcus aureus	1:10 dil	15 sec
Escherichia coli	1:10 dil	30 sec

volume of the liquid chemical (solute) is diluted in a larger volume of solvent to achieve a certain ratio. For example, a common laboratory phenolic disinfectant such as Lysol is usually diluted 1:200; that is, one part of chemical has been added to 200 parts of water by volume. Solutions such as chlorine that are effective in very diluted concentrations are expressed in parts per million (ppm). In percent solutions, the solute is added to water by weight or volume to achieve a certain percentage in the solution. Alcohol, for instance, is used in percentages ranging from 50% to 95%. In general, solutions of low dilution or high percentage have more of the active chemical (are more concentrated) and tend to be more germicidal, but expense and potential toxicity can necessitate using the minimum strength that is effective.

Another factor that contributes to germicidal effectiveness is the length of exposure. Most compounds require adequate contact time to allow the chemical to penetrate and to act on the microbes present. The composition of the material being treated must also be considered. Smooth, solid objects are more reliably disinfected than are those with pores or pockets that can trap soil. An item contaminated with common biological matter such as serum, blood, saliva, pus, fecal material, or urine presents a problem in disinfection. Large amounts of organic material can hinder the penetration of a disinfectant and, in some cases, can form bonds that reduce its activity. Adequate cleaning of instruments and other reusable materials ensures that the germicide or sterilant will better accomplish the job for which it was chosen.

IN THE NEWS (Continued from page 315)

The disinfection of the hospital likely failed because of the difficulty with disinfecting rough surfaces, such as concrete block walls, porous mats, and brushed concrete floors.

The problem with *Salmonella* was brought under control by several measures. First, the concrete block walls of the stalls were painted with a special epoxy product, making them smooth and easy to clean. Second, the more porous mats were discarded and replaced with smooth, solid rubber mats. The walls and mats were then cleaned with detergent containing a quaternary ammonium product, and then effectively disinfected with sodium hypochlorite (bleach). Surfaces must be completely cleaned of organic material before the disinfection process is begun.

Hand scrubbing of the surfaces—including the walls, mats, brushed concrete floors and drains—was found to be more effective than power washing. Hand cleaning and scrubbing likely increased the amount of time that the *Salmonella* was exposed to the detergent and quaternary ammonium disinfectant.

Additional hand-washing stations were installed, and faculty, staff, and students were provided education on infection control measures.

See: Tillotson, K. et al. 1997. Outbreak of Salmonella infantis infection in a large animal veterinary teaching hospital. J. Am. Vet. Med. Assoc. (12):1554–1557.

of exposure, and the strength and chemical action of the germicide **(table 11.6).** Standardized procedures for testing the effectiveness of germicides are summarized in appendix C. The modes of action of most germicides are to attack the cellular targets discussed earlier: proteins, nucleic acids, the cell wall, and the cell membrane.

A chemical's strength or concentration is expressed in various ways, depending upon convention and the method of preparation. The content of many chemical agents can be expressed by more than one notation. In dilutions, a small

Germicidal Categories According to Chemical Group

Several general groups of chemical compounds are widely used for antimicrobial purposes in medicine and commerce (see table 11.5). Prominent agents include halogens, heavy metals, alcohols, phenolic compounds, oxidizers, aldehydes, detergents, and gases. These groups will be surveyed in the following section from the standpoint of each agent's specific forms, modes of action, indications for use, and limitations.

The Halogen Antimicrobial Chemicals

The **halogens** are fluorine, bromine, chlorine, and iodine, a group of nonmetallic elements, all of which are found in group VII of the periodic table. Although they can exist in either the ionic (halide) or nonionic state, most halogens exert their antimicrobial effect primarily in the nonionic state, not the halide state (chloride, iodide, for example). Because fluorine and bromine are difficult and dangerous to handle, and are no more effective than chlorine and iodine, only the latter two are used routinely in germicidal preparations. These elements are highly effective components of disinfectants and antiseptics because they are microbicidal and not just microbistatic, and they are sporicidal with longer exposure. For these reasons, halogens are the active ingredients in nearly one-third of all antimicrobial chemicals currently marketed.

Chlorine and Its Compounds Chlorine has been used for disinfection and antisepsis for approximately 200 years. The major forms used in microbial control are liquid and gaseous chlorine (Cl_2), hypochlorites (OCl), and chloramines (NH_2Cl). In solution, these compounds combine with water and release hypochlorous acid (HOCl), which oxidizes the sulfhydryl (S—H) group on the amino acid cysteine and interferes with disulfide (S—S) bridges on numerous enzymes. The resulting denaturation of the enzymes is permanent and suspends metabolic reactions. Chlorine kills not only bacteria and endospores but also fungi and viruses. Chlorine compounds are less effective and relatively unstable, if exposed to light, alkaline pH, and excess organic matter.

Chlorine Compounds in Disinfection and Antisepsis Gaseous and liquid chlorine are used almost exclusively for large-scale disinfection of drinking water, sewage, and wastewater from such sources as agriculture and industry. Chlorination to a concentration of 0.6 to 1.0 parts of chlorine per million parts of water will usually ensure that water is safe to drink. This treatment rids the water of most pathogenic vegetative microorganisms without unduly affecting its taste (some persons may debate this). In chapter 22, however, you will learn about pathogenic organisms that can survive water chlorination.

Hypochlorites are perhaps the most extensively used of all chlorine compounds. The scope of applications is broad, including sanitization and disinfection of food equipment in dairies, restaurants, and canneries and treatment of swimming pools, spas, drinking water, and even fresh foods.

Hypochlorites are used in the allied health areas to treat wounds and to disinfect equipment, bedding, and instruments. Common household bleach is a weak solution (5%) of sodium hypochlorite that serves as an all-around disinfectant, deodorizer, and stain remover.

Chloramines (dichloramine, halazone) are being employed more frequently as an alternative to pure chlorine in treating water supplies. Because standard chlorination of water is now believed to produce unsafe levels of cancer-causing substances such as trihalomethanes, some water districts have been directed by federal agencies to adopt chloramine treatment of water supplies. Chloramines also serve as sanitizers and disinfectants and for treating wounds and skin surfaces.

Iodine and Its Compounds Iodine is a pungent black chemical that forms brown-colored solutions when dissolved in water or alcohol. The two primary iodine preparations are *free iodine* in solution (I_2) and *iodophors*. Iodine rapidly penetrates the cells of microorganisms, where it apparently disturbs a variety of metabolic functions by interfering with the hydrogen and disulfide bonding of proteins (a mode of action similar to chlorine). All classes of microorganisms are killed by iodine if proper concentrations and exposure times are used. Iodine activity is not as adversely affected by organic matter and pH as chlorine is.

Applications of Iodine Solutions Aqueous iodine contains 2% iodine and 2.4% sodium iodide; it is used as a topical antiseptic before surgery and occasionally as a treatment for burned and infected skin. A stronger iodine solution (5% iodine and 10% potassium iodide) is used primarily as a disinfectant for plastic items, rubber instruments, cutting blades, thermometers, and other inanimate items. Iodine tincture is a 2% solution of iodine and sodium iodide in 70% alcohol that can be used in skin antisepsis. Because iodine can be extremely irritating to the skin and toxic when absorbed, strong aqueous solutions and tinctures (5–7%) are no longer considered safe for routine antisepsis. Iodine tablets are available for disinfecting water during emergencies or destroying pathogens in impure water supplies.

Iodophors are complexes of iodine and a neutral polymer such as a polyvinylalcohol. This formulation allows the slow release of free iodine and increases its degree of penetration. These compounds have largely replaced free iodine solutions in medical antisepsis because they are less prone to staining or irritating tissues. Common iodophor products marketed as Betadine, Povidone (PVP), and Isodine contain 2% to 10% of available iodine. They are used to prepare skin and mucous membranes for surgery and injections, in surgical handscrubs, to treat burns, and to disinfect equipment and surfaces. Although pure iodine is toxic to the eye, a recent study showed that Betadine solution is an effective means of preventing eye infections in newborn infants, and it may replace antibiotics and silver nitrate as the method of choice.

Phenol and Its Derivatives

Phenol (carbolic acid) is an acrid, poisonous compound derived from the distillation of coal tar. First adopted by Joseph Lister in 1867 as a surgical germicide, phenol was the major antimicrobial chemical until other phenolics with fewer toxic and irritating effects were developed. Solutions of phenol are now used only in certain limited cases, but it remains one standard against which other phenolic disinfectants are rated. The *phenol coefficient* quantitatively compares a chemical's antimicrobic properties to those of phenol. Substances chemically related to phenol are often referred to as phenolics. Hundreds of these chemicals are now available.

Phenolics consist of one or more aromatic carbon rings with added functional groups **(figure 11.12)**. Among the most important are alkylated phenols (cresols), chlorinated phenols, and bisphenols. In high concentrations, they are cellular poisons, rapidly disrupting cell walls and membranes and precipitating proteins; in lower concentrations, they inactivate certain critical enzyme systems. The phenolics are strongly microbicidal and will destroy vegetative bacteria (including the tuberculosis bacterium), fungi, and most viruses (not hepatitis B), but they are not reliably sporicidal. Their continued activity in the presence of organic matter and their detergent actions contribute to their usefulness. Unfortunately, the toxicity of many of the phenolics makes them too dangerous to use as antiseptics.

Applications of Phenolics

Phenol itself is still used for general disinfection of drains, cesspools, and animal quarters, but it is seldom applied as a medical germicide. The cresols are simple phenolic derivatives that are combined with soap for intermediate or low levels of disinfection in the hospital. Lysol and creolin, in a 1% to 3% emulsion, are common household versions of this type.

The bisphenols are also widely employed in commerce, clinics, and the home. One type, orthophenyl phenol, is the major ingredient in disinfectant aerosol sprays. This same phenolic is also found in some proprietary compounds (Lysol) often used in hospital and laboratory disinfection. One particular bisphenol, hexachlorophene, was once a common additive of cleansing soaps (pHisoHex) used in the hospital and home. When hexachlorophene was found to be absorbed through the skin and a cause of neurological damage, it was no longer available without a prescription. It is occasionally used to control outbreaks of skin infections.

Perhaps the most widely used phenolic is *triclosan*, chemically known as dichlorophenoxyphenol (see Insight 11.2). It is the antibacterial compound added to dozens of products, from soaps to kitty litter. It acts as both a disinfectant and antiseptic and is broad-spectrum in its effects.

Chlorhexidine

The compound chlorhexidine (Hibiclens, Hibitane) is a complex organic base containing chlorine and two phenolic rings. Its mode of action targets both cell membranes (lowering sur-

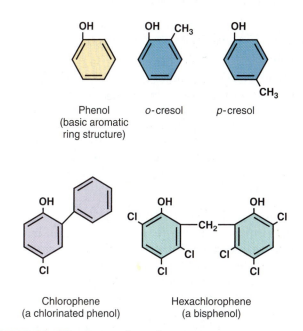

FIGURE 11.12 **Some phenolics.**
All contain a basic aromatic ring, but they differ in the types of additional compounds such as Cl and CH$_3$.

face tension until selective permeability is lost) and protein structure (causing denaturation). At moderate to high concentrations, it is bactericidal for both gram-positive and gram-negative bacteria but inactive against spores. Its effects on viruses and fungi vary. It possesses distinct advantages over many other antiseptics because of its mildness, low toxicity, and rapid action, and it is not absorbed into deeper tissues to any extent. Alcoholic or aqueous solutions of chlorhexidine are now commonly used for handscrubbing, preparing skin sites for surgical incisions and injections, and whole body washing. Chlorhexidine solution also serves as an obstetric antiseptic, a neonatal wash, a wound degermer, a mucous membrane irrigant, and a preservative for eye solutions.

Alcohols as Antimicrobial Agents

Alcohols are colorless hydrocarbons with one or more —OH functional groups. Of several alcohols available, only ethyl and isopropyl are suitable for microbial control. Methyl alcohol is not particularly microbicidal, and more complex alcohols are either poorly soluble in water or too expensive for routine use. Alcohols are employed alone in aqueous solutions or as solvents for tinctures (iodine, for example).

Alcohol's mechanism of action depends in part upon its concentration. Concentrations of 50% and higher dissolve membrane lipids, disrupt cell surface tension, and compromise membrane integrity. Alcohol that has entered the protoplasm denatures proteins through coagulation, but only in alcohol-water solutions of 50% to 95%. Alcohol is the exception to the rule that higher concentrations of an antimicrobial chemical have greater microbicidal activity. Because water is needed for proteins to coagulate, alcohol shows a greater microbicidal activity at 70% concentration (that is, 30% water) than at 100% (0% water). Absolute alcohol (100%) dehydrates cells and inhibits their growth but is generally not a protein coagulant.

Although useful in intermediate- to low-level germicidal applications, alcohol does not destroy bacterial spores at room temperature. Alcohol can, however, destroy resistant vegetative forms, including tuberculosis bacteria and fungal spores, provided the time of exposure is adequate. Alcohol is generally more effective in inactivating enveloped viruses than the more resistant nonenveloped viruses such as poliovirus and hepatitis A virus.

Applications of Alcohols   Ethyl alcohol, also called ethanol or grain alcohol, is known for being germicidal, nonirritating, and inexpensive. Solutions of 70% to 95% are routinely used as skin degerming agents because the surfactant action removes skin oil, soil, and some microbes sheltered in deeper skin layers. One limitation to its effectiveness is the rate at which it evaporates. Ethyl alcohol is occasionally used to disinfect electrodes, face masks, and thermometers, which are first cleaned and then soaked in alcohol for 15 to 20 minutes. Isopropyl alcohol, sold as rubbing alcohol, is even more microbicidal and less expensive than ethanol, but these benefits must be weighed against its toxicity. It must be used with caution in disinfection or skin cleansing, because inhalation of its vapors can adversely affect the nervous system.

Hydrogen Peroxide and Related Germicides

Hydrogen peroxide (H_2O_2) is a colorless, caustic liquid that decomposes in the presence of light, metals, or catalase into water and oxygen gas. Early formulations were unstable and inhibited by organic matter, but manufacturing methods now permit synthesis of H_2O_2 so stable that even dilute solutions retain activity through several months of storage.

The germicidal effects of hydrogen peroxide are due to the direct and indirect actions of oxygen. Oxygen forms hydroxyl free radicals (—OH), which, like the superoxide radical (see chapter 7), are highly toxic and reactive to cells. Although most microbial cells produce catalase to inactivate the metabolic hydrogen peroxide, it cannot neutralize the amount of hydrogen peroxide entering the cell during disinfection and antisepsis. Hydrogen peroxide is bactericidal, virucidal, and fungicidal and, in higher concentrations, sporicidal.

Applications of Hydrogen Peroxide   As an antiseptic, 3% hydrogen peroxide serves a variety of needs, including skin and wound cleansing, bedsore care, and mouthwashing. It is especially useful in treating infections by anaerobic bacteria because of the lethal effects of oxygen on these forms. Hydrogen peroxide is also a versatile disinfectant for soft contact lenses, surgical implants, plastic equipment, utensils, bedding, and room interiors.

A number of clinical procedures involve delicate reusable instruments such as endoscopes and dental handpieces. Because these devices can become heavily contaminated by tissues and fluids, they need to undergo sterilization, not just

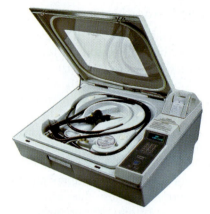

A cabinet for rapid (within 30 minutes) sterile processing of endoscopes and other microsurgical instruments

FIGURE 11.13   Sterile processing of invasive equipment protects patients.

disinfection, between patients to prevent transmission of infections such as hepatitis, tuberculosis, and genital warts. These very effective and costly diagnostic tools (a colonoscope may cost up to $30,000) have created another dilemma. They may trap infectious agents where they cannot be easily removed, and they are delicate, complex, and difficult to clean. Traditional methods are either too harsh (heat) to protect the instruments from damage or too slow (ethylene oxide) to sterilize them in a timely fashion between patients. The need for effective rapid sterilization has led to the development of low-temperature sterilizing cabinets that contain liquid chemical sterilants (figure 11.13). The major types of chemical sterilants used in these machines are powerful oxidizing agents such as hydrogen peroxide (35%) and peracetic acid (35%) that penetrate into delicate machinery, kill the most resistant microbes, and do not corrode or damage the working parts.

Vaporized hydrogen peroxide is currently being used as a sterilant in enclosed areas. Hydrogen peroxide plasma sterilizers exist for those applications involving small industrial or medical items. For larger enclosed spaces, such as isolators and passthrough rooms, peroxide generators can be used to fill a room with hydrogen peroxide vapors at concentrations high enough to be sporicidal.

Another compound with effects similar to those of hydrogen peroxide is ozone (O_3), used to disinfect air, water, and industrial air conditioners and cooling towers.

Chemicals with Surface Action: Detergents

Detergents are polar molecules that act as surfactants. Most anionic detergents have limited microbicidal power. This includes most soaps. Much more effective are positively charged (cationic) detergents, particularly the quaternary ammonium compounds (usually shortened to *quats*).

The activity of cationic detergents arises from the amphipathic (two-headed) nature of the molecule. The positively

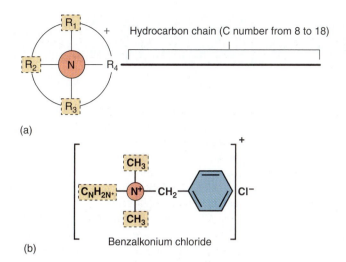

(a)

(b)

FIGURE 11.14 **The structure of detergents.**
(a) In general, detergents are polar molecules with a positively charged head and at least one long, uncharged hydrocarbon chain. The head contains a central nitrogen nucleus with various alkyl (R) groups attached. **(b)** A common quaternary ammonium detergent, benzalkonium chloride.

charged end binds well with the predominantly negatively charged bacterial surface proteins while the long, uncharged hydrocarbon chain allows the detergent to disrupt the cell membrane **(figure 11.14)**. Eventually the cell membrane loses selective permeability, leading to the death of the cell. Several other effects are seen but the loss of integrity of the cell membrane is most important.

The effects of detergents are varied. When used at high enough concentrations, quaternary ammonium compounds are effective against some gram-positive bacteria, viruses, fungi, and algae. In low concentrations they exhibit only microbistatic effects. Drawbacks to the quats include their ineffectiveness against the tuberculosis bacterium, hepatitis virus, *Pseudomonas*, and spores at any concentration. Furthermore, their activity is greatly reduced in the presence of organic matter, and they function best in alkaline solutions. As a result of these limitations, quats are rated only for low-level disinfection in the clinical setting.

Applications of Detergents and Soaps   Quaternary ammonium compounds **(quats)** include benzalkonium chloride, Zephiran, and cetylpyridinium chloride (Ceepryn). In dilutions ranging from 1:100 to 1:1,000, quats are mixed with cleaning agents to simultaneously disinfect and sanitize floors, furniture, equipment surfaces, and restrooms. They are used to clean restaurant eating utensils, food-processing equipment, dairy equipment, and clothing. They are common preservatives for ophthalmic solutions and cosmetics. Their level of disinfection is far too low for disinfecting medical instruments.

Soaps are alkaline compounds made by combining the fatty acids in oils with sodium or potassium salts. In usual practice, soaps are only weak microbicides, and they destroy only highly sensitive forms such as the agents of gonorrhea,

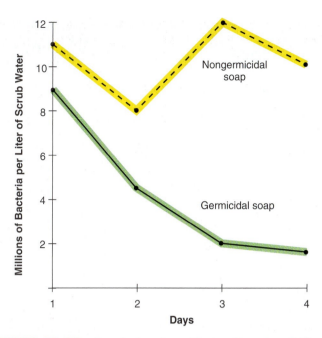

FIGURE 11.15 **Graph showing effects of handscrubbing.**
Comparison of scrubbing over several days with a nongermicidal soap versus a germicidal soap. Germicidal soap has persistent effects on skin over time, keeping the microbial count low. Without germicide, soap does not show this sustained effect.

meningitis, and syphilis. The common hospital pathogen *Pseudomonas* is so resistant to soap that various species grow abundantly in soap dishes.

Soaps function primarily as cleansing agents and sanitizers in industry and the home. The superior sudsing and wetting properties of soaps help to mechanically remove large amounts of surface soil, greases, and other debris that contains microorganisms. Soaps gain greater germicidal value when mixed with agents such as chlorhexidine or iodine. They can be used for cleaning instruments before heat sterilization, degerming patients' skin, routine handwashing by medical and dental personnel, and preoperative handscrubbing. Vigorously brushing the hands with germicidal soap over a 15-second period is an effective way to remove dirt, oil, and surface contaminants as well as some resident microbes, but it will never sterilize the skin **(Insight 11.3 and figure 11.15)**.

Heavy Metal Compounds

Various forms of the metallic elements mercury, silver, gold, copper, arsenic, and zinc have been applied in microbial control over several centuries. These are often referred to as heavy metals because of their relatively high atomic weight. However, from this list, only preparations containing mercury and silver still have any significance as germicides. Although some metals (zinc, iron) are actually needed in small concentrations as cofactors on enzymes, the higher molecular weight metals (mercury, silver, gold) can be very toxic, even in minute quantities (parts per million). This property of having antimicrobial effects in exceedingly small amounts

INSIGHT 11.3

The Quest for Sterile Skin

More than a hundred years ago, before sterile gloves were a routine part of medical procedures, the hands remained bare during surgery. Realizing the danger from microbes, medical practitioners attempted to sterilize the hands of surgeons and their assistants to prevent surgical infections. Several stringent (and probably very painful) techniques involving strong chemical germicides and vigorous scrubbing were practiced. Here are a few examples.

In Schatz's method, the hands and forearms were first cleansed by brisk scrubbing with liquid soap for 3 to 5 minutes, then soaked in a saturated solution of permanganate at a temperature of 110°F until they turned a deep mahogany brown. Next, the limbs were immersed in saturated oxalic acid until the skin became decolorized. Then, as if this were not enough, the hands and arms were rinsed with sterile limewater and washed in warm bichloride of mercury for 1 minute.

Or, there was Park's method (more like a torture). First, the surfaces of the hands and arms were rubbed completely with a mixture of cornmeal and green soap to remove loose dirt and superficial skin. Next, a paste of water and mustard flour was applied to the skin until it began to sting. This potion was rinsed off in sterile water, and the hands and arms were then soaked in hot bichloride of mercury for a few minutes, during which the solution was rubbed into the skin.

Another method once earnestly suggested for getting rid of microorganisms was to expose the hands to a hot-air cabinet to "sweat the germs" out of skin glands. Pasteur himself advocated a quick flaming of the hands to maintain asepsis.

The old dream of sterilizing the skin was finally reduced to some basic realities: The microbes entrenched in the epidermis and skin glands cannot be completely eradicated even with the most intense efforts, and the skin cannot be sterilized without also seriously damaging it. Because this is true for both medical personnel and their patients, the chance always exists that infectious agents can be introduced during invasive medical procedures. Safe surgery had to wait until 1890, when rubber gloves were first made available for placing a sterile barrier around the hands. Of course, this did not mean that skin cleansing and antiseptic procedures were abandoned or downplayed. A thorough scrubbing of the skin, followed by application of an antiseptic, is still needed to remove the most dangerous source of infections—the superficial contaminants constantly picked up from whatever we touch.

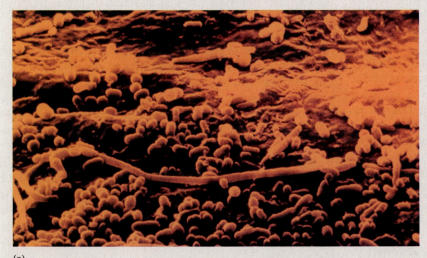

(a)

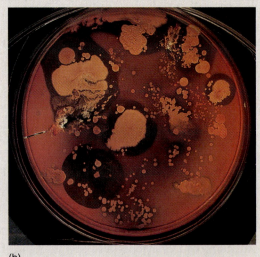

(b)

Microbes on normal unwashed hands. **(a)** A scanning electron micrograph of a piece of skin from a fingertip shows clusters of bacteria perched atop a fingerprint ridge (47,000×). **(b)** Heavy growth of microbial colonies on a plate of blood agar. This culture was prepared by passing an open sterile plate around a classroom of 30 students and having each one touch its surface. After incubation, a mixed population of bacteria and fungi appeared.

is called an **oligodynamic** (ol″-ih-goh-dy-nam′-ik) **action** (**figure 11.16**). Heavy metal germicides contain either an inorganic or an organic metallic salt, and they come in the form of aqueous solutions, tinctures, ointments, or soaps.

Mercury, silver, and most other metals exert microbicidal effects by binding onto functional groups of proteins and inactivating them, rapidly bringing metabolism to a standstill (see **figure 11.4c**). This mode of action can destroy many types of microbes, including vegetative bacteria, fungal cells and spores, algae, protozoa, and viruses (but not endospores).

Unfortunately, there are several drawbacks to using metals in microbial control:

1. metals are very toxic to humans if ingested, inhaled, or absorbed through the skin, even in small quantities, for the same reasons that they are toxic to microbial cells;
2. they commonly cause allergic reactions;
3. large quantities of biological fluids and wastes neutralize their actions; and
4. microbes can develop resistance to metals.

Health and environmental considerations have dramatically reduced the use of metallic antimicrobial compounds in medicine, dentistry, commerce, and agriculture.

Applications of Heavy Metals Weak (0.001–0.2%) organic mercury tinctures such as thimerosal (Merthiolate) and nitromersol (Metaphen) are fairly effective antiseptics and infection preventives, but they should never be used on broken skin because they are harmful and can delay healing. The organic mercurials also serve as preservatives in cosmetics and ophthalmic solutions. Mercurochrome, that old staple of the medicine cabinet, is now considered among the poorest of antiseptics.

A silver compound with several applications is silver nitrate (AgNO₃) solution. German professor of obstetrics Carl Siegmund Franz Credé introduced it in the late nineteenth century for preventing gonococcal infections in the eyes of newborn infants who had been exposed to an infected birth canal (described in chapter 23). This preparation is not used as often now because many pathogens are resistant to it. It has been replaced by antibiotics in most instances. Solutions of silver nitrate (1–2%) can also be used as topical germicides on mouth ulcers and occasionally root canals. Silver sulfadiazine ointment, when added to dressings, effectively prevents infection in second- and third-degree burn patients, and pure silver is now incorporated into catheters to prevent urinary tract infections in the hospital. Colloidal silver preparations are mild germicidal ointments or rinses for the mouth, nose, eyes, and vagina. Silver ions are increasingly incorporated into many hard surfaces, such as plastics and steel, as a way to control microbial growth on items such as toilet seats, stethoscopes, and even refrigerator doors.

Aldehydes As Germicides

Organic substances bearing a —CHO functional group (a strong reducing group) on the terminal carbon are called aldehydes. Several common substances such as sugars and

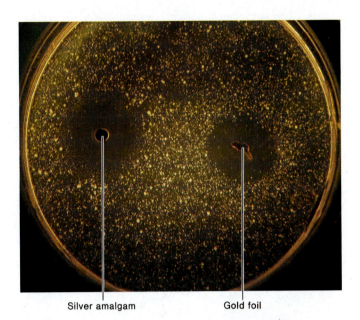

Silver amalgam   Gold foil

FIGURE 11.16 **Demonstration of the oligodynamic action of heavy metals.**
A pour plate inoculated with saliva has small fragments of heavy metals pressed lightly into it. During incubation, clear zones indicating growth inhibition developed around both fragments. The slightly larger zone surrounding the amalgam (used in tooth fillings) probably reflects the synergistic effect of the silver and mercury it contains.

some fats are technically aldehydes. The two aldehydes used most often in microbial control are *glutaraldehyde* and *formaldehyde*.

Glutaraldehyde is a yellow acidic liquid with a mild odor. The molecule's two aldehyde groups favor the formation of polymers. The mechanism of activity involves cross-linking protein molecules on the cell surface. In this process, amino acids are alkylated, meaning that a hydrogen atom on an amino acid is replaced by the glutaraldehyde molecule itself (**figure 11.17**). It can also irreversibly disrupt the activity of enzymes within the cell. Glutaraldehyde is rapid and broad-spectrum, and is one of the few chemicals officially accepted as a sterilant and high-level disinfectant. It kills spores in 3 hours and fungi and vegetative bacteria (even *Mycobacterium* and *Pseudomonas*) in a few minutes. Viruses, including the most resistant forms, appear to be inactivated after relatively short exposure times. Glutaraldehyde retains its potency even in the presence of organic matter, is noncorrosive, does not damage plastics, and is less toxic or irritating than formaldehyde. Its principal disadvantage is that it is somewhat unstable, especially with increased pH and temperature.

Formaldehyde is a sharp, irritating gas that readily dissolves in water to form an aqueous solution called **formalin**. Pure formalin is a 37% solution of formaldehyde gas dissolved in water. The chemical is microbicidal through its attachment to nucleic acids and functional groups of amino acids. Formalin is an intermediate- to high-level disinfectant,

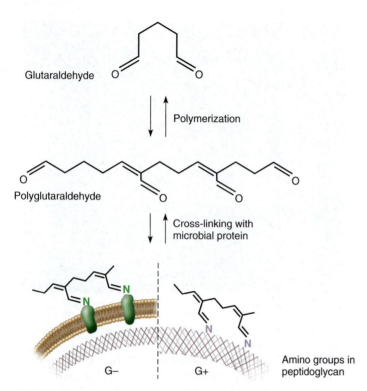

FIGURE 11.17 **Actions of glutaraldehyde.**
The molecule polymerizes easily. When these alkylating polymers react with amino acids, they cross-link and inactivate proteins.

although it acts more slowly than glutaraldehyde. Formaldehyde's extreme toxicity (it is classified as a carcinogen) and irritating effects on the skin and mucous membranes greatly limit its clinical usefulness.

A third aldehyde, ortho-phthalaldehyde (OPA) has recently been registered by the EPA as a high-level disinfectant. OPA is pale blue liquid with a barely detectable odor and can be most directly compared to glutaraldehyde. It has a mechanism of action similar to glutaraldehyde, is stable, nonirritating to the eyes and nasal passages and, for most uses, is much faster acting than glutaraldehyde. It is effective against vegetative bacteria, including *Mycobacterium* and *Pseudomonas*, fungi, and viruses. Chief among its disadvantages is an inability to reliably destroy spores and, on a more practical note, its tendency to stain proteins, including those in human skin.

Applications of the Aldehydes   Glutaraldehyde is a milder chemical for sterilizing materials that are damaged by heat. Commercial products (Cidex, Sporicidin) diluted to 2% are used to sterilize respiratory therapy equipment, hemostats, fiberoptic endoscopes (laparoscopes, arthroscopes), and kidney dialysis equipment. Glutaraldehyde is employed on dental instruments (usually in combination with autoclaving) to inactivate hepatitis B and other blood-borne viruses. It also serves to preserve vaccines, sanitize poultry carcasses, and degerm cows' teats.

Formalin tincture (8%) has limited use as a disinfectant for surgical instruments, and formalin solutions have appli-

cations in aquaculture to kill fish parasites and control growth of algae and fungi. Any object that is intended to come into intimate contact with the body must be thoroughly rinsed to neutralize the formalin residue. It is, after all, one of the active ingredients in embalming fluid.

Gaseous Sterilants and Disinfectants

Processing inanimate substances with chemical vapors, gases, and aerosols provides a versatile alternative to heat or liquid chemicals. Currently, those vapors and aerosols having the broadest applications are ethylene oxide (ETO), propylene oxide, and chlorine dioxide.

Ethylene oxide is a colorless substance that exists as a gas at room temperature. It is very explosive in air, a feature that can be eliminated by combining it with a high percentage of carbon dioxide or fluorocarbon. Like the aldehydes, ETO is a very strong alkylating agent, and it reacts vigorously with functional groups of DNA and proteins. Through these actions, it blocks both DNA replication and enzymatic actions. Ethylene oxide is one of a very few gases generally accepted for chemical sterilization because, when employed according to strict procedures, it is a sporicide. A specially designed ETO sterilizer called a *chemiclave*, a variation on the autoclave, is equipped with a chamber, gas ports, and temperature, pressure, and humidity controls (**figure 11.18**). Ethylene oxide is rather penetrating but relatively slow-acting, requiring from 90 minutes to 3 hours. Some items absorb ETO residues and must be aerated with sterile air for several hours after exposure to ensure dissipation of as much residual gas as possible. For all of its effectiveness, ETO has some unfortunate features. Its explosiveness makes it dangerous to handle; it can damage the lungs, eyes, and mucous membranes if contacted directly; and it is rated as a carcinogen by the government.

Chlorine dioxide is another gas that has of late been used as a sterilant. Despite the name, chlorine dioxide works in a completely different way from the chlorine compounds discussed earlier in the chapter. It is a strong alkylating agent, which disrupts proteins and is effective against vegetative bacteria, fungi, viruses, and endospores. Although chlorine dioxide is used for the treatment of drinking water, wastewater, food processing equipment, and medical waste, its most well known use was in the decontamination of the Senate offices after the anthrax attack of 2001 (see **Insight 11.4**).

Applications of Gases and Aerosols   Ethylene oxide (carboxide, cryoxide) is an effective way to sterilize and disinfect plastic materials and delicate instruments in hospitals and industries. It can safely sterilize prepackaged medical devices, surgical supplies, syringes, and disposable petri plates. Ethylene oxide has been used extensively to disinfect sugar, spices, dried foods, and drugs.

Propylene oxide is a close relative of ETO, with similar physical properties and mode of action, although it is less toxic. Because it breaks down into a relatively harmless substance, it is safer than ETO for sterilization of foods (nuts, powders, starches, spices).

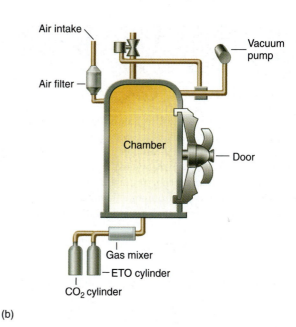

FIGURE 11.18 **Sterilization using gas.**
(a) An automatic ethylene oxide sterilizer. **(b)** The machine is equipped with gas canisters containing ethylene oxide (ETO) and carbon dioxide, a chamber to hold items, and mechanisms that evacuate gas and introduce air.

Dyes As Antimicrobial Agents

Dyes are important in staining techniques and as selective and differential agents in media; they are also a primary source of certain drugs used in chemotherapy. Because aniline dyes such as crystal violet and malachite green are very active against gram-positive species of bacteria and various fungi, they are incorporated into solutions and ointments to treat skin infections (ringworm, for example). The yellow acridine dyes, acriflavine and proflavine, are sometimes utilized for antisepsis and wound treatment in medical and veterinary clinics. For the most part, dyes will continue to have limited applications because they stain and have a narrow spectrum of activity.

Acids and Alkalis

Conditions of very low or high pH can destroy or inhibit microbial cells; but they are limited in applications due to their corrosive, caustic, and hazardous nature. Aqueous solutions of ammonium hydroxide remain a common component of detergents, cleansers, and deodorizers. Organic acids are widely used in food preservation because they prevent spore germination and bacterial and fungal growth and because they are generally regarded as safe to eat. Acetic acid (in the form of vinegar) is a pickling agent that inhibits bacterial growth; propionic acid is commonly incorporated into breads and cakes to retard molds; lactic acid is added to sauerkraut and olives to prevent growth of anaerobic bacteria (especially the clostridia), and benzoic and sorbic acids are added to beverages, syrups, and margarine to inhibit yeasts.

For a look at the antimicrobial chemicals found in some common household products, see **table 11.7.**

☑ CHECKPOINT

- Chemical agents of microbial control are classified by their physical state and chemical nature.
- Chemical agents can be either microbicidal or microbistatic. They are also classified as high-, medium-, or low-level germicides.
- Factors that determine the effectiveness of a chemical agent include the type and numbers of microbes involved, the material involved, the strength of the agent, and the exposure time.
- Halogens are effective chemical agents at both microbicidal and microbistatic levels. Chlorine compounds disinfect water, food, and industrial equipment. Iodine is used as either free iodine or iodophor to disinfect water and equipment. Iodophors are also used as antiseptic agents.
- Phenols are strongly microbicidal agents used in general disinfection. Milder phenol compounds, the bisphenols, are also used as antiseptics.
- Alcohols dissolve membrane lipids and destroy cell proteins. Their action depends upon their concentration, but they are generally only microbistatic.
- Hydrogen peroxide is a versatile microbicide that can be used as an antiseptic for wounds and a disinfectant for utensils. A high concentration is an effective sporicide.
- Surfactants are of two types: detergents and soaps. They reduce cell membrane surface tension, causing membrane rupture. Cationic detergents, or quats, are low-level germicides limited by the amount of organic matter present and the microbial load.
- Aldehydes are potent sterilizing agents and high-level disinfectants that irreversibly disrupt microbial enzymes.
- Ethylene oxide and chlorine dioxide are gaseous sterilants that work by alkylating protein and DNA.

INSIGHT 11.4

Microbiology

Decontaminating Congress

Choosing a microbial control technique is usually a straightforward process: cultures are autoclaved, milk is pasteurized, and medical supplies may be irradiated. When letters containing spores of *Bacillus anthracis* were opened in the Hart Office Building of the U.S. Senate, the process got just a bit trickier. Among the many concerns of the Environmental Protection Agency, which was charged with the building's remediation, were these:

- Anthrax is lethal disease.
- *Bacillus anthracis* is a spore-forming bacterium, making eradication difficult.
- The Hart Office Building is populated by thousands of people, who could quickly and easily spread endospores from office to office.
- The area to be decontaminated included heating and air-conditioning vents, carpeting, furniture, office equipment, sensitive papers, artwork, and the various belongings of quickly evacuated workers.

The goal of the project could be simply stated: Detect and remove all traces of a lethal, highly infectious, spore-forming bacterium from an enormous space filled with all manner of easily damaged material. Easier said than done.

With this goal in mind, the EPA first set to work to determine the extent of contamination. Samples were taken from 25 buildings on Capitol Hill. Nonporous surfaces were swabbed; furniture and carpets were vacuumed into a HEPA filter; air was pumped through a filter to remove any airborne spores. Samples were placed in sterile vials and double bagged before being transferred to the laboratory. Analysis of the samples revealed the presence of spores in many areas of the Hart Office Building, including the mail-processing areas of 11 senators. Because spores were not found in the entrances to these offices, it was theorized that the spores spread primarily through the mail. The heaviest contamination was found in the office of Senator Tom Daschle, to whom the original anthrax-containing letter was addressed. Additional contamination was found in a conference room, stairwell, elevator, and restroom.

With the scope of the problem identified, the EPA could set out to devise a strategy for remediation, keeping in mind the difficulty of killing spore-forming organisms and the myriad contents of the building, much of it delicate, expensive, and in many cases irreplaceable. It was decided that most areas would be cleaned by a combination of HEPA vacuuming followed by either treatment with liquid chlorine dioxide or Sandia decontamination foam, an antibacterial foam that combines surfactants with oxidizing agents. For the most heavily contaminated areas (Senator Daschle's office and parts of the heating and air-conditioning system), gaseous chlorine dioxide would be used. Chlorine dioxide has been accepted as a sterilant since 1988 and is used for, among other things, treatment of medical waste. Its use in this project was based primarily on the facts that it would both work and be unlikely to harm the contents of the building. (It was even tested to ensure that it would not damage the ink making up the signature on a document.)

Before fumigation could begin, however, a method of evaluating the success of the remediation effort needed to be devised. Borrowing from a common laboratory technique, 3,000 small slips of paper covered with spores from the organism *Bacillus stearothermophilus* (which is generally considered harder to kill than *B. anthracis*) were dispersed throughout the building. If after the treatment was complete these spores were unable to germinate, then the fumigation could be considered a success.

On December 1, 2001, technicians prepared the 3,000-square-foot office for fumigation. This included constructing barriers to seal off the portion of building being fumigated and raising the humidity in the offices to approximately 75%, to enable the gas to adhere to any lingering spores. Office machines (computers, copiers, for example) were turned on so that the fans inside the equipment would aid in spreading the gas. Finally, at 3:15 A.M. fumigation began. It ended 20 hours later and, after the gas was neutralized and ventilated from the building, technicians entered to collect the test strips. Analyzing the results, it was clear that trace amounts of spores were still present but only in those areas that originally had the worst contamination (these areas were again cleaned with liquid chlorine dioxide). Senator Daschle's office got a makeover with new carpeting, paint, and furniture, and shortly thereafter the building was reoccupied. The senator also had his office equipment replaced as it was deemed too contaminated to be used. Just as antibiotics are useless against viruses, computer antivirus software doesn't work against bacteria.

Workers in protective garments prepare the Hart Office Building for decontamination.

TABLE 11.7 Active Ingredients of Various Commercial Antimicrobial Products

Product	Specific Chemical Agent	Antimicrobial Category
Lysol Sanitizing Wipes	Dimethyl benzyl ammonium chloride	Detergent (quat)
Clorox Disinfecting Wipes	Dimethyl benzyl ammonium chloride	Detergent (quat)
Tilex Mildew Remover	Sodium hypochlorites	Halogen
Lysol Mildew Remover	Sodium hypochlorites	Halogen
Ajax Antibacterial Hand Soap	Triclosan	Phenolic
Dawn Antibacterial Hand Soap	Triclosan	Phenolic
Dial Antibacterial Hand Soap	Triclosan	Phenolic
Lysol Disinfecting Spray	Alkyl dimethyl benzyl ammonium saccharinate/ethanol	Detergent (quats)/alcohol
ReNu Contact Lens Solution	Polyaminopropyl biguanide	Chlorhexidine
Wet Ones Antibacterial Moist Towelettes	Benzethonium chloride	Detergents (quat)
Noxzema Triple Clean	Triclosan	Phenolic
Scope Mouthwash	Ethanol	Alcohol
Purell Instant Hand Sanitizer	Ethanol	Alcohol
Pine-Sol	Phenolics and surfactant	Mixed
Allergan Eye Drops	Sodium chlorite	Halogen

Chapter Summary With Key Terms

11.1 Controlling Microorganisms

A. *Decontamination* procedures involve the destruction or removal of contaminants. Contaminants are defined as microbes present at a given place and time that are undesirable or unwanted.

B. **Sterilization** is a process that destroys or removes all microbes, including viruses.
1. A **bacteri*cide*** is a chemical that destroys bacteria.
2. **Bacterio*static*** agents inhibit or prevent the growth of bacteria on tissues or on other objects in the environment.
3. A **germicide** is a chemical that will kill any pathogenic microorganisms.

C. **Disinfection** refers to a physical or chemical process that destroys vegetative pathogens but not bacterial endospores.

D. **Sanitization** is any chemical technique that removes microorganisms to "safe levels or standards."

E. Several factors influence the rate at which antimicrobial agents work. These factors are:
1. Exposure time to the agent.
2. Numbers of microbes present.
3. Relative resistance of microbes (for example, endospores vs. vegetative forms).
4. Activity of the agent (microbicidal vs. microbistatic).

F. How antimicrobial agents work: their *modes of action.* Agents affect cell wall synthesis, membrane permeability, and protein and nucleic acid synthesis and function.

11.2 Methods of Physical Control

A. Moist heat denatures proteins and DNA while destroying membranes.
1. *Sterilization:* **Autoclaves** utilize steam under pressure to sterilize heat-resistant materials, whereas intermittent sterilization can be used to sterilize more delicate items.
2. *Disinfection:* Pasteurization subjects liquids to temperatures below 100°C and is used to lower the microbial load in liquids. Boiling water can be used to destroy vegetative pathogens in the home.

B. Dry heat, using higher temperatures than moist heat, can also be used to sterilize.
1. **Incineration** can be carried out using a Bunsen burner or incinerator. Temperatures range between 600°C and 1800°C.
2. *Dry ovens* coagulate proteins at temperatures of 15°C to 180°C.

C. Cold temperatures are microbistatic, with refrigeration (0°C to 15°C) and freezing (below 0°C) commonly used to preserve food, media, and cultures.

D. Drying and desiccation lead to (often temporary) metabolic inhibition by reducing water in the cell.

E. *Radiation:* Energy in the form of radiation is a method of *cold sterilization*, which works by introducing mutations into the DNA of target cells.
1. **Ionizing radiation,** such as gamma rays and X-rays, has deep penetrating power and works by causing breaks in the DNA of target organisms.
2. **Nonionizing radiation** uses **ultraviolet** waves with very little penetrating power and works by creating dimers between adjacent pyrimidines, which interferes with replication.

F. Filtration involves the physical removal of microbes by passing a gas or liquid through a fine filter, and can be used to sterilize air as well as heat-sensitive liquids.

11.3 Chemical Agents in Microbial Control

A. Chemicals are divided into disinfectants, **antiseptics,** sterilants, **sanitizers,** and **degermers** based on their level of effectiveness and the surfaces to which they are applied.

B. Antimicrobial chemicals are found as solids, gases, and liquids. Liquids can be either aqueous (water based) or tinctures (alcohol based).

C. Halogens are chemicals based on elements from group VII of the periodic table.
   1. Chlorine is used as chlorine gas, hypochlorites, and chloramines. All work by disrupting disulfide bonds and, given adequate time, are sporicidal.
   2. Iodine is found both as free iodine (I_2) and iodophors (iodine bound to organic polymers such as soaps). Iodine has a mode of action similar to chlorine and is also sporicidal, given enough time.
D. Phenolics are chemicals based on phenol, which work by disrupting cell membranes and precipitating proteins. They are bactericidal, fungicidal, and viricidal, but not sporicidal.
E. Chlorhexidine (Hibiclens, Hibitane) is a surfactant and protein denaturant with broad microbicidal properties, although it is not sporicidal. Solutions of chlorhexidine are used as skin degerming agents for preoperative scrubs, skin cleaning, and burns.
F. Ethyl and isopropyl alcohol, in concentrations of 50% to 90%, are useful for microbial control. Alcohols act as **surfactants,** dissolving membrane lipids and coagulating proteins of vegetative bacterial cells and fungi. They are not sporicidal.
G. Hydrogen peroxide produces highly reactive hydroxyl-free radicals that damage protein and DNA while also decomposing to O_2 gas, which is toxic to anaerobes. Strong solutions of H_2O_2 are sporicidal.
H. Detergents and soaps
   1. Cationic detergents known as quaternary ammonium compounds **(quats)** act as surfactants that alter the membrane permeability of some bacteria and fungi. They are not sporicidal.
   2. Soaps have little microbicidal activity but rather function by removing grease and soil that contain microbes.
I. Heavy metals: Solutions of silver and mercury kill vegetative cells (but not spores) in exceedingly low concentrations **(oligodynamic action)** by inactivating proteins.
J. Aldehydes such as glutaraldehyde and formaldehyde kill microbes by alkylating protein and DNA molecules.
K. Gases and aerosols such as **ethylene oxide (ETO),** propylene oxide, and chlorine dioxide are strong alkylating agents, all of which are sporicidal.
L. Dyes, acids, and alkalis can also inhibit or destroy microbes.

Multiple-Choice Questions

1. A microbicidal agent has what effect?
   a. sterilizes
   b. inhibits microorganisms
   c. is toxic to human cells
   d. destroys microorganisms

2. Microbial control methods that kill _____ are able to sterilize.
   a. viruses
   b. the tubercle bacillus
   c. endospores
   d. cysts

3. Any process that destroys the non-spore-forming contaminants on inanimate objects is
   a. antisepsis
   b. disinfection
   c. sterilization
   d. degermation

4. Sanitization is a process by which
   a. the microbial load on objects is reduced
   b. objects are made sterile with chemicals
   c. utensils are scrubbed
   d. skin is debrided

5. An example of an agent that lowers the surface tension of cells is
   a. phenol
   b. chlorine
   c. alcohol
   d. formalin

6. High temperatures _____ and low temperatures _____.
   a. sterilize, disinfect
   b. kill cells, inhibit cell growth
   c. denature proteins, burst cells
   d. speed up metabolism, slow down metabolism

7. The temperature-pressure combination for an autoclave is
   a. 100°C and 4 psi
   b. 121°C and 15 psi
   c. 131°C and 9 psi
   d. 115°C and 3 psi

8. Microbe(s) that is/are the target(s) of pasteurization include:
   a. *Clostridium botulinum*
   b. *Mycobacterium* species
   c. *Salmonella* species
   d. both b and c

9. Ionizing radiation removes _____ from atoms.
   a. protons
   b. waves
   c. electrons
   d. ions

10. The primary mode of action of nonionizing radiation is to
    a. produce superoxide ions
    b. make pyrimidine dimers
    c. denature proteins
    d. break disulfide bonds

11. The most versatile method of sterilizing heat-sensitive liquids is
    a. UV radiation
    b. exposure to ozone
    c. beta propiolactone
    d. filtration

12. _____ is the iodine antiseptic of choice for wound treatment.
    a. Eight percent tincture
    b. Five percent aqueous
    c. Iodophor
    d. Potassium iodide solution

13. A chemical with sporicidal properties is
    a. phenol
    b. alcohol
    c. quaternary ammonium compound
    d. glutaraldehyde

14. Silver nitrate is used
    a. in antisepsis of burns
    b. as a mouthwash
    c. to treat genital gonorrhea
    d. to disinfect water

15. Detergents are
    a. high-level germicides
    b. low-level germicides
    c. excellent antiseptics
    d. used in disinfecting surgical instruments

16. Which of the following is an approved sterilant?
    a. chlorhexidine
    b. betadyne
    c. ethylene oxide
    d. ethyl alcohol

Concept Questions

These questions are suggested as a *writing-to-learn* experience. For each question, compose a one- or two-paragraph answer that includes the factual information needed to completely address the question.

1. Compare sterilization with disinfection and sanitization. Describe the relationship of the concepts of sepsis, asepsis, and antisepsis.

2. a. Briefly explain how the type of microorganisms present will influence the effectiveness of exposure to antimicrobial agents.
   b. Explain how the numbers of contaminants can influence the measures used to control them.

3. a. Precisely what is microbial death?
   b. Why does a population of microbes not die instantaneously when exposed to an antimicrobial agent?

4. Why are antimicrobial processes inhibited in the presence of extraneous organic matter?

5. Describe four modes of action of antimicrobial agents, and give a specific example of how each works.

6. a. Summarize the nature, mode of action, and effectiveness of moist and dry heat.
   b. Compare the effects of moist and dry heat on vegetative cells and spores.
   c. Explain the concepts of TDT and TDP, using examples. What are the minimum TDTs for vegetative cells and endospores?

7. How can the temperature of steam be raised above 100°C? Explain the relationship involved.

8. a. What are several microbial targets of pasteurization?
   b. What are the primary purposes of pasteurization?
   c. What is ultrapasteurization?

9. Explain why desiccation and cold are not reliable methods of disinfection.

10. a. What are some advantages of ionizing radiation as a method of control?
    b. Some disadvantages?

11. a. What is the precise mode of action of ultraviolet radiation?
    b. What are some disadvantages to its use?

12. What are the superior characteristics of iodophors over free iodine solutions?

13. a. Name one chemical for which the general rule that a higher concentration is more effective is *not* true.
    b. What is a sterilant?
    c. Name the principal sporicidal chemical agents.

14. Why is hydrogen peroxide solution so effective against anaerobes?

15. Give the uses and disadvantages of the heavy metal chemical agents, glutaraldehyde, and the sterilizing gases.

16. What does it mean to say that a chemical has an oligodynamic action?

Critical Thinking Questions

Critical thinking is the ability to reason and solve problems using facts and concepts. These questions can be approached from a number of angles, and in most cases, they do not have a single correct answer.

1. What is wrong with this statement: "Prior to vaccination, the patient's skin was sterilized with alcohol"? What would be the more correct wording?

2. For each item on the following list, give a reasonable method of sterilization. You cannot use the same method more than three times; the method must sterilize, not just disinfect; and the method must not destroy the item or render it useless unless there is no other choice. After considering a workable method, think of a method that would not work. Note: Where an object containing something is given, you must sterilize everything (for example, both the jar and the Vaseline in it). Some examples of methods are autoclave, ethylene oxide gas, dry oven, and ionizing radiation.

room air	carcasses of cows with
blood in a syringe	"mad cow" disease
serum	inside of a refrigerator
a pot of soil	wine
plastic petri plates	a jar of Vaseline
heat-sensitive drugs	fruit in plastic bags
cloth dressings	talcum powder
leather shoes from a thrift shop	milk
a cheese sandwich	orchid seeds
human hair (for wigs)	metal instruments
a flask of nutrient agar	mail contaminated with
an entire room (walls, floor, etc.)	anthrax spores
rubber gloves	
disposable syringes	

3. a. Graph the data on tables 11.3 and 11.4, plotting the time on the Y axis and the temperature on the X axis for three different organisms.
   b. Using pasteurization techniques as a model, compare the TDTs and explain the relationships between temperature and length of exposure.
   c. Is there any difference between the graph for a sporeformer and the graph for a non-sporeformer? Explain.

4. Can you think of situations in which the same microbe would be considered a serious contaminant in one case and completely harmless in another?

5. A supermarket/drugstore assignment: Look at the labels of 10 different products used to control microbes, and make a list of their active ingredients, their suggested uses, and information on toxicity and precautions.

6. Devise an experiment that will differentiate between bactericidal and bacteristatic effects.

7. There is quite a bit of concern that chlorine used as a water purification chemical presents serious dangers. What alternative methods covered by this chapter could be used to purify water supplies and yet keep them safe from contamination?

8. The shelf-life and keeping qualities of fruit and other perishable foods are greatly enhanced through irradiation, saving industry and consumers billions of dollars. How would you personally feel about eating a piece of irradiated fruit? How about spices that had been sterilized with ETO?

9. Can you think of some innovations the health care community can use to deal with medical waste and its disposal that prevent infection but are ecologically sound?

Internet Search Topics

1. Look for information on the Internet concerning triclosan-resistant bacteria. Based on what you find, do you think the widespread use of this antimicrobial is a good idea? Why or why not?

2. Find websites that discuss problems with sterilizing reusable medical instruments such as endoscopes and the types of diseases that can be transmitted with them.

3. Research "prion decontamination." What would you do if you were caring for a patient in the hospital with Creutzfeldt-Jakob disease (a prion disease) and there was a blood splatter on the floor?

4. Go to the Online Learning Center for chapter 11 of this text at http://www.mhhe.com/cowan1. Access the URLs listed under Internet Search Topics.

   Handwashing is one of our main protections to ensure good health and hygiene. Log on to one of the websites listed to research information, statistics, and other aspects of handwashing.

Drugs, Microbes, Host—
The Elements of Chemotherapy

IN THE NEWS

Dana, a 21-year-old college student, was suffering from a persistent cough, fever, and constant fatigue. She was asked whether she had these symptoms in the past and if she had been treated for them at that time. She reported that she had gotten sick with similar symptoms before but never received any treatment. Since she had mild fever, and her chest X rays were normal, her doctor prescribed streptomycin and cough medicine. Her symptoms improved drastically within a week and, consequently, she decided to stop taking her medication.

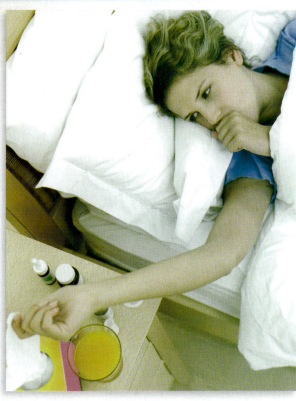

After 2 months, she felt ill again. Her symptoms included coughing, loss of appetite, high fever, and night sweats. She said the cough, which had lasted for the past 3 weeks, was accompanied by a blood-stained sputum.

A preliminary diagnosis was made based on the symptoms. Chest X rays, Mantoux PPD test, and an acid-fast stain from a sputum sample were later performed. She also was tested for HIV. Based on the results of the tests and her symptoms, a cocktail of antibiotics, including isoniazid and rifampin, was prescribed. She was instructed to take the antibiotics for 9 months.

▶ *What disease is described here?*

▶ *Why might Dana's symptoms have resurfaced despite the original streptomycin treatment?*

▶ *Why was it necessary to place Dana on a multidrug treatment?*

CHAPTER OVERVIEW

▶ Antimicrobial chemotherapy involves the use of chemicals to prevent and treat infectious diseases. These chemicals include antibiotics, which are derived from the natural metabolic processes of bacteria and fungi, as well as synthetic drugs.

▶ To be effective, antimicrobial therapy must disrupt a necessary component of a microbe's structure or metabolism to the extent that microbes are killed or their growth is inhibited. At the same time, the drug must be safe for humans, while not being overly toxic, causing allergies or disrupting normal flora.

▶ Drugs are available to treat infections by all manner of organisms, including viruses. Drug choice depends on the microbe's sensitivity, the drug's toxicity, and the health of the patient.

▶ Adverse side effects of drugs include damage to organs, allergies, and disruption of the host's normal flora, which can lead to other infections.

▶ A major problem in modern chemotherapy is the development of drug resistance. Many microbes have undergone genetic changes that allow them to circumvent the effects of a drug. The misapplication of drugs and antibiotics has worsened this problem.

12.1 Principles of Antimicrobial Therapy

A hundred years ago in the United States, one out of three children was expected to die of an infectious disease before the age of 5. Early death or severe lifelong debilitation from scarlet fever, diphtheria, tuberculosis, meningitis, and many other bacterial diseases was a fearsome yet undeniable fact of life to most of the world's population. The introduction of modern drugs to control infections in the 1930s was a medical revolution that has added significantly to the lifespan and health of humans. It is no wonder that for many years, antibiotics in particular were regarded as miracle drugs. In later discussions, we will evaluate this misconception in light of the shortcomings of chemotherapy. Although antimicrobial drugs have greatly reduced the incidence of certain infections, they have definitely not eradicated infectious disease and probably never will. In fact, in some parts of the world, mortality rates from infectious diseases are as high as before the arrival of antimicrobial drugs. Nevertheless, humans have been taking medicines to try to control diseases for thousands of years **(Insight 12.1).**

The goal of antimicrobial chemotherapy is deceptively simple: administer a drug to an infected person that destroys the infective agent without harming the host's cells. In actuality, this goal is rather difficult to achieve, because many (often contradictory) factors must be taken into account. The ideal drug should be easily administered, yet be able to reach the infectious agent anywhere in the body; be absolutely toxic to the infectious agent while simultaneously being nontoxic to the host; and remain active in the body as long as needed, yet be safely and easily broken down and excreted. In short, the perfect drug does not exist, but by balancing drug characteristics against one another, a satisfactory compromise can be achieved **(table 12.1).**

Chemotherapeutic agents are described with regard to their origin, range of effectiveness, and whether they are naturally produced or chemically synthesized. A few of the more important terms you will encounter are found in **table 12.2.**

In this chapter, we describe different types of antibiotic drugs, their mechanism of action, and the types of microbes on which they are effective. The organ system chapters 18 through 23 list specific disease agents and the drugs used to treat them.

TABLE 12.1	Characteristics of the Ideal Antimicrobial Drug

- Selectively toxic to the microbe but nontoxic to host cells
- Microbicidal rather than microbistatic
- Relatively soluble; functions even when highly diluted in body fluids
- Remains potent long enough to act and is not broken down or excreted prematurely
- Doesn't lead to the development of antimicrobial resistance
- Complements or assists the activities of the host's defenses
- Remains active in tissues and body fluids
- Readily delivered to the site of infection
- Reasonably priced
- Does not disrupt the host's health by causing allergies or predisposing the host to other infections

TABLE 12.2	Terminology of Chemotherapy
Chemotherapeutic drug	Any chemical used in the treatment, relief, or prophylaxis of a disease
Prophylaxis	Use of a drug to prevent imminent infection of a person at risk
Antimicrobial chemotherapy	The use of chemotherapeutic drugs to control infection
Antimicrobials	All-inclusive term for any antimicrobial drug, regardless of its origin
Antibiotics	Substances produced by the natural metabolic processes of some microorganisms that can inhibit or destroy other microorganisms
Semisynthetic drugs	Drugs which are chemically modified in the laboratory after being isolated from natural sources
Synthetic drugs	The use of chemical reactions to synthesize antimicrobial compounds in the laboratory
Narrow spectrum (limited spectrum)	Antimicrobials effective against a limited array of microbial types—for example, a drug effective mainly on gram-positive bacteria
Broad spectrum (extended spectrum)	Antimicrobials effective against a wide variety of microbial types—for example, a drug effective against both gram-positive and gram-negative bacteria

INSIGHT 12.1

From Witchcraft to Wonder Drugs

Early human cultures relied on various types of primitive medications such as potions, poultices, and mudplasters. Many were concocted from plant, animal, and mineral products that had been found, usually through trial and error or accident, to have some curative effect upon ailments and complaints. In one ancient Chinese folk remedy, a fermented soybean curd was applied to skin infections. The Greeks used wine and plant resins (myrrh and frankincense), rotting wood, and various mineral salts to treat diseases. Many folk medicines were effective, but some of them either had no effect or were even harmful. It is interesting that the Greek word *pharmakeutikos* originally meant the practice of witchcraft. These ancient remedies were handed down from generation to generation, but it was not until the Middle Ages that a specific disease was first treated with a specific chemical. Dosing syphilitic patients with inorganic arsenic and mercury compounds may have proved the ancient axiom: *Graviora quaedum sunt remedia periculus.* ("Some remedies are worse than the disease.")

An enormous breakthrough in the science of drug therapy came with the germ theory of infection by Robert Koch (see chapter 1). This allowed disease treatment to focus on a particular microbe, which in turn opened the way for Paul Ehrlich to formulate the first theoretical concepts in chemotherapy in the late 1800s. Ehrlich had observed that certain dyes affixed themselves to specific microorganisms and not to animal tissues. This observation led to the profound idea that if a drug was properly selective in its actions, it would zero in on and destroy a microbial target and leave human cells unaffected. His first discovery was an arsenic-based drug that was very toxic to the spirochete of syphilis but, unfortunately, to humans as well. Ehrlich systematically altered this parent molecule, creating numerous derivatives. Finally, on the 606th try, he arrived at a compound he called *salvarsan*. This drug had some therapeutic merit and was used for a few years, but it eventually had to be discontinued because it was still not selective enough in its toxicity. Ehrlich's work had laid important foundations for many of the developments to come.

Another pathfinder in early drug research was Gerhard Domagk, whose discoveries in the 1930s launched a breakthrough in therapy that marked the true beginning of broad-scale usage of antimicrobial drugs. Domagk showed that the red dye prontosil was chemically changed by the body into a compound with specific activity against bacteria. This substance was sulfonamide—the first *sulfa* drug. In a short time, the structure of this drug was determined, and it became possible to synthesize it on a wide scale and to develop scores of other sulfonamide drugs. Although these drugs had immediate applications in therapy (and still do), still another fortunate discovery was needed before the golden age of antibiotics could really blossom.

The discovery of antibiotics dramatically demonstrates how developments in science and medicine often occur through a combination of accident, persistence, collaboration, and vision. In 1928 in the London laboratory of Sir Alexander Fleming, a plate of *Staphylococcus aureus* became contaminated with the mold *Penicillium notatum*. Observing these plates, Fleming noted that the colonies of *Staphylococcus* were evidently being destroyed by some activity of the nearby mold colonies. Struck by this curious phenomenon, he extracted from the fungus a compound he called penicillin and showed that it was responsible for the inhibitory effects.

Although Fleming understood the potential for penicillin, he was unable to develop it. A decade after his discovery, English chemists Howard Florey and Ernst Chain worked out methods for industrial production of penicillin to help in the war effort. Clinical trials conducted in 1941 ultimately proved its effectiveness, and cultures of the mold were brought to the United States for an even larger-scale effort. When penicillin was made available to the world's population, at first it seemed a godsend. By the 1950s, the pharmaceutical industry had entered an era of drug research and development that soon made penicillin only one of a large assortment of antimicrobial drugs. But in time, because of extreme overuse and misunderstanding of its capabilities, it also became the model for one of the most serious drug problems—namely, drug resistance.

The father of modern antibiotics. A Scottish physician, Sir Alexander Fleming, accidently discovered penicillin when his keen eye noticed that colonies of bacteria were being lysed by a fungal contaminant. He studied the active ingredient and set the scene for the development of the drug 10 years later.

The Origins of Antimicrobial Drugs

Nature is a prolific producer of antimicrobial drugs. Antibiotics, after all, are common metabolic products of aerobic bacteria and fungi. By inhibiting the growth of other microorganisms in the same habitat (antagonism), antibiotic producers presumably enjoy less competition for nutrients and space. The greatest numbers of antibiotics are derived from bacteria in the genera *Streptomyces* and *Bacillus* and molds in the genera *Penicillium* and *Cephalosporium*. Not only have chemists created new drugs by altering the structure of naturally occurring antibiotics, they are actively searching for metabolic compounds with antimicrobial effects in species other than bacteria and fungi (discussed later).

✔ CHECKPOINT

- Antimicrobial chemotherapy involves the use of drugs to control infection on or in the body.
- Antimicrobial drugs are produced either synthetically or from natural sources. They inhibit or destroy microbial growth in the infected host. Antibiotics are the subset of antimicrobials produced by the natural metabolic processes of microorganisms.
- Antimicrobial drugs are classified by their range of effectiveness. Broad-spectrum antimicrobials are effective against many types of microbes. Narrow-spectrum antimicrobials are effective against a limited group of microbes.
- Antimicrobial therapy involves three interacting factors: the microbe's sensitivity, the drug's toxicity, and the health of the patient.
- Bacteria and fungi are the primary sources of most antibiotics. The molecular structures of these compounds can be chemically altered to form additional semisynthetic antimicrobials.

12.2 Interactions Between Drug and Microbe

The goal of antimicrobial drugs is either to disrupt the cell processes or structures of bacteria, fungi, and protozoa or to inhibit virus replication. Most of the drugs used in chemotherapy interfere with the function of enzymes required to synthesize or assemble macromolecules, or they destroy structures already formed in the cell. Above all, drugs should be **selectively toxic,** which means they should kill or inhibit microbial cells without simultaneously damaging host tissues. This concept of selective toxicity is central to chemotherapy, and the best drugs are those that block the actions or synthesis of molecules in microorganisms but not in vertebrate cells. Examples of drugs with excellent selective toxicity are those that block the synthesis of the cell wall in bacteria (penicillins). They have low toxicity and few direct effects on human cells because human cells lack the chemical peptidoglycan and are thus unaffected by this action of the antibiotic. Among the most toxic to human cells are drugs that act upon a structure common to both the infective agent and the host cell, such as the cell membrane (for example, amphotericin B used to treat fungal infections). As the characteristics of the infectious agent become more and more similar to those of the host cell, selective toxicity becomes more difficult to achieve, and more and greater side effects are seen. The previous example briefly illustrates this concept. We will examine the subject in more detail in a later section.

Mechanisms of Drug Action

If the goal of chemotherapy is to disrupt the structure or function of an organism to the point where it can no longer survive, then the first step toward this goal is to identify the structural and metabolic needs of a living cell. Once the requirements of a living cell have been determined, methods of removing, disrupting, or interfering with these requirements can be employed as potential chemotherapeutic strategies. The metabolism of an actively dividing cell is marked by the production of new cell wall components (in most cells), DNA, RNA, proteins, and cell membrane. Consequently, antimicrobial drugs are divided into categories based on which of these metabolic targets they affect. These categories are outlined in **figure 12.1** and include:

1. inhibition of cell wall synthesis,
2. inhibition of nucleic acid (RNA and DNA) structure and function,

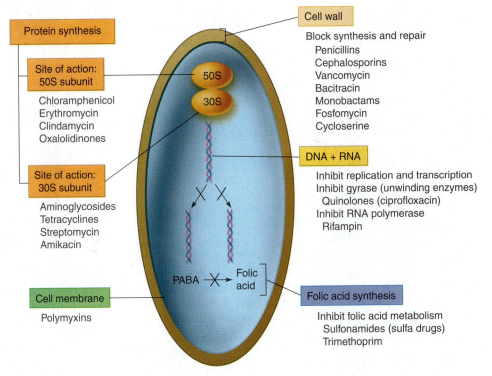

FIGURE 12.1 Primary sites of action of antimicrobial drugs on bacterial cells.

Protein synthesis

Site of action: 50S subunit

Chloramphenicol
Erythromycin
Clindamycin
Oxalolidinones

Site of action: 30S subunit

Aminoglycosides
Tetracyclines
Streptomycin
Amikacin

Cell membrane

Polymyxins

50S
30S

PABA — Folic acid

Cell wall

Block synthesis and repair
Penicillins
Cephalosporins
Vancomycin
Bacitracin
Monobactams
Fosfomycin
Cycloserine

DNA + RNA

Inhibit replication and transcription
Inhibit gyrase (unwinding enzymes)
Quinolones (ciprofloxacin)
Inhibit RNA polymerase
Rifampin

Folic acid synthesis

Inhibit folic acid metabolism
Sulfonamides (sulfa drugs)
Trimethoprim

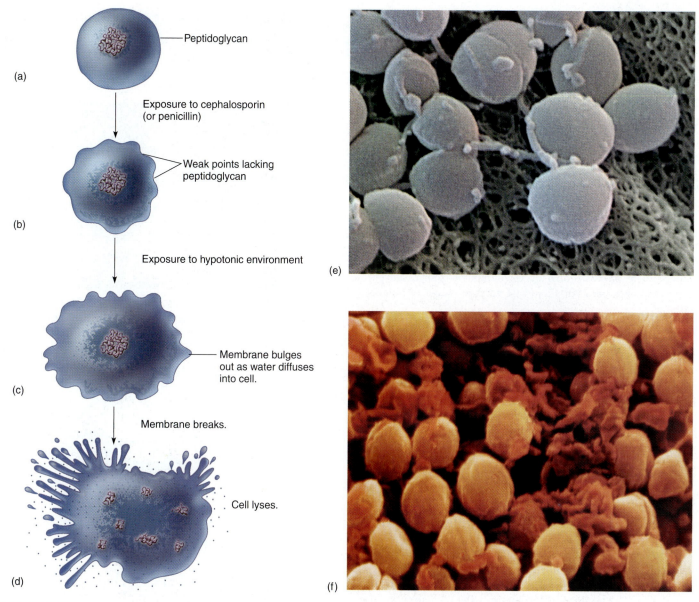

FIGURE 12.2 **The consequences of exposing a growing cell to antibiotics that prevent cell wall synthesis.**
(a)–(d) Diagram of effect of cephalosporin on a *Staphylococcus*. **(e)** Scanning electron micrograph of bacterial cells in their normal state (10,000×). **(f)** Scanning electron micrograph of the same cells in the drug-affected state, showing deformation and disintegration (10,000×).

3. inhibition of protein synthesis,
4. interference with cell membrane structure or function, and
5. inhibition of folic acid synthesis.

As you will see, these categories are not completely discrete, and some effects can overlap.

Antimicrobial Drugs That Affect the Bacterial Cell Wall

The cell walls of most bacteria contain a rigid girdle of peptidoglycan, which protects the cell against rupture from hypotonic environments. Active cells must constantly synthesize new peptidoglycan and transport it to its proper place in the cell envelope. Drugs such as penicillins and cephalosporins react with one or more of the enzymes required to complete this process, causing the cell to develop weak points at growth sites and to become osmotically fragile **(figure 12.2).** Antibiotics that produce this effect are considered bactericidal, because the weakened cell is subject to lysis. It is essential to note that most of these antibiotics are active only in young, growing cells, because old, inactive, or dormant cells do not synthesize peptidoglycan. (One exception is a new class of antibiotics called the "-penems.")

Cycloserine inhibits the formation of the basic peptidoglycan subunits, and vancomycin hinders the elongation of the peptidoglycan. Penicillins and cephalosporins bind and block peptidases that cross-link the glycan molecules, thereby

INSIGHT 12.2   *Discovery*

A Modern Quest for Designer Drugs

Once the first significant drug was developed, the world immediately witnessed a scientific scramble to find more antibiotics.
This search was advanced on several fronts. Hundreds of investigators began the laborious task of screening samples from soil,
dust, muddy lake sediments, rivers, estuaries, oceans, plant surfaces, compost heaps, sewage, skin, and even the hair and skin of
animals for antibiotic-producing bacteria and fungi. This intense
effort has paid off over the past 50 years, because more than
10,000 antibiotics have been discovered (although surprisingly,
only a relatively small number have actually been used clinically). Finding a new antimicrobial substance is only a first step.
The complete pathway of drug development from discovery
to therapy takes between 12 and 24 years at a cost of billions
of dollars.

Antibiotics are products of fermentation pathways that occur
in many spore-forming bacteria and fungi. The role of antibiotics
in the lives of these microbes must be important since the genes for
antibiotic production are preserved in evolution. Some experts
theorize that antibiotic-releasing microorganisms can inhibit or
destroy nearby competitors or predators; others propose that antibiotics play a part in spore formation. Whatever benefit the microbes derive, these compounds have been extremely profitable
for humans. Every year, the pharmaceutical industry farms vast
quantities of microorganisms and harvests their products to treat
diseases caused by other microorganisms. Researchers have facilitated the work of nature by selecting mutant species that yield
more abundant or useful products, by varying the growth medium,
or by altering the procedures for large-scale industrial production
(see chapter 24).

Another approach in the drug quest is to chemically manipulate molecules by adding or removing functional groups. Drugs
produced in this way are designed to have advantages over other,
related drugs. Using this **semisynthetic** method, a natural product of the microorganism is joined with various preselected functional groups. The antibiotic is reduced to its basic molecular
framework (called the nucleus), and to this nucleus specially selected side chains (R groups) are added. A case in point is the
metamorphosis of the semisynthetic penicillins. The nucleus is an
inactive penicillin derivative called aminopenicillanic acid, which
has an opening on the number 6 carbon for addition of R groups.
A particular carboxylic acid (R group) added to this nucleus can
"fine-tune" the penicillin, giving it special characteristics. For instance, some R groups will make the product resistant to penicillinase (methicillin), some confer a broader activity spectrum
(ampicillin), and others make the product acid-resistant (penicillin V). Cephalosporins and tetracyclines also exist in several
semisynthetic versions. The potential for using bioengineering
techniques to design drugs seems almost limitless, and, indeed,
several drugs have already been produced by manipulating the
genes of antibiotic producers.

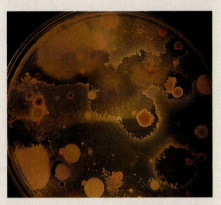

A plate with several discrete colonies of soil bacteria was sprayed with a
culture of *Escherichia coli* and incubated. Zones of inhibition (clear areas
with no growth) surrounding several colonies indicate species that
produce antibiotics.

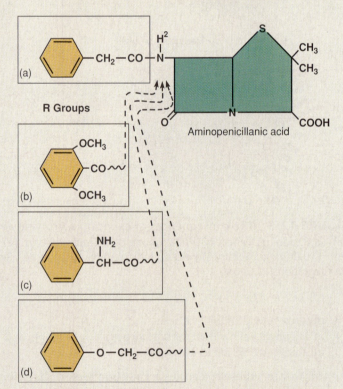

Synthesizing penicillins. **(a)** The original penicillin G molecule is a
fermentation product of *Penicillium chrysogenum* that appears somewhat
like a house with a removable patio on the left. This house without the
patio is the basic nucleus called aminopenicillanic acid. **(b–d)** Various
new fixtures (R groups) can be added to change the properties of the
drug. These R groups will produce different penicillins: **(b)** methicillin;
(c) ampicillin; and **(d)** penicillin V.

FIGURE 12.3 The mode of action of penicillins and cephalosporins on the bacterial cell wall.
(a) Intact peptidoglycan has chains of NAM (*N*-acetyl muramic acid) and NAG (*N*-acetyl glucosamine) glycans cross-linked by peptide bridges.
(b) These two drugs block the peptidases that link the cross-bridges between NAMs, thereby greatly weakening the cell wall meshwork.

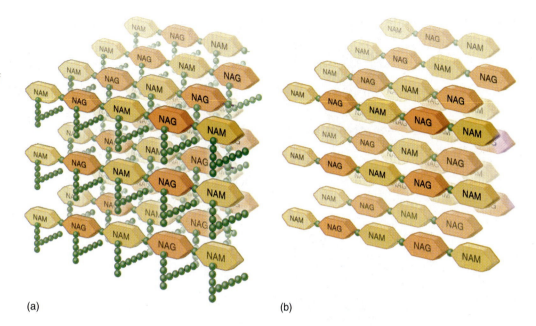

(a)   (b)

interrupting the completion of the cell wall **(figure 12.3).** Penicillins that do not penetrate the outer membrane are less effective against gram-negative bacteria, but broad-spectrum penicillins and cephalosporins, such as carbenicillin or ceftriaxone, can access the cell walls of gram-negative species.

Antimicrobial Drugs That Affect Nucleic Acid Synthesis

As you learned in chapter 9, the metabolic pathway that generates DNA and RNA is a long, enzyme-catalyzed series of reactions. Like any complicated process, it is subject to breakdown at many different points along the way, and inhibition at any point in the sequence can block subsequent events. Antimicrobial drugs interfere with nucleic acid synthesis by blocking synthesis of nucleotides, inhibiting replication, or stopping transcription. Because functioning DNA and RNA are required for proper translation as well, the effects on protein metabolism can be far-reaching.

Other antimicrobials inhibit DNA synthesis. Chloroquine (an antimalarial drug) binds and cross-links the double helix. The newer broad-spectrum quinolones inhibit DNA unwinding enzymes or helicases, thereby stopping DNA transcription. Antiviral drugs that are analogs of purines and pyrimidines, including azidothymidine (AZT) and acyclovir, insert in the viral nucleic acid and block further replication.

Antimicrobial Drugs That Block Protein Synthesis

Most inhibitors of translation, or protein synthesis, react with the ribosome-mRNA complex. Although human cells also have ribosomes, the ribosomes of eucaryotes are different in size and structure from those of procaryotes, so these antimicrobials usually have a selective action against bacteria. One potential therapeutic consequence of drugs that bind to the procaryotic ribosome is the damage they can do to eucaryotic mitochondria, which contain a procaryotic type of ribosome. Two possible targets of ribosomal inhibition are the 30S

subunit and the 50S subunit **(figure 12.4).** Aminoglycosides (streptomycin, gentamicin, for example) insert on sites on the 30S subunit and cause the misreading of the mRNA, leading to abnormal proteins. Tetracyclines block the attachment of tRNA on the A acceptor site and effectively stop further synthesis. Other antibiotics attach to sites on the 50S subunit in a way that prevents the formation of peptide bonds (chloramphenicol) or inhibits translocation of the subunit during translation (erythromycin).

Antimicrobial Drugs That Disrupt Cell Membrane Function

A cell with a damaged membrane invariably dies from disruption in metabolism or lysis and does not even have to be actively dividing to be destroyed. The antibiotic classes that damage cell membranes have specificity for particular microbial groups, based on differences in the types of lipids in their cell membranes.

Polymyxins interact with membrane phospholipids, distort the cell surface, and cause leakage of proteins and nitrogen bases, particularly in gram-negative bacteria. The polyene antifungal antibiotics (amphotericin B and nystatin) form complexes with the sterols on fungal membranes; these complexes cause abnormal openings and seepage of small ions. Unfortunately, this selectivity is not exact, and the universal presence of membranes in microbial and animal cells alike means that most of these antibiotics can be quite toxic to humans.

Antimicrobial Drugs That Inhibit Folic Acid Synthesis

Sulfonamides and trimethoprim are drugs that act by mimicking the normal substrate of an enzyme in a process called **competitive inhibition.** They are supplied to the cell in high concentrations to ensure that a needed enzyme is constantly occupied with the **metabolic analog** rather than the true substrate of the enzyme. As the enzyme is no longer able to

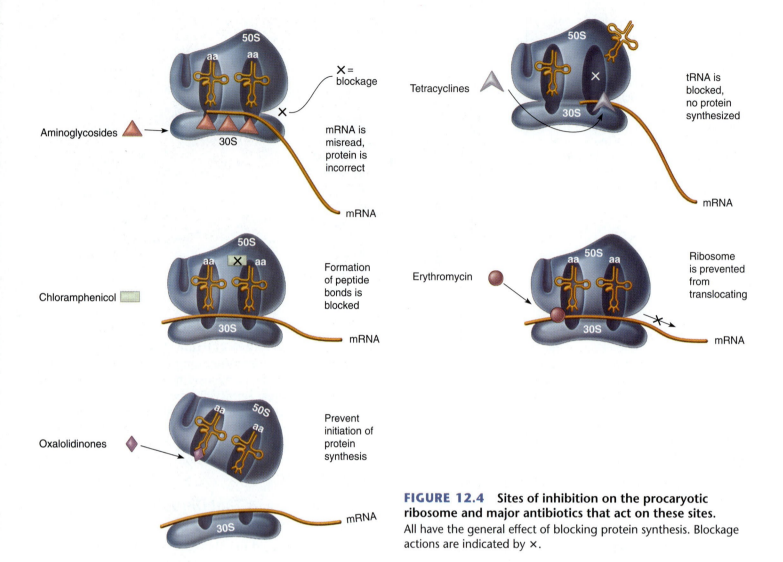

FIGURE 12.4 **Sites of inhibition on the procaryotic ribosome and major antibiotics that act on these sites.** All have the general effect of blocking protein synthesis. Blockage actions are indicated by ×.

produce a needed product, cellular metabolism slows or stops. Sulfonamides and trimethoprim interfere with folate metabolism by blocking enzymes required for the synthesis of tetrahydrofolate, which is needed by bacterial cells for the synthesis of folic acid and the eventual production of DNA and RNA and amino acids. Trimethoprim and sulfonamides are often given simultaneously to achieve a *synergistic effect*, which, in pharmacological terms, refers to an additive effect achieved by multiple drugs working together, thus requiring a lower dose of each. **Figure 12.5** illustrates sulfonamides' competition with PABA for the active site of the enzyme pteridine synthetase.

The selective toxicity of these compounds is explained by the fact that mammals derive folic acid from their diet and so do not possess this enzyme system. Therefore, inhibition of bacterial and protozoan parasites, which must synthesize folic acid, is easily accomplished while leaving the human host unaffected.

12.3 Survey of Major Antimicrobial Drug Groups

Scores of antimicrobial drugs are marketed in the United States. Although the medical and pharmaceutical literature contains a wide array of names for antimicrobials, most of them are variants of a small number of drug families. About 260 different antimicrobial drugs are currently classified in 20 drug families. Drug reference books may give the impression that there are 10 times that many because various drug companies assign different trade names to the very same generic drug. Ampicillin, for instance, is available under 50 different names. Most antibiotics are useful in controlling bacterial infections, although we shall also consider a number of antifungal, antiviral, and antiprotozoan drugs. **Table 12.3** summarizes some major infectious agents, the diseases they cause, and the drugs indicated to treat them (drugs of choice).

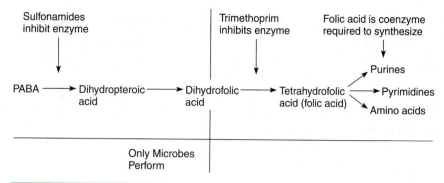

(a) Normal metabolic pathway

(b) Normal folic acid synthesis

(c) Inhibition of folic acid synthesis by sulfa drug

FIGURE 12.5 **The mode of action of sulfa drugs.**
(a) The metabolic pathway needed to synthesize tetrahydrofolic acid (THFA) contains two enzymes that are chemotherapeutic targets.
(b) Under normal circumstances, PABA acts as a substrate for the enzyme pteridine synthetase, the product of which is needed for the eventual production of folic acid. **(c)** Sulfonamides act as a competitive inhibitor, constantly filling the active site of the enzyme and preventing PABA from entering the site and being converted to the folic acid molecule.

✶ Remember at least 1 drug from each

TABLE 12.3 Selected Survey of Chemotherapeutic Agents in Infectious Diseases

Infectious Agent	Typical Infection	Drugs of Choice*
Bacteria		
Gram-Positive Cocci		
Staphylococcus aureus	Abscess, skin infections, toxic shock	Penicillins, vancomycin, cephalosporin
Streptococcus pyogenes	Strep throat, erysipelas, rheumatic fever	Penicillin, cephalosporin, erythromycin
Streptococcus pneumoniae	Pneumonia	Penicillin, if sensitive, cephalosporin, erythromycin
Gram-Positive Rods		
Bacillus	Anthrax	Penicillin, doxycycline
Clostridium	Gas gangrene, tetanus	Penicillin, cephalosporin, clindamycin
Corynebacterium	Diphtheria	Erythromycin, penicillin
Acid-Fast Rods		
Mycobacterium tuberculosis	Tuberculosis	(Isoniazid, rifampin, pyrazinamide),* ethambutol, streptomycin
Mycobacterium leprae	Leprosy	(Dapsone, rifampin, clofazimine)**
Nocardia	Nocardiosis	Sulfamethoxazole-trimethoprim, cephalosporin
Gram-Negative Cocci		
Neisseria gonorrhoeae	Gonorrhea	Ceftriaxone, spectinomycin, quinolones
Neisseria meningitidis	Meningitis	Penicillin G, ceftriaxone, ampicillin
Gram-Negative Rods		
Bordetella	Pertussis	Erythromycin
Brucella	Brucellosis	Tetracycline, rifampin, gentamicin
Escherichia coli	Sepsis, diarrhea, urinary tract infection	Cephalosporin, quinolone, sulfamethoxazole-trimethoprim
Haemophilus influenzae	Meningitis	Cefotaxime, cephtriaxone
Legionella	Legionnaires disease	Erythromycin, rifampin
Pseudomonas	Opportunistic lung and burn infections	Ticarcillin, aminoglycoside
Salmonella	Typhoid fever, salmonellosis	Cephalosporin, quinolone
Shigella	Dysentery	Quinolone, sulfamethoxazole-trimethoprim, ampicillin
Vibrio cholerae	Cholera	Tetracyclines, sulfamethoxazole-trimethoprim
Yersinia pestis	Plague	Streptomycin, gentamicin
Spirochetes		
Borrelia	Lyme disease	Ceftriaxone, doxycycline
Treponema pallidum	Syphilis	Penicillin, tetracyclines
Mycoplasma	Pneumonia, urinary infections	Erythromycin, azithromycin, clarithromycin
Rickettsia	Rocky Mountain spotted fever	Tetracyclines, chloramphenicol
Chlamydia	Urethritis, vaginitis	Azithromycin, doxycycline, erythromycin, quinolones
Fungi		
Systemic Mycoses		
Aspergillus	Aspergillosis	Amphotericin B, azoles, flucytosine
Blastomyces	Blastomycosis	Ketoconazole, amphotericin B
Candida albicans	Candidiasis	Amphotericin B, fluconazole
Coccidioides immitis	Valley fever	Amphotericin B, azoles
Cryptococcus neoformans	Cryptococcosis	Amphotericin B, fluconazole, flucytosine
Pneumocystis (carinii) jiroveci	Pneumonia (PCP)	Trimethoprim-sulfamethoxazole
Sporothrix schenckii	Sporotrichosis	Iodides, itraconazole
Protozoa		
Entamoeba histolytica	Amoebiasis	Metronidazole, fiodoquinol
Giardia lamblia	Giardiasis	Quinacrine, metronidazole
Plasmodium	Malaria	Chloroquine, quinine, mefloquine
Toxoplasma gondii	Toxoplasmosis	Pyrimethamine, sulfadiazine
Trichomonas vaginalis	Trichomoniasis	Metronidazole
Trypanosoma brucei	Sleeping sickness	Suramin,*** pentamidine

continued

TABLE 12.3	continued	
Infectious Agent	**Typical Infection**	**Drugs of Choice***
Helminths		
Ascaris	Ascariosis	Mebendazole/pyrantel, piperazine
Cestodes	Tapeworm	Niclosamide, praziquantel
Schistosoma	Schistosomiasis	Praziquantel
Various fluke infections		Praziquantel
Various roundworm infections		Mebendazole, piperazine
Viruses		
Herpesvirus	Genital herpes, oral herpes, shingles	Acyclovir, ganciclovir
HIV	AIDS	(AZT, protease inhibitors), ddI, ddC, d4T
Orthomyxovirus	Type A influenza	Amantadine, rimantidine

*More or less in order of preference.

**( ) Usually given in combination.

***Available in the United States only from the Drug Service of the Centers for Disease Control and Prevention.

Antibacterial Drugs

Penicillin and Its Relatives

The **penicillin** group of antibiotics, named for the parent compound, is a large, diverse group of compounds, most of which end in the suffix *-cillin*. Although penicillins could be completely synthesized in the laboratory from simple raw materials, it is more practical and economical to obtain natural penicillin through microbial fermentation. The natural product can then be used either in unmodified form or to make semisynthetic derivatives. *Penicillium chrysogenum* is the major source of the drug. All penicillins consist of three parts: a thiazolidine ring, a *beta-lactam* (bey'-tuh-lak'-tam) ring, and a variable side chain that dictates its microbicidal activity **(figure 12.6).**

Subgroups and Uses of Penicillins The characteristics of certain penicillin drugs are shown in **table 12.4.** Penicillins G and V are the most important natural forms. Penicillin is considered the drug of choice for infections by known sensitive, gram-positive cocci (most streptococci) and some gram-negative bacteria (meningococci and the syphilis spirochete).

Certain semisynthetic penicillins such as ampicillin, carbenicillin, and amoxicillin have broader spectra and thus can be used to treat infections by gram-negative enteric rods. Many bacteria produce enzymes that are capable of destroying the beta-lactam ring of penicillin. The enzymes are referred to as **penicillinases** or *beta-lactamases*, and they make the bacteria that possess them resistant to many penicillins. Penicillinase-resistant penicillins such as methicillin, nafcillin, and cloxacillin are useful in treating infections caused by some penicillinase-producing bacteria. Mezlocillin and azlocillin have such an extended spectrum that they can be substituted for combinations of antibiotics. All of the "-cillin" drugs are relatively mild and well tolerated because of their specific

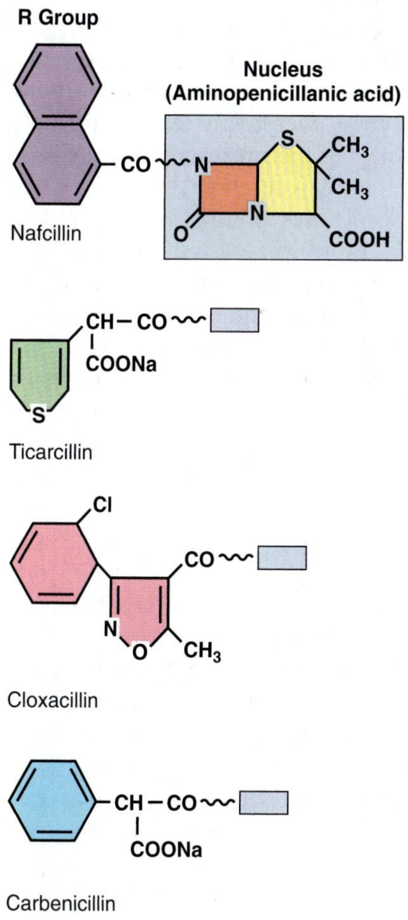

FIGURE 12.6 **Chemical structure of penicillins.**
All penicillins contain a thiazolidine ring (yellow) and a beta-lactam ring (red), but each differs in the nature of the side chain (R group), which is also responsible for differences in biological activity.

TABLE 12.4 Characteristics of Selected Penicillin Drugs

Name	Spectrum of Action	Uses, Advantages	Disadvantages
Penicillin G	Narrow	Best drug of choice when bacteria are sensitive; low cost; low toxicity	Can be hydrolyzed by penicillinase; allergies occur; requires injection
Penicillin V	Narrow	Good absorption from intestine; otherwise, similar to penicillin G	Hydrolysis by penicillinase; allergies
Oxacillin, dicloxacillin	Narrow	Not susceptible to penicillinase; good absorption	Allergies; expensive
Methicillin, nafcillin	Narrow	Not usually susceptible to penicillinase	Poor absorption; allergies; growing resistance
Ampicillin	Broad	Works on gram-negative bacilli	Can be hydrolyzed by penicillinase; allergies; only fair absorption
Amoxicillin	Broad	Gram-negative infections; good absorption	Hydrolysis by penicillinase; allergies
Carbenicillin	Broad	Same as ampicillin	Poor absorption; used only parenterally
Azlocillin, mezlocillin ticarcillin	Very broad	Effective against *Pseudomonas* species; low toxicity compared with aminoglycosides	Allergies, susceptible to many beta-lactamases

mode of action on cell walls (which humans lack). The primary problems in therapy include allergy, which is altogether different than toxicity, and resistant strains of pathogens. Clavulanic acid is a chemical that inhibits beta-lactamase enzymes, thereby increasing the longevity of beta-lactam antibiotics in the presence of penicillinase-producing bacteria. For this reason, clavulanic acid is often added to semisynthetic penicillins to augment their effectiveness. For example, clavamox is a combination of amoxicillin and clavulanate and is marketed under the trade name Augmentin. Zosyn is a similar combination of tazobactum, a beta-lactamase inhibitor, and piperacillin that is used for a wide variety of systemic infections.

The Cephalosporin Group of Drugs

The **cephalosporins** are a newer group of antibiotics that currently account for a majority of all antibiotics administered. The first compounds in this group were isolated in the late 1940s from the mold *Cephalosporium acremonium*. Cephalosporins are similar to penicillins; they have a beta-lactam structure that can be synthetically altered **(figure 12.7)** and have a similar mode of action. The generic names of these compounds are often recognized by the presence of the root *cef, ceph,* or *kef* in their names.

Subgroups and Uses of Cephalosporins

The cephalosporins are versatile. They are relatively broad-spectrum, resistant

to most penicillinases, and cause fewer allergic reactions than penicillins. Although some cephalosporins are given orally, many are poorly absorbed from the intestine and must be administered **parenterally** (par-ehn'-tur-ah-lee), by injection into a muscle or a vein.

Three generations of cephalosporins exist, based upon their antibacterial activity. First-generation cephalosporins

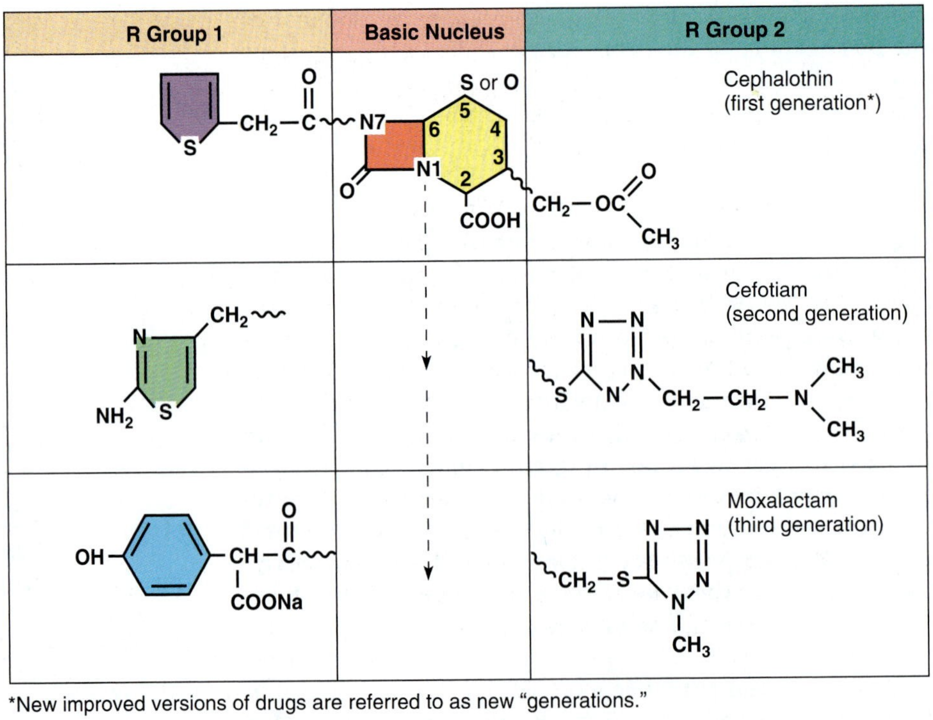

*New improved versions of drugs are referred to as new "generations."

FIGURE 12.7 The structure of cephalosporins.
Like penicillin, they have a beta-lactam ring (red), but they have a different main ring (yellow). However, unlike penicillins, they have two sites for placement of R groups (at positions 3 and 7). This makes possible several generations of molecules with greater versatility in function and complexity in structure.

such as cephalothin and cefazolin are most effective against gram-positive cocci and a few gram-negative bacteria. Second-generation forms include cefaclor and cefonacid, which are more effective than the first-generation forms in treating infections by gram-negative bacteria such as *Enterobacter, Proteus,* and *Haemophilus.* Third-generation cephalosporins, such as cephalexin (Keflex) and cefotaxime, are broad-spectrum with especially well-developed activity against enteric bacteria that produce beta-lactamases. Ceftriaxone (rocephin) is a newer semisynthetic broad-spectrum drug for treating a wide variety of respiratory, skin, urinary, and nervous system infections.

Other Beta-Lactam Antibiotics

Related antibiotics include imipenem, a broad-spectrum drug for infections with aerobic and anaerobic pathogens. It is active in very low concentrations and can be taken by mouth with few side effects. Aztreonam, isolated from the bacterium *Chromobacterium violaceum,* is a newer narrow-spectrum drug for treating pneumonia, septicemia, and urinary tract infections by gram-negative aerobic bacilli. Aztreonam is especially useful when treating persons who are allergic to penicillin. Because of similarities in their chemical structure, allergies to penicillin often are accompanied by allergies to cephalosporins and carboxypenems (of which imipenem is a member). The structure of aztreonam is chemically distinct so that persons with allergies to penicillin are not usually adversely affected by treatment with aztreonam.

The Aminoglycoside Drugs

Antibiotics composed of one or more amino sugars and an aminocyclitol (6-carbon) ring are referred to as **aminoglycosides (figure 12.8).** These complex compounds are exclusively the products of various species of soil **actinomycetes** in the genera *Streptomyces* **(figure 12.9)** and *Micromonospora.*

Subgroups and Uses of Aminoglycosides   The aminoglycosides have a relatively broad antimicrobial spectrum because they inhibit protein synthesis. They are especially useful in treating infections caused by aerobic gram-negative rods and certain gram-positive bacteria. Streptomycin is among the oldest of the drugs and has gradually been replaced by newer forms with less mammalian toxicity. It is still the antibiotic of choice for treating bubonic plague and tularemia and is considered a good antituberculosis agent. Gentamicin is less toxic and is widely administered for infections caused by gram-negative rods (*Escherichia, Pseudomonas, Salmonella,* and *Shigella*). Two relatively new aminoglycosides, tobramycin and amikacin, are also used for gram-negative infections and have largely replaced kanamycin.

Tetracycline Antibiotics

In 1948, a colony of *Streptomyces* isolated from a soil sample gave off a substance, aureomycin, with strong antimicrobial properties. This antibiotic was used to synthesize its relatives terramycin and tetracycline. These natural parent compounds and semisynthetic derivatives are known as the

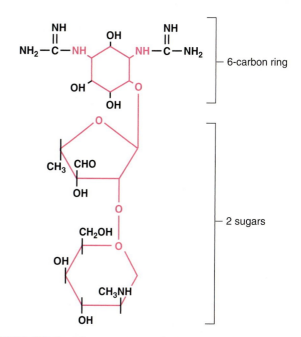

FIGURE 12.8   **The structure of streptomycin.**
Colored portions of the molecule show the general arrangement of an aminoglycoside.

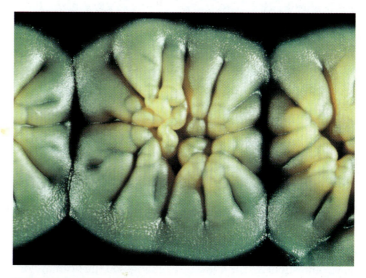

FIGURE 12.9   **A colony of *Streptomyces,* one of nature's most prolific antibiotic producers.**

tetracyclines **(figure 12.10a).** Their action of binding to ribosomes and blocking protein synthesis accounts for the broad-spectrum effects in the group.

Subgroups and Uses of Tetracyclines   The scope of microorganisms inhibited by tetracyclines includes gram-positive and gram-negative rods and cocci, aerobic and anaerobic bacteria, mycoplasmas, rickettsias, and spirochetes. Tetracycline compounds such as doxycycline and minocycline are administered orally to treat several sexually transmitted diseases, Rocky Mountain spotted fever, Lyme disease, typhus, *Mycoplasma* pneumonia, cholera, leptospirosis, acne, and even

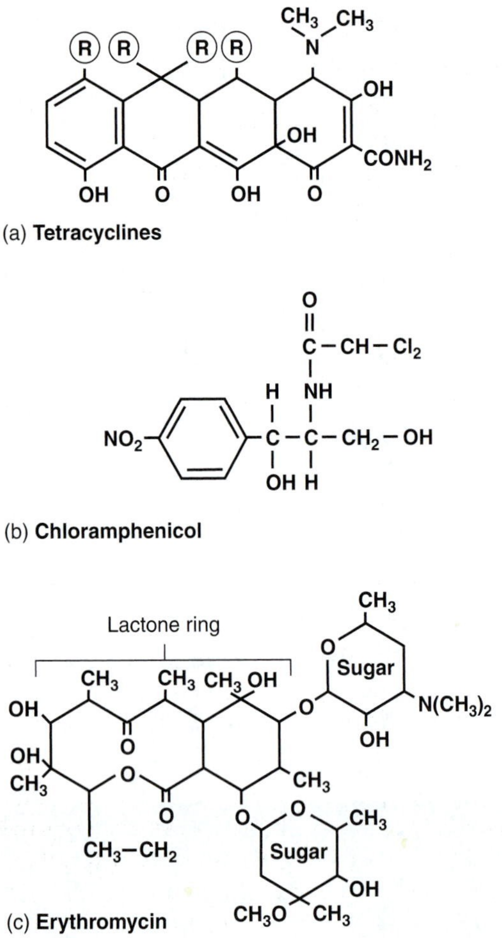

(a) Tetracyclines

(b) Chloramphenicol

(c) Erythromycin

FIGURE 12.10 **Structures of three broad-spectrum antibiotics.**
(a) Tetracyclines. These are named for their regular group of four rings. The several types vary in structure and activity by substitution at the four R groups. (b) Chloramphenicol. (c) Erythromycin, an example of a macrolide drug. Its central feature is a large lactone ring to which two hexose sugars are attached.

some protozoan infections. Although generic tetracycline is low in cost and easy to administer, its side effects—namely, gastrointestinal disruption due to changes in the normal flora of the gastrointestinal tract and deposition in hard tissues—can limit its use (see table 12.6).

Chloramphenicol

Originally isolated in the late 1940s from *Streptomyces venezuelae*, chloramphenicol is a potent broad-spectrum antibiotic with a unique nitrobenzene structure (**figure 12.10b**). Its primary effect on cells is to block peptide bond formation and protein synthesis. It is one type of antibiotic that is no longer derived from the natural source but is entirely synthesized through chemical processes. Although this drug is fully as broad-spectrum as the tetracyclines, it is so toxic to human cells that its uses are restricted. A small number of people undergoing long-term therapy with this drug incur irreversible damage to the bone marrow that usually results in a fatal

form of aplastic anemia.[1] Its administration is now limited to typhoid fever, brain abscesses, and rickettsial and chlamydial infections for which an alternative therapy is not available. Chloramphenicol should never be given in large doses repeatedly over a long time period, and the patient's blood must be monitored during therapy.

Other *Streptomyces* Antibiotics: Erythromycin, Clindamycin, Vancomycin, Rifamycin

Erythromycin is a macrolide[2] antibiotic first isolated in 1952 from a strain of *Streptomyces*. Its structure consists of a large lactone ring with sugars attached (**figure 12.10c**). This drug is relatively broad-spectrum and of fairly low toxicity. Its mode of action is to block protein synthesis by attaching to the ribosome. It is administered orally for *Mycoplasma* pneumonia, legionellosis, *Chlamydia* infections, pertussis, and diphtheria, and as a prophylactic drug prior to intestinal surgery. It also offers a useful substitute for dealing with penicillin-resistant streptococci and gonococci and for treating syphilis and acne. Newer semisynthetic macrolides include *clarithromycin* and *azithromycin*. Both drugs are useful for middle ear, respiratory, and skin infections and have also been approved for *Mycobacterium* (MAC) infections in AIDS patients. Clarithromycin has additional applications in controlling infectious stomach ulcers (see chapter 22).

Clindamycin is a broad-spectrum antibiotic derived from the less effective lincomycin. The tendency of clindamycin to cause adverse reactions in the gastrointestinal tract limits its applications to

1. serious infections in the large intestine and abdomen due to anaerobic bacteria (*Bacteroides* and *Clostridium*) that are unresponsive to other antibiotics,
2. infections with penicillin-resistant staphylococci, and
3. acne medications applied to the skin.

In 2004 the FDA approved the first representative of a new class of drugs called ketolides which are similar to macrolides like erythromycin but have a different ring structure. The new drug, called Ketek, is to be used for respiratory tract infections that are suspected to be caused by antibiotic-resistant bacteria such as *Streptococcus pneumoniae*.

Vancomycin is a narrow-spectrum antibiotic most effective in treating staphylococcal infections in cases of penicillin and methicillin resistance or in patients with an allergy to penicillins. It has also been chosen to treat *Clostridium* infections in children and endocarditis (infection of the lining of the heart) caused by *Enterococcus faecalis*. Because it is very toxic and hard to administer, vancomycin is usually restricted to the most serious, life-threatening conditions.

Another product of the genus *Streptomyces* is rifamycin, which is altered chemically into rifampin. It is somewhat

---

1. A failure of the blood-producing tissue that results in very low levels of blood cells.

2. Macrolide antibiotics get their name from the type of chemical ring structure (macrolide ring) they all possess.

limited in spectrum because the molecule cannot pass through the cell envelope of many gram-negative bacilli. It is mainly used to treat infections by several gram-positive rods and cocci and a few gram-negative bacteria. Rifampin figures most prominently in treating mycobacterial infections, especially tuberculosis and leprosy, but it is usually given in combination with other drugs to prevent development of resistance. Rifampin is also recommended for prophylaxis in *Neisseria meningitidis* carriers and their contacts, and it is occasionally used to treat *Legionella, Brucella,* and *Staphylococcus* infections.

The Bacillus Antibiotics: Bacitracin and Polymyxin

Bacitracin is a narrow-spectrum peptide antibiotic produced by a strain of the bacterium *Bacillus subtilis.* Since it was first isolated, its greatest claim to fame has been as a major ingredient in a common drugstore antibiotic ointment (Neosporin) for combating superficial skin infections by streptococci and staphylococci. For this purpose, it is usually combined with neomycin (an aminoglycoside) and polymyxin.

Bacillus polymyxa is the source of the **polymyxins,** narrow-spectrum peptide antibiotics with a unique fatty acid component that contributes to their detergent activity. Only two polymyxins—B and E (also known as colistin)—have any routine applications, and even these are limited by their toxicity to the kidney. Either drug can be indicated to treat drug-resistant *Pseudomonas aeruginosa* and severe urinary tract infections caused by other gram-negative rods.

New Classes of Antibiotics

Most new antibiotics are formulated from the traditional drug classes. Three recent additions that are completely new drug types are fosfomycin, synercid, and daptomycin. Fosfomycin trimethamine is a phosphoric acid agent effective as alternate treatment for urinary tract infections caused by enteric bacteria. It works by inhibiting an enzyme necessary for cell wall synthesis. Synercid is a combined antibiotic from the streptogramin group of drugs. It is effective against *Staphylococcus* and *Enterococcus* species that cause endocarditis and surgical infections, and against resistant strains of *Streptococcus.* It is one of the main choices when other drugs are ineffective due to resistance. Synercid works by binding to sites on the 50S ribosome, inhibiting translation. Daptomycin is a lipopeptide made by *Streptomyces.* It is most active against gram-positive bacteria, acting to disrupt multiple aspects of membrane function. Many experts are urging physicians to use these medications only when no other drugs are available to slow the development of drug resistance.

Synthetic Antibacterial Drugs

The synthetic antimicrobials as a group do not originate from bacterial or fungal fermentations. Some were developed from aniline dyes, and others were originally isolated from plants. Although they have been largely supplanted by antibiotics, several types are still useful.

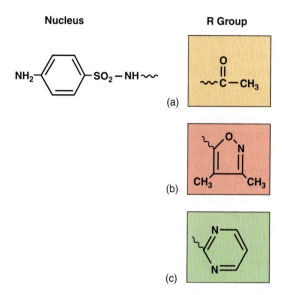

FIGURE 12.11 **The structures of some sulfonamides.** **(a)** Sulfacetamide, **(b)** sulfadiazine, and **(c)** sulfisoxazole.

The Sulfonamides, Trimethoprim, and Sulfones

The very first modern antimicrobial drugs were the **sulfonamides,** or sulfa drugs, named for sulfanilamide, an early form of the drug **(figure 12.11).** Although thousands of sulfonamides have been formulated, only a few have gained any importance in chemotherapy. Because of its solubility, sulfisoxazole is the best agent for treating shigellosis, acute urinary tract infections, and certain protozoan infections. Silver sulfadiazine ointment and solution are prescribed for treatment of burns and eye infections. Another drug, trimethoprim (Septra, Bactrim), inhibits the enzymatic step immediately following the step inhibited by sulfonamides in the synthesis of folic acid. Because of this, trimethoprim is often given in combination with sulfamethoxazole to take advantage of the synergistic effect of the two drugs. This combination is one of the primary treatments for *Pneumocystis (carinii) jiroveci* pneumonia (PCP) in AIDS patients.

Sulfones are compounds chemically related to the sulfonamides but lacking their broad-spectrum effects. This lack does not diminish their importance as key drugs in treating Hansen's disease (leprosy). The most active form is dapsone, usually given in combination with rifampin and clofazamine (an antibacterial dye) over long periods.

Miscellaneous Antibacterial Agents   Isoniazid (INH) is bactericidal to *Mycobacterium tuberculosis,* but only against growing cells. Although still in use, it has been largely supplanted by rifampicin. Oral doses are indicated for both active tuberculosis and prophylaxis in cases of a positive TB test. Ethambutol, a closely related compound, is effective in treating the early stages of tuberculosis.

A new class of synthetic antibacterial drugs, oxazolidinones, was recently developed and the first member of that class, linezolid, was approved for use in 2000 by the FDA. These drugs work by a completely novel mechanism, inhibiting the initiation of protein synthesis. Because this class of drug is not found in nature, it is hoped that resistance among bacteria will be slow to develop. Linezolid (under the name Zyvox) is used to treat infections caused by two of the most difficult clinical pathogens, methicillin-resistant *Staphylococcus aureus* (MRSA) and vancomycin-resistant *Enterococcus* (VRE).

Much excitement was generated by a new class of synthetic drugs chemically related to quinine called **fluoroquinolones**. These drugs exhibit several ideal traits, including high potency and broad spectrum. Even in minimal concentrations, quinolones inhibit a wide variety of gram-positive and gram-negative bacterial species. In addition, they are readily absorbed from the intestine. The principal quinolones, norfloxacin and ciprofloxacin, have been successful in therapy for urinary tract infections, sexually transmitted diseases, gastrointestinal infections, osteomyelitis, respiratory infections, and soft tissue infections. Newer drugs in this category are sparfloxacin and levofloxacin. These agents are especially recommended for pneumonia, bronchitis, and sinusitis. Ciprofloxacin received a great deal of publicity recently as the drug of choice for treating anthrax. Shortly after the first attacks, however, the Centers for Disease Control and Prevention changed their recommendation from ciprofloxacin to doxycycline. The change in preferred drug was made not because ciprofloxacin was ineffective but rather to prevent the emergence of ciprofloxacin-resistance bacteria (see Insight 12.4). Side effects that limit the use of quinolones include seizures and other brain disturbances.

Agents to Treat Fungal Infections

Because the cells of fungi are eucaryotic, they present special problems in chemotherapy. For one, the great majority of chemotherapeutic drugs are designed to act on bacteria and are generally ineffective in combating fungal infections. For another, the similarities between fungal and human cells often mean that drugs toxic to fungal cells are also capable of harming human tissues. A few agents with special antifungal properties have been developed for treating systemic and superficial fungal infections. Four main drug groups currently in use are the macrolide polyene antibiotics, griseofulvin, synthetic azoles, and flucytosine **(figure 12.12)**.

Polyenes bind to fungal membranes and cause loss of selective permeability. They are specific for fungal membranes because fungal membranes contain a particular sterol component called ergosterol, while human membranes do not. The toxicity of polyenes is not completely selective, however, because mammalian cell membranes contain compounds similar to ergosterol that bind polyenes to a small extent.

Macrolide polyenes, represented by amphotericin B (named for its acidic and basic—amphoteric—properties) and nystatin (for New York State, where it was discovered), have a structure that mimics the lipids in some cell membranes. Amphotericin B is by far the most versatile and effective of all antifungals. Not only does it work on most fungal infections, including skin and mucous membrane lesions caused by *Candida albicans*, but it is one of the few drugs that

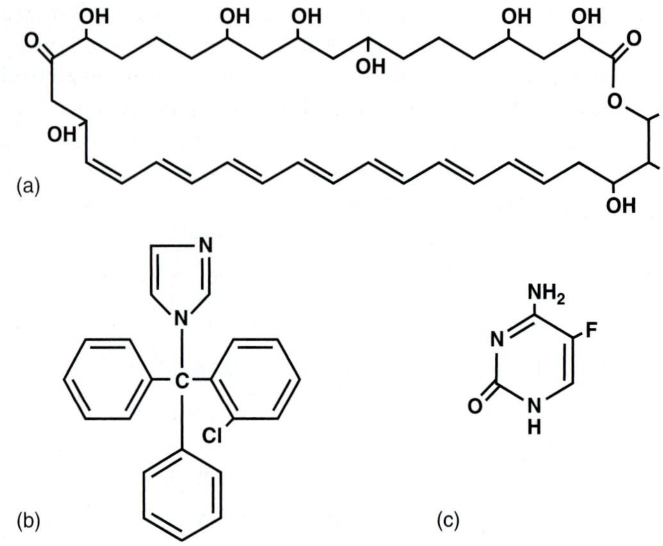

FIGURE 12.12 **Some antifungal drug structures.**
(a) Polyenes. The example shown is amphotericin B, a complex steroidal antibiotic that inserts into fungal cell membranes.
(b) Clotrimazole, one of the azoles. **(c)** Flucytosine, a structural analog of cystosine that contains fluoride.

can be injected to treat systemic fungal infections such as histoplasmosis and cryptococcus meningitis. Nystatin is used only topically or orally to treat candidiasis of the skin and mucous membranes, but it is not useful for subcutaneous or systemic fungal infections or for ringworm.

Griseofulvin is an antifungal product especially active in certain dermatophyte infections such as athlete's foot. The drug is deposited in the epidermis, nails, and hair, where it inhibits fungal growth. Because complete eradication requires several months and griseofulvin is relatively nephrotoxic, this therapy is given only for the most stubborn cases.

The **azoles** are broad-spectrum antifungal agents with a complex ringed structure. The most effective drugs are ketoconazole, fluconazole, clotrimazole, and miconazole. Ketoconazole is used orally and topically for cutaneous mycoses, vaginal and oral candidiasis, and some systemic mycoses. Fluconazole can be used in selected patients for AIDS-related mycoses such as aspergillosis and cryptococcus meningitis. Clotrimazole and miconazole are used mainly as topical ointments for infections in the skin, mouth, and vagina.

Flucytosine is an analog of the nucelotide cytosine that has antifungal properties. It is rapidly absorbed after oral therapy, and it is readily dissolved in the blood and cerebrospinal fluid. Alone, it can be used to treat certain cutaneous mycoses. Many fungi are resistant to flucytosine, so it is usually combined with amphotericin B to effectively treat systemic mycoses.

Antiparasitic Chemotherapy

The enormous diversity among protozoan and helminth parasites and their corresponding therapies reaches far beyond the scope of this textbook; however, a few of the more common drugs will be surveyed here and described again for particular diseases in the organ systems chapters. Presently, a small number of approved and experimental drugs are used to treat malaria, leishmaniasis, trypanosomiasis, amoebic dysentery, and helminth infections, but the need for new and better drugs has spurred considerable research in this area.

Antimalarial Drugs: Quinine and Its Relatives

Quinine, extracted from the bark of the cinchona tree, was the principal treatment for malaria for hundreds of years, but it has been replaced by the synthesized quinolines, mainly chloroquine and primaquine, which have less toxicity to humans. Because there are several species of *Plasmodium* (the malaria parasite) and many stages in its life cycle, no single drug is universally effective for every species and stage, and each drug is restricted in application. For instance, primaquine eliminates the liver phase of infection, and chloroquine suppresses acute attacks associated with infection of red blood cells. Chloroquine is taken alone for prophylaxis and suppression of acute forms of malaria. Primiquine is administered to patients with relapsing cases of malaria. Mefloquine is a semisynthetic analog of quinine used to treat infections caused by chloroquine-resistant strains of *Plasmodium*.

Chemotherapy for Other Protozoan Infections

A widely used amoebicide, metronidazole (Flagyl), is effective in treating mild and severe intestinal infections and hepatic disease caused by *Entamoeba histolytica*. Given orally, it also has applications for infections by *Giardia lamblia* and *Trichomonas vaginalis* (described in chapters 22 and 23, respectively). Other drugs with antiprotozoan activities are quinicrine (a quinine-based drug), sulfonamides, and tetracyclines.

Antihelminthic Drug Therapy

Treating helminthic infections has been one of the most difficult and challenging of all chemotherapeutic tasks. Flukes, tapeworms, and roundworms are much larger parasites than other microorganisms and, being animals, have greater similarities to human physiology. Also, the usual strategy of using drugs to block their reproduction is usually not successful in eradicating the adult worms. The most effective drugs immobilize, disintegrate, or inhibit the metabolism of all stages of the life cycle.

Mebendazole and thiabendazole are broad-spectrum antiparasitic drugs used in several roundworm and tapeworm intestinal infestations. These drugs work locally in the intestine to inhibit the function of the microtubules of worms, eggs, and larvae, which interferes with their glucose utilization and disables them. The compounds pyrantel and piperazine paralyze the muscles of intestinal roundworms. Niclosamide destroys the scolex and the adjoining proglottids of tapeworms, thereby loosening the worm's holdfast. In these forms of therapy, the worms are unable to maintain their grip on the intestinal wall and are expelled along with the feces by the normal peristaltic action of the bowel. Two newer antihelminthic drugs are praziquantel, a treatment for various tapeworm and fluke infections, and ivermectin, a veterinary drug now used for strongyloidiasis and oncocercosis in humans. Helminthic diseases are described in chapter 22 because these organisms spend at least some part of their life cycles in the digestive tract.

Antiviral Chemotherapeutic Agents

The chemotherapeutic treatment of viral infections presents unique problems. Throughout our discussion of infections by bacteria, fungi, protozoa, and helminths, we've relied on differences in structure and metabolism to guide the choice of drug. With viruses, we are dealing with an infectious agent that relies upon the host cell for the vast majority of its metabolic functions. Disrupting viral metabolism requires that we disrupt the metabolism of the host cell to a much greater extent than is desirable. Put another way, selective toxicity with regard to viral infection is almost impossible to achieve because a single metabolic system is responsible for the well-being of both virus and host. Although viral diseases such as measles, mumps, and hepatitis are routinely prevented by the use of effective vaccinations, epidemics of AIDS, influenza, and even the common cold attest to the need for more effective medications for the treatment of viral pathogens.

Over the last few years, several antiviral drugs have been developed that target specific points in the infectious cycle of viruses. Although antiviral drugs are certainly a welcome discovery, the chemotherapeutic treatment of viruses is still in its infancy and most antiviral compounds are quite limited in their usefulness.

Most compounds have their effects on the completion of the virus cycle. Three major modes of action are (table 12.5):

1. barring penetration of the virus into the host cell,
2. blocking the transcription and translation of viral molecules, and
3. preventing the maturation of viral particles.

Although antiviral drugs protect uninfected cells by keeping viruses from being synthesized and released, most are unable to destroy extracellular viruses or those in a latent state.

Several antiviral agents mimic the structure of nucleotides and compete for sites on replicating DNA. The incorporation of these synthetic nucleotides inhibits further DNA synthesis. **Acyclovir** (Zovirax) and its relatives are synthetic purine compounds that block DNA synthesis in a small group of viruses, particularly the herpesviruses (see chapters 18 and 23). In the topical form, it is most effective in controlling the primary attack of facial or genital herpes. Intravenous or oral acyclovir therapy can reduce the severity of primary and recurrent genital herpes episodes. Some newer relatives (valacyclovir) are more effective and require fewer doses. Famciclovir is used to treat shingles and chickenpox caused by the herpes zoster virus, and ganciclovir is approved to treat cytomegalovirus infections of the

TABLE 12.5	Actions of Selected Antiviral Drugs	
Category	**Example of Drug and Its Structure**	**Effects of Drug**

Inhibition of Virus Entry: receptor/fusion/uncoating inhibitors

Fuzeon

Polypeptide of 36 amino acids (trade secret)

Amantidine and relatives

Example: Amantidine

Tamiflu, Relenza

Example: Tamiflu

(1) Blocks HIV infection by preventing the binding of viral GP-41 receptors to cell receptor, thereby preventing fusion of virus with cell

(2) Blocks entry of influenza virus by interfering with fusion of virus with cell membrane (also release)

(3) Stops the actions of influenza neuraminidase, required for entry of virus into cell (also assembly)

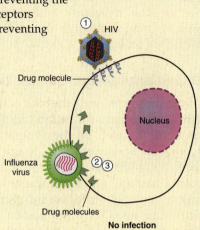

Inhibition of Nucleic Acid Synthesis

Acyclovir, other "cyclovirs"

Example: Acyclovir

Nucleotide analog reverse transcriptase (RT) inhibitors

Example: Zidovudine (AZT)

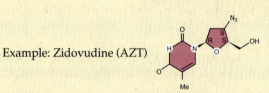

Non-nucleoside reverse transcriptase inhibitors

Example: Nevirapine

(4) Inactivates viral DNA polymerase and terminates DNA replication in herpesviruses

(5) Stops the action of reverse transcriptase in HIV, blocking viral DNA production

(6) Attaches to HIV RT binding site, stopping its action

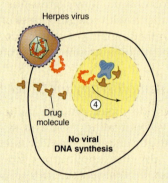

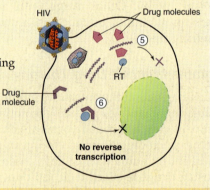

Inhibition of Viral Assembly/Release

Protease inhibitors

Example: Saquinavir

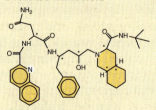

Amantidine and relatives; Tamiflu (see above)

(7) Inserts into HIV protease, stopping its action and resulting in inactive noninfectious viruses

(8) Thought to interfere with influenza virus assembly or budding

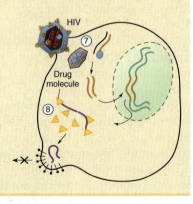

INSIGHT 12.3   *Discovery*

Household Remedies—From Apples to Zinc

Who would have thought that drinking a glass of apple juice, swallowing a clove of garlic, or eating yogurt could be a reliable therapy for infections? A series of research findings from the past few years seems to point to a possible role for these as medicinal aids. Apple juice, fruit juices, and even tea contain natural antiviral substances, thought to be organic acids, that kill the poliovirus and coxsackievirus. Drinking beverages that contain these substances can help prevent the passage of those viruses into the intestine (their usual site of entry). Could this be a reason that "an apple a day keeps the doctor away"?

Another common home remedy has been drinking cranberry juice to reduce the symptoms of urinary tract infections. New research indicates that multiple chemicals in the juice are excreted through the kidney and can block the attachment of pili by pathogens such as *Escherichia coli*. Controlled studies now support the benefits of zinc ions to control the common cold. It has been suggested that the zinc attaches to cell receptors and blocks the attachment of cold viruses. Various tablets and lozenges are now sold over the counter as cold deterrents.

The therapeutic benefits of certain foods are often surprising. Yogurt made with live cultures has been shown in controlled studies to inhibit *Staphylococcus* infections. Further examination demonstrated that some compound given off by *Lactobacillus* in yogurt prevents the adhesion of pathogens to tissues. This may explain the benefits of eating yogurt to control infections of the gastrointestinal tract and vagina. Research indicates that compounds in garlic extract inactivate viruses and destroy bacteria.

---

eye. An analog of adenine, *vidarabine,* is also effective against the herpesviruses. Ribavirin is a guanine analog used in aerosol form to treat life-threatening infections by the respiratory syncytial virus (RSV) in infants and some types of viral hemorrhagic fever. An interesting aspect of some of these antiviral agents (specifically valacyclovir and famciclovir) is that they are activated by an enzyme encoded by the virus itself, activating the drug only in virally infected cells. The enzyme thymidine kinase is used by the virus to process nucleosides before incorporating them into viral RNA or DNA. When the inactive drug enters a virally infected cell, it is activated by the virus' thymidine kinase to produce a working antiviral agent. In cells without viruses, the drug is never activated and DNA replication is allowed to continue unabated.

HIV is classified as a retrovirus, meaning it carries its genetic information in the form of RNA rather than DNA (HIV and AIDS are discussed in chapter 20). Upon infection, the RNA genome is used as a template by the enzyme **reverse transcriptase** to produce a DNA copy of the virus' genetic information. Because this particular reaction is not seen outside of the retroviruses, it offers two ideal targets for chemotherapy. The first is interfering with the synthesis of the new DNA strand, which is accomplished using *nucleoside reverse transcriptase inhibitors* (nucleotide analogs), while the second involves interfering with the action of the enzyme responsible for the synthesis, which is accomplished using *non-nucleoside reverse transcriptase inhibitors.*

Azidothymidine (AZT or zidovudine) is a thymine analog that exerts its effect by incorporating itself into the growing DNA chain of HIV and terminating synthesis, in a manner analogous to that seen with acyclovir. AZT is used at all stages of HIV infection, including prophylactically with people accidentally exposed to blood or other body fluids.

Other approved anti-HIV drugs that act as nucleotide analogs are abacavir, lamivudine (3T3), didanosine (ddI), zalcitabine (ddC), and stavudine (d4T).

Non-nucleoside reverse transcriptase inhibitors accomplish the same goal (preventing reverse transcription of the HIV genome) by binding to the reverse transcriptase enzyme itself, inhibiting its ability to synthesize DNA. Drugs with this mode of action include nevirapine, efavirenz, and delaviridine.

Assembly and release of mature viral particles is also targeted in HIV through the use of protease inhibitors. These drugs (indinavir, saquinavir, nelfinavir, crixivan), usually used in combination with nucleotide analogs and reverse transcriptase inhibitors, have been shown to reduce the HIV load to undetectable levels by preventing the maturation of virus particles in the cell. Refer to table 12.5 for a summary of HIV drug mechanisms and see chapter 20 for further coverage of this topic.

Amantadine and its relative, rimantidine, are restricted almost exclusively to treating infections by influenza A virus. Relenza and Tamiflu medications are effective treatments for influenza A and B and useful prophylactics as well. Because one action of these drugs is to inhibit the fusion and uncoating of the virus, they must be given rather early in an infection (table 12.5). In addition to these antiviral drugs, dozens of other agents are under investigation. For a novel approach for controlling viruses see **Insight 12.3.**

A sensible alternative to artificial drugs has been a human-based substance, **interferon (IFN).** Interferon is a glycoprotein produced primarily by fibroblasts and leukocytes in response to various immune stimuli. It has numerous biological activities, including antiviral and anticancer properties. Studies have shown that it is a versatile part of animal host defenses, having a major role in natural immunities. (Its mechanism is discussed in chapter 14.)

The first investigations of interferon's antiviral activity were limited by the extremely minute quantities that could be extracted from human blood. Several types of interferon are currently produced by the recombinant DNA technology techniques outlined in chapter 10. Extensive clinical trials have tested its effectiveness in viral infections and cancer. Some of the known therapeutic benefits of interferon include:

1. reducing the time of healing and some of the complications in certain infections (mainly of herpesviruses);
2. preventing or reducing some symptoms of cold and papillomaviruses (warts);
3. slowing the progress of certain cancers, including bone cancer and cervical cancer, and certain leukemias and lymphomas; and
4. treating a rare cancer called hairy-cell leukemia, hepatitis C (a viral liver infection), genital warts, and Kaposi's sarcoma in AIDS patients.

✔ CHECKPOINT

- Antimicrobials are classified into +/−20 major drug families, based on their chemical composition, source or origin, and their site of action.
- The majority of antimicrobials are effective against bacteria, but a limited number are effective against protozoa, helminths, fungi, and viruses.
- Penicillins, cephalosporins, bacitracin, vancomycin, and cycloserines block cell wall synthesis, primarily in gram-positive bacteria.
- Aminoglycosides and tetracyclines block protein synthesis in procaryotes.
- Sulfonamides, trimethoprim, isoniazid, nitrofurantoin, and the fluoroquinolones are synthetic antimicrobials effective against a broad range of microorganisms. They block steps in the synthesis of nucleic acids.
- Fungal antimicrobials, such as macrolide polyenes, griseofulvin, azoles, and flucytosine, must be monitored carefully because of the potential toxicity to the infected host. They promote lysis of cell membranes.
- There are fewer antiparasitic drugs than antibacterial drugs because parasites are eucaryotes like their human hosts and they have several life stages, some of which can be resistant to the drug.
- Antihelminthic drugs immobilize or disintegrate infesting helminths or inhibit their metabolism in some manner.
- Antiviral drugs interfere with viral replication by blocking viral entry into cells, blocking the replication process, or preventing the assembly of viral subunits into complete virions.
- Many antiviral agents are analogs of nucleotides. They inactivate the replication process when incorporated into viral nucleic acids. HIV antivirals interfere with reverse transcriptase or proteases to prevent the maturation of viral particles.
- Although interferon is effective *in vivo* against certain viral infections, commercial interferon is not currently effective as a broad-spectrum antiviral agent.

Interactions Between Microbes and Drugs: The Acquisition of Drug Resistance

One unfortunate outcome of the use of antimicrobials is the development of microbial **drug resistance,** an adaptive response in which microorganisms begin to tolerate an amount of drug that would ordinarily be inhibitory. The development of mechanisms for circumventing or inactivating antimicrobial drugs is due largely to the genetic versatility and adaptability of microbial populations. The property of drug resistance can be intrinsic as well as acquired. Intrinsic drug resistance can best be exemplified by the fact that bacteria must, of course, be resistant to any antibiotic that they produce. This type of resistance is limited, however, to a small group of organisms and is generally not a problem with regard to antimicrobial chemotherapy. Of much greater importance is the acquisition of resistance to a drug by a microbe that was previously sensitive to the drug. In our context, the term drug resistance will refer to this last type of acquired resistance.

How Does Drug Resistance Develop?

Contrary to popular belief, antibiotic resistance is not a recent phenomenon. Resistance to penicillin developed in some bacteria as early as 1940, three years before the drug was even approved for public use. The scope of the problem became apparent in the 1980s and 1990s, when scientists and physicians observed treatment failures on a large scale.

Microbes become newly resistant to a drug after one of the following events occurs: (1) spontaneous mutations in critical chromosomal genes, or (2) acquisition of entire new genes or sets of genes via transfer from another species. Chromosomal drug resistance usually results from spontaneous random mutations in bacterial populations. The chance that such a mutation will be advantageous is minimal, and the chance that it will confer resistance to a specific drug is lower still. Nevertheless, given the huge numbers of microorganisms in any population and the constant rate of mutation, such mutations do occur. The end result varies from slight changes in microbial sensitivity, which can be overcome by larger doses of the drug, to complete loss of sensitivity.

Resistance occurring through intermicrobial transfer originates from plasmids called **resistance (R) factors** that are transferred through conjugation, transformation, or transduction. Studies have shown that plasmids encoded with drug resistance are naturally present in microorganisms before they have been exposed to the drug. Such traits are "lying in wait" for an opportunity to be expressed and to confer adaptability on the species. Many bacteria also maintain transposable drug resistance sequences (transposons) that are duplicated and inserted from one plasmid to another or from a plasmid to the chromosome. Chromosomal genes and plasmids containing codes for drug resistance are faithfully replicated and inherited by all subsequent

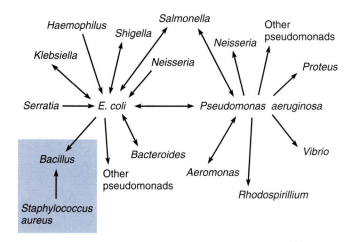

FIGURE 12.13 **Spread of resistance factors.**
This figure traces documented evidence of known cases in which R factors have been transferred among pathogens. Such promiscuous exchange of drug resistance occurs primarily by conjugation and transduction. Most of the bacteria are gram negative, but *Bacillus* and *Staphylococcus* are unrelated gram-positive genera. This phenomenon is responsible for the rapid spread of drug-resistant microbes.
Source: Data from Young and Mayer, *Review of Infectious Diseases,* 1:55, 1979.

progeny. This sharing of resistance genes accounts for the rapid proliferation of drug-resistant species **(figure 12.13)**. A growing body of evidence points to the ease and frequency of gene transfers in nature, from totally unrelated bacteria living in the body's normal flora and the environment.

Specific Mechanisms of Drug Resistance

Inside a bacterial cell, the net effect of the two events described earlier is one of the following, which actually causes the bacterium to be resistant:

1. New enzymes are synthesized; these inactivate the drug (only occurs when new genes are acquired).
2. Permeability or uptake of drug into bacterium is decreased or eliminated (can occur via mutation or acquisition of new genes).
3. Binding sites for drug are decreased in number or affinity (can occur via mutation or acquisition of new genes).
4. An affected metabolic pathway is shut down or an alternate pathway is used (occurs due to mutation of original enzyme(s)).

Some bacteria can become resistant indirectly by lapsing into dormancy, or, in the case of penicillin, by converting to a cell-wall-deficient form (L form) that penicillin cannot affect.

Drug Inactivation Mechanisms Microbes inactivate drugs by producing enzymes that permanently alter drug structure. One example, bacterial exoenzymes called **beta-lactamases,** hydrolyze the beta-lactam ring structure of some penicillins

and cephalosporins rendering the drugs inactive. Two beta-lactamases—penicillinase and cephalosporinase—disrupt the structure of certain penicillin or cephalosporin molecules so their activity is lost **(figure 12.14)**. So many strains of *Staphylococcus aureus* produce penicillinase that regular penicillin is rarely a possible therapeutic choice. Now that some strains of *Neisseria gonorrhoeae,* called PPNG,[3] have also acquired penicillinase, alternative drugs are required to treat gonorrhea (see chapter 23). A large number of other gram-negative species are inherently resistant to some of the penicillins and cephalosporins because of naturally occurring beta-lactamases.

Decreased Drug Permeability or Increased Drug Elimination
The resistance of some bacteria can be due to a mechanism that prevents the drug from entering the cell and acting on its target. For example, the outer membrane of the cell wall of certain gram-negative bacteria is a natural blockade for some of the penicillin drugs. Resistance to the tetracyclines can arise from plasmid-encoded proteins that pump the drug out of the cell. Aminoglycoside resistance can arise in multiple ways; one of these is change in drug permeability caused by point mutations in the transport system or LPS machinery (figure 12.14).

Many bacteria possess multidrug resistant (MDR) pumps that actively transport drugs and other chemicals out of cells. These pumps are proteins encoded by plasmids or chromosomes. They are stationed in the cell membrane and expel molecules by a proton-motive force similar to ATP synthesis (figure 12.14). They confer drug resistance on many gram-positive pathogens (*Staphylococcus, Streptococcus*) and gram-negative pathogens (*Pseudomonas, E. coli*). Because they lack selectivity, one type of pump can expel a broad array of antimicrobial drugs, detergents, and other toxic substances.

Change of Drug Receptors Because most drugs act on a specific target such as protein, RNA, DNA, or membrane structure, microbes can circumvent drugs by altering the nature of this target. Bacteria can become resistant to aminoglycosides when point mutations in ribosomal proteins arise (figure 12.14). Erythromycin and clindamycin resistance is associated with an alteration on the 50S ribosomal binding site. Penicillin resistance in *Streptococcus pneumoniae* and methicillin resistance in *Staphylococcus aureus* are related to an alteration in the binding proteins in the cell wall. Several species of enterococci have acquired resistance to vancomycin through a similar alteration of cell wall proteins. Fungi can become resistant by decreasing their synthesis of ergosterol, the principal receptor for certain antifungal drugs.

Changes in Metabolic Patterns The action of antimetabolites can be circumvented if a microbe develops an alternative metabolic pathway or enzyme (figure 12.14). Sulfonamide and trimethoprim resistance develops when microbes deviate from the usual patterns of folic acid synthesis. Fungi

---

3. Penicillinase-producing *Neisseria gonorrhoeae.*

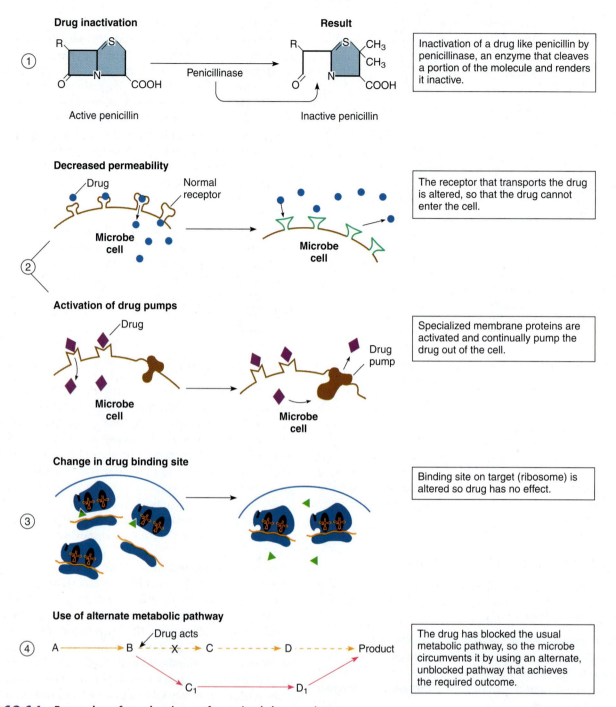

FIGURE 12.14   Examples of mechanisms of acquired drug resistance.

can acquire resistance to flucytosine by completely shutting off certain metabolic activities.

Natural Selection and Drug Resistance

So far, we have been considering drug resistance at the cellular and molecular levels, but its full impact is felt only if this resistance occurs throughout the cell population. Let us examine how this might happen and its long-term therapeutic consequences.

Any large population of microbes is likely to contain a few individual cells that are already drug-resistant because of prior mutations or transfer of plasmids **(figure 12.15a).** As long as the drug is not present in the habitat, the numbers of these resistant forms will remain low because they have no particular growth advantage (and often are disadvantaged relative to their nonmutated counterparts). But if the population is subsequently exposed to this drug **(figure 12.15b),** sensitive individuals are inhibited or destroyed, and resistant forms survive and proliferate. During subsequent population growth, offspring of these resistant microbes will inherit this drug resistance. In time, the replacement population will have a preponderance of the drug-resistant forms and can eventually become completely resistant **(figure 12.15c).** In

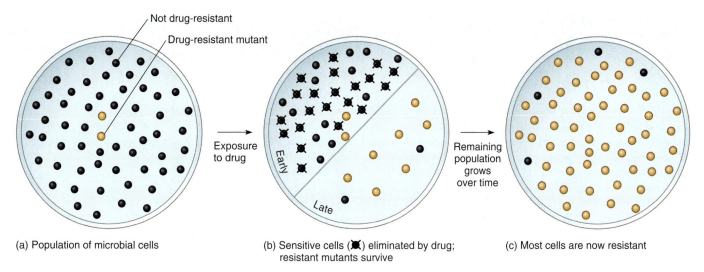

FIGURE 12.15 **The events in natural selection for drug resistance.**
(a) Populations of microbes can harbor some members with a prior mutation that confers drug resistance. **(b)** Environmental pressure (here, the presence of the drug) selects for survival of these mutants. **(c)** They eventually become the dominant members of the population.

ecological terms, the environmental factor (in this case, the drug) has put selection pressure on the population, allowing the more "fit" microbe (the drug-resistant one) to survive, and the population has evolved to a condition of drug resistance.

Natural selection for drug-resistant forms is apparently a common phenomenon. It takes place most frequently in various natural habitats, laboratories, and medical environments, and it also can occur within the bodies of humans and animals during drug therapy **(Insight 12.4).**

IN THE NEWS *(Continued from page 347)*

Dana, the college student described on the first page of chapter 12, was sick with tuberculosis (TB). The nature of the symptoms plus positive chest X rays and Mantoux PPD (skin test) provided strong pieces of evidence that the tuberculosis microbe, *Mycobacterium tuberculosis,* was the causative agent.

Tuberculosis is a contagious disease, with a high mortality rate worldwide. It can remain dormant in a person for many years, and become reactivated if their immune system becomes weakened. Dana's illness, with its seeming "recurrence" 2 months after the initial episode, was not a case of reactivation tuberculosis; the original case had never actually been cured. Since the physician did not recognize the infection as TB, inadequate antimicrobial therapy was prescribed. To make matters worse, Dana stopped taking her antibiotics as soon as she felt better.

Mycobacterium tuberculosis is particularly prone to becoming antibiotic resistant. Prescribing a single antibiotic to treat it often leads to situations like Dana's, because mutant bacteria that aren't affected by the drug eventually dominate the population in the lungs. TB infections are always treated with multiple drugs simultaneously for this reason. It is highly improbable that a single bacterial cell would mutate in the multiple genetic locations needed to render it resistant to more than one drug.

New Approaches to Antimicrobial Therapy

Researchers are constantly "pushing the envelope" of antimicrobial research. Often the quest focuses on finding new targets in the bacterial cell, and custom-designing drugs that aim for them. There are many interesting new strategies that have not yet resulted in a marketable drug—for example, (1) targeting iron-scavenging capabilities of bacteria, and (2) targeting a genetic control mechanism in bacteria referred to as riboswitches.

The best example of a strategy aimed at iron-scavenging capabilities is recent work with *Staphylococcus aureus*. Scientists have found that this bacterium has a special pathway involving several proteins that punch holes in red blood cells and then "reach in" to bind the heme, strip it of its iron and use it for their own purposes. Researchers are currently investigating inhibitory substances that block the iron-collecting pathway, which would result in inevitable bacterial death. This approach may prove effective in wiping out antibiotic-resistant strains of *S. aureus*.

Riboswitches are areas of bacterial RNA that are used to control translation of mRNA. They act by binding substances in the bacterial cell, prohibiting translation processes. Since riboswitches appear to be ubiquitous in bacteria, finding a drug that blocks their action could be a useful antimicrobial.

Other novel approaches to controlling infections include the use of **probiotics** and **prebiotics.** Probiotics are preparations of live microorganisms that are fed to animals and humans to improve the intestinal flora. This can serve to replace microbes lost during antimicrobial therapy or simply to augment the flora that is already there. This is a slightly more sophisticated application of methods that have long been used in an empiric fashion, for instance, by people who consume yogurt because of the beneficial microbes it contains. A probiotic approach is also sometimes used in female patients who have recurring urinary tract infections. These patients receive vaginal inserts or perineal swabs of *Lactobacillus* in an

INSIGHT 12.4

Medical

The Rise of Drug Resistance

Many people unrealistically assume that science will come to the rescue and solve the problem of drug resistance. If drug companies just keep making more and better antimicrobials, soon infectious diseases will be vanquished. This unfortunate attitude has vastly underestimated the extreme versatility and adaptability of microorganisms and the complexity of the task. It is a fact of nature that if a large number of microbes are exposed to a variety of drugs, there will always be some genetically favored individuals that survive and thrive. The AIDS virus (HIV) is so prone to drug resistance that it can become resistant during the first few weeks of therapy in a single individual. Because HIV mutates so rapidly, in most cases, it will eventually become resistant to all drugs that have been developed so far.

Ironically, thousands of patients die every year in the United States from infections that lack effective drugs, and 60% of hospital infections are caused by drug-resistant microbes. For many years, concerned observers reported the gradual development of drug resistance in staphylococci, *Salmonella*, and gonococci. But during the past decade, the scope of the problem has escalated. It is now a common event to discover microbes that have become resistant to relatively new drugs in a very short time. In fact, many strains of pathogens have multiple drug resistance, and a few are resistant to all drugs. **Table 12.A** lists some notable examples.

The Hospital Factor

The clinical setting is a prolific source of drug-resistant strains of bacteria. This environment continually exposes pathogens to a variety of drugs. The hospital also maintains patients with weakened defenses, making them highly susceptible to pathogens. A classic example occurred with *Staphylococcus aureus* and penicillin. In the 1950s, hospital strains began to show resistance to this drug, and because of indiscriminate use, these strains became nearly 100% resistant in 30 years. In a short time, MRSA (methicillin-resistant *S. aureus*) strains appeared, which can tolerate nearly all antibiotics. Up until recently, MRSA has been sensitive to the drug of last resort, vancomycin. In 2002, the first cases of complete resistance to this drug were reported (VRSA). To complicate this problem, strains of MRSA are now being spread into the community.

Drugs in Animal Feeds

Another practice that has contributed significantly to growing drug resistance is the addition of antimicrobials to livestock feed, with the idea of decreasing infections and thereby improving animal health and size. This practice has had serious impact in both the United States and Europe. Enteric bacteria such as *Salmonella*,

TABLE 12.A				
		Year and Prevalence of Resistance		
Organism	**Drug**			
Pneumococci		**1989**	**2000**	**2002**
	Penicillin	Rare	30%	40%
	Erythromycin	0	15%	40%
	Cephalosporin	0	14%	27%
		1990	**2000**	**2002**
Gonococci	Penicillin	10	70%	74%
	Fluoroquinolone	1.4%	10%	12%
Enterococci	Vancomycin	Rare	50%	52%
		1995	**2000**	**2002**
Campylobacter	Fluoroquinolones	Rare	10%	21%

Escherichia coli, and enterococci that live as normal intestinal flora of these animals readily share resistance plasmids and are constantly selected and amplified by exposure to drugs. These pathogens subsequently "jump" to humans and cause drug-resistant infections, oftentimes at epidemic proportions. In a deadly outbreak of *Salmonella* infection in Denmark, the pathogen was found to be resistant to seven different antimicrobials. In the United States, a strain of fluoroquinolone-resistant *Campylobacter* from chickens caused over 5,000 cases of food infection in the late 1990s. The opportunistic pathogen called VRE (vancomycin-resistant enterococcus) has been traced to the use of a vancomycin-like drug in cattle feed. It is now one of the most tenacious of hospital-acquired infections for which there are few drug choices. To attempt to curb this source of resistance, Europe and the United States have begun to ban the use of human drugs in animal feeds.

The move seems to be working. Denmark banned all agricultural antibiotic use for growth-promotion in 1998. Resistance to the drugs declined dramatically among bacteria isolated from the farm animals without significant reductions in animal size or health. Scientists are confident that this will lead to a reduction in human carriage of antibiotic-resistant bacteria.

Worldwide Drug Resistance

The drug dilemma has become a widespread problem, affecting all countries and socioeconomic groups. In general, the majority of infectious diseases, whether bacterial, fungal, protozoan, or viral, are showing increases in drug resistance. In parts of India, the main drugs used to treat cholera (furazolidone, ampicillin) have gone from being highly effective to essentially useless in 10 years. In Southeast Asia, 98% of gonococcus infections are multidrug resistant. Malaria, tuberculosis, and typhoid fever pathogens are gaining in resistance, with few alternative drugs to control them. To add to the problem, global travel and globalization of food products means that drug resistance can be rapidly exported.

In countries with adequate money to pay for antimicrobials, most infections will be treated, but at some expense. In the United States alone, the extra cost for treating the drug-resistant variety is around $10 billion per year. In many developing countries, drugs are mishandled by overuse and underuse, either of which can contribute to drug resistance. Many countries that do not regulate the sale of prescription drugs make them readily available to purchase over the counter. For example, the antituberculosis drug INH (isoniazid) is sometimes used as a "lung vitamin" to improve health, and antibiotics are taken in the wrong dose and wrong time for undiagnosed conditions. These countries serve as breeding grounds for drug resistance that can eventually be carried to other countries.

It is clear that we are in a race with microbes and we are falling behind. If the trend is not contained, the world may return to a time when there are few effective drugs left. We simply cannot develop them as rapidly as microbes can develop resistance. In this light, it is essential to fight the battle on more than one front. **Table 12.B** summarizes the several critical strategies to give us an edge in controlling drug resistance.

TABLE 12.B	**Strategies to Limit Drug Resistance by Microorganisms**

Drug Usage
- Physicians have the responsibility for making an accurate diagnosis and prescribing the correct drug therapy.
- Patients must comply with and carefully follow the physician's guidelines. It is important for the patient to take the correct dosage, by the best route, for the appropriate period. This diminishes the selection for mutants that can resist low drug levels, and ensures elimination of the pathogen.
- Administration of two or more drugs together increases the chances that at least one of the drugs will be effective and that a resistant strain of either drug will not be able to persist. The basis for this combined therapy method lies in the unlikelihood of simultaneous resistance to several drugs.

Drug Research
- Research focuses on developing shorter-term, higher-dose antimicrobials that are more effective, less expensive, and have fewer side effects.
- Pharmaceutical companies continue to seek new antimicrobial drugs with structures that are not readily inactivated by microbial enzymes or drugs with modes of action that are not readily circumvented.

Long-Term Strategies
- Proposals to reduce the abuse of antibiotics range from educational programs for health workers to requiring written justification from the physician on all antibiotics prescribed.
- Especially valuable antimicrobials may be restricted in their use to only one or two types of infections.
- The addition of antimicrobials to animal feeds must be curtailed worldwide.
- Government programs that make effective therapy available to low-income populations should be increased.
- Vaccines should be used whenever possible to provide alternative protection.

attempt to restore a healthy acidic environment to the region, and to displace disease-causing microorganisms. Probiotics are thought to be useful for influencing the development of food allergies; their role in the stimulation of mucosal immunity is also being investigated.

Prebiotics are nutrients that encourage the growth of beneficial microbes in the intestine. For instance, certain sugars such as fructans are thought to encourage the growth of *Bifidobacterium* in the large intestine, and to discourage the growth of potential pathogens. You can be sure that you will hear more about prebiotics and probiotics as the concepts become increasingly well studied by scientists. Clearly, the use of these agents is a different type of antimicrobial strategy than we are used to, but it may have its place in a future in which traditional antibiotics are more problematic.

Another category of antimicrobial is the **lantibiotics.** Lantibiotics are short peptides produced by bacteria that inhibit the growth of other bacteria. They are distinguished by the presence of unusual amino acids that are not seen elsewhere in the cell. They exert their antimicrobial activity either by puncturing cell membranes or by inhibiting bacterial enzymes. The most well-known lantibiotic is *nisin*. Lantibiotics have long been used in food preservation, in veterinary medicine, and more recently in personal care products such as deodorants; they may soon find use in human chemotherapy. Several lantibiotics have been found to be effective against human pathogens such as *Propionibacterium acne*, methicillin-resistant *S. aureus*, and *Helicobacter pylori*.

So far virtually all of the antimicrobials used in human infections have been derived from other microorganisms, or artificially synthesized in the laboratory. But many scientists are investigating antimicrobial substances derived from plants and from animals (called "natural products"). Plant-derived antimicrobials are by no means an original idea; indigenous cultures have been using plants as medicines for centuries, as pointed out in Insight 12.1. In the fifth century B.C., Hippocrates mentioned 300 to 400 medicinal plants. But phyto- (plant) chemicals have been overlooked by modern infectious disease scientists until recently. (Many widely used chemotherapeutic agents, such as Coumadin, a drug that prevents blood clots, are derived from plants, but virtually none of them are anti-infectives.) Now modern techniques have proven the antimicrobial efficacy of a wide variety of extracts from botanical sources, and the search for new ones has intensified. In addition, lantibiotic-like substances have been found in amphibians and other animals.

✔ CHECKPOINT

- Microorganisms are termed drug resistant when they are no longer inhibited by an antimicrobial to which they were previously sensitive.
- Drug resistance is genetic; microbes acquire genes that code for methods of inactivating or escaping the antimicrobial, or acquire mutations that affect the drug's impact. Resistance is selected for in environments where antimicrobials are present in high concentrations, such as in hospitals.

- Microbial drug resistance develops through the selection of preexisting random mutations and through acquisition of resistance genes from other microorganisms.
- Varieties of microbial drug resistance include drug inactivation, decreased drug uptake, decreased drug receptor sites, and modification of metabolic pathways formerly attacked by the drug.
- Widespread indiscriminate use of antimicrobials has resulted in an explosion of microorganisms resistant to all common drugs.

12.4 Interaction Between Drug and Host

Until now this chapter has focused on the interaction between antimicrobials and the microorganisms they target. During an infection the microbe is living in or on a host; therefore the drug is administered to the host though its target is the microbe. Therefore, the effect of the drug on the host must always be considered.

Although selective antimicrobial toxicity is the ideal constantly being sought, chemotherapy by its very nature involves contact with foreign chemicals that can harm human tissues. In fact, estimates indicate that at least 5% of all persons taking an antimicrobial drug experience some type of serious adverse reaction to it. The major side effects of drugs fall into one of three categories: direct damage to tissues through toxicity, allergic reactions, and disruption in the balance of normal microbial flora. The damage incurred by antimicrobial drugs can be short-term and reversible or permanent, and it ranges in severity from cosmetic to lethal. **Table 12.6** summarizes drug groups and their major side effects.

Toxicity to Organs

Drugs can adversely affect the following organs: the liver (hepatotoxic), kidneys (nephrotoxic), gastrointestinal tract, cardiovascular system and blood-forming tissue (hemotoxic), nervous system (neurotoxic), respiratory tract, skin, bones, and teeth.

Because the liver is responsible for metabolizing and detoxifying foreign chemicals in the blood, it can be damaged by a drug or its metabolic products. Injury to liver cells can result in enzymatic abnormalities, fatty liver deposits, hepatitis, and liver failure. The kidney is involved in excreting drugs and their metabolites. Some drugs irritate the nephron tubules, creating changes that interfere with their filtration abilities. Drugs such as sulfonamides can crystallize in the kidney and form stones that can obstruct the flow of urine.

The most common complaint associated with oral antimicrobial therapy is diarrhea, which can progress to severe intestinal irritation or colitis. Although some drugs directly irritate the intestinal lining, the usual gastrointestinal complaints are caused by disruption of the intestinal microflora (discussed in a subsequent section).

TABLE 12.6 Major Adverse Toxic Reactions to Common Drug Groups

Antimicrobial Drug	Primary Tissue Affected	Primary Damage or Abnormality Produced
Antibacterials		
Penicillin G	Skin	Rash
Carbenicillin	Platelets	Abnormal bleeding
Ampicillin	GI tract	Diarrhea and enterocolitis
Cephalosporins	Platelet function	Inhibition of prothrombin synthesis
	White blood cells	Decreased circulation
	Kidney	Nephritis
Tetracyclines	GI tract	Diarrhea and enterocolitis
	Teeth, bones	Discoloration of tooth enamel
	Skin	Reactions to sunlight (photosensitization)
Chloramphenicol	Bone marrow	Injury to red and white blood cell precursors
Aminoglycosides (streptomycin, gentamicin, amikacin)	GI tract, hair cells in cochlea, vestibular cells, neuromuscular, kidney tubules	Diarrhea and enterocolitis; malabsorption; loss of hearing, dizziness, kidney damage
Isoniazid	Liver	Hepatitis
	Brain	Seizures
	Skin	Dermatitis
Sulfonamides	Kidney	Formation of crystals; blockage of urine flow
	Red blood cells	Hemolysis
	Platelets	Reduction in number
Polymyxin	Kidney	Damage to membranes of tubule cells
	Neuromuscular system	Weakened muscular responses
Quinolones (ciprofloxacin, norfloxacin)	Nervous system, bones, GI tract	Headache, dizziness, tremors, GI distress
Rifampin	Liver	Damage to hepatic cells
	Skin	Dermatitis
Antifungals		
Amphotericin B	Kidney	Disruption of tubular filtration
Flucytosine	White blood cells	Decreased number
Antiprotozoan Drugs		
Metronidazole	GI tract	Nausea, vomiting
Chloroquine	GI tract	Vomiting
	Brain	Headache
	Skin	Itching
Antihelminthics		
Niclosamide	GI tract	Nausea, abdominal pain
Pyrantel	GI tract	Irritation
	Brain	Headache, dizziness
Antivirals		
Acyclovir	Brain	Seizures, confusion
	Skin	Rash
Amantadine	Brain	Nervousness, light-headedness
	GI tract	Nausea
AZT	Bone marrow	Immunosuppression, anemia

Many drugs given for parasitic infections are toxic to the heart, causing irregular heartbeats and even cardiac arrest in extreme cases. Chloramphenicol can severely depress blood-forming cells in the bone marrow, resulting in either a reversible or a permanent (fatal) anemia. Some drugs hemolyze the red blood cells, others reduce white blood cell counts, and still others damage platelets or interfere with their formation, thereby inhibiting blood clotting.

Certain antimicrobials act directly on the brain and cause seizures. Others, such as aminoglycosides, damage nerves (very commonly, the 8th cranial nerve), leading to dizziness, deafness, or motor and sensory disturbances. When drugs block the transmission of impulses to the diaphragm, respiratory failure can result.

The skin is a frequent target of drug-induced side effects. The skin response can be a symptom of drug allergy or a direct toxic effect. Some drugs interact with sunlight to cause photodermatitis, a skin inflammation. Tetracyclines are contraindicated (not advisable) for children from birth to 8 years of age because they bind to the enamel of the teeth, creating

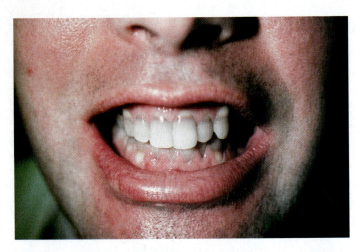

FIGURE 12.16 **Drug-induced side effect.**
An adverse effect of tetracycline given to young children is the permanent discoloration of tooth enamel.

a permanent gray to brown discoloration **(figure 12.16).** Pregnant women should avoid tetracyclines because they cross the placenta and can be deposited in the developing fetal bones and teeth.

Allergic Responses to Drugs

One of the most frequent drug reactions is heightened sensitivity, or **allergy.** This reaction occurs because the drug acts as an antigen (a foreign material capable of stimulating the immune system) and stimulates an allergic response. This response can be provoked by the intact drug molecule or by substances that develop from the body's metabolic alteration of the drug. In the case of penicillin, for instance, it is not the penicillin molecule itself that causes the allergic response but a product, *benzlpenicilloyl.* Allergic reactions have been reported for every major type of antimicrobial drug, but the penicillins account for the greatest number of antimicrobial allergies, followed by the sulfonamides.

People who are allergic to a drug become sensitized to it during the first contact, usually without symptoms. Once the immune system is sensitized, a second exposure to the drug can lead to a reaction such as a skin rash (hives), respiratory inflammation, and, rarely, anaphylaxis, an acute, overwhelming allergic response that develops rapidly and can be fatal. (This topic is discussed in greater detail in chapter 16.)

Suppression and Alteration of the Microflora by Antimicrobials

Most normal, healthy body surfaces, such as the skin, large intestine, outer openings of the urogenital tract, and oral cavity, provide numerous habitats for a virtual "garden" of microorganisms. These normal colonists or residents, called the **flora** or microflora, consist mostly of harmless or beneficial bacteria, but a small number can potentially be pathogens. Although we shall defer a more detailed discussion of this topic to chapter 13 and later chapters, here we focus on the general effects of drugs on this population.

If a broad-spectrum antimicrobial is introduced into a host to treat infection, it will destroy microbes regardless of their roles in the balance, affecting not only the targeted infectious agent but also many others in sites far removed from the original infection **(figure 12.17).** When this therapy destroys beneficial resident species, the microbes that were once in small numbers begin to overgrow and cause disease. This complication is called a **superinfection.**

Some common examples demonstrate how a disturbance in microbial flora leads to replacement flora and superinfection. A broad-spectrum cephalosporin used to treat a urinary tract infection by *Escherichia coli* will cure the infection, but it will also destroy the lactobacilli in the vagina that normally maintain a protective acidic environment there. The drug has no effect, however, on *Candida albicans,* a yeast that also resides in normal vaginas. Released from the inhibitory environment provided by lactobacilli, the yeasts proliferate and cause symptoms. *Candida* can cause similar superinfections of the oropharynx (thrush) and the large intestine.

Oral therapy with tetracyclines, clindamycin, and broad-spectrum penicillins and cephalosporins is associated with a serious and potentially fatal condition known as *antibiotic-associated colitis* (pseudomembranous colitis). This condition is due to the overgrowth in the bowel of *Clostridium difficile,* an endospore-forming bacterium that is resistant to the antibiotic. It invades the intestinal lining and releases toxins that induce diarrhea, fever, and abdominal pain. (You'll learn more about infectious diseases of the gastrointestinal tract, including *C. difficile,* in chapter 22.)

> ✔ **CHECKPOINT**
>
> ■ The three major side effects of antimicrobials are toxicity to organs, allergic reactions, and problems resulting from suppression or alteration of normal flora.
>
> ■ Antimicrobials that destroy most but not all normal flora allow the unaffected normal flora to overgrow, causing a superinfection.

12.5 Considerations in Selecting an Antimicrobial Drug

Before actual antimicrobial therapy can begin, it is important that at least three factors be known:

1. the nature of the microorganism causing the infection,
2. the degree of the microorganism's susceptibility (also called sensitivity) to various drugs, and
3. the overall medical condition of the patient.

Identifying the Agent

Identification of infectious agents from body specimens should be attempted as soon as possible. It is especially important that such specimens be taken before any antimicrobial drug is given, just in case the drug eliminates the infectious agent. Direct examination of body fluids, sputum, or stool is a rapid initial method for detecting and perhaps even identifying

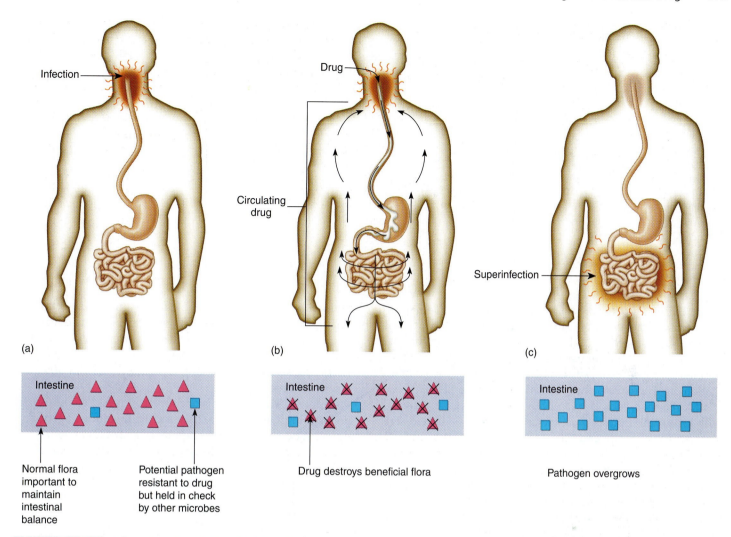

FIGURE 12.17   **The role of antimicrobials in disrupting microbial flora and causing superinfections.**
(a) A primary infection in the throat is treated with an oral antibiotic. **(b)** The drug is carried to the intestine and is absorbed into the circulation. **(c)** The primary infection is cured, but drug-resistant pathogens have survived and create an intestinal superinfection.

bacteria or fungi. A doctor often begins the therapy on the basis of such immediate findings. The choice of drug will be based on experience with drugs that are known to be effective against the microbe; this is called the "informed best guess." For instance, if a sore throat appears to be caused by *Streptococcus pyogenes,* the physician might prescribe penicillin, because this species seems to be almost universally sensitive to it so far. If the infectious agent is not or cannot be isolated, epidemiologic statistics may be required to predict the most likely agent in a given infection. For example, *Streptococcus pneumoniae* accounts for the majority of cases of meningitis in children, followed by *Neisseria meningitidis* (discussed in detail in chapter 19).

Testing for the Drug Susceptibility of Microorganisms

Testing is essential in those groups of bacteria commonly showing resistance, primarily *Staphylococcus* species, *Neisseria gonorrhoeae, Streptococcus pneumoniae,* and *Enterococcus faecalis,*

and the aerobic gram-negative enteric bacilli. However, not all infectious agents require antimicrobial sensitivity testing. Drug testing in fungal or protozoan infections is difficult and is often unnecessary. When certain groups, such as group A streptococci and all anaerobes (except *Bacteroides*), are known to be uniformly susceptible to penicillin G, testing may not be necessary unless the patient is allergic to penicillin.

Selection of a proper antimicrobial agent begins by demonstrating the *in vitro* activity of several drugs against the infectious agent by means of standardized methods. In general, these tests involve exposing a pure culture of the bacterium to several different drugs and observing the effects of the drugs on growth.

The *Kirby-Bauer* technique is an agar diffusion test that provides useful data on antimicrobial susceptibility. In this test, the surface of a plate of special medium is spread with the test bacterium, and small discs containing a premeasured amount of antimicrobial are dispensed onto the bacterial lawn. After incubation, the zone of inhibition surrounding the discs is measured and compared with a standard for each drug

TABLE 12.7	Results of a Sample Kirby-Bauer Test			
	Zone Sites (mm) Required For:		Actual Result (mm) For:	
Drug	Susceptibility (S)	Resistance (R)	*Staphylococcus aureus*	Evaluation
Bacitracin	>13	<8	15	S
Chloramphenicol	>18	<12	20	S
Erythromycin	>18	<13	15	I
Gentamicin	>13	<12	16	S
Kanamycin	>18	<13	20	S
Neomycin	>17	<12	12	R
Penicillin G	>29	<20	10	R
Polymyxin B	>12	<8	10	R
Streptomycin	>15	<11	11	R
Vancomycin	>12	<9	15	S
Tetracycline	>19	<14	25	S

R = resistant, I = intermediate, S = sensitive

(table 12.7 and figure 12.18). The profile of antimicrobial sensitivity, or *antibiogram,* provides data for drug selection. The Kirby-Bauer procedure is less effective for bacteria that are anaerobic, highly fastidious, or slow-growing (*Mycobacterium*). An alternative diffusion system that provides additional information on drug effectiveness is the E-test (figure 12.19).

More sensitive and quantitative results can be obtained with tube dilution tests. First the antimicrobial is diluted serially in tubes of broth, and then each tube is inoculated with a small uniform sample of pure culture, incubated, and examined for growth (turbidity). The smallest concentration (highest dilution) of drug that visibly inhibits growth is called the minimum inhibitory concentration, or MIC. The MIC is useful in determining the smallest effective dosage of a drug and in providing a comparative index against other antimicrobials (figure 12.20 and table 12.8). In many clinical laboratories, these antimicrobial testing procedures are performed in automated machines that can test dozens of drugs simultaneously.

The MIC and Therapeutic Index

The results of antimicrobial sensitivity tests guide the physician's choice of a suitable drug. If therapy has already commenced, it is imperative to determine if the tests bear out the use of that particular drug. Once therapy has begun, it is important to observe the patient's clinical response, because the *in vitro* activity of the drug is not always correlated with its *in vivo* effect. When antimicrobial treatment fails, the failure is due to

1. the inability of the drug to diffuse into that body compartment (the brain, joints, skin);
2. a few resistant cells in the culture that did not appear in the sensitivity test; or
3. an infection caused by more than one pathogen (mixed), some of which are resistant to the drug.

If therapy does fail, a different drug, combined therapy, or a different method of administration must be considered.

Many factors influence the choice of an antimicrobial drug besides microbial sensitivity to it. The nature and spectrum of the drug, its potential adverse effects, and the condition of the patient can be critically important. When several antimicrobial drugs are available for treating an infection, final drug selection advances to a new series of considerations. In general, it is better to choose the narrowest-spectrum drug of those that are effective if the causative agent is known. This decreases the potential for superinfections and other adverse reactions.

Because drug toxicity is of concern, it is best to choose the one with high selective toxicity for the infectious agent and low human toxicity. The therapeutic index (TI) is defined as the ratio of the dose of the drug that is toxic to humans as compared to its minimum effective (therapeutic) dose. The closer these two figures are (the smaller the ratio), the greater is the potential for toxic drug reactions. For example, a drug that has a therapeutic index of:

$$\frac{10 \ \mu g/ml: \text{toxic dose}}{9 \ \mu g/\mu l \ (MIC)} \quad \boxed{TI = 1.1}$$

is a riskier choice than one with a therapeutic index of:

$$\frac{10 \ \mu g/ml}{1 \ \mu g/ml} \quad \boxed{TI = 10}$$

Drug companies recommend dosages that will inhibit the microbes but not adversely affect patient cells. When a series of drugs being considered for therapy have similar MICs, the drug with the highest therapeutic index usually has the widest margin of safety.

The physician must also take a careful history of the patient to discover any preexisting medical conditions that will influence the activity of the drug or the response of the patient. A history of allergy to a certain class of drugs should preclude the administration of that drug and any drugs related to it. Underlying liver or kidney disease will ordinarily necessitate the modification of drug therapy, because these organs play such an important part in metabolizing or excreting the drug. Infants, the elderly, and pregnant women require special precautions. For example, age can diminish gastrointestinal absorption and organ function, and most antimicrobial drugs cross the placenta and could affect fetal development.

The intake of other drugs must be carefully scrutinized, because incompatibilities can result in increased toxicity or failure

Kirby-Bauer Disk Diffusion Test*

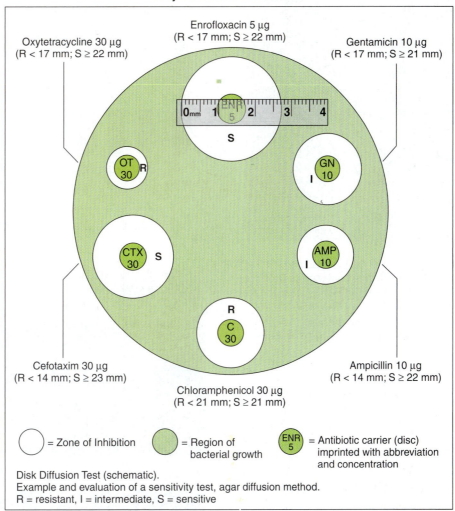

Enrofloxacin 5 μg
(R < 17 mm; S ≥ 22 mm)

Oxytetracycline 30 μg
(R < 17 mm; S ≥ 22 mm)

Gentamicin 10 μg
(R < 17 mm; S ≥ 21 mm)

Cefotaxim 30 μg
(R < 14 mm; S ≥ 23 mm)

Ampicillin 10 μg
(R < 14 mm; S ≥ 22 mm)

Chloramphenicol 30 μg
(R < 21 mm; S ≥ 21 mm)

◯ = Zone of Inhibition   ◯ = Region of bacterial growth   ⬤ENR 5 = Antibiotic carrier (disc) imprinted with abbreviation and concentration

Disk Diffusion Test (schematic).
Example and evaluation of a sensitivity test, agar diffusion method.
R = resistant, I = intermediate, S = sensitive

*R and S values differ from Table 12.7 due to differing concentrations of the antimicrobials.

(a)

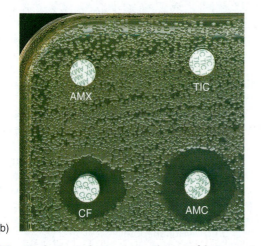

(b)

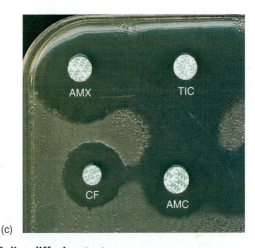

(c)

FIGURE 12.18   Technique for preparation and interpretation of disc diffusion tests.
(a) Standardized methods are used to seed a lawn of bacteria over the medium. A dispenser delivers several drugs onto a plate, followed by incubation. Interpretation of results: During incubation, antimicrobials become increasingly diluted as they diffuse out of the disc into the medium. If the test bacterium is sensitive to a drug, a zone of inhibition develops around its disc. The larger the size of this zone, the greater is the bacterium's sensitivity to the drug. The diameter of each zone is measured in millimeters and evaluated for susceptibility or resistance by means of a comparative standard (see table 12.7). **(b)** Antibiogram of *Escherichia coli* showing resistance to amoxicillin (AMX) and ticarcillin (TIC); when clavulanic acid is combined with amoxicillin (AMC), the result is sensitivity (lower right) due to the combined action of the two drugs. **(c)** Results of test with *Escherichia hermannii* indicate a synergistic effect between ticarcillin (TIC) and AMC (note the expended zone between these two drugs).

FIGURE 12.19   Alternative to the Kirby-Bauer procedure.
Another diffusion test is the E-test, which uses a strip to produce the zone of inhibition. The advantage of the E-test is that the strip contains a gradient of drug calibrated in μg. This way, the MIC can be measured by observing the mark on the strip that corresponds to the edge of the zone of inhibition. (IP = imipenem and TZ = tazobactam)

of one or more of the drugs. For example, the combination of aminoglycosides and cephalosporins increases nephrotoxic effects; antacids reduce the absorption of isoniazid; and the interaction of tetracycline or rifampin with oral contraceptives can abolish the contraceptive's effect. Some drugs (penicillin with certain aminoglycosides, or amphotericin B with flucytosine) act synergistically, so that reduced doses of each can be used in combined therapy. Other concerns in choosing drugs include any genetic or metabolic abnormalities in the patient, the site of infection, the route of administration, and the cost of the drug.

The Art and Science of Choosing an Antimicrobial Drug

Even when all the information is in, the final choice of a drug is not always easy or straightforward. Consider the case of an elderly alcoholic patient with pneumonia caused by *Klebsiella* and complicated by diminished liver and kidney function. All drugs must be given parenterally because of prior damage to the gastrointestinal lining and poor absorption. Drug tests show that the infectious agent is sensitive to third-generation cephalosporins, gentamicin, imipenem and azlocillin. The patient's history shows previous allergy to the penicillins, so these would be ruled out. Drug interactions occur between alcohol and the cephalosporins, which are also associated with serious bleeding in elderly patients, so this may not be a good choice. Aminoglycosides such as gentamicin are nephrotoxic and poorly cleared by damaged kidneys. Imipenem causes intestinal discomfort, but it has less toxicity and would be a viable choice.

In the case of a cancer patient with severe systemic *Candida* infection, there will be fewer criteria to weigh. Intravenous amphotericin B or fluconazole are the only possible choices, despite drug toxicity and other possible adverse side

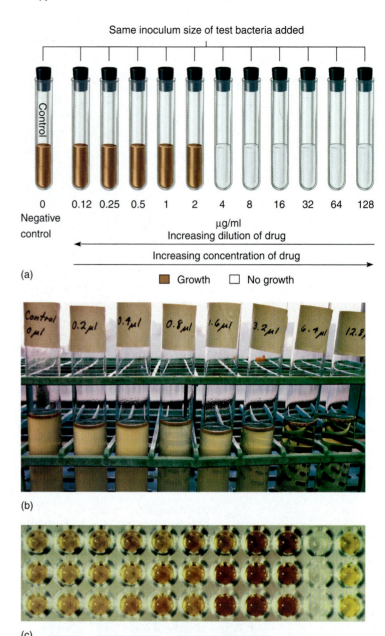

(a)

■ Growth   □ No growth

(b)

(c)

FIGURE 12.20   Tube dilution test for determining the minimum inhibitory concentration (MIC).
(a) The antibiotic is diluted serially through tubes of liquid nutrient from right to left. All tubes are inoculated with an identical amount of a test bacterium and then incubated. The first tube on the left is a control that lacks the drug and shows maximum growth. The dilution of the first tube in the series that shows no growth (no turbidity) is the MIC. **(b)** Photograph of MIC tube test using *Escherichia coli* and tetracycline. **(c)** Multiwell plate with an array of three tests. This system can be automated to read the MICs of several drugs simultaneously.

effects. In a life-threatening situation in which a dangerous chemotherapy is perhaps the only chance for survival, the choices are reduced and the priorities are different.

An Antimicrobial Drug Dilemma

We began this chapter with a view of the exciting strides made in chemotherapy during the past few years, but we

TABLE 12.8	Comparitive MICs (µg/ml) for Common Drugs and Pathogens				
Bacterium	**Penicillin G**	**Ampicillin**	**Sulfamethoxazole**	**Tetracycline**	**Cefaclor**
Staphylococcus aureus	4.0	0.05	3.0	0.3	4.0
Enterococcus faecalis	3.6	1.6	100.0	0.3	60.0
Neisseria gonorrhoeae	0.5	0.5	5.0	0.8	2.0
Escherichia coli	100.0	12.0	3.0	6–50.0	3.0
Pseudomonas aeruginosa	>500.0	>200.0	NA	>100.0	NA
Salmonella species	12.0	6.0	10.0	1.0	0.8
Clostridium	0.16	NA	NA	3.0	12.0

NA = not available

must end it on a note of qualification and caution. There is now a worldwide problem in the management of antimicrobial drugs, which rank second only to some nervous system drugs in overall usage. The remarkable progress in treating many infectious diseases has spawned a view of antimicrobials as a "cure-all" for infections as diverse as the common cold and acne. And, although it is true that nothing is as dramatic as curing an infectious disease with the correct antimicrobial drug, in many instances, drugs have no effect or can be harmful. The depth of this problem can perhaps be appreciated better with a few statistics:

1. Roughly 200 million prescriptions for antimicrobials are written in the United States every year. A recent study disclosed that 75% of antimicrobial prescriptions are for pharyngeal, sinus, lung, and upper respiratory infections. A fairly high percentage of these are viral in origin and will have little or no benefit from antibacterial drugs.

   Many drugs are also misprescribed as to type, dosage, or length of therapy. Such overuse of antimicrobials is known to increase the development of antimicrobial resistance, harm the patient, and waste billions of dollars.

2. Drugs are often prescribed without benefit of culture or susceptibility testing, even when such testing is clearly warranted.

3. Some physicians tend to use a "shotgun" antimicrobial therapy for minor infections, which involves administering a broad-spectrum drug instead of a more specific narrow-spectrum one. This practice can lead to superinfections and other adverse reactions. Tetracyclines and chloramphenicol are still prescribed routinely for infections that would be treated more effectively with narrower spectrum, less toxic drugs.

4. More expensive newer drugs are chosen when a less costly older one would be just as effective. Among the most expensive drugs are cephalosporins and the longer-acting tetracyclines, yet these are among the most commonly prescribed antibiotics.

5. Tons of excess antimicrobial drugs produced in this country are exported to other countries, where controls are not as strict. Nearly 200 different antibiotics are sold over the counter in Latin America and Asian countries. It is common for people in these countries to self-medicate without understanding the correct medical indication. Drugs used in this way are largely ineffectual but, worse yet, they are known to be responsible for emergence of drug-resistant bacteria that subsequently cause epidemics.

The medical community recognizes that most physicians are motivated by important and prudent practical concerns, such as the need for immediate therapy to protect a sick patient and for defensive medicine to provide the very best care possible. But in the final analysis, every allied health professional should be critically aware not only of the admirable and utilitarian nature of antimicrobials, but also of their limitations.

✔ CHECKPOINT

- The three major considerations necessary to choose an effective antimicrobial are the nature of the infecting microbe, the microbe's sensitivity to available drugs, and the overall medical status of the infected host.
- The Kirby-Bauer test identifies antimicrobials that are effective against a specific infectious bacterial isolate.
- The MIC (minimum inhibitory concentration) identifies the smallest effective dose of an antimicrobial toxic to the infecting microbe.
- The therapeutic index is a ratio of the amount of drug toxic to the infected host and the MIC. The smaller the ratio, the greater the potential for toxic host-drug reactions.
- The effectiveness of antimicrobial drugs is being compromised by several alarming trends: inappropriate prescription, use of broad-spectrum instead of narrow-spectrum drugs, use of higher-cost drugs, sale of over-the-counter antimicrobials in other countries, and lack of sufficient testing before prescription.

Chapter Summary With Key Terms

12.1 Principles of Antimicrobial Therapy
Chemotherapeutic drugs
A. Used to control microorganisms in the body. Depending on their source, these drugs are described as **antibiotics**, **semisynthetic** or synthetic.

B. Based on their mode and spectrum of action, they are described as **broad spectrum** or **narrow spectrum** and microbistatic or microbicidal.

12.2 Interactions Between Drug and Microbe

A. The ideal antimicrobial is **selectively toxic,** highly potent, stable, and soluble in the body's tissues and fluids, does not disrupt the immune system or microflora of the host and is exempt from drug resistance.

B. Strategic approaches to the use of chemotherapeutics include
   1. **Prophylaxis,** where drugs are administered to *prevent* infection in susceptible people.
   2. Combined therapy, where two or more drugs are given simultaneously, either to prevent the emergence of resistant species or achieve synergism.

C. The inappropriate use of drugs on a worldwide basis has led to numerous medical and economic problems.

12.3 Survey of Major Antimicrobial Drug Groups

A. **Penicillins** are beta-lactam-based drugs originally isolated from the mold *Penicillium chrysogenum.* The natural form is penicillin G, although various semisynthetic forms, such as ampicillin and methicillin, exist which vary in their spectrum and applications.
   1. Penicillin is bactericidal, blocking completion of the cell wall, which causes eventual rupture of the cell.
   2. Major problems encountered in penicillin therapy include allergic reactions and bacterial resistance to the drug through **beta-lactamase.**

B. **Cephalosporins** include both natural and semisynthetic forms initially isolated from the mold *Cephalosporium.* Cephalosporins inhibit peptidoglycan synthesis (like penicillin) but have a much broader spectrum.

C. **Aminoglycosides** include several narrow-spectrum drugs isolated from unique bacteria found in the genera *Streptomyces.* Examples include streptomycin, gentamicin, tobramycin, and amikacin.

D. **Tetracyclines** and chloramphenicol are very broad-spectrum drugs isolated from *Streptomyces.* Both interfere with translation but their use is limited by adverse effects.

E. Erythromycin, clindamycin, vancomycin, and rifampin are all isolated from *Streptomyces.*
   1. Erythromycin and clindamycin both affect protein synthesis. Erythromycin is a broad-spectrum alternative for use with penicillin-resistant bacteria, while clindamycin is used primarily for intestinal infection by anaerobes. Ketek is a new ketolide antibiotic related to erythromycin.
   2. Vancomycin interferes with the early stages of cell wall synthesis and is used for life-threatening, methicillin-resistant staphylococcal infection.
   3. Rifampin interferes with RNA polymerase (thereby affecting transcription) and is primarily used for tuberculosis and leprosy infections.
   4. Resistant bacteria can be found for each of the above drugs.

F. Bacitracin and **polymyxin** are narrow-spectrum antibiotics isolated from the bacterial genera *Bacillus.*
   1. Bacitracin prevents cell wall synthesis in gram-positive organisms. It is used in antibacterial skin ointments, often in combination with neomycin.
   2. Polymyxin is also found in skin ointments and can be used to treat *Pseudomonas* infections.

G. Fosfomycin, Synercid, and daptomycin are newly developed antibiotics that are useful when bacteria have developed resistance to more traditional drugs.

H. *Synthetic Antibacterial Drugs*
   1. **Sulfonamides** are broad-spectrum drugs that act as **metabolic analogs, competitively inhibiting** enzymes needed for nucleic acid synthesis.
   2. Trimethoprim, dapsone, isoniazid, fluoroquinolones, and oxazolidinones are all synthetic antibacterial drugs.
      a. Trimethoprim is often used in combination with sulfa drugs.
      b. Dapsone is a narrow-spectrum drug related to the sulfonamides. It is used (often in combination with rifampin) to treat leprosy.
      c. Isoniazid (INH) blocks the synthesis of cell wall components in *Mycobacteria.* It is used in the treatment of tuberculosis.
      d. Fluoroquinolones (ciprofloxacin) are a new class of broad-spectrum synthetic drug.
      e. Oxazolidinones are another new class of synthetic drug that work by inhibiting the start of protein synthesis. They are useful for treatment of staphylococci that are resistant to other drugs (MRSA and VRE).

I. *Drugs for Fungal Infection*
   1. Amphoteracin B and nystatin disrupt fungal membranes by detergent action.
   2. **Azoles** are synthetic drugs that interfere with membrane synthesis. They include ketoconazole, fluconazole, and miconazole.
   3. Flucytosine inhibits DNA synthesis. Because many fungi are now resistant, it must usually be used in conjunction with amphotericin.

J. *Drugs for Protozoan Infections*
   1. Quinine, or the related compounds chloraquine, primaquine, or mefloquine are used to treat infections by the malarial parasite *Plasmodium.*
   2. Other drugs including metronidazole, suramin, melarsopral, and nitrifurimox are used for other protozoan infections.

K. *Drugs for Helminth Infections* include mebendazole, thiabendazole, praziquantel, pyrantel, piperizine, and niclosamide, which are all used for antihelminthic chemotherapy.

L. *Drugs for Viral Infection* usually act by inhibiting viral penetration, multiplication, or assembly. Because viral and host metabolism are so closely related, toxicity is a potential adverse reaction with all antiviral drugs.
   1. **Acyclovir,** valacyclovir, famciclovir, and ribavirin act as nucleoside analogs, inhibiting viral DNA replication, especially in herpesviruses.
   2. Amantadine acts early in the cycle of influenza A to prevent viral uncoating.
   3. AZT and protease inhibitors such as indinavir are used as anti-HIV drugs.
   4. Fuzeon prevents the fusion of HIV to its host receptor, thereby blocking virus entry.
   5. Interferon is a naturally occurring protein with both antiviral and anticancer properties.

M. *The Acquisition of Drug Resistance*
   1. Microbes can lose their sensitivity to a drug through the acquisition of **resistance factors.** Drug resistance takes the form of:
      a. Drug inactivation.
      b. Decreased permeability to drug or increased elimination of drug from the cell.

c. Change in drug receptors.
d. Change of metabolic patterns.
2. New approaches to antimicrobial therapy include the use of **probiotics** and **prebiotics**.

12.4 Interaction Between Drug and Host
Side effects of chemotherapy include organ toxicity, allergic responses, alteration of microflora.

12.5 Considerations in Selecting an Antimicrobial Drug
A. Rapid identification of the infectious agent is important.
B. The pathogenic microbe should be tested for its susceptibility to different antimicrobial agents. This involves using standardized methods such as the *Kirby-Bauer* and **minimum inhibitory concentration (MIC)** techniques.

Multiple-Choice Questions

1. A compound synthesized by bacteria or fungi that destroys or inhibits the growth of other microbes is a/an
   a. synthetic drug
   b. antibiotic
   c. antimicrobial drug
   d. competitive inhibitor

2. Which statement is *not* an aim in the use of drugs in antimicrobial chemotherapy? The drug should:
   a. have selective toxicity
   b. be active even in high dilutions
   c. be broken down and excreted rapidly
   d. be microbicidal

3. Drugs that prevent the formation of the bacterial cell wall are
   a. quinolones
   b. beta-lactams
   c. tetracyclines
   d. aminoglycosides

4. Sulfonamide drugs initially disrupt which process?
   a. folic acid synthesis
   b. transcription
   c. PABA synthesis
   d. protein synthesis

5. Microbial resistance to drugs is acquired through
   a. conjugation
   b. transformation
   c. transduction
   d. all of these

6. R factors are _____ that contain a code for _____.
   a. genes, replication
   b. plasmids, drug resistance
   c. transposons, interferon
   d. plasmids, conjugation

7. When a patient's immune system becomes reactive to a drug, this is an example of
   a. superinfection
   b. drug resistance
   c. allergy
   d. toxicity

8. An antibiotic that disrupts the normal flora can cause
   a. the teeth to turn brown
   b. aplastic anemia
   c. a superinfection
   d. hepatotoxicity

9. Most antihelminthic drugs function by
   a. weakening the worms so they can be flushed out by the intestine
   b. inhibiting worm metabolism
   c. blocking the absorption of nutrients
   d. inhibiting egg production

10. An example of an antiviral drug that can prevent a viral nucleic acid from being replicated is
    a. azidothymidine
    b. acyclovir
    c. amantadine
    d. both a and b

11. Which of the following effects do antiviral drugs *not* have?
    a. killing extracellular viruses
    b. stopping virus synthesis
    c. inhibiting virus maturation
    d. blocking virus receptors

12. Which of the following modes of action would be most selectively toxic?
    a. interrupting ribosomal function
    b. dissolving the cell membrane
    c. preventing cell wall synthesis
    d. inhibiting DNA replication

13. The MIC is the _____ of a drug that is required to inhibit growth of a microbe.
    a. largest concentration
    b. standard dose
    c. smallest concentration
    d. lowest dilution

14. An antimicrobial drug with a _____ therapeutic index is a better choice than one with a _____ therapeutic index.
    a. low, high
    b. high, low

Concept Questions

These questions are suggested as a *writing-to-learn* experience. For each question, compose a one- or two-paragraph answer that includes the factual information needed to completely address the question.

1. Differentiate between antibiotics and synthetic drugs.

2. a. Differentiate between narrow-spectrum and broad-spectrum antibiotics.
   b. Can you determine why some drugs have narrower spectra than others? (Hint: Look at their mode of action.)
   c. How might one determine whether a particular antimicrobial is broad- or narrow-spectrum?

3. a. What is the major source of antibiotics?
   b. What appears to be the natural function of antibiotics?

4. a. Using the diagram at right as a guide, briefly explain how the three factors in drug therapy interact.
   b. What drug characteristics will make treatment most effective?
   c. Why is it better for a drug to be microbicidal than microbistatic?

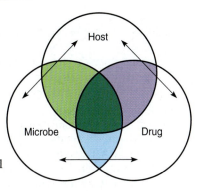

5. a. Explain the major modes of action of antimicrobial drugs, and give an example of each.
   b. What is competitive inhibition?
   c. What is the basic reason that a metabolic analog molecule can inhibit metabolism?
   d. Why does the penicillin group of drugs have milder toxicity than other antibiotics?
   e. What are the long-term effects of drugs that block transcription?
   f. Why would a drug that blocks translation on the ribosomes of bacteria also affect human cells?
   g. Why do drugs that act on bacterial and fungal membranes generally have high toxicity?

6. Construct a chart that summarizes the modes of action and applications of the major groups of antibacterial drugs (antibiotics and synthetics), antifungal drugs, antiparasitic drugs, and antiviral drugs.

7. a. Explain why there are fewer antifungal, antiparasitic, and antiviral drugs than antibacterial drugs.
   b. What effect do nitrogen-base analogs have upon viruses?
   c. Summarize the origins and biological actions of interferon.

8. Explain the phenomenon of drug resistance from the standpoint of microbial genetics (include a description of R factors).

Critical Thinking Questions

Critical thinking is the ability to reason and solve problems using facts and concepts. These questions can be approached from a number of angles, and in most cases, they do not have a single correct answer.

1. Occasionally, one will hear the expression that a microbe has become "immune" to a drug.
   a. What is a better way to explain what is happening?
   b. Explain a simple test one could do to determine if drug resistance was developing in a culture.

2. a. Can you think of additional ways that drug resistance can be prevented?
   b. What can health care workers do?
   c. What can one do on a personal level?

3. Drugs are often given to surgical patients, to dental patients with heart disease, or to healthy family members exposed to contagious infections.
   a. What word would you use to describe this use of drugs?
   b. What is the purpose of this form of treatment?
   c. Explain some potential undesired effects of this form of therapy.

4. a. Your pregnant neighbor has been prescribed a daily dose of oral tetracycline for acne. Do you think this therapy is advisable for her? Why or why not?
   b. A woman has been prescribed a broad-spectrum oral cephalosporin for a strep throat. What are some possible consequences in addition to cure of the infected throat?
   c. A man has a severe case of sinusitis that is negative for bacterial pathogens. A physician prescribes an oral antibacterial drug in treatment. What are your opinions of this therapy?

5. You have been directed to take a sample from a growth-free portion of the zone of inhibition in the Kirby-Bauer test and inoculate it onto a plate of nonselective medium.
   a. What does it mean if growth occurs on the new plate?
   b. What if there is no growth?

6. In cases in which it is not possible to culture or drug test an infectious agent (such as middle ear infection), how would the appropriate drug be chosen?

7. Using the results in tables 12.7 and 12.8 and reviewing drug characteristics, choose an antimicrobial for each of the following situations (explain your choice):
   a. for an adult patient suffering from *Mycoplasma* pneumonia
   b. for a child with meningitis (drug must enter into cerebrospinal fluid)
   c. for a patient with allergy to erythromycin
   d. for a urinary tract infection by *Enterococcus*
   e. for gonorrhea

8. What factors can play a part in drug synergism?

9. How would you personally feel about being told by a physician that your infection cannot be cured by an antibiotic, and that the best thing to do is go home, drink a lot of fluids, and take aspirin or other symptom-relieving drugs?

10. a. Refer to figure 12.18*a* and interpret the results.
    b. Give the MICs for the tests in figure 12.20*a*.

11. a. Explain the basis for combined therapy.
    b. Give reasons why it could be helpful to use combined therapy in treating HIV infection.

Internet Search Topics

1. Locate information on new types of antibacterial drugs called linezolid, daptomycin, and Zyvox. Determine their mode of action and indications for use.

2. Go to the Online Learning Center for chapter 12 of this text at http://www.mhhe.com/cowan1. Access the URLs listed under Internet Search Topics and research the following:
   a. Go to the AIDS website listed. Look up examples of several types of anti-HIV drugs, comparing information on costs, side effects, and problems in therapy.
   b. Go to the Department of Health website as listed. Use the information provided there to discuss the problems of antibiotic resistance with family and friends.

Microbe–Human Interactions

Infection and Disease

In July of 1999, Karen visited her family physician with a large itchy red area on her left calf. She had first noticed a small red bump on her calf about a week before. Over the next 7 days, the redness spread into a slightly raised circular rash about 4 inches in diameter with a central clear area. Karen also reported a headache, muscle aches, and fatigue. A physical exam revealed a fever and some lymph node swelling. No one else in the family had similar symptoms. Karen's mother, however, had recently been complaining of arthritis pain in her knees, and her knee joints were swollen and stiff.

The characteristic rash led the physician to a probable diagnosis, which was confirmed by the presence of antibodies in Karen's blood.

Oral antibiotic treatment resulted in complete recovery. Karen's mother was diagnosed with a more advanced form of the same disease. She was hospitalized for intravenous antibiotic treatment. After several courses of antibiotics, her arthritis symptoms also disappeared.

The disease that Karen and her mother had is not contagious. Ticks that feed on infected deer or mice transmit the infection to humans. Karen's family had recently moved into a new subdivision on the outskirts of their Wisconsin town. Karen had enjoyed walking her dog on the nearby wooded trails, watching for the occasional white-tailed deer. Her mother spent hours in their new, larger yard planting a vegetable garden. Immature ticks called nymphs, less than 2 mm in size and nearly colorless, infected Karen and her mother as they took a blood meal. During the summer of 1999, the local health department confirmed 29 cases of the disease, 24 of them in Karen's neighborhood.

▶ *What was the cause of Karen's rash and her mother's arthritis?*

▶ *How are the rash and the swollen knee joints related?*

CHAPTER OVERVIEW

▶ The normal flora of humans includes bacteria, fungi, and protozoa that live on the body without causing disease. These microbes can be found in areas exposed to the outside environment, such as the gastrointestinal tract, skin, and respiratory tract, and are generally beneficial to humans.

▶ Pathogens are those microbes that infect the body and cause disease. Disease results when an adequate number of pathogenic cells enter the body through a specific route, grow, and disrupt tissues.

▶ Pathogens produce virulence factors such as toxins and enzymes that help them invade and damage the cells of their host. The effects of infection and disease are seen in the host as signs and symptoms, which may include both short- and long-term damage.

▶ Pathogens may be spread by direct or indirect means involving overtly infected people, carriers, vectors, and vehicles. A significant source of infection is exposure to the hospital environment.

▶ Pathogens may be found residing in humans, animals, food, soil, and water.

▶ The field of epidemiology is concerned with the patterns of disease occurrence in a population.

13.1 The Human Host

The human body exists in a state of dynamic equilibrium with microorganisms. In the healthy individual, this balance is maintained as a peaceful coexistence and lack of disease. But on occasion, the balance tips in favor of the microorganism, and an infection or disease results. In this chapter, we explore each component of the host-parasite relationship, beginning with the nature and function of normal flora, moving to the stages of infection and disease, and closing with a study of epidemiology and the patterns of disease in populations. These topics will set the scene for chapters 14 and 15, which deal with the ways the host defends itself against assault by microorganisms, and also for chapters 18 through 23, which examine diseases affecting different organ systems.

In chapter 7, several of the basic interrelationships between humans and microorganisms were considered. Most of the microbes inhabiting the human body benefit from the nutrients and protective habitat it provides. From the human point of view, these relationships run the gamut from mutualism to commensalism to parasitism and can have beneficial, neutral, or harmful effects. A common characteristic of all microbe-human relationships, regardless of where they lead, is that they begin with contact.

Contact, Infection, Disease—A Continuum

The body surfaces are constantly exposed to microbes. Some microbes become implanted there as colonists (normal flora), some are rapidly lost (transients), and others invade the tissues. Such intimate contact with microbes inevitably leads to **infection,** a condition in which pathogenic microorganisms penetrate the host defenses, enter the tissues, and multiply. When the cumulative effects of the infection damage or disrupt tissues and organs, the **pathologic** state that results is known as a disease. A disease is defined as any deviation from health. There are hundreds of different diseases caused by such factors as infections, diet, genetics, and aging. In this chapter, however, we will discuss only **infectious disease—** the disruption of a tissue or organ caused by microbes or their products.

The pattern of the host-parasite relationship can be viewed as a series of stages that begins with contact, pro-

gresses to infection, and ends in disease. Because of numerous factors relating to host resistance and degree of pathogenicity, not all contacts lead to infection and not all infections lead to disease. In fact, contamination without infection and infection without disease are the rule. Before we consider further details of infection and disease, let us examine what happens when microbes colonize the body, thereby establishing a long-term, usually beneficial, relationship **(figure 13.1).**

Resident Flora: The Human As a Habitat

With its constant source of nourishment and moisture, relatively stable pH and temperature, and extensive surfaces upon which to settle, the human body provides a favorable habitat for an abundance of microorganisms. In fact, it is so favorable that, cell-for-cell, microbes on the human body outnumber human cells ten to one! The large and mixed collection of microbes adapted to the body has been variously called the **normal (resident) flora,** or indigenous flora, though some microbiologists prefer to use the terms *microflora* and *commensals*. The normal residents include an array of bacteria, fungi, protozoa, and, to a certain extent, viruses and arthropods.

Acquiring Resident Flora

The human body offers a seemingly endless variety of environmental niches, with wide variations in temperature, pH, nutrients, and oxygen tension occurring from one area to another. Because the body provides such a range of habitats, it should not be surprising that the body supports a wide range of microbes. As shown in **tables 13.1** and **13.2,** most areas of the body in contact with the outside environment harbor resident microorganisms, while internal organs and tissue, along with the fluids they contain, are generally microbe-free.

It should be mentioned that several groups of scientists have recently reported detecting bacteria in blood from healthy humans and animals. They are bacteria that cannot be cultivated on laboratory media and were only detected using microscopy and molecular identification techniques. While these results are still somewhat controversial, they raise the possibility that microbes do exist in what we have long thought of as sterile body compartments, but that they

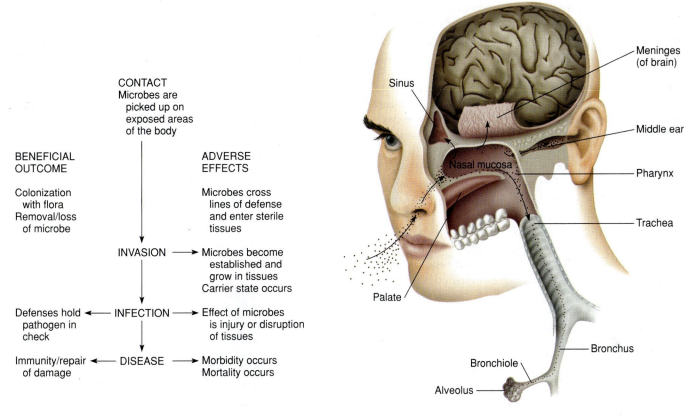

CONTACT
Microbes are
picked up on
exposed areas
of the body

BENEFICIAL
OUTCOME

Colonization
with flora
Removal/loss
of microbe

ADVERSE
EFFECTS

Microbes cross
lines of defense
and enter sterile
tissues

INVASION → Microbes become
established and
grow in tissues
Carrier state occurs

Defenses hold ← INFECTION → Effect of microbes
pathogen in   is injury or disruption
check   of tissues

Immunity/repair ← DISEASE → Morbidity occurs
of damage   Mortality occurs

FIGURE 13.1 **Associations between microbes and humans.**
Effects of exposure can progress in a variety of directions, ranging from no effect to colonization, and from infection to immunity to disease. The example shown here follows the possible events in the case of contact with a pathogen such as *Streptococcus pneumoniae* (the pneumococcus). This bacterium can be harbored harmlessly in the upper respiratory tract, but it may also invade and infect various sites in the cranium and respiratory tract.

TABLE 13.1   Sites That Harbor a Normal Flora
• Skin and its contiguous mucous membranes
• Upper respiratory tract
• Gastrointestinal tract (various parts)
• Outer opening of urethra
• External genitalia
• Vagina
• External ear canal
• External eye (lids, conjunctiva)

TABLE 13.2   Sterile (Microbe-Free) Anatomical Sites and Fluids
All Internal Tissues and Organs
Heart and circulatory system
Liver
Kidneys and bladder
Lungs
Brain and spinal cord
Muscles
Bones
Ovaries/testes
Glands (pancreas, salivary, thyroid)
Sinuses
Middle and inner ear
Internal eye
Fluids Within an Organ or Tissue
Blood
Urine in kidneys, ureters, bladder
Cerebrospinal fluid
Saliva prior to entering the oral cavity
Semen prior to entering the urethra
Amniotic fluid surrounding the embryo and fetus

are fundamentally different from the types of microbes with which we are familiar. For the purposes of this book we will consider the blood and other sites listed in table 13.2 to be sterile. Discoveries such as these remind us that science is a continuously changing body of knowledge, and no scientific truth is ever "final."

The vast majority of microbes that come in contact with the body are removed or destroyed by the host's defenses long before they are able to colonize a particular area. Of those microbes able to establish an ongoing presence, an even smaller number are able to remain without attracting the unwanted attention of the body's defenses. This last group of organisms have evolved, along with their human hosts, to produce a complex relationship where the effects of normal flora are generally not deleterious to the host.

Although generally stable, the flora can fluctuate to a limited extent with general health, age, variations in diet, hygiene, hormones, and drug therapy. In many cases, bacterial flora actually benefits the human host by preventing the overgrowth of harmful microorganisms. A common example is the fermentation of glycogen by lactobacilli, which keep the pH in the vagina quite acidic and prevent the overgrowth of the yeast *Candida albicans*. A second example is seen in the large intestine where a protein produced by *E. coli* prevents the growth of pathogenic bacteria such as *Salmonella* and *Shigella*.

The generally antagonistic effect "good" microbes have against intruder microorganisms is called **microbial antagonism.** Normal flora exist in a steady established relationship with the host and are unlikely to be displaced by incoming microbes. This antagonistic protection may simply be a result of a limited number of attachment sites in the host site, all of which are stably occupied by normal flora. Antagonism may also result from the chemical or physiological environment created by the resident flora, which is hostile to other microbes.

In experiments performed with mice, scientists discovered that a species of the intestinal bacterium *Bacteroides* controlled the production of a host chemical that suppressed other microbes in the area. This is an extreme case of microbial antagonism, but the general phenomenon is of great importance to human health.

Characterizing the normal flora as beneficial or, at worst, commensal to the host presupposes that the host is in good health, with a fully functioning immune system, and that the flora is present only in its natural microhabitat within the body. Hosts with compromised immune systems could very easily be infected by their own flora (see table 13.4). This outcome is seen when AIDS patients contract pneumococcal pneumonia, the causative agent of which (*Streptococcus pneumoniae*) is often carried as normal flora in the nasopharynx. **Endogenous** infections can also occur when normal flora is introduced to a site that was previously sterile, as when *E. coli* enters the bladder, resulting in a urinary tract infection.

Initial Colonization of the Newborn

The uterus and its contents are normally sterile during embryonic and fetal development and remain essentially germ-free until just before birth. The event that first exposes the infant to microbes is the breaking of the fetal membranes, at which time microbes from the mother's vagina can enter the womb. Comprehensive exposure occurs during the birth process itself, when the baby unavoidably comes into intimate contact with the birth canal **(figure 13.2).** Within 8 to 12 hours after delivery, the newborn typically has been colonized by bacteria such as streptococci, staphylococci, and lactobacilli, acquired primarily from its mother. The nature of the flora initially colo-

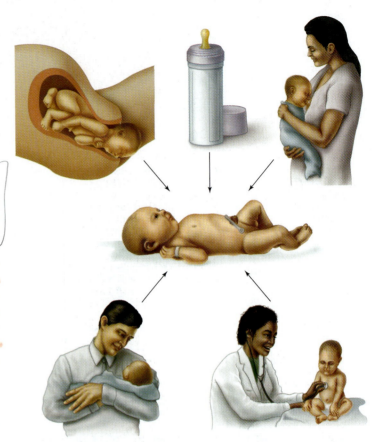

FIGURE 13.2 The origins of flora in newborns.

nizing the large intestine depends upon whether the baby is bottle- or breast-fed. Bottle-fed infants (receiving milk or a milk-based formula) tend to acquire a mixed population of coliforms, lactobacilli, enteric streptococci, and staphylococci. In contrast, the intestinal flora of breast-fed infants consists primarily of *Bifidobacterium* species whose growth is favored by a growth factor from the milk. This bacterium metabolizes sugars into acids that protect the infant from infection by certain intestinal pathogens. The skin, gastrointestinal tract, and portions of the respiratory and genitourinary tract all continue to be colonized as contact continues with family members, health care personnel, the environment, and food.

Milestones that contribute to development of the adult pattern of flora are eruption of teeth, weaning, and introduction of the first solid food. Although exposure to microbes is unavoidable and even necessary for the maturation of the infant's flora, contact with pathogens is dangerous, because the neonate is not yet protected by a full complement of flora, and owing to its immature immune defenses, is extremely susceptible to infection.

Indigenous Flora of Specific Regions

Although we tend to speak of the flora as a single unit, it is a complex mixture of hundreds of species, differing somewhat in

TABLE 13.3	Life on Humans: Sites Containing Well-Established Flora and Representative Examples	
Anatomic Sites	**Common Genera**	**Remarks**
Skin	**Bacteria:** *Staphylococcus, Micrococcus, Corynebacterium, Propionibacterium, Streptococcus*	Microbes live only in upper dead layers of epidermis, glands, and follicles; dermis and layers below are sterile.
	Fungi: *Candida, Pityrosporum*	Dependent on skin lipids for growth.
	Arthropods: *Demodix* mite	Present in sebaceous glands and hair follicles.
Gastrointestinal Tract		
Oral cavity	**Bacteria:** *Streptococcus, Neisseria, Veillonella, Fusobacterium, Lactobacillus, Bacteroides, Actinomyces, Eikenella, Treponema, Haemophilus*	Colonize the epidermal layer of cheeks, gingiva, pharynx; surface of teeth; found in saliva in huge numbers.
	Fungi: *Candida* species	Can cause thrush.
	Protozoa: *Entamoeba gingivalis*	Inhabit the gingiva of persons with poor oral hygiene.
Large intestine and rectum	**Bacteria:** *Bacteroides, Fusobacterium, Bifidobacterium, Clostridium,* fecal streptococci, *Lactobacillus,* coliforms (*Escherichia, Enterobacter*)	Areas of lower gastrointestinal tract other than large intestine and rectum have sparse or nonexistent flora. Flora consists predominantly of strict anaerobes; other microbes are aerotolerant or facultative.
	Fungi: *Candida*	Intestinal thrush
	Protozoa: *Entamoeba coli, Trichomonas hominis*	Feed on waste materials in the large intestine.
Upper Respiratory Tract	Microbial population exists in the nasal passages, throat, and pharynx; owing to proximity, flora is similar to that of oral cavity	Trachea and bronchi have a sparse population; bronchioles and alveoli have no normal flora and are essentially sterile.
Genital Tract	**Bacteria:** *Lactobacillus, Streptococcus, Corynebacterium, Escherichia*	In females, flora occupies the external genitalia and vaginal and cervical surfaces; internal reproductive structures normally remain sterile. Flora responds to hormonal changes during life.
	Fungi: *Candida*	Cause of yeast infections.
Urinary Tract	**Bacteria:** *Staphylococcus, Streptococcus, Corynebacterium, Lactobacillus*	In females, flora exists only in the first portion of the urethral mucosa; the remainder of the tract is sterile. In males, the entire reproductive and urinary tract is sterile except for a short portion of the anterior urethra.

↑External Ear Canal

←Eye (eyelids)

quality and quantity from one individual to another. Studies of the flora have shown that most people harbor certain specially adapted bacteria, fungi, and protozoa. The normal, indigenous flora present in specific body sites is presented in detail in chapters 18 through 23. **Table 13.3** provides an overview.

For a look into the laboratory study of resident flora see **Insight 13.1.**

13.2 The Progress of an Infection

A microbe whose relationship with its host is parasitic and results in infection and disease is termed a **pathogen.** The type and severity of infection depend both on the pathogenicity of the organism and the condition of the host **(figure 13.3).** Pathogenicity, you will recall, is a broad concept that describes an organism's potential to cause infection or disease, and is used to divide pathogenic microbes into one of two groups. **True pathogens** (primary pathogens) are capable of causing disease in healthy persons with normal immune defenses. They are generally associated with a specific, recognizable disease, which may vary in severity from mild (colds) to severe (malarial) to fatal (rabies). Examples of true pathogens include influenza virus, plague bacillus, and malarial protozoan.

Opportunistic pathogens cause disease when the host's defenses are compromised[1] or when they become established

---
1. People with weakened immunity are often termed *immunocompromised.*

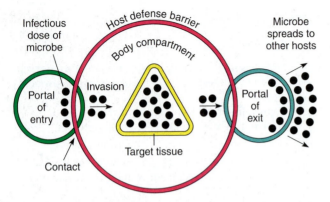

FIGURE 13.3 **An overview of the events in infection.**
An adequate dose of an infectious agent overcomes a defense barrier and enters the tissues of a host through one of several portals of entry. From here, it moves or is carried to a specific organ or tissue, called the target tissue, where it multiplies and usually causes some degree of damage. The exit of the pathogen through the same or another portal can facilitate its transmission to another host.

TABLE 13.4	Factors That Weaken Host Defenses and Increase Susceptibility to Infection*

- Old age and extreme youth (infancy, prematurity)
- Genetic defects in immunity and acquired defects in immunity (AIDS)
- Surgery and organ transplants
- Organic disease: cancer, liver malfunction, diabetes
- Chemotherapy/immunosuppressive drugs
- Physical and mental stress
- Other infections

*These conditions compromise defense barriers or immune responses.

in a part of the body that is not natural to them. Opportunists are not considered pathogenic to a normal healthy person and, unlike primary pathogens, do not generally possess well-developed virulence properties. Examples of opportunistic pathogens include *Pseudomonas* species and *Candida albicans*. Factors that greatly predispose a person to infections, both primary and opportunistic, are shown in **table 13.4.**

The relative severity of the disease caused by a particular microorganism depends on the **virulence** of the microbe. Although the terms *pathogenicity* and *virulence* are often used interchangeably, virulence is the accurate term for describing the degree of pathogenicity. The virulence of a microbe is determined by its ability to

1. establish itself in the host, and
2. cause damage

There is much involved in both of these steps. To establish themselves in a host microbes must enter the host, attach firmly to host tissues, and survive the host defenses. To cause damage microbes produce toxins or induce a host response that is actually injurious to the host. Any characteristic or structure of the microbe that contributes to the above activities is called a **virulence factor.** Virulence can be due to single or multiple factors. In some microbes, the causes of virulence are clearly established, but in others they are not. In the following section, we examine the effects of virulence factors while simultaneously outlining the stages in the progress of an infection.

Recognizing that classifying pathogens into only two categories may be unduly restrictive, the Centers for Disease Control and Prevention has adopted a system of biosafety categories for pathogens based on their degree of pathogenicity and the relative danger in handling them. This system assigns microbes to one of four levels or classes. Microbes not known to cause disease in humans are assigned to level 1, and highly contagious viruses that pose an extreme risk to humans are classified as level 4. Microbes of intermediate virulence are assigned to levels 2 or 3. This system is explained in more detail in **Insight 13.2.**

Becoming Established: Step One—Portals of Entry

To initiate an infection, a microbe enters the tissues of the body by a characteristic route, the **portal of entry,** usually a cutaneous or membranous boundary. The source of the infectious agent can be **exogenous,** originating from a source outside the body (the environment or another person or animal), or endogenous, already existing on or in the body (normal flora or latent infection).

For the most part, the portals of entry are the same anatomical regions that also support normal flora: the skin, gastrointestinal tract, respiratory tract, and urogenital tract. The majority of pathogens have adapted to a specific portal of entry, one that provides a habitat for further growth and spread. This adaptation can be so restrictive that if certain pathogens enter the "wrong" portal, they will not be infectious. For instance, inoculation of the nasal mucosa with the influenza virus invariably gives rise to the flu, but if this virus contacts only the skin, no infection will result. Likewise, contact with athlete's foot fungi in small cracks in the toe webs can induce an infection, but inhaling the fungus spores will not infect a healthy individual. Occasionally, an infective agent can enter by more than one portal. For instance, *Mycobacterium tuberculosis* enters through both the respiratory and gastrointestinal tracts, and pathogens in the genera *Streptococcus* and *Staphylococcus* have adapted to invasion through several portals of entry such as the skin, urogenital tract, and respiratory tract.

Infectious Agents That Enter the Skin

The skin is a very common portal of entry. The actual sites of entry are usually nicks, abrasions, and punctures (many of which are tiny and inapparent) rather than smooth, unbroken skin. *Staphylococcus aureus* (the cause of boils), *Streptococcus pyogenes* (an agent of impetigo), the fungal dermatophytes, and agents of gangrene and tetanus gain

INSIGHT 13.1   *Discovery*

Life Without Flora

For years, questions lingered about how essential the microbial flora is to normal life and what functions various members of the flora might serve. The need for animal models to further investigate these questions led eventually to development of laboratory strains of *germ-free,* or **axenic,** mammals and birds. The techniques and facilities required for producing and maintaining a germ-free colony are exceptionally rigorous. After the young mammals are taken from the mother aseptically by cesarian section, they are immediately transferred to a sterile isolator or incubator. The newborns must be fed by hand through gloved ports in the isolator until they can eat on their own, and all materials entering their chamber must be sterile. Rats, mice, rabbits, guinea pigs, monkeys, dogs, hamsters, and cats are some of the mammals raised in the germ-free state.

Experiments with germ-free animals are of two basic varieties: (1) general studies on how the lack of normal microbial flora influences the nutrition, metabolism, and anatomy of the animal, and (2) **gnotobiotic** (noh″-toh-by-ah′-tik) studies, in which the germ-free subject is inoculated either with a single type of microbe to determine its individual effect or with several known microbes to determine interrelationships. Results are validated by comparing the germ-free group with a conventional, normal control group. **Table 13.A** summarizes some major conclusions arising from studies with germ-free animals.

A dramatic characteristic of germ-free animals is that they live longer and have fewer diseases than normal controls, as long as they remain in a sterile environment. From this standpoint, it is clear that the flora is not needed for survival and may even be the source of infectious agents. At the same time, it is also clear that axenic life is highly impractical. Additional studies have revealed important facts about the effect of the flora on various organs and systems. For example, the flora contributes significantly to the development of the immune system. When germ-free animals are placed in contact with normal control animals, they gradually

TABLE 13.A	Effects of the Germ-Free State	
Germ-Free Animals Display	**Significance**
Enlargement of the cecum; other degenerative diseases of the intestinal tract of rats, rabbits, chickens	Microbes are needed for normal intestinal development
Vitamin deficiency in rats	Microbes are a significant nutritional source of vitamins
Underdevelopment of immune system in most animals	Microbes are needed to stimulate development of certain host defenses
Absence of dental caries and periodontal disease in dogs, rats, hamsters	Microbes are essential in caries formation and gum disease
Heightened sensitivity to enteric pathogens (*Shigella, Salmonella, Vibrio cholerae*) and to fungal infections	Normal flora are antagonistic against pathogens
Lessened susceptibility to amoebic dysentery	Normal flora facilitate the completion of the life cycle of the amoeba in the gut

develop a flora similar to that of the controls. However, germ-free subjects are less tolerant of microorganisms and can die from infections by relatively harmless species. This susceptibility is due to the immature character of the immune system of germ-free animals. These animals have a reduced number of certain types of white blood cells and slower antibody response.

Gnotobiotic experiments have clarified the dynamics of several infectious diseases. Perhaps the most striking discoveries were made in the case of oral diseases. For years, the precise involvement of microbes in dental caries had been ambiguous. Studies with germ-free rats, hamsters, and beagles confirmed that caries development is influenced by heredity, a diet high in sugars, and poor oral hygiene. Even when all these predisposing factors are present, however, germ-free animals still remain free of caries unless they have been inoculated with specific bacteria. Further discussion on dental diseases is found in chapter 22.

The ability of known pathogens to cause infection can also be influenced by normal flora, sometimes in opposing ways. Studies have indicated that germ-free animals are highly susceptible to experimental infection by the enteric pathogens *Shigella* and *Vibrio,* whereas normal animals are less susceptible, presumably because of their protective flora. In marked contrast, *Entamoeba histolytica* (the agent of amoebic dysentery) is more pathogenic in the normal animal than in the germ-free animal. One explanation for this phenomenon is that *E. histolytica* must feed on intestinal bacteria to complete its life cycle.

Sterile enclosure for rearing and handling germ-free laboratory animals.

access through damaged skin. The viral agent of cold sores (herpes simplex, type 1) enters through the mucous membranes near the lips.

Some infectious agents create their own passageways into the skin using digestive enzymes. For example, certain helminth worms burrow through the skin directly to gain access to the tissues. Other infectious agents enter through bites. The bites of insects, ticks, and other animals offer an avenue to a variety of viruses, rickettsias, and protozoa. An artificial means for breaching the skin barrier is contaminated hypodermic needles by intravenous drug abusers. Users who inject drugs are predisposed to a disturbing list of well-known diseases: hepatitis, AIDS, tetanus, tuberculosis, osteomyelitis, and malaria. A resurgence of some of these infections is directly traceable to drug use. Contaminated needles often contain bacteria from the skin or environment that induce heart disease (endocarditis), lung abscesses, and chronic infections at the injection site.

Although the conjunctiva, the outer protective covering of the eye, is ordinarily a relatively good barrier to infection, bacteria such as *Haemophilus aegyptius* (pinkeye), *Chlamydia trachomatis* (trachoma), and *Neisseria gonorrhoeae* have a special affinity for this membrane.

The Gastrointestinal Tract as Portal

The gastrointestinal tract is the portal of entry for pathogens contained in food, drink, and other ingested substances. They are adapted to survive digestive enzymes and abrupt pH changes. Most enteric pathogens possess specialized mechanisms for entering and localizing in the mucosa of the small or large intestine. The best-known enteric agents of disease are gram-negative rods in the genera *Salmonella, Shigella, Vibrio,* and certain strains of *Escherichia coli.* Viruses that enter through the gut are poliovirus, hepatitis A virus, echovirus, and rotavirus. Important enteric protozoans are *Entamoeba histolytica* (amoebiasis) and *Giardia lamblia* (giardiasis). Although the anus is not a typical portal of entry, it becomes one in people who practice anal sex. See chapter 22 for details of these diseases.

The Respiratory Portal of Entry

The oral and nasal cavities are also the gateways to the respiratory tract, the portal of entry for the greatest number of pathogens. Because there is a continuous mucous membrane surface covering the upper respiratory tract, the sinuses, and the auditory tubes, microbes are often transferred from one site to another. The extent to which an agent is carried into the respiratory tree is based primarily on its size. In general, small cells and particles are inhaled more deeply than larger ones. Infectious agents with this portal of entry include the bacteria of streptococcal sore throat, meningitis, diphtheria, and whooping cough and the viruses of influenza, measles, mumps, rubella, chickenpox, and the common cold. Pathogens that are inhaled into the lower regions of the respiratory tract (bronchioles and lungs) can cause **pneumonia,** an inflammatory condition of the lung. Bacteria (*Streptococcus pneumoniae, Klebsiella, Mycoplasma*) and fungi (*Cryptococcus* and *Pneumocystis*) are a few of the agents involved in pneumonias. All types of pneumonia are on the increase owing to the greater susceptibility of AIDS patients and other immunocompromised hosts to them. Other agents causing unique recognizable lung diseases are *Mycobacterium tuberculosis* and fungal pathogens such as *Histoplasma*. Chapter 21 describes infections of the respiratory system.

Urogenital Portals of Entry

The urogenital tract is the portal of entry for pathogens that are contracted by sexual means (intercourse or intimate direct contact). In the past, a judgmental attitude toward women and a somewhat Victorian attitude toward sex led to these diseases being referred to as venereal, from Venus, the Latin name of the goddess of love. The more contemporary description, **sexually transmitted disease (STD),** reflects the mode of transmission of these diseases more accurately. STDs account for an estimated 4% of infections worldwide, with approximately 13 million new cases occurring in the United States each year. The most recent available statistics (2002) for the estimated incidence of common STDs is provided in **table 13.5.**

The microbes of STDs enter the skin or mucosa of the penis, external genitalia, vagina, cervix, and urethra. Some can penetrate an unbroken surface; others require a cut or abrasion. The once predominant sexual diseases syphilis and gonorrhea have been supplanted by a large and growing list of STDs led by genital warts, chlamydia, and herpes. Evolving sexual practices have increased the incidence of STDs that were once uncommon, and diseases that were not originally considered STDs are now so classified.[2] Other common sexually transmitted agents are HIV (AIDS virus), *Trichomonas* (a protozoan), *Candida albicans* (a yeast), and hepatitis B virus. STDs are described in detail in chapter 23, with the exception of HIV (chapter 20) and hepatitis B (chapter 22).

---

2. Amoebic dysentery, scabies, salmonellosis, and *Strongyloides* worms are examples.

TABLE 13.5	Incidence of Common Sexually Transmitted Diseases
STD	**Estimated Number of New Cases per Year in U.S.**
Human papillomavirus	5,500,000
Trichomoniasis	5,000,000
Herpes simplex	1,000,000
Chlamydiosis	783,000
Gonorrhea	361,000
Hepatitis B	77,000
AIDS	41,002
Syphilis	32,200

INSIGHT 13.2

Medical

Laboratory Biosafety Levels and Classes of Pathogens

Personnel handling infectious agents in the laboratory must be protected from possible infection through special risk management or containment procedures. These involve:

1. carefully observing standard laboratory aseptic and sterile procedures while handling cultures and infectious samples;
2. using large-scale sterilization and disinfection procedures;
3. restricting eating, drinking, and smoking; and
4. wearing personal protective items such as gloves, masks, safety glasses, laboratory coats, boots, and headgear.

Some circumstances also require additional protective equipment such as biological safety cabinets for inoculations and specially engineered facilities to control materials entering and leaving the laboratory in the air and on personnel. **Table 13.B** summarizes the primary biosafety levels and agents of disease as characterized by the Centers for Disease Control and Prevention.

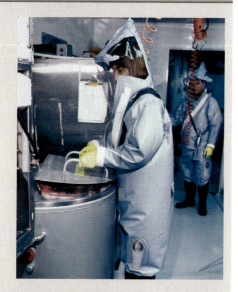

TABLE 13.B	Primary Biosafety Levels and Agents of Disease	
Biosafety Level	**Facilities and Practices**	**Risk of Infection and Class of Pathogens**
1	Standard, open bench, no special facilities needed; typical of most microbiology teaching labs; access may be restricted.	Low infection hazard; microbes not generally considered pathogens and will not colonize the bodies of healthy persons; *Micrococcus luteus, Bacillus megaterium, Lactobacillus, Saccharomyces.*
2	At least Level 1 facilities and practices; plus personnel must be trained in handling pathogens; lab coats and gloves required; safety cabinets may be needed; biohazard signs posted; access restricted.	Agents with moderate potential to infect; class 2 pathogens can cause disease in healthy people but can be contained with proper facilities; most pathogens belong to class 2; includes *Staphylococcus aureus, Escherichia coli, Salmonella* spp., *Corynebacterium diphtheriae*; pathogenic helminths; hepatitis A, B, and rabies viruses; *Cryptococcus* and *Blastomyces.*
3	Minimum of Level 2 facilities and practices; plus all manipulation performed in safety cabinets; lab designed with special containment features; only personnel with special clothing can enter; no unsterilized materials can leave the lab; personnel warned, monitored, and vaccinated against infection dangers.	Agents can cause severe or lethal disease especially when inhaled; class 3 microbes include *Mycobacterium tuberculosis, Francisella tularensis, Yersinia pestis, Brucella* spp., *Coxiella burnetii, Coccidioides immitis,* and yellow fever, WEE, and AIDS viruses.
4	Minimum of Level 3 facilities and practices; plus facilities must be isolated with very controlled access; clothing changes and showers required for all people entering and leaving; materials must be autoclaved or fumigated prior to entering and leaving lab.	Agents being handled are highly virulent microbes that pose extreme risk for morbidity and mortality when inhaled in droplet or aerosol form; most are exotic flaviviruses; arenaviruses, including Lassa fever virus; or filoviruses, including Ebola and Marburg viruses.

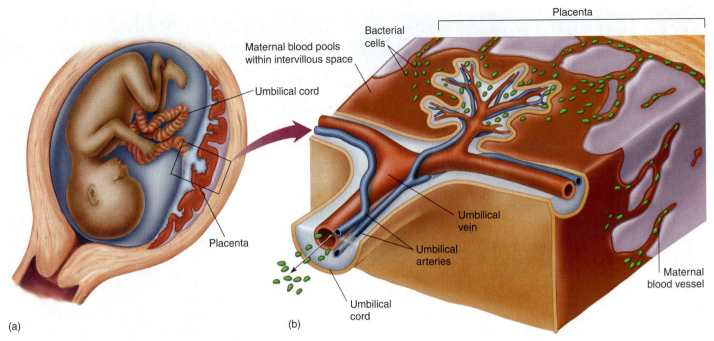

FIGURE 13.4   Transplacental infection of the fetus.
(a) Fetus in the womb. **(b)** In a closer view, microbes are shown penetrating the maternal blood vessels and entering the blood pool of the placenta. They then invade the fetal circulation by way of the umbilical vein.

Not all urogenital infections are STDs. Some of these infections are caused by displaced organisms (as when normal flora from the gastrointestinal tract cause urinary tract infections) or by opportunistic overgrowth of normal flora ("yeast infections").

Pathogens That Infect During Pregnancy and Birth

The placenta is an exchange organ formed by maternal and fetal tissues that separates the blood of the developing fetus from that of the mother, yet permits diffusion of dissolved nutrients and gases to the fetus. The placenta is ordinarily an effective barrier against microorganisms in the maternal circulation. However, a few microbes such as the syphilis spirochete can cross the placenta, enter the umbilical vein, and spread by the fetal circulation into the fetal tissues **(figure 13.4).**

Other infections, such as herpes simplex, occur perinatally when the child is contaminated by the birth canal. The common infections of fetus and neonate are grouped together in a unified cluster, known by the acronym **STORCH,** that medical personnel must monitor. STORCH stands for **s**yphilis, **t**oxoplasmosis, **o**ther diseases (hepatitis B, AIDS, and chlamydia), **r**ubella, **c**ytomegalovirus, and **h**erpes simplex virus. The most serious complications of STORCH infections are spontaneous abortion, congenital abnormalities, brain damage, prematurity, and stillbirths.

The Size of the Inoculum

Another factor crucial to the course of an infection is the quantity of microbes in the inoculating dose. For most agents, infection will proceed only if a minimum number, called the *infectious dose* (ID), is present. This number has been determined experimentally for many microbes. In general, microorganisms with smaller infectious doses have greater virulence. On the low end of the scale, the ID for rickettsia, the causative agent of Q fever, is only a single cell, and it is only about 10 infectious cells in tuberculosis, giardiasis, and coccidioidomycosis. The ID is 1,000 bacteria for gonorrhea and 10,000 bacteria for typhoid fever, in contrast to 1,000,000,000 bacteria in cholera. Numbers below an infectious dose will generally not result in an infection. But if the quantity is far in excess of the ID, the onset of disease can be extremely rapid. Even weakly pathogenic species can be rendered more virulent with a large inoculum.

Becoming Established: Step Two— Attaching to the Host

How Pathogens Attach

Adhesion is a process by which microbes gain a more stable foothold at the portal of entry. Because adhesion is dependent on binding between specific molecules on both the host and pathogen, a particular pathogen is limited to only those cells (and organisms) to which it can bind. Once attached, the pathogen is poised advantageously to invade the body compartments. Bacterial, fungal, and protozoal pathogens attach most often by mechanisms such as fimbriae (pili), surface proteins, and adhesive slimes or capsules; viruses attach by means of specialized receptors **(figure 13.5).** In addition, parasitic worms are mechanically fastened to the portal of entry by suckers, hooks, and barbs. Adhesion

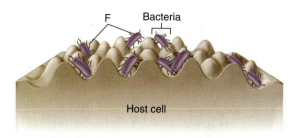

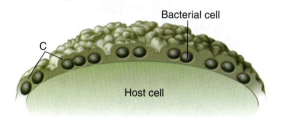

(a) **Fimbriae**

(b) **Capsules**

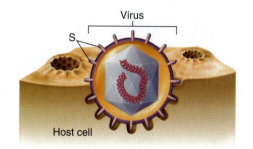

(c) **Spikes**

FIGURE 13.5 **Mechanisms of adhesion by pathogens.**
(a) Fimbriae (F), minute bristlelike appendages. **(b)** Adherent
extracellular capsules (C) made of slime or other sticky substances.
(c) Viral envelope spikes (S). See table 13.6 for specific examples.

TABLE 13.6	Adhesion Properties of Microbes	
Microbe	**Disease**	**Adhesion Mechanism**
Neisseria gonorrhoeae	Gonorrhea	Fimbriae attach to genital epithelium
Escherichia coli	Diarrhea	Well-developed fimbrial adhesin
Shigella	Dysentery	Fimbriae can attach to intestinal epithelium
Vibrio	Cholera	Glycocalyx anchors microbe to intestinal epithelium
Treponema	Syphilis	Tapered hook embeds in host cell
Mycoplasma	Pneumonia	Specialized tip at ends of bacteria fuse tightly to lung epithelium
Pseudomonas aeruginosa	Burn, lung infections	Fimbriae and slime layer
Streptococcus pyogenes	Pharyngitis, impetigo	Lipotechoic acid and capsule anchor cocci to epithelium
Streptococcus mutants, *S. sobrinus*	Dental caries	Dextran slime layer glues cocci to tooth surface
Influenza virus	Influenza	Viral spikes react with receptor on cell surface
Poliovirus	Polio	Capsid proteins attach to receptors on susceptible cells
HIV	AIDS	Viral spikes adhere to white blood cell receptors
Giardia lamblia (protozoan)	Giardiasis	Small suction disc on underside attaches to intestinal surface

methods of various microbes and the diseases they lead to
are shown in **table 13.6.** Firm attachment to host tissues is
almost always a prerequisite for causing disease since the
body has so many mechanisms for flushing microbes and
foreign materials from its tissues.

Becoming Established: Step Three—Surviving Host Defenses

Microbes that aren't established in a normal flora relationship in a particular body site in a host are likely to encounter resistance from host defenses when first entering,
especially from certain white blood cells called **phagocytes.** These cells ordinarily engulf and destroy pathogens
by means of enzymes and antimicrobial chemicals (see
chapter 14).

Antiphagocytic factors are a type of virulence factor used by
some pathogens to avoid phagocytes. The antiphagocytic factors of resistant microorganisms help them to circumvent
some part of the phagocytic process (see **figure 13.6c**). The
most aggressive strategy involves bacteria that kill phagocytes outright. Species of both *Streptococcus* and *Staphylococcus* produce **leukocidins,** substances that are toxic to
white blood cells. Some microorganisms secrete an extracel-

lular surface layer (slime or capsule) that makes it physically
difficult for the phagocyte to engulf them. *Streptococcus pneumoniae, Salmonella typhi, Neisseria meningitidis,* and *Cryptococcus neoformans* are notable examples. Some bacteria are well
adapted to survival inside phagocytes after ingestion. For instance, pathogenic species of *Legionella, Mycobacterium,* and
many rickettsias are readily engulfed but are capable of
avoiding further destruction. The ability to survive intracellularly in phagocytes has special significance because it provides a place for the microbes to hide, grow, and be spread
throughout the body.

✓ CHECKPOINT

- Microbial infections result when a microorganism penetrates
host defenses, multiplies, and damages host tissue. The **pathogenicity** of a microbe refers to its ability to cause infection
or disease. The *virulence* of a pathogen refers to the degree of
damage it inflicts on the host tissues.

continued

- *True pathogens* cause infectious disease in healthy hosts, whereas *opportunistic pathogens* become infectious only when the host immune system is compromised in some way.
- The site at which a microorganism first contacts host tissue is called the *portal of entry*. Most pathogens have one preferred portal of entry, although some have more than one.
- The respiratory system is the portal of entry for the greatest number of pathogens.
- The *infectious dose*, or ID, refers to the minimum number of microbial cells required to initiate infection in the host. The ID varies widely among microbial species.
- Fimbriae, hooks, and adhesive capsules are types of adherence factors by which pathogens physically attach to host tissues.
- Antiphagocytic factors produced by microorganisms include leukocidins, capsules, and factors that resist digestion by white blood cells.

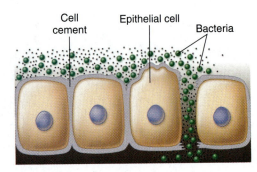

(a) **Exoenzymes**

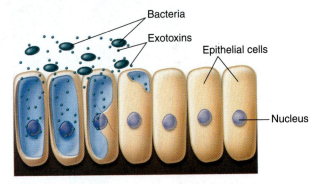

(b) **Toxins**

Causing Disease

How Virulence Factors Contribute to Tissue Damage

Virulence factors from a microbe's perspective are simply adaptations it uses to invade and establish itself in the host. These same factors determine the degree of tissue damage that occurs. The effects of a pathogen's virulence factors on tissues vary greatly. Cold viruses, for example, invade and multiply but cause relatively little damage to their host. At the other end of the spectrum, pathogens such as *Clostridium tetani* or HIV severely damage or kill their host. Microorganisms either inflict direct damage on hosts through the use of exoenzymes or toxins **(figure 13.6a,b),** or they cause damage indirectly when their presence causes an excessive or inappropriate host response **(figure 13.6c).** For convenience, we will divide the "directly damaging" virulence factors into exoenzymes and toxins. Although this distinction is useful, there is often a very fine line between enzymes and toxins because many substances called toxins actually function as enzymes.

Microbial virulence factors are often responsible for inducing the host to cause damage, as well. The capsule of *Streptococcus pneumoniae* is a good example. Its presence prevents the bacterium from being cleared from the lungs by phagocytic cells, leading to a continuous influx of fluids into the lung spaces, and the condition we know as pneumonia.

Extracellular Enzymes Many pathogenic bacteria, fungi, protozoa, and worms secrete **exoenzymes** that break down and inflict damage on tissues. Other enzymes dissolve the host's defense barriers and promote the spread of microbes to deeper tissues.

Examples of enzymes are:

1. mucinase, which digests the protective coating on mucous membranes and is a factor in amoebic dysentery;
2. keratinase, which digests the principal component of skin and hair, and is secreted by fungi that cause ringworm;

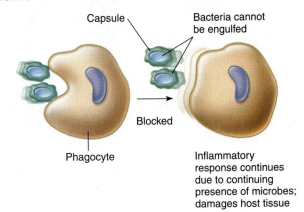

(c) **Induction of host response**

FIGURE 13.6 **Three ways microbes damage the host.** **(a)** Exoenzymes. Bacteria produce extracellular enzymes that dissolve extracellular barriers and penetrate through or between cells to underlying tissues. **(b)** Toxins (primarily exotoxins) secreted by bacteria diffuse to target cells, which are poisoned and disrupted. **(c)** Bacterium has a property that enables it to escape phagocytosis and remain as an "irritant" to host defenses, which are deployed excessively.

3. collagenase, which digests the principal fiber of connective tissue and is an invasive factor of *Clostridium* species and certain worms; and
4. hyaluronidase, which digests hyaluronic acid, the ground substance that cements animal cells together. This enzyme is an important virulence factor in staphylococci, clostridia, streptococci, and pneumococci.

Some enzymes react with components of the blood. Coagulase, an enzyme produced by pathogenic staphylococci, causes clotting of blood or plasma. By contrast, the bacterial

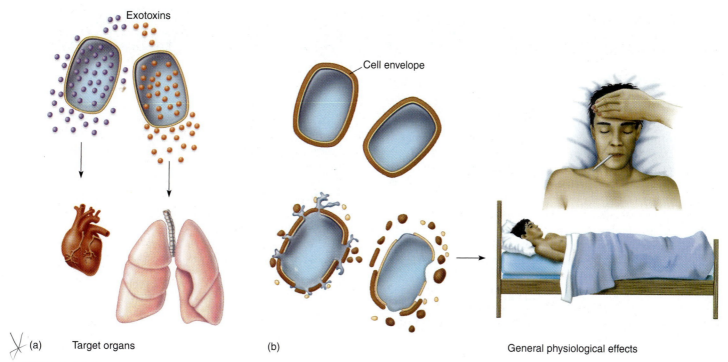

Exotoxins

Cell envelope

(a)   Target organs

(b)

General physiological effects

FIGURE 13.7 **The origins and effects of circulating exotoxins and endotoxin.**
(a) Exotoxins, given off by live cells, have highly specific targets and physiological effects. **(b)** Endotoxin, given off when the cell wall of gram-negative bacteria disintegrates, has more generalized physiological effects.

kinases (streptokinase, staphylokinase) do just the opposite, dissolving fibrin clots and expediting the invasion of damaged tissues. In fact, one form of streptokinase (streptase) is marketed as a therapy to dissolve blood clots in patients with problems with thrombi and emboli.[3]

Bacterial Toxins: A Potent Source of Cellular Damage   A **toxin** is a specific chemical product of microbes, plants, and some animals that is poisonous to other organisms. **Toxigenicity,** the power to produce toxins, is a genetically controlled characteristic of many species and is responsible for the adverse effects of a variety of diseases generally called **toxinoses.** Toxinoses in which the toxin is spread by the blood from the site of infection are called **toxemias** (tetanus and diphtheria, for example), whereas those caused by ingestion of toxins are **intoxications** (botulism). A toxin is named according to its specific target of action: Neurotoxins act on the nervous system; enterotoxins act on the intestine; hemotoxins lyse red blood cells; and nephrotoxins damage the kidneys.

   A more traditional scheme classifies toxins according to their origins **(figure 13.7).** A toxin molecule secreted by a living bacterial cell into the infected tissues is an **exotoxin.** A toxin that is not secreted but is released only after the cell is damaged or lysed is an **endotoxin.** Other important differences between the two groups are summarized in **table 13.7.**

TABLE 13.7	Differential Characteristics of Bacterial Exotoxins and Endotoxin	
Characteristic	**Exotoxins**	**Endotoxin**
Toxicity	Toxic in minute amounts	Toxic in high doses
Effects on the Body	Specific to a cell type (blood, liver, nerve)	Systemic: fever, inflammation
Chemical Composition	Small proteins	Lipopolysaccharide of cell wall
Heat Denaturation at 60°C	Unstable	Stable
Toxoid Formation	Can be converted to toxoid*	Cannot be converted to toxoid
Immune Response	Stimulate antitoxins**	Does not stimulate antitoxins
Fever Stimulation	Usually not	Yes
Manner of Release	Secreted from live cell	Released by cell during lysis
Typical Sources	A few gram-positive and gram-negative	All gram-negative bacteria

*A toxoid is an inactivated toxin used in vaccines.

**An antitoxin is an antibody that reacts specifically with a toxin.

---

3. These conditions are intravascular blood clots that can cause circulatory obstructions.

The Classic Stages of Clinical Infections

As the body of the host responds to the invasive and toxigenic activities of a parasite, it passes through four distinct phases of infection and disease: the incubation period, the prodrome, the period of invasion, and the convalescent period.

The **incubation period** is the time from initial contact with the infectious agent (at the portal of entry) to the appearance of the first symptoms. During the incubation period, the agent is multiplying at the portal of entry but has not yet caused enough damage to elicit symptoms. Although this period is relatively well defined and predictable for each microorganism, it does vary according to host resistance, degree of virulence, and distance between the target organ and the portal of entry (the farther apart, the longer the incubation period). Overall, an incubation period can range from several hours in pneumonic plague to several years in leprosy. The majority of infections, however, have incubation periods ranging between 2 and 30 days.

The earliest notable symptoms of infection appear as a vague feeling of discomfort, such as head and muscle aches, fatigue, upset stomach, and general malaise. This short period (1–2 days) is known as the **prodromal stage.** The infectious agent next enters a **period of invasion,** during which it multiplies at high levels, exhibits its greatest toxicity, and becomes well established in its target tissue. This period is often marked by fever and other prominent and more specific signs and symptoms, which can include cough, rashes, diarrhea, loss of muscle control, swelling, jaundice, discharge of exudates, or severe pain, depending on the particular infection. The length of this period is extremely variable.

As the patient begins to respond to the infection, the symptoms decline—sometimes dramatically, other times slowly. During the recovery that follows, called the **convalescent period,** the patient's strength and health gradually return owing to the healing nature of the immune response. An infection that results in death is called terminal.

The transmissibility of the microbe during these four stages must be considered on an individual basis. A few agents are released mostly during incubation (measles, for example); many are released during the invasive period (*Shigella*); and others can be transmitted during all of these periods (hepatitis B).

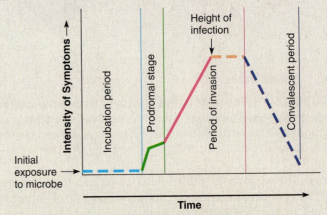

Stages in the course of infection and disease: Dashed lines represent periods with a variable length.

Exotoxins are proteins with a strong specificity for a target cell and extremely powerful, sometimes deadly, effects. They generally affect cells by damaging the cell membrane and initiating lysis or by disrupting intracellular function. **Hemolysins** (hee-mahl'-uh-sinz) are a class of bacterial exotoxin that disrupts the cell membrane of red blood cells (and some other cells too). This damage causes the red blood cells to **hemolyze**—to burst and release hemoglobin pigment. Hemolysins that increase pathogenicity include the streptolysins of *Streptococcus pyogenes* and the alpha (α) and beta (β) toxins of *Staphylococcus aureus.* When colonies of bacteria growing on blood agar produce hemolysin, distinct zones appear around the colony. The pattern of hemolysis is often used to identify bacteria and determine their degree of pathogenicity.

The exotoxins of diphtheria, tetanus, and botulism, among others, attach to a particular target cell, become internalized, and interrupt an essential cell pathway. The consequences of cell disruption depend upon the target. One toxin of *Clostridium tetani* blocks the action of certain spinal neu-

rons; the toxin of *Clostridium botulinum* prevents the transmission of nerve-muscle stimuli; pertussis toxin inactivates the respiratory cilia; and cholera toxin provokes profuse salt and water loss from intestinal cells. More details of the pathology of exotoxins are found in later chapters on specific diseases.

In contrast to the category *exotoxin,* which contains many specific examples, the word *endotoxin* refers to a single substance. Endotoxin is actually a chemical called lipopolysaccharide (LPS), which is part of the outer membrane of gram-negative cell walls. When gram-negative bacteria cause infections, some of them eventually lyse and release these LPS molecules into the infection site or into the circulation. Endotoxin differs from exotoxins in having a variety of systemic effects on tissues and organs. Depending upon the amounts present, endotoxin can cause fever, inflammation, hemorrhage, and diarrhea. Blood infection by gram-negative bacteria such as *Salmonella, Shigella, Neisseria meningitidis,* and *Escherichia coli* are particularly dangerous, in that it can lead to fatal endotoxic shock.

Inducing an injurious host response   Despite the extensive discussion on direct virulence factors, such as enzymes and toxins, it is probably the case that more microbial diseases are the result of indirect damage, or the host's excessive or inappropriate response to a microorganism. This is an extremely important point, since it means that pathogenicity is not a trait inherent in microorganisms, but is really a consequence of the interplay between microbe and host.

Of course, it is easier to study and characterize the microbes that cause direct damage through toxins or enzymes. For this reason, these true pathogens were the first to be fully understood as the science of microbiology progressed. But in the last 15 to 20 years, microbiologists have come to appreciate exactly how important the relationship between microbe and host is, and this has greatly expanded our understanding of infectious diseases.

The various ways in which a microbe/host interaction causes damage will be described in the chapters describing specific diseases (chapters 18–23).

The Process of Infection and Disease

Establishment, Spread, and Pathologic Effects

Aided by virulence factors, microbes eventually settle in a particular target organ and continue to cause damage at the site. The type and scope of injuries inflicted during this process account for the typical stages of an infection **(Insight 13.3),** the patterns of the infectious disease, and its manifestations in the body.

In addition to the adverse effects of enzymes, toxins, and other factors, multiplication by a pathogen frequently weakens host tissues. Pathogens can obstruct tubular structures such as blood vessels, lymphatic channels, fallopian tubes, and bile ducts. Accumulated damage can lead to cell and tissue death, a condition called **necrosis.** Although viruses do not produce toxins or destructive enzymes, they destroy cells by multiplying in and lysing them. Many of the cytopathic effects of viral infection arise from the impaired metabolism and death of cells (see chapter 6).

Patterns of Infection   Patterns of infection are many and varied. In the simplest situation, a **localized infection,** the microbe enters the body and remains confined to a specific tissue **(figure 13.8a).** Examples of localized infections are boils, fungal skin infections, and warts.

Many infectious agents do not remain localized but spread from the initial site of entry to other tissues. In fact, spreading is necessary for pathogens such as rabies and hepatitis A virus, whose target tissue is some distance from the site of entry. The rabies virus travels from a bite wound along nerve tracts to its target in the brain, and the hepatitis A virus moves from the intestine to the liver via the circulatory system.

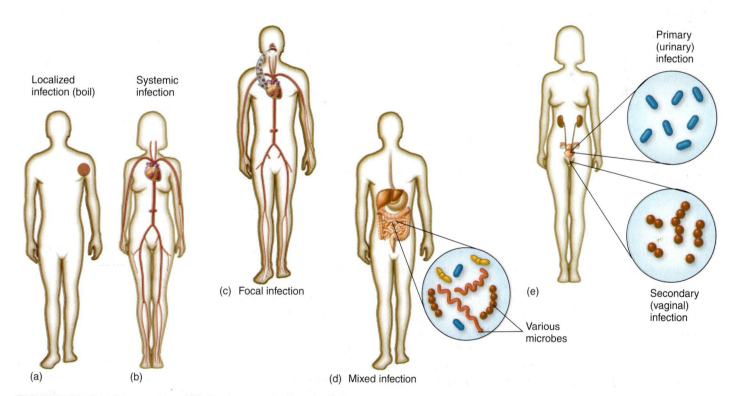

FIGURE 13.8   **The occurrence of infections with regard to location, type of microbe, and length of time.**
(a) A localized infection, in which the pathogen is restricted to one specific site. **(b)** Systemic infection, in which the pathogen spreads through circulation to many sites. **(c)** A focal infection occurs initially as a local infection, but circumstances cause the microbe to be carried to other sites systemically. **(d)** A mixed infection, in which the same site is infected with several microbes at the same time. **(e)** In a primary-secondary infection, an initial infection is complicated by a second one in the same or a different location and caused by a different microbe.

INSIGHT 13.4   *Medical*

A Quick Guide to the Terminology of Infection and Disease

Words in medicine have great power and economy. A single technical term can often replace a whole phrase or sentence, thereby saving time and space in patient charting. The beginning student may feel overwhelmed by what seems like a mountain of new words. However, having a grasp of a few root words and a fair amount of anatomy can help you learn many of these words and even deduce the meaning of unfamiliar ones. Some examples of medical shorthand follow.

The suffix *-itis* means an inflammation and, when affixed to the end of an anatomical term, indicates an inflammatory condition in that location. Thus, meningitis is an inflammation of the meninges surrounding the brain; encephalitis is an inflammation of the brain itself; hepatitis involves the liver; vaginitis, the vagina; gastroenteritis, the intestine; and otitis media, the middle ear. Although not all inflammatory conditions are caused by infections, many infectious diseases inflame their target organs.

The suffix *-emia* is derived from the Greek word *haeima*, meaning blood. When added to a word, it means "associated with the blood." Thus, septicemia means sepsis (infection) of the blood; bacteremia, bacteria in the blood; viremia, viruses in the blood; and fungemia, fungi in the blood. It is also applicable to specific conditions such as toxemia, gonococcemia, and spirochetemia.

The suffix *-osis* means "a disease or morbid process." It is frequently added to the names of pathogens to indicate the disease they cause: for example, listeriosis, histoplasmosis, toxoplasmosis, shigellosis, salmonellosis, and borreliosis. A variation of this suffix is *-iasis*, as in trichomoniasis and candidiasis.

The suffix *-oma* comes from the Greek word *onkomas* (swelling) and means tumor. Although the root is often used to describe cancers (sarcoma, melanoma), it is also applied in some infectious diseases that cause masses or swellings (tuberculoma, leproma).

When an infection spreads to several sites and tissue fluids, usually in the bloodstream, it is called a **systemic infection (figure 13.8b).** Examples of systemic infections are viral diseases (measles, rubella, chickenpox, and AIDS); bacterial diseases (brucellosis, anthrax, typhoid fever, and syphilis); and fungal diseases (histoplasmosis and cryptococcosis). Infectious agents can also travel to their targets by means of nerves (as in rabies) or cerebrospinal fluid (as in meningitis).

A **focal infection** is said to exist when the infectious agent breaks loose from a local infection and is carried into other tissues **(figure 13.8c).** This pattern is exhibited by tuberculosis or by streptococcal pharyngitis, which gives rise to scarlet fever. In the condition called toxemia,[4] the infection itself remains localized at the portal of entry, but the toxins produced by the pathogens are carried by the blood to the actual target tissue. In this way, the target of the bacterial cells can be different from the target of their toxin.

An infection is not always caused by a single microbe. In a **mixed infection,** several agents establish themselves simultaneously at the infection site **(figure 13.8d).** In some mixed or synergistic infections, the microbes cooperate in breaking down a tissue. In other mixed infections, one microbe creates an environment that enables another microbe to invade. Gas gangrene, wound infections, dental caries, and human bite infections tend to be mixed. These are sometimes called **polymicrobial** diseases.

Some diseases are described according to a sequence of infection. When an initial, or **primary, infection** is compli-

cated by another infection caused by a different microbe, the second infection is termed a **secondary infection (figure 13.8e).** This pattern often occurs in a child with chickenpox (primary infection) who may scratch his pox and infect them with *Staphylococcus aureus* (secondary infection). The secondary infection need not be in the same site as the primary infection, and it usually indicates altered host defenses.

Infections that come on rapidly, with severe but short-lived effects, are called **acute infections.** Infections that progress and persist over a long period of time are **chronic infections. Insight 13.4** illustrates other common terminology used to describe infectious diseases.

Signs and Symptoms: Warning Signals of Disease

When an infection causes pathologic changes leading to disease, it is often accompanied by a variety of signs and symptoms. A **sign** is any objective evidence of disease as noted by an observer; a **symptom** is the subjective evidence of disease as sensed by the patient. In general, signs are more precise than symptoms, though both can have the same underlying cause. For example, an infection of the brain might present with the sign of bacteria in the spinal fluid and symptom of headache. Or a streptococcal infection might produce a sore throat (symptom) and inflamed pharynx (sign). Disease indicators that can be sensed and observed can qualify as either a sign or a symptom. When a disease can be identified or defined by a certain complex of signs and symptoms, it is termed a **syndrome.** Signs and symptoms with considerable importance in diagnosing

---

4. Not to be confused with toxemia of pregnancy, which is a metabolic disturbance and not an infection.

infectious diseases are shown in **table 13.8.** Specific signs and symptoms for particular infectious diseases are covered in chapters 18 through 23.

Signs and Symptoms of Inflammation

The earliest symptoms of disease result from the activation of the body defense process called **inflammation.** The inflammatory response includes cells and chemicals that respond nonspecifically to disruptions in the tissue. This subject is discussed in greater detail in chapter 14, but as noted earlier, many signs and symptoms of infection are caused by the mobilization of this system. Some common symptoms of inflammation include fever, pain, soreness, and swelling. Signs of inflammation include **edema,** the accumulation of fluid in an afflicted tissue; **granulomas** and **abscesses,** walled-off collections of inflammatory cells and microbes in the tissues; and **lymphadenitis,** swollen lymph nodes.

Rashes and other skin eruptions are common symptoms and signs in many diseases, and because they tend to mimic each other, it can be difficult to differentiate among diseases on this basis alone. The general term for the site of infection or disease is **lesion.** Skin lesions can be restricted to the epidermis and its glands and follicles, or they can extend into the dermis and subcutaneous regions. The lesions of some infections undergo characteristic changes in appearance during the course of disease and thus fit more than one category. (See Insight 18.3.)

Signs of Infection in the Blood

Changes in the number of circulating white blood cells, as determined by special counts, are considered to be signs of possible infection. **Leukocytosis** (loo″-koh′-sy-toh′-sis) is an increase in the level of white blood cells, whereas **leukopenia** (loo″-koh-pee′-nee-uh) is a decrease. Other signs of infection revolve around the occurrence of a microbe or its products in the blood. The clinical term for blood infection, **septicemia,** refers to a general state in which microorganisms are multiplying in the blood and are present in large numbers. When small numbers of bacteria or viruses are found in the blood, the correct terminology is **bacteremia** or **viremia,** which means that these microbes are present in the blood but are not necessarily multiplying.

During infection, a normal host will invariably show signs of an immune response in the form of antibodies in the serum or some type of sensitivity to the microbe. This fact is the basis for several serological tests used in diagnosing infectious diseases such as AIDS or syphilis. Such specific immune reactions indicate the body's attempt to develop specific immunities against pathogens. We will concentrate on this role of the host defenses in chapters 14 and 15.

Infections That Go Unnoticed

It is rather common for an infection to produce no noticeable symptoms, even though the microbe is active in the host tissue. In other words, although infected, the host

TABLE 13.8	Common Signs and Symptoms of Infectious Diseases
Signs	**Symptoms**
Fever	Chills
Septicemia	Pain, ache, soreness, irritation
Microbes in tissue fluids	Nausea
Chest sounds	Malaise, fatigue
Skin eruptions	Chest tightness
Leukocytosis	Itching
Leukopenia	Headache
Swollen lymph nodes	Nausea
Abscesses	Abdominal cramps
Tachycardia (increased heart rate)	Anorexia (lack of appetite)
Antibodies in serum	Sore throat

does not manifest the disease. Infections of this nature are known as **asymptomatic, subclinical,** or *inapparent* because the patient experiences no symptoms or disease and does not seek medical attention. However, it is important to note that most infections are attended by some sort of sign. In the section on epidemiology, we further address the significance of subclinical infections in the transmission of infectious agents.

The Portal of Exit: Vacating the Host

Earlier, we introduced the idea that a parasite is considered *unsuccessful* if it does not have a provision for leaving its host and moving to other susceptible hosts. With few exceptions, pathogens depart by a specific avenue called the **portal of exit (figure 13.9).** In most cases, the pathogen is shed or released from the body through secretion, excretion, discharge, or sloughed tissue. The usually very high number of infectious agents in these materials increases both virulence and the likelihood that the pathogen will reach other hosts. In many cases, the portal of exit is the same as the portal of entry, but a few pathogens use a different route. As we see in the next section, the portal of exit concerns epidemiologists because it greatly influences the dissemination of infection in a population.

Respiratory and Salivary Portals

Mucus, sputum, nasal drainage, and other moist secretions are the media of escape for the pathogens that infect the lower or upper respiratory tract. The most effective means of releasing these secretions are coughing and sneezing (see figure 13.15), although they can also be released during talking and laughing. Tiny particles of liquid released into the air form aerosols or droplets that can spread the infectious agent to other people. The agents of tuberculosis, influenza, measles, and chickenpox most often leave the host through airborne droplets. Droplets of saliva are the exit route for several viruses, including those of mumps, rabies, and infectious mononucleosis.

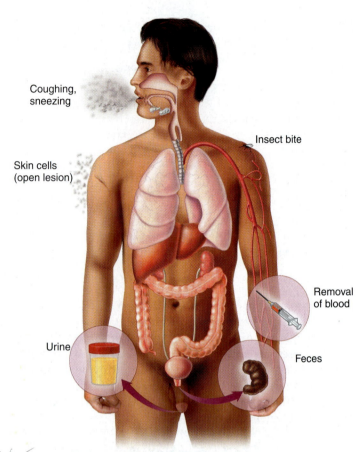

FIGURE 13.9 Major portals of exit of infectious diseases.

Skin Scales

The outer layer of the skin and scalp are constantly being shed into the environment. A large proportion of household dust is actually composed of skin cells. A single person can shed several billion skin cells a day, and some persons, called shedders, disseminate massive numbers of bacteria into their immediate surroundings. Skin lesions and their exudates can serve as portals of exit in warts, fungal infections, boils, herpes simplex, smallpox, and syphilis.

Fecal Exit

Feces is a very common portal of exit. Some intestinal pathogens grow in the intestinal mucosa and create an inflammation that increases the motility of the bowel. This increased motility speeds up peristalsis, resulting in diarrhea, and the more fluid stool provides a rapid exit for the pathogen. A number of helminth worms release cysts and eggs through the feces (see chapter 22). Feces containing pathogens are a public health problem when allowed to contaminate drinking water or when used to fertilize crops.

Urogenital Tract

A number of agents involved in sexually transmitted infections leave the host in vaginal discharge or semen. This is also the source of neonatal infections such as herpes simplex, *Chlamydia*, and *Candida albicans*, which infect the infant as it passes through the birth canal. Less commonly, certain

pathogens that infect the kidney are discharged in the urine: for instance, the agents of leptospirosis, typhoid fever, tuberculosis, and schistosomiasis.

Removal of Blood or Bleeding

Although the blood does not have a direct route to the outside, it can serve as a portal of exit when it is removed or released through a vascular puncture made by natural or artificial means. Blood-feeding animals such as ticks and fleas are common transmitters of pathogens (see Insight 20.2). The AIDS and hepatitis viruses are transmitted by shared needles or through small gashes in a mucous membrane caused by sexual intercourse. Blood donation is also a means for certain microbes to leave the host, though this means of exit is now unusual because of close monitoring of the donor population and blood used for transfusions.

The Persistence of Microbes and Pathologic Conditions

The apparent recovery of the host does not always mean that the microbe has been completely removed or destroyed by the host defenses. After the initial symptoms in certain chronic infectious diseases, the infectious agent retreats into a dormant state called **latency.** Throughout this latent state, the microbe can periodically become active and produce a recurrent disease. The viral agents of herpes simplex, herpes zoster, hepatitis B, AIDS, and Epstein-Barr can persist in the host for long periods. The agents of syphilis, typhoid fever, tuberculosis, and malaria also enter into latent stages. The person harboring a persistent infectious agent may or may not shed it during the latent stage. If it is shed, such persons are chronic carriers who serve as sources of infection for the rest of the population.

Some diseases leave **sequelae** in the form of long-term or permanent damage to tissues or organs. For example, meningitis can result in deafness, a strep throat can lead to rheumatic heart disease, Lyme disease can cause arthritis, and polio can produce paralysis.

- Infectious diseases that are asymptomatic or subclinical nevertheless often produce clinical signs.
- The portal of exit by which a pathogen leaves its host is often but not always the same as the portal of entry.
- The portals of exit and entry determine how pathogens spread in a population.
- Some pathogens persist in the body in a latent state; others cause long-term diseases called *sequelae.*

13.3 Epidemiology: The Study of Disease in Populations

So far, our discussion has revolved primarily around the impact of an infectious disease in a single individual. Let us now turn our attention to the effects of diseases on the community—the realm of **epidemiology.** By definition, this term involves the study of the frequency and distribution of disease and other health-related factors in defined human populations. It involves many disciplines—not only microbiology, but anatomy, physiology, immunology, medicine, psychology, sociology, ecology, and statistics—and it considers many diseases other than infectious ones, including heart disease, cancer, drug addiction, and mental illness. The epidemiologist is a medical sleuth who collects clues on the causative agent, pathology, sources and modes of transmission, and tracks the numbers and distribution of cases of disease in the community. In fulfilling these demands, the epidemiologist asks who, when, where, how, why, and what about diseases. The outcome of these studies helps public health departments develop prevention and treatment programs and establish a basis for predictions.

Who, When, and Where? Tracking Disease in the Population

Epidemiologists are concerned with all of the factors covered earlier in this chapter: virulence, portals of entry and exit, and the course of disease. But they are also interested in surveillance—that is, collecting, analyzing, and reporting data on the rates of occurrence, mortality, morbidity, and transmission of infections. Surveillance involves keeping data for a large number of diseases seen by the medical community and reported to public health authorities. By law, certain **reportable,** or notifiable, diseases must be reported to authorities; others are reported on a voluntary basis.

A well-developed network of individuals and agencies at the local, district, state, national, and international levels keeps track of infectious diseases. Physicians and hospitals report all notifiable diseases that are brought to their attention. Case reporting can focus on a single individual or collectively on group data.

Local public health agencies first receive the case data and determine how they will be handled. In most cases, health officers investigate the history and movements of patients to trace their prior contacts and to control the further spread of the infection as soon as possible through drug therapy, immunization, and education. In sexually transmitted diseases, patients are asked to name their partners so that these persons can be notified, examined, and treated. It is very important to maintain the confidentiality of the persons in these reports. The principal government agency responsible for keeping track of infectious diseases nationwide is the Centers for Disease Control and Prevention (CDC) in Atlanta, Georgia, which is a part of the United States Public Health Service. The CDC publishes a weekly notice of diseases (the *Morbidity and Mortality Report*) that provides weekly and cumulative summaries of the case rates and deaths for about 50 notifiable diseases, highlights important and unusual diseases, and presents data concerning disease occurrence in the major regions of the United States. It is available to anyone at www.cdc.gov/mmwr/. Ultimately, the CDC shares its statistics on disease with the World Health Organization (WHO) for worldwide tabulation and control.

Epidemiologic Statistics: Frequency of Cases

The **prevalence** of a disease is the total number of existing cases with respect to the entire population. It is often thought of as a snapshot, and is usually reported as the percentage of the population having a particular disease at any given time. Disease **incidence** measures the number of new cases over a certain time period, as compared with the general healthy population. This statistic, also called the case, or morbidity, rate, indicates both the rate and the risk of infection. The equations used to figure these rates are:

$$\text{Prevalence} = \frac{\text{Total number of cases in population}}{\text{Total number of persons in population}} \times 100 = \%$$

$$\text{Incidence} = \frac{\text{Number of new cases}}{\text{Total number of susceptible persons}} \quad \text{(Usually reported per 100,000 persons)}$$

As an example, let us use a classroom of 50 students exposed to a new strain of influenza. Before exposure, the prevalence and incidence in this population are both zero (0/50). If in one week, 5 out of the 50 people contract the disease, the prevalence is 5/50 = 10%, and the incidence is 1 in 10 (10 is used instead of 100,000 due to smaller population). If after 1 week, 5 more students contract the flu, the prevalence becomes 5 + 5 = 10/50 = 20%, and the incidence becomes 2 in 10. The 20% figure for prevalence assumes that all 10 patients were still sick in the second week. If the first 5 patients were well by the time the second group of five became ill, the prevalence at that second time point would be 5/50, or 10%. But the incidence for the 2-week period would still be 2 in 10 (10 cases per 50).

The changes in incidence and prevalence are usually followed over a seasonal, yearly, and long-term basis and are helpful in predicting trends **(figure 13.10).** Statistics of concern to the epidemiologist are the rates of disease with regard to sex, race, or geographic region. Also of importance is the

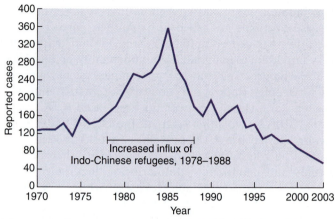

(a) **Leprosy — Reported cases by year, United States, 1970–2003**

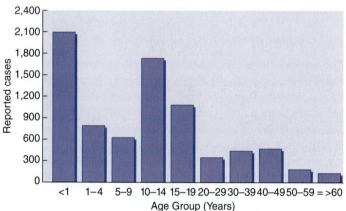

(b) **Pertussis — Reported cases by age group, United States, 2000**

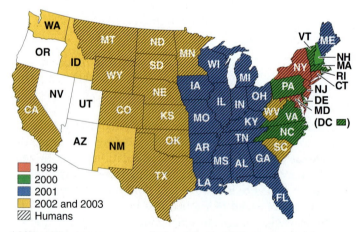

(c) West Nile virus cases were first reported in 1999 among a variety of animals and humans. Colors track its rapid progress across the United States in just four years.

FIGURE 13.10 **Graphical representation of epidemiological data.**
The Centers for Disease Control and Prevention collects epidemiological data which is analyzed with regard to **(a)** time frame, **(b)** age, and **(c)** geographic region. See also the inside back cover.
Source: Centers for Disease Control and Prevention. *Morbidity and Mortality Weekly Report.* United States, 2003.

mortality rate, which measures the total number of deaths in a population due to a certain disease. Over the past century, the overall death rate from infectious diseases has dropped, although the number of persons afflicted with infectious diseases (the **morbidity rate**) has remained relatively high.

Monitoring statistics also makes it possible to define the frequency of a disease in the population. An infectious disease that exhibits a relatively steady frequency over a long time period in a particular geographic locale is **endemic** (**figure 13.11***a*). For example, Lyme disease is endemic to certain areas of the United States where the tick vector is found. A certain number of new cases are expected in these areas every year. When a disease is **sporadic,** occasional cases are reported at irregular intervals in random locales (**figure 13.11***b*). Tetanus and diphtheria are reported sporadically in the United States (fewer than 50 cases a year).

When statistics indicate that the prevalence of an endemic or sporadic disease is increasing beyond what is expected for that population, the pattern is described as an **epidemic** (**figure 13.11***c*). An epidemic exists when an increasing trend is observed in a particular population. The time period is not defined—it can range from hours in food poisoning to years in syphilis—nor is an exact percentage of increase needed before an outbreak can qualify as an epidemic. Several epidemics occur every year in the United States, most recently among STDs such as chlamydia and gonorrhea. The spread of an epidemic across continents is a **pandemic,** as exemplified by AIDS and influenza (**figure 13.11***d*).

One important epidemiologic truism might be called the "iceberg effect," which refers to the fact that only a small portion of an iceberg is visible above the surface of the ocean, with a much more massive part lingering unseen below the surface. Regardless of case reporting and public health screening, a large number of cases of infection in the community go undiagnosed and unreported. (For a list of reportable diseases in the United States, see **table 13.9.**) In the instance of salmonellosis, approximately 40,000 cases are reported each year. Epidemiologists estimate that the actual number is more likely somewhere between 400,000 and 4,000,000. The iceberg effect can be even more lopsided for sexually transmitted diseases or for infections that are not brought to the attention of reporting agencies.

Investigative Strategies of the Epidemiologist

Initial evidence of a new disease or an epidemic in the community is fragmentary. A few sporadic cases are seen by physicians and are eventually reported to authorities. However, it can take several reports before any alarm is registered. Epidemiologists and public health departments must piece together odds and ends of data from a series of apparently unrelated cases and work backward to reconstruct the epidemic pattern. A completely new disease requires even greater preliminary investigation, because the infectious agent must be isolated and linked directly to the disease (see the discussion of Koch's postulates in a later section of this

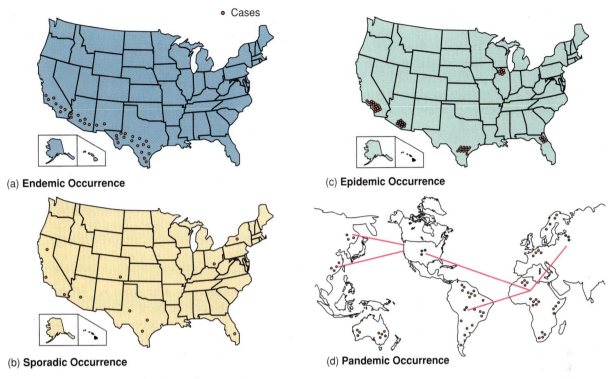

• Cases

(a) **Endemic Occurrence**

(c) **Epidemic Occurrence**

(b) **Sporadic Occurrence**

(d) **Pandemic Occurrence**

FIGURE 13.11 Patterns of infectious disease occurrence.

(a) In endemic occurrence, cases are concentrated in one area at a relatively stable rate. (b) In sporadic occurrence, a few cases occur randomly over a wide area. (c) An epidemic is an increased number of cases that often appear in geographic clusters. The clusters may be local, as in the case of a restaurant-related food-borne epidemic, or nationwide, as is the case with *Chlamydia*. (d) Pandemic occurrence means that an epidemic ranges over more that one continent.

TABLE 13.9	**Reportable Diseases in the United States***

- Acquired immunodeficiency syndrome (AIDS)
- Anthrax
- Botulism
- Brucellosis
- Chancroid
- *Chlamydia trachomatis* genital infections
- Cholera
- Coccidioidomycosis
- Cryptosporidiosis
- Cyclosporiasis
- Diphtheria
- Ehrlichiosis
- Encephalitis/meningitis, arboviral
  - Encephalitis/meningitis, California serogroup viral
  - Encephalitis/meningitis, eastern equine
  - Encephalitis/meningitis, Powassan
  - Encephalitis/meningitis, St. Louis

- Encephalitis/meningitis, western equine
- Encephalitis/meningitis, West Nile
- Enterohemorrhagic *Escherichia coli*
- Giardiasis
- Gonorrhea
- *Haemophilus influenzae* invasive disease
- Hansen's disease (leprosy)
- Hantavirus pulmonary syndrome
- Hemolytic uremic syndrome
- Hepatitis, viral, acute
  - Hepatitis A, acute
  - Hepatitis B, acute
  - Hepatitis B virus, perinatal infection
  - Hepatitis C, acute
- Hepatitis, viral, chronic
  - Chronic hepatitis B
  - Hepatitis C virus infection (past or present)

- HIV infection
  - HIV infection, adult (>13 years)
  - HIV infection, pediatric (<13 years)
- Legionellosis
- Listeriosis
- Lyme disease
- Malaria
- Measles
- Meningococcal disease
- Mumps
- Pertussis
- Plague
- Poliomyelitis, paralytic
- Psittacosis
- Q fever
- Rabies
  - Rabies, animal
  - Rabies, human
- Rocky Mountain spotted fever
- Rubella
- Rubella, congenital syndrome
- Salmonellosis

- Severe acute respiratory syndrome–associated coronavirus (SARS-CoV) disease
- Shigellosis
- Streptococcal disease, invasive, group A
- Streptococcal toxic shock syndrome
- *Streptococcus pneumoniae*, drug-resistant, invasive disease
- *Streptococcus pneumoniae*, invasive in children <5 years
- Syphilis
- Syphilis, congenital
- Tetanus
- Toxic shock syndrome
- Trichinosis
- Tuberculosis
- Tularemia
- Typhoid fever
- Varicella
- Yellow fever

*Other diseases may be reportable to State Departments of Health.

Source: Centers for Disease Control and Prevention, 2003.

chapter). All factors possibly impinging on the disease are scrutinized. Investigators search for *clusters* of cases indicating spread between persons or a public (common) source of infection; they also look at possible contact with animals, contaminated food, water, and public facilities, and at human interrelationships or changes in community structure. Out of this maze of case information, the investigators hope to recognize a pattern that indicates the source of infection so that they can quickly move to control it.

Reservoirs: Where Pathogens Persist

In order for an infectious agent to continue to exist and be spread, it must have a permanent place to reside. The **reservoir** is the primary habitat in the natural world from which a pathogen originates. Often it is a human or animal carrier, although soil, water, and plants are also reservoirs. The reservoir can be distinguished from the infection **source**, which is the individual or object from which an infection is actually acquired. In diseases such as syphilis, the reservoir and the source are the same (the human body). In the case of hepatitis A, the reservoir (a human carrier) is usually different from the source of infection (contaminated food).

Living Reservoirs

Many pathogens continue to exist and spread because they are harbored by members of a host population. Persons or animals with frank symptomatic infection are obvious sources of infection, but a **carrier** is, by definition, an individual who *inconspicuously* shelters a pathogen and spreads it to others without any notice. Although human carriers are occasionally detected through routine screening (blood tests, cultures) and other epidemiologic devices, they are unfortunately very difficult to discover and control. As long as a pathogenic reservoir is maintained by the carrier state, the disease will continue to exist in that population, and the potential for epidemics will be a constant threat. The duration of the carrier state can be short- or long-term, and the carrier may or may not have experienced disease due to the microbe.

Several situations can produce the carrier state. **Asymptomatic** (apparently healthy) **carriers** are infected, but as previously indicated, they show no symptoms **(figure 13.12a)**. A few asymptomatic infections (gonorrhea and genital warts, for instance) can carry out their entire course without overt manifestations. **Figure 13.12b** demonstrates three types of carriers who have had or will have the disease but do not at the time they transmit the organism. *Incubation carriers* spread the infectious agent during the incubation period. For example, AIDS patients can harbor and spread the virus for months and years before their first symptoms appear. Recuperating patients without symptoms are considered *convalescent carriers* when they continue to shed viable microbes and convey the infection to others. Diphtheria patients, for example, spread the microbe for up to 30 days after the disease has subsided.

An individual who shelters the infectious agent for a long period after recovery because of the latency of the infectious agent is a *chronic carrier*. Patients who have recovered from tuberculosis, hepatitis, and herpes infections frequently carry the agent chronically. About one in 20 victims of typhoid fever continues to harbor *Salmonella typhi* in the gallbladder for several years, and sometimes for life. The most infamous of these was "Typhoid Mary," a cook who spread the infection to hundreds of victims in the early 1900s. (*Salmonella* infection is described in chapter 22.)

The **passive carrier** state is of great concern during patient care (see a later section on nosocomial infections). Medical and dental personnel who must constantly handle materials that are heavily contaminated with patient secretions and blood risk picking up pathogens mechanically and accidently transferring them to other patients **(figure 13.12c)**. Proper handwashing, handling of contaminated materials, and aseptic techniques greatly reduce this likelihood.

Animals As Reservoirs and Sources Up to now, we have lumped animals with humans in discussing living reservoirs or carriers, but animals deserve special consideration as vectors of infections. The word **vector** is used by epidemiologists to indicate a live animal that transmits an infectious agent from one host to another. (The term is sometimes misused to include any object that spreads disease.) The majority of vectors are arthropods such as fleas, mosquitoes, flies, and ticks, although larger animals can also spread infection—for example, mammals (rabies), birds (psittacosis), or lower vertebrates (salmonellosis).

By tradition, vectors are placed into one of two categories, depending upon the animal's relationship with the microbe **(figure 13.13)**. A **biological vector** actively participates in a pathogen's life cycle, serving as a site in which it can multiply or complete its life cycle. A biological vector communicates the infectious agent to the human host by biting, aerosol formation, or touch. In the case of biting vectors, the animal can

1. inject infected saliva into the blood (the mosquito) **(figure 13.13a)**,
2. defecate around the bite wound (the flea), or
3. regurgitate blood into the wound (the tsetse fly).

A detailed discussion of arthropod vectors is found in Insight 20.2.

Mechanical vectors are not necessary to the life cycle of an infectious agent and merely transport it without being infected. The external body parts of these animals become contaminated when they come into physical contact with a source of pathogens. The agent is subsequently transferred to humans indirectly by an intermediate such as food or, occasionally, by direct contact (as in certain eye infections). Houseflies **(figure 13.13b)** are noxious mechanical vectors. They feed on decaying garbage and feces, and while they are feeding, their feet and mouthparts easily become contaminated. They also regurgitate juices onto food to soften and digest it. Flies spread more than 20 bacterial, viral, protozoan, and worm infections. Other nonbiting flies transmit tropical ulcers, yaws, and trachoma. Cockroaches, which have similar unsavory habits,

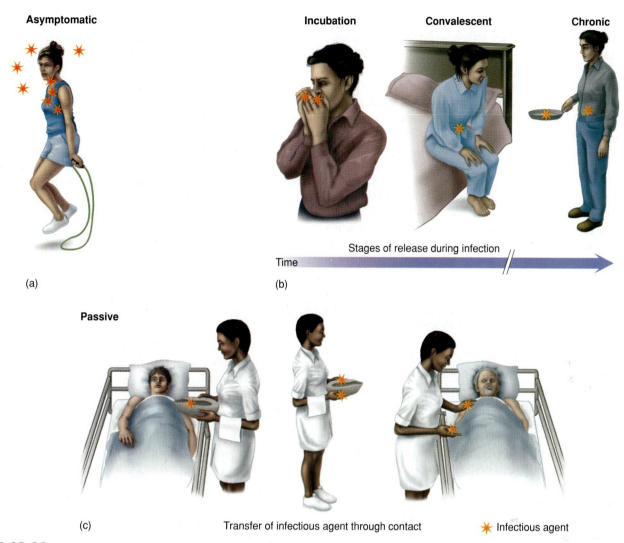

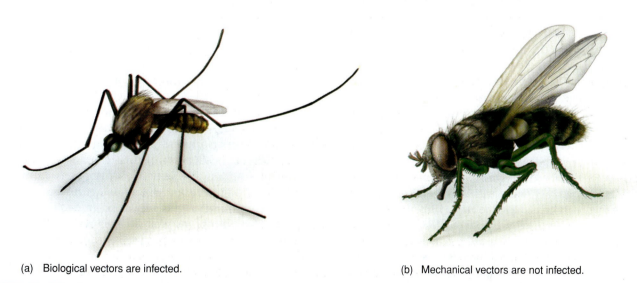

Asymptomatic

(a)

Incubation   **Convalescent**   **Chronic**

Stages of release during infection

Time

(b)

Passive

(c)   Transfer of infectious agent through contact   ✳ Infectious agent

FIGURE 13.12

(a) An asymptomatic carrier is infected without symptoms. **(b)** Incubation, convalescent, and chronic carriers can transmit the infection either before or after the period of symptoms. **(c)** A passive carrier is contaminated but not infected.

(a)   Biological vectors are infected.

(b)   Mechanical vectors are not infected.

FIGURE 13.13   Two types of vectors (magnified views).

(a) Biological vectors serve as hosts during pathogen development. One example is the mosquito, a carrier of malaria. **(b)** Mechanical vectors such as the housefly transport pathogens on their feet and mouthparts.

play a role in the mechanical transmission of fecal pathogens as well as contributing to allergy attacks in asthmatic children.

Many vectors and animal reservoirs spread their own infections to humans. An infection indigenous to animals but naturally transmissible to humans is a **zoonosis** (zoh"-uh-noh'-sis). In these types of infections, the human is essentially a dead-end host and does not contribute to the natural persistence of the microbe. Some zoonotic infections (rabies, for instance) can have multihost involvement, and others can have very complex cycles in the wild (see plague in chapter 20). Zoonotic spread of disease is promoted by close associations of humans with animals, and people in animal-oriented or outdoor professions are at greatest risk. At least 150 zoonoses exist worldwide; the most common ones are listed in **table 13.10**. Zoonoses make up a full 70% of all new emerging diseases worldwide. It is worth noting that zoonotic infections are impossible to completely eradicate without also eradicating the animal reservoirs. Attempts have been made to eradicate mosquitoes and certain rodents, and in 2004 China slaughtered tens of thousands of civet cats who were thought to be a source of SARS.

One technique that can provide an early warning signal for the occurrence of certain mosquito-borne zoonoses has been the use of *sentinel animals.* These are usually domestic animals (most often chickens or horses) that can serve as hosts for diseases such as West Nile fever, various viral encephalitides, and malaria. Sentinel animals are placed at various sites throughout the community, and their blood is monitored periodically for antibodies to the infectious agents that would indicate a recent infection by means of a mosquito bite. The presence of infected animals provides useful data on the potential for human exposure, and it also helps establish the epidemiologic pattern of the zoonosis including where it may have spread.

IN THE NEWS (Continued from page 383)

Karen and her mother had Lyme disease, a zoonosis caused by the spirochete *Borrelia burgdorferi*. The reservoir hosts have *B. burgdorferi* in their blood but show no signs of disease. *Borrelia burgdorferi* enters the skin as the tick vector feeds and multiplies at the site of the bite, initially causing a localized infection. The host's immune system responds with inflammation, resulting in the classic bull's-eye rash. If the infection is not controlled, it can spread to the spleen, kidneys, liver, brain, and joints. Inflammation at these sites results in a wide variety of symptoms, including heart damage, vomiting and diarrhea, mood swings and memory loss, and arthritis.

Lyme disease was named for Lyme, Connecticut, where a cluster of arthritis cases occurred in 1975. Since Dr. Willy Burgdorfer identified the spirochete that bears his name in the early 1980s, the number of reported U.S. cases has increased from 491 in 1982 to more than 18,000 in 2002. The incidence is highest in northeastern, mid-Atlantic, and midwestern states, and in northern California. Humans moving into the reservoir host and tick habitat are at highest risk of acquiring Lyme disease. Other risk factors are camping, hiking, and outdoor occupations such as wildlife biologist and park ranger. Wearing long pants and long-sleeved shirts and using DEET-containing insect repellent on exposed skin are the best ways to prevent Lyme disease.

See: Barbour, A. G. 1996. Lyme disease: The cause, the cure, the controversy. Baltimore: The Johns Hopkins University Press.

Benach, J. L. et al. 1983. Spirochetes isolated from the blood of two patients with Lyme disease. N. Eng. J. Med. 308:740–742.

Burgdorfer, W. 1993. How the discovery of Borrelia burgdorferi *came about. Clin. Dermatol. 11:335–338.*

TABLE 13.10	Common Zoonotic Infections
Disease	**Primary Animal Reservoirs**
Viruses	
Rabies	All mammals
Yellow fever	Wild birds, mammals, mosquitoes
Viral fevers	Wild mammals
Hantavirus	Rodents
Influenza	Chickens, swine
West Nile virus	Wild birds, mosquitoes
Bacteria	
Rocky Mountain spotted fever	Dogs, ticks
Psittacosis	Birds
Leptospirosis	Domestic animals
Anthrax	Domestic animals
Brucellosis	Cattle, sheep, pigs
Plague	Rodents, fleas
Salmonellosis	Variety of mammals, birds, and rodents
Tularemia	Rodents, birds, arthropods
Miscellaneous	
Ringworm	Domestic mammals
Toxoplasmosis	Cats, rodents, birds
Trypanosomiasis	Domestic and wild mammals
Trichinosis	Swine, bears
Tapeworm	Cattle, swine, fish
Scabies	Domestic animals

Nonliving Reservoirs

Clearly, microorganisms have adapted to nearly every habitat in the biosphere. They thrive in soil and water and often find their way into the air. Although most of these microbes are saprobic and cause little harm and considerable benefit to humans, some are opportunists and a few are regular pathogens. Because human hosts are in regular contact with these environmental sources, acquisition of pathogens from natural habitats is of diagnostic and epidemiologic importance.

Soil harbors the vegetative forms of bacteria, protozoa, helminths, and fungi, as well as their resistant or developmental stages such as spores, cysts, ova, and larvae. Bacterial pathogens include the anthrax bacillus and species of *Clostridium* that are responsible for gas gangrene, botulism, and tetanus. Pathogenic fungi in the genera *Coccidioides* and *Blastomyces* are spread by spores in the soil and dust. The invasive stages of the hookworm *Necator* occur in the soil. Natural bodies of water carry fewer nutrients than soil does but still support pathogenic species such as *Legionella*, *Cryptosporidium*, and *Giardia*.

How and Why? The Acquisition and Transmission of Infectious Agents

Infectious diseases can be categorized on the basis of how they are acquired. A disease is **communicable** when an infected host can transmit the infectious agent to another host and establish infection in that host. (Although this terminology is standard, one must realize that it is not the disease that is communicated, but the microbe. Also be aware that the word *infectious* is sometimes used interchangeably with the word *communicable*, but this is not precise usage.) The transmission of the agent can be direct or indirect, and the ease with which the disease is transmitted varies considerably from one agent to another. If the agent is highly communicable, especially through direct contact, the disease is **contagious.** Influenza and measles move readily from host to host and thus are contagious, whereas leprosy is only weakly communicable. Because they can be spread through the population, communicable diseases will be our main focus in the following sections.

In contrast, a **non-communicable** infectious disease does *not* arise through transmission of the infectious agent from host to host. The infection and disease are acquired through some other, special circumstance. Non-communicable infections occur primarily when a compromised person is invaded by his or her own microflora (as with certain pneumonias, for example) or when an individual has accidental contact with a facultative parasite that exists in a nonliving reservoir such as soil. Some examples are certain mycoses, acquired through inhalation of fungal spores, and tetanus, in which *Clostridium tetani* spores from a soiled object enter a cut or wound. Persons thus infected do not become a source of disease to others.

Patterns of Transmission in Communicable Diseases

The routes or patterns of disease transmission are many and varied. The spread of diseases is by direct or indirect contact with animate or inanimate objects and can be horizontal or vertical. The term *horizontal* means the disease is spread through a population from one infected individual to another; *vertical* signifies transmission from parent to offspring via the ovum, sperm, placenta, or milk. The extreme complexity of transmission patterns among microorganisms makes it very difficult to generalize. However, for easier organization, we will divide microorganisms into two major groups, as shown in **figure 13.14**: transmission by contact or transmission by indirect routes, in which some vehicle is involved.

Modes of Contact Transmission   In order for microbes to be directly transferred, some type of contact must occur between the skin or mucous membranes of the infected person and that of the infectee. It may help to think of this route as the

portal of exit meeting the portal of entry without the involvement of an intermediate object or substance. Included in this category are fine droplets sprayed directly upon a person during sneezing or coughing (as distinguished from droplet nuclei that are transmitted some distance by air). Most sexually transmitted diseases are spread directly. In addition, infections that result from kissing, or bites by biological vectors are direct. Most obligate parasites are far too sensitive to survive for long outside the host and can be transmitted only through direct contact. Diseases transmitted vertically from mother to baby fit in this contact category also.

Routes of Indirect Transmission   For microbes to be indirectly transmitted, the infectious agent must pass from an infected host to an intermediate conveyor and from there to another host. This form of communication is especially pronounced when the infected individuals contaminate inanimate objects, food, or air through their activities. The transmitter of the infectious agent can be either openly infected or a carrier.

Indirect Spread by Vehicles: Contaminated Materials   The term **vehicle** specifies any inanimate material commonly used by humans that can transmit infectious agents. A *common vehicle* is a single material that serves as the source of infection for many individuals. Some specific types of vehicles are food, water, various biological products (such as blood, serum, and tissue), and fomites. A **fomite** is an inanimate object that harbors and transmits pathogens. The list of possible fomites is as long as your imagination allows. Probably highest on the list would be objects commonly in contact with the public such as doorknobs, telephones, push buttons, and faucet handles that are readily contaminated by touching. Shared bed linens, handkerchiefs, toilet seats, toys, eating utensils, clothing, personal articles, and syringes are other examples. Although paper money is impregnated with a disinfectant to inhibit microbes, pathogens are still isolated from bills as well as coins.

Outbreaks of food poisoning often result from the role of food as a common vehicle. The source of the agent can be soil, the handler, or a mechanical vector. Because milk provides a rich growth medium for microbes, it is a significant means of transmitting pathogens from diseased animals, infected milk handlers, and environmental sources of contamination. The agents of brucellosis, tuberculosis, Q fever, salmonellosis, and listeriosis are transmitted by contaminated milk. Water that has been contaminated by feces or urine can carry *Salmonella*, *Vibrio* (cholera) viruses (hepatitis A, polio), and pathogenic protozoans (*Giardia*, *Cryptosporidium*).

In the type of transmission termed the *oral-fecal route*, a fecal carrier with inadequate personal hygiene contaminates food during handling, and an unsuspecting person ingests it. Hepatitis A, amoebic dysentery, shigellosis, and typhoid fever are often transmitted this way. Oral-fecal transmission can also involve contaminated materials such as toys and diapers. It is really a special category of indirect transmission, which specifies the way in which the vehicle became contaminated was through contact with fecal material, and that it found its way to someone's mouth.

TABLE 13.12 Levels of Isolation Used in Clinical Settings

Type of Isolation*	Protective Measures**	To Prevent Spread of
Enteric Precautions	Gowns and gloves must be worn by all persons having direct contact with patient; masks not required; special precautions taken for disposing of feces and urine	Diarrheal diseases; *Shigella, Salmonella,* and *Escherichia coli* gastroenteritis; cholera; hepatitis A; rotavirus; and giardiasis
Respiratory Precautions	Private room with closed door is necessary; gowns and gloves not required; masks usually indicated; items contaminated with secretions must be disinfected	Tuberculosis, measles, mumps, meningitis, pertussis, rubella, chickenpox
Drainage and Secretion Precautions	Gowns and gloves required for all persons; masks not needed; contaminated instruments and dressings require special precautions	Staphylococcal and streptococcal infections; gas gangrene; herpes zoster; burn infections
Strict Isolation	Private room with closed door required; gowns, masks, and gloves must be worn by all persons; contaminated items must be wrapped and sent to central supply for decontamination	Mostly highly virulent or contagious microbes; includes diphtheria, some types of pneumonia, extensive skin and burn infections, disseminated herpes simplex and zoster
Reverse Isolation (Also Called Protective Isolation)	Same guidelines as for strict isolation; room may be ventilated by unidirectional or laminar airflow filtered through a high-efficiency particulate air (HEPA) filter that removes most airborne pathogens; infected persons must be barred	Used to protect patients extremely immunocompromised by cancer therapy, surgery, genetic defects, burns, prematurity, or AIDS and therefore vulnerable to opportunistic pathogens

*Precautions are based upon the primary portal of entry and communicability of the pathogen.

**In all cases, visitors to the patient's room must report to the nurses' station before entering the room; all visitors and personnel must wash their hands upon entering and leaving the room.

The potential seriousness and impact of nosocomial infections have required hospitals to develop committees that monitor infectious outbreaks and develop guidelines for infection control and aseptic procedures.

Medical asepsis includes practices that lower the microbial load in patients, caregivers, and the hospital environment. These practices include proper handwashing, disinfection, and sanitization, as well as patient isolation (**table 13.12** summarizes guidelines for the major types of isolation). The goal of these procedures is to limit the spread of infectious agents from person to person. An even higher level of stringency is seen with *surgical asepsis*, which involves all of the strategies listed previously plus ensuring that all surgical procedures are conducted under sterile conditions. This includes sterilization of surgical instruments, dressings, sponges and the like, as well as clothing personnel in sterile garments and scrupulously disinfecting the room surfaces and air.

Hospitals generally employ an *infection control officer* who not only implements proper practices and procedures throughout the hospital but is also charged with tracking potential outbreaks, identifying breaches in asepsis, and training other health care workers in aseptic technique. Among those most in need of this training are nurses and other caregivers whose work, by its very nature, exposes them to needlesticks, infectious secretions, blood, and physical contact with the patient. The same practices that interrupt the routes of infection in the patient can also protect the health care worker. It is for this reason that most hospitals have adopted universal precautions that recognize that all secretions from all persons in the clinical setting are potentially infectious and that transmission can occur in either direction.

Universal Blood and Body Fluid Precautions

Medical and dental settings require stringent measures to prevent the spread of nosocomial infections from patient to patient, from patient to worker, and from worker to patient. But even with precautions, the rate of such infections is rather high. Recent evidence indicates that more than one-third of nosocomial infections could be prevented by consistent and rigorous infection control methods.

Previously, control guidelines were disease-specific, and clearly identified infections were managed with particular restrictions and techniques. With this arrangement, personnel tended to handle materials labeled *infectious* with much greater care than those that were not so labeled. The AIDS epidemic spurred a reexamination of that policy. Because of the potential for increased numbers of undiagnosed

HIV-infected patients, the Centers for Disease Control and Prevention laid down more-stringent guidelines for handling patients and body substances. These guidelines have been termed **universal precautions (UPs),** because they are based on the assumption that all patient specimens could harbor infectious agents and so must be treated with the same degree of care. They also include body substance isolation (BSI) techniques to be used in known cases of infection.

It is worth mentioning that these precautions are designed to protect all individuals in the clinical setting—patients, workers, and the public alike. In general, they include techniques designed to prevent contact with pathogens and contamination and, if prevention is not possible, to take purposeful measures to decontaminate potentially infectious materials.

The universal precautions recommended for all health care settings are:

1. Barrier precautions, including masks and gloves, should be taken to prevent contact of skin and mucous membranes with patients' blood or other body fluids. Because gloves can develop small invisible tears, double gloving decreases the risk further. For protection during surgery, venipuncture, or emergency procedures, gowns, aprons, and other body coverings should be worn. Dental workers should wear eyewear and face shields to protect against splattered blood and saliva.

2. More than 10% of health care personnel are pierced each year by sharp (and usually contaminated) instruments. These accidents carry risks not only for AIDS but also for hepatitis B, hepatitis C, and other diseases. Preventing inoculation infection requires vigilant observance of proper techniques. All disposable needles, scalpels, or sharp devices from invasive procedures must immediately be placed in puncture-proof containers for sterilization and final discard. Under no circumstances should a worker attempt to recap a syringe, remove a needle from a syringe, or leave unprotected used syringes where they pose a risk to others. Reusable needles or other sharp devices must be heat-sterilized in a puncture-proof holder before they are handled. If a needlestick or other injury occurs, immediate attention to the wound, such as thorough degermation and application of strong antiseptics, can prevent infection.

3. Dental handpieces should be sterilized between patients, but if this is not possible, they should be thoroughly disinfected with a high-level disinfectant (peroxide, hypochlorite). Blood and saliva should be removed completely from all contaminated dental instruments and intraoral devices prior to sterilization.

4. Hands and other skin surfaces that have been accidently contaminated with blood or other fluids should be scrubbed immediately with a germicidal soap. Hands should likewise be washed after removing rubber gloves, masks, or other barrier devices.

5. Because saliva can be a source of some types of infections, barriers should be used in all mouth-to-mouth resuscitations.

6. Health care workers with active, draining skin or mucous membrane lesions must refrain from handling patients or equipment that will come into contact with other patients. Pregnant health care workers risk infecting their fetuses and must pay special attention to these guidelines. Personnel should be protected by vaccination whenever possible.

Isolation procedures for known or suspected infections should still be instituted on a case-by-case basis.

Which Agent Is the Cause? Using Koch's Postulates to Determine Etiology

An essential aim in the study of infection and disease is determining the precise **etiologic,** or causative, agent. In our modern technological age, we take for granted that a certain infection is caused by a certain microbe, but such has not always been the case. More than a century ago, Robert Koch realized that in order to prove the germ theory of disease he would have to develop a standard for determining causation that would stand the test of scientific scrutiny. Out of his experimental observations on the transmission of anthrax in cows came a series of proofs, called **Koch's postulates,** that established the principal criteria for etiologic studies **(figure 13.17).** These postulates direct an investigator to

1. find evidence of a particular microbe in every case of a disease,
2. isolate that microbe from an infected subject and cultivate it artificially in the laboratory,
3. inoculate a susceptible healthy subject with the laboratory isolate and observe the same resultant disease, and
4. reisolate the agent from this subject.

Valid application of Koch's postulates requires attention to several critical details. Each isolated culture must be pure, observed microscopically, and identified by means of characteristic tests; the first and second isolate must be identical; and the pathologic effects, signs, and symptoms of the disease in the first and second subject must be the same. Once established, these postulates were rapidly put to the test, and within a short time, they had helped determine the causative agents of tuberculosis, diphtheria, and plague. Today, most infectious diseases have been directly linked to a known infectious agent.

Koch's postulates continue to play an essential role in modern epidemiology. Every decade, new diseases challenge the scientific community and require application of the postulates. Prominent examples are toxic shock syndrome,

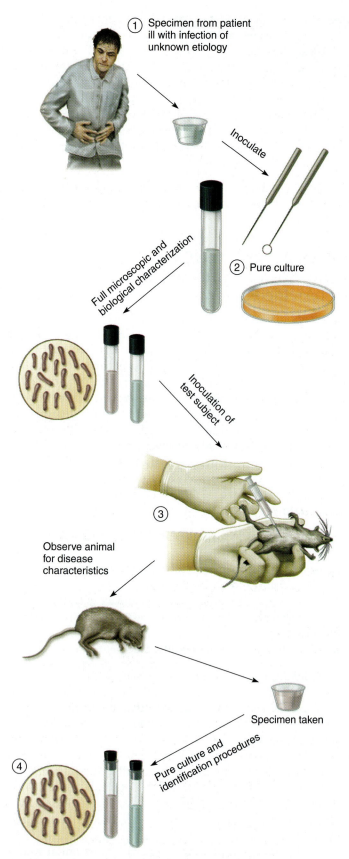

① Specimen from patient ill with infection of unknown etiology

Inoculate

Full microscopic and biological characterization

② Pure culture

Inoculation of test subject

③

Observe animal for disease characteristics

Specimen taken

Pure culture and identification procedures

④

FIGURE 13.17 Koch's postulates: Is this the etiologic agent?
The microbe in the initial and second isolations and the disease in the patient and experimental animal must be identical for the postulates to be satisfied.

AIDS, Lyme disease, and Legionnaires disease (named for the American Legion members who first contracted a mysterious lung infection in Philadelphia).

Koch's postulates are reliable for most infectious diseases, but they cannot be completely fulfilled in certain situations. For example, some infectious agents are not readily isolated or grown in the laboratory. If one cannot elicit a similar infection by inoculating it into an animal, it is very difficult to prove the etiology. In the past, scientists have attempted to circumvent this problem by using human subjects **(Insight 13.5)**.

A small but vocal group of critics has claimed that the postulates have not been adequately carried out for AIDS, and thus that HIV cannot be claimed as the causative agent, despite overwhelming evidence from observing infected humans and primates that it is. Cases of accidental infection through exposure to blood have yielded much proof. In all cases, subjects developed an early virus syndrome, carried high virus levels, and later developed AIDS.

✓ CHECKPOINT

- Epidemiology is the study of the determinants and distribution of all diseases in populations. The study of infectious disease in populations is just one aspect of this field.
- Data on specific, reportable diseases is collected by local, national, and worldwide agencies.
- The *prevalence* of a disease is the percentage of existing cases in a given population. The disease *incidence*, or *morbidity rate*, is the number of newly infected members in a population during a specified time period.
- Disease frequency is described as sporadic, epidemic, pandemic, or endemic.
- The primary habitat of a pathogen is called its reservoir. A human reservoir is also called a carrier.
- Animals can be either reservoirs or vectors of pathogens. An infected animal is a biological vector. Uninfected animals, especially insects, that transmit pathogens mechanically are called mechanical vectors.
- Soil and water are nonliving reservoirs for pathogenic bacteria, protozoa, fungi, and worms.
- A communicable disease can be transmitted from an infected host to others, but not all infectious diseases are communicable.
- The spread of infectious disease from person to person is called horizontal transmission. The spread of infectious disease from parent to offspring is called vertical transmission.
- Infectious diseases are spread by either contact or indirect routes of transmission. Vehicles of indirect transmission include soil, water, food, air, and fomites (inanimate objects).
- Nosocomial infections are acquired in a hospital from surgical procedures, equipment, personnel, and exposure to drug-resistant microorganisms.
- Causative agents of infectious disease must be isolated and identified according to Koch's postulates.

INSIGHT 13.5   *Historical*

The History of Human Guinea Pigs

These days, human beings are not used as subjects for determining the cause of infectious disease, but in earlier times they were. A long tradition of human experimentation dates well back into the eighteenth century, with the subject frequently the experimenter himself. In some studies, mycologists inoculated their own skin and even that of family members with scrapings from fungal lesions to demonstrate that the disease was transmissible. In the early days of parasitology, it was not uncommon for a brave researcher to swallow worm eggs in order to study the course of his infection and the life cycle of the worm.

In a sort of reverse test, a German colleague of Koch's named Max von Petenkofer believed so strongly that cholera was *not* caused by a bacterium that he and his assistant swallowed cultures of *Vibrio cholerae*. Fortunately for them, they acquired only a mild form of the disease. Many self-experimenters have not been so fortunate.

One of the most famous cases is that of Jesse Lazear, a Cuban physician who worked with Walter Reed on the etiology of yellow fever in 1900. Dr. Lazear was convinced that mosquitoes were directly involved in the spread of yellow fever, and by way of proof, he allowed himself and two volunteers to be bitten by mosquitoes infected with the blood of yellow fever patients. Although all three became ill as a result of this exposure, Dr. Lazear's sacrifice was the ultimate one—he died of yellow fever. Years later, paid volunteers were used to completely fulfill the postulates.

Dr. Lazear was not the first martyr in this type of cause. Fifteen years previously, a young Peruvian medical student, Daniel Carrion, attempted to prove that a severe blood infection, Oroya fever, had the same etiology as *verucca peruana*, an ancient disfiguring skin disease. After inoculating himself with fluid from a skin lesion, he developed the severe form and died, becoming a national hero. In his honor, the disease now identified as bartonellosis is also sometimes called Carrion's disease. Eventually, the microbe (*Bartonella bacilliformis*) was isolated, a monkey model was developed, and the sand fly was shown to be the vector for this disease.

The incentive for self-experimentation has continued, but present-day researchers are more likely to test experimental vaccines than dangerous microbes. One exception to this was Dr. J. Robert Warren and Barry Marshall who tested the effects of *Helicobacter pylori* by swallowing a culture. Although they didn't get ulcers, they helped to establish the pathogenicity of the microbe. Even more recently, a parasitologist fed several of his patients live worm eggs to treat their inflammatory bowel disease, and it actually improved their condition significantly (described in Insight 22.4).

Chapter Summary With Key Terms

13.1 The Human Host
A. The human body is in constant contact with microbes, some of which invade the body and multiply. Microbes are classified as either part of the normal flora of the body or as **pathogens,** depending on whether or not their colonization of the body results in infection.
B. *Resident Flora: The Human as Habitat*
1. **Resident flora** or microflora consists of a huge and varied population of microbes that reside permanently on all surfaces of the body exposed to the environment.
2. Colonization begins just prior to birth and continues throughout life. Variations in flora occur in response to changes in an individual's age, diet, hygiene, and health.
3. Flora often provide some benefit to the host but are also capable of causing disease, especially when the immune system of the host is compromised or when flora invades a normally sterile area of the body.
4. An important benefit provided by normal flora is **microbial antagonism.**
C. *Indigenous Flora of Specific Regions*
Resident normal flora includes bacteria, fungi, and protozoa. These normal flora occupy the skin, mouth, gastrointestinal tract, large intestine, respiratory tract, and genitourinary tract.

13.2 The Progress of an Infection
A. **Pathogenicity** is the ability of microorganisms to cause infection and disease; **virulence** refers to the degree to which a microbe can invade and damage host tissues.
1. **True pathogens** are able to cause disease in a normal healthy host with intact immune defenses.
2. **Opportunistic pathogens** can cause disease only in persons whose host defenses are compromised by *predisposing conditions.*
B. *Becoming Established: Step One—Portals of Entry*
1. Microbes typically enter the body through a specific **portal of entry.** Portals of entry are generally the same as those areas of the body that harbor microflora.
2. The minimum number of microbes needed to produce an infection is termed the infectious dose, and these pathogens may be classified as **exogenous** or **endogenous,** based on their source.
C. *Becoming Established: Step Two—Attaching to the Host*
Microbes attach to the host cell, a process known as **adhesion,** by means of fimbriae, capsules, or receptors.
D. *Becoming Established: Step Three—Surviving Host Defenses*
Microbes that persist in the human host have developed mechanisms of disabling early defensive strategies, especially phagocytes.
a. **Exoenzymes** digest epithelial tissues and permit invasion of pathogens.
b. **Toxigenicity** refers to a microbe's capacity to produce **toxins** at the site of multiplication. Toxins are divided into **endotoxins** and **exotoxins,** and

diseases caused by toxins are classified as **intoxications, toxemias, or toxinoses.**
   c. Microbes have a variety of ways of causing the host to damage itself.
E. *Causing Disease*
   Virulence factors are structures or properties of a microbe that lead to pathologic effects on the host. While some virulence factors are those that enable entry or adhesion, many virulence factors lead directly to damage. These fall in three categories: enzymes, toxins, and the induction of a damaging host response.
F. *The Process of Infection and Disease*
   1. **Infectious diseases** follow a typical pattern consisting of an *incubation period, prodromal stage, period of invasion*, and *convalescent period*. Each stage is marked by symptoms of a specific intensity.
   2. Infections are classified by:
      a. whether they remain localized at the site of inoculation **(localized infection, systemic infection, focal infection).**
      b. number of microbes involved in an infection, and the order in which they infect the body **(mixed infection, primary infection, secondary infection).**
      c. persistence of the infection **(acute infection, chronic infection,** subacute infection).
   3. **Signs** of infection refer to *objective* evidence of infection, and **symptoms** refer to *subjective* evidence. A **syndrome** is a disease that manifests as a predictable complex of symptoms.
   4. After the initial infection pathogens may remain in the body in a **latent** state and may later cause recurrent infections. Long-term damage to host cells is referred to as **sequelae.**
   5. Microbes are released from the body through a specific **portal of exit** that allows them access to another host.

13.3 **Epidemiology : The Study of Disease in Populations**
   A. Epidemiologists are involved in the surveillance of reportable diseases in populations. Diseases are tracked with regard to their **prevalence** and **incidence,** while populations are tracked with regard to **morbidity** and **mortality.**
   B. Diseases are described as being **endemic, sporadic, epidemic,** or **pandemic** based on the frequency of their occurrence in a population.
   C. The **reservoir** of a microbe refers to its natural habitat; a pathogen's **source** refers to the immediate origin of an infectious agent.
   D. **Carriers** are individuals who inconspicuously shelter a pathogen and spread it to others. Carriers are divided into **asymptomatic carriers,** incubation carriers, convalescent carriers, chronic carriers, and **passive carriers.**
   E. **Vectors** refer to animals that transmit pathogens.
      1. **Biological vectors** are animals that are involved in the life cycle of the pathogen they transmit; **mechanical vectors** transmit pathogens without being involved in the life cycle.
      2. **Zoonoses** are diseases that can be transmitted to humans but are normally found in animal populations.
   F. Agents of disease are either **communicable** or **noncommunicable,** with highly communicable diseases referred to as **contagious.**
      1. Direct transmission of a disease occurs when a portal of exit meets a portal of entrance.
      2. Indirect transmission occurs when an intermediary such as a **vehicle, fomite,** or **droplet nuclei** connect portals of entrance and exit.
   G. **Nosocomial** infections are infectious diseases acquired in a hospital setting. Special vigilance is required to reduce their occurrence.
   H. **Koch's postulates** are a series of proofs used to determine the **etiologic** agent of a specific disease.

Multiple-Choice Questions

1. The best descriptive term for the resident flora is:
   a. commensals   c. pathogens
   b. parasites   d. mutualists

2. Resident flora is commonly found in the
   a. liver   c. salivary glands
   b. kidney   d. urethra

3. Resident flora is absent from the
   a. pharynx   c. intestine
   b. lungs   d. hair follicles

4. Virulence factors include
   a. toxins   c. capsules
   b. enzymes   d. all of these

5. The specific action of hemolysins is to
   a. damage white blood cells
   b. cause fever
   c. damage red blood cells
   d. cause leukocytosis

6. The _____ is the time that lapses between encounter with a pathogen and the first symptoms.
   a. prodrome   c. period of convalescence
   b. period of invasion d. period of incubation

7. A short period early in a disease that manifests with general malaise and achiness is the
   a. period of incubation c. sequela
   b. prodrome   d. period of invasion

8. The presence of a few bacteria in the blood is termed
   a. septicemia   c. bacteremia
   b. toxemia   d. a secondary infection

9. A _____ infection is acquired in a hospital.
   a. subclinical   c. nosocomial
   b. focal   d. zoonosis

10. A/an _____ is a passive animal transporter of pathogens.
   a. zoonosis   c. mechanical vector
   b. biological vector d. asymptomatic carrier

11. An example of a non-communicable infection is:
   a. measles   c. tuberculosis
   b. leprosy   d. tetanus

12. A general term that refers to an increased white blood cell count is:
   a. leukopenia   c. leukocytosis
   b. inflammation   d. leukemia

13. A positive antibody test for HIV would be a ____ of infection.
    a. sign
    b. symptom
    c. syndrome
    d. sequela

14. Which of the following would *not* be a portal of entry?
    a. the meninges
    b. the placenta
    c. skin
    d. small intestine

15. Which of the following is *not* a condition of Koch's postulates?
    a. isolate the causative agent of a disease
    b. cultivate the microbe in a lab
    c. inoculate a test animal to observe the disease
    d. test the effects of a pathogen on humans

Concept Questions

These questions are suggested as a *writing-to-learn* experience. For each question (except #16), compose a one- or two-paragraph answer that includes the factual information needed to completely address the question.

1. Differentiate between contamination, infection, and disease. What are the possible outcomes in each?

2. How are infectious diseases different from other diseases?

3. Name the general body areas that are sterile. Why is the inside of the intestine not sterile like many other organs?

4. What causes variations in the flora of the newborn intestine?

5. Why must axenic young be delivered by cesarian section?

6. Explain several ways that true pathogens differ from opportunistic pathogens.

7. a. Distinguish between pathogenicity and virulence.
   b. Define virulence factors, and give examples of them in gram-positive and gram-negative bacteria, viruses, and parasites.

8. Describe the course of infection from contact with the pathogen to its exit from the host.

9. a. Explain why most microbes are limited to a single portal of entry.
   b. For each portal of entry, give a vehicle that carries the pathogen and the course it must travel to invade the tissues.
   c. Explain how the portal of entry could differ from the site of infection.

10. Differentiate between exogenous and endogenous infections.

11. a. What factors possibly affect the size of the infectious dose?
    b. Name four factors involved in microbial adhesion.

12. Which body cells or tissues are affected by hemolysins, leukocidins, hyaluronidase, kinases, tetanus toxin, pertussis toxin, and enterotoxin?

13. Compare and contrast: systemic versus local infections; primary versus secondary infections; infection versus intoxication.

14. What are the differences between signs and symptoms? (First put yourself in the place of a patient with an infection and then in the place of a physician examining you. Describe what you would feel and what the physician would detect upon examining the affected area.)

15. a. What are some important considerations about the portal of exit?
    b. Name some examples of infections and their portals of exit.

16. Complete the table:

	Exotoxins	Endotoxins
Chemical makeup		
General source		
Degree of toxicity		
Effects on cells		
Symptoms in disease		
Examples		

17. a. Explain what it means to be a carrier of infectious disease.
    b. Describe four ways that humans can be carriers.
    c. What is epidemiologically and medically important about carriers in the population?

18. a. Outline the science of epidemiology and the work of an epidemiologist.
    b. Using the following statistics, based on number of reported cases, determine which show endemic, sporadic, or epidemic patterns. Explain how you can determine each type.

United States Region	Cases	
Chlamydiosis	**2000**	**2001**
Northeast	93,116	115,467
Midwest	168,332	184,111
South	286,136	307,405
West	161,868	176,259
Total cases	709,452	783,242
Lyme disease		
Northeast	14,932	14,435
Midwest	1,343	1,260
South	1,319	1,198
West	136	136
Total cases	17,730	17,029
Rubella		
Northeast	27	9
Midwest	3	5
South	123	5
West	15	4
Total cases	168	23

   c. Explain what would have to occur for these diseases to have a pandemic distribution.

19. a. Explain the precise difference between communicable and non-communicable infectious diseases.
    b. Between contact and indirect modes of transmission.
    c. Between vectors and vehicles as modes of transmission.

The respiratory tract is constantly guarded from infection by elaborate and highly effective adaptations. Nasal hair traps larger particles. Rhinitis, the copious flow of mucus and fluids that occurs in allergy and colds, exerts a flushing action. In the respiratory tree (primarily the trachea and bronchi), a ciliated epithelium (called the ciliary escalator) conveys foreign particles entrapped in mucus toward the pharynx to be removed **(figure 14.3).** Irritation of the nasal passage reflexively initiates a sneeze, which expels a large volume of air at high velocity. Similarly, the acute sensitivity of the bronchi, trachea, and larynx to foreign matter triggers coughing, which ejects irritants.

The genitourinary tract derives partial protection from the continuous trickle of urine through the ureters and from periodic bladder emptying that flushes the urethra.

The composition of flora and its protective effect were discussed in chapter 13. Even though the resident flora does not constitute an anatomical barrier, its presence can block the access of pathogens to epithelial surfaces and can create an unfavorable environment for pathogens by competing for limited nutrients or by altering the local pH.

Nonspecific Chemical Defenses

The skin and mucous membranes offer a variety of chemical defenses. Sebaceous secretions exert an antimicrobial effect, and specialized glands such as the meibomian glands of the eyelids lubricate the conjunctiva with an antimicrobial secre-

tion. An additional defense in tears and saliva is **lysozyme,** an enzyme that hydrolyzes the peptidoglycan in the cell wall of bacteria. The high lactic acid and electrolyte concentrations of sweat and the skin's acidic pH and fatty acid content are also inhibitory to many microbes. Likewise, the hydrochloric acid in the stomach renders protection against many pathogens that are swallowed, and the intestine's digestive juices and bile are potentially destructive to microbes. Even semen contains an antimicrobial chemical that inhibits bacteria, and the vagina has a protective acidic pH maintained by normal flora.

Genetic Defenses

Some hosts are genetically immune to the diseases of other hosts. One explanation for this phenomenon is that some pathogens have such great specificity for one host species that they are incapable of infecting other species. One way of putting it is: "Humans can't acquire distemper from cats, and cats can't get mumps from humans." This specificity is particularly true of viruses, which can invade only by attaching to a specific host receptor. But it does not hold true for zoonotic infectious agents that attack a broad spectrum of animals. Genetic differences in susceptibility can also exist within members of one species. Humans carrying a gene or genes for sickle-cell anemia are resistant to malaria. Genetic differences also exist in susceptibility to tuberculosis, leprosy, and certain systemic fungal infections.

The vital contribution of barriers is clearly demonstrated in people who have lost them or never had them. Patients with severe skin damage due to burns are extremely susceptible to infections; those with blockages in the salivary glands, tear ducts, intestine, and urinary tract are also at

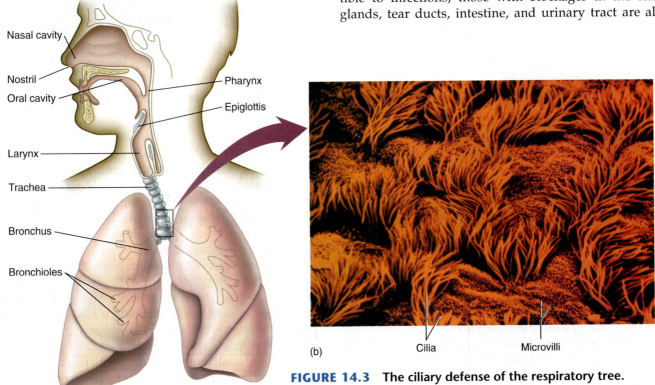

(a)   Right lung   Left lung

(b)   Cilia   Microvilli

FIGURE 14.3 **The ciliary defense of the respiratory tree.**
(a) The epithelial lining of the airways contains a brush border of cilia to entrap and propel particles upward toward the pharynx. **(b)** Tracheal mucosa (5,000×).

Nasal cavity
Nostril
Oral cavity
Pharynx
Epiglottis
Larynx
Trachea
Bronchus
Bronchioles

greater risk for infection. But as important as it is, the first line of defense alone is not sufficient to protect against infection. Because many pathogens find a way to circumvent the barriers by using their virulence factors (discussed in chapter 13), a whole new set of defenses—inflammation, phagocytosis, specific immune responses—are brought into play.

> ✔ **CHECKPOINT**
>
> - The multilevel, interconnecting network of host protection against microbial invasion is organized into three lines of defense.
> - The first line consists of physical and chemical barricades provided by the skin and mucous membranes.
> - The second line encompasses all the nonspecific cells and chemicals found in the tissues and blood.
> - The third line, the specific immune response, is customized to react to specific antigens of a microbial invader. This response immobilizes and destroys the invader every time it appears in the host.

14.2 The Second and Third Lines of Defense: An Overview

Immunology encompasses the study of all features of the body's second and third lines of defense. Although this chapter is concerned, not surprisingly, with infectious microbial agents, be aware that immunology is central to the study of fields as diverse as cancer (at least partly the result of an underactive immune system) and allergy (an overactive immune system). In chapter 17 you will see that many of the most powerful aspects of immunology have been developed for use in health care, laboratory, or commercial settings.

In the body, the mandate of the immune system can be easily stated. A healthy functioning immune system is responsible for

1. surveillance of the body,
2. recognition of foreign material, and
3. destruction of entities deemed to be foreign **(figure 14.4).**

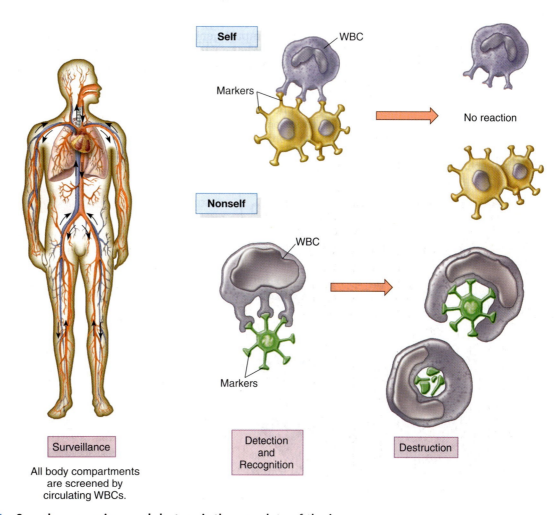

FIGURE 14.4 **Search, recognize, and destroy is the mandate of the immune system.**
White blood cells are equipped with a very sensitive sense of "touch." As they sort through the tissues, they feel surface markers that help them determine what is self and what is not. When self markers are recognized, no response occurs. However, when nonself is detected, a reaction to destroy it is mounted.

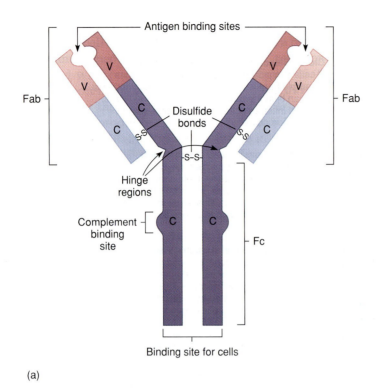

(a)

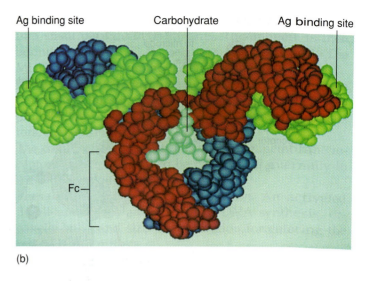

(b)

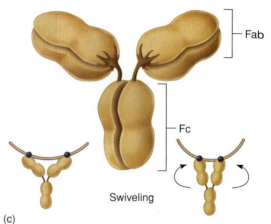

(c)

FIGURE 15.11   Working models of antibody structure.
(a) Diagrammatic view of IgG depicts the principal functional areas (Fabs and Fc) of the molecule. **(b)** Realistic model of immunoglobulin shows the tertiary and quaternary structure achieved by additional intrachain and interchain bonds and the position of the carbohydrate component. **(c)** The "peanut" model of IgG helps illustrate swiveling of Fabs relative to one another and to Fc.

that vary slightly in distance and position. The Fc fragment is involved in binding to various cells and molecules of the immune system itself. **Figure 15.11** shows three views of antibody structure.

Antibody-Antigen Interactions and the Function of the Fab

The site on the antibody where the antigenic determinant binds is composed of a *hypervariable region* whose amino acid content can be extremely varied. Antibodies differ somewhat in the exactness of this groove for antigen, but a certain complementary fit is necessary for the antigen to be held effectively **(figure 15.12)**. The specificity of antigen binding sites for antigens is very similar to enzymes and substrates (in fact, some antibodies are used as enzymes, as you learned in Insight 8.2). So specific are some immunoglobulins for antigen that they can distinguish between a single functional group of a few atoms. Because the specificity of

the Fab sites is identical, an Ig molecule can bind antigenic determinants on the same cell or on two separate cells and thereby link them.

The principal activity of an antibody is to unite with, immobilize, call attention to, or neutralize the antigen for which it was formed **(figure 15.13)**. Antibodies called opsonins stimulate **opsonization,** (ahp"-son-uh-zay'-shun) a process in which microorganisms or other particles are coated with specific antibodies so that they will be more readily recognized by phagocytes, which dispose of them. Opsonization has been likened to putting handles on a slippery object to provide phagocytes a better grip. The capacity for antibodies to aggregate, or **agglutinate,** antigens is the consequence of their cross-linking cells or particles into large clumps. Agglutination renders microbes immobile and enhances their phagocytosis. This is a principle behind certain immune tests discussed in chapter 17. The interaction of an antibody with complement can result in the specific rupturing of cells and some viruses. In **neutralization**

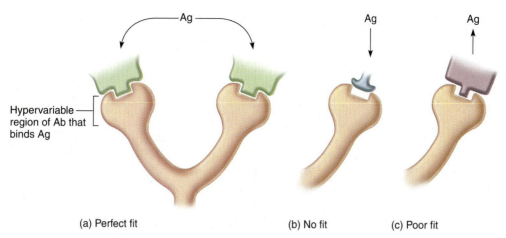

FIGURE 15.12 **Antigen-antibody binding.**
The union of antibody (Ab) and antigen (Ag) is characterized by a certain degree of fit and is supported by a multitude of weak linkages, especially hydrogen bonds and electrostatic attraction. The better the fit (i.e., antigen in **(a)** vs antigen in **(c)**), the stronger the stimulation of the lymphocyte during the activation stage.

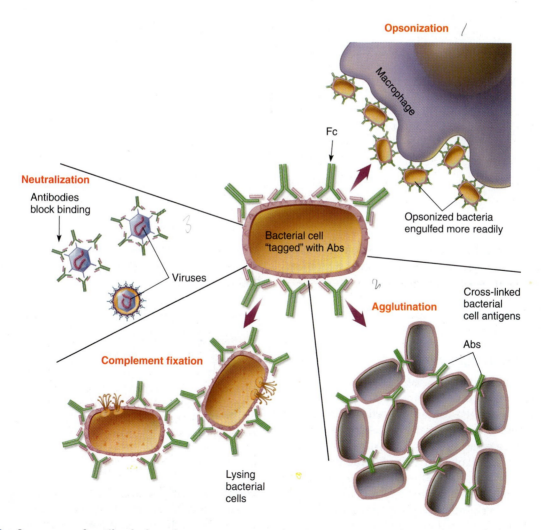

FIGURE 15.13 **Summary of antibody functions.**

TABLE 15.2   Characteristics of the Immunoglobulin (Ig) Classes

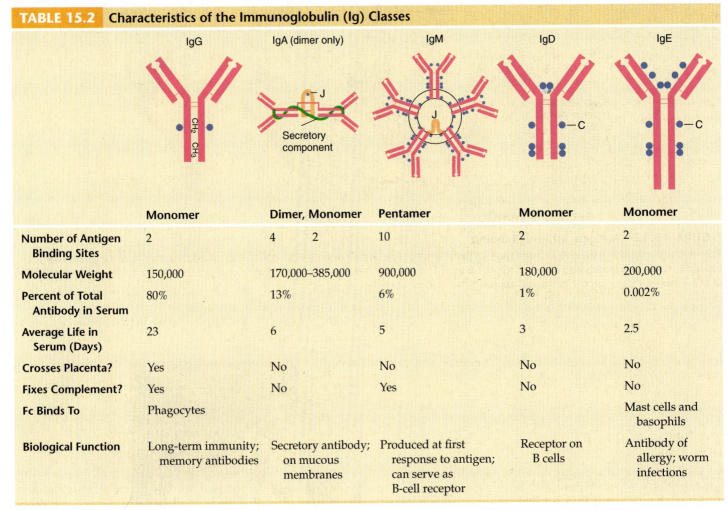

	IgG	IgA (dimer only)		IgM	IgD	IgE
	Monomer	Dimer, Monomer		Pentamer	Monomer	Monomer
Number of Antigen Binding Sites	2	4	2	10	2	2
Molecular Weight	150,000	170,000–385,000		900,000	180,000	200,000
Percent of Total Antibody in Serum	80%	13%		6%	1%	0.002%
Average Life in Serum (Days)	23	6		5	3	2.5
Crosses Placenta?	Yes	No		No	No	No
Fixes Complement?	Yes	No		Yes	No	No
Fc Binds To	Phagocytes					Mast cells and basophils
Biological Function	Long-term immunity; memory antibodies	Secretory antibody; on mucous membranes		Produced at first response to antigen; can serve as B-cell receptor	Receptor on B cells	Antibody of allergy; worm infections

C = carbohydrate.

J = J chain.

reactions, antibodies fill the surface receptors on a virus or the active site on a microbial enzyme to prevent it from attaching normally. **Antitoxins** are a special type of antibody that neutralize bacterial exotoxins. It should be noted that not all antibodies are protective; some neither benefit nor harm, and a few actually cause diseases.

Functions of the Fc Fragment

Although the Fab fragments bind antigen, the Fc fragment has a different binding function. In most classes of immunoglobulin, the end of Fc contains an effector molecule that can bind to the membrane of cells, such as macrophages, neutrophils, eosinophils, mast cells, basophils, and lymphocytes. The effect of an antibody's Fc fragment binding to a cell depends upon that cell's role. In the case of opsonization, the attachment of antibody to foreign cells and viruses exposes the Fc fragments to phagocytes. Certain antibodies have regions on the Fc portion for fixing complement, and in some immune reactions, the binding of Fc causes the release of cytokines. For example, the Fc end of the antibody of allergy (IgE) binds to basophils and mast cells, which causes the release of allergic mediators such as

histamine. The size and amino acid composition of Fc also determine an antibody's permeability, its distribution in the body, and its class.

Accessory Molecules on Immunoglobulins

All antibodies contain molecules in addition to the basic polypeptides. Varying amounts of carbohydrates are affixed to the constant regions in most instances (table 15.2). Two additional accessory molecules are the *J chain* that joins the monomers of IgA and IgM, and the *secretory component*, which helps move IgA across mucous membranes. These proteins occur only in certain immunoglobulin classes.

The Classes of Immunoglobulins

Immunoglobulins exist as structural and functional classes called *isotypes* (compared and contrasted in table 15.2). The differences in these classes are due primarily to variations in the Fc fragment. The classes are differentiated with shorthand names (Ig, followed by a letter: IgG, IgA, IgM, IgD, IgE).

The structure of IgG has already been presented. It is a monomer produced by plasma cell in a primary response and by memory cells responding the second time to a given antigenic stimulus. It is by far the most prevalent antibody circulating throughout the tissue fluids and blood. It has numerous functions: It neutralizes toxins, opsonizes, and fixes complement, and it is the only antibody capable of crossing the placenta.

The two forms of IgA are: (1) a monomer that circulates in small amounts in the blood and (2) a dimer that is a significant component of the mucous and serous secretions of the salivary glands, intestine, nasal membrane, breast, lung, and genitourinary tract. The dimer, called secretory IgA, is formed in a plasma cell by two monomers held together by a J chain. To facilitate the transport of IgA across membranes, a secretory piece is later added by the epithelial cells of the mucosa. IgA coats the surface of these membranes and appears free in saliva, tears, colostrum, and mucus. It confers the most important specific local immunity to enteric, respiratory, and genitourinary pathogens. Its contribution in protecting newborns who derive it passively from nursing is mentioned in Insight 15.2.

IgM is a huge molecule composed of five monomers (making it a pentamer) attached by the Fc portions to a central J chain. With its 10 binding sites, this molecule has tremendous capacity for binding antigen. IgM is the first class synthesized following the host's first encounter with antigen. Its complement-fixing qualities make it an important antibody in many immune reactions. It circulates mainly in the blood and is far too large to cross the placental barrier.

IgD is a monomer found in minuscule amounts in the serum, and it does not fix complement, opsonize, or cross the placenta. Its main function is to serve as a receptor for antigen on B cells, usually along with IgM. It seems to be the triggering molecule for B-cell activation.

IgE is also an uncommon blood component unless one is allergic or has a parasitic worm infection. Its Fc region interacts with receptors on mast cells and basophils. It biological significance is to stimulate an inflammatory response through the release of potent physiological substances by the basophils and mast cells. Because inflammation would enlist blood cells such as eosinophils and lymphocytes to the site of infection, it would certainly be one defense against parasites. Unfortunately, IgE has another, more insidious effect—that of mediating anaphylaxis, asthma, and certain other allergies.

Evidence of Antibodies in Serum

Regardless of the site where antibodies are first secreted, a large quantity eventually ends up in the blood by way of the body's communicating networks. If one subjects a sample of **antiserum** (serum containing specific antibodies) to electrophoresis, the major groups of proteins migrate in a pattern consistent with their mobility and size (figure 15.14). The albumins show up in one band, and the globulins in four bands called alpha-1 (α_1), alpha-2 (α_2), beta (β), and gamma (γ) globulins. Most of the globulins represent antibodies, which explains how the term *immunoglobulin* was derived.

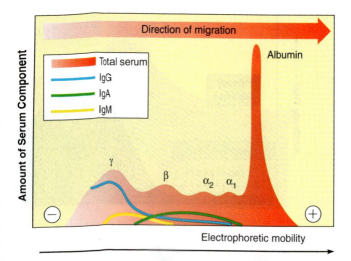

FIGURE 15.14 Pattern of human serum after electrophoresis.

When antiserum is subjected to electrical current, the various proteinaceous components are separated into bands. This is a means of separating the different antibodies and serum proteins as well as quantifying them.

Gamma globulin is composed primarily of IgG, whereas β and α_2 globulins are a mixture of IgG, IgA, and IgM.

Monitoring Antibody Production over Time: Primary and Secondary Responses to Antigens

We can learn a great deal about how the immune system reacts to an antigen by studying the levels of antibodies in serum over time (figure 15.15). This level is expressed quantitatively as the **titer** (ty'-tur), or concentration of antibodies. Upon the first exposure to an antigen, the system undergoes a **primary response.** The earliest part of this response, the *latent period,* is marked by a lack of antibodies for that antigen, but much activity is occurring. During this time, the antigen is being concentrated in lymphoid tissue, and is being processed by the correct clones of B lymphocytes. As plasma cells synthesize antibodies, the serum titer increases to a certain plateau and then tapers off to a low level over a few weeks or months. When the class of antibodies produced during this response is tested, an important characteristic of the response is uncovered. It turns out that, early in the primary response, most of the antibodies are the IgM type, which is the first class to be secreted by plasma cells. Later, the class of the antibodies (but not their specificity) is switched to IgG or some other class (IgA or IgE).

When the immune system is exposed again to the same immunogen within weeks, months, or even years, a **secondary response** occurs. The rate of antibody synthesis, the peak titer, and the length of antibody persistence are greatly increased over the primary response. The rapidity and amplification seen in this response are attributable to the memory B cells that were formed during the primary response. Because of its association with recall, the secondary response is also called the **anamnestic response.** The

Monoclonal Antibodies: Variety Without Limit

The value of antibodies as tools for locating or identifying antigens is well established. For many years, antiserum extracted from human or animal blood was the main source of antibodies for tests and therapy, but most antiserum has a basic problem. It contains **polyclonal antibodies,** meaning that it is a mixture of different antibodies because it reflects dozens of immune reactions from a wide variety of B-cell clones. This characteristic is to be expected, because several immune reactions may be occurring simultaneously, and even a single species of microbe can stimulate several different types of antibodies. Certain applications in immunology require a pure preparation of **monoclonal antibodies (MAbs)** that originate from a single clone and have a single specificity for antigen.

The technology for producing monoclonal antibodies is possible by hybridizing cancer cells and activated B cells *in vitro*. This technique began with the discovery that tumors isolated from multiple myelomas in mice consist of identical plasma cells. These monoclonal plasma cells secrete a strikingly pure form of antibodies with a single specificity and continue to divide indefinitely. Immunologists recognized the potential in these plasma cells and devised a **hybridoma** approach to creating MAb. The basic idea behind this approach is to hybridize or fuse a myeloma cell with a normal plasma cell from a mouse spleen to create an immortal cell that secretes a supply of functional antibodies with a single specificity.

The introduction of this technology has the potential for numerous biomedical applications. Monoclonal antibodies have provided immunologists with excellent standardized tools for studying the immune system and for expanding disease diagnosis and treatment. Most of the successful applications thus far use MAbs in *in vitro* diagnostic testing and research. Although injecting monoclonal antibodies to treat human disease is an exciting prospect, so far this therapy has been stymied because most MAbs are of mouse origin, and many humans will develop hypersensitivity to them. The development of human MAbs and other novel approaches using genetic engineering is currently under way.

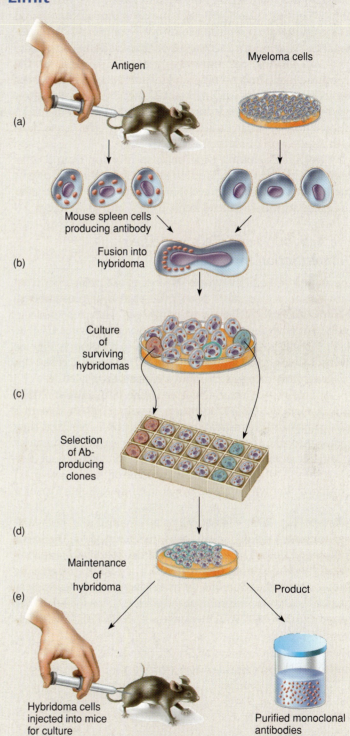

Summary of the technique for producing monoclonal antibodies by hybridizing myeloma tumor cells with normal plasma cells. **(a)** A normal mouse is inoculated with an antigen having the desired specificity, and activated cells are isolated from its spleen. A special strain of mouse provides the myeloma cells. **(b)** The two cell populations are mixed with polyethylene glycol, which causes some cells in the mixture to fuse and form hybridomas. **(c)** Surviving cells are cultured and separated into individual wells. **(d)** Tests are performed on each hybridoma to determine the specificity of the antibody (Ab) it secretes. **(e)** A hybridoma with the desired specificity is grown in tissue culture; antibody product is then isolated and purified. The hybridoma is maintained in a susceptible mouse for future use.

15.6 T-Cell Response
Cell-Mediated Immunity (CMI)

During the time that B cells have been actively responding to antigens, the T-cell limb of the system has been similarly engaged. The responses of T cells, however, are **cell-mediated immunities**, which require the direct involvement of T lymphocytes throughout the course of the reaction. These reactions are among the most complex and diverse in the immune system and involve several subsets of T cells whose particular actions are dictated by CD receptors. T cells are restricted; that is, they require some type of MHC (self) recognition before they can be activated, and all produce cytokines with a spectrum of biological effects.

T cells have notable differences in function from B cells. Rather than making antibodies to control foreign antigens, the whole T cell acts directly in contact with the antigen. They also stimulate other T cells, B cells, and phagocytes.

The Activation of T Cells and Their Differentiation Into Subsets

The mature T cells in lymphoid organs are primed to react with antigens that have been processed and presented to them by dendritic cells and macrophages. They recognize an antigen only when it is presented in association with an MHC carrier (see figure 15.9). T cells with CD4 receptors recognize endocytosed peptides presented on MHC II and T cells with CD8 receptors recognize peptides presented on MHC I.

A T cell is initially sensitized when an antigen/MHC complex is bound to its receptors. As with B cells, activated T cells transform in preparation for mitotic divisions, and they divide into one of the subsets of effector cells and memory cells that can interact with the antigen upon subsequent contact **(table 15.3)**. Memory T cells are some of the longest-lived blood cells known (70 years in one well-documented case).

T Helper (T_H) Cells Helper cells play a central role in regulating immune reactions to antigens, including those of B cells and other T cells. They are also involved in activating macrophages and improving opsonization. They do this directly by receptor contact and indirectly by releasing cytokines such as interleukin-2, which stimulates the primary growth and activation of B and T cells, and interleukins-4, -5, and -6, which stimulate various activities of B cells. T helper cells are the most prevalent type of T cell in the blood and lymphoid organs, making up about 65% of this population. The severe depression of this class of T cells (with CD4 receptors) by HIV is what largely accounts for the immunopathology of AIDS.

When T helper (CD4) cells are stimulated by antigen/MHC complex, they differentiate into either T helper 1 (T_H1) cells, or T helper 2 (T_H2) cells, probably depending on what type of cytokines the antigen-presenting cells secrete. It is thought that if the dendritic (APC) cell secretes IL-2 and/or interferon gamma, the T cell will become a T_H1 cell. A T_H1 cell will activate more T cells and is also involved in delayed hypersensitivity. (Delayed hypersensitivity is a type of response to allergens, distinct from immediate allergies such as hay fever and anaphylaxis. Both of these reactions will be discussed in chapter 16.)

If the APC secretes another set of cytokines (some think IL-4 is involved), the T cell will differentiate into a T_H2 cell. These cells have the function of secreting substances that influence B-cell differentiation, and enhancing the antibody response. They may also be able to dampen the response of T_H1 cells when necessary.

Cytotoxic T (T_C) Cells: Cells That Kill Other Cells When CD8 cells are stimulated by antigen/MHC complex, they differentiate into T cytotoxic cells (T_C cells). **Cytotoxicity** is the capacity of certain T cells to kill a specific target cell. It is a fascinating and powerful property that accounts for much of our immunity to foreign cells and cancer, and yet, under some circumstances, it can lead to disease. For a **killer T cell** to become activated, it must recognize a foreign peptide complexed with self MHC I presented to it, and mount a direct attack upon the target cell. After activation, the T_C cell severely injures the target cell **(figure 15.16)**. This process involves the secretion of **perforins**[2] and **granzymes**. Perforins are proteins that can punch holes in the membranes of target cells. Granzymes are enzymes that attack proteins of target cells. The action of the perforins causes ions to leak out of target cells and creates a passageway for granzymes to enter. These events are usually followed by target cell death through *apoptosis*.

---

2. **perforin** From the term *perforate* or to penetrate with holes.

TABLE 15.3	Characteristics of Subsets of T Cells	
Types	**Primary Receptors on T Cell**	**Functions/Important Features**
T helper cell 1 (T_H1)	CD4	Activates the cell-mediated immunity pathway, secrete tumor necrosis factor and interferon gamma, also responsible for delayed hypersensitivity (allergy occurring several hours or days after contact)
T helper cell 2 (T_H2)	CD4	Drives B-cell proliferation, secrete IL-4, IL-5, IL-6, IL-10; can dampen T_H1 activity
T cytotoxic cell (T_C)	CD8	Destroys a target foreign cell by lysis; important in destruction of complex microbes, cancer cells, virus-infected cells; graft rejection; requires MHC I for function

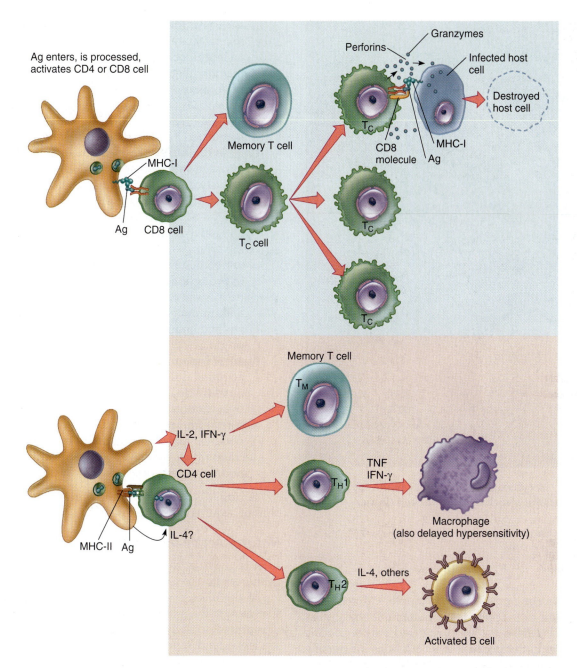

FIGURE 15.16 Overall scheme of T-cell activation and differentiation into different types of T cells.
Antigen-presenting cells present antigenic peptides to T cells bearing either CD4 or CD8 markers. Upon binding antigen/MHC-I complex, CD8 cells become T cytotoxic cells, which bind host cells displaying antigen/MHC-I complexes, release perforins, and lead to the apoptosis of those cells. CD4 cells bind antigen/MHC-II complexes on APCs and, depending on the type of cytokine released by the APC, become either T_H1 or T_H2 cells. T_H1 cells influence macrophages to destroy ingested microbes or to become more active. T_H2 cells secrete cytokines that enhance B-cell activation.

Target Cells That T_C Cells Can Destroy Include the Following:

• Virally infected cells (figure 15.16). Cytotoxic cells recognize these because of telltale virus peptides expressed on their surface. Cytotoxic defenses are an essential protection against viruses.
• Cancer cells. T cells constantly survey the tissues and immediately attack any abnormal cells they encounter (figure 15.17). The importance of this function is clearly demonstrated in the susceptibility of T-cell-deficient people to cancer (chapter 16).
• Cells from other animals and humans. Cytotoxic CMI is the most important factor in graft rejection. In this instance, the T_C cells attack the foreign tissues that have been implanted into a recipient's body.

Other Types of Killer Cells Natural killer (NK) cells are a type of lymphocyte related to T cells that lack specificity for antigens. They circulate through the spleen, blood, and lungs

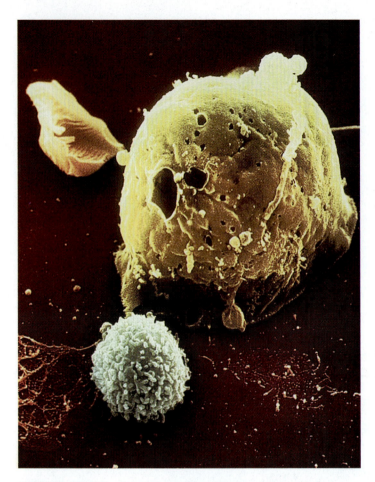

FIGURE 15.17 A cytotoxic T cell (lower blue cell) has mounted a successful attack on a tumor cell (larger yellow cell).
These small killer cells perforate their cellular targets with holes that lead to lysis and death.

and are probably the first killer cells to attack cancer cells and virus-infected cells. They destroy such cells by similar mechanisms as T cells. They are considered part of specific cell-mediated immunity because their activities are acutely sensitive to cytokines such as interleukin-12 and interferon.

As you can see, the T-cell system is very complex. In summary, T cells differentiate into three different types of cells, each of which contributes to the orchestrated immune response, under the influence of a multitude of cytokines.

15.7 A Practical Scheme for Classifying Specific Immunities

The means by which humans acquire immunities can be conveniently encapsulated within four interrelated categories: active, passive, natural, and artificial.

Active immunity occurs when an individual receives an immune stimulus (antigen) that activates the B and T cells, causing the body to produce immune substances such as antibodies. Active immunity is marked by several characteristics: (1) It is an essential attribute of an immunocompetent individual; (2) it creates a memory that renders the person ready for quick action upon reexposure to that same antigen; (3) it requires several days to develop; and (4) it lasts for a relatively long time, sometimes for life. Active immunity can be stimulated by natural or artificial means.

Passive immunity occurs when an individual receives immune substances (antibodies) that were produced actively in the body of another human or animal donor. The recipient is protected for a time even though he or she has not had prior exposure to the antigen. It is characterized by: (1) lack of memory for the original antigen, (2) lack of production of new antibodies against that disease, (3) immediate onset of protection, and (4) short-term effectiveness, because antibodies have a limited period of function, and ultimately, the recipient's body disposes of them. Passive immunity can also be natural or artificial in origin.

Natural immunity encompasses any immunity that is acquired during the normal biological experiences of an individual rather than through medical intervention.

Artificial immunity is protection from infection obtained through medical procedures. This type of immunity is induced by immunization with vaccines and immune serum.

Figure 15.18 illustrates the various possible combinations of acquired immunities.

Natural Active Immunity: Getting the Infection

After recovering from infectious disease, a person may be actively resistant to reinfection for a period that varies according to the disease. In the case of childhood viral infections such as measles, mumps, and rubella, this natural active stimulus provides nearly lifelong immunity. Other diseases result in a less extended immunity of a few months to years (such as pneumococcal pneumonia and shigellosis), and reinfection is possible. Even a subclinical infection can stimulate natural active immunity. This probably accounts for the fact that some people are immune to an infectious agent without ever having been noticeably infected with or vaccinated for it.

Natural Passive Immunity: Mother to Child

Natural, passively acquired immunity occurs only as a result of the prenatal and postnatal, mother-child relationship. During fetal life, IgG antibodies circulating in the maternal bloodstream are small enough to pass or be actively transported across the placenta. Antibodies against tetanus, diphtheria, pertussis, and several viruses regularly cross the placenta. This natural mechanism provides an infant with a mixture of many maternal antibodies that can protect it for the first few critical months outside the womb, while its own immune system is gradually developing active immunity. Depending upon the

Acquired Immunity

Natural Immunity
is acquired through the normal life experiences of
a human and is not induced through medical means.

Artificial Immunity
is that produced purposefully through
medical procedures (also called immunization).

Active Immunity
is the consequence of
a person developing his
own immune response
to a microbe.

Passive Immunity
is the consequence of
one person receiving
preformed immunity
made by another person.

Active Immunity
is the consequence of a
person developing his
own immune response
to a microbe.

Passive Immunity
is the consequence
of one person receiving
preformed immunity
made by another person.

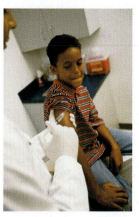
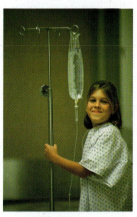

FIGURE 15.18   Categories of acquired immunities.
Natural immunities, which occur during the normal course of life, are either active (acquired from an infection and then recovering) or passive (antibodies donated by the mother to her child). Artificial immunities are acquired through medical practices and can be active (vaccinations with antigen, to stimulate an immune response) or passive (immune therapy with a serum containing antibodies).

microbe, passive protection lasts anywhere from a few months to a year. But eventually, the infant's body clears the antibody. Most childhood vaccinations are timed so that there is no lapse in protection against common childhood infections.

Another source of natural passive immunity comes to the baby by way of mother's milk **(Insight 15.2).** Although the human infant acquires 99% of natural passive immunity in utero and only about 1% through nursing, the milk-borne antibodies provide a special type of intestinal protection that is not available from transplacental antibodies.

Artificial Immunity: Immunization

Immunization is any clinical process that produces immunity in a subject. Because it is often used to give advance protection against infection, it is also called *immunoprophylaxis.* The use of these terms is sometimes imprecise, thus it should be stressed that active immunization, in which a person is administered antigen, is synonymous with vaccination, and that passive immunization, in which a person is given antibodies, is a type of immune therapy.

Vaccination: Artificial Active Immunization

The term *vaccination* originated from the Latin word *vacca* (cow), because the cowpox virus was used in the first preparation for active immunization against smallpox (see Insight 15.3). Vaccination exposes a person to a specially prepared microbial (antigenic) stimulus, which then triggers the im-

mune system to produce antibodies and lymphocytes to protect the person upon future exposure to that microbe. As with natural active immunity, the degree and length of protection vary. Commercial vaccines are currently available for many diseases.

Immunotherapy: Artificial Passive Immunization

In immunotherapy, a patient at risk for acquiring a particular infection is administered a preparation that contains specific antibodies against that infectious agent. In the past, these therapeutic substances were obtained by vaccinating animals (horses in particular), then taking blood and extracting the serum. However, horse serum is now used only in limited situations because of the potential for hypersensitivity to it. Pooled human serum from donor blood (gamma globulin) and immune serum globulins containing high quantities of antibodies are more frequently used. Immune serum globulins are used to protect people who have been exposed to hepatitis, measles, and rubella. More specific immune serum, obtained from patients recovering from a recent infection, is useful in preventing and treating hepatitis B, rabies, pertussis, and tetanus.

An outline summarizing the system of host defenses covered in chapters 14 and 15 was presented in figure 14.1. You may want to use this resource to review major aspects of immunity and to guide you in answering certain questions (see concept question 15).

INSIGHT 15.2 *Historical*

Breast Feeding: The Gift of Antibodies

An advertising slogan from the past claims that cow's milk is "nature's most nearly perfect food." One could go a step further and assert that human milk is nature's *perfect* food for young humans. Clearly, it is loaded with essential nutrients, not to mention being available on demand from a readily portable, hygienic container that does not require refrigeration or warming. But there is another and perhaps even greater benefit. During lactation, the breast becomes a site for the proliferation of lymphocytes that produce IgA, a special class of antibody that protects the mucosal surfaces from local invasion by microbes. The very earliest secretion of the breast, a thin, yellow milk called **colostrum,** is very high in IgA. These antibodies form a protective coating in the gastrointestinal tract of a nursing infant that guards against infection by a number of enteric pathogens (*Escherichia coli, Salmonella,* poliovirus, rotavirus). Protection at this level is especially critical because an infant's own IgA and natural intestinal barriers are not yet developed. As with immunity in utero, the necessary antibodies will be donated only if the mother herself has active immunity to the microbe through a prior infection or vaccination.

In recent times, the ready availability of artificial formulas and the changing life-styles of women have reduced the incidence of breast feeding. Where adequate hygiene and medical care prevail, bottle-fed infants get through the critical period with few problems, because the foods given them are relatively sterile and they have received protection against some childhood infections in utero. Mothers in developing countries with untreated water supplies or poor medical services are strongly discouraged from using prepared formulas, because they can actually inoculate the baby's intestine with pathogens from the formula. Millions of neonates suffer from severe and life-threatening diarrhea that could have been prevented by the hygienic practice of nursing.

In the mid-1900s, baby formula manufacturers tried to introduce the widespread use of prepared formula to developing countries. They provided free samples of formula, and once babies were weaned from breast milk, mothers were forced to buy formula. The health effects were so damaging that in the 1980s the World Health Organization issued an "International Code on the Marketing of Breast Milk Substitutes," discouraging formula use in the developing world.

✔ CHECKPOINT

- T cells do not produce antibodies. Instead, they produce different cytokines that play diverse roles in the immune response. Each subset of T cell produces a distinct set of cytokines that stimulate lymphocytes or destroys foreign cells.

- Active immunity means that your body produces antibodies to a disease agent. If you contract the disease, you can develop natural active immunity. If you are vaccinated, your body will produce artificial active immunity.

- In passive immunity, you receive antibodies from another person. Natural passive immunity comes from the mother. Artificial passive immunity is administered medically.

15.8 Immunization: Methods of Manipulating Immunity for Therapeutic Purposes

Methods that actively or passively immunize people are widely used in disease prevention and treatment. In the case of passive immunization, a patient is given preformed antibodies, which is actually a form of **immunotherapy.** In the case of active immunization, a patient is vaccinated with a microbe or its antigens, providing a form of advance protection.

Passive Immunization

As mentioned earlier, the first attempts at passive immunization involved the transfusion of horse serum containing antitoxins to prevent tetanus and to treat patients exposed to diphtheria. Since then, antisera from animals have been replaced with products of human origin that function with various degrees of specificity. Immune serum globulin (ISG), sometimes called *gamma globulin*, contains immunoglobulin extracted from the pooled blood of at least 1,000 human donors. The method of processing ISG concentrates the antibodies to increase potency and eliminates potential pathogens (such as the hepatitis B and HIV viruses). It is a treatment of choice in preventing measles and hepatitis A and in replacing antibodies in immunodeficient patients. Most forms of ISG are injected intramuscularly to minimize adverse reactions, and the protection it provides lasts 2 to 3 months.

A preparation called specific immune globulin (SIG) is derived from a more defined group of donors. Companies that prepare SIG obtain serum from patients who are convalescing and in a hyperimmune state after such infections as pertussis, tetanus, chickenpox, and hepatitis B. These globulins are preferable to ISG because they contain higher titers of specific antibodies obtained from a smaller pool of patients. Although useful for prophylaxis in persons who have been exposed or may be exposed to infectious agents, these sera are often limited in availability.

INSIGHT 15.3   *Historical*

The Lively History of Active immunization

The basic notion of immunization has existed for thousands of years. It probably stemmed from the observation that persons who had recovered from certain communicable diseases rarely if ever got a second case. Undoubtedly, the earliest crude attempts involved bringing a susceptible person into contact with a diseased person or animal. The first recorded attempt at immunization occurred in sixth century China. It consisted of drying and grinding up smallpox scabs and blowing them with a straw into the nostrils of vulnerable family members. By the tenth century, this practice had changed to the deliberate inoculation of dried pus from the smallpox pustules of one patient into the arm of a healthy person, a technique later called **variolation** (variola is the smallpox virus). This method was used in parts of the Far East for centuries before Lady Mary Montagu brought it to England in 1721. Although the principles of the technique had some merit, unfortunately many recipients and their contacts died of smallpox. This outcome vividly demonstrates a cardinal rule for a workable vaccine: It must contain an antigen that will provide protection but not cause the disease. Variolation was so controversial that any English practitioner caught doing it was charged with a felony.

Eventually, this human experimentation paved the way for the first really effective vaccine, developed by the English physician Edward Jenner in 1796. Jenner conducted the first scientifically controlled study, one that had a tremendous impact on the advance of medicine. His work gave rise to the words **vaccine** and *vaccination* (from L., *vacca*, cow), which now apply to any immunity obtained by inoculation with selected antigens. Jenner was inspired by the case of a dairymaid who had been infected by a pustular infection called cowpox. This related virus afflicts cattle but causes a milder condition in humans. She explained that she and other milkmaids had remained free of smallpox. Other residents of the region expressed a similar confidence in the cross-protection of cowpox. To test the effectiveness of this new vaccine, Jenner prepared material from human cowpox lesions and inoculated a young boy. When challenged 2 months later with an injection of crusts from a smallpox patient, the boy proved immune.

Jenner's discovery—that a less pathogenic agent could confer protection against a more pathogenic one—is especially remarkable in view of the fact that microscopy was still in its infancy and the nature of viruses was unknown. At first, the use of the vaccine was regarded with some fear and skepticism (see illustration). When Jenner's method proved successful and word of its significance spread, it was eventually adopted in many other countries. Eventually, the original virus mutated into a unique strain (*vaccinia* virus) that became the basis of the current vaccine. In 1973 the World Health Organization declared that smallpox had been eradicated. As a result, smallpox vaccination had been discontinued until recently, due to the threat of bioterrorism.

Other historical developments in vaccination included using heat-killed bacteria in vaccines for typhoid fever, cholera, and plague, and techniques for using neutralized toxins for diphtheria and tetanus. Throughout the history of vaccination, there have been vocal opponents and minimizers, but numbers do not lie. Whenever a vaccine has been introduced, the prevalence of that disease has declined dramatically.

Detail from "The Cowpock," an 1808 etching that caricatured the worst fears of the English public concerning Edward Jenner's smallpox vaccine.

When a human immune globulin is not available, antisera and antitoxins of animal origin can be used. Sera produced in horses are available for diphtheria, botulism, and spider and snake bites. Unfortunately, the presence of horse antigens can stimulate allergies such as serum sickness or anaphylaxis (see chapter 16). Although donated immunities only last a relatively short time, they act immediately and can protect patients for whom no other useful medication or vaccine exists.

Artificial Active Immunity: Vaccination

Active immunity can be conferred artificially by **vaccination**—exposing a person to material that is antigenic but not pathogenic. The discovery of vaccination was one of the farthest reaching and most important developments in medical science **(Insight 15.3).** The basic principle behind vaccination is to stimulate a primary and secondary anamnestic response that primes the immune system for future exposure to a virulent pathogen. If this pathogen enters the body, the immune response will be immediate, powerful, and sustained.

Vaccines have profoundly reduced the prevalence and impact of many infectious diseases that were once common and often deadly. In this section, we survey the principles of vaccine preparation and important considerations surrounding vaccine indication and safety. (Vaccines are also given specific consideration in later chapters on infectious diseases and organ systems.)

Principles of Vaccine Preparation   A vaccine must be considered from the standpoints of antigen selection, effectiveness, ease in administration, safety, and cost. In natural immunity, an infectious agent stimulates a relatively long-term protective response. In artificial active immunity, the objective is to obtain this same response with a modified version of the microbe or its components. Qualities of an effective vaccine are listed in **table 15.4.** Most vaccine preparations are based on one of the following antigen preparations **(figure 15.19):**

1. killed whole cells or inactivated viruses;
2. live, attenuated cells or viruses;
3. antigenic molecules derived from bacterial cells or viruses, or
4. genetically engineered microbes or microbial antigens.

A survey of the major licensed vaccines and their indications is presented in **table 15.5.**

Large, complex antigens such as whole cells or viruses are very effective immunogens. Depending on the vaccine, these are either killed or attenuated. **Killed** or **inactivated vaccines** are prepared by cultivating the desired strain or strains of a bacterium or virus and treating them with formalin, radiation, heat, or some other agent that does not destroy antigenicity. One type of vaccine for the bacterial disease cholera is of this type (see chapter 22). The Salk polio vaccine and one form of influenza vaccine contain inactivated viruses. Because the microbe does not multiply, killed vaccines often require a larger dose and more boosters to be effective.

A number of vaccines are prepared from live, **attenuated** microbes. Attenuation is any process that substantially lessens or negates the virulence of viruses or bacteria. It is usually achieved by modifying the growth conditions or manipulating microbial genes in a way that eliminates virulence factors. Attenuation methods include long-term cultivation, selection of mutant strains that grow at colder temperatures (cold mutants), passage of the microbe through unnatural hosts or tissue culture, and removal of virulence genes. The vaccine for tuberculosis (BCG) was obtained after 13 years of subculturing the agent of bovine tuberculosis. Vaccines for measles, mumps, polio (Sabin), and rubella contain live, nonvirulent viruses. The advantages of live preparations are:

1. Viable microorganisms can multiply and produce infection (but not disease) like the natural organism.

TABLE 15.4	Checklist of Requirements for an Effective Vaccine

- It should have a low level of adverse side effects or toxicity and not cause serious harm.
- It should protect against exposure to natural, wild forms of pathogen.
- It should stimulate both antibody (B-cell) response and cell-mediated (T-cell) response.
- It should have long-term, lasting effects (produce memory).
- It should not require numerous doses or boosters.
- It should be inexpensive, have a relatively long shelf life, and be easy to administer.

2. They confer long-lasting protection.
3. They usually require fewer doses and boosters than other types of vaccines.

Disadvantages of using live microbes in vaccines are that they require special storage facilities, can be transmitted to other people, and can conceivably mutate back to a virulent strain.

If the exact antigenic determinants that stimulate immunity are known, it is possible to produce a vaccine based on a selected component of a microorganism. These vaccines for bacteria are called **acellular** or **subcellular vaccines.** For viruses, they are called **subunit vaccines.** The antigen used in these vaccines may be taken from cultures of the microbes, produced by genetic engineering or synthesized chemically.

Examples of component antigens currently in use are the capsules of the pneumococcus and meningococcus, the protein surface antigen of anthrax, and the surface proteins of hepatitis B virus. A special type of vaccine is the **toxoid,** which consists of a purified bacterial exotoxin that has been chemically denatured. By eliciting the production of antitoxins that can neutralize the natural toxin, toxoid vaccines provide protection against toxinoses such as diphtheria and tetanus.

Development of New Vaccines

Despite considerable successes, dozens of bacterial, viral, protozoan, and fungal diseases still remain without a functional vaccine. At the present time, no reliable vaccines are available for malaria, HIV/AIDS, various diarrheal diseases (*E. coli, Shigella*), respiratory diseases, and worm infections that affect over 200 million people per year worldwide. Of all of the challenges facing vaccine specialists, probably the most difficult has been choosing a vaccine antigen that is safe and that properly stimulates immunity. Currently, much attention is being focused on newer strategies for vaccine preparation that employ antigen synthesis, recombinant DNA, and gene cloning technology.

When the exact composition of an antigenic determinant is known, it is sometimes possible to artificially synthesize it.

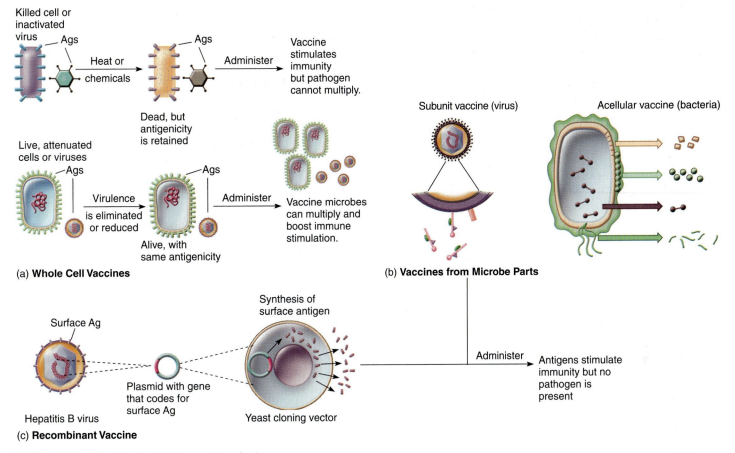

FIGURE 15.19 **Strategies in vaccine design.**
(a) Whole cells or viruses, killed or attenuated. **(b)** Acellular or subunit vaccines are made by disrupting the microbe to release various molecules or cell parts that can be isolated and purified. **(c)** Recombinant vaccines are made by isolating a gene for the antigen from the pathogen (here a hepatitis virus) and splicing it into a plasmid. Insertion of the recombinant plasmid into a cloning host (yeast) results in the production of large amounts of viral surface antigen to use in vaccine preparation.

This ability permits preservation of antigenicity while greatly increasing antigen purity and concentration. The malaria vaccine currently being used in areas of South America and Africa is composed of three synthetic peptides from the parasite. Several biotechnology companies are using plants to mass produce vaccine antigens. Tests are underway to grow tomatoes, potatoes, and bananas that synthesize proteins from cholera, hepatitis, papillomavirus, and *E. coli* pathogens. This strategy aims to deliver vaccines to populations that otherwise would not have access to them, by making them part of the food supply.

Genetically Engineered Vaccines Some of the genetic engineering concepts introduced in chapter 10 offer novel approaches to vaccine development. These methods are particularly effective in designing vaccines for obligate parasites that are difficult or expensive to culture, such as the syphilis spirochete or the malaria parasite. This technology provides a means of isolating the genes that encode various

microbial antigens, inserting them into plasmid vectors, and cloning them in appropriate hosts. The outcome of recombination can be varied as desired. For instance, the cloning host can be stimulated to synthesize and secrete a protein product (antigen), which is then harvested and purified **(figure 15.19c).** Certain vaccines for hepatitis are currently being prepared in this way. Antigens from the agents of syphilis, *Schistosoma,* and influenza have been similarly isolated and cloned and are currently being considered as potential vaccine material.

Another ingenious technique using genetic recombination has been nicknamed the *Trojan horse* vaccine. The term derives from an ancient legend in which the Greeks sneaked soldiers into the fortress of their Trojan enemies by hiding them inside a large, mobile wooden horse. In the microbial equivalent, genetic material from a selected infectious agent is inserted into a live carrier microbe that is nonpathogenic. In theory, the recombinant microbe will multiply and express the foreign genes, and the vaccine recipient will be immunized

TABLE 15.5 Currently Approved Vaccines

Disease/Vaccine Preparation	Route of Administration	Recommended Usage/Comments
Contain Killed Whole Bacteria		
Cholera	Subcutaneous (SQ) injection	For travelers; effect not long-term
Plague	SQ	For exposed individuals and animal workers; variable protection
Contain Live, Attenuated Bacteria		
Tuberculosis (BCG)	Intradermal (ID) injection	For high-risk occupations only; protection variable
Typhoid	Oral	For travelers only; low rate of effectiveness
Acellular Vaccines (Capsular Polysaccharides or Proteins)		
Anthrax	SQ	For protection in military recruits, occupationally exposed
Meningitis (meningococcal)	SQ	For protection in high-risk infants, military recruits; short duration
Meningitis (*Haemophilus influenzae*)	IM	For infants and children; may be administered with DTaP
Pneumococcal pneumonia	IM or SQ	Important for people at high risk: the young, elderly, and immunocompromised; moderate protection
Pertussis (aP)	IM	For newborns and children; contains recombinant protein antigens
Toxoids (Formaldehyde-Inactivated Bacterial Exotoxins)		
Diphtheria	IM	A routine childhood vaccination; highly effective in systemic protection
Tetanus	IM	A routine childhood vaccination; highly effective
Pertussis	IM	A routine childhood vaccination; highly effective
Botulism	IM	Only for exposed individuals such as laboratory personnel
Contain Inactivated Whole Viruses		
Poliomyelitis (Salk)	IM	Routine childhood vaccine; now used as first choice
Rabies	IM	For victims of animal bites or otherwise exposed; effective
Influenza	IM	For high-risk populations; requires constant updating for new strains; immunity not durable
Japanese encephalitis	SQ	For those residing in endemic areas, lab workers
Hepatitis A	IM	Protection for travelers, institutionalized people
Contain Live, Attenuated Viruses		
Adenovirus infection	Oral	For immunizing military recruits
Measles (rubeola)	SQ	Routine childhood vaccine; very effective
Mumps (parotitis)	SQ	Routine childhood vaccine; very effective
Poliomyelitis (Sabin)	Oral	Routine childhood vaccine; very effective, but can cause polio
Rubella	SQ	Routine childhood vaccine; very effective
Chickenpox (varicella)	SQ	Routine childhood vaccine; immunity can diminish over time
Smallpox (live vaccinia virus, not attenuated variola)	MP*	Since 2003 offered on voluntary basis for health care workers, some military
Yellow fever	SQ	Travelers, military personnel in endemic areas
Influenza	Inhaled	Same as for inactivated
Subunit Viral Vaccines		
Hepatitis B	IM	Recommended for all children, starting at birth; also for health workers and others at risk
Influenza	IM	See influenza, inactivated, above
Recombinant Vaccines		
Hepatitis B	IM	Used more often than subunit, but for same groups
Pertussis	IM	See acellular above

*MP = multiple puncture method.

against the microbial antigens. Vaccinia, the virus originally used to vaccinate for smallpox, and adenoviruses have proved practical agents for this technique. Vaccinia is used as the carrier in one of the experimental vaccines for AIDS, herpes simplex 2, leprosy, and tuberculosis.

DNA vaccines are being hailed as the most promising of all of the newer approaches to immunization. The technique in these formulations is very similar to gene therapy as described in chapter 10, except in this case, microbial (not human) DNA is inserted into a plasmid vector and inoculated into a recipient (figure 15.20). The expectation is that the human cells will take up some of the plasmids and express the microbial DNA in the form of proteins. Because these proteins are foreign, they will be recognized during immune surveillance and cause B and T cells to be sensitized and form memory cells.

Experiments with animals have shown that these vaccines are very safe and that only a small amount of the foreign antigen need be expressed to produce effective immunity. Another advantage to this method is that any number of potential microbial proteins can be expressed, making the antigenic stimulus more complex and improving the likelihood that it will stimulate both antibody and cell mediated immunity. At the present time, over 30 DNA-based vaccines are being tested in animals. Vaccines for Lyme disease, hepatitis C, herpes simplex, influenza, tuberculosis, papillomavirus, malaria, and SARS are undergoing animal trials, most with encouraging results.

Concerns about potential terrorist release of Class A bioterrorism agents have created pressure to re-introduce or improve vaccines for smallpox, anthrax, botulism, plague, and tularemia. Smallpox vaccination is now being given again to first responders (healthcare providers, the military). Earlier vaccines already exist for anthrax, tularemia, botulism, and plague, but they have been difficult to give or not effective. New vaccines are being developed and tested to immunize military and other personnel in the event of a disaster.

Route of Administration and Side Effects of Vaccines

Most vaccines are injected by subcutaneous, intramuscular, or intradermal routes. Oral vaccines are available for only three diseases (see table 15.5), but they have some distinct advantages. An oral dose of a vaccine can stimulate protection (IgA) on the mucous membrane of the portal of entry. Oral vaccines are also easier to give, more readily accepted, and well tolerated.

Some vaccines require the addition of a special binding substance, or **adjuvant** (ad'-joo-vunt). An adjuvant is any compound that enhances immunogenicity and prolongs antigen retention at the injection site. The adjuvant precipitates the antigen and holds it in the tissues so that it will be released gradually. Its gradual release presumably facilitates contact with antigen-presenting cells and lymphocytes. Common adjuvants are alum (aluminum hydroxide salts), Freund's adjuvant (emulsion of mineral oil, water, and extracts of mycobacteria), and beeswax.

Vaccines must go through many years of trials in experimental animals and human volunteers before they are licensed for general use. Even after they have been approved, like all therapeutic products, they are not without complications. The most common of these are local reactions at the

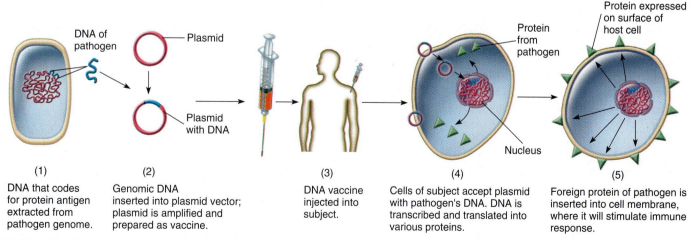

(1)	(2)	(3)	(4)	(5)
DNA that codes for protein antigen extracted from pathogen genome.	Genomic DNA inserted into plasmid vector; plasmid is amplified and prepared as vaccine.	DNA vaccine injected into subject.	Cells of subject accept plasmid with pathogen's DNA. DNA is transcribed and translated into various proteins.	Foreign protein of pathogen is inserted into cell membrane, where it will stimulate immune response.

FIGURE 15.20 **DNA vaccine preparation.**
DNA vaccines contain all or part of the pathogen's DNA, which is used to "infect" a recipient's cells. Processing of the DNA leads to production of an antigen protein that can stimulate a specific response against that pathogen.

injection site, fever, allergies, and other adverse reactions. Relatively rare reactions (about 1 case out of 220,000 vaccinations) are panencephalitis (from measles vaccine), back-mutation to a virulent strain (from polio vaccine), disease due to contamination with dangerous viruses or chemicals, and neurological effects of unknown cause (from pertussis and swine flu vaccines). Some patients experience allergic reactions to the medium (eggs or tissue culture) rather than to vaccine antigens. Some recent studies have attempted to link childhood vaccinations to later development of diabetes, asthma, and autism. After thorough examination of records, epidemiologists have found no convincing evidence for a vaccine connection to these diseases.

When known or suspected adverse effects have been detected, vaccines are altered or withdrawn. Recently, the whole-cell pertussis vaccine was replaced by the acellular capsule (aP) form when it was associated with adverse neurological effects. The live oral rotavirus vaccine had to be withdrawn when children experienced intestinal blockage. Polio vaccine was switched from live, oral to inactivated when too many cases of paralytic disease occurred from back-mutated vaccine stocks. Vaccine companies have also phased out certain preservatives, such as thimerosal, that are thought to cause allergies and other potential side effects.

Professionals involved in giving vaccinations must understand their inherent risks but also realize that the risks from the infectious disease almost always outweigh the chance of an adverse vaccine reaction. The greatest caution must be exercised in giving live vaccines to immunocompromised or pregnant patients, the latter because of possible risk to the fetus.

To Vaccinate: Why, Whom, and When?

Vaccination confers long-lasting, sometimes lifetime, protection in the individual, but an equally important effect is to protect the public health. Vaccination is an effective method of establishing **herd immunity** in the population. According to this concept, individuals immune to a communicable infectious disease will not harbor it, thus reducing the occurrence of that pathogen. With a larger number of immune individuals in a population (herd), it will be less likely that an unimmunized member of the population will encounter the agent. In effect, collective immunity through mass immunization confers indirect protection on the nonimmune members (such as children). Herd immunity maintained through immunization is an important force in preventing epidemics.

Until recently, vaccination was recommended for all typical childhood diseases for which a vaccine is available and

for adults only in certain special circumstances (health workers, travelers, military personnel). It has become apparent to public health officials that vaccination of adults is often needed in order to boost an older immunization, protect against "adult" infections (such as pneumonia in the elderly), or provide special protection in people with certain medical conditions.

In **table 15.6** the current recommended schedule for childhood immunizations is provided, with complete footnotes from the Centers for Disease Control and Prevention. **Table 15.7** contains the recommended adult immunization schedule, categorized by age group and by medical condition. Some vaccines are mixtures of antigens from several pathogens, notably Pediatrix (DTaP, IPV, and HB). It is also common for several vaccines to be given simultaneously, as occurs in military recruits who receive as many as 15 injections within a few minutes and children who receive boosters for DTaP and polio at the same time they receive the MMR vaccine. Experts doubt that immune interference (inhibition of one immune response by another) is a significant problem in these instances, and the mixed vaccines are carefully balanced to prevent this eventuality. The main problem with simultaneous administration is that side effects can be amplified.

In July of 2004 the Centers for Disease Control and Prevention reported that 79 percent of children in the United States were being vaccinated on time. While this represents a record high, it means that at least one million children in this country have not received adequate immunization.

✔ CHECKPOINT

- Knowledge of the specific immune response has a practical application: commercial production of antisera and vaccines.
- Artificial passive immunity usually involves administration of antiserum, and occasionally B and T cells. Antibodies collected from donors (human or otherwise) are injected into people who need protection immediately. Examples include ISG (immune serum globulin) and SIG (specific immune globulin).
- Artificial active agents are vaccines that provoke a protective immune response in the recipient but do not cause the actual disease. Vaccination is the process of challenging the immune system with a specially selected antigen. Examples are (1) killed or inactivated microbes; (2) live, attenuated microbes; (3) subunits of microbes; and (4) genetically engineered microbes or microbial parts.
- Vaccination programs seek to protect the individual directly through raising the antibody titer and indirectly through the development of herd immunity.

	Range of Recommended Ages			Catch-up Immunization				Preadolescent Assessment			
Vaccine ▼ Age ▶	Birth	1 mo	2 mos	4 mos	6 mos	12 mos	15 mos	18 mos	24 mos	4–6 yrs	11–12 yrs	13–18 yrs
Hepatitis B[1]	HepB #1 only if mother HBsAg (-)		HepB #2			HepB #3					HepB series	
Diphtheria, Tetanus, Pertussis[2]			DTaP	DTaP	DTaP		DTaP			DTaP	Td	Td
***Haemophilus influenzae* Type b[3]**			Hib	Hib	Hib[3]	Hib						
Inactivated Poliovirus			IPV	IPV		IPV				IPV		
Measles, Mumps, Rubella[4]						MMR #1				MMR #2		MMR #2
Varicella[5]						Varicella				Varicella		
Pneumococcal[6]			PCV	PCV	PCV	PCV				PCV	PPV	
Hepatitis A[7]										Hepatitis A series		
Influenza[8]						Influenza (yearly)						

Vaccines below this line are for selected populations

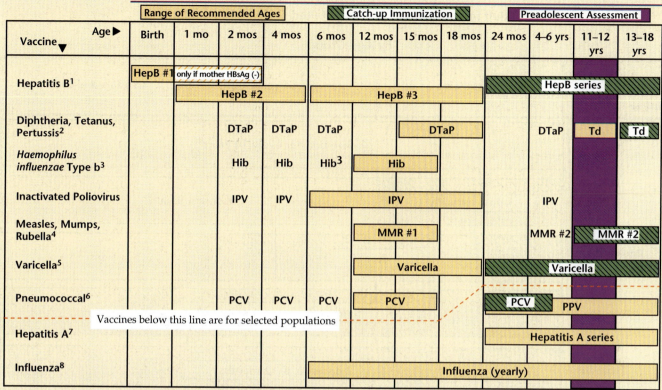

This schedule indicates the recommended ages for routine administration of currently licensed childhood vaccines, as of December 1, 2003, for children through age 18 years. Any dose not given at the recommended age should be given at any subsequent visit when indicated and feasible. ▨▨ Indicates age groups that warrant special effort to administer those vaccines not previously given. Additional vaccines may be licensed and recommended during the year. Licensed combination vaccines may be used whenever any components of the combination are indicated and the vaccine's other components are not contraindicated. Providers should consult the manufacturers' package inserts for detailed recommendations. Clinically significant adverse events that follow immunization should be reported to the Vaccine Adverse Event Reporting System (VAERS). Guidance about how to obtain and complete a VAERS form can be found on the Internet: http://www.vaers.org/ or by calling 1-800-822-7967.

1. **Hepatitis B (HepB) vaccine.** All infants should receive the first dose of hepatitis B vaccine soon after birth and before hospital discharge; the first dose may also be given by age 2 months if the infant's mother is hepatitis B surface antigen (HBsAg) negative. Only monovalent HepB can be used for the birth dose. Monovalent or combination vaccine containing HepB may be used to complete the series. Four doses of vaccine may be administered when a birth dose is given. The second dose should be given at least 4 weeks after the first dose, except for combination vaccines which cannot be administered before age 6 weeks. The third dose should be given at least 16 weeks after the first dose and at least 8 weeks after the second dose. The last dose in the vaccination series (third or fourth dose) should not be administered before age 24 weeks.

   Infants born to HBsAg-positive mothers should receive HepB and 0.5 mL of Hepatitis B Immune Globulin (HBIG) within 12 hours of birth at separate sites. The second dose is recommended at age 1 to 2 months. The last dose in the immunization series should not be administered before age 24 weeks. These infants should be tested for HBsAg and antibody to HBsAg (anti-HBs) at age 9 to 15 months.

   Infants born to mothers whose HBsAg status is unknown should receive the first dose of the HepB series within 12 hours of birth. Maternal blood should be drawn as soon as possible to determine the mother's HBsAg status; if the HBsAg test is positive, the infant should receive HBIG as soon as possible (no later than age 1 week). The second dose is recommended at age 1 to 2 months. The last dose in the immunization series should not be administered before age 24 weeks.

2. **Diphtheria and tetanus toxoids and acellular pertussis (DTaP) vaccine.** The fourth dose of DTaP may be administered as early as age 12 months, provided 6 months have elapsed since the third dose and the child is unlikely to return at age 15 to 18 months. The final dose in the series should be given at age ≥4 years. **Tetanus and diphtheria toxoids (Td)** is recommended at age 11 to 12 years if at least 5 years have elapsed since the last dose of tetanus and diphtheria toxoid-containing vaccine. Subsequent routine Td boosters are recommended every 10 years.

3. ***Haemophilus influenzae* type b (Hib) conjugate vaccine.** Three Hib conjugate vaccines are licensed for infant use. If PRP-OMP (PedvaxHIB or ComVax [Merck]) is administered at ages 2 and 4 months, a dose at age 6 months is not required. DTaP/Hib combination products should not be used for primary immunization in infants at ages 2, 4 or 6 months but can be used as boosters following any Hib vaccine. The final dose in the series should be given at age ≥12 months.

4. **Measles, mumps, and rubella vaccine (MMR).** The second dose of MMR is recommended routinely at age 4 to 6 years but may be administered during any visit, provided at least 4 weeks have elapsed since the first dose and both doses are administered beginning at or after age 12 months. Those who have not previously received the second dose should complete the schedule by the 11- to 12-year-old visit.

5. **Varicella vaccine.** Varicella vaccine is recommended at any visit at or after age 12 months for susceptible children (i.e., those who lack a reliable history of chickenpox). Susceptible persons age ≥13 years should receive 2 doses, given at least 4 weeks apart.

6. **Pneumococcal vaccine.** The heptavalent **pneumococcal conjugate vaccine (PCV)** is recommended for all children age 2 to 23 months. It is also recommended for certain children age 24 to 59 months. The final dose in the series should be given at age ≥12 months. **Pneumococcal polysaccharide vaccine (PPV)** is recommended in addition to PCV for certain high-risk groups. See *MMWR* 2000;49(RR-9): 1-38.

7. **Hepatitis A vaccine.** Hepatitis A vaccine is recommended for children and adolescents in selected states and regions and for certain high-risk groups; consult your local public health authority. Children and adolescents in these states, regions, and high-risk groups who have not been immunized against hepatitis A can begin the hepatitis A immunization series during any visit. The 2 doses in the series should be administered at least 6 months apart. See *MMWR* 1999;48(RR-12): 1-37.

8. **Influenza vaccine.** Influenza vaccine is recommended annually for children age ≥6 months with certain risk factors (including but not limited to children with asthma, cardiac disease, sickle cell disease, human immunodeficiency virus infection, and diabetes; and household members of persons in high-risk groups [see *MMWR* 2003;52(RR-8): 1–36]) and can be administered to all others wishing to obtain immunity. In addition, healthy children age 6 to 23 months are encouraged to receive influenza vaccine if feasible, because children in this age group are at substantially increased risk of influenza-related hospitalizations. For healthy persons age 5 to 49 years, the intranasally administered live-attenuated influenza vaccine (LAIV) is an acceptable alternative to the intramuscular trivalent inactivated influenza vaccine (TIV). See *MMWR* 2003;52(RR-13): 1-8. Children receiving TIV should be administered a dosage appropriate for their age (0.25 mL if age 6 to 35 months or 0.5 mL if age ≥3 years). Children age ≤8 years who are receiving influenza vaccine for the first time should receive 2 doses (separated by at least 4 weeks for TIV and at least 6 weeks for LAIV).

For additional information about vaccines, including precautions and contraindications for immunization and vaccine shortages, please visit the National Immunization Program Web site at www.cdc.gov/nip/ or call the National Immunization Information Hotline at 800-232-2522 (English) or 800-232-0233 (Spanish).

Approved by the Advisory Committee on Immunization Practices (www.cdc.gov/nip/acip), the American Academy of Pediatrics (www.aap.org), and the American Academy of Family Physicians (www.aafp.org).

Source: www.cdc.gov/nip/recs/child-schedule.pdf.

TABLE 15.7   Recommended Adult Immunization Schedule, United States, 2003–2004

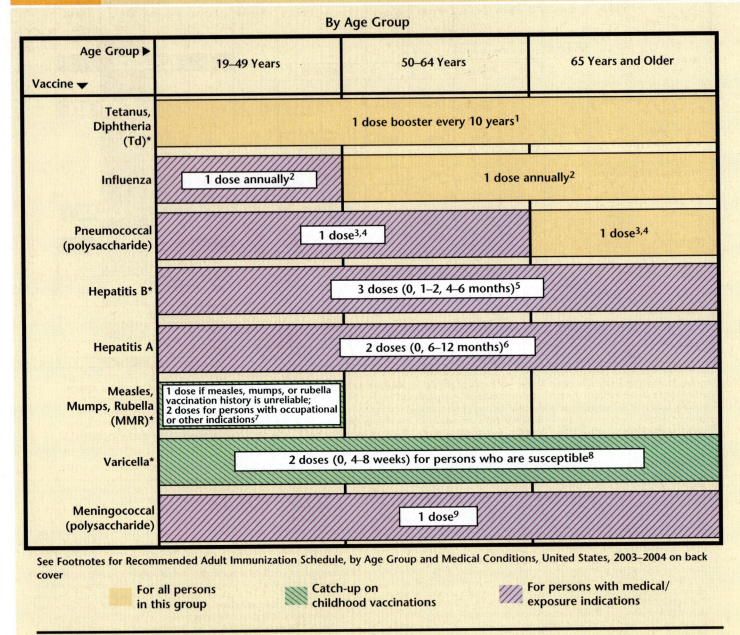

By Age Group

Age Group ▶ Vaccine ▼	19–49 Years	50–64 Years	65 Years and Older
Tetanus, Diphtheria (Td)*	1 dose booster every 10 years[1]		
Influenza	1 dose annually[2]	1 dose annually[2]	
Pneumococcal (polysaccharide)	1 dose[3,4]		1 dose[3,4]
Hepatitis B*	3 doses (0, 1–2, 4–6 months)[5]		
Hepatitis A	2 doses (0, 6–12 months)[6]		
Measles, Mumps, Rubella (MMR)*	1 dose if measles, mumps, or rubella vaccination history is unreliable; 2 doses for persons with occupational or other indications[7]		
Varicella*	2 doses (0, 4–8 weeks) for persons who are susceptible[8]		
Meningococcal (polysaccharide)	1 dose[9]		

See Footnotes for Recommended Adult Immunization Schedule, by Age Group and Medical Conditions, United States, 2003–2004 on back cover

For all persons in this group	Catch-up on childhood vaccinations	For persons with medical/ exposure indications

---

*Covered by the Vaccine Injury Compensation Program. For information on how to file a claim call 800-338-2382. Please also visit www.hrsa.gov/osp/vicp To file a claim for vaccine injury contact: U.S. Court of Federal Claims, 717 Madison Place, N.W., Washington, D.C. 20005, 202-219-9657.

This schedule indicates the recommended age groups for routine administration of currently licensed vaccines for persons 19 years of age and older. Licensed combination vaccines may be used whenever any components of the combination are indicated and the vaccine's other components are not contraindicated. Providers should consult the manufacturers' package inserts for detailed recommendations.

Report all clinically significant post-vaccination reactions to the Vaccine Adverse Event Reporting System (VAERS). Reporting forms and instructions on filing a VAERS report are available by calling 800-822-7967 or from the VAERS website at www.vaers.org.

For additional information about the vaccines listed above and contraindications for immunization, visit the National Immunization Program Website at www.cdc.gov/nip/ or call the National Immunization Hotline at 800-232-2522 (English) or 800-232-0233 (Spanish).

Approved by the Advisory Committee on Immunization Practices (ACIP), and accepted by the American College of Obstetricians and Gynecologists (ACOG) and the American Academy of Family Physicians (AAFP)

Source: www.cdc.gov/nip/recs/adult-schedule-2page.pdf

By Medical Conditions

Medical Conditions ▼ Vaccine ▶	Tetanus-Diphtheria (Td)*,[1]	Influenza[2]	Pneumo-coccal (polysac-charide)[3,4]	Hepatitis B*,[5]	Hepatitis A[6]	Measles, Mumps, Rubella (MMR)*,[7]	Varicella*,[8]
Pregnancy		A					
Diabetes, heart disease, chronic pulmonary disease, chronic liver disease, including chronic alcoholism		B	C		D		
Congenital Immunodeficiency, leukemia, lymphoma, generalized malignancy, therapy with alkylating agents, antimetabolites, radiation or large amounts of corticosteroids			E				F
Renal failure/end stage renal disease, recipients of hemodialysis or clotting factor concentrates			E	G			
Asplenia including elective splenectomy and terminal complement component deficiencies		H	E, I, J				
HIV infection			E, K			L	

See Special Notes for Medical Conditions below—also see Footnotes for Recommended Adult Immunization Schedule, by Age Group and Medical Conditions, United States, 2003–2004 on back cover

- For all persons in this group
- Catch-up on childhood vaccinations
- For persons with medical/exposure indications
- Contraindicated

Special Notes for Medical Conditions

A. For women without chronic diseases/conditions, vaccinate if pregnancy will be at 2nd or 3rd trimester during influenza season. For women with chronic diseases/conditions, vaccinate at any time during the pregnancy.

B. Although chronic liver disease and alcoholism are not indicator conditions for influenza vaccination, give 1 dose annually if the patient is age 50 years or older, has other indications for influenza vaccine, or if the patient requests vaccination.

C. Asthma is an indicator condition for influenza but not for pneumococcal vaccination.

D. For all persons with chronic liver disease.

E. For persons < 65 years, revaccinate once after 5 years or more have elapsed since initial vaccination.

F. Persons with impaired humoral immunity but intact cellular immunity may be vaccinated. *MMWR* 1999;48 (RR-06):1-5.

G. Hemodialysis patients: Use special formulation of vaccine (40 μg/mL) or two 1.0 mL 20 μg doses given at one site. Vaccinate early in the course of renal disease. Assess antibody titers to hep B surface antigen (anti-HBs) levels annually. Administer additional doses if anti-HBs levels decline to <10 milli international units (mIU)/mL.

H. There are no data specifically on risk of severe or complicated influenza infections among persons with asplenia. However, influenza is a risk factor for secondary bacterial infections that may cause severe disease in asplenics.

I. Administer meningococcal vaccine and consider Hib vaccine.

J. Elective splenectomy: vaccinate at least 2 weeks before surgery.

K. Vaccinate as close to diagnosis as possible when CD4 cell counts are highest.

L. Withhold MMR or other measles containing vaccines from HIV-infected persons with evidence of severe immunosuppression. *MMWR* 1998; 47 (RR-8):21-22; *MMWR* 2002;51 (RR-02):22-24.

Footnotes for
Recommended Adult Immunization Schedule by Age Group and Medical Conditions, United States, 2003–2004

1. **Tetanus and diphtheria (Td) toxoids**—Adults including pregnant women with uncertain histories of a complete primary vaccination series should receive a primary series of Td. A primary series for adults is 3 doses: the first 2 doses given at least 4 weeks apart and the 3rd dose, 6–12 months after the second. Administer 1 dose if the person had received the primary series and the last vaccination was 10 years ago or longer. Consult *MMWR* 1991; 40 (RR-10): 1-21 for administering Td as prophylaxis in wound management. The ACP Task Force on Adult Immunization supports a second option for Td use in adults: a single Td booster at age 50 years for persons who have completed the full pediatric series, including the teenage/young adult booster. *Guide for Adult Immunization.* 3rd ed. ACP 1994: 20.

2. **Influenza vaccination**—Medical indications: chronic disorders of the cardiovascular or pulmonary systems including asthma; chronic metabolic diseases including diabetes mellitus, renal dysfunction, hemoglobin-opathies, or immunosuppression (including immunosuppression caused by medications or by human immunodeficiency virus [HIV]), requiring regular medical follow-up or hospitalization during the preceding year; women who will be in the second or third trimester of pregnancy during the influenza season. Occupational indications: health-care workers. Other indications: residents of nursing homes and other long-term care facilities; persons likely to transmit influenza to persons at high-risk (in-home care givers to persons with medical indications, household contacts and out-of-home caregivers of children birth to 23 months of age, or children with asthma or other indicator conditions for influenza vaccination, household members and care givers of elderly and adults with high-risk conditions); and anyone who wishes to be vaccinated. For healthy persons aged 5–49 years without high risk conditions, either the inactivated vaccine or the intranasally administered influenza vaccine (Flumist) may be given. *MMWR* 2003; 52 (RR-8): 1-36; *MMWR* 2003; 53 (RR-13): 1-8.

3. **Pneumococcal polysaccharide vaccination**—Medical indications: chronic disorders of the pulmonary system (excluding asthma), cardiovascular diseases, diabetes mellitus, chronic liver diseases including liver disease as a result of alcohol abuse (e.g., cirrhosis), chronic renal failure or nephrotic syndrome, functional or anatomic asplenia (e.g., sickle cell disease or splenectomy), immunosuppressive conditions (e.g., congenital immunodeficiency, HIV infection, leukemia, lymphoma, multiple myeloma, Hodgkins disease, generalized malignancy, organ or bone marrow transplantation), chemotherapy with alkylating agents, anti-metabolites, or long-term systemic corticosteroids. Geographic/other indications: Alaskan Natives and certain American Indian populations. Other indications: residents of nursing homes and other long-term care facilities. *MMWR* 1997; 46 (RR-8): 1–24.

4. **Revaccination with pneumococcal polysaccharide vaccine**—One time revaccination after 5 years for persons with chronic renal failure or nephrotic syndrome, functional or anatomic asplenia (e.g., sickle cell disease or splenectomy), immunosuppressive conditions (e.g., congenital immunodeficiency, HIV infection, leukemia, lymphoma, multiple myeloma, Hodgkins disease, generalized malignancy, organ or bone marrow transplantation), chemotherapy with alkylating agents, anti-metabolites, or long-term systemic corticosteroids. For persons 65 and older, one-time revaccination if they were vaccinated 5 or more years previously and were aged less than 65 years at the time of primary vaccination. *MMWR* 1997; 46 (RR-8): 1-24.

5. **Hepatitis B vaccination**—Medical indications: hemodialysis patients, patients who receive clotting-factor concentrates. Occupational indications: health-care workers and public-safety workers who have exposure to blood in the workplace, persons in training in schools of medicine, dentistry, nursing, laboratory technology, and other allied health professions. Behavioral indications: injecting drug users, persons with more than one sex partner in the previous 6 months, persons with a recently acquired sexually-transmitted disease (STD), all clients in STD clinics, men who have sex with men. Other indications: household contacts and sex partners of persons with chronic HBV infection, clients and staff of institutions for the developmentally disabled, international travelers who will be in countries with high or intermediate prevalence of chronic HBV infection for more than 6 months, inmates of correctional facilities. *MMWR* 1991; 40 (RR-13): 1-19. (www.cdc.gov/travel/diseases/hbv.htm).

6. **Hepatitis A vaccination**—For the combined HepA-HepB vaccine use 3 doses at 0, 1, 6 months). Medical indications: persons with clotting-factor disorders or chronic liver disease. Behavioral indications: men who have sex with men, users of injecting and noninjecting illegal drugs. Occupational indications: persons working with HAV-infected primates or with HAV in a research laboratory setting. Other indications: persons traveling to or working in countries that have high or intermediate endemicity of hepatitis A. *MMWR* 1999; 48 (RR-12): 1-37. (www.cdc.gov/travel/diseases/hav.htm).

7. **Measles, Mumps, Rubella vaccination (MMR)**—Measles component: Adults born before 1957 may be considered immune to measles. Adults born in or after 1957 should receive at least one dose of MMR unless they have a medical contraindication, documentation of at least one dose or other acceptable evidence of immunity. A second dose of MMR is recommended for adults who:

- are recently exposed to measles or in an outbreak setting
- were previously vaccinated with killed measles vaccine
- were vaccinated with an unknown vaccine between 1963 and 1967
- are students in post-secondary educational institutions
- work in health care facilities
- plan to travel internationally

Mumps component: 1 dose of MMR should be adequate for protection. Rubella component: Give 1 dose of MMR to women whose rubella vaccination history is unreliable and counsel women to avoid becoming pregnant for 4 weeks after vaccination. For women of child-bearing age, regardless of birth year, routinely determine rubella immunity and counsel women regarding congenital rubella syndrome. Do not vaccinate pregnant women or those planning to become pregnant in the next 4 weeks. If pregnant and susceptible, vaccinate as early in postpartum period as possible. *MMWR* 1998; 47 (RR-8): 1-57; *MMWR* 2001; 50: 1117.

8. **Varicella vaccination**—Recommended for all persons who do not have reliable clinical history of varicella infection, or serological evidence of varicella zoster virus (VZV) infection who may be at high risk for exposure or transmission. This includes, health-care workers and family contacts of immunocompromised persons, those who live or work in environments where transmission is likely (e.g., teachers of young children, day care employees, and residents and staff members in institutional settings), persons who live or work in environments where VZV transmission can occur (e.g., college students, inmates and staff members of correctional institutions, and military personnel), adolescents and adults living in households with children, women who are not pregnant but who may become pregnant in the future, international travelers who are not immune to infection. Note: Greater than 95% of U.S. born adults are immune to VZV. Do not vaccinate pregnant women or those planning to become pregnant in the next 4 weeks. If pregnant and susceptible, vaccinate as early in postpartum period as possible. *MMWR* 1996; 45 (RR-11): 1-36; *MMWR* 1999; 48 (RR-6): 1-5.

9. **Meningococcal vaccination (quadrivalent polysaccharide vaccine for serogroups A, C, Y, and W-135)**—Consider vaccination for persons with medical indications: adults with terminal complement component deficiencies, with anatomic or functional asplenia. Other indications: travelers to countries in which disease is hyperendemic or epidemic ("meningitis belt" of sub-Saharan Africa, Mecca, Saudi Arabia for Hajj). Revaccination at 3–5 years may be indicated for persons at high risk for infection (e.g., persons residing in areas in which disease is epidemic). Counsel college freshmen, especially those who live in dormitories, regarding meningococcal disease and the vaccine so that they can make an educated decision about receiving the vaccination. *MMWR* 2000; 49 (RR-7): 1-20.

Note: The AAFP recommends that colleges should take the lead on providing education on meningococcal infection and vaccination and offer it to those who are interested. Physicians need not initiate discussion of the meningococcal quadrivalent polysaccharide vaccine as part of routine medical care.

Chapter Summary With Key Terms

15.1 Specific Immunity: The Third and Final Line of Defense Development of Lymphocyte Specificity/Receptors

Acquired immunity involves the reactions of B and T lymphocytes to foreign molecules, or **antigens.** Before they can react, each lymphocyte must undergo differentiation into its final functional type by developing protein receptors for antigen, the specificity of which is randomly generated and unique for each lymphocyte.

15.2 An Overview of Specific Immune Responses

A. Genetic recombination and mutation during embryonic and fetal development produce billions of different lymphocytes, each bearing a different receptor.

B. *Tolerance to self* occurs during this time.

C. The receptors on B cells are **immunoglobulin (Ig)** molecules, and receptors on T cells are smaller unrelated glycoprotein molecules.

D. T-cell receptors bind antigenic determinants on MHC molecules, which in humans are called **human leukocyte** antigens (HLA).

15.3 The Lymphocyte Response System in Depth

An antigen (Ag) is any substance that stimulates an immune response.

A. Requirements for **antigenicity** include foreignness (recognition as nonself), large size, and complexity of cell or molecule.

B. Foreign molecules less than 1,000 MW **(haptens)** are not antigenic unless attached to a larger carrier molecule.

C. The epitope is the small molecular group of the foreign substance that is recognized by lymphocytes. Cells, viruses, and large molecules can have numerous antigenic determinants.

15.4 Cooperation in Immune Reactions to Antigens

A. T-cell-dependent antigens must be processed by large phagocytes such as dendritic cells or macrophages called **antigen-presenting cells (APCs).**

B. The presentation of a single antigen involves a direct collaboration between an APC, a T helper (T_H), and an antigen specific B or T cell.

C. Cytokines involved are **interleukin-1** from the APC which activates the T_H cells, and **interleukin-2** produced by the T_H cell, which activates B and other T cells.

15.5 B-Cell Response

A. Once B cells process the antigen, interact with T_H cells, and are stimulated by B-cell growth and differentiation factors, they enter the cell cycle in preparation for mitosis and clonal expansion.

B. Divisions give rise to **plasma cells** that secrete antibodies and **memory cells** that can react to that same antigen later.

1. *Nature of Antibodies* (Immunoglobulins): A single immunoglobulin molecule (monomer) is a large Y-shaped protein molecule consisting of four polypeptide chains. It contains two identical regions **(Fab)** with ends that form the active site that binds with a unique specificity to an antigen. The **Fc** region determines the location and the function of the antibody molecule.

2. *Antigen-Antibody (Ag-Ab) Reactions* include **opsonization, neutralization, agglutination,** and complement fixation.

3. The five **antibody classes,** which differ in size and function, are **IgG, IgA** (secretory Ab), **IgM, IgD,** and **IgE.**

4. *Antibodies in Serum* **(Antiserum):**

   a. The first introduction of an Ag to the immune system produces a **primary response,** with a gradual increase in Ab titer.

   b. The second contact with the same Ag produces a **secondary,** or **memory response,** due to memory cells produced during initial response.

15.6 T-Cell Response

A. T_H cells (CD4) secrete cytokines that activate macrophages to kill phagocytosed antigens and activate cytotoxic (T_C) cells to kill infected cells and to become memory T_C cells. These cytokines suppress humoral immunity (antibody production).

B. T helper 2 (T_H2) CD4 cells activate B cells to make IgG, IgM and IgE antibodies and become memory B cells. T_H2 cells suppress cellular immunity.

C. Cytotoxic (T_C) CD8 cells recognize and kill infected cells, tumor cells, and foreign transplant cells.

15.7 A Practical Scheme for Classifying Specific Immunities

Immunities acquired through B and T lymphocytes can be classified by a simple system.

A. Natural immunity is acquired as part of normal life experiences.

B. Artificial immunity is acquired through medical procedures such as immunization.

C. **Active immunity** results when a person is challenged with antigen that stimulates production of antibodies. It creates memory, takes time, and is lasting.

D. In **passive immunity,** preformed antibodies are donated to an individual. It does not create memory, acts immediately, and is short term.

15.8 Immunization: Methods of Manipulating Immunity for Therapeutic Purposes

A. Passive immunotherapy includes administering immune serum globulin and specific immune globulins pooled from donated serum to prevent infection and disease in those at risk; antisera and antitoxins from animals are occasionally used.

B. Active immunization is synonymous with **vaccination;** provides an antigenic stimulus that does not cause disease but can produce long-lasting, protective immunity. **Vaccines** are made with:

   1. **Killed** whole cells or **inactivated viruses** that do not reproduce but are antigenic.

   2. Live, **attenuated** cells or viruses that are able to reproduce but have lost virulence.

   3. **Acellular** or **subunit** components of microbes such as surface antigen or neutralized toxins **(toxoids).**

   4. Genetic engineering techniques, including cloning of antigens, recombinant attenuated microbes, and **DNA**-based **vaccines.**

C. Boosters (additional doses) are often required.

D. Vaccination increases herd immunity, protection provided by mass immunity in a population.

Multiple-Choice Questions

1. The primary B-cell receptor is
   a. IgD
   b. IgA
   c. IgE
   d. IgG

2. In humans, B cells mature in the _____, and T cells mature in the _____.
   a. GALT, liver
   b. bursa, thymus
   c. bone marrow, thymus
   d. lymph nodes, spleen

3. Small, simple molecules are _____ antigens.
   a. poor
   b. never
   c. good
   d. heterophilic

4. Which type of cell actually secretes antibodies?
   a. T cell
   b. macrophage
   c. plasma cell
   d. monocyte

5. The cross-linkage of antigens by antibodies is known as
   a. opsonization
   b. a cross-reaction
   c. agglutination
   d. complement fixation

6. The greatest concentration of antibodies is found in the _____ fraction of the serum.
   a. gamma globulin
   b. albumin
   c. beta globulin
   d. alpha globulin

7. T _____ cells assist in the functions of certain B cells and other T cells.
   a. Sensitized
   b. Cytotoxic
   c. Helper
   d. Natural killer

8. T$_C$ cells are important in controlling
   a. virus infections
   b. allergy
   c. autoimmunity
   d. all of these

9. Vaccination is synonymous with _____ immunity.
   a. natural active
   b. artificial passive
   c. artificial active
   d. natural passive

10. Which of the following can serve as antigen-presenting cells (APCs)?
    a. T cells
    b. B cells
    c. macrophages
    d. dendritic cells
    e. b, c, and d

11. **Multiple matching.** Place all possible matches in the space at the left.
    _____ IgG   b f g h
    _____ IgA   a b c
    _____ IgD   b
    _____ IgE   b i
    _____ IgM   d e h
    a. found in mucous secretions
    b. a monomer
    c. a dimer
    d. has greatest number of Fabs
    e. major Ig of primary response to Ag
    f. major Ig of secondary response to Ag
    g. crosses the placenta
    h. fixes complement
    i. involved in allergic reactions

12. A living microbe with reduced virulence that is used for vaccination is considered
    a. a toxoid
    b. attenuated
    c. denatured
    d. an adjuvant

13. A vaccine that contains parts of viruses is called
    a. acellular
    b. recombinant
    c. subunit
    d. attenuated

14. Widespread immunity that protects the population from the spread of disease is called
    a. seropositivity
    b. cross-reactivity
    c. epidemic prophylaxis
    d. herd immunity

15. DNA vaccines contain _____ DNA that stimulates cells to make _____ antigens.
    a. human, RNA
    b. microbial, protein
    c. human, protein
    d. microbial, polysaccharide

16. What is the purpose of an adjuvant?
    a. to kill the microbe
    b. to stop allergic reactions
    c. to improve the contact between the antigen and lymphocytes
    d. to make the antigen more soluble in the tissues

Concept Questions

These questions are suggested as a *writing-to-learn* experience. For each question (except #15), compose a one- or two-paragraph answer that includes the factual information needed to completely address the question.

1. a. What function do receptors play in specific immune responses?
   b. How can receptors be made to vary so widely?

2. Describe the major histocompatibility complex, and explain how it participates in immune reactions.

3. a. Explain the clonal selection theory of antibody specificity and diversity.
   b. Why must the body develop tolerance to self?

4. a. Trace the origin and development of B lymphocytes; of T lymphocytes.
   b. What is happening during lymphocyte maturation?

5. Describe three ways that B cells and T cells are similar and at least five major ways in which they are different.

6. a. What is an antigen or immunogen?
   b. What is an epitope?
   c. How do foreignness, size, and complexity contribute to antigenicity?

7. a. Describe the actions of an antigen-presenting cell.
   b. What is the difference between a T-cell-dependent and T-cell-independent response?

8. a. Trace the immune response system, beginning with the entry of a T-cell-dependent antigen, antigen processing, presentation, the cooperative response among the macrophage and lymphocytes, and the reactions of activated B and T cells.
   b. What are the actions of interleukins-1 and -2?

9. a. On what basis is a particular B-cell clone selected?
   b. How are B cells activated, and what events are involved in this process?
   c. What happens after B cells are activated?
   d. What are the functions of plasma cells, clonal expansion, and memory cells?

10. a. Describe the structure of immunoglobulin.
    b. What are the functions of the Fab and Fc portions?
    c. Describe four or five ways that antibodies function in immunity.
    d. Describe the attachment of Abs to Ags. (What eventually happens to the Ags?)

11. a. Contrast the primary and secondary response to Ag.
    b. Explain the type, order of appearance, and amount of immunoglobulin in each response and the reasons for them.
    c. What causes the latent period? The anamnestic response?

12. a. Explain how monoclonal and polyclonal antibodies are different.
    b. Describe several possible applications of monoclonals in medicine.

13. a. Why are the immunities involving T cells called cell-mediated?
    b. How do T cells become sensitized?
    c. How do cytotoxic cells kill their target?

14. a. Contrast active and passive immunity in terms of how each is acquired, how long it lasts, whether memory is triggered, how soon it becomes effective, and what immune cells and substances are involved.
    b. Name at least two major ways that natural and artificial immunities are different.

15. **Multiple matching.** (Summarizes information from chapters 14 and 15 [see figure 14.1].) In the blanks on the left, place the letters of all of the host defenses and immune responses in the right column that can fit the description.

    _____ vaccination for tetanus
    _____ lysozyme in tears
    _____ immunization with horse serum
    _____ in utero transfer of antibodies
    _____ booster injection for diphtheria
    _____ recovery from a case of mumps
    _____ colostrum
    _____ interferon
    _____ action of neutrophils
    _____ injection of gamma globulin
    _____ recovery from a case of mumps
    _____ edema
    _____ humans having protection from canine distemper virus
    _____ stomach acid
    _____ cilia in trachea
    _____ asymptomatic chickenpox
    _____ complement

    a. active
    b. passive
    c. natural
    d. artificial
    e. acquired
    f. innate, inborn
    g. chemical barrier
    h. mechanical barrier
    i. genetic barrier
    j. specific
    k. nonspecific
    l. inflammatory response
    m. second line of defense
    n. none of these

16. a. What are the advantages and disadvantages of a killed vaccine; a live, attenuated vaccine; a subunit vaccine; a recombinant vaccine; and a DNA vaccine?
    b. Use an outline to explain how an inoculation with tetanus toxoid will protect a person the next time he or she steps on a dirty piece of glass.

17. a. Describe the concept of herd immunity.
    b. How does vaccination contribute to its development in a community?
    c. Give some possible explanations for recent epidemics of diphtheria and whooping cough.

Critical Thinking Questions

Critical thinking is the ability to reason and solve problems using facts and concepts. These questions can be approached from a number of angles, and in most cases, they do not have a single correct answer.

1. What is the advantage of having lymphatic organs screen the body fluids, directly and indirectly?

2. Cells contain built-in suicide genes to self-destruct by apoptosis under certain conditions. Can you explain why development of the immune system might depend in part on this sort of adaptation?

3. a. Give some possible explanations for the need to have immune tolerance.
   b. Why would it be necessary for the T cells to bind both antigen and self (MHC) receptors?

4. Double-stranded DNA is a large, complex molecule, but it is not generally immunogenic unless it is associated with proteins or carbohydrates. Can you think why this might be so? (Hint: How universal is DNA?)

5. Explain how it is possible for people to give a false-positive reaction in blood tests for syphilis, AIDS, and infectious mononucleosis.

6. Describe the cellular/microscopic pathology in the immune system of AIDS patients that results in opportunistic infections and cancers.

7. Explain why most immune reactions result in a polyclonal collection of antibodies.

8. a. Combine information on the functions of different classes of Ig to explain the exact mechanisms of natural passive immunity (both transplacental and colostrum-induced).
   b. Why are these sorts of immunity short-term?

9. Use a football game or warfare analogy to produce a scenario depicting the major activities of the immune system from this chapter and from chapter 14.

10. Using words and arrows, complete a flow outline of an immune response, beginning with entrance of antigen; include processing, cell interaction, involvement of cytokines, and the end results for B and T cells.

11. Describe the relationship between an antitoxin, a toxin, and a toxoid.

▶ Autoimmune diseases, such as rheumatoid arthritis and multiple sclerosis, are due to B and T cells that are abnormally sensitized to react with the body's natural molecules and thus can damage cells and tissues.

▶ T-cell responses to certain allergens and foreign molecules cause the diseases known as delayed-type hypersensitivities and graft rejection.

▶ Immunodeficiencies occur when B and T cells and other immune cells are missing or destroyed. They may be inborn and genetic or acquired.

▶ The primary outcome of immunodeficiencies is manifest in recurrent infections and lack of immune competence.

16.1 The Immune Response: A Two-Sided Coin

Humans possess a powerful and intricate system of defense, which by its very nature also carries the potential to cause injury and disease. In most instances, a defect in immune function is expressed in commonplace, but miserable, symptoms such as those of hay fever and dermatitis. But abnormal or undesirable immune functions are also actively involved in debilitating or life-threatening diseases such as asthma, anaphylaxis, rheumatoid arthritis, and graft rejection.

With few exceptions, our previous discussions of the immune response have centered around its numerous beneficial effects. The precisely coordinated system that seeks out, recognizes, and destroys an unending array of foreign materials is clearly protective, but it also presents another side—a side that promotes rather than prevents disease. In this chapter, we will survey **immunopathology,** the study of disease states associated with overreactivity or underreactivity of the immune response **(figure 16.1).** In the cases of allergies and *autoimmunity,* the tissues are innocent bystanders attacked by immunologic functions that can't distinguish one's own tissues from those expressing foreign material. In **grafts** and **transfusions,** a recipient reacts to the foreign tissues and cells of another individual. In **immunodeficiency** diseases, immune function is incompletely developed, suppressed, or destroyed. Cancer falls into a special category, because it is both a cause and an effect of immune dysfunction. As we shall see, one fascinating by-product of studies of immune disorders has been our increased understanding of the basic workings of the immune system.

Overreactions to Antigens: Allergy/Hypersensitivity

The term **allergy** means a condition of altered reactivity or exaggerated immune response that is manifested by inflammation. Although it is sometimes used interchangeably with hypersensitivity, some experts refer to immediate reactions such as hay fever as allergies and to delayed reactions as hypersensitivities. Allergic individuals are acutely sensitive to repeated contact with antigens, called **allergens,** that do not noticeably affect nonallergic individuals. Although the general effects of hypersensitivity are detrimental, we must be aware that it involves the very same types of immune reactions as those at work in protective immunities. These in-

clude humoral and cell-mediated actions, the inflammatory response, phagocytosis, and complement. Such an association means that all humans have the potential to develop hypersensitivity under particular circumstances.

Originally, allergies were defined as either immediate or delayed, depending upon the time lapse between contact with the allergen and onset of symptoms. Subsequently, they were differentiated as humoral versus cell-mediated. But as information on the nature of the allergic immune response accumulated, it became evident that, although useful, these schemes oversimplified what is really a very complex spectrum of reactions. The most widely accepted classification, first introduced by immunologists P. Gell and R. Coombs, includes four major categories: type I ("common" allergy and anaphylaxis), type II (IgG- and IgM-mediated cell damage), type III (immune complex), and type IV (delayed hypersensitivity) **(table 16.1).** In general, types I, II, and III involve a B-cell–immunoglobulin response, and type IV involves a T-cell response (figure 16.1). The antigens that elicit these reactions can be exogenous, originating from outside the body (microbes, pollen grains, and foreign cells and proteins), or endogenous, arising from self tissue (autoimmunities).

One of the reasons allergies are easily mistaken for infections is that both involve damage to the tissues and thus trigger the inflammatory response, as described in chapter 14. Many symptoms and signs of inflammation (redness, heat, skin eruptions, edema, and granuloma) are prominent features of allergies.

✔ CHECKPOINT

■ Immunopathology is the study of diseases associated with excesses and deficiencies of the immune response. Such diseases include allergies, autoimmunity, grafts, transfusions, immunodeficiency disease, and cancer.

■ An allergy or hypersensitivity is an exaggerated immune response that injures or inflames tissues.

■ There are four categories of hypersensitivity reactions: type I (allergy and anaphylaxis), type II (IgG and IgM tissue destruction), type III (immune complex reactions), and type IV (delayed hypersensitivity reactions).

■ Antigens that trigger hypersensitivity reactions are allergens. They can be either exogenous (originate outside the host) or endogenous (involve the host's own tissue).

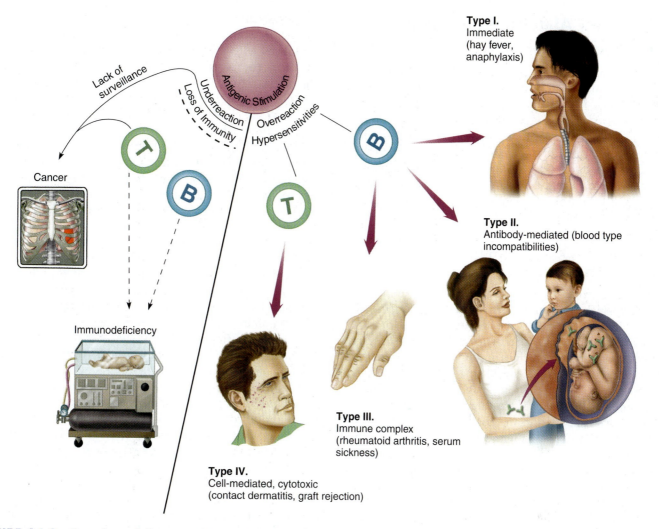

FIGURE 16.1   **Overview of diseases of the immune system.**
Just as the system of T cells and B cells provides necessary protection against infection and disease, the same system can cause serious and debilitating conditions by overreacting or underreacting to immune stimuli.

TABLE 16.1	Hypersensitivity States	
Type	**Systems and Mechanisms Involved**	**Examples**
I. Immediate hypersensitivity	IgE-mediated; involves mast cells, basophils, and allergic mediators	Anaphylaxis, allergies such as hay fever, asthma
II. Antibody mediated	IgG, IgM antibodies act upon cells with complement and cause cell lysis; includes some autoimmune diseases	Blood group incompatibility, pernicious anemia; myasthenia gravis
III. Immune complex mediated	Antibody-mediated inflammation; circulating IgG complexes deposited in basement membranes of target organs; includes some autoimmune diseases	Systemic lupus erythematosus; rheumatoid arthritis; serum sickness; rheumatic fever
IV. T-cell mediated	Delayed hypersensitivity and cytotoxic reactions in tissues	Infection reactions; contact dermatitis; graft rejection; some types of autoimmunity

16.2 Type I Allergic Reactions: Atopy and Anaphylaxis

All type I allergies share a similar physiological mechanism, are immediate in onset, and are associated with exposure to specific antigens. However, it is convenient to recognize two levels of severity: **Atopy** is any chronic local allergy such as hay fever or asthma; **anaphylaxis** (an"-uh-fih-lax'-us) is a systemic, sometimes fatal reaction that involves airway obstruction and circulatory collapse. In the following sections, we will consider the epidemiology of type I allergies, allergens and routes of inoculation, mechanisms of disease, and specific syndromes.

Epidemiology and Modes of Contact with Allergens

Allergies exert profound medical and economic impact. Allergists (physicians who specialize in treating allergies) estimate that about 10% to 30% of the population is prone to atopic allergy. It is generally acknowledged that self-treatment with over-the-counter medicines accounts for significant underreporting of cases. The 35 million people afflicted by hay fever (15–20% of the population) spend about half a billion dollars annually for medical treatment. The monetary loss due to employee debilitation and absenteeism is immeasurable. The majority of type I allergies are relatively mild, but certain forms such as asthma and anaphylaxis may require hospitalization and can cause death. Millions of people in the United States suffer from asthma.

The predisposition for type I allergies has a strong familial association. Be aware that what is hereditary is a generalized *susceptibility*, not the allergy to a specific substance. For example, a parent who is allergic to ragweed pollen can have a child who is allergic to cat hair. The prospect of a child's developing atopic allergy is at least 25% if one parent is atopic, increasing up to 50% if grandparents or siblings are also afflicted. The actual basis for atopy appears to be a genetic program that favors allergic antibody (IgE) production, increased reactivity of mast cells, and increased susceptibility of target tissue to allergic mediators. Allergic persons often exhibit a combination of syndromes, such as hay fever, eczema, and asthma.

Other factors that affect the presence of allergy are age, infection, and geographic locale. New allergies tend to crop up throughout an allergic person's life, especially as new exposures occur after moving or changing lifestyle. In some persons, atopic allergies last for a lifetime; others "outgrow" them, and still others suddenly develop them later in life. Some features of allergy are not yet completely explained.

The Nature of Allergens and Their Portals of Entry

As with other antigens, allergens have certain immunogenic characteristics. Not unexpectedly, proteins are more allergenic than carbohydrates, fats, or nucleic acids. Some allergens are haptens, nonproteinaceous substances with a molecular weight of less than 1,000 that can form complexes with carrier molecules in the body (shown in figure 15.8). Organic and inorganic chemicals found in industrial and household products, cosmetics, food, and drugs are commonly of this type. **Table 16.2** lists a number of common allergenic substances.

Allergens typically enter through epithelial portals in the respiratory tract, gastrointestinal tract, and skin. The mucosal surfaces of the gut and respiratory system present a thin, moist surface that is normally quite penetrable. The dry, tough keratin coating of skin is less permeable, but access still occurs through tiny breaks, glands, and hair follicles. It is worth noting that the organ of allergic expression may or may not be the same as the portal of entry.

Airborne environmental allergens such as pollen, house dust, dander (shed skin scales), or fungal spores are termed *inhalants*. Each geographic region harbors a particular combination of airborne substances that varies with the season and humidity **(figure 16.2a)**. Pollen, the most common offender, is given off seasonally by the reproductive structures of pines and flowering plants (weeds, trees, and grasses). Unlike pollen, mold spores are released throughout the year and are especially profuse in moist areas of the home and garden. Airborne animal hair and dander (skin flakes), feathers, and the saliva of dogs and cats are common sources of allergens. The component of house dust that appears to account for most dust allergies is not soil or other debris, but the decomposed bodies and feces of tiny mites that commonly live in this dust **(figure 16.2b)**. Some people are allergic to their work, in the sense that they are exposed to allergens on the job. Examples include florists, woodworkers, farmers, drug processors, welders, and plastics manufacturers whose work can aggravate inhalant and contact allergies.

Allergens that enter by mouth, called *ingestants*, often cause food allergies: *Injectant* allergies are an important adverse side effect of drugs or other substances used in diagnosing, treating, or preventing disease. A natural source of injectants is venom from stings by hymenopterans, a family of

TABLE 16.2	Common Allergens, Classified by Portal of Entry		
Inhalants	**Ingestants**	**Injectants**	**Contactants**
Pollen	Food	Hymenopteran	Drugs
Dust	(milk, peanuts,	venom (bee,	Cosmetics
Mold spores	wheat, shellfish,	wasp)	Heavy metals
Dander	soybeans,	Drugs	Detergents
Animal hair	nuts, eggs,	Vaccines	Formalin
Insect parts	fruits)	Serum	Rubber
Formalin	Food additives	Enzymes	Glue
Drugs	Drugs (aspirin,	Hormones	Solvents
Enzymes	penicillin)		Dyes

**National Allergy Bureau
Pollen and Mold Report**

Location: Sacramento, CA   Date: June 04, 2003
Counting Station: Allergy Medical Group of the North Area

Trees	Moderate severity	Total count: 41 / m³
Weeds	High severity	Total count: 64 / m³
Grass	High severity	Total count: 60 / m³
Mold	Low severity	Total count: 4,219 / m³

(a)

(b)

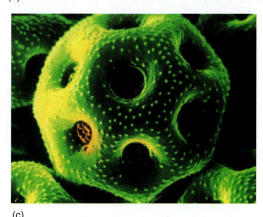

(c)

FIGURE 16.2   Monitoring airborne allergens.
(a) The air in heavily vegetated places with a mild climate is especially laden with allergens such as pollen and mold spores. These counts vary seasonally. **(b)** Because the dust mite *Dermatophagoides* feeds primarily on human skin cells in house dust, these mites are found in abundance in bedding and carpets. Airborne mite feces and particles from their bodies are an important source of allergies. **(c)** Scanning electron micrograph of a single pollen grain from a rose (6,000x). Millions of these are released from a single flower.

insects that includes honey-bees and wasps. *Contactants* are allergens that enter through the skin. Many contact allergies are of the type IV, delayed variety discussed later in this chapter.

Mechanisms of Type I Allergy: Sensitization and Provocation

What causes some people to sneeze and wheeze every time they step out into the spring air, while others suffer no ill effects? In order to answer this question, we must examine what occurs in the tissues of the allergic individual that does not occur in the normal person. In general, type I allergies develop in stages **(figure 16.3)**. The initial encounter with an allergen provides a **sensitizing dose** that primes the immune system for a subsequent encounter with that allergen but generally elicits no signs or symptoms. The memory cells and immunoglobulin are then ready to react with a subsequent provocative dose of the same allergen. It is this dose that precipitates the signs and symptoms of allergy. Despite numerous anecdotal reports of people showing an allergy upon first contact with an allergen, it is generally believed that these individuals unknowingly had contact at some previous time. Fetal exposure to allergens from the mother's bloodstream is one possibility, and foods can be a prime source of "hidden" allergens such as penicillin.

The Physiology of IgE-Mediated Allergies

During primary contact and sensitization, the allergen penetrates the portal of entry **(figure 16.3a)**. When large particles such as pollen grains, hair, and spores encounter a moist membrane, they release molecules of allergen that pass into the tissue fluids and lymphatics. The lymphatics then carry the allergen to the lymph nodes, where specific clones of B cells recognize it, are activated, and proliferate into plasma cells. These plasma cells produce immunoglobulin E (IgE), the antibody of allergy. IgE is different from other immunoglobulins in having an Fc region with great affinity for mast cells and basophils. The binding of IgE to these cells in the tissues sets the scene for the reactions that occur upon repeated exposure to the same allergen **(figure 16.3b)**.

The Role of Mast Cells and Basophils

The most important characteristics of mast cells and basophils relating to their roles in allergy are:

1. Their ubiquitous location in tissues. Mast cells are located in the connective tissue of virtually all organs, but particularly high concentrations exist in the lungs, skin, gastrointestinal tract, and genitourinary tract. Basophils circulate in the blood but migrate readily into tissues.

(a) Sensitization/IgE Production   **(b) Subsequent Exposure to Allergen**

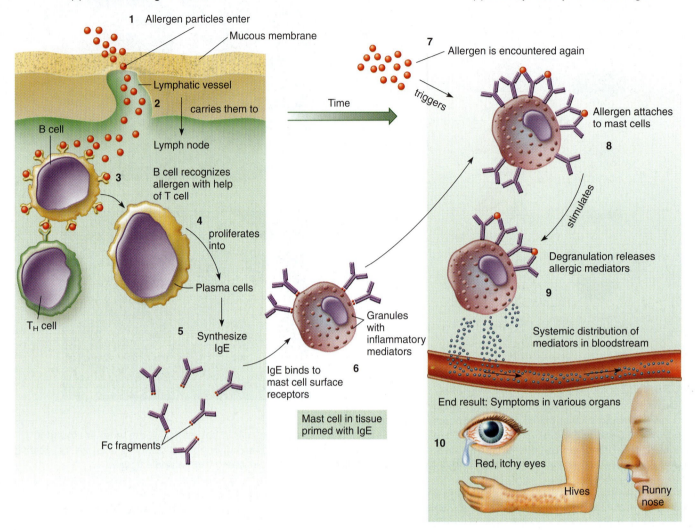

FIGURE 16.3   A schematic view of cellular reactions during the type I allergic response.
(a) Sensitization (initial contact with sensitizing dose), 1–6. **(b)** Provocation (later contacts with provocative dose), 7–10.

2. Their capacity to bind IgE during sensitization (figure 16.3). Each cell carries 30,000 to 100,000 cell receptors that bind 10,000 to 40,000 IgE antibodies.
3. Their cytoplasmic granules (secretory vesicles), which contain physiologically active cytokines (histamine, serotonin—introduced in chapter 14).
4. Their tendency to **degranulate** (figures 16.3*b* and 16.4), or release the contents of the granules into the tissues when triggered by a specific allergen through the IgE bound to them.

Let us now see what occurs when sensitized cells are challenged with allergen a second time.

The Second Contact with Allergen

After sensitization, the IgE-primed mast cells can remain in the tissues for years. Even after long periods without contact, a person can retain the capacity to react immediately upon re-

exposure. The next time allergen molecules contact these sensitized cells, they bind across adjacent receptors and stimulate degranulation. As chemical mediators are released, they diffuse into the tissues and bloodstream. Cytokines give rise to numerous local and systemic reactions, many of which appear quite rapidly (figure 16.3*b*). The symptoms of allergy are not caused by the direct action of allergen on tissues but by the physiological effects of mast cell mediators on target organs.

Cytokines, Target Organs, and Allergic Symptoms

Numerous substances involved in mediating allergy (and inflammation) have been identified. The principal chemical mediators produced by mast cells and basophils are histamine, serotonin, leukotriene, platelet-activating factor, prostaglandins, and bradykinin **(figure 16.4)**. These chemicals, acting alone or in combination, account for the tremen-

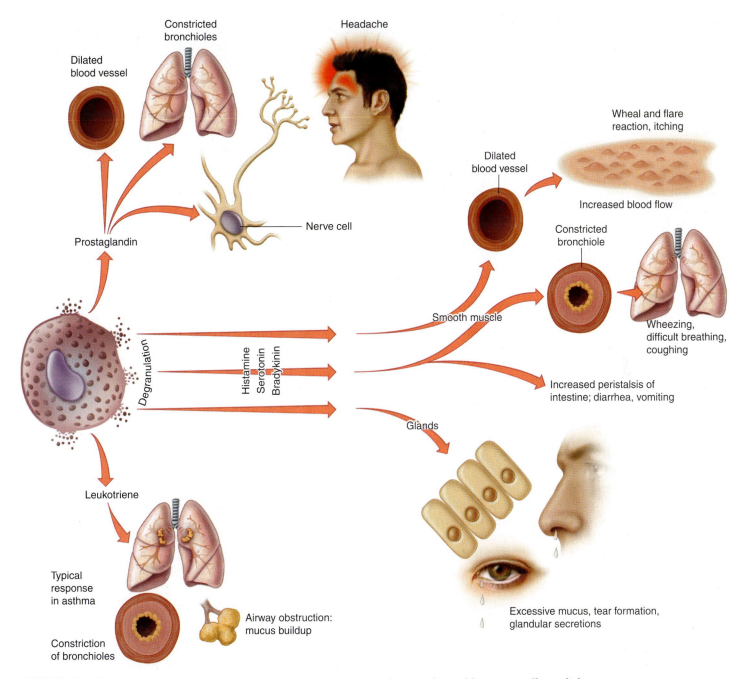

FIGURE 16.4 **The spectrum of reactions to inflammatory cytokines released by mast cells and the common symptoms they elicit in target tissues and organs.**
Note the extensive overlapping effects.

dous scope of allergic symptoms. For some theories pertaining to this function of the allergic response, see **Insight 16.1.** Targets of these mediators include the skin, upper respiratory tract, gastrointestinal tract, and conjunctiva. The general responses of these organs include rashes, itching, redness, rhinitis, sneezing, diarrhea, and shedding of tears. Systemic targets include smooth muscle, mucous glands, and nervous tissue. Because smooth muscle is responsible for regulating the size of blood vessels and respiratory passageways, changes in its activity can profoundly alter blood flow, blood

pressure, and respiration. Pain, anxiety, agitation, and lethargy are also attributable to the effects of mediators on the nervous system.

Histamine is the most profuse and fastest-acting allergic mediator. It is a potent stimulator of smooth muscle, glands, and eosinophils. Histamine's actions on smooth muscle vary with location. It *constricts* the smooth muscle layers of the small bronchi and intestine, thereby causing labored breathing and increased intestinal motility. In contrast, histamine *relaxes* vascular smooth muscle and dilates arterioles and venules. It

INSIGHT 16.1 *Medical*

Of What Value Is Allergy?

Why would humans and other mammals evolve an allergic response that is capable of doing so much harm and even causing death? It is unlikely that this limb of immunity exists merely to make people miserable; it must have a role in protection and survival. What are the underlying biological functions of IgE, mast cells, and the array of potent cytokines? Analysis has revealed that, although allergic persons have high levels of IgE, trace quantities are present even in the sera of nonallergic individuals, just as mast cells and inflammatory chemicals are also part of normal human physiology. It is generally believed that one important function of this system is to defend against helminth worms and other multicellular organisms that are ubiquitous human parasites. In chapter 14, you learned that inflammatory mediators serve valuable functions, such as increasing blood flow and vascular permeability to summon essential immune components to an injured site. They are also responsible for increased mucous secretion, gastric motility, sneezing, and coughing, which help expel noxious agents. The difference is that, in allergic persons, the quantity and quality of these reactions are excessive and uncontrolled.

is responsible for the *wheal and flare* reaction in the skin (see figure 16.6), pruritis (itching), and headache. More severe reactions such as anaphylaxis can be accompanied by edema and vascular dilation, which lead to hypotension, tachycardia, circulatory failure, and, frequently, shock. Salivary, lacrimal, mucous, and gastric glands are also histamine targets.

Although the role of **serotonin** in human allergy is uncertain, its effects appear to complement those of histamine. In experimental animals, serotonin increases vascular permeability, capillary dilation, smooth muscle contraction, intestinal peristalsis, and respiratory rate, but it diminishes central nervous system activity.

Before the specific types were identified, **leukotriene** (loo"-koh-try'-een) was known as the "slow-reacting substance of anaphylaxis" for its property of inducing gradual contraction of smooth muscle. This type of leukotriene is responsible for the prolonged bronchospasm, vascular permeability, and mucous secretion of the asthmatic individual. Other leukotrienes stimulate the activities of polymorphonuclear leukocytes.

Platelet-activating factor is a lipid released by basophils, neutrophils, monocytes, and macrophages. The physiological response to stimulation by this factor is similar to that of histamine, including increased vascular permeability, pulmonary smooth muscle contraction, pulmonary edema, hypotension, and a wheal and flare response in the skin.

Prostaglandins are a group of powerful inflammatory agents. Normally, these substances regulate smooth muscle contraction (for example, they stimulate uterine contractions during delivery). In allergic reactions, they are responsible for vasodilation, increased vascular permeability, increased sensitivity to pain, and bronchoconstriction. Certain anti-inflammatory drugs work by preventing the actions of prostaglandins.

Bradykinin is related to a group of plasma and tissue peptides known as kinins that participate in blood clotting and chemotaxis. In allergy, it causes prolonged smooth muscle contraction of the bronchioles, dilatation of peripheral arterioles, increased capillary permeability, and increased mucous secretion.

Specific Diseases Associated with IgE- and Mast Cell–Mediated Allergy

The mechanisms just described are basic to hay fever, allergic asthma, food allergy, drug allergy, eczema, and anaphylaxis. In this section, we cover the main characteristics of these conditions, followed by methods of detection and treatment.

Atopic Diseases

Hay fever is a generic term for **allergic rhinitis,** a seasonal reaction to inhaled plant pollen or molds, or a chronic, year-round reaction to a wide spectrum of airborne allergens or inhalants (see table 16.2). The targets are typically respiratory membranes, and the symptoms include nasal congestion; sneezing; coughing; profuse mucous secretion; itchy, red, and teary eyes; and mild bronchoconstriction.

Asthma is a respiratory disease characterized by episodes of impaired breathing due to severe bronchoconstriction. The airways of asthmatic people are exquisitely responsive to minute amounts of inhalant allergens, food, or other stimuli, such as infectious agents. The symptoms of asthma range from occasional, annoying bouts of difficult breathing to fatal suffocation. Labored breathing, shortness of breath, wheezing, cough, and ventilatory **rales** are present to one degree or another. The respiratory tract of an asthmatic person is chronically inflamed and severely over-reactive to allergy chemicals, especially leukotrienes and serotonin from pulmonary mast cells. Other pathologic components are thick mucous plugs in the air sacs and lung damage that can result in long-term respiratory compromise. An imbalance in the nervous control of the respiratory smooth muscles is apparently involved in asthma, and the episodes are influenced by the psychological state of the person, which strongly supports a neurological connection.

The number of asthma sufferers in the United States is estimated at more than 10 million, with nearly one-third of them children. For reasons that are not completely understood, asthma is on the increase, and deaths from it have doubled since 1982, even though effective agents to control it are more available now than they have ever been before. It has been suggested that more highly insulated buildings, mandated by energy efficiency regulations, have created indoor air conditions that harbor higher concentrations of contaminants, including insect remains and ozone.

Atopic dermatitis is an intensely itchy inflammatory condition of the skin, sometimes also called **eczema.** Sensitization occurs through ingestion, inhalation, and, occasionally, skin contact with allergens. It usually begins in infancy with reddened, vesicular, weeping, encrusted skin lesions **(figure 16.5).** It then progresses in childhood and adulthood to a dry, scaly, thickened skin condition. Lesions can occur on the face, scalp, neck, and inner surfaces of the limbs and trunk. The itchy, painful lesions cause considerable discomfort, and they are often predisposed to secondary bacterial infections. An anonymous writer once aptly described eczema as "the itch that rashes" or "one scratch is too many but one thousand is not enough."

Food Allergy

The ordinary diet contains a vast variety of compounds that are potentially allergenic. Although the mode of entry is intestinal, food allergies can also affect the skin and respiratory tract. Gastrointestinal symptoms include vomiting, diarrhea, and abdominal pain. In severe cases, nutrients are poorly absorbed, leading to growth retardation and failure to thrive in young children. Other manifestations of food allergies include eczema, hives, rhinitis, asthma, and occasionally, anaphylaxis. Classic food hypersensitivity involves IgE and degranulation of mast cells, but not all reactions involve this mechanism. The most common food allergens come from peanuts, fish, cow's milk, eggs, shellfish, and soybeans.[1]

Drug Allergy

Modern chemotherapy has been responsible for many medical advances. Unfortunately, it has also been hampered by the fact that drugs are foreign compounds capable of stimulating allergic reactions. In fact, allergy to drugs is one of the most common side effects of treatment (present in 5–10% of hospitalized patients). Depending upon the allergen, route of entry, and individual sensitivities, virtually any tissue of the body can be affected, and reactions range from mild atopy to fatal anaphylaxis. Compounds implicated most often are antibiotics (penicillin is number one in prevalence), synthetic antimicrobials (sulfa drugs), aspirin, opiates, and contrast dye used in X rays. The actual allergen is not the intact drug itself but a hapten given off when the liver processes the drug. Some forms of penicillin sensitivity are due to the presence of small amounts of the drug in meat, milk, and other foods and to exposure to *Penicillium* mold in the environment.

---

1. Do not confuse food allergy with food intolerance. Many people are lactose intolerant, for example due to a deficiency in the enzyme that degrades the milk sugar.

FIGURE 16.5 Atopic dermatitis, or eczema.
Vesicular, encrusted lesions are typical in afflicted infants. This condition is prevalent enough to account for 1% of pediatric care.

Anaphylaxis: An Overpowering Systemic Reaction

The term **anaphylaxis,** or anaphylactic shock, was first used to denote a reaction of animals injected with a foreign protein. Although the animals showed no response during the first contact, upon reinoculation with the same protein at a later time, they exhibited acute symptoms—itching, sneezing, difficult breathing, prostration, and convulsions—and many died in a few minutes. Two clinical types of anaphylaxis are seen in humans. *Cutaneous anaphylaxis* is the wheal and flare inflammatory reaction to the local injection of allergen. *Systemic anaphylaxis*, on the other hand, is characterized by sudden respiratory and circulatory disruption that can be fatal in a few minutes. In humans, the allergen and route of entry are variable, though bee stings and injections of antibiotics or serum are implicated most often. Bee venom is a complex material containing several allergens and enzymes that can create a sensitivity that can last for decades after exposure.

The underlying physiological events in systemic anaphylaxis parallel those of atopy, but the concentration of chemical mediators and the strength of the response are greatly amplified. The immune system of a sensitized person exposed to a provocative dose of allergen responds with a sudden, massive release of chemicals into the tissues and blood, which act rapidly on the target organs. Anaphylactic persons have been known to die in 15 minutes from complete airway blockage.

IN THE NEWS *(Continued from page 483)*

Further investigation by health care personnel revealed that the woman was allergic to wasp stings; a sting mark was discovered on her fingertip, and a nest of yellow jackets was found in the ground below the hedge she had been trimming.

Her death was caused by a severe, systemic allergic reaction to protein in the wasp venom, which resulted in a condition called anaphylaxis or anaphylactic shock. Anaphylactic shock initiates a host of conditions simultaneously throughout the body, causing airway blockage and circulatory collapse within a matter of minutes.

Approximately 1% to 3% of the population has some type of allergy to insect venom, usually to the venom produced by bees and wasps. Reactions can range from minor allergic reactions resulting in hives, headache, and stomach upset, to severe allergic reactions that cause anaphylactic shock. People with known allergies to insect venom should wear a medical ID bracelet stating their allergy, as well as talk to their doctors about getting an insect sting allergy kit to carry with them at all times.

See: Li, J. T., and Yunginger, J. W. 1992. Management of insect sting hypersensitivity. Mayo Clin. Proc. 67:188–194.

Muellman, R. L., Lindzon, R. D., and Silvers, N. S. 1998. Allergy, hypersensitivity and anaphylaxis. In P. Rosen, editor. Emergency medicine, concepts and clinical practice. 4th ed. St. Louis: Mosby Year Book.

Pumphrey, R. S., and Roberts, I. S. 2000. Postmortem findings after fatal anaphylactic reactions. J. Clin. Pathol. 53:273–276.

Diagnosis of Allergy

Because allergy mimics infection and other conditions, it is important to determine if a person is actually allergic. If possible or necessary, it is also helpful to identify the specific allergen or allergens. Allergy diagnosis involves several levels of tests, including nonspecific, specific, *in vitro*, and *in vivo* methods.

A new test that can distinguish whether a patient has experienced an allergic attack measures elevated blood levels of tryptase, an enzyme released by mast cells that increases during an allergic response. Several types of specific *in vitro* tests can determine the allergic potential of a patient's blood sample. A differential blood cell count can indicate the levels of basophils and eosinophils—a higher level of these indicates allergy. The leukocyte histamine-release test measures the amount of histamine released from the patient's basophils when exposed to a specific allergen. Serological tests that use radioimmune assays (see chapter 17) to reveal the quantity and type of IgE are also clinically helpful.

Skin Testing

A useful *in vivo* method to detect precise atopic or anaphylactic sensitivities is skin testing. With this technique, a patient's skin is injected, scratched, or pricked with a small amount of a pure allergen extract. There are hundreds of these allergen extracts containing common airborne allergens (plant and mold pollen) and more unusual allergens (mule dander, theater dust, bird feathers). Unfortunately, skin tests for food allergies using food extracts are unreliable in most cases. In patients with numerous allergies, the allergist maps the skin on the inner aspect of the forearms or back and injects the allergens intradermally according to this predetermined pattern **(figure 16.6a)**. Approximately 20 minutes after antigenic challenge, each site is appraised for a wheal response indicative of histamine release. The diameter of the wheal is measured and rated on a scale of 0 (no reaction) to 4+ (greater than 15 mm). **Figure 16.6b** shows skin test results for a person with extreme inhalant allergies.

Treatment and Prevention of Allergy

In general, the methods of treating and preventing type I allergy involve

1. avoiding the allergen, although this may be very difficult in many instances;
2. taking drugs that block the action of lymphocytes, mast cells, or chemical mediators; and
3. undergoing desensitization therapy.

It is not possible to completely prevent initial sensitization, since there is no way to tell in advance if a person will develop an allergy to a particular substance. The practice of delaying the introduction of solid foods apparently has some merit in preventing food allergies in children, although even breast milk can contain allergens ingested by the mother. Rigorous cleaning and air conditioning can reduce contact with airborne allergens, but it is not feasible to isolate a person from all allergens, which is the reason drugs are so important in control.

Therapy to Counteract Allergies

The aim of antiallergy medication is to block the progress of the allergic response somewhere along the route between IgE production and the appearance of symptoms **(figure 16.7)**. Oral anti-inflammatory drugs such as corticosteroids inhibit the activity of lymphocytes and thereby reduce the production of IgE, but they also have dangerous side effects and should not be taken for prolonged periods. Some drugs block the degranulation of mast cells and reduce the levels of inflammatory cytokines. The most effective of these are diethylcarbamazine and cromolyn. Asthma and rhinitis sufferers can find relief with a drug that blocks synthesis of leukotriene and a monoclonal antibody that inactivates IgE (Xolair).

Widely used medications for preventing symptoms of atopic allergy are **antihistamines,** the active ingredients in most over-the-counter allergy-control drugs. Antihistamines interfere with histamine activity by binding to histamine receptors on target organs. Most of them have major side

FIGURE 16.6   A method for conducting an allergy skin test.

The forearm (or back) is mapped and then injected with a selection of allergen extracts. The allergist must be very aware of potential anaphylaxis attacks triggered by these injections. **(a)** Close-up of skin wheals showing a number of positive reactions (dark lines are measurer's marks). **(b)** An actual skin test record for some common environmental allergens [not related to **(a)**.]

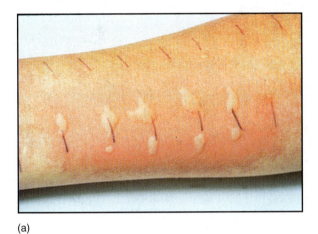

(a)

(b)

Environmental Allergens

No. 1 Standard Series — ID 8/85

+++	1. Acacia gum
+++	2. Cat dander
++++	3. Chicken feathers
++++	4. Cotton lint
++	5. Dog dander
+	6. Duck feathers
+	7. Glue, animal
++	8. Horse dander
✕	9. Horse serum
+++	10. House dust #1
+	11. Kapok
+	12. Mohair (goat)
+	13. Paper
++++	14. Pyrethrum
+++	15. Rug pad, ozite
+	16. Silk dust
+	17. Tobacco dust
+	18. Tragacanth gum
+++++	19. Upholstery dust
+++	20. Wool

No. 2 Airborne Particles — ID 8/85

+++	1. Ant
+++++	2. Aphid
++++	3. Bee
++++	4. Housefly
✕	5. House mite
+++	6. Mosquito
++++	7. Moth
+++	8. Roach
++	9. Wasp
0	10. Yellow jacket

Airborne mold spores

++	11. *Alternaria*
+++	12. *Aspergillus*
++	13. *Cladosporium*
+++	14. *Hormodendrum*
0	15. *Penicillium*
+	16. *Phoma*
+++	17. *Rhizopus*
	18. ....

✕ - not done   ++ - mild reaction
0 - no reaction   +++ - moderate reaction
+ - slight reaction   ++++ - severe reaction

effects, however, such as drowsiness. Newer antihistamines lack this side effect because they do not cross the blood-brain barrier. Other drugs that relieve inflammatory symptoms are aspirin and acetaminophen, which reduce pain by interfering with prostaglandin, and theophylline, a bronchodilator that reverses spasms in the respiratory smooth muscles. Persons who suffer from anaphylactic attacks are urged to carry at all times injectable epinephrine (adrenaline) and an identification tag indicating their sensitivity. An aerosol inhaler containing epinephrine can also provide rapid relief. Epinephrine reverses constriction of the airways and slows the release of allergic mediators.

Approximately 70% of allergic patients benefit from controlled injections of specific allergens as determined by skin tests. This technique, called **desensitization** or **hyposensitization,** is a therapeutic way to prevent reactions between allergen, IgE, and mast cells. The allergen preparations contain pure, preserved suspensions of plant antigens, venoms, dust mites, dander, and molds (but so far, hyposensitization for foods has not proved very effective). The immunologic basis of this treatment is open to differences in interpretation. One theory suggests that injected allergens stimulate the formation of high levels of allergen-specific IgG (**figure 16.8**) instead of IgE. It has been proposed that these IgG **blocking antibodies** remove allergen from the system before it can bind to IgE, thus preventing the degranulation of mast cells. It is also possible that allergen delivered in this fashion combines with the IgE itself and takes it from circulation before it can react with the mast cells.

✔ CHECKPOINT

- Type I hypersensitivity reactions result from excessive IgE production in response to an exogenous antigen.
- The two kinds of type I hypersensitivities are atopy, a chronic, local allergy, and anaphylaxis, a systemic, potentially fatal allergic response.
- The predisposition to type I hypersensitivities is inherited, but age, geographic locale, and infection also influence allergic response.
- Type I allergens include inhalants, ingestants, injectants, and contactants.
- The portals of entry for type I antigens are the skin, respiratory tract, gastrointestinal tract, and genitourinary tract.
- Type I hypersensitivities are set up by a sensitizing dose of allergen and expressed when a second provocative dose triggers the allergic response. The time interval between the two can be many years.
- The primary participants in type I hypersensitivities are IgE, basophils, mast cells, and agents of the inflammatory response.
- Allergies are diagnosed by a variety of *in vitro* and *in vivo* tests that assay specific cells, IgE, and local reactions.
- Allergies are treated by medications that interrupt the allergic response at certain points. Allergic reactions can often be prevented by desensitization therapy.

FIGURE 16.7   Strategies for circumventing allergic attacks.

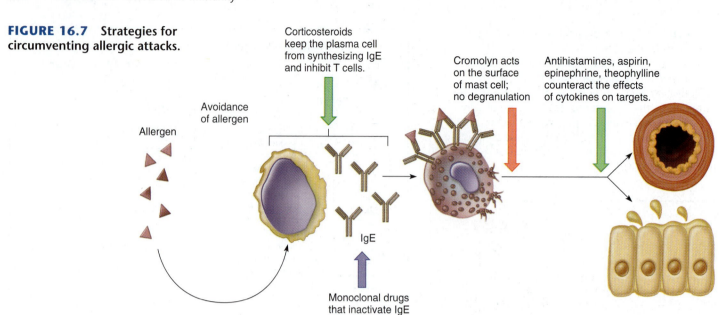

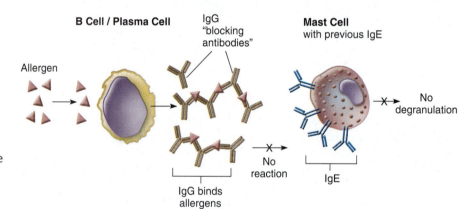

FIGURE 16.8   The blocking antibody theory for allergic desensitization.
An injection of allergen causes IgG antibodies to be formed instead of IgE; these blocking antibodies cross-link and effectively remove the allergen before it can react with the IgE in the mast cell.

16.3 Type II Hypersensitivities: Reactions That Lyse Foreign Cells

The diseases termed type II hypersensitivities are a complex group of syndromes that involve complement-assisted destruction (lysis) of cells by antibodies (IgG and IgM) directed against those cells' surface antigens. This category includes transfusion reactions and some types of autoimmunities (discussed in a later section). The cells targeted for destruction are often red blood cells, but other cells can be involved.

Chapters 14 and 15 described the functions of unique surface markers on cell membranes. Ordinarily, these molecules play essential roles in transport, recognition, and development, but they become medically important when the tissues of one person are placed into the body of another person. Blood transfusions and organ donations introduce alloantigens (molecules that differ in the same species) on donor cells that are recognized by the lymphocytes of the recipient. These reactions are not really immune dysfunctions as allergy and autoimmunity are. The immune sys-

tem is in fact working normally, but it is not equipped to distinguish between the desirable foreign cells of a transplanted tissue and the undesirable ones of a microbe.

The Basis of Human ABO Antigens and Blood Types

The existence of human blood types was first demonstrated by an Austrian pathologist, Karl Landsteiner, in 1904. While studying incompatibilities in blood transfusions, he found that the serum of one person could clump the red blood cells of another. Landsteiner identified four distinct types, subsequently called the **ABO blood groups.**

Like the MHC antigens on white blood cells, the ABO antigen markers on red blood cells are genetically determined and composed of glycoproteins. These ABO antigens are inherited as two (one from each parent) of three alternative **alleles:** A, B, or O. A and B alleles are dominant over O and codominant with one another. As **table 16.3** indicates, this mode of inheritance gives rise to four blood types (phe-

				Incidence of Type in United States		
Genotype	Blood Type	Antigen Present on Erythrocyte Membranes	Antibody in Plasma	Among Whites (%)	Among Asians (%)	Among Those of African and Caribbean Descent (%)
AA, AO	A	A	Anti-b	41	28	27
BB, BO	B	B	Anti-a	10	27	20
AB	AB	A and B	Neither anti-a nor anti-b	4	5	7
OO	O	Neither A nor B	Anti-a and anti-b	45	40	46

TABLE 16.3   Characteristics of ABO Blood Groups

notypes), depending on the particular combination of genes. Thus, a person with an *AA* or *AO* genotype has type A blood; genotype *BB* or *BO* gives type B; genotype *AB* produces type AB; and genotype *OO* produces type O. Some important points about the blood types are:

1. They are named for the dominant antigen(s);
2. the RBCs of type O persons have antigens, but not A and B antigens; and
3. tissues other than RBCs carry A and B antigens.

A diagram of the AB antigens and blood types is shown in **figure 16.9.** The A and B genes each code for an enzyme that adds a terminal carbohydrate to RBC surface molecules during maturation. RBCs of type A contain an enzyme that adds *N*-acetylgalactosamine to the molecule; RBCs of type B have an enzyme that adds D-galactose; RBCs of type AB contain both enzymes that add both carbohydrates; and RBCs of type O lack the genes and enzymes to add a terminal molecule.

Antibodies Against A and B Antigens

Although an individual does not normally produce antibodies in response to his or her own RBC antigens, the serum can contain antibodies that react with blood of another antigenic type, even though contact with this other blood type has *never* occurred. These preformed antibodies account for the immediate and intense quality of transfusion reactions. As a rule, type A blood contains antibodies (anti-b) that react against the B antigens on type B and AB red blood cells. Type B blood contains antibodies (anti-a) that react with A antigen on type A and AB red blood cells. Type O blood contains antibodies against both A and B antigens. Type AB blood does not contain antibodies against either A or B antigens[2] (table 16.3). What is the source of these anti-a and anti-b antibodies? It appears that they develop in early infancy because of exposure to certain antigens that are widely distributed in nature. These antigens are surface molecules on bacteria and plant cells that mimic the structure of A and B antigens. Ex-

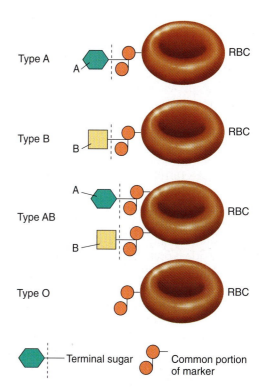

FIGURE 16.9   The genetic/molecular basis for the A and B antigens (receptors) on red blood cells.
In general, persons with blood types A, B, and AB inherit a gene for the enzyme that adds a certain terminal sugar to the basic RBC receptor. Type O persons do not have such an enzyme and lack the terminal sugar.

posure to these sources stimulates the production of corresponding antibodies.

Clinical Concerns in Transfusions

The presence of ABO antigens and a, b antibodies underlie several clinical concerns in giving blood transfusions. First, the individual blood types of donor and recipient must be determined. By use of a standard technique, drops of blood are mixed with antisera that contain antibodies against the A

---

2. Why would this be true? The answer lies in the first sentence of the paragraph.

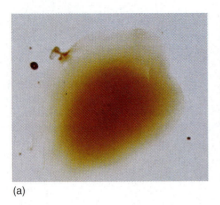 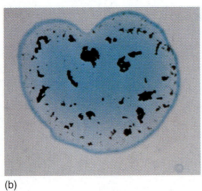
(a)   (b)

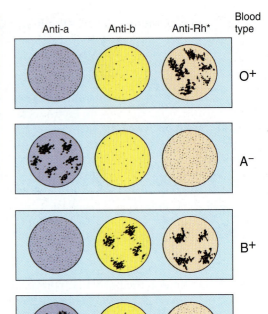

FIGURE 16.10   Interpretation of blood typing.
In this test, a drop of blood is mixed with a specially prepared antiserum known to contain antibodies against the A, B, or Rh antigens. **(a)** If that particular antigen is not present, the red blood cells in that droplet do not agglutinate, and form an even suspension. **(b)** If that antigen is present, agglutination occurs and the RBCs form visible clumps. **(c)** Several patterns and their interpretations. Anti-a, anti-b, and anti-Rh are shorthand for the antiserum applied to the drops. (In general, O$^+$ is the most common blood type, and AB$^-$ is the rarest.)

and B antigens and are then observed for the evidence of agglutination **(figure 16.10).**

Knowing the blood types involved makes it possible to determine which transfusions are safe to do. The general rule of compatibility is that the RBC antigens of the donor must not be agglutinated by antibodies in the recipient's blood **(figure 16.11).** The ideal practice is to transfuse blood that is a perfect match (A to A, B to B). But even in this event, blood samples must be cross-matched before the transfusion because other blood group incompatibilities can exist. This test involves mixing the blood of the donor with the serum of the recipient to check for agglutination.

Under certain circumstances (emergencies, the battlefield), the concept of universal transfusions can be used. To appreciate how this works, we must apply the rule stated in the previous paragraph. Type O blood lacks A and B antigens and will not be agglutinated by other blood types, so it could theoretically be used in any transfusion. Hence, a person with this blood type is called a **universal donor.** Because type AB blood lacks agglutinating antibodies, an individual with this blood could conceivably receive any type of blood. Type AB persons are consequently called *universal recipients.* Although both types of transfusions involve antigen-antibody incompatibilities, these are of less concern because of the dilution of the donor's blood in the body of the recipient. Additional RBC markers that can be significant in transfusions are the Rh, MN, and Kell antigens (see next sections).

Transfusion of the wrong blood type causes differing degrees of adverse reaction. The severest reaction is massive hemolysis when the donated red blood cells react with recipient antibody and trigger the complement cascade (figure 16.11). The resultant destruction of red cells leads to sys-

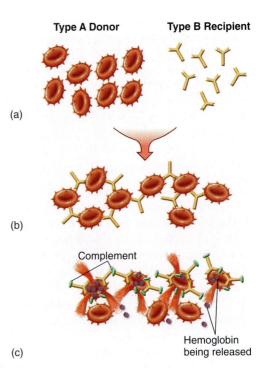

FIGURE 16.11   Microscopic view of a transfusion reaction.
(a) Incompatible blood. The red blood cells of the type A donor contain antigen A, while the serum of the type B recipient contains anti-a antibodies that can agglutinate donor cells. **(b)** Agglutination complexes can block the circulation in vital organs. **(c)** Activation of the complement by antibody on the RBCs can cause hemolysis and anemia. This sort of incorrect transfusion is very rare because of the great care taken by blood banks to ensure a correct match.

INSIGHT 16.2 *Medical*

Why Doesn't a Mother Reject Her Fetus?

Think of it: Even though mother and child are genetically related, the father's genetic contribution guarantees that the fetus will contain molecules that are antigenic to the mother. In fact, with the recent practice of implanting one woman with the fertilized egg of another woman, the surrogate mother is carrying a fetus that has no genetic relationship to her. Yet, even with this essentially foreign body inside the mother, dangerous immunologic reactions such as Rh incompatibility are rather rare. In what ways do fetuses avoid the surveillance of the mother's immune system? The answer appears to lie in the placenta and embryonic tissues. The fetal components that contribute to these tissues are not strongly antigenic, and they form a barrier that keeps the fetus isolated in its own antigen-free environment. The placenta is surrounded by a dense, many-layered envelope that prevents the passage of maternal cells, and it actively absorbs, removes, and inactivates circulating antigens.

temic shock and kidney failure brought on by the blockage of glomeruli (blood-filtering apparatus) by cell debris. Death is a common outcome. Other reactions caused by RBC destruction are fever, anemia, and jaundice. A transfusion reaction is managed by immediately halting the transfusion, administering drugs to remove hemoglobin from the blood, and beginning another transfusion with red blood cells of the correct type.

The Rh Factor and Its Clinical Importance

Another RBC antigen of major clinical concern is the **Rh factor** (or D antigen). This factor was first discovered in experiments exploring the genetic relationships among animals. Rabbits inoculated with the RBCs of rhesus monkeys produced an antibody that also reacted with human RBCs. Further tests showed that this monkey antigen (termed Rh for rhesus) was present in about 85% of humans and absent in the other 15%. The details of Rh inheritance are more complicated than those of ABO, but in simplest terms, a person's Rh type results from a combination of two possible alleles—a dominant one that codes for the factor and a recessive one that does not. A person inheriting at least one Rh gene will be Rh^+; only those persons inheriting two recessive genes are Rh^-. The "+" or "−" appearing after a blood type refers to the Rh status of the person, as in O^+ or AB^- (see **figure 16.10***c*). However, unlike the ABO antigens, exposure to environmental antigens does not sensitize Rh^- persons to the Rh factor. The only ways one can develop antibodies against this factor are through placental sensitization or transfusion.

Hemolytic Disease of the Newborn and Rh Incompatibility

The potential for placental sensitization occurs when a mother is Rh^- and her unborn child is Rh^+. The obvious intimacy between mother and fetus makes it possible for fetal RBCs to leak into the mother's circulation during childbirth, when the detachment of the placenta creates avenues for fetal blood to enter the maternal circulation. The mother's immune system detects the foreign Rh factors on the fetal RBCs and is sensi-

tized to them by producing antibodies and memory B cells. The first Rh^+ child is usually not affected because the process begins so late in pregnancy that the child is born before maternal sensitization is completed. However, the mother's immune system has been strongly primed for a second contact with this factor in a subsequent pregnancy **(figure 16.12***a*).

In the next pregnancy with an Rh^+ fetus, fetal blood cells escape into the maternal circulation late in pregnancy and elicit a memory response. The fetus is at risk when the maternal anti-Rh antibodies cross the placenta into the fetal circulation, where they affix to fetal RBCs and cause complement-mediated lysis. The outcome is a potentially fatal **hemolytic disease of the newborn (HDN)** called *erythroblastosis fetalis* (eh-rith″-roh-blas-toh′-sis fee-tal′-is). This term is derived from the presence of immature nucleated RBCs called erythroblasts in the blood. They are released into the infant's circulation to compensate for the massive destruction of RBCs stimulated by maternal antibodies. Additional symptoms are severe anemia, jaundice, and enlarged spleen and liver.

Maternal-fetal incompatibilities are also possible in the ABO blood group, but adverse reactions occur less frequently than with Rh sensitization because the antibodies to these blood group antigens are IgM rather than IgG and are unable to cross the placenta in large numbers. In fact, the maternal-fetal relationship is a fascinating instance of foreign tissue not being rejected, despite the extensive potential for contact **(Insight 16.2).**

Preventing Hemolytic Disease of the Newborn

Once sensitization of the mother to Rh factor has occurred, all other Rh^+ fetuses will be at risk for hemolytic disease of the newborn. Prevention requires a careful family history of an Rh^- pregnant woman. It can predict the likelihood that she is already sensitized or is carrying an Rh^+ fetus. It must take into account other children she has had, their Rh types, and the Rh status of the father. If the father is also Rh^-, the child will be Rh^- and free of risk, but if the father is Rh^+, the probability that the child will be Rh^+ is 50% or 100%, depending on the exact genetic makeup of the father. If there is any

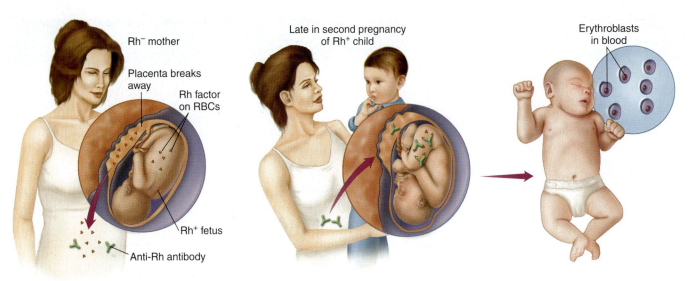

(a) **The development and aftermath of Rh sensitization**
Initial sensitization of the maternal immune system to fetal Rh⁺ factor occurs when fetal cells leak into the
Rh⁻ mother's circulation late in pregnancy, or during delivery, when the placenta tears away. The child will
escape hemolytic disease in most instances, but the mother, now sensitized, will be capable of an
immediate reaction to a second Rh⁺ fetus and its Rh-factor antigen. At that time, the mother's anti-Rh
antibodies pass into the fetal circulation and elicit severe hemolysis in the fetus and neonate.

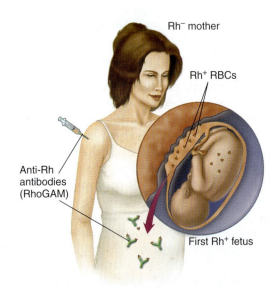

(b) **Prevention of erythroblastosis fetalis with anti-Rh immune globulin (RhoGAM)**
Injecting a mother who is at risk with RhoGAM during her first Rh⁺ pregnancy helps to
inactivate and remove the fetal Rh-positive cells before her immune system can react
and develop sensitivity.

FIGURE 16.12   Development and control of Rh incompatibility.

possibility that the fetus is Rh⁺, the mother must be passively immunized with antiserum containing antibodies against the Rh factor (Rh_0 [D] *immune globulin*, or RhoGAM[3]). This antiserum, injected at 28 to 32 weeks and again immediately after delivery, reacts with any fetal RBCs that have escaped into the maternal circulation, thereby preventing the sensitization of the mother's immune system to Rh factor **(figure 16.12b)**. Anti-Rh antibody must be given with each pregnancy that involves an Rh⁺ fetus. It is ineffective if the mother has already been sensitized by a prior Rh⁺ fetus or an incorrect blood transfusion, which can be determined by a serological test.

As in ABO blood types, the Rh factor should be matched for a transfusion, although it is acceptable to transfuse Rh⁻ blood if the Rh type is not known.

3. RhoGAM: Immunoglobulin fraction of human anti-Rh serum, prepared from pooled human sera.

Other RBC Antigens

Although the ABO and Rh systems are of greatest medical significance, about 20 other red blood cell antigen groups have been discovered. Examples are the *MN, Ss, Kell,* and *P* blood groups. Because of incompatibilities that these blood groups present, transfused blood is screened to prevent possible cross-reactions. The study of these blood antigens (as well as ABO and Rh) has given rise to other useful applications. For example, they can be useful in forensic medicine (crime detection), studying ethnic ancestry, and tracing prehistoric migrations in anthropology. Many blood cell antigens are remarkably hardy and can be detected in dried blood stains, semen, and saliva. Even the 3,300-year-old mummy of King Tutankhamen has been typed A_2MN!

✔ CHECKPOINT

- Type II hypersensitivity reactions occur when preformed antibodies react with foreign cell-bound antigens. The most common type II reactions occur when transfused blood is mismatched to the recipient's ABO type. IgG or IgM antibodies attach to the foreign cells, resulting in complement fixation. The resultant formation of membrane attack complexes lyses the donor cells.
- Complement, IgG, and IgM antibodies are the primary mediators of type II hypersensitivities.
- The concepts of universal donor (type O) and universal recipient (type AB) apply only under emergency circumstances. Cross-matching donor and recipient blood is necessary to determine which transfusions are safe to perform.
- Type II hypersensitivities can also occur when Rh^- mothers are sensitized to Rh^+ RBCs of their unborn babies and the mother's anti-Rh antibodies cross the placenta, causing hemolysis of the newborn's RBCs. This is called hemolytic disease of the newborn, or erythroblastosis fetalis.

16.4 Type III Hypersensitivities: Immune Complex Reactions

Type III hypersensitivity involves the reaction of soluble antigen with antibody and the deposition of the resulting complexes in basement membranes of epithelial tissue. It is similar to type II, because it involves the production of IgG and IgM antibodies after repeated exposure to antigens and the activation of complement. Type III differs from type II because its antigens are not attached to the surface of a cell. The interaction of these antigens with antibodies produces free-floating complexes that can be deposited in the tissues, causing an **immune complex reaction** or disease. This category includes therapy-related disorders (serum sickness and the Arthus reaction) and a number of autoimmune diseases (such as glomerulonephritis and lupus erythematosus).

Mechanisms of Immune Complex Disease

After initial exposure to a profuse amount of antigen, the immune system produces large quantities of antibodies that circulate in the fluid compartments. When this antigen enters the system a second time, it reacts with the antibodies to form antigen-antibody complexes **(figure 16.13).** These complexes summon various inflammatory components such as complement and neutrophils, which would ordinarily eliminate Ag-Ab complexes as part of the normal immune response. In an immune complex disease, however, these complexes are so abundant that they deposit in the **basement membranes**[4] of epithelial tissues and become inaccessible. In response to these events, neutrophils release lysosomal granules that digest tissues and cause a destructive inflammatory condition. The symptoms of type III hypersensitivities are due in great measure to this pathologic state.

Types of Immune Complex Disease

During the early tests of immunotherapy using animals, hypersensitivity reactions to serum and vaccines were common. In addition to anaphylaxis, two syndromes, the **Arthus reaction**[5] and **serum sickness,** were identified. These syndromes are associated with certain types of passive immunization (especially with animal serum).

Serum sickness and the Arthus reaction are like anaphylaxis in requiring sensitization and preformed antibodies. Characteristics that set them apart from anaphylaxis are:

1. They depend upon IgG, IgM, or IgA (precipitating antibodies) rather than IgE;
2. They require large doses of antigen (not a minuscule dose as in anaphylaxis); and
3. Their symptoms are delayed (a few hours to days).

The Arthus reaction and serum sickness differ from each other in some important ways. The Arthus reaction is a *localized* dermal injury due to inflamed blood vessels in the vicinity of any injected antigen. Serum sickness is a *systemic* injury initiated by antigen-antibody complexes that circulate in the blood and settle into membranes at various sites.

The Arthus Reaction

The Arthus reaction is usually an acute response to a second injection of vaccines (boosters) or drugs at the same site as the first injection. In a few hours, the area becomes red, hot to the touch, swollen, and very painful. These symptoms are mainly due to the destruction of tissues in and around the blood vessels and the release of histamine from mast cells and basophils. Although the reaction is usually self-limiting and rapidly cleared, intravascular blood clotting can occasionally cause necrosis and loss of tissue.

---

4. **Basement membranes** are basal partitions of epithelia that normally filter out circulating antigen-antibody complexes.

5. Named after Maurice Arthus, the physiologist who first identified this localized inflammatory response.

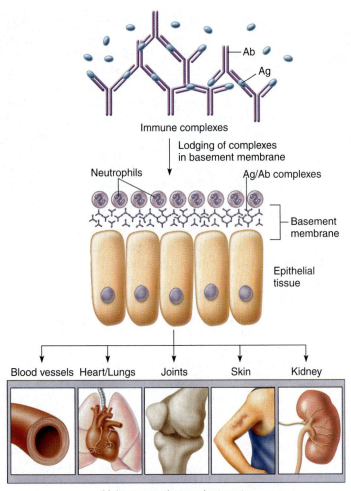

Immune complexes

Lodging of complexes
in basement membrane

Neutrophils

Ag/Ab complexes

Basement
membrane

Epithelial
tissue

Blood vessels   Heart/Lungs   Joints   Skin   Kidney

Major organs that can be targets
of immune complex deposition

Steps:

1. Antibody combines with excess soluble antigen, forming large quantities of Ab/Ag complexes.

2. Circulating immune complexes become lodged in the basement membrane of epithelia in sites such as kidney, lungs, joints, skin.

3. Fragments of complement cause release of histamine and other mediator substances.

4. Neutrophils migrate to the site of immune complex deposition and release enzymes that cause severe damage in the tissues and organs involved.

FIGURE 16.13   **Pathogenesis of immune complex disease.**

Serum Sickness

Serum sickness was named for a condition that appeared in soldiers after repeated injections of horse serum to treat tetanus. It can also be caused by injections of animal hormones and drugs. The immune complexes enter the circulation, are carried throughout the body, and are eventually deposited in blood vessels of the kidney, heart, skin, and joints (figure 16.13). The condition can become chronic, causing symptoms such as enlarged lymph nodes, rashes, painful joints, swelling, fever, and renal dysfunction.

■ The mediators of type III hypersensitivity reactions include soluble IgA, IgG, or IgM, and agents of the inflammatory response.

■ Two kinds of type III hypersensitivities are localized (Arthus) reactions and systemic (serum sickness). Arthus reactions occur at the site of injected drugs or booster immunizations. Systemic reactions occur when repeated antigen challenges cause systemic distribution of the immune complexes and subsequent inflammation of joints, lymph nodes, and kidney tubules.

✔ **CHECKPOINT**

■ Type III hypersensitivities are induced when a profuse amount of antigen enters the system and results in large quantities of antibody formation.

■ Type III hypersensitivity reactions occur when large quantities of antigen react with host antibody to form small, soluble immune complexes that settle in tissue cell membranes, causing chronic destructive inflammation. The reactions appear hours or days after the antigen challenge.

16.5 Type IV Hypersensitivities: Cell-Mediated (Delayed) Reactions

The adverse immune responses we have covered so far are explained primarily by B-cell involvement and antibodies. A notable difference exists in type IV hypersensitivity, which involves primarily the T-cell branch of the immune system. Type IV immune dysfunction has traditionally been known as delayed hypersensitivity because the symptoms arise one

INSIGHT 16.3   *Medical*

Pretty, Pesky, Poisonous Plants

As a cause of allergic contact dermatitis (affecting about 10 million people a year), nothing can compare with a single family of plants belonging to the genus *Toxicodendron*. At least one of these plants—either poison ivy, poison oak, or poison sumac—flourishes in the forests, woodlands, or along the trails of most regions of America. The allergen in these plants, an oil called urushiol, has such extreme potency that a pinhead-sized amount could spur symptoms in 500 people, and it is so long-lasting that botanists must be careful when handling 100-year-old plant specimens. Although degrees of sensitivity vary among individuals, it is estimated that 85% of all Americans are potentially hypersensitive to this compound. Some people are so acutely sensitive that even the most minuscule contact, such as handling pets or clothes that have touched the plant, or breathing vaporized urushiol, can trigger an attack.

Humans first become sensitized by contact during childhood. Individuals at great risk (firefighters, hikers) are advised to determine their degree of sensitivity using a skin test, so that they can be adequately cautious and prepared. Some odd remedies include skin potions containing bleach, buttermilk, ammonia, hair spray, and meat tenderizer. Commercial products are available for blocking or washing away the urushiol. Allergy researchers are currently testing oral vaccines containing a form of urushiol, which seem to desensitize experimental animals. An effective method using poison ivy desensitization injection is currently available to people with extreme sensitivity.

Poison oak

Poison sumac

Poison ivy

Learning to identify these common plants can prevent exposure and sensitivity. One old saying that might help warns, "Leaves of three, let it be; berries white, run with fright."

to several days following the second contact with an antigen. In general, type IV diseases result when T cells respond to antigens displayed on self tissues or transplanted foreign cells. Examples of type IV hypersensitivity include delayed allergic reactions to infectious agents, contact dermatitis, and graft rejection.

Delayed-Type Hypersensitivity

Infectious Allergy

A classic example of a delayed-type hypersensitivity occurs when a person sensitized by tuberculosis infection is injected with an extract (tuberculin) of the bacterium *Mycobacterium tuberculosis*. The so-called tuberculin reaction is an acute skin inflammation at the injection site appearing within 24 to 48 hours. So useful and diagnostic is this technique for detecting present or prior tuberculosis that it is the chosen screening device (see chapter 21). Other infections that use similar skin testing are leprosy, syphilis, histoplasmosis, toxoplasmosis, and candidiasis. This form of hypersensitivity arises from time-consuming cellular events involving a specific class of T cells (T_H1) that receive the processed allergens from dendritic cells. Activated T_H cells release cytokines that recruit various inflammatory cells such as macrophages, neutrophils, and eosinophils. The build-up of fluid and cells at the site gives rise to a red papule (for example, see **figure 16.14**). In a chronic infection (tertiary syphilis, for example), extensive damage to organs can occur through granuloma formation.

Contact Dermatitis

The most common delayed allergic reaction, contact dermatitis, is caused by exposure to resins in poison ivy or poison oak **(Insight 16.3),** to simple haptens in household and personal articles (jewelry, cosmetics, elasticized undergarments), and to certain drugs. Like immediate atopic dermatitis, the reaction to these allergens requires a sensitizing and a

The Origins of Autoimmune Disease

Since otherwise healthy individuals show (very low levels of) autoantibodies, it is suspected that there is a function for them. A moderate, regulated amount of autoimmunity is probably required to dispose of old cells and cellular debris. Disease apparently arises when this regulatory or recognition apparatus goes awry. Attempts to explain the origin of autoimmunity include the following theories.

The *sequestered antigen theory* explains that during embryonic growth, some tissues are immunologically privileged; that is, they are sequestered behind anatomical barriers and cannot be scanned by the immune system. Examples of these sites are regions of the central nervous system, which are shielded by the meninges and blood-brain barrier; the lens of the eye, which is enclosed by a thick sheath; and antigens in the thyroid and testes, which are sequestered behind an epithelial barrier. Eventually the antigen becomes exposed by means of infection, trauma, or deterioration, and is perceived by the immune system as a foreign substance.

According to the **clonal selection theory,** the immune system of a fetus develops tolerance by eradicating all self-reacting lymphocyte clones, called *forbidden clones,* while retaining only those clones that react to foreign antigens. Some of these forbidden clones may survive, and since they have not been subjected to this tolerance process, they can attack tissues with self antigens.

The *theory of immune deficiency* proposes that mutations in the receptor genes of some lymphocytes render them reactive to self or that a general breakdown in the normal T-suppressor function sets the scene for inappropriate immune responses.

Inappropriate expression of MHC II markers on cells that don't normally express them has been found to cause abnormal immune reactions to self. In a related phenomenon T-cell activation may incorrectly "turn on" B cells that can react with self antigens. This phenomenon is called the *bystander effect.*

Some autoimmune diseases appear to be caused by *molecular mimicry,* in which microbial antigens bear molecular determinants similar to normal human cells. An infection could cause formation of antibodies that can cross-react with tissues. This is one purported explanation for the pathology of rheumatic fever. Another probable example of mimicry leading to autoimmune disease is the skin condition psoriasis. Although the etiology of this condition is complex, and involves the inheritance of certain types of MHC alleles, infection with group A streptococci also plays a role. Scientists report that T cells primed to react with streptococcal surface proteins also react with keratin cells in the skin, causing them to proliferate. For this reason, psoriasis patients often report flare-ups after a strep throat infection.

Autoimmune disorders such as type I diabetes and multiple sclerosis are likely triggered by *viral infection.* Viruses can noticeably alter cell receptors, thereby causing immune cells to attack the tissues bearing viral receptors.

Some researchers believe that many, if not most, autoimmune diseases will someday be discovered to have an underlying microbial etiology, either through molecular mimicry or viral alteration of host antigens (just described) or due to the presence of as yet undetectable microbes in the sites affected by autoimmunity.

Examples of Autoimmune Disease

Systemic Autoimmunities

One of the most severe chronic autoimmune diseases is systemic lupus erythematosus (SLE, or lupus). This name originated from the characteristic butterfly-shaped rash that drapes across the nose and cheeks **(figure 16.17a).** Apparently ancient physicians thought the rash resembled a wolf

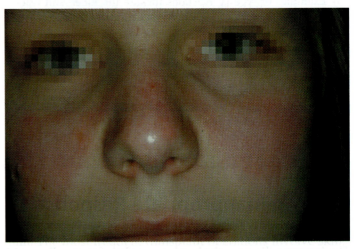

(a)

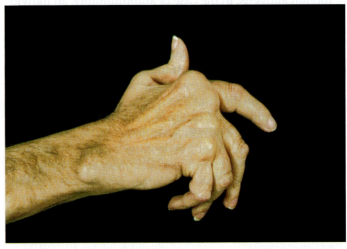

(b)

FIGURE 16.17 **Common autoimmune diseases.**
(a) Systemic lupus erythematosus. One symptom is a prominent rash across the bridge of the nose and on the cheeks. These papules and blotches can also occur on the chest and limbs.
(b) Rheumatoid arthritis commonly targets the synovial membrane of joints. Over time, chronic inflammation causes thickening of this membrane, erosion of the articular cartilage, and fusion of the joint. These effects severely limit motion and can eventually swell and distort the joints.

bite (*lupus* in Latin for wolf). Although the manifestations of the disease vary considerably, all patients produce autoantibodies against a great variety of organs and tissues. The organs most involved are the kidneys, bone marrow, skin, nervous system, joints, muscles, heart, and GI tract. Antibodies to intracellular materials such as the nucleoprotein of the nucleus and mitochondria are also common.

In SLE, autoantibody-autoantigen complexes appear to be deposited in the basement membranes of various organs. Kidney failure, blood abnormalities, lung inflammation, myocarditis, and skin lesions are the predominant symptoms. One form of chronic lupus (called discoid) is influenced by exposure to the sun and primarily afflicts the skin. The etiology of lupus is still a puzzle. It is not known how such a generalized loss of self-tolerance arises, though viral infection or loss of T-cell suppressor function are suspected. The fact that women of childbearing years account for 90% of cases indicates that hormones may be involved. The diagnosis of SLE can usually be made with blood tests. Antibodies against the nucleus and various tissues (detected by indirect fluorescent antibody or radioimmune assay techniques) are common, and a positive test for the lupus factor (an antinuclear factor) is also very indicative of the disease.

Rheumatoid arthritis, another systemic autoimmune disease, incurs progressive, debilitating damage to the joints. In some patients, the lung, eye, skin, and nervous system are also involved. In the joint form of the disease, autoantibodies form immune complexes that bind to the synovial membrane of the joints and activate phagocytes and stimulate release of cytokines. Chronic inflammation leads to scar tissue and joint destruction. The joints in the hands and feet are affected first, followed by the knee and hip joints **(figure 16.17b)**. The precipitating cause in rheumatoid arthritis is not known, though infectious agents such as Epstein-Barr virus have been suspected. The most common feature of the disease is the presence of an IgM antibody, called rheumatoid factor (RF), directed against other antibodies. This does not cause the disease but is used mainly in diagnosis. The symptoms are complicated by a type IV delayed hypersensitivity response.

Autoimmunities of the Endocrine Glands

On occasion, the thyroid gland is the target of autoimmunity. The underlying cause of **Graves disease** is the attachment of autoantibodies to receptors on the follicle cells that secrete the hormone thyroxin. The abnormal stimulation of these cells causes the overproduction of this hormone and the symptoms of hyperthyroidism. In **Hashimoto's thyroiditis**, both autoantibodies and T cells are reactive to the thyroid gland, but in this instance, they reduce the levels of thyroxin by destroying follicle cells and by inactivating the hormone. As a result of these reactions, the patient suffers from hypothyroidism.

The pancreas and its hormone, insulin, are other autoimmune targets. Insulin, secreted by the beta cells in the pancreas, regulates and is essential to the utilization of glucose by cells. **Diabetes mellitus** is caused by a dysfunction in

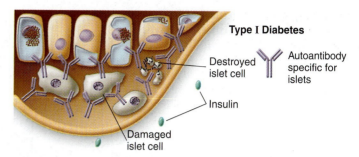

FIGURE 16.18 **The autoimmune component in diabetes mellitus, type I.**
Autoantibodies produced against the beta cells of the pancreas destroy the cells and greatly reduce insulin synthesis.

insulin production or utilization **(figure 16.18)**. Type I diabetes is associated with autoantibodies and sensitized T cells that damage the beta cells. A complex inflammatory reaction leading to lysis of these cells greatly reduces the amount of insulin secreted.

Neuromuscular Autoimmunities

Myasthenia gravis is named for the pronounced muscle weakness that is its principal symptom. Although the disease afflicts all skeletal muscle, the first effects are usually felt in the muscles of the eyes and throat. Eventually, it can progress to complete loss of muscle function and death. The classic syndrome is caused by autoantibodies binding to the receptors for acetylcholine, a chemical required to transmit a nerve impulse across the synaptic junction to a muscle **(figure 16.19)**. The immune attack so severely damages the muscle cell membrane that transmission is blocked and paralysis ensues. Current treatment usually includes immunosuppressive drugs and therapy to remove the autoantibodies from the circulation. Experimental therapy using immunotoxins to destroy lymphocytes that produce autoantibodies shows some promise.

Multiple sclerosis (MS) is a paralyzing neuromuscular disease associated with lesions in the insulating myelin sheath that surrounds neurons in the white matter of the central nervous system. The underlying pathology involves damage to the sheath by both T cells and autoantibodies that severely compromise the capacity of neurons to send impulses. The principal motor and sensory symptoms are muscular weakness and tremors, difficulties in speech and vision, and some degree of paralysis. Most MS patients first experience symptoms as young adults, and they tend to experience remissions (periods of relief) alternating with recurrences of disease throughout their lives. Convincing evidence from studies of the brain tissue of MS patients points to a strong connection between the disease and infection with human herpesvirus 6. The disease can be treated passively with monoclonal antibodies that target T cells, and a vaccine containing the myelin protein has shown beneficial effects. Immunosuppressants such as cortisone and interferon beta may also alleviate symptoms.

FIGURE 16.19 **Mechanism for involvement of autoantibodies in myasthenia gravis.**
Antibodies developed against receptors on the postsynaptic membrane block them so that acetylcholine cannot bind and muscle contraction is inhibited.

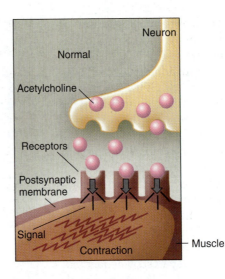
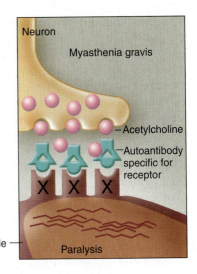

16.7 Immunodeficiency Diseases: Hyposensitivity of the Immune System

It is a marvel that development and function of the immune system proceed as normally as they do. On occasion, however, an error occurs and a person is born with or develops weakened immune responses. In many cases, these "experiments" of nature have provided penetrating insights into the exact functions of certain cells, tissues, and organs because of the specific signs and symptoms shown by the immunodeficient individuals. The predominant consequences of immunodeficiencies are recurrent, overwhelming infections, often with opportunistic microbes. Immunodeficiencies fall into two general categories: *primary diseases,* present at birth (congenital) and usually stemming from genetic errors, and *secondary diseases,* acquired after birth and caused by natural or artificial agents **(table 16.5).**

Primary Immunodeficiency Diseases

Deficiencies affect both specific immunities such as antibody production and less-specific ones such as phagocytosis. Consult **figure 16.20** to survey the places in the normal sequential development of lymphocytes where defects can occur and the possible consequences. In many cases, the deficiency is due to an inherited abnormality, though the exact nature of the abnormality is not known for a number of diseases. Because the development of B cells and T cells departs at some point, an individual can lack one or both cell lines. It must be emphasized, however, that some deficiencies affect other cell functions. For example a T-cell deficiency can affect B-cell function because of the role of T helper cells. In some deficiencies, the lymphocyte in question is completely absent or is present at very low levels, whereas in others, lymphocytes are present but do not function normally.

Clinical Deficiencies in B-Cell Development or Expression

Genetic deficiencies in B cells usually appear as an abnormality in immunoglobulin expression. In some instances, only certain immunoglobulin classes are absent; in others, the levels of all types of immunoglobulins (Ig) are reduced. A significant number of B-cell deficiencies are X-linked (also called sex-linked) recessive traits, meaning that the gene occurs on the X chromosome and the disease appears primarily in male children.

The term **agammaglobulinemia** literally means the absence of gamma globulin, the fraction of serum that contains immunoglobulins. Because it is very rare for Ig to be completely absent, some physicians prefer the term **hypogammaglobulinemia.** T-cell function in these patients is usually normal. The symptoms of recurrent, serious bacterial infections usually appear about 6 months after birth. The bacteria most often implicated are pyogenic cocci, *Pseudomonas,* and *Haemophilus influenzae,* and the most common infection sites are the lungs, sinuses, meninges, and blood. Many Ig-deficient patients can have recurrent infections with viruses

TABLE 16.5	General Categories of Immunodeficiency Diseases with Selected Examples	
Primary Immune Deficiencies (Genetic)		**Secondary Immune Deficiencies (Acquired)**

Primary Immune Deficiencies (Genetic)

B-Cell Defects (Low Levels of B Cells and Antibodies)
Agammaglobulinemia (X-linked, non-sex-linked)
Hypogammaglobulinemia
Selective immunoglobulin deficiencies

T-Cell Defects (Lack of All Classes of T Cells)
Thymic aplasia (DiGeorge syndrome)
Chronic mucocutaneous candidiasis

Combined B-Cell and T-Cell Defects (Usually Caused by Lack or Abnormality of Lymphoid Stem Cell)
Severe combined immunodeficiency disease (SCID)
X-SCIDI due to an interleukin defect
Adenosine deaminase (ADA) deficiency
Wiskott-Aldrich syndrome
Ataxia-telangiectasia

Phagocyte Defects
Chédiak-Higashi syndrome
Chronic granulomatous disease of children (see In The News, chapter 14)
Lack of surface adhesion molecules

Complement Defects
Lacking one of C components
Hereditary angioedema
Associated with rheumatoid diseases

Secondary Immune Deficiencies (Acquired)

From Natural Causes
Infection: AIDS, leprosy, tuberculosis, measles
Other disease: cancer, diabetes
Nutrition deficiencies
Stress
Pregnancy
Aging

From Immunosuppressive Agents
Irradiation
Severe burns
Steroids (cortisones)
Drugs to treat graft rejection and cancer
Removal of spleen

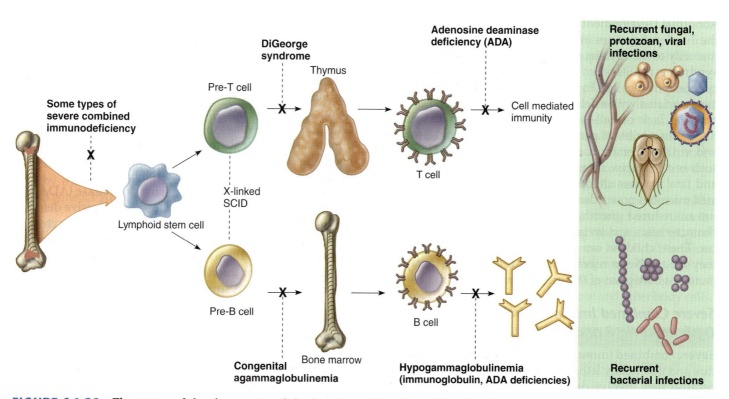

FIGURE 16.20 The stages of development and the functions of B cells and T cells, whose failure causes immunodeficiencies. Dotted lines represent the phases in development where breakdown can occur.

Chapter Summary With Key Terms

16.1 The Immune Response: A Two-Sided Coin
The study of disease states involving the malfunction of the immune system is called **immunopathology.**
A. **Allergy,** or hypersensitivity, is an exaggerated, misdirected expression of certain immune responses.
B. Autoimmunity involves abnormal responses to self antigens. A deficiency or loss in immune function is called **immunodeficiency.**

16.2 Type I Allergic Reactions: Atopy and Anaphylaxis
A. Immediate-onset allergies involve contact with **allergens,** antigens that affect certain people; susceptibility is inherited.
B. On first contact with allergen, specific B cells react with allergen and form a special antibody class called IgE, which affixes by its Fc receptor to mast cells and basophils.
  1. This **sensitizing dose** primes the allergic response system.
  2. Upon subsequent exposure with a provocative dose, the same allergen binds to the IgE-mast cell complex.
  3. This causes **degranulation,** release of intracellular granules containing mediators **(histamine, serotonin, leukotriene, prostaglandin),** with physiological effects such as vasodilation and bronchoconstriction.
  4. Symptoms are rash, itching, redness, increased mucous discharge, pain, swelling, and difficulty in breathing.
C. Diagnosis of allergy can be made by a histamine release test on basophils; serological assays for IgE; and skin testing, which injects allergen into the skin and mirrors the degree of reaction.
D. Control of allergy involves drugs to interfere with the action of histamine, inflammation, and release of cytokines from mast cells. **Desensitization** therapy involves the administration of purified allergens.

16.3 Type II Hypersensitivities: Reactions That Lyse Foreign Cells
A. Type II reactions involve the interaction of antibodies, foreign cells, and complement, leading to lysis of the foreign cells. In transfusion reactions, humans may become sensitized to special antigens on the surface of the red blood cells of other humans.
B. The **ABO blood groups** are genetically controlled: Type A blood has A antigens on the RBCs; type B has B antigens; type AB has both A and B antigens; and type O has neither antigen. People produce antibodies against A or B antigens if they lack these antigens. Antibodies can react with antigens if the wrong blood type is transfused.
C. **Rh factor** is another RBC antigen that becomes a problem if an Rh⁻ mother is sensitized by an Rh⁺ fetus. A second fetus can receive antibodies she has made against the factor and develop **hemolytic disease of the newborn.** Prevention involves therapy with Rh immune globulin.

16.4 Type III Hypersensitivities: Immune Complex Reactions
Exposure to a large quantity of soluble foreign antigens (serum, drugs) stimulates antibodies that produce small, soluble Ag-Ab complexes. These **immune complexes** are trapped in various organs and tissues, which incites a damaging inflammatory response.

16.5 Type IV Hypersensitivities: Cell-Mediated (Delayed) Reactions
A delayed response to antigen involving the activation of and damage by T cells.
A. *Delayed allergic response:* Skin response to allergens, including infectious agents. Example is tuberculin reaction; contact dermatitis is caused by exposure to plants (ivy, oak) and simple environmental molecules (metals, cosmetics); cytotoxic T cells acting on allergen elicit a skin reaction.
B. *Graft rejection:* Reaction of cytotoxic T cells directed against foreign cells of a grafted tissue; involves recognition of foreign HLA by T cells and rejection of tissue.
  1. Host may reject graft; graft may reject host.
  2. Types of grafts include: **autograft,** from one part of body to another, **isograft,** grafting between identical twins; **allograft,** between two members of same species; **xenograft,** between two different species.
  3. All major organs may be successfully transplanted.
  4. Allografts require tissue match (HLA antigens must correspond); rejection is controlled with drugs.

16.6 An Inappropriate Response Against Self, or Autoimmunity
A. In certain type II and III hypersensitivities, the immune system has lost tolerance to self molecules (autoantigens) and forms **autoantibodies** and sensitized T cells against them. Disruption of function can be systemic or organ specific.
B. **Autoimmune diseases** are genetically determined and more common in females.

16.7 Immunodeficiency Diseases: Hyposensitivity of the Immune System
Components of the immune response system are absent. Deficiencies involve B and T cells, phagocytes, and complement.
A. Primary immunodeficiency is genetically based, congenital; defect in inheritance leads to lack of B-cell activity, T-cell activity, or both.
B. B-cell defect is called **agammaglobulinemia;** patient lacks antibodies; serious recurrent bacterial infections result. In Ig deficiency, one of the classes of antibodies is missing or deficient.
C. In T-cell defects, the thymus is missing or abnormal. In **DiGeorge syndrome,** the thymus fails to develop; afflicted children experience recurrent infections with eucaryotic pathogens and viruses; immune response is generally underdeveloped.
D. In **severe combined immunodeficiency (SCID),** both limbs of the lymphocyte system are missing or defective; no adaptive immune response exists; fatal without replacement of bone marrow or other therapies.
E. Secondary (acquired) immunodeficiency is due to damage after birth (infections, drugs, radiation). **AIDS** is the most common of these; T helper cells are main target; deficiency manifests in numerous opportunistic infections and cancers.

Multiple-Choice Questions

1. Pollen is which type of allergen?
   a. contactant
   b. ingestant
   c. injectant
   d. inhalant

2. B cells are responsible for which allergies?
   a. asthma
   b. anaphylaxis
   c. tuberculin reactions
   d. both a and b

3. Which allergies are T-cell mediated?
   a. type I
   b. type II
   c. type III
   d. type IV

4. The contact with allergen that results in symptoms is called the
   a. sensitizing dose
   b. degranulation dose
   c. provocative dose
   d. desensitizing dose

5. Production of IgE and degranulation of mast cells are involved in
   a. contact dermatitis
   b. anaphylaxis
   c. Arthus reaction
   d. both a and b

6. The direct, immediate cause of allergic symptoms is the action of
   a. the allergen directly on smooth muscle
   b. the allergen on B lymphocytes
   c. allergic mediators released from mast cells and basophils
   d. IgE on smooth muscle

7. Theoretically, type ___ blood can be donated to all persons because it lacks ___.
   a. AB, antibodies
   b. O, antigens
   c. AB, antigens
   d. O, antibodies

8. An example of a type III immune complex disease is
   a. serum sickness
   b. contact dermatitis
   c. graft rejection
   d. atopy

9. Type II hypersensitivities are due to
   a. IgE reacting with mast cells
   b. activation of cytotoxic T cells
   c. IgG-allergen complexes that clog epithelial tissues
   d. complement-induced lysis of cells in the presence of antibodies

10. Production of autoantibodies may be due to
    a. emergence of forbidden clones of B cells
    b. production of antibodies against sequestered tissues
    c. infection-induced change in receptors
    d. all of these are possible

11. Rheumatoid arthritis is an ___ that affects the ___ .
    a. immunodeficiency disease, muscles
    b. autoimmune disease, nerves
    c. allergy, cartilage
    d. autoimmune disease, joints

12. A positive tuberculin skin test is an example of
    a. a delayed-type allergy
    b. acute contact dermatitis
    c. autoimmunity
    d. eczema

13. Contact dermatitis can be caused by
    a. pollen grains
    b. chemicals absorbed by the skin
    c. microbes
    d. proteins found in foods

14. Which disease would be most similar to AIDS in its pathology?
    a. X-linked agammaglobulinemia
    b. SCID
    c. ADA deficiency
    d. DiGeorge syndrome

Concept Questions

These questions are suggested as a *writing-to-learn* experience. For each question, compose a one- or two-paragraph answer that includes the factual information needed to completely address the question.

1. a. Define allergy and hypersensitivity.
   b. What accounts for the reactions that occur in these conditions?
   c. What does it mean when a reaction is immediate or delayed?
   d. Give examples of each type.

2. Describe several factors that influence types and severity of allergic responses.

3. a. How are atopic allergies similar to anaphylaxis?
   b. How are they different?

4. a. How do allergens gain access to the body?
   b. What are some examples of allergens that enter by these portals?

5. a. Trace the course of a pollen grain through sensitization and provocation in type I allergies.
   b. Include in the discussion the role of mast cells, basophils, IgE, and allergic mediators.
   c. Outline the target organs and symptoms of the principal atopic diseases and their diagnosis and treatment.

6. a. Describe the allergic response that leads to anaphylaxis. Include its usual causes, how it is diagnosed and treated, and two effective physiological targets for treatment.
   b. Explain how hyposensitization is achieved and suggest two mechanisms by which it might work.

7. a. What is the mechanism of type II hypersensitivity?
   b. Why are the tissues of some people antigenic to others?
   c. Would we be concerned about this problem if it were not for transfusions?
   d. What is the actual basis of the four ABO and Rh blood groups?
   e. Where do we derive our natural hypersensitivities to the A or B antigens that we do not possess?
   f. How does a person become sensitized to Rh factor? List consequences.

8. Explain the rules of transfusion. Illustrate what will happen if type A blood is accidently transfused into a type B person.

9. a. Contrast type II and type III hypersensitivities with respect to type of antigen, antibody, and manifestations of disease.
   b. What is immune complex disease?

CHAPTER OVERVIEW

▶ The ability to identify microbes that are responsible for a patient's symptoms is central to infectious disease microbiology. Diagnosis might be considered an art and a science, involving multiple health care providers and clinical personnel.

▶ Accurate specimen collection is the cornerstone of accurate diagnosis.

▶ Phenotypic identification methods assess a microbe's appearance, growth characteristics, and/or arsenal of chemicals and enzymes. Direct examination of specimens, cultivation of specimens, and biochemical testing are tools used for phenotypic methods of diagnosis.

▶ Genotypic methods examine the genetic content of a microbe, by analyzing its G + C content, its DNA or rRNA sequence. Polymerase chain reaction is used to increase the amount of DNA in a sample so that it can be analyzed.

▶ Immunological methods of identification can be used to probe the antigenic makeup of a microbe, or to identify the presence of antibodies to a microbe in a patient's blood. There are many variations on immunological methods, including introducing antigens into a patient to detect the presence of an immune response. These methods are often referred to as *serological* tests.

▶ Accurate diagnosis depends on the ability to differentiate the causative organism from normal flora or contaminating microbes.

17.1 Preparation for the Survey of Microbial Diseases

In chapters 18 through 22, the most clinically significant bacterial, fungal, parasitic, and viral diseases will be covered. The chapters will survey the most prevalent infectious conditions and the organisms that cause them. This chapter gets us started with an introduction to the how-to of diagnosing the infections.

For many students (and professionals), the most pressing topic in microbiology is *how to identify unknown bacteria* in patient specimens or in samples from nature. Methods microbiologists use to identify bacteria to the level of genus and species fall into three main categories: *phenotypic*, which includes a consideration of morphology (microscopic and macroscopic) as well as bacterial physiology or biochemistry; *immunological*, which entails serological analysis; and *genotypic* (or genetic) techniques. Data from a cross section of such tests can produce a unique profile of each bacterium. Increasingly, genetic means of identification are being used as a sole resource for identifying bacteria. As universally used databases become more complete, because of submissions from scientists and medical personnel worldwide, genetic analyses provide a more accurate and speedy way of identifying microbes than was possible even a decade ago. There are still many organisms, however, that must be identified in the "old-fashioned" way—via biochemical, serological, and morphological means. Serology is so reliable for some diseases that it may never be replaced. All of these methods—phenotypic, genotypic, and serological—will be described in this chapter.

Phenotypic Methods

Microscopic Morphology

Traits that can be valuable aids to identification are combinations of cell shape and size, Gram stain reaction, acid-fast reaction, and special structures, including endospores, granules, and capsules. Electron microscope studies can pinpoint additional structural features (such as the cell wall, flagella, pili, and fimbriae).

Macroscopic Morphology

Traits that can be assessed with the naked eye are also useful in diagnosis. These include the appearance of colonies, including texture, size, shape, pigment, speed of growth, and patterns of growth in broth and gelatin media.

Physiological/Biochemical Characteristics

These have been the traditional mainstay of bacterial identification. Enzymes and other biochemical properties of bacteria are fairly reliable and stable expressions of the chemical identity of each species. Dozens of diagnostic tests exist for determining the presence of specific enzymes and to assess nutritional and metabolic activities. Examples include tests for fermentation of sugars; capacity to digest or metabolize complex polymers such as proteins and polysaccharides; production of gas; presence of enzymes such as catalase, oxidase, and decarboxylases; and sensitivity to antimicrobic drugs. Special rapid identification test systems that record the major biochemical reactions of a culture have streamlined data collection.

Chemical Analysis

This involves analyzing the types of specific structural substances that the microorganism contains, such as the chemical composition of peptides in the cell wall and lipids in membranes.

Genotypic Methods

Examining the genetic material itself has revolutionized the identification and classification of bacteria. There are many advantages of genotypic methods over phenotypic methods,

INSIGHT 17.1 — Discovery

The Uncultured

By the 1990s it was clear to microbiologists that culture-based (phenotypic) methods for identifying bacteria were becoming inadequate. This was first confirmed by environmental researchers, who came to believe that at most 1% (and in some environments it was 0.001%) of microbes present in lakes, soil, and saltwater environments could be grown in laboratories and, therefore, were unknown and unstudied. These microbes are termed **viable nonculturable, or VNC.**

Although it took microbiologists many years to come to this realization, once they did, it made sense. Scientists had spent several decades (since microbes could first routinely be grown) concocting recipes for media and having great success in growing all kinds of bacteria from all kinds of environments. They had plenty to do, just in identifying and studying those. By the 1990s the advent of non-culture-dependent tools, such as gene probing and PCR, revealed vast numbers of species that had never before turned up on a culture dish. That this vast zoo of microbes was revealed in environmental samples was not surprising due to the huge array of microenvironments that would have had to have been reproduced in media for them to be grown in the lab.

But medical microbiologists felt fairly confident that they could culture microbes from a human, since the "environment" of human tissues is well understood. Although some human-inhabiting microbes cannot be grown in culture, we never suspected that we were missing a large proportion of them. In 1999 three Stanford University scientists applied PCR techniques to a

collection of subgingival plaque harvested from one of their own mouths. They used a wide library of DNA fragments as probes, essentially "fishing" for new isolates. Oral biologists had previously recovered about 500 bacterial strains from this site; the Stanford scientists found 30 species that had never before been cultured or described. This discovery shook the medical world and led to increased investigation of "normal" human flora, using non-culture-based methods to find VNCs in the human body.

The new realization that our bodies are hosts to a wide variety of microbes about which we know nothing has several implications. As evolutionary microbiologist Paul Ewald has said, "What are all those microbes doing in there?" He points out that many oral microbes previously assumed to be innocuous are now associated with cancer and heart disease. Many of the diseases that we currently think of as noninfectious will likely be found to have an infectious cause once we learn to look for VNCs. Another question is: *How did the microbes get there?* One organism found in the oral cavity was a metal-oxidizing soil bacterium. The researchers speculate that it may have been obtained from drinking water. Ewald suggests that many of our oral residents (and gastrointestinal flora), including those that may turn out to be stealthily pathogenic, have been obtained from a very common activity—kissing. This is an activity that may not be as innocuous as previously thought.

---

when they are available. The primary advantage is that actually culturing the microorganisms is not always necessary. In recent decades scientists have come to realize that there are many more microorganisms that we can't grow in the lab compared with those that we can **(Insight 17.1).** Another advantage is that genotypic methods are increasingly automated, and results are obtained very quickly, often with more precision than with phenotypic methods.

Immunological Methods

Bacteria and other microbes have surface and other molecules called antigens that are recognized by the immune system. One immune response to antigens is the production of molecules called antibodies that are designed to bind tightly to the antigens. The nature of the antibody response is also exploited for diagnosis when a patient's blood (or other tissue) is tested for the presence of specific antibodies to a suspected pathogen. This is often easier than testing for the

microbe itself, especially in the case of viral infections. Most HIV testing entails examination of a person's blood for presence of antibody to the virus. Laboratory kits based on this technique are available for immediate identification of a number of pathogens.

17.2 On the Track of the Infectious Agent: Specimen Collection

Regardless of the method of diagnosis, specimen collection is the common point that guides the health care decisions of every member of a clinical team. Indeed, the success of identification and treatment depends on how specimens are collected, handled, and stored. Specimens can be taken by a clinical laboratory scientist or medical technologist, nurse, physician, or even by the patient himself. However, it is imperative that general aseptic procedures be used, including sterile sample containers and other tools to prevent

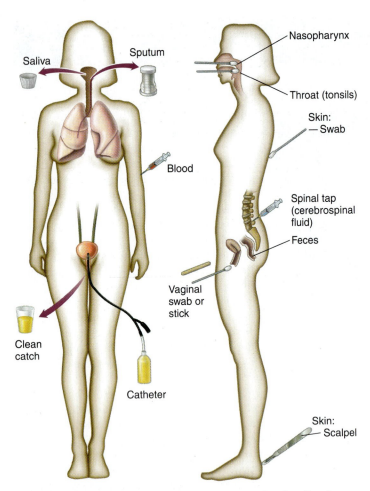

FIGURE 17.1   Sampling sites and methods of collection for clinical laboratories.

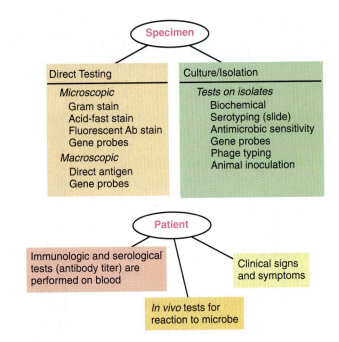

FIGURE 17.2   A scheme of specimen isolation and identification.

contamination from the environment or the patient. **Figure 17.1** delineates the most common sampling sites and procedures.

In sites that normally contain resident microflora, care should be taken to sample only the infected site and not surrounding areas. For example, throat and nasopharyngeal swabs should not touch the tongue, cheeks, or saliva. Saliva is an especially undesirable contaminant because it contains millions of bacteria per milliliter, most of which are normal flora. Saliva samples are occasionally taken for dental diagnosis by having the patient expectorate into a container. Depending on the nature of the lesion, skin can be swabbed or scraped with a scalpel to expose deeper layers. The mucous lining of the vagina, cervix, or urethra can be sampled with a swab or applicator stick.

Urine is taken aseptically from the bladder with a thin tube called a catheter. Another method, called a "clean catch," is taken by washing the external urethra and collecting the urine midstream. The latter method inevitably incorporates a few normal flora into the sample, but these can usually be differentiated from pathogens in an actual infection. Sometimes diagnostic techniques require first-voided "dirty catch" urine. Sputum, the mucous secretion that coats the lower respiratory surfaces, especially the lungs, is discharged by coughing or taken by catheterization to avoid

contamination with saliva. Sterile materials such as blood, cerebrospinal fluid, and tissue fluids must be taken by sterile needle aspiration. Antisepsis of the puncture site is extremely important in these cases. Additional sources of specimens are the vagina, eye, ear canal, nasal cavity (all by swab), and diseased tissue that has been surgically removed (biopsied).

After proper collection, the specimen is promptly transported to a lab and stored appropriately (usually refrigerated) if it must be held for a time. Nonsterile samples in particular, such as urine, feces, and sputum, are especially prone to deterioration at room temperature. Special swab and transport systems are designed to collect the specimen and maintain it in stable condition for several hours. These devices contain nonnutritive maintenance media (so the microbes do not grow), a buffering system, and an anaerobic environment to prevent possible destruction of oxygen-sensitive bacteria.

Overview of Laboratory Techniques

The routes taken in specimen analysis are the following: (1) direct tests using microscopic, immunological, or genetic methods that provide immediate clues as to the identity of the microbe or microbes in the sample and (2) cultivation, isolation, and identification of pathogens using a wide variety of general and specific tests (**figure 17.2**). Most test results fall into two categories: presumptive data, which place the isolated microbe (isolate) in a preliminary category such as a genus, and more specific, confirmatory data, which provide more definitive evidence of a species. Some tests are more important for some groups of bacteria than for others. The total time required for analysis ranges from a

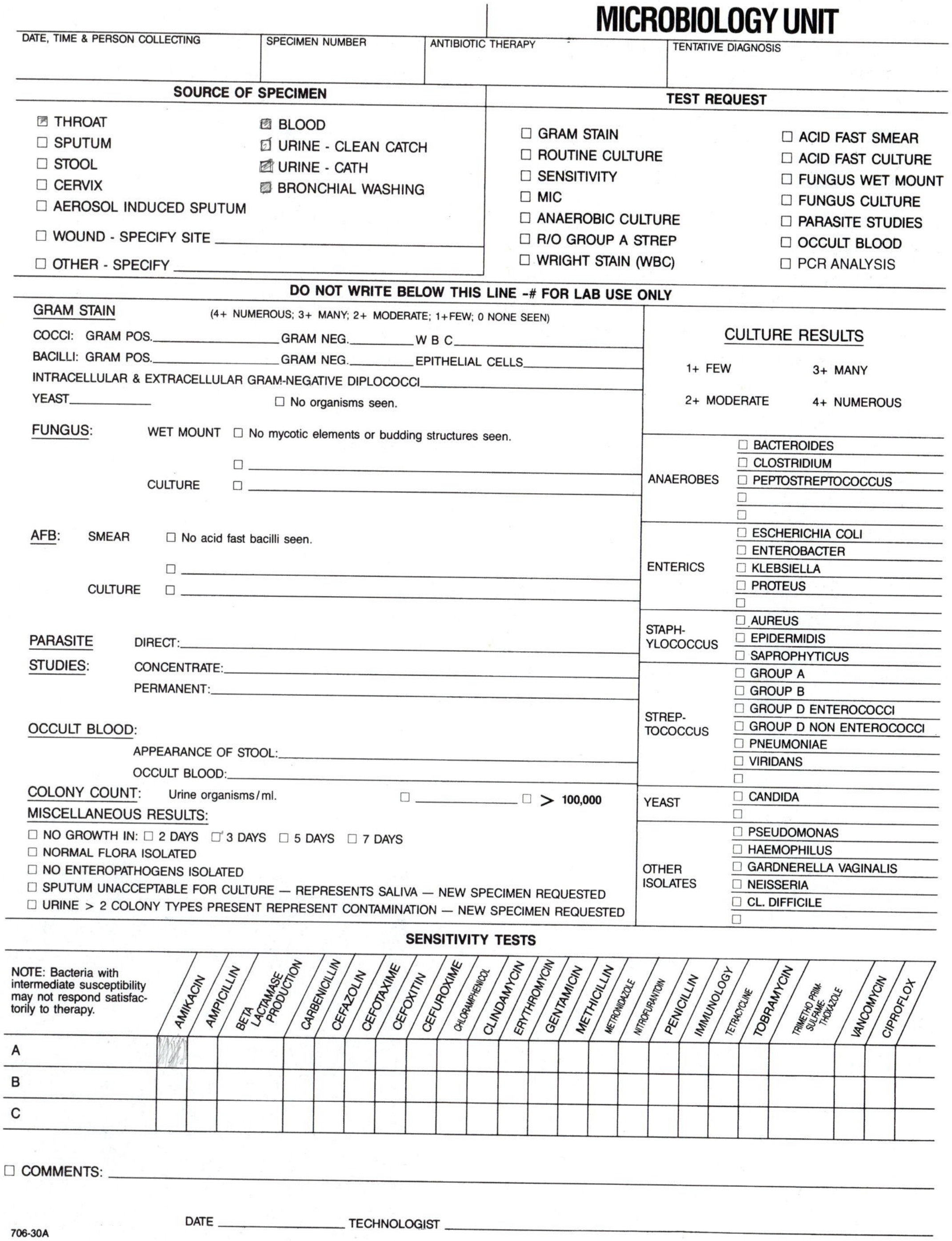

MICROBIOLOGY UNIT

DATE, TIME & PERSON COLLECTING	SPECIMEN NUMBER	ANTIBIOTIC THERAPY	TENTATIVE DIAGNOSIS

SOURCE OF SPECIMEN

☑ THROAT   ☑ BLOOD
☐ SPUTUM   ☑ URINE - CLEAN CATCH
☐ STOOL   ☑ URINE - CATH
☐ CERVIX   ☑ BRONCHIAL WASHING
☐ AEROSOL INDUCED SPUTUM

☐ WOUND - SPECIFY SITE _____

☐ OTHER - SPECIFY _____

TEST REQUEST

☐ GRAM STAIN   ☐ ACID FAST SMEAR
☐ ROUTINE CULTURE   ☐ ACID FAST CULTURE
☐ SENSITIVITY   ☐ FUNGUS WET MOUNT
☐ MIC   ☐ FUNGUS CULTURE
☐ ANAEROBIC CULTURE   ☐ PARASITE STUDIES
☐ R/O GROUP A STREP   ☐ OCCULT BLOOD
☐ WRIGHT STAIN (WBC) ☐ PCR ANALYSIS

DO NOT WRITE BELOW THIS LINE -# FOR LAB USE ONLY

GRAM STAIN   (4+ NUMEROUS; 3+ MANY; 2+ MODERATE; 1+FEW; 0 NONE SEEN)

COCCI: GRAM POS._____ GRAM NEG._____ W B C_____

BACILLI: GRAM POS._____ GRAM NEG._____ EPITHELIAL CELLS_____

INTRACELLULAR & EXTRACELLULAR GRAM-NEGATIVE DIPLOCOCCI_____

YEAST_____   ☐ No organisms seen.

FUNGUS: WET MOUNT ☐ No mycotic elements or budding structures seen.

                         ☐ _____

            CULTURE   ☐ _____

AFB: SMEAR   ☐ No acid fast bacilli seen.

                      ☐ _____

         CULTURE   ☐ _____

PARASITE   DIRECT:_____

STUDIES: CONCENTRATE:_____

              PERMANENT:_____

OCCULT BLOOD:

         APPEARANCE OF STOOL:_____

         OCCULT BLOOD:_____

COLONY COUNT: Urine organisms / ml.   ☐ _____ ☐ > 100,000

MISCELLANEOUS RESULTS:

☐ NO GROWTH IN: ☐ 2 DAYS ☐ 3 DAYS ☐ 5 DAYS ☐ 7 DAYS
☐ NORMAL FLORA ISOLATED
☐ NO ENTEROPATHOGENS ISOLATED
☐ SPUTUM UNACCEPTABLE FOR CULTURE — REPRESENTS SALIVA — NEW SPECIMEN REQUESTED
☐ URINE > 2 COLONY TYPES PRESENT REPRESENT CONTAMINATION — NEW SPECIMEN REQUESTED

CULTURE RESULTS

1+ FEW   3+ MANY

2+ MODERATE   4+ NUMEROUS

ANAEROBES	☐ BACTEROIDES
	☐ CLOSTRIDIUM
	☐ PEPTOSTREPTOCOCCUS
	☐
	☐
ENTERICS	☐ ESCHERICHIA COLI
	☐ ENTEROBACTER
	☐ KLEBSIELLA
	☐ PROTEUS
	☐
STAPH-YLOCOCCUS	☐ AUREUS
	☐ EPIDERMIDIS
	☐ SAPROPHYTICUS
STREP-TOCOCCUS	☐ GROUP A
	☐ GROUP B
	☐ GROUP D ENTEROCOCCI
	☐ GROUP D NON ENTEROCOCCI
	☐ PNEUMONIAE
	☐ VIRIDANS
	☐
YEAST	☐ CANDIDA
	☐
OTHER ISOLATES	☐ PSEUDOMONAS
	☐ HAEMOPHILUS
	☐ GARDNERELLA VAGINALIS
	☐ NEISSERIA
	☐ CL. DIFFICILE

SENSITIVITY TESTS

NOTE: Bacteria with intermediate susceptibility may not respond satisfactorily to therapy.

Columns: AMIKACIN, AMPICILLIN, BETA LACTAMASE PRODUCTION, CARBENICILLIN, CEFAZOLIN, CEFOTAXIME, CEFOXITIN, CEFUROXIME, CHLORAMPHENICOL, CLINDAMYCIN, ERYTHROMYCIN, GENTAMICIN, METHICILLIN, METRONIDAZOLE, NITROFURANTOIN, PENICILLIN, IMMUNOLOGY, TETRACYCLINE, TOBRAMYCIN, TRIMETHO PRIM-SULFAME-THOXAZOLE, VANCOMYCIN, CIPROFLOX

A																						
B																						
C																						

☐ COMMENTS: _____

706-30A   DATE _____ TECHNOLOGIST _____

FIGURE 17.3 Example of a clinical form used to report data on a patient's specimens.

few minutes in a streptococcal sore throat to several weeks in tuberculosis.

Results of specimen analysis are entered in a summary patient chart (**figure 17.3**) that can be used in assessment and treatment regimens. The type of antimicrobial drugs chosen for testing varies with the type of microorganism isolated.

Some diseases are diagnosed without the need to identify microbes from specimens. Serological tests on a patient's serum can detect signs of an antibody response. One method that clarifies whether a positive test indicates current or prior infection is to take two samples several days apart to see if the antibody titer is rising. Skin testing can pinpoint a delayed allergic reaction to a microorganism. These tests are also important in screening the general population for exposure to an infectious agent such as rubella or tuberculosis.

Because diagnosis is both a science and an art, the ability of the practitioner to interpret signs and symptoms of disease can be very important. AIDS, for example, is usually diagnosed by serological tests and a complex of signs and symptoms without ever isolating the virus. Some diseases (athlete's foot, for example) are diagnosed purely by the typical presenting symptoms and may require no lab tests at all.

17.3 Phenotypic Methods

Immediate Direct Examination of Specimen

Direct microscopic observation of a fresh or stained specimen is one of the most rapid methods of determining presumptive and sometimes confirmatory characteristics. Stains most often employed for bacteria are the Gram stain (see Insight 4.2) and the acid-fast stain (see figure 21.16). For many species these ordinary stains are useful, but they do not work with certain organisms. Direct fluorescence antibody (DFA) tests can highlight the presence of the microbe in patient specimens by means of labeled antibodies (figure 17.4). DFA tests are particularly useful for bacteria, such as the syphilis spirochete, that are not readily cultivated in the laboratory or if rapid diagnosis is essential for the survival of the patient.

Another way that specimens can be analyzed is through *direct antigen testing*, a technique similar to direct fluorescence in that known antibodies are used to identify antigens on the surface of bacterial isolates. But in direct antigen testing, the reactions can be seen with the naked eye. Quick test kits that greatly speed clinical diagnosis are available for *Staphylococcus aureus*, *Streptococcus pyogenes*, *Neisseria gonorrhoeae*, *Haemophilus influenzae*, and *Neisseria meningitidis*.

However, when the microbe is very sparse in the specimen, direct testing is like looking for a needle in a haystack, and more sensitive methods are necessary.

Cultivation of Specimen

Isolation Media

Such a wide variety of media exist for microbial isolation that a certain amount of preselection must occur, based on the nature of the specimen. In cases where the suspected pathogen is present in small numbers or is easily overgrown, the specimen can be initially enriched with specialized media. In specimens such as urine and feces that have high bacterial counts and a diversity of species, selective media are used. In most cases, specimens are also inoculated into differential media that define such characteristics as reactions in blood (blood agar) and fermentation patterns (mannitol salt and MacConkey agar). A patient's blood is usually cultured in a special bottle of broth that can be periodically sampled for growth. Numerous other examples of isolation, differential, and biochemical media were presented in chapter 3. So that subsequent steps in identification will be as accurate as possible, all work must be done from isolated colonies or pure cultures, because working with a mixed or contaminated culture gives misleading and inaccurate results. From such isolates, clinical microbiologists obtain information about a pathogen's microscopic morphology and staining reactions, cultural appearance, motility, oxygen requirements, and biochemical characteristics.

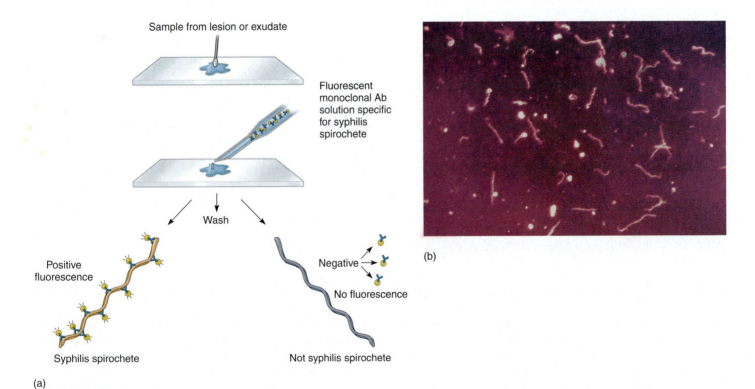

FIGURE 17.4 Direct fluorescence antigen test.
(a) Results for *Treponema pallidum*, the syphilis spirochete, and an unrelated spirochete. **(b)** Photomicrograph of this technique used on a blood sample from a syphilitic patient.

Biochemical Testing

The physiological reactions of bacteria to nutrients and other substrates provide excellent indirect evidence of the types of enzyme systems present in a particular species. Many of these tests are based on the following scheme:

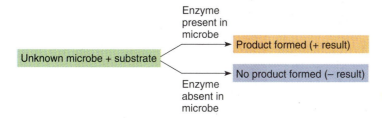

The microbe is cultured in a medium with a special substrate and then tested for a particular end product. The presence of the end product indicates that the enzyme is expressed in that species; its absence means it lacks the enzyme for utilizing the substrate in that particular way. These types of reactions are particularly meaningful in bacteria, which are haploid and generally express their genes for utilizing a given nutrient.

Among the prominent biochemical tests are carbohydrate fermentation (acid and/or gas); hydrolysis of gelatin, starch, and other polymers; enzyme actions such as catalase, oxidase, and coagulase; and various by-products of metabolism. Many are presently performed with rapid, miniaturized systems that can simultaneously determine up to 23 characteristics in small individual cups or spaces (**figure 17.5**). An important plus, given the complexity of biochemical profiles, is that such systems are readily adapted to computerized analysis.

Common schemes for identifying bacteria are somewhat artificial but convenient. They are based on easily recognizable characteristics such as motility, oxygen requirements, Gram stain reactions, shape, spore formation, and various biochemical reactions. Schemes can be set up as flowcharts (**figure 17.6**) or keys that trace a route of identification by offering pairs of opposing characteristics (positive versus negative, for example) from which to select. Eventually, an endpoint is reached, and the name of a genus or species that fits that particular combination of characteristics appears. Diagnostic tables that provide more complete information are preferred by many laboratories because variations from the general characteristics used on the flowchart can be misleading. Both systems are used in this text.

Miscellaneous Tests

When morphological and biochemical tests are insufficient to complete identification, other tests come into play.

Bacteria host viruses called bacteriophages that are very species- and strain-specific. Such selection by a virus for its host is useful in typing some bacteria, primarily *Staphylococcus* and *Salmonella*. The technique of phage typing involves inoculating a lawn of cells onto a Petri dish, mapping it off into blocks, and applying a different phage to each block. Cleared areas corresponding to lysed cells indicate sensitivity to that phage. Phage typing is chiefly used for tracing strains of bacteria in epidemics.

Animals must be inoculated to cultivate bacteria such as *Mycobacterium leprae* and *Treponema pallidum*, whereas avian embryos and cell cultures are used to grow rickettsias, chlamydias, and viruses. Animal inoculation is also occasionally used to test bacterial or fungal virulence.

Antimicrobial sensitivity tests are not only important in determining the drugs to be used in treatment (see figure 12.20), but the patterns of sensitivity can also be used in presumptive identification of some species of *Streptococcus*, *Pseudomonas*, and *Clostridium*. Antimicrobials are also used as selective agents in many media.

Determining Clinical Significance of Cultures

Questions that can be difficult but necessary to answer in this era of debilitated patients and opportunists are: Is an isolate clinically important, and how do you decide whether it is a contaminant or just part of the normal flora? The number of microbes in a sample is one useful criterion. For example, a few colonies of *Escherichia coli* in a urine sample can simply indicate normal flora, whereas several hundred can mean

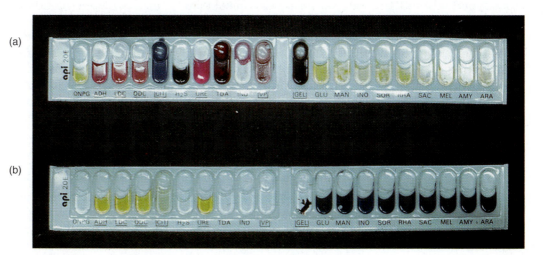

FIGURE 17.5 Rapid tests.
The API 20E manual biochemical system for microbial identification. **(a)** Positive and **(b)** negative results.

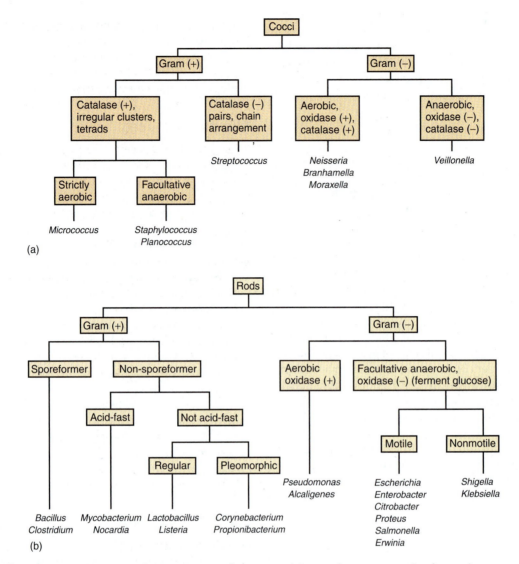

FIGURE 17.6 **Flowchart to separate primary genera of gram-positive and gram-negative bacteria.** **(a)** Cocci and **(b)** rods commonly involved in human diseases.

active infection. In contrast, the presence of a single colony of a true pathogen such as *Mycobacterium tuberculosis* in sputum or an opportunist in sterile sites such as cerebrospinal fluid or blood is highly suggestive of its role in disease. Furthermore, the repeated isolation of a relatively pure culture of any microorganism can mean it is an agent of disease, though care must be taken in this diagnosis. Another problem facing clinical laboratory personnel is that of differentiating a pathogen from species in the normal flora that are similar in morphology from their more virulent relatives.

17.4 Genotypic Methods
DNA Analysis Using Genetic Probes

The exact order and arrangement of the DNA code is unique to each organism. With a technique called *hybridization*, it is possible to identify a bacterial species by analyzing segments of its DNA. This requires small fragments of single-stranded DNA (or RNA) called **probes** that are known to be complementary to the specific sequences of DNA from a particular microbe. The test is conducted by extracting unknown test DNA from cells in specimens or cultures and binding it to special blotter paper. After several different probes have been added to the blotter, it is observed for visible signs that the probes have become fixed (hybridized) to the test DNA. The binding of probes onto several areas of the test DNA indicates close correspondence and makes positive identification possible **(figure 17.7)**.

Nucleic Acid Sequencing and rRNA Analysis

One of the most viable indicators of evolutionary relatedness and affiliation is comparison of the sequence of nitrogen bases in ribosomal RNA, a major component of ribosomes. Ribosomes have the same function (protein synthesis) in all

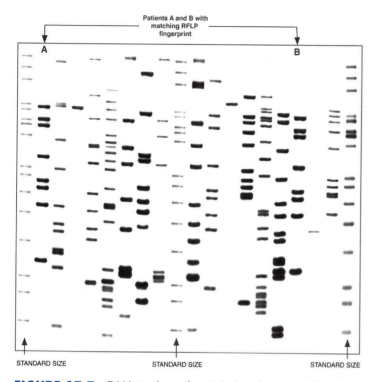

Patients A and B with matching RFLP fingerprint

A   B

STANDARD SIZE   STANDARD SIZE   STANDARD SIZE

FIGURE 17.7 **DNA typing of restriction fragment length polymorphisms (RFLPs) for *Mycobacterium tuberculosis*.** The pattern shows results for various strains isolated from 17 patients. Bands were developed by DNA hybridization using probes specific to genes from several *M. tuberculosis* strains. Lanes 1, 10, and 20 provide size markers for reference. Patients A and B are infected with the same common strain of the pathogen.

and other morphological characteristics, *Escherichia* has a G + C base composition of 48% to 52% and *Pseudomonas* has a composition of 58% to 70%, indicating that they probably are not closely related. This technique is most applicable for clarifying the taxonomic position of a bacterium, but it is too nonspecific to be applicable as a precise identification tool.

✔ CHECKPOINT

- Microbiologists use three categories of techniques to diagnose infections: *phenotypic, genotypic,* and *immunological.*
- The first step in clinical diagnosis (after observing the patient) is obtaining a sample. If this step is not performed correctly, the test will not be accurate no matter how "sensitive."
- The main phenotypic methods include the direct examination of specimens, observing the growth of specimen cultures on special media, and biochemical testing of specimen cultures.
- The use of genotypic methods has been increasing rapidly. These include genetic probing, nucleic acid sequencing, rRNA analysis, PCR-based methods, and determination of G + C composition.

cells, and they tend to remain more or less stable in their nucleic acid content over long periods. Thus, any major differences in the sequence, or "signature," of the rRNA is likely to indicate some distance in ancestry. This technique is powerful at two levels: It is effective for differentiating general group differences (it was used to separate the three superkingdoms of life discussed in chapter 1), and it can be fine-tuned to identify at the species level (for example in *Mycobacterium* and *Legionella*).

Polymerase Chain Reaction

Many nucleic acid assays use the polymerase chain reaction (PCR). This method can amplify DNA present in samples even in tiny amounts, which greatly improves the sensitivity of the test (see figure 10.6). PCR tests are being used or developed for a wide variety of bacteria, viruses, protozoa, and fungi.

G + C Base Composition

The overall percentage of guanine and cytosine (the G + C content as compared with A + T content) in DNA is a general indicator of relatedness because it is a trait that does not change rapidly. Bacteria with a significant difference in G + C percentage are likely to be genetically distinct species or genera. For example, although superficially similar in Gram reaction, shape,

17.5 Immunological Methods

The antibodies formed during an immune reaction are important in combating infection, but they hold additional practical value. Characteristics of antibodies (such as their quantity or specificity) can reveal the history of a patient's contact with microorganisms or other antigens. This is the underlying basis of serological testing. **Serology** is the branch of immunology that traditionally deals with *in vitro* diagnostic testing of serum. Serological testing is based on the familiar concept that antibodies have extreme specificity for antigens, so when a particular antigen is exposed to its specific antibody, it will fit like a hand in a glove. The ability to visualize this interaction by some means provides a powerful tool for detecting, identifying, and quantifying antibodies—or for that matter, antigens. The scheme works both ways, depending on the situation. One can detect or identify an unknown antibody using a known antigen, or one can use an antibody of known specificity to help detect or identify an unknown antigen **(figure 17.8).** Modern serological testing has grown into a field that tests more than just serum. Urine, cerebrospinal fluid, whole tissues, and saliva can also be used to determine the immunological status of patients. These and other immune tests are helpful in confirming a suspected diagnosis or in screening a certain population for disease.

General Features of Immune Testing

The strategies of immunological tests are diverse, and they underline some of the brilliant and imaginative ways that antibodies and antigens can be used as tools. We will summarize them under the headings of agglutination, precipitation,

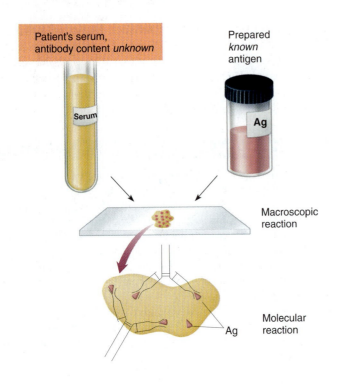

Patient's serum, antibody content *unknown*

Serum

Prepared *known* antigen

Ag

Macroscopic reaction

Molecular reaction

Ag

(a) In serological diagnosis of disease, a blood sample is scanned for the presence of antibody using an antigen of known specificity. A positive reaction is usually evident as some visible sign, such as color change or clumping, that indicates a specific interaction between antibody and antigen. (The reaction at the molecular level is rarely observed.)

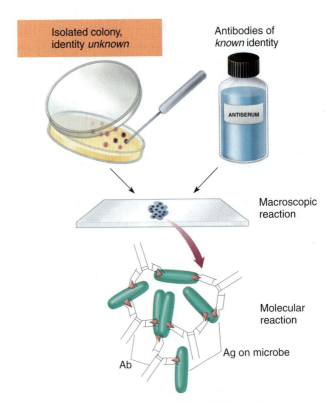

Isolated colony, identity *unknown*

Antibodies of *known* identity

ANTISERUM

Macroscopic reaction

Molecular reaction

Ag on microbe

Ab

(b) An unknown microbe is mixed with serum containing antibodies of known specificity, a procedure known as serotyping. Microscopically or macroscopically observable reactions indicate a correct match between antibody and antigen and permit identification of the microbe.

FIGURE 17.8 Basic principles of serological testing using antibodies and antigens.

immunodiffusion, complement fixation, fluorescent antibody tests, and immunoassay tests. First we will overview the general characteristics of immune testing, and we will then look at each type separately.

The most effective serological tests have a high degree of specificity and sensitivity **(figure 17.9)**. Specificity is the property of a test to focus upon only a certain antibody or antigen and not to react with unrelated or distantly related ones. Sensitivity means that the test can detect even very small amounts of antibodies or antigens that are the targets of the test. New systems using monoclonal antibodies have greatly improved specificity, and those using radioactivity, enzymes, and electronics have improved sensitivity.

Visualizing Antigen-Antibody Interactions

The primary basis of most tests is the binding of an antibody (Ab) to a specific molecular site on an antigen (Ag). Because this reaction cannot be readily seen without an electron microscope, tests involve some type of endpoint reaction visible to the naked eye or with regular magnification that tells whether the result is positive or negative. In the case of large antigens such as cells, Ab binds to Ag and creates large clumps or aggregates that are visible macroscopically or microscopically **(figure 17.10a)**. Smaller Ag-Ab complexes that

do not result in readily observable changes will require special indicators in order to be visualized. Endpoints are often revealed by dyes or fluorescent reagents that can tag molecules of interest. Similarly, radioactive isotopes incorporated into antigens or antibodies constitute sensitive tracers that are detectable with photographic film.

An antigen-antibody reaction can be used to read a **titer,** or the quantity of antibodies in the serum. Titer is determined by serially diluting a sample in tubes or in a multiple-welled microtiter plate and mixing it with antigen **(figure 17.10b)**. It is expressed as the highest dilution of serum that produces a visible reaction with an antigen. The more a sample can be diluted and yet still react with antigen, the greater is the concentration of antibodies in that sample and the higher is its titer. Interpretation of testing results is discussed in **Insight 17.2.**

Agglutination and Precipitation Reactions

The essential differences between agglutination and precipitation are in size, solubility, and location of the antigen. In agglutination, the antigens are whole cells such as red blood cells or bacteria with determinant groups on the surface. In precipitation, the antigen is a soluble molecule. In both instances, when Ag and Ab are optimally combined so that

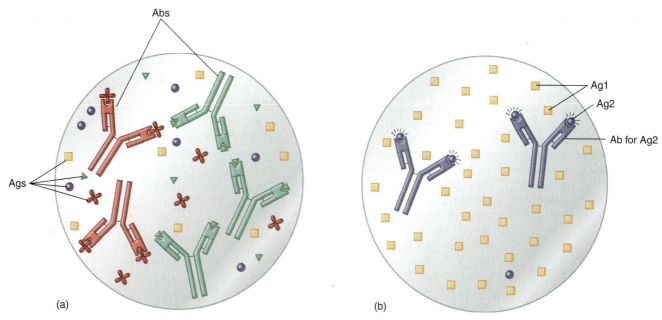

(a)

(b)

FIGURE 17.9   Specificity and sensitivity in immune testing.
(a) This test shows specificity in which an antibody (Ab) attaches with great exactness with only one type of antigen (Ag). **(b)** Sensitivity is demonstrated by the fact that Ab can pick up antigens even when the antigen is greatly diluted.

Agglutination

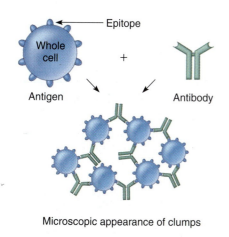

Precipitation

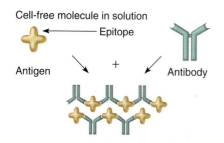

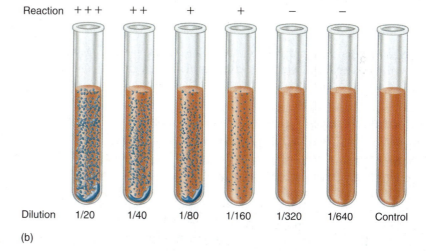

FIGURE 17.10   Cellular/molecular view of agglutination and precipitation reactions that produce visible antigen-antibody complexes.
Although IgG is shown as the Ab, IgM is also involved in these reactions. **(a)** Agglutination involves clumping of whole cells; precipitation is the formation of antigen-antibody complexes in cell free solution. **(b)** The tube agglutination test. A sample of patient's serum is serially diluted with saline. The dilution is made in a way that halves the number of antibodies in each subsequent tube. An equal amount of the antigen (here, blue bacterial cells) is added to each tube. The control tube has antigen, but no serum. After incubation and centrifugation, each tube is examined for agglutination clumps as compared with the control, which will be cloudy and clump-free. The titer is defined as the dilution of the last tube in the series that shows agglutination.

INSIGHT 17.2
Medical

When Positive Is Negative: How to Interpret Serological Test Results

What if a patient's serum gives a positive reaction—is **seropositive**—in a serological test? In most situations, it means that antibodies specific for a particular microbe have been detected in the sample. But one must be cautious in proceeding to the next level of interpretation. The mere presence of antibodies does not necessarily indicate that the patient has a disease, but only that he or she has possibly had contact with a microbe or its antigens through infection or vaccination. In screening tests for determining a patient's history (rubella, for instance), knowing that a certain titer of antibodies is present can be significant, because it shows that the person has some protection.

When the test is being used to diagnose current disease, however, a series of tests to show a rising titer of antibodies is necessary. The accompanying figure indicates how such a test can be used to diagnose patients who have nonspecific symptoms that could fit several diseases. Lyme disease, for instance, can be mistaken for arthritis or viral infections. In the first group, note that the antibody titer against *Borrelia burgdorferi*, the causative agent, increased steadily over a 6-week period. A control group that shared similar symptoms did not exhibit a rise in titer for antibodies to this microbe. Clinicians call samples collected early and late in an infection *acute* and *convalescent* sera.

Another important consideration in testing is the occasional appearance of biological false positives. These are results in which

a patient's serum shows a positive reaction, even though, in reality, he is not or has not been infected by the microbe. False positives, such as those in syphilis and AIDS testing, arise when antibodies or other substances present in the serum cross-react with the test reagents, producing a positive result. Such false results may require retesting by a method that greatly minimizes cross-reactions.

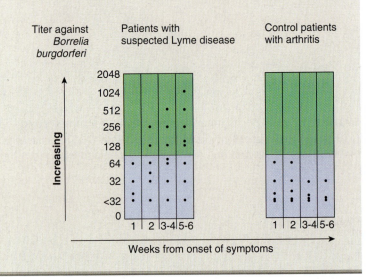

neither is in excess, one antigen is interlinked by several antibodies to form an insoluble, three-dimensional aggregate so large that it cannot remain suspended and it settles out.

Agglutination Testing

Agglutination is discernible because the antibodies cross-link the antigens to form visible clumps. **Agglutination** tests are performed routinely by blood banks to determine ABO and Rh (Rhesus) blood types in preparation for transfusions. In this type of test, antisera containing antibodies against the blood group antigens on red blood cells are mixed with a small sample of blood and read for the presence or absence of clumping. The **Widal test** is an example of a tube agglutination test for diagnosing salmonelloses and undulant fever. In addition to detecting specific antibody, it also gives the serum titer.

Numerous variations of agglutination testing exist. The rapid plasma reagin (RPR) test is one of several tests commonly used to test for antibodies to syphilis. The cold agglutinin test, named for antibodies that react only at lower temperatures (4°C to 20°C), was developed to diagnose *Mycoplasma* pneumonia. The *Weil-Felix reaction* is an agglutination test sometimes used in diagnosing rickettsial infections.

In some tests, special agglutinogens have been prepared by affixing antigen to the surface of an inert particle. In *latex*

agglutination tests, the inert particles are tiny latex beads. Kits using latex beads are available for assaying pregnancy hormone in the urine, identifying *Candida* yeasts and bacteria (staphylococci, streptococci, and gonococci), and diagnosing rheumatoid arthritis.

Precipitation Tests

In precipitation reactions, the soluble antigen is precipitated (made insoluble) by an antibody. This reaction is observable in a test tube in which antiserum has been carefully laid over an antigen solution. At the point of contact, a cloudy or opaque zone forms.

One example of this technique is the VDRL (Veneral Disease Research Lab) test that also detects antibodies to syphilis. Although it is a good screening test, it contains a heterophilic antigen (cardiolipin) that may give rise to false positive results. Although precipitation is a useful detection tool, the precipitates are so easily disrupted in liquid media that most precipitation reactions are carried out in agar gels. These substrates are sufficiently soft to allow the reactants (Ab and Ag) to freely diffuse, yet firm enough to hold the Ag-Ab precipitate in place. One technique with applications in microbial identification and diagnosis of disease is the double diffusion (Ouchterlony) method **(figure 17.11).** It is called

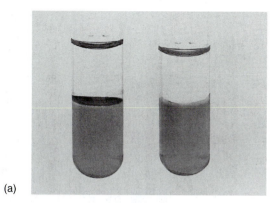

(a)

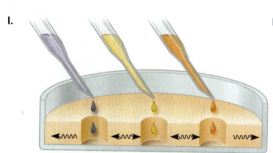

Side view

I. In one method of setting up a double-diffusion test, wells are punctured in soft agar, and antibodies (Ab) and antigens (Ag) are added in a pattern. As the contents of the wells diffuse toward each other, a number of reactions can result, depending on whether antibodies meet and precipitate antigens.

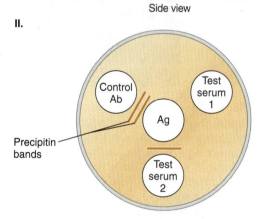

II. Example of test pattern and results. Antigen (Ag) is placed in the center well and antibody (Ab) samples are placed in outer wells. The control contains known Abs to the test Ag. Note bands that form where Ab/Ag meet. The other wells (1, 2) contain unknown test sera. One is positive and the other is negative. Double bands indicate more than one antigen and antibody that can react.

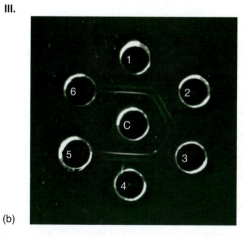

(b)

III. Actual test results for detecting infection with the fungal pathogen *Histoplasma*. Numbers 1 and 4 are controls and 2, 3, 5, and 6 are patient test sera. Can you determine which patients have the infection and which do not?

FIGURE 17.11 **Precipitation reactions.**

(a) A tube precipitation test for streptococcal group antigens. Specific antiserum has been placed in the bottom of tubes and antigen solution carefully overlaid to form a zone of contact. The left-hand tube has developed a band of precipitate; the right-hand tube is negative. **(b)** Double-diffusion (Ouchterlony) tests in a semisolid matrix.

Step 1   Serum sample separated next to trough

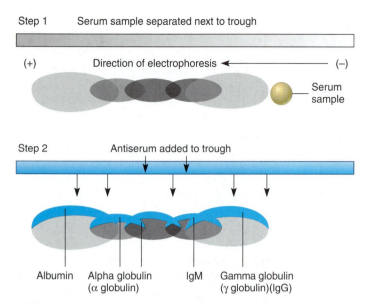

FIGURE 17.12   Immunoelectrophoresis of normal human serum.
Step 1. Proteins are separated by electrophoresis on a gel.
Step 2. To identify the bands and increase visibility, antiserum containing antibodies specific for serum proteins is placed in a trough and allowed to diffuse toward the bands. This diffusion produces a pattern of numerous arcs representing major serum components.

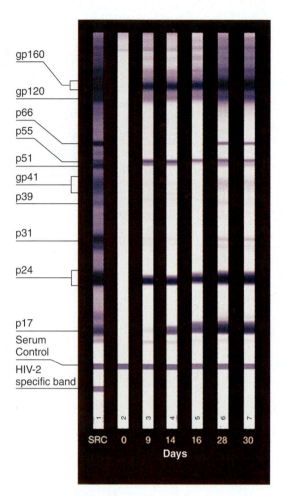

FIGURE 17.13   The Western blot procedure.
The example shown here tests one patient's blood for antibodies to specific HIV antigens. Samples were taken at six different time periods after suspected exposure. The test strips are prepared by electrophoresing several of the major HIV surface and core antigens and then blotting them onto special filters. The test strips are incubated with a patient's serum and developed with a radioactive or colorimetric label. Sites where HIV antigens have bound antibodies show up as bands. Patients' sera are then compared with a positive control strip (SRC) containing antibodies for all HIV antigens. Certain criteria must be met to consider the result positive.

Interpretation of Bands

Labels correspond with glycoproteins (GP) or proteins (P) that are part of HIV-1 antigen structure.

- The test is considered positive if bands occur at two locations: gp 160 or gp 120 and g31 or g24.
- The test is considered negative if no bands are present for any HIV antigen.
- The test is considered indeterminate if bands are present, but not at the criteria locations. This result may require retesting at a later date.

double diffusion because it involves diffusion of both antigens and antibodies. The test is performed by punching a pattern of small wells into an agar medium and filling them with test antigens and antibodies. A band forming between two wells indicates that antibodies from one well have met and reacted with antigens from the other well. Variations on this technique provide a means of identifying unknown antibodies or antigens.

Immunoelectrophoresis constitutes yet another refinement of diffusion and precipitation in agar. With this method, a serum sample is first electrophoresed to separate the serum proteins as previously described in chapter 15. Antibodies that react with specific serum proteins are placed in a trough parallel to the direction of migration, forming reaction arcs specific for each protein **(figure 17.12)**. This test is widely used to detect disorders in the production of antibodies.

The Western Blot for Detecting Proteins

The **Western blot** test is somewhat similar to the previous tests because it involves the electrophoretic separation of proteins, followed by an immunoassay to detect these proteins. This test is a counterpart of the Southern blot test for identifying DNA, described in chapter 10. It is a highly specific and sensitive way to identify or verify a particular protein (antibody or antigen) in a sample **(figure 17.13)**.

First, the test material is electrophoresed in a gel to separate out particular bands. The gel is then transferred to a special blotter that binds the reactants in place. The blot is developed by incubating it with a solution of antigen or antibody that has been labeled with radioactive, fluorescent, or luminescent labels. Sites of specific binding will appear as a pattern of bands that can be compared with known positive and negative samples. This is currently the second (verification) test for people who are antibody-positive for HIV in the ELISA test (described in a later section), because it tests more types of antibodies and is less subject to misinterpretation than are other antibody tests. The technique has significant applications for detecting microbes and their antigens in specimens.

✔ CHECKPOINT

- Serological tests can test for either antigens or antibodies. Most are *in vitro* assessments of antigen-antibody reactivity from a variety of body fluids. The basis of these tests is an antigen-antibody reaction made visible through the processes of agglutination, precipitation, immunodiffusion, complement fixation, fluorescent antibody, and immunoassay techniques.

- One measurement is the *titer*, described as the concentration of antibody in serum. It is the highest dilution of serum that gives a visible antigen-antibody reaction. The higher the titer, the greater the level of antibody present.

- Agglutination reactions occur between antibody and antigens bound to cells. This results in visible clumps caused by large antibody-antigen complexes. In viral hemagglutination testing, the antibody reacts with the antigen and inhibits it from agglutinating red blood cells.

- In precipitation reactions, soluble antigen and antibody react to form insoluble, visible precipitates. Precipitation reactions can also be visualized by adding radioactive or enzyme markers to the antigen-antibody complex.

- In immunoelectrophoresis techniques such as the Western blot, proteins that have been separated by electrical current are identified by labeled antibodies. HIV infections are verified with this method.

Complement Fixation

An antibody that requires complement to complete the lysis of its antigenic target cell is termed a **lysin** or cytolysin. When lysins act in conjunction with the intrinsic complement system on red blood cells, the cells hemolyze (lyse and release their hemoglobin). This lysin-mediated hemolysis is the basis of a group of tests called complement fixation, or CF **(figure 17.14).**

Complement fixation testing uses four components—antibody, antigen, complement, and sensitized sheep red blood cells—and it is conducted in two stages. In the first stage, the test antigen is allowed to react with the test antibody (at least one must be of known identity) in the absence of complement. If the Ab-Ag are specific for each other, they form complexes. To this mixture, purified complement proteins from guinea pig blood are added. If antibody and antigen have complexed during the previous step, they attach, or fix, the complement to them, thus preventing it from participating in further reactions. The extent of this complement fixation is determined in the second stage by means of sheep RBCs with surface lysin molecules. The sheep RBCs serve as an indicator complex that can also fix complement. Contents of the stage 1 tube are mixed with the stage 2 tube and observed for hemolysis, which can be observed with the naked eye as a clearing of the solution. If hemolysis *does not* occur, it means that the complement was used up by the first stage Ab-Ag complex and that the unknown antigen or antibody was indeed present. This result is considered positive. If hemolysis *does* occur, it means that unfixed complement from tube 1 reacted with the RBC complex instead, thereby causing lysis of the sheep RBCs. This result is negative for the antigen or antibody that was the target of the test. Complement fixation tests are invaluable in diagnosing influenza, polio, and various fungi.

The antistreptolysin O (ASO) titer test measures the levels of antibody against the streptolysin toxin, an important hemolysin of group A streptococci. It employs a technique related to complement fixation. A serum sample is exposed to known suspensions of streptolysin and then allowed to incubate with RBCs. Lack of hemolysis indicates antistreptolysin antibodies in the patient's serum that have neutralized the streptolysin and prevented hemolysis. This is an important verification procedure for scarlet fever, rheumatic fever, and other related streptococcal syndromes (see chapter 21).

Miscellaneous Serological Tests

A test that relies on changes in cellular activity as seen microscopically is the *Treponema pallidum immobilization* (TPI) test for syphilis. The impairment or loss of motility of the *Treponema* spirochete in the presence of test serum and complement indicates that the serum contains anti-*Treponema pallidum* antibodies. In *toxin neutralization* tests, a test serum is incubated with the microbe that produces the toxin. If the serum inhibits the growth of the microbe, one can conclude that antitoxins are present.

Serotyping is an antigen-antibody technique for identifying, classifying, and subgrouping certain bacteria into categories called serotypes, using antisera for cell antigens such as the capsule, flagellum, and cell wall. It is widely used in typing *Salmonella* species and strains and is the basis for identifying the numerous serotypes of streptococci. The Quellung test, which identifies serotypes of the pneumococcus, involves a precipitation reaction in which antibodies react with the capsular polysaccharide. Although the reaction makes the capsule seem to swell, it is actually creating a zone of Ab-Ag complex on the cell's surface.

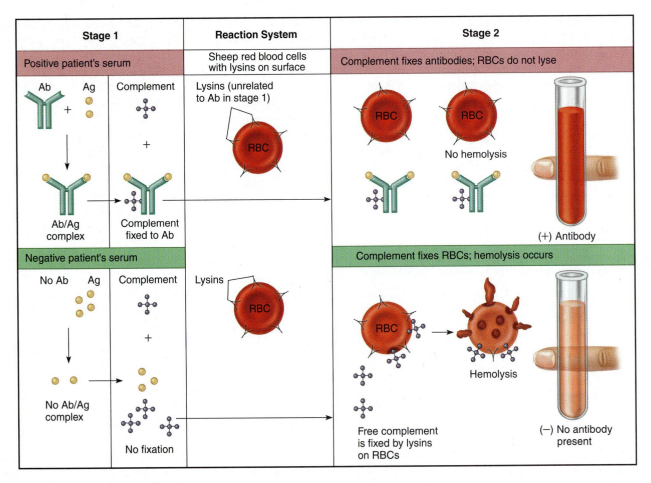

Stage 1			Reaction System	Stage 2	
Positive patient's serum			Sheep red blood cells with lysins on surface	**Complement fixes antibodies; RBCs do not lyse**	
Ab	Ag	Complement	Lysins (unrelated to Ab in stage 1)	RBC RBC	No hemolysis
Ab/Ag complex		Complement fixed to Ab	RBC		(+) Antibody
Negative patient's serum				**Complement fixes RBCs; hemolysis occurs**	
No Ab	Ag	Complement	Lysins	RBC	Hemolysis
No Ab/Ag complex		No fixation	RBC	Free complement is fixed by lysins on RBCs	(−) No antibody present

FIGURE 17.14 Complement fixation test.
In this example, two serum samples are being tested for antibodies to a certain infectious agent. In reading this test one observes the cloudiness of the tube. If it is cloudy, the RBCs are not hemolyzed and the test is positive. If it is clear and pink, the RBCs are hemolyzed and the test is negative.

Fluorescent Antibodies and Immunofluorescence Testing

The property of dyes such as fluorescein and rhodamine to emit visible light in response to ultraviolet radiation was discussed in chapter 3. This property of fluorescence has found numerous applications in diagnostic immunology. The fundamental tool in immunofluorescence testing is a fluorescent antibody—a monoclonal antibody labeled by a fluorescent dye (fluorochrome).

The two ways that fluorescent antibodies (FABs) can be used are shown in **figure 17.15.** In *direct testing*, an unknown test specimen or antigen is fixed to a slide and exposed to a fluorescent antibody solution of known composition. If the antibodies are complementary to antigens in the material, they will bind to it. After the slide is rinsed to remove unattached antibodies, it is observed with the fluorescent microscope. Fluorescing cells or specks indicate the presence of Ab-Ag complexes and a positive result. These tests are valuable for identifying and locating antigens on the surfaces of cells or in tissues and in identifying the disease agents of syphilis, gonorrhea, chlamydiosis, whooping cough, Legionnaires' disease, plague, trichomoniasis, meningitis, and listeriosis.

In *indirect testing* methods, the fluorescent antibodies are antibodies made to react with the Fc region of another antibody (remember that antibodies can be antigenic). In this scheme, an antigen of known character (a bacterial cell, for example) is combined with a test serum of unknown antibody content. The fluorescent antibody solution that can react with the unknown antibody is applied and rinsed off to visualize whether the serum contains antibodies that have affixed to the antigen. A positive test shows fluorescing aggregates or cells, indicating that the fluorescent antibodies have combined with the unlabeled antibodies. In a negative test, no fluorescent complexes will appear. This technique is frequently used to diagnose syphilis (FTA-ABS) and various viral infections.

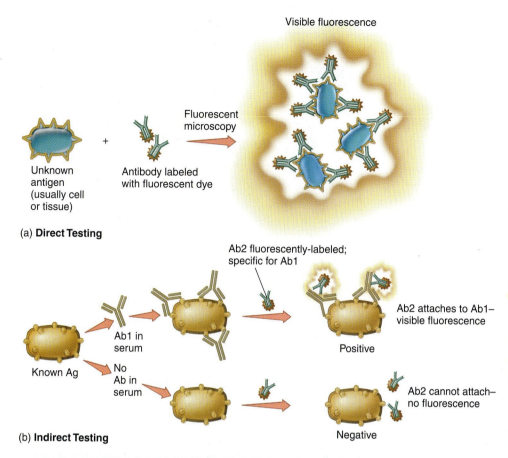

Visible fluorescence

Fluorescent microscopy

Unknown antigen (usually cell or tissue)

Antibody labeled with fluorescent dye

(a) **Direct Testing**

Ab2 fluorescently-labeled; specific for Ab1

Ab1 in serum

Known Ag

No Ab in serum

Ab2 attaches to Ab1– visible fluorescence

Positive

Ab2 cannot attach– no fluorescence

Negative

(b) **Indirect Testing**

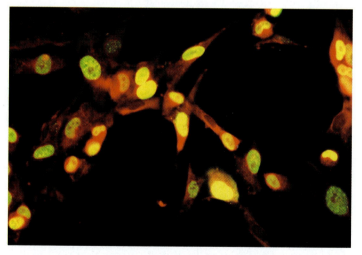

(c) **Indirect Immunofluorescence Testing**

FIGURE 17.15 Immunofluorescence testing.
(a) Direct: Unidentified antigen (Ag) is directly tagged with fluorescent Ab. **(b)** Indirect: Ag of known identity is used to assay unknown Ab; a positive reaction occurs when the second Ab (with fluorescent dye) affixes to the first Ab. **(c)** An indirect immunofluorescent stain of cells infected with two different viruses. Cells fluorescing green contain cytomegalovirus; cells fluorescing yellow contain adenovirus.

Immunoassays: Tests of Great Sensitivity

The elegant tools of the microbiologist and immunologist are being used increasingly in athletics, criminology, government, and business to test for trace amounts of substances such as hormones, metabolites, and drugs. But traditional techniques in serology are not refined enough to detect a few molecules of these chemicals. Extremely sensitive alternative methods that permit rapid and accurate measurement of trace antigen or antibody are called **immunoassays.** Examples of the technology for detecting an antigen or antibody in minute quantities include radioactive isotope labels, enzyme labels, and sensitive electronic sensors. Many of these tests are based on specifically formulated monoclonal antibodies.

Radioimmunoassay (RIA)

Antibodies or antigens labeled with a radioactive isotope can be used to pinpoint minute amounts of a corresponding antigen or antibody. Although very complex in practice, these assays compare the amount of radioactivity present in a sample before and after incubation with a known, labeled antigen or antibody. The labeled substance competes with its natural, nonlabeled partner for a reaction site. Large amounts of a bound radioactive component indicate that the unknown test substance was not present. The amount of radioactivity is measured with an isotope counter or a photographic emulsion (autoradiograph). Radioimmunoassay has been employed to measure the levels of insulin and other hormones and to diagnose allergies, chiefly by the radioimmunosorbent test (RIST) for measurement of IgE in allergic patients and the radioallergosorbent test (RAST) to standardize allergenic extracts.

Enzyme-Linked Immunosorbent Assay (ELISA)

The **ELISA test**, also known as enzyme immunoassay (EIA), contains an enzyme-antibody complex that can be used as a color tracer for antigen-antibody reactions. The enzymes used most often are horseradish peroxidase and alkaline phosphatase, both of which release a dye (chromogen) when exposed to their substrate. This technique also relies on a solid support such as a plastic microtiter plate that can *adsorb* (attract on its surface) the reactants **(figure 17.16).**

(a) **Indirect ELISA,** comparing a positive vs. negative reaction. This is the basis for HIV screening tests.

Well A   Well B

Known antigen is adsorbed to well.

Serum samples with unknown antibodies.

A

B

Sample A   Sample B

Well is rinsed to remove unbound (nonreactive) antibodies.

Indicator antibody linked to enzyme attaches to any bound antibody.

Wells are rinsed to remove unbound indicator antibody. A colorless substrate for enzyme is added.

Enzymes linked to indicator Ab hydrolyze the substrate, which releases a dye. Wells that develop color are positive for the antibody; colorless wells are negative.

(+)   (−)

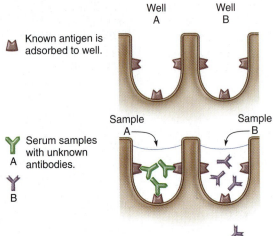
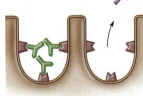

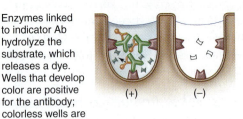

(b) **Microtiter ELISA Plate with 96 Tests for HIV Antibodies.** Colored wells indicate a positive reaction.

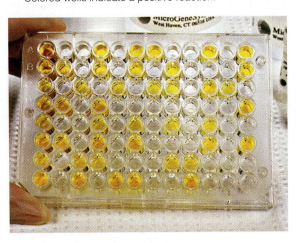

(c) **Capture or Antibody Sandwich ELISA Method.** Note that an antigen is trapped between two antibodies. This test is used to detect hantavirus and measles virus.

Antibody is adsorbed to well.

Test antigen is added; if complimentary, antigen binds to antibody.

Enzyme

Enzyme-linked antibody specific for test antigen then binds to another antigen, forming a sandwich.

Enzyme's substrate (□) is added, and reaction produces a visible color change (●).

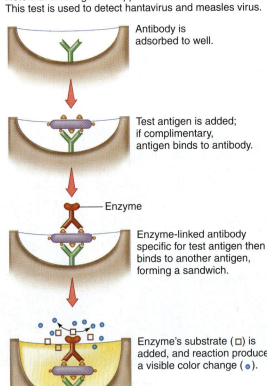

FIGURE 17.16   Methods of ELISA testing.

The *indirect ELISA* test can detect antibodies in a serum sample. As with other indirect tests, the final positive reaction is achieved by means of an antibody-antibody reaction. The indicator antibody is complexed to an enzyme that produces a color change with positive serum samples (**figure 17.16a,b**). The starting reactant is a known antigen that is adsorbed to the surface of a well. To this, an unknown serum is added. After rinsing, an enzyme-Ab reagent that can react with the unknown test antibody is placed in the well. The substrate for the enzyme is then added, and the wells are scanned for color changes. Color development indicates that all the components reacted and that the antibody was present in the patient's serum. This is the common screening test for the antibodies to HIV, various rickettsial species, hepatitis A and C, the cholera vibrio, and *Helicobacter,* a cause of gastric ulcers. Because false positives can occur, a verification test may be necessary (such as Western blot for HIV).

In *capture ELISA* (or sandwich) tests, a known antibody is adsorbed to the bottom of a well and incubated with a solution containing unknown antigen (**figure 17.16c**). After excess unbound components have been rinsed off, an enzyme-antibody indicator that can react with the antigen is added. If antigen is present, it will attract the indicator-antibody and hold it in place. Next, the substrate for the enzyme is placed in the wells and incubated. Enzymes affixed to the antigen will hydrolyze the substrate and release a colored dye. Thus, any color developing in the wells is a positive result. Lack of color means that the antigen was not present and that the subsequent rinsing removed the enzyme-antibody complex. The capture technique is used to detect antibodies to hantavirus, rubella virus, and *Toxoplasma.*

A newer technology uses electronic monitors that directly read out antibody-antigen reactions. Without belaboring the technical aspects, these systems contain computer chips that sense the minute changes in electrical current given off when an antibody binds to antigen. The potential for sensitivity is extreme; it is thought that amounts as small as 12 molecules of a substance can be detected in a sample. In another procedure, antibody substrate molecules are incubated with sample and then exposed to the enzyme alkaline phosphatase. If the antibody is bound, the enzyme reacts with the substrate and causes visible light to be emitted. The light can be detected by machines or photographic films.

IN THE NEWS (Continued from page 515)

The nurse described at the beginning of the chapter was tested for HIV infection. Initially, she tested positive by the ELISA screening test; however, by the confirmatory test she was shown to be HIV negative. As a result, the ELISA test is referred to as a false positive. Although the ELISA test is very sensitive, its specificity is much lower than many of the confirmatory tests. Almost 9 out of 10 low-risk individuals (such as blood donor volunteers) that test positive for HIV by the ELISA method are shown to be HIV negative by the confirmatory test methods. Patients that have delivered multiple babies, or those with certain autoimmune diseases, can test positive by the ELISA method in the absence of HIV infection.

Up to 800,000 U.S. health care workers annually experience needlestick or other "sharps" injuries. Based on data analysis, only 0.3% of those exposed to HIV-contaminated blood, after the skin had been pierced through a sharps injury, became productively infected with HIV.

Interestingly, most ELISA tests will detect antibodies to both HIV-1 and HIV-2, whereas the Western blot method is initially performed with HIV-1, since the virus is more common in the United States. An HIV-2 Western blot can be performed if the HIV-1 Western blot yields negative results.

CDC. 1989. *Interpretation and use of the Western blot assay for serodiagnosis of human immunodeficiency virus type 1 infections.* MMWR; *38:S4–S6.*

CDC. 2001. *Revised guidelines for HIV counseling, testing, and referral.* MMWR; *50(No. RR-19).*

CDC. 2001. *Updated U.S. Public Health Service guidelines for the management of occupational exposures to HBV, HCV, and HIV and recommendations for post-exposure prophylaxis.* MMWR; *50(No. RR-11).*

Tests That Differentiate T Cells and B Cells

So far we have concentrated on tests that identify antigens and antibodies in samples, but techniques also exist that differentiate between B cells and T cells and can quantify subsets of each. Information on the types and numbers of lymphocytes in blood and other samples is a common way to evaluate immune dysfunctions such as those in AIDS, immunodeficiencies, and cancer. A simple method for identifying T cells is to mix them with untreated sheep red blood cells. Receptors on the T cells bind the RBCs into a flowerlike cluster called a **rosette formation (figure 17.17a).** Rosetting can also occur in B cells if one uses Ig-coated bovine RBCs or mouse erythrocytes.

For routine identification of lymphocytes, rosette formation has been replaced by fluorescent techniques. These techniques can differentiate between T cells and B cells as well as to subgroup them (**figure 17.17b**). These subgroup tests utilize monoclonal antibodies produced in response to specific cell markers. B-cell tests categorize different stages in B-cell development and are very useful in characterizing B-cell cancers. Tests that can help differentiate the CD4, CD8, and other T-cell subsets are important in monitoring AIDS and other immunodeficiency diseases.

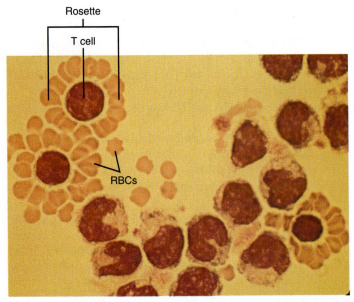

(a)

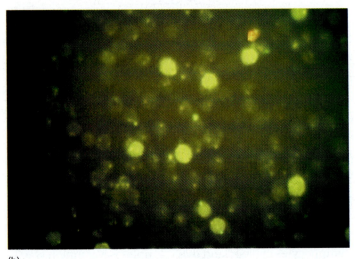

(b)

FIGURE 17.17 **Tests for characterizing T cells and B cells.**
(a) Photomicrograph of rosette formation that identifies T cells.
(b) Plasma cells highlighted by fluorescent antibodies.

In Vivo Testing

Probably the first immunologic tests were performed not in a test tube but on the body itself. A classic example of one such technique is the **tuberculin test,** which uses a small amount of purified protein derivative (PPD) from *Mycobacterium tuberculosis* injected into the skin. The appearance of a red,

raised, thickened lesion in 48 to 72 hours can indicate previous exposure to tuberculosis (shown in figure 16.14). In practice, *in vivo* tests employ principles similar to serological tests, except in this case an antigen or an antibody is introduced into a patient to elicit some sort of visible reaction. Like the tuberculin test, some of these diagnostic skin tests are useful for evaluating infections due to fungi (coccidioidin and histoplasmin tests, for example) or allergens. Allergic reactions and other immune system disorders are the topics of chapter 16.

A Viral Example

All of the methods discussed so far—phenotypic, genotypic, and immunological—are applicable to the different types of microorganisms. Viruses sometimes present special difficulties since they are not cells and they are more labor intensive to culture in the laboratory. **Figure 17.18** presents an overview of various techniques that might be used to diagnose viral infections. It provides one example of the variety of methods that can be employed for many infections regardless of their cause.

> ### ✔ CHECKPOINT
>
> - Complement fixation involves a two-part procedure in which complement fixes to a specific antibody if present, or to red blood cell antigens, if antibody is absent. Lack of RBC hemolysis is indicative of a positive test.
> - Serological tests can measure the degree to which host antibody binds directly to disease agents or toxins. This is the principle behind tests for syphilis and rheumatic fever.
> - Direct fluorescent antibody tests indicate presence of an antigen and are useful in identifying infectious agents. Indirect fluorescent tests indicate the presence of a particular antibody and can diagnose infection.
> - Immunoassays can detect very small quantities of antigen, antibody, or other substances. Radioimmunoassay uses radioisotopes to detect trace amounts of biological substances.
> - The ELISA test uses enzymes and dyes to detect antigen-antibody complexes. It is widely used to detect viruses, bacteria, and antibodies in HIV infection.
> - Technicians use precise assays to differentiate between B and T cells and to identify subgroups of these cells for disease diagnosis.
> - *In vivo* serological testing, such as the tuberculin test, involves subcutaneous injection of antigen to elicit a visible immune response in the host.

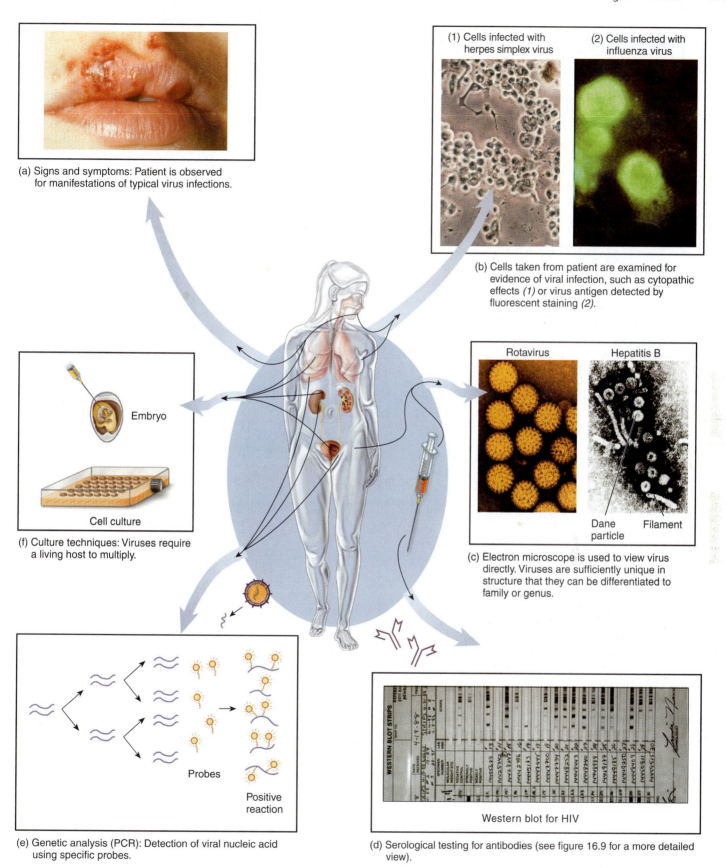

(a) Signs and symptoms: Patient is observed for manifestations of typical virus infections.

(1) Cells infected with herpes simplex virus

(2) Cells infected with influenza virus

(b) Cells taken from patient are examined for evidence of viral infection, such as cytopathic effects (1) or virus antigen detected by fluorescent staining (2).

Embryo

Cell culture

(f) Culture techniques: Viruses require a living host to multiply.

Rotavirus

Hepatitis B

Dane particle   Filament

(c) Electron microscope is used to view virus directly. Viruses are sufficiently unique in structure that they can be differentiated to family or genus.

Probes

Positive reaction

(e) Genetic analysis (PCR): Detection of viral nucleic acid using specific probes.

Western blot for HIV

(d) Serological testing for antibodies (see figure 16.9 for a more detailed view).

FIGURE 17.18 Summary of methods used to diagnose viral infections.

Chapter Summary With Key Terms

17.1 Preparation for the Survey of Microbial Diseases

A. Phenotypic methods are those that assess microscopic morphology, macroscopic morphology, physiological and biochemical characteristic, and chemical composition.

B. Genotypic methods examine the genetic composition of a microorganism.

C. Immunological methods exploit the host's antibody reaction to microbial antigens for purposes of diagnosis.

17.2 On the Track of the Infectious Agent: Specimen Collection

A. Diagnosis begins with accurate specimen collection, which means the sampling of body sites or fluids that are suspected to contain the infective agent.

B. Laboratory methods yield either presumptive data or confirmatory data. Sometimes pathogens are diagnosed based solely on signs and symptoms in the patient.

17.3 Phenotypic Methods

A. The most obvious phenotypic characteristic of a microbe is what it looks like. Many infections can be diagnosed by microscopic examination of fresh or stained microorganisms from specimens. Direct antigen or antibody testing can also be performed on fresh specimens.

B. Cultivation of specimens allows the examination of colony morphology and/or growth characteristics, which are useful for identifying microbes.

C. Biochemical tests determine whether microbes possess particular enzymes or biochemical pathways which can serve to identify them.

D. A variety of other tests are available to test specific phenotypic characteristics of particular suspected microorganisms.

E. It is of utmost importance to consider whether an identified microbe is actually causing the disease or is simply a bystander, or member of the normal flora.

17.4 Genotypic Methods

Genetic methods of identification are being used increasingly in diagnostic microbiology. Tests can assess the proportion of G + C nucleotides relative to the A + T content, or determine DNA or rRNA sequences. An increasingly useful procedure is the use of **polymerase chain reaction** to amplify DNA present in a sample in order to detect small quantities of microbial sequences.

17.5 Immunological Methods

A. Serology is a science that attempts to detect signs of infection in a patient's serum such as antibodies specific for a microbe.

B. The basis of serological tests is that Abs specifically bind to Ag *in vitro*. An Ag of known identity will react with antibodies in an unknown serum sample. The reverse is also true; known antibodies can be used to detect and type antigens.

C. These Ag-Ab reactions are visible in the form of obvious clumps and precipitates, color changes, or the release of radioactivity. Test results are read as positive or negative.

D. Desirable properties of tests are high specificity and sensitivity.

E. Types of Tests

1. In **agglutination** tests, antibody cross-links whole-cell antigens, forming complexes that settle out and form visible clumps in the test chamber; examples are tests for blood type, some bacterial diseases, and viral diseases.

2. Double diffusion precipitation tests involve the diffusion of Ags and Abs in a soft agar gel, forming zones of precipitation where they meet.

3. In immunoelectrophoresis, migration of serum proteins in gel is combined with precipitation by antibodies.

4. The **Western blot** test separates antigen into bands. After the gel is affixed to a blotter, it is reacted with a test specimen and developed by radioactivity or with dyes.

5. Complement fixation tests detect *lysins*—antibodies that fix complement and can lyse target cells. It involves first mixing test Ag and Ab with complement and then with sensitized sheep RBCs. If the complement is fixed by the Ag-Ab, the RBCs remain intact, and the test is positive. If RBCs are hemolyzed, specific antibodies are lacking.

6. In direct assays, known marked Ab is used to detect unknown Ag (microbe).

   a. In indirect testing, known Ag reacts with unknown Ab, and the reaction is made visible by a second Ab that can affix to and identify the unknown Ab.

   b. Immunofluorescence testing uses fluorescent antibodies (FABs tagged with fluorescent dye) either directly or indirectly to visualize cells or cell aggregates that have reacted with the FABs.

7. **Immunoassays** are highly sensitive tests for Ag and Ab.

   a. In *radioimmunoassay*, Ags or Abs are labeled with radioactive isotopes and traced.

   b. The **enzyme-linked immunosorbent assay (ELISA)** can detect unknown Ag or Ab by direct or indirect means. A positive result is visualized when a colored product is released by an enzyme-substrate reaction.

   c. Tests are also available to differentiate B cells from T cells and their subtypes.

8. With *in vivo* testing, Ags are introduced into the body directly to determine the patient's immunologic history.

Multiple-Choice Questions

1. **Multiple Matching.** Evaluate and match each of the following culture results to the most likely interpretation.
   a. probable infection
   b. normal flora
   c. contamination

   ____ 1. Isolation of two colonies of *E. coli* on a plate streaked from the urine sample

   ____ 2. Isolation of 50 colonies of *Streptococcus pneumoniae* on a plate streaked with sputum

   ____ 3. A mixture of 80 colonies of various streptococci on a culture from a throat swab

   ____ 4. Colonies of black bread mold on selective media used to isolate bacteria from stool

   ____ 5. Blood culture bottle with heavy growth

2. Which of the following methods can identify different strains of a microbe?
   a. microscopic examination
   b. radioimmunoassay
   c. DNA typing
   d. agglutination test

3. In agglutination reactions, the antigen is a ____; in precipitation reactions, it is a ____.
   a. soluble molecule, whole cell
   b. whole cell, soluble molecule
   c. bacterium, virus
   d. protein, carbohydrate

4. Which reaction requires complement?
   a. hemagglutination
   b. precipitation
   c. hemolysis
   d. toxin neutralization

5. A patient with a ____ titer of antibodies to an infectious agent generally has greater protection than a patient with a ____ titer.
   a. high, low
   b. low, high
   c. negative, positive
   d. old, new

6. Direct immunofluorescence tests use a labeled antibody to identify ____.
   a. an unknown microbe
   b. an unknown antibody
   c. fixed complement
   d. agglutinated antigens

7. The Western blot test can be used to identify
   a. unknown antibodies
   b. unknown antigens
   c. specific DNA
   d. both a and b

8. An example of an *in vivo* serological test is
   a. indirect immunofluorescence
   b. radioimmunoassay
   c. tuberculin test
   d. complement fixation

Concept Questions

These questions are suggested as a *writing-to-learn* experience. For each question, compose a one- or two-paragraph answer that includes the factual information needed to completely address the question.

1. Why do specimens need to be taken aseptically even when nonsterile sites are being sampled and selective media are to be used?

2. Explain the general principles in specimen collection.

3. a. What is involved in direct specimen testing?
   b. In presumptive and confirmatory tests?
   c. In cultivating and isolating the pathogen?
   d. In biochemical testing?
   e. In gene probes?

4. Differentiate between the serological tests used to identify isolated cultures of pathogens and those used to diagnose disease from patients' serum.

5. Why is it important to prevent microbes from growing in specimens?

6. Why is speed so important in the clinical laboratory?

7. Summarize the important points in determining if a clinical isolate is involved in infection.

8. In figure 17.18, which of the methods (a) through (f) is a phenotypic method? Genotypic? Immunological?

9. a. What is the basis of serology and serological testing?
   b. Differentiate between specificity and sensitivity.
   c. Describe several general ways that Ag-Ab reactions are detected.

10. a. What does seropositivity mean?
    b. What is a false positive test result and what are some possible causes?
    c. What is meant by a false negative result and what might account for it?
    d. What does the titer of serum tell us about the immune status?

11. a. Explain how agglutination and precipitation reactions are alike.
    b. In what ways are they different?
    c. Make a drawing of the manner in which antibodies cross-link the antigens in agglutination and precipitation reactions.
    d. Give examples of several tests that employ the two reactions.

12. a. What is meant by complement fixation? What are cytolysins?
    b. What is the purpose of using sheep red blood cells in this test?

13. a. Explain the differences between direct and indirect procedures in serological or immunoassay tests.
    b. How is fluorescence detected?
    c. How is the reaction in a radioimmunoassay detected?
    d. How does a positive reaction in an ELISA test appear? How many wells are positive in figure 17.16*b*?

14. Briefly describe the principles and give an example of the use of a specific test using immunoelectrophoresis, Western blot, complement fixation, fluorescence testing (direct and indirect), and immunoassays (direct and indirect ELISA).

Critical Thinking Questions

Critical thinking is the ability to reason and solve problems using facts and concepts. These questions can be approached from a number of angles, and in most cases, they do not have a single correct answer.

1. See Insight 17.1. How would you explain to a junior high biology class that in the next decade some diseases currently thought to be non-infectious will probably be found to be caused by microbes?

2. In what way could the extreme sensitivity of the PCR method be a problem when working with clinical specimens?

3. Why do some tests for antibody in serum (such as for HIV and syphilis) require backup verification with additional tests at a later date?

4. a. Look at figure 17.10*b*. What is the titer as shown?
   b. If the titer had been 1:40, what interpretation would be made as to the immune status of the patient?

   c. What would it mean if a test 2 weeks later revealed a titer of 1:1280?
   d. What would it mean if no agglutination had occurred in any tube?

5. Why do we interpret positive hemolysis in the complement fixation test to mean negative for the test substance?

6. Observe figure 17.16 and make note of the several steps in the indirect ELISA test. What four essential events are necessary to develop a positive reaction (besides having antibody A)? Hint: what would happen without rinses?

7. Using the criteria for band interpretation in figure 17.13, does this patient test positive for HIV? Why or why not?

8. Explain how an immunoassay method could use monoclonal antibodies to differentiate between B and T cells and between different subsets of T cells.

Internet Search Topics

Go to the Online Learning Center for chapter 17 of this text at http://www.mhhe.com/cowan1. Access the URLs listed under Internet Search Topics and research the following:

1. Access the websites listed for excellent graphics and explanations of immune tests.

Infectious Diseases Affecting the Skin and Eyes

IN THE NEWS

In February of 2001, a 10-month-old baby in Texas became ill with fever, conjunctivitis, and a maculopapular rash. The baby's parents reported that the fever had begun during their recent flight from China, where they had adopted the child. The hospital where the child was taken conducted a thorough physical exam, making note of low-grade fever, cough, runny nose, and conjunctivitis. Raised bumps were apparent on the mucosal lining inside the cheeks.

A presumptive diagnosis was made based on the symptoms; blood was drawn and the diagnosis was confirmed when high levels of a specific IgM were found. The diagnosis set off a flurry of activity because the disease is relatively serious and highly contagious. Furthermore, the baby could have potentially exposed hundreds of people during the incubation period preceding the flight home from China. First, of course, there were the other children and staff at the orphanage where the child lived. Also potentially exposed were 63 families that had traveled together to and from China on this trip and dozens of staff members at the Chinese hospital where medical exams were performed, as well as staff at the U.S. Consulate and passengers and crew on the two flights (from China to Los Angeles and from Los Angeles to Houston).

The Centers for Disease Control and Prevention worked together with the Central China Adoption Agency to identify contacts, isolate additional cases, and supply vaccinations where needed. Eventually 14 cases of this disease were found in the United States, most of them babies from a single orphanage in China, or their U.S. contacts. Officials at the orphanage acknowledged that recent arrivals at the orphanage had not been adequately immunized against the disease. A vaccine against this disease has been widely used in the United States since the 1960s, but less developed countries may have low vaccination rates.

▶ *What is the disease described here?*

▶ *If a vaccine has been widely used in the United States for 40 years, why did American contacts of the child also become infected?*

CHAPTER OVERVIEW

▶ The skin is organized in layers, from the deepest layer, called the stratum basale, to the uppermost layer, which is the epidermis. The epidermis is composed of cells packed with the protein keratin, which protects the skin and "waterproofs" it. Keratin also provides protection against microbial invasion. Other defenses include low pH, high-salt, and low-moisture conditions. Both the skin and the eye are protected by lysozyme.

▶ The surfaces of the eye exposed to microbial infection are the conjunctiva and the cornea. The flushing action of the tears, which contain lysozyme and lactoferrin, is the major protective feature of the eye.

▶ The skin has a diverse array of microbes as its normal flora. Gram-positive bacteria are most common. Inhabitants of the skin must be tolerant of high-salt, low-moisture conditions. The eye is also home to an array of (mostly gram-positive) bacteria, but in lower numbers than on the skin.

▶ Infectious diseases with their most visible manifestations on the surface of the body (or on the skin) range from acne and athletes foot to leprosy and gangrene.

▶ Infectious diseases with their major manifestations in the eye include neonatal eye infections as well as conditions afflicting all ages, such as keratitis and trachoma.

18.1 The Skin and Eyes

The skin makes contact directly with the environment—not only with solid objects but also with water and other fluids and with the atmosphere. As a result, the skin is a vulnerable site for microbial infection. What's more, many infectious diseases include skin eruptions or lesions as part of the course of illness and often as a major symptom, even if the infective agent does not enter via the skin. Prior to more sophisticated diagnostic methods, the appearance of a skin rash was often the best clue to the type of disease being experienced by a patient. This is still true in many instances.

The eye surface, like the skin, is also exposed constantly to the environment. For this reason, we include diseases of both organ systems in this chapter.

18.2 The Skin and Its Defenses

The organ under consideration in this chapter is the boundary between the organism and the environment. The skin, together with the hair, nails, and sweat and oil glands, forms the **integument.** The skin has a total surface area of 1.5 to 2 square meters. Its thickness varies from 1.5 mm at places such as the eyelids to 4 mm on the soles of the feet. Several distinct layers can be found in this thickness, and we will summarize them here. Follow **figure 18.1** as you read.

The outermost portion of the skin is the epidermis, which is further subdivided into four or five distinct layers. On top is a thick layer of epithelial cells called the stratum corneum, about 25 cells thick. The cells in this layer are dead and have migrated from the deeper layers during the normal course of cell division. They are packed with a protein called keratin, which the cells have been producing ever since they arose from the deepest level of the epidermis. Because this process is continuous, the entire epidermis is replaced every 25 to 45 days. Keratin gives the cells their ability to withstand damage, abrasion, and water penetration; the surface of the skin is termed *keratinized* for this reason. Below the stratum corneum are three or four more layers of epithelial cells. Notice in figure 18.1 that there are no nerve endings or blood vessels in the stratum corneum or any other portion of the epidermis. The lowest layer, the stratum basale, or basal layer, is attached to the underlying dermis and is the source for all of the cells that make up the epidermis.

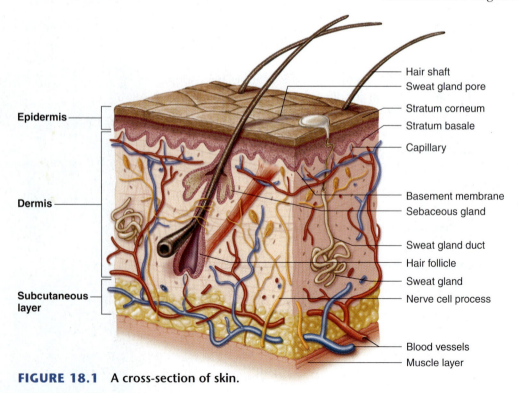

Epidermis

Dermis

Subcutaneous layer

Hair shaft
Sweat gland pore
Stratum corneum
Stratum basale
Capillary
Basement membrane
Sebaceous gland
Sweat gland duct
Hair follicle
Sweat gland
Nerve cell process
Blood vessels
Muscle layer

FIGURE 18.1   A cross-section of skin.

The dermis, underneath the epidermis, is composed of connective tissue instead of epithelium. This means that it is a rich matrix of fibroblast cells and fibers such as collagen, and it contains macrophages and mast cells. The dermis also harbors a dense network of nerves, blood vessels, and lymphatic vessels. Damage to the epidermis generally does not result in bleeding, whereas damage deep enough to penetrate the dermis results in broken blood vessels. Blister formation, the result of friction trauma or burns, causes a separation between the dermis and epidermis.

The "roots" of hairs, called follicles, are in the dermis. Sebaceous (oil) glands and scent glands are associated with the hair follicle. Separate sweat glands are also found in this tissue. All of these glands have openings on the surface of the skin, so they pass through the epidermis as well.

It could be said that the skin is its own defense—in other words, the very nature of its keratinized surface prevents most microorganisms from penetrating into sensitive deeper tissues. Millions of cells from the stratum corneum slough off every day, and attached microorganisms slough off with them. The skin is also brimming with antimicrobial substances. The sebaceous glands' secretion, called **sebum,** has a low pH, which makes the skin inhospitable to most microorganisms. Sebum is oily due to its high concentration of lipids. The lipids can serve as nutrients for normal microbiota, but breakdown of the fatty acids contained in lipids leads to toxic by-products that inhibit the growth of microorganisms not adapted to the skin environment. This mechanism helps control the growth of potentially pathogenic bacteria. Sweat is also inhibitory to microorganisms, because of both its low pH and its high salt concentration. **Lysozyme** is an enzyme found in sweat (and tears and saliva) that specifically breaks down peptidoglycan, which you learned in chapter 4 is a unique component of eubacterial cell walls.

18.3 Normal Flora of the Skin

Microbes that live on the skin surface as normal flora must be capable of living in the dry, salty conditions they find there. Microbes are rather sparsely distributed over dry, flat areas of the body such as on the back, but they can grow into dense populations in moist areas and skin folds, such as the underarm and groin areas. The normal microflora also live in the protected environment of the hair follicles and glandular ducts.

Three main categories of microorganisms reside on the skin: the diphtheroids, the micrococci (including staphylococci and streptococci), and yeasts. The diphtheroids are club-shaped bacteria that resemble *Corynebacterium diptheriae*. They are gram positive and can be aerobic, aerotolerant, or anaerobic. Unlike *C. diptheriae* they are not considered virulent. One very prominent member of this group is *Propionibacterium acnes*, which is aerotolerant or even anaerobic. It lives on healthy skin, but its metabolic activities can contribute to the development of acne (discussed later in the chapter).

The micrococcus group includes the genus *Staphylococcus* as well as the genus *Micrococcus*. It has been said that *S. epidermidis* is the bacterial species most well adapted to life

on the human body. It is present on the skin of every human. *S. aureus*, by contrast, could be called the bacterium best adapted to damage its host. It is found living on the skin of at least 20% of the human population, but it can hardly be called "normal flora" because it is such a potentially dangerous pathogen. Sometimes it is referred to as "abnormal flora," indicating that although it can inhabit skin without causing disease, it can also cause dangerous infections (see Insight 18.1).

S. epidermidis is sometimes referred to as "CNS" or coagulase-negative staphylococcus, highlighting its lack of the enzyme **coagulase,** which is indeed found in *S. aureus*. The staphylococci are particularly well adapted to life on the skin since they can tolerate high salt concentrations. In fact, a common culturing method to select for staphylococci employs a high-salt agar, which inhibits the growth of most nonstaphylococci.

Alpha-hemolytic and nonhemolytic streptococci are also found on the skin.

Various yeast groups grow in low numbers on the skin; they can cause opportunistic disease. These groups include familiar species, such as *Candida albicans*, and less familiar types, such as *Pityrosporum*.

18.4 The Surface of the Eye and Its Defenses

The eye is a complex organ with many different tissue types, but for the purposes of this chapter we will consider only its exposed surfaces, the *conjunctiva* and the *cornea* **(figure 18.2).** The **conjunctiva** is a very thin membranelike tissue that covers the eye (except for the cornea) and lines the eyelids. It secretes an oil- and mucous-containing fluid that lubricates

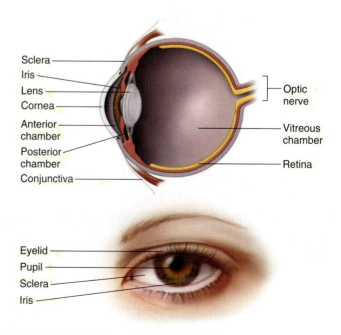

FIGURE 18.2   The anatomy of the eye.

and protects the eye surface. The **cornea** is the dome-shaped central portion of the eye, lying over the iris (the colored part of the eye). It has five to six layers of epithelial cells that can regenerate quickly if they are superficially damaged. It has been called "the windshield of the eye."

The eye's best defense is the film of tears, which consists of an aqueous fluid, oil, and mucous. The tears are formed in the lacrimal gland at the outer and upper corner of each eye **(figure 18.3),** and they drain into the lacrimal duct at the inner corner. The aqueous portion of tears contains sugars, lysozyme, and lactoferrin. These last two substances have antimicrobial properties. The mucous layer contains proteins and sugars and plays a protective role. And of course, the flow of the tear film prevents the attachment of microorganisms to the eye surface.

Because the eye's primary function is vision, anything that hinders vision would be counterproductive. For that reason, inflammation does not occur in the eye as readily as it does elsewhere in the body. Flooding the eye with fluid containing a large number of light-diffracting objects such as lymphocytes and phagocytes in response to every irritant would mean almost constantly blurred vision. So even though the eyes are relatively vulnerable to infection (not being covered by keratinized epithelium), the evolution of the vertebrate eye has of necessity favored reduced innate immunity. This characteristic is sometimes known as *immune privilege.*

The specific immune response, involving B and T cells, is also somewhat restricted in the eye. The anterior chamber (figure 18.2) is largely cut off from the blood supply. Lymphocytes that do gain access to this area are generally less active than lymphocytes elsewhere in the body.

18.5 Normal Flora of the Eye

The normal flora of the eye is generally sparse. When people are tested, up to 20% have no recoverable bacteria in their eyes. The few bacteria that are found resemble the normal flora of the skin—namely, diphtheroids, coagulase-negative staphylococci, *Micrococcus,* nonhemolytic streptococci, and some yeast. *Neisseria* species can also live on the surface of the eye.

A NOTE ABOUT THE CHAPTER ORGANIZATION

Beginning in this chapter we discuss all the conditions caused by microbial infection. The chapter organization mirrors the clinical experience. Patients present themselves to health care practitioners with a set of symptoms, and the health care team makes an "anatomical" diagnosis—such as a *generalized vesicular rash.* The anatomical diagnosis allows practitioners to narrow down the list of possible causes to microorganisms that are known to be capable of creating such a condition. Then the proper tests can be performed to arrive at an etiological diagnosis (that is, determining the exact microbial cause). So the order of events is (1) anatomical diagnosis based on signs and symptoms; (2) consideration of a number of agents that are known to cause disease in that anatomical location (often called the differential diagnosis); followed by (3) the etiological diagnosis. In practice this process may be shortened. For instance, if a patient has a disease such as Hansen's disease (leprosy), the distinctive signs and symptoms of that disease may allow the practitioner to make the anatomical and the etiological diagnosis at the same time, followed by confirmation of the etiology through laboratory methods, if necessary. In other cases, such as the common cold, the physician may consider only the anatomical diagnosis and never advance to the etiological diagnosis because a cold is a mild self-limiting disease.

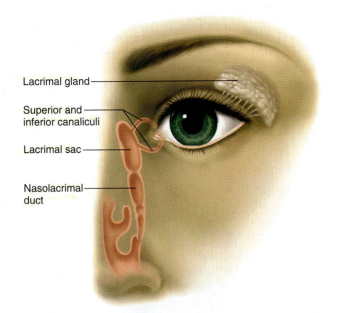

Lacrimal gland

Superior and inferior canaliculi

Lacrimal sac

Nasolacrimal duct

FIGURE 18.3 **The lacrimal apparatus of the eye.**

Defenses and Normal Flora of the Skin and Eyes		
	Defenses	**Normal Flora**
Skin	Keratinized surface, sloughing, low pH, high salt, lysozyme	*Corynebacterium, Propionibacterium, Staphylooccus epidermidis* and *S. aureus, Micrococcus,* α- and nonhemolytic streptococci, *Candida,* and *Pityrosporum*
Eyes	Mucus in conjunctiva and in tears, lysozyme and lactoferrin in tears	Sparse but similar to skin

The chapters are organized by anatomical diagnosis (for example, Microbial Diseases of the Skin and Eyes). Specific diseases and the microorganisms that cause them then are detailed. Some diseases are the result of infection by a single type of microorganism. Hansen's disease is an example of this type of disease. In other cases, single diseases or conditions can be caused by many different microorganisms, including bacteria, viruses, and so on. The classic examples are pneumonia in the respiratory tract, meningitis in the central nervous system, and diarrhea in the gastrointestinal system. In this chapter, for instance, a maculopapular rash may be caused by the measles virus, the rubella virus, or a parvovirus. A table at the end of each disease/condition makes it clear whether a single microorganism or multiple agents are to be considered in the diagnosis.

18.6 Skin Diseases Caused by Microorganisms

Acne

The term *acne* encompasses all follicle-associated lesions, from the isolated pimple to severe widespread acne. Normally, the sebaceous glands associated with hair follicles (figure 18.1) are a self-contained system for protecting, softening, and lubricating the skin. As hair and skin grow, dead epidermal cells and sebum work their way upward and are discharged from the pore to the skin surface.

Skin prone to pimples and acne has a structure that traps the mass of sebum and dead cells, clogging the pores. An exaggerated process of keratinization occurs in skin cells in and around the follicle, which also helps to block the pore. An added factor is overproduction of sebum when the sebaceous gland is stimulated by hormones (especially male). *Propionibacterium acnes* present in the follicle releases lipases to digest this surplus of oil. The combination of digestive products (fatty acids) and bacterial antigens stimulates an intense local inflammation that eventually can burst the follicle. In time the lesion can erupt on the surface.

Different types of lesions are associated with this process. When the skin initially swells over the pore leading out of a hair follicle, it is called a *comedo*. If the pore is closed, this comedo is commonly called a whitehead. If the pore remains open to the surface but is blocked with a dark plug of sebum, it appears as a blackhead. When the lesion erupts on the surface, it is called a pustule or papule. At this point the lesion contains sebum and pus, a collection of bacteria, dead skin cells, and white blood cells from the inflammatory reaction. Pustules that come to involve deeper layers of skin are called cysts, and they can be quite painful. Widespread lesions of this type are called cystic acne.

Causative Agent

Propionibacterium acnes is the bacterium associated with acne, but the "cause" of acne is multifactorial, requiring other conditions, just described, to be just right before the presence of this otherwise benign bacterium results in acne.

The bacterium is an anaerobic or aerotolerant gram-positive rod arranged in short chains or clumps. It releases a variety of enzymes that contribute to its virulence. The most important of these appears to be lipase, although it also releases proteases, neuraminidase, and a hyaluronidase. In addition, it secretes a low molecular weight protein that is a strong attractant for white blood cells (contributing to inflammation).

The complete genome sequence of the bacterium was published in 2004. This will allow researchers to identify additional virulence factors and to design more precise therapies for it.

Transmission and Epidemiology

As already noted, *P. acnes* is normal flora on human skin, so it is not a transmissible infection. The epidemiology of a condition refers to its distribution in populations, and usually takes into consideration the mode of transmission of the microorganism, the degree of susceptibility of different hosts, environmental parameters such as climate and geography, and even the behavior of current and potential hosts. When speaking of the epidemiology of acne, we are really considering what groups have the combination of factors that can result in acne, rather than the distribution of *P. acnes*. Almost 100% of adolescents and young adults experience acne of some degree at some time in their lives. More severe forms of adolescent acne are more common in males than females, probably because male hormones, or androgens, aggravate the condition. Females produce male hormones as well, but during adolescence males have a higher incidence of moderate to severe acne. Evidence exists that acne extending into adulthood, or beginning in adulthood, is more common in women.

Prevention and Treatment

There is no effective prevention of acne; it is not the result of poor hygiene or even of eating the wrong foods. For many years the only treatment options were (1) topical agents that enhanced the sloughing of skin cells, which could help to prevent comedo formation or to keep comedos from becoming pustules or papules; and (2) either topical or oral antibiotics, such as erythromycin or tetracycline. It has become apparent that such long-term use of antibiotics causes a high rate of antibiotic resistance in skin bacteria (in the case of topical application) or in whole-body normal flora. This result should have been predicted since oral antibiotics are typically given for long periods of time in low, sublethal doses—the perfect set of conditions for creating antibiotic resistance in bacteria. It has ~ shown that live-in family members of peopl~ for their acne also eventually harbo~ the same antibiotic the acne pa~ tant bacteria can spread beyona~ ~up

Recently, females have been ~ ~tigo, pills (containing estrogen) to trea~ ~t fever, controversial because of the da~ ~er in this patients with severe acne, and f~ ~streptococ- options have failed, isotretinoin (Acc~

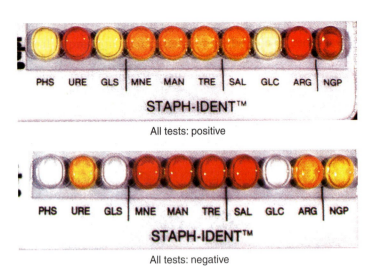

FIGURE 18.7 **Miniaturized test system used in further identification of *Staphylococcus* isolates.**
A single isolate is used to inoculate all the cupules on a strip. The cupules contain substrates that detect phosphatase production (PHS), urea hydrolysis (URE), glucosidase production (GLS), mannose fermentation (MNE), mannitol fermentation (MAN), trehalose fermentation (TRE), salicin fermentation (SAL), glucuronidase production (GLC), arginine hydrolysis (ARG), and galactosidase production (NGP).

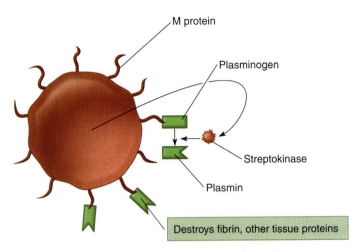

FIGURE 18.8 **Plasmin activation by *S. pyogenes*.**
The bacterium binds host plasminogen, then secretes an enzyme (streptokinase) that cleaves it, creating plasmin, which has tissue-degrading power. In the figure bacterial components are red, and host components are green.

✔ CHECKPOINT 18.2   Impetigo

	Staphylococcus aureus	*Streptococcus pyogenes*
Causative Organism(s)	*Staphylococcus aureus*	*Streptococcus pyogenes*
Most Common Modes of Transmission	Direct contact, indirect contact	Direct contact, indirect contact
Virulence Factors	Exfoliative toxin A, coagulase, other enzymes	Streptokinase, plasminogen-binding ability, hyaluronidase, M protein
Culture/Diagnosis	Routinely based on clinical signs, when necessary, culture and Gram stain, coagulase and catalase tests, multitest systems, PCR	Routinely based on clinical signs, when necessary, culture and Gram stain, coagulase and catalase tests, multitest systems, PCR
Prevention	Hygiene practices	Hygiene practices
Treatment	Topical mupirocin, oral cephalexin	Topical mupirocin, oral cephalexin
Distinguishing Features	Seen more often in older children, adults	Seen more often in newborns; may have some involvement in all impetigo (preceding *S. aureus* in staphylococcal impetigo)

If the precise etiological agent must be identified, there are well-established methods for identifying group A streptococci. Refer to chapter 21.

Pathogenesis and Virulence Factors The symptoms of *S. pyogenes* impetigo are indistinguishable from that caused by *S. aureus*. Like *S. aureus*, this bacterium possesses a huge arsenal of enzymes and toxins. As mentioned earlier, it anchors itself to surfaces (including skin), using a variety of adhesive elements on its surface (LTA, M protein and other proteins, and a hyaluronic acid capsule). M protein also protects it from phagocytosis. And like *S. aureus* it possesses hyaluronidase.

S. pyogenes has a clever system to exploit host factors to increase its ability to spread in tissues. The bacterium's M protein has a high-affinity binding site for plasminogen—a host plasma protein which, when activated (cleaved) becomes plasmin **(figure 18.8)**. Plasmin itself is a protein-splitting enzyme that digests fibrin and other tissue proteins. But *S. pyogenes* doesn't wait for the normal course of events in which another host protein activates the plasminogen. Instead it secretes an enzyme called streptokinase, which is a plasminogen activator. So the bacterium coats itself with host plasminogen, then uses its own enzyme (streptokinase) to activate it. It turns itself into a tissue degrader.

Rarely, impetigo caused by *S. pyogenes* can be followed by acute poststreptococcal glomerulonephritis (see chapter 21). The strains that cause impetigo never cause rheumatic fever, however.

Transmission and Epidemiology of Impetigo

Impetigo, whether it is caused by *S. pyogenes*, *S. aureus*, or both, is highly contagious and transmitted through direct contact, but also via fomites and mechanical vector transmission. It affects mostly preschool children, but all ages can acquire the disease. The peak incidence is in the summer and fall. *S. pyogenes* is more often the cause of impetigo in newborns, and *S. aureus* is more often the cause of impetigo in older children, but both can cause infection in either age group.

Prevention

The only current prevention for impetigo is good hygiene. Vaccines are in development for both of the etiological agents, but none are currently available.

Treatment

Impetigo is usually treated with a drug that will kill either bacterium, *S. pyogenes* or *S. aureus*, eliminating the need to determine the exact etiological agent. The drug of choice is topical mupirocin (brand name Bactroban), a protein synthesis inhibitor. In cases of widespread skin involvement, oral antibiotics such as cephalexin may be used.

The treatment of nonimpetigo *S. pyogenes* infections is usually straightforward because this organism is sensitive to penicillin. *S. aureus* infections are another story. Although *S. aureus* impetigo is usually easily treated, other *S. aureus* diseases can be very difficult to treat effectively. We'll discuss these difficulties later in this chapter.

Cellulitis

Cellulitis is a condition caused by a fast-spreading infection in the dermis and in the subcutaneous tissues below. It causes pain, tenderness, swelling, and warmth. Fever and swelling of the lymph nodes draining the area may also occur. Frequently red lines leading away from the area are visible (a phenomenon called *lymphangitis*); this symptom is the result of microbes and inflammatory products being carried by the lymphatic system. Bacteremia could develop with this disease, but uncomplicated cellulitis has a good prognosis.

Cellulitis generally follows introduction of bacteria or fungi into the dermis, either through trauma or by subtle means (with no obvious break in the skin). Symptoms take several days to develop. The most common causes of the condition in healthy people are *Staphylococcus aureus* and *Streptococcus pyogenes*, although almost any bacterium and some fungi can cause this condition in an immunocompromised patient. In infants, group B streptococci are a frequent cause (see chapter 23).

People who are immunocompromised, or who have cardiac insufficiency, are at higher risk for this condition than are healthy persons. They also risk complications, such as spread to the bloodstream, rapid spreading through adjacent tissues, and, especially in children, meningitis. Occasionally, cellulitis is a complication of varicella (chickenpox) infections.

Mild cellulitis responds well to oral antibiotics chosen to be effective against both *S. aureus* and *S. pyogenes*. More involved infections and infections in immunocompromised people require intravenous antibiotics. If there are extensive areas of tissue damage, surgical debridement (duh-breed'-munt) is warranted **(Checkpoint 18.3).**

Staphylococcal Scalded Skin Syndrome (SSSS)

This syndrome is another **dermolytic** condition caused by *Staphylococcus aureus*. It affects mostly newborns and babies, although children and adults can experience the infection. Newborns are susceptible when sharing a nursery with another newborn who is colonized with *S. aureus*. Transmission may occur when caregivers carry the bacterium from one baby to another. Adults in the nursery can also directly transfer *S. aureus* because approximately 30% of adults are asymptomatic carriers. Carriers can harbor the bacteria in the nasopharynx, axilla, perineum, and even the vagina. (Fortunately, only about 5% of *S. aureus* strains are lysogenized by the type of phage that codes for the toxins responsible this disease.)

✔ CHECKPOINT 18.3 Cellulitis

Causative Organism(s)	*Staphylococcus aureus*	*Streptococcus pyogenes*	Other bacteria or fungi
Most Common Modes of Transmission	Parenteral implantation	Parenteral implantation	Parenteral implantation
Virulence Factors	Exfoliative toxin A, coagulase, other enzymes	Streptokinase, plasminogen-binding ability, hyaluronidase, M protein	–
Culture/Diagnosis	Based on clinical signs	Based on clinical signs	Based on clinical signs
Prevention	–	–	
Treatment	Aggressive treatment with oral or IV antibiotic (cephalexin); surgery sometimes necessary	Aggressive treatment with oral or IV antibiotic (cephalexin); surgery sometimes necessary	Aggressive treatment with oral or IV antibiotic (cephalexin); surgery sometimes necessary
Distinguishing Features	–	–	More common in immunocompromised

This condition can be thought of as a systemic form of impetigo. Like impetigo, it is an exotoxin-mediated disease. The phage-encoded exfoliative toxins A and B are responsible for the damage. Unlike impetigo, the toxins enter the bloodstream from some focus of infection (the throat, the eye, or sometimes an impetigo infection) and then travel to the skin throughout the body. These toxins cause **bullous** lesions, which often appear first around the umbilical cord (in neonates), or in the diaper or axilla area. The lesions begin as red areas, take on the appearance of wrinkled tissue paper, and then form very large blisters. Fever may precede the skin manifestations. Eventually the top layers of epidermis peel off completely. The split occurs in the epidermal tissue layers just above the stratum basale (see figure 18.1). Widespread **desquamation** of the skin follows, leading to the burned appearance referred to in the name **(figure 18.9).**

At this point, the protective keratinized layer is gone, and the patient is vulnerable to secondary infections, cellulitis, and bacteremia. In the absence of these complications, young patients nearly always recover if treated promptly. Adult patients have a higher mortality rate—as high as 50%. Once a tentative diagnosis of SSSS is made, immediate antibiotic therapy should be instituted, using cloxacillin or cephalexin.

It is important, however, to differentiate this disease from a similar skin condition called *toxic epidermal necrolysis (TEN)*, which is caused by a reaction to antibiotics, barbiturates, or other drugs. TEN has a significant mortality rate. The treatments for the two diseases are very different, so it is important to distinguish between them before instituting therapy. In TEN, the split in skin tissue occurs *between* the dermis and the epidermis, not within the epidermis as is the case with SSSS. Histological examination of tissue from a lesion is usually a better way to diagnose the disease than reliance on culture. Because SSSS is caused by the dissemination of exotoxin, *S. aureus* may not be found in lesions. Nevertheless, culture should be attempted so that antibiotic sensitivities can be established.

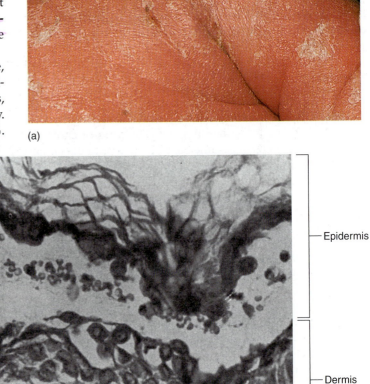

(a)

(b)

FIGURE 18.9 **Staphylococcal scalded skin syndrome (SSSS) in a newborn child.**
(a) Exfoliative toxin produced in local infections causes blistering and peeling away of the outer layer of skin. **(b)** Photomicrograph of a segment of skin affected with SSSS. The point of epidermal shedding, or desquamation, is in the epidermis. The lesions will heal weil because the level of separation is so superficial.

✔ CHECKPOINT 18.4	Scalded Skin Syndrome
Causative Organism(s)	*Staphylococcus aureus*
Most Common Modes of Transmission	Direct contact, droplet contact
Virulence Factors	Exfoliative toxins A and B
Culture/Diagnosis	Histological sections; culture performed but false negatives common
Prevention	Eliminate carriers in contact with neonates
Treatment	Immediate systemic antibiotics (cloxacillin or cephalexin)
Distinguishing Features	Split in skin occurs *within* epidermis

Gas Gangrene

Clostridium perfringens, a gram-positive endospore-forming bacterium, as well as some related species, cause a serious condition called **gas gangrene**, or clostridial **myonecrosis** (my"-oh-neh-kro'-sis). The spores of these species can be found in soil, on human skin, and in the human intestine and vagina. The bacteria are mainly anaerobic, and they require anaerobic conditions to manufacture and release the exotoxins that mediate the damage in the disease.

Signs and Symptoms

Two forms of gas gangrene have been indentified. In anaerobic cellulitis, the bacteria spread within damaged necrotic muscle tissue, producing toxin and gas, but the infection remains localized and does not spread into healthy tissue. The pathology of true myonecrosis is more destructive. Toxins produced in large muscles, such as the thigh, shoulder, and buttocks, diffuse into nearby healthy tissue and cause local necrosis there. This damaged tissue then serves as a focus for continued clostridial growth, toxin formation, and gas production. The disease can progress through an entire limb or body area, destroying tissues as it goes **(figure 18.10)**. Initial symptoms of pain, edema, and a bloody exudate in the lesion are followed by fever, tachycardia, and blackened necrotic tissue filled with bubbles of gas. Gangrenous infections of the uterus due to septic abortions, and clostridal septicemia, are particularly serious complications. If treatment is not begun early, the disease is invariably fatal.

Pathogenesis and Virulence Factors

Because clostridia are not highly invasive, infection requires damaged or dead tissue that supplies growth factors, and an anaerobic environment. The low-oxygen environment results from an interrupted blood supply and the presence of aerobic bacteria that deplete oxygen. Such conditions stimulate spore germination, rapid vegetative growth in the dead tissue, and release of exotoxins. *C. perfringens* produces several physiologically active exotoxins; the most potent one, *alpha toxin,* causes red blood cell rupture, edema, and tissue destruction **(figure 18.11)**. Additional virulence factors that enhance tissue destruction are collagenase, hyaluronidase, and DNase. The gas formed in tissues, resulting from fermentation of muscle carbohydrates, can also destroy muscle structure.

Transmission and Epidemiology

The conditions that may predispose a person to gangrene are surgical incisions, compound fractures, diabetic ulcers, septic abortions, puncture and gunshot wounds, and crushing

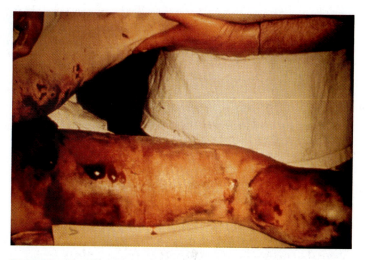

FIGURE 18.10   The clinical appearance of myonecrosis in a compound fracture of the leg.
Necrosis has traveled from the main site of the break to other areas of the leg. Note the severe degree of involvement, with blackening, general tissue destruction, and bubbles on the skin caused by gas formation in underlying tissue.

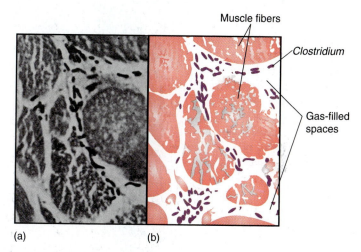

FIGURE 18.11   Growth of *Clostridium perfringens* (plump rods), causing gas formation and separation of the fibers.
(a) A microscopic analysis of clostridial myonecrosis, showing a histological section of gangrenous skeletal muscle. **(b)** A schematic drawing of the same section.
(a) From N.A. Boyd et al., *Journal of Medical Microbiology,* 5:459, 1972. Reprinted by permission of Longman Group, Ltd.

✔ CHECKPOINT 18.5	Gas Gangrene
Causative Organism(s)	*Clostridium perfringens,* other species
Most Common Modes of Transmission	Vehicle (soil), endogenous transfer from skin, GI tract, reproductive tract
Virulence Factors	Alpha toxin, other exotoxins, enzymes, gas formation
Culture/Diagnosis	Gram stain, CT scans (abdominal infections), X ray, clinical picture
Prevention	Clean wounds, debride dead tissue
Treatment	Cephalosporin, surgical removal, oxygen therapy

INSIGHT 18.1   *Medical*

The Skin Predators: *Staphlyococcus* and *Streptococcus*

The relatively hostile environment of the skin makes it difficult for many microorganisms to set up shop there. But two genera of gram-positive bacteria, *Staphylococcus* and *Streptococcus*, are uniquely suited to living there and sometimes cause disease there. Many of these diseases are described in this chapter. Here we discuss some very common skin conditions caused by the two bacteria.

Staphylococcus aureus: Folliculitis, Furuncles, and Carbuncles

Currently, 31 species have been placed in the genus *Staphylococcus*. Of these, the most important human pathogen is probably *S. aureus*. Elsewhere in this chapter we have discussed some of the other staphylococcal skin conditions (impetigo, cellulitis, and scalded skin syndrome), but other common skin diseases have *S. aureus* as a cause. **Folliculitis** is a mild, superficial inflammation of hair follicles or glands. Although these lesions are usually resolved with no complications, they can lead to infections of subcutaneous tissues. An *abscess* is a more serious localized staphylococcal skin infection, which appears as an inflamed, fibrous lesion enclosing a core of pus. There are two types: furuncles and carbuncles. A **furuncle** results when the inflammation of a single hair follicle or sebaceous gland progresses into a large, red, and extremely tender abscess or pustule. Furuncles often oc-

cur in clusters on parts of the body such as the buttocks, axillae, and back of the neck, where skin rubs against other skin or clothing. They are also commonly called *boils*. A **carbuncle** is a larger and deeper lesion, sometimes as big as a baseball, created by aggregation and interconnection of a cluster of furuncles. It is usually found in areas of thick, tough skin such as on the back of the neck. Carbuncles are extremely painful and can even be fatal in elderly patients when they give rise to systemic disease.

Streptococcus pyogenes: Erysipelas

In addition to impetigo and cellulitis, at least two other important *S. pyogenes* diseases begin on the skin. One fairly invasive manifestation is **erysipelas.** The pathogen usually enters through a small wound or incision on the face or extremities and eventually spreads to the dermis and subcutaneous tissues. Early symptoms are edema

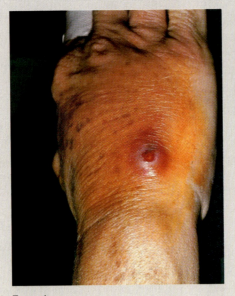

Furuncle

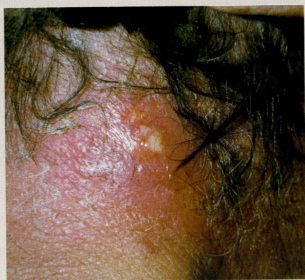

Carbuncle

injuries contaminated by spores from the body or the environment.

Prevention and Treatment

One of the most effective ways to prevent clostridial wound infections is immediate and rigorous cleansing and surgical repair of deep wounds, decubitus ulcers (bedsores), compound fractures, and infected incisions. Debridement of diseased tissue eliminates the conditions that promote the spread of gangrenous infection. This procedure is most difficult in the

intestine or body cavity, where only limited amounts of tissue can be removed. Surgery is supplemented by large doses of a broad-spectrum cephalosporin (cefoxitin) or penicillin to control infection. Hyperbaric oxygen therapy, in which the affected part is exposed to an increased oxygen mix in a pressurized chamber, can also lessen the severity of infection.

Extensive myonecrosis of a limb may call for amputation. Because there are so many different antigenic subtypes in this bacterial group, active immunization is not possible **(Checkpoint 18.5).**

and redness of the skin near the portal of entry, and fever and chills. The lesion begins to spread outward, producing a slightly elevated edge that is noticeably red and hot. Depending on the depth of the lesion and how the infection progresses, cutaneous lesions can remain superficial or can produce long-term systemic complications. Severe cases involving large areas of skin are occasionally fatal.

Both *Streptococcus pyogenes* and *Staphylococcus aureus:* Necrotizing Fasciitis

As you read early in this chapter, both *S. pyogenes* and *S. aureus* can lead to impetigo and cellulitis. Often, both of the pathogens are isolated from lesions. The same is true for a very invasive infection called necrotizing fasciitis. The disease has been known for hundreds of years, but in recent years small outbreaks of the disease have received heavy publicity as the "flesh-eating disease." Cases of

this disease are rather rare, but its potential for harm is high. It can begin with an innocuous cut in the skin and spread rapidly into nearby tissue, causing severe disfigurement and even death.

There is really no mystery to the pathogenesis of necrotizing fasciitis. It begins very much like impetigo and other skin infections: Streptococci and/or staphylococci on the skin are readily introduced into small abrasions or cuts, where they begin to grow rapidly. The particular strains of bacteria that cause this condition have great toxigenicity and invasiveness, because of special enzymes and toxins. The enzymes digest the connective tissue in skin, and the toxins poison the epidermal and dermal tissue. As the flesh is killed, it separates and sloughs off, forming a pathway for the bacteria to spread into deeper tissues such as muscle. More dangerous cases involve polymicrobial infections that can include anaerobic bacteria and the systemic spread of the toxin to other organs. Some patients have lost parts of their limbs and faces, and others have suffered amputation, but early diagnosis and treatment can prevent these complications.

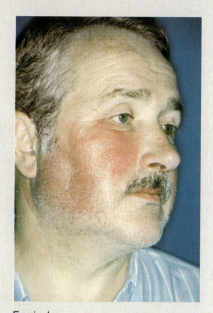

Erysipelas

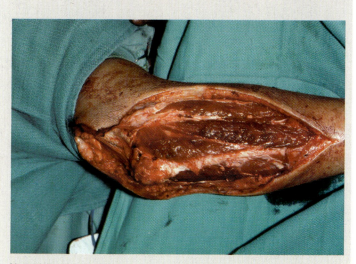

Necrotizing fasciitis

Hansen's Disease (Leprosy)

Leprosy is a chronic, progressive disease of the skin and nerves known for its extensive medical and cultural ramifications. From ancient times, leprosy victims were stigmatized because of the severe disfigurement of the disease and the belief that it was a divine curse. Leprosy patients not only suffered the torture of the disease, but they also endured terrible brutalities, including imprisonment under the most gruesome conditions. The modern view of leprosy is more enlightened. We know that it is caused by a bac-

terium, *Mycobacterium leprae*, that it is not readily communicated and that it should not be accompanied by social banishment. Because of the unfortunate connotations associated with the word *leprosy*, the preferred name today is Hansen's disease, named for the person who first detected the bacterium.

Signs and Symptoms

Tuberculoid leprosy, the most superficial form, is characterized by asymmetrical, shallow skin lesions containing very

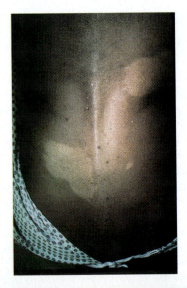

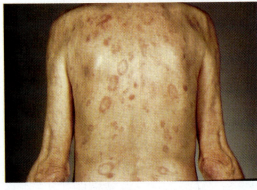

FIGURE 18.12 **Leprosy lesions.**
Both views are of the tuberculoid form, with shallow, painless lesions. **(a)** Infection in dark-skinned persons manifests as hypopigmented patches or macules. **(b)** In light-skinned individuals, it appears as reddish patches or papules.

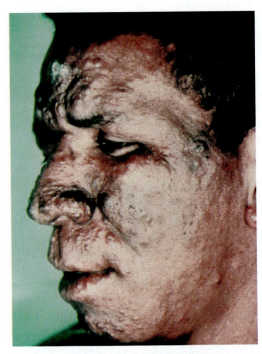

FIGURE 18.13 **A clinical picture of lepromatous leprosy (LL).**
Infection of the nose, lips, chin, and brows produces moderate facial deformation, typical of lepromas.

few bacteria **(figure 18.12).** Microscopically, the lesions appear as thin granulomas and enlarged dermal nerves. Damage to these nerves usually results in local loss of pain reception and feeling. This form has fewer complications and is more easily treated than other types of Hansen's disease.

Lepromatous leprosy (LL) is responsible for the disfigurement commonly associated with the disease. It is marked by chronicity and severe complications due to widespread dissemination of the bacteria. Leprosy bacteria grow primarily in macrophages in cooler regions of the body, including the nose, ears, eyebrows, chin, and testes. As growth proceeds, the face of the afflicted person develops folds and granulomous thickenings, called **lepromas,** which are caused by massive intracellular overgrowth of the bacterium **(figure 18.13).** Advanced LL causes a loss of sensitivity that predisposes the patient to unrecognized injury, secondary infections, blindness, and kidney or respiratory failure.

Borderline leprosy patients can progress in either direction along the scale, depending upon their treatment and immunologic competence. The most severe effect of intermediate forms of Hansen's disease is early damage to nerves that control the muscles of the hands and feet. The subsequent wasting of the muscles and loss of control produces gross deformities in the extremities.

Because the bacterium causing this disease grows extremely slowly, early symptoms may not appear for 1 to 7 years, and in some cases 40 years, after infection.

Causative Agent

Mycobacterium leprae, the causative agent of Hansen's disease, is in the same genus as *M. tuberculosis.* The general morphology and staining characteristics of the leprosy bacillus are similar to those of other mycobacteria (slender rod with an acid-fast wall), but it is exceptional in two ways: (1) it is a strict parasite that has not been grown in artificial medium or human tissue culture, and (2) it is the slowest growing of all the species. *M. leprae* multiplies within host cells in large packets called *globi* at an optimum temperature of 30°C.

Thirty years ago, Eleanor Storrs discovered that *M. leprae* could be grown in the footpads of armadillos, opening the door for research on the bacterium and the production of antigens and other bacterial components for diagnosis and eventual immunization.

When the genome of *M. leprae* was sequenced in 2001, researchers discovered that it had far fewer genes than *M. tuberculosis* or any other *Mycobacterium* species. Scientists suspect that this is the result of *reductive evolution*, a gradual loss of gene functions that become unnecessary when a pathogen adapts itself to a strictly intracellular life-style.

Pathogenesis and Virulence Factors

M. leprae is considered to be a pathogen of low virulence, mainly because of the very slow progression of disease. Because of the ultimate disfiguring and disabling effects of the condition, it is hard to think of it as having low virulence. Nevertheless, few virulence factors have been identified. Its ability to survive inside macrophages certainly contributes to its virulence. Recently a surface protein from *M. leprae* was shown to bind avidly to receptors on Schwann cells, which form the myelin coating around peripheral nerves. This capability likely gives rise to the bacterium's ability to cause the progressive nerve damage characteristic of the disease.

Transmission and Epidemiology

The incidence of Hansen's disease has recently been declining due to a worldwide control effort. The World Health Organization (WHO) estimates of the current disease incidence range from 500,000 to 1 million cases, mostly in areas of Asia, Africa, Central and South America, and the Pacific islands. Hansen's disease is not restricted to warm climates, however, since it also occurs in Siberia, Korea, and northern China. The disease is also reported in the United States, especially in Hawaii, Texas, Louisiana, Florida, and California. The total number of new cases reported nationwide is from 300 to 500 per year, mostly among recent immigrants from endemic areas.

The mechanism of transmission among humans has yet to be verified. Theories proposed that the bacterium is directly inoculated into the skin through contact with an infected person, that mechanical vectors are involved, or that inhalation of droplet nuclei is a factor. Although the human body was long considered the sole host and reservoir of the bacterium, it is now clear that armadillos naturally harbor a mycobacterial species genetically identical to *M. leprae* and may develop a granulomatous disease similar to leprosy. Whether humans can acquire the disease from armadillos is not yet known, but it is tempting to speculate that the infection is actually a zoonosis.

Most people who come into contact with *M. leprae* do not develop clinical disease. As with tuberculosis, it appears that health and living conditions influence susceptibility and the course of the disease. Some have called it the least contagious of the communicable diseases, which is quite ironic considering that in ancient times, infected persons were required to wear bells around their necks in public and to ring them constantly to warn people to stay away from them. It has been suggested that people who do acquire the infection have some defect in the regulation of T cells. Long-term household contact with an infected person, poor nutrition, and crowded conditions increase the risks of infection. Many people become infected as children and harbor the microbe through adulthood.

Culture and Diagnosis

Diagnosing the disease, at least in the early stages, is difficult because of the varying clinical presentations and the fact that the bacteria are not always detectable in "lesions." Serological testing is complicated by the fact that antibodies may be present in people who have the infection but don't have the disease. In any case, diagnosis starts with assessing the symptomology and is complemented by microscopic examination of lesions and by patient history. Biopsies of affected tissues are examined for acid-fast bacteria. PCR has recently been made available for *M. leprae*, but a negative result is not always reliable since bacteria may not be present.

Prevention and Treatment

Although vaccine trials were undertaken in the late 1990s, no vaccine has yet become available. Preventing Hansen's disease requires constant surveillance of high-risk populations to discover early cases, chemoprophylaxis of healthy persons in close contact with infected people, and isolation of the infected.

Hansen's disease can be treated with drugs, and if caught early cases can be cured within 12 to 24 months, if the treatment regimen can be maintained. Long-term drug regimens are difficult in many developing countries. Because of an increase in resistant strains, multidrug therapy is necessary. Tuberculoid leprosy can be managed with rifampin and dapsone. Lepromatous leprosy requires a combination of rifampin, dapsone, and clofazimine until the numbers of bacteria in skin lesions has been substantially reduced. Then dapsone can be taken alone for an indeterminate period.

✓ CHECKPOINT 18.6	Leprosy
Causative Organism(s)	*Mycobacterium leprae*
Most Common Modes of Transmission	Not clear, possibly direct or droplet contact, mechanical vector
Virulence Factors	Binding to Schwann cells, ability to survive within macrophages
Culture/Diagnosis	Clinical signs, microscopy, biopsy, PCR, patient history
Prevention	Isolation of infected people, chemoprophylaxis of contacts
Treatment	Multidrug treatment including rifampin and dapsone; varies with form of leprosy

Vesicular or Pustular Rash Diseases

There are two diseases that present as generalized "rashes" over the body, in which the individual lesions contain fluid. The lesions are often called *pox*, and the two diseases are chickenpox and smallpox. Chickenpox is very common and mostly benign, but even a single case of smallpox constitutes a public health emergency. Both are viral diseases.

to the bloodstream, causing generalized illness. Once vaccination began in December 2002, several people developed cardiac symptoms, so people at risk for heart problems were advised not to take the vaccine. Since vaccinations began, there have been a handful of cases of inadvertent transmission of the vaccinia strain, resulting in localized symptoms in persons contacting those who had been immunized. The status of smallpox vaccination is changing continuously; the most current information is available at the Centers for Disease Control and Prevention website, at www.cdc.gov.

Vaccination is also useful for postexposure prophylaxis, meaning that it can prevent or lessen the effects of the disease after you have already been infected with it.

Another chapter was added to the smallpox story in 2003, when dozens of people came down with a disease called monkeypox, caused by the monkey variant of the smallpox virus. They had apparently caught the disease from their pet prairie dogs, which had seemingly caught it from an exotic species of African rat. Both of these animals were imported to the United States as part of the exotic pet trade. The U.S. government recommended that people exposed to infected animals be vaccinated with the vaccinia vaccine.

Treatment   There is no treatment for smallpox. Antiviral treatment is not effective. If lesions become infected secondarily with bacteria, antibiotics can be used for that complication **(Checkpoint 18.7).**

Maculopapular Rash Diseases

Insight 18.3 contains a description of the different infectious conditions that can result in a rash of some sort on the skin. The infectious conditions described in this section are those with their major manifestations on the skin. (Meningococcal meningitis, for instance, can result in a diffuse rash on the skin, but its major manifestations are in the central nervous system, so it is discussed in chapter 19.) In this section we examine measles, rubella, "fifth disease," and roseola. They all cause skin eruptions classified as maculopapular.

Measles

Most of us living in the United States don't think twice about measles. It is just another vaccination we get when we are children. But every year approximately 1 million children in the developing world die from this disease, even though an extremely effective vaccine has been available since 1964. Health campaigns all over the world seek to make measles vaccine available to all, but there is much work to be done. Many scientists and public health advocates hope that once polio is eradicated (see chapter 19), measles will be the next disease targeted globally for eradication.

Measles is also known as **rubeola.** Be very careful not to confuse it with the next maculopapular rash disease, rubella.

Signs and Symptoms   The initial symptoms of measles are sore throat, dry cough, headache, conjunctivitis, lymphadeni-

---

✔ CHECKPOINT 18.7   Vesicular/Pustular Rash Diseases

Disease	Chickenpox	Smallpox
Causative Organism(s)	Human herpesvirus 3 (varicella-zoster virus)	Variola virus
Most Common Modes of Transmission	Droplet contact, inhalation of aerosolized lesion fluid	Droplet contact, indirect contact
Virulence Factors	Ability to fuse cells, ability to remain latent in ganglia	Ability to dampen, avoid immune response
Culture/Diagnosis	Based largely on clinical appearance	Based largely on clinical appearance
Prevention	Live attenuated vaccine	Live virus vaccine (vaccinia virus)
Treatment	None in uncomplicated cases; acyclovir for high risk	–
Distinguishing Features	No fever prodrome; lesions are superficial; in centripetal distribution (more in center of body)	Fever precedes rash, lesions are deep and in centrifugal distribution (more on extremities)
Appearance of Lesion		

INSIGHT 18.3 *Medical*

Naming Skin Lesions

There seems to be no end to the types of lesions or irregularities that can occur on the skin. Dermatology, the study of the skin, is a branch of medicine that relies heavily on visual characteristics for initial diagnoses. The many types of skin lesions or irregularities have been given specific descriptive names for this purpose. None of these names points to an exact etiological cause, but because certain infectious agents generally cause distinctive types of lesions, the list of possible causes can be narrowed considerably once the "style" of irregularity is identified and named. **Table 18.3A** contains a list of the more common descriptors of skin bumps, lesions, and irregularities.

TABLE 18.3A	Skin Terms	
Descriptive Name	**Appearance**	**Examples**
Macule	Flat, well-demarcated lesion characterized mainly by color change	Freckle, tinea versicolor (fungus infection)
Papule	Small elevated, solid bump	Warts, cutaneous leishmaniasis
Maculopapular Rash	Flat to slightly raised colored bump	Measles, rubella, fifth disease, roseola
Plaque	Elevated flat-topped lesion larger than 1 cm (i.e., a wider papule)	Psoriasis
Vesicle	Elevated lesion filled with clear fluid	Chickenpox
Bulla	Large (wide) vesicle	Blister, gas blisters in gangrene
Pustule	Small elevated lesion filled with purulent fluid (pus)	Acne, smallpox, mucocutaneous leishmanisasis, cutaneous anthrax
Cyst	Raised, encapsulated lesion, usually solid or semisolid when palpated	Severe acne
Purpura	Reddish-purple discoloration due to blood in small areas of tissue; does not blanch when pressed	Meningococcal bloodstream infection (see chapter 19)
Petechiae	Small purpura	Meningococcal bloodstream infection
Scale	Flaky portions of skin separated from deeper portions	Ringworm of body and scalp, athlete's foot

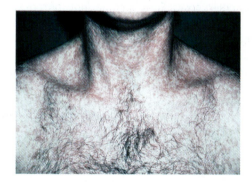

FIGURE 18.16 The rash of measles.

tis, and fever. In a short time, unusual oral lesions called *Koplik's spots* appear as a prelude to the characteristic red maculopapular **exanthum** (eg-zan'-thum) that erupts on the head and then progresses to the trunk and extremities, until most of the body is covered **(figure 18.16)**. The rash gradually coalesces into red patches that fade to brown.

In a small number of cases, children develop laryngitis, bronchopneumonia, and bacterial secondary infections such as ear and sinus infections. Children afflicted with leukemia or thymic deficiency are especially predisposed to pneumonia because of their lack of the natural T-cell defense. Undernourished children may experience severe diarrhea and abdominal discomfort that adds to their debilitation.

In a small percentage of cases, the virus can cause pneumonia. Affected patients are very ill and often have a characteristic dusky skin color from lack of oxygen. Occasionally (1 in 100 cases) measles progresses to encephalitis, resulting in various CNS changes ranging from disorientation to coma. Permanent brain damage or epilepsy can result.

A large number of measles patients experience secondary bacterial infections with *Haemophilus influenzae, Streptococcus pneumoniae*, or other streptococci or staphylococci. These can also lead to pneumonia or upper respiratory tract complications.

The most serious complication is **subacute sclerosing pan-encephalitis (SSPE),** a progressive neurological degeneration of the cerebral cortex, white matter, and brain stem. Its incidence is approximately one case in a million measles infections, and it afflicts primarily male children and adolescents. The pathogenesis of SSPE appears to involve a defective virus, one that has lost its ability to form a capsid and be released from an infected cell. Instead, it spreads unchecked through the brain by cell fusion, gradually destroying neurons and accessory cells, and breaking down myelin. The disease is known for profound intellectual and neurological impairment. The course of the disease invariably leads to coma and death in a matter of months or years.

Measles during pregnancy has been associated with spontaneous miscarriage and low-birthweight babies, but severe birth defects have not been reported.

Causative Agent   The measles virus is a member of the *Morbillivirus* genus. It is a single-stranded enveloped RNA virus in the Paramyxovirus family.

Pathogenesis and Virulence Factors   The virus implants in the respiratory mucosa and infects the tracheal and bronchial cells. From there it travels to the lymphatic system, where it multiplies and then enters the bloodstream. Viremia carries the virus to the skin and to various organs.

The measles virus induces the cell membranes of adjacent host cells to fuse into large **syncytia** (sin-sish'-uh), giant cells with many nuclei. These cells no longer perform their proper function. The virus seems proficient at disabling many aspects of the host immune response, especially cell-mediated immunity and delayed-type hypersensitivity. The host may be left vulnerable for many weeks after infection; this immune response disruption is one of the reasons that secondary bacterial infections are so common.

Transmission and Epidemiology   Measles is one of the most contagious infectious diseases, transmitted principally by respiratory droplets. Epidemic spread is favored by crowding, low levels of herd immunity, malnutrition, and inadequate medical care. Outbreaks of measles occasionally have been linked to the lack of immunization in children or the failure of a single dose of vaccine in many children. There is no reservoir other than humans, and a person is infectious during the periods of incubation, prodrome phase, and the skin rash, but usually not during convalescence. Only relatively large, dense populations of susceptible individuals can sustain the continuous chain necessary for transmission. In the United States, the incidence of measles is sporadic, usually less than 100 cases per year. In the 1980s and 1990s, a significant proportion of U.S. measles cases occurred among college students, perhaps because of communal living conditions and perhaps due to a waning of their childhood immunity. Now at least 32 states have laws requiring that students present proof of two measles immunizations before they can enroll in college.

Culture and Diagnosis   The disease can be diagnosed on clinical presentation alone, but if further identification is required an ELISA test is available that tests for patient IgM to measles antigen, indicating a current infection. For best results, blood should be drawn on the third day of onset or later, because before that time titers of IgM may not be high enough to be detected by the test. Also, the method of comparing acute and convalescent sera may be used to confirm a measles infection after the fact. As you may recall from chapter 17, much higher IgG titers 14 days after onset when compared to titers at day 1 or 2 are a clear indication of current or recent infection. This knowledge allows health care providers to be on the lookout for complications, and to be ahead of the game if a person who has had contact with the patient presents with similar symptoms.

Prevention   The MMR vaccine (for measles, mumps, and rubella) contains live attenuated measles virus, which confers protection for about 20 years. Measles immunization is recommended for all healthy children at the age of 12 to 15 months, with a booster before the child enters school. Failing that, the preadolescent health check serves as a good time to get the second dose of measles vaccine.

Treatment   Treatment relies on reducing fever, suppressing cough, and replacing lost fluid. Complications require additional remedies to relieve neurological and respiratory symptoms and to sustain nutrient, electrolyte, and fluid levels. Therapy includes antibiotics for bacterial complications and doses of immune globulin. Vitamin A supplements are recommended by some physicians; they have been found effective in reducing the symptoms and decreasing the rate of complications.

IN THE NEWS   *(Continued from page 539)*

The baby from China described at the beginning of the chapter was sick with measles. Hospital staff looked inside the baby's mouth for Koplik's spots. Finding these spots provides some assurance that, among all the maculopapular diseases, measles is the likely diagnosis.

The U.S. incidence of measles is low (usually less than 100 per year), but sporadic outbreaks do occur, even among people who were fully vaccinated as children, probably due to waning of their artificially acquired active immunity. This case highlights the need for constant vigilance and continued immunization, especially as the world "shrinks," and we come in contact with people from other parts of the world where different levels of immunization are achieved.

This story provides an example of some unintended consequences of an otherwise joyous event: international adoptions. In 1992, about 6,000 babies from foreign countries were brought to the United States and adopted. In 2001, nearly 19,000 infants were adopted by happy U.S. families, and the trend is likely to continue.

See: CDC. 2003. Measles outbreak among internationally adopted children arriving in the United States, February–March 2001. MMWR 51:1115–1116.

Rubella

This disease is also known as German measles. Rubella is derived from the Latin for "little red," and that's a good way to remember it because it causes a relatively minor rash disease with few complications. Sometimes it is called the three-day measles. The only exception to this mild course of events is when a fetus is exposed to the virus while in its mother's womb (in utero). Serious damage can occur, and for that reason women of childbearing years must be sure to have been vaccinated well before they plan to conceive.

Signs and Symptoms   The two clinical forms of rubella are referred to as postnatal infection, which develops in children or adults, and **congenital** (prenatal) infection of the fetus, expressed in the newborn as various types of birth defects.

Postnatal Rubella   During an incubation period of 2 to 3 weeks, the rubella virus multiplies in the respiratory epithelium, infiltrates local lymphoid tissue, and enters the bloodstream. Early symptoms include malaise, mild fever, sore throat, and lymphadenopathy. The rash of pink macules and papules first appears on the face and progresses down the trunk and toward the extremities, advancing and resolving in about 3 days. The rash is milder looking than the measles rash (Checkpoint Table 18.8). Adult rubella is often accompanied by joint inflammation and pain rather than a rash. Very occasionally complications such as arthralgia/arthritis, or even encephalitis, can occur but more often in adults than in children.

Congenital Rubella   Rubella is a strongly **teratogenic** (ter-at'-oh-jen"-ik) virus. Transmission of the rubella virus to a fetus in utero can result in a serious complication called **congenital rubella (figure 18.17).** The mother is able to transmit the virus even if she is asymptomatic. Fetal injury varies according to the time of infection. It is generally accepted that infection in the first trimester is most likely to induce miscarriage or multiple permanent defects in the newborn. The most common of these is deafness, and may be the only defect seen in some babies. Other babies may experience cardiac abnormalities, ocular lesions, deafness, and mental and physical retardation in varying combinations. Less drastic sequelae that usually resolve in time are anemia, hepatitis, pneumonia, carditis, and bone infection.

Causative Agent   The rubella virus is a *Rubivirus,* in the family Togavirus. It is a nonsegmented single-stranded RNA virus with a loose lipid envelope. There is only one known serotype of the virus, and humans are the only natural host. Its envelope contains two different viral proteins.

Pathogenesis and Virulence Factors   The course of disease in postnatal rubella is mostly unremarkable. But when exposed to a fetus, the virus creates havoc. It has the ability to stop mitotis, which is an important process in a rapidly developing embryo and fetus. It also induces apoptosis of normal tissue cells. This inappropriate cell death can do irreversible harm to organs it affects. And last, the virus

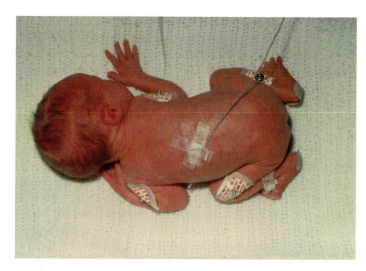

FIGURE 18.17   **An infant born with congenital rubella can manifest a papular and purpurotic rash.**
Courtesy Kenneth Schiffer, from *AJDC* 118:25, July 1969. © American Medical Association.

damages vascular endothelium, leading to poor development of many organs. Studies have shown that the earlier in gestation that the infection process begins, the more devastating its effects.

Transmission and Epidemiology   Rubella is an endemic disease with worldwide distribution. Infection is initiated through contact with respiratory secretions and occasionally urine. The virus is shed during the prodromal phase and up to a week after the rash appears. Congenitally infected infants are contagious for a much longer period of time. Because the virus is only moderately communicable, close living conditions are required for its spread. Although epidemics and pandemics of rubella once regularly occurred in 6- to 9-year cycles, the introduction of vaccination has essentially stopped this pattern in the United States. Most cases are reported among adolescents and young adults in military training camps, colleges, and summer camps. The greatest concern is that nonimmune women of childbearing age might be caught up in this cycle, raising the prospect of congenital rubella.

Culture and Diagnosis   Diagnosing rubella relies on the same twin techniques discussed earlier for measles. Because it mimics other diseases, rubella should not be diagnosed on clinical grounds alone. IgM antibody to rubella virus can be detected early using an ELISA technique or a latex-agglutination card. Other conditions and infections can lead to false positives, however, and the IgM test should be augmented by an acute and convalescent measurement of IgG antibody. It is important to know whether the infection is indeed rubella, especially in women, because if so, they will be immune to reinfection.

Prevention   The attenuated rubella virus vaccine is usually given to children in the combined form (MMR vaccination) at 12 to 15 months and a booster at 4 or 6 years of age. The vaccine for rubella can be administered on its own, without the measles and mumps components. Some estimates suggest that from 10% to 20% of the U.S. population is nonimmune to rubella, so the possibility for outbreaks, and for the accompanying congenital rubella syndrome, is still significant.

Many health care providers recommend screening adult women of childbearing age for antibodies to rubella, which would indicate either that they had had the infection or that they had been immunized. The current recommendation for nonpregnant, antibody-negative women is immediate immunization. Because the vaccine contains live virus, and because a teratogenic effect is theoretically possible, the vaccine is administered on the condition that the patient not become pregnant for 3 months afterward. The vaccine is not given to pregnant women.

Treatment   Postnatal rubella is generally benign and requires only symptomatic treatment. No specific treatment is available for the congenital manifestations.

Fifth Disease

This disease, more precisely called *erythema infectiosum*, is so named because about 100 years ago it was the fifth of the diseases recognized by doctors to cause rashes in children. The first four were scarlet fever (see chapter 21), measles, rubella, and roseola (coming up next in this chapter). Fifth disease is a very mild disease that often results in a characteristic "slapped-cheek" appearance because of a confluent reddish rash that begins on the face. Within 2 days the rash spreads on the body but is most prominent on the arms, legs and trunk. The rash is maculopapular and the blotches tend to run together rather than to appear as distinct bumps. The illness is rather mild, featuring low-grade fever and malaise and lasting 5 to 10 days. The rash may persist for days to weeks, and it tends to recur under stress or with exposure to sunlight. As with almost any infectious agent, it can cause more serious disease in people with underlying immune disease.

The causative agent is parvovirus B19. You may have heard of "parvo" as a disease of dogs, but strains of this virus group infect humans as well. Fifth disease is usually diagnosed by the clinical presentation, but sometimes it is helpful to rule out rubella by testing for IgM against rubella. Specific serological tests for fifth disease are available if they are considered necessary.

This infection is very contagious. It is transmitted through respiratory droplets or even direct contact. It can be transmitted through the placenta, with a range of possible effects, from no symptoms to stillbirth. There is no vaccine and no treatment for this usually mild disease.

Roseola

This disease is common in young children and babies. It sometimes results in a maculopapular rash, but a high percentage (up to 70%) of cases proceed without the rash stage. Children sick with this disease exhibit a high fever (up to 41°C,

or 105°F) that comes on quickly and lasts for up to 3 days. Seizures may occur during this period, but other than that patients remain alert and do not act terribly ill. On the fourth day, the fever disappears, and it is at this point that a rash can appear, first on the chest and trunk and less prominently on the face and limbs. By the time the rash appears, the disease is almost over.

Roseola is caused by a human herpesvirus called HHV-6, and sometimes by HHV-7. Like all herpesviruses, it can remain latent in its host indefinitely after the disease has cleared. Very occasionally the virus reactivates in childhood or adulthood, leading to mononucleosis-like or hepatitis-like symptoms. Immunocompetent hosts generally do not experience reactivation. It is thought that 100% of the U.S. population is infected with this virus by adulthood. Some people experienced the disease roseola when they became infected, and some of them did not. The suggestion has been made that this virus causes other disease conditions later in life, such as multiple sclerosis or chronic fatigue syndrome, but so far no convincing connection has been demonstrated. No vaccine and no treatment exists for roseola.

The two HHV viruses can cause severe disseminated disease in AIDS patients and other people with compromised immunity **(Checkpoint 18.8).**

Wartlike Eruptions

All types of warts are caused by viruses. Most common warts you have seen on yourself and others are probably caused by one of more than 80 human papillomaviruses, or HPVs. HPVs are also the cause of genital warts, described in chapter 23. Another virus in the poxvirus family causes a condition called **molluscum contagiosum,** which causes bumps that may look like warts.

Warts

Warts, also known as **papillomas,** afflict nearly everyone. Children seem to get them more frequently than adults, and there is speculation that people gradually build up immunity to the various HPVs that they encounter over time, as is the case with the viruses that cause the common cold.

The warts are benign, squamous epithelial growths. Some HPVs can infect mucous membranes, others invade skin. The appearance and seriousness of the infection varies somewhat from one anatomical region to another. Painless, elevated, rough growths on the fingers and occasionally on other body parts are called common, or seed, warts **(Checkpoint 18.9).** These growths commonly occur in children and young adults. Just as certain types of HPVs are associated with particular outcomes in the genital area, common warts are most often caused by HPV 2, 4, 27, and 29. **Plantar warts** are often caused by HPV 1. They are deep, painful papillomas on the soles of the feet. Flat warts (HPV types 3, 10, 28, and 49) are smooth, skin-colored lesions that develop on the face, trunk, elbows, and knees.

The warts contain variable amounts of virus. Transmission occurs through direct contact, and often warts are transmitted

✔ CHECKPOINT 18.8 Maculopapular Rash Diseases

Disease	Measles	Rubella	Fifth Disease	Roseola
Causative Organism(s)	Measles virus	Rubella virus	Parvovirus B19	Human herpesvirus 6 or 7
Most Common Modes of Transmission	Droplet contact	Droplet contact	Droplet contact, direct contact	?
Virulence Factors	Syncytium formation, ability to suppress CMI	In fetuses: inhibition of mitosis, induction of apoptosis, and damage to vascular endothelium	–	Ability to remain latent
Culture/Diagnosis	ELISA for IgM, acute/convalescent IgG	Acute IgM, acute/convalescent IgG	Usually diagnosed clinically	Usually diagnosed clinically
Prevention	Live attenuated vaccine (MMR)	Live attenuated vaccine (MMR)	–	–
Treatment	No antivirals; vitamin A, antibiotics for secondary bacterial infections	–	–	–
Distinguishing Features of the Rashes	Starts on head, spreads to whole body, lasts over a week	Milder red rash, lasts approximately 3 days	"Slapped-face" rash first, spreads to limbs and trunk, tends to be confluent rather than distinct bumps	High fever precedes rash stage—rash not always present
Appearance of Lesions				

from one part of the body to another by autoinoculation. Because the viruses are fairly stable in the environment, they can also be transmitted indirectly, from towels or from a shower stall, where they persist inside the protective covering of sloughed-off keratinized skin cells. The incubation period can be from 1 to 8 months. Almost all nongenital warts are harmless, and they tend to resolve themselves over time. Rarely, a wart can become malignant, when caused by a particular type of HPV.

The warts caused by papillomaviruses are usually distinctive enough to permit reliable clinical diagnosis without much difficulty. However, a biopsy and histological examination can help clarify ambiguous cases. Warts disappear on their own 60% to 70% of the time, usually over the course of 2 to 3 years. Physicians do approve of home remedies for resolving warts. These include nonprescription salicylic acid preparations, as well as the use of adhesive tape. Yes, you read that right: well-controlled medical studies have shown that adhesive tape (even duct tape!) can cause warts to disappear, presumably because the tape creates an airtight atmosphere that stops virus reproduction. But a psychological component, similar to a placebo effect, cannot be ruled out. (Neither of these treatments should be used for genital warts; see chapter 23.) Physicians have other techniques for removing warts, including a number of drugs and/or cryosurgery. No treatment guarantees that the viruses are eliminated; therefore warts can always grow back.

Molluscum Contagiosum

This disease is distributed throughout the world, with highest incidence occurring on certain Pacific islands, although its incidence in North America has been increasing since the 1980s. Skin lesions take the form of smooth, waxy nodules on the face, trunk, and limbs. The firm nodules may be indented in the middle (Checkpoint Table 18.9), and they contain a milky fluid containing epidermal cells filled with viruses in intracytoplasmic inclusion bodies. This condition is common in children, where it most often causes nodules on the face, arms, legs, and trunk. In adults it appears mostly in the genital areas. In immunocompromised patients, the lesions can be more disfiguring and more widespread on the body. It is particularly common in AIDS patients and often presents as facial lesions.

The molluscum contagiosum virus is a poxvirus, containing double-stranded DNA and possessing an envelope. It is spread via direct contact and also through fomites. Adults who acquire this infection usually acquire it through sexual contact. Autoinoculation can spread the virus from existing lesions to new places on the body, resulting in new nodules.

☑ CHECKPOINT 18.9 Wart and Wartlike Eruptions

Disease	Warts	Molluscum contagiosum
Causative Organism(s)	Human papillomaviruses	Molluscum contagiosum viruses
Most Common Modes of Transmission	Direct contact, autoinoculation, indirect contact	Direct contact, including sexual contact, autoinoculation
Virulence Factors	–	–
Culture/Diagnosis	Clinical diagnosis, also histology, microscopy, PCR	Clinical diagnosis, also histology, microscopy, PCR
Prevention	Avoid contact	Avoid contact
Treatment	Home treatments, cryosurgery (virus not eliminated)	Usually none, although mechanical removal can be performed (virus not eliminated)
Appearance of Lesions		

The condition may be diagnosed on clinical appearance alone, or a skin biopsy may be performed and histological analysis undertaken. A clinician can perform a more simple "squash procedure," in which fluid from the lesion is extracted onto a microscope slide, squashed by another microscope slide, stained, and examined for the presence of the characteristic inclusion bodies in the epithelial cells. PCR can also be used to detect the virus in skin lesions. In most cases no treatment is indicated, although a physician may remove the lesions or treat them with a topical chemical. Treatment of lesions does not ensure elimination of the virus (Checkpoint 18.9).

Larger Pustular Skin Lesions

Leishmaniasis

Two infections that result in large lesions (greater than a few millimeters across) deserve mention in this chapter on skin infections. The first is leishmaniasis, a zoonosis transmitted among various mammalian hosts by female sand flies. This infection can express itself in several different forms, depending on which species of the protozoan *Leishmania* is involved. Cutaneous leishmaniasis is a localized infection of the capillaries of the skin caused by *L. tropica*, found in Mediterranean, African, and Indian regions. A form of mucocutaneous leishmaniasis called espundia is caused by *L. brasiliensis*, endemic to parts of Central and South America. It affects both the skin and mucous membranes. Another form of this infection is systemic leishmaniasis.

Leishmania is transmitted to the mammalian host by the sand fly when it ingests the host's blood. The disease is endemic to equatorial regions that provide favorable conditions for the sand fly. Numerous wild and domesticated animals, especially dogs, serve as reservoirs for the protozoan. Although humans are usually accidental hosts, the flies freely feed on them. At particular risk are travelers or immigrants who have never had contact with the protozoan and lack specific immunity.

Leishmania infection begins when an infected fly injects the motile forms of the protozoan into the host while feeding. After being engulfed by macrophages, the parasite converts to a nonmotile reproductive form and multiplies in the macrophage. The manifestations of the disease vary with the fate of the macrophages. If they remain fixed, the infection stays localized in the skin or mucous membranes, but if the infected macrophages migrate, systemic disease occurs.

In cutaneous leishmaniasis, a small red papule occurs at the site of the bite and spreads laterally into a large ulcer **(Checkpoint 18.10)**. The edges of the ulcer are raised and the base is moist. It can be filled with a serous/purulent exudate or covered with a crust. Satellite lesions may occur. Mucocutaneous leishmaniasis usually begins with a skin lesion on the head or face and then progresses to single or multiple lesions, usually in the mouth and nose. Lesions can be quite extensive, eventually involving and disfiguring the hard palate, the nasal septum, and the lips.

There is no vaccine; avoiding the sand fly is the only prevention. The disease can be treated with chemicals such as an antimony compound called Pentastam; other antimicrobials may be indicated for secondary infections of the lesions.

Cutaneous Anthrax

This form of anthrax is the most common and least dangerous version of infection with *Bacillus anthracis*. (The spectrum of anthrax disease is discussed fully in chapter 20.) It is caused by endospores entering the skin through small cuts or abrasions. Germination and growth of the pathogen in the skin are marked by the production of a papule that becomes increasingly necrotic and later ruptures to form a painless, black, **eschar** (ess'-kar) (Checkpoint Table 18.10). In the fall of 2001, 11 cases of cutaneous anthrax occurred in the United

✓ CHECKPOINT 18.10 | Large Pustular Skin Lesions

Disease	Leishmaniasis	Cutaneous Anthrax
Causative Organism(s)	*Leishmania* spp.	*Bacillus anthracis*
Most Common Modes of Transmission	Biological vector	Direct contact with endospores
Virulence Factors	Multiplication within macrophages	Endospore formation; capsule, lethal factor, edema factor (see chapter 20)
Culture/Diagnosis	Culture of protozoa, microscopic visualization	Culture on blood agar; serology, PCR performed by CDC
Prevention	Avoiding sand fly	Avoid contact; vaccine available but not widely used
Treatment	Pentastam	Ciprofloxacin, doxycycline, penicillin
Distinguishing Features	Mucocutaneous and systemic forms	Can be fatal
Appearance of Lesions		

States as a result of bioterrorism (along with 11 cases of inhalational anthrax). Mail workers and others contracted the infection when endospores were sent through the mail. The infection can be naturally transmitted by contact with hides of infected animals (especially goats).

Left untreated, even the cutaneous form of anthrax is fatal approximately 20% of the time. A vaccine exists, but is recommended only for high-risk persons and the military. Upon suspicion of cutaneous anthrax, ciprofloxacin and/or doxycycline should be used initially. If the isolate is found to be sensitive to penicillin, patients can be switched to that drug **(Checkpoint 18.10)**.

Ringworm (Cutaneous Mycoses)

A group of fungi that is collectively termed **dermatophytes** cause a constellation of integument conditions. These mycoses are strictly confined to the nonliving epidermal tissues (stratum corneum) and its derivatives (hair and nails). All these conditions have different names that begin with the word **tinea** (tin'-ee-ah), which derives from the erroneous belief that they were caused by worms. That misconception is also the reason these diseases are often called *ringworm*—ringworm of the scalp (tinea capitis), beard (tinea barbae), body (tinea corporis), groin (tinea cruris), foot (tinea pedis), and hand (tinea manuum). (Don't confuse these "tinea" terms with genus and species names. It is simply an old practice for naming conditions.) Most of these conditions are caused by one of three different dermatophytes, which will be discussed here.

One fungal infection is even more superficial than the others; it infects only the most superficial layers of the stra-

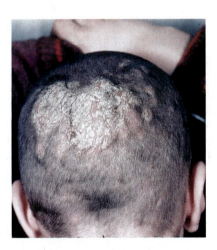

FIGURE 18.18 **Ringworm of the scalp.**
Hair loss can accompany these lesions.

tum corneum and causes a condition called **tinea versicolor.** It is not a ringworm but is nevertheless included at the end of this section.

Signs and Symptoms of the Cutaneous Mycoses

Ringworm of the Scalp (Tinea Capitis) This mycosis results from the fungal invasion of the scalp and the hair of the head, eyebrows, and eyelashes **(figure 18.18)**. Very common in children, tinea capitis is acquired from other children and adults or from domestic animals. Manifestations range from small scaly patches, to a severe inflammatory reaction, to destruction of the hair follicle and temporary or permanent hair loss.

Ringworm of the Beard (Tinea Barbae) This tinea, also called *barber's itch*, affects the chin and beard of adult males. Although once a common aftereffect of unhygienic barbering, it is now contracted mainly from animals.

Ringworm of the Body (Tinea Corporis) This extremely prevalent infection of humans can appear nearly anywhere on the body's glabrous (smooth and bare) skin. The principal sources are other humans, animals, and soil, and it is transmitted primarily by direct contact and fomites (clothing, bedding). The infection usually appears as one or more scaly reddish rings on the trunk, hip, arm, neck, or face **(figure 18.19)**. The ringed pattern is formed when the infection radiates from the original site of invasion into the surrounding skin. Depending on the causal species, and the health and hygiene of the patient, lesions vary from mild and diffuse to florid and pustular.

Ringworm of the Groin (Tinea Cruris) Sometimes known as *jock itch*, crural ringworm occurs mainly in males on the groin, perianal skin, scrotum, and occasionally, the penis. The fungus thrives under conditions of moisture and humidity created by sweating. It is transmitted primarily from human to human and is pervasive among athletes and persons living in close quarters (ships, military installations).

Ringworm of the Foot (Tinea Pedis) Tinea pedis has more colorful names as well, including athlete's foot and jungle rot. The disease is clearly connected to wearing shoes because it is uncommon in cultures where people customarily go barefoot. Conditions that encase the feet in a closed, warm, moist environment increase the possibility of infection. Tinea pedis is a known hazard in shared facilities such as shower stalls, public floors, and locker rooms. Infections begin with blisters between the toes that burst, crust over, and can spread to the rest of the foot and nails **(figure 18.20a)**.

Ringworm of the Hand (Tinea Manuum) Infection of the hand by dermatophytes is nearly always associated with concurrent infection of the foot. Lesions usually occur on the fingers and palms of one hand, and they vary from white and patchy to deep and fissured.

Ringworm of the Nail (Tinea Unguium) Fingernails and toenails, being masses of keratin, are often sites for persistent fungus colonization. The first symptoms are usually superficial white patches in the nail bed. A more invasive form causes thickening, distortion, and darkening of the nail **(figure 18.20b)**. Nail problems caused by dermatophytes are on the rise as more women wear artificial fingernails, which can provide a portal of entry into the nail bed.

Causative Agents

There are about 39 species in the genera *Trichophyton, Microsporum,* and *Epidermophyton* that can cause the preceding conditions. The causative agent of a given type of ringworm varies from one geographic location to another and is not restricted to

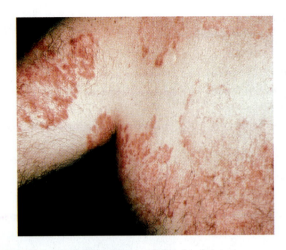

FIGURE 18.19 **Ringworm of the body.**

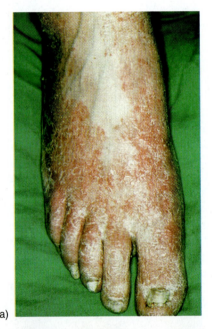

(a)

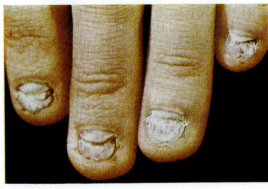

(b)

FIGURE 18.20 **Ringworm of the extremities.**
(a) *Trichophyton* infection spreading over the foot in a "moccasin" pattern. The chronicity of tinea pedis is attributed to the lack of fatty-acid-forming glands in the feet. **(b)** Ringworm of the nails. Invasion of the nail bed causes some degree of thickening, accumulation of debris, cracking, and discoloration; nails can be separated from underlying structures as shown here.

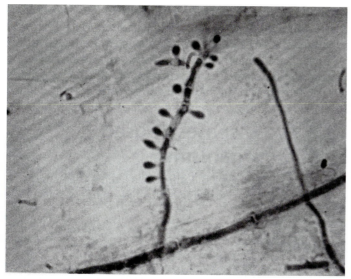

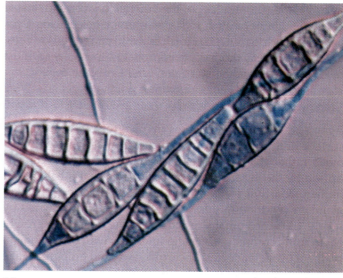

(a)

(b)

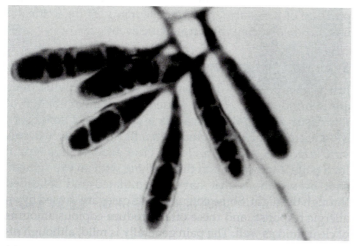

(c)

FIGURE 18.21 **Examples of dermatophyte spores.**
(a) Regular, numerous microconidia of *Trichophyton*.
(b) Macroconidia of *Microsporum canis*, a cause of ringworm in
cats, dogs, and humans. **(c)** Smooth-surfaced macroconidia in
clusters characteristic of *Epidermophyton*.

a particular genus and species. These fungi are so closely related, and morphologically similar, that they can be difficult to
differentiate. Various species exhibit unique macroconidia, microconidia, and unusual types of hyphae. In general, *Trichophyton* produces thin-walled, smooth macroconidia and numerous
microconidia **(figure 18.21a)**; *Microsporidium* produces thick-
walled, rough macroconidia and sparser microconidia **(figure
18.21b)**; and *Epidermophyton* has ovoid, smooth, clustered
macroconidia and no microconidia **(figure 18.21c)**.

The presenting symptoms of a cutaneous mycosis occasionally are so dramatic and suggestive of these genera that
no further testing is necessary. In most cases, however, direct
microscopic examination and culturing are required. Diagnosis of tinea of the scalp caused by some species is aided by
use of a long-wave ultraviolet lamp that causes infected hairs
to fluoresce. Samples of hair, skin scrapings, and nail debris
treated with heated potassium hydroxide (KOH) show a
thin, branching fungal mycelium if infection is present.

Pathogenesis and Virulence Factors

The dermatophytes have the ability to invade and digest keratin, which is naturally abundant in the cells of the stratum
corneum. The fungi do not invade deeper epidermal layers.
Important factors that promote infection are the hardiness of
the dermatophyte spores (they can last for years on fomites);
presence of abraded skin; and intimate contact. Most infections exhibit a long incubation period (months), followed by
localized inflammation and allergic reactions to fungal proteins. As a general rule, infections acquired from animals and
soil cause more severe reactions than do infections acquired
from other humans, and infections eliciting stronger immune
reactions are resolved faster.

Transmission and Epidemiology

Transmission of the fungi that cause these diseases is direct
and indirect contact, with other humans or with infected animals. Some of these fungi can be acquired from the soil.

Prevention and Treatment

The only way to prevent these infections is to avoid contact
with the dermatophytes, which is impractical. Keeping susceptible skin areas dry is helpful. Treatment of ringworm is
based on the knowledge that the dermatophyte is feeding on
dead epidermal tissues. These regions undergo constant replacement from living cells deep in the epidermis, so if multiplication of the fungus can be blocked, the fungus will
eventually be sloughed off along with the skin or nail. Unfortunately, this takes time. By far the most satisfactory choice for
therapy is a topical antifungal agent. Ointments containing
tolnaftate, miconazole, itraconazole, terbinafine, or thiaben-
dazine are applied regularly for several weeks. Some drugs

Trachoma

Ocular trachoma is a chronic *Chlamydia trachomatis* infection of the epithelial cells of the eye. It is an ancient disease and a major cause of blindness in certain parts of the world. Although a few cases occur annually in the United States, several million cases occur endemically in parts of Africa and Asia. Transmission is favored by contaminated fingers, fomites, fleas, and a hot, dry climate. It is caused by a different *C. trachomatis* strain than that which causes simple conjunctivitis. Ongoing infection or many recurrent infections with this strain eventually lead to chronic inflammatory damage and scarring.

The first signs of infection are a mild conjunctival discharge and slight inflammation of the conjunctiva. These symptoms are followed by marked infiltration of lymphocytes and macrophages into the infected area. As these cells build up, they impart a pebbled (rough) appearance to the inner aspect of the upper eyelid **(figure 18.24)**. In time, a vascular pseudomembrane of exudates and inflammatory leukocytes forms over the cornea, a condition called *pannus*, which lasts a few weeks. Chronic and secondary infections can lead to corneal damage and impaired vision. Early treatment of this disease with tetracycline or sulfa drugs is highly effective and prevents all of the complications. It is a tragedy that in this day of sophisticated preventive medicine, millions of children worldwide will develop blindness for lack of a few dollars' worth of antibiotics.

FIGURE 18.24   **Ocular trachoma caused by *C. trachomatis*.**

CHECKPOINT 18.13	Trachoma
Causative Organism(s)	*C. trachomatis* serovars A–C
Most Common Modes of Transmission	Indirect contact, mechanical vector
Virulence Factors	Intracellular growth
Culture/Diagnosis	Detection of inclusion bodies in stained preparations
Prevention	Hygiene, vector control, prompt treatment of initial infection
Treatment	Oral doxycycline or topical erythromycin

Keratitis

Keratitis is a more serious eye infection than conjunctivitis. Invasion of deeper eye tissues occurs and can lead to complete corneal destruction. Any microorganism can cause this condition, especially after trauma to the eye, but this section will focus on one of the more common causes: herpes simplex virus. It can cause keratitis in the absence of predisposing trauma.

The usual cause of herpetic keratitis is a "misdirected" reactivation of (oral) herpes simplex virus type 1 (HSV-1). The virus, upon reactivation, travels into the ophthalmic rather than the mandibular branch of the trigeminal nerve. Infections with HSV-2 can also occur, as a result of a sexual encounter with the virus or transfer of the virus from the genital to eye area, or if an individual has a recurrent oral infection with HSV-2. Preliminary symptoms are a gritty feeling in the eye, conjunctivitis, sharp pain, and sensitivity to light. Some patients develop characteristic branched or opaque corneal lesions as well. In 25% to 50% of cases, this keratitis is recurrent and chronic and can interfere with vision. Blindness due to herpes is the leading infectious cause of blindness in the United States.

The viral condition is treated with topical vidarabine or oral acyclovir or both. Keratitis resulting from trauma and subsequent bacterial infection is treated with appropriate antibiotics. Most physicians will prescribe antibiotics for prophylactic reasons when there is damage to the eye, even if the original cause is viral **(Checkpoint 18.14)**.

River Blindness

River blindness is a chronic parasitic (helminthic) infection. It is endemic in dozens of countries in Latin America, Africa, Asia, and the Middle East. At any given time, tens of millions

✔ CHECKPOINT 18.14	Keratitis	
Causative Organism(s)	Herpes simplex virus	Miscellaneous microorganisms
Most Common Modes of Transmission	Reactivation of latent virus, although primary infections can occur in the eye	Often traumatic introduction (parenteral)
Virulence Factors	Latency	Various
Culture/Diagnosis	Usually clinical diagnosis; viral culture or PCR if needed	Various
Prevention	–	–
Treatment	Topical vidarabine and/or oral acyclovir	Specific antimicrobials

of people are infected with the worm called *Onchocerca volvulus* (ong″-koh′ser′-kah′ volv′-yoo′lus.) This organism is a filarial (threadlike) helminthic worm transmitted by small biting vectors called *black flies*. These voracious flies often attack in large numbers, and it is not uncommon in endemic areas to be bitten several hundred times a day. The disease gets its name from the habitat where these flies are most often found, rural settlements along rivers bordered with overhanging vegetation.

The *Onchocerca* larvae are deposited into a bite wound and develop into adults in the immediate subcutaneous tissues, where disfiguring nodules form within 1 to 2 years after initial contact. Microfilariae given off by the adult female migrate via the bloodstream to many locations, but especially to the eyes. While the worms are in the blood they can be transmitted to other feeding black flies.

Some cases of onchocerciasis result in a severe itchy rash that can last for years. It was previously thought that condition was caused by degeneration of the worms and the inflammation and granulomatous lesion formation that result from the release of their antigens. It is in fact the case that the worms eventually invade the entire eye, producing much inflammation and permanent damage to the retina and optic nerve. In 1999 researchers first discovered large colonies of bacteria called *Wolbachia* living *inside* the *Onchocerca* worms. By 2002 scientists felt very strongly that the damage caused to human tissues was induced by the bacteria rather than by the worms. Of course, the worms serve as the delivery system to the human as it does not appear that the bacteria can infect humans on their own. These bacteria enjoy a mutualistic relationship with their hosts; they are essential for normal *Onchocerca* development.

In regions of high prevalence, it is not unusual for an ophthalmologist to see microfilariae wiggling in the anterior chamber during a routine eye checkup. Microfilariae die in several months, but adults can exist for up to 15 years in skin nodules.

River blindness has been a serious problem in many areas of Africa. In some villages, nearly half of the residents are affected by the disease. A campaign to eradicate onchocerciasis by 2007 is currently underway, supported by the Carter Center, an organization run by former U.S. President Jimmy Carter. The approach is to treat people with *ivermectin,* a potent antifilarial drug and to use insecticides to control the black flies. This approach need not be changed because of the new information about *Wolbachia;* eliminating the protozoan will still eliminate the disease. The drug company that manufactures ivermectin has promised to provide the drug for free for as long as the need for it exists.

✔ CHECKPOINT 18.15	River Blindness
Causative Organism(s)	*Wolbachia* plus *Onchocerca volvulus*
Most Common Modes of Transmission	Biological vector
Virulence Factors	Induction of inflammatory response
Culture/Diagnosis	"Skin snips": small piece of skin in NaCl solution examined under microscope and microfilariae counted
Prevention	Avoiding black fly
Treatment	Ivermectin
Distinguishing Features	Worms often visible in eye

Taxonomic Organization of Microorganisms Causing Diseases of the Skin and Eyes

Microorganism	Disease	Chapter Location
Gram-Positive Bacteria		
Propionibacterium acnes	Acne	Acne, p. 543
Staphylococcus aureus	Impetigo, cellulitis, scalded skin syndrome, folliculitis, abscesses (furuncles and carbuncles), necrotizing fasciitis	Impetigo, p. 544 Cellulitis, p. 547 Scalded Skin Syndrome, p. 547, Insight 18.1, p. 550
Streptococcus pyogenes	Impetigo, cellulitis, erysipelas, necrotizing fasciitis	Impetigo, p. 545 Cellulitis, p. 547, Insight 18.1, p. 550
Clostridium perfringens	Gas gangrene	Gas gangrene, p. 548
Bacillus anthracis	Cutaneous anthrax	Large pustular skin lesions, p. 564
Gram-Negative Bacteria		
*Mycobacterium leprae**	Leprosy	Leprosy, p. 551
Neisseria gonorrhoeae	Neonatal conjunctivitis	Conjunctivitis, p. 568
Chlamydia trachomatis	Neonatal conjunctivitis, trachoma	Conjunctivitis, p. 568 Trachoma, p. 570
Wolbachia (in combination with *Onchocerca*)	River blindness	River blindness, p. 570
DNA Viruses		
Human herpesvirus 3 (varicella) virus	Chickenpox	Vesicular or pustular rash diseases, p. 554
Variola virus	Smallpox	Vesicular or pustular rash diseases, p. 556
Parvovirus B 19	Fifth disease	Maculopapular rash diseases, p. 562
Human herpesvirus 6 and 7	Roseola	Maculopapular rash diseases, p. 562
Human papillomavirus	Warts	Warts and wartlike eruptions, p. 562
Molluscum contagiosum virus	Molluscum contagiosum	Warts and wartlike eruptions, p. 563
Herpes simplex virus	Keratitis	Keratitis, p. 570
RNA Viruses		
Measles virus	Measles	Maculopapular rash diseases, p. 558
Rubella virus	Rubella	Maculopapular rash diseases, p. 561
Fungi		
Trichophyton	Ringworm	Ringworm, p. 565
Microsporum	Ringworm	Ringworm, p. 565
Epidermophyton	Ringworm	Ringworm, p. 565
Malassezia furfur	Superfical mycosis	Superficial mycoses, p. 565
Protozoa		
Leishmania spp.	Leishmaniasis	Large pustular skin lesions, p. 564
Helminths		
Onchocerca volvulus (in combination with *Wolbachia*)	River blindness	River blindness, p. 570

*There is some debate about the gram status of the genus *Mycobacterium*; it is generally not considered gram positive or gram negative.

Infectious Diseases Affecting the Skin and Eyes

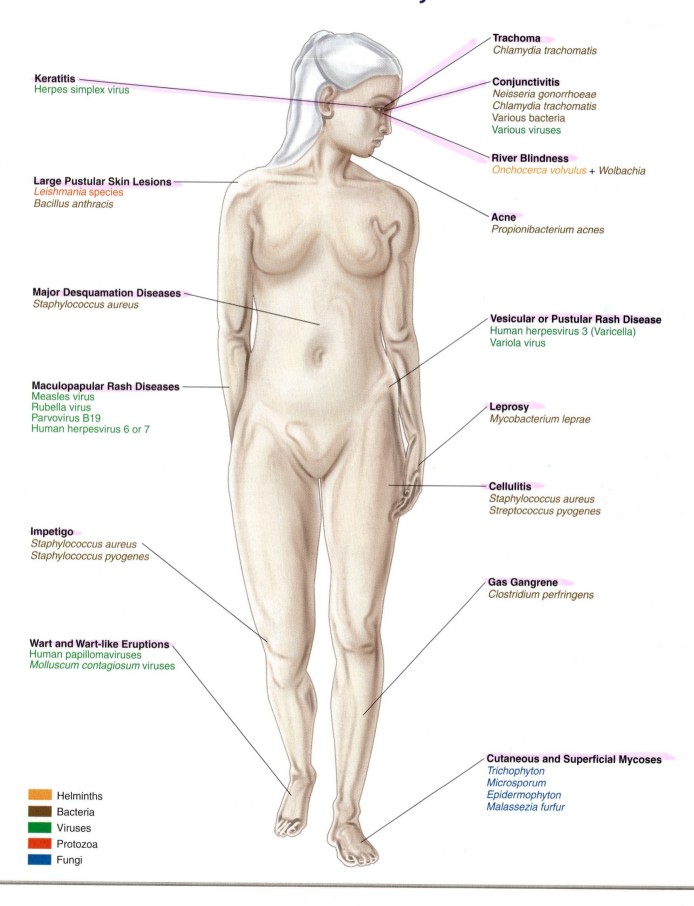

Keratitis
Herpes simplex virus

Large Pustular Skin Lesions
Leishmania species
Bacillus anthracis

Major Desquamation Diseases
Staphylococcus aureus

Maculopapular Rash Diseases
Measles virus
Rubella virus
Parvovirus B19
Human herpesvirus 6 or 7

Impetigo
Staphylococcus aureus
Staphylococcus pyogenes

Wart and Wart-like Eruptions
Human papillomaviruses
Molluscum contagiosum viruses

Trachoma
Chlamydia trachomatis

Conjunctivitis
Neisseria gonorrhoeae
Chlamydia trachomatis
Various bacteria
Various viruses

River Blindness
Onchocerca volvulus + *Wolbachia*

Acne
Propionibacterium acnes

Vesicular or Pustular Rash Disease
Human herpesvirus 3 (Varicella)
Variola virus

Leprosy
Mycobacterium leprae

Cellulitis
Staphylococcus aureus
Streptococcus pyogenes

Gas Gangrene
Clostridium perfringens

Cutaneous and Superficial Mycoses
Trichophyton
Microsporum
Epidermophyton
Malassezia furfur

Helminths
Bacteria
Viruses
Protozoa
Fungi

Chapter Summary With Key Terms

18.1 The Skin and Eyes
The skin is organized in layers, from the deepest layer called the stratum basale, to the uppermost epidermis.

18.2 The Skin and Its Defenses
A. The epidermal cells contain the protein keratin, which "waterproofs" the skin and protects it from microbial invasion.
B. Other defenses include, low pH sebum, and high salt and lysozyme in sweat.

18.3 Normal Flora of the Skin
The skin has a diverse array of microbes as its normal flora, including the diphtheroids, micrococci, and yeasts.

18.4 The Surface of the Eye and Its Defenses
The flushing action of the tears, which contain lysozyme and lactoferrin, is the major protective feature of the eye.

18.5 Normal Flora of the Eye
The eye has similar microbes as the skin, but in lower numbers.

18.6 Skin Diseases Caused by Microorganisms
A. **Acne:** These are follicle-associated lesions caused by microbial digestion of excess sebum trapped in the pores of the skin. *Propionibacterium acnes* is the main causative agent.
B. **Impetigo:** A highly contagious superficial bacterial infection that can cause the skin to peel or flake off that is transmitted both by direct contact and via fomites and mechanical vectors. Causative organisms can be either *Staphylococcus aureus* or *Streptococcus pyogenes* or both.
C. **Cellulitis:** This condition results from a fast-spreading infection of the dermis and subcutaneous tissue below. Most commonly it is caused by the introduction of *S. aureus* or *S. pyogenes* into the dermis.
D. **Staphylococcal Scalded Skin Syndrome (SSSS):** This dermolytic condition is caused by *S. aureus*. It affects mostly newborns and babies and is similar to a systemic form of impetigo. *Toxic epidermal necrolysis (TEN)* is a similar manifestation caused by a reaction to antibiotics, barbiturates, or other drugs.
E. **Gas Gangrene:** Also called clostridial myonecrosis, this disease can be manifested in two forms: anaerobic cellulitis or myonecrosis. The spore-forming anaerobe, *Clostridium perfringens,* is the most common causative organism.
F. **Hansen's Disease (Leprosy):** Leprosy is a chronic, progressive disease of the skin and nerves caused by a bacterium, *Mycobacterium leprae.* Tuberculoid leprosy is the most superficial form; **Lepromatous leprosy** is a chronic manifestation that is responsible for the disfigurement commonly associated with the disease.
G. **Vesicular or Pustular Rash Diseases**
  1. **Chickenpox:** Skin lesions progress quickly from macules and papules to itchy vesicles, filled with a clear fluid. Patients are considered contagious until all of the lesions have crusted over.
  2. **Shingles:** Recuperation from chickenpox is associated with the virus becoming latent in the ganglia and may reemerge as **shingles.** Human herpesvirus 3, an enveloped DNA virus, causes chickenpox, as well as herpes zoster or shingles. The virus is sometimes referred to as the varicella-zoster virus (VZV).
  3. **Smallpox:** Naturally occurring smallpox has been eradicated from the world. This infection manifests as a rash in the pharynx, spreads to the face, and progresses to the extremities. Variola major is a highly virulent form of smallpox that causes toxemia, shock, and intravascular coagulation. A patient with variola minor has a milder form of the disease. The causative agent of smallpox, the variola virus, is an orthopoxvirus, an enveloped DNA virus.
H. **Maculopapular Rash Diseases**
  1. **Measles:** Measles or **rubeola** results in oral lesions called *Koplik's spots* and characteristic red maculopapular **exanthemum** that erupts on the head and then progresses to the trunk and extremities, until most of the body is covered. The most serious complication is **subacute sclerosing panencephalitis (SSPE),** a progressive neurological degeneration of the cerebral cortex, white matter, and brain stem. The measles virus is a member of the *Morbillivirus* genus. It is a single-stranded enveloped RNA virus in the Paramyxovirus family. The MMR vaccine (measles, mumps, and rubella) contains live attenuated measles virus.
  2. **Rubella:** This disease is also known as German measles, and can appear in two forms: postnatal and **congenital** (prenatal) infection of the fetus. The rubella virus is a nonsegmented single-stranded RNA virus with a loose lipid envelope called *Rubivirus,* in the family Togavirus. The MMR vaccination contains protection from rubella.
  3. **Fifth Disease:** Also called *erythema infectiosum,* fifth disease is a very mild, but highly contagious, disease that often results in a characteristic "slapped-cheek" appearance because of a confluent reddish rash that begins on the face. The causative agent is parvovirus B19.
  4. **Roseola:** This disease can result in a maculopapular rash, and is caused by a human herpesvirus called HHV-6, and sometimes by HHV-7.
I. **Wartlike Eruptions:** Viruses cause virtually all warts. Most common warts are caused by human papillomavirus, or a poxvirus, **molluscum contagiosum,** which causes bumps that may look like warts. Warts, or **papillomas,** are benign, squamous epithelial growths. Virus can infect mucous membranes or invade skin. Rarely, a wart can become malignant, when caused by a particular type of HPV.
J. **Larger Pustular Skin Lesions**
  1. **Leishmaniasis:** This is a zoonosis transmitted by female sand fly when it ingests the host's blood. A protozoan causes this equatorial disease, and the infection can either be localized in the skin or mucous membranes, or systemic.
  2. **Cutaneous Anthrax:** This form of anthrax is the most common and least dangerous version of infection with *Bacillus anthracis*. The skin shows formation of a papule that becomes necrotic and later ruptures to form a painless black **eschar.**

K. **Ringworm (Cutaneous Mycoses):** A group of fungi that are collectively termed **dermatophytes** cause mycoses that are confined to the nonliving epidermal tissues, hair and nails. These diseases are often called "ringworm"— ringworm of the scalp (tinea capitis), beard (tinea barbae), body (tinea corporis), groin (tinea cruris), foot (tinea pedis), and hand (tinea manuum). Species in the genera *Trichophyton, Microsporum,* and *Epidermophyton* cause all of the cutaneous mycoses.

L. **Superficial Mycosis:** Agents of **superficial mycoses** involve the outer epidermis. **Tinea versicolor** is caused by the yeast *Malassezia furfur,* a normal inhabitant of human skin that feeds on the high oil content of the skin glands.

18.7 Eye Diseases Caused by Microorganisms

A. **Conjunctivitis:** Infection of the conjunctiva (commonly called pinkeye) has many different clinical presentations. Neonatal eye infection is usually associated with *Neisseria gonorrhoeae* or *Chlamydia trachomatis;* they are transmitted vertically via a genital tract infection in the mother. Bacterial conjunctivitis in other age groups is most commonly caused by *Staphylococcus epidermidis* or by *Streptococcus pyogenes, Streptococcus pneumoniae, Haemophilus influenzae,* or *Moraxella* species. Viral conjunctivitis is commonly caused by adenoviruses. Both bacterial and viral conjunctivitis are highly contagious.

B. **Trachoma:** Ocular trachoma is a chronic *Chlamydia trachomatis* infection of the epithelial cells of the eye and a major cause of blindness in certain parts of the world. Trachoma and simple conjunctivitis are caused by different strains of *C. trachomatis.*

C. **Keratitis:** Keratitis is a more serious eye infection than conjunctivitis. Herpes simplex viruses (HSV-1 and HSV-2) have been implicated in this condition.

D. **River Blindness:** River blindness is a chronic parasitic helminth infection that is endemic in dozens of countries in Latin America, Africa, Asia, and the Middle East. The condition is caused by a symbiotic pair, the bacterium *Wolbachia* living inside the helminth *Onchocerca.* The worm is transmitted to humans by small biting black flies.

Multiple-Choice Questions

1. An effective treatment for a cutaneous mycosis like tinea pedis would be
   a. penicillin
   b. miconazole
   c. griseofulvin
   d. doxycycline

2. What is the antimicrobial enzyme found in sweat, tears, and saliva that can specifically break down peptidoglycan?
   a. lysozyme
   b. β-lactamase
   c. catalase
   d. coagulase

3. Which of the following have been used as treatments for acne?
   a. erythromycin
   b. tetracycline
   c. oral contraceptives
   d. Accutane
   e. all of the above

4. Name the organism(s) most commonly associated with cellulitis.
   a. *Staphylococcus aureus*
   b. *Propionibacterium acnes*
   c. *Streptococcus pyogenes*
   d. both a and b
   e. both a and c

5. Which of the following is *not* a characteristic of *Staphylococcus aureus*?
   a. coagulase positive
   b. gram-negative coccus
   c. catalase positive
   d. surface protein A

6. Due to a highly successful vaccination program, the WHO has managed the worldwide eradication of the naturally occurring disease:
   a. chickenpox
   b. leprosy
   c. smallpox
   d. German measles

7. Measles can potentially be eradicated because
   a. humans are the only reservoir.
   b. it is not very contagious.
   c. it is not particularly hazardous to any group.
   d. it is easily controlled by antibiotics.
   e. most people have developed resistance to it.

8. Warts are caused by
   a. human herpesvirus 3
   b. papillomavirus virus
   c. herpes simplex virus
   d. morbillivirus

9. Which of the following is *not* a characteristic of *Streptococcus pyogenes*?
   a. Lancefield group A
   b. gram-positive coccus
   c. coagulase positive
   d. β-hemolytic on blood agar

10. Herpesviruses can cause all of the following diseases, except
    a. chickenpox
    b. shingles
    c. keratitis
    d. smallpox
    e. roseola

11. What is the treatment for someone infected with fifth disease?
    a. acyclovir
    b. tetracycline
    c. vaccine
    d. no treatment

12. Which disease is incorrectly matched with the causative agent?
    a. viral conjunctivitis—adenovirus
    b. river blindness—*Onchocerca volvulus*
    c. smallpox—variola virus
    d. gas gangrene—*Staphylococcus aureus*

13. Dermatophytes are fungi that infect the epidermal tissue by invading and attacking
    a. collagen
    b. keratin
    c. fibroblasts
    d. sebaceous glands

14. The dermolytic conditions seen in impetigo and scalded skin syndrome are primarily due to the following *Staphylococcus aureus*–produced product(s)
    a. exfoliative toxins A and B
    b. M protein
    c. alpha toxin
    d. none of the above
    e. all of the above

▶ Infections in the nervous system generally affect the brain, the meninges, and/or the peripheral nerves. Brain infections are called encephalitis; meningeal infections are called meningitis. Some infections affect multiple sites, as in meningoencephalitis.

▶ Infections in the central nervous system are often very serious, because the tissues affected have lowered defenses and can be permanently damaged by inflammation.

19.1 The Nervous System and Its Defenses

The nervous system can be thought of as having two component parts: the central nervous system (CNS), consisting of the brain and spinal cord, and the peripheral nervous system (PNS), which contains the nerves that emanate from the brain and spinal cord to sense organs and to the periphery of the body **(figure 19.1).** The nervous system performs three important functions—sensory, integrative, and motor. The sensory function is fulfilled by sensory receptors at the ends of peripheral nerves. They generate nerve impulses that are transmitted to the central nervous system. There the impulses are translated, or integrated, into sensation or thought, which in turn drive the motor function. The motor function necessarily involves structures outside of the nervous system, such as muscles and glands.

The brain and the spinal cord are dense structures made up of cells called *neurons.* They are both surrounded by bone. The brain is situated inside the skull, and the spinal cord lies within the spinal column **(figure 19.2),** which is composed of a stack of interconnected bones called vertebrae. The soft tissue of the brain and spinal cord is encased within a tough casing of three membranes called the **meninges.** The layers of membranes, from outer to inner, are the dura mater, the arachnoid mater, and the pia mater. Between the arachnoid mater and pia mater is the subarachnoid space (that is, the space under the arachnoid mater). The subarachnoid space is filled with a clear serumlike fluid called cerebrospinal fluid (CSF). The CSF provides nutrition to the CNS, while also providing a liquid cushion for the sensitive brain and spinal cord. The meninges are a common site of infection, and microorganisms can often be found in the CSF when meningeal infection **(meningitis)** occurs.

The PNS consists of cranial and spinal nerves (figure 19.1). Nerves, or neurons, are bundles of cellular fibers, in the form of axons and dendrites that receive and transmit nerve signals. The axons and dendrites of adjacent neurons communicate with each other over a very small space, called a synapse. Chemicals called neurotransmitters are released from one cell and act on the next cell in the synapse.

The defenses of the nervous system are mainly structural. The bony casings of the brain and spinal cord protect them from traumatic injury. The cushion of surrounding CSF also serves a protective function. The entire nervous system is served by the vascular system, but the interface between the blood vessels serving the brain and the brain itself is different from that of other areas of the body, and provide a third structural protection. The cells that make up the walls of the blood vessels allow very few molecules to pass through. In other parts of the body there is freer passage of ions, sugars, and other metabolites through the walls of blood vessels. The restricted permeability of blood vessels in the brain is called the **blood-brain barrier,** and it prohibits most microorganisms from passing into the central nervous system. The drawback of this phenomenon is that drugs and antibiotics are difficult to introduce into the CNS when needed.

The CNS is considered an "immunologically privileged" site. These sites are able to mount only a partial, or at least a different, immune response when exposed to immunologic challenge. The functions of the CNS are so vital for the life of an organism that even temporary damage that could potentially result from "normal" immune responses would be very detrimental. The uterus and parts of the eye are other immunologically privileged sites. Cells in the CNS express lower levels of MHC antigens. Complement proteins are also in much lower quantities in the CNS. However, specialized cells in the central nervous system perform defensive functions. Microglia are a type of nerve cell having phagocytic capabilities, and brain

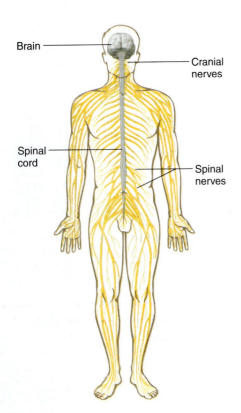

FIGURE 19.1 **Nervous system.**
The central nervous system and the peripheral nerves.

Brain

Cranial nerves

Spinal cord

Spinal nerves

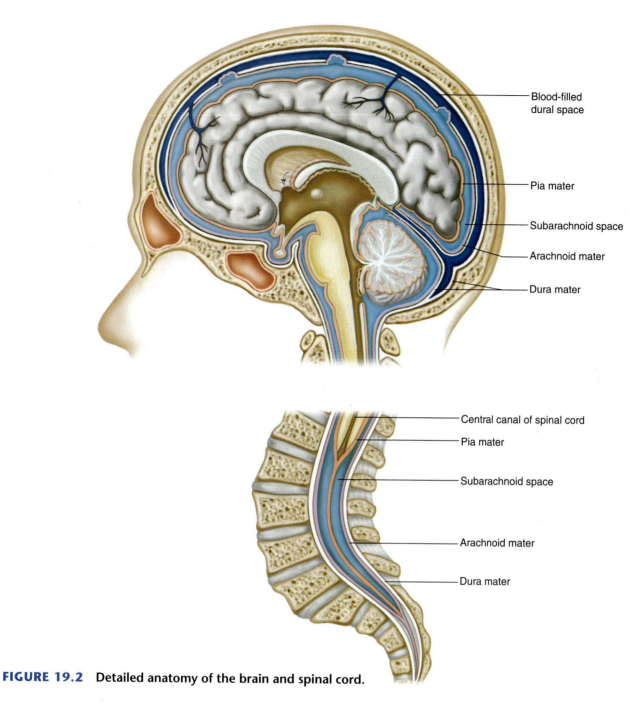

FIGURE 19.2 **Detailed anatomy of the brain and spinal cord.**

macrophages also exist in the CNS, although the activity of both of these types of cells is thought to be reduced when compared with phagocytic cells elsewhere in the body.

19.2 Normal Flora of the Nervous System

There is no normal flora in either the CNS or PNS. Finding microorganisms of any type in these tissues represents a deviation from the healthy state. Viruses such as herpes simplex live in a dormant state in the nervous system between episodes of acute disease, but they are not considered normal flora.

Nervous System Defenses and Normal Flora		
	Defenses	**Normal Flora**
Nervous System	Bony structures, blood-brain barrier, microglial cells, and macrophages	None

19.3 Nervous System Diseases Caused by Microorganisms

Meningitis

Meningitis, an inflammation of the meninges, is an excellent example of an anatomical syndrome. Many different microorganisms can cause an infection of the meninges, and they produce a similar constellation of symptoms. Noninfectious causes of meningitis exist as well, but they are much less common than the infections listed here.

The more serious forms of acute meningitis are caused by bacteria, but it is thought that their entrance to the CNS is often facilitated by coinfection or previous infection with respiratory viruses. Meningitis in neonates is most often caused by different microorganisms, and therefore it is described separately in the following section.

Whenever meningitis is suspected, lumbar puncture (spinal tap) is performed to obtain cerebrospinal fluid (CSF), which is then examined by Gram stain and/or culture. Most physicians will begin treatment with a broad-spectrum antibiotic immediately, and shift treatment if necessary after a diagnosis has been confirmed.

Signs and Symptoms

No matter the cause, meningitis results in these typical symptoms: headache, painful or stiff neck, fever, and usually an increased number of white blood cells in the CSF. Specific microorganisms may cause additional, and sometimes characteristic, symptoms, which are described in the individual sections that follow.

Like many other infectious diseases, meningitis can manifest as acute or chronic disease. Some microorganisms are more likely to cause acute meningitis, and others are more likely to cause chronic disease.

In a normal healthy patient, it is very difficult for microorganisms to gain access to the nervous system. Those that are successful usually have specific virulence factors.

Neisseria meningitidis

This bacterium appears as gram-negative diplococci lined up side by side **(figure 19.3)** and is commonly known as the meningococcus. It is often associated with epidemic forms of meningitis. This organism causes the most serious form of acute meningitis, and it is responsible for about 25% of all meningitis cases, most of them in children younger than age 2. Although 12 different strains of capsular antigens exist, serotypes A, B, and C are responsible for most cases of infection.

Pathogenesis and Virulence Factors Bacteria entering the blood vessels rapidly permeate the meninges and produce symptoms of meningitis. Meningitis is marked by fever, sore throat, headache, stiff neck, convulsions, and vomiting. The most serious complications of meningococcal infection are due to meningococcemia **(figure 19.4),** which can accompany meningitis but can also occur on its own. The pathogen releases endotoxin into the generalized circulation, which is a potent stimulus for certain white blood cells. Damage to the

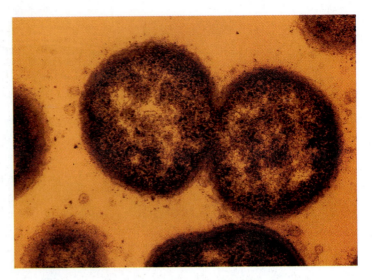

FIGURE 19.3 **Transmission electron micrograph of Neisseria (52,000×).**
This cross-section makes the bacteria appear more spherical than usual.

blood vessels caused by cytokines released by the white blood cells leads to vascular collapse, hemorrhage, and crops of lesions called **petechiae** (pee-tee'-kee-ay) on the trunk and appendages.

In a small number of cases, meningococcemia becomes a fulminant disease with a high mortality rate. Recent evidence suggests that persons who experience meningitis, rather than mild infection, have a genetic predisposition to it. These patients have changes in the genes that encode toll-like receptors (see chapter 14). The changes make it less likely that the host will initiate an early defensive response to the bacterium.

The disease has a sudden onset, marked by fever higher than 40°C, chills, delirium, severe widespread ecchymosis (ek"-ih'moh'seez) (areas of bleeding under the skin), shock, and coma. Generalized intravascular clotting, cardiac failure, damage to the adrenal glands, and death can occur within a few hours. The bacterium has an IgA protease and a capsule, both of which counter the body's defenses.

Transmission and Epidemiology Because meningococci do not survive long in the environment, these bacteria are usually acquired through close contact with secretions or droplets. Upon reaching its portal of entry in the nasopharynx, the meningococci attach there using pili. In many people, this can result in simple asymptomatic colonization. In the more vulnerable individual, however, the meningococci are engulfed by epithelial cells of the mucosa and penetrate into the nearby blood vessels, along the way damaging the epithelium and causing pharyngitis.

Meningococcal meningitis has a sporadic or epidemic incidence in late winter or early spring. The continuing reservoir of infection is humans who harbor the pathogen in the nasopharynx. The carriage state, which can last from a few days to several months, exists in 3% to 30% of the adult population and can exceed 50% in institutional settings. The scene is set for transmission when carriers live in close quarters with

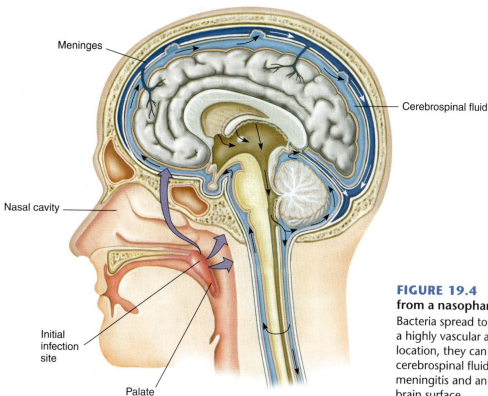

Meninges

Cerebrospinal fluid

Nasal cavity

Initial infection site

Palate

FIGURE 19.4 **Dissemination of the meningococcus from a nasopharyngeal infection.**
Bacteria spread to the roof of the nasal cavity, which borders a highly vascular area at the base of the brain. From this location, they can enter the blood and escape into the cerebrospinal fluid. Infection of the meninges leads to meningitis and an inflammatory purulent exudate over the brain surface.

nonimmune individuals, as might be expected in families, day care facilities, college dormitories, and military barracks. The highest risk groups are young children (6 to 36 months old) and older children and young adults (10 to 20 years old).

Culture and Diagnosis Suspicion of bacterial meningitis constitutes a medical emergency, and differential diagnosis must be done with great haste and accuracy. It is most important to confirm (or rule out) meningococcal meningitis, since it can be rapidly fatal. Treatment (described in the following section) is usually begun with this bacterium in mind until it can be ruled out. Cerebrospinal fluid, blood, or nasopharyngeal samples are stained and observed directly for the typical gram-negative diplococci. Cultivation may be necessary to differentiate the bacterium from other species. Specific rapid tests are also available for detecting the capsular polysaccharide or the cells directly from specimens without culturing.

It is usually necessary to differentiate this species from normal *Neisseria* that also live in the human body and can be present in infectious fluids. Immediately after collection, specimens are streaked on Modified Thayer-Martin medium (MTM) or chocolate agar and incubated in a high CO_2 atmosphere. Presumptive identification of the genus is obtained by a Gram stain and oxidase testing on isolated colonies **(figure 19.5)**. Further testing may be necessary to differentiate *N. meningitidis* and *N. gonorrhoeae* from one another, from other oxidase-positive species, and from normal flora of the oropharynx that can be confused with the pathogens. Several rapid-method identification kits have been developed for this purpose.

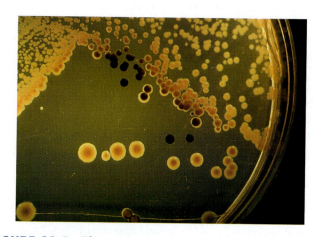

FIGURE 19.5 **The oxidase test.**
A drop of oxidase reagent is placed on a suspected *Neisseria* or *Branhamella* colony. If the colony reacts with the chemical to produce a purple to black color, it is oxidase-positive; those that remain white to tan are oxidase-negative. Because several species of gram-negative rods are also oxidase-positive, this test is presumptive for these two genera only if a Gram stain has verified the presence of gram-negative cocci.

Prevention and Treatment The infection rate in most populations is about 1%, so well-developed natural immunity to the meningococcus appears to be the rule. A sort of natural immunization occurs during the early years of life as one is exposed to the meningococcus and its close relatives. Resistance is due to opsonizing antibodies that develop against the capsular polysaccharides in groups A and C and against membrane antigens to group B. Because even treated meningococcemial disease has a mortality rate of up to 15%, it is vital that chemotherapy begin as soon as possible with one or more drugs. Penicillin G is the most potent of the drugs available for meningococcal infections; it is generally given in high doses intravenously. Patients may also require treatment for shock and intravascular clotting.

When family members, medical personnel, or children in day care or school have come in close contact with infected people, preventive therapy with rifampin or tetracycline may be warranted. Meningococcal vaccines that contain specific purified capsular antigens are available to protect high-risk groups, especially during epidemics. Group A vaccine protects all ages, but group C vaccine is useful only for individuals over 2 years of age, and a group B vaccine is not yet available.

Streptococcus pneumoniae

Because this bacterium causes the majority of bacterial pneumonias (see chapter 21), it is also referred to as the **pneumococcus.** Pneumococcal meningitis is also caused by this bacterium; indeed, it is the most frequent cause of community-acquired meningitis and is also very severe. It does not cause the petechiae associated with meningococcal meningitis, and that difference is useful diagnostically. As many as 25% of pneumococcal meningitis patients will also have pneumococcal pneumonia. Pneumococcal meningitis is most likely to occur in patients with underlying susceptibility, such as alcoholics and patients with sickle-cell disease or those with absent or defective spleen function.

This bacterium is covered thoroughly in chapter 21, because it is a common cause of ear infections and pneumonia. It obviously has the potential to be highly pathogenic, while at the same time appearing as normal flora in many people. It can penetrate the respiratory mucosa, gain access to the bloodstream and then, under certain conditions, enter the meninges.

Like the meningococcus, this bacterium has a polysaccharide capsule that protects it against phagocytosis. It also produces an α-hemolysin and hydrogen peroxide, both of which have been shown to induce damage in the CNS. It also appears capable of inducing brain cell apoptosis (for a discussion of apoptosis, see chapter 14).

The bacterium is a small gram-positive flattened coccus that appears in end-to-end pairs. It has a distinctive appearance in a Gram stain of cerebrospinal fluid. Staining or culturing the nasopharynx is not useful since it is often normal flora there. It is also α-hemolytic on blood agar. Treatment requires a drug to which the bacterium is not resistant; penicillin is therefore not a good choice. Ceftriaxone is often used, but drug susceptibilities must always be tested.

As mentioned in chapter 21, two vaccines are available for *S. pneumoniae*: a seven-valent conjugated vaccine (Prevnar), which is now recommended as part of the childhood immunization schedule, and a 23-valent polysaccharide vaccine (Pneumovax) is available for adults.

Haemophilus influenzae

This bacterium was originally named when it was isolated from patients with "flu" about 100 years ago. For over 40 years, it was erroneously proclaimed the causative agent until the real agent, the influenza virus, was discovered. This species eventually was shown to be an agent of acute bacterial meningitis in humans. *Haemophilus* cells are tiny (0.5×0.8 mm) gram-negative pleomorphic rods sometimes confused with the genus *Neisseria* in clinical samples. The members of this group tend to be fastidious and are sensitive to drying, temperature extremes, and disinfectants. Even though their name means blood-loving, none of these organisms can be grown on blood agar alone without special techniques. They require certain factors from blood, namely Factor X (hemin), a necessary component of cytochromes, catalase, and peroxidase, and Factor V, nicotinamide adenine dinucleotide (NAD or NADP), an important coenzyme. Media such as chocolate agar (a form of cooked blood agar) and Fildes medium provide these factors.

The meningitis caused by this bacterium is severe. The disease is caused primarily by the b serotype and was once most common in children between 3 months and 5 years of age. The case rates have declined because of an increased emphasis on vaccination. In contrast to meningococcal meningitis, *Haemophilus* meningitis is not associated with epidemics in the general population, but tends to occur as sporadic outbreaks in day care and family settings. It is transmitted by close contact and nose and throat discharges. Healthy adult carriers are the usual reservoirs of the bacterium.

Haemophilus meningitis is very similar to meningococcal meningitis, with symptoms of fever, vomiting, stiff neck, and neurological impairment. Untreated cases have a fatality rate of nearly 90%, and even with prompt diagnosis and aggressive treatment, about 33% of children sustain residual damage.

Haemophilus infections are usually treated with a ceftriaxone. Outbreaks of disease in families and day care centers may necessitate rifampin prophylaxis for all contacts. Routine vaccination with a subunit vaccine (Hib containing capsular polysaccharide conjugated to a protein is recommended for all children, beginning at age 2 months, with three follow-up boosters).

Listeria monocytogenes

Listeria monocytogenes is a gram-positive bacterium that ranges in morphology from coccobacilli to long filaments in palisades formation **(figure 19.6).** Cells do not produce capsules or spores and have from one to four flagella. *Listeria* is not fastidious and is resistant to cold, heat, salt, pH extremes, and bile. It grows inside host cells.

Listeriosis in normal adults is often a mild or subclinical infection with nonspecific symptoms of fever, diarrhea, and sore throat. However, listeriosis in elderly or imunocompromised patients, fetuses, and neonates (described later) usually affects the brain and meninges and results in septicemia. The death rate is around 20%. Pregnant women are especially

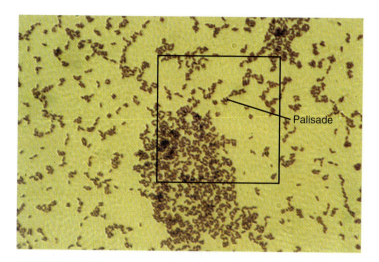

FIGURE 19.6 *Listeria monocytogenes.*
The bacterium is generally rod shaped. In Gram stains individual cells tend to stack up in structures called palisades.

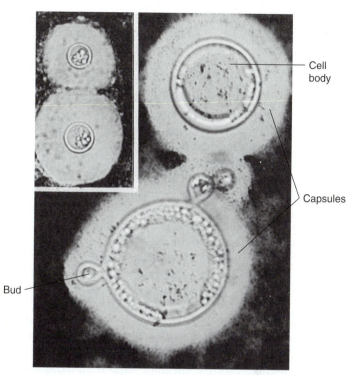

FIGURE 19.7 *Cryptococcus neoformans* **from infected spinal fluid stained negatively with India ink.**
Halos around the large spherical yeast cells are thick capsules. Also note the buds forming on one cell. Encapsulation is a useful diagnostic sign for cryptococcosis, although the capsule is fragile and may not show up in some preparations (150×).

susceptible to infection, which can be transmitted to the infant prenatally when the microbe crosses the placenta or postnatally through the birth canal. Intrauterine infections are widely systemic and usually result in premature abortion and fetal death.

The distribution of *L. monocytogenes* is so broad that its reservoir has been difficult to determine. It has been isolated all over the world from water, soil, plant materials, and the intestines of healthy mammals (including humans), birds, fish, and invertebrates. Apparently, the primary reservoir is soil and water, and animals, plants, and food are secondary sources of infection. Most cases of listeriosis are associated with ingesting contaminated dairy products, poultry, and meat. Recent epidemics have spurred an in-depth investigation into the prevalence of *L. monocytogenes* in these sources. A 2003 U.S. government report concluded that consumers are exposed to low to moderate levels of *L. monocytogenes* on a regular basis. The pathogen has been isolated in 10% to 15% of ground beef and in 25% to 30% of chicken and turkey carcasses, and is also present in 5% to 10% of luncheon meats, hot dogs, and cheeses. Aged cheeses made from raw milk are of special concern because *Listeria* readily survives long storage and can grow during refrigeration.

In late 2002, *Listeria* contamination of a poultry processing plant in Pennsylvania led to the recall of 27.4 million pounds of processed chicken and turkey, the largest meat recall in U.S. history.

Except in cases of pregnancy, human-to-human transmission is probably not a significant factor. A predisposing factor in listeriosis seems to be the weakened condition of host defenses in the intestinal mucosa, since studies have shown that immunocompetent individuals are somewhat resistant to infection.

Diagnosing listeriosis is hampered by the difficulty in isolating it. The chances of isolation, however, can be improved by using a procedure called *cold enrichment,* in which the specimen is held at 4°C and periodically plated onto media, but this procedure can take 4 weeks. Rapid diagnostic kits using ELISA, immunofluorescence, and gene probe technology are now available for direct testing of dairy products and cultures. Antibiotic therapy should be started as soon as listeriosis is suspected. Ampicillin and trimethoprim-sulfamethoxazole are the first choices, followed by erythromycin. Prevention can be improved by adequate pasteurization temperatures and by proper washing, refrigeration, and cooking of foods that are suspected of being contaminated with animal manure or sewage. Pregnant women are cautioned by the U.S. Food and Drug Administration not to eat soft, unpasteurized cheeses.

Cryptococcus neoformans

The fungus *Cryptococcus neoformans* causes a more chronic form of meningitis with a more gradual onset of symptoms, although in AIDS patients the onset may be fast and the course of the disease more acute. It is sometimes classified as a meningoencephalitis. Headache is the most common symptom, but nausea and neck stiffness are very common. This fungus is a widespread resident of human habitats. It has a spherical to ovoid shape, with small, constricted buds and a large capsule that is important in its pathogenesis **(figure 19.7).**

Transmission and Epidemiology   The primary ecological niche of *C. neoformans* is the bird population. It is prevalent in urban areas where pigeons congregate, and it proliferates in the high-nitrogen environment of droppings that accumulate on pigeon roosts. Masses of dried yeast cells are readily scattered into the air and dust. Its role as an opportunist is supported by evidence that healthy humans have strong resistance to it and that frank infection occurs primarily in debilitated patients. Most cryptococcal infections cause symptoms in the respiratory and central nervous systems.

By far the highest rates of cryptococcal meningitis occur among patients with AIDS. This meningitis is frequently fatal. Other conditions that predispose individuals to infection are steroid treatment, diabetes, and cancer. It is not considered communicable among humans.

The primary portal of entry for *C. neoformans* is the respiratory tract, but most lung infections are subclinical and rapidly resolved.

Pathogenesis and Virulence Factors   The escape of the yeasts into the blood is intensified by weakened host defenses and results in severe complication. *Cryptococcus* shows an extreme affinity for the meninges and brain. The tumor-like masses formed in these locations can cause headache, mental changes, coma, paralysis, eye disturbances, and seizures. In some cases the infection disseminates into the skin, bones, and viscera **(figure 19.8).**

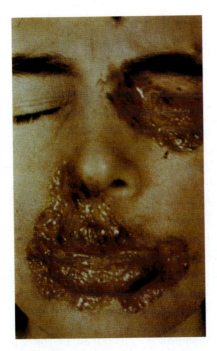

FIGURE 19.8   Cryptococcosis.
A late disseminated case of cutaneous cryptococcosis in which fungal growth produces a gelatinous exudate. The texture is due to the capsules surrounding the yeast cells.

Culture and/or Diagnosis   The first step in diagnosis of cryptococcosis is negative staining of specimens to detect encapsulated budding yeast cells that do not occur as pseudohyphae. Isolated colonies can be used to perform screening tests that presumptively differentiate *C. neoformans* from other cryptococcal species. Confirmatory results include a negative nitrate assimilation, pigmentation on bird-seed agar, and fluorescent antibody tests. Cryptococcal antigen can be detected in a specimen by means of serological tests, and DNA probes can make a positive genetic identification.

Prevention and Treatment   Systemic cryptococcosis requires immediate treatment with amphotericin B and fluconazole over a period of weeks or months. There is no prevention.

Coccidioides immitis

Although this fungus has probably lived in soil for millions of years, human encounters with it are relatively recent and coincide with increased exposure to soil and encroachment of humans into its habitat, through, for example, expanded agricultural practices. The morphology of *Coccidioides immitis* is very distinctive. At 25°C, it forms a moist white to brown colony with abundant, branching, septate hyphae. These hyphae fragment into thick-walled, blocklike **arthroconidia** (arthrospores) at maturity **(figure 19.9).** On special media incubated at 37°C to 40°C, an arthrospore germinates into the parasitic phase, a small, spherical cell called a spherule, which can be found in infected tissues as well. This structure swells into a giant sporangium that cleaves internally to form numerous endospores that look like bacterial endospores but lack their resistance traits.

Pathogenesis and Virulence Factors   This is a true systemic fungal infection of high virulence, as opposed to an opportunistic infection. It usually begins with pulmonary infection but can disseminate quickly throughout the body. Coccidioidomycosis of the meninges is the most serious manifestation. All persons inhaling the arthrospores probably develop some degree of infection, but certain groups have a genetic susceptibility that gives rise to more serious disease.

Transmission and Epidemiology   *C. immitis* occurs endemically in various natural reservoirs and casually in areas where it has been carried by wind and animals. Conditions favoring its settlement include high carbon and salt content and a semiarid, relatively hot climate. The fungus has been isolated from soils, plants, and a large number of vertebrates. The natural history of *C. immitis* follows a cyclic pattern—a period of dormancy in winter and spring, followed by growth in summer and fall. Growth and spread are greatly increased by cycles of drought and heavy rains, followed by rainstorms.

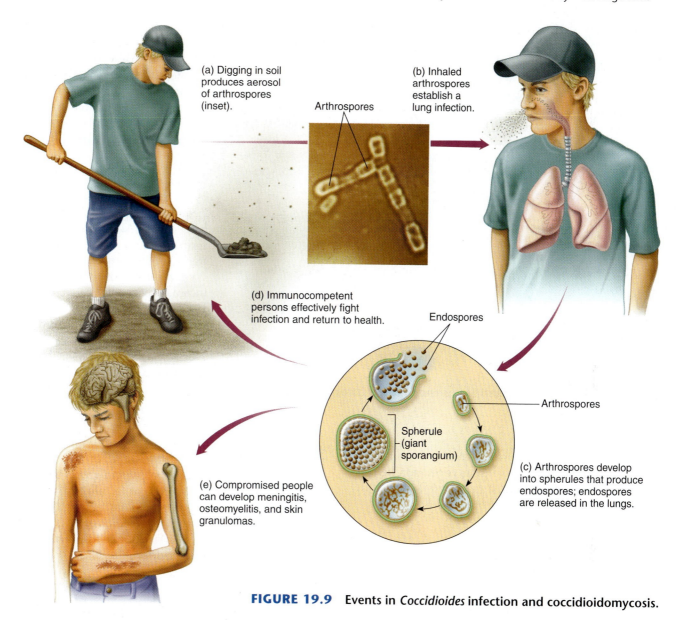

(a) Digging in soil produces aerosol of arthrospores (inset).

(b) Inhaled arthrospores establish a lung infection.

Arthrospores

(d) Immunocompetent persons effectively fight infection and return to health.

Endospores

Arthrospores

Spherule (giant sporangium)

(c) Arthrospores develop into spherules that produce endospores; endospores are released in the lungs.

(e) Compromised people can develop meningitis, osteomyelitis, and skin granulomas.

FIGURE 19.9   Events in *Coccidioides* infection and coccidioidomycosis.

Skin testing has disclosed that the highest incidence of coccidiomycosis, estimated at 100,000 cases per year, occurs in the southwestern United States **(figure 19.10),** although it also occurs in Mexico and parts of Central and South America. Especially concentrated reservoirs exist in the San Joaquin Valley of California and in southern Arizona. Outbreaks are usually associated with farming activity, archeological digs, construction, and mining. A highly unusual outbreak of coccidiomycosis was traced to the Northridge, California earthquake. Clouds of dust bearing loosened spores were given off by landslides, and local winds then carried the dust into the outlying residential areas.

Culture and Diagnosis   Diagnosis of coccidioidomycosis is straightforward when the highly distinctive spherules are found in sputum, spinal fluid, and biopsies. This finding is further supported by isolation of typical mycelia and arthrospores on Sabouraud's agar. Newer specific antigen tests have been effective tools to identify and differentiate *Coccidioides* from other fungi. All cultures must be grown in closed tubes or bottles and opened in a biological containment hood to prevent laboratory infections. Immunodiffusion and latex agglutination tests on serum samples are excellent screens for detecting early infection. Skin tests using an extract of the fungi are of primary importance in epidemiological studies.

Prevention and Treatment   The majority of patients do not require treatment. In people with disseminated disease, however, amphotericin B is administered intravenously; alternatively, oral fluconazole is used. Minimizing contact with

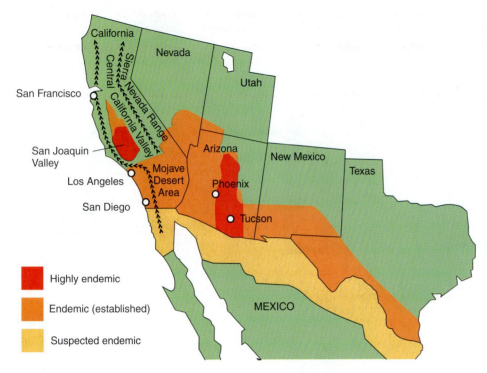

FIGURE 19.10   Areas in the United States endemic for *Coccidioides immitis.*

Highly endemic

Endemic (established)

Suspected endemic

the fungus in its natural habitat has been of some value. For example, oiling dirt roads and planting vegetation help reduce spore aerosols, and using dust masks while excavating soil prevents workers from inhaling spores.

Viruses

A wide variety of viruses can cause meningitis. Because no bacteria or fungi are found in the CSF in viral meningitis, the condition is often called *aseptic meningitis,* although the microorganisms previously described can occasionally cause aseptic meningitis as well when no microbes can be found in the CSF. This may happen when bacterial or fungal meningitis is incompletely treated with antimicrobials; aseptic meningitis may also have noninfectious causes.

By far the majority of cases of viral meningitis occur in children, and 90% are caused by enteroviruses. But many other viruses also gain access to the central nervous system on occasion. An initial infection with herpes simplex type 2 is sometimes known to cause meningitis. Also, other herpesviruses such as HHV-6 and HHV-7, and also HHV-3 (the chickenpox virus) and cytomegalovirus (CMV) can infect the meninges, resulting in symptoms. Arboviruses, arenaviruses, and adenoviruses have also been found in meningitis cases. Finally, HIV infection may manifest as meningitis.

Viral meningitis is generally milder than bacterial or fungal meningitis, and it is usually resolved within 2 weeks. The mortality rate is less than 1%. Diagnosis begins with the failure to find bacteria, fungi, or protozoa in CSF, and can be confirmed, depending on the virus, by viral culture or specific antigen tests. In most cases no treatment is indicated. Acy-

clovir can be used when the causative agent is a herpesvirus, and, of course, if HIV is the cause the entire HIV antiviral regimen is called for (HIV is discussed in chapter 20). **Checkpoint 19.1** summarizes the agents causing meningitis.

Neonatal Meningitis

Meningitis in newborns is almost always a result of infection transmitted by the mother, either in utero or (more frequently) during passage through the birth canal. (Although **Insight 19.1** describes a troubling exception to this trend.) As more premature babies survive, the rates of neonatal meningitis increase, because the condition is favored in patients with immature immune systems. The two most common causes are *Streptococcus agalactiae* and *Escherichia coli. Listeria monocytogenes* is also found frequently in neonates. It has already been covered here, but is included in Checkpoint 19.2 as a reminder that it can cause neonatal cases as well.

Streptococcus agalactiae

This species of *Streptococcus* belongs to group B of the streptococci. It colonizes 10% to 30% of female genital tracts, and is the most frequent cause of neonatal meningitis (for details about this condition in women, see chapter 23). The treatment for neonatal disease is penicillin G, sometimes supplemented with an aminoglycoside.

Escherichia coli

The K1 strain of *Escherichia coli* is the second most common cause of neonatal meningitis. Most babies who suffer from this infection are premature, and their prognosis is poor. Twenty percent of them die, even with aggressive antibiotic treatment, and those who survive often have brain damage.

The bacterium is usually transmitted from the mother's birth canal. It causes no disease in the mothers but can infect the vulnerable tissues of a neonate. It seems to have a predilection for the tissues of the central nervous system. Ceftriaxone is usually administered intravenously, in combination with aminoglycosides **(Checkpoint 19.2).**

Meningoencephalitis

Up to this point, we have described microorganisms causing meningitis (inflammation of the meninges). Next we will discuss microorganisms that cause **encephalitis,** inflammation of the brain. Because the brain and the spinal cord (and the meninges) are so closely connected, infections of one of these structures may also involve the other.

✔ CHECKPOINT 19.1 — Meningitis

Causative Organism(s)	*Neisseria meningitidis*	*Streptococcus pneumoniae*	*Haemophilus influenzae*	*Listeria monocytogenes*	*Cryptococcus neoformans*	*Coccidioides immitis*	Viruses
Most Common Modes of Transmission	Droplet contact	Droplet contact	Droplet contact	Vehicle (food)	Vehicle (air, dust)	Vehicle (air, dust, soil)	Droplet contact
Virulence Factors	Capsule, endotoxin, IgA protease	Capsule, induction of apoptosis, hemolysin and hydrogen peroxide production	Capsule	Intracellular growth	Capsule, melanin production	Granuloma (spherule) formation	Lytic infection of host cells
Culture/ Diagnosis	Gram stain/ culture of CSF, blood, rapid antigenic tests	Gram stain/ culture of CSF	Culture on chocolate agar	Cold enrichment, rapid methods	Negative staining, biochemical tests, DNA probes	Identification of spherules, cultivation on Sabouraud's agar	Initially, absence of bacteria/ fungi/ protozoa, followed by viral culture or antigen tests
Prevention	Vaccine used in some situations; rifampin or tetracycline used to protect contacts	Two vaccines: Prevnar (children), and Pneumovax (adults)	Hib vaccine	Cooking food, avoiding unpasteurized dairy products	–	Avoiding airborne spores	–
Treatment	Penicillin G or third-generation cephalosporins	Ceftriaxone— check for resistance	Ceftriaxone	Ampicillin, trimethoprim-sulfamethoxazole	Amphotericin B and fluconazole	Amphotericin B or oral fluconazole	Usually none unless specific virus identified and specific antiviral exists
Distinctive Features	Petechiae, meningococcemia	Serious, acute, most common meningitis in adults	Serious, acute, less common since vaccine became available	Asymptomatic in healthy adults, meningitis in neonates, elderly and immunocompromised	Acute or chronic, most common in AIDS patients	Almost exclusively in endemic regions	Generally milder than bacterial or fungal

✔ CHECKPOINT 19.2 — Neonatal Meningitis

Causative Organism(s)	*Streptococcus agalactiae*	*Escherichia coli*, strain K1	*Listeria monocytogenes*
Most Common Modes of Transmission	Vertical (during birth)	Vertical (during birth)	Vertical
Virulence Factors	Capsule	–	Intracellular growth
Culture/Diagnosis	Culture mother's genital tract on blood agar; CSF culture of neonate	CSF Gram stain/culture	Cold enrichment, rapid methods
Prevention	Culture and treatment of mother	–	Cooking food, avoiding unpasteurized dairy products
Treatment	Penicillin G plus aminoglycosides	Ceftriaxone plus aminoglycoside	Ampicillin, trimethoprim-sulfamethoxazole
Distinctive Features	Most common; positive culture of mother confirms diagnosis	Suspected if infant is premature	Suspected if infant is premature

INSIGHT 19.1   *Discovery*

Baby Food and Meningitis

It will come as no surprise to you that there are potentially dangerous bacteria in a wide variety of foods we consume. Cases of *E. coli* O157:H7 disease associated with hamburgers and hepatitis contracted by customers of a Mexican restaurant get a lot of media attention. It may surprise you to learn that pathogenic bacteria are found in dried infant formula and dried baby food, as well.

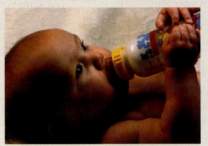

In 2001 an outbreak of meningitis in a neonatal intensive care unit in Tennessee was traced to a batch of powdered formula. The manufacturer recalled the product after the Centers for Disease Control and Prevention issued a warning. In 2004 scientists in England investigated 110 different types of baby foods. Ten percent of the powdered formula samples and 25% of the dried infant foods contained intestinal bacteria. In many of the samples they found a bacterium called *Enterobacter sakazakii*, an intestinal bacterium that has been linked to several fatal outbreaks of meningitis in children's hospitals.

The disease is rare but almost always associated with infant foods, and has a 33% fatality rate, with up to 80% of infants suffering permanent neurological damage. So what is to be done? In this case, it is "consumer beware." Manufacturers have never claimed that their formulas and foods are sterile. The scientists who conducted the infant food study also investigated ideal conditions for preparing and storing the products. They noted that the bacterial doubling time in prepared formula is 10 hours when refrigerated, and 30 minutes at room temperature. This means that leaving prepared formula in your diaper bag or on the kitchen counter for even a few hours could lead to high levels of bacteria in the bottle.

Powdered formula is made by manufacturing the nutritious liquid and then freeze-drying it. It is sterile as a liquid but bacteria can be introduced during the freeze-drying and packaging phases. Since the outbreaks, the FDA has recommended that the powder be reconstituted with boiling water. The CDC has not supported this recommendation because of many problems with it including the risk of destroying important nutrients and the lack of data that boiling it would be sufficient to kill *E. sakazakii*. Hospitals are advised to use ready-to-feed or concentrated liquid formulas.

But two microorganisms cause a distinct disease called *meningoencephalitis*, and they are both amoebas. *Naegleria fowleri* and *Acanthamoeba* are accidental parasites that invade the body only under unusual circumstances.

Naegleria fowleri

The trophozoite of *Naegleria* is a small, flask-shaped amoeba that moves by means of a single, broad pseudopod **(figure 19.11)**. It forms a rounded, thick-walled, uninucleate cyst that is resistant to temperature extremes and mild chlorination.

Most cases of *Naegleria* infection reported worldwide occur in people who have been swimming in warm, natural bodies of fresh water. One epidemic in Australia was due to contamination of a public water supply, and in Belgium, an outbreak was reported among people who had bathed in polluted canal water. Infection can begin when amoebas are forced into human nasal passages as a result of swimming, diving, or other aquatic activities. Once the amoeba is inoculated into the favorable habitat of the nasal mucosa, it burrows in, multiplies, and subsequently migrates into the brain and surrounding structures. The result is primary amoebic meningoencephalitis (PAM), a rapid, massive destruction of brain and spinal tissue that causes hemorrhage and coma and invariably ends in death within a week or so. We should note that this organism is very common—children often carry the amoeba as harmless flora, especially during the

FIGURE 19.11   **Scanning electron micrograph of** *Naegleria fowleri.*
The "eyes" and "mouth" of its facelike appearance are its attachment and feeding structures.

✔ CHECKPOINT 19.3   Meningoencephalitis

	Primary Amoebic Meningoencephalitis	Granulomatous Amoebic Meningoencephalitis
Causative Organism(s)	*Naegleria fowleri*	*Acanthamoeba*
Most Common Modes of Transmission	Vehicle (exposure while swimming in water)	Direct contact
Virulence Factors	Invasiveness	Invasiveness
Culture/Diagnosis	Examination of CSF; brain imaging, biopsy	Examination of CSF; brain imaging, biopsy
Prevention	Avoid warm fresh water	–
Treatment	Mostly ineffective	Surgical excision of granulomas; amphotericin may help

summer months, and the series of events leading to disease is exceedingly rare.

Unfortunately, *Naegleria* meningoencephalitis advances so rapidly that treatment usually proves futile. Studies have indicated that early therapy with amphotericin B, sulfadiazine, or tetracycline in some combination can be of some benefit. Because of the wide distribution of the amoeba and its hardiness, no general means of control exists. Public swimming pools and baths must be adequately chlorinated and checked periodically for the amoeba.

Acanthamoeba

This protozoan has a large, amoeboid trophozoite with spiny pseudopods and a double-walled cyst. It differs from *Naegleria* in its portal of entry; it invades broken skin, the conjunctiva, and occasionally the lungs and urogenital epithelia. Although it causes a meningoencephalitis somewhat similar to that of *Naegleria,* the course of infection is lengthier. The disease is called granulomatous amoebic meningoencephalitis (GAM). At special risk for infection are people with traumatic eye injuries, contact lens wearers, and AIDS patients exposed to contaminated water. Ocular infections can be avoided by carefully tending to injured eyes and by using sterile solutions to store and clean contact lenses. Cutaneous and CNS infections with this organism are occasional complications in AIDS **(Checkpoint 19.3).**

Acute Encephalitis

Encephalitis can present as acute or **subacute.** It is always a serious condition, as the tissues of the brain are extremely sensitive to damage by inflammatory processes. Acute encephalitis is almost always caused by viral infection. One category of viral encephalitis is caused by viruses borne by insects (arboviruses), including West Nile virus. Alternatively, other viruses, such as members of the herpes family, are causative agents. Bacteria such as those covered under meningitis can also cause encephalitis, but the symptoms are almost always more pronounced in the meninges than in the brain.

The signs and symptoms of encephalitis vary, but they may include behavior changes or confusion because of

inflammation. Decreased consciousness and seizures frequently occur. Symptoms of meningitis are often also present. Few of these agents have specific treatments, but because swift initiation of acyclovir therapy can save the life of a patient suffering from herpesvirus encephalitis, most physicians will begin empiric therapy with acyclovir in all seriously ill neonates and most other patients showing evidence of encephalitis. Treatment will, in any case, do no harm in patients who are infected with other agents.

Arboviruses

Wherever there are arthropods, there are also arboviruses, so collectively, their distribution is worldwide. The vectors and viruses tend to be clustered in the tropics and subtropics, but many temperate zones report periodic epidemics. A given arbovirus type may have very restricted distribution, even to a single isolated region, but some types range over several continents, and others can spread along with their vectors **(figure 19.12).**

Most arthropods that serve as infectious disease vectors feed on the blood of hosts, a process that infects them for varying time periods. Infections show a peak incidence when the arthropod is actively feeding and reproducing, usually from late spring through early fall. Warm-blooded vertebrates also maintain the virus during the cold and dry seasons. Humans can serve as dead-end, accidental hosts, as in equine encephalitis, or they can be a maintenance reservoir, as in yellow fever (discussed in chapter 20).

Arboviral diseases have a great impact on humans. Although exact statistics are unavailable, it is believed that millions of people acquire infections each year and thousands of them die. One common outcome of arboviral infection is an acute fever, often accompanied by rash. Viruses that primarily cause these symptoms are covered in chapter 20.

The arboviruses discussed in this chapter can cause encephalitis, and we will consider them as a group because the symptoms and management are similar. The transmission and epidemiology of individual viruses are different, however, and are discussed for each virus. **Insight 19.2** discusses West Nile virus, an arbovirus that has spread across North America in recent years.

INSIGHT 19.2   *Medical*

A Long Way from Egypt: West Nile Virus in the United States

In 1999 the first cases of West Nile encephalitis were seen in several northeastern states. Over the next few years, West Nile virus spread westward across the country, resulting in at least 4,000 confirmed infections and over 250 deaths by early 2003. In the summer of 2003, the Centers for Disease Control and Prevention reported that they had detected the virus in 600 blood donors in the United States. Government officials were made aware of the need to test for the virus after 23 patients acquired West Nile from blood transfusions in 2002.

The arrival of a deadly new disease, especially at a time of public nervousness concerning potential biological warfare, led to a great deal of media attention concerning West Nile virus, much of it sensational, even if not entirely accurate.

West Nile virus is an arbovirus commonly found in Africa, the Middle East, and parts of Asia, but until mid-1999 had not been detected in the Americas. The virus is known to infect a host of mammals (including humans), as well as birds and mosquitoes. Mosquitoes generally become infected when they feed on birds infected with the virus, and they can then bite and transmit the

virus to humans. If infection results, the illness is generally characterized by flulike symptoms that last just a few days and have no long-term consequences. Less than 1% of infected persons will suffer the potentially lethal inflammation of the brain known as West Nile encephalitis.

Because transmission of the virus depends on the presence of mosquitoes to act as vectors, viral control is synonymous with vector control. Insect repellant, long-sleeved shirts and long pants, and control of mosquito breeding grounds (primarily stagnant water) all decrease the spread of the virus. Because the virus is blood-borne, person-to-person transmission is unlikely. There has been one confirmed case, however, in which the virus was transmitted from mother to fetus, as well as another case in which an infected organ donor passed on the virus to four organ recipients.

Epidemiologists are also closely monitoring the magnitude of the disease in animals. So far, the virus has been isolated in nearly 200 bird species as well as cats, dogs, rodents, horses, and even alligators.

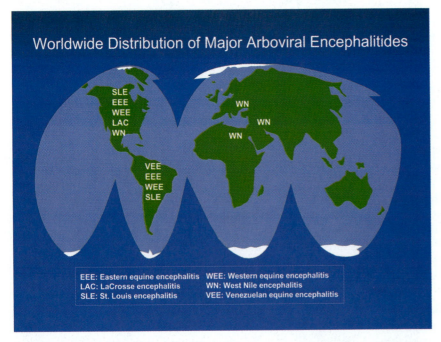

FIGURE 19.12   Worldwide distribution of major arboviral encephalitides.

memory deficits, changes in speech and personality, and heart disorders. In some cases, survivors experience some degree of permanent brain damage. Young children and the elderly are most sensitive to injury by arboviral encephalitis.

The virulence of these viruses is not well understood, but much research has focused on proteins that the virus uses to attach to host tissues, or to induce fusion with host cell membranes. Both of these functions facilitate invasion of the virus.

Culture and Diagnosis   Except during epidemics, detecting arboviral infections can be difficult. The patient's history of travel to endemic areas or contact with vectors, along with serum analysis, are highly supportive of a diagnosis. Rapid serological tests are available for some of the viruses.

Prevention and Treatment   No satisfactory treatment exists for any of the arboviral encephalitides (plural of *encephalitis*). As mentioned earlier, empiric acyclovir treatment may be begun in case the infection is actually caused by either herpes simplex virus or varicella zoster. Treatment of the other infections relies entirely on support measures to control fever, convulsions, dehydration, shock, and edema.

Pathogenesis and Virulence Factors   Arboviral encephalitis begins with an arthropod bite, the release of the virus into tissues, and its replication in nearby lymphatic tissues. Prolonged viremia establishes the virus in the brain, where inflammation can cause swelling and damage to the brain, nerves, and meninges. Symptoms are extremely variable and can include coma, convulsions, paralysis, tremor, loss of coordination,

Most of the control safeguards for arbovirus disease are aimed at the arthropod vectors. Mosquito abatement by eliminating breeding sites and by broadcasting insecticides has been highly effective in restricted urban settings. Birds play a role as reservoirs of the virus, but direct transmission between birds and humans does not occur.

Live attenuated vaccines are available for Western equine encephalitis and Eastern equine encephalitis and are administered to laboratory workers, veterinarians, ranchers, and horses.

Western Equine Encephalitis (WEE) This disease occurs sporadically in the western United States and Canada, appearing first in horses and later in humans. The mosquito that carries the virus emerges in the early summer when irrigation begins in rural areas and breeding sites are abundant. The disease is extremely dangerous to infants and small children, with a case fatality rate of 3% to 7%.

Eastern Equine Encephalitis (EEE) EEE is endemic to an area along the eastern coast of North America and Canada. The usual pattern is sporadic, but occasional epidemics can occur in humans and horses. High periods of rainfall in the late summer increase the chance of an outbreak, and disease usually appears first in horses and caged birds. The case fatality rate can be very high (70%).

California Encephalitis This condition may be caused by two different viral strains. The California strain occurs occasionally in the western United States and has little impact on humans. The LaCrosse strain is widely distributed in the eastern United States and Canada and is a prevalent cause of viral encephalitis in North America. Children living in rural areas are the primary target group, and most of them exhibit mild, transient symptoms. Fatalities are rare.

St. Louis Encephalitis (SLE) St. Louis encephalitis may be the most common of all American viral encephalitides. Cases appear throughout North and South America, but epidemics in the United States occur most often in the Midwest and South. Inapparent infection is very common, and the total number of cases is probably thousands of times greater than the 50 to 100 reported each year. The seasons of peak activity are spring and summer, depending on the region and species of mosquito. In the eastern United States, mosquitoes breed in stagnant or polluted water in urban and suburban areas during the summer. In the West, mosquitoes frequent spring floodwaters in rural areas.

West Nile Encephalitis The West Nile virus is a close relative of the SLE virus. It emerged in the United States in 1999, and by the beginning of 2003 more than 4,000 people had been infected. See Insight 19.2 for details.

IN THE NEWS *(Continued from page 577)*

The encephalitis victims in Queens suffered from West Nile virus infection. It was not initially suspected because until that time, it had never been seen in North America. And most arboviruses are confined to specific geographical areas because their insect vectors are confined to those areas. West Nile virus had been seen in France in 1962 and in Romania in 1996, but never in North America before 1999.

The virus has exhibited other changes besides its geographical distribution. In its "native" locale of Africa and the Middle East, West Nile virus doesn't seem to cause overt disease in birds, although it infects large percentages of them. Something has changed in the virus that we are seeing in the United States, since it is proving lethal to birds as well as to some of the humans it infects.

See: CDC. 1999. Outbreak of West Nile-like viral encephalitis—New York, 1999. MMWR 48:890–892.

Herpes Simplex Virus

Herpes simplex type 1 and 2 viruses can cause encephalitis in newborns born to HSV-positive mothers. In this case the virus is disseminated and the prognosis is poor. Older children and young adults (ages 5 to 30), as well as older adults (over 50 years old) are also susceptible to herpes simplex encephalitis caused most commonly by HSV-1. In these cases the HSV encephalitis represents a reactivation of dormant HSV from the trigeminal ganglion.

It should be noted the varicella-zoster virus (see chapter 18) can also reactivate from the dormant state, and it is responsible for rare cases of encephalitis.

JC Virus

The **JC virus (JCV)** gets its name from the initials of the patient in whom it was first diagnosed as the cause of illness. Serological studies indicate that infection with this polyoma virus is commonplace. In patients with immune dysfunction, especially in those with AIDS, it can cause a condition called **progressive multifocal leukoencephalopathy** (loo"-koh-en-sef"uh-lop'-uh-thee) **(PML).** This uncommon but generally fatal infection is a result of JC virus attack of accessory brain cells. The infection demyelinizes certain parts of the cerebrum. This virus should be considered when encephalitis symptoms are observed in AIDS patients. Recently a few deaths from this condition have been prevented with high doses of zidovudine.

Other Virus-Associated Encephalitides

Infection with measles and other childhood rash diseases can result, 1 to 2 weeks later, in an inappropriate immune response with consequences in the CNS. The condition is called postinfection encephalitis (PIE), and it is thought to be a result of immune system action and not of direct viral invasion of neural tissue. Very rarely PIE can occur after immunization with live attenuated vaccines against viral infections. Note that PIE is distinct from another possible sequela of measles virus infection called SSPE (discussed later in this chapter) **(Checkpoint 19.4).**

Subacute Encephalitis

When encephalitis symptoms take longer to show up, and when the symptoms are less striking, the condition is termed **subacute encephalitis.** The most common cause of subacute encephalitis is the protozoan *Toxoplasma.* Another form of

✓ CHECKPOINT 19.4   Encephalitis

Causative Organism(s)	Arboviruses (viruses causing WEE, EEE, California encephalitis, SLE, West Nile encephalitis)	Herpes simplex 1 or 2	JC virus	Immunological reaction to other viral infections
Most Common Modes of Transmission	Vector (arthropod bites)	Vertical or reactivation of latent infection	? Ubiquitous	Sequelae of measles, other viral infections and occasionally, vaccination
Virulence Factors	Attachment, fusion, invasion capabilities	–	–	–
Culture/Diagnosis	History, rapid serological tests	Clinical presentation, PCR, Ab tests, growth of virus in cell culture	PCR of cerebrospinal fluid	History of viral infection or vaccination
Prevention	Insect control, vaccines for WEE and EEE available	Maternal screening for HSV	None	–
Treatment	None	Acyclovir	Zidovudine or other antivirals	Steroids, anti-inflammatory agents
Distinctive Features	History of exposure to insect important	In infants, disseminated disease present; rare between 30–50 yrs	In severely immunocompromised, especially AIDS	History of virus/vaccine exposure critical

subacute encephalitis can be caused by persistent measles virus as many as 7 to 15 years after the initial infection. Finally, a class of infectious agent known as prions can cause a condition called spongiform encephalopathy.

Toxoplasma gondii

Toxoplasma gondii is a flagellated parasite with such extensive cosmopolitan distribution that some experts estimate it affects the majority of the world's population at some time in their lives. In most of these cases, toxoplasmosis goes unnoticed, but disease in the fetus and in immunodeficient people, especially those with AIDS, is severe and often fatal. *T. gondii* is a very successful parasite with so little host specificity that it can attack at least 200 species of birds and mammals. However, its primary reservoir and hosts are members of the feline family, both domestic and wild.

Signs and Symptoms   As just mentioned, most cases of toxoplasmosis are asymptomatic or marked by mild symptoms such as sore throat, lymph node enlargement, and low-grade fever. In patients whose immunity is suppressed by infection, cancer, or drugs, the outlook may be grim. The infection causes a more chronic or subacute form of encephalitis than do most viruses, often producing extensive brain lesions and fatal disruptions of the heart and lungs. A pregnant woman with toxoplasmosis has a 33% chance of transmitting the infection to her fetus. Congenital infection occurring in the first or second trimester is associated with stillbirth and severe abnormalities such as liver and spleen enlargement, liver failure, hydrocephalus, convulsions, and damage to the retina that can result in blindness.

Pathogenesis and Virulence Factors   Little is known about the actual mechanism of pathogenesis for *Toxoplasma*, but it is an obligate intracellular parasite, making its ability to invade host cells an important factor for virulence.

Transmission and Epidemiology   To follow the transmission of toxoplasmosis, we must first look at the general stages of the *Toxoplasma* life cycle in the cat (**figure 19.13a**). The parasite undergoes a sexual phase in the intestine and is then released in feces, where it becomes an infective *oocyst* that survives in moist soil for several months. Ingested oocysts release an invasive asexual tissue phase called a *tachyzoite* that infects many different tissues and often causes disease in the cat. These forms eventually enter an asexual cyst state in tissues, called a *pseudocyst*. Most of the time, the parasite does not cycle in cats alone and is spread by oocysts to intermediate hosts, usually rodents and birds. The cycle returns to cats when they eat these infected prey animals.

Other vertebrates become a part of this transmission cycle (**figure 19.13b**). Herbivorous animals such as cattle and sheep ingest oocysts that persist in the soil of grazing areas and then develop pseudocysts in their muscles and other organs. Carnivores such as canines are infected by eating pseudocysts in the tissues of carrier animals.

Humans appear to be constantly exposed to the pathogen. The rate of prior infections, as detected through serological tests, can be as high as 90% in some populations. Many cases are caused by ingesting pseudocysts in contaminated meats. A common source is raw or undercooked meat. The grooming habits of cats spread fecal oocysts on their body surfaces, and unhygienic handling of them presents an

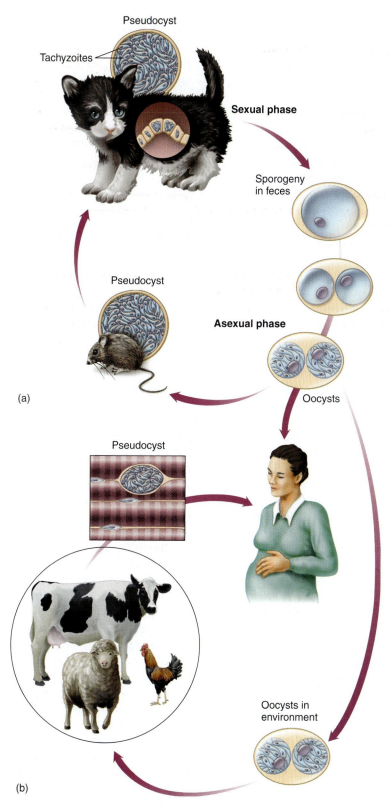

(a)

(b)

FIGURE 19.13 **The life cycle and morphological forms of** *Toxoplasma gondii.*
(a) The cycle in cats and their prey. **(b)** The cycle in other animal hosts. The zoonosis has a large animal reservoir (domestic and wild) that becomes infected through contact with oocysts in the soil. Humans can be infected through contact with cats or ingestion of pseudocysts in animal flesh. Infection in pregnant women is a serious complication because of the potential damage to the fetus.

opportunity to ingest oocysts. Infection can also occur when oocysts are inhaled in air or dust contaminated with cat droppings and when tachyzoites cross the placenta to the fetus.

Culture and Diagnosis   This infection can be differentiated from viral encephalitides by means of serological tests that detect antitoxoplasma antibodies, especially those for IgM, which appears early in infection. Disease can also be diagnosed by culture and histological analysis.

Prevention and Treatment   The most effective drugs are pyrimethamine and sulfadiazine alone or in combination. Because these drugs do not destroy the cyst stage, they must be given for long periods to prevent recurrent infection.

In view of the fact that the oocysts are so widespread and resistant, hygiene is of paramount importance in controlling toxoplasmosis. There is no such thing as a safe form of raw meat, even salted or spiced. Adequate cooking or freezing below −20°C destroys both oocysts and tissue cysts. Oocysts can also be avoided by washing the hands after handling cats or soil possibly contaminated with cat feces, especially sandboxes and litter boxes. Pregnant women should be especially attentive to these rules and should never clean the cat's litter box.

Measles Virus: Subacute Sclerosing Panencephalitis

Subacute sclerosing panencephalitis (SSPE) is sometimes called a "slow virus infection." It occurs years after an initial measles episode and is different from immune-mediated postinfectious encephalitis, described earlier. SSPE seems to be caused by direct viral invasion of neural tissue. It is not clear what factors lead to persistence of the virus in some people. See chapter 18 for more details. SSPE's important features are listed in **Checkpoint 19.5.**

Prions

As you read in chapter 6, prions are *proteinaceous infectious particles* containing, apparently, no genetic material. They are known to cause diseases called **transmissible spongiform encephalopathies (TSEs),** neurodegenerative diseases with long incubation periods but rapid progressions once they begin. The human TSEs are **Creutzfeldt-Jakob disease (CJD),** Gerstmann-Strussler-Scheinker disease, and fatal familial insomnia. TSEs are also found in animals and include a disease called **scrapie** in sheep and goats, transmissible mink encephalopathy, and bovine spongiform encephalopathy (BSE). This last disease is commonly known as mad cow disease and has been in the headlines in recent years due to its apparent link to a variant form of CJD human disease in Great Britain.

☑ **CHECKPOINT 19.5**	**Subacute Encephalitis**		
Causative Organism(s)	*Toxoplasma gondii*	Subacute sclerosing panencephalitis	Prions
Most Common Modes of Transmission	Vehicle (meat) or fecal-oral	Persistence of measles virus	CJD = direct/parenteral contact with infected tissue; or inherited vCJD = vehicle (meat)
Virulence Factors	Intracellular growth	Cell fusion, evasion of immune system	Avoidance of host immune response
Culture/Diagnosis	Serological detection of IgM, culture, histology	EEGs, MRI, serology (Ab vs. measles virus)	Biopsy, image of brain
Prevention	Personal hygiene, food hygiene	None	Avoiding tissue
Treatment	Pyrimethamine and/or sulfadiazine	None	None
Distinctive Features	Subacute, slower development of disease	History of measles	Long incubation period; fast progression once it begins

Signs and Symptoms of CJD Symptoms of CJD include altered behavior, dementia, memory loss, impaired senses, delirium, and premature senility. Uncontrollable muscle contractions continue until death, which usually occurs within 1 year of diagnosis.

Causative Agent of CJD The transmissible agent in CJD is a prion. It seems that a normal host protein (called PrP) found in mammalian brains has undergone a mutation that alters its structure. Once this happens, the abnormal PrP itself becomes catalytic and able to spontaneously convert other normal human PrP proteins into the abnormal form. This becomes a self-propagating chain reaction that creates a massive accumulation of altered PrP, leading to plaques, spongiform damage (that is, holes in the brain), and severe loss of brain function.

Using the term *transmissible agent* may be a bit misleading, however, as scientists believe that some cases of CJD arise through genetic mutation of the PrP gene, which can be a heritable trait. So it seems that although one can acquire a defective PrP protein via transmission, one can also have an altered PrP gene passed on through heredity. It is thought that 10% to 15% of CJD cases are inherited in this way.

Prions are incredibly hardy "pathogens." They are highly resistant to chemicals, radiation, and heat. They can withstand prolonged autoclaving.

Pathogenesis and Virulence Factors Autopsies of the brain of CJD patients reveal spongiform lesions as well as tangled protein fibers (neurofibrillary tangles) and enlarged astroglial cells (figure 19.14). These changes affect the gray matter of the CNS and seem to be caused by the massive accumulation of altered PrP, which may be toxic to neurons. The altered PrPs apparently stimulate no host immune response.

Transmission and Epidemiology CJD is not a communicable disease, in that ordinary contact with infected people will not allow transmission of the disease. Direct or indirect contact with infected brain tissue or cerebrospinal fluid may

result in prion transmission. In the classical presentation of CJD, a certain proportion of cases are considered to have been transmitted, whereas another group seem to have been inherited. This classical form is considered endemic, and occurs at the rate of 1 case per million persons in the United States per year. It mainly affects the elderly.

In the late 1990s, it became apparent that humans were contracting a variant form of CJD (vCJD) after ingesting meat from cattle that had been afflicted by bovine spongiform encephalopathy. Presumably meat products had been contaminated with fluid or tissues infected with the prion, although the exact food responsible for the transmission has not been pinpointed. When experimenters purposely infect cattle with prions, they subsequently detect the agent in the retina, dorsal root ganglia, parts of the digestive tract, and the bone marrow of the animals. Even so, the risk of contracting vCJD from the ingestion of meat is extremely small, even in countries such as Great Britain, where a significant number of livestock has been found to have BSE. There the risk of infection is estimated to be one case per 10 billion meat servings. In May 2003, the first North American case of mad cow disease appeared in Canada; in December of that year a single cow in the United States was found to have BSE. As of spring of 2002, a total of 125 cases of vCJD had occurred worldwide. The median age at death of patients with vCJD is 28 years. In contrast, the median age at death of patients with classical CJD is 68 years.

Health care professionals should be aware of the possibility of CJD in patients, especially when surgical procedures are performed, as cases have been reported of transmission of CJD via contaminated surgical instruments. Due to the heat and chemical resistance of prions, normal disinfection and sterilization procedures are usually not sufficient to eliminate them from instruments and surfaces. The latest CDC guidelines for handling of CJD patients in a health care environment should be consulted. CJD has also been transmitted through corneal grafts and administration of contaminated human growth hormone. Currently no cases have been documented of transmission through blood products,

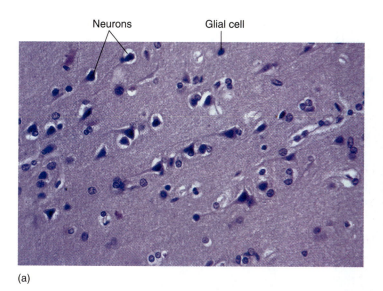

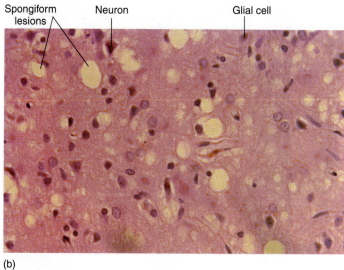

(a)   (b)

FIGURE 19.14 **The microscopic effects of spongiform encephalopathy.**
(a) Normal cerebral cortex section, showing neurons and glial cells. **(b)** Section c cortex in CJD patient shows numerous round holes, producing a "spongy" appearance. This destroys brain architecture and causes massive loss of neurons and glial cells.

although scientists have shown that it is possible in laboratory experiments. In 2003 a British patient died of CJD after receiving a blood transfusion in 1996 from a donor who had CJD. While it is possible that the second patient could have contracted CJD independently of the blood transfusion, health officials took this death as a sign that blood-borne transmission might well be possible. Experiments suggest that vCJD seems to be more transmissible through blood than is classic CJD. For that reason, blood donation programs screen for possible exposure to BSE by asking about travel and residence history.

Culture and Diagnosis   It is very difficult to diagnose CJD. Definitive diagnosis requires examination of biopsied brain or nervous tissue, and this procedure is usually considered too risky, because of both the trauma induced in the patient and the undesirability of contaminating surgical instruments and operating rooms. Electroencephalograms and magnetic resonance imaging can provide important clues. A new test has been developed that can detect prion proteins in cerebrospinal fluid, but it is not yet widely used because it has a relatively high error rate.

Prevention and/or Treatment   Prevention of this disease relies on avoiding infected tissues. Avoiding vCJD entails not ingesting tainted meats. No known treatment exists for either form of CJD, and patients inevitably die. Medical intervention focuses on easing symptoms and making the patient as comfortable as possible **(Checkpoint 19.5)**.

Rabies

Rabies is a slow, progressive zoonotic disease characterized by a fatal encephalitis. It is so distinctive in its pathogenesis and its symptoms that we discuss it separately from the other encephalitides. It is distributed nearly worldwide, except for 34 countries that have remained rabies-free by practicing rigorous animal control.

Signs and Symptoms

The average incubation period of rabies is 1 to 2 months or more, depending upon the wound site, its severity, and the inoculation dose. The incubation period is shorter in facial, scalp, or neck wounds because of closer proximity to the brain. The prodromal phase begins with fever, nausea, vomiting, headache, fatigue, and other nonspecific symptoms.

In the form of rabies termed *furious,* the first acute signs of neurological involvement are periods of agitation, disorientation, seizures, and twitching. Spasms in the neck and pharyngeal muscles lead to severe pain upon swallowing, leading to a symptom known as *hydrophobia* (fear of water). Throughout this phase, the patient is fully coherent and alert. With the *dumb* form of rabies, a patient is not hyperactive but is paralyzed, disoriented, and stuporous. Ultimately, both forms progress to the coma phase, resulting in death from cardiac or respiratory arrest. Until recently, humans were never known to survive rabies. But a handful of patients have recovered in recent years after receiving intensive, long-term treatment.

Causative Agent

The rabies virus is in the family Rhabdoviridiae, genus *Lyssavirus.* The particles of this virus have a distinctive bulletlike appearance, round on one end and flat on the other. Additional features are a helical nucleocapsid and spikes that protrude through the envelope **(figure 19.15)**. The family contains about 60 different viruses, but only the rabies *Lyssavirus* infects humans.

there is no sensory or mental lapse. Although nausea and vomiting can occur at an early stage, they are not common. Later symptoms are descending muscular paralysis and respiratory compromise. In the past, death resulted from respiratory arrest, but mechanical respirators have reduced the fatality rate to about 10%.

Causative Agent

Clostridium botulinum, like *Clostridium tetani,* is a spore-forming anaerobe that does its damage through the release of an exotoxin. *C. botulinum* commonly inhabits soil and water and occasionally the intestinal tract of animals. It is distributed worldwide but occurs most often in the Northern Hemisphere. The species has eight distinctly different types (designated A, B, C_α, C_β, D, E, F, and G), which vary in distribution among animals, regions of the world, and types of exotoxin. Human disease is usually associated with types A, B, E, and F, and animal disease with types A, B, C, D, and E.

Both *C. tetani* and *C. botulinum* produce neurotoxins; but tetanospasmin, the toxin made by *C. tetani,* results in spastic paralysis (uncontrolled muscle contraction). In contrast, botulin, the *C. botulinum* neurotoxin, results in flaccid paralysis, a loss of ability to contract the muscles.

Pathogenesis and Virulence Factors

As just described, the symptoms are caused entirely by the exotoxin botulin.

Transmission and Epidemiology of Food-Borne Botulism in Children and Adults

There is a high correlation between cultural dietary preferences and food-borne botulism. In the United States, the disease is often associated with low-acid vegetables (green beans, corn), fruits, and occasionally meats, fish, and dairy products. Most botulism outbreaks occur in home-processed foods, including canned vegetables, smoked meats, and cheese spreads. The demand for prepackaged convenience foods, such as vacuum-packed cooked vegetables and meats, has created a new source of risk, but most commercially canned foods are held to very high standards of preservation and are only rarely a source of botulism.

Several factors in food processing can lead to botulism. Spores are present on the vegetables or meat at the time of gathering and are difficult to remove completely. When contaminated food is put in jars and steamed in a pressure cooker that does not reach reliable pressure and temperature, some spores survive (botulinum spores are highly heat-resistant). At the same time, the pressure is sufficient to evacuate the air and create anaerobic conditions. Storage of the jars at room temperature favors spore germination and vegetative growth, and one of the products of the cell's metabolism is botulin, the most potent microbial toxin known.

Bacterial growth may not be evident in the appearance of the jar or can or in the food's taste or texture, and only minute amounts of toxin may be present. Botulism is never transmitted person-to-person. Of the more than 100 cases of botulism every year in the United States, about one-quarter are food-borne.

Transmission and Epidemiology of Infant Botulism

Infant botulism was first described in the late 1970s in children between the ages of 2 weeks and 6 months who had ingested spores. It is currently the most common type of botulism in the United States, with approximately 80 cases reported annually. The exact food source is not always known, although raw honey has been implicated in some cases, and the spores are common in dust and soil. Apparently, the immature state of the neonatal intestine and microbial flora allows the spores to gain a foothold, germinate, and give off neurotoxin. As in adults, babies exhibit flaccid paralysis, usually manifested as a weak sucking response, generalized loss of tone (the "floppy baby syndrome"), and respiratory complications. Although adults can also ingest botulinum spores in contaminated vegetables and other foods, the adult intestinal tract normally inhibits this sort of infection.

Transmission and Epidemiology of Wound Botulism

Perhaps three or four cases of wound botulism occur each year in the United States. In this form of the disease, spores enter a wound or puncture, much as in tetanus, but the symptoms are similar to those of food-borne botulism. Increased cases of this form of botulism are being reported in intravenous drug users as a result of needle puncture.

Culture and Diagnosis

Diagnostic standards are slightly different for the three different presentations of botulism. In food-borne botulism, some laboratories attempt to identify the toxin in the offending food. Alternatively, if multiple patients present with the same symptoms after ingesting the same food, a presumptive diagnosis can be made. The cultivation of *C. botulinum* in feces is considered confirmation of the diagnosis since the carrier rate is very low.

In infant botulism, finding the toxin or the organism in the feces confirms the diagnosis. In wound botulism, the toxin should be demonstrated in the serum, or the organism should be grown from the wound. Because minute amounts of the toxin are highly dangerous, laboratory testing should only be performed by experienced personnel. A suspected case of botulism should trigger a phone call to the state health department or the CDC before proceeding with diagnosis or treatment.

Prevention and Treatment

The CDC maintains a supply of type A, B, and E trivalent horse antitoxin, which, when administered early, can prevent the worst outcomes of the disease. Patients are also managed with respiratory and cardiac support systems. Antitoxin therapy is

INSIGHT 19.4   Discovery

Botox: No Wrinkles. No Headaches. No Worries?

In 2002, nearly 2 million people paid good money (and lots of it) to have one of the most potent toxins on earth injected into their faces. The toxin, of course, is Botox, short for botulinum toxin, and the story of how these injections came to be the most popular cosmetic procedure in the United States is a fascinating one.

Scientists have long known that death from *Clostridium botulinum* infection results from paralysis of the respiratory muscles. In fact, researchers had even determined that botulinum toxin causes death by interfering with the release of acetylcholine, a neurotransmitter that causes the contraction of skeletal muscles. The trick was finding a practical application for this knowledge.

In 1989 Botox was first approved to treat cross-eyes and uncontrollable blinking, two conditions resulting from the inappropriate contracting of muscles around the eye. Success in this first arena led to Botox treatment for a variety of neurological disorders that cause painful contraction of neck and shoulder muscles. A much wider use of Botox occurred in so-called "off label" uses, as doctors found that injecting facial muscles with the toxin inhibited contraction of these muscles, and consequent wrinkling of the overlying skin. The "lunch-hour facelift" went over exactly as most people would imagine, becoming the most popular cosmetic procedure even before winning official FDA approval (which it did in 2002). In a surprise twist, patients undergoing Botox treatment for wrinkles reported fewer headaches, especially migraines.

Clinical trials have shown this result to be widespread and reproducible, but the exact mechanism by which Botox works to prevent headaches is unknown.

So then, Botox is perfect and we need never have, as George Orwell once said, ". . . the face he (or she) deserves"? Not so fast. In the rush to embrace the admittedly dramatic results of Botox injections, most patients have paid scant attention to potential problems that can arise from the use of a potent paralytic agent. The most common problem arising from Botox treatment is excessive paralysis of facial muscles, resulting from poorly targeted injections. Depending on the site of the injection, results such as drooping eyelids, facial paralysis, slurred speech, and drooling are possible. Even if the treatment works perfectly, the wrinkle-free visage is a result of muscle paralysis, meaning that patients are often unable to move their eyebrows or in some cases to frown or squint.

Finally, Botox is not a permanent solution; as the effects of the toxin wear off, the wrinkles (or headaches, as the case may be) return. Every 4 to 6 months, the treatment must be repeated. This last fact has been a boon to doctors for two reasons. The first is obvious—namely, that a permanent solution to wrinkles would rule out any repeat clientele. The second advantage has to do with malpractice lawsuits. By the time a patient has consulted a lawyer and made it through the legal system, any consequences of a botched treatment may have already worn off.

---

generally not administered to infants with botulism; supportive care is primary. In all cases, hospitalization is required and recovery takes weeks. There is an overall 5% mortality rate.

✔ CHECKPOINT 19.9	Botulism
Causative Organism(s)	*Clostridium botulinum*
Most Common Modes of Transmission	Vehicle (food-borne toxin, airborne organism); direct contact (wound); parenteral (injection)
Virulence Factors	Botulinum exotoxin
Culture/Diagnosis	Culture of organism; demonstration of toxin
Prevention	Food hygiene; toxoid immunization available for laboratory professionals
Treatment	Antitoxin, supportive care

African Sleeping Sickness

This condition is caused by *Trypanosoma brucei*, a member of the protozoan group known as hemoflagellates because of their propensity to live in the blood and tissues of the human host. The disease, also called **trypanosomiasis**, has greatly affected the living conditions of Africans since ancient times. Today at least 50 million people are at risk, and 30,000 to 40,000 new cases occur each year. It imposes an additional hardship when it attacks domestic and wild mammals.

Signs and Symptoms

Trypanosomiasis affects the lymphatics and areas surrounding blood vessels. Usually a long asymptomatic period precedes onset of symptoms. Symptoms include intermittent fever, enlarged spleen, swollen lymph nodes, and joint pain. There are two variants of the disease, caused by two different subspecies of the protozoan. In both forms, the central nervous system is affected, the initial signs being personality and behavioral changes that progress to lassitude and sleep disturbances. The disease is commonly called *sleeping sickness*, but in fact, uncontrollable sleepiness occurs primarily in the day and is followed by sleeplessness at night. Signs of advancing neurological deterioration are muscular tremors, shuffling gait, slurred speech, seizures, and local paralysis. Death results from coma, secondary infections, or heart damage.

Causative Agent

Trypanosoma brucei is a flagellated protozoan, an obligate parasite that is spread by a blood-sucking insect called the tsetse fly, which serves as its intermediate host. It shares a complicated life cycle with other hemoflagellates. In chapter 5, we first described the trypanosome life cycle using the example of *T. cruzi*, the agent that causes Chagas disease.

Transmission and Epidemiology

The cycle begins when a tsetse fly becomes infected after feeding on an infected reservoir host, such as a wild animal (antelope, pig, lion, hyena), domestic animal (cow, goat), or human **(figure 19.25).** In the fly's gut, the trypanosome multiplies, migrates to the salivary glands, and develops into the infec-

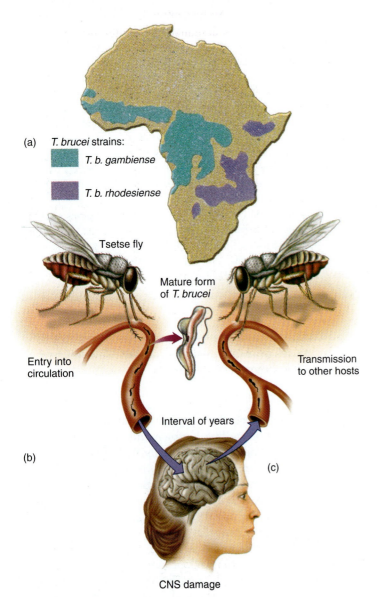

(a) *T. brucei* strains:

■ *T. b. gambiense*

■ *T. b. rhodesiense*

Tsetse fly

Mature form of *T. brucei*

Entry into circulation

Transmission to other hosts

Interval of years

(b)

(c)

CNS damage

FIGURE 19.25 **The generalized cycle between humans and the tsetse fly vector.**
(a) The distribution of African trypanosomiasis. **(b)** The saliva of a fly infected with *T. brucei* inoculates the human bloodstream. The parasite matures and invades various organs. In time, its cumulative effects cause central nervous system (CNS) damage. **(c)** The trypanosome is spread to other hosts through another fly in whose alimentary tract the parasite completes a series of developmental stages.

tious stage. When the fly bites a new host, it releases the large, fully formed stage of the parasite into the wound. At this site, the trypanosome multiplies and produces a sore called the *primary chancre*. From there, the pathogen moves into the lymphatics and the blood (figure 19.25). The trypanosome can also cross the placenta and damage a developing fetus.

Two variants of sleeping sickness are the Gambian (West African) strain, caused by the subspecies *Trypanosoma brucei gambiense,* and the Rhodesian (East African) strain, caused by *T. b. rhodesiense* (figure 19.25). These geographically isolated types are associated with different ecological niches of the principal tsetse fly vectors. In the West African form, the fly inhabits the dense vegetation along rivers and forests typical of that region, whereas the East African form is adapted to savanna woodlands and lakefront thickets.

African sleeping sickness occurs only in Sub-Saharan Africa. Tsetse flies exist elsewhere, and it is not known why they do not support *Trypanosoma* in other regions. In some parts of equatorial Africa, *T. b. gambiense* has recently undergone a resurgence. Epidemics are reported in areas of the Sudan and the Democratic Republic of the Congo, where a significant segment of the population is infected. These outbreaks were partly due to a civil war and an accompanying disruption in medical services.

Pathogenesis and Virulence Factors

The protozoan manages to flourish in the blood even though it stimulates a strong immune response. The immune response is counteracted by an unusual adaptation of the trypanosome. As soon as the host begins manufacturing IgM antibodies to the trypanosome, surviving organisms change the structure of their surface glycoprotein antigens. This change in specificity (sometimes referred to as an *antigenic shift*) renders the existing IgM ineffective, so that the parasite eludes control and multiplies in the blood. The host responds by producing IgM of a new specificity, but the protozoan changes its antigens again. The host eventually becomes exhausted and overwhelmed by repeated efforts to catch up with this trypanosome masquerade. This cycle has tremendous impact on the pathology and control of the disease. The presence of the trypanosome in the blood, and the severity of symptoms, follow a wavelike pattern.

Culture and Diagnosis

Sleeping sickness may be suspected if the patient has been bitten by a distinctive fly while living or traveling in an endemic area. Trypanosomes are readily demonstrated in blood smears, as well as in spinal fluid or lymph nodes.

Prevention and Treatment

Control of trypanosomiasis in western Africa, where humans are the main reservoir hosts, involves eliminating tsetse flies by applying insecticides, trapping flies, or destroying the shelter and breeding sites. In eastern regions, where cattle herds and large wildlife populations are reservoir hosts, control is less practical because large mammals are the hosts, and flies are less concentrated in specific sites. The antigenic shifting practiced by the trypanosome makes the development of a vaccine very difficult.

Chemotherapy is most successful if administered prior to nervous system involvement. Two different drugs are available for the early stages of the disease. Suramin works against *T. b. rhodesiense,* and pentamidine is used for *T. b. gambiense.* Brain infection must be treated with drugs that can cross the blood-brain barrier. One of these is a highly toxic arsenic-based drug called melarsoprol. It causes nervous system symptoms itself, sometimes permanent, and is itself fatal in a small percentage of cases. An alternative was developed in 1990, but the company stopped making it and gave the license to the World Health Organization. The WHO is seeking a new company to make the drug.

✔ CHECKPOINT 19.10	African Sleeping Sickness
Causative Organism(s)	*Trypanosoma brucei* subspecies *gambiense* or *rhodesiense*
Most Common Modes of Transmission	Vector, vertical
Virulence Factors	Immune evasion by antigen shifting
Culture/Diagnosis	Microscopic examination of blood, CSF
Prevention	Vector control
Treatment	Suramin or pentamidine (early), melarsoprol (late)

Taxonomic Organization of Microorganisms Causing Disease in the Nervous System

Microorganism	Disease	Chapter Location
Gram-Positive Endospore-Forming Bacteria		
Clostridium botulinum	Botulism	Botulism, p. 603
Clostridium tetani	Tetanus	Tetanus, p. 601
Gram-Positive Bacteria		
Streptococcus agalactiae	Neonatal meningitis	Neonatal meningitis, p. 586
Streptococcus pneumoniae	Meningitis	Meningitis, p. 582
Listeria monocytogenes	Meningitis, neonatal meningitis	Meningitis, p. 582 Neonatal meningitis, p. 586
Gram-Negative Bacteria		
Escherichia coli	Neonatal meningitis	Neonatal meningitis, p. 586
Haemophilus influenzae	Meningitis	Meningitis, p. 582
Neisseria meningitidis	Meningococcal meningitis	Meningitis, p. 580
DNA Viruses		
Herpes simplex virus 1 and 2	Encephalitis	Encephalitis, p. 591
JC virus	Progressive multifocal leukoencephalopathy	Encephalitis, p. 591
RNA Viruses		
Arboviruses		
Western equine encephalitis virus, Eastern equine encephalitis virus, California encephalitis virus (California and LaCrosse strains), St. Louis encephalitis virus, West Nile virus	Encephalitis	Encephalitis, p. 589
Measles virus	Subacute sclerosing panencephalitis	Subacute encephalitis, p. 593
Poliovirus	Poliomyelitis	Poliomyelitis, p. 597
Rabies virus	Rabies	Rabies, p. 595
Fungi		
Cryptococcus neoformans	Meningitis	Meningitis, p. 583
Coccidioides immitis	Meningitis	Meningitis, p. 584
Prions		
Creutzfeldt-Jakob prion	Creutzfeldt-Jakob disease	Subacute encephalitis, p. 593
Protozoa		
Acanthamoeba	Meningoencephalitis	Meningoencephalitis, p. 589
Naegleria fowleri	Meningoencephalitis	Meningoencephalitis, p. 588
Toxoplasma gondii	Subacute encephalitis	Subacute encephalitis, p. 591
Trypanosoma brucei subspecies *gambiense* and *rhodesiense*	African sleeping sickness	African sleeping sickness, p. 605

Infectious Diseases Affecting the Nervous System

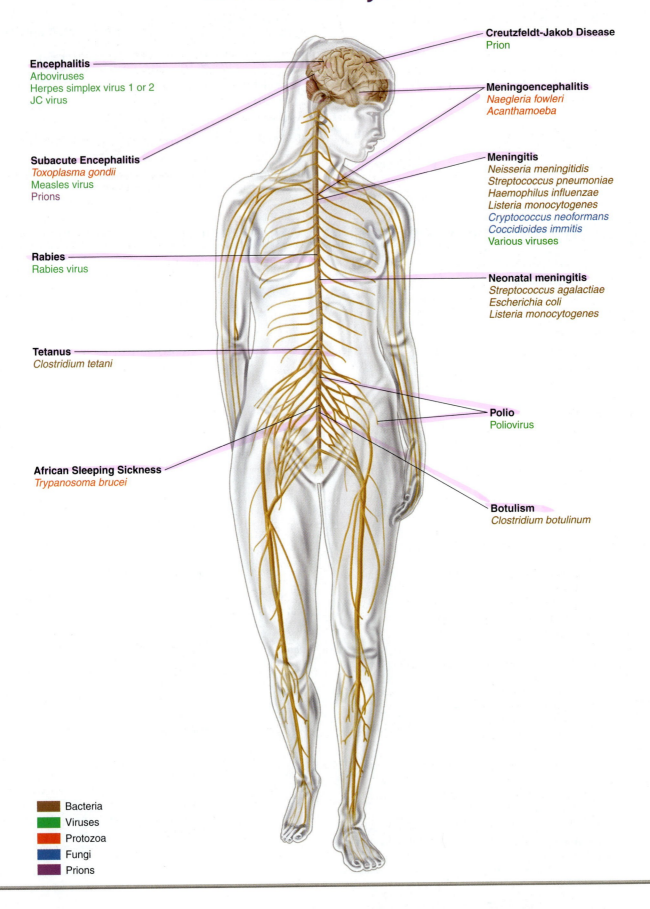

Creutzfeldt-Jakob Disease
Prion

Encephalitis
Arboviruses
Herpes simplex virus 1 or 2
JC virus

Meningoencephalitis
Naegleria fowleri
Acanthamoeba

Subacute Encephalitis
Toxoplasma gondii
Measles virus
Prions

Meningitis
Neisseria meningitidis
Streptococcus pneumoniae
Haemophilus influenzae
Listeria monocytogenes
Cryptococcus neoformans
Coccidioides immitis
Various viruses

Rabies
Rabies virus

Neonatal meningitis
Streptococcus agalactiae
Escherichia coli
Listeria monocytogenes

Tetanus
Clostridium tetani

Polio
Poliovirus

African Sleeping Sickness
Trypanosoma brucei

Botulism
Clostridium botulinum

Bacteria
Viruses
Protozoa
Fungi
Prions

Chapter Summary With Key Terms

19.1 The Nervous System and Its Defenses

A. The nervous system has two parts: the central nervous system (the brain and spinal cord), and the peripheral nervous system (spinal and cranial nerves). The soft tissue of the brain and spinal cord is encased within a tough casing of three membranes called the **meninges.** The subarachnoid space (under the arachnoid mater) is filled with a clear serumlike fluid called cerebrospinal fluid (CSF). The meninges are a common site of infection, and microorganisms can often be found in the CSF when meningeal infection **(meningitis)** occurs.

B. The nervous system is protected by the **blood-brain barrier,** which limits the passage of substances from the bloodstream to the brain.

19.2 Normal Flora of the Nervous System

There is no normal flora in either the CNS or PNS.

19.3 Nervous System Diseases Caused by Microorganisms

A. **Meningitis** is an inflammation of the meninges. Symptoms include headache, painful or stiff neck, fever, and usually an increased number of white blood cells in the CSF. The more serious forms of acute meningitis are caused by bacteria, often facilitated by coinfection or previous infection with respiratory viruses.

1. *Neisseria meningitidis:* This gram-negative diplococcus is commonly known as the meningococcus and causes the most serious form of acute meningitis. The most serious complications of meningococcal infection are due to meningococcemia.

2. *Streptococcus pneumoniae:* This **pneumococcus** is the most frequent cause of community-acquired pneumococcal meningitis.

3. *Haemophilus influenzae:* This bacterium is a gram-negative pleomorphic rod and an agent of acute bacterial meningitis in humans. The disease is caused primarily by the b serotype and was once most common in children between 3 months and 5 years of age, although these rates have declined because of vaccination.

4. *Listeria monocytogenes:* L. monocytogenes is a gram-positive bacterium that ranges in morphology from coccobacilli to long filaments in palisades formation. Pregnant women are especially susceptible to infection, which can result in premature abortion and fetal death. Most cases of listeriosis are associated with ingesting contaminated dairy products, poultry, and meat.

5. *Cryptococcus neoformans:* The fungus C. neoformans causes a more chronic form of meningitis with a more gradual onset of symptoms. The primary ecological niche of C. neoformans is the bird population. Most cryptococcal infections cause symptoms in the respiratory and central nervous systems, with the highest rates of cryptococcal meningitis occurring among patients with AIDS; frequently it is fatal.

6. *Coccidioides immitis:* This is a true systemic fungal infection that begins with pulmonary infection but can disseminate quickly throughout the body and can lead to coccidioidomycosis of the meninges. The highest incidence of coccidioidomycosis occurs in the southwestern United States, Mexico, and parts of Central and South America.

7. *Viruses:* A wide variety of viruses can cause meningitis, particularly in children, and 90% are caused by enteroviruses.

B. **Neonatal Meningitis:** Meningitis in newborns is usually transmitted by the mother, either in utero or during passage through the birth canal.

1. *Streptococcus agalactiae:* This species of *Streptococcus* belongs to the group B streptococci and is the most frequent cause of neonatal meningitis.

2. *Escherichia coli:* The K1 strain of *E. coli* is the second most common cause of neonatal meningitis. Most babies who suffer from this infection are premature, and their prognosis is poor.

C. **Meningoencephalitis:** Because the brain and the spinal cord (and the meninges) are so closely connected, infections of one of these structures may also involve the other. Two amoebas, *Naegleria fowleri* and *Acanthamoeba,* are parasites that cause meningoencephalitis.

D. **Acute encephalitis** is almost always caused by viral infection.

1. Arboviruses: Arthropods serve as infectious disease vectors for many arboviruses. Arboviral encephalitis begins with an arthropod bite, the release of the virus into tissues, and its replication in nearby lymphatic tissues.

   a. **Western equine encephalitis (WEE)** occurs sporadically in the western United States and Canada and is carried by a mosquito.

   b. **Eastern equine encephalitis (EEE)** is endemic to an area along the eastern coast of North America and Canada.

   c. **California encephalitis** may be caused by two different viral strains. The California strain occurs occasionally in the western United States and has little impact on humans. The LaCrosse strain is widely distributed in the eastern United States and Canada.

   d. **St. Louis encephalitis (SLE)** may be the most common of all American viral encephalitides. Cases appear throughout North and South America, but epidemics occur most often in the Midwest and South.

   e. **West Nile encephalitis:** The West Nile virus is a close relative of the SLE virus. It emerged in the United States in 1999, and by the beginning of 2003 more than 4,000 people had been infected.

2. Herpes simplex virus: Herpes simplex type 1 and 2 viruses can cause encephalitis in newborns born to HSV-positive mothers, as well as older children and young adults (ages 5 to 30), and even older adults (over 50 years old).

3. JC virus: The JC virus (JCV) can cause a condition called **progressive multifocal leukoencephalopathy (PML),** particularly in immunocompromised individuals. This is a fatal infection.

E. **Subacute Encephalitis:** When encephalitis symptoms take longer to show up, it is termed subacute encephalitis.
 1. *Toxoplasma gondii* is a protozoan that causes toxoplasmosis, the most common form of subacute encephalitis. Although relatively asymptomatic in the general population, the disease in pregnant women and in immunodeficient people, is severe and often fatal. *T. gondii* has primary reservoir and hosts in members of the feline family, both domestic and wild.
 2. Measles virus can produce **subacute sclerosing panencephalitis (SSPE),** which occurs years after an initial measles infection.
 3. Prions are proteinaceous infectious particles containing no genetic material. They cause diseases called transmissible spongiform encephalopathies (TSEs), neurodegenerative diseases with long incubation periods but rapid progressions once they begin. The human TSEs are **Creutzfeldt-Jakob disease (CJD),** Gerstmann-Strussler-Scheinker disease, and fatal familial insomnia. Prions are very resistant to chemicals, radiation heat, and autoclaving.

F. **Rabies** is a slow, progressive zoonotic disease characterized by fatal encephalitis. The rabies virus is in the family Rhabdoviridiae, genus *Lyssavirus.* The particles of this virus have a distinctive bulletlike appearance, round on one end and flat on the other. The primary reservoirs of the virus are wild mammals. An effective vaccine regimen is available.

G. **Poliomyelitis:** Polio is an acute enterovirus infection of the spinal cord that can cause neuromuscular paralysis. Because it often affects small children, in the past it was called infantile paralysis. Prevention is by vaccination using one of two forms of vaccine currently in use— inactivated Salk poliovirus vaccine (IPV), and oral Sabin poliovirus vaccine (OPV).

H. **Tetanus** is a neuromuscular disease, also called lockjaw, and is caused by *Clostridium tetani,* a gram-positive, spore-forming rod. *C. tetani* releases a powerful neurotoxin, **tetanospasmin,** which binds to target sites on spinal neurons and blocks the inhibition of muscle contraction. Without inhibition of contraction, the muscles contract uncontrollably, which can be fatal.

I. **Botulism** is often an *intoxication* (that is, caused by an exotoxin) associated with eating poorly preserved foods, although it can also occur as a true infection. There are three major forms of botulism: food-borne botulism (in children and adults), infant botulism, and wound botulism. The causative agent is *Clostridium botulinum,* a spore-forming anaerobe that does its damage through the release of an exotoxin.

J. **African sleeping sickness** is caused by a protozoan, *Trypanosoma brucei.* Trypanosomiasis affects the central nervous system, leading to neurological deterioration: muscular tremors, shuffling gait, slurred speech, seizures, and local paralysis. Death results from coma, secondary infections, or heart damage.

Multiple-Choice Questions

1. Which of the following organisms does *not* cause meningitis?
 a. *Haemophilus influenzae*
 b. *Streptococcus pneumoniae*
 c. *Neisseria meningitidis*
 d. *Clostridium tetani*

2. The first choice antibiotic for bacterial meningitis is the broad-spectrum
 a. cephalosporin   c. ampicillin
 b. penicillin   d. vancomycin

3. Meningococcal meningitis is caused by
 a. *Haemophilus influenzae*
 b. *Streptococcus pneumoniae*
 c. *Neisseria meningitidis*
 d. *Listeria monocytogenes*

4. In order to diagnose bacterial meningitis, a sample of ____ is taken.
 a. brain   c. urine
 b. cerebrospinal fluid   d. sputum

5. Which of the following neurological diseases is not caused by a prion?
 a. Creutzfeldt-Jakob disease
 b. scrapie
 c. mad cow disease
 d. St. Louis encephalitis

6. *Cryptococcus neoformans* is primarily transmitted by
 a. direct contact
 b. bird droppings
 c. fomites
 d. sexual activity

7. Which of the following is *not* caused by an arbovirus?
 a. St. Louis encephalitis
 b. Eastern equine encephalitis
 c. West Nile encephalitis
 d. PAM

8. CJD is caused by a(n)
   a. arbovirus
   b. prion
   c. protozoan
   d. bacterium

9. *Toxoplasma gondii* is a(n)
   a. arbovirus
   b. prion
   c. protozoan
   d. bacterium

10. Which of the following is a common reservoir for the rabies virus?
    a. pigeons
    b. humans
    c. raccoons
    d. mosquitoes

11. What food should you avoid feeding a child under 1 year old because of potential botulism?
    a. honey
    b. milk

c. apple juice
d. applesauce

12. *Naegleria fowleri* meningoencephalitis is commonly acquired by
    a. bird droppings
    b. swimming in ponds and streams
    c. mosquito bites
    d. chickens

13. Which organism is responsible for progressive multifocal leukoencephalopathy?
    a. JC virus
    b. herpesvirus
    c. *E. coli*
    d. *Haemophilus influenzae*

14. What statement regarding coccidiomycosis is untrue?
    a. It is caused by a fungus.
    b. Fluconazole treatment is indicated.
    c. Penicillin G is the first line of treatment.
    d. It commonly occurs in the western and southwestern part of the United States.

Concept Questions

These questions are suggested as a *writing-to-learn* experience. For each question, compose a one- or two-paragraph answer that includes the factual information needed to completely address the question.

1. Describe the components of the human nervous system.

2. a. What is meningitis?
   b. Describe the symptoms of this condition.
   c. Define petechiae.
   d. What is the most frequent cause of community-acquired meningitis?
   e. Name the clinical symptom that can distinguish between this organism and the agent of meningococcal meningitis.

3. What is the common mode(s) of transmission in neonatal meningitis?

4. What is the difference between meningitis and encephalitis?

5. What sterilization methods are most effective against prions?

6. Discuss the transmission of CJD.

7. Name the infectious viral disease for which postexposure passive and active immunization is indicated. Describe the procedure.

8. a. Name and describe the three major forms of botulism.
   b. What is the causative agent of this disease?

Critical Thinking Questions

Critical thinking is the ability to reason and solve problems using facts and concepts. These questions can be approached from a number of angles, and in most cases, they do not have a single correct answer.

1. Why is there no normal flora associated with the nervous system?

2. What organisms cause aseptic meningitis? Why is this condition referred to as aseptic meningitis?

3. Discuss the roles of arthropods and birds in arbovirus infections.

4. How did West Nile encephalitis spread to the United States?

5. Why should pregnant women be careful around cats and their litter boxes?

6. Why is the Sabin oral polio vaccine no longer recommended for childhood vaccinations?

7. Why is botulism associated with preserved canned foods?

8. How should trypanosomiasis (African sleeping sickness) be controlled or eradicated?

9. Why are young children so susceptible to infant botulism?

INSIGHT 20.1 — *Medical*

Atherosclerosis

Atherosclerosis is a condition you may not associate with infection. In atherosclerosis, plaques form on the inner endothelium of the arteries, decreasing the flexibility of the arterial walls and possibly leading to obstruction. The well-established cause of plaque formation seems to be an elevation in low-density lipoproteins (LDLs) in the plasma (originating in the diet) accompanied by chronic endothelial injury events. Injury to the endothelium results in adhesion of platelets to the surface, accumulation of other blood components, and release of growth factors that cause the proliferation of smooth muscle cells. This begs the question: What causes the chronic endothelial injury? The answer is complex, and can include such things as nicotine in the bloodstream from cigarette smoke, or high blood levels of insulin as seen in diabetics. But recent findings suggest that another of the causes may be chronic bloodstream infection with the bacterium *Chlamydia pneumoniae* or even various viruses.

Other researchers have found that reproductive tract infections with *Chlamydia trachomatis* may predispose people to heart disease. These studies found that the immune response to proteins on the surface of *C. trachomatis* can cross-react with myosin, a protein in heart muscle. This phenomenon is called molecular mimicry, referring to the similarity between *Chlamydia* proteins and heart proteins. The misdirected immune response may injure the heart muscle.

Obviously much remains to be discovered about heart disease, atherosclerosis, and the role infection may play.

---

types of white blood cells include the lymphocytes, responsible for specific immunity, and the phagocytes, which are so critical to nonspecific as well as specific immune responses. Very few microbes can survive in the blood with so many defensive elements in it. That said, a handful of infectious agents have nonetheless evolved exquisite mechanisms for avoiding blood-borne defenses. The same defenses are present in the lymphatic system, whose very existence is centered on host immunity.

Medical conditions involving the blood often have the suffix *-emia.* For instance, viruses that cause meningitis can travel to the nervous system via the bloodstream. Their presence in the blood is called **viremia.** When fungi are in the blood, the condition is termed **fungemia,** and bacterial presence is called **bacteremia,** a general term denoting only their *presence.* Although the blood contains no normal flora (see next section), bacteria frequently are introduced into the bloodstream during the course of daily living. Brushing your teeth or tearing a hangnail can introduce bacteria from the mouth or skin into the bloodstream; this situation is usually temporary. But when bacteria flourish and grow in the bloodstream, the condition is termed **septicemia.** Septicemia can very quickly lead to cascading immune responses, resulting in decreased systemic blood pressure that can lead to **septic shock,** a life-threatening condition.

20.2 Normal Flora of the Cardiovascular and Lymphatic Systems

Like the nervous system, the cardiovascular and lymphatic systems are "closed" systems with no normal access to the external environment. Therefore they possess no normal flora. In the absence of disease, microorganisms may be transiently present in either system as just described. The lymphatic system serves to filter microbes and their products out of tissues. Thus, in the healthy state no microorganisms *colonize* either the lymphatic or cardiovascular systems. Of course, this is biology, and it is never quite that simple. Recent studies have suggested that the bloodstream is not completely sterile, even during periods of apparent health. It is tempting to speculate that these low-level microbial "infections" may contribute to diseases for which no etiology has previously been found, or for conditions currently thought to be noninfectious (see **Insight 20.1**).

Cardiovascular and Lymphatic Systems Defenses and Normal Flora		
	Defenses	**Normal Flora**
Cardiovascular System	Blood-borne components of nonspecific and specific immunity—including phagocytosis, specific immunity	None
Lymphatic System	Numerous immune defenses reside here	None

20.3 Cardiovascular and Lymphatic System Diseases Caused by Microorganisms

Categorizing cardiovascular and lymphatic infections according to clinical presentation is somewhat difficult because most of these conditions are systemic, with effects on multiple organ systems. We begin with infections involving the heart and the blood in general, and then discuss conditions with more specific causes.

Endocarditis

Endocarditis is an inflammation of the endocardium, or inner lining of the heart. Most of the time endocarditis refers to an infection of the valves of the heart, often the mitral or aortic valve. Two variations of infectious endocarditis have been described: acute and subacute. Each has distinct groups of possible causative agents. Rarely, endocarditis can also be caused by vascular trauma or by circulating immune complexes in the absence of infectious agents.

The surgical innovation of prosthetic valves presents a new hazard for development of endocarditis. Patients with prosthetic valves can acquire acute endocarditis if bacteria are introduced during the surgical procedure; alternatively, the prosthetic valves can serve as infection sites for the subacute form of endocarditis long after the surgical procedure. Because the symptoms and the diagnostic procedures are similar for both forms of endocarditis, they will be discussed first; then the specific aspects of acute and subacute endocarditis will be addressed.

Signs and Symptoms

The signs and symptoms are similar for both types of endocarditis, except that in the subacute condition they develop more slowly and are less pronounced than with the acute disease. Symptoms include fever, anemia, abnormal heartbeat, and sometimes symptoms similar to myocardial infarction (heart attack). Abdominal or side pain is sometimes reported. The patient may look very ill and may have petechiae (small red-to-purple discolorations) over the upper half of the body and under the fingernails. In subacute cases, an enlarged spleen may have developed over time; cases of extremely long duration can lead to clubbed fingers and toes.

Culture and Diagnosis

The diagnostic procedures for the two forms of endocarditis are essentially the same. One of the most important diagnostic tools is a high index of suspicion. A history of risk factors, or behaviors, such as abnormal valves, intravenous drug use, recent surgery, or bloodstream infections, should lead one to consider endocarditis when the symptoms just described are observed. Blood cultures, if positive, are the gold standard for diagnosis, but negative blood cultures do not rule out endocarditis. If it is possible to obtain the agent, it is very important to determine its antimicrobial susceptibilities.

In acute endocarditis, the symptoms may be magnified. The patient may also display central nervous system symptoms suggestive of meningitis, such as stiff neck or headache.

Acute Endocarditis

Acute endocarditis is most often the result of an overwhelming bloodstream challenge with bacteria. Certain of these bacteria seem to have the ability to colonize normal heart valves. Accumulations of bacteria on the valves (vegetations) hamper their function and can lead directly to cardiac malfunction and death. Alternatively, pieces of the bacterial vegetation can break off and create emboli (blockages) in vital organs. The bacterial colonies can also provide a constant source of blood-borne bacteria, with the accompanying systemic inflammatory response and shock. Bacteria that are attached to surfaces bathed by blood (such as heart valves) quickly become covered with a mesh of fibrin and platelets that protects them from the immune components in the blood.

Causative Agents   The acute form of endocarditis is most often caused by *Staphylococcus aureus*. Other agents that cause it are *Streptococcus pyogenes, Streptococcus pneumoniae,* and *Neisseria gonorrhoeae,* as well as a host of other bacteria. Each of these bacteria is described elsewhere in this book; all are pathogenic.

Transmission and Epidemiology   The most common route of transmission for acute endocarditis is parenteral—that is, via direct entry into the body. Intravenous or subcutaneous drug users have been a growing risk group for the condition. Traumatic injuries and surgical procedures can also introduce the large number of bacteria required for the acute form of endocarditis.

Prevention and Treatment   Prevention is based on avoiding the introduction of bacteria into the bloodstream during surgical procedures or injections. Untreated, this condition is invariably fatal. Treatment depends on the identity and the antimicrobial susceptibility of the causative agent. In the case of gram-positive cocci, the drug of choice is penicillin for susceptible strains. Otherwise vancomycin plus an aminoglycoside may be used. High, continuous blood levels of antibiotics are required to resolve the infection because the bacteria exist in biofilm vegetations. In addition to the decreased access of antibiotics to bacteria deep in the biofilm, these bacteria often express a phenotype of lower susceptibility to antibiotics. Surgical debridement of the valves, accompanied by antibiotic therapy, is sometimes required.

Subacute Endocarditis

Subacute forms of this condition are almost always preceded by some form of damage to the heart valves, or by congenital malformation. Irregularities in the valves encourage the attachment of bacteria, which then form biofilms and impede normal function, as well as provide an ongoing source of bacteria to the bloodstream. People who have suffered rheumatic fever, and the accompanying damage to heart valves, are particularly susceptible to this condition (see chapter 21 for a complete discussion of rheumatic fever).

Causative Agents   Most commonly, subacute endocarditis is caused by bacteria of low pathogenicity, often originating in the oral cavity. Alpha-hemolytic streptococci, such as *Streptococcus sanguis, S. oralis, and S. mutans,* are most often responsible, although normal flora from the skin and other bacteria can also colonize abnormal valves and lead to this condition.

Transmission and Epidemiology   Minor disruptions in the skin or mucous membranes, such as those induced by vigorous toothbrushing, dental procedures, or relatively minor cuts and

✔ CHECKPOINT 20.1	Endocarditis	
Disease	**Acute Endocarditis**	**Subacute Endocarditis**
Causative Organism(s)	*Staphylococcus aureus, Streptococcus pyogenes, S. pneumoniae, Neisseria gonorrhoeae,* others	Alpha-hemolytic streptococci, others
Most Common Modes of Transmission	Parenteral	Endogenous transfer of normal flora to bloodstream
Virulence Factors	Attachment	Attachment
Culture/Diagnosis	Blood culture	Blood culture
Prevention	Aseptic surgery, injections	Prophylactic antibiotics before invasive procedures
Treatment	Penicillin, or vancomycin plus aminoglycoside; surgery may be necessary	Penicillin, or vancomycin plus aminoglycoside; surgery may be necessary
Distinctive Features	Acute onset, high fatality rate	Slower onset

lacerations, can introduce bacteria into the bloodstream and lead to valve colonization. The bacteria are not, therefore, transmitted from other people or from the environment. The average age of onset for subacute endocarditis has increased in recent decades from the mid-twenties to the mid-fifties. Males are slightly more likely to experience it than females.

Prevention and Treatment The practice of prophylactic antibiotic therapy in advance of surgical and dental procedures on patients with underlying valve irregularities has decreased the incidence of this infection. When it occurs, treatment is similar to treatment for the acute form of the disease, described earlier **(Checkpoint 20.1).**

Septicemias

Septicemia occurs when organisms are actively multiplying in the blood. Many different bacteria (and a few fungi) can cause this condition. Patients suffering from these infections are sometimes described as "septic." One infection that should be considered in cases of aggressive septicemia, especially if respiratory symptoms are also present, is anthrax.

Signs and Symptoms

Fever is a prominent feature of septicemia. The patient appears very ill, and may have an altered mental state, shaking chills, and gastrointestinal symptoms. Often an increased breathing rate is exhibited, accompanied by respiratory alkalosis (increased tissue pH due to breathing disorder). Low blood pressure is a hallmark of this condition and is caused by the inflammatory response to infectious agents in the bloodstream, which leads to a loss of fluid from the vasculature. This condition is the most dangerous feature of the disease, often culminating in death.

Causative Agents

The vast majority of septicemias are caused by bacteria, and they are approximately evenly divided between gram-

positives and gram-negatives. Perhaps 10% are caused by fungal infections. Polymicrobic bloodstream infections increasingly are being identified, in which more than one microorganism is causing the infection.

Pathogenesis and Virulence Factors

Gram-negative bacteria multiplying in the blood release large amounts of endotoxin into the bloodstream, stimulating a massive inflammatory response mediated by a host of cytokines. This response invariably leads to a drastic drop in blood pressure, a condition called **endotoxic shock.** Gram-positive bacteria can instigate a similar cascade of events when fragments of their cell walls are released into the blood.

Transmission and Epidemiology

In many cases, septicemias can be traced to parenteral introduction of the microorganisms via intravenous lines or surgical procedures. Other infections may arise from serious urinary tract infections or from renal, prostatic, pancreatic, or gallbladder abscesses. Patients with underlying spleen malfunction may be predisposed to multiplication of microbes in the bloodstream. Meningeal infections or pneumonia occasionally can lead to sepsis. Approximately half a million cases occur each year in the United States, resulting in more than 100,000 deaths.

Culture and Diagnosis

Because the infection is in the bloodstream, a blood culture is the obvious route to diagnosis. A full regimen of media should be inoculated to ensure isolation of the causative microorganism. Antibiotic susceptibilities should be assessed. Empiric therapy should be started immediately, before culture and susceptibility results are available. The choice of antimicrobial agent should be informed by knowledge of any suspected source of the infection, such as an intravenous catheter (in which case, skin flora should be considered), urinary tract infections (in which case, gram-negatives and *Streptococci* should be considered), and so forth.

Prevention and Treatment

Empiric therapy, which is begun immediately after blood cultures are taken, often begins with a broad-spectrum antibiotic. Once the organism is identified, and its antibiotic susceptibility is known, treatment can be adjusted accordingly.

✔ CHECKPOINT 20.2	Septicemia
Causative Organism(s)	Bacteria or fungi
Most Common Modes of Transmission	Parenteral, endogenous transfer
Virulence Factors	Cell wall or membrane components
Culture/Diagnosis	Blood culture
Prevention	–
Treatment	Broad-spectrum antibiotic until identification and susceptibilities tested

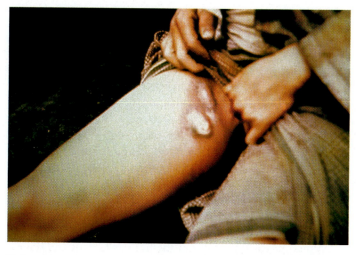

FIGURE 20.3 **A classic inguinal bubo of bubonic plague.** This hard nodule is very painful and can rupture onto the surface.

Plague

The word **plague**[1] conjures up visions of death and morbidity unlike any other infectious disease. Although pandemics of plague have probably occurred since antiquity, the first one that was reliably chronicled killed an estimated 100 million people in the sixth century A.D. The last great pandemic occurred in the late 1800s and was transmitted around the world, primarily by rat-infested ships. The disease was brought to the United States through the port of San Francisco around 1906. Infected rats eventually mingled with native populations of rodents and gradually spread the disease throughout the West and Midwest.

Signs and Symptoms

Three possible manifestations of infection occur with the bacterium causing plague. **Pneumonic plague** is a respiratory disease, described in chapter 21. In **bubonic plague,** the bacterium, which is injected by the bite of a flea, enters the lymph and is filtered by a local lymph node. Infection causes inflammation and necrosis of the node, resulting in a swollen lesion called a **bubo,** usually in the groin or axilla **(figure 20.3).** The incubation period lasts 2 to 8 days, ending abruptly with the onset of fever, chills, headache, nausea, weakness, and tenderness of the bubo. Mortality rates, even with treatment, are greater than 15%.

These cases often progress to massive bacterial growth in the blood termed septicemic plague. The presence of the bacteria in the blood results in disseminated intravascular coagulation, subcutaneous hemorrhage, and purpura that may degenerate into necrosis and gangrene. Mortality rates, once the disease has progressed to this point, are 30% to 50% with treatment, and 100% without treatment. Because of the visible

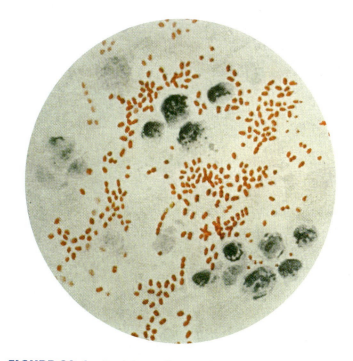

FIGURE 20.4 *Yersinia pestis.* Note the more darkly stained poles of the bacterium, lending it a "safety pin" appearance.

darkening of the skin, the plague has often been called the "black death."

Causative Agent

The cause of this dreadful disease is a tiny, harmless-looking gram-negative rod, *Yersinia pestis,* a member of the Family Enterobacteriaceae. Other species members are *Y. enterocolitica* and *Y. pseudotuberculosis.* Those species cause gastrointestinal tract diseases in humans. *Y. pestis* displays unusual bipolar staining that makes it look like a safety pin **(figure 20.4).**

---

1. From the Latin *plaga,* meaning to strike, infect, or afflict with disease, calamity, or some other evil.

The girl in the chapter opener was suffering from septicemic plague. But wait! you say—the original Gram stain revealed gram-*positive* diplococci, later identified as *Streptococcus pneumonia*, and as a result she was treated for gram-positive septicemia.

This story is an excellent example of the importance of using multiple indicators for presumptive diagnosis. It is not clear why *S. pneumoniae* was present in large numbers in the patient's blood, but apparently its presence indicated an infection secondary to plague. Any patient with clinical signs of sepsis and a history that suggests possible plague exposure should immediately be treated empirically with antibiotics known to be highly effective against *Y. pestis*, such as streptomycin or gentamycin.

Testing revealed that four of the five family dogs and the cat with the abscess were all positive for *Y. pestis*. They had presumably been infected with fleas shared with the local prairie dog colony. The victim had cared for the sick family cat in recent days. In this case, patient history (place of residence, recent exposure to a sick animal) was a better indicator than laboratory tests.

The presenting symptom of pain in the arm, and especially in the axilla, was probably an early sign of *Y. pestis* infection and multiplication in the axillary lymph nodes. The earlier trampoline injury only served to deflect attention away from the true cause of the axillary pain.

See: CDC. 1997. Fatal human plague—Arizona and Colorado, 1996. MMWR 46:617–620.

Pathogenesis and Virulence Factors

The number of bacteria required to initiate a plague infection is small—perhaps only 3 to 50 cells. Much research has been conducted on the differences between the two *Yersinia* species that cause GI tract disease and this *Yersinia*, since it has such different effects on the host. Scientists have discovered that *Y. pestis* carries three plasmids: One is common to all three *Yersinia* species; two are unique to *Y. pestis*. All three plasmids carry genes important for pathogenesis. The plasmid all *Yersinia* carry contains genes for a system called the Yop system, a series of proteins the bacteria use to attach to host cells and inject proteins into them that short circuit the immune response. The two plasmids unique to *Y. pestis* carry genes that help it to cause disease in mice and to survive in the flea vector. Examples of these genes include a gene for capsule formation and a gene for plasminogen activation (similar to the streptokinase expressed by *S. pyogenes*) (see chapter 18). Plasminogen activation leads to clotting, which helps the microbe resist phagocytosis.

Transmission and Epidemiology

The principal agents in the transmission of the plague bacterium are fleas. These tiny, bloodsucking insects have a special relationship with the bacterium. After a flea ingests a blood meal from an infected animal, the bacteria multiply in its gut. In fleas that effectively transmit the bacterium, the esophagus becomes blocked due to coagulation factors produced by the pathogen. Being unable to feed properly, the ravenous flea jumps from animal to animal in a futile attempt to get nourishment. During this process, regurgitated infectious material is inoculated into the bite wound.

The plague bacterium exists naturally in many animal hosts, and its distribution is extensive. Although the incidence of disease has been reduced in the developed world, it has actually been increasing in Africa and other parts of the world. Plague still exists endemically in large areas of Africa, South America, the Mideast, Asia, and the former Soviet Union, and it sometimes erupts into epidemics such as the outbreak in India in the 1990s that infected hundreds of residents. This new surge of cases was attributed to increased populations of rats following the monsoon floods. In the United States, sporadic cases (usually less than 10 per year) occur as a result of contact with wild and domestic animals. This disease is considered endemic in U.S. western and southwestern states. Persons most at risk for developing plague are veterinarians and people living and working near woodlands and forests. Dogs and cats can be infected with the plague, often from contact with infected wild animals such as prairie dogs. Human cases have been traced to a chain of events involving a flea from a prairie dog moving to a domestic cat, and then a flea from cat moving to a human.

The epidemiology of plague is among the most complex of all diseases. It involves several different types of vertebrate hosts and flea vectors, and its exact cycle varies from one region to another. A general scheme of the cycle is presented in **figure 20.5.** Humans can develop plague through contact with the fleas of wild or domestic or semidomestic animals. Contact with infected body fluids can also spread the disease. If a person has breaks in the skin on his or her hands, handling infected animals or animal skins is a possible means of transmission. (Persons with the pneumonic form of the disease can spread *Y. pestis* through respiratory droplets.)

The Animal Reservoirs The plague bacillus occurs in 200 different species of mammals. The primary long-term *endemic reservoirs* are various rodents, such as mice and voles, that harbor the organism but do not develop the disease. These hosts spread the disease to other mammals called *amplifying hosts* that become infected with the bacterium and experience massive die-offs during epidemics. These hosts, including rats, ground squirrels, chipmunks, and rabbits, are the usual sources of human plague. The particular mammal that is most important in this process depends on the area of the world. Other mammals (camels, sheep, coyotes, deer, dogs, and cats) can also be involved in the transmission cycle.

Culture and Diagnosis

Because death can ensue as quickly as 2 to 4 days after the appearance of symptoms, prompt diagnosis and treatment of plague are imperative. The patient's history, including recent

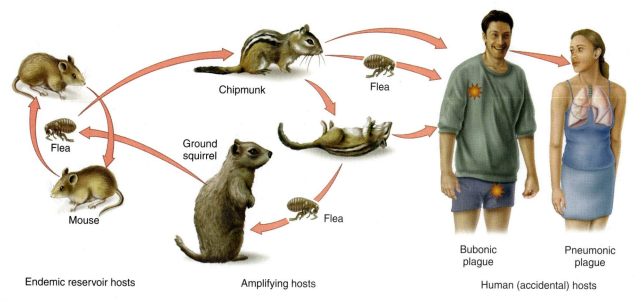

Endemic reservoir hosts   Amplifying hosts   Human (accidental) hosts

FIGURE 20.5 The infection cycle of *Yersinia pestis* (simplified for clarity).

travel to endemic regions, can help establish a diagnosis. Culture of the organism is the definitive method of diagnosis, although a Gram stain of aspirate from buboes often reveals the presence of the safety-pin-shaped bacteria.

Prevention and Treatment

Plague is one of a handful of internationally quarantinable diseases—(others are cholera and yellow fever). In addition to quarantine during epidemics, plague is controlled by trapping rodents and by poisoning their burrows with insecticide to kill fleas. These methods, however, cannot begin to control the reservoir hosts, so the potential for plague will always be present in endemic areas, especially as humans encroach into rodent habitats. A killed or attenuated vaccine that protects against the disease for a few months is given to military personnel, veterinarians, and laboratory workers.

Streptomycin or gentamycin are the drugs of choice.

✔ **CHECKPOINT 20.3**	**Plague**
Causative Organism(s)	*Yersinia pestis*
Most Common Modes of Transmission	Vector, biological; also droplet contact (pneumonic) and direct contact with body fluids
Virulence Factors	Capsule, Yop system, plasminogen activator
Culture/Diagnosis	Culture or Gram stain of blood or bubo aspirate
Prevention	Flea and or animal control; vaccine available for high-risk individuals
Treatment	Streptomycin or gentamycin

Tularemia

The causative agent of tularemia is a facultative intracellular gram-negative bacterium called *Francisella tularensis*. It has several characteristics in common with *Yersinia pestis*, and the two species were previously often included in a single genus called *Pasteurella*. It is a zoonotic disease of assorted mammals endemic to the Northern Hemisphere. Because it has been associated with outbreaks of disease in wild rabbits, it is sometimes called rabbit fever. It is currently listed as a pathogen of concern on the lists of bioterrorism agents (see Insight 21.2 for details).

Tularemia is abundantly distributed through numerous animal reservoirs and vectors in northern Europe, Asia, and North America, but not in the tropics. This disease is noteworthy for its complex epidemiology and spectrum of symptoms. Although rabbits and rodents (muskrats and ground squirrels) are the chief reservoirs, other wild animals (skunks, beavers, foxes, opossums) and some domestic animals are implicated as well. The chief route of transmission in the past had been through the activity of skinning rabbits, but with the decline of rabbit hunting, transmission via tick bites is more common. Ticks are the most frequent arthropod vector, followed by biting flies, mites, and mosquitoes.

Tularemia is strikingly varied in its portals of entry and disease manifestations. Although bites by a vector are the most common source of infection, in many cases infection results when the skin or eye is inoculated through contact with infected animals, animal products, contaminated water, and dust. Pulmonary forms of the infection can result from aerosolized soils or animal fluids and also from spread of the bacterium in the bloodstream. The disease is not communicated from human to human. With an estimated infective dose of between 10 and 50 organisms, *F. tularensis* is often considered one of the most infectious of all bacteria. The term "lawnmower" tularemia refers to tularemia acquired while performing grass-mowing or brush-cutting chores. Cases of tularemia have appeared in people who have accidentally run over dead rabbits while lawn mowing, presumably from inhaling aerosolized bacteria.

After an incubation period ranging from a few days to 3 weeks, acute symptoms of headache, backache, fever, chills, malaise, and weakness appear. Further clinical manifestations are tied to the portal of entry. They include ulcerative skin lesions, swollen lymph glands, conjunctival inflammation, sore throat, intestinal disruption, and pulmonary involvement. The death rate in the most serious forms of disease is 10%, but proper treatment with gentamycin or tetracycline reduces mortality to almost zero. Because the intracellular persistence of *F. tularensis* can lead to relapses, antimicrobial therapy must not be discontinued prematurely. Protection is available in the form of a live attenuated vaccine. Laboratory workers and other occupationally exposed personnel must wear gloves, masks, and eyewear.

✔ CHECKPOINT 20.4	Tularemia
Causative Organism(s)	*Francisella tularensis*
Most Common Modes of Transmission	Vector, biological; also direct contact with body fluids from infected animal; airborne
Virulence Factors	Intracellular growth
Culture/Diagnosis	Culture dangerous to lab workers and not reliable; serology most often used
Prevention	Live attenuated vaccine for high-risk individuals
Treatment	Gentamycin or tetracycline

Infectious Mononucleosis

This lymphatic system disease, which is often simply called "mono" or the "kissing disease," can be caused by a number of bacteria or viruses, but the vast majority of cases are caused by the **Epstein-Barr virus (EBV),** and most of the remainder are caused by cytomegalovirus (CMV). Both of these viruses are in the herpes family.

Signs and Symptoms

The symptoms of mononucleosis are sore throat, high fever, and cervical lymphadenopathy, which develop after a long incubation period (30 to 50 days). Many patients also have a gray-white exudate in the throat, a skin rash, and enlarged spleen and liver. A notable sign of mononucleosis is sudden leukocytosis, consisting initially of infected B cells and later T cells. Fatigue is a hallmark of the disease. Patients remain fatigued for a period of weeks. During that time, they are advised not to engage in strenuous activity due to the possibility of injuring their enlarged spleen (or liver).

Eventually, the strong, cell-mediated immune response is decisive in controlling the infection and preventing complications. But after recovery, people usually remain chronically infected with EBV and CMV.

Epstein-Barr Virus

Although "mono" was first described more than a century ago, its most frequent cause was finally discovered through a series of accidental events starting in 1958, when Michael Burkitt discovered an unusual malignant tumor in African children (Burkitt's lymphoma) that appeared to be infectious. Later, Michael Epstein and Yvonne Barr cultured a virus from tumors that showed typical herpesvirus morphology. Evidence that the two diseases had a common cause was provided when a laboratory technician accidentally acquired mononucleosis while working with the Burkitt's lymphoma virus. The Epstein-Barr virus shares morphological and antigenic features with other herpesviruses, and in addition, it contains a circular form of DNA that is readily spliced into the host cell DNA.

Scientists have long suspected a link between chronic EBV infection and illnesses such as chronic fatigue syndrome, but the connection is still controversial. In 2003, a report in the *New England Journal of Medicine* presented strong evidence that chronic EBV infection was necessary, although probably not sufficient, to cause certain forms of Hodgkin's lymphoma.

Pathogenesis and Virulence Factors The latency of the virus, and its ability to splice its DNA into host cell DNA, make it an extremely versatile virus that can avoid the host's immune response.

Transmission and Epidemiology More than 90% of the world's population is infected with EBV. In general, the virus causes no noticeable symptoms, but the time of life when the virus is first encountered seems to matter. In the case of EBV, infection during the teen years seems to result in disease, whereas infection before or after this period is usually asymptomatic. You will soon see that infection with CMV during the fetal period can lead to severe disease.

Direct oral contact and contamination with saliva are the principal modes of transmission, although transfer through blood transfusions, sexual contact, and organ transplants is possible.

Culture and Diagnosis A differential blood count that shows excess lymphocytes, reduced neutrophils, and large, atypical lymphocytes with lobulated nuclei and vacuolated cytoplasm is suggestive of EBV infection (**figure 20.6**). A test called the "Monospot test" detects *heterophile antibodies*—which are antibodies that are not directed against EBV but are seen when a person has an EBV infection. This test is not reliable in children younger than age 4, in which case a specific EBV antigen/antibody test is conducted.

Prevention and Treatment   The usual treatments for infectious mononucleosis are directed at symptomatic relief of fever and sore throat. Hospitalization is rarely needed. Occasionally, rupture of the spleen necessitates immediate surgery to remove it.

Cytomegalovirus

Cytomegalovirus (CMV) is also a herpesvirus. It is generally distinguished by its ability to produce giant (megalo) cells (cyto) with nuclear and cytoplasmic inclusion bodies. Like other herpesviruses, both EBV and CMV have a tendency to become latent in host cells. Infections are likely to be permanent. The viruses do not reemerge the way herpes simplex viruses do, unless a patient becomes severely immunocompromised. (CMV ocular symptoms are a common complication of AIDS—affecting up to 40% of all AIDS patients.)

Pathogenesis and Virulence Factors   The ability of the virus to fuse cells and its latency both contribute to its virulence.

Transmission and Epidemiology   Like EBV, CMV is ubiquitous in humans. Unlike EBV, CMV generally causes disease only in fetuses, newborns, and immunodeficient adults. Although not covered here, CMV infection of fetuses affects up to 5,000 babies a year and can cause long-term neurological and sensory disturbances.

CMV is transmitted in saliva, respiratory mucus, milk, urine, semen, cervical secretions, and feces. Transmission usually involves intimate contact such as sex, vaginal birth, transplacental infection, blood transfusion, and organ transplantation.

Culture and Diagnosis   During CMV mononucleosis, the virus can be isolated from virtually all organs as well as from epithelial tissue. Cell enlargement and prominent inclusions in the cytoplasm and nucleus are suggestive of CMV. The virus can be cultured and tested with monoclonal antibody against a CMV protein called *early nuclear antigen*. Direct ELISA tests and DNA probe analysis are also useful in diagnosis. Testing serum for antibodies may fail to diagnose infection in neonates and in the immunocompromised, so it is less reliable.

Prevention and Treatment   Drug therapy is generally reserved for serious disease in immunosuppressed patients, and not for CMV mononucleosis. The three main drugs are ganciclovir, valacyclovir, and foscarnet, which have toxic side effects and cannot be administered for long periods. The development of a vaccine is hampered by the lack of an animal than can be infected with human cytomegalovirus. One crucial concern is whether vaccine-stimulated antibodies would be protective, since patients already seropositive can become naturally reinfected. Despite these odds, clinical trials began in late 2003 for an experimental CMV vaccine.

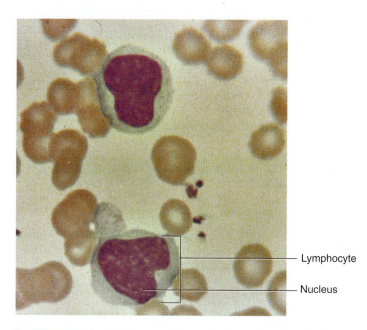

Lymphocyte

Nucleus

FIGURE 20.6   **Evidence of Epstein-Barr infection in the blood smear of a patient with infectious mononucleosis.** Note the abnormally large lymphocytes containing indented nuclei with light discolorations.

✔ CHECKPOINT 20.5	Infectious Mononucleosis	
Causative Organism(s)	Epstein-Barr virus (EBV)	Cytomegalovirus (CMV)
Most Common Modes of Transmission	Direct, indirect contact, parenteral	Direct, indirect contact, parenteral, vertical
Virulence Factors	Latency, ability to incorporate into host DNA	Latency, ability to fuse cells
Culture/Diagnosis	Differential blood count, Monospot test for heterophile antibody, specific ELISA	Virus isolation and growth, ELISA or PCR tests
Prevention	–	Vaccine in trials
Treatment	Supportive	Only for immunosuppressed patients, not usually for mononucleosis
Distinctive Features	Most common in teens	More common in adults, dangerous to fetus

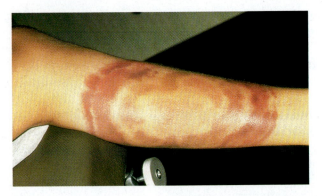

FIGURE 20.7 **Lesions of Lyme disease on the lower leg.**
Note the flat, reddened rings in the form of a bull's-eye.

Lyme disease

In the 1970s, an enigmatic cluster of arthritis cases appeared in the town of Old Lyme, Connecticut. The phenomenon caught the attention of nonprofessionals and professionals alike, whose persistence and detective work ultimately disclosed the unusual nature and epidemiology of Lyme disease. The process of discovery began in the home of Polly Murray, who, along with her family, was beset for years by recurrent bouts of stiff neck, swollen joints, malaise, and fatigue that seemed vaguely to follow a rash from tick bites. When Mrs. Murray's son was diagnosed as having juvenile rheumatoid arthritis, she became skeptical. Conducting her own literature research, she began to discover inconsistencies. Rheumatoid arthritis was described as a rare, noninfectious disease, yet over an 8-year period, she found that 30 of her neighbors had experienced similar illnesses. Ultimately this cluster of cases, and others, were reported to state health authorities. Eventually Lyme disease was shown to be caused by *Borrelia burgdorferi*. It is now recognized that Lyme disease has been around for centuries.

Signs and Symptoms

Lyme disease is nonfatal, but it often evolves into a slowly progressive syndrome that mimics neuromuscular and rheumatoid conditions. An early symptom in 70% of cases is a rash at the site of a tick bite. The lesion, called *erythema migrans*, looks something like a bull's-eye, with a raised erythematous (reddish) ring that gradually spreads outward and a pale central region **(figure 20.7)**. Other early symptoms are fever, headache, stiff neck, and dizziness. If not treated or if treated too late, the disease can advance to the second stage, during which cardiac and neurological symptoms, such as facial palsy, can develop. After several weeks or months, a crippling polyarthritis can attack joints. Some people acquire chronic neurological complications that are severely disabling.

Causative Agent

Borrelia burgdorferi was discovered in 1981 by Dr. Willy Burgdorfer, although he did not realize at that time its con-

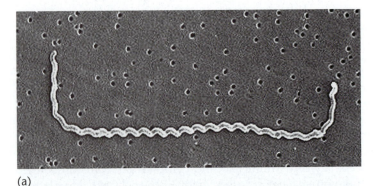

(a)

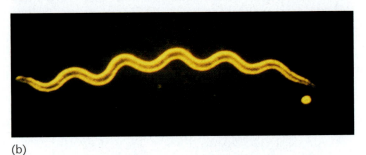

(b)

FIGURE 20.8 **Spirochetes.**
(a) *Leptospira* has numerous fine, regular coils and one or both ends curved. (b) *Borrelia* has 3 to 10 loose, irregular coils.

nection with disease. Borrelia are spirochetes, but they are morphologically distinct from other pathogenic spirochetes. They are comparatively larger, ranging from 0.2 to 0.5 μm in width and from 10 to 20 μm in length, and they contain 3 to 10 irregularly spaced and loose coils **(figure 20.8)**. The nutritional requirements of *Borrelia* are so complex that the bacterium can be grown in artificial media only with difficulty.

Pathogenesis and Virulence Factors

The bacterium is a master of immune evasion. It changes its surface antigens while it is in the tick, and again after it has been transmitted to a mammalian host. It provokes a strong humoral and cellular immune response, but this response is mainly ineffective, perhaps because of the bacterium's ability to switch its antigens. Indeed, it is possible that the immune response contributes to the pathology of the infection.

B. burgdorferi also has multiple proteins for attachment to host cells; these are considered virulence factors as well.

Transmission and Epidemiology

B. burgdorferi is transmitted primarily by hard ticks of the genus *Ixodes*. (See **Insight 20.2** for a discussion of ticks and other arthropod vectors of diseases.) In the northeastern part of the United States, *Ixodes scapularis* (the black-legged deer tick) passes through a complex 2-year cycle that involves two principal hosts **(figure 20.9)**. As a larva or nymph, it feeds on the white-footed mouse, where it picks up the infectious agent. The nymph is relatively nonspecific and will try to feed on nearly any type of vertebrate—thus, it is the form most likely to bite humans. The adult tick reproductive phase

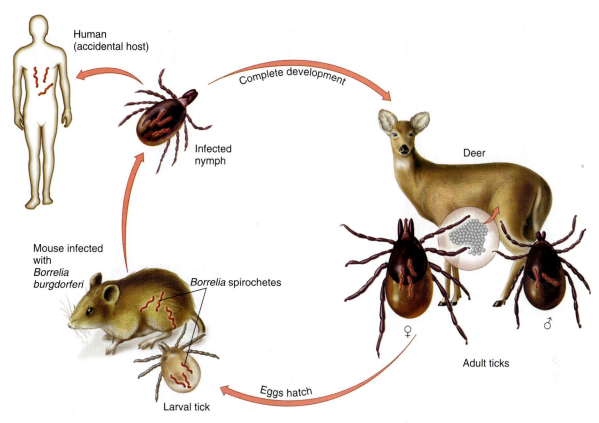

FIGURE 20.9 **The cycle of Lyme disease in the northeastern United States.**
The exact reservoir hosts vary from region to region in the United States and worldwide.

of the cycle is completed on deer. In California, the transmission cycle involves *Ixodes pacificus* and the dusky-footed woodrat as reservoir.

The incidence of Lyme disease is showing a gradual upward trend from about 10,000 cases per year in 1991 to 18,000 in 2002. This increase may be partly due to improved diagnosis, but it also reflects changes in the numbers of hosts and vectors. The greatest concentrations of Lyme disease are found in areas having high mouse and deer populations. Most of the cases have occurred in New York, Pennsylvania, Connecticut, New Jersey, Rhode Island, and Maryland, but the numbers in the Midwest and West are growing. Highest risk groups include hikers, backpackers, and people living in newly developed communities near woodlands and forests. Peak seasons are the summer and early fall.

Culture and Diagnosis

Diagnosis of Lyme disease can be difficult because of the range of symptoms it presents. Most suggestive are the ring-shaped lesions, isolation of spirochetes from the patient, and serological testing with an ELISA method that tracks a rising antibody titer. Tests for spirochetal DNA in specimens is especially helpful for late-stage diagnosis.

Prevention and Treatment

A vaccine for Lyme disease was available for a brief period of time, but it was withdrawn from the market in early 2002 be-

cause of controversy over its possible side effects. Other vaccines are in development. Because dogs can also acquire the disease, there is a vaccine for them. Anyone involved in outdoor activities should wear protective clothing, boots, leggings, and insect repellant containing DEET.[2] Individuals exposed to heavy infestation should routinely inspect their bodies for ticks and remove ticks gently without crushing, preferably with forceps or fingers protected with gloves, because it is possible to become infected by tick feces or body fluids.

Early, prolonged (3 to 4 weeks) treatment with doxycycline and amoxicillin is effective, and other antibiotics such as ceftriaxone and penicillin are used in late Lyme disease therapy.

✔ CHECKPOINT 20.6	Lyme Disease
Causative Organism(s)	*Borrelia burgdorferi*
Most Common Modes of Transmission	Vector, biological
Virulence Factors	Antigenic shifting, adhesins
Culture/Diagnosis	ELISA for Ab, PCR
Prevention	Tick avoidance
Treatment	Doxycycline and/or amoxicillin (3–4 weeks), also ceftriaxone and penicillin

2. N,N-Diethyl-M-toluamide—the active ingredient in OFF! and Cutter brand insect repellants.

The Arthropod Vectors of Infectious Disease

Many bacterial pathogens have evolved with, and made complex adaptations to, the bodies of arthropods, particularly insects and arachnids. In their role as biological vectors, they are an important source of zoonotic infections in humans. In this chapter alone, you learn about the flea transmitting plague, lice transmitting trench fever, and the very busy tick transmitting tularemia, Lyme disease, and ehrlichioses to humans, while playing a part in keeping Q fever cycling among animal species. Here we describe some main groups implicated in disease.

Ticks

There are over 810 species of ticks throughout the world. About 100 of them are vectors of infectious disease. Ticks are arachnids, as compared with fleas and lice, which are insects.

Hard (ixodid) ticks have adapted to a wide-ranging life-style, hitchhiking along as their hosts wander through forest, savanna, or desert regions. Depending upon the species, ticks feed during larval, nymph, and adult metamorphic stages. The longevity of ticks is formidable; metamorphosis can extend for 2 years, and adults can survive for 4 years away from a host without feeding. The tiny unengorged ticks humans pick up from vegetation crawl on the body, embed their mouthparts in the skin, and fill with blood, expanding to hundreds of times in size. Ixodid ticks are implicated in Rocky Mountain spotted fever and Q fever, as well as the ehrlichioses.

Fleas

Fleas are laterally flattened, wingless insects with well-developed jumping legs and a prominent proboscis for piercing the skin of warm-blooded animals. They are known for their extreme longevity and resistance, and many are notorious in their nonspecificity, passing with ease from wild or domesticated mammals to humans. In response to mechanical stimulation and warmth, fleas jump onto their targets and crawl about, feeding as they go. A well-known example is the oriental rat flea that transmits *Rickettsia typhi*, the cause of murine typhus. The flea harbors the pathogen in its gut and period-

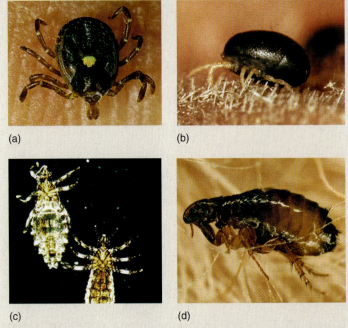

(a)   (b)

(c)   (d)

Arthropod vectors.
(a) Hard (ixodid) tick. **(b)** An engorged soft tick. **(c)** The body louse. **(d)** Cat flea.

ically contaminates the environment with virulent bacteria by defecating. This same flea is involved in the transmission of plague.

Lice

Lice (singular, louse) are small, flat insects equipped with biting or sucking mouthparts. The lice of humans usually occupy head and body hair, or pubic, chest, and axillary hair. They feed by gently piercing the skin and sucking blood and tissue fluid. Infection develops when the louse (or its feces) is inadvertently squashed and rubbed into wounds, skin, eyes, or mucous membranes. See the discussion on trench fever to read about a disease transmitted by lice.

Hemorrhagic Fever Diseases

A number of agents that infect the blood and lymphatics cause extreme fevers, some of which are accompanied by internal hemorrhaging. The diseases are grouped into the category of "hemorrhagic fevers" and are covered in this section. The following section deals with diseases in which the main symptom is fever—without the hemorrhagic part.

All hemorrhagic fever diseases described here are caused by viruses in one of three families: Arenaviridae, Filoviridae, and Flaviviridae. Bunyaviridae is a fourth family with members that cause hemorrhagic fevers, but we will not discuss examples of these here. All of these viruses are RNA enveloped viruses, the distribution of which is restricted to their natural host's distribution.

Yellow Fever

This disease is caused by an arbovirus, a single-stranded RNA flavivirus that is generally called the yellow fever virus. It currently occurs only in parts of Africa and South America. Two patterns of transmission are seen in nature. One is an urban cycle between humans and the mosquito *Aedes aegypti*, which reproduces in standing water in cities. The other is a sylvan (forest) cycle, maintained between forest monkeys and mosquitoes.

The presence of the virus in the bloodstream causes capillary fragility and disrupts the blood clotting system, which can lead to localized bleeding and shock. Infection begins acutely with fever, headache, and muscle pain. In some patients, the disease progresses to oral hemorrhage, nosebleed, vomiting, jaundice, and liver and kidney damage with significant mortality rates. Most cases occur during the rainy season.

Dengue Fever

Dengue fever is caused by a single-stranded RNA flavivirus and is also carried by *Aedes* mosquitoes. Although mild infection is the usual pattern, a form called dengue hemorrhagic shock syndrome can be lethal. Dengue fever is also called "breakbone fever" because of the severe pain it induces in muscles and joints (it does not actually cause fractures). The illness is endemic to Southeast Asia and India, and several epidemics have occurred in South America and Central America, the Caribbean, and Mexico. The Pan American Health Organization has reported an ongoing epidemic of dengue fever in the Americas that has increased from 390,000 cases in 1984 to more than 1 million cases in 2002.

Researchers in Thailand, where dengue fever is one of the leading causes of child mortality, have developed a live attenuated vaccine, which is being tested in clinical trials.

Ebola and Marburg

Unlike the two viruses causing yellow fever and dengue fever, the Ebola and Marburg viruses are filoviruses (Family Filoviridae). The two viruses are related and cause similar symptoms, although Ebola has received the greatest share of media attention. Its gruesome symptoms are extreme manifestations of the same kind of hemorrhagic events described for yellow fever and dengue fever. The virus in the bloodstream leads to extensive capillary fragility and disruption of clotting. Patients bleed from their orifices, even from their mucous membranes, and experience massive internal and external hemorrhage. Very often they manifest a rash on their trunk in early stages of the disease. The mortality rate is between 25% and 100%, and there is no effective treatment.

It is not known how humans acquire these viruses. They are both indigenous to Africa. Their natural reservoir is also unknown, although many scientists suspect that it might be non-

human primates. Direct contact with an infected person or with their body fluids will transmit the virus. Hospital workers caring for Ebola patients are at high risk of becoming infected.

In 2002 and 2003, Ebola caused a catastrophic epidemic among lowland gorillas in Central Africa. During that time dozens of humans also contracted the disease, probably from handling dead gorilla carcasses or from using primate meat for food. The disease was then transmitted from human to human. Although the extent of the epidemic is still unknown, researchers suspect that thousands of gorillas and hundreds of people have died over the past several years in this outbreak.

Outbreaks with Marburg virus are very rare, but individuals have been infected sporadically since it was first recognized in 1967. Symptoms are similar to Ebola virus infection.

Lassa Fever

The Lassa fever virus is an arenavirus. Several related arenaviruses cause the diseases Argentine hemorrhagic fever, Bolivian hemorrhagic fever, and lymphocytic choriomeningitis (an infection of the brain and meninges). Lassa fever virus is found in West Africa. In most cases, infection with this virus is asymptomatic, but in 20% of the cases, a severe hemorrhagic syndrome develops. The syndrome includes chest pain, hemorrhaging, sore throat, back pain, vomiting, diarrhea, and sometimes encephalitis. Patients who recover suffer from deafness at a significant rate.

The reservoir of the virus is a rodent found in Africa called the multimammate rat. It is spread to humans through aerosolization of rat droppings, urine, hair, and so forth. Eating food contaminated by rat excretions also transmits the virus. Infected persons can spread it to other people through their own secretions. Vertical transmission also occurs, and the disease leads to spontaneous abortions in 95% of infected pregnant women.

This hemorrhagic fever has been shown to respond to the antiviral agent ribavirin, especially if administered in the early stages of infection. There is no vaccine.

✔ CHECKPOINT 20.7 Hemorrhagic Fevers

Disease	Yellow Fever	Dengue Fever	Ebola and/or Marburg	Lassa Fever
Causative Organism(s)	Yellow fever virus	Dengue fever virus	Ebola virus, Marburg virus	Lassa fever virus
Most Common Modes of Transmission	Biological vector	Biological vector	Direct contact, body fluids	Droplet contact (aerosolized rodent excretions), direct contact with infected fluids
Virulence Factors	Disruption of clotting factors	Disruption of clotting factors	Disruption of clotting factors	Disruption of clotting factors
Culture/Diagnosis	ELISA, PCR	Rise in IgM titers	PCR, viral culture (conducted at CDC)	ELISA
Prevention	Live attenuated vaccine available	Live attenuated vaccine being tested	–	Avoiding rats, safe food storage
Treatment	Supportive	Supportive	Supportive	Ribavirin
Distinctive Features	Accompanied by jaundice	"Breakbone fever"—so named due to severe pain	Massive hemorrhage; rash sometimes present	Chest pain, deafness as long-term sequelae

Nonhemorrhagic Fever Diseases

In this section, we examine some infectious diseases that result in a syndrome characterized by high fever, but without the capillary fragility that leads to hemorrhagic symptoms. All of the diseases in this section are caused by bacteria.

Brucellosis

This disease goes by several different names (besides brucellosis): Malta fever, undulant fever, and Bang's disease.[3] It is on the CDC list of possible bioterror agents, though it is not designated as being "of highest concern."

Signs and Symptoms   The *Brucella* bacteria responsible for this disease live in phagocytic cells. These cells carry the bacteria into the bloodstream, creating focal lesions in the liver, spleen, bone marrow, and kidney. The cardinal manifestation of human brucellosis is a fluctuating pattern of fever, which is the origin of the common name, undulant fever **(figure 20.10)**. It is also accompanied by chills, profuse sweating, headache, muscle pain and weakness, and weight loss. Fatalities are not common, although the syndrome can last for a few weeks to a year, even with treatment.

Causative Agent   The bacterial genus *Brucella* contains tiny, aerobic gram-negative coccobacilli. Two species can cause this disease in humans: *B. abortus* (common in cattle), and *B. suis* (from pigs). Humans can become infected with either of these bacteria and experience severe disease. Even though a principal manifestation of the disease in animals is an infection of the placenta and fetus, human placentas do not become infected.

Pathogenesis and Virulence Factors   *Brucella* enters through damaged skin or via mucous membranes of the digestive tract, conjunctiva, and respiratory tract. From there it is taken up by phagocytic cells. Because it is able to avoid destruction in the phagocytes, the bacterium is transported easily through the bloodstream and to various organs, such as the liver, kidney, breast tissue, or joints. Scientists suspect that the up-and-down nature of the fever is related to unusual properties of the bacterial lipopolysaccharide.

Transmission and Epidemiology   Brucellosis occurs worldwide, with concentrations in Europe, Africa, India, and Latin America. It is associated predominantly with occupational contact in slaughterhouses, livestock handling, and the veterinary profession. Infection takes place through contact with blood, urine, placentas, and through consumption of raw milk and cheese. Human-to-human transmission is rare. Needlesticks are one of the more common modes of transmission in the United States. Inhalation of aerosolized bacteria occurs with some frequency as well.

Brucellosis is also a common disease of wild herds of bison and elk. Cattle that share grazing land with these wild herds often suffer severe outbreaks of the placental infections (called Bang's disease).

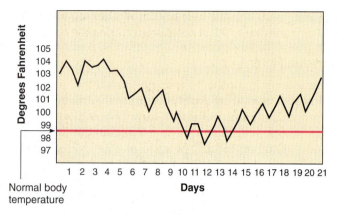

FIGURE 20.10   **The temperature cycle in classic brucellosis.**
Body temperature undulates between day and night and between fever, normal, and subnormal.
Source: A. Smith, Principles of Microbiology, *10th ed., 1985.*

Culture and Diagnosis   The patient's history can be very helpful in diagnosis, as are serological tests of the patient's blood and blood culture of the pathogen. In areas where *Brucella* is endemic, serology is of limited use since significant proportions of the population already display antibodies to the bacterium. Blood culture is positive in less than 40% of cases; Gram staining of biopsy material from lymph nodes or bone marrow (from the sternum) is considered more reliable.

Prevention and Treatment   Prevention is effectively achieved by testing and elimination of infected animals, quarantine of imported animals, and pasteurization of milk. Although several types of animal vaccines are available, those developed so far for humans are ineffective or unsafe. The status of this pathogen as a potential germ warfare agent makes a reliable vaccine even more urgent.

A combination of tetracycline and rifampin or streptomycin taken for 3 to 6 weeks is usually effective in controlling infection.

Q Fever

The name of this disease arose from the frustration created by not being able to identify its cause. The Q stands for "query." Its cause, a bacterium called *Coxiella burnetii*, was finally identified in the mid-1900s. The clinical manifestations of Q fever are abrupt onset of fever, chills, head and muscle ache, and occasionally, a rash. The disease is sometimes complicated by pneumonitis (30% of cases), hepatitis, and endocarditis.

C. burnetii is a very small pleomorphic gram-negative bacterium, and for a time it was considered a rickettsia. It is an intracellular parasite, but it is much more resistant to environmental pressures because it produces an unusual type of endospore-like structure **(figure 20.11)**. *C. burnetii* is apparently harbored by a wide assortment of vertebrates and arthropods, especially ticks, which play an essential role in transmission

---

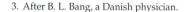

3. After B. L. Bang, a Danish physician.

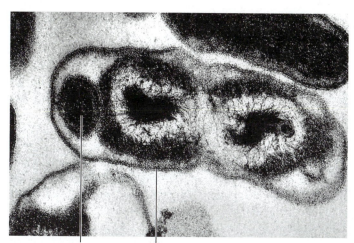

Endospore   Vegetative cell

FIGURE 20.11   **The agent of Q fever.**
The vegetative cells of *Coxiella burnetii* produce unique endospores that are released when the cell disintegrates. Free spores survive outside the host and are important in transmission.

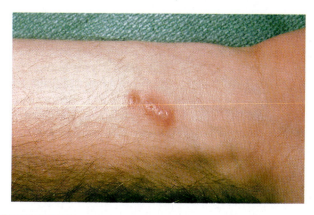

FIGURE 20.12   **Cat-scratch disease.**
A primary nodule appears at the site of the scratch in about 21 days. In time, large quantities of pus collect, and the regional lymph nodes swell.

between wild and domestic animals. Ticks do not transmit the disease to humans, however. Humans acquire infection largely by means of environmental contamination and airborne spread. Birth products, such as placentas, of infected domestic animals contain large numbers of bacteria. Other sources of infectious material include urine, feces, milk, and airborne particles from infected animals. The primary portals of entry are the lungs, skin, conjunctiva, and gastrointestinal tract.

C. burnetii has been isolated from most regions of the world. California and Texas have the highest case rates in the United States, although most cases probably go undetected. People at highest risk are farm workers, meat cutters, veterinarians, laboratory technicians, and consumers of raw milk products.

Mild or subclinical cases resolve spontaneously, and more severe cases respond to tetracycline therapy. A vaccine is available in many parts of the world and is used for U.S. military and laboratory workers. Q fever is of potential concern as a bioterror agent because it is very resistant to heat and drying, can be inhaled, and even a single bacterium is enough to cause disease. It is an organism that the U.S. military worked with during the period when potential biowarfare agents were being developed in this country (the 1950s and 1960s).

Cat-Scratch Disease

This disease is one of a pair of diseases caused by different species of the small gram-negative rod *Bartonella*. *Bartonella* species are considered to be emerging pathogens. They are fastidious but not obligate intracellular parasites, so they will grow on blood agar. In addition to cat-scratch disease and trench fever, discussed next, *Bartonella* species cause a particularly nasty cutaneous and systemic infection in AIDS patients called bacillary angiomatosis.

Bartonella henselae is the agent of cat-scratch disease (CSD), an infection connected with being clawed or bitten by a cat.

The pathogen is present in over 40% of cats, especially kittens. There are approximately 25,000 cases per year in the United States, 80% of them in children 2 to 14 years old. The symptoms start after 1 to 2 weeks, with a cluster of small papules at the site of inoculation **(figure 20.12)**. In a few weeks, the lymph nodes along the lymphatic drainage swell and can become pus-filled. Only about one-third of patients experience high fever. Most infections remain localized and resolve in a few weeks, but drugs such as tetracycline, erythromycin, and rifampin can be effective therapies. The disease can be prevented by thorough antiseptic cleansing of a cat bite or scratch.

Trench Fever

This disease has a long history. Trench fever was once a common condition of soldiers in battle. The causative agent, *Bartonella quintana*, is carried by lice. Most cases occur in endemic regions of Europe, Africa, and Asia, although the disease is beginning to show up in poverty-stricken areas of large cities in the developed world. This version of the disease is called "urban trench fever." Highly variable symptoms can include a 5- to 6-day fever (the species epithet, *Quintana*, refers to a 5-day fever). Symptoms also include leg pains, especially in the tibial region (the disease is sometime called "shinbone fever"), headache, chills, and muscle aches. A macular rash can also occur. (See Insight 18.3 for definitions of skin lesions.) Endocarditis can develop, especially in the urban version of the disease. The microbe can persist in the blood long after convalescence and is responsible for later relapses.

Trench fever may be treated with doxycycline or erythromycin.

Ehrlichioses

"Ehrlichioses" is plural because there are four tick-borne, fever-producing diseases caused by members of the genus *Ehrlichia*. All of the human *Ehrlichia* diseases are newly described (since the 1980s), although animal infections have been recognized for years.

Members of the genus *Ehrlichia* are small intracellular bacteria, and like *Coxiella*, they share many characteristics with rickettsia, including a strict parasitic existence and association with ticks. *Ehrlichia chaffeensis* causes human monocytic ehrlichiosis (HME). *Ehrlichia phagocytophila* causes human granulocytic ehrlichiosis (HGE). Another species, *Ehrlichia ewingii*, can cause either syndrome. The diseases are sometimes referred to as "spotless" Rocky Mountain spotted fever. *Ehrlichia sennetsu* causes a disease resembling infectious mononucleosis, but only in Japan.

Since 1986, approximately 1,300 cases of HME have been diagnosed and traced to contact with the Lone Star tick (*Ixodes scapularis*). The incidence of human ehrlichiosis since 1993 has been about 100 to 150 cases a year. Both types of ehrlichiosis are showing increased incidence, probably due to improved diagnosis.

The signs and symptoms of HGE and HME are similar: an acute febrile state manifesting headache, muscle pain, and rigors. Most patients recover rapidly with no lasting effects, but around 5% of older chronically ill patients die from disseminated infection. Rapid diagnosis is enabled by PCR tests and indirect fluorescent antibody tests. It can be critical to differentiate or detect coinfection with Lyme disease *Borrelia*, which is carried by the same tick. Doxycycline will clear up most infections within 7 to 10 days.

Rocky Mountain Spotted Fever (RMSF)

This disease is named for the region in which it was first detected in the United States—the Rocky Mountains of Montana and Idaho. In spite of its name, the disease occurs infrequently in the western United States. The majority of cases are concentrated in the Southeast and eastern seaboard regions **(figure 20.13)**. It also occurs in Canada and Central and South America. Infections occur most frequently in the spring and summer, when the tick vector is most active. The yearly rate of RMSF is 20 to 40 cases per 10,000 population, with fluctuations coinciding with weather and tick infestations.

RMSF is caused by a bacterium called *Rickettsia rickettsii* transmitted by hard ticks such as the wood tick (*Dermacentor andersoni*), the American dog tick (*D. variabilis*, among others), and the Lone Star tick (*Ambylomma americanum*). The dog tick is probably most responsible for transmission to humans because it is the major vector in the southeastern United States.

After 2 to 4 days of incubation, the first symptoms are sustained fever, chills, headache, and muscular pain. A distinctive spotted rash usually comes on within 2 to 4 days after the prodrome **(figure 20.14)**. Early lesions are slightly mottled like measles, but later ones are macular, maculopapular, and even petechial. In the most severe untreated cases, the enlarged lesions merge and can become necrotic, predisposing to gangrene of the toes or fingertips.

Although the spots are the most obvious symptom of the disease, the most grave manifestations are cardiovascular disruption, including hypotension, thrombosis, and hemorrhage. Conditions of restlessness, delirium, convulsions, tremor, and coma are signs of the often overwhelming effects on the central nervous system. Fatalities occur in an average of 20% of untreated cases and 5% to 10% of treated cases.

Suspected cases of RMSF require immediate treatment even before laboratory confirmation. A recent aid to early diagnosis is a method for staining rickettsias directly in a tissue biopsy using fluorescent antibodies. Isolating rickettsias from the patient's blood or tissues is desirable, but it is

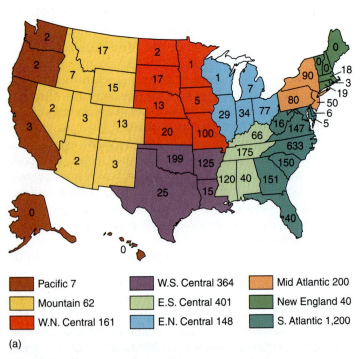

Pacific 7	W.S. Central 364	Mid Atlantic 200
Mountain 62	E.S. Central 401	New England 40
W.N. Central 161	E.N. Central 148	S. Atlantic 1,200

(a)

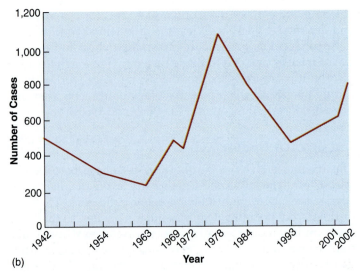

(b)

FIGURE 20.13 **Trends in infection for Rocky Mountain spotted fever.**

(a) A summary map showing the distribution of Rocky Mountain spotted fever cases reported over a 4-year period. **(b)** Reported cases of Rocky Mountain spotted fever in the United States, 1942–2002. The spike in cases in the late 1970s may be due to increased reporting.

✔ CHECKPOINT 20.8 Nonhemorrhagic Fever Diseases

Disease	Brucellosis	Q fever	Cat-Scratch Disease	Trench Fever	Ehrlichioses	Rocky Mountain Spotted Fever
Causative Organism(s)	*Brucella abortus* or *B. suis*	*Coxiella burnetii*	*Bartonella henselae*	*Bartonella quintana*	*Ehrlichia* species	*Rickettsia rickettsii*
Most Common Modes of Transmission	Direct contact, airborne, parenteral (needlesticks)	Airborne, direct contact, food-borne	Parenteral (cat scratch or bite)	Biological vector (lice)	Biological vector (tick)	Biological vector (tick)
Virulence Factors	Intracellular growth; avoidance of destruction by phagocytes	Endospore-like structure	Endotoxin	Endotoxin	–	Induces apoptosis in cells lining blood vessels
Culture/ Diagnosis	Gram stain of biopsy material	Serological tests for antibody	Biopsy of lymph nodes plus Gram staining; ELISA (performed by CDC)	ELISA (performed by CDC)	PCR, indirect antibody test	Fluorescent antibody, PCR
Prevention	Animal control, pasteurization of milk	Vaccine for high-risk population	Clean wound sites	Avoid lice	Avoid ticks	Avoid ticks
Treatment	Tetracycline plus (rifampin or streptomycin)	Tetracycline	Tetracycline, erythromycin, or rifampin	Doxycycline or erythromycin	Doxycycline	Doxycycline or Tetracycline
Distinguishing Characteristics	Undulating fever, muscle aches	Airborne route of transmission, pneumonia also sometimes present	History of cat bite or scratch; fever not always present	Endocarditis common, 5-day fever	Seasonal occurrence (April–Oct.)	Most common in east and southeast U.S.

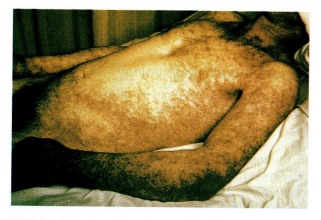

FIGURE 20.14 **Late generalized rash of Rocky Mountain spotted fever.**
In some cases, lesions become hemorrhagic and may become gangrenous in the extremities.

expensive and requires specially qualified lab personnel and lab facilities. Specimens taken from the rash lesions are suitable for PCR assay, which is very specific and sensitive and can circumvent the need for culture.

The drug of choice for suspected and known cases is tetracycline or doxycycline administered for 1 week. Other preventive measures parallel those for Lyme disease: wearing protective clothing, using insect sprays, and fastidiously removing ticks (**Checkpoint 20.8**).

Malaria

Throughout human history, including prehistoric times, malaria has been one of the greatest afflictions, in the same rank as bubonic plague, influenza, and tuberculosis. Even now, as the dominant protozoan disease, it threatens 40% of the world's population every year. The origin of the name is from the Italian words *mal*, bad, and *aria*, air. The superstitions of the Middle Ages alleged that evil spirits or mists and vapors arising from swamps caused malaria, because many victims came down with the disease after this sort of exposure. We now know that a swamp was mainly involved as a habitat for the mosquito vector.

Signs and Symptoms

After a 10- to 16-day incubation period, the first symptoms are malaise, fatigue, vague aches, and nausea with or without diarrhea, followed by bouts of chills, fever, and sweating. These symptoms occur at 48- or 72-hour intervals, as a result of the synchronous rupturing of red blood cells. The interval, length, and regularity of symptoms reflect the type of malaria (described next). Patients with falciparum malaria,

because it protects the bacterium or the host directly. It helps the edema factor get to its target site. The third exotoxin is called *lethal factor*. It also uses enzymatic action to inhibit important cellular processes. The end result of lethal factor action is massive inflammation and initiation of shock.

The *B. anthracis* exotoxin complex is like other bacterial "A-B toxins," which are described in detail in chapter 21. Most A-B toxins have two components: a "B" component that binds to host cells, and an "A," or active, component that enters the cell and exerts some toxic effect. *B. anthracis* is a bit different; its protective antigen is the B component, and both lethal factor and edema factor are A components. The bacteria that cause cholera, shigellosis, pertussis, and diphtheria all use A-B exotoxins.

Additional virulence factors for *B. anthracis* include hemolysins and other enzymes that damage host membranes.

Transmission and Epidemiology

The anthrax bacillus is a facultative parasite that undergoes its cycle of vegetative growth and sporulation in the soil. Animals become infected while grazing on grass contaminated with spores. When the pathogen is returned to the soil in animal excrement or carcasses, it can sporulate and become a long-term reservoir of infection for the animal population. The majority of natural anthrax cases are reported in livestock from Africa, Asia, and the Middle East. Most recent (natural) cases in the United States have occurred in textile workers handling imported animal hair or hide, or products made from them. Because of effective control procedures, the number of cases in the United States is extremely low (fewer than 10 per year).

As a result of the terrorist attacks of 2001, anthrax has dominated the public consciousness as never before. The anthrax attack aimed at two senators and several media outlets focused a great deal of attention on the threat of bioterrorism. During that attack, 22 people acquired anthrax and 5 people died. The attacks led to permanent and drastic changes to our public health response system.

Culture and Diagnosis

Diagnosis requires a high index of suspicion. This means that anthrax must be present as a possibility in the clinician's mind or it is likely not to be diagnosed, since it is such a rare disease in the developed world and, because, in all of its manifestations, it can mimic other infections that are not so rare. (A very astute public health clinician in Florida first suspected anthrax in the attacks of 2001 and called for the proper tests.) First-level (presumptive) diagnosis begins with culturing the bacterium on blood agar and performing a Gram stain. Further tests can be performed to provide evidence regarding presence of *B. anthracis* as opposed to other *Bacillus* species. These tests include motility (*B. anthracis* is nonmotile) and a lack of hemolysis on blood agar. Ultimately, samples should be handled by the Centers for Disease Control and Prevention, which will perform confirmatory tests, usually involving direct fluorescent antibody testing and phage lysis tests.

Prevention and Treatment

A vaccine containing live spores and a toxoid prepared from a special strain of *B. anthracis* are used to protect livestock in areas of the world where anthrax is endemic. Humans should be vaccinated with the purified toxoid if they have occupational contact with livestock or products such as hides and bone, or if they are members of the military. Effective vaccination requires six inoculations given over 1.5 years, with yearly boosters. The cumbersome nature of vaccination has spurred research and development of more manageable vaccines. Persons who are suspected of being exposed to the bacterium are given prophylactic antibiotics, which seems to be effective at preventing disease even after exposure.

Carcasses of animals that have died from anthrax must be burned or chemically decontaminated before burial to prevent establishing the microbe in the soil. Imported items containing animal hides, hair, and bone should be gas sterilized.

The recommended treatment for anthrax is penicillin, doxycycline, or ciprofloxacin. During the attacks in 2001, initial treatment of exposed and sick persons was with ciprofloxacin, because of fear that the *B. anthracis* strains used in the attacks could have been penicillin-resistant, either through intentional genetic engineering or due to the natural presence of β-lactamase genes in the bacterium. Ciprofloxacin treatment continued for the course of the 2001 incident. The CDC is now recommending the use of doxycycline instead of ciprofloxacin, because ciprofloxacin is often used for empirical treatment of all types of infections of unknown etiology. More frequent use of ciprofloxacin could lead to higher levels of antibiotic resistance in bacteria in the U.S. population, which would render ciprofloxacin less effective as an empirical agent.

✔ CHECKPOINT 20.10   Anthrax

Causative Organism(s)	*Bacillus anthracis*
Most Common Modes of Transmission	Vehicle (air, soil) indirect contact (animal hides), vehicle (food)
Virulence Factors	Triple exotoxin, capsule
Culture/Diagnosis	Culture, direct fluorescent antibody tests
Prevention	Vaccine for high-risk population, postexposure antibiotic prophylaxis
Treatment	Doxycycline, ciprofloxacin, penicillin

HIV Infection and AIDS

The sudden emergence of AIDS in the early 1980s focused an enormous amount of public attention, research studies, and financial resources on the virus and its disease.

The first cases of AIDS were seen by physicians in Los Angeles, San Francisco, and New York City. They observed clusters of young male patients with one or more of a complex of symptoms: severe pneumonia caused by *Pneumocystis (carinii) jiroveci* (ordinarily a harmless fungus); a rare vascular cancer called Kaposi's sarcoma; sudden weight loss; swollen lymph nodes; and general loss of immune function. Another common feature was that all of these young men were homosexuals. Early hypotheses attempted to explain the disease as a consequence of a "homosexual lifestyle" or as a result of immune suppression by chronic drug abuse or infections. Soon, however, cases were reported in nonhomosexual patients who had been transfused with blood or blood products. Eventually, virologists at the Pasteur Institute in France isolated a novel retrovirus, later named the **human immunodeficiency virus (HIV).** This cluster of symptoms was therefore clearly a communicable infectious disease, and the medical community termed it **acquired immunodeficiency syndrome,** or **AIDS.**

One important question about HIV seems to have been answered: *Where did it come from?* Researchers have been comparing the genetics of HIV with the various African monkey viruses, called simian immunodeficiency viruses, or SIVs. The genetic sequences in these various viruses led them to conclude that HIV is a hybrid virus, with genetic sequences from two separate monkey SIVs. One of the SIVs has as its natural host the greater spot-nosed monkey and the other infects red-capped mangabeys. Apparently, one or more chimpanzees became coinfected with the two viruses after making a meal of both of the smaller monkeys. Within the chimpanzee, a third type of virus emerged that contained genetic sequences from both SIVs. This new type of SIV was probably transmitted to humans when they captured chimps, butchered them and used them for food. So humans originally acquired HIV from eating or skinning chimps; chimps got SIV from eating monkeys. The crossover into humans probably occurred in the early part of the 1900s; the earliest record we have of human infection is a blood sample preserved from an African man who died in 1959.

HIV probably remained in small isolated villages, causing sporadic cases and mutating into more virulent strains that were readily transmitted from human to human. When this pattern was combined with changing social and sexual practices and increased immigration and travel, a pathway was opened up for rapid spread of the virus to the rest of the world.

Signs and Symptoms

A spectrum of clinical disease is associated with HIV infection. To understand the progression, follow **figure 20.19** closely. Symptoms in HIV infection are directly tied to two things: the level of virus in the blood, and the level of T cells in the blood. (The figure shows two different lines that correspond to virus and T cells.) Note also that the figure depicts the course of HIV infection in the absence of medical intervention or chemotherapy.

Initial infection is often attended by vague, mononucleosis-like symptoms that soon disappear. This phase corresponds

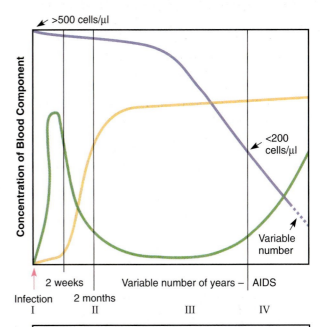

Virus levels are high during the initial acute infection and decrease until the later phases of HIV disease and AIDS. Antibody levels gradually rise and remain relatively high throughout phases III and IV. T-cell numbers remain relatively normal until the later phases of HIV disease and full-blown AIDS.

FIGURE 20.19 Dynamics of virus antigen, antibody, and T cells in circulation.

to the initial high levels of virus (the green line in the figure), and the subsequent drop in virus load. Note that antibody levels (the orange line) rise at the same time that virus load is dropping; the immune response is responsible for the decreasing numbers of virus in the blood. This initial state is followed by a period of (mostly) asymptomatic infection that varies in length from 2 to 15 years (the average is 10). The lymphadenopathy that attended the initial infection usually persists throughout the entire infection. Note that during the mid-to-late asymptomatic periods, the number of T cells in the blood is steadily decreasing (purple line). Once the T cells reach low enough levels, symptoms of AIDS ensue.

Initial symptoms may be fatigue, diarrhea, weight loss, and neurological changes, but most patients first notice this phase of infection because of one or more opportunistic infections or neoplasms. These are detailed in **Insight 20.3.** Other disease-related symptoms appear to accompany severe immune deregulation, hormone imbalances, and metabolic disturbances. Pronounced wasting of body mass is a consequence of weight loss, diarrhea, and poor nutrient absorption. Protracted fever, fatigue, sore throat, and night sweats are significant and debilitating. Both a rash and

TABLE 20.1	Regional HIV/AIDS Statistics and Features, 2002		
	Estimated Adults and Children Living with HIV/AIDS	Estimated Adults and Children Newly Infected with HIV	Percent of HIV-positive Adults Who Are Women
Sub-Saharan Africa	29.4 million	3.5 million	58
North Africa and Middle East	550,000	83,000	55
South and Southeast Asia	6.0 million	700,000	36
East Asia and Pacific	1.2 million	270,000	24
Latin America	1.5 million	150,000	30
Caribbean	440,000	60,000	50
Eastern Europe and Central Asia	1.2 million	250,000	27
Western Europe	570,000	30,000	25
North America	980,000	45,000	20
Australia and New Zealand	15,000	500	7
TOTAL	42 million	5 million	50

not yet begun to show symptoms because they are in the latent phase of the disease.

AIDS first became a notifiable disease at the national level in 1984, and it has continued in an epidemic pattern, although the number of AIDS cases in the United States has decreased since 1994. This decrease is occurring at the same time that new HIV infections are being reported at the rate of 40,000 a year. This situation is due to the advent of effective therapies that prevent the progression to AIDS. But in developing countries, which are hardest hit by the HIV epidemic, access to these life-saving drugs is still very limited. And even in the United States, despite treatment advances, HIV infection and AIDS is the sixth most common cause of death among people aged 25 to 44, although it has fallen out of the top ten list for causes of death overall.

Figure 20.23 depicts the numbers of new infections broken down by gender, risk group, and race. Men still account for 70% of new infections. Forty-two percent of all new infections are acquired through male homosexual activity. Men having sex with other men (a group labeled MSM) have increased susceptibility because of the practice of anal sex, which is known to lacerate the rectal mucosa, and can provide an entrance for viruses from semen into the blood. The receptive partner (whether male or female) is the more likely of the two to become infected. In addition, bisexual men provide an avenue for the virus to infect females.

In large metropolitan areas especially, as many as 60% of intravenous drug users (IDUs) can be HIV carriers. Infection from contaminated needles is growing more rapidly than any other mode of transmission, and it is another significant factor in the spread of HIV to the heterosexual population.

In most parts of the world, heterosexual intercourse is the primary mode of transmission. In the industrialized world, the overall rate of heterosexual infection has increased dramatically in the past several years, especially in adolescent and young adult women. In the United States, about 33% of HIV infections arise from unprotected sexual intercourse with an infected partner of the opposite sex.

Now that donated blood is routinely tested for antibodies to the AIDS virus, transfusions are no longer considered a serious risk. Because there can be a lag period of a few weeks to several months before antibodies appear in an infected person, it is remotely possible to be infected through donated blood. Rarely, organ transplants can carry HIV, so they too should be tested. Other blood products (serum, coagulation factors) were once implicated in AIDS. Thousands of hemophiliacs died from the disease in the 1980s and 1990s. It is now standard practice to heat-treat any therapeutic blood products to destroy all viruses.

A small percentage of AIDS cases (less than 8%) occur in people without apparent risk factors. This does not mean that some other unknown route of spread exists. Factors such as patient denial, unavailability of history, death, or uncooperativeness make it impossible to explain every case.

We should note that not everyone who becomes infected or is antibody-positive develops AIDS. About 5% of people who are antibody-positive remain free of disease, indicating that functioning immunity to the virus can develop. Any person who remains healthy despite HIV infection is termed a *nonprogressor*. These people are the object of intense scientific study. Some have been found to lack the cytokine receptors that HIV requires. Others are infected by a weakened virus mutant.

Treatment of HIV-infected mothers with AZT has dramatically decreased the rate of maternal to infant transmission of HIV during pregnancy. Current treatment regimens result in a transmission rate of approximately 11%, with some studies of multidrug regimens claiming rates as low as 5%. Evidence suggests that giving mothers protease inhibitors can reduce the transmission rate to around 1%. (Untreated mothers pass the virus to their babies at the rate

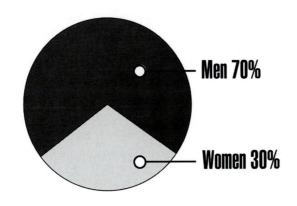

(a)

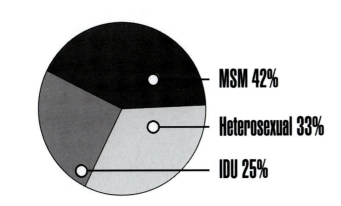

(b)

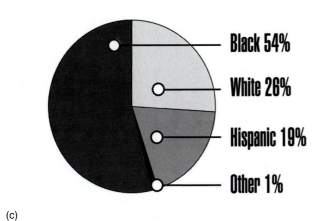

(c)

FIGURE 20.23 **New HIV infections each year in the United States.**

(a) Percentage of new infections by gender. (b) Percentage of new infections by risk. (c) Percentage of new infections by race.

of 33%.) The cost of perinatal prevention strategies (approximately $1,000 per pregnancy) and the scarcity of medical counseling in underserved areas has led to an increase in maternal transmission of HIV in developing parts of the world, at the same time that the developed world has seen a marked decrease.

Medical and dental personnel are not considered a high-risk group, although several hundred medical and dental workers are known to have acquired HIV or become antibody-positive as a result of clinical accidents. A health care worker involved in an accident in which gross inoculation with contaminated blood occurs (as in the case of a needle-stick) has a less than 1 in 1,000 chance of becoming infected. We should emphasize that transmission of HIV will not occur through casual contact or routine patient care procedures, and that universal precautions for infection control (see chapter 13) were designed to give full protection for both worker and patient.

Culture and Diagnosis

First, let's define some terms. A person is diagnosed as having HIV infection if they have tested positive for the human immunodeficiency virus. This diagnosis is not the same as having AIDS.

Most viral testing is based on detection of antibodies specific to the virus in serum or other fluids, which allows for the rapid, inexpensive screening of large numbers of samples. Testing usually proceeds at two levels. The initial screening tests include the older ELISA and newer latex agglutination and rapid antibody tests. The favored initial screening test is an "oral swab test" that provides results within minutes. It is licensed only for use by health care professionals. Even though saliva can be tested with this kit, blood is more often used in the test because of increased viral load in blood relative to saliva. The advantage of this test is that results are available within minutes instead of the days to weeks previously required for the return of results from ELISA testing. One kit has been licensed by the Food and Drug Administration for home use; in this test, a blood spot is applied to a card and then sent to a testing laboratory to perform the actual procedure. The client can call an automated phone service, enter his or her anonymous testing code, and receive results over the phone. Other home kits have been devised to provide results immediately in the client's home, much like a home pregnancy test, but none of these is yet approved by the FDA, and their accuracy is questionable.

Although the approved tests just described are largely accurate, around 1% may of results are false positives, and they always require follow-up with a more specific test called *Western blot* analysis (see p. 528). This test detects several different anti-HIV antibodies and can usually rule out false positive results.

Another inaccuracy can be false negative results that occur when testing is performed before the onset of detectable antibody production. To rule out this possibility, persons who test negative, but feel they may have been exposed, should be tested a second time 3 to 6 months later.

Blood and blood products are sometimes tested for HIV antigens (rather than for HIV antibodies) to close the window of time between infection and detectable levels of antibodies during which contamination could be missed by antibody tests. The American Red Cross is currently participating in a program to test its blood supply with a DNA probe for HIV.

In the United States, a person is diagnosed with AIDS if they meet the following criteria: (1) they are positive for the virus, *and* (2) they fulfill one of these additional criteria:

- They have a CD4 (helper T cell) count of fewer than 200 cells per milliliter of blood.
- Their CD4 cells account for fewer than 14% of all lymphocytes.
- They experience one or more of a CDC-provided list of AIDS-defining illnesses (ADIs).

The list of ADIs is long and includes opportunistic infections such as *Pneumocystis (carinii) jiroveci* pneumonia and *Cryptosporidium* diarrhea; neoplasms such as Kaposi's sarcoma and invasive cervical cancer; and other conditions such as wasting syndrome and neuropathy (see Insight 20.3).

Prevention

Avoidance of sexual contact with infected persons is a cornerstone of HIV prevention. Abstaining from sex is an obvious prevention method, although those who are sexually active can also take steps to decrease their risk. Epidemiologists cannot overemphasize the need to screen prospective sex partners and to follow a monogamous sexual life-style. And monogamous or not, a sexually active person should consider every partner to be infected unless proven otherwise. This may sound harsh, but it is the only sure way to avoid infection during sexual encounters. Barrier protection (condoms) should be used when having sex with anyone whose HIV status is not known with certainty to be negative. Although avoiding intravenous drugs is an obvious deterrent, many drug addicts do not, or cannot, choose this option. In such cases, risk can be decreased by not sharing syringes or needles, or by cleaning needles with bleach and then rinsing before another use.

From the very first years of the AIDS epidemic, the potential for creating a vaccine has been regarded as slim, because the virus presents many seemingly insurmountable problems. Among them, HIV becomes latent in cells; its cell surface antigens mutate rapidly; and although it does elicit immune responses, it is apparently not completely controlled by them. In view of the great need for a vaccine, however, none of those facts has stopped the medical community from moving ahead.

Currently hundreds of potential HIV vaccines are in clinical trials. To protect against as many strains as possible, most vaccine designers have concentrated their efforts on proteins that seems to be highly conserved (show very few changes) between strains. Currently only one vaccine is in Phase III trials—the last step before possible FDA approval. While earlier trials investigate vaccine safety and whether the vaccine causes antibody production, Phase III trials examine the abil-

ity of the vaccine to protect humans from becoming infected. The trial is being conducted in Thailand, through a joint venture between the United States and Thailand.

Treatment

It must be clearly stated: There is no cure for HIV. None of the therapies do more than prolong life or diminish symptoms.

Clear-cut guidelines exist for treating people who test HIV-positive. These guidelines are updated regularly, and they differ depending on whether a person is completely asymptomatic or experiencing some of the manifestation of HIV infection, and also whether a person has previously been treated with antiretroviral agents. A person diagnosed with AIDS receives treatment for the HIV infection, and also a wide array of drugs to prevent or treat a variety of opportunistic infections, and other ADIs such as wasting disease. These treatment regimens vary according to each patient's profile and needs.

The first effective drugs developed were the synthetic nucleoside analogs (reverse transcriptase inhibitors) azidothymidine (AZT), didanosine (ddI), Epivir (3TC), and stavudine (d4T). They interrupt the HIV multiplication cycle by mimicking the structure of actual nucleosides and being added to viral DNA by reverse transcriptase. Because these drugs lack all of the correct binding sites for further DNA synthesis, viral replication and the viral cycle are terminated **(figure 20.24***a***).** Other reverse transcriptase inhibitors that are not nucleosides are nevirapine and sustiva, both of which bind to the enzyme and restructure it. Another important class of drugs is the protease inhibitors **(figure 20.24***b***)** that block the action of the HIV enzyme (protease) involved in the final assembly and maturation of the virus. Examples of these drugs include Crixivan, Norvir, and Agenerase. Integrase inhibitors are currently under investigation as a means to stop virus multiplication **(figure 20.24***c***).**

The latest addition to the arsenal is Fuzeon, a drug classified as a fusion inhibitor. It prevents the virus from fusing with the membrane of target cells, thereby stopping infection altogether.

A regimen that has proved to be extremely effective in controlling AIDS and inevitable drug resistance is **HAART,** short for *highly active anti-retroviral therapy*. By combining two reverse transcriptase inhibitors and one protease inhibitor in a "cocktail," the virus is interrupted in two different phases of its cycle. This therapy has been successful in reducing viral load to undetectable levels and facilitating the improvement of immune function. It has also reduced the incidence of viral drug resistance, because the virus would have to undergo at least two separate mutations simultaneously, at nearly impossible odds. Patients who are HIV-positive but asymptomatic can remain healthy with this therapy as well. The primary drawbacks are high cost, toxic side effects, drug failure due to patient noncompliance, and an inability to completely eradicate the virus.

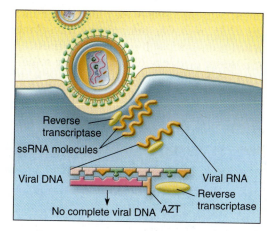

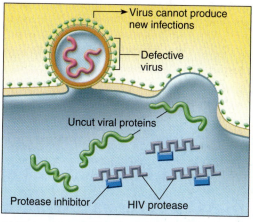

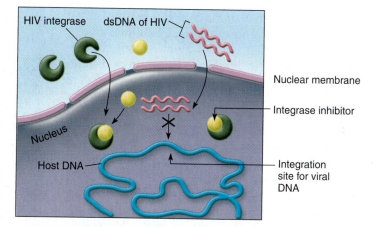

Location of reaction

- ☐ External to cell (yellow)
- ☐ Cytoplasm (blue)
- ☐ Nucleus (purple)

(a) A prominent group of drugs (AZT, ddI, 3TC) are nucleoside analogs that inhibit reverse transcriptase. They are inserted in place of the natural nucleotide by reverse transcriptase but block further action of the enzyme and synthesis of viral DNA.

(b) Protease inhibitors plug into the active sites on HIV protease. This enzyme is necessary to cut elongate HIV protein strands and produce functioning smaller protein units. Because the enzyme is blocked, the proteins remain uncut, and abnormal defective viruses are formed.

(c) Integrase inhibitors are a new class of experimental drugs that attach to the enzyme required to splice the dsDNA from HIV into the host genome. This will prevent formation of the provirus and block future virus multiplication in that cell.

FIGURE 20.24   Mechanisms of action of anti-HIV drugs.

✔ CHECKPOINT 20.11	HIV Infection and AIDS
Causative Organism(s)	Human immunodeficiency virus 1 or 2
Most Common Modes of Transmission	Direct contact (sexual), parenteral (blood-borne), vertical (perinatal and via breast milk)
Virulence Factors	Attachment, syncytia formation, reverse transcriptase, high mutation rate
Culture/Diagnosis	Initial screening for antibody followed by Western blot confirmation of antibody
Prevention	Avoidance of contact with infected sex partner, contaminated blood, breast milk
Treatment	HAART (reverse transcriptase inhibitors plus protease inhibitors), Fuzeon, nonnucleoside RT inhibitors

Adult T-Cell Leukemia and Hairy-Cell Leukemia

Leukemia is the general name for at least four different malignant diseases of the white blood cell forming elements originating in the bone marrow. Leukemias are all acquired, rather than being inherited; some forms are acute and others are chronic. Leukemias have many causes, only two of which are thought to be viral. The retrovirus HTLV-I is associated with a form of leukemia called adult T-cell leukemia. HTLV-II is thought by some to cause hairy-cell leukemia.

The signs and symptoms of all leukemias are similar, and include easy bruising or bleeding, paleness, fatigue, and recurring minor infections. These symptoms are associated with the underlying pathologies of anemia, platelet deficiency, and immune dysfunction brought about by the disturbed lymphocyte ratio and function. In some cases of adult T-cell leukemia, cutaneous T-cell lymphoma is the prime clinical manifestation, accompanied by dermatitis, with thickened, scaly, ulcerative, or tumorous skin lesions. Other complications are lymphadenopathy and dissemination of the tumors to the lung, spleen, and liver.

The possible mechanisms by which retroviruses stimulate cancer are not entirely clear. One hypothesis is that the virus carries an oncogene that, when spliced into a host's chromosome and triggered by various carcinogens, can immortalize the cell and deregulate the cell division cycle. One of HTLV's genetic targets seems to be the gene and receptor for interleukin-2, a potent stimulator of T cells.

Adult T-cell leukemia was first described by physicians working with a cluster of patients in southern Japan. Later, a similar clinical disease was described in Caribbean immigrants. In time, it was shown that these two diseases were the same. Although more common in Japan, Europe, and the Caribbean, a small number of cases occur in the United States. The disease is not highly transmissible; studies among families show that repeated close or intimate contact is required. Because the virus is thought to be transferred in infected blood cells, blood transfusions and blood products are potential agents of transmission. Intravenous drug users could spread it through needle sharing.

Hairy-cell leukemia is a rare form of cancer that may be caused by HTLV-II. The disease derives its name from the appearance of the afflicted lymphocytes, which have fine cytoplasmic projections that resemble hairs. The disease is more common in males than in females and is probably spread through blood and shared syringes and needles. Unlike adult T-cell leukemia, this form of leukemia exhibits overall leukopenia, but with an increased number of neoplastic B lymphocytes.

Treatment of these diseases may include a number of antineoplastic drugs, radiation therapy, and transplants. Alpha-interferon has been used with some effectiveness.

There is some evidence to suggest that some forms of non-Hodgkin's lymphoma may also be caused by HTLV-I.

✔ CHECKPOINT 20.12	Adult T-Cell Leukemia and Hairy-Cell Leukemia	
Disease	**Adult T-Cell Leukemia**	**Hairy-Cell Leukemia**
Causative Organism(s)	HTLV-I	(possibly) HTLV-II
Most Common Modes of Transmission	Unclear— blood-borne transmission implicated	Unclear— blood-borne transmission implicated
Virulence Factors	Induction of malignant state	Induction of malignant state
Culture/Diagnosis	Differential blood count followed by histological examination of excised lymph node tissue	Differential blood count followed by histological examination of excised lymph node tissue
Prevention	–	–
Treatment	Antineoplastic drugs, interferon alpha	Antineoplastic drugs, interferon alpha

Taxonomic Organization of Microorganisms Causing Disease in the Cardiovascular and Lymphatic System

Microorganism	Disease	Chapter Location
Gram-Positive Endospore-Forming Bacteria		
Bacillus anthracis	Anthrax	Anthrax, p. 635
Gram-Positive Bacteria		
Staphylococcus aureus	Acute endocarditis	Endocarditis, p. 617
Streptococcus pyogenes	Acute endocarditis	Endocarditis, p. 617
Streptococcus pneumoniae	Acute endocarditis	Endocarditis, p. 617
Gram-Negative Bacteria		
Yersinia pestis	Plague	Plague, p. 619
Francisella tularensis	Tularemia	Tularemia, p. 621
Borellia burgdorferi	Lyme disease	Lyme disease, p. 624
Brucella abortus, B. suis	Brucellosis	Nonhemorrhagic fever diseases, p. 628
Coxiella burnetii	Q fever	Nonhemorrhagic fever diseases, p. 628
Bartonella henselae	Cat-scratch disease	Nonhemorrhagic fever diseases, p. 629
Bartonella quintana	Trench fever	Nonhemorrhagic fever diseases, p. 629
Ehrlichia chaffeensis, E. phagocytophila, E. ewingii	Ehrlichiosis	Nonhemorrhagic fever diseases, p. 629
Neisseria gonorrhoeae	Acute endocarditis	Endocarditis, p. 617
Rickettsia rickettsii	Rocky Mountain spotted fever	Nonhemorrhagic fever diseases, p. 630
DNA Viruses		
Epstein-Barr virus	Infectious mononucleosis	Infectious mononucleosis, p. 622
Cytomegalovirus	Infectious mononucleosis	Infectious mononucleosis, p. 622
RNA Viruses		
Yellow fever virus	Yellow fever	Hemorrhagic fevers, p. 626
Dengue fever virus	Dengue fever	Hemorrhagic fevers, p. 627
Ebola and Marburg viruses	Ebola and Marburg hemorrhagic fevers	Hemorrhagic fevers, p. 627
Lassa fever virus	Lassa fever	Hemorrhagic fevers, p. 627
Retroviruses		
Human immunodeficiency virus 1 and 2	HIV infection and AIDS	HIV infection and AIDS, p. 636
Human T-cell lymphotropic virus I	Adult T-cell leukemia	Leukemias, p. 645
Human T-cell lymphotropic virus II	Hairy-cell leukemia (?)	Leukemias, p. 645
Protozoa		
Plasmodium falciparum, P. vivax, P. ovale, P. malariae	Malaria	Malaria, p. 631

Infectious Diseases Affecting
the Cardiovascular and Lymphatic Systems

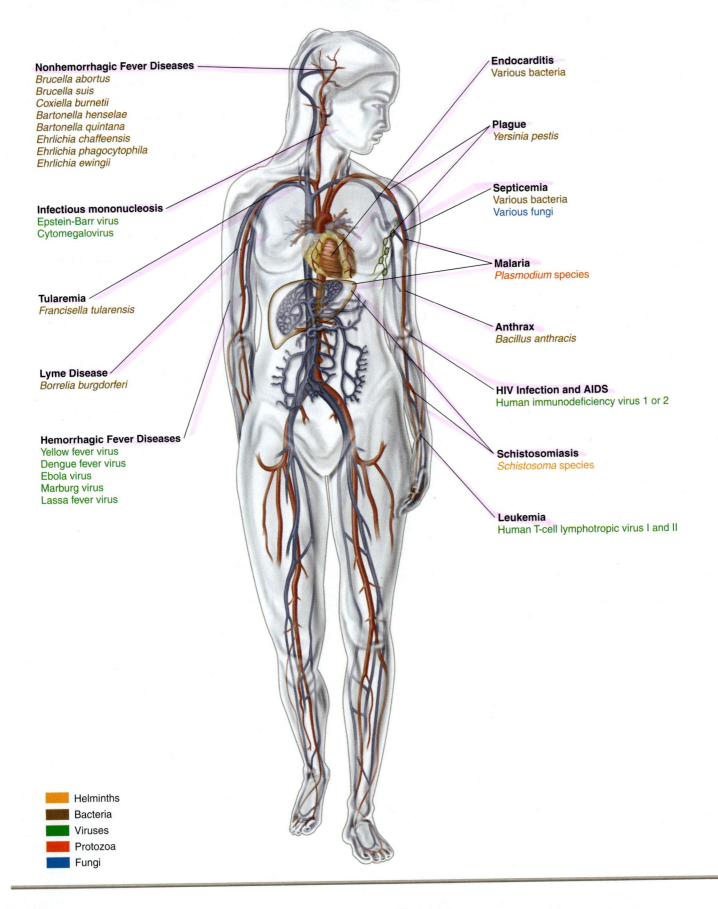

Nonhemorrhagic Fever Diseases
Brucella abortus
Brucella suis
Coxiella burnetii
Bartonella henselae
Bartonella quintana
Ehrlichia chaffeensis
Ehrlichia phagocytophila
Ehrlichia ewingii

Infectious mononucleosis
Epstein-Barr virus
Cytomegalovirus

Tularemia
Francisella tularensis

Lyme Disease
Borrelia burgdorferi

Hemorrhagic Fever Diseases
Yellow fever virus
Dengue fever virus
Ebola virus
Marburg virus
Lassa fever virus

Endocarditis
Various bacteria

Plague
Yersinia pestis

Septicemia
Various bacteria
Various fungi

Malaria
Plasmodium species

Anthrax
Bacillus anthracis

HIV Infection and AIDS
Human immunodeficiency virus 1 or 2

Schistosomiasis
Schistosoma species

Leukemia
Human T-cell lymphotropic virus I and II

Helminths
Bacteria
Viruses
Protozoa
Fungi

Chapter Summary With Key Terms

20.1 The Cardiovascular and Lymphatic Systems and Their Defenses

A. *The cardiovascular system* is composed of the blood vessels, which carry blood to and from all regions of the body, and the heart, which pumps the blood. The cardiovascular system provides tissues with oxygen and nutrients, and carries away carbon dioxide and waste products. The blood vessels consist of *arteries, veins,* and *capillaries.* The heart is a fist-sized muscular organ that pumps blood around the body. The entire organ is encased in a fibrous covering, the pericardium, which is an occasional site of infection. The actual wall of the heart has three layers: from outer to inner they are the epicardium, the myocardium, and the endocardium. The endocardium also covers the valves of the heart, and it is a relatively common target of microbial infection.

B. *The Lymphatic System:* A major source of immune cells and fluids, the **lymphatic system** serves as a one-way passage, returning fluid from the tissues to the cardiovascular system. It collects fluid that has left the blood vessels and entered tissues, filters it of impurities and infectious agents, and returns it to the blood.

C. *Defenses of the Cardiovascular and Lymphatic Systems:* The cardiovascular system is highly protected from microbial infection.

1. Microbes that successfully invade the system give rise to **systemic infections.** The various types of white blood cells include the lymphocytes, responsible for specific immunity, and the phagocytes, essential in defending the body from microbial invasion. The same defenses are present in the lymphatic system, whose very existence is centered on host immunity.

2. Viruses present in the blood lead to **viremia.** When fungi are in the blood, the condition is termed fungemia, and bacterial presence is called **bacteremia,** but when bacteria flourish and grow in the bloodstream, the condition is termed **septicemia.**

20.2 Normal Flora of the Cardiovascular and Lymphatic Systems

The cardiovascular and lymphatic systems are "closed" systems with no normal flora.

20.3 Cardiovascular and Lymphatic System Diseases Caused by Microorganisms

A. **Endocarditis** is an inflammation of the endocardium, usually due to an infection of the valves of the heart. Two variations of infectious endocarditis have been described: acute and subacute.

1. **Acute endocarditis** is most often caused by *Staphylococcus aureus,* as well as group A streptococci, *Streptococcus pneumoniae,* and *Neisseria gonorrhoeae.* The most common route of transmission for acute endocarditis is parenteral.

2. **Subacute forms of endocarditis** are almost always preceded by some form of damage to the heart valves, or by congenital malformation. Alpha-hemolytic streptococci, such as *Streptococcus sanguis, S. oralis,* and *S. mutans,* are most often responsible, although normal flora from the skin and other bacteria can also colonize abnormal valves and lead to this condition. Transmission can be due vigorous tooth brushing, dental procedures, or relatively minor cuts and lacerations.

B. **Septicemias** occur when organisms are actively multiplying in the blood. Most are caused by bacteria, and to a lesser extent fungi. Gram-negative bacteria multiplying in the blood release large amounts of endotoxin into the bloodstream, resulting in **endotoxic shock.** Gram-positive bacteria can instigate a similar cascade of events when fragments of their cell walls are released into the blood.

C. **Plague** can manifest in three different ways: **Pneumonic plague** is a respiratory disease; **bubonic plague** causes inflammation and necrosis of the lymph nodes, resulting in a swollen lesion called a **bubo; septicemic plague** is the result of multiplication of bacteria in the blood. *Yersinia pestis* is the causative organism—a tiny, gram-negative rod. The principal agents in the transmission of the plague bacterium are fleas.

D. **Tularemia's** causative agent is a facultative intracellular gram-negative bacterium called *Francisella tularensis.* This disease is often called rabbit fever, since it is associated with outbreaks of disease in wild rabbits.

E. **Infectious mononucleosis:** The vast majority of cases are caused by the herpesviruses **Epstein-Barr virus (EBV)** and **cytomegalovirus (CMV).** A strong, cell-mediated immune response can control the infection, but people usually remain chronically infected with EBV and CMV.

1. **Epstein-Barr virus** is also associated with Burkitt's lymphoma, and due to its tendency for latency, it is very effective at avoiding the host's immune response.

2. **Cytomegalovirus (CMV)** is transmitted in saliva, respiratory mucus, milk, urine, semen, cervical secretions, and feces. Transmission usually involves intimate contact such as sex, vaginal birth, transplacental infection, blood transfusion, and organ transplantation.

F. **Lyme disease** is caused by *Borrelia burgdorferi.* This nonfatal syndrome mimics neuromuscular and rheumatoid conditions. An early symptom is a bull's-eye rash at the site of a tick bite. *B. burgdorferi* is a unique spirochete that is transmitted primarily by *Ixodes* ticks.

G. **Hemorrhagic fever diseases** are extreme fevers which are often accompanied by internal hemorrhaging. All hemorrhagic fever diseases described here are caused by RNA enveloped viruses in one of three families: Arenaviridae, Filoviridae and Flaviviridae.

1. **Yellow fever** is caused by an arbovirus, a single-stranded RNA flavivirus that is transmitted by the mosquito, *Aedes aegypti.* Infection begins acutely with fever, headache, and muscle pain.

2. **Dengue fever** is caused by a single-stranded RNA flavivirus and is also carried by *Aedes* mosquitoes. Although mild infection is the usual pattern, a form called dengue hemorrhagic shock syndrome can be lethal.

3. *Ebola and Marburg* viruses are filoviruses (Family Filoviridae) endemic to Central Africa. The virus in the bloodstream leads to extensive capillary fragility and disruption of clotting. Marburg outbreaks are less common than Ebola fever.

4. The *Lassa fever* virus is an arenavirus found in West Africa. The reservoir of the virus is a rodent found in Africa called the multimammate rat.

H. **Nonhemorrhagic Fever Diseases:** These infectious diseases are characterized by high fever, but without the capillary fragility that leads to hemorrhagic symptoms.

1. **Brucellosis** is also called Malta fever, undulant fever, and Bang's disease. The bacterial genus *Brucella* contains tiny, aerobic gram-negative coccobacilli. Two species can cause this disease in humans: *B. abortus* (common in cattle), and *B. suis* (from pigs).

2. **Q fever** is caused by *Coxiella burnetii*, is a very small pleomorphic gram-negative bacterium, and an intracellular parasite. *C. burnetii* is apparently harbored by a wide assortment of vertebrates and arthropods, especially ticks. Ticks, however, do not transmit the disease to humans, who acquire infection mainly by environmental contamination and airborne transmission.

3. **Cat-Scratch Disease:** *Bartonella henselae* is the agent of **cat-scratch disease (CSD),** an infection connected with being clawed or bitten by a cat. The pathogen is present in over 40% of cats, especially kittens.

4. **Trench fever's** causative agent, *Bartonella quintana,* is carried by lice. Highly variable symptoms can include a 5- to 6-day fever, leg pains, headache, chills, and muscle aches.

5. **Ehrlichioses:** There are four tick-borne, fever-producing diseases caused by members of the genus *Ehrlichia.* Members of the genus *Ehrlichia* are small intracellular bacteria and are strict parasites. *Ehrlichia chaffeensis* causes human monocytic ehrlichiosis (HME). *Ehrlichia phagocytophila* causes human granulocytic ehrlichiosis (HGE). Another species, *Ehrlichia ewingii* can cause either syndrome. The diseases are sometimes referred to as "spotless" Rocky Mountain spotted fever.

I. **Malaria:** This protozoan disease exhibits symptoms of malaise, fatigue, vague aches, and nausea, followed by bouts of chills, fever, and sweating. These symptoms occur at 48- or 72-hour intervals, as a result of the synchronous rupturing of red blood cells. The causative organisms are *Plasmodium* species: *P. malariae, P. vivax, P. falciparum,* and *P. ovale.* Humans and some other primates are the primary vertebrate hosts for the species. The *sexual phase* of the protozoan is carried out in the *Anopheles* mosquito.

J. **Anthrax** infection can exhibit its primary symptoms in various locations of the body: on the skin (cutaneous anthrax), in the lungs (pulmonary anthrax), in the gastrointestinal tract (acquired through ingestion of contaminated foods), and in the central nervous system (anthrax meningitis). The cutaneous and pulmonary forms of the disease are the most common. In all of these forms, the anthrax bacterium gains access to the bloodstream, and death, if it occurs, is usually a result of an overwhelming septicemia. *Bacillus anthracis* is a gram-positive endospore-forming rod that is found in the soil.

K. **HIV Infection and AIDS: Human immunodeficiency virus (HIV)** is the organism responsible for the cluster of symptoms known as **acquired immunodeficiency syndrome,** or **AIDS.** Symptoms in HIV infection are directly tied to two things: the level of virus in the blood and the level of T-cells in the blood.

1. Initial infection is often attended by vague, mononucleosis-like symptoms that soon disappear. This phase corresponds to the initial high levels of virus. This initial state is followed by a period of latent infection that varies in length from 2 to 15 years. The antibody response becomes ineffective once the virus has become latent. During the mid-to-latent periods, the number of T cells in the blood is steadily decreasing. Once the T cells reach low enough levels, symptoms of AIDS ensue.

2. HIV is a retrovirus, in the genus lentivirus. It contains **reverse transcriptase** that catalyzes the replication of double-stranded DNA from single-stranded RNA. Retroviral DNA is incorporated into the host genome as a provirus that can be passed on to progeny cells in a latent state. There are two types of HIV: HIV-1, which is the dominant form in most of the world, and HIV-2.

3. The primary effects of HIV infection—those directly due to viral action—are harm to T cells and the central nervous system. The death of T cells and other white blood cells results in extreme **leukopenia** and loss of essential T4 memory clones and stem cells. The destruction of T4 lymphocytes paves the way for invasion by opportunistic agents and malignant cells. The secondary effects of HIV infection are the opportunistic infections and malignancies that occur as the immune system becomes progressively weakened by virus multiplication.

4. HIV transmission occurs mainly through sexual intercourse and transfer of blood or blood products.

5. A regimen that has proved to be extremely effective in controlling AIDS and inevitable drug resistance is **HAART,** short for *highly active anti-retroviral therapy.*

L. **Adult T-Cell Leukemia and Hairy-Cell Leukemia:** Leukemia is the general name for at least four different malignant diseases of the white blood cell forming elements originating in the bone marrow. The retrovirus HTLV-I is associated with a form of leukemia called adult T-cell leukemia. HTLV-II is associated with hairy-cell leukemia.

Multiple-Choice Questions

1. When bacteria flourish and grow in the bloodstream, this is referred to as
   a. viremia
   b. bacteremia
   c. septicemia
   d. fungemia

2. Which of the following is *not* a symptom of septicemia?
   a. fever
   b. respiratory alkalosis
   c. shaking chills
   d. high blood pressure

3. The plague bacterium, *Yersinia pestis*, is transmitted mainly by
   a. mosquitoes
   b. fleas
   c. dogs
   d. birds

4. Rabbit fever is caused by
   a. *Yersinia pestis*
   b. *Francisella tularensis*
   c. *Borrellia burgdorferi*
   d. *Chlamydia bunnyensis*

5. A distinctive bull's-eye rash results from a tick bite transmitting
   a. Lyme disease
   b. tularemia
   c. Q fever
   d. Rocky Mountain spotted fever

6. Lyme disease is caused by
   a. *Yersinia pestis*
   b. *Francisella tularensis*
   c. *Borrellia burgdorferi*
   d. *Chlamydia bunnyensis*

7. Cat-scratch disease is caused by
   a. *Coxiella burnetii*
   b. *Bartonella henselae*
   c. *Bartonella quintana*
   d. *Brucella abortus*

8. Q fever is caused by the intracellular parasite
   a. *Coxiella burnetii*
   b. *Bartonella henselae*
   c. *Bartonella quintana*
   d. *Brucella abortus*

9. The bite of the Lone Star tick, *Ixodes scapularis*, can cause
   a. ehrlichioses
   b. Lyme disease
   c. trench fever
   d. both a and b
   e. both b and c

10. Cat-scratch disease is effectively treated with
    a. rifampin
    b. penicillin
    c. amoxicillin
    d. acyclovir

11. Brucellosis can be transmitted to humans by
    a. sexual activity
    b. mosquitoes
    c. inhalation of spores
    d. drinking contaminated milk

12. Wool-sorter's disease is caused by
    a. *Brucella abortus*
    b. *Bacillus anthracis*
    c. *Coxiella burnetii*
    d. *Rabies virus*

13. Which of the following is *not* a hemorrhagic fever?
    a. Lassa fever
    b. Marburg fever
    c. Ebola fever
    d. Trench fever

14. Which characteristic of yellow fever is not true?
    a. single-stranded RNA flavivirus
    b. jaundice
    c. transmitted by fleas
    d. mosquito borne

15. Which of the following respiratory tract infections are considered to be AIDS-defining conditions?
    a. candidiasis
    b. *Pneumocystis (carinii) jiroveci* pneumonia
    c. *Mycobacterium tuberculosis*
    d. herpes simplex bronchitis
    e. all of the above

Concept Questions

These questions are suggested as a *writing-to-learn* experience. For each question, compose a one- or two-paragraph answer that includes the factual information needed to completely address the question.

1. What is endotoxic shock?

2. a. Name the agent(s) responsible for infectious mononucleosis.
   b. What other diseases are associated with these agents?

3. Describe the infectious cycle of HIV.

4. Name the four virus families associate with hemorrhagic fevers. Do they have any common characteristics?

5. a. What is the causative organism of Q fever?
   b. What characteristic(s) of this organism labels it as an "atypical" bacterium?

6. Describe the life cycle of the malarial parasite, including the significant events of sexual and asexual reproduction.

7. What criteria are used in the United States to diagnose a person with AIDS?

8. Describe some of the chemotherapeutic strategies for treating HIV-positive patients.

9. a. What are retroviruses? Where does the name come from?
   b. Name some retroviruses implicated in human diseases.

10. a. What are the different locations in the human body that anthrax infection can be exhibited?
    b. Which of these are the most common forms of the disease?
    c. What organism(s) cause this disease?

Critical Thinking Questions

Critical thinking is the ability to reason and solve problems using facts and concepts. These questions can be approached from a number of angles, and in most cases, they do not have a single correct answer.

1. What is the significance of biofilms in subacute bacterial endocarditis?

2. Why is the bubonic plague called the "black death"?

3. A Nigerian tourist is hospitalized with fever and chills that have cycled for the last 48 hours. A blood smear reveals circular rings within the erythrocytes. What is the treatment and why?

▶ Sore throats, ear infections, sinusitis are common URT infections that can be caused by multiple microorganisms.

▶ Diphtheria is a serious URT that has a single causative organism; it is controlled by the DTaP vaccination.

▶ Whooping cough, respiratory syncytial virus infection and influenza affect the lower URT and the lower respiratory tract (LRT).

▶ Tuberculosis and pneumonia are two major LRT infections.

▶ Many different organisms can cause pneumonia, including the newly discovered SARS virus.

21.1 The Respiratory Tract and Its Defenses

The respiratory tract is the most common place for infectious agents to gain access to the body. We breathe 24 hours a day, and anything in the air we breathe passes at least temporarily into this organ system.

The structure of the system is illustrated in **figure 21.1a.** Most clinicians divide the system into two parts, the *upper* and *lower respiratory tracts.* The upper respiratory tract includes the mouth, the nose, nasal cavity and sinuses above it, the throat or pharynx, and the epiglottis and larynx. The lower respiratory tract begins with the trachea, which feeds into the bronchi and bronchioles in the lungs. Attached to the bronchioles are small balloonlike structures called alveoli, which inflate and deflate with inhalation and exhalation. These are the site of oxygen exchange in the lungs.

Several anatomical features of the respiratory system protect it from infection. As described in chapter 14, nasal hair serves to trap particles and cilia **(figure 21.1b)** on the epithelium of the trachea, and bronchi (the ciliary escalator) propel particles upward and out of the respiratory tract. Mucus on the surface of the mucous membranes lining the respiratory tract is a natural trap for invading microorganisms. Once the microorganisms are trapped, involuntary responses such as coughing, sneezing, and swallowing can move them out of sensitive areas.

The second and third lines of defense also help protect the respiratory tract. Macrophages inhabit the alveoli of the lungs and the clusters of lymphoid tissue (tonsils) in the throat. Secretory IgA against specific pathogens can be found in the mucous secretions as well.

21.2 Normal Flora of the Respiratory Tract

Because of its constant contact with the external environment, the respiratory system harbors a large number of commensal microorganisms. The normal flora is generally limited to the upper respiratory tract, and gram-positive bacteria such as streptococci and staphylococci are very common. It is important to note that some bacteria that can cause serious disease are frequently present in the upper respiratory tract as "normal" flora; these include *Streptococcus pyogenes, Haemophilus influenzae, Streptococcus pneumoniae, Neisseria meningitidis,* and *Staphylococcus aureus.* These bacteria can potentially cause disease if their host becomes immunocompromised for some reason, and they can cause disease in other hosts when they are innocently transferred to them. Other normal flora bacteria include nonhemolytic and α-hemolytic streptococci, *Moraxella* species, and *Corynebacterium* species (often called diphtheroids). Yeasts, especially *Candida albicans,* also colonize the mucosal surfaces of the mouth.

In the respiratory system, as in some other organ systems, the normal flora performs the important function of microbial antagonism (see chapter 13), which reduces the chances of pathogens establishing themselves in the same area by competing with them for resources and space.

Respiratory Tract Defenses and Normal Flora		
	Defenses	**Normal Flora**
Upper Respiratory Tract	Nasal hair, ciliary escalator, mucus, involuntary responses such as coughing, secretory IgA	*Moraxella,* nonhemolytic and α-hemolytic streptococci, *Corynebacterium* and other diphtheroids, *Candida albicans* Note: *Streptococcus pyogenes, Streptococcus pneumoniae, Haemophilus influenzae, Neisseria meningitidis, Staphylococcus aureus* often present as "normal" flora
Lower Respiratory Tract	Mucus, alveolar macrophages, secretory IgA	None

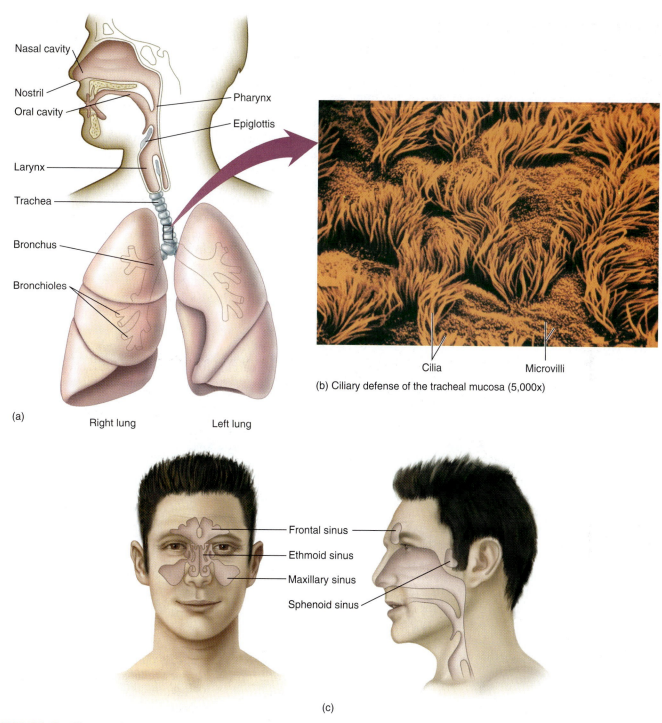

Nasal cavity

Nostril

Oral cavity

Pharynx

Epiglottis

Larynx

Trachea

Bronchus

Bronchioles

(a)

Right lung   Left lung

Cilia   Microvilli

(b) Ciliary defense of the tracheal mucosa (5,000x)

Frontal sinus

Ethmoid sinus

Maxillary sinus

Sphenoid sinus

(c)

FIGURE 21.1   The respiratory tract.
(a) Important structures in the upper and lower respiratory tract. (b) Ciliary defense of the respiratory tract. (c) The four pairs of sinuses in the face and skull.

21.3 Upper Respiratory Tract Diseases Caused by Microorganisms

Rhinitis, or the Common Cold

In the course of a year, people in the United States suffer from about 1 billion colds, called rhinitis because *rhin-* means nose and *-itis* means inflammation. Many people have several episodes a year. Economists estimate that this fairly innocuous infection costs the United States $40 billion a year in trips to the doctors, medications, and lost work time.

Signs and Symptoms

Everyone is familiar with the symptoms of rhinitis: sneezing, scratchy throat, and runny nose, which usually begin 2 or 3 days after infection. An uncomplicated cold generally is not accompanied by fever, although children can experience low fevers (less than 102°F). The incubation period is usually 2 to 5 days.

Causative Agents

The common cold is caused by one of over 200 different kinds of viruses. The particular virus is almost never identified, and the symptoms and handling of the infection are the same no matter which of the viruses is responsible.

The most common type of virus leading to rhinitis is the group called rhinoviruses. Coronaviruses probably are in second place. Most viruses causing the common cold never lead to any serious consequences, but some of them can be serious for some patients. For instance, the respiratory syncytial virus (RSV) causes colds in most people, but in some, especially children, they can lead to more serious respiratory tract symptoms. (RSV is discussed later in the chapter.) In this section we consider all cold-causing viruses together as a group because they are treated similarly.

Viral infection of the upper respiratory tract can predispose a patient to secondary infections by other microorganisms, such as bacteria. Secondary infections may explain why some people report that their colds improved when they were given antibiotics. The cold was caused by viruses; bacterial infection may have followed.

Pathogenesis and Virulence Factors

Viruses that induce rhinitis do not have many virulence mechanisms. They must penetrate the mucus that coats the respiratory tract and then find firm attachment points. Once they are attached, they use host cells to produce more copies of themselves (see chapter 6). The symptoms we experience as the common cold are mainly the result of our body fighting back against the viral invaders. Virus-infected cells in the upper respiratory tract release chemicals that attract certain types of white blood cells to the site, and these cells release cytokines and other inflammatory mediators, as described earlier in chapters 14 and 16. These mediators generate a localized inflammatory reaction, characterized by swelling and inflammation of the nasal mucosa, leakage of fluid from capillaries and lymph vessels, and the increased production of mucus. The similarity of these symptoms to those of inhalant allergies illustrates that the same immune reactions are involved in both conditions.

Transmission and Epidemiology

Cold viruses are transmitted by droplet contact, but indirect transmission may be more common, such as when a healthy person touches a fomite and then touches one of their own vulnerable surfaces, such as the mouth, nose, or an eye. In some cases, the viruses can remain airborne in droplet nuclei and aerosols and can be transmitted in that way.

The epidemiology of the common cold is fairly simple: Practically everybody gets them, and fairly frequently. Children have more frequent infections than adults, probably because nearly every virus they encounter is a new one, and they have no secondary immunity to it. People can acquire some degree of immunity to a cold virus that they have encountered before, but because there are more than 200 viruses, this immunity doesn't provide much overall protection.

Prevention

There is no vaccine for rhinitis. A traditional vaccine would need to contain antigens from 200-plus viruses to provide complete protection. Researchers are studying novel types of immunization strategies, however. Because most of the viruses causing rhinitis use only a few different chemicals on host epithelium for their attachment site, some scientists have proposed developing a vaccine that would stimulate antibody to the docking site on the host. Other approaches include inducing antibody to the sites of action for the inflammatory mediators. But for now, the best prevention is to stop the transmission between hosts. The best way to prevent transmission is frequent handwashing, followed closely by stopping droplets from traveling away from the mouth and nose by covering them when sneezing or coughing. It is better to do this by covering the face with the crook of the arm, rather than the hand, because subsequent contact with surfaces is less likely.

Treatment

No chemotherapeutic agents cure the common cold. A wide variety of over-the-counter agents are available that improve symptoms by blocking inflammatory mediators and their action. The use of these agents may also cut down on transmission to new hosts, because fewer virus-loaded secretions are produced.

✔ CHECKPOINT 21.1	Rhinitis
Causative Organism(s)	200-plus viruses
Most Common Modes of Transmission	Indirect contact, droplet contact
Virulence Factors	Adhesins; most symptoms induced by host response
Culture/Diagnosis	Not necessary
Prevention	Hygiene practices
Treatment	For symptoms only

Sinusitis

Commonly called a *sinus infection,* this inflammatory condition of any of the four pairs of sinuses in the skull **(figure 21.1c)** can actually be caused by allergy (most common), infections, or simply by structural problems such as narrow passageways or a deviated nasal septum. The infectious agents that may be responsible for the condition commonly include a variety of viruses or bacteria, and less commonly, fungi. Infections of the sinuses generally follow a bout with the common cold. But the inflammatory symptoms produce a large amount of fluid and mucus and when trapped in the sinuses, these secretions provide an excellent growth medium for bacteria or fungi. So viral rhinitis is frequently followed by sinusitis caused by bacteria or fungi.

Signs and Symptoms

A person suffering from any form of sinusitis experiences nasal congestion, pressure above the nose or in the forehead, and sometimes the feeling of a headache or a toothache. Facial swelling and tenderness are common. Discharge from the nose and mouth appears opaque and has a green or yellow color in the case of bacterial infections. Discharge caused by an allergy is usually clear, and the symptoms may be accompanied by itchy, watery eyes.

Causative Agents

Bacteria Any number of bacteria that are normal flora in the upper respiratory tract may cause sinus infections. The causative organism is usually not identified, but treatment is begun empirically, based on the symptoms.

The bacteria that cause these infections are most often normal flora in the host and don't have an arsenal of virulence factors that lead to their ability to cause disease. The pathogenesis of this condition is brought about by the confluence of several factors: predisposition to infection because of underlying (often viral) infection; buildup of fluids, providing a rich environment for bacterial multiplication; and sometimes the anatomy of the sinuses, which can contribute to entrapment of mucus and bacterial growth.

Bacterial sinusitis is not a communicable disease. Of course, the virus originally causing rhinitis is transmissible, but the host takes it from there by creating the conditions favorable for respiratory tract microorganisms to multiply in the sinus spaces, which normally do not harbor microorganisms to any significant extent.

Sinusitis is extremely common, resulting in approximately 11.5 million office visits a year in the United States. A large proportion of these cases are allergic sinusitis episodes, but approximately 30% of them are caused by bacterial overgrowth in the sinuses. Women and residents of the southern United States have slightly higher rates. As with many upper respiratory tract infections, smokers have higher rates of infection than nonsmokers. Children who are exposed to large amounts of secondhand smoke are also more susceptible.

Broad-spectrum antibiotics may be prescribed when the physician feels that the sinusitis is bacterial in origin (that is, when allergic and fungal sinusitis are ruled out).

Fungi Fungal sinusitis is rare, but it is often recognized when antibacterial drugs fail to alleviate symptoms. Simple fungal infections may normally be found in the maxillary sinuses and are noninvasive in nature. These colonies are generally not treated with antifungal agents but instead simply mechanically removed by a physician. *Aspergillus fumigatus* is a common fungus involved in this type of infection. The growth of fungi in this type of sinusitis may be encouraged by trauma to the area.

More serious invasive fungal infections of the sinuses may be found in severely immunocompromised patients. Fungi such as *Aspergillus* and *Mucor* species may invade the bony structures in the sinuses and even travel to the brain or eye. These infections are treated aggressively with a combination of surgical removal of the fungus and intravenous antifungal therapy **(Checkpoint 21.2).**

Acute Otitis Media (Ear Infection)

This condition is another common sequela of rhinitis, or the common cold, and for reasons similar to the ones described for sinusitis. Viral infections of the upper respiratory tract lead to inflammation of the eustachian tubes and the buildup of fluid in the middle ear, which can lead to bacterial multiplication in those fluids. Although the middle ear normally

✔ CHECKPOINT 21.2	Sinusitis	
Causative Organism(s)	Various bacteria, often mixed infection	Various fungi
Most Common Modes of Transmission	Endogenous (opportunism)	Introduction by trauma *or* opportunistic overgrowth
Virulence Factors	–	–
Culture/Diagnosis	Culture not usually performed; diagnosis based on clinical presentation, occasionally X rays or other imaging technique used	Same
Prevention	–	–
Treatment	Broad-spectrum antibiotics	Physical removal of fungus; in severe cases antifungals used
Distinctive Features	Much more common than fungal	Suspect in immunocompromised patients

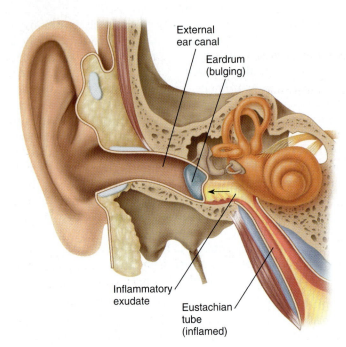

External
ear canal

Eardrum
(bulging)

Inflammatory
exudate

Eustachian
tube
(inflamed)

FIGURE 21.2 **An infected middle ear.**

has no flora, bacteria can migrate along the eustachian tube from the upper respiratory tract **(figure 21.2).** When bacteria encounter mucus and fluid buildup in the middle ear, they multiply rapidly. Their presence increases the inflammatory response, leading to pus production and continued fluid secretion, referred to as *effusion.*

Another condition, known as chronic otitis media, occurs when fluid remains in the middle ear for indefinite periods of time. Until recently, physicians considered it to be the result of a noninfectious immune reaction because they could not culture bacteria from the site, and because antibiotics were not effective. New data suggest that this form of otitis media is caused by a mixed biofilm of bacteria that are attached to the membrane of the inner ear. Biofilm bacteria generally are less susceptible to antibiotics (as discussed in chapter 4), and their presence in biofilm form would explain the inability to culture them from ear fluids.

Signs and Symptoms

Otitis media may be accompanied by a sensation of fullness or pain in the ear and loss of hearing. Younger children may exhibit irritability, fussiness, and difficulty in sleeping, eating, or hearing. Severe or untreated infections can lead to rupture of the eardrum because of pressure of pus buildup, or to internal breakthrough of these infected fluids, which can lead to more serious conditions such as mastoiditis, meningitis, or intracranial abscess.

Causative Agents

Many different viruses and bacteria can cause acute otitis media, but the most common cause is *Streptococcus pneumoniae* (also discussed in the section on pneumonia later in this chapter). *Haemophilus influenzae* is another common cause of this condition; however, the incidence of all types of infec-

tions with this bacterium was significantly reduced with the introduction of a childhood vaccine against it in the 1980s.

Streptococcus pneumoniae appears as pairs of elongated gram-positive cocci joined end to end. It is often called by the familiar name *pneumococcus,* and diseases caused by it are termed *pneumococcal.*

Transmission and Epidemiology

Otitis media is a sequela of upper respiratory tract infection and is not communicable, although the upper respiratory infection preceding it is. Children are particularly susceptible, and boys have a slightly higher incidence than do girls.

Prevention

A vaccine against *S. pneumoniae* has been a part of the recommended childhood vaccination schedule since 2000. The vaccine (Prevnar) is a seven-valent conjugated vaccine (see chapter 15). It contains polysaccharide capsular material from seven different strains of the bacterium, complexed with a chemical that makes it more antigenic. It is distinct from another vaccine for the same bacterium (Pneumovax), which is primarily targeted to the older population to prevent pneumococcal pneumonia.

Treatment

Until the late 1990s, broad-spectrum antibiotics were routinely prescribed for otitis media. When it became clear that frequently treating children with these drugs was producing a bacterial flora with high rates of antibiotic resistance, the treatment regimen was reexamined.

The current recommendation for uncomplicated acute otitis media with a fever below 104°F is "watchful waiting" for 72 hours to allow the body to clear the infection, avoiding the use of antibiotics.

Children who experience frequent recurrences of ear infections sometimes have small tubes placed through the tympanic membranes into their middle ears, to provide a means of keeping fluid out of the site when inflammation occurs **(Checkpoint 21.3).**

Pharyngitis
Signs and Symptoms

The name says it all—this is an inflammation of the throat, which the host experiences as pain and swelling. The severity of pain can range from moderate to severe, depending on the causative agent. Viral sore throats are generally mild and sometimes lead to hoarseness. Sore throats caused by group A streptococci are generally more painful than those caused by viruses, and they are more likely to be accompanied by fever, headache, and nausea.

Clinical signs of a sore throat are reddened mucosa, swollen tonsils, and sometimes white packets of inflammatory products visible on the walls of the throat, especially in streptococcal disease **(figure 21.3).** The mucous membranes may be swollen, affecting speech and swallowing. Often

✓ CHECKPOINT 21.3   Otitis Media

	Streptococcus pneumoniae	*Haemophilus influenzae*	Other bacteria
Causative Organism(s)	*Streptococcus pneumoniae*	*Haemophilus influenzae*	Other bacteria
Most Common Modes of Transmission	Endogenous (may follow upper respiratory tract infection by *S. pneumoniae* or other microorganisms)	Endogenous (follows upper respiratory tract infection)	Endogenous
Virulence Factors	Capsule, hemolysin	Capsule, fimbriae	–
Culture/Diagnosis	Usually relies on clinical symptoms and failure to resolve within 72 hours	Same	Same
Prevention	Pneumococcal conjugate vaccine (heptavalent)	Hib vaccine	None
Treatment	Wait for resolution; if needed, amoxicillin (are high rates of resistance) or trimethoprim/ sulfamethoxazole	Wait for resolution; if needed, ceftriaxone or ampicillin if isolate is sensitive	Wait for resolution; if needed, a broad-spectrum antibiotic (azithromycin) might be used in absence of etiological diagnosis
Distinctive Features	–	–	Suspect if fully vaccinated against other two

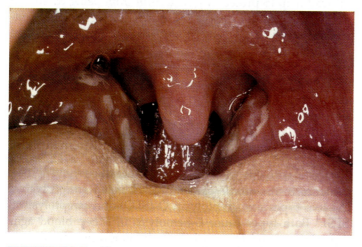

FIGURE 21.3 **The appearance of the throat in pharyngitis and tonsillitis.**
The pharynx and tonsils become bright red and suppurative. Whitish pus nodules may also appear on the tonsils.

pharyngitis results in foul-smelling breath. The incubation period for most sore throats is generally 2 to 5 days.

Causative Agents

A sore throat is most commonly caused by the same viruses causing the common cold. It can also accompany other diseases, such as infectious mononucleosis (described in chapter 20). Pharyngitis may simply be the result of mechanical irritation from prolonged shouting or from drainage of an infected sinus cavity. The most serious cause of pharyngitis is *Streptococcus pyogenes*. The possibility of this bacterium being involved is the reason that even moderately severe sore throats should be monitored and diagnosed.

S. pyogenes is a gram-positive coccus that grows in chains. It does not form spores, is nonmotile, and forms capsules and slime layers. *S. pyogenes* is a facultative anaerobe that ferments a variety of sugars. It does not form catalase, but it does have a peroxidase system for inactivating hydrogen peroxide, which allows its survival in the presence of oxygen.

Pathogenesis

Untreated streptococcal throat infections occasionally can result in serious complications, either right away or days to weeks after the throat symptoms subside. These complications include scarlet fever, rheumatic fever, and glomerulonephritis. More rarely, invasive and deadly conditions such as necrotizing fasciitis can result from infection by *S. pyogenes*. These invasive conditions are described in chapter 18.

Scarlet Fever   Scarlet fever is the result of infection with an *S. pyogenes* strain that is itself infected with a bacteriophage. This virus confers on the streptococcus the ability to produce erythrogenic toxin, described in the section on virulence. Scarlet fever is characterized by a sandpaper-like rash, most often on the neck, chest, elbows, and inner surfaces of the thighs. High fever accompanies the rash. It most often affects school-age children, and was a source of great suffering in the United States in the early part of the twentieth century. In epidemic form, the disease can have a fatality rate of up to 95%. Most cases seen today are mild. They are easily recognizable and amenable to antibiotic therapy. Because of the fear elicited by the name "scarlet fever," the disease is often called scarlatina in North America.

Rheumatic Fever   Rheumatic fever is thought to be due to an immunological cross-reaction between the streptococcal M protein and heart muscle. It tends to occur approximately 3 weeks after pharyngitis has subsided. It can result in permanent damage to heart valves **(figure 21.4)**. Other symptoms include arthritis in multiple joints and the appearance

✔ CHECKPOINT 21.4 Pharyngitis

	Streptococcus pyogenes	Viruses
Causative Organism(s)	*Streptococcus pyogenes*	Viruses
Most Common Modes of Transmission	Droplet or direct contact	All forms of contact
Virulence Factors	LTA, M protein, hyaluronic acid capsule, SLS and SLO, superantigens	–
Culture/Diagnosis	β-hemolytic on blood agar, sensitive to bacitracin, rapid antigen tests	Goal is to rule out *S. pyogenes*, further diagnosis usually not performed
Prevention	Hygiene practices	Hygiene practices
Treatment	Penicillin, cephalexin in penicillin-allergic	Symptom relief only
Distinctive Features	Generally more severe than viral pharyngitis	Hoarseness frequently accompanies viral pharyngitis

(figure 21.7b). If the pharyngitis is caused by a virus, the blood agar dish will show a variety of colony types, representing the normal bacterial flora. Active infection with *S. pyogenes* will yield a plate with a majority of β-hemolytic colonies. Group A streptococci are by far the most common β-hemolytic isolates in human diseases, but lately an increased number of infections by group B streptococci (also β-hemolytic), as well as the existence of β-hemolytic enterococci, have made it important to use differentiation tests. A positive bacitracin disc test (figure 21.7b) provides additional evidence for group A.

Prevention

No vaccine exists for group A streptococci, although many researchers are working on the problem. A vaccine against this bacterium would also be a vaccine against rheumatic fever, and thus it is in great demand. In the meantime, infection can be prevented by good handwashing, especially after coughing and sneezing and before preparing foods or eating.

Treatment

The antibiotic of choice for *S. pyogenes* is penicillin; many group A streptococci have become resistant to erythromycin, a macrolide antibiotic. In patients with penicillin allergies, a first-generation cephalosporin, such as cephalexin, is indicated **(Checkpoint 21.4).**

Diphtheria

For hundreds of years, diphtheria was a significant cause of morbidity and mortality, but in the last 50 years, both the number of cases and the fatality rate have steadily declined throughout the world. In the United States in recent years, only one or two cases have been reported each year. But when healthy people are screened for the presence of the bacterium, it is found in a significant percentage of them, indicating that the lack of cases is due to the protection afforded by immunization with the diphtheria toxoid, which is part of the childhood immunization series.

Indeed, during the 1990s a diphtheria epidemic occurred in the former Soviet Union in which 157,000 people became

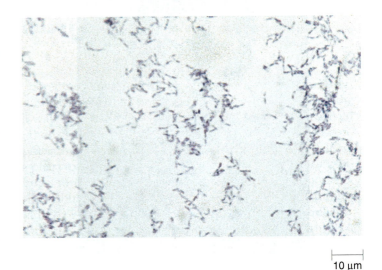

10 μm

FIGURE 21.8 *Corynebacterium diphtheriae.*

ill with diphtheria and 5,000 people died. This upsurge of cases was attributed to a breakdown in immunization practices and production of vaccine, which followed the breakup of the Soviet Union. These examples emphasize the importance of maintaining vaccination, even for diseases that have long been kept under control.

Signs, Symptoms, and Causative Organism

The disease is caused by *Corynebacterium diphtheriae*, a non-spore-forming, gram-positive club-shaped bacterium **(figure 21.8).** The symptoms of diphtheria are experienced initially in the upper respiratory tract. At first the patient experiences a sore throat, lack of appetite, and low-grade fever. A characteristic membrane forms on the tonsils or pharynx **(figure 21.9).** The membrane is formed by the bacteria and it may be quite extensive. It adheres to tissues and cannot easily be removed. It may eventually completely block respiration. The patient may recover after this crisis. Alternatively, exotoxin manufactured by the bacterium may penetrate the bloodstream and travel throughout the body.

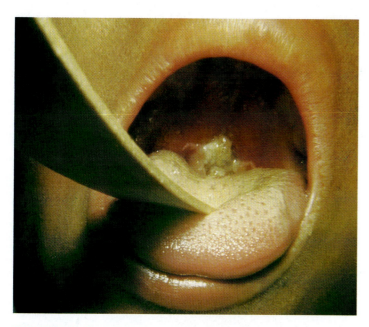

FIGURE 21.9   Diagnosing diphtheria.
The clinical appearance in diphtheria infection includes gross inflammation of the pharynx and tonsils marked by grayish patches (a pseudomembrane) and swelling over the entire area.

Pathogenesis and Virulence Factors

The exotoxin is encoded by a bacteriophage of *C. diphtheriae.* Strains of the bacterium that are not lysogenized by this bacterium do not cause serious disease. The exotoxin is of a type called **A-B toxin.** It is illustrated in **figure 21.10** and explained briefly here. A-B toxins are so named because they consist of two parts, an A (active) component and a B (binding) component. The B component binds to a receptor molecule on the surface of the host cell. The next step is for the A component to be moved across the host cell membrane. The A components of most A-B toxins then catalyze a reaction by which they remove a sugar derivative called the ADP-ribosyl group from the coenzyme NAD and attach it to one host cell protein or another. This process is called *ADP-ribosylation.* This process disrupts the normal function of that host protein, resulting in some type of symptom for the patient.

The release of diphtheria toxin in the blood leads to complications in distant organs, especially myocarditis and neuritis. Myocarditis can cause abnormal cardiac rhythms and in the worst cases can lead to heart failure. Neuritis affects motor nerves and may result in temporary paralysis of limbs, the soft palate, and even the diaphragm, a condition that can predispose a patient to other lower respiratory tract infections.

Prevention and Treatment

Diphtheria can easily be prevented by a series of vaccinations with toxoid, usually given as part of a mixed vaccine against tetanus and pertussis, as well, called the *DTaP* (for diphtheria, tetanus, and acellular pertussis). If a patient has diphtheria,

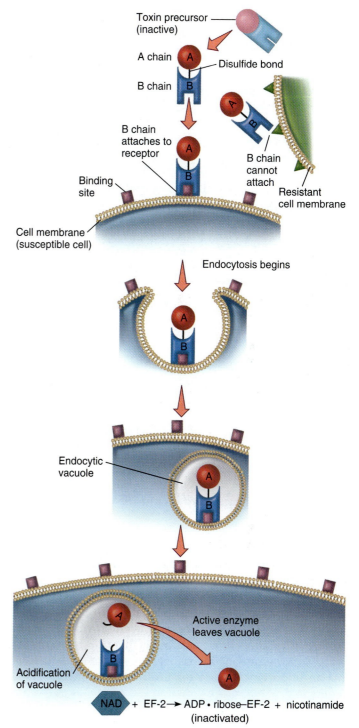

FIGURE 21.10   A-B toxin of *Corynebacterium diphtheriae.*
The B chain attaches to host cell membrane, then the toxin enters the cell. The two chains separate and the A chain enters the cytoplasm as an active enzyme that ADP-ribosylates a protein (EF-2) needed for protein synthesis. Cell death follows.

and it has progressed to the bloodstream, the adverse effects of toxemia are treated with diphtheria antitoxin derived from horses. Prior to injection, the patient must be tested for allergy to horse serum and be desensitized if necessary. The infection itself may be treated with antibiotics from the penicillin or erythromycin family. Bed rest, heart medication, and tracheostomy or bronchoscopy to remove the membrane (sometimes called a pseudomembrane) may be indicated.

✔ **CHECKPOINT 21.5**	**Diphtheria**
Causative Organism(s)	*Corynebacterium diphtheriae*
Most Common Modes of Transmission	Droplet contact, direct contact or indirect contact with contaminated fomites
Virulence Factors	Exotoxin: diphtheria toxin
Culture/Diagnosis	Tellurite medium—gray/black colonies, club-shaped morphology on Gram stain; *treatment begun before definitive identification*
Prevention	Diphtheria toxoid vaccine (part of DTaP)
Treatment	Antitoxin plus penicillin or erythromycin

21.4 Diseases Caused by Microorganisms Affecting the Upper and Lower Respiratory Tract

A number of infectious agents affect both the upper and lower respiratory tract regions. We discuss the more well-known diseases in this section; specifically, they are whooping cough, respiratory syncytial virus (RSV), and influenza.

Whooping Cough

Whooping cough is also known as *pertussis* (the suffix *-tussis* is Latin for cough). A vaccine for this potentially serious infection has been available since 1926. The disease is still troubling to the public health community because its incidence is increasing in the United States, despite improvements in the vaccine. In addition, in the recent past there has been concern over the vaccine among the general public. For these reasons, it is an important disease for health care professionals to understand.

Signs and Symptoms

The disease has two distinct symptom phases called the catarrhal and paroxysmal stages, which are followed by a long recovery (or convalescent) phase, during which a patient is particularly susceptible to other respiratory infections. After an incubation period of from 3 to 21 days, the **catarrhal stage** begins when bacteria present in the respiratory tract cause what appear to be cold symptoms, most notably a runny nose. This stage lasts 1 to 2 weeks. The disease worsens in the second **(paroxysmal)** stage, which is characterized by severe and uncontrollable coughing (a *paroxysm* can be thought of as a convulsive attack). The common name for the

disease comes from the whooping sound a patient makes as he or she tries to grab a breath between uncontrollable bouts of coughing. The violent coughing spasms can result in burst blood vessels in the eyes or even vomiting. In the worst cases, seizures result from small hemorrhages in the brain.

As in any disease, the **convalescent phase** is the time when numbers of bacteria are decreasing and no longer cause ongoing symptoms. But the active stages of the disease damage the cilia on respiratory tract epithelial cells, and complete recovery of these surfaces requires weeks or even months. During this time, other microorganisms can more easily colonize and cause secondary infection.

Causative Agent

Bordetella pertussis is a very small gram-negative rod. Sometimes it looks like a coccobacillus. It is strictly aerobic and fastidious, having specific nutritional requirements for successful culture.

IN THE NEWS *(Continued from page 653)*

The outbreak described at the beginning of the chapter was caused by *B. pertussis*. That might surprise you, if you think of whooping cough as a disease of childhood, or if you assume that it has been eliminated by vaccination. But the CDC reports that pertussis is the only disease among those for which children are routinely vaccinated that has *increased* in incidence during the past 20 years. As many as 8,200 cases occurred in the United States in 2002; the incidence rate in adults increased 400% during the 1990s.

The Illinois refinery outbreak described here was noticed because many of the adults experienced full-blown disease. Scientists suspect that adults often have mild or unrecognized infections, creating a reservoir for new infections. It was once believed that childhood immunization conferred lifelong protection, but most scientists now believe that immunity wanes during adolescence and early adulthood. There is growing interest in selectively reimmunizing young adults with acellular pertussis vaccine.

See: CDC. 2003. Pertussis outbreak among adults at an oil refinery—Illinois, August–October 2002. MMWR 52:1–4.

Pathogenesis and Virulence Factors

The progress of this disease can be clearly traced to the virulence mechanisms of the bacterium. It is absolutely essential for the bacterium to attach firmly to the epithelial cells of the mouth and throat, and it does so using specific adhesive molecular structures on its surface. One of these structures is called *filamentous hemagglutinin (FHA)*. It is a fibrous structure that surrounds the bacterium like a capsule and is also secreted in soluble form. In that form it can act as a bridge between the bacterium and the epithelial cell.

Once the bacteria are attached in large numbers, production of mucus increases and localized inflammation ensues, resulting in the early stages of the disease. Then the real damage begins: The bacteria release multiple exotoxins that damage ciliated respiratory epithelial cells and cripple other components of the host defense, including phagocytic cells.

The two most important exotoxins are *pertussis toxin* and *tracheal cytotoxin*. Pertussis toxin is a classic A-B toxin, like the diphtheria toxin illustrated in figure 21.10. In the case of pertussis toxin, the host protein affected by the process of ADP-ribosylation is one that normally limits the production of cyclic AMP. Cyclic AMP is a critical molecule that regulates numerous functions inside host cells. The excessive amounts of cyclic AMP result in copious production of mucus and a variety of other effects in the respiratory tract and the immune system.

Tracheal cytotoxin results in more direct destruction of ciliated cells. Another important contributor to the pathology of the disease is *B. pertussis* endotoxin. As always with endotoxins, its release leads to the production of a host of cytokines that have direct and indirect effects on physiological processes and on the host response.

Transmission and Epidemiology

B. pertussis is transmitted via respiratory droplets. It is highly contagious during both the catarrhal and paroxysmal stages. The disease manifestations are most serious in infants. Twenty-five percent of infections occur in older children and adults, who generally have milder symptoms. The disease results in 300,000 to 500,000 deaths annually worldwide.

Pertussis outbreaks continue to occur in the United States and elsewhere. Even though it is estimated that approximately 85% of U.S. children are vaccinated against pertussis, it continues to be spread, perhaps by adults whose own immunity has dwindled. These adults may experience mild, unrecognized disease, and unwittingly pass it to others. It has also been found that fully vaccinated children can experience the disease, possibly due to antigenic changes in the bacterium.

Culture and Diagnosis

This disease is often diagnosed based solely on its symptoms, since they are so distinctive. When culture confirmation is desired, nasopharyngeal swabs can be inoculated on specific media—Bordet-Gengou (B-G) medium, charcoal agar, or potato-glycerol agar.

Prevention

The current vaccine for pertussis is an acellular formulation of important *B. pertussis* antigens. It results in far fewer side effects than the previous whole cell vaccine, which was used until the mid-1990s. It is generally given in the form of the DTaP vaccine.

A second prevention strategy is the administration of antibiotics to contacts of people who have been diagnosed with the disease. Erythromycin or trimethoprim-sulfamethoxazole is given for 14 days to prevent disease in those who may have been infected.

Treatment

Treating someone who is already ill with pertussis is focused on supportive care; antibiotics may or may not shorten the course of the disease, which is often the case when major symptoms of a condition are the result of exotoxin secretion. Antibiotics (erythromycin) are sometimes administered because they do decrease the contagiousness of the patient.

✔ CHECKPOINT 21.6	Pertussis (Whooping Cough)
Causative Organism(s)	*Bordetella pertussis*
Most Common Modes of Transmission	Droplet contact
Virulence Factors	FHA (adhesion), pertussis toxin and tracheal cytotoxin, endotoxin
Culture/Diagnosis	Grown on B-G, charcoal or potato-glycerol agar; diagnosis can be made on symptoms
Prevention	Acellular vaccine (DTaP), erythromycin or trimethoprim; sulfamethoxazole for contacts
Treatment	Mainly supportive; erythromycin to decrease communicability

Respiratory Syncytial Virus Infection

As its name indicates, respiratory syncytial virus (RSV) infects the respiratory tract and produces giant multinucleated cells (syncytia). Outbreaks of droplet-spread RSV disease occur regularly throughout the world, with peak incidence in the winter and early spring. Children 6 months of age or younger, as well as premature babies, are especially susceptible to serious disease caused by this virus. RSV is the most prevalent cause of respiratory infection in the newborn age group, and nearly all children have experienced it by age 2. An estimated 100,000 children are hospitalized with RSV infection each year in the United States. The mortality rate is highest for children with complications such as prematurity, congenital disease, and immunodeficiency. Infection in older children and adults usually manifests as a cold.

The first symptoms are fever that lasts for approximately 3 days, rhinitis, pharyngitis, and otitis. More serious infections progress to the bronchial tree and lung parenchyma, giving rise to symptoms of croup that include acute bouts of coughing, wheezing, difficulty in breathing (called **dyspnea**), and abnormal breathing sounds (called rales). (Note: This condition is often called croup and also bronchiolitis; be aware that both of these terms are clinical descriptions of diseases caused by a variety of viruses [in addition to RSV] and sometimes by bacteria.)

The virus is highly contagious and is transmitted through droplet contact, but also through fomite contamination. Diagnosis of RSV infection is more critical in babies than in older children or adults. The afflicted child is conspicuously ill, with signs typical of pneumonia and bronchitis. The best diagnostic procedures are those that demonstrate the viral antigen directly from specimens (direct and indirect fluorescent staining, ELISA, and DNA probes).

There is no RSV vaccine available yet, but an effective passive antibody preparation is used as prevention in high-risk children. Ribavirin, an antiviral drug, can be administered as an inhaled aerosol to very sick children, although the clinical benefit is uncertain.

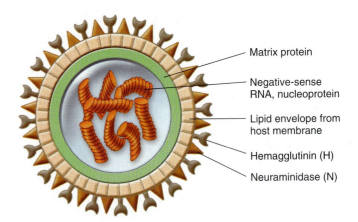

FIGURE 21.11 Schematic drawing of influenza virus.

✔ CHECKPOINT 21.7	RSV Disease
Causative Organism(s)	Respiratory syncytial virus (RSV)
Most Common Modes of Transmission	Droplet and indirect contact
Virulence Factors	Syncytia formation
Culture/Diagnosis	Direct antigen testing
Prevention	Passive antibody
Treatment	Ribavirin in severe cases

Influenza

The "flu" is a very important disease to study for several reasons. First of all, everyone is familiar with the cyclical increase of influenza infections occurring during the winter months in the United States. Second, many conditions are erroneously termed the "flu," while in fact only diseases caused by influenza viruses are actually the flu. Third, the way that influenza viruses behave provides an excellent illustration of the way other viruses can, and do, change to cause more serious diseases than they did previously.

Signs and Symptoms

Influenza begins in the upper respiratory tract but in serious cases may also affect the lower respiratory tract. There is a 1- to 4-day incubation period, after which symptoms begin very quickly. These include headache, chills, dry cough, body aches, fever, stuffy nose, and sore throat. Even the sum of all these symptoms can't describe how a person actually feels: lousy. The flu is known to "knock you off your feet." Extreme fatigue can last for a few days or even a few weeks. An infection with influenza can leave patients vulnerable to secondary infections, often bacterial. Influenza infection alone occasionally leads to a pneumonia that can cause rapid death, even in young healthy adults.

Patients with emphysema or cardiopulmonary disease, along with very young, elderly, or pregnant patients are more susceptible to serious complications.

Causative Agent

Influenza is caused by one of three influenza viruses: A, B, or C. They belong to the family Orthomyxoviridae. They are spherical particles with an average diameter of 80 to 120 nm. Each virion is covered with a lipoprotein envelope that is studded with glycoprotein spikes acquired during viral maturation **(figure 21.11)**. The two glycoproteins that make up the spikes of the envelope and contribute to virulence are called hemagglutinin (H) and neuraminidase (N). The name hemagglutinin is derived from this glycoprotein's agglutinating action on red blood cells, which is the basis for viral assays used to identify the viruses. Hemagglutinin contributes to infectivity by binding to host cell receptors of the respiratory mucosa, a process that facilitates viral penetration. Neuraminidase breaks down the protective mucous coating of the respiratory tract, assists in viral budding and release, keeps viruses from sticking together, and participates in host cell fusion.

The ssRNA genome of the influenza virus is known for its extreme variability. It is subject to constant genetic changes that alter the structure of its envelope glycoproteins. Research has shown that genetic changes are very frequent in the area of the glycoproteins recognized by the host immune response, but very rare in the areas of the glycoproteins used for attachment to the host cell **(figure 21.12)**. In this way, the virus can continue to attach to host cells while managing to decrease the effectiveness of the host response to its presence. This constant mutation of the glycoproteins is called **antigenic drift**—the antigens gradually change their amino acid composition, resulting in decreased ability of host memory cells to recognize them.

An even more serious phenomenon is known as **antigenic shift.** The genome of the virus consists of just 10 genes, encoded on 8 separate RNA strands. Antigenic shift is the swapping out of one of those genes or strands with a gene or strand from a different influenza virus. Some explanation is in order. First, we know that certain influenza viruses infect both humans and swine. Other influenza viruses infect birds (or ducks) and swine. All of these viruses have 10 genes coding for the same important influenza proteins (including H and N)—but the actual sequence of the genes is different in the different types of viruses. Second, when the two viruses just described infect a single swine host, with both virus types infecting the same host cell, the viral packaging step can accidentally produce a human influenza virus that contains 7 human influenza virus RNA strands plus a single duck influenza virus RNA strand (**figure 21.13**). When that virus infects a human, no immunologic recognition of the protein that came from the duck virus occurs. Experts have traced the flu pandemics of 1918, 1957, 1968, and 1977 to strains of a virus that came from pigs (swine flu). Influenza A viruses are named according to the different types of H and N spikes they display on their surfaces. For instance, in 2004 the most common circulating subtypes of influenza A viruses were H1N1 and H3N2. Influenza B viruses are not divided into subtypes since they are thought only to undergo antigenic drift and not antigenic shift. Influenza C viruses are thought to cause only minor respiratory disease and are probably not involved in epidemics.

Scientists have also recently found that antigenic drift and shift are not even required to make an influenza virus deadly. It appears that a minor genetic alteration in another influenza virus gene, one that seems to produce an enzyme used to manufacture new viruses in the host cell, can make the difference between a somewhat pathogenic influenza virus and a lethal one. It is still not clear exactly how many of these minor changes can lead to pandemic levels of infection and a catastrophe for the public health.

Pathogenesis and Virulence Factors

The influenza virus binds primarily to ciliated cells of the respiratory mucosa. Infection causes the rapid shedding of these cells along with a load of viruses. Stripping the respiratory epithelium to the basal layer eliminates protective ciliary clearance and leads to severe inflammation and irritation. The illness is further aggravated by fever, headache, and the other symptoms just described. The viruses tend to remain in the respiratory tract rather than spread to the bloodstream. As the

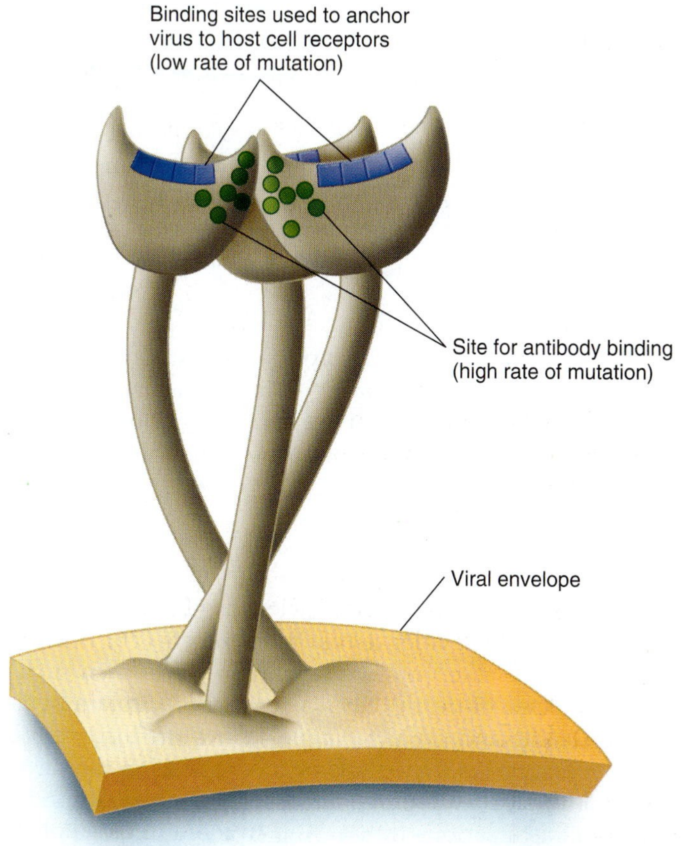

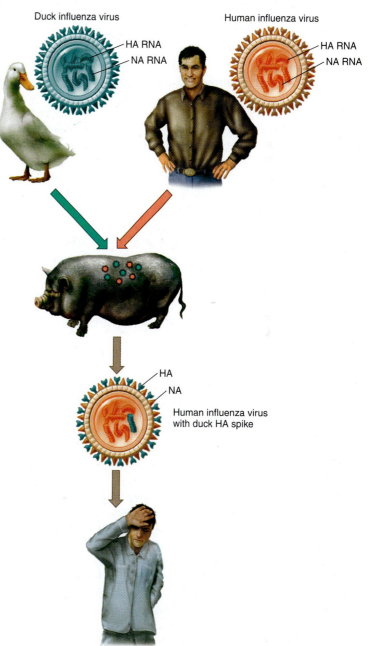

FIGURE 21.13   **Antigenic shift event.**
Where ducks and swine and humans live close together, the swine can serve as a melting pot for creating "hybrid" influenza viruses that are not recognized by the human immune system.

FIGURE 21.12   **Schematic drawing of hemagglutinin (HA) of influenza virus.**
Blue boxes depict site used to attach virus to host cells; green circles depict sites for anti-influenza antibody binding.

normal ciliated epithelium is restored in a week or two, the symptoms subside.

As just noted, the glycoproteins and their structure are important virulence determinants. First of all, they mediate the adhesion of the virus to host cells. Second, they change gradually and sometimes suddenly, evading immune recognition.

Transmission and Epidemiology

Inhalation of virus-laden aerosols and droplets constitutes the major route of influenza infection, although fomites can play a secondary role. Transmission is greatly facilitated by crowding and poor ventilation in classrooms, barracks, nursing homes, dormitories, and military installations in the late fall and winter. It is highly contagious and affects people of all ages. Annually there are approximately 36,000 U.S. deaths from influenza and its complications, mainly among the very young and the very old. The winter of 2003–2004 was a particularly bad influenza season.

Culture and Diagnosis

Very often physicians will diagnose influenza based on symptoms alone. But there is a wide variety of culture-based and nonculture-based methods to diagnose the infection. Rapid influenza tests (such as PCR, ELISA-type assays, or immunofluorescence) provide results within 24 hours; viral culture provides results in 3 to 10 days. Cultures are not typically performed at the point of care; they must be sent to diagnostic laboratories, and they require up to 10 days for results. Despite these disadvantages, culture can be useful to identify which subtype of influenza is causing infections, which is important for public health authorities to know. Also, nonculture-based test results are not terribly accurate; they may produce as many as 30% false negatives.

Prevention

Preventing influenza infections and epidemics is one of the top priorities for public health officials. The standard vaccine contains dead viruses grown in embryonated eggs. It has an overall effectiveness of 70% to 90%. The vaccine consists of three different influenza viruses (usually two influenza A and one influenza B) that have been judged to most resemble the virus variants likely to cause infections in the coming flu season. Because of the changing nature of the antigens on the viral surface, annual vaccination is considered the best way to avoid infection. Anyone over the age of 6 months can take the vaccine, and it is recommended for anyone in a high-risk group or for people who have a high degree of contact with the public.

A new vaccine called FluMist is a nasal mist vaccine consisting of the three strains of influenza virus in live attenuated form. It is designed to stimulate secretory immunity in the upper respiratory tract. Its safety and efficacy have so far been demonstrated only for persons between the ages of 5 and 49. It is not advised for immunocompromised individuals, and it is significantly more expensive than the injected vaccine.

Several of the anti-influenza drugs listed in the following section can be used for preventive purposes, especially in epidemics.

Treatment

Influenza is one of the first viral diseases for which effective antiviral drugs became available. The drugs must be taken early in the infection, preferably by the second day. This requirement is an inherent difficulty because most people do not realize until later that they may have the flu. Amantadine and rimantadine can be used to treat and prevent influenza type A infections, but they do not work against influenza type B viruses.

Zanamivir (Relenza) is an inhaled drug that works against influenza A and B. Oseltamivir (Tamiflu) is available in capsules or as a powdered mix to be made into a drink. It can also be used for prevention of influenza A and B.

✔ CHECKPOINT 21.8	Influenza
Causative Organism(s)	Influenza A, B, and C viruses
Most Common Modes of Transmission	Droplet contact, direct contact, some indirect contact
Virulence Factors	Glycoprotein spikes, overall ability to change genetically
Culture/Diagnosis	Viral culture (3–10 days) or rapid antigen-based tests
Prevention	Killed injected vaccine or inhaled live attenuated vaccine—taken annually
Treatment	Amantadine, rimantadine, zanamivir, or oseltamivir

21.5 Lower Respiratory Tract Diseases Caused by Microorganisms

In this section, we consider microbial diseases that affect the lower respiratory tract primarily—namely, the bronchi, bronchioles, and lungs, with minimal involvement of the upper respiratory tract. Our discussion focuses on tuberculosis and pneumonia.

Tuberculosis

Mummies from the Stone Age, ancient Egypt, and Peru provide unmistakable evidence that tuberculosis (TB) is an ancient human disease. In fact, historically it has been such a prevalent cause of death that it was called "Captain of the Men of Death" and "White Plague." After the discovery of streptomycin in 1943, the rates of tuberculosis in the developed world declined rapidly. But since the mid-1980s, it has reemerged as a serious threat. And in many regions of the world, the rates of TB are so high that the World Health Organization has requested emergency aid. The cause of tuberculosis is primarily the bacterial species *Mycobacterium tuberculosis,* informally called the tubercle bacillus.

Fungal Lung Diseases

Increasingly, the microorganisms that cause pulmonary infections are fungi. Although still much rarer than bacterial lung infections, fungal pneumonias have shown a remarkable rise in incidence. One hospital in the Midwest reported an overall 20-fold increase in fungal infections (of all types) in the 10 years between the late 1970s and the late 1980s. And a great many of those infections occur in the lungs. As you read in chapter 5, two broad categories of fungi cause human infections: those considered to be *primary pathogens*, which readily cause disease even in healthy hosts, and *opportunists*, which cause disease primarily in hosts that are weakened due to underlying illness, advanced age, immune deficiency, or chemotherapy of some sort.

The primary pathogens usually have restricted geographic distributions. **Table 21.A** describes major characteristics of these fungi. As you can imagine, when primary pathogens invade people with weakened immune systems, the results can be disastrous.

In contrast to the primary pathogens, the opportunists are more likely to be ubiquitous, and can affect weakened patients indiscriminately. **Table 21.B** lists some of the most common oppor-

tunistic fungal infections of the lungs. These opportunistic fungal infections are the ones increasing at a steady rate in the modern era, for several reasons:

- Fungi and their spores are everywhere. They constantly enter our respiratory tracts. They live in our GI tracts and on our skin.
- Antibiotic use decreases the bacterial count in our bodies, leaving fungi unhindered and able to flourish.
- More invasive procedures are being employed in hospitals and for outpatient procedures, opening pathways for fungi to access "sterile" areas of the body.
- The number of patients who are immunosuppressed (or otherwise "weakened") is constantly increasing.

For these reasons, health care professionals should be particularly vigilant for symptoms of fungal diseases in patients who are hospitalized, are HIV-positive, or have other underlying health problems. Invasive fungal infections are extremely difficult to treat effectively; there is a significant mortality rate for patients suffering from opportunistic fungal infections in the lungs.

TABLE 21.A	Primary Fungal Pathogens of the Lungs	
Pathogen	**Geographic Distribution**	**Disease and Symptoms**
Histoplasma capsulatum	All continents except Australia; highest rates in U.S. Ohio Valley	Histoplasmosis (see p. 679); aches, pains, and coughing; more severe symptoms include fever, night sweats, and weight loss
Blastomyces dermatitidis	Forest soils, areas of decaying wood and organic matter; worldwide distribution, in U.S. most common on East Coast and in Midwest	Blastomycosis—cough, chest pain, hoarseness, fever; severe cases involve skin and other organs; lung abscesses resemble malignant tumors; skin nodules, bone infections, involvement of central nervous system possible
Coccidioides immitis	Semiarid, hot climates; Mexico, Central and South America; SW U.S., especially California and southern Arizona	Coccidioidomycosis—fever, chest pain, headaches, malaise, chronic infection can lead to pulmonary nodular growths and cavity formation in lungs
Paracoccidioides brasiliensis	Tropical and semitropical regions of South and Central America	Paracoccidioidomycosis—infections of lung and skin; in severe cases, fungus can invade lungs, skin, and lymphatic organs

TABLE 21.B	Opportunistic Fungi in the Lungs
Fungus	**Disease**
Pneumocystis (carinii) jiroveci	"PCP" pneumonia (see p. 681); cough, fever, shallow respiration, and cyanosis
Aspergillus spp.	Aspergillosis; fungus balls form in the lungs and other tissues, necrotic pneumonia, dissemination to the brain, heart, skin
Geotrichum candidum	Geotrichosis; secondary infections in tuberculosis or very ill patients
*Cryptococcus neoformans**	Cryptococcosis; lung infections followed often by brain and meninges involvement
Candida albicans	Candidal lung infections; in HIV-positive and lung transplant patients

*Cryptococcus could fit in either category—primary or opportunistic pathogen—but its array of virulence factors are (individually) less potent than most of those expressed by primary pathogens. However, it often causes disease in otherwise healthy patients.

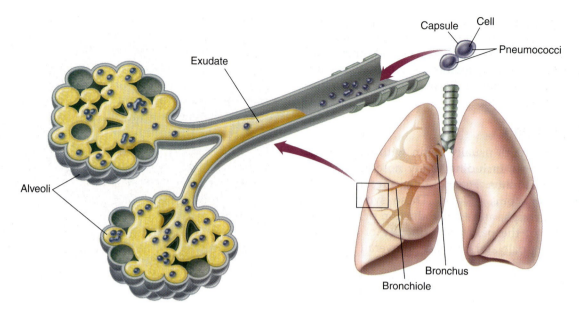

FIGURE 21.21 **The course of bacterial pneumonia.**
As the pneumococcus traces a pathway down the respiratory tree, it provokes intense inflammation and exudate formation. The blocking of the bronchioles and alveoli by consolidation of inflammatory cells and products is evident.

In infants and the elderly, the areas of infection are usually spottier and centered more in the bronchi than in the alveoli (bronchial pneumonia). Systemic complications of pneumonia are pleuritis and endocarditis, but pneumococcal bacteremia and meningitis are the greatest danger to the patient.

Because the pneumococcus is such a frequent cause of pneumonia in older adults, this population is encouraged to seek immunization with the older pneumococcal polysaccharide vaccine, which stimulates immunity to the capsular polysaccharides of 23 different strains of the bacterium. Active disease is treated with antibiotics, but the choice of antibiotic is often difficult. Many isolates of *S. pneumoniae* are resistant to penicillin and its derivatives, prompting the use of macrolide antibiotics such as erythromycin. Rising resistance to the macrolides has been noted, so often trimethoprim-sulfamethoxazole is now prescribed. A new drug called ketek, in a new class of drugs called ketolides, has recently been developed for drug-resistant *S. pneumoniae*. This bacterium is clearly capable of rapid development of resistance, and effective treatment requires that the practitioner be familiar with local resistance trends.

Legionella pneumophila *Legionella* is a weakly gram-negative bacterium that has a range of shapes, from coccus to filaments. Several species or subtypes have been characterized, but *L. pneumophila* (lung-loving) is the one most frequently isolated from infections.

Although the organisms were originally described in the late 1940s, they were not clearly associated with human disease until 1976. The incident that brought them to the attention of medical microbiologists was a sudden and mysterious epidemic of pneumonia that afflicted 200 American Legion members attending a convention in Philadelphia and killed 29 of them. After 6 months of painstaking analysis, epidemiologists isolated the pathogen and traced its

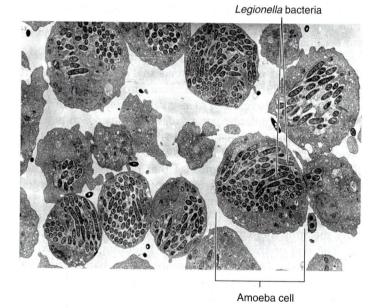

FIGURE 21.22 *Legionella* **living intracellularly in the amoeba** *Hartmanella.*
Amoebas inhabiting natural waters appear to be the reservoir for this pathogen and a means for it to survive in rather hostile environments. The pathogenesis of *Legionella* in humans is likewise dependent on its uptake by and survival in phagocytes.

source to contaminated air-conditioning vents in the Legionnaires' hotel.

Legionella's ability to survive and persist in natural habitats has been something of a mystery, yet it appears to be widely distributed in aqueous habitats as diverse as tap water, cooling towers, spas, ponds, and other fresh waters. The bacteria can live in close association with free-living amoebas **(figure 21.22)**. It is released during aerosol formation and can be carried for long distances. Cases have been traced to supermarket vegetable sprayers, hotel fountains, and even the fallout from the Mount St. Helens volcano eruption in 1980.

Although this bacterium can cause another disease called Pontiac fever, pneumonia is the more serious disease, with a fatality rate of 3% to 30%. *Legionella* pneumonia is thought of as an opportunistic disease, usually affecting the elderly and rarely being seen in children and healthy adults. It is difficult to diagnose, even with specific antibody tests. Curiously, urine is often used for antigen testing with this microorganism.

Mycoplasma pneumoniae

Mycoplasmas, as you learned in chapter 4, are among the smallest known self-replicating microorganisms. They naturally lack a cell wall and are therefore irregularly shaped. They may resemble cocci, filaments, doughnuts, clubs, or helices. They are free-living but fastidious, requiring complex medium to grow in the lab. (This genus should not be confused with *Mycobacterium*.)

Pneumonias caused by *Mycoplasma* (as well as those caused by *Chlamydia* and some other microorganisms) are often called atypical pneumonia—atypical in the sense that the symptoms do not resemble those of pneumococcal or other severe pneumonias. *Mycoplasma* pneumonia is transmitted by aerosol droplets among people confined in close living quarters, especially families, students, and the military.

The bacterium binds very tightly to specific receptors of the respiratory epithelium and inhibits ciliary action. Gradual spread of the bacteria over the next 2 to 3 weeks disrupts the cilia and damages the respiratory epithelium. The first symptoms—fever, malaise, sore throat, and headache—are not suggestive of pneumonia. A cough is not a prominent early symptom, and when it does appear it is mostly unproductive. As the disease progresses, nasal symptoms, chest pain, and earache can develop. The lack of acute illness in most patients has given rise to the name "walking pneumonia." For some reason, there is an increase in *Mycoplasma* pneumonias every 3 to 6 years in the United States.

Diagnosis of *Mycoplasma* may begin with ruling out other bacteria or viral agents. Serological or PCR tests confirm the diagnosis. These bacteria do not stain with Gram's stain and are not visible in direct smears of sputum.

Hantavirus

In 1993 hantavirus suddenly burst into the American consciousness. A cluster of unusual cases of severe lung edema among healthy young adults arose in the Four Corners area of New Mexico. Most of the patients died within a few days. They were later found to have been infected with hantavirus, an agent that had previously only been known to cause severe kidney disease and hemorrhagic fevers in other parts of the world. The new condition was named hantavirus pulmonary syndrome (HPS). Since 1993 the disease has occurred sporadically, but it has a mortality rate of at least 33%. It is considered an emerging disease.

Symptoms, Pathogenesis, and Virulence Factors Common features of the prodromal phase of this infection include fever, chills, myalgias, headache, nausea, vomiting, and di-

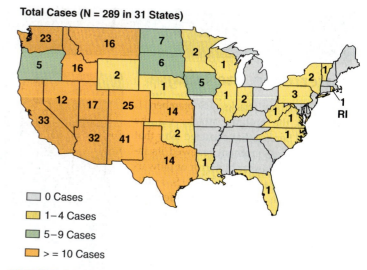

Total Cases (N = 289 in 31 States)

- ☐ 0 Cases
- ☐ 1–4 Cases
- ☐ 5–9 Cases
- ☐ >= 10 Cases

FIGURE 21.23 Hantavirus pulmonary syndrome cases, United States.
Cumulative data through January 2002.

arrhea or a combination of these symptoms. A cough is common but is not a prominent early feature. Initial symptoms resemble those of other common viral infections. Soon a severe pulmonary edema occurs, and causes acute respiratory distress (ARDS, or acute respiratory distress syndrome, has many microbial and nonmicrobial causes; this is but one of them).

The acute lung symptoms appear to be due to the presence of large amounts of hantavirus antigen, which becomes disseminated throughout the bloodstream (including the capillaries surrounding the alveoli of the lung). Massive amounts of fluid leave the blood vessels and flood the alveolar spaces in response to the inflammatory stimulus, causing severe breathing difficulties and a drop in blood pressure. The propensity to cause a massive inflammatory response could be considered a virulence factor for this organism.

Transmission and Epidemiology Very soon after the initial cases in 1993, it became clear that the virus was associated with the presence of mice in close proximity to the victims. Investigators eventually determined that the virus is transmitted via airborne dust contaminated with the urine, feces, or saliva of infected rodents. Deer mice and other rodents can carry the virus with few apparent symptoms. Small outbreaks of the disease are usually correlated with increases in the local rodent population. Epidemiologists suspect that rodents have been infected with this pathogen for centuries. It has no doubt been the cause of sporadic cases of unexplained pneumonia in humans for decades, but the incidence seems to be increasing, especially in areas of the United States west of the Mississippi River **(figure 21.23)**.

Treatment and Prevention The diagnosis is established by detection of IgM to hantavirus in the patient's blood, or by using PCR techniques to find hantavirus genetic material in

aware of the possibility of *E. coli*–contaminated hamburgers or *Salmonella*–contaminated ice cream. New food safety measures are being implemented all the time, but it is still necessary for the consumer to be aware and to practice good food handling. As just mentioned, the increased use of day care centers has also led to increased transmission of diarrheal agents.

For a disease that exacts such a high price, there is relatively little consensus on how to manage a patient with acute diarrhea. Although most diarrhea episodes are self-limiting, and therefore do not require treatment, others (such as *E. coli* O157:H7) can have devastating effects. In most diarrheal illnesses, antimicrobial treatment is contraindicated (inadvisable), but some, such as shigellosis, call for quick treatment with antibiotics. For public health reasons, it is important to know which agents are causing diarrhea in the community, but in most cases identification of the agent is not performed.

In this section we describe acute diarrhea having infectious agents as the cause. In the sections following this one, we discuss acute diarrhea and vomiting caused by toxins, commonly known as food poisoning, and chronic diarrhea and its causes.

Salmonella

It is estimated that one of every three chickens destined for human consumption is contaminated with *Salmonella*, and other poultry such as ducks and turkeys are also affected. Eggs are a particular problem because the bacteria may actually enter the egg while the shell is being formed in the chicken. *Salmonella* is a very large genus of bacteria, but only one species is of interest to us: *S. enterica* is divided into many serovars, based on variation in the major surface antigens.

As mentioned in chapter 4, serotype or serovar analysis aids in bacterial identification. Many gram-negative enteric bacteria are named and designated according to the following antigens: **H,** the flagellar antigen; **K,** the capsular antigen; and **O,** the cell wall antigen. Not all enteric bacteria carry the H and K antigens, but all have O, the polysaccharide portion of the lipopolysaccharide implicated in endotoxic shock (see chapter 20). Most species of gram-negative enterics exhibit a variety of subspecies, serovars, or serotypes caused by slight variations in the chemical structure of the HKO antigens. Some bacteria in this chapter (for example, *E. coli* O157:H7) are named according to their surface antigens; however, we will use Latin serovar names for *Salmonella*.

Salmonellae are motile; they ferment glucose with acid and sometimes gas; and most of them produce hydrogen sulfide (H_2S), but not urease. They grow readily on most laboratory media and can survive outside the host in inhospitable environments such as fresh water and freezing temperatures. These pathogens are resistant to chemicals such as bile and dyes, which are the basis for isolation on selective media.

Signs and Symptoms The genus *Salmonella* causes a variety of illnesses in the GI tract and beyond. Until fairly recently its most severe manifestation was typhoid fever, which will

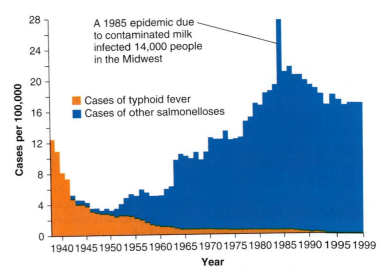

FIGURE 22.10 **Data on the prevalence of typhoid fever and other salmonelloses from 1940 to 1999.**
Nontyphoidal salmonelloses did occur before 1940, but the statistics are not available.
Source: Data from *Morbidity and Mortality Weekly Report,* January 9, 1998, Vol. 46. Centers for Disease Control and Prevention, Atlanta, GA.

be discussed shortly. Since the mid-1900s, a milder disease usually called *salmonellosis* has been much more common **(figure 22.10).** Sometimes the condition is also called enteric fever or gastroenteritis. Whereas typhoid fever is caused by the *typhi* serovar, gastroenteritises are generally caused by the serovars known as *paratyphi, hirschfeldii,* and *typhimurium.* Another serovar, which is sometimes called *Arizona hinshawii* (even though it is still a *Salmonella*) is a pathogen found in the intestines of reptiles. Most of these strains come from animals, unlike the typhi strain, which infects humans exclusively. *Salmonella* bacteria are normal intestinal flora in cattle, poultry, rodents, and reptiles.

Salmonellosis can be relatively severe, with an elevated body temperature and septicemia as more prominent features than GI tract disturbance. But it can also be fairly mild, with gastroenteritis—vomiting, diarrhea, and mucosal irritation—as its major feature. Blood can appear in the stool. In otherwise healthy adults, symptoms spontaneously subside after 2 to 5 days; death is infrequent except in debilitated persons.

Typhoid fever is so named because it bears a superficial resemblance to typhus, a rickettsial disease, even though the two diseases are otherwise very different. In the United States, the incidence of typhoid fever has remained at a steady rate for the last 30 years, appearing sporadically (figure 22.10). Of the 50 to 100 cases reported annually, roughly half are imported from endemic regions. In other parts of the world, typhoid fever is still a serious health problem, responsible for 25,000 deaths each year and probably millions of cases.

Typhoid fever, caused by the typhi serovar of *S. enterica*, is characterized by a progressive, invasive infection that leads eventually to septicemia. Symptoms are fever, diarrhea, and abdominal pain. The bacterium infiltrates the mesenteric lymph nodes and the phagocytes of the liver and spleen. In some people, the small intestine develops areas of ulceration that are vulnerable to hemorrhage, perforation, and peritonitis. Its presence in the circulatory system may lead to nodules or abscesses in the liver or urinary tract.

Because it is so rare compared with the less severe salmonellosis, the rest of this section refers mainly to salmonellosis and not to typhoid fever.

Pathogenesis and Virulence Factors   The ability of *Salmonella* to cause disease seems to be highly dependent on its ability to adhere effectively to the gut mucosa. Recent research has uncovered an "island" of genes in *Salmonella* that seems to confer virulence on the bacterium. This island was discovered when those genes were inactivated, and the bacterium was no longer capable of causing disease in an experimental model. Researchers weren't sure what the functions of those genes were, but when they *injected* the inactivated strain or the wild-type strain into experimental animals rather than transmitted the bacteria orally, both were still capable of causing disease. This result indicated to the researchers that the "virulence genes" were most important for entry, adhesion, or invasion into the host. It is also believed that endotoxin is an important virulence factor for *Salmonella*.

Transmission and Epidemiology   Animal products such as meat and milk can be readily contaminated with *Salmonella* during slaughter, collection, and processing. Inherent risks are involved in eating poorly cooked beef or unpasteurized fresh or dried milk, ice cream, and cheese. A 2001 U.S. outbreak was traced to green grapes. A particular concern is the contamination of foods by rodent feces. Several outbreaks of infection have been traced to unclean food storage or to food-processing plants infested with rats and mice.

Most cases are traceable to a common food source such as milk or eggs. Some cases may be due to poor sanitation. In one outbreak, about 60 people became infected after visiting the Komodo dragon exhibit at the Denver zoo. They picked up the infection by handling the rails and fence of the dragon's cage. In 2002 two people apparently acquired salmonellosis from a blood transfusion, and one of them died. The blood donor, who had an asymptomatic infection with *Salmonella*, had contracted the infection from his pet snake.

In recent years many cancer patients and HIV-positive people have become deathly ill from ingesting a folk remedy called "rattlesnake pill," sometimes known as Pulvo de Vibora. It is particularly common in California, the Southwest, and in Mexico. The CDC has investigated these pills and found that they can contain the *Arizona hinshawii* serovar of *Salmonella* (found in the intestines of reptiles) as well as many other pathogens. Extreme care should be taken by immunocompromised persons when considering such unlicensed alternative therapies, and health care professionals should be alert to the possibility of such exposures when they see unusual infections in patients.

Prevention and Treatment   The only prevention for salmonellosis is avoiding contact with the bacterium. In 1998 a vaccine was approved for use in poultry, making it the first "food safety" vaccine. A vaccine for humans is undergoing testing, as well.

Uncomplicated cases of salmonellosis are treated with fluid and electrolyte replacement; if the patient has underlying immunocompromise or if the disease is severe, trimethoprim-sulfamethoxazole is recommended.

Typhoid fever, by contrast, is always treated with antibiotics, in part to clear the patient of the typhi strain, which has a tendency to be shed for weeks after recovery. A small number of people chronically carry the bacterium for longer periods in the gallbladder; from this site, the bacteria are constantly released into the intestine and feces. In some people gallbladder removal is necessary to stop the shedding. Two vaccines are available for the typhi strain, and are recommended for people traveling to endemic areas.

Shigella

The *Shigella* are gram-negative straight rods, nonmotile and non-spore-forming. They do not produce urease or hydrogen sulfide, traits that help in its identification. They are primarily human parasites, though they can infect apes. All produce a similar disease that can vary in intensity. These bacteria resemble some types of pathogenic *E. coli* very closely. Diagnosis is complicated by the fact that several alternative candidates can cause bloody diarrhea, such as *E. coli* and others. Isolation and identification follow the usual protocols for enterics. Stool culture is still the gold standard for identification in the case of *Shigella* infections **(Insight 22.1)**.

Although *Shigella dysenteriae* causes the most severe form of dysentery, it is uncommon in the United States and occurs primarily in the Eastern Hemisphere. In the past decade, the prevalent agents in the United States have been *Shigella sonnei* and *Shigella flexneri*, which cause approximately 20,000 to 25,000 cases each year, half of them in children.

Signs and Symptoms   The symptoms of shigellosis include frequent, watery stools, as well as fever, and often intense abdominal pain. Nausea and vomiting are common. Stools often contain obvious blood, and even more often are found to have occult (not visible to the naked eye) blood. Diarrhea containing blood is also called **dysentery**. Mucus from the GI tract will also be present in the stools.

Pathogenesis and Virulence Factors   Shigellosis is different from many GI tract infections in that *Shigella* invades the villus cells of the large intestine, rather than the small intestine. In addition, it is not as invasive as *Salmonella* and does not perforate the intestine or invade the blood. It enters the intestinal mucosa by means of lymphoid cells in Peyer's patches. Once in the mucosa, *Shigella* instigates an inflammatory response that causes extensive tissue destruction. The release of endotoxin causes fever. **Enterotoxin,** an exotoxin that affects the enteric (or GI) tract, damages the mucosa and

INSIGHT 22.1   *Medical*

Stools: To Culture or Not to Culture?

The practice of diagnosing GI tract infections is really at a cross-roads in the early twenty-first century. For decades, clinical microbiologists have relied on stool cultures complemented with a battery of biochemical tests to try to tease out the single pathogenic bacterium among the multitude of normal strains that reside in the intestinal tract. Now many physicians feel that stool cultures are not necessary except in certain circumstances. Indeed, some studies show that as few as 2% of routinely ordered stool cultures come back positive for anything. When we consider that some of these cultures can cost as much as $1,000, it is easy to see their point.

It seems that the best guideline to use is this: *Will the results of the culture change the therapy?* Physicians generally agree that when fever is present, when there is blood in the stools or pain suggesting appendicitis, or if the patient gives a history that suggests possible exposure to *E. coli* O157:H7, stool cultures should be ordered. In other cases physicians should use a variety of other indicators, both clinical and epidemiological, to diagnose diarrhea and other gastrointestinal disorders. Newer technologies that can test for specific pathogens without culturing, such as ELISA and PCR tests, may eventually make costly and slow culture techniques obsolete.

villi. Local areas of erosion give rise to bleeding and heavy secretion of mucus **(figure 22.11).** *Shigella dysenteriae* (and perhaps some of the other species) produces a heat-labile exotoxin called **shiga toxin,** which seems to be responsible for the more serious damage to the intestine as well as any systemic effects, including injury to nerve cells. It is an A-B toxin (see figure 21.10). To review, the B portion of the toxin attaches to host cells, and the whole toxin is internalized. Once inside, the A portion of the toxin exerts its effect. In the case of the shiga toxin, the A portion of the toxin binds to ribosomes, interrupting protein synthesis and leading to the damage just described. You'll encounter shiga toxin again when we discuss *E. coli* O157:H7.

Transmission and Epidemiology   In addition to the usual oral route, shigellosis is also acquired through direct person-to-person contact, largely because of the small infectious dose required (from 10 to 200 bacteria). The disease is mostly associated with lax sanitation, malnutrition, and crowding, and it is spread epidemically in day care centers, prisons, mental institutions, nursing homes, and military camps. *Shigella* was responsible for some cruise ship outbreaks in the mid-1990s (later cruise ship outbreaks were caused by viruses). As in

other enteric infections, *Shigella* can establish a chronic carrier condition in some people that lasts several months.

Prevention and Treatment   The only prevention of this and most other diarrheal diseases is good hygiene and avoiding contact with infected persons. Although some experts say that bloody diarrhea in this country should not be treated with antibiotics (which is generally accepted for *E. coli* O157:H7 infections), most physicians recommend prompt treatment of shigellosis with trimethoprim-sulfamethoxazole (TMP-SMZ).

E. coli O157:H7 (EHEC)

In January of 1993, this awkwardly named bacterium burst into the public's consciousness when three children died after eating undercooked hamburgers at a fast-food restaurant in Washington State. The cause of their illness was determined to be this particular strain of *E. coli*, which had actually been recognized since the 1980s. Since then, it has led to approximately 73,000 illnesses and about 50 deaths each year in the United States. It is considered an emerging pathogen.

Dozens of different strains of *E. coli* exist, many of which cause no disease at all. A handful of them cause various

FIGURE 22.11   **The appearance of the large intestinal mucosa in** *Shigella* **dysentery.**
Note the patches of blood and mucus, the erosion of the lining, and the absence of perforation.

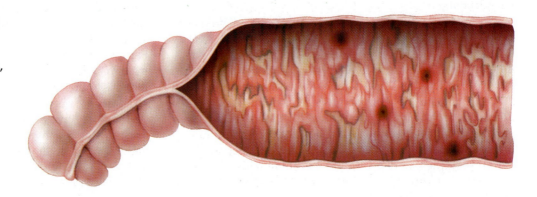

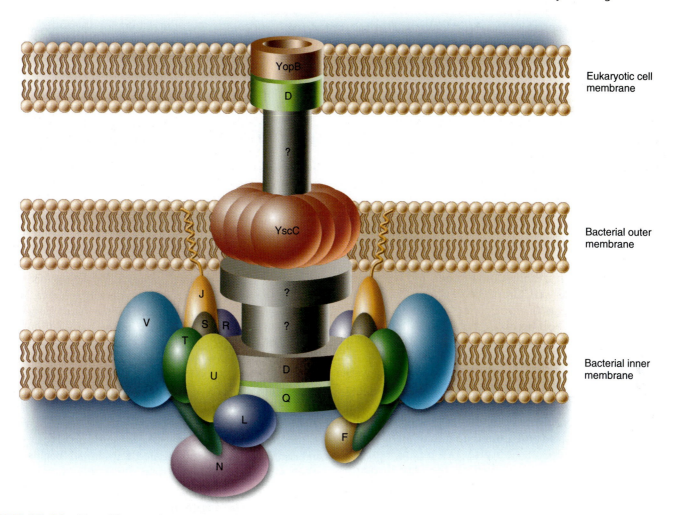

FIGURE 22.12   **Type III secretion system.**
This multiprotein "pipeline" is situated in the cytoplasmic and outer membranes of *E. coli*. Each of the differently shaped objects is a different protein.

degrees of intestinal symptoms as described in this and the following section. Some of them cause urinary tract infections (see chapter 23). *E. coli* O157:H7 and its close relatives are the most virulent of them all. The group of *E. coli* of which this strain is the most famous representative is generally referred to as **enterohemorrhagic *E. coli*, or EHEC.**

Signs and Symptoms   *E. coli* O157:H7 is the agent of a spectrum of conditions, ranging from mild gastroenteritis with fever to bloody diarrhea. About 10% of patients develop **hemolytic uremic syndrome (HUS),** a severe hemolytic anemia that can cause kidney damage and failure. Neurologic symptoms such as blindness, seizure, and stroke (and long-term debilitation) are also possible. These serious manifestations are most likely to occur in children younger than 5 and in the elderly.

Pathogenesis and Virulence Factors   This bacterium owes much of its virulence to shiga toxins (so named because they

are identical to the shiga exotoxin secreted by virulent *Shigella* species). Sometimes this *E. coli* is referred to as STEC (shiga toxin–producing *E. coli*). For simplicity, EHEC is used here. The shiga toxin genes are present on bacteriophage in *E. coli*, but are on the chromosome of *Shigella dysenteriae*, suggesting that the *E. coli* acquired the virulence factor through phage-mediated transfer. As described earlier for *Shigella*, the shiga toxin interrupts protein synthesis in its target cells. It seems to be responsible especially for the systemic effects of this infection.

Another important virulence determinant for EHEC is the ability to efface (rub out or destroy) enterocytes, which are gut epithelial cells. This is accomplished with a set of bacterial proteins, one of which is called *intimin*—used for "intimate" attachment to host cells. Another set of proteins enables the bacterium to construct a complex bridging system between *E. coli* and host cell membranes, which allows *E. coli* to insert its products into the host cell. This system is called the *Type III secretion system* (**figure 22.12**).

The bacterium also produces a set of proteins that are actually passed through the apparatus in figure 22.12, including the protein that does the damage to host cells. The most startling discovery, however, has been that one of the products sent through the Type III "pipeline" was a protein that the bacterium inserts into the host cell membrane, so that it will become a receptor for the bacterial intimin protein. Essentially, the bacterium is sending over the lock into which it can insert its key—ensuring a very tight bond indeed.

The net effect of the action of these products is a lesion in the gut (effacement), usually in the large intestine. The microvilli are lost from the gut epithelium, and the lesions produce bloody diarrhea.

Transmission and Epidemiology   The most common mode of transmission for EHEC is the ingestion of contaminated and undercooked beef, although other foods and beverages can be contaminated as well. Ground beef is more dangerous than steaks or other cuts of meat, for several reasons. Consider the way that the beef becomes contaminated in the first place. The bacterium is a natural inhabitant of the GI tracts of cattle. Contamination occurs when intestinal contents contact the animal carcass, so bacteria are confined to the surface of meats. Because high heat destroys this bacterium, even a brief trip under the broiler is usually sufficient to kill *E. coli* on the surface of steaks or roasts. But in ground beef, the "surface" of meat is mixed and ground up throughout a batch, meaning any bacteria are mixed in also. This mixing explains why hamburgers should be cooked all the way through. Hamburger is also a common vehicle because meat processing plants tend to grind meats from several cattle sources together, thereby contaminating large amounts of hamburger with meat from one animal carrier.

Other farm products may also become contaminated by cattle feces. Products that are eaten raw, such as lettuce, vegetables, and apples used in unpasteurized cider are particularly problematic. The disease can also be spread via the fecal-oral route of transmission, especially among young children in group situations. Even touching surfaces contaminated with cattle feces can cause disease, since ingesting as few as 10 organisms has been found to be sufficient to initiate this disease.

Culture and Diagnosis   Infection with this type of *E. coli* should be confirmed with stool culture, or with newer techniques such as ELISA or PCR.

Prevention and Treatment   The best prevention for this disease is never to eat raw or even rare hamburger. The shiga toxin is heat-labile and the *E. coli* is killed by heat as well. If you are thinking "I used to be able to eat rare hamburgers," you are correct, but things have changed. The emergence of this pathogen in the early 1980s, probably resulting from a regular *E. coli* picking up the shiga toxin from *Shigella*, has changed the rules.

No vaccine exists for *E. coli* O157:H7. A great deal of research is directed at vaccinating livestock to break the chain of transmission to humans.

Antibiotics are contraindicated for this infection. Even with severe disease manifestations, antibiotics have been found to be of no help, and they may increase the pathology. It is also recommended that antimotility drugs (to limit the diarrhea) not be used. Supportive therapy is the only option.

IN THE NEWS   *(Continued from page 687)*

In the case of the dairy-farm-associated illness, investigators found that the patients had had close contact with calves at the farm, and most did not wash their hands immediately afterward. These two actions put them at high risk to contract *E. coli* O157:H7 illness. Most of the victims were young children (median age: 4 years old). Investigators recovered organisms from rectal swabs of a high percentage of the young cattle, as well as from surfaces such as fence railings. The dairy farm allowed patrons to purchase food and beverages and to consume them in the petting zoo area, which no doubt increased the chances that contaminated fingers would transfer microbes to the children's mouths.

Sixteen of the 51 patients were hospitalized, all of them children, and one of them developed end-stage renal failure (a consequence of hemolytic uremic syndrome) and required a kidney transplant. All patients eventually recovered, but this incident and others like it have led health officials to warn that all cattle should be handled as though they are colonized with this dangerous bacterium. Handwashing stations have started appearing at state fairs and petting zoos, and food-related activities are beginning to be clearly separated from animal areas.

See: Crump, J. A. et al. 2002. An outbreak of Escherichia coli *O157:H7 infections among visitors to a dairy farm. N. Eng. J. Med. 347:555*

Other *E. coli*

At least four other categories of *E. coli* can cause diarrheal diseases. Scientists call these **enterotoxigenic** *E. coli*, **enteroinvasive** *E. coli*, **enteropathogenic** *E. coli*, and **enteroaggregative** *E. coli*. In clinical practice, most physicians are interested in differentiating shiga toxin–producing *E. coli* (EHEC) from all the others. Each of these will be considered separately and briefly here; in Checkpoint Table 22.5, the non-shiga toxin–producing *E. coli* will be grouped together in one column.

Enterotoxigenic *E. coli* (ETEC)   The presentation varies depending on which type of *E. coli* is causing the disease. **Traveler's diarrhea,** characterized by watery diarrhea, low-grade fever, nausea, and vomiting, is usually caused by enterotoxigenic *E. coli* (ETEC). These strains also cause a great deal of illness in infants in developing countries.

The bacterium is transmitted through the fecal-oral route or via contaminated vehicles or even fomites (such as a dirty glass). Travelers are susceptible to these strains because they are likely to be new to their immune systems. People living in endemic areas probably encounter the bacteria as infants. As the name suggests, the virulence of the bacterium derives from its ability to secrete two types of exotoxins that act on the enteric tract (enterotoxin). One toxin is a heat-labile A-B toxin, and it acts like the cholera toxin, described later. Another toxin, actually a group of toxins, is heat-stable. These toxins are very small proteins that alter host cell function in order to cause large amounts of fluid secretion into the intestinal tract. The bacterium mainly affects the small intestine.

Most infections with ETEC are self-limiting, however miserable they make you feel. They are treated only with fluid replacement. In infants, ETEC can be life-threatening, and fluid replacement is vital to survival.

Enteroinvasive *E. coli* (EIEC) These strains cause a disease that is very similar to *Shigella* dysentery. The bacteria invade gut mucosa and cause widespread destruction. Blood and pus will be found in the stool. Significant fever is often present. EIEC does not produce the heat-labile or heat-stable exotoxins just described and does not have a shiga toxin, despite the clinical similarity to *Shigella* disease. EIEC does seem to have a protein that is expressed inside host cells, which leads to its destruction.

Disease caused by this bacterium is more common in developing countries. It is transmitted primarily through contaminated food and water. Treatment is supportive (including rehydration).

Enteropathogenic *E. coli* (EPEC) These strains result in a profuse, watery diarrhea. Fever and vomiting are also common. The EPEC bacteria are very similar to the EHEC *E. coli* described earlier—they produce effacement of gut surfaces. The important difference between EPEC and EHEC is that EPEC does not produce a shiga toxin, and therefore does not produce the systemic symptoms characteristic of those bacteria.

EPEC has been known to cause outbreaks in hospital nurseries in this country but is more notorious for causing diarrhea in infants in developing countries.

Most disease is self-limiting. As with any other diarrhea, however, it can be life-threatening in young babies. Rehydration is the main treatment.

Enteroaggregative *E. coli* (EAEC) These bacteria are most notable for their ability to cause chronic diarrhea, most notably in young children and in AIDS patients. EAEC will be considered in the section on chronic diarrhea.

Campylobacter

Although you may never have heard of *Campylobacter*, it is considered to be the most common bacterial cause of diarrhea in the United States. It probably causes more diarrhea than *Salmonella* and *Shigella* combined, with 2 million cases of diarrhea credited to it per year.

The symptoms of campylobacteriosis are frequent watery stools, fever, vomiting, headaches, and severe abdominal pain. The symptoms may last longer than most acute diarrheal episodes, sometimes extending beyond 2 weeks. They may subside, and then recur over a period of weeks.

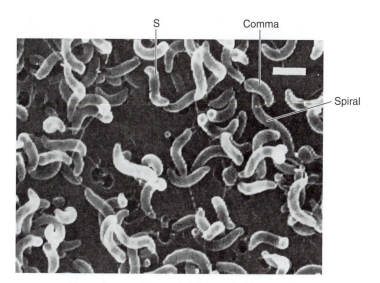

FIGURE 22.13 Scanning micrograph of *Campylobacter jejuni*, showing comma, S, and spiral forms.

Campylobacter jejuni is the most common cause, although there are other *Campylobacter* species. Campylobacters are slender, curved or spiral gram-negative bacteria propelled by polar flagella at one or both poles, often appearing in S-shaped or gull-winged pairs **(figure 22.13)**. These bacteria tend to be microaerophilic inhabitants of the intestinal tract, genitourinary tract, and oral cavity of humans and animals. A close relative, *Helicobacter pylori*, is the causative agent of most stomach ulcers (described earlier). Transmission of this pathogen takes place via the ingestion of contaminated beverages and food, especially water, milk, meat, and chicken.

Once ingested, *C. jejuni* cells reach the mucosa at the last segment of the small intestine (ileum) near its junction with the colon; they adhere, burrow through the mucus, and multiply. Symptoms commence after an incubation period of 1 to 7 days. The mechanisms of pathology appear to involve a heat-labile enterotoxin that stimulates a secretory diarrhea like that of cholera. In a small number of cases, infection with this bacterium can lead to a serious neuromuscular paralysis called Guillain-Barré syndrome.

Guillain-Barré syndrome (GBS) is the leading cause of acute paralysis in the United States since the eradication of polio here. The good news is that many patients recover completely from this paralysis. The condition is still mysterious in many ways, but it seems to be an autoimmune reaction that can be brought on by infection with viruses and bacteria, by vaccination in rare cases, and even by surgery. The single most common precipitating event for the onset of GBS is *Campylobacter* infection. Twenty to forty percent of GBS cases are preceded by infection with *Campylobacter*. The reasons for this are not clear. (Note that even though 20% to 40% of GBS cases are preceded by *Campylobacter* infection, only about 1 in 1,000 cases of *Campylobacter* infection result in GBS.)

Diagnosis of *C. jejuni* enteritis requires isolation of the bacterium from stool samples and occasionally from blood samples. More rapid presumptive diagnosis can be obtained from direct examination of feces with a dark-field microscope, which accentuates the characteristic curved rods and darting motility. This procedure is difficult to perform and not often used except in specialized labs. Resolution of infection occurs in most instances with simple, nonspecific rehydration and electrolyte balance therapy. In more severely affected patients, it may be necessary to administer erythromycin. Antibiotic resistance is growing in these bacteria. Because vaccines are yet to be developed, prevention depends on rigid sanitary control of water and milk supplies and care in food preparation.

Yersinia Species

Yersinia is a genus of gram-negative bacteria that includes the infamous plague bacterium, *Yersina pestis* (discussed in chapter 20). There are two species that cause GI tract disease: *Y. enterocolitica* and *Y. pseudotuberculosis*. The infections are most notable for the high degree of abdominal pain they cause. This symptom is accompanied by fever. Often the symptoms are mistaken for appendicitis.

The disease is uncommon in the United States, but outbreaks do occasionally occur. Food and beverages can become contaminated with these bacteria, which inhabit the intestines of farm animals, pets, and wild animals. Transmission also occurs when people handle raw food and then touch fomites such as toys or baby bottles without washing their hands.

The bacteria invade the small intestinal mucosa, and some enter the lymphatics and are harbored intracellularly in phagocytes. Inflammation of the ileum and mesenteric lymph nodes gives rise to severe abdominal pain. The infection occasionally spreads to the bloodstream, but systemic effects are rare. Two to three percent of patients experience joint pain a month following the diarrhea episode. This symptom resolves spontaneously within a few months. Infections with *Y. pseudotuberculosis* tend to be milder than those with *Y. enterocolitica*, and center on lymph node inflammation rather than mucosal involvement.

Simple rules of food hygiene are usually sufficient to prevent the spread of this infection. Antibiotics are not usually prescribed for this disease, unless bacteremia is documented. In that case, doxycycline or TMP-SMZ is used.

Clostridium difficile

Clostridium difficile is a gram-positive endospore-forming rod found as normal flora in the intestine. It was once considered relatively harmless, but now is known to cause a condition called pseudomembranous colitis. It is also sometimes called antibiotic-associated colitis. In most cases, this infection is precipitated by therapy with broad-spectrum antibiotics such as ampicillin, clindamycin, or cephalosporins. It is a major cause of diarrhea in hospitals. Although *C. difficile* is relatively noninvasive, it is able to superinfect the large intestine when drugs have disrupted the normal flora. It produces two enterotoxins, toxins A and B, that cause areas of necrosis in the wall of the intestine. The predominant symptom is diar-

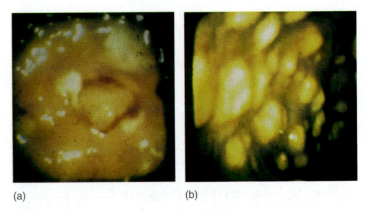

(a)   (b)

FIGURE 22.14 **Antibiotic-associated colitis.**
(a) A mild form with diffuse, inflammatory patches. **(b)** Heavy yellow plaques, or pseudomembranes, typical of more severe cases. Photographs were made by a sigmoidoscope, an instrument capable of photographing the interior of the colon.

rhea commencing late in therapy or even after therapy has stopped. More severe cases exhibit abdominal cramps, fever, and leukocytosis. The colon is inflamed and gradually sloughs off loose, membrane-like patches called pseudomembranes consisting of fibrin and cells **(figure 22.14)**. If the condition is not stopped, perforation of the cecum and death can result.

Mild, uncomplicated cases respond to withdrawal of antibiotics and replacement therapy for lost fluids and electrolytes. More severe infections are treated with oral vancomycin or metronidazole for several weeks until the intestinal flora returns to normal. Because infected persons often shed large numbers of spores in their stools, increased precautions are necessary to prevent spread of the agent to other patients who may be on antimicrobial therapy. Some new techniques on the horizon are vaccination with *C. difficile* toxoid and restoration of normal flora by ingestion of a mixed culture of lactobacilli and yeasts.

Vibrio cholerae

Cholera has been a devastating disease for centuries. It is not an exaggeration to say that the disease has shaped a good deal of human history in Asia and Latin America, where it has been endemic. These days we have come to expect outbreaks of cholera to occur after natural disasters, war, or large refugee movements, especially in underdeveloped parts of the world.

Vibrios are comma-shaped rods with a single polar flagellum. They belong to the family Vibrionaceae. A freshly isolated specimen of *Vibrio cholerae* reveals quick, darting cells that slightly resemble a cooked hot dog or a comma **(figure 22.15)**. *Vibrio* shares many cultural and physiological characteristics with members of the Enterobacteriaceae, a closely related family. Vibrios are fermentative and grow on ordinary or selective media containing bile at 37°C. They possess unique O and H antigens and membrane receptor antigens that provide some basis for classifying members of the family. There are two major biotypes, called classic and *El Tor*.

FIGURE 22.15 *Vibrio cholerae.*
Note the characteristic curved shape and single polar flagellum.

Signs and Symptoms   After an incubation period of a few hours to a few days, symptoms begin abruptly with vomiting, followed by copious watery feces called secretory diarrhea. The intestinal contents are lost very quickly, leaving only secreted fluids. This voided fluid contains flecks of mucus, hence the description "rice-water stool." Fluid losses of nearly 1 liter per hour have been reported in severe cases, and an untreated patient can lose up to 50% of body weight during the course of this disease. The diarrhea causes loss of blood volume, acidosis from bicarbonate loss, and potassium depletion that manifest in muscle cramps, severe thirst, flaccid skin, sunken eyes, and in young children, coma and convulsions. Secondary circulatory consequences can include hypotension, tachycardia, cyanosis, and collapse from shock within 18 to 24 hours. If cholera is left untreated, death can occur in less than 48 hours, and the mortality rate approaches 55%.

Pathogenesis and Virulence Factors   After being ingested with food or water, *V. cholerae* encounters the potentially destructive acidity of the stomach. This hostile environment influences the size of the infectious dose (10^8 cells), although certain types of food shelter the pathogen more readily than others. At the junction of the duodenum and jejunum, the vibrios penetrate the mucous barrier using their flagella, adhere to the microvilli of the epithelial cells, and multiply there. The bacteria never enter the host cells or invade the mucosa. The virulence of *V. cholerae* is due entirely to an enterotoxin called cholera toxin (CT), which disrupts the normal physiology of intestinal cells. It is a typical A-B type toxin as previously described for *Shigella.* When this toxin binds to specific intestinal receptors, a secondary signaling system is activated. Under the influence of this system, the cells shed large amounts of electrolytes into the intestine, an event accompanied by profuse water loss. Most cases of cholera are mild or self-limited, but in children and weakened individuals, the disease can be deadly.

Transmission and Epidemiology   Although the human intestinal tract was once thought to be the primary reservoir, it is now known that the parasite is free-living in certain endemic regions. The pattern of cholera transmission, and the onset of epidemics, are greatly influenced by the season of the year and the climate. Cold, acidic, dry environments inhibit the migration and survival of *Vibrio,* whereas warm, monsoon, alkaline, and saline conditions favor them. The bacteria survive in water sources for long periods of time. The disease has persisted in a pandemic pattern since 1961, when the *El Tor* biotype began to prevail worldwide. This strain survives longer in the environment, infects a higher number of people, and is more likely to be chronically carried than any other strain. Recent outbreaks in several parts of the world have been traced to giant cargo ships that pick up ballast water in one port and empty it in another elsewhere in the world. Cholera ranks among the top seven causes of morbidity and mortality, affecting several million people in endemic regions of Asia and Africa.

In nonendemic areas such as the United States, the microbe is spread by water and food contaminated by asymptomatic carriers, but it is relatively uncommon. Sporadic outbreaks occur along the Gulf of Mexico, and *V. cholerae* is sometimes isolated from shellfish in that region.

Culture and Diagnosis   During epidemics of this disease, clinical evidence is usually sufficient to diagnose cholera. But confirmation of the disease is often required for epidemiological studies and detection of sporadic cases. *V. cholerae* can be readily isolated and identified in the laboratory from stool samples. Direct dark-field microscopic observation reveals characteristic curved cells with brisk, darting motility as confirmatory evidence. Immobilization or fluorescent staining of feces with group-specific antisera is supportive as well. Difficult or elusive cases can be traced by detecting a rising antitoxin titer in the serum.

Prevention and Treatment   Effective prevention is contingent upon proper sewage treatment and water purification. Detecting and treating carriers with mild or asymptomatic cholera is a serious goal, but it is difficult to accomplish because of inadequate medical provisions in those countries where cholera is endemic. Vaccines are available for travelers and people living in endemic regions. One contains killed *V. cholerae* but protects for only 6 months or less. An oral vaccine containing live, attenuated bacteria was developed to be a more effective alternative, but evidence suggests it also confers only short-term immunity.

The key to cholera therapy is prompt replacement of water and electrolytes, since their loss accounts for the severe morbidity and mortality. This therapy can be accomplished by various rehydration techniques that replace the lost fluid and electrolytes. One of these, oral rehydration therapy (ORT), is described in **Insight 22.2.**

INSIGHT 22.2   *Discovery*

A Little Water, Some Sugar and Salt, Save Millions of Lives

In 1970 a clinical trial was conducted on a very low-tech solution to the devastating problem of death from diarrhea, especially among children in the developing world. Until that time, the treatment, if a child could get it, was rehydration through an IV drip. This treatment usually required traveling to the nearest clinic, often miles or days away. Most children received no treatment at all, and 3 million of them died every year. Then scientists tested a simple sugar-salt solution that patients could drink. They tested it first in India, where cholera was rampant, and found that mortality rates were greatly decreased. After more testing in Bangladesh, Turkey, the Philippines, and the United States, oral-rehydration therapy (ORT) became the treatment of choice for diarrhea from all causes. The WHO and UNICEF began providing packages of the sugar and salt mixture, and instructions for mixing it with boiled water, to dozens of countries. They also oversaw training of individuals who could in turn teach townspeople and villagers about ORT.

Volunteers in front of an Oral Rehydration Clinic in the Philippines. ORT clinics are commonplace in developing countries.

The relatively simple solution, developed by the WHO, consists of a mixture of the electrolytes sodium chloride, sodium bicarbonate, potassium chloride, and glucose or sucrose dissolved in water. When administered early in amounts ranging from 100 to 400 milliliters per hour, the solution can restore patients in 4 hours, often bringing them literally back from the brink of death. Infants and small children who once would have died now survive so often that the mortality rate for treated cases of cholera is near zero. This therapy has several advantages, especially for countries with few resources. It does not require medical

facilities, high-technology equipment, or complex medication protocols. It also eliminates the need for clean needles, which is a pressing issue in many parts of the world.

In 1978 the British Medical journal *The Lancet* called ORT "potentially the most important medical advance this century." With estimates of at least a million lives saved every year since its introduction, this statement seems to have been proven correct.

Cases in which the patient is unconscious or has complications from severe dehydration require intravenous replenishment as well. Oral antibiotics such as tetracycline and drugs such as trimethoprim-sulfamethoxazole can terminate the diarrhea in 48 hours. They also diminish the period of vibrio excretion.

Cryptosporidium

Cryptosporidium is an intestinal protozoan of the apicomplexan type (see chapter 5) that infects a variety of mammals, birds, and reptiles. For many years, cryptosporidiosis was considered an intestinal ailment exclusive to calves, pigs, chickens, and other poultry, but it is clearly a zoonosis as well. The organism's life cycle includes a hardy intestinal oocyst as well as a tissue phase. Humans accidentally ingest the oocysts with water or food that has been contaminated by feces from infected animals. The oocyst "excysts" once it reaches the intestines, and releases sporozoites that attach to the epithelium of the small intestine **(figure 22.16).** The organism penetrates the intestinal

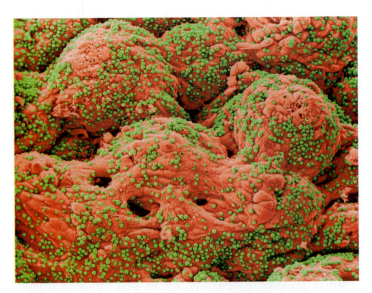

FIGURE 22.16 Scanning electron micrograph of *Cryptosporidium* attached to the intestinal epithelium.

cells and lives intracellularly in them. It undergoes asexual and sexual reproduction in the gut and produces more oocysts, which are excreted from the host and after a short time become infective again. The oocysts are highly infectious and extremely resistant to treatment with chlorine and other disinfectants.

The prominent symptoms mimic other types of gastroenteritis, with headache, sweating, vomiting, severe abdominal cramps, and diarrhea. AIDS patients may experience chronic persistent cryptosporidial diarrhea that can be used as a criterion to help diagnose AIDS. The agent can be detected in fecal samples with indirect immunofluorescence and by acid-fast staining of biopsy tissues **(figure 22.17)**. Stool cultures should be performed to rule out other (bacterial) causes of infection.

Cryptosporidiosis has a cosmopolitan distribution. Its highest prevalence is in areas with unreliable water and food sanitation. The carrier state occurs in 3% to 30% of the population in developing countries. The susceptibility of the general public to this pathogen has been amply demonstrated by several large-scale epidemics. In 1993, 370,000 people developed *Cryptosporidium* gastroenteritis from the municipal water supply in Milwaukee, Wisconsin. Other mass outbreaks of this sort have been traced to contamination of the local water reservoir by livestock wastes. Other studies revealed that at least one-third of all fresh surface waters harbor this parasite. Because chlorination is not entirely successful in eradicating the cysts, most treatment plants use filtration to remove them, but even this method can fail.

Treatment is not usually required for otherwise healthy patients. Antidiarrheal agents (antimotility drugs) may be used. Although no curative antimicrobial agent exists for *Cryptosporidium*, physicians will often try paromomycin, an aminoglycoside that can be effective against protozoa.

Rotavirus

Rotavirus is a member of the *Reovirus* group, which consists of an unusual double-stranded RNA genome with both an inner and an outer capsid. Globally, rotavirus is the primary viral cause of morbidity and mortality resulting from diarrhea, accounting for nearly 50% of all cases. It is estimated that there are 1 million cases of rotavirus infection in the United States every year, leading to 70,000 hospitalizations. Peak occurrences of this infection are seasonal; in the U.S. Southwest the peak is often in the late fall and in the Northeast the peak comes in the spring.

Diagnosis of rotavirus infections is usually not performed, as it is treated symptomatically. Nevertheless, studies are often conducted so that public health officials can maintain surveillance of how prevalent the infection is. Stool samples from infected persons contain large amounts of virus, which is readily visible using an electron microscope **(figure 22.18)**. The virus gets its name from its physical appearance, which is said to resemble a "spoked wheel." An ELISA test is also available.

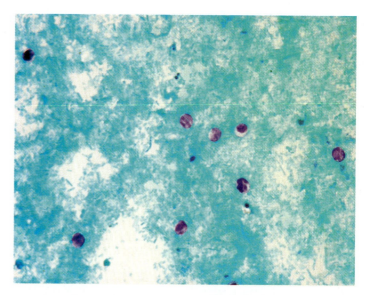

FIGURE 22.17 **Acid-fast stain of *Cryptosporidium*.** Oocysts of *Cryptosporidium* stain bright red or purple.

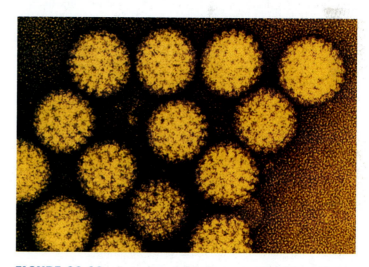

FIGURE 22.18 ***Rotavirus* visible in a sample of feces from a child with gastroenteritis.** Note the unique "spoked-wheel" morphology of the virus.

The virus is transmitted by the fecal-oral route, including through contaminated food, water, and fomites. For this reason, disease is most prevalent in areas of the world with poor sanitation. In the United States, rotavirus infection is relatively common, but its course is generally mild.

The effects of infection vary with the age, nutritional state, general health, and living conditions of the patient. Babies from 6 to 24 months of age lacking maternal antibodies have the greatest risk for fatal disease. These children present symptoms of watery diarrhea, fever, vomiting, dehydration, and shock. The intestinal mucosa can be damaged in a way that chronically compromises nutrition, and long-term or repeated infections can retard growth. Newborns seem to be

✓ CHECKPOINT 22.5 Acute Diarrhea

	Bacterial Causes				
Causative Organism(s)	*Salmonella*	*Shigella*	Shiga-toxin-producing *E. coli* (O157:H7), (EHEC)	Other *E. coli* (non-shiga-toxin-producing)	*Campylobacter*
Most Common Modes of Transmission	Vehicle (food, beverage), fecal-oral	Fecal-oral, direct contact	Vehicle (food, beverage), fecal-oral	Vehicle, fecal-oral	Vehicle (food, water), fecal-oral
Virulence Factors	Adhesins, endotoxin	Endotoxin, enterotoxin, shiga toxins in some strains	Shiga toxins; proteins for attachment, secretion, effacement	Various: proteins for attachment, secretion, effacement; heat-labile and/or heat-stable exotoxins; invasiveness	Adhesins, exotoxin, induction of autoimmunity
Culture/ Diagnosis	Stool culture, not usually necessary	Stool culture; antigen testing for shiga toxin	Stool culture, antigen testing for shiga toxin	Stool culture not usually necessary in absence of blood, fever	Stool culture not usually necessary; dark-field microscopy
Prevention	Food hygiene and personal hygiene	Food hygiene and personal hygiene	Avoid live *E. coli* (cook meat and clean vegetables)	Food and personal hygiene	Food and personal hygiene
Treatment	Rehydration; no antibiotic for uncomplicated disease	TMP-SMZ, rehydration	Antibiotics contraindicated, supportive measures	Rehydration	Rehydration, erythromycin in severe cases (antibiotic resistance rising)
Fever Present	Usually	Often	Often	Sometimes	Usually
Blood in Stool	Sometimes	Often	Usually	Sometimes	No
Distinctive Features	Often associated with chickens, reptiles	Very low ID$_{50}$	Hemolytic uremic syndrome	EIEC, ETEC, EPEC	Guillain-Barré syndrome

protected by maternal antibodies. Adults can also acquire this infection, but it is generally mild and self-limiting.

Children are treated with oral replacement fluid and electrolytes. A vaccine was introduced in 1998 but was withdrawn 9 months later because of a side effect called intussusception, a form of intestinal blockage that seemed to be associated with immunization.

Other Viruses

A bewildering array of viruses can cause gastroenteritis, including adenoviruses, noroviruses (sometimes known as Norwalk viruses), and astroviruses. They are extremely common in the United States and around the world. They are usually "diagnosed" when no other agent (such as those just described) is identified.

Transmission is fecal-oral or via contamination of food and water. Viruses generally cause a profuse, watery diarrhea of 3 to 5 days duration. Vomiting may accompany the disease, especially in the early phases. Mild fever is often seen.

In 2002, a series of gastroenteritis outbreaks occurred on cruise ships, most of which were ascribed to viruses other than rotavirus.

Treatment of these infections always focuses on rehydration (Checkpoint 22.5).

			Nonbacterial Causes		
Yersinia	*Clostridium difficile*	*Vibrio cholerae*	*Cryptosporidium*	Rotavirus	Other viruses
Vehicle (food, water), fecal-oral, indirect contact	Endogenous (normal flora)	Vehicle (water and some foods), fecal-oral	Vehicle (water, food), fecal-oral	Fecal-oral, vehicle, fomite	Fecal-oral, vehicle
Intracellular growth	Enterotoxins A and B	Cholera toxin (CT)	Intracellular growth	–	–
Cold-enrichment stool culture	Stool culture, PCR, ELISA demonstration of toxins in stool	Clinical diagnosis, microscopic techniques, serological detection of antitoxin	Acid-fast staining, ruling out bacteria	Usually not performed	Usually not performed
Food and personal hygiene	–	Water hygiene	Water treatment, proper food handling	Hygiene	Hygiene
None in most cases, doxycycline or TMP-SMZ for bacteremia	Withdrawal of antibiotic, in severe cases metronidazole or vancomycin	Rehydration, in severe cases tetracycline, TMP-SMZ	None, paromomycin used sometimes	Rehydration	Rehydration
Usually	Sometimes	No	Often	Often	Sometimes
Occasionally	Not usually; mucus prominent	No	Not usually	No	No
Severe abdominal pain	Antibiotic-associated diarrhea	Rice-water stools	Resistant to chlorine disinfection	Severe in babies	–

Acute Diarrhea with Vomiting (Food Poisoning)

If a patient presents with severe nausea, frequent vomiting accompanied by diarrhea, and reports that companions with whom he or she shared a recent meal (within the last 1 to 6 hours) are suffering the same fate, food poisoning should be suspected. **Food poisoning** refers to symptoms in the gut that are caused by a preformed toxin of some sort. In many cases the toxin comes from *Staphylococcus aureus*. In others, the source of the toxin is *Bacillus cereus* or *Clostridium perfringens*. The toxin occasionally comes from nonmicrobial

sources such as fish, shellfish, or mushrooms. In any case, if the symptoms are violent and the incubation period is very short, *intoxication* (the effects of a toxin) rather than *infection* should be considered. (See **Insight 22.3** for information about outbreak investigations in general).

Staphylococcus aureus Exotoxin

This illness is associated with eating foods such as custards, sauces, cream pastries, processed meats, chicken salad, or ham that have been contaminated by handling and then left unrefrigerated for a few hours. Because of the high salt tolerance of *S. aureus*, even foods containing salt as a preservative are not exempt. The toxins produced by the multiplying bacteria do not noticeably alter the food's taste or smell. The exotoxin (which is an enterotoxin) is heat-stable; inactivation requires 100°C for at least 30 minutes. Thus, heating the food after toxin production may not prevent disease. The ingested toxin acts upon the gastrointestinal epithelium and stimulates nerves, with acute symptoms of cramping, nausea, vomiting, and diarrhea. Recovery is also rapid, usually within 24 hours. The disease is not transmissible person to person. Often a single source will contaminate several people, leading to a mini-outbreak.

The illness is caused by the toxin and does not require *S. aureus* to be present, or alive, in the contaminated food. If the bacterium is allowed to multiply in the food, it produces its exotoxin. Even if the bacteria are subsequently destroyed by heating, the preformed toxin will act quickly once it is ingested.

As you learned earlier, many diarrheal diseases have symptoms caused by bacterial exotoxins. In most cases, the bacteria take up temporary residence in the gut and then start producing exotoxin, so the incubation period is longer than the 1 to 6 hours seen with *S. aureus* food poisoning. Because this toxin is heat-stable, mishandling of food, such as allowing bacteria to multiply and then heating or reheating, can provide the perfect conditions for food poisoning to occur.

This condition is almost always self-limiting, and antibiotics are definitely not warranted.

Bacillus cereus Exotoxin

Bacillus cereus is a sporulating gram-positive bacterium that is naturally present in soil. As a result, it is a common resident on vegetables and other products in close contact with soil. It produces two exotoxins, one of which causes a diarrheal-type disease, the other of which causes an **emetic** (ee-met'-ik) or vomiting disease. The type of disease that takes place is influenced by the type of food that is contaminated by the bacterium. The emetic form is most frequently linked to fried rice, especially when it has been cooked and kept warm for long periods of time. These conditions are apparently ideal for the expression of the low-molecular-weight, heat-stable exotoxin having an emetic effect. Outbreaks are often associated with Chinese restaurants, although a notable outbreak occurred at two day care centers in 1993.

The diarrheal form of the disease is usually associated with cooked meats or vegetables that are held at a warm temperature for long periods of time. These conditions apparently favor the production of the high-molecular-weight, heat-labile exotoxin. The symptom in these cases is a watery, profuse diarrhea that lasts only for about 24 hours.

Diagnosis of the emetic form of the disease is accomplished by finding the bacterium in the implicated food source. Microscopic examination of stool samples is used to diagnose the diarrheal form of the disease. Of course, in everyday practice, diagnosis as well as treatment is not performed because of the short duration of the disease.

In both cases, the only prevention is the proper handling of food.

Clostridum perfringens Exotoxin

Another sporulating gram-positive bacterium that causes intestinal symptoms is *Clostridium perfringens*. You first read about this bacterium as the causative agent of gas gangrene in chapter 18. Endospores from *C. perfringens* can also contaminate many kinds of foods. Those most frequently implicated in disease are animal flesh (meat, fish) and vegetables such as beans that have not been cooked thoroughly enough to destroy endospores. When these foods are cooled, spores germinate, and the germinated cells multiply, especially if the food is left unrefrigerated. If the food is eaten without adequate reheating, live *C. perfringens* cells enter the small intestine and release exotoxin. The toxin, acting upon epithelial cells, initiates acute abdominal pain, diarrhea, and nausea in 8 to 16 hours. Recovery is rapid, and deaths are extremely rare.

C. perfringens also causes an enterocolitis infection similar to that caused by *C. difficile*. This infectious type of diarrhea is acquired from contaminated food, or it may be transmissible by inanimate objects **(Checkpoint 22.6)**.

Chronic Diarrhea

Chronic diarrhea is defined as lasting longer than 14 days. It can have infectious causes or can reflect noninfectious conditions. Most of us are familiar with diseases that present a constellation of bowel syndromes, such as irritable bowel syndrome, Crohn's disease, and ulcerative colitis, none of which are directly caused by a microorganism as far as we know. They may indeed represent an overreaction to the presence of an infectious agent or another irritant, but the host response seems to be responsible for the pathology. When the presence of an infectious agent is ruled out by a negative stool culture or other tests, these conditions are suspected.

People suffering from AIDS almost universally suffer from chronic diarrhea. Most of the patients who are not taking antiretroviral drugs have diarrhea caused by a variety of opportunistic microorganisms, including *Cryptosporidium*, *Mycobacterium avium*, and so forth. Recently, investigators have found that patients who are aggressively treating their HIV infection with the cocktail of drugs known as HAART (see chapter 20) still suffer from chronic diarrhea at a high rate. The causes for this diarrhea are not completely understood. A patient's HIV status should be considered if he or she presents with chronic diarrhea.

✔ CHECKPOINT 22.6	Acute Diarrhea with Vomiting (Food Poisoning)		
Causative Organism(s)	*Staphylococcus aureus* exotoxin	*Bacillus cereus*	*Clostridium perfringens*
Most Common Modes of Transmission	Vehicle (food)	Vehicle (food)	Vehicle (food)
Virulence Factors	Heat-stable exotoxin	Heat-stable toxin, heat-labile toxin	Heat-labile toxin
Culture/Diagnosis	Usually based on epidemiological evidence	Microscopic analysis of food or stool	Detection of toxin in stool
Prevention	Proper food handling	Proper food handling	Proper food handling
Treatment	None	None	None
Fever Present	Not usually	Not usually	Not usually
Blood in Stool	No	No	No
Distinctive Features	Suspect in foods with high salt or sugar content	Two forms: emetic and diarrheal	Acute abdominal pain

Next we examine a few of the microbes that can be responsible for chronic diarrhea in otherwise healthy people. Keep in mind that practically any disease of the intestinal tract has a sexual mode of transmission in addition to the ones that are commonly stated. For example, any kind of oral-anal sexual contact efficiently transfers pathogens to the "oral" partner. This mode is more commonly seen in cases of chronic illness than it is in patients experiencing acute diarrhea, for obvious reasons.

Enteroaggregative E. coli (EAEC)

In the section on acute diarrhea, you read about the various categories of *E. coli* that can cause disease in the gut. One type, the enteroaggregative *E. coli* (EAEC), is particularly associated with chronic disease, especially in children. This bacterium was first recognized in 1987. It secretes neither the heat-stable nor heat-labile exotoxins previously described for enterotoxigenic *E. coli* (ETEC). It is distinguished by its ability to adhere to human cells in aggregates, rather than as single cells **(figure 22.19)**. Its presence appears to stimulate secretion of large amounts of mucus in the gut, which may be part of its role in causing chronic diarrhea. The bacterium also seems capable of exerting toxic effects on the gut epithelium, although the mechanisms are not well understood.

Transmission of the bacterium is through contaminated food and water. It is difficult to diagnose in a clinical lab because EAEC is not easy to distinguish from other *E. coli*, including normal flora. And the designation EAEC is not actually a serotype, but is functionally defined as an *E. coli* that adheres in an aggregative pattern.

This bacterium seems to be associated with chronic diarrhea in people who are malnourished. It is not exactly clear whether the malnutrition predisposes patients to this infec-

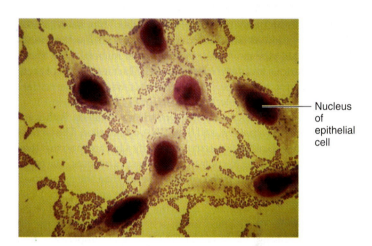

Nucleus of epithelial cell

FIGURE 22.19 Enteroaggregative *E. coli* adhering to epithelial cells.

tion, or whether this infection contributes to malnutrition. Probably both possibilities are operating in patients, who are usually children in developing countries. More recently, the bacterium has been associated with acute diarrhea in industrialized countries, perhaps providing a clue to this question. It may be that in well-nourished hosts the bacterium produces acute, self-limiting disease.

Cyclospora

Cyclospora cayetanensis is an emerging protozoan pathogen. Since the first occurrence in 1979, hundreds of outbreaks have been reported in the United States and Canada. Its mode of transmission is fecal-oral, and most cases have been associated with consumption of fresh produce and water, presumably contaminated with feces. This disease occurs

unfortunately, the amount of chlorine used in municipal water supplies does not destroy the cysts.

Treatment is with quinacrine or metronidazole.

Entamoeba

Amoebas are widely distributed in aqueous habitats and are frequent parasites of animals, but only a small number of them have the necessary virulence to invade tissues and cause serious pathology. One of the most significant pathogenic amoebas is *Entamoeba histolytica* (en"-tah-mee'-bah his"-toh-lit'-ihkuh). The relatively simple life cycle of this parasite alternates between a large trophozoite that is motile by means of pseudopods and a smaller, compact, nonmotile cyst **(figure 22.22).** The trophozoite lacks most of the organelles of other eukaryotes, and it has a large single nucleus that contains a prominent nucleolus called a *karyosome.* Amoebas from fresh specimens are often packed with food vacuoles containing host cells and bacteria. The mature cyst is encased in a thin yet tough wall and contains four nuclei as well as distinctive cigar-shaped bodies called *chromatoidal bodies,* which are actually dense clusters of ribosomes.

Signs and Symptoms As hinted by its species name, tissue damage is one of the formidable characteristics of untreated *E. histolytica* infection. Clinical amoebiasis exists in intestinal and extraintestinal forms. The initial targets of intestinal amoebiasis are the cecum, appendix, colon, and rectum. The amoeba secretes enzymes that dissolve tissues, and it actively penetrates deeper layers of the mucosa, leaving erosive ulcerations **(figure 22.23).** This phase is marked by dysentery (bloody, mucus-filled stools), abdominal pain, fever, diarrhea, and weight loss. The most life-threatening manifestations of intestinal infection are hemorrhage, perforation, appendicitis, and tumorlike growths called amoebomas. Lesions in the mucosa of the colon have a characteristic flask-like shape.

Extraintestinal infection occurs when amoebas invade the viscera of the peritoneal cavity. The most common site of invasion is the liver. Here, abscesses containing necrotic tissue and trophozoites develop and cause amoebic hepatitis. Another rarer complication is pulmonary amoebiasis. Other infrequent targets of infection are the spleen, adrenals, kidney, skin, and brain. Severe forms of the disease result in about a 10% fatality rate.

Pathogenesis and Virulence Factors Amoebiasis begins when viable cysts are swallowed and arrive in the small intestine, where the alkaline pH and digestive juices of this environment stimulate excystment. Each cyst releases four trophozoites, which are swept into the cecum and large intes-

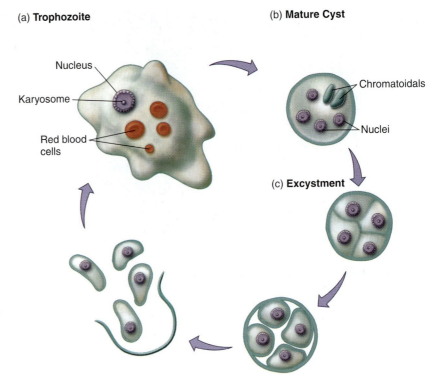

FIGURE 22.22 **Cellular forms of *Entamoeba histolytica.***
(a) A trophozoite containing a single nucleus, a karyosome, and red blood cells. **(b)** A mature cyst with four nuclei and two blocky chromatoidals. **(c)** Stages in excystment. Divisions in the cyst create four separate cells, or metacysts, that differentiate into trophozoites and are released.

tine. There the trophozoites attach by fine pseudopods **(figure 22.24),** multiply, actively move about, and feed. In about 90% of patients, infection is asymptomatic or very mild, and the trophozoites do not invade beyond the most superficial layer. The severity of the infection can vary with the strain of the parasite, inoculum size, diet, and host resistance.

The secretion of lytic enzymes by the amoeba seems to induce apoptosis of host cells. This means that the host is contributing to the process by destroying its own tissues on cue from the protozoan. The invasiveness of the amoeba is also a clear contributor to its pathogenicity.

Transmission and Epidemiology of Amoebiasis *Entamoeba* is harbored by chronic carriers whose intestines favor the encystment stage of the life cycle. Cyst formation cannot occur in active dysentery because the feces are so rapidly flushed from the body; but after recuperation, cysts are continuously shed in feces.

Humans are the primary hosts of *E. histolytica.* Infection is usually acquired by ingesting food or drink contaminated with cysts released by an asymptomatic carrier. The amoeba is thought to be carried in the intestines of one-tenth of the world's population, and it kills up to 100,000 people a year. Its geographic distribution is partly due to local sewage disposal and fertilization practices. Occurrence is highest in tropical

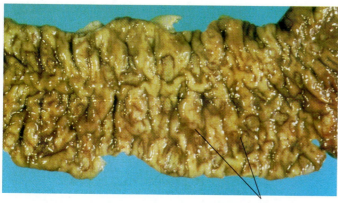

FIGURE 22.23 **Intestinal amoebiasis and dysentery of the cecum.**
Red patches are sites of amoebic damage to the intestinal mucosa.

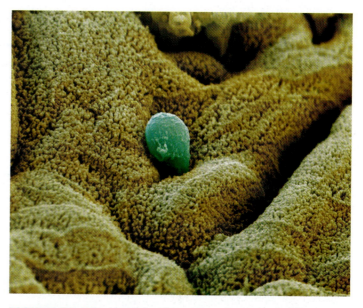

FIGURE 22.24 **Trophozoite of *Entamoeba histolytica*.**
Note the fringe of very fine pseudopods it uses to invade and feed on tissue.

regions (Africa, Asia, and Latin America), where night soil (human excrement) or untreated sewage is used to fertilize crops, and sanitation of water and food can be substandard. Although the prevalence of the disease is lower in the United States, as many as 10 million people could harbor the agent.

Epidemics of amoebiasis are infrequent but have been documented in prisons, hospitals, juvenile care institutions, and communities where water supplies are polluted. Amoebic infections can also be transmitted by anal-oral sexual contact.

Culture and Diagnosis   Diagnosis of this protozoal infection relies on a combination of tests, including microscopic examination of stool for the characteristic cysts or trophozoites, ELISA tests of stool for *E. histolytica* antigens, and serological testing for the presence of antibodies to the pathogen. PCR testing is currently being refined. It is important to differentiate *E. histolytica* from the similar *Entamoeba coli* and *Entamoeba dispar,* which occur as normal flora.

Prevention and Treatment   No vaccine yet exists for *E. histolytica*, although several are in development. Prevention of the disease therefore relies on purification of water. Because regular chlorination of water supplies does not kill cysts, more rigorous methods such as boiling or iodine are required.

Effective treatment usually involves the use of drugs such as iodoquinol, which acts in the feces, and metronidazole or chloroquine, which work in the tissues. Flagyl is used as well. Dehydroemetine is used to control symptoms, but it will not cure the disease. Other drugs are given to relieve diarrhea and cramps, while lost fluid and electrolytes are replaced by oral or intravenous therapy. Infection with *E. histolytica* provokes antibody formation against several antigens, but permanent immunity is unlikely and reinfection can occur **(Checkpoint 22.7).**

Hepatitis

When certain viruses infect the liver, they cause **hepatitis,** an inflammatory disease marked by necrosis of hepatocytes and a mononuclear response that swells and disrupts the liver architecture. This pathologic change interferes with the liver's excretion of bile pigments such as bilirubin into the intestine. When bilirubin, a greenish-yellow pigment, accumulates in the blood and tissues, it causes **jaundice,** a yellow tinge in the skin and eyes. The condition can be caused by a variety of different viruses. They are all named hepatitis viruses, but only because they all can cause this inflammatory condition in the liver.

We should note that noninfectious conditions can also cause inflammation and disease in the liver, including some autoimmune conditions, drugs, and alcohol overuse.

Hepatitis A Virus

Hepatitis A virus (HAV) is a nonenveloped, single-stranded RNA enterovirus. It belongs to the family Picornaviridae. In general, HAV disease is far milder and shorter term than the other forms.

Signs and Symptoms   Most infections by this virus are either subclinical or accompanied by vague, flu-like symptoms. In more overt cases, the presenting symptoms may include jaundice and swollen liver. Darkened urine is often seen in this and other hepatitises. Jaundice is present in only about 10% of the cases. Hepatitis A occasionally occurs as a fulminating disease and causes liver damage, but this manifestation is quite rare. The virus is not **oncogenic**

✓ CHECKPOINT 22.7   Chronic Diarrhea

	Enteroaggregative *E. coli* (EAEC)	*Cyclospora cayetanensis*	*Giardia lamblia*	*Entamoeba histolytica*
Causative Organism(s)				
Most Common Modes of Transmission	Vehicle (food, water), fecal-oral	Fecal-oral, vehicle	Vehicle, fecal-oral, direct and indirect contact	Vehicle, fecal-oral
Virulence Factors	?	Invasiveness	Attachment to intestines alters mucosa	Lytic enzymes, induction of apoptosis, invasiveness
Culture/Diagnosis	Difficult to distinguish from other *E. coli*	Stool examination, PCR	Stool examination, ELISA	Stool examination, ELISA, serology
Prevention	?	Washing, cooking food, personal hygiene	Water hygiene, personal hygiene	Water hygiene, personal hygiene
Treatment	None	TMP-SMZ	Quinacrine, metronidazole	Iodoquinol plus metronidazole or chloroquine, Flagyl
Fever Present	No	Usually	Not usually	Yes
Blood in Stool	Sometimes, mucus also	No	No, mucus present (greasy and malodorous)	Yes
Distinctive Features	Chronic in the malnourished	–	Frequently occurs in backpackers, campers	–

(cancer causing), and complete uncomplicated recovery results.

Pathogenesis and Virulence Factors   The hepatitis A virus is generally of low virulence. Most of the pathogenic effects are thought to be the result of host response to the presence of virus in the liver.

Transmission and Epidemiology   There is an important distinction between this virus and hepatitis B and C viruses: Hepatitis A virus is spread through the fecal-oral route (and is sometimes known as infectious hepatitis). In general, the disease is associated with deficient personal hygiene and lack of public health measures. In countries with inadequate sewage control, most outbreaks are associated with fecally contaminated water and food. The United States has a yearly reported incidence of 15,000 to 20,000 cases. Most of these result from close institutional contact, unhygienic food handling, eating shellfish, sexual transmission, or travel to other countries. In 2003 the largest single hepatitis A outbreak to date in the United States was traced to contaminated green onions used in salsa dips at a Mexican restaurant. At least 600 people who had eaten at the restaurant fell ill with hepatitis A.

Hepatitis A occasionally can be spread by blood or blood products, but this is the exception rather than the rule. In developing countries, children are the most common victims, because exposure to the virus tends to occur early in life, whereas in North America and Europe, more cases appear in adults. Because the virus is not carried chronically, the principal reservoirs are asymptomatic, short-term carriers (often children) or people with clinical disease.

Culture and Diagnosis   Diagnosis of the disease is aided by detection of anti-HAV IgM antibodies produced early in the infection and by tests to identify HA antigen or virus directly in stool samples.

Prevention and Treatment   Prevention of hepatitis A is based primarily on immunization. An inactivated viral vaccine (Havrix) is currently approved, and an oral vaccine based on an attenuated strain of virus is in development. Short-term protection can be conferred by passive immune globulin. This treatment is useful for people who have come in contact with HAV-infected individuals, or who have eaten at a restaurant that was the source of a recent outbreak. In the 2003 green onion outbreak, 9,000 patrons of the Mexican restaurant received passive immunization as a precaution. A combined hepatitis A/hepatitis B vaccine, called Twinrix, is recommended for people who may be at risk for both diseases, such as people with chronic liver disfunction, intravenous drug users, and men who have sex with men. Travelers to areas with high rates of both diseases should obtain vaccine coverage as well.

No specific medicine is available for hepatitis A once the symptoms begin. Drinking lots of fluids and avoiding liver irritants, such as aspirin or alcohol, will speed recovery. Patients who receive immune globulin early in the disease usually experience milder symptoms than patients who do not receive it.

A Note About Hepatitis E Another RNA virus, called hepatitis E, causes a type of hepatitis very similar to that caused by hepatitis A. It is transmitted by the fecal-oral route, although it does not seem to be transmitted person to person. It is usually self-limiting, except in the case of pregnant women, for whom the fatality rate is 15% to 25%. It is more common in developing countries, and almost all of the cases reported in the United States occur in people who have traveled to these regions. There is currently no vaccine.

Hepatitis B Virus

Hepatitis B virus (HBV) is an enveloped DNA virus in the family Hepadnaviridae. Intact viruses are often called Dane particles. An antigen of clinical and immunological significance is the surface (or S) antigen. The genome is partly double-stranded, and partly single-stranded.

Signs and Symptoms In addition to the direct damage to liver cells just outlined, the spectrum of hepatitis disease may include fever, chills, malaise, anorexia, abdominal discomfort, diarrhea, and nausea. Rashes may appear and arthritis may occur. Hepatitis B infection can be very serious, even life-threatening. A small number of patients develop glomerulonephritis and arterial inflammation. Complete liver regeneration and restored function occur in most patients; however, a small number of patients develop chronic liver disease in the form of necrosis or **cirrhosis** (permanent liver scarring and loss of tissue). In some cases, chronic HBV infection can lead to a malignant condition.

Patients who become infected as children have significantly higher risks of long-term infection and disease. In fact, 90% of neonates infected at birth develop chronic infection, as do 30% of children infected between the ages of 1 and 5, but only 6% of persons infected after the age of 5. This finding is one of the major justifications for the routine vaccination of children. The mortality rate is 15% to 25% for people with chronic infection.

The association of HBV with **hepatocellular carcinoma** is based on these observations:

1. Certain hepatitis B antigens are found in malignant cells and are often detected as integrated components of the host genome.
2. Persistent carriers of the virus are more likely to develop this cancer.
3. People from areas of the world with a high incidence of hepatitis B (Africa and the Far East) are more frequently affected by liver cancer.

In addition, investigators have found that mass vaccination against HBV in Taiwan, begun 18 years ago, has resulted in a significant decrease in liver cancer in that country. (Taiwan previously had one of the highest rates of this cancer.) It is speculated that cancer is probably a result of infection early in life and the long-term carrier state. In general, people with chronic hepatitis are 200 times more likely to develop liver cancer, though the exact role of the virus is still the object of molecular analysis.

Some patients infected with hepatitis B are coinfected with a particle called the delta agent, sometimes also called a hepatitis D virus. This agent seems to be a defective RNA virus that cannot produce infection unless a cell is also infected with HBV. Hepatitis D virus invades host cells by "borrowing" the outer receptors of HBV. When HBV infection is accompanied by the delta agent, the disease becomes more severe and is more likely to progress to permanent liver damage.

Pathogenesis and Virulence Factors The hepatitis B virus enters the body through a break in the skin or mucous membrane, or by injection into the bloodstream. Eventually, it reaches the liver cells (hepatocytes) where it multiplies and releases viruses into the blood during an incubation period of 4 to 24 weeks (7 weeks average). Surprisingly, the majority of those infected exhibit few overt symptoms and eventually develop an immunity to HBV, but some people experience the symptoms described earlier. The precise mechanisms of virulence are not clear. The ability of HBV to remain latent in some patients contributes to its pathogenesis.

Transmission and Epidemiology An important factor in the transmission pattern of hepatitis B virus is that it multiplies exclusively in the liver, which continuously seeds the blood with viruses. Electron microscopic studies have revealed up to 10^7 virions per milliliter of infected blood. Even a minute amount of blood (a *millionth* of a milliliter) can transmit infection. The abundance of circulating virions is so high, and the minimal dose so low, that such simple practices as sharing a toothbrush or a razor can transmit the infection. Over the past 10 years, HBV has also been detected in semen and vaginal secretions, and it can be transmitted by these fluids. Spread of the virus by means of close contact in families or institutions is also well documented. Vertical transmission is possible, and it predisposes the child to development of the carrier state and increased risk of liver cancer. It is sometimes known as *serum hepatitis*.

Hepatitis B is an ancient disease that has been found in all populations, although the incidence and risk are highest among people living under crowded conditions, drug addicts, the sexually promiscuous, and those in certain occupations, including people who conduct medical procedures involving blood or blood products.

This virus is one of the major infectious concerns for health care workers. Needle sticks can easily transmit the virus, and therefore most workers are required to have the full series of HBV vaccinations. Unlike the more notorious HIV, HBV remains infective for days in dried blood, for months when stored in serum at room temperature, and for decades if frozen. Although it is not inactivated after 4 hours of exposure to 60°C, boiling for the same period can destroy it. Disinfectants containing chlorine, iodine, and glutaraldehyde show potent anti–hepatitis B activity.

Cosmetic manipulation such as tattooing and ear or body piercing can expose a person to infection if the instruments are not properly sterilized. The only reliable method for destroying HBV on reusable instruments is autoclaving.

Culture and Diagnosis Serological tests can detect either virus antigen or antibodies. Radioimmunoassay and ELISA testing permit detection of the important surface antigen of HBV very early in infection. These same tests are essential for screening blood destined for transfusions, semen in sperm banks, and organs intended for transplant. Antibody tests are most valuable in patients who are negative for the antigen.

Prevention and Treatment Since 1981, the primary prevention for HBV infection is vaccination. The most widely used vaccines are recombinant, containing the pure surface antigen cloned in yeast cells. Vaccines are given in three doses over 18 months, with occasional boosters. Vaccination is a must for medical and dental workers and students, patients receiving multiple transfusions, immunodeficient persons, and cancer patients. The vaccine is also now strongly recommended for all newborns as part of a routine immunization schedule. As just mentioned, a combined vaccine for HAV/HBV may be appropriate for certain people.

Passive immunization with hepatitis B immune globulin (HBIG) gives significant immediate protection to people who have been exposed to the virus through needle puncture, broken blood containers, or skin and mucosal contact with blood. Another group for whom passive immunization is highly recommended is neonates born to infected mothers.

Mild cases of hepatitis B are managed by symptomatic treatment and supportive care. Chronic infection can be controlled with recombinant human interferon, a drug called adefovir dipivoxil (a nucleotide analog), or lamivudine (another nucleotide analog best known for its use in HIV patients). All of these can help to stop virus multiplication and prevent liver damage in many, but not all, patients. None of the drugs are considered curative.

Hepatitis C Virus

Hepatitis C is sometimes referred to as the "silent epidemic" because more than 4 million Americans are infected with the virus, but it takes many years to cause noticeable symptoms. In the United States at least 35,000 new infections occur every year. Liver failure from hepatitis C is one of the most common reasons for liver transplants in this country. Hepatitis C is an RNA virus in the Flaviviridae family. It used to be known as "non-A non-B" virus. It is usually diagnosed with a blood test for antibodies to the virus.

Signs and Symptoms People have widely varying experiences with this infection. It shares many characteristics of hepatitis B disease, but it is much more likely to become chronic. Of those infected, 75% to 85% will remain infected indefinitely. (In contrast, only about 6% of persons who acquire hepatitis B after the age of 5 will be chronically infected.) With HCV infection it is possible to have severe symptoms without

✔ CHECKPOINT 22.8	Hepatitis		
	Hepatitis A or E virus	Hepatitis B virus	Hepatitis C virus
Causative Organism(s)	Hepatitis A or E virus	Hepatitis B virus	Hepatitis C virus
Most Common Modes of Transmission	Fecal-oral, vehicle	Parenteral (blood contact), direct contact (especially sexual), vertical	Parenteral (blood contact), vertical
Virulence Factors	–	Latency	Core protein suppresses immune function?
Culture/Diagnosis	IgM serology	Serology (ELISA, radioimmunoassay)	Serology
Prevention	Hepatitis A vaccine or combined HAV/HBV vaccine	HBV recombinant vaccine	–
Treatment	Immune globulin	Interferon, nucleoside analogs	(Pegylated) interferon with or without ribavirin
Long-Term Consequences	None	Chronic infection, liver cancer, death	Chronic infection and liver disease very common; cancer, death
Incubation Period	2–7 weeks	1–6 months	2–8 weeks

permanent liver damage; but it is more common to have chronic liver disease, even if there are no overt symptoms. Cancer may also result from chronic HCV infection.

Pathogenesis and Virulence Factors
The virus is so adept at establishing chronic infections that researchers are studying the ways that it evades immunological detection and destruction. The virus's core protein seems to play a role in the suppression of cell-mediated immunity as well as in the production of various cytokines.

Transmission and Epidemiology   This virus is acquired in similar ways to HBV. It is more commonly transmitted through blood contact (both "sanctioned," such as in blood transfusions, and "unsanctioned," such as needle sharing by injecting drug users) than through transfer of other body fluids. Vertical transmission is also possible.

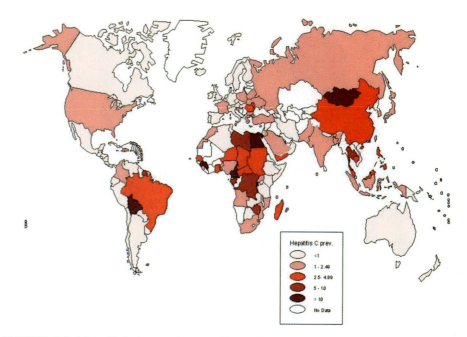

FIGURE 22.25   Global prevalence of hepatitis C as of June 1999.

Before a test was available to test blood products for this virus, it seems to have been frequently transmitted through blood transfusions. Hemophiliacs who were treated with clotting factor prior to 1985 were infected at a high rate with HCV. Once blood began to be tested for HIV (in 1985) and screened for so-called "non-A non-B" hepatitis, the risk of contracting HCV from blood was greatly reduced. The current risk for transfusion-associated HCV is thought to be 1 in 100,000 units transfused.

Because HCV was not recognized sooner, a relatively large percentage of the population is infected. Eighty percent of the 4 million affected in this country are suspected to have no symptoms. It has a very high prevalence in parts of South America, Central Africa, and in China **(figure 22.25).**

Prevention and Treatment   There is currently no vaccine for hepatitis C. Various treatment regimens have been attempted; most include the use of therapeutic interferon, and a more effective derivative of interferon called pegylated interferon. Some clinicians also prescribe ribavirin to try to suppress viral multiplication. The treatments are not curative, but they may prevent or lessen damage to the liver **(Checkpoint 22.8).**

Helminthic Intestinal Infections

Helminths that parasitize humans are amazingly diverse, ranging from barely visible roundworms (0.3 mm) to huge tapeworms (25 m long). In the introduction to these organisms in chapter 5, we grouped them into three categories: nema-todes (roundworms), trematodes (flukes), and cestodes (tapeworms), and we discussed basic characteristics of each group. You may wish to review those sections before continuing. In this section, we examine the intestinal diseases caused by helminths. Although they can cause symptoms that might be mistaken for some of the diseases discussed elsewhere in this chapter, helminthic diseases are usually accompanied by an additional set of symptoms that arises from the host response to helminths. Worm infection usually provokes an increase in granular leukocytes called eosinophils, which have a specialized capacity to destroy worms. This increase, termed **eosinophilia,** is a hallmark of helminth infection, and is detectable in blood counts. If the following symptoms occur coupled with eosinophilia, helminthic infection should be suspected.

Helminthic infections may be acquired through the fecal-oral route or through penetration of the skin, but most of them spend part of their lives in the intestinal tract. (**Figure 22.26** depicts the four different types of life cycles of the helminths.) While the worms are in the intestines, they can produce a gamut of intestinal symptoms. Some of them also produce symptoms outside of the intestines; they will be considered in separate categories.

General Clinical Considerations

This section on helminthic intestinal infections is organized a bit differently: We will talk about diagnosis, pathogenesis and prevention, and treatment of the helminths as a group in the next subsections. Each type of infection is then described in the sections that follow.

Cycle A

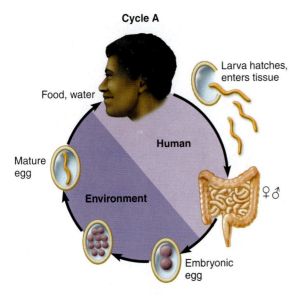

In cycle A, the worm develops in intestine; egg is released with feces into environment; eggs are ingested by new host and hatch in intestine (examples: *Ascaris, Trichuris*).

Cycle B

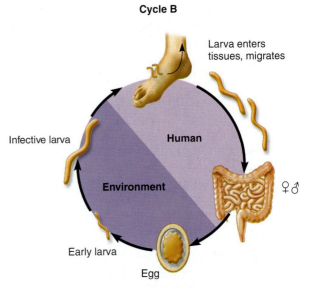

In cycle B, the worm matures in intestine; eggs are released with feces; larvae hatch and develop in environment; infection occurs through skin penetration by larvae (example: hookworms).

Cycle C

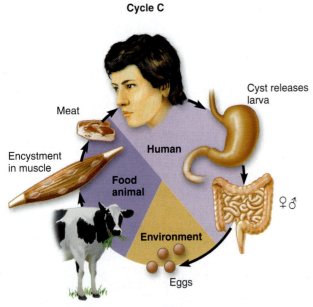

In cycle C, the adult matures in human intestine; eggs are released into environment; eggs are eaten by grazing animals; larval forms encyst in tissue; humans eating animal flesh are infected (example: *Taenia*).

Cycle D

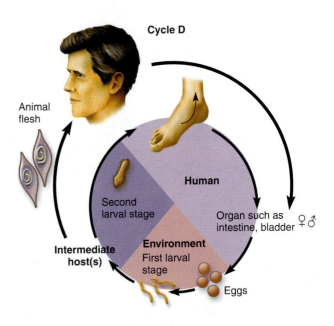

In cycle D, eggs are released from human; humans are infected through ingestion or direct penetration by larval phase (examples: *Opisthorchis* and *Schistosoma*).

FIGURE 22.26 **Four basic helminth life and transmission cycles.**

INSIGHT 22.4   *Discovery*

Treating Inflammatory Bowel Disease with Worms?

Probably every one of us knows someone who suffers from an inflammatory bowel condition, such as Crohn's disease or ulcerative colitis. These are not thought to be caused by microorganisms, so how do they occur? The answer might surprise you.

Many recent epidemiological investigations have revealed that inflammatory bowel disease (IBD) is most common in Western industrialized countries and is very rare in developing countries. More specifically, the prevalence of IBD in any given country is inversely proportional to the prevalence of helminthic infections in that country. Looking at the picture in this country, the incidence of helminthic infections decreased dramatically between the 1930s and the 1950s; the incidence of IBD began its continuous rise in the 1950s. Scientists suspect a connection here: that the *absence* of exposure to helminth infection predisposes a person to IBD.

These researchers have developed a hypothesis that the parts of the immune system that are activated during helminthic infection begin to "malfunction" when left idle, eventually resulting in damage to host tissue. Researchers wondered whether they could "treat" IBD by exposing patients to an intestinal helminth infection. The first studies were conducted in mice, and the results looked promising. Then researchers at the University of Iowa conducted studies in human volunteers. They selected eight patients with either Crohn's disease or ulcerative colitis and administered to them Gatorade containing 2,500 eggs of the pig whipworm *Trichuris suis*. They chose this worm because it colonizes the intestines for a few weeks, and then is completely eliminated without treatment. It does not invade tissues, and the eggs that are shed in the stools are not infective.

The researchers found marked improvement in the inflammatory bowel conditions in all of the patients. They determined that the effects were of short duration and asked several of the patients to continue in the study, receiving fresh doses of *T. suis* every 3 weeks. All of these patients experienced significant and longlasting remission of their IBD symptoms. What's more, they indicated that they would be willing to continue the treatments indefinitely. It seems that occasional contact with helminths keeps the complicated network of immunoregulatory mechanisms in good working order. This story serves to remind us about the intimate association between humans and their parasites.

Pathogenesis and Virulence Factors in General In most cases, helminths that infect humans do not have sophisticated virulence factors. They do have numerous adaptations that allow them to survive in their hosts. They have specialized mouthparts for attaching to tissues and for feeding, enzymes with which they liquefy and penetrate tissues, and a cuticle or other covering to protect them from host defenses. In addition, their organ systems are usually reduced to the essentials: getting food and processing it, moving, and reproducing. The damage they cause in the host is very often the result of the host's response to the presence of the invader.

Many helminths have more than one host during their lifetimes. If this is the case, the host in which the adult worm is found is called the **definitive host** (usually a vertebrate). Sometimes the actual definitive host is not the host usually used by the parasite, but an accidental bystander. Humans often become the accidental definitive hosts for helminths whose normal definitive host is a cow, pig, or fish. Larval stages of helminths are found in intermediate hosts. Humans can serve as intermediate hosts, too. Helminths may require no intermediate host at all, or may need one or more intermediate hosts for its entire life cycle.

Diagnosis in General Diagnosis of almost all helminthic infections follows a similar series of steps. A differential blood count showing eosinophilia and serological tests indicating sensitivity to helminthic antigens both provide indirect evidence of worm infection. A history of travel to the tropics or immigration from those regions is also helpful, even if it occurred years ago, because some flukes and nematodes persist for decades. The most definitive evidence, however, is the discovery of eggs, larvae, or adult worms in stools or other tissues. The worms are sufficiently distinct in morphology that positive identification can be based on any stage, including eggs. That said, not all of these diseases result in eggs or larval stages that can easily be found in stool.

Prevention and Treatment in General Preventive measures are aimed at minimizing human contact with the parasite or interrupting its life cycle. In areas where the worm is transmitted by fecally contaminated soil and water, disease rates are significantly reduced through proper sewage disposal, using sanitary latrines, avoiding human feces as fertilizer, and disinfection of the water supply. In cases where the larvae invade through the skin, people should avoid direct contact with infested water and soil. Food-borne disease can be avoided by thoroughly washing and cooking vegetable and meats. Also, because adult worms, larvae, and eggs are sensitive to cold, freezing foods is a highly satisfactory preventive measure. These methods work best if humans are the sole host of the parasite; if they are not, control of reservoirs or vector populations may be necessary.

Although several useful antihelminthic medications exist, the cellular physiology of the eukaryotic parasites resembles that of humans, and drugs toxic to them can also be toxic to us. Some antihelminthic drugs suppress a metabolic process that is more important to the worm than to the human. Others inhibit the worm's movement and prevent it

TABLE 22.1	Antihelminthic Therapeutic Agents and Their Effects
Drug	**Effect**
Piperazine	Paralyzes worm so it can be expelled in feces
Pyrantel	Paralyzes worm so it can be expelled in feces
Mebendazole	Blocks key step in worm metabolism
Thiabendazole	Blocks key step in worm metabolism
Praziquantel	Interferes with worm metabolism
Niclosamide	Inhibits ATP formation in worm; destroys proglottids but not eggs

from maintaining its position in a certain organ. Therapy is also based on a drug's greater toxicity to the more vulnerable helminths or on the local effects of oral drugs in the intestine. Antihelminthic drugs of choice and their effects are given in **table 22.1.** We should note that some helminths have developed resistance to the drugs used to treat them. In some cases, surgery may be necessary to remove worms or larvae, although this procedure can be difficult if the parasite load is high or is not confined to one area.

Intestinal Distress as the Primary Symptom

Both tapeworms and roundworms can infect the intestinal tract in such a way as to cause primary symptoms there. The pork tapeworm *(Taenia solium)* and the fish tapeworm *(Diphyllobothrium latum)* are highlighted, as well as two nematodes (roundworms): the whipworm *Trichuris trichiura* and the pinworm *Enterobius vermicularis.* Both of the roundworms are deposited in the small intestine and migrate to the large intestine. We'll start with these.

Trichuris trichiura The common name for this nematode— whipworm—refers to its likeness to a miniature buggy whip. Its life cycle and transmission is of the cycle A type (figure 22.26). Humans are the sole host. Trichuriasis has its highest incidence in areas of the tropics and subtropics that have poor sanitation. Embryonic eggs deposited in the soil are not immediately infective and continue development for 3 to 6 weeks in this habitat. Ingested eggs hatch in the small intestine, where the larvae attach, penetrate the outer wall, and go through several molts. The mature adults move to the large intestine and gain a hold with their long, thin heads, while the thicker tail dangles free in the intestinal lumen. Following sexual maturation and fertilization, the females eventually lay 3,000 to 5,000 eggs daily into the bowel. The entire cycle requires about 90 days, and untreated infection can last up to 2 years.

Symptoms of this infection may include localized hemorrhage of the bowel, caused by worms burrowing and piercing intestinal mucosa. This can also provide a portal of entry for secondary bacterial infection. Heavier infections can cause dysentery, loss of muscle tone, and rectal prolapse, which can prove fatal in children.

Enterobius vermicularis This nematode is often called the pinworm, or seatworm. It is the most common worm disease of children in temperate zones. Some estimates put the prevalence of this infection in the United States at 5% to 15%, although most experts feel that this has declined in recent years. The transmission of this roundworm is of the cycle A type. Freshly deposited eggs have a sticky coating that causes them to lodge beneath the fingernails and to adhere to fomites. Upon drying, the eggs become airborne and settle in house dust. Worms are ingested from contaminated food or drink, and from self-inoculation from one's own fingers. Eggs hatch in the small intestine and release larvae that migrate to the large intestine. There the larvae mature into adult worms and mate.

The symptoms of this condition are pronounced anal itching when the mature female emerges from the anus and lays eggs. Although infection is not fatal and most cases are asymptomatic, the afflicted child can suffer from disrupted sleep and sometimes nausea, abdominal discomfort, and diarrhea. A simple rapid test can be performed by pressing a piece of transparent adhesive tape against the anal skin and then applying it to a slide for microscopic examination. When one member of the family is diagnosed, the entire family should be tested and/or treated since it is likely that multiple members are infected.

Taenia solium In contrast to the last two helminths, this one is a tapeworm. Adult worms are usually around 5 m long and have a scolex with hooklets and suckers to attach to the intestine (**figure 22.27**). Taeniasis caused by the *T. solium* (the pig tapeworm) is distributed worldwide but is mainly concentrated in areas where humans live in close proximity with pigs, or eat undercooked pork. In pigs, the eggs hatch in the small intestine, and the released larvae migrate throughout the organs. Ultimately, they encyst in the muscles, becoming *cysticerci,* young tapeworms that are the infective stage for humans. When humans ingest a live cysticercus in pork, the coat is digested and the organism is flushed into the intestine, where it firmly attaches by the scolex and develops into an adult tapeworm. Infection with *T. solium* can take another form, when humans ingest the tapeworm eggs rather than cysticerci. Although humans are not the usual intermediate hosts, the eggs can still hatch in the intestine, releasing tapeworm larvae that migrate to all tissues. They form bladderlike sacs throughout the body that can cause serious damage. This transmission and life cycle is cycle C in figure 22.26. The pork tapeworm is not the same as the more commonly known pork helminthic infection, *trichinosis.* It is discussed in a later section.

For such a large organism, it is remarkable how few symptoms a tapeworm causes. Occasionally, a patient discovers proglottids in his or her stool, and some patients complain of vague abdominal pain and nausea.

Other tapeworms of the genus *Taenia* infect humans. One of them is the beef tapeworm, *Taenia saginata.* It usually causes similar general symptoms of helminth infection. But humans are not known to acquire *T. saginata* infection by ingesting the eggs.

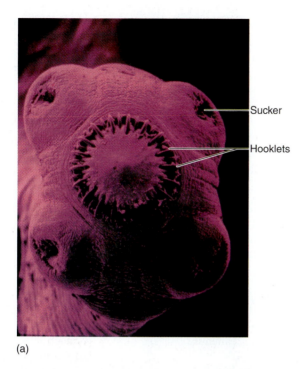

(a)

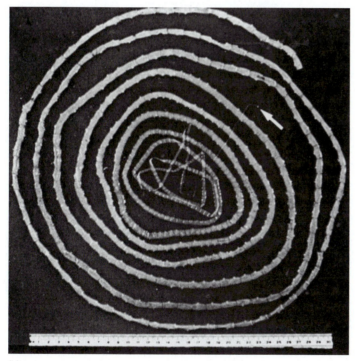

(b)

FIGURE 22.27   Tapeworm characteristics.
(a) Tapeworm scolex showing sucker and hooklets. **(b)** Adult
Taenia saginata. The arrow points to the scolex; the remainder of
the tape, called the strobila, has a total length of 5 meters.

Diphyllobothrium latum   This tapeworm has an intermediate
host in fish. It is common in the Great Lakes, Alaska, and
Canada. Humans are its definitive host. It develops in the intestine and can cause long-term symptoms. It can be transmitted in raw food such as sushi and sashimi made from

salmon. (Reputable sushi restaurants employ authentic sushi
chefs who are trained to carefully examine fish for larvae and
other signs of infection.)

As is the case with most tapeworms, symptoms are minor and usually vague, and include possible abdominal discomfort or nausea. The tapeworm seems to have the ability
to absorb and use the vitamin B12, making it unavailable to
its human host. Anemia is therefore sometimes reported with
this infection.

***Hymenolepis* species**   These relatively small tapeworms are
the most common tapeworm infections in the world. There
are two species: *H. nana*, known as the dwarf tapeworm since
it is only 15 to 40 mm in length, and *H. diminuta*, the rat tapeworm, which is usually 20 to 60 cm in length as an adult.

The life cycle of these tapeworms often involves insects
as well as the definitive host, which may be a rodent or a
human. When eggs are passed in the feces of a rodent or human, they can be ingested by various insects, which are in
turn accidentally ingested by humans (in cereals or other
foods). Alternatively, eggs in the environment can be directly
ingested by humans. Tapeworms become established in the
small intestine and eggs can be released after proglottids
break off from the attached worms.

Symptoms are mild, and the treatment of choice is praziquantel **(Checkpoint 22.9).**

Intestinal Distress Accompanied by Migratory Symptoms

A diverse group of helminths enter the body as larvae or
eggs, mature to the worm stage in the intestine, and then migrate into the circulatory and lymphatic systems, after which
they travel to the heart and lungs, migrate up the respiratory
tree to the throat, and are swallowed. This journey returns
the mature worms to the intestinal tract where they then take
up residence. All of these conditions, in addition to causing
symptoms in the digestive tract, may induce inflammatory reactions along their migratory routes, resulting in eosinophilia
and, during their lung stage, pneumonia. Three different examples of this type of infection follow.

Ascaris lumbricoides   *Ascaris lumbricoides* is a giant intestinal
roundworm (up to 300 mm long) that probably accounts for
the greatest number of worm infections (estimated at 1 billion cases worldwide). Most reported cases in the United
States occur in the southeastern states. *Ascaris* spends its larval and adult stages in humans and releases embryonic eggs
in feces, which are then spread to other humans through
food, drink, or contaminated objects placed in the mouth.
The eggs thrive in warm, moist soils and resist cold and
chemical disinfectants, but they are sensitive to sunlight,
high temperatures, and drying. After ingested eggs hatch in
the human intestine, the larvae embark upon an odyssey in the
tissues. First, they penetrate the intestinal wall and enter
the lymphatic and circulatory systems. They are swept into
the heart and eventually arrive at the capillaries of the

✔ CHECKPOINT 22.9 Intestinal Distress

Causative Organism(s)	*Trichuris trichiura* (whipworm)	*Enterobius vermicularis* (pinworm)	*Taenia solium* (pork tapeworm)	*Diphyllobothrium latum* (fish tapeworm)	*Hymenolepis nana* and *H. diminuta*
Most Common Modes of Transmission	Cycle A: vehicle (soil)/fecal-oral	Cycle A: vehicle (food, water), fomites, self-inoculation	Cycle C: vehicle (pork)—also fecal-oral	Cycle C: vehicle (seafood)	Cycle C: vehicle (ingesting insects)—also fecal-oral
Virulence Factors	Burrowing and invasiveness	–	–	Vitamin B12 usage	–
Culture/Diagnosis	Blood count, serology, egg or worm detection	Adhesive tape method	Blood count, serology, egg or worm detection	Blood count, serology, egg or worm detection	Blood count, serology, egg or worm detection
Prevention	Hygiene, sanitation	Hygiene	Cook meat, avoid pig feces	Cook meat	Hygienic environment
Treatment	Mebendazole	Piperazine, pyrantel	Praziquantel, Niclosamide	Praziquantel, Niclosamide	Praziquantel
Distinctive Features	Humans sole host	Common in U.S.	Tapeworm; intermediate host is pigs	Large tapeworm; anemia	Most common tapeworm infection

lungs. From this point, the larvae migrate up the respiratory tree to the glottis. Worms entering the throat are swallowed and returned to the small intestine, where they reach adulthood and reproduce, producing up to 200,000 fertilized eggs a day.

Even as adults, male and female worms are not attached to the intestine and retain some of their exploratory ways. They are known to invade the biliary channels of the liver and gallbladder, and on occasion the worms emerge from the nose and mouth. Severe inflammatory reactions mark the migratory route, and allergic reactions such as bronchospasm, asthma, or skin rash can occur. Heavy worm loads can retard the physical and mental development of children. One possibility with intestinal worm infections is self-reinoculation due to poor personal hygiene.

Necator americanus and Ancylostoma duodenale These two different nematodes are called by the common name hookworm. *Necator americanus* (nee-kay'-tor ah-mer"-ih-cah'-nus) is endemic to the New World, and *Ancylostoma duodenale* (an'-kih-los'-toh-mah doo-oh-den-ah'-lee) is endemic to the Old World, although the two species overlap in parts of Latin America. Otherwise, with respect to transmission, life cycle, and pathology, they are usually lumped together. The *hook* refers to the adult's oral cutting plates by which it anchors to the intestinal villi, and its curved anterior end **(figure 22.28)**.

Unlike other intestinal worms, hookworm larvae hatch outside the body and infect by penetrating the skin. Hookworm transmission is described by cycle B (figure 22.26). Ordinarily, the parasite is present in soil contaminated with human feces. It enters sites on bare feet such as hair follicles, abrasions, or the soft skin between the toes, but cases have

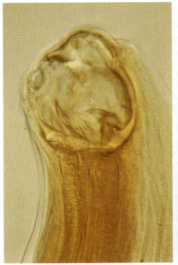 

FIGURE 22.28 Cutting teeth on the mouths of (left) *Necator americanus* and (right) *Ancylostoma duodenale*.

occurred via mud that was splattered on the ankles of people wearing shoes. Infection has even been reported in people handling soiled laundry.

On contact the hookworm larvae actively burrow into the skin. After several hours, they reach the lymphatic or blood circulation and are immediately carried into the heart and lungs. The larvae proceed up the bronchi and trachea to the throat. Most of the larvae are swallowed with sputum and arrive in the small intestine, where they anchor, feed on blood, and mature. Eggs first appear in the stool about

6 weeks after the time of entry, and the untreated infection can last about 5 years.

Symptoms from these infections follow the progress of the worm in the body. A localized dermatitis called *ground itch* may be caused by the initial penetration of larvae. The transit of the larvae to the lungs is ordinarily brief, but it can cause symptoms of pneumonia and eosinophilia. The potential for injury is greatest during the intestinal phase, when heavy worm burdens can cause nausea, vomiting, cramps, and bloody diarrhea. Because blood loss is significant, iron-deficient anemia develops, and infants are especially susceptible to hemorrhagic shock. Chronic fatigue, listlessness, apathy, and anemia worsen with chronic and repeated infections.

Hookworm infections are treated with antihelminthic drugs, but frequent reinfection is a problem. U.S. and Brazilian researchers are testing a vaccine against *Necator americanus.* In 2000, the Bill and Melinda Gates Foundation, recognizing the impact of worldwide hookworm infections, contributed $18 million to the development of a hookworm vaccine.

Strongyloides stercoralis The agent of strongyloidiasis, or threadworm infection, is *Strongyloides stercoralis* (stron′-jih-loy-deez ster″-kor-ah′-lis). This nematode is exceptional because of its minute size and its capacity to complete its life cycle either within the human body or outside in moist soil. It shares a similar distribution and life cycle to hookworms and afflicts an estimated 100 to 200 million people worldwide. Infection occurs when soil larvae penetrate the skin (cycle B in figure 22.26). The worm then enters the circulation, is carried to the respiratory tract and swallowed, and then enters the small intestine to complete development. Although adult *S. stercoralis* lays eggs in the gut just as hookworms do, the eggs hatch into larvae in the colon and can remain entirely in the host's body to complete the cycle. The larval form of the

organism can likewise exit with feces and go through an environmental cycle. These numerous alternative life cycles greatly increase the chance of transmission and the likelihood for chronic infection.

The first symptom of threadworm infection is usually a red, intensely itchy skin rash at the site of entry. Mild migratory activity in an otherwise normal person can escape notice, but heavy worm loads can cause symptoms of pneumonitis and eosinophilia. The nematode activities in the intestine produce bloody diarrhea, liver enlargement, and malabsorption. In immunocompromised patients, there is a risk of disseminated infection involving numerous organs **(figure 22.29).** Hardest hit are AIDS patients, transplant patients on immunosuppressant drugs, and cancer patients receiving irradiation therapy, who can die if not treated promptly.

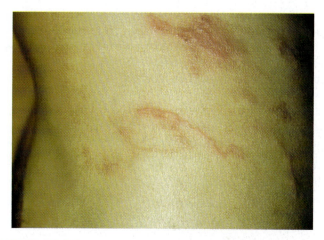

FIGURE 22.29 A patient with disseminated *Strongyloides* infection.
Trails under the skin indicate the migration tracks of the worms.

✔ CHECKPOINT 22.10	Intestinal Distress plus Migratory Symptoms		
Causative Organism(s)	*Ascaris lumbricoides* (intestinal roundworm)	*Necator americanus* and *Ancylostoma duodenale* (hookworms)	*Strongyloides stercoralis* (threadworm)
Most Common Modes of Transmission	Cycle A: vehicle (soil/fecal-oral), fomites, self-inoculation	Cycle B: vehicle (soil), fomite	Cycle B: vehicle (soil), fomite
Virulence Factors	Induction of hypersensitivity, adult worm migration, and abdominal obstruction	Induction of hypersensitivity, adult worm migration, and abdominal obstruction	Induction of hypersensitivity, adult worm migration, and abdominal obstruction
Culture/Diagnosis	Blood count, serology, egg or worm detection	Blood count, serology, egg or worm detection	Blood count, serology, egg or worm detection
Prevention	Hygiene	Sanitation	Sanitation
Treatment	Piperazine, pyrantel, mebendazole	Piperazine, pyrantel, mebendazole, thiabendazole	Thiabendazole
Distinctive Features	Roundworm; 1 billion persons infected	Penetrates skin, serious intestinal symptoms	Penetrates skin, severe for immunocompromised

Liver and Intestinal Disease

One group of worms that lands in the intestines has a particular affinity for the liver. Two of these worms are trematodes (flatworms), and they are categorized as liver flukes.

Opisthorchis sinensis* and *Clonorchis sinensis   *Opisthorchis sinensis* and *Clonorchis sinensis* are two worms known as Chinese liver flukes. They complete their sexual development in mammals such as humans, cats, dogs, and swine. Their intermediate development occurs in snail and fish hosts. Humans ingest cercariae in inadequately cooked or raw freshwater fish (see cycle D in figure 22.26). Larvae hatch and crawl into the bile duct, where they mature and shed eggs into the intestinal tract. Feces containing eggs are passed into standing water that harbors the intermediate snail host. The cycle is complete when infected snails release cercariae that invade fish living in the same water.

Symptoms of *Opisthorchis* and *Clonorchis* infection are slow to develop but include thickening of the lining of the bile duct, and possible granuloma formation in areas of the liver if eggs enter the stroma of the liver. If the infection is heavy, the bile duct could be blocked.

Fasciola hepatica   This liver fluke is a common parasite in sheep, cattle, goats, and other mammals and is occasionally transmitted to humans **(figure 22.30)**. Periodic outbreaks in temperate regions of Europe and South America are associated with eating wild watercress. The life cycle is very complex, involving the mammal as the definitive host, the release of eggs in the feces, the hatching of eggs in the water into *miracidia*, invasion of freshwater snails, development and release of cercariae, encystment of cercariae on a water plant, and ingestion of the cyst by a mammalian host eating the plant. The cysts release young flukes into the intestine that wander to the liver, lodge in the gallbladder, and develop into adults. Humans develop symptoms of vomiting, diarrhea, hepatomegaly, and bile obstruction only if they are chronically infected by a large number of flukes.

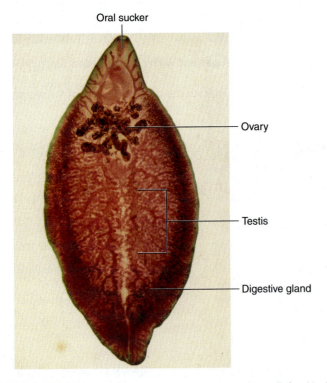

FIGURE 22.30   *Fasciola hepatica,* **the sheep liver fluke (2×).**

Oral sucker

Ovary

Testis

Digestive gland

✔ CHECKPOINT 22.11   Liver and Intestinal Disease

	Opisthorchis sinensis, *Clonorchis sinensis*	*Fasciola hepatica*
Causative Organism(s)	*Opisthorchis sinensis,* *Clonorchis sinensis*	*Fasciola hepatica*
Most Common Modes of Transmission	Cycle D: vehicle (fish or crustaceans)	Cycle D: vehicle (water and water plants)
Virulence Factors	–	–
Culture/Diagnosis	Blood count, serology, egg or worm detection	Blood count, serology, egg or worm detection
Prevention	Cook food, sanitation of water	Sanitation of water
Treatment	Praziquantel	Praziquantel
Distinctive Features	Live in bile duct	Live in liver and gallbladder

Muscle and Neurological Symptoms

Trichinosis is an infection transmitted by eating pork (and sometimes other wildlife) that have the cysts of *Trichinella* species embedded in the meat. The life cycle of this nematode is spent entirely within the body of a mammalian host such as a pig, bear, cat, dog, or rat. In nature the parasite is maintained in an encapsulated (encysted) larval form in the muscles of these animal reservoirs and is transmitted when other animals prey upon them. The disease cannot be transmitted from one human to another except in the case of cannibalism.

Because all wild and domesticated mammals appear to be susceptible to *Trichinella* species, one might expect human trichinosis to be common worldwide. But in reality, it is more common in the United States and in Europe than in the rest of the world. This distribution appears to be related to regional or ethnic customs of eating raw or rare pork dishes or wild animal meats. Bear meat is the source of up to one-third of the cases in the United States. Home or small-scale butchering enterprises that do not carefully inspect pork can spread the parasite, although commercial pork can also be a source. Practices such as tasting raw homemade pork sausage or serving rare pork or pork-beef mixtures have been responsible for sporadic outbreaks.

The cyst envelope is digested in the stomach and small intestine, which liberates the larvae. After burrowing into the intestinal mucosa, the larvae reach adulthood and mate. The larvae that result from this union penetrate the intestine and enter the lymphatic channels and blood. All tissues are at risk for invasion, but final development occurs when the coiled larvae are encysted in the skeletal muscle. At maturity, the cyst is about 1 mm long and can be observed by careful inspection of meat. Although larvae can deteriorate over time, they have also been known to survive for years.

Symptoms may be unnoticeable or they could be life-threatening, depending on how many larvae were ingested in the tainted meat. The first symptoms, when present, mimic influenza or viral fevers, with diarrhea, nausea, abdominal pains, fever, and sweating. The second phase, brought on by the mass migration of larvae and their entrance into muscle, produces puffiness around the eyes, intense muscle and joint pain, shortness of breath, and pronounced eosinophilia. The most serious life-threatening manifestations are heart and brain involvement. Although the symptoms eventually subside, a cure is not available once the larvae have encysted in muscles.

The most effective preventive measures for trichinosis are to adequately store and cook pork and wild meats.

✔ CHECKPOINT 22.12	Muscle and Neurological Symptoms
Causative Organism(s)	*Trichinella* species
Most Common Modes of Transmission	Vehicle (food)
Virulence Factors	–
Culture/Diagnosis	Serology combined with clinical picture; muscle biopsy
Prevention	Cook meat
Treatment	Pyrantel, mebendazole, thiabendazole
Distinctive Features	Brain and heart involvement can be fatal

Liver Disease

When liver swelling or malfunction is accompanied by eosinophilia, **schistosomiasis** should be suspected. Schistosomiases has afflicted humans for thousands of years. The disease is caused by the blood flukes *Schistosoma mansoni*, or *S. japonicum*, species that are morphologically and geograph-ically distinct but share similar life cycles, transmission methods, and general disease manifestations. It is one of the few infectious agents that can invade intact skin.

Signs and Symptoms The first symptoms of infection are itchiness in the area where the worm enters the body, followed by fever, chills, diarrhea, and cough. The most severe consequences, associated with chronic infection, are hepatomegaly and liver disease and splenomegaly. Other serious conditions caused by a different schistosome occur in the urinary tract—bladder obstruction and blood in the urine. This condition will be discussed in chapter 23 (genitourinary tract diseases). Occasionally, eggs from the worms are carried into the central nervous system and heart and create a severe granulomatous response. Adult flukes can live for many years, and by eluding the immune defenses, cause a chronic affliction.

Causative Agent Schistosomes are trematodes, or flukes (see chapter 5), but they are more cylindrical than flat **(figure 22.31b).** They are often called *blood flukes*. Flukes have digestive, excretory, neuromuscular, and reproductive systems, but they lack circulatory and respiratory systems. Humans are the definitive hosts for the blood fluke, and snails are the intermediate host.

Pathogenesis and Virulence Factors This parasite is clever indeed. Once inside the host, it coats its outer surface with proteins from the host's bloodstream, basically "cloaking" itself from the host defense system. This coat reduces its surface antigenicity and allows it to remain in the host indefinitely.

Other virulence attributes are the organism's ability to invade intact skin and attach to vascular endothelium, to sequester iron from the bloodstream, and to induce a granulomatous response.

Transmission and Epidemiology The life cycle of the schistosome is complex (figure 22.31). The cycle begins when infected humans release eggs into irrigated fields or ponds, either by deliberate fertilization with excreta or by defecating or urinating directly into the water. The egg hatches in the water and gives off an actively swimming ciliated larva called a **miracidium (figure 22.31a)** which instinctively swims to a snail and burrows into a vulnerable site, shedding its ciliated covering in the process. In the body of the snail, the miracidium multiplies into a larger, fork-tailed swimming larva called a **cercaria** (figure 22.31b). Cercariae are given off by the thousands into the water by infected snails.

Upon contact with a human wading or bathing in water, cercariae attach themselves to the skin by ventral suckers and penetrate into hair follicles. They pass into small blood and

9. *Cryptosporidium* is an intestinal waterborne protozoan that infects a variety of mammals, birds, and reptiles. AIDS patients may experience chronic persistent cryptosporidial diarrhea that can be used as a criterion to help diagnose AIDS.

10. Rotavirus is the primary viral cause of morbidity and mortality resulting from diarrhea, accounting for nearly 50% of all cases. The virus is transmitted by the fecal-oral route, including through contaminated food, water, and fomites.

G. **Acute Diarrhea with Vomiting: Food poisoning** refers to symptoms in the gut that are caused by a preformed toxin.

1. *Staphylococcus aureus* exotoxin: The heat-stable enterotoxin requires 100°C for at least 30 minutes to achieve inactivation. Thus, heating the food after toxin production may not prevent disease. The ingested toxin acts upon the gastrointestinal epithelium and stimulates nerves, with acute symptoms of cramping, nausea, vomiting, and diarrhea.

2. *Bacillus cereus* exotoxin: *Bacillus cereus* is a common resident on vegetables and other products in close contact with soil. It produces two exotoxins, one of which causes a diarrheal-type disease, the other of which causes an **emetic** disease. The emetic form is most frequently linked to fried rice.

3. *Clostridium perfringens* exotoxin: Another sporulating gram-positive bacterium that contaminates animal flesh and vegetables such as beans that have not been cooked thoroughly enough to destroy endospores. The toxin, acting upon epithelial cells, initiates acute abdominal pain, diarrhea, and nausea in 8 to 16 hours.

H. **Chronic Diarrhea**

1. Enteroaggregative *E. coli* (EAEC) is particularly associated with chronic disease, especially in children. Transmission of the bacterium is through contaminated food and water, and is associated with people who are malnourished.

2. *Cyclospora cayetanensis* is an emerging protozoan pathogen that is transmitted via the fecal-oral route, and has been associated with consumption of fresh produce and water.

3. *Giardia lamblia* is a protozoan that can cause diarrhea of long duration, abdominal pain, and flatulence. Freshwater supplies are common vehicles of infection.

4. *Entamoeba histolytica* is a freshwater parasite that causes intestinal amoebiasis, which targets the cecum, appendix, colon, and rectum, leading to dysentery, abdominal pain, fever, diarrhea, and weight loss.

I. **Hepatitis** is an inflammatory disease marked by necrosis of hepatocytes and a mononuclear response that swells and disrupts the liver architecture, causing **jaundice**, a yellow tinge in the skin and eyes. The condition can be caused by a variety of different viruses.

1. **Hepatitis A virus (HAV)** is a nonenveloped, single-stranded RNA enterovirus of low virulence. Most of the pathogenic effects are thought to be the result of host response to the presence of virus in the liver. Hepatitis A virus is spread through the fecal-oral route. An inactivated viral vaccine (Havrix) is currently approved, and an oral vaccine based on an attenuated strain of virus is in development.

2. **Hepatitis B virus (HBV)** is an enveloped DNA virus in the family Hepadnaviridae. Hepatitis B infection can be very serious, even life-threatening; some patients develop chronic liver disease in the form of necrosis or cirrhosis. HBV is also associated with **hepatocellular carcinoma.** Some patients infected with hepatitis B are coinfected with a particle called the delta agent, sometimes also called a hepatitis D virus. HBV is transmitted by blood and other bodily fluids. Thus, this virus is one of the major infectious concerns for health care workers.

3. *Hepatitis C virus:* Hepatitis C is an RNA virus in the Flaviviridae family. It shares many characteristics of hepatitis B disease, but it is much more likely to become chronic. It is more commonly transmitted through blood contact than through transfer of other body fluids.

J. **Helminthic intestinal infections:** intestinal distress as the primary symptom—both tapeworms and roundworms can infect the intestinal tract in such a way as to cause primary symptoms there.

1. *Trichuris trichiura:* Humans are the sole host for this tropical and subtropical parasite. Symptoms of infection may include localized hemorrhage of the bowel, caused by worms burrowing and piercing intestinal mucosa.

2. *Enterobius vermicularis:* This pinworm is the most common worm disease of children in temperate zones. The transmission of this roundworm is by the fecal-oral route. Infection is not fatal and most cases are asymptomatic.

3. *Taenia solium:* This tapeworm is transmitted to humans by the consumption of raw or undercooked pork. Other tapeworms of the genus *Taenia* infect humans. One of them is the beef tapeworm, *Taenia saginata.*

4. *Diphyllobothrium latum:* The intermediate host for this tapeworm is fish, and it can be transmitted in raw food such as sushi and sashimi made from salmon.

K. **Helminthic intestinal infections:** intestinal distress accompanied by migratory symptoms.

1. *Ascaris lumbricoides* is an intestinal roundworm that releases eggs in feces, which are then spread to other humans through fecal-oral routes.

2. *Necator americanus* and *Ancylostoma duodenale:* These two different nematodes are called by the common name "hookworm." Hookworm larvae hatch outside the body in soil contaminated with feces, and infect by penetrating the skin.

3. *Strongyloides stercoralis:* This nematode infection occurs when soil larvae penetrate the skin, similar to hookworm infestations. The most susceptible are AIDS patients, transplant patients on immunosuppressant drugs, and cancer patients receiving radiation therapy.

L. **Liver and intestinal disease:** One group of worms that appear in the intestines has a particular affinity for the liver—**liver flukes.**

1. *Opisthorchis sinensis* and *Clonorchis sinensis* complete their sexual development in mammals such as cats, dogs, and swine. Their intermediate development occurs in snail and fish hosts. Humans are infested by eating inadequately cooked or raw freshwater fish and crustaceans.

2. *Fasciola hepatica:* This liver fluke is a common parasite in sheep, cattle, goats, and other mammals and is occasionally transmitted to humans. Humans develop symptoms only if they are chronically infected by a large number of flukes.

M. **Muscle and Neurological Symptoms**

1. **Trichinosis** is an infection transmitted by eating undercooked pork that has the cysts of *Trichinella* species embedded in the meat. All tissues are at risk for invasion, but final development occurs when the coiled larvae are encysted in the skeletal muscle.

2. **Schistosomiasis** in the intestines is caused by the blood flukes *Schistosoma mansoni* and *S. japonicum* species. Symptoms of infection include fever, chills, diarrhea, hepatomegaly and liver disease, and splenomegaly. Humans are the definitive hosts for the blood fluke, and snails are the intermediate host.

Multiple-Choice Questions

1. Food moves down the GI tract through the action of
   a. cilia
   b. peristalsis
   c. gravity
   d. microorganisms

2. The microorganism(s) most associated with acute necrotizing ulcerative periodontitis (ANUP) is (are):
   a. *Treponema vincentii*
   b. *Prevotella intermedia*
   c. *Fusobacterium*
   d. all of the above

3. Gastric ulcers are caused by
   a. *Treponema vincentii*
   b. *Prevotella intermedia*
   c. *Helicobacter pylori*
   d. all of the above

4. Virus family Paramyxoviridae contains viruses that cause which of the following diseases?
   a. measles
   b. mumps
   c. influenza
   d. both a and b
   e. both b and c

5. Which of these microorganisms is considered the most common cause of diarrhea in the United States?
   a. *E. coli*
   b. *Salmonella*
   c. *Campylobacter*
   d. *Shigella*

6. Which of these microorganisms is associated with Guillain-Barré syndrome?
   a. *E. coli*
   b. *Salmonella*
   c. *Campylobacter*
   d. *Shigella*

7. Besides humans, other natural hosts for the mumps virus include
   a. dogs
   b. monkeys
   c. gophers
   d. all of the above
   e. none of the above

8. Pseudomembranous colitis or antibiotic-associated colitis is caused by
   a. *Vibrio cholerae*
   b. *Clostridium difficile*
   c. *Campylobacter jejuni*
   d. *Shigella*

9. This microorganism can thrive even in salt-preserved foods, causing food poisoning.
   a. *Bacillus cereus*
   b. *Clostridium perfringens*
   c. *Shigella*
   d. *Staphylococcus aureus*

10. This microorganism is commonly associated with fried rice and produces an emetic (vomiting) toxin.
    a. *Bacillus cereus*
    b. *Clostridium perfringens*
    c. *Shigella*
    d. *Staphylococcus aureus*

11. This sporeformer contaminates meats as well as vegetables and is also the causative agent of gas gangrene.
    a. *Bacillus cereus*
    b. *Clostridium perfringens*
    c. *Shigella*
    d. *Staphylococcus aureus*

12. This flagellated protozoan is a water-borne source of chronic diarrhea.
    a. *Cyclospora cayetanensis*
    b. *Giardia lamblia*
    c. *Entamoeba histolytica*
    d. hepatitis A

13. This hepatic disease, transmitted by the fecal-oral route, is caused by
    a. hepatitis A virus
    b. hepatitis B virus
    c. hepatitis C virus
    d. none of the above

14. This hepatitis virus is an enveloped DNA virus.
    a. hepatitis A virus
    b. hepatitis B virus
    c. hepatitis C virus
    d. hepatitis E virus

15. This pinworm is very common among young children in the United States.
    a. *Necator americanus*
    b. *Ascaris lumbricoides*
    c. *Enterobius vermicularis*
    d. *Trichinella spiralis*

Concept Questions

These questions are suggested as a *writing-to-learn* experience. For each question, compose a one- or two-paragraph answer that includes the factual information needed to completely address the question.

1. a. Which microorganism(s) is (are) the major culprit(s) associated with tooth decay?
   b. How do these microorganisms facilitate tooth decay?

2. a. What is the main preventive measure against contracting the mumps?
   b. Besides mumps, what two other diseases are also prevented by this treatment?

3. a. What is the cause of antibiotic-associated colitis?
   b. How is this treated?

4. a. What is food poisoning?
   b. What are some likely microbial culprits associated with food poisoning?
   c. List some nonmicrobial sources of toxins involved in food poisoning.

5. *Entamoeba histolytica* can cause three different forms of amoebiasis. Discuss them.

6. How can hepatitis A infections be prevented?

Causative Agent

The urinary manifestations occur if a host is infected with a particular species of schistosome, *Schistosoma haematobium*. It is found throughout Africa, the Caribbean, and the Middle East. (*S. mansoni* and *S. japonicum* are the species responsible for liver manifestations.) *Schistosomes* are trematodes, or flukes (illustrated in figure 22.31). Humans are the definitive hosts for schistosomes, and snails are the intermediate host.

Pathogenesis and Virulence Factors

Like the other species, *S. haematobium* is able to invade intact skin and attach to vascular endothelium. It engages in the same antigenic cloaking behavior as the other two species. The disease manifestations occur when the eggs in the bladder induce a massive granulomatous response that leads to leakage in the blood vessels and blood in the urine. Significant portions of the bladder eventually can be filled with granulomatous tissue and scar tissue. Function of the bladder is decreased or halted altogether. Chronic infection with *S. haematobium* can also lead to bladder cancer.

Transmission and Epidemiology

The life cycle of the schistosome is described completely in chapter 22. After the worms pass into small blood and lymphatic vessels, they are carried to the liver. Eventually *S. haematobium* enters the venous plexus of the bladder. While attached to these intravascular sites, the worms feed upon blood, and the female lays eggs that are eventually voided in urine.

Culture and Diagnosis

Diagnosis depends on identifying the eggs in urine.

Prevention and Treatment

The cycle of infection cannot be broken as long as people are exposed to untreated sewage in their environment. It is quite common for people to be cured and then to be reinfected because their village has no sewage treatment. A vaccine would provide widespread control of the disease, but so far none is licensed. More than one vaccine is in development, however.

Praziquantel is the drug treatment of choice and is quite effective at eliminating the worms.

✔ CHECKPOINT 23.3	Urinary Schistosomiasis
Causative Organism(s)	*Schistosoma haematobium*
Most Common Modes of Transmission	Vehicle (contaminated water)
Virulence Factors	Antigenic "cloaking," induction of granulomatous response
Culture/Diagnosis	Identification of eggs in urine
Prevention	Avoiding contaminated vehicles
Treatment	Praziquantel

23.4 Reproductive Tract Diseases Caused by Microorganisms

We saw earlier that reproductive tract diseases in men almost always involve the urinary tract as well, and this is sometimes, but not always, the case with women. We should note that although many of the infectious diseases of the reproductive tract are transmitted through sexual contact, not all of them are.

We begin this section with a discussion of infections that are symptomatic primarily in women: *vaginitis* and *vaginosis*. Men may also harbor these infections with or without symptoms. We next consider three broad categories of **sexually transmitted diseases (STDs):** *discharge diseases* in which increased fluid is released in male and female reproductive tracts; *ulcer diseases* in which microbes cause distinct open lesions; and the *wart diseases*. The section concludes with a neonatal disease caused by group B *Streptococcus* colonization.

Vaginitis and Vaginosis
Signs and Symptoms

Vaginitis, an inflammation of the vagina, is a condition characterized by some degree of vaginal itching, depending on the etiological agent. Symptoms may also include burning, and sometimes a discharge, which may take different forms as well. From the name it is obvious that vaginitis only affects women, but most of the agents can also colonize the male reproductive tract.

Causative Agents

The most common cause of vaginitis is *Candida albicans*. The vaginal condition caused by this fungus is known as a *yeast infection*. Most women experience this condition one or multiple times during their lives. Other causes can be bacterial, as in the case of *Gardnerella*, or even protozoal, as in the case of *Trichomonas*. We describe each of these agents here.

Candida albicans   *C. albicans* is a dimorphic fungus that is normal flora in from 50% to 100% of humans, living in low numbers on many mucosal surfaces such as the mouth, gastrointestinal tract, vagina, etc. The vaginal condition it causes is often called *vulvovaginal candidiasis*. The yeast is easily detectable on a wet prep or a Gram stain of material obtained during a pelvic exam **(figure 23.5).** The presence of pseudohyphae in the smear is a clear indication that the yeast is growing rapidly and causing a yeast infection.

Pathogenesis and Virulence Factors   The fungus grows in thick curdlike colonies on the walls of the vagina. The colony debris contributes to a white vaginal discharge. In otherwise healthy people, the fungus is not invasive and limits itself to this surface infection. Please note, however, that *Candida* infections of the bloodstream do occur, and they have high mortality rates. They do not normally stem from vaginal infections with the fungus, however.

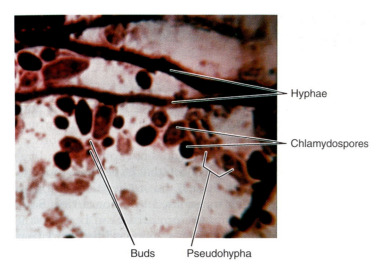

FIGURE 23.5   Gram stain of *Candida albicans* in a vaginal smear.

Transmission and Epidemiology   Vaginal infections with this organism are nearly always opportunistic. Disruptions of the normal bacterial flora, or even minor damage to the mucosal epithelium in the vagina, can lead to overgrowth by this fungus. Disruptions may be mechanical, such as wearing very tight pants, or they may be chemical, as when broad-spectrum antibiotics taken for some other purpose temporarily diminish the vaginal bacterial population. Diabetics and pregnant women are also predisposed to vaginal yeast overgrowths. Some women are prone to this condition during menstruation. The term "infection" is really a misnomer—because this condition is not the result of a new infection, but rather an increased rate of growth of a member of the normal flora.

It is probably possible to transmit this yeast through sexual contact, especially if a woman is experiencing an overgrowth of it. The recipient's immune system may well subdue the yeast so that it acts as normal flora in them. But the yeast may be passed back to the original partner during further sexual contact after treatment. By that time, the circumstances that led to it becoming dominant in the vagina may have returned to normal, and its growth would be limited by the normal bacterial flora. So the sexual route of transmission is difficult to assess. Nevertheless it is recommended that a patient's sexual partner also be treated to short-circuit the possibility of retransmission. The important thing to remember is that *Candida* is an opportunistic fungus.

Women with HIV infection experience frequently recurring yeast infections. Also, a small percentage of women with no underlying immune disease experience chronic or recurrent vaginal infection with *Candida* for reasons that are not clear.

Prevention and Treatment   No vaccine is available for *C. albicans*. Topical and oral azole drugs are used to treat vaginal candidiasis, and some of them are now available over the counter. If infections recur frequently, or fail to resolve, it is important to see a physician for evaluation.

Gardnerella **Species**   The bacterium *Gardnerella* is associated with a particularly common condition in women in their childbearing years. This condition is usually called vaginosis rather than vaginitis because it doesn't appear to induce inflammation in the vagina. It is also known as BV, or bacterial vaginosis. Despite the absence of an inflammatory response, a vaginal discharge is associated with the condition, which is said to have a very fishy odor, especially after sex. Itching is common. But it is also true that many women have this condition with no noticeable symptoms.

Vaginosis is most likely a result of a shift from a predominance of "good bacteria" (lactobacilli) in the vagina to a predominance of "bad bacteria," and one of those is *Gardnerella vaginalis*. This genus of bacteria is aerotolerant and gram-positive, although in a Gram stain it usually appears gram-negative. Probably a mixed infection leads to the condition, however. Anaerobic streptococci and other bacteria, particularly a genus known as *Mobiluncus*, that are normally found in low numbers in a healthy vagina can also often be found in high numbers in this condition. The often-mentioned fishy odor comes from the metabolic by-products of anaerobic metabolism by these bacteria.

Pathogenesis and Virulence Factors   The mechanism of damage in this disease is not well understood. But some of the outcomes are. Besides the symptoms just mentioned, vaginosis can lead to complications such as **pelvic inflammatory disease (PID;** to be discussed later in the chapter), infertility, and more rarely, ectopic pregnancies. Babies born to some mothers with vaginosis have low birth weights.

Transmission and Epidemiology   This mixed infection is not considered to be sexually transmitted, although women who have never had sex rarely develop the condition. It is very common in sexually active women. It may be that the condition is *associated* with sex but not transmitted by it. This situation could occur if the act of penetration or the presence of semen (or saliva) causes changes in the vaginal epithelium, or in the vaginal flora. We do not know exactly what causes the increased numbers of *Gardnerella* and other normally rare flora. The low pH typical of the vagina is usually higher in vaginosis. It is not clear whether this causes, or is caused by, the change in bacterial flora.

Culture and Diagnosis   The condition can be diagnosed by a variety of methods. Sometimes a simple stain of vaginal secretions is used to examine sloughed vaginal epithelial cells. In vaginosis some cells will appear to be nearly covered with adherent bacteria. In normal times, vaginal epithelial cells are sparsely covered with bacteria. These cells are called clue cells and are a helpful diagnostic indicator **(figure 23.6).** They can also be found on Pap smears.

Prevention and Treatment   No known prevention exists. Asymptomatic cases are generally not treated. Women who find the condition uncomfortable, or who are planning on becoming pregnant, should be treated. Women who use intrauterine devices (IUDs) for contraception should also be treated because IUDs can provide a passageway for the

Infectious Diseases Affecting the Genitourinary System

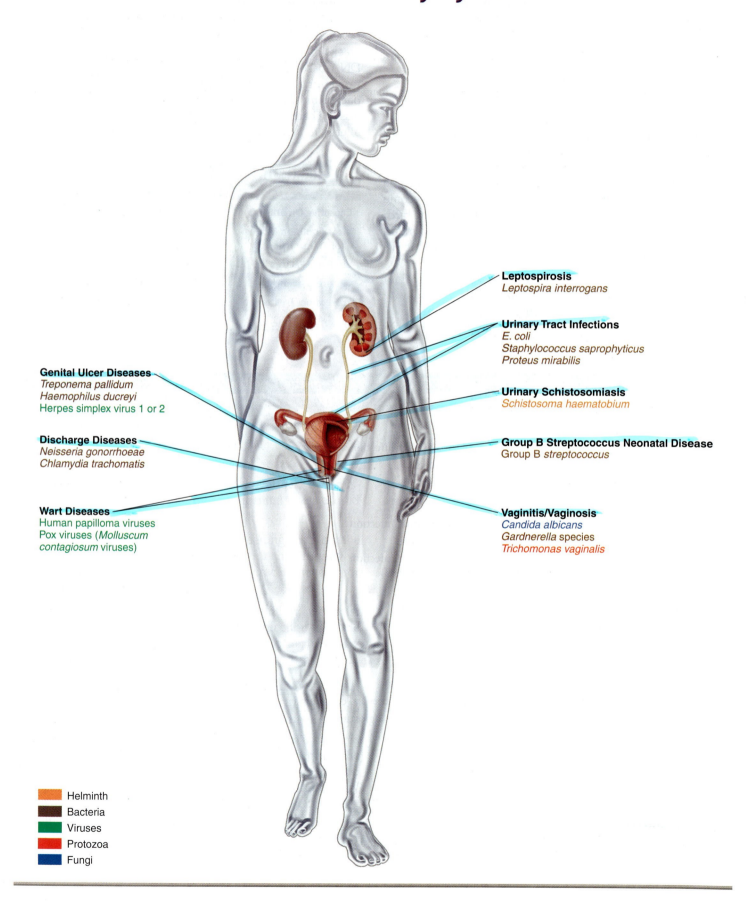

Leptospirosis
Leptospira interrogans

Urinary Tract Infections
E. coli
Staphylococcus saprophyticus
Proteus mirabilis

Urinary Schistosomiasis
Schistosoma haematobium

Group B Streptococcus Neonatal Disease
Group B *streptococcus*

Vaginitis/Vaginosis
Candida albicans
Gardnerella species
Trichomonas vaginalis

Genital Ulcer Diseases
Treponema pallidum
Haemophilus ducreyi
Herpes simplex virus 1 or 2

Discharge Diseases
Neisseria gonorrhoeae
Chlamydia trachomatis

Wart Diseases
Human papilloma viruses
Pox viruses (*Molluscum contagiosum* viruses)

- Helminth
- Bacteria
- Viruses
- Protozoa
- Fungi

Chapter Summary With Key Terms

23.1 The Genitourinary Tract and Its Defenses
This "system" is really two systems, the reproductive system and the urinary system.
A. The reproductive tract in males and females is composed of structures and substances which allow for sexual intercourse and the creation of a new fetus; it is protected by normal mucosal defenses as well as specialized features (such as the low pH of the adult female reproductive tract).
B. The urinary system allows the excretion of fluid and wastes from the body. It has mechanical as well as chemical defense mechanisms.

23.2 Normal Flora of the Genitourinary Tract
Both the genital and the urinary systems have normal flora only in their most distal regions. Normal flora in the male reproductive and urinary systems are found in the distal part of the urethra and resemble skin flora. The same is generally true for the female urinary system. The normal flora in the female reproductive tract changes over the course of a woman's lifetime.

23.3 Urinary Tract Diseases Caused by Microorganisms
A. **Urinary Tract Infections (UTIs):** Infection can occur at a number of sites; the bladder (*cystitis*), the kidneys (*pyelonephritis*), and the urethra (*urethritis*). In most cystitis and pyelonephritis cases, the cause is bacteria that are normal flora in the gastrointestinal tract—most commonly, *Escherichia coli, Staphylococcus saprophyticus,* and *Proteus mirabilis.* Community-acquired UTIs are most often transmitted from the GI tract to the urinary system. UTIs are the most common of nosocomial infections. Patients of both sexes who have urinary catheters are susceptible to infections with a variety of microorganisms.
B. **Leptospirosis** is a zoonosis associated with wild animals that can affect the kidneys, liver, brain, and eyes. The causative agent is the *Leptospira interrogans* spirochete.
C. **Urinary Schistosomiasis:** This form of schistosomiasis is caused by *S. haematobium.* The bladder is damaged by trematode eggs and the granulomatous response they induce.

23.4 Reproductive Tract Diseases Caused by Microorganisms
A. **Vaginitis and Vaginosis**
1. Vaginitis is an inflammation of the vagina, most commonly caused by *Candida albicans.* This vulvovaginal candidiasis is nearly always an opportunistic infection. Women with HIV infection experience frequently recurring yeast infections. Topical and oral azole drugs are used to treat vaginal candidiasis.
2. The bacterium *Gardnerella* is associated with vaginosis that has a discharge but no inflammation in the vagina. Vaginosis could lead to complications such as **pelvic inflammatory disease (PID).**
3. *Trichomonas vaginalis* causes mostly asymptomatic infections in females and males. *Trichomonas* is a flagellated protozoan and is easily transmitted through sexual contact.

B. **Discharge Diseases with Major Manifestation in Genitourinary Tract:** Discharge diseases are those in which the infectious agent causes an increase in fluid discharge in the male and female reproductive tracts.
1. **Gonorrhea** elicits *urethritis* in males, but many cases are asymptomatic. In females, both the urinary and genital tracts will be infected during sexual intercourse. Major complications occur when the infection reaches uterus and fallopian tubes. One disease resulting from this progression is **salpingitis,** which can lead to pelvic inflammatory disease (PID). The causative agent, *Neisseria gonorrhoeae,* is a gram-negative diploccoccus that is most infectious when transferred to a suitable mucous membrane. Hence, sexual activity is the most common transmission method.
2. **Chlamydia:** Genital chlamydial infection is the most common reportable infectious disease in the United States. In males, the bacterium causes an inflammation of the urethra (NGU). Females have cervicitis, a discharge, salpingitis, and frequently PID as a result of a *Chlamydia* infection.
   Certain strains of *Chlamydia trachomatis* can invade the lymphatic tissues, resulting in another condition called lymphogranuloma venereum. *C. trachomatis* is a very small bacterium, gram-negative, and is an obligate intracellular parasite. All *Chlamydia* species alternate between two distinct stages: an elementary body and a reticulate body. Both forms are necessary for a complete life cycle to occur.

C. **Genital Ulcer Diseases**
1. **Syphilis** is caused by *Treponema pallidum,* a spirochete; it is a thin, regularly coiled cell with a gram-negative cell wall. There are three distinct clinical stages, designated as *primary, secondary,* and *tertiary syphilis,* with a latent period of quiescence. The spirochete appears in the lesions and blood during the primary and secondary stages and thus is transmissible at these times. During the early latency period between secondary and tertiary syphilis, it is also transmissible. Syphilis is largely nontransmissible during the "late latent" and tertiary stages.
   a. *Primary Syphilis:* Early syphilis infection exhibits a hard **chancre** at the site of entry. These painless ulcers contain live spirochetes. The chancre then heals without scarring, but by this time the spirochete has escaped into the circulation.
   b. *Secondary Syphilis:* About 3 weeks to 6 months after the chancre heals, the secondary stage appears. Initial symptoms are fever, headache, and sore throat, followed by lymphadenopathy and a peculiar red or brown rash that breaks out on all skin surfaces, including the palms of the hands and the soles of the feet. The lesions contain viable spirochetes and disappear spontaneously in a few weeks.
   c. *Latency and Tertiary Syphilis:* After resolution of secondary syphilis, some infections enter a latent

period that can last for 20 years or longer. The final stage of the disease, tertiary syphilis, is very damaging. Cardiovascular syphilis results from damage to the small arteries in the aortic wall. Painful, swollen syphilitic tumors called **gummas** can develop in the liver, skin, bone, and cartilage. Neurosyphilis involves the blood vessels in the brain, cranial nerves, and dorsal roots of the spinal cord.

    d. *Congenital Syphilis:* The syphilis bacterium can pass from a pregnant woman's circulation into the placenta and can be carried throughout the fetal tissues, leading to **congenital syphilis.** The pathogen inhibits fetal growth and disrupts critical periods of development, which can lead to spontaneous miscarriage or stillbirth.

2. **Chancroid:** This ulcerative disease caused by a **pleomorphic** gram-negative rod called *Haemophilus ducreyi*. Chancroid is transmitted exclusively through direct—mainly sexual—contact.

3. **Genital herpes** is caused by **herpes simplex viruses (HSVs).** Two types of HSV have been identified, HSV-1 and HSV-2. After infection, there may be no symptoms, or there may be fluid-filled, painful vesicles on the genitalia, perineum, thigh, and buttocks. In severe cases, meningitis or encephalitis can develop. Patients that recover remain asymptomatic or experience recurrent "surface" infections indefinitely. HSV infections in the neonate and the fetus can be fatal.

    HSV-1 and HSV-2 are DNA viruses with icosahedral capsids and envelopes containing glycoprotein spikes. Herpesviruses can become latent, most likely by the incorporation of viral nucleic acid into the host genome. The virus becomes latent in the nerve cells of the body. People with active lesions are the most contagious.

D. **Wart Diseases**

1. *Human Papillomaviruses* are the causative agents of genital warts. Certain types of the virus infect cells on the female cervix that can eventually result in malignancies of the cervix. Males can also get cancer from infection with these viruses.

    The human papillomaviruses are a group of nonenveloped DNA viruses belonging to the Papoviridae family. Two types, HPV-16 and HPV-18, appear to be very closely associated with development of cervical cancer. The major virulence factors for cancer-causing HPVs are **oncogenes,** which code for proteins that interfere with normal host cell function, resulting in uncontrolled growth. The mode of transmission is direct contact.

    Infection with any HPV is incurable. Genital warts can be removed, but the virus will remain. Treatment of cancerous cell changes is an important part of HPV therapy, and it can only be instituted if the changes are detected through Pap smears. A vaccine designed to protect against HPV-16 is being tested in clinical trials.

2. A pox family virus causes a condition called **molluscum contagiosum.** This disease can take the form of skin lesions from wartlike growths in the membranes of the genitalia, and it can also be transmitted sexually.

E. **Group B Streptococcus "Colonization"—Neonatal Disease:** Asymptomatic colonization of women by a β-hemolytic *Streptococcus* in Lancefield group B is very common. But when these women become pregnant and give birth, about half of their infants become colonized by the bacterium during passage through the birth canal, or by ascension of the bacteria through ruptured membranes; some infected infants experience life-threatening bloodstream infections, meningitis, or pneumonia. In 2002, the CDC recommended that all pregnant women be screened for group B *Streptococcus* colonization and treated with antibiotics prior to childbirth.

Multiple-Choice Questions

1. Cystitis is an infection of the
   a. bladder   c. kidney
   b. urethra   d. vagina

2. Nongonococcal urethritis (NGU) is caused by
   a. *Neisseria gonorrhoeae*
   b. *Chlamydia trachomatis*
   c. *Treponema pallidum*
   d. *Trichomonas vaginalis*

3. Leptospirosis is transmitted to humans by
   a. person to person
   b. fomites
   c. mosquitoes
   d. contaminated soil or water

4. Syphilis is caused by
   a. *Treponema pallidum*
   b. *Neisseria gonorrhoeae*
   c. *Trichomonas vaginalis*
   d. *Haemophilus ducreyi*

5. What diagnostic method is used to detect the syphilis microorganism?
   a. dark-field microscopy
   b. immunofluorescence
   c. Wasserman test
   d. FTA-ABS
   e. all of the above

6. Bacterial vaginosis is commonly associated with the following organism:
   a. *Candida albicans*
   b. *Gardnerella*
   c. *Trichomonas*
   d. all of the above
   e. none of the above

7. Genital herpes can be treated with
   a. acyclovir   c. herpes vaccine
   b. penicillin   d. both a and b

8. This dimorphic fungus is a common cause of vaginitis.
   a. *Candida albicans*
   b. *Gardnerella*
   c. *Trichomonas*
   d. all of the above

9. Chancroid is caused by
   a. *Treponema pallidum*
   b. *Neisseria gonorrhoeae*
   c. *Trichomonas vaginalis*
   d. *Haemophilus ducreyi*

10. There are estimates that approximately ____% of adult Americans have genital herpes.
    a. 2   c. 20
    b. 10   d. 50

11. The majority of cervical cancers are caused by
    a. genital herpes infections
    b. chancroids
    c. syphilis
    d. human papillomavirus

12. Warts are caused by
    a. human papillomavirus
    b. molluscum contagiosum
    c. toads
    d. both a and b
    e. both a and c

13. Genital herpes transmission can be reduced or prevented by all of the following except
    a. condom
    b. abstinence
    c. contraceptive pill
    d. female condom

14. This protozoan can be treated with the drug Flagyl.
    a. *Neisseria gonorrhoeae*
    b. *Chlamydia trachomatis*
    c. *Treponema pallidum*
    d. *Trichomonas vaginalis*

Concept Questions

These questions are suggested as a writing-to-learn experience. For each question, compose a one- or two-paragraph answer that includes the factual information needed to completely address the question.

1. Besides *E. coli,* name two other microorganisms associated with cystitis and pyelonephritis.

2. Describe the symptoms of Weil's syndrome.

3. Describe the common treatments for gonorrhea.

4. a. What is PID?
   b. What are the two most common microorganisms associated with this disease?
   c. Describe the long-term consequences of untreated PID.

5. Describe the life cycle of *Chlamydia.*

6. What are some of the stimuli that can trigger reactivation of a latent herpesvirus infection? Speculate on why.

7. a. Human papillomavirus is associated with what condition?
   b. Name some of the different sites on the body that can be affected by this virus.

8. What are the clinical stages of syphilis?

9. a. What is the standard screening for cervical cancer?
   b. In this screening technique, cervical cells are screened for abnormalities. What are some of the terms used to describe these abnormalities?

Critical Thinking Questions

Critical thinking is the ability to reason and solve problems using facts and concepts. These questions can be approached from a number of angles, and in most cases, they do not have a single correct answer.

1. What characteristics of *Neisseria gonorrhoeae* allow its effective sexual transmission?

2. What is the concern regarding sexually transmitted diseases such as genital herpes and syphilis and an increased risk of HIV infections?

3. What is a reportable disease?

4. Why is *Chlamydia* considered an "atypical" bacterium?

5. What has "malaria therapy" had to do with curing syphilis?

6. It has been stated that the actual number of people in the United States who have genital herpes may be a lot higher than official statistics depict. What are some possible reasons for this discrepancy?

7. Why is herpes of the newborn of particular concern, and what can be done to prevent this type of transmission?

8. Why are urinary tract infections such common nosocomial infections?

Food Web

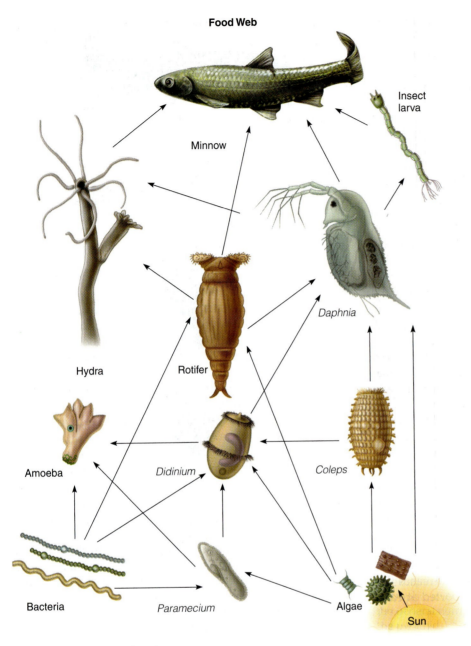

FIGURE 24.5 **Food web.**
More complex trophic patterns are accurately depicted by a food web, which traces the multiple feeding options that exist for most organisms. Note: Arrows point toward the consumers. Compare this pattern of feeding with the chain in figure 24.4 (organisms not to scale).

The most basic image of a feeding pathway can be provided by a food chain. Although it is a somewhat simplistic way to describe feeding relationships, a food chain helps identify the types of organisms that are present at a given trophic level in a natural setting (figure 24.4). Feeding relationships in communities are more accurately represented by a multichannel food chain, or a **food web** (figure 24.5). A food web reflects the actual nutritional structure of a community. It can help to identify feeding patterns typical of herbivores (plant eaters),

carnivores (flesh eaters), and omnivores (feed on both plants and flesh).

Ecological Interactions Between Organisms in a Community

Whenever complex mixtures of organisms associate, they develop various dynamic interrelationships based on nutrition and shared habitat. These relationships, some of which were described in earlier chapters, include mutualism, commensalism, parasitism, competition, synergism (cross-feeding), predation, and scavenging.

Mutually beneficial associations (mutualism), such as that of protozoans living in the termite intestine, are so well evolved that the two members require each other for survival. In contrast, **commensalism** is one-sided and independent. (These terms describing relationships between organisms echo the terms we described in chapter 13.) Although the action of one microbe favorably affects another, the first microbe receives no benefit. Many commensal unions involve *co-metabolism*, meaning that the waste products of the first microbe are useful nutrients for the second one. In **synergism**, two organisms that are usually independent cooperate to break down a nutrient neither one could have metabolized alone. *Parasitism* is an intimate relationship whereby a parasite derives its nutrients and habitat from a host that is usually harmed in the process. In *competition*, one microbe gives off antagonistic substances that inhibit or kill susceptible species sharing its habitat. A *predator* is a form of consumer that actively seeks out and ingests live prey (protozoa that prey on algae and bacteria). *Scavengers* are nutritional jacks-of-all-trades; they feed on a variety of food sources, ranging from live cells to dead cells and wastes.

- A living community is composed of populations that show a pattern of energy and nutritional relationships called a food web. Microorganisms are essential producers and decomposers in any ecosystem.
- The relationships between populations in a community are described according to the degree of benefit or harm they pose to one another. These relationships include mutualism, commensalism, predation, parasitism, synergism, scavenging, and competition.

The Natural Recycling of Bioelements

Environmental ecosystems are exposed to the sun, which constantly infuses them with a renewable source of energy. In contrast, the bioelements and nutrients that are essential components of protoplasm are supplied exclusively by sources somewhere in the biosphere and are not being continually replenished from outside the earth. In fact, the lack of a required nutrient in the immediate habitat is one of the chief factors limiting organismic and population growth. Because of the finite source of life's building blocks, the long-term sustenance of the biosphere requires continuous **recycling** of elements and nutrients. Essential elements such as carbon, nitrogen, sulfur, phosphorus, oxygen, and iron are cycled through biological, geologic, and chemical mechanisms called **biogeochemical cycles**. Although these cycles vary in certain specific characteristics, they share several general qualities, as summarized in the following list:

- All elements ultimately originate from a nonliving, long-term reservoir in the atmosphere, sedimentary rocks, or water. They cycle in pure form (N_2) or as compounds (PO_4).
- Elements make the rounds between the abiotic environment and the biotic environment.
- Recycling maintains a necessary balance of nutrients in the biosphere so that they do not build up or become unavailable.
- Cycles are complex systems that rely upon the interplay of producers, consumers, and decomposers. Often the waste products of one organism become a source of energy or building material for another.
- All organisms participate directly in recycling, but only certain categories of microorganisms have the metabolic pathways for converting inorganic compounds from one nutritional form to another.

The English biologist James Lovelock has postulated a concept called the **Gaia** (guy'-uh) **Theory**, after the mythical Greek goddess of earth. This hypothesis proposes that the biosphere contains a diversity of habitats and niches favorable to life because living things have made it that way. Not only does the earth shape the character of living things, but living things shape the character of the earth. After all, we know that the chemical compositions of the aquatic environment, the atmosphere, and even the soil would not exist as

they do without the actions of living things. Organisms are also very active in evaporation and precipitation cycles, formation of mineral deposits, and rock weathering.

For billions of years, microbes have played prominent roles in the formation and maintenance of the earth's crust, the development of rocks and minerals, and the formation of fossil fuels. This revolution in understanding the biological involvement in geologic processes has given rise to a new field called *geomicrobiology*.

In the next several sections we examine how, jointly and over a period of time, the varied microbial activities affect and are themselves affected by the abiotic environment.

Atmospheric Cycles

The Carbon Cycle

Because carbon is the fundamental atom in all biomolecules and accounts for at least one-half of the dry weight of protoplasm, the **carbon cycle** is more intimately associated with the energy transfers and trophic patterns in the biosphere than are other elements. Besides the enormous organic reservoir in the bodies of organisms, carbon also exists in the gaseous state as carbon dioxide (CO_2) and methane (CH_4) and in the mineral state as carbonate (CO_3). In general, carbon is recycled through ecosystems via photosynthesis (carbon fixation), respiration, and fermentation of organic molecules, limestone decomposition, and methane production. A convenient starting point from which to trace the movement of carbon is with carbon dioxide, which occupies a central position in the cycle and represents a large common pool that diffuses into all parts of the ecosystem **(figure 24.6)**. As a general rule, the cycles of oxygen and hydrogen are closely allied to the carbon cycle.

The principal users of the atmospheric carbon dioxide pool are photosynthetic autotrophs (phototrophs) such as plants, algae, and cyanobacteria. An estimated 165 billion tons of organic material per year are produced by terrestrial and aquatic photosynthesis. A smaller amount of CO_2 is used by chemosynthetic autotrophs such as methane-producing bacteria. A review of the general equation for photosynthesis in **figure 24.7** reveals that phototrophs use energy from the sun to fix CO_2 into organic compounds such as glucose that can be used in synthesis and respiration. Photosynthesis is also the primary means by which the atmospheric supply of O_2 is regenerated.

Just as photosynthesis removes CO_2 from the atmosphere, other modes of generating energy, such as respiration and fermentation, return it. As you may recall from the discussion of aerobic respiration in chapter 8, in the presence of O_2, organic compounds such as glucose are degraded completely to CO_2 and H_2O, with the release of energy. Carbon dioxide is also released by anaerobic respiration, and by certain types of fermentation reactions.

A small but important phase of the carbon cycle involves certain limestone deposits composed primarily of calcium carbonate ($CaCO_3$). Limestone is produced when marine

H. **Aquatic Microbiology**
1. The surface water, atmospheric moisture, and groundwater are linked through a **hydrologic cycle** that involves evaporation and precipitation. Living things contribute to the cycle through respiration and transpiration.
2. The diversity and distribution of water communities are related to sunlight, temperature, aeration, and dissolved nutrients. Phytoplankton and zooplankton drifting in the uppermost zone constitute a microbial community that supports the aquatic ecosystem.
3. Water Management
   a. Providing **potable water** is central to prevention of water-borne disease.
   b. Water is constantly surveyed for certain **indicator bacteria** (coliforms and enterococci) that signal fecal contamination.
   c. Assays for possible water contamination include the **standard plate count** and **membrane filter tests** to enumerate coliforms.
4. Water and sewage treatment:
   a. Drinking water is rendered safe by a purification process that involves storage, sedimentation, settling, aeration, filtration, and disinfection.
   b. Sewage or used wastewater can be processed to remove solid matter, dangerous chemicals, and microorganisms.
   c. Microbes biodegrade the waste material or sludge. Solid wastes are further processed in **anaerobic digesters.**

24.2 Applied Microbiology and Biotechnology
Biotechnology is the practical application of microbiology in the manufacture of food, industrial chemicals, drugs, and other products. Many of these processes use mass, controlled microbial **fermentations** and bioengineered microorganisms.

24.3 Microorganisms and Food
A. Microbes and humans compete for the rich nutrients in food. Many microbes are present on food as harmless contaminants; some are used to create flavors and nutrients; others may produce unfavorable reactions.
B. Fermentations in foods: Microbes can impart desirable aroma, flavor, or texture to foods. Bread, alcoholic beverages, some vegetables, and some dairy products are infused with pure microbial strains to yield the necessary fermentation products.

24.4 General Concepts in Industrial Microbiology
A. Industrial microbiology involves the large-scale commercial production of organic compounds such as antibiotics, vitamins, amino acids, enzymes, and hormones using specific microbes in carefully controlled fermentation settings.
B. Microbes are chosen for their production of a desired **metabolite;** several different species can be used to biotransform raw materials in a stepwise series of metabolic reactions.
C. Fermentations are conducted in massive culture devices called **fermentors** that have special mechanisms for adding nutrients, stirring, oxygenating, altering pH, cooling, monitoring, and harvesting product.

Multiple-Choice Questions

1. Which of the following is *not* a major subdivision of the biosphere?
   a. hydrosphere
   b. lithosphere
   c. stratosphere
   d. atmosphere

2. A/an ____ is defined as a collection of populations sharing a given habitat.
   a. biosphere
   b. community
   c. biome
   d. ecosystem

3. The quantity of available nutrients ____ from the lower levels of the energy pyramid to the higher ones.
   a. increases
   b. decreases
   c. remains stable
   d. cycles

4. Photosynthetic organisms convert the energy of ____ into chemical energy.
   a. electrons
   b. protons
   c. photons
   d. hydrogen atoms

5. Which of the following is considered a greenhouse gas?
   a. CO_2
   b. CH_4
   c. N_2O
   d. all of these

6. Root nodules contain ____, which can ____.
   a. *Azotobacter*, fix N_2
   b. *Nitrosomonas*, nitrify NH_3
   c. rhizobia, fix N_2
   d. *Bacillus*, denitrify NO_3

7. Which element(s) has/have an inorganic reservoir that exists primarily in sedimentary deposits?
   a. nitrogen
   b. phosphorus
   c. sulfur
   d. both b and c

8. Which of the following bacteria would be the most accurate indicator of fecal contamination?
   a. *Enterobacter*
   b. *Thiobacillus*
   c. *Escherichia*
   d. *Staphylococcus*

9. Milk is usually pasteurized by
   a. the high-temperature short-time method
   b. ultrapasteurization
   c. batch method
   d. electrical currents

10. The dried, presprouted grain that is soaked to activate enzymes for beer is
    a. hops
    b. malt
    c. wort
    d. mash

11. Substances given off by yeasts during fermentation are
    a. alcohol
    b. carbon dioxide
    c. organic acids
    d. all of these

12. Which of the following is added to facilitate milk curdling during cheese making?
    a. lactic acid
    b. salt
    c. *Lactobacillus*
    d. rennin

13. Secondary metabolites of microbes are formed during the _____ phase of growth.
    a. exponential
    c. death
    b. stationary
    d. lag

14. The large tanks used in industrial production of antibiotics are termed
    a. digesters
    c. spargers
    b. fermentors
    d. biotransformers

Concept Questions

These questions are suggested as a writing-to-learn experience. For each question, compose a one- or two-paragraph answer that includes the factual information needed to completely address the question.

1. a. Present in outline form the levels of organization in the biosphere. Define the term *biome*.
   b. Compare autotrophs and heterotrophs; producers and consumers.
   c. Where in the energy and trophic schemes do decomposers enter?
   d. Compare the concepts of habitat and niche using *Chlamydomonas* (figure 24.2) as an example.

2. a. Using figures 24.4 and 24.5, point out specific examples of producers; primary, secondary, and tertiary consumers; herbivores; primary, secondary, and tertiary carnivores; and omnivores.
   b. What is mineralization, and which organisms are responsible for it?

3. a. Outline the general characteristics of a biogeochemical cycle.
   b. What are the major sources of carbon, nitrogen, phosphorus, and sulfur?

4. a. In what major forms is carbon found? Name three ways carbon is returned to the atmosphere.
   b. Name a way it is fixed into organic compounds.
   c. What form is the least available for the majority of living things?

5. a. Describe nitrogen fixation, ammonification, nitrification, and denitrification.
   b. What form of nitrogen is required by plants? By animals?

6. a. Outline the phosphorus and sulfur cycles.
   b. What are the most important inorganic and organic phosphorus compounds? How is phosphorus made available to plants?
   c. What are the most important inorganic and organic sulfur compounds?
   d. What are the roles of microorganisms in these cycles?

7. a. Describe the structure of the soil and the rhizosphere.
   b. What is humus?
   c. Compare and contrast root nodules with mycorrhizae.

8. a. Outline the modes of cycling water through the lithosphere, hydrosphere, and atmosphere.
   b. What are the roles of precipitation, condensation, respiration, transpiration, surface water, and aquifers?

9. a. What causes the formation of the epilimnion, hypolimnion, and thermocline?
   b. What is upwelling?
   c. In what ways are red tides and eutrophic algal blooms similar and different?

10. a. Why must water be subjected to microbiological analysis?
    b. What are the characteristics of good indicator organisms, and why are they monitored rather than pathogens?
    c. Give specific examples of indicator organisms and water-borne pathogens.
    d. Describe two methods of water analysis.
    e. What are the principles behind the most probable number test?
    f. Describe the three phases of sewage treatment.
    g. What is activated sludge?

11. a. Explain the meaning of fermentation from the standpoint of industrial microbiology.
    b. Describe five types of fermentations.

12. a. Which microbes are used as starter cultures in bread, beer, wine, cheeses, and sauerkraut?
    b. Outline the steps in beer making.
    c. List the steps in wine making.
    d. What are curds and whey, and what causes them?

13. a. Describe the aims of industrial microbiology.
    b. Differentiate between primary and secondary metabolites.
    c. Describe a fermentor.
    d. How is it scaled up for industrial use?
    e. What are specific examples of products produced by these processes?

Critical Thinking Questions

Critical thinking is the ability to reason and solve problems using facts and concepts. These questions can be approached from a number of angles, and in most cases, they do not have a single correct answer.

1. a. What factors cause energy to decrease with each trophic level?
   b. How is it possible for energy to be lost and the ecosystem to still run efficiently?

   c. Are the nutrients on the earth a renewable resource? Why, or why not?

2. Give specific examples from biogeochemical cycles that support the Gaia Theory.

3. Biologists can set up an ecosystem in a small, sealed aquarium that continues to function without maintenance for years. Describe the minimum biotic and abiotic components it must contain to remain balanced and stable.

4. a. Is the greenhouse effect harmful under ordinary circumstances?
   b. What occurrence has made it dangerous to the global ecosystem?
   c. What could each person do on a daily basis to cut down on the potential for disrupting the delicate balance of the earth?

5. a. If we are to rely on microorganisms to biodegrade wastes in landfills, aquatic habitats, and soil, list some ways that this process could be made more efficient.
   b. Since elemental poisons (heavy metals) cannot be further degraded even by microbes, what is a possible fate of these metals?
   c. Provide some possible solutions for this form of pollution.

6. Why are organisms in the abyssal zone of the ocean necessarily halophilic, psychrophilic, barophilic, and anaerobic?

7. a. What eventually happens to the nutrients that run off into the ocean with sewage and other effluents?
   b. Why can high mountain communities usually dispense with water treatment?

8. Every year supposedly safe municipal water supplies cause outbreaks of enteric illness.
   a. How in the course of water analysis and treatment might these pathogens be missed?
   b. What kinds of microbes are they most likely to be?
   c. Why is there less tolerance for a fecal coliform in drinking or recreational water than other bacteria?

9. Describe four food-preparation and food-maintenance practices in your own kitchen that could expose people to food poisoning and explain how to prevent them.

10. a. What is the purpose of boiling the wort in beer preparation?
    b. What are hops used for?
    c. If fermentation of sugars to produce alcohol in wine is anaerobic, why do wine makers make sure that the early phase of yeast growth is aerobic?

11. Predict the differences in the outcome of raw milk that has been incubated for 48 hours versus pasteurized milk that has been incubated for the same length of time.

12. Explain the ways that co-metabolism and biotransformations of microorganisms are harnessed in industrial microbiology.

13. Review chapter 10 and describe several ways that recombinant DNA technology can be used in biotechnology processes.

Internet Search Topics

1. Go to websites that contain information on microbial food-borne illness. List the 15 most common pathogens in order of prevalence, and determine if they are food infections or intoxications.

2. Look up information on techniques for testing water. Explain how several of the tests work and their uses.

3. Find information on red tide outbreaks and illness in humans.

4. Research the subject of bioremediation. What sorts of toxic substances are being cleaned up and what types of microbes are involved?

5. Go to websites to research wine making. How are different types of wines made and how do they vary in color, flavor, alcohol content, and other features?

Exponents

Dealing with concepts such as microbial growth often requires working with numbers in the billions, trillions, and even greater. A mathematical shorthand for expressing such numbers is with exponents. The exponent of a number indicates how many times (designated by a superscript) that number is multiplied by itself. These exponents are also called common *logarithms,* or logs. The following chart, based on multiples of 10, summarizes this system.

Exponential Notation for Base 10

Number	Quantity	Exponential Notation*	Number Arrived at By:	One Followed By:
1	One	10^0	Numbers raised to zero power are equal to one	No zeros
10	Ten	10^{1}**	10×1	One zero
100	Hundred	10^2	10×10	Two zeros
1,000	Thousand	10^3	$10 \times 10 \times 10$	Three zeros
10,000	Ten thousand	10^4	$10 \times 10 \times 10 \times 10$	Four zeros
100,000	Hundred thousand	10^5	$10 \times 10 \times 10 \times 10 \times 10$	Five zeros
1,000,000	Million	10^6	10 times itself 6 times	Six zeros
1,000,000,000	Billion	10^9	10 times itself 9 times	Nine zeros
1,000,000,000,000	Trillion	10^{12}	10 times itself 12 times	Twelve zeros
1,000,000,000,000,000	Quadrillion	10^{15}	10 times itself 15 times	Fifteen zeros
1,000,000,000,000,000,000	Quintillion	10^{18}	10 times itself 18 times	Eighteen zeros

Other large numbers are sextillion (10^{21}), septillion (10^{24}), and octillion (10^{27}).

*The proper way to say the numbers in this column is 10 raised to the *nth* power, where *n* is the exponent. The numbers in this column can also be represented as 1×10^n, but for brevity, the $1 \times$ can be omitted.

**The exponent 1 is usually omitted.

Converting Numbers to Exponent Form

As the chart shows, using exponents to express numbers can be very economical. When simple multiples of 10 are used, the exponent is always equal to the number of zeros that follow the 1, but this rule will not work with numbers that are more varied. Other large whole numbers can be converted to exponent form by the following operation: First, move the decimal (which we assume to be at the end of the number) to the left until it sits just behind the first number in the series (example: 3568. = 3.568). Then count the number of spaces (digits) the decimal has moved; that number will be the exponent. (The decimal has moved from 8. to 3., or 3 spaces.) In final notation, the converted number is multiplied by 10 with its appropriate exponent: 3568 is now 3.568×10^3.

Rounding Off Numbers

The notation in the previous example has not actually been shortened, but it can be reduced further by rounding off the decimal fraction to the nearest thousandth (three digits), hundredth (two digits), or tenth (one digit). To round off a number, drop its last digit and either increase the one next to it or leave it as it is. If the number dropped is 5, 6, 7, 8, or 9, the subsequent digit is increased by one (rounded up); if it is 0, 1, 2, 3, or 4, the subsequent digit remains as is. Using the example of 3.528, removing the 8 rounds off the 2 to a 3 and produces 3.53 (two digits). If further rounding is desired, the same rule of thumb applies, and the number becomes 3.5 (one digit). Other examples of exponential conversions are shown on the next page.

Number	Is the Same As	Rounded Off, Placed in Exponent Form
16,825.	$1.6825 \times 10 \times 10 \times 10 \times 10$	1.7×10^4
957,654.	$9.57654 \times 10 \times 10 \times 10 \times 10 \times 10$	9.58×10^5
2,855,000.	$2.855000 \times 10 \times 10 \times 10 \times 10 \times 10 \times 10$	2.86×10^6

Negative Exponents

The numbers we have been using so far are greater than 1 and are represented by positive exponents. But the correct notation for numbers less than 1 involves negative exponents (10 raised to a negative power, or 10^{-n}). A negative exponent says that the number has been divided by a certain power of 10 (10, 100, 1,000). This usage is handy when working with concepts such as pH that are based on very small numbers otherwise needing to be represented by large decimal fractions—for example, 0.003528. Converting this and other such numbers to exponential notation is basically similar to converting positive numbers, except that you work from left to right and the exponent is negative. Using the example of 0.003528, first convert the number to a whole integer followed by a decimal fraction and keep track of the number of spaces the decimal point moves (example: 0.003528 = 3.528). The decimal has moved three spaces from its original position, so the finished product is 3.528×10^{-3}. Other examples are:

Number	Is the Same As	Rounded Off, Express with Exponents
0.0005923	$\dfrac{5.923}{10 \times 10 \times 10 \times 10}$	5.92×10^{-4}
0.00007295	$\dfrac{7.295}{10 \times 10 \times 10 \times 10 \times 10}$	7.3×10^{-5}

Significant Events in Microbiology

Date	Discovery/People Involved	Date	Discovery/People Involved
1546	Italian physician Girolamo Fracastoro suggests that invisible organisms may be involved in disease.	1869	Johann Miescher, a Swiss pathologist, discovers in the cell nucleus the presence of complex acids, which he terms nuclein (DNA, RNA).
1590	Zaccharias Janssen, a Dutch spectacle maker, invents the first compound microscope.	1876–1877	German bacteriologist Robert Koch* studies anthrax in cattle and implicates the bacterium *Bacillus anthracis* as its causative agent.
1660	Englishman Robert Hooke explores various living and nonliving matter with a compound microscope that uses reflected light.	1881	Pasteur develops a vaccine for anthrax in animals.
			Koch introduces the use of pure culture techniques for handling bacteria in the laboratory.
1668	Francesco Redi, an Italian naturalist, conducts experiments that demonstrate the fallacies in the spontaneous generation theory.		Walther and Fanny Hesse introduce agar-agar as a solidifying gel for culture media.
1676	Antonie van Leeuwenhoek, a Dutch linen merchant, uses a simple microscope of his own design to observe bacteria and protozoa.	1882	Koch identifies the causative agent of tuberculosis.
		1884	Koch outlines his postulates.
1776	An Italian anatomist, Lazzaro Spallanzani, conducts further convincing experiments that dispute spontaneous generation.		Elie Metchnikoff,* a Russian zoologist, lays groundwork for the science of immunology by discovering phagocytic cells.
1796	English surgeon Edward Jenner introduces a vaccination for smallpox.		The Danish physician Hans Christian Gram devises the Gram stain technique for differentiating bacteria.
1838	Phillipe Ricord, a French physician, inoculates 2,500 human subjects to demonstrate that syphilis and gonorrhea are two separate diseases.	1885	Pasteur develops a special vaccine for rabies.
		1887	Julius Petri, a German bacteriologist, adapts two plates to form a container for holding media and culturing microbes.
1839	Theodor Schwann, a German zoologist, and Matthias Schleiden, a botanist, formalize the theory that all living things are composed of cells.	1890	A German, Emil von Behring,* and a Japanese, Shibasaburo Kitasato, demonstrate the presence of antibodies in serum that neutralize the toxins of diphtheria and tetanus.
1847–1850	The Hungarian physician Ignaz Semmelweis substantiates his theory that childbed fever is a contagious disease transmitted to women by their physicians during childbirth.	1892	A Russian, D. Ivanovski, is the first to isolate a virus (the tobacco mosaic virus) and show that it could be transmitted in a cell-free filtrate.
1853–1854	John Snow, a London physician, demonstrates the epidemic spread of cholera through a water supply contaminated with human sewage.	1895	Jules Bordet,* a Belgian bacteriologist, discovers the antimicrobial powers of complement.
1857	French bacteriologist Louis Pasteur shows that fermentations are due to microorganisms and originates the process now known as pasteurization.	1898	R. Ross* and G. Grassi demonstrate that malaria is transmitted by the bite of female mosquitoes.
			Germans Friedrich Loeffler and P. Frosch discover that "filterable viruses" cause foot-and-mouth disease in animals.
1858	Rudolf Virchow, a German pathologist, introduces the concept that all cells originate from preexisting cells.	1899	Dutch microbiologist Martinus Beijerinck further elucidates the viral agent of tobacco mosaic disease and postulates that viruses have many of the properties of living cells and that they reproduce within cells.
1861	Louis Pasteur completes the definitive experiments that finally lay to rest the theory of spontaneous generation.		
1867	The English surgeon Joseph Lister publishes the first work on antiseptic surgery, beginning the trend toward modern aseptic techniques in medicine.		*continued*

*These scientists were awarded Nobel prizes for their contributions to the field.

Date	Discovery/People Involved	Date	Discovery/People Involved
1900	The American physician Walter Reed and his colleagues clarify the role of mosquitoes in transmitting yellow fever.	1954	Jonas Salk develops the first polio vaccine.
	An Austrian pathologist, Karl Landsteiner,* discovers the ABO blood groups.	1957	Alick Isaacs and Jean Lindenmann discover the natural antiviral substance interferon.
1903	American pathologist James Wright and others demonstrate the presence of antibodies in the blood of immunized animals.		D. Carleton Gajdusek* discovers the underlying cause of slow virus diseases.
1905	Syphilis is shown to be caused by *Treponema pallidum*, through the work of German bacteriologists Fritz Schaudinn and E. Hoffman.	1959–1960	Gerald Edelman* and Rodney Porter* determine the structure of antibodies.
1906	August Wasserman, a German bacteriologist, develops the first serologic test for syphilis.	1972	Paul Berg* develops the first recombinant DNA in a test tube.
	Howard Ricketts, an American pathologist, links the transmission of Rocky Mountain spotted fever to ticks.	1973	Herb Boyer and Stanley Cohen clone the first DNA using plasmids.
1908	The German Paul Ehrlich* becomes the pioneer of modern chemotherapy by developing salvarsan to treat syphilis.	1975	A technique for making monoclonal antibodies is developed by Cesar Milstein, Georges Kohler, and Niels Kai Jerne.
1910	An American pathologist, Francis Rous,* discovers viruses that can induce cancer.	1979	Genetically engineered insulin is first synthesized by bacteria.
1915–1917	British scientist F. Twort and French scientist F. D'Herelle independently discover bacterial viruses.	1982	Development of first hepatitis B vaccine using virus isolated from human blood.
1928	Frederick Griffith lays the foundation for modern molecular genetics by his discovery of transformation in bacteria.	1983	Isolation and characterization of human immunodeficiency virus (HIV) by Luc Montagnier of France and Robert Gallo of the United States.
1929	A Scottish bacteriologist, Alexander Fleming,* discovers and describes the properties of the first antibiotic, penicillin.		The polymerase chain reaction is invented by Kary Mullis.*
1933–1938	Germans Ernst Ruska* and B. von Borries develop the first electron microscope.	1987	The molecular genetics of antibody genes is worked out by Susumu Tonegawa.*
1935	Gerhard Domagk,* a German physician, discovers the first sulfa drug and paves the way for the era of antimicrobic chemotherapy.		First release of recombinant strain of *Pseudomonas* to prevent frost formation on strawberry plants.
	Wendell Stanley* is successful in inducing tobacco mosaic viruses to form crystals that still retain their infectiousness.	1989	Cancer-causing genes called oncogenes are characterized by J. Michael Bishop, Robert Huber, Hartmut Michel, and Harold Varmus.
1941	Australian Howard Florey* and Englishman Ernst Chain* develop commercial methods for producing penicillin; this first antibiotic is tested and put into widespread use.	1990	First clinical trials in gene therapy testing.
			Vaccine for *Haemophilus influenzae*, a cause of meningitis, is introduced.
1944	Oswald Avery, Colin MacLeod, and Maclyn McCarty show that DNA is the genetic material.	1991	Development of transgenic animals to synthesize human hemoglobin.
	Joshua Lederberg* and E. L. Tatum* discover conjugation in bacteria.	1994	Human breast cancer gene isolated.
	The Russian Selman Waksman* and his colleagues discover the antibiotic streptomycin.	1995	First bacterial genome fully sequenced, for *Haemophilus influenzae*.
1953	James Watson,* Francis Crick,* Rosalind Franklin, and Maurice Wilkins* determine the structure of DNA.	1997	Medical researchers at Case Western University construct human artificial chromosomes (HACs).
			Scottish researchers clone first mammal (a sheep) from adult nuclei.
		1999	Heat-loving bacteria discovered 2 miles beneath earth in African gold mine.
		2000	A rough version of the human genome is mapped.
			250-million-year-old bacterium unearthed by Pennsylvania team.
		2001	Mailed anthrax spores cause major bioterrorism event.
		2002	New York virologists create polio virus in a test tube.
		2003	New roles for small nuclear RNAs discovered.

Answers to Multiple-Choice Questions and Selected Matching Questions

Chapter 1
1. d
2. c
3. d
4. c
5. d
6. a
7. c
8. b
9. d
10. c
11. c
12. b
13. d
14. 1st col:
   3, 7, 4, 2
   2nd col:
   8, 5, 6, 1
15. c
16. c

Chapter 2
1. c
2. c
3. b
4. c
5. a
6. d
7. c
8. a
9. c
10. b
11. b
12. d
13. b
14. c
15. c
16. c
17. a
18. d
19. d
20. b

Chapter 3
1. c
2. b
3. c
4. d

5. b
6. d
7. b
8. b
9. c
10. c
11. a
12. b
13. c
14. abf, df,
   abf, ef,
   af, bef,
   ac, bef
15. d
16. b

Chapter 4
1. d
2. a
3. c
4. a
5. c
6. b
7. d
8. b
9. d
10. c
11. c
12. b
13. c

Chapter 5
1. b
2. d
3. d
4. a
5. b
6. c
7. c
8. b
9. d
10. a
11. d
12. b
13. c
14. Matching:
   b, e, c, h, g,
   j, i, d, a, f

15. d
16. b

Chapter 6
1. c
2. d
3. d
4. b
5. d
6. a
7. a
8. d
9. b
10. b
11. c
12. d
13. a

Chapter 7
1. c
2. a
3. a
4. c
5. c
6. b
7. a
8. a
9. b
10. b
11. c
12. c
13. c
14. b

Chapter 8
1. b
2. a
3. c
4. d
5. d
6. b
7. b
8. b
9. b
10. a
11. c
12. a
13. d
14. b

15. c
16. c
17. c
18. c
19. Matching:
   c, a, b, c,
   b, a, c, c

Chapter 9
1. b
2. e
3. b
4. b
5. c
6. b
7. c
8. a
9. b
10. a
11. b
12. d
13. b
14. d
15. b
16. Matching:
   e/h, f, b, g,
   e, a, i, c/e/h

Chapter 10
1. c
2. c
3. d
4. c
5. a
6. b
7. c
8. c
9. c
10. d
11. d
12. Matching:
   h, c, f, a,
   g, b, e, d

Chapter 11
1. d
2. c
3. b
4. a

5. c
6. b
7. b
8. d
9. c
10. b
11. d
12. c
13. d
14. a
15. b
16. c

Chapter 12
1. b
2. c
3. b
4. a
5. d
6. b
7. c
8. c
9. a
10. d
11. a
12. c
13. c
14. b

Chapter 13
1. a
2. d
3. b
4. d
5. c
6. d
7. b
8. c
9. c
10. c
11. d
12. c
13. a
14. a
15. d

Chapter 14
1. b
2. b

3. d
4. b
5. b
6. c
7. b
8. a
9. c
10. d
11. c
12. d
13. c

Chapter 15
1. a
2. c
3. a
4. c
5. c
6. a
7. c
8. d
9. c
10. c
11. IgG b, f, g, h
   IgA a, b, c
   IgD b
   IgE b, i
   IgM d, e, h
12. b
13. c
14. d
15. b
16. c

Chapter 16
1. d
2. d
3. d
4. c
5. b
6. c
7. b
8. a
9. d
10. d
11. d
12. a
13. b
14. d

Chapter 17
1. a, b, c, a
2. c
3. b
4. c
5. a
6. a
7. d
8. c

Chapter 18
1. b
2. a
3. e
4. e
5. b
6. c
7. a
8. b
9. c
10. d
11. d
12. d
13. b
14. a

Chapter 19
1. d
2. a
3. c
4. b
5. d
6. b
7. d
8. b
9. c
10. c
11. a
12. b
13. a
14. c

Chapter 20
1. c
2. d
3. b
4. b
5. a
6. c

7. b
8. a
9. d
10. a
11. d
12. b
13. d
14. c
15. e

Chapter 21
1. d
2. d
3. b
4. b
5. b
6. d
7. c
8. d
9. c
10. b
11. b
12. c
13. d
14. a
15. c

Chapter 22
1. b
2. d
3. c
4. d
5. c
6. c
7. e
8. b
9. d
10. a
11. b
12. b
13. a
14. b
15. c

Chapter 23
1. a
2. b
3. d
4. a

5. e
6. b
7. a
8. a
9. d
10. c
11. d
12. d
13. c
14. d

Chapter 24
1. c
2. b
3. b
4. c
5. d
6. c
7. d
8. c
9. a
10. b
11. d
12. d
13. b
14. b

Glossary

A

abiogenesis The belief in spontaneous generation as a source of life.

abiotic Nonliving factors such as soil, water, temperature, and light that are studied when looking at an ecosystem.

ABO blood group system Developed by Karl Landsteiner in 1904; the identification of different blood groups based on differing isoantigen markers characteristic of each blood type.

abscess An inflamed, fibrous lesion enclosing a core of pus.

A-B toxin A class of bacterial exotoxin consisting of two components: a binding (B) component and an active (A) or enzymatic component.

abyssal zone The deepest region of the ocean; a sunless, high-pressure, cold, anaerobic habitat.

acellular vaccine A vaccine preparation that contains specific antigens such as the capsule or toxin from a pathogen and not the whole microbe. Acellular (without a cell).

acid-fast A term referring to the property of mycobacteria to retain carbol fuchsin even in the presence of acid alcohol. The staining procedure is used to diagnose tuberculosis.

acidic A solution with a pH value below 7 on the pH scale.

acidic fermentation An anaerobic degradation of pyruvic acid that results in organic acid production.

acquired immunodeficiency syndrome See *AIDS*.

actin Long filaments of protein arranged like ribbons under the cell membrane of some bacteria; contribute to cell shape.

actinomycetes A group of filamentous, fungus-like bacteria.

active immunity Immunity acquired through direct stimulation of the immune system by antigen.

active site The specific region on an apoenzyme that binds substrate. The site for reaction catalysis.

active transport Nutrient transport method that requires carrier proteins in the membranes of the living cells and the expenditure of energy.

acute Characterized by rapid onset and short duration.

acyclovir A synthetic purine analog that blocks DNA synthesis in certain viruses, particularly the herpes simplex viruses.

adenine (A) One of the nitrogen bases found in DNA and RNA, with a purine form.

adenosine deaminase (ADA) deficiency An immunodeficiency disorder and one type of SCIDS that is caused by an inborn error in the metabolism of adenine. The accumulation of adenine destroys both B and T lymphocytes.

adenosine triphosphate (ATP) A nucleotide that is the primary source of energy to cells.

adhesion The process by which microbes gain a more stable foothold at the portal of entry; often involves a specific interaction between the molecules on the microbial surface and the receptors on the host cell.

adjuvant In immunology, a chemical vehicle that enhances antigenicity, presumably by prolonging antigen retention at the injection site.

adsorption A process of adhering one molecule onto the surface of another molecule.

aerobe A microorganism that lives and grows in the presence of free gaseous oxygen (O_2).

aerobic respiration Respiration in which the final electron acceptor in the electron transport chain is oxygen (O_2).

aerosols Suspensions of fine dust or moisture particles in the air that contain live pathogens.

aflatoxin From *Aspergillus flavus* toxin, a mycotoxin that typically poisons moldy animal feed and can cause liver cancer in humans and other animals.

agammaglobulinemia Also called hypogammaglobulinemia. The absence of or severely reduced levels of antibodies in serum.

agar A polysaccharide found in seaweed and commonly used to prepare solid culture media.

agglutination The aggregation by antibodies of suspended cells or similar-sized particles (agglutinogens) into clumps that settle.

agranulocyte One form of leukocyte (white blood cells), having globular, nonlobed nuclei and lacking prominent cytoplasmic granules.

AIDS Acquired immunodeficiency syndrome. The complex of signs and symptoms characteristic of the late phase of human immunodeficiency virus (HIV) infection.

alcoholic fermentation An anaerobic degradation of pyruvic acid that results in alcohol production.

algae Photosynthetic, plant-like organisms which generally lack the complex structure of plants; they may be single-celled or multicellular, and inhabit diverse habitats such as marine and freshwater environments, glaciers, and hot springs.

allele A gene that occupies the same location as other alternative (allelic) genes on paired chromosomes.

allergen A substance that provokes an allergic response.

allergy The altered, usually exaggerated, immune response to an allergen. Also called hypersensitivity.

alloantigen An antigen that is present in some but not all members of the same species.

allograft Relatively compatible tissue exchange between nonidentical members of the same species. Also called homograft.

allosteric Pertaining to the altered activity of an enzyme due to the binding of a molecule to a region other than the enzyme's active site.

Ames test A method for detecting mutagenic and potentially carcinogenic agents based upon the genetic alteration of nutritionally defective bacteria.

amination The addition of an amine ($-NH_2$) group to a molecule.

amino acids The building blocks of protein. Amino acids exist in 20 naturally occurring forms that impart different characteristics to the various proteins they compose.

aminoglycoside A complex group of drugs derived from soil actinomycetes that impairs ribosome function and has antibiotic potential. Example: streptomycin.

ammonification Phase of the nitrogen cycle in which ammonia is released from decomposing organic material.

amphibolism Pertaining to the metabolic pathways that serve multiple functions in the breakdown, synthesis, and conversion of metabolites.

amphipathic Relating to a compound that has contrasting characteristics, such as hydrophilic-hydrophobic or acid-base.

amphitrichous Having a single flagellum or a tuft of flagella at opposite poles of a microbial cell.

amplicon DNA strand that has been primed for replication during polymerase chain reaction.

anabolism The energy-consuming process of incorporating nutrients into protoplasm through biosynthesis.

anaerobe A microorganism that grows best, or exclusively, in the absence of oxygen.

anaerobic digesters Closed chambers used in a microbial process that converts organic sludge from waste treatment plants into useful fuels such as methane and hydrogen gases. Also called bioreactors.

anaerobic respiration Respiration in which the final electron acceptor in the electron transport chain is an inorganic molecule containing sulfate, nitrate, nitrite, carbonate, etc.

analog In chemistry, a compound that closely resembles another in structure.

anamnestic In immunology, an augmented response or memory related to a prior stimulation of the immune system by antigen. It boosts the levels of immune substances.

anaphylaxis The unusual or exaggerated allergic reaction to antigen that leads to severe respiratory and cardiac complications.

anion A negatively charged ion.

antagonism Relationship in which microorganisms compete for survival in a common environment by taking actions that inhibit or destroy another organism.

antibiotic A chemical substance from one microorganism that can inhibit or kill another microbe even in minute amounts.

antibody A large protein molecule evoked in response to an antigen that interacts specifically with that antigen.

anticodon The trinucleotide sequence of transfer RNA that is complementary to the trinucleotide sequence of messenger RNA (the codon).

antigen Any cell, particle, or chemical that induces a specific immune response by B cells or T cells and can stimulate resistance to an infection or a toxin. See *immunogen.*

antigen binding site Specific region at the ends of the antibody molecule that recognize specific antigens. These sites have numerous shapes to fit a wide variety of antigens.

antigen-presenting cell (APC) A macrophage or dendritic cell that ingests and degrades an antigen and subsequently places the antigenic determinant molecules on its surface for recognition by CD4 T lymphocytes.

antigenic drift Minor antigenic changes in the influenza A virus due to mutations in the spikes' genes.

antigenic shift Major changes in the influenza A virus due to recombination of viral strains from two different host species.

antigenicity The property of a substance to stimulate a specific immune response such as antibody formation.

antihistamine A drug that counters the action of histamine and is useful in allergy treatment.

antimetabolite A substance such as a drug that competes with, substitutes for, or interferes with a normal metabolite.

antimicrobial A special class of compounds capable of destroying or inhibiting microorganisms.

antisense DNA A type of gene therapy which utilizes an oligonucleotide to bind to the sense strand of a specific piece of DNA, thereby inhibiting transcription.

antisepsis Chemical treatments to kill or inhibit the growth of all vegetative microorganisms on body surfaces.

antiseptic A growth-inhibiting agent used on tissues to prevent infection.

antiserum Antibody-rich serum derived from the blood of animals (deliberately immunized against infectious or toxic antigen) or from people who have recovered from specific infections.

antitoxin Globulin fraction of serum that neutralizes a specific toxin. Also refers to the specific antitoxin antibody itself.

apoenzyme The protein part of an enzyme, as opposed to the nonprotein or inorganic cofactors.

apoptosis The genetically programmed death of cells that is both a natural process of development and the body's means of destroying abnormal or infected cells.

appendages Accessory structures that sprout from the surface of bacteria. They can be divided into two major groups: those that provide motility and those that enable adhesion.

aquifer A subterranean water-bearing stratum of permeable rock, sand, or gravel.

archaea Procaryotic single-celled organisms of primitive origin that have unusual anatomy, physiology and genetics, and live in harsh habitats; when capitalized **(Archaea)** the term refers to one of the three domains of living organisms as proposed by Woese.

Arthus reaction An immune complex phenomenon that develops after repeat injection. This localized inflammation results from aggregates of antigen and antibody that bind, complement, and attract neutrophils.

ascospore A spore formed within a saclike cell (ascus) of Ascomycota following nuclear fusion and meiosis.

ascus Special fungal sac in which haploid spores are created.

asepsis A condition free of viable pathogenic microorganisms.

aseptic technique Methods of handling microbial cultures, patient specimens, and other sources of microbes in a way that prevents infection of the handler and others who may be exposed.

assembly (viral) The step in viral multiplication in which capsids and genetic material are packaged into virions.

asymptomatic An infection that produces no noticeable symptoms even though the microbe is active in the host tissue.

asymptomatic carrier A person with an inapparent infection who shows no symptoms of being infected yet is able to pass the disease agent on to others.

atmosphere That part of the biosphere that includes the gaseous envelope up to 14 miles above the earth's surface. It contains gases such as carbon dioxide, nitrogen, and oxygen.

atom The smallest particle of an element to retain all the properties of that element.

atomic number (AN) A measurement that reflects the number of protons in an atom of a particular element.

atomic weight The average of the mass numbers of all the isotopic forms for a particular element.

atopy Allergic reaction classified as type I, with a strong familial relationship; caused by allergens such as pollen, insect venom, food, and dander; involves IgE antibody; includes symptoms of hay fever, asthma, and skin rash.

ATP synthase A unique enzyme located in the mitochondrial cristae and chloroplast grana that harnesses the flux of hydrogen ions to the synthesis of ATP.

attenuate To reduce the virulence of a pathogenic bacterium or virus by passing it through a non-native host or by long-term subculture.

AUG (start codon) The codon that signals the point at which translation of a messenger RNA molecule is to begin.

autoantibody An "anti-self" antibody having an affinity for tissue antigens of the subject in which it is formed.

autoclave A sterilization chamber which allows the use of steam under pressure to sterilize materials. The most common temperature/pressure combination for an autoclave is 121°C and 15 psi.

autograft Tissue or organ surgically transplanted to another site on the same subject.

autoimmune disease The pathologic condition arising from the production of antibodies against autoantigens. Example: rheumatoid arthritis. Also called autoimmunity.

autotroph A microorganism that requires only inorganic nutrients and whose sole source of carbon is carbon dioxide.

axenic A sterile state such as a pure culture. An axenic animal is born and raised in a germ-free environment. See *gnotobiotic.*

axial filament A type of flagellum (called an endoflagellum) that lies in the periplasmic space of spirochetes and is responsible for locomotion. Also called periplasmic flagellum.

azole Five-membered heterocyclic compounds typical of histidine, which are used in antifungal therapy.

B

bacillus Bacterial cell shape that is cylindrical (longer than it is wide).

back-mutation A mutation which counteracts an earlier mutation, resulting in the restoration of the original DNA sequence.

bacteremia The presence of viable bacteria in circulating blood.

bacteremic Bacteria present in the bloodstream.

Bacteria When capitalized can refer to one of the three domains of living organisms proposed by Woese, containing all non-archaea procaryotes.

bacteria (plural of bacterium) Category of procaryotes with peptidoglycan in their cell walls and circular chromosome(s). This group of small cells is widely distributed in the earth's habitats.

bacterial chromosome A circular body in bacteria that contains the primary genetic material. Also called nucleoid.

bactericide An agent that kills bacteria.

bacteriocin Proteins produced by certain bacteria that are lethal against closely related bacteria and are narrow spectrum compared with antibiotics; these proteins are coded and transferred in plasmids.

bacteriophage A virus that specifically infects bacteria.

bacteriostatic Any process or agent that inhibits bacterial growth.

bacterium A tiny unicellular procaryotic organism that usually reproduces by binary fission and usually has a peptidoglycan cell wall, has various shapes, and can be found in virtually any environment.

barophile A microorganism that thrives under high (usually hydrostatic) pressure.

basement membrane A thin layer (1–6 μm) of protein and polysaccharide found at the base of epithelial tissues.

basic A solution with a pH value above 7 on the pH scale.

basidiospore A sexual spore that arises from a basidium. Found in basidiomycota fungi.

basidium A reproductive cell created when the swollen terminal cell of a hypha develops filaments (sterigmata) that form spores.

basophil A motile polymorphonuclear leukocyte that binds IgE. The basophilic cytoplasmic granules contain mediators of anaphylaxis and atopy.

beta-lactamase An enzyme secreted by certain bacteria that cleaves the beta-lactam ring of penicillin and cephalosporin and thus provides for resistance against the antibiotic. See *penicillinase*.

beta oxidation The degradation of long-chain fatty acids. Two-carbon fragments are formed as a result of enzymatic attack directed against the second or beta carbon of the hydrocarbon chain. Aided by coenzyme A, the fragments enter the tricarboxylic acid (TCA) cycle and are processed for ATP synthesis.

binary fission The formation of two new cells of approximately equal size as the result of parent cell division.

binomial system Scientific method of assigning names to organisms that employs two names to identify every organism—genus name plus species name.

biochemistry The study of organic compounds produced by (or components of) living things. The four main categories of biochemicals are carbohydrates, lipids, proteins, and nucleic acid.

bioenergetics The study of the production and use of energy by cells.

bioethics The study of biological issues and how they relate to human conduct and moral judgment.

biofilm A complex association that arises from a mixture of microorganisms growing together on the surface of a habitat.

biogenesis Belief that living things can only arise from others of the same kind.

biogeochemical cycle A process by which matter is converted from organic to inorganic form and returned to various nonliving reservoirs on earth (air, rocks, and water) where it becomes available for reuse by living

things. Elements such as carbon, nitrogen, and phosphorus are constantly cycled in this manner.

biological vector An animal which not only transports an infectious agent but plays a role in the life cycle of the pathogen, serving as a site in which it can multiply or complete its life cycle. It is usually an alternate host to the pathogen.

biomes Particular climate regions in a terrestrial realm.

bioremediation The use of microbes to reduce or degrade pollutants, industrial wastes, and household garbage.

biosphere Habitable regions comprising the aquatic (hydrospheric), soil-rock (lithospheric), and air (atmospheric) environments.

biotechnology The use of microbes or their products in the commercial or industrial realm.

biotic Living factors such as parasites, food substrates, or other living or once-living organisms that are studied when looking at an ecosystem.

blast cell An immature precursor cell of B and T lymphocytes. Also called a lymphoblast.

blocking antibody The IgG class of immunoglobulins that competes with IgE antibody for allergens, thus blocking the degranulation of basophils and mast cells.

blood-brain barrier Decreased permeability of the walls of blood vessels in the brain, restricting access to that compartment.

blood cells Cellular components of the blood consisting of red blood cells, primarily responsible for the transport of oxygen and carbon dioxide, and white blood cells, primarily responsible for host defense and immune reactions.

B lymphocyte (B cell) A white blood cell that gives rise to plasma cells and antibodies.

botulin *Clostridium botulinum* toxin. Ingestion of this potent exotoxin leads to flaccid paralysis.

bradykinin An active polypeptide that is a potent vasodilator released from IgE-coated mast cells during anaphylaxis.

Brownian movement The passive, erratic, nondirectional motion exhibited by microscopic particles. The jostling comes from being randomly bumped by submicroscopic particles, usually water molecules, in which the visible particles are suspended.

brucellosis A zoonosis transmitted to humans from infected animals or animal products; causes a fluctuating pattern of severe fever in humans as well as muscle pain, weakness, headache, weight loss, and profuse sweating. Also called undulant fever.

bubo The swelling of one or more lymph nodes due to inflammation.

bubonic plague The form of plague in which bacterial growth is primarily restricted to the lymph and is characterized by the appearance of a swollen lymph node referred to as a bubo.

budding See *exocytosis*.

bulbar poliomyelitis Complication of polio infection in which the brain stem, medulla, or cranial nerves are affected. Leads to loss of respiratory control and paralysis of the trunk and limbs.

C

calculus Dental deposit formed when plaque becomes mineralized with calcium and phosphate crystals. Also called tartar.

cancer Any malignant neoplasm that invades surrounding tissue and can metastasize to other locations. A carcinoma is derived from epithelial tissue, and a sarcoma arises from proliferating mesodermal cells of connective tissue.

capsid The protein covering of a virus's nucleic acid core. Capsids exhibit symmetry due to the regular arrangement of subunits called capsomers. See *icosahedron*.

capsomer A subunit of the virus capsid shaped as a triangle or disc.

capsule In bacteria, the loose, gel-like covering or slime made chiefly of polysaccharides. This layer is protective and can be associated with virulence.

carbohydrate A compound containing primarily carbon, hydrogen, and oxygen in a 1:2:1 ratio.

carbon cycle That pathway taken by carbon from its abiotic source to its use by producers to form organic compounds (biotic), followed by the breakdown of biotic compounds and their release to a nonliving reservoir in the environment (mostly carbon dioxide in the atmosphere).

carbon fixation Reactions in photosynthesis that incorporate inorganic carbon dioxide into organic compounds such as sugars. This occurs during the Calvin cycle and uses energy generated by the light reactions. This process is the source of all production on earth.

carbuncle A deep staphylococcal abscess joining several neighboring hair follicles.

carotenoid Yellow, orange, or red photosynthetic pigments.

carrier A person who harbors infections and inconspicuously spreads them to others. Also, a chemical agent that can accept an atom, chemical radical, or subatomic particle from one compound and pass it on to another.

caseous lesion Necrotic area of lung tubercle superficially resembling cheese. Typical of tuberculosis.

catabolism The chemical breakdown of complex compounds into simpler units to be used in cell metabolism.

catalyst A substance that alters the rate of a reaction without being consumed or permanently changed by it. In cells, enzymes are catalysts.

catalytic site The niche in an enzyme where the substrate is converted to the product (also active site).

catarrhal A term referring to the secretion of mucus or fluids; term for the first stage of pertussis.

cation A positively charged ion.

cell An individual membrane-bound living entity; the smallest unit capable of an independent existence.

cell-mediated The type of immune responses brought about by T cells, such as cytotoxic and helper effects.

cellulitis The spread of bacteria within necrotic tissue.

cellulose A long, fibrous polymer composed of β-glucose; one of the most common substances on earth.

cephalosporins A group of broad-spectrum antibiotics isolated from the fungus *Cephalosporium*.

cercaria The free-swimming larva of the schistosome trematode that emerges from the snail host and can penetrate human skin, causing schistosomiasis.

cestode The common name for tapeworms that parasitize humans and domestic animals.

chancre The primary sore of syphilis that forms at the site of penetration by *Treponema pallidum*. It begins as a hard, dull red, painless papule that erodes from the center.

chancroid A lesion that resembles a chancre but is soft and is caused by *Haemophilus ducreyi*.

chemical bond A link formed between molecules when two or more atoms share, donate, or accept electrons.

chemical mediators Small molecules that are released during inflammation and specific immune reactions that allow communication between the cells of the immune system and facilitate surveillance, recognition and attack.

chemiosmosis The generation of a concentration gradient of hydrogen ions (called the proton motive force) by the pumping of hydrogen ions to the outer side of the membrane during electron transport.

chemoautotroph An organism that relies upon inorganic chemicals for its energy and carbon dioxide for its carbon. Also called a chemolithotroph.

chemoheterotroph Microorganisms that derive their nutritional needs from organic compounds.

chemokine Chemical mediators (cytokines) that stimulate the movement and migration of white blood cells.

chemostat A growth chamber with an outflow that is equal to the continuous inflow of nutrient media. This steady-state growth device is used to study such events as cell division, mutation rates, and enzyme regulation.

chemotactic factors Chemical mediators that stimulate the movement of white blood cells. See *chemokines*.

chemotaxis The tendency of organisms to move in response to a chemical gradient (toward an attractant or to avoid adverse stimuli).

chemotherapy The use of chemical substances or drugs to treat or prevent disease.

chemotroph Organism that oxidizes compounds to feed on nutrients.

chitin A polysaccharide similar to cellulose in chemical structure. This polymer makes up the horny substance of the exoskeletons of arthropods and certain fungi.

chlorophyll A group of mostly green pigments that are used by photosynthetic eucaryotic organisms and cyanobacteria to trap light energy to use in making chemical bonds.

chloroplast An organelle containing chlorophyll that is found in photosynthetic eucaryotes.

cholesterol Best-known member of a group of lipids called steroids. Cholesterol is commonly found in cell membranes and animal hormones.

chromatin The genetic material of the nucleus. Chromatin is made up of nucleic acid and stains readily with certain dyes.

chromosome The tightly coiled bodies in cells that are the primary sites of genes.

chronic Any process or disease that persists over a long duration.

cilium (plural: *cilia*) Eucaryotic structure similar to flagella that propels a protozoan through the environment.

class In the levels of classification, the division of organisms that follows phylum.

classical pathway Pathway of complement activation initiated by a specific antigen-antibody interaction.

clonal selection theory A conceptual explanation for the development of lymphocyte specificity and variety during immune maturation.

clone A colony of cells (or group of organisms) derived from a single cell (or single organism) by asexual reproduction. All units share identical characteristics. Also used as a verb to refer to the process of producing a genetically identical population of cells or genes.

cloning host An organism such as a bacterium or a yeast that receives and replicates a foreign piece of DNA inserted during a genetic engineering experiment.

coagulase A plasma-clotting enzyme secreted by *Staphylococcus aureus*. It contributes to virulence and is involved in forming a fibrin wall that surrounds staphylococcal lesions.

coccobacillus An elongated coccus; a short, thick, oval-shaped bacterial rod.

coccus A spherical-shaped bacterial cell.

codon A specific sequence of three nucleotides in mRNA (or the sense strand of DNA) that constitutes the genetic code for a particular amino acid.

coenzyme A complex organic molecule, several of which are derived from vitamins (e.g., nicotinamide, riboflavin). A coenzyme operates in conjunction with an enzyme. Coenzymes serve as transient carriers of specific atoms or functional groups during metabolic reactions.

cofactor An enzyme accessory. It can be organic, such as coenzymes, or inorganic, such as Fe^{+2}, Mn^{+2}, or Zn^{+2} ions.

cold sterilization The use of nonheating methods such as radiation or filtration to sterilize materials.

coliform A collective term that includes normal enteric bacteria that are gram-negative and lactose-fermenting.

colony A macroscopic cluster of cells appearing on a solid medium, each arising from the multiplication of a single cell.

colostrum The clear yellow early product of breast milk that is very high in secretory antibodies. Provides passive intestinal protection.

commensalism An unequal relationship in which one species derives benefit without harming the other.

communicable infection Capable of being transmitted from one individual to another.

community The interacting mixture of populations in a given habitat.

competitive inhibition Control process that relies on the ability of metabolic analogs to control microbial growth by successfully competing with a necessary enzyme to halt the growth of bacterial cells.

complement In immunology, serum protein components that act in a definite sequence when set in motion either by an antigen-antibody complex or by factors of the alternative (properdin) pathway.

complementary DNA (cDNA) DNA created by using reverse transcriptase to synthesize DNA from RNA templates.

compounds Molecules that are a combination of two or more different elements.

concentration The expression of the amount of a solute dissolved in a certain amount of solvent. It may be defined by weight, volume, or percentage.

condylomata acuminata Extensive, branched masses of genital warts caused by infection with human papillomavirus.

congenital Transmission of an infection from mother to a fetus.

congenital rubella Transmission of the rubella virus to a fetus in utero. Injury to the fetus is generally much more serious than it is to the mother.

congenital syphilis A syphilis infection of the fetus or newborn acquired from maternal infection in utero.

conidia Asexual fungal spores shed as free units from the tips of fertile hyphae.

conjugation In bacteria, the contact between donor and recipient cells associated with the transfer of genetic material such as plasmids. Can involve special (sex) pili. Also a form of sexual recombination in ciliated protozoans.

conjunctiva The thin fluid-secreting tissue that covers the eye and lines the eyelid.

constitutive enzyme An enzyme present in bacterial cells in constant amounts, regardless of the presence of substrate. Enzymes of the central catabolic pathways are typical examples.

consumer An organism that feeds on producers or other consumers. It gets all nutrients and energy from other organisms (also called heterotroph). May exist at several levels, such

as primary (feeds on producers), secondary (feeds on primary consumers).

contagious Communicable; transmissible by direct contact with infected people and their fresh secretions or excretions.

contaminant An impurity; any undesirable material or organism.

contaminated culture A medium that once held a pure (single or mixed) culture but now contains unwanted microorganisms.

convalescence Recovery; the period between the end of a disease and the complete restoration of health in a patient.

corepressor A molecule that combines with inactive repressor to form active repressor, which attaches to the operator gene site and inhibits the activity of structural genes subordinate to the operator.

covalent A type of chemical bond that involves the sharing of electrons between two atoms.

covalent bond A chemical bond formed by the sharing of electrons between two atoms.

Creutzfeldt-Jakob disease A spongiform encephalopathy caused by infection with a prion. The disease is marked by dementia, impaired senses and uncontrollable muscle contractions.

crista The infolded inner membrane of a mitochondrion that is the site of the respiratory chain and oxidative phosphorylation.

culture The visible accumulation of microorganisms in or on a nutrient medium. Also, the propagation of microorganisms with various media.

curd The coagulated milk protein used in cheese making.

cutaneous Second level of skin, including the stratum corneum and occasionally the upper dermis.

cyanosis Blue discoloration of the skin or mucous membranes indicative of decreased oxygen concentration in blood.

cyst The resistant, dormant, but infectious form of protozoans. Can be important in spread of infectious agents such as *Entamoeba histolytica* and *Giardia lamblia*.

cystine An amino acid, HOOC—CH(NH$_2$)—CH$_2$—S—S—CH$_2$—CH(NH$_2$)COOH. An oxidation product of two cysteine molecules in which the —SH (sulfhydryl) groups form a disulfide union. Also called dicysteine.

cytochrome A group of heme protein compounds whose chief role is in electron and/or hydrogen transport occurring in the last phase of aerobic respiration.

cytokine A chemical substance produced by white blood cells and tissue cells that regulates development, inflammation, and immunity.

cytopathic effect The degenerative changes in cells associated with virus infection. Examples: the formation of multinucleate giant cells (Negri bodies), the prominent cytoplasmic inclusions of nerve cells infected by rabies virus.

cytoplasm Dense fluid encased by the cell membrane; the site of many of the cell's biochemical and synthetic activities.

cytosine (C) One of the nitrogen bases found in DNA and RNA, with a pyrimidine form.

D

deamination The removal of an amino group from an amino acid.

death phase End of the cell growth due to lack of nutrition, depletion of environment, and accumulation of wastes. Population of cells begins to die.

debridement Trimming away devitalized tissue and foreign matter from a wound.

decomposer A consumer that feeds on organic matter from the bodies of dead organisms. These microorganisms feed from all levels of the food pyramid and are responsible for recycling elements (also called saprobes).

decomposition The breakdown of dead matter and wastes into simple compounds, that can be directed back into the natural cycle of living things.

decontamination The removal or neutralization of an infectious, poisonous, or injurious agent from a site.

deduction Problem-solving process in which an individual constructs a hypothesis, tests its validity by outlining particular events that are predicted by the hypothesis, and then performs experiments to test for those events.

definitive host The organism in which a parasite develops into its adult or sexually mature stage. Also called the final host.

degerm To physically remove surface oils, debris, and soil from skin to reduce the microbial load.

degranulation The release of cytoplasmic granules, as when cytokines are secreted from mast cell granules.

dehydration synthesis During the formation of a carbohydrate bond, the step in which one carbon molecule gives up its OH group and the other loses the H from its OH group, thereby producing a water molecule. This process is common to all polymerization reactions.

denaturation The loss of normal characteristics resulting from some molecular alteration. Usually in reference to the action of heat or chemicals on proteins whose function depends upon an unaltered tertiary structure.

dendritic cell A large, antigen-processing cell characterized by long, branchlike extensions of the cell membrane.

denitrification The end of the nitrogen cycle when nitrogen compounds are returned to the reservoir in the air.

dental caries A mixed infection of the tooth surface that gradually destroys the enamel and may lead to destruction of the deeper tissue.

deoxyribonucleic acid (DNA) The nucleic acid often referred to as the "double helix." DNA carries the master plan for an organism's heredity.

deoxyribose A 5-carbon sugar that is an important component of DNA.

dermatophytes A group of fungi that cause infections of the skin and other integument components. They survive by metabolizing keratin.

desensitization See *hyposensitization*.

desiccation To dry thoroughly. To preserve by drying.

desquamate To shed the cuticle in scales; to peel off the outer layer of a surface.

diabetes mellitus A disease involving compromise in insulin function. In one form, the pancreatic cells that produce insulin are destroyed by autoantibodies, and in another, the pancreas does not produce sufficient insulin.

diapedesis The migration of intact blood cells between endothelial cells of a blood vessel such as a venule.

differential medium A single substrate that discriminates between groups of microorganisms on the basis of differences in their appearance due to different chemical reactions.

differential stain A technique that utilizes two dyes to distinguish between different microbial groups or cell parts by color reaction.

diffusion The dispersal of molecules, ions, or microscopic particles propelled down a concentration gradient by spontaneous random motion to achieve a uniform distribution.

DiGeorge syndrome A birth defect usually caused by a missing or incomplete thymus gland that results in abnormally low or absent T-cells and other developmental abnormalities.

dimorphic In mycology, the tendency of some pathogens to alter their growth form from mold to yeast in response to rising temperature.

diplococcus Spherical or oval-shaped bacteria, typically found in pairs.

direct, or total cell count 1. Counting total numbers of individual cells being viewed with magnification. 2. Counting isolated colonies of organisms growing on a plate of media as a way to determine population size.

disaccharide A sugar containing two monosaccharides. Examples: sucrose (fructose + glucose).

disease Any deviation from health, as when the effects of microbial infection damage or disrupt tissues and organs.

disinfection The destruction of pathogenic nonsporulating microbes or their toxins, usually on inanimate surfaces.

division In the levels of classification, an alternate term for phylum.

DNA See *deoxyribonucleic acid*.

DNA fingerprint A pattern of restriction enzyme fragments which is unique for an individual organism.

DNA polymerase Enzyme responsible for the replication of DNA. Several versions of the enzyme exist, each completing a unique portion of the replication process.

DNA sequencing Determining the exact order of nucleotides in a fragment of DNA. Most commonly done using the Sanger dideoxy sequencing method.

DNA vaccine A newer vaccine preparation based on inserting DNA from pathogens into host cells to encourage them to express the foreign protein and stimulate immunity.

domain In the levels of classification, the broadest general category to which an organism is assigned. Members of a domain share only one or a few general characteristics.

droplet nuclei The dried residue of fine droplets produced by mucus and saliva sprayed while sneezing and coughing. Droplet nuclei are less than 5 μm in diameter (large enough to bear a single bacterium and small enough to remain airborne for a long time) and can be carried by air currents. Droplet nuclei are drawn deep into the air passages.

drug resistance An adaptive response in which microorganisms begin to tolerate an amount of drug that would ordinarily be inhibitory.

dysentery Diarrheal illness in which stools contain blood and/or mucus.

dyspnea Difficulty in breathing.

E

ecosystem A collection of organisms together with its surrounding physical and chemical factors.

ectoplasm The outer, more viscous region of the cytoplasm of a phagocytic cell such as an amoeba. It contains microtubules, but not granules or organelles.

eczema An acute or chronic allergy of the skin associated with itching and burning sensations. Typically, red, edematous, vesicular lesions erupt, leaving the skin scaly and sometimes hyperpigmented.

edema The accumulation of excess fluid in cells, tissues, or serous cavities. Also called swelling.

electrolyte Any compound that ionizes in solution and conducts current in an electrical field.

electron A negatively charged subatomic particle that is distributed around the nucleus in an atom.

electrophoresis The separation of molecules by size and charge through exposure to an electrical current.

electrostatic Relating to the attraction of opposite charges and the repulsion of like charges. Electrical charge remains stationary as opposed to electrical flow or current.

element A substance comprising only one kind of atom that cannot be degraded into two or more substances without losing its chemical characteristics.

ELISA Abbreviation for **e**nzyme-**l**inked **i**mmuno**s**orbent **a**ssay, a very sensitive serological test used to detect antibodies in diseases such as AIDS.

emerging disease Newly identified diseases that are becoming more prominent.

emetic Inducing to vomit.

encephalitis An inflammation of the brain, usually caused by infection.

endemic disease A native disease that prevails continuously in a geographic region.

endergonic reaction A chemical reaction that occurs with the absorption and storage of surrounding energy. Antonym: exergonic.

endocytosis The process whereby solid and liquid materials are taken into the cell through membrane invagination and engulfment into a vesicle.

endoenzyme An intracellular enzyme, as opposed to enzymes that are secreted.

endogenous Originating or produced within an organism or one of its parts.

endoplasmic reticulum An intracellular network of flattened sacs or tubules with or without ribosomes on their surfaces.

endospore A small, dormant, resistant derivative of a bacterial cell that germinates under favorable growth conditions into a vegetative cell. The bacterial genera *Bacillus* and *Clostridium* are typical sporeformers.

endosymbiosis Relationship in which a microorganism resides within a host cell and provides a benefit to the host cell.

endotoxic shock A massive drop in blood pressure caused by the release of endotoxin from gram-negative bacteria multiplying in the bloodstream.

endotoxin A bacterial toxin that is not ordinarily released (as is exotoxin). Endotoxin is composed of a phospholipid-polysaccharide complex that is an integral part of gram-negative bacterial cell walls. Endotoxins can cause severe shock and fever.

energy of activation The minimum energy input necessary for reactants to form products in a chemical reaction.

energy pyramid An ecological model that shows the energy flow among the organisms in a community. It is structured like the food pyramid, but shows how energy is reduced from one trophic level to another.

enriched medium A nutrient medium supplemented with blood, serum, or some growth factor to promote the multiplication of fastidious microorganisms.

enteric Pertaining to the intestine.

enteroaggregative The term used to describe certain types of intestinal bacteria that tend to stick to each other in large clumps.

enteroinvasive Predisposed to invade the intestinal tissues.

enteropathogenic Pathogenic to the alimentary canal.

enterotoxigenic Having the capacity to produce toxins that act on the intestinal tract.

enterotoxin A bacterial toxin that specifically targets intestinal mucous membrane cells. Enterotoxigenic strains of *Escherichia coli* and *Staphylococcus aureus* are typical sources.

enveloped virus A virus whose nucleocapsid is enclosed by a membrane derived in part from the host cell. It usually contains exposed glycoprotein spikes specific for the virus.

enzyme A protein biocatalyst that facilitates metabolic reactions.

enzyme induction One of the controls on enzyme synthesis. This occurs when enzymes appear only when suitable substrates are present.

enzyme repression The inhibition of enzyme synthesis by the end product of a catabolic pathway.

eosinophil A leukocyte whose cytoplasmic granules readily stain with red eosin dye.

epidemic A sudden and simultaneous outbreak or increase in the number of cases of disease in a community.

epidemiology The study of the factors affecting the prevalence and spread of disease within a community.

epitope The precise molecular group of an antigen that defines its specificity and triggers the immune response.

Epstein-Barr virus (EBV) Herpesvirus linked to infectious mononucleosis, Burkitt's lymphoma and nasopharyngeal carcinoma.

erysipelas An acute, sharply defined inflammatory disease specifically caused by hemolytic *Streptococcus*. The eruption is limited to the skin but can be complicated by serious systemic symptoms.

erythroblastosis fetalis Hemolytic anemia of the newborn. The anemia comes from hemolysis of Rh-positive fetal erythrocytes by anti-Rh maternal antibodies. Erythroblasts are immature red blood cells prematurely released from the bone marrow.

erythrocytes (red blood cells) Blood cells involved in the transport of oxygen and carbon dioxide.

erythrogenic toxin An exotoxin produced by lysogenized group A strains of β-hemolytic streptococci that is responsible for the severe fever and rash of scarlet fever in the nonimmune individual. Also called a pyrogenic toxin.

eschar A dark, sloughing scab that is the lesion of anthrax and certain rickettsioses.

essential nutrient Any ingredient such as a certain amino acid, fatty acid, vitamin, or mineral that cannot be formed by an organism and must be supplied in the diet. A growth factor.

ester bond A covalent bond formed by reacting carboxylic acid with an OH group:

$$(R—\overset{\overset{\textstyle O}{||}}{C}—O—R')$$

Olive and corn oils, lard, and butter fat are examples of triacylglycerols—esters formed between glycerol and three fatty acids.

ethylene oxide A potent, highly water-soluble gas invaluable for gaseous sterilization of heat-sensitive objects such as plastics, surgical and diagnostic appliances, and spices.

etiologic agent The microbial cause of disease; the pathogen.

eubacteria Term used for non-archaea prokaryotes, means "true bacteria."

eucaryotic cell A cell that differs from a procaryotic cell chiefly by having a nuclear membrane (a well-defined nucleus), membrane-bounded subcellular organelles, and mitotic cell division.

Eukarya One of the three domains (sometimes called superkingdoms) of living organisms, as proposed by Woese; contains all eucaryotic organisms.

eutrophication The process whereby dissolved nutrients resulting from natural seasonal enrichment or industrial pollution of water cause overgrowth of algae and cyanobacteria to the detriment of fish and other large aquatic inhabitants.

evolution Scientific principle that states that living things change gradually through hundreds of millions of years, and these changes are expressed in structural and functional adaptations in each organism. Evolution presumes that those traits which favor survival are preserved and passed on to following generations, and those traits which do not favor survival are lost.

exanthem An eruption or rash of the skin.

exergonic A chemical reaction associated with the release of energy to the surroundings. Antonym: endergonic.

exfoliative toxin A poisonous substance that causes superficial cells of an epithelium to detach and be shed. Example: staphylococcal exfoliatin. Also called an epidermolytic toxin.

exocytosis The process that releases enveloped viruses from the membrane of the host's cytoplasm.

exoenzyme An extracellular enzyme chiefly for hydrolysis of nutrient macromolecules that are otherwise impervious to the cell membrane. It functions in saprobic decomposition of organic debris and can be a factor in invasiveness of pathogens.

exogenous Originating outside the body.

exon A stretch of eucaryotic DNA coding for a corresponding portion of mRNA that is translated into peptides. Intervening stretches of DNA that are not expressed are called introns. During transcription, exons are separated from introns and are spliced together into a continuous mRNA transcript.

exotoxin A toxin (usually protein) that is secreted and acts upon a specific cellular target. Examples: botulin, tetanospasmin, diphtheria toxin, and erythrogenic toxin.

exponential Pertaining to the use of exponents, numbers that are typically written as a superscript to indicate how many times a factor is to be multiplied. Exponents are used in scientific notation to render large, cumbersome numbers into small workable quantities.

exponential growth phase The period of maximum growth rate in a growth curve. Cell population increases logarithmically.

extrapulmonary tuberculosis A condition in which tuberculosis bacilli have spread to organs other than the lungs.

extremophiles Organisms capable of living in harsh environments, such as extreme heat or cold.

F

facilitated diffusion The passive movement of a substance across a plasma membrane from an area of higher concentration to an area of lower concentration utilizing specialized carrier proteins.

facultative Pertaining to the capacity of microbes to adapt or adjust to variations; not obligate. Example: the presence of oxygen is not obligatory for a facultative anaerobe to grow. See *obligate.*

family In the levels of classification, a mid-level division of organisms that groups more closely related organisms than previous levels. An order is divided into families.

fastidious Requiring special nutritional or environmental conditions for growth. Said of bacteria.

fecal coliforms Any species of gram negative lactose positive bacteria (primarily *Escherichia coli*) that live primarily in the intestinal tract and not the environment. Finding evidence of these bacteria in a water or food sample is substantial evidence of fecal contamination and potential for infection (see *coliform*).

feedback inhibition Temporary end to enzyme action caused by an end product molecule binding to the regulatory site and preventing the enzyme's active site from binding to its substrate.

fermentation The extraction of energy through anaerobic degradation of substrates into simpler, reduced metabolites. In large industrial processes, fermentation can mean any use of microbial metabolism to manufacture organic chemicals or other products.

fermentor A large tank used in industrial microbiology to grow mass quantities of microbes that can synthesize desired products. These devices are equipped with means to stir, monitor and harvest products such as drugs, enzymes, and proteins in very large quantities.

fertility (F') factor Donor plasmid that allows synthesis of a pilus in bacterial conjugation. Presence of the factor is indicated by F^+, and lack of the factor is indicated by F^-.

filament A helical structure composed of proteins that is part of bacterial flagella.

fimbria A short, numerous surface appendage on some bacteria that provides adhesion but not locomotion.

Firmicutes Taxonomic category of bacteria that have gram-positive cell envelopes.

flagellum A structure that is used to propel the organism through a fluid environment.

flora Beneficial or harmless resident bacteria commonly found on and/or in the human body.

fluid mosaic model A conceptualization of the molecular architecture of cellular membranes as a bilipid layer containing proteins. Membrane proteins are embedded to some degree in this bilayer, where they float freely about.

fluorescence The property possessed by certain minerals and dyes to emit visible light when excited by ultraviolet radiation. A fluorescent dye combined with specific antibody provides a sensitive test for the presence of antigen.

fluoroquinolones Synthetic antimicrobial drugs chemically related to quinine. They are broad-spectrum and easily adsorbed from the intestine.

focal infection Occurs when an infectious agent breaks loose from a localized infection and is carried by the circulation to other tissues.

folliculitis An inflammatory reaction involving the formation of papules or pustules in clusters of hair follicles.

fomite Virtually any inanimate object an infected individual has contact with that can serve as a vehicle for the spread of disease.

food chain A simple straight-line feeding sequence among organisms in a community.

food fermentations Addition to and growth of known cultures of microorganisms in foods to produce desirable flavors, smells, or textures. Includes cheeses, breads, alcoholic beverages, and pickles.

food web A complex network that traces all feeding interactions among organisms in a community (see *food chain*). This is considered to be a more accurate picture of food relationships in a community than a food chain.

formalin A 37% aqueous solution of formaldehyde gas; a potent chemical fixative and microbicide.

frameshift mutation An insertion or deletion mutation which changes the codon reading frame from the point of the mutation to the final codon. Almost always leads to a nonfunctional protein.

fructose One of the carbohydrates commonly referred to as sugars. Fructose is commonly fruit sugars.

functional group In chemistry, a particular molecular combination that reacts in predictable ways and confers particular properties on a compound. Examples: —COOH, —OH, —CHO.

fungi Macroscopic and microscopic heterotrophic eucaryotic organisms that can be uni- or multicellular.

fungus Heterotrophic unicellular or multicellular eucaryotic organism which may take the form of a larger macroscopic organism, as in the case of mushrooms, or a smaller microscopic organism, as in the case of yeasts and molds.

furuncle A boil; a localized pyogenic infection arising from a hair follicle.

G

Gaia Theory The concept that biotic and abiotic factors sustain suitable conditions for one another simply by their interactions. Named after the mythical Greek goddess of earth.

gamma globulin The fraction of plasma proteins high in immunoglobulins (antibodies). Preparations from pooled human plasma containing normal antibodies make useful passive immunizing agents against pertussis, polio, measles, and several other diseases.

gas gangrene Disease caused by a clostridial infection of soft tissue or wound. The name refers to the gas produced by the bacteria growing in the tissue. Unless treated early, it is fatal. Also called myonecrosis.

gastritis Pain and/or nausea, usually experienced after eating; result of inflammation of the lining of the stomach.

gel electrophoresis A laboratory technique for separating DNA fragments according to length by employing electricity to force the DNA through a gel-like matrix typically made of agarose. Smaller DNA fragments move more quickly through the gel, thereby moving farther than larger fragments during the same period of time.

gene A site on a chromosome that provides information for a certain cell function. A specific segment of DNA that contains the necessary code to make a protein or RNA molecule.

gene probe Short strands of single-stranded nucleic acid that hybridize specifically with complementary stretches of nucleotides on test samples and thereby serve as a tagging and identification device.

generation time Time required for a complete fission cycle—from parent cell to two new daughter cells. Also called doubling time.

gene therapy The introduction of normal functional genes into people with genetic diseases such as sickle cell anemia and cystic fibrosis. This is usually accomplished by a virus vector.

genetic engineering A field involving deliberate alterations (recombinations) of the genomes of microbes, plants, and animals through special technological processes.

genetics The science of heredity.

genital warts A prevalent STD linked to some forms of cancer of the reproductive organs. Caused by infection with human papillomavirus.

genome The complete set of chromosomes and genes in an organism.

genotype The genetic makeup of an organism. The genotype is ultimately responsible for an organism's phenotype, or expressed characteristics.

genus In the levels of classification, the second most specific level. A family is divided into several genera.

germ free See *axenic.*

germicide An agent lethal to non-endospore-forming pathogens.

germ theory of disease A theory first originating in the 1800s which proposed that microorganisms can be the cause of diseases. The concept is actually so well established in the present time that it is considered a fact.

giardiasis Infection by the *Giardia* flagellate. Most common mode of transmission is contaminated food and water. Symptoms include diarrhea, abdominal pain, and flatulence.

gingivitis Inflammation of the gum tissue in contact with the roots of the teeth.

gluconeogenesis The formation of glucose (or glycogen) from noncarbohydrate sources such as protein or fat. Also called glyconeogenesis.

glucose One of the carbohydrates commonly referred to as sugars. Glucose is characterized by its 6-carbon structure.

glycerol A 3-carbon alcohol, with three OH groups that serve as binding sites.

glycocalyx A filamentous network of carbohydrate-rich molecules that coats cells.

glycogen A glucose polymer stored by cells.

glycolysis The energy-yielding breakdown (fermentation) of glucose to pyruvic or lactic acid. It is often called anaerobic glycolysis because no molecular oxygen is consumed in the degradation.

glycosidic bond A bond that joins monosaccharides to form disaccharides and polymers.

gnotobiotic Referring to experiments performed on germ-free animals.

Golgi apparatus An organelle of eucaryotes that participates in packaging and secretion of molecules.

gonococcus Common name for *Neisseria gonorrhoeae,* the agent of gonorrhea.

Gracilicutes Taxonomic category of bacteria that have gram-negative envelopes.

graft Live tissue taken from a donor and transplanted into a recipient to replace damaged or missing tissues such as skin, bone, blood vessels.

graft vs. host disease (GVHD) A condition associated with a bone marrow transplant in which T cells in the transplanted tissue mount an immune response against the recipient's (host) normal tissues.

Gram stain A differential stain for bacteria useful in identification and taxonomy. Gram-positive organisms appear purple from crystal violet-mordant retention, whereas gram-negative organisms appear red after loss of crystal violet and absorbance of the safranin counterstain.

grana Discrete stacks of chlorophyll-containing thylakoids within chloroplasts.

granulocyte A mature leukocyte that contains noticeable granules in a Wright stain. Examples: neutrophils, eosinophils, and basophils.

granuloma A solid mass or nodule of inflammatory tissue containing modified macrophages and lymphocytes. Usually a chronic pathologic process of diseases such as tuberculosis or syphilis.

Grave's disease A malfunction of the thyroid gland in which autoantibodies directed at thyroid cells stimulate an overproduction of thyroid hormone (hyperthyroidism).

greenhouse effect The capacity to retain solar energy by a blanket of atmospheric gases that redirects heat waves back toward the earth.

group translocation A form of active transport in which the substance being transported is altered during transfer across a plasma membrane.

growth curve A graphical representation of the change in population size over time. This graph has four periods known as lag phase, exponential or log phase, stationary phase, and death phase.

growth factor An organic compound such as a vitamin or amino acid that must be provided in the diet to facilitate growth. An essential nutrient.

guanine (G) One of the nitrogen bases found in DNA and RNA in the purine form.

Guillain-Barré syndrome A neurological complication of infection or vaccination.

gumma A nodular, infectious granuloma characteristic of tertiary syphilis.

gut-associated lymphoid tissue (GALT) A collection of lymphoid tissue in the gastrointestinal tract which includes the appendix, the lacteals, and Peyer's patches.

gyrase The enzyme responsible for supercoiling DNA into tight bundles, a type of topoisomerase.

H

habitat The environment to which an organism is adapted.

halogens A group of related chemicals with antimicrobial applications. The halogens most often used in disinfectants and antiseptics are chlorine and iodine.

halophile A microbe whose growth is either stimulated by salt or requires a high concentration of salt for growth.

Hansen's disease A chronic, progressive disease of the skin and nerves caused by infection by a mycobacterium that is a slow-growing, strict parasite. Hansen's disease is the preferred name for leprosy.

hapten An incomplete or partial antigen. Although it constitutes the determinative group and can bind antigen, hapten cannot stimulate a full immune response without being carried by a larger protein molecule.

Hashimoto's thyroiditis An autoimmune disease of the thyroid gland that damages the thyroid follicle cells and results in decreased production of thyroid hormone (hypothyroidism).

hay fever A form of atopic allergy marked by seasonal acute inflammation of the conjunctiva and mucous membranes of the respiratory passages. Symptoms are irritative itching and rhinitis.

helical Having a spiral or coiled shape. Said of certain virus capsids and bacteria.

helminth A term that designates all parasitic worms.

helper T cell A class of thymus-stimulated lymphocytes that facilitate various immune activities such as assisting B cells and macrophages. Also called a T helper cell.

hemagglutinin A molecule that causes red blood cells to clump or agglutinate. Often found on the surfaces of viruses.

hemolysin Any biological agent that is capable of destroying red blood cells and causing the release of hemoglobin. Many bacterial pathogens produce exotoxins that act as hemolysins.

hemolytic disease Incompatible Rh factor between mother and fetus causes maternal antibodies to attack the fetus and trigger complement-mediated lysis in the fetus.

hemolytic uremic syndrome Severe hemolytic anemia, leading to kidney damage or failure; can accompany *E. coli* O157:H7 intestinal infection.

hemolyze When red blood cells burst and release hemoglobin pigment.

hemopoiesis The process by which the various types of blood cells are formed, such as in the bone marrow.

hepatitis Inflammation and necrosis of the liver, often the result of viral infection.

hepatitis A virus (HAV) Enterovirus spread by contaminated food responsible for short-term (infectious) hepatitis.

hepatitis B virus (HBV) DNA virus that is the causative agent of serum hepatitis.

hepatocellular carcinoma A liver cancer associated with infection with hepatitis B virus.

herd immunity The status of collective acquired immunity in a population that reduces the likelihood that nonimmune individuals will contract and spread infection. One aim of vaccination is to induce herd immunity.

heredity Genetic inheritance.

hermaphroditic Containing the sex organs for both male and female in one individual.

herpes zoster A recurrent infection caused by latent chickenpox virus. Its manifestation on the skin tends to correspond to dermatomes and to occur in patches that "girdle" the trunk. Also called shingles.

heterotroph An organism that relies upon organic compounds for its carbon and energy needs.

hexose A 6-carbon sugar such as glucose and fructose.

hierarchies Levels of power. Arrangement in order of rank.

histamine A cytokine released when mast cells and basophils release their granules. An important mediator of allergy, its effects include smooth muscle contraction, increased vascular permeability, and increased mucus secretion.

histiocyte Another term for macrophage.

histone Proteins associated with eucaryotic DNA. These simple proteins serve as winding spools to compact and condense the chromosomes.

HLA An abbreviation for **h**uman **l**eukocyte **a**ntigens. This closely linked cluster of genes programs for cell surface glycoproteins that control immune interactions between cells and is involved in rejection of allografts. Also called the major histocompatibility complex (MHC).

holoenzyme An enzyme complete with its apoenzyme and cofactors.

hops The ripe, dried fruits of the hop vine (*Humulus lupulus*) that is added to beer wort for flavoring.

host Organism in which smaller organisms or viruses live, feed, and reproduce.

host range The limitation imposed by the characteristics of the host cell on the type of virus that can successfully invade it.

human diploid cell vaccine A vaccine made using cell culture that is currently the vaccine of choice for preventing infection by rabies virus.

human immunodeficiency virus (HIV) A retro virus that causes acquired immunodeficiency syndrome (AIDS).

human papillomavirus (HPV) A group of DNA viruses whose members are responsible for common, plantar and genital warts.

humoral immunity Protective molecules (mostly B lymphocytes) carried in the fluids of the body.

hybridization A process that matches complementary strands of nucleic acid (DNA-DNA, RNA-DNA, RNA-RNA). Used for locating specific sites or types of nucleic acids.

hybridoma An artificial cell line that produces monoclonal antibodies. It is formed by fusing (hybridizing) a normal antibody-producing cell with a cancer cell, and it can produce pure antibody indefinitely.

hydration The addition of water as in the coating of ions with water molecules as ions enter into aqueous solution.

hydrogen bond A weak chemical bond formed by the attraction of forces between molecules or atoms—in this case, hydrogen and either oxygen or nitrogen. In this type of bond, electrons are not shared, lost, or gained.

hydrologic cycle The continual circulation of water between hydrosphere, atmosphere, and lithosphere.

hydrolysis A process in which water is used to break bonds in molecules. Usually occurs in conjunction with an enzyme.

hydrophilic The property of attracting water. Molecules that attract water to their surface are called hydrophilic.

hydrophobic The property of repelling water. Molecules that repel water are called hydrophobic.

hydrosphere That part of the biosphere which encompasses water-containing environments such as oceans, lakes, rivers.

hypertonic Having a greater osmotic pressure than a reference solution.

hyphae The tubular threads that make up filamentous fungi (molds). This web of branched and intertwining fibers is called a mycelium.

hypogammaglobulinemia An inborn disease in which the gamma globulin (antibody) fraction of serum is greatly reduced. The condition is associated with a high susceptibility to pyogenic infections.

hyposensitization A therapeutic exposure to known allergens designed to build tolerance and eventually prevent allergic reaction.

hypothesis A tentative explanation of what has been observed or measured.

hypotonic Having a lower osmotic pressure than a reference solution.

I

icosahedron A regular geometric figure having 20 surfaces that meet to form 12 corners. Some virions have capsids that resemble icosahedral crystals.

immune complex reaction Type III hypersensitivity of the immune system. It is characterized by the reaction of soluble antigen with antibody, and the deposition of the resulting complexes in basement membranes of epithelial tissue.

immunity An acquired resistance to an infectious agent due to prior contact with that agent.

immunoassays Extremely sensitive tests that permit rapid and accurate measurement of trace antigen or antibody.

immunocompetence The ability of the body to recognize and react with multiple foreign substances.

immunodeficiency Immune function is incompletely developed, suppressed, or destroyed.

immunodeficiency disease A form of immunopathology in which white blood cells are unable to mount a complete, effective immune response, which results in recurrent infections. Examples would be AIDS and agammaglobulinemia.

immunogen Any substance that induces a state of sensitivity or resistance after processing by the immune system of the body.

immunoglobulin The chemical class of proteins to which antibodies belong.

immunology The study of the system of body defenses that protect against infection.

immunopathology The study of disease states associated with overreactivity or underreactivity of the immune response.

immunotherapy Preventing or treating infectious diseases by administering substances that produce artificial immunity. May be active or passive.

incidence In epidemiology, the number of new cases of a disease occurring during a period.

incineration Destruction of microbes by subjecting them to extremes of dry heat. Microbes are reduced to ashes and gas by this process.

inclusion A relatively inert body in the cytoplasm such as storage granules, glycogen, fat, or some other aggregated metabolic product.

incubate To isolate a sample culture in a temperature-controlled environment to encourage growth.

incubation period The period from the initial contact with an infectious agent to the appearance of the first symptoms.

indicator bacteria In water analysis, any easily cultured bacteria that may be found in the intestine and can be used as an index of fecal contamination. The category includes coliforms and enterococci. Discovery of these bacteria in a sample means that pathogens may also be present.

induced mutation Any alteration in DNA that occurs as a consequence of exposure to chemical or physical mutagens.

inducible enzyme An enzyme that increases in amount in direct proportion to the amount of substrate present.

inducible operon An operon that under normal circumstances is not transcribed. The presence of a specific inducer molecule can cause transcription of the operon to begin.

infection The entry, establishment, and multiplication of pathogenic organisms within a host.

infectious disease The state of damage or toxicity in the body caused by an infectious agent.

inflammation A natural, nonspecific response to tissue injury that protects the host from further damage. It stimulates immune reactivity and blocks the spread of an infectious agent.

inoculation The implantation of microorganisms into or upon culture media.

inorganic chemicals Molecules that lack the basic framework of the elements of carbon and hydrogen.

integument The outer surfaces of the body: skin, hair, nails, sweat glands, and oil glands.

interferon Naturally occurring polypeptides produced by fibroblasts and lymphocytes that can block viral replication and regulate a variety of immune reactions.

interferon gamma A protein produced by a virally infected cell that induces production of antiviral substances in neighboring cells. This defense prevents the production and maturation of viruses and thus terminates the viral infection.

interleukin A class of chemicals released from host cells that have potent effects on immunity.

intoxication Poisoning that results from the introduction of a toxin into body tissues through ingestion or injection.

intron The segments on split genes of eucaryotes that do not code for polypeptide. They can have regulatory functions. See *exon.*

in utero Literally means "in the uterus"; pertains to events or developments occurring before birth.

in vitro Literally means "in glass," signifying a process or reaction occurring in an artificial environment, as in a test tube or culture medium.

in vivo Literally means "in a living being," signifying a process or reaction occurring in a living thing.

iodophor A combination of iodine and an organic carrier that is a moderate-level disinfectant and antiseptic.

ion An unattached, charged particle.

ionic bond A chemical bond in which electrons are transferred and not shared between atoms.

ionization The aqueous dissociation of an electrolyte into ions.

ionizing radiation Radiant energy consisting of short-wave electromagnetic rays (X ray) or high-speed electrons that cause dislodgment of electrons on target molecules and create ions.

irradiation The application of radiant energy for diagnosis, therapy, disinfection, or sterilization.

irritability Capacity of cells to respond to chemical, mechanical, or light stimuli. This property helps cells adapt to the environment and obtain nutrients.

isograft Transplanted tissue from one monozygotic twin to the other; transplants between highly inbred animals that are genetically identical.

isolation The separation of microbial cells by serial dilution or mechanical dispersion on solid media to create discrete colonies.

isotonic Two solutions having the same osmotic pressure such that, when separated by a semipermeable membrane, there is no net movement of solvent in either direction.

isotope A version of an element that is virtually identical in all chemical properties to another version except that their atoms have slightly different atomic masses.

J

jaundice The yellowish pigmentation of skin, mucous membranes, sclera, deeper tissues, and excretions due to abnormal deposition of bile pigments. Jaundice is associated with liver infection, as with hepatitis B virus and leptospirosis.

K

Kaposi sarcoma A malignant or benign neoplasm that appears as multiple hemorrhagic sites on the skin, lymph nodes, and viscera and apparently involves the metastasis of abnormal blood vessel cells. It is a clinical feature of AIDS.

killed or inactivated vaccine A whole cell or intact virus preparation in which the microbes are dead or preserved and cannot multiply, but are still capable of conferring immunity.

killer T cells A T lymphocyte programmed to directly affix cells and kill them. See *cytotoxic.*

kingdom In the levels of classification, the second division from more general to more specific. Each domain is divided into kingdoms.

Koch's postulates A procedure to establish the specific cause of disease. In all cases of infection: (1) The agent must be found; (2) inoculations of a pure culture must reproduce the same disease in animals; (3) the agent must again be present in the experimental animal; and (4) a pure culture must again be obtained.

Koplik's spots Tiny red blisters with central white specks on the mucosal lining of the cheeks. Symptomatic of measles.

L

labile In chemistry, molecules or compounds that are chemically unstable in the presence of environmental changes.

lactose One of the carbohydrates commonly referred to as sugars. Lactose is commonly found in milk.

lactose (*lac*) operon Control system that manages the regulation of lactose metabolism. It is composed of three DNA segments, including a regulator, a control locus, and a structural locus.

lager The maturation process of beer, which is allowed to take place in large vats at a reduced temperature.

lagging strand The newly forming 5' DNA strand that is discontinuously replicated in segments (Okazaki fragments).

lag phase The early phase of population growth during which no signs of growth occur.

lantibiotics Short peptides produced by bacteria that inhibit the growth of other bacteria.

latency The state of being inactive. Example: a latent virus or latent infection.

leading strand The newly forming 3' DNA strand that is replicated in a continuous fashion without segments.

leaven To lighten food material by entrapping gas generated within it. Example: the rising of bread from the CO_2 produced by yeast or baking powder.

Legionnaire's disease Infection by *Legionella* bacterium. Weakly gram-negative rods are able to survive in aquatic habitats. Some forms may be fatal.

lepromas Skin nodules seen on the face of persons suffering from lepromatous leprosy. The skin folds and thickenings are caused by the overgrowth of *Mycobacterium leprae.*

lepromatous leprosy Severe, disfiguring leprosy characterized by widespread dissemination of the leprosy bacillus in deeper lesions.

leprosy See *Hansen's disease.*

lesion A wound, injury, or some other pathologic change in tissues.

leukocidin A heat-labile substance formed by some pyogenic cocci that impairs and sometimes lyses leukocytes.

leukocytes White blood cells. The primary infection-fighting blood cells.

leukocytosis An abnormally large number of leukocytes in the blood, which can be indicative of acute infection.

leukopenia A lower than normal leukocyte count in the blood that can be indicative of blood infection or disease.

leukotriene An unsaturated fatty acid derivative of arachidonic acid. Leukotriene functions in chemotactic activity, smooth muscle contractility, mucous secretion, and capillary permeability.

L form L-phase variants; wall-less forms of some bacteria that are induced by drugs or chemicals. These forms can be involved in infections.

ligase An enzyme required to seal the sticky ends of DNA pieces after splicing.

light-dependent reactions The series of reactions in photosynthesis that are driven by the light energy (photons) absorbed by chlorophyll. They involve splitting of water into hydrogens and oxygen, transport of electrons by NADP, and ATP synthesis.

light-independent reactions The series of reactions in photosynthesis that can proceed with or without light. It is a cyclic system that uses ATP from the light reactions to incorporate or fix carbon dioxide into organic compounds, leading to the production of glucose and other carbohydrates (also called the Calvin cycle).

lipase A fat-splitting enzyme. Example: triacylglycerol lipase separates the fatty acid chains from the glycerol backbone of triglycerides.

lipid A term used to describe a variety of substances that are not soluble in polar solvents such as water, but will dissolve in nonpolar solvents such as benzene and chloroform. Lipids include triglycerides, phospholipids, steroids, and waxes.

lipopolysaccharide A molecular complex of lipid and carbohydrate found in the bacterial cell wall. The lipopolysaccharide (LPS) of gram-negative bacteria is an endotoxin with generalized pathologic effects such as fever.

lithoautotroph Bacteria that rely on inorganic minerals to supply their nutritional needs. Sometimes referred to as chemoautotrophs.

lithosphere That part of the biosphere which encompasses the earth's crust, including rocks and minerals.

lithotroph An autotrophic microbe that derives energy from reduced inorganic compounds such as N_2S.

lobar pneumonia Infection involving whole segments (lobes) of the lungs, which may lead to consolidation and plugging of the alveoli and extreme difficulty in breathing.

localized infection Occurs when a microbe enters a specific tissue, infects it, and remains confined there.

loci A site on a chromosome occupied by a gene.

log phase Maximum rate of cell division during which growth is geometric in its rate of increase. Also called exponential growth phase.

lophotrichous Describing bacteria having a tuft of flagella at one or both poles.

lumen The cavity within a tubular organ.

lymphadenitis Inflammation of one or more lymph nodes. Also called lymphadenopathy.

lymphatic system A system of vessels and organs that serve as sites for development of immune cells and immune reactions. It includes the spleen, thymus, lymph nodes, and GALT.

lymphocyte The second most common form of white blood cells.

lyophilization A method for preserving microorganisms (and other substances) by freezing and then drying them directly from the frozen state.

lyse To burst.

lysin A complement-fixing antibody that destroys specific targeted cells. Examples: hemolysin and bacteriolysin.

lysis The physical rupture or deterioration of a cell.

lysogenic conversion A bacterium acquires a new genetic trait due to the presence of genetic material from an infecting phage.

lysogeny The indefinite persistence of bacteriophage DNA in a host without bringing about the production of virions.

lysosome A cytoplasmic organelle containing lysozyme and other hydrolytic enzymes.

lysozyme An enzyme found in sweat, tears, and saliva that breaks down bacterial peptidoglycan.

M

macromolecules Large, molecular compounds assembled from smaller subunits, most notably biochemicals.

macronutrient A chemical substance required in large quantities (phosphate, for example).

macrophage A white blood cell derived from a monocyte that leaves the circulation and enters tissues. These cells are important in nonspecific phagocytosis and in regulating, stimulating, and cleaning up after immune responses.

macroscopic Visible to the naked eye.

malt The grain, usually barley, that is sprouted to obtain digestive enzymes and dried for making beer.

maltose One of the carbohydrates referred to as sugars. A fermentable sugar formed from starch.

Mantoux test An intradermal screening test for tuberculin hypersensitivity. A red, firm patch of skin at the injection site greater than 10 mm in diameter after 48 hours is a positive result that indicates current or prior exposure to the TB bacillus.

mapping Determining the location of loci and other qualities of genomic DNA.

marker Any trait or factor of a cell, virus, or molecule that makes it distinct and recognizable. Example: a genetic marker.

mash In making beer, the malt grain is steeped in warm water, ground up, and fortified with carbohydrates to form mash.

mass number (MN) Measurement that reflects the number of protons and neutrons in an atom of a particular element.

mast cell A nonmotile connective tissue cell implanted along capillaries, especially in the lungs, skin, gastrointestinal tract, and genitourinary tract. Like a basophil, its granules store mediators of allergy.

matrix The dense ground substance between the cristae of a mitochondrion that serves as a site for metabolic reactions.

matter All tangible materials that occupy space and have mass.

maximum temperature The highest temperature at which an organism will grow.

mechanical vector An animal which transports an infectious agent but is not infected by it, such as houseflies whose feet become contaminated with feces.

medium (plural, *media*) A nutrient used to grow organisms outside of their natural habitats.

meiosis The type of cell division necessary for producing gametes in diploid organisms. Two nuclear divisions in rapid succession produce four gametocytes, each containing a haploid number of chromosomes.

membrane In a single cell, a thin double-layered sheet composed of lipids such as phospholipids and sterols and proteins.

memory (immunologic memory) The capacity of the immune system to recognize and act against an antigen upon second and subsequent encounters.

memory cell The long-lived progeny of a sensitized lymphocyte that remains in circulation and is genetically programmed to react rapidly with its antigen.

Mendosicutes Taxonomic category of bacteria that have unusual cell walls; archaea.

meninges The tough tri-layer membrane covering the brain and spinal cord. Consists of the dura mater, arachnoid mater, and pia mater.

meningitis An inflammation of the membranes (meninges) that surround and protect the brain. It is often caused by bacteria such as *Neisseria meningitidis* (the meningococcus) and *Haemophilus influenzae*.

merozoite The motile, infective stage of an apicomplexan parasite that comes from a liver or red blood cell undergoing multiple fission.

mesophile Microorganisms that grow at intermediate temperatures.

messenger RNA A single-stranded transcript that is a copy of the DNA template that corresponds to a gene.

metabolic analog Enzyme that mimics the natural substrate of an enzyme and vies for its active site.

metabolism A general term for the totality of chemical and physical processes occurring in a cell.

metabolites Small organic molecules that are intermediates in the stepwise biosynthesis or breakdown of macromolecules.

metachromatic Exhibiting a color other than that of the dye used to stain it.

methanogens Methane producers.

MHC Major histocompatibility complex. See *HLA*.

MIC Abbreviation for **m**inimum **i**nhibitory **c**oncentration. The lowest concentration of antibiotic needed to inhibit bacterial growth in a test system.

microaerophile An aerobic bacterium that requires oxygen at a concentration less than that in the atmosphere.

microbe See *microorganism*.

microbial ecology The study of microbes in their natural habitats.

microbiology A specialized area of biology that deals with living things ordinarily too small to be seen without magnification, including bacteria, archaea, fungi, protozoa and viruses.

microfilaments Cellular cytoskeletal element formed by thin protein strands that attach to cell membrane and form a network through the cytoplasm. Responsible for movement of cytoplasm.

micronutrient A chemical substance required in small quantities (trace metals, for example).

microorganism A living thing ordinarily too small to be seen without magnification; an organism of microscopic size.

microscopic Invisible to the naked eye.

microscopy Science that studies structure, magnification, lenses, and techniques related to use of a microscope.

microtubules Long hollow tubes in eucaryotic cells; maintain the shape of the cell and transport substances from one part of cell to another; involved in separating chromosomes in mitosis.

miliary tuberculosis Rapidly fatal tuberculosis due to dissemination of mycobacteria in the blood and formation of tiny granules in various organs and tissues. The term *miliary* means resembling a millet seed.

mineralization The process by which decomposers (bacteria and fungi) convert organic debris into inorganic and elemental form. It is part of the recycling process.

minimum inhibitory concentration (MIC) The smallest concentration of drug needed to visibly control microbial growth.

minimum temperature The lowest temperature at which an organism will grow.

miracidium The ciliated first-stage larva of a trematode. This form is infective for a corresponding intermediate host snail.

missense mutation A mutation in which a change in the DNA sequence results in a different amino acid being incorporated into a protein, with varying results.

mitochondrion A double-membrane organelle of eucaryotes that is the main site for aerobic respiration.

mitosis Somatic cell division that preserves the somatic chromosome number.

mixed acid fermentation An anaerobic degradation of pyruvic acid that results in more than one organic acid being produced (e.g. acetic acid, lactic acid, succinic acid).

mixed culture A container growing two or more different, known species of microbes.

mixed infection Occurs when several different pathogens interact simultaneously to produce an infection. Also called a synergistic infection.

molecule A distinct chemical substance that results from the combination of two or more atoms.

molluscum contagiosum Poxvirus-caused disease which manifests itself by the appearance of small lesions on the face, trunk, and limbs. Can be associated with sexual transmission.

monoclonal antibody An antibody produced by a clone of lymphocytes that respond to a particular antigenic determinant and generate identical antibodies only to that determinant. See *hybridoma*.

monocyte A large mononuclear leukocyte normally found in the lymph nodes, spleen, bone marrow, and loose connective tissue. This type of cell makes up 3% to 7% of circulating leukocytes.

monomer A simple molecule that can be linked by chemical bonds to form larger molecules.

mononuclear phagocyte system A collection of monocytes and macrophages scattered throughout the extracellular spaces that function to engulf and degrade foreign molecules.

monosaccharide A simple sugar such as glucose that is a basic building block for more complex carbohydrates.

monotrichous Describing a microorganism that bears a single flagellum.

morbidity A diseased condition.

mordant A chemical that fixes a dye in or on cells by forming an insoluble compound and thereby promoting retention of that dye. Example: Gram's iodine in the Gram stain.

morphology The study of organismic structure.

mortality rate Total number of deaths in a population attributable to a particular disease.

most probable number (MPN) Test used to detect the concentration of contaminants in water and other fluids.

motility Self-propulsion.

mumps Viral disease characterized by inflammation of the parotid glands.

must Juices expressed from crushed fruits that are used in fermentation for wine.

mutagen Any agent that induces genetic mutation. Examples: certain chemical substances, ultraviolet light, radioactivity.

mutant strain A subspecies of microorganism which has undergone a mutation, causing expression of a trait that differs from other members of that species.

mutation A permanent inheritable alteration in the DNA sequence or content of a cell.

mutualism Organisms living in an obligatory, but mutually beneficial, relationship.

mycelium The filamentous mass that makes up a mold. Composed of hyphae.

mycorrhizae Various species of fungi adapted in an intimate, mutualistic relationship to plant roots.

mycosis Any disease caused by a fungus.

N

NAD/NADH Abbreviations for the oxidized/reduced forms of nicotinamide adenine dinucleotide, an electron carrier. Also known as the vitamin niacin.

nanobes Cell-like particles, found in sediments and other geologic deposits, that some scientists speculate are the smallest bacteria. Short for nanobacteria.

narrow-spectrum Denotes drugs that are selective and limited in their effects. For example, they inhibit either gram-negative or gram-positive bacteria, but not both.

natural selection A process in which the environment places pressure on organisms to adapt and survive changing conditions. Only the survivors will be around to continue the life cycle and contribute their genes to future generations. This is considered a major factor in evolution of species.

necrosis A pathologic process in which cells and tissues die and disintegrate.

negative stain A staining technique that renders the background opaque or colored and leaves the object unstained so that it is outlined as a colorless area.

nematode A common name for helminths called roundworms.

neurotropic Having an affinity for the nervous system. Most likely to affect the spinal cord.

neutralization The process of combining an acid and a base until they reach a balanced proportion, with a pH value close to 7.

neutron An electrically neutral particle in the nuclei of all atoms except hydrogen.

neutrophil A mature granulocyte present in peripheral circulation, exhibiting a multilobular nucleus and numerous cytoplasmic granules that retain a neutral stain. The neutrophil is an active phagocytic cell in bacterial infection.

niche In ecology, an organism's biological role in or contribution to its community.

nitrification Phase of the nitrogen cycle in which ammonium is oxidized.

nitrogen base A ringed compound of which pyrimidines and purines are types.

nitrogen cycle The pathway followed by the element nitrogen as it circulates from inorganic sources in the nonliving environment to living things and back to the nonliving environment. The longtime reservoir is nitrogen gas in the atmosphere.

nitrogen fixation A process occurring in certain bacteria in which atmospheric N_2 gas is converted to a form (NH_4) usable by plants.

nitrogenous base A nitrogen-containing molecule found in DNA and RNA that provides the basis for the genetic code. Adenine, guanine, and cytosine are found in both DNA and RNA while thymine is found exclusively in DNA and uracil is found exclusively in RNA.

nomenclature A set system for scientifically naming organisms, enzymes, anatomical structures, etc.

non-communicable An infectious disease that does not arrive through transmission of an infectious agent from host to host.

noncompetitive inhibition Form of enzyme inhibition that involves binding of a regulatory molecule to a site other than the active site.

nonionizing radiation Method of microbial control, best exemplified by ultraviolet light, that causes the formation of abnormal bonds within the DNA of microbes, increasing the rate of mutation. The primary limitation of nonionizing radiation is its inability to penetrate beyond the surface of an object.

nonpolar A term used to describe an electrically neutral molecule formed by covalent bonds between atoms that have the same or similar electronegativity.

non-self Molecules recognized by the immune system as containing foreign markers, indicating a need for immune response.

nonsense codon A triplet of mRNA bases that does not specify an amino acid but signals the end of a polypeptide chain.

nonsense mutation A mutation that changes an amino acid-producing codon into a stop codon, leading to premature termination of a protein.

normal flora The native microbial forms that an individual harbors.

nosocomial infection An infection not present upon admission to a hospital but incurred while being treated there.

nucleocapsid In viruses, the close physical combination of the nucleic acid with its protective covering.

nucleoid The basophilic nuclear region or nuclear body that contains the bacterial chromosome.

nucleolus A granular mass containing RNA that is contained within the nucleus of a eucaryotic cell.

nucleosome Structure in the packaging of DNA. Formed by the DNA strands wrapping around the histone protein to form nucleus bodies arranged like beads on a chain.

nucleotide The basic structural unit of DNA and RNA; each nucleotide consists of a phosphate, a sugar (ribose in RNA, deoxyribose in DNA), and a nitrogenous base such as adenine, guanine, cytosine, thymine (DNA only) or uracil (RNA only).

numerical aperture In microscopy, the amount of light passing from the object and into the object in order to maximize optical clarity and resolution.

nutrient Any chemical substance that must be provided to a cell for normal metabolism and growth. Macronutrients are required in large amounts, and micronutrients in small amounts.

nutrition The acquisition of chemical substances by a cell or organism for use as an energy source or as building blocks of cellular structures.

O

obligate Without alternative; restricted to a particular characteristic. Example: an obligate parasite survives and grows only in a host; an obligate aerobe must have oxygen to grow; an obligate anaerobe is destroyed by oxygen.

Okazaki fragment In replication of DNA, a segment formed on the lagging strand in which biosynthesis is conducted in a discontinuous manner dictated by the $5' \rightarrow 3'$ DNA polymerase orientation.

oligodynamic action A chemical having antimicrobial activity in minuscule amounts. Example: certain heavy metals are effective in a few parts per billion.

oligonucleotides Short pieces of DNA or RNA that are easier to handle than long segments.

oligotrophic Nutrient-deficient ecosystem.

oncogene A naturally occurring type of gene that when activated can transform a normal cell into a cancer cell.

oncovirus Mammalian virus capable of causing malignant tumors.

oocyst The encysted form of a fertilized macrogamete or zygote; typical in the life cycles of apicomplexan parasites.

operator In an operon sequence, the DNA segment where transcription of structural genes is initiated.

operon A genetic operational unit that regulates metabolism by controlling mRNA production. In sequence, the unit consists of a regulatory gene, inducer or repressor control sites, and structural genes.

opportunistic In infection, ordinarily nonpathogenic or weakly pathogenic microbes that cause disease primarily in an immunologically compromised host.

opsonization The process of stimulating phagocytosis by affixing molecules (opsonins such as antibodies and complement) to the surfaces of foreign cells or particles.

optimum temperature The temperature at which a species shows the most rapid growth rate.

orbitals The pathways of electrons as they rotate around the nucleus of an atom.

order In the levels of classification, the division of organisms that follows class. Increasing similarity may be noticed among organisms assigned to the same order.

organelle A small component of eucaryotic cells that is bounded by a membrane and specialized in function.

organic chemicals Molecules that contain the basic framework of the elements carbon and hydrogen.

osmophile A microorganism that thrives in a medium having high osmotic pressure.

osmosis The diffusion of water across a selectively permeable membrane in the direction of lower water concentration.

osteomyelitis A focal infection of the internal structures of long bones, leading to pain and inflammation. Often caused by *Staphylococcus aureus*.

oxidation In chemical reactions, the loss of electrons by one reactant.

oxidation-reduction Redox reactions, in which paired sets of molecules participate in electron transfers.

oxidative phosphorylation The synthesis of ATP using energy given off during the electron transport phase of respiration.

oxygenic Any reaction that gives off oxygen; usually in reference to the result of photosynthesis in eucaryotes and cyanobacteria.

P

palindrome A word, verse, number, or sentence that reads the same forward or backward. Palindromes of nitrogen bases in DNA have genetic significance as transposable elements, as regulatory protein targets, and in DNA splicing.

palisades The characteristic arrangement of *Corynebacterium* cells resembling a row of fence posts and created by snapping.

pandemic A disease afflicting an increased proportion of the population over a wide geographic area (often worldwide).

papilloma Benign, squamous epithelial growth commonly referred to as a wart.

parasite An organism that lives on or within another organism (the host), from which it obtains nutrients and enjoys protection. The parasite produces some degree of harm in the host.

parasitism A relationship between two organisms in which the host is harmed in some way while the colonizer benefits.

parenteral Administering a substance into a body compartment other than through the gastrointestinal tract, such as via intravenous, subcutaneous, intramuscular, or intramedullary injection.

paroxysmal Events characterized by sharp spasms or convulsions; sudden onset of a symptom such as fever and chills.

passive carrier Persons who mechanically transfer a pathogen without ever being infected by it. For example, a health care worker who doesn't wash his/her hands adequately between patients.

passive immunity Specific resistance that is acquired indirectly by donation of preformed immune substances (antibodies) produced in the body of another individual.

passive transport Nutrient transport method that follows basic physical laws and does not require direct energy input from the cell.

pasteurization Heat treatment of perishable fluids such as milk, fruit juices, or wine to destroy heat-sensitive vegetative cells, followed by rapid chilling to inhibit growth of survivors and germination of spores. It prevents infection and spoilage.

pathogen Any agent (usually a virus, bacterium, fungus, protozoan, or helminth) that causes disease.

pathogenicity The capacity of microbes to cause disease.

pathognomic Distinctive and particular to a single disease, suggestive of a diagnosis.

pathologic Capable of inducing physical damage on the host.

pathology The structural and physiological effects of disease on the body.

pellicle A membranous cover; a thin skin, film, or scum on a liquid surface; a thin film of salivary glycoproteins that forms over newly cleaned tooth enamel when exposed to saliva.

pelvic inflammatory disease (PID) An infection of the uterus and fallopian tubes that has ascended from the lower reproductive tract. Caused by gonococci and chlamydias.

penetration (viral) The step in viral multiplication in which virus enters the host cell.

penicillinase An enzyme that hydrolyzes penicillin; found in penicillin-resistant strains of bacteria.

penicillins A large group of naturally occurring and synthetic antibiotics produced by *Penicillium* mold and active against the cell wall of bacteria.

pentose A monosaccharide with five carbon atoms per molecule. Examples: arabinose, ribose, xylose.

peptide Molecule composed of short chains of amino acids, such as a dipeptide (two amino acids), a tripeptide (three), and a tetrapeptide (four).

peptide bond The covalent union between two amino acids that forms between the amine group of one and the carboxyl group of the other. The basic bond of proteins.

peptidoglycan A network of polysaccharide chains cross-linked by short peptides that forms the rigid part of bacterial cell walls. Gram-negative bacteria have a smaller amount of this rigid structure than do gram-positive bacteria.

perforin Proteins released by cytotoxic T cells that produce pores in target cells.

perinatal In childbirth, occurring before, during, or after delivery.

period of invasion The period during a clinical infection when the infectious agent multiplies at high levels, exhibits its greatest toxicity and becomes well established in the target tissues.

periodontal Involving the structures that surround the tooth.

periplasmic space The region between the cell wall and cell membrane of the cell envelopes of gram-negative bacteria.

peritrichous In bacterial morphology, having flagella distributed over the entire cell.

petechiae Minute hemorrhagic spots in the skin that range from pinpoint- to pinhead-sized.

Peyer's patches Oblong lymphoid aggregates of the gut located chiefly in the wall of the terminal and small intestine. Along with the tonsils and appendix, Peyer's patches make up the gut-associated lymphoid tissue that responds to local invasion by infectious agents.

pH The symbol for the negative logarithm of the H ion concentration; p (power) or $[H^+]_{10}$. A system for rating acidity and alkalinity.

phage A bacteriophage; a virus that specifically parasitizes bacteria.

phagocyte A class of white blood cells capable of engulfing other cells and particles.

phagocytosis A type of endocytosis in which the cell membrane actively engulfs large particles or cells into vesicles.

phagolysosome A body formed in a phagocyte, consisting of a union between a vesicle containing the ingested particle (the phagosome) and a vacuole of hydrolytic enzymes (the lysosome).

phenotype The observable characteristics of an organism produced by the interaction between its genetic potential (genotype) and the environment.

phosphate An acidic salt containing phosphorus and oxygen that is an essential inorganic component of DNA, RNA, and ATP.

phospholipid A class of lipids that compose a major structural component of cell membranes.

phosphorylation Process in which inorganic phosphate is added to a compound.

photoactivation (light repair) A mechanism for repairing DNA with ultraviolet light-induced mutations using an enzyme (photolyase) that is activated by visible light.

photoautotroph An organism that utilizes light for its energy and carbon dioxide chiefly for its carbon needs.

photon A subatomic particle released by electromagnetic sources such as radiant energy (sunlight). Photons are the ultimate source of energy for photosynthesis.

photophosphorylation The process of electron transport during photosynthesis that results in the synthesis of ATP from ADP.

photosynthesis A process occurring in plants, algae, and some bacteria that traps the sun's energy and converts it to ATP in the cell. This energy is used to fix CO_2 into organic compounds.

phototrophs Microbes that use photosynthesis to feed.

phylum In the levels of classification, the third level of classification from general to more specific. Each kingdom is divided into numerous phyla. Sometimes referred to as a division.

physiology The study of the function of an organism.

phytoplankton The collection of photosynthetic microorganisms (mainly algae and cyanobacteria) that float in the upper layers of aquatic habitats where sun penetrates. These microbes are the basis of aquatic food pyramids and, together with zooplankton, make up the plankton.

pili Small, stiff filamentous appendages in gram-negative bacteria that function in DNA exchange during bacterial conjugation.

pilus A hollow appendage used to bring two bacterial cells together to transfer DNA.

pinocytosis The engulfment, or endocytosis, of liquids by extensions of the cell membrane.

plague Zoonotic disease caused by infection with *Yersinia pestis*. The pathogen is spread by flea vectors and harbored by various rodents.

plankton Minute animals (zooplankton) or plants (phytoplankton) that float and drift in the limnetic zone of bodies of water.

plantar warts Deep, painful warts on the soles of the feet as a result of infection by human papillomavirus.

plaque In virus propagation methods, the clear zone of lysed cells in tissue culture or chick embryo membrane that corresponds to the area containing viruses. In dental application, the filamentous mass of microbes that adheres tenaciously to the tooth and predisposes to caries, calculus, or inflammation.

plasma The carrier fluid element of blood.

plasma cell A progeny of an activated B cell that actively produces and secretes antibodies.

plasmids Extrachromosomal genetic units characterized by several features. A plasmid is a double-stranded DNA that is smaller than and replicates independently of the cell chromosome; it bears genes that are not essential for cell growth; it can bear genes that code for adaptive traits; and it is transmissible to other bacteria.

platelet-activating factor A substance released from basophils that causes release of allergic mediators and the aggregation of platelets.

platelets Formed elements in the blood which develop when megakaryocytes disintegrate. Platelets are involved in hemostasis and blood clotting.

pleomorphism Normal variability of cell shapes in a single species.

pluripotential Stem cells having the developmental plasticity to give rise to more than one type. Example: undifferentiated blood cells in the bone marrow.

pneumococcus Common name for *Streptococcus pneumoniae*, the major cause of bacterial pneumonia.

pneumonia An inflammation of the lung leading to accumulation of fluid and respiratory compromise.

pneumonic plague The acute, frequently fatal form of pneumonia caused by *Yersinia pestis*.

point mutation A change that involves the loss, substitution, or addition of one or a few nucleotides.

polar Term to describe a molecule with an asymmetrical distribution of charges. Such a molecule has a negative pole and a positive pole.

poliomyelitis An acute enteroviral infection of the spinal cord that can cause neuromuscular paralysis.

polyclonal In reference to a collection of antibodies with mixed specificities that arose from more than one clone of B cells.

polymer A macromolecule made up of a chain of repeating units. Examples: starch, protein, DNA.

polymerase An enzyme that produces polymers through catalyzing bond formation between building blocks (polymerization).

polymerase chain reaction (PCR) A technique that amplifies segments of DNA for testing. Using denaturation, primers, and heat-resistant DNA polymerase, the number can be increased several million-fold.

polymorphonuclear leukocytes (PMNLs) White blood cells with variously shaped nuclei. Although this term commonly denotes all granulocytes, it is used especially for the neutrophils.

polymyxin A mixture of antibiotic polypeptides from *Bacillus polymyxa* that are particularly effective against gram-negative bacteria.

polypeptide A relatively large chain of amino acids linked by peptide bonds.

polyribosomal complex An assembly line for mass production of proteins composed of a chain of ribosomes involved in mRNA transcription.

polysaccharide A carbohydrate that can be hydrolyzed into a number of monosaccharides. Examples: cellulose, starch, glycogen.

population A group of organisms of the same species living simultaneously in the same habitat. A group of different populations living together constitutes the community level.

porin Transmembrane proteins of the outer membrane of gram-negative cells that permit transport of small molecules into the periplasmic space but bar the penetration of larger molecules.

portal of entry Route of entry for an infectious agent; typically a cutaneous or membranous route.

portal of exit Route through which a pathogen departs from the host organism.

positive stain A method for coloring microbial specimens that involves a chemical that sticks to the specimen to give it color.

potable Describing water that is relatively clear, odor-free, and safe to drink.

PPNG Penicillinase-producing *Neisseria gonorrhoeae*.

prevalence The total number of cases of a disease in a certain area and time period.

primary infection An initial infection in a previously healthy individual that is later complicated by an additional (secondary) infection.

primary response The first response of the immune system when exposed to an antigen.

primary structure Initial protein organization described by type, number, and order of amino acids in the chain. The primary structure varies extensively from protein to protein.

primers Synthetic oligonucleotides of known sequence that serve as landmarks to indicate where DNA amplification will begin.

prion A concocted word to denote "proteinaceous infectious agent"; a cytopathic protein associated with the slow-virus spongiform encephalopathies of humans and animals.

probes Small fragments of single-stranded DNA (RNA) that are known to be complementary to the specific sequence of DNA being studied.

probiotics Preparations of live microbes used as a preventive or therapeutic measure to displace or compete with potential pathogens.

procaryotic cell Small cells, lacking special structures such as a nucleus and organelles. All procaryotes are microorganisms.

prodromal stage A short period of mild symptoms occurring at the end of the period of incubation. It indicates the onset of disease.

producer An organism that synthesizes complex organic compounds from simple inorganic molecules. Examples would be photosynthetic microbes and plants. These organisms are solely responsible for originating food pyramids and are the basis for life on earth (also called autotroph).

proglottid The egg-generating segment of a tapeworm that contains both male and female organs.

progressive multifocal leukoencephalopathy An uncommon, fatal complication of infection with JC virus (polyoma virus).

promastigote A morphological variation of the trypanosome parasite responsible for leishmaniasis.

promoter Part of an operon sequence. The DNA segment that is recognized by RNA polymerase as the starting site for transcription.

promoter region The site composed of a short signaling DNA sequence that RNA polymerase recognizes and binds to commence transcription.

prophage A lysogenized bacteriophage; a phage that is latently incorporated into the host chromosome instead of undergoing viral replication and lysis.

prophylactic Any device, method, or substance used to prevent disease.

prostaglandin A hormonelike substance that regulates many body functions. Prostaglandin comes from a family of organic acids containing 5-carbon rings that are essential to the human diet.

protease inhibitors Drugs that act to prevent the assembly of functioning viral particles.

protein Predominant organic molecule in cells, formed by long chains of amino acids.

proton An elementary particle that carries a positive charge. It is identical to the nucleus of the hydrogen atom.

protoplast A bacterial cell whose cell wall is completely lacking and that is vulnerable to osmotic lysis.

protozoa A group of single-celled, eucaryotic organisms.

pseudohypha A chain of easily separated, spherical to sausage-shaped yeast cells partitioned by constrictions rather than by septa.

pseudomembrane A tenacious, noncellular mucous exudate containing cellular debris that tightly blankets the mucosal surface in infections such as diphtheria and pseudomembranous enterocolitis.

pseudopodium A temporary extension of the protoplasm of an ameboid cell. It serves both in ameboid motion and for food gathering (phagocytosis).

pseudopods Protozoan appendage responsible for motility. Also called "false feet."

psychrophile A microorganism that thrives at low temperature (0°–20°C), with a temperature optimum of 0°–15°C.

pulmonary Occurring in the lungs. Examples include pulmonary anthrax and pulmonary nocardiosis.

pure culture A container growing a single species of microbe whose identity is known.

purine A nitrogen base that is an important encoding component of DNA and RNA. The two most common purines are adenine and guanine.

pus The viscous, opaque, usually yellowish matter formed by an inflammatory infection. It consists of serum exudate, tissue debris, leukocytes, and microorganisms.

pyogenic Pertains to pus formers, especially the pyogenic cocci: pneumococci, streptococci, staphylococci, and neisseriae.

pyrimidine Nitrogen bases that help form the genetic code on DNA and RNA. Uracil, thymine, and cytosine are the most important pyrimidines.

pyrimidine dimer The union of two adjacent pyrimidines on the same DNA strand, brought about by exposure to ultraviolet light. It is a form of mutation.

pyrogen A substance that causes a rise in body temperature. It can come from pyrogenic microorganisms or from polymorphonuclear leukocytes (endogenous pyrogens).

Q

quaternary structure Most complex protein structure characterized by the formation of large, multiunit proteins by more than one of the polypeptides. This structure is typical of antibodies and some enzymes that act in cell synthesis.

quats A word that pertains to a family of surfactants called quaternary ammonium compounds. These detergents are only weakly microbicidal and are used as sanitizers and preservatives.

quinolone A class of synthetic antimicrobic drugs with broad-spectrum effects.

R

rabies The only rhabdovirus that infects humans. Zoonotic disease characterized by fatal meningoencephalitis.

radiation Electromagnetic waves or rays, such as those of light given off from an energy source.

radioactive isotopes Unstable isotopes whose nuclei emit particles of radiation. This emission is called radioactivity or radioactive decay. Three naturally occurring emissions are alpha, beta, and gamma radiation.

reactants Molecules entering or starting a chemical reaction.

real image An image formed at the focal plane of a convex lens. In the compound light microscope, it is the image created by the objective lens.

receptor Cell surface molecules involved in recognition, binding, and intracellular signaling.

recombinant DNA A technology, also known as genetic engineering, that deliberately modifies the genetic structure of an organism to create novel products, microbes, animals, plants, and viruses.

recombination A type of genetic transfer in which DNA from one organism is donated to another.

recycling A process which converts unusable organic matter from dead organisms back into their essential inorganic elements and returns them to their nonliving reservoirs to make them available again for living organisms. This is a common term that means the same as mineralization and decomposition.

redox Denoting an oxidation-reduction reaction.

reduction In chemistry, the gain of electrons.

redundancy The property of the genetic code which allows an amino acid to be specified by several different codons.

refraction In optics, the bending of light as it passes from one medium to another with a different index of refraction.

regulated enzymes Enzymes whose extent of transcription or translation are influenced by changes in the environment.

regulator DNA segment that codes for a protein capable of repressing an operon.

regulatory site The location on an enzyme where a certain substance can bind and block the enzyme's activity.

rennin The enzyme casein coagulase, which is used to produce curd in the processing of milk and cheese.

replication In DNA synthesis, the semiconservative mechanisms that ensure precise duplication of the parent DNA strands.

replication fork The Y-shaped point on a replicating DNA molecule where the DNA polymerase is synthesizing new strands of DNA.

reportable disease Those diseases that must be reported to health authorities by law.

repressible operon An operon that under normal circumstances is transcribed. The buildup of the operon's amino acid product causes transcription of the operon to stop.

repressor The protein product of a repressor gene that combines with the operator and arrests the transcription and translation of structural genes.

reservoir In disease communication, the natural host or habitat of a pathogen.

resident flora The deeper, more stable microflora that inhabit the skin and exposed mucous membranes, as opposed to the superficial, variable, transient population.

resistance (R) factor Plasmids, typically shared among bacteria by conjugation, that provide resistance to the effects of antibiotics.

resolving power The capacity of a microscope lens system to accurately distinguish between two separate entities that lie close to each other. Also called resolution.

respiratory chain A series of enzymes that transfer electrons from one to another, resulting in the formation of ATP. It is also known as the electron transport chain. The chain is located in the cell membrane of bacteria and in the inner mitochondrial membrane of eucaryotes.

respiratory syncytial virus (RSV) An RNA virus that infects the respiratory tract. RSV is the most prevalent cause of respiratory infection in newborns.

restriction endonuclease An enzyme present naturally in cells that cleaves specific locations on DNA. It is an important means of inactivating viral genomes, and it is also used to splice genes in genetic engineering.

reticuloendothelial system Also known as the mononuclear phagocyte system, it pertains to a network of fibers and phagocytic cells (macrophages) that permeates the tissues of all organs. Examples: Kupffer cells in liver sinusoids, alveolar phagocytes in the lung, microglia in nervous tissue.

retrovirus A group of RNA viruses (including HIV) that have the mechanisms for converting their genome into a double strand of DNA that can be inserted on a host's chromosome.

reverse transcriptase The enzyme possessed by retroviruses that carries out the reversion of RNA to DNA—a form of reverse transcription.

Reye's syndrome A sudden, usually fatal neurological condition that occurs in children after a viral infection. Autopsy shows cerebral edema and marked fatty change in the liver and renal tubules.

Rh factor An isoantigen that can trigger hemolytic disease in newborns due to incompatibility between maternal and infant blood factors.

rhizobia Bacteria that live in plant roots and supply supplemental nitrogen that boosts plant growth.

rhizosphere The zone of soil, complete with microbial inhabitants, in the immediate vicinity of plant roots.

ribonucleic acid (RNA) The nucleic acid responsible for carrying out the hereditary program transmitted by an organism's DNA.

ribose A 5-carbon monosaccharide found in RNA.

ribosome A bilobed macromolecular complex of ribonucleoprotein that coordinates the codons of mRNA with tRNA anticodons and, in so doing, constitutes the peptide assembly site.

ribozyme A part of an RNA-containing enzyme in eucaryotes that removes intervening sequences of RNA called introns and splices together the true coding sequences (exons) to form a mature messenger RNA.

rickettsias Medically important family of bacteria, commonly carried by ticks, lice, and fleas. Significant cause of important emerging diseases.

ringworm A superficial mycosis caused by various dermatophytic fungi. This common name is actually a misnomer.

RNA polymerase Enzyme process that translates the code of DNA to RNA.

rolling circle An intermediate stage in viral replication of circular DNA into linear DNA.

root nodules Small growths on the roots of legume plants that arise from a symbiotic association between the plant tissues and bacteria (Rhizobia). This association allows fixation of nitrogen gas from the air into a usable nitrogen source for the plant.

rosette formation A technique for distinguishing surface receptors on T cells by reacting them with sensitized indicator sheep red blood cells. The cluster of red cells around the central white blood cell resembles a little rose blossom and is indicative of the type of receptor.

rough endoplasmic reticulum (RER) Microscopic series of tunnels that originates in the outer membrane of the nuclear envelope and is used in transport and storage. Large numbers of ribosomes, partly attached to the membrane, give the rough appearance.

rubeola (red measles) Acute disease caused by infection with Morbillivirus.

S

saccharide Scientific term for sugar. Refers to a simple carbohydrate with a sweet taste.

salpingitis Inflammation of the fallopian tubes.

sanitize To clean inanimate objects using soap and degerming agents so that they are safe and free of high levels of microorganisms.

saprobe A microbe that decomposes organic remains from dead organisms. Also known as a saprophyte or saprotroph.

sarcina A cubical packet of 8, 16, or more cells; the cellular arrangement of the genus *Sarcina* in the family Micrococcaceae.

saturation The complete occupation of the active site of a carrier protein or enzyme by the substrate.

schistosomiasis Infection by blood fluke, often as a result of contact with contaminated water in rivers and streams. Symptoms appear in liver, spleen, or urinary system depending on species of *Schistosoma*. Infection may be chronic.

schizogony A process of multiple fission whereby first the nucleus divides several times, and subsequently the cytoplasm is subdivided for each new nucleus during cell division.

scientific method Principles and procedures for the systematic pursuit of knowledge, involving the recognition and formulation of a problem, the collection of data through observation and experimentation, and the formulation and testing of a hypothesis.

scolex The anterior end of a tapeworm characterized by hooks and/or suckers for attachment to the host.

sebaceous glands The sebum- (oily, fatty) secreting glands of the skin.

secondary infection An infection that compounds a preexisting one.

secondary response The rapid rise in antibody titer following a repeat exposure to an antigen that has been recognized from a previous exposure. This response is brought about by memory cells produced as a result of the primary exposure.

secondary structure Protein structure that occurs when the functional groups on the outer surface of the molecule interact by forming hydrogen bonds. These bonds cause the amino acid chain to either twist, forming a helix, or to pleat into an accordion pattern called a β-pleated sheet.

secretory antibody The immunoglobulin (IgA) that is found in secretions of mucous membranes and serves as a local immediate protection against infection.

selectively toxic Property of an antimicrobial agent to be highly toxic against its target microbe while being far less toxic to other cells, particularly those of the host organism.

selective media Nutrient media designed to favor the growth of certain microbes and to inhibit undesirable competitors.

self Natural markers of the body that are recognized by the immune system.

self-limited Applies to an infection that runs its course without disease or residual effects.

semiconservative replication In DNA replication, the synthesis of paired daughter strands, each retaining a parent strand template.

semisolid media Nutrient media with a firmness midway between that of a broth (a liquid medium) and an ordinary solid medium; motility media.

semisynthetic Drugs which, after being naturally produced by bacteria, fungi, or other living sources, are chemically modified in the laboratory.

sensitizing dose The initial effective exposure to an antigen or an allergen that stimulates an immune response. Often applies to allergies.

sepsis The state of putrefaction; the presence of pathogenic organisms or their toxins in tissue or blood.

septicemia Systemic infection associated with microorganisms multiplying in circulating blood.

septic shock Blood infection resulting in a pathological state of low blood pressure accompanied by a reduced amount of blood circulating to vital organs. Endotoxins of all gram-negative bacteria can cause shock, but most clinical cases are due to gram-negative enteric rods.

septum A partition or cellular cross wall, as in certain fungal hyphae.

sequela A morbid complication that follows a disease.

sequencing Determining the actual order and types of bases in a segment of DNA.

serology The branch of immunology that deals with *in vitro* diagnostic testing of serum.

seropositive Showing the presence of specific antibody in a serological test. Indicates ongoing infection.

serotonin A vasoconstrictor that inhibits gastric secretion and stimulates smooth muscle.

serotyping The subdivision of a species or subspecies into an immunologic type, based upon antigenic characteristics.

serum The clear fluid expressed from clotted blood that contains dissolved nutrients, antibodies, and hormones but not cells or clotting factors.

serum sickness A type of immune complex disease in which immune complexes enter circulation, are carried throughout the body, and are deposited in the blood vessels of the kidney, heart, skin, and joints. The condition may become chronic.

severe acute respiratory syndrome (SARS) A severe respiratory disease caused by infection with a newly described coronavirus.

severe combined immunodeficiencies A collection of syndromes occurring in newborns caused by a genetic defect that knocks out both B and T cell types of immunity. There are several versions of this disease, termed SCIDS for short.

sex pilus A conjugative pilus.

sexually transmitted disease (STD) Infections resulting from pathogens that enter the body via sexual intercourse or intimate, direct contact.

shingles Lesions produced by reactivated human herpesvirus 3 (chickenpox) infection; also known as herpes zoster.

sign Any abnormality uncovered upon physical diagnosis that indicates the presence of disease. A sign is an objective assessment of disease, as opposed to a symptom, which is the subjective assessment perceived by the patient.

silent mutation A mutation that, because of the degeneracy of the genetic code, results in a nucleotide change in both the DNA and mRNA but not the resultant amino acid and thus, not the protein.

simple stain Type of positive staining technique that uses a single dye to add color to cells so that they are easier to see. This technique tends to color all cells the same color.

smooth endoplasmic reticulum (SER) A microscopic series of tunnels lacking ribosomes that functions in the nutrient processing function of a cell.

solute A substance that is uniformly dispersed in a dissolving medium or solvent.

solution A mixture of one or more substances (solutes) that cannot be separated by filtration or ordinary settling.

solvent A dissolving medium.

somatic (O or cell wall antigen) One of the three major antigens commonly used to differentiate gram-negative enteric bacteria.

source The person or item from which an infection is directly acquired. See *reservoir*.

Southern blot A technique that separates fragments of DNA using electrophoresis and identifies them by hybridization.

species In the levels of classification, the most specific level of organization.

specificity Limited to a single, precise characteristic or action.

spheroplast A gram-negative cell whose peptidoglycan, when digested by lysozyme, remains intact but is osmotically vulnerable.

spike A receptor on the surface of certain enveloped viruses that facilitates specific attachment to the host cell.

spirillum A type of bacterial cell with a rigid spiral shape and external flagella.

spirochete A coiled, spiral-shaped bacterium that has endoflagella and flexes as it moves.

spontaneous generation Early belief that living things arose from vital forces present in nonliving, or decomposing, matter.

spontaneous mutation A mutation in DNA caused by random mistakes in replication and not known to be influenced by any mutagenic agent. These mutations give rise to an organism's natural, or background, rate of mutation.

sporadic Description of a disease which exhibits new cases at irregular intervals in unpredictable geographic locales.

sporangium A fungal cell in which asexual spores are formed by multiple cell cleavage.

spore A differentiated, specialized cell form that can be used for dissemination, for survival in times of adverse conditions, and/or for reproduction. Spores are usually unicellular and may develop into gametes or vegetative organisms.

sporicide A chemical agent capable of destroying bacterial endospores.

sporozoite One of many minute elongated bodies generated by multiple division of the oocyst. It is the infectious form of the malarial parasite that is harbored in the salivary gland of the mosquito and inoculated into the victim during feeding.

sporulation The process of spore formation.

start codon The nucleotide triplet AUG that codes for the first amino acid in protein sequences.

starter culture The sizeable inoculation of pure bacterial, mold, or yeast sample for bulk processing, as in the preparation of fermented foods, beverages, and pharmaceuticals.

stasis A state of rest or inactivity; applied to nongrowing microbial cultures. Also called microbistasis.

stationary growth phase Survival mode in which cells either stop growing or grow very slowly.

stem cells Pluripotent, undifferentiated cells.

sterile Completely free of all life forms, including spores and viruses.

sterilization Any process that completely removes or destroys all viable microorganisms, including viruses, from an object or habitat. Material so treated is sterile.

STORCH Acronym for common infections of the fetus and neonate. Storch stands for **s**yphilis, **t**oxoplasmosis, **o**ther diseases (hepatitis B, AIDS and chlamydiosis), **r**ubella, **c**ytomegalovirus, and **h**erpes simplex virus.

strain In microbiology, a set of descendants cloned from a common ancestor that retain the original characteristics. Any deviation from the original is a different strain.

streptolysin A hemolysin produced by streptococci.

strict, or obligate anaerobe An organism which does not use oxygen gas in metabolism and cannot survive in oxygen's presence.

stroma The matrix of the chloroplast that is the site of the dark reactions.

structural gene A gene that codes for the amino acid sequence (peptide structure) of a protein.

subacute Indicates an intermediate status between acute and chronic disease.

subacute sclerosing panencephalitis (SSPE) A complication of measles infection in which progressive neurological degeneration of the cerebral cortex invariably leads to coma and death.

subclinical A period of inapparent manifestations that occurs before symptoms and signs of disease appear.

subculture To make a second-generation culture from a well-established colony of organisms.

subcutaneous The deepest level of the skin structure.

substrate The specific molecule upon which an enzyme acts.

subunit vaccine A vaccine preparation that contains only antigenic fragments such as surface receptors from the microbe. Usually in reference to virus vaccines.

sucrose One of the carbohydrates commonly referred to as sugars. Common table or cane sugar.

sulfonamide Antimicrobial drugs that interfere with the essential metabolic process of bacteria and some fungi.

superantigens Bacterial toxins that are potent stimuli for T cells and can be a factor in diseases such as toxic shock.

superficial mycosis A fungal infection located in hair, nails, and the epidermis of the skin.

superinfection An infection occurring during antimicrobial therapy that is caused by an overgrowth of drug-resistant microorganisms.

superoxide A toxic derivative of oxygen; (O_2^-).

surfactant A surface-active agent that forms a water-soluble interface. Examples: detergents, wetting agents, dispersing agents, and surface tension depressants.

sylvatic Denotes the natural presence of disease among wild animal populations. Examples: sylvatic (sylvan) plague, rabies.

symbiosis An intimate association between individuals from two species; used as a synonym for mutualism.

symptom The subjective evidence of infection and disease as perceived by the patient.

syncytium A multinucleated protoplasmic mass formed by consolidation of individual cells.

syndrome The collection of signs and symptoms that, taken together, paint a portrait of the disease.

synergism The coordinated or correlated action by two or more drugs or microbes that results in a heightened response or greater activity.

syngamy Conjugation of the gametes in fertilization.

synthesis (viral) The step in viral multiplication in which viral genetic material and proteins are made through replication and transcription/translation.

syphilis A sexually transmitted bacterial disease caused by the spirochete *Treponema pallidum*.

systemic Occurring throughout the body; said of infections that invade many compartments and organs via the circulation.

T

Taq polymerase DNA polymerase from the thermophilic bacterium *Thermus aquaticus* that enables high-temperature replication of DNA required for the polymerase chain reaction.

tartar See *calculus*.

taxa Taxonomic categories.

taxonomy The formal system for organizing, classifying, and naming living things.

temperate phage A bacteriophage that enters into a less virulent state by becoming incorporated into the host genome as a prophage instead of in the vegetative or lytic form that eventually destroys the cell.

template The strand in a double stranded DNA molecule which is used as a model to synthesize a complementary strand of DNA or RNA during replication or transcription.

Tenericutes Taxonomic category of bacteria that lack cell walls.

teratogenic Causing abnormal fetal development.

tertiary structure Protein structure that results from additional bonds forming between functional groups in a secondary structure, creating a three-dimensional mass.

tetanospasmin The neurotoxin of *Clostridium tetani*, the agent of tetanus. Its chief action is directed upon the inhibitory synapses of the anterior horn motor neurons.

tetracyclines A group of broad-spectrum antibiotics with a complex 4-ring structure.

tetrads Groups of four.

theory A collection of statements, propositions, or concepts that explains or accounts for a natural event.

therapeutic index The ratio of the toxic dose to the effective therapeutic dose that is used to assess the safety and reliability of the drug.

thermal death point The lowest temperature that achieves sterilization in a given quantity of broth culture upon a 10-minute exposure. Examples: 55°C for *Escherichia coli*, 60°C for *Mycobacterium tuberculosis*, and 120°C for spores.

thermal death time The least time required to kill all cells of a culture at a specified temperature.

thermocline A temperature buffer zone in a large body of water that separates the warmer water (the epilimnion) from the colder water (the hypolimnion).

thermoduric Resistant to the harmful effects of high temperature.

thermophile A microorganism that thrives at a temperature of 50°C or higher.

thrush *Candida albicans* infection of the oral cavity.

thylakoid Vesicles of a chloroplast formed by elaborate folding of the inner membrane to form "discs." Solar energy trapped in the thylakoids is used in photosynthesis.

thymine (T) One of the nitrogen bases found in DNA, but not in RNA. Thymine is in a pyrimidine form.

thymus Butterfly-shaped organ near the tip of the sternum that is the site of T-cell maturation.

tincture A medicinal substance dissolved in an alcoholic solvent.

tinea Ringworm; a fungal infection of the hair, skin, or nails.

tinea versicolor A condition of the skin appearing as mottled and discolored skin pigmentation as a result of infection by the yeast *Malassezia furfur*.

titer In immunochemistry, a measure of antibody level in a patient, determined by agglutination methods.

T lymphocyte (T cell) A white blood cell that is processed in the thymus gland and is involved in cell-mediated immunity.

tonsils A ring of lymphoid tissue in the pharynx which acts as a repository for lymphocytes.

topoisomerases Enzymes that can add or remove DNA twists and thus regulate the degree of supercoiling.

toxigenicity The tendency for a pathogen to produce toxins. It is an important factor in bacterial virulence.

toxin A specific chemical product of microbes, plants, and some animals that is poisonous to other organisms.

toxinosis Disease whose adverse effects are primarily due to the production and release of toxins.

toxoid A toxin that has been rendered nontoxic but is still capable of eliciting the formation of protective antitoxin antibodies; used in vaccines.

trace elements Micronutrients (zinc, nickel, and manganese) that occur in small amounts, and are involved in enzyme function and maintenance of protein structure.

transamination The transfer of an amino group from an amino acid to a carbohydrate fragment.

transcript A newly transcribed RNA molecule.

transcription mRNA synthesis; the process by which a strand of RNA is produced against a DNA template.

transduction The transfer of genetic material from one bacterium to another by means of a bacteriophage vector.

transfer RNA (tRNA) A transcript of DNA that specializes in converting RNA language into protein language.

transformation In microbial genetics, the transfer of genetic material contained in "naked" DNA fragments from a donor cell to a competent recipient cell.

transfusion Infusion of whole blood, red blood cells, or platelets directly into a patient's circulation.

translation Protein synthesis; the process of decoding the messenger RNA code into a polypeptide.

transposon A DNA segment with an insertion sequence at each end, enabling it to migrate to another plasmid, to the bacterial chromosome, or to a bacteriophage.

traveler's diarrhea A type of gastroenteritis typically caused by infection with enterotoxigenic strains of *E. coli* that are ingested through contaminated food and water.

tricarboxylic acid cycle (TCA or Krebs cycle) The second pathway of the three pathways that complete the process of primary catabolism. Also called the citric acid cycle.

trichinosis Infection by the *Trichinella spiralis* parasite, usually caused by eating the meat of an infected animal. Early symptoms include fever, diarrhea, nausea, and abdominal pain that progress to intense muscle and joint pain and shortness of breath. In the final stages, heart and brain function are at risk, and death is possible.

trichomoniasis Sexually transmitted disease caused by infection by the trichomonads, a group of protozoa. Symptoms include urinary pain and frequency, and foul-smelling vaginal discharge in females or recurring urethritis, with a thin milky discharge, in males.

triglyceride A type of lipid composed of a glycerol molecule bound to three fatty acids.

triplet See *codon.*

trophozoite A vegetative protozoan (feeding form) as opposed to a resting (cyst) form.

true pathogen A microbe capable of causing infection and disease in healthy persons with normal immune defenses.

trypomastigote The infective morphological stage transmitted by the tsetse fly or the reduviid bug in African trypanosomiasis and Chagas disease.

tubercle In tuberculosis, the granulomatous well-defined lung lesion that can serve as a focus for latent infection.

tuberculin A glycerinated broth culture of *Mycobacterium tuberculosis* that is evaporated and filtered. Formerly used to treat tuberculosis, tuberculin is now used chiefly for diagnostic tests.

tuberculoid leprosy A superficial form of leprosy characterized by asymmetrical, shallow skin lesions containing few bacterial cells.

turbid Cloudy appearance of nutrient solution in a test tube due to growth of microbe population.

tyndallization Fractional (discontinuous, intermittent) sterilization designed to destroy spores indirectly. A preparation is exposed to flowing steam for an hour, and then the mineral is allowed to incubate to permit spore germination. The resultant vegetative cells are destroyed by repeated steaming and incubation.

typhoid fever Form of salmonellosis. It is highly contagious. Primary symptoms include fever, diarrhea, and abdominal pain. Typhoid fever can be fatal if untreated.

U

ubiquitous Present everywhere at the same time.

ultraviolet radiation Radiation with an effective wavelength from 240 nm to 260 nm. UV radiation induces mutations readily but has very poor penetrating power.

uncoating The process of removal of the viral coat and release of the viral genome by its newly invaded host cell.

undulant fever See *brucellosis.*

universal donor In blood grouping and transfusion, a group O individual whose erythrocytes bear neither agglutinogen A nor B.

universal precautions (UP) Centers for Disease Control and Prevention guidelines for health care workers regarding the prevention of disease transmission when handling patients and body substances.

uracil (U) One of the nitrogen bases in RNA, but not in DNA. Uracil is in a pyrimidine form.

urinary tract infection (UTI) Invasion and infection of the urethra and bladder by bacterial residents, most often *E. coli.*

V

vaccination Exposing a person to the antigenic components of a microbe without its pathogenic effects for the purpose of inducing a future protective response.

vaccine Originally used in reference to inoculation with the cowpox or vaccinia virus to protect against smallpox. In general, the term now pertains to injection of whole microbes (killed or attenuated), toxoids, or parts of microbes as a prevention or cure for disease.

vacuoles In the cell, membrane-bounded sacs containing fluids or solid particles to be digested, excreted, or stored.

valence The combining power of an atom based upon the number of electrons it can either take on or give up.

variable region The antigen binding fragment of an immunoglobulin molecule, consisting of a combination of heavy and light chains whose molecular conformation is specific for the antigen.

variolation A hazardous, outmoded process of deliberately introducing smallpox material scraped from a victim into the nonimmune subject in the hope of inducing resistance.

vector An animal that transmits infectious agents from one host to another, usually a biting or piercing arthropod like the tick, mosquito, or fly. Infectious agents can be conveyed mechanically by simple contact or biologically whereby the parasite develops in the vector. A genetic element such as a plasmid or a bacteriophage used to introduce genetic material into a cloning host during recombinant DNA experiments.

vegetative In describing microbial developmental stages, a metabolically active feeding and dividing form, as opposed to a dormant, seemingly inert, nondividing form. Examples: a bacterial cell versus its spore; a protozoan trophozoite versus its cyst.

vehicle An inanimate material (solid object, liquid, or air) that serves as a transmission agent for pathogens.

vesicle A blister characterized by a thin-skinned, elevated, superficial pocket filled with serum.

viable nonculturable (VNC) Describes microbes that cannot be cultivated in the laboratory but that maintain metabolic activity (i.e., are alive).

vibrio A curved, rod-shaped bacterial cell.

viremia The presence of viruses in the bloodstream.

virion An elementary virus particle in its complete morphological and thus infectious form. A virion consists of the nucleic acid core surrounded by a capsid, which can be enclosed in an envelope.

viroid An infectious agent that, unlike a virion, lacks a capsid and consists of a closed circular RNA molecule. Although known viroids are all plant pathogens, it is conceivable that animal versions exist.

virtual image In optics, an image formed by diverging light rays; in the compound light microscope, the second, magnified visual impression formed by the ocular from the real image formed by the objective.

virucide A chemical agent which inactivates viruses, especially on living tissue.

virulence In infection, the relative capacity of a pathogen to invade and harm host cells.

virulence factors A microbe's structures or capabilities that allow it to establish itself in a host and cause damage.

virus Microscopic, acellular agent composed of nucleic acid surrounded by a protein coat.

vitamins A component of coenzymes critical to nutrition and the metabolic function of coenzyme complexes.

W

wart An epidermal tumor caused by papillomaviruses. Also called a verruca.

Western blot test A procedure for separating and identifying antigen or antibody mixtures by two-dimensional electrophoresis in polyacrylamide gel, followed by immune labeling.

wheal A welt; a marked, slightly red, usually itchy area of the skin that changes in size and shape as it extends to adjacent area. The reaction is triggered by cutaneous contact or intradermal injection of allergens in sensitive individuals.

whey The residual fluid from milk coagulation that separates from the solidified curd.

whitlow A deep inflammation of the finger or toe, especially near the tip or around the nail. Whitlow is a painful herpes simplex virus infection that can last several weeks and is most common among health care personnel who come in contact with the virus in patients.

whole blood A liquid connective tissue consisting of blood cells suspended in plasma.

Widal test An agglutination test for diagnosing typhoid.

wild type The natural, nonmutated form of a genetic trait.

wort The clear fluid derived from soaked mash that is fermented for beer.

X

xenograft The transfer of a tissue or an organ from an animal of one species to a recipient of another species.

Z

zoonosis An infectious disease indigenous to animals that humans can acquire through direct or indirect contact with infected animals.

zooplankton The collection of non-photosynthetic microorganisms (protozoa, tiny animals) that float in the upper regions of aquatic habitat, and together with phytoplankton comprise the plankton.

zygospore A thick-walled sexual spore produced by the zygomycete fungi. It develops from the union of two hyphae, each bearing nuclei of opposite mating types.

Credits

Photographs

Front Matter

Kelly Cowan author photo, page iii
Courtesy of Michael Williams, Miami University Middletown.

Chapter 1

Opener: © AFP PHOTO/Peter PARKS/Getty; **1.2a:** © Doug Sokell/Tom Stack & Associates; **1.2b:** © Tom Volk; **1.3a:** © Corale L. Brierley/Visuals Unlimited; **1.3b:** © Science VU/SIM, NBS/Visuals Unlimited; **1.3c:** GE Global Research; **Insight 1.1a:** National Institutes of Health (NIH)/U.S. National Library of Medicine; **Insight 1.1b:** © Ron Edmonds/AP Photo; **1.6a:** © Janice Carr/Public Health Image Library; **1.6b:** © Tom Volk; **1.6c:** © T.E. Adams/Visuals Unlimited; **1.6 d and f:** Public Health Image Library; **1.6e:** © Carolina Biological Supply/Phototake; **1.8:** © Bettmann/Corbis **1.9 a and inset:** © Kathy Park Talaro/Visuals Unlimited; **1.9b:** © Science VU/Visuals Unlimited; **1.11:** © AKG/Photo Researchers; **1.12:** © Bettmann/Corbis; **Insight 1.3 a, b and c:** © Brian Smale.

Chapter 2

Opener: © Royalty-Free/Corbis; **2.6d:** © Kathy Park Talaro; **2.10 a, b and c:** © John W. Hole; **Insight 2.3:** © Don Facett/Visuals Unlimited; **2.22d:** From A.S. Moffat, "Nitrogenase Structure Revealed," *Science,* 250:1513, 12/14/90. Photo by M.M. Georgiadis and D.C. Rees, Caltech.

Chapter 3

Opener: © Royalty-Free/Corbis/Vol. 52; **3.3 b, d and f:** © Kathy Park Talaro; **3.4b:** © Kathy Park Talaro; **3.5a:** © Fundamental Photographs; **3.5b:** © Kathy Park Talaro; **Insight 3.1:** Charles River Lab; **3.6b:** © Kathy Park Talaro; **3.7 a and b, 3.9 a and b, 3.10 a and b, 3.12 a and b, 3.12c:** © Kathy Park Talaro; **3.11:** Harold J. Benson; **3.13:** Kathy Park Talaro/Visuals Unlimited; **3.14:** Leica Microsystems Inc.; **3.19a:** © Carolina Biological Supply/Phototake; **3.19 b and c, 3.20b:** © Abbey/Visuals Unlimited; **3.20a:** © George J. Wilder/Visuals Unlimited; **3.21:** © Molecular Probes, Inc.; **3.22:** Anne Fleury; **3.23:** © William Ormerod/Visuals Unlimited; **3.24a:** © Billy Curran, Department of Veterinary Science. Queen's University Belfast; **3.24b:** J.P. Dubley et al., Clinical Microbiology Reviews, © ASM, April 1998, Vol. II, #2, 281. Image courtesy of Dr. Jitender P. Dubey; **3.25:** © Dennis Kunkel/CNRI/Phototake; **Insight 3.2:** Courtesy of IBM Corporation. Almaden Research Center. Unauthorized use not permitted; **3.26a1:** © Kathy Park Talaro; **3.26a2:** Harold J. Benson; **3.26b1:** © Jack Bostrack/Visuals Unlimited; **3.26b2:** © Jack Bostrack/Visuals Unlimited; **3.26b3:** © Manfred Kage/Peter Arnold, Inc.; **3.26c1:** © A.M. Siegelman/Visuals Unlimited; **3.26c2:** © David Frankhauser.

Chapter 4

Opener: © Royalty-Free/Corbis; **4.3a:** Dr. Jeffrey C. Burnham; **4.3b:** From Reichelt and Baumann, *Arch. Microbiol.* 94:283-330. © Springer-Verlag, 1973; **4.3c:** From Noel R. Krieg in *Bacteriological Reviews,* March 1976, Vol. 40(1):87 fig 7; **4.3d:** From Preer et al., *Bacteriological Reviews,* June 1974, 38(2):121, fig 7. © ASM; **4.6:** Stanley F. Hayes, Rocky Mountain Laboratories, NIAID, NIH; **4.7a:** © Eye of Science/Photo Researchers, Inc.; **4.7b:** Dr. S. Knutton from D.R. Lloyd and S. Knurron, *Infection and Immunity,* January 1987, p. 86–92. © ASM; **4.8:** © L. Caro/SPL/Photo Researchers, Inc.; **4.10:** © John D. Cunningham/Visuals Unlimited; **4.11:** © Science VU-Charles W. Stratton/Visuals Unlimited; **4.13a:** © S.C. Holt/Biological Photo Service; **4.13b:** © T. J. Beveridge/Biological Photo Service; **4.15:** © David M. Phillips/Visuals Unlimited; **4.17:** © E.S. Anderson/Photo Researchers, Inc.; **4.19:** © Paul W. Johnson/Biological Photo Service; **4.20:** © Rut CARBALLIDO-LOPEZ/I.N.R.A. Jouyen-Josas, Laboratoire de Génétique Microbienne; **4.21:** Dr. Peter Lewis; **Table 4.1a:** Kit Pogliano and Marc Sharp/UCSD; **Table 4.1b:** © Lee D. Simon/Photo Researchers, Inc.; **4.23 a and b:** © David M. Phillips/Visuals Unlimited; **4.23c:** From *Microbiological Reviews,* 55(1):25, fig 2b, March 1991. Courtesy of Jorge Benach; **4.23d:** © R.G. Kessel-G. Shih/Visuals Unlimited; **4.24:** © A.M. Siegelman/Visuals Unlimited; **4.26:** Baca and Paretsky, *Microbiological Reviews,* 47(20):133, fig 16, June 1983 © ASM; **4.27a:** John Waterbury, Woods Hole Oceanographic Institute; **4.27 b and c:** © T.E. Adams/Visuals Unlimited; **Insight 4.3:** Heide N. Schulz/Max Planck Institute for Marine Microbiology; **4.28:** From *ASM News,* 53(2), Feb. 1987. © ASM, H. Kaltwasser; **4.29:** GBF-German Research Center for Biotechnology, Braunschweig, Germany; **4.30a:** © Wayne P. Armstrong, Palomar College; **4.30b:** Dr. Mike Dyall-Smith, University of Melbourne.

Chapter 5

Opener: © Royalty-Free/Corbis/Vol. 9; **5.1 a and b:** © Andrew Knoll; **5.3 a and b:** © RMF/FDF/Visuals Unlimited; **5.5:** R.G. Garrison Ph.D. from Fungal Dimorphism with Emphasis on Fungi Pathogenic for Human, P.J. Szaniszlo (ed). Reprinted with permission of Plenum Publishing Corp.; **5.6, 5.12:** © Don Facett/Visuals Unlimited; **5.7b:** © Science VU/Visuals Unlimited; **5.15 a and b:** Dr. Judy A. Murphy, San Joaquin Delta College, Department of Microscopy, Stockton, CA; **5.16:** © David M. Phillips/Visuals Unlimited; **5.17a, 5.23:** © Kathy Park Talaro; **5.17b:** © Everett S. Beneke/Visuals Unlimited; **Insight 5.2:** Gregory M. Filip; **5.24a:** © A.M. Siegelman/Visuals Unlimited; **5.24b:** © John D. Cunningham/Visuals Unlimited; **5.25:** Neil Carlson, Department of Environmental Health and Safety; **5.26a:** © T.E. Adams/Visuals Unlimited; **5.26b:** © Jan Hinsch/Photo Researchers, Inc.; **5.26c:** Center for Applied Aquatic Ecology, Department of Botany, North Carolina State University; **5.28:** © David M. Phillips/Visuals Unlimited; **5.30a:** © BioMEDIA ASSOCIATES; **5.30b:** © Yuuji Tsukii, Protist Information Server, http://protist.i.hosei.ac.jp/protist_menuE.html; **5.31b:** Michael Riggs et al., Infection and Immunity, Vol. 62, #5, May 1994, p. 1931 © ASM.

Chapter 6

Opener: © Royalty-Free/Corbis/Vol. 9; **6.2a:** © K.G. Murti/Visuals Unlimited; **6.2b:** © CDC/Phototake; **6.2c:** © A.B. Dowsette/SPL/Photo Researchers; **6.3a:** Schaffer et al., Proceedings of the National Academy of Science, 41:1020, 1955; **6.3b:** © Omnikron/Photo Researchers; **Insight 6.1:** © Science VU/Wayside/Visuals Unlimited; **6.6b:** © Dennis Kunkel/CNRI/Phototake; **6.6d:** © K.G. Murti/Visuals Unlimited; **6.8a:** © Lennart Nilsson/Boehringer Ingelheim International GMBH; **6.8b:** © Kathy Park Talaro; **6.10b:** © Harold Fisher; **6.14:** © K.G. Murti/Visuals Unlimited; **6.15b:** © Chris Bjornberg/Photo Researchers, Inc.; **6.16a:** © Patricia Barber/Custom Medical Stock; **6.16b:** Massimo Battaglia, INeMM CNR, Rome, Italy; **6.18b:** © Lee D. Simon/Photo Researchers, Inc.; **6.19:** © K.G. Murti/Visuals Unlimited; **6.21a:** Ted Heald, State of Iowa Hygienic Laboratory; **6.22a:** © E.S. Chan/Visuals Unlimited; **6.22 b and c:** © Jack W. Frankel; **Insight 6.3:** © Dr. Dennis Kunkel/Visuals Unlimited.

Chapter 7

Opener: © Royalty-Free/Corbis/Vol. 52; **Insight 7.2a:** E.E. Adams, Montana State University/National Science Foundation, Office of Polar Programs; **Insight 7.2b, 7.1a:** © Ralph Robinson/Visuals Unlimited; **7.1b:** Jack Jones, US EPA; **7.9a:** © Pat Armstrong/Visuals Unlimited; **7.9b:** © Philip Sze/Visuals Unlimited; **Insight 7.4:** © Michael Milstein; **7.10a:** Sheldon Manufacturing, Inc.; **7.11:** © Terese M. Barta, Ph.D.; **Insight 7.5:** © Mike Abbey/Visuals Unlimited; **7.12:** © Science VU-Fred Marsik/Visuals Unlimited; **7.16a:** © Kathy Park Talaro/Visuals Unlimited.

Chapter 8

Opener: © Custom Medical Stock Photo, Inc.

Chapter 9

Opener: © Jean Claude Revy-ISM/Phototake—All rights reserved.; **9.1:** PhotoDisc VO6, Nature, Wildlife and the Environment; **9.3:** © K.G. Murti/Visuals Unlimited; **Insight 9.2a:** © A. Barrington Brown/Photo Researchers Inc.; **Insight 9.2b:** © Lawrence Livermore Laboratory/SPL/Custom Medical Stock Photo; **9.17b:** Steven McKnight and Oscar L. Miller, Department of Biology, University of Virginia.

Chapter 10:

Opener © Royalty-Free/Corbis; **10.2b:** © Kathy Park Talaro; **Insight 10.1:** © AFP/Getty; **10.7a:** © Koester Axel/Corbis; **10.7b:** © Milton P. Gordon, Department of Biochemistry, University of Washington; **10.7c:** © Andrew Brookes/Corbis; **Insight 10.2:** © Marilyn Humphries; **Table 10.3a:** David M. Stalker; **Table 10.3b:** Richard Shade, Purdue University; **Table 10.3c:** Peter Beyer, UNI-Freiburg; **10.12:** Brigid Hogan, Howard Hughes Medical Institute, Vanderbilt University; **Table 10.4a:** R.L. Brinster, School of Veterinary Medicine, University of Pennsylvania; **Table 10.4b:** © Karen Kasmauski/National Geographic Image Collection; **10.15c:** Dr. Michael Baird; **10.17:** Tyson Clark, University of California, Santa Cruz.

Chapter 11

Opener: © Anthony Reynolds; Cordaiy Photo Library Ltd./CORBIS; **Insight 11.1:** © Bettmann/Corbis; **11.5a:** © Science VU/Visuals Unlimited; **11.6:** © Raymond B. Otero/Visuals Unlimited; **11.8a:** © Science VU/Nordion International/Visuals Unlimited; **11.10:** © Tom Pantages; **11.11b:** © Fred Hossler/Visuals Unlimited; **11.13a:** STERIS Corporation. System 1 ® is a registered trademark of STERIS Corporation; **Insight 11.3a:** © David M. Phillips/Visuals Unlimited; **Insight 11.3b:** © Kathy Park Talaro; **11.16:** © Kathy Park Talaro/Visuals Unlimited; **11.18a:** Anderson Products, www.anpro.com; **Insight 11.4:** © AFP/Getty.

Chapter 12

Opener: © Photosearch; **Insight 12.1:** © Bettmann/Corbis; **12.2e:** © David Scharf/Peter Arnold; **12.2f:** © CNRI/PHOTO RESEARCHERS, INC.; **Insight 12.2:** © Kathy Park Talaro/Visuals Unlimited; **12.9:** © Cabisco/Visuals Unlimited; **12.16:** © Kenneth E. Greer/Visuals Unlimited; **12.18 b and c, 12.20c:** © Alain Philippon; **12.19:** Etest® is a registered trademark belonging to AB BIODISK, Sweden, and the product and underlying technologies are patented by AB BIODISK in all major markets; **12.20b:** © Kathy Park Talaro.

Chapter 13

Opener: © David Young-Wolff/PhotoEdit; **Insight 13.1:** © Eurelios/Phototake; **Insight 13.2:** © PHIL image 1965; **13.15:** Marshall W. Jennison, Massachusetts Institute of Technology, 1940.

Chapter 14

Opener: © David Grossman/Photo Researchers, Inc.; **14.3b:** © Ellen R. Dirksen/Visuals Unlimited; **14.16c:** Steve Kunkel; **14.18a:** © David M. Phillips/Visuals Unlimited; **14.21e:** Reproduced from *The Journal of Experimental Medicine*, 1966, Vol. 123, p. 969 by copyright permission of The Rockefeller University Press.

Chapter 15

Opener: © Royalty-Free/CORBIS; **15.11b:** © R. Feldman/Rainbow; **15.17:** © Lennart Nilsson, "The Body Victorious," Bonnier Fakta; **15.18a:** © Photodisc/Getty; **15.18b:** © Photodisc/Getty; **15.18c:** © Photodisc/Getty; **15.18d:** © Creatas/PictureQuest; **Insight 15.3:** James Gillroy, British, 1757–1815. "The CowPock," engraving, 1802, William McCallin McKee Memorial Collection, 1928. 1407. © 1991, The Art Institute of Chicago. All rights reserved.

Chapter 16

Opener: © PhotoLink/Getty; **16.2b:** © SPL/Photo Researchers; **16.2c:** © David M. Phillips/Visuals Unlimited; **16.5:** © Kenneth E. Greer/Visuals Unlimited; **16.6a:** © STU/Custom Medical Stock; **16.10 a and b:** © Stuart I. Fox; **Insight 16.3a:** © Renee Lynn/Photo Researchers; **Insight 16.3b:** © Walter H. Hodge/Peter Arnold; **Insight 16.3c:** © Runk/Schoenberger/Grant Heilman Photography; **16.14:** © Kathy Park Talaro; **16.15b:** © Kenneth E. Greer/Visuals Unlimited; **16.17a:** © Diepgen T.L., Yihume G. et al. Dermatology Online Atlas published online at: www.dermis.net. Reprinted with permission.; **16.17b:** © SIU/Visuals Unlimited; **16.21:** Reprinted from R. Kretchmer, *New England Journal of Medicine*, 279:1295, 1968 Massachusetts Medical Society. All rights reserved; **Insight 16.5:** Baylor College of Medicine, Public Affairs.

Chapter 17

Opener: © Keith Brofsky/Getty; **Insight 17.1:** © Photodisc/Getty; **17.4b:** © Fred Marsik/Visuals Unlimited; **17.5:** Analytab Products, a division of Sherwood Medical; **17.7:** Wadsworth Center, NYS Department of Public Health; **17.11b:** Immuno-Mycologics, Inc.; **17.13:** Genelabs Diagnostics Pte Ltd.; **17.15c:** CHEMICON® International, Inc.; **17.16b:** © Hank Morgan/Science Source/Photo Researchers; **17.17a:** © PhotoTake; **17.17b:** © Custom Medical Stock Photo, Inc.; **17.18b1:** © A.M. Siegelman/Visuals Unlimited; **17.18b2:** © Science VU/CDC/Visuals Unlimited.

Chapter 18

Openers: © Lynsey Addario/Corbis; **18.4:** Farrar W.E., Woods M.J., Innes J.A.: Infectious Diseases: Text and Color Atlas, ed. 2. London, Mosby Europe, 1993; **18.5a:** © David M. Phillips/Visuals Unlimited; **18.5b:** © Kathy Park Talaro/Visuals Unlimited; **18.9a:** National Institute Slide Bank/The Welcome Centre for Medical Sciences; **18.9b:** Braude, *Infections, Diseases, and Medical Microbiology*, 2/e, Fig. 3, pg. 1320. With permission from Elsevier; **18.10:** © Science VU-Charles W. Stratton/Visuals Unlimited; **18.11:** M.A. Boyd et al., *Journal of Medical Microbiology*, 5:459, 1972. Reprinted by permission of Longman Group, Ltd. © Pathological Society of Great Britain and Ireland; **Insight 18.1 a and b:** © Carroll H. Weiss/Camera M.D. Studios; **Insight 18.1c:** © ISM/Phototake; **Insight 18.1d:** © Biomedical Communications/Custom Medical Stock Photo; **18.12a, 18.13:** © Kenneth E. Greer/Visuals Unlimited; **18.12b:** © Science VU/Visuals Unlimited; **18.14 a and b:** © Centers for Disease Control; **18.15:** © Logical Images/Custom Medical Stock Photo; **Checkpoint 18.7:** © World Health Org./Peter Arnold; **Checkpoint 18.7:** © Phil Degginger; **18.16:** © Kenneth E. Greer/Visuals Unlimited; **18.17:** © James Stevenson/Photo Researchers, Inc.; **Checkpoint 18.8a:** © Lennart Nilsson/Boehringer Ingelheim International GMBH; **Checkpoint 18.8b, Checkpoint 18.10a:** © Centers for Disease Control; **Checkpoint 18.8c:** © Jack Ballard/Visuals Unlimited; **Checkpoint 18.8d:** © Custom Medical Stock Photo, Inc.; **Checkpoint 18.9a:** © Kenneth E. Greer/Visuals Unlimited; **Checkpoint 18.9b:** © Charles Stoer/Camera M.D. Studios; **Checkpoint 18.10b:** © Science VU-Charles W. Stratton/Visuals Unlimited; **18.18:** © Everett S. Beneke/Visuals Unlimited; **18.19, 18.20a:** © Kenneth E. Greer/Visuals Unlimited; **18.20b:** Reprinted from J. Walter Wilson, *Fungous Diseases of Man*, Plate 42 (middle right), © 1965, The Regents of the University of California; **18.21 a and c:** From Elmer W. Koneman and Roberts, *Practical Laboratory Mycology*, 1985, pages 133, 134 © Williams and Wilkins Co., Baltimore, MD; **18.21b:** © A. M. Siegelman/Visuals Unlimited; **18.22:** © Carroll H. Weiss/Camera M.D. Studios; **18.23:** © Science VU-Bascom Palmer Institute/Visuals Unlimited; **18.24:** Armed Forces Institue of Pathology.

Chapter 19

Opener: © Reuters/CORBIS; **19.5:** © Kathy Park Talaro/Visuals Unlimited; **19.6:** © Louis De Vos; **19.7:** © Gordon Love, M.D. VA, North CA Healthcare System, Martinez, CA; **19.8:** Reprinted from J. Walter Wilson, *Fungous Diseases of Man*, Plate 21, © 1965, The Regents of the University of California; **Insight 19.1:** © Royalty-Free/CORBIS/Vol. 124; **19.11:** © Science VU-David John/Visuals Unlimited; **19.12:** © Centers for Disease Control; **19.14a:** © M. Abbey/Photo Researchers, Inc.; **19.14b:** © Pr. J.J. Hauw/ISM/Phototake; **19.15:** © Lennart Nilsson/Boehringer Ingelheim International GMBH; **19.16:** © Science VU-AFIP/Visuals Unlimited; **19.17a:** © Cabisco/Visuals Unlimited; **19.17b:** © Science VU-Charles W. Stratton/Visuals Unlimited; **19.18:** From I. Katayarma, C.Y. Li, and L.T. Yam, "Ultrastructure Characteristics of the Hairy Cells of Leukemic Reticuloenadotheliosis," *American Journal of Pathology*, 67:361, 1972. Reprinted by permission of the American Society for Investigative Pathology; **Insight 19.3:** Marching Mothers ® Photo courtesy of Ontario March of Dimes; **19.21:** © John D. Cunningham/Visuals Unlimited; **19.22:** Dr. T.F. Sellers, Jr.

Chapter 20

Opener: © Javier Pierini/CORBIS; **20.3, 20.4:** © Centers for Disease Control; **20.6:** Barbara O'Connor; **20.7:** Centers for Disease Control/Peter Arnold; **20.8a:** CDC/NCID/HIP/Janice Carr; **20.8b:** © Science VU-Charles W. Stratton/Visuals Unlimited; **Insight 20.2a:** © Dwight Kuhn; **Insight 20.2b:** © Science VU/Visuals Unlimited; **Insight 20.2c:** © A.M. Siegelman/Visuals Unlimited; **Insight 20.2d:** © George D. Lepp/CORBIS; **20.11:** McCaul and Williams, "Development Cycle of C. Burnetii," *Journal of Bacteriology*, 147:1063, 1981. Reprinted with permission of American Society for Microbiology; **20.12:** © Kenneth E. Greer/Visuals Unlimited; **20.14:** Department of Health and Human Resources, Courtesy of Dr. W. Burgdorfer; **20.16:** Stephen B. Aley, Ph.D., University of Texas at El Paso; **20.17:** © Roll Back Malaria Partnership; **20.18:** © A.M. Siegelman/Visuals Unlimited; **Insight 20.3:** © Science VU/Visuals Unlimited; **20.23 a, b and c:** Centers for Disease Control.

Chapter 21

Opener: © Richard Melloul/CORBIS SYGMA; **21.1b:** © Ellen R. Dirksen/Visuals Unlimited; **21.3:** Farrar W.E., Woods M.J., Innes J.A.: Infectious Diseases: Text and Color Atlas, ed. 2. London, Mosby Europe, 1993; **21.6:** Courtesy of Wellesley College Archives; **21.7a:** Diagnostic Products Corporation; **21.7b:** © Dr. David Schlaes/John D. Cunningham/Visuals Unlimited; **21.8:** From Nester et al., *Microbiology: A Human Perspective*, 4th ed. © Evans Roberts; **21.9:** Centers for Disease Control; **21.14a:** © John D. Cunningham/Visuals Unlimited; **21.16:** © Elmer Koneman/Visuals Unlimited; **21.17:** Gillies and Dodds, *Bacteriology Illustrated*, 5th ed., fig 25, p. 58. Reprinted with permission of Churchill Livingstone; **21.18:** © CNRI/SPL/Photo Researchers, Inc.; **21.19:** © Dr. Leonid Heifets, National Jewish Medical Research Center; **21.20a:** From Nester et al., *Microbiology: A Human Perspective*, 4th ed. © Evans Roberts; **21.20b:** © L.M. Pope and D.R. Grote/Biological Photo Service; **Insight 21.2a:** Centers for Disease Control; **Insight 21.2b:** © JAMA; **21.24:** © Tom Volk.

Chapter 22

Opener: © Royalty-Free/CORBIS; **22.4a:** © R. Gottsegen/Peter Arnold, Inc.; **22.4b:** © Stanley Flegler/Visuals Unlimited; **22.6:** © Science VU-Max A. Listgarten/Visuals Unlimited, Inc.; **22.7:** © Biophoto Associates/Photo Researchers, Inc.; **22.8:** Exeen M. Morgan and Fred Rapp, "Measles Virus and Its Associated Disease," *Bacteriological Reviews*, 41(3):636–666, 1977. Reprinted by permission of American Society for Microbiology; **22.9a:** © PhotoTake; **22.13:** R.R. Colwell and D.M. Rollins, "Viable but Nonculturable Stage of *Campylobacter jejuni* and Its Role in Survival in the Natural Aquatic Environment," *Applied and Environmental Microbiology*, 52(3):531–538, 1986. Reprinted with permission of American Society for Microbiology; **22.14a:** Fred Pittman; **22.14b:** Farrar and Lambert: *Pocket Guide for Nurses: Infectious Diseases*. © 1984, Williams and Wilkins, Baltimore, MD; **22.15:** Centers for Disease Control; **Insight 22.2:** © Kathleen Jagger; **22.16:** © Moredun Animal Health Ltd./Photo Researchers, Inc.; **22.17:** Original image from DPDx-Identification and Diagnosis of Parasites of Public Health Concern; **22.18:** © K.G. Murti/Visuals Unlimited; **Insight 22.3:** © Tom Pantages; **22.19:** © Iruka Okeke; **22.20:** © Ynes R. Ortega; **22.23:** © Science VU-Charles W. Stratton/Visuals Unlimited; **22.24:** © Eye of Science/Photo Researchers, Inc.; **22.25:** *WHO Weekly Epidemiological Record*, Vol. 75, No. 3, 2000; **22.27a:** © Stanley Flegler/Visuals Unlimited; **22.27b:** Katz et al., "Parasitic Diseases," © Springer-Verlag; **22.28a:** © R. Calentine/Visuals Unlimited; **22.28b:** © Science VU-Fred Marsik/Visuals Unlimited; **22.29:** © Carroll H. Weiss/Camera M.D. Studios; **22.30:** © A.M.

Siegelman/Visuals Unlimited; **22.31a:** © Cabisco/Visuals Unlimited; **22.31b:** Harvey Blankespoor; **22.31c:** © Science VU/Visuals Unlimited.

Chapter 23

Opener: © Annie Griffiths Belt/CORBIS; **23.4:** Science VU/Fred Marsik/Visuals Unlimited; **23.5:** © Raymond B. Otero/Visuals Unlimited; **23.6:** © Mary Stallone/Medical Images, Inc.; **23.7:** © David M. Phillips/The Population Council/Photo Researchers, Inc.; **23.9:** James Bingham, *Pocket Guide for Clinical Medicine.* © 1984 Williams and Wilkins Co., Baltimore, MD; **23.10:** © George J. Wilder/Visuals Unlimited; **23.12:** Courtesy Morris D. Cooper, Ph.D., Professor of Medical Microbiology, Southern Illinois University School of Medicine, Springfield, IL; **23.14:** © Science VU/Visuals Unlimited; **23.15:** Kenneth E. Greer/Visuals Unlimited; **23.16 a and b:** © Science VU/CDC/Visuals Unlimited; **23.17:** © Custom Medical Stock Photo, Inc.; **23.18:** © Science VU/CDC/Visuals Unlimited; **23.19:** © Kenneth E. Greer/Visuals Unlimited; **23.20:** © Carroll H. Weiss/Camera M.D. Studios; **23.21:** Public Health Image Library; **23.22:** SLACK Incorporated **Checkpoint 23.6 (all):** © Carroll H. Weiss/Camera M.D. Studios; **23.23a:** © CHOR SOKUNTHEA/Reuters/Corbis; **23.23b:** © Tatiana Markow/Sygma/CORBIS; **Checkpoint 23.7:** © Kenneth E. Greer/Visuals Unlimited; **Checkpoint 23.7:** © Charles Stoer/Camera M.D. Studios.

Chapter 24

Opener: © Vanessa Vick/Photo Researchers; **24.1:** Reprinted cover image from December 1, 2000 *Science* with permission from Jillian Banfield, Vol. 290, 12/1/2000. © 2000 American Association for the Advancement of Science; Image courtesy of Jillian Banfield; **Insight 24.1:** U.S. National Climatic Data Center, 2001; **24.9b:** © John D. Cunningham/Visuals Unlimited; **24.10 a and b:** © Sylvan Wittwer/Visuals Unlimited; **Insight 24.2a:** © Kevin Schafer/Peter Arnold Inc.; **Insight 24.2b:** © Gorm Kallestad/AP Photo; **24.13:** © John D. Cunningham/Visuals Unlimited; **Insight 24.3 (both):** © Carl Oppenheimer; **24.16b:** © Carleton Ray/Photo Researchers, Inc.; **24.17:** © John D. Cunningham/Visuals Unlimited; **24.18a:** © Kathy Park Talaro; **24.18 b and c:** Reprinted from EPA Method 1604 (EPA-821-R-02-024) courtesy of Dr. Kristen Brenner from the Microbial Exposure Research Branch, Microbiological and Chemical Exposure Assessment Research Division, National Exposure Research Laboratory, Office of Research and Development, U.S. Environmental Protection Agency; **24.21 a and b:** Sanitation Districts of Los Angeles County; **24.22:** © John D. Cunningham/Visuals Unlimited; **24.23b:** © Kevin Schafer/Peter Arnold Inc.; **24.24a:** © Kathy Park Talaro/Visuals Unlimited; **24.25:** © Joe Munroe/Photo Researchers; **Insight 24.5 a and b:** © Kathy Park Talaro; **24.27:** © Kathy Park Talaro; **24.3:** © J.T. MacMillan.

Line Art

Chapter 3

3.23 a and b: William A. Jensen and Roderic B. Park, *Cell Ultrastructure,* © 1967 Wadsworth Publishing Company. Diagram A-1: comparison between the components of the light and electron microscopes, p. 56. Courtesy William A. Jensen.

Chapter 6

6.10: From Westwood et al., *Journal of Microbiology,* 34:67, 1964. Reprinted by permission of The Society for General Microbiology, United Kingdom.

Chapter 11

11.5b: From John J. Perkins, *Principles and Methods of Sterilization in Health Sciences,* 2/e, 1969. Courtesy of Charles C. Thomas Publisher, Ltd., Springfield, Illinois; **11.15:** From Nolte, et al., *Oral Microbiology,* 4e. © 1982 Mosby.

Chapter 15

15.15: From Joseph A. Bellanti, MD, *Immunology III.* (Philadelphia, PA: W.B. Saunders, 1985). Reprinted by permission of Joseph A. Bellanti, MD.

Chapter 21

21.23: From CDC Special Pathogens Branch. All About Hantavirus. http://www.cdc.gov/ncidod/diseases/hanta/hps. Click "Case Information," "Maps."

Chapter 22

22.25: From *WHO Weekly Epidemiological Record,* Vol. 75, No. 3, 2000.

Chapter 23

23.22: Image from www.infectiousdiseasenews.com/200007/alexander1aCREAM.gif. Reprinted by permission.

Inside back cover: All graphics courtesy of Centers for Disease Control and Prevention. Summary of notifiable diseases—United States, 2002. Published April 30, 2004, for *MMWR* 2002; 51(No. 53):[36–49].

Index

Note: In this index, page numbers followed by a *t* designate tables; page numbers followed by an *f* refer to figures; page numbers followed by an *n* refer to footnotes; page numbers set in *italics* refer to definitions of terms or introductory discussions.

Displaying Disease Statistics

Infectious disease specialists use a number of different methods to visually represent the numbers of disease cases or deaths.

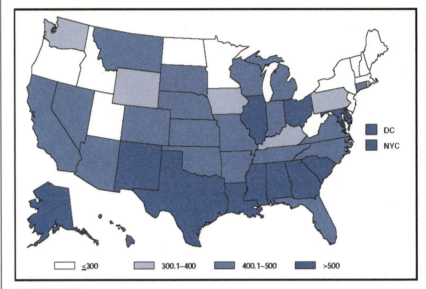

CHLAMYDIA. Reported cases among women per 100,000 female population — United States, 2002

DC
NYC

≤300 | 300.1–400 | 400.1–500 | >500

Chlamydia refers to genital infections caused by *Chlamydia trachomatis*. In 2002, the chlamydia rate among women was 455.37 cases/100,000 population. Rates for men are not given because reporting for men is limited.

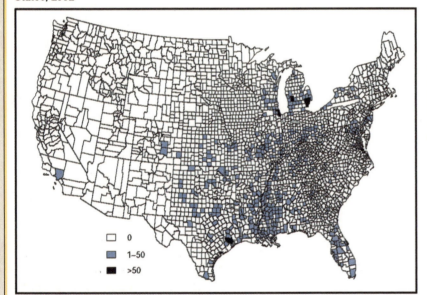

ENCEPHALITIS/MENINGITIS, WEST NILE. Reported cases, by county — United States, 2002

0
1–50
>50

In 2002, 36 states and the District of Columbia reported 2,146 West Nile virus (WNV), through the Arbonet surveillance system, neuroinvasive cases (i.e., encephalitis or meningitis) compared with a total of 64 cases from 10 states in 2001. Since WNV was first discovered during an encephalitis outbreak in New York City, a median of 61.5 (average: 572; range: 21–2,146) cases were reported per year in the United States.

Some methods emphasize the geographical distribution of disease.